Elements

Element	Symbol	Atomic number	Relative atomic mass / g mol⁻¹
Actinium	Ac	89	227.03
Aluminium	Al	13	26.98
Americium	Am	95	241.06
Antimony	Sb	51	121.75
Argon	Ar	18	39.95
Arsenic	As	33	74.92
Astatine	At	85	210
Barium	Ba	56	137.34
Berkelium	Bk	97	249.08
Beryllium	Be	4	9.01
Bismuth	Bi	83	208.98
Boron	B	5	10.81
Bromine	Br	35	79.91
Cadmium	Cd	48	112.40
Caesium	Cs	55	132.91
Calcium	Ca	20	40.08
Californium	Cf	98	252.08
Carbon	C	6	12.01
Cerium	Ce	58	140.12
Chlorine	Cl	17	35.45
Chromium	Cr	24	52.01
Cobalt	Co	27	58.93
Copper	Cu	29	63.54
Curium	Cm	96	244.07
Dysprosium	Dy	66	162.50
Einsteinium	Es	99	252.09
Erbium	Er	68	167.26
Europium	Eu	63	151.96
Fermium	Fm	100	257.10
Fluorine	F	9	19.00
Francium	Fr	87	223
Gadolinium	Gd	64	157.25
Gallium	Ga	31	69.72
Germanium	Ge	32	72.59
Gold	Au	79	196.97
Hafnium	Hf	72	178.49
Helium	He	2	4.00
Holmium	Ho	67	164.93
Hydrogen	H	1	1.008
Indium	In	49	114.82
Iodine	I	53	126.90
Iridium	Ir	77	192.2
Iron	Fe	26	55.85
Krypton	Kr	36	83.80
Lanthanum	La	57	138.91
Lawrencium	Lr	103	262
Lead	Pb	82	207.19
Lithium	Li	3	6.94
Lutetium	Lu	71	174.97
Magnesium	Mg	12	24.31
Manganese	Mn	25	54.94
Mendelevium	Md	101	258.10
Mercury	Hg	80	200.59
Molybdenum	Mo	42	95.94
Neodymium	Nd	60	144.24
Neon	Ne	10	20.18
Neptunium	Np	93	237.05
Nickel	Ni	28	58.69
Niobium	Nb	41	92.91
Nitrogen	N	7	14.01
Nobelium	No	102	259
Osmium	Os	76	190.2
Oxygen	O	8	16.00
Palladium	Pd	46	106.4
Phosphorus	P	15	30.97
Platinum	Pt	78	195.08
Plutonium	Pu	94	239.05
Polonium	Po	84	210
Potassium	K	19	39.10
Praseodymium	Pr	59	140.91
Promethium	Pm	61	146.92
Protactinium	Pa	91	231.04
Radium	Ra	88	226.03
Radon	Rn	86	222
Rhenium	Re	75	186.2
Rhodium	Rh	45	102.91
Rubidium	Rb	37	85.47
Ruthenium	Ru	44	101.07
Samarium	Sm	62	150.35
Scandium	Sc	21	44.96
Selenium	Se	34	78.96
Silicon	Si	14	28.09
Silver	Ag	47	107.87
Sodium	Na	11	22.99
Strontium	Sr	38	87.62
Sulfur	S	16	32.06
Tantalum	Ta	73	180.95
Technetium	Tc	43	98.91
Tellurium	Te	52	127.60
Terbium	Tb	65	158.92
Thallium	Tl	81	204.37
Thorium	Th	90	232.04
Thulium	Tm	69	168.93
Tin	Sn	50	118.71
Titanium	Ti	22	47.90
Tungsten	W	74	183.85
Uranium	U	92	238.03
Vanadium	V	23	50.94
Xenon	Xe	54	131.30
Ytterbium	Yb	70	173.04
Yttrium	Y	39	88.91
Zinc	Zn	30	65.37
Zirconium	Zr	40	91.22

INORGANIC
CHEMISTRY

Catherine E. Housecroft and Alan G. Sharpe

Prentice
Hall

An imprint of **Pearson Education**

Harlow, England · London · New York · Reading, Massachusetts · San Francisco · Toronto · Don Mills, Ontario · Sydney
Tokyo · Singapore · Hong Kong · Seoul · Taipei · Cape Town · Madrid · Mexico City · Amsterdam · Munich · Paris · Milan

Pearson Education Limited
Edinburgh Gate
Harlow
Essex CM20 2JE
England

and Associated Companies throughout the world.

Visit us on the World Wide Web at:
www.pearsoneduc.com

First edition 2001

© Pearson Education Limited 2001

ISBN 0582-31080-6

British Library Cataloguing-in-Publication Data
A catalogue record for this book can be obtained from the British Library

10 9 8 7 6
06 05 04 03 02

Typeset in $9\frac{1}{2}/12$ pt Times by 60
Printed by Ashford Colour Press Ltd., Gosport

Contents

22 *d*-Block metal chemistry: the second and third row metals   534

Preface

Inorganic Chemistry has developed from the three editions of Alan Sharpe's *Inorganic Chemistry* and builds upon the success of this text. The aim of the two books is the same: to provide a single volume that gives a critical introduction to modern inorganic chemistry. However, in making the transition, the book has undergone a complete overhaul, not only in a complete rewriting of the text, but also in the general format, pedagogical features and illustrations. These changes give *Inorganic Chemistry* a more modern feel while retaining the original characteristic approach to the discussions, in particular of general principles of inorganic chemistry. *Inorganic Chemistry* provides students with numerous fully-worked examples of calculations, extensive end-of-chapter problems, and 'boxed' material relating to chemical and theoretical background, chemical resources, the effects of chemicals on the environment and applications of inorganic chemicals. The book contains chapters on physical inorganic chemistry and descriptive chemistry of the elements. Descriptive chapters build upon the foundations laid in the earlier chapters. The material is presented in a logical order but navigation through the text is aided by comprehensive cross-references. The book is completed by four 'topic' chapters covering inorganic kinetics, catalysis, aspects of the solid state and bioinorganic chemistry. Each chapter in the book ends with a summary and a checklist of new chemical terms. The reading lists contain suggestions both for books and articles in the current literature. Additional information about websites of interest to readers of this book can be accessed via: http://www.booksites.net/housecroft

The content of all descriptive chemistry chapters contains up-to-date information and takes into account the results of the latest research; in particular, the chapters on organometallic chemistry of the *s*- and *p*-block and *d*-block elements reflect a surge in research interest in this area of chemistry. Another major development from Alan Sharpe's original text has been to extend the discussion of molecular orbital theory, with an aim not only of introducing the topic but also showing how an objective (and cautious) approach can provide insight into particular bonding features of molecular species. Greater emphasis on the use of multinuclear NMR spectroscopy has been included; case studies introduce $I > \frac{1}{2}$ nuclei and the observation of satellite peaks and applications of NMR spectroscopy are discussed where appropriate throughout the text. Appendices are included and are a feature of the book; they provide tables of physical data, selected character tables, and a list of abbreviations.

Answers to non-descriptive problems are included in *Inorganic Chemistry*, but a separate *Solutions Manual* has been written by Catherine Housecroft, and this gives detailed answers or essay plans for all end of chapter problems.

Most of the 3D-structural diagrams in the book have been drawn using Chem3D Pro. with coordinates accessed from the Cambridge Crystallographic Data Base and implemented through the ETH in Zürich. The protein structures in Chapter 28 have been drawn using Rasmol with data from the Protein Data Bank (http://www/rcsb.org/pdb).

Suggestions passed on by readers of Alan Sharpe's *Inorganic Chemistry* have helped us to identify 'holes' and, in particular, we thank Professor Derek Corbridge. We gratefully acknowledge comments made on the manuscript by members of the panel of reviewers (from the UK, the Netherlands and the US) set up by Pearson Education. A number of colleagues have read chapters of the manuscript and their suggestions and criticisms have been invaluable: special thanks go to Professors Steve Chapman, Edwin Constable, Michael Davies and Georg Süss-Fink, and Dr Malcolm Gerloch. We should also like to thank Dr Paul Bowyer for information on sulfur dioxide in wine production, and Dr Bo Sundman for providing data for the iron phase-diagram. A text of this type cannot become reality without dedicated work from the publisher: from among those at Pearson Education who have seen this project develop from infancy and provided us with support, particular thanks go to Lynn Brandon, Pauline Gillett, Julie Knight, Paul Nash, Alex Seabrook and Ros Woodward, and to Bridget Allen and Kevin Ancient for tireless and dedicated work on the design and artwork.

One of us must express sincere thanks to her husband, Edwin Constable, for endless discussions and critique. Thanks again to two very special feline companions, Philby and Isis, who have sat, slept and played by the Macintosh through every minute of the writing of this edition – they are not always patient, but their love and affection is an integral part of writing.

Catherine E. Housecroft
Alan G. Sharpe
June 2000

The publishers would like to thank the following for reviewing this text:

Thomas P. Fehlner, Department of Chemistry and Biochemistry, University of Notre Dame, US
Professor Jan Reedijk, Leiden Institute of Chemistry, Leiden University, Netherlands
Dr Philip Bailey, Department of Chemistry, University of Edinburgh, UK
Dr Mark Winter, Department of Chemistry, Sheffield University, UK

The publishers would like to thank the following for permission to use material in this book:

Professor B.N. Figgis for Figure 20.20 from Figgis, B.N. (1966) *Introduction to Ligand Fields*, New York: Interscience.

Though every effort has been made to trace the owners of copyright material, we take this opportunity to apologise to any copyright holders whose rights may have been unwittingly infringed.

A Companion Web Site accompanies

INORGANIC CHEMISTRY

by Catherine E. Housecroft and Alan G. Sharpe

Visit the Inorganic Chemistry Companion Web Site at
www.booksites.net/housecroft
to find valuable teaching and learning material including:

For Students:
- Chapter-by-chapter annotated web links
- Multiple choice practice questions
- Questions to help you test your understanding

For Lecturers:
- A secure, password protected site with teaching guidance
- Chem3D figures in the book are available for viewing in Chime

Chapter 1   Some basic concepts

TOPICS

- Fundamental particles
- Atomic number, mass number and isotopes
- An overview of quantum theory
- Orbitals of the hydrogen atom and quantum numbers
- The multi-electron atom, the *aufbau* principle and electronic configurations
- The periodic table
- Ionization energies and electron affinities
- Lewis structures

- Valence bond theory
- Fundamentals of molecular orbital theory
- The octet rule
- Electronegativity
- Dipole moments
- MO theory: heteronuclear diatomic molecules
- Isoelectronic molecules
- Molecular shape and the VSEPR model
- Geometrical isomerism

1.1 Introduction

Inorganic chemistry: it is not an isolated branch of chemistry

If organic chemistry is considered to be the 'chemistry of carbon', then inorganic chemistry is the chemistry of all elements except carbon. In its broadest sense, this is true, but of course there are overlaps between branches of chemistry. A topical example is the chemistry of the *fullerenes* (see *Section 13.4*) including C_{60} (see *Figure 13.3*) and C_{70}; this was the subject of the award of the 1996 Nobel Prize in chemistry to Professors Sir Harry Kroto, Richard Smalley and Richard Curl. An understanding of such molecules and related species called *nanotubes* involves studies by organic, inorganic and physical chemists as well as by physicists and materials scientists.

Inorganic chemistry is not simply the study of elements and compounds; it is also the study of physical principles. For example, in order to understand why some compounds are soluble in a given solvent and others are not, we apply laws of thermodynamics. If our aim is to propose details of a reaction mechanism, then a knowledge of reaction kinetics is needed. Overlap between physical and inorganic chemistry is also significant in the study of molecular structure. In the solid state, X-ray diffraction methods are routinely used to obtain pictures of the spatial arrangements of atoms in a molecule or molecular ion. To interpret the behaviour of molecules in solution, we use physical techniques such as nuclear magnetic resonance (NMR) spectroscopy; the equivalence or not of particular nuclei on a spectroscopic timescale may indicate whether a molecule is static or undergoing a dynamic process (see *Section 2.11*). In this text, we describe the *results* of such experiments but we will not, in general, discuss underlying theories; several texts which cover experimental details of such techniques are listed at the end of Chapter 1.

The aims of Chapter 1

In this chapter, we outline some concepts fundamental to an understanding of inorganic chemistry. We have assumed that readers are to some extent familiar with most of these concepts and our aim is to give a point of reference for review purposes.

1.2 Fundamental particles of an atom

An *atom* is the smallest unit quantity of an element that is capable of existence, either alone or in chemical combination with other atoms of the same or another element. The fundamental particles of which atoms are composed are the *proton*, *electron* and *neutron*.

Table 1.1 Properties of the proton, electron and neutron.

	Proton	Electron	Neutron
Charge / C	$+1.602 \times 10^{-19}$	-1.602×10^{-19}	0
Charge number (relative charge)	1	-1	0
Rest mass / kg	1.673×10^{-27}	9.109×10^{-31}	1.675×10^{-27}
Relative mass	1837	1	1839

A neutron and a proton have approximately the same mass and, relative to these, an electron has negligible mass (Table 1.1). The charge on a proton is positive and of equal magnitude, but opposite sign, to that on a negatively charged electron; a neutron has no charge. In an atom of any element, there are equal numbers of protons and electrons and so an atom is neutral. The *nucleus* of an atom consists of protons and (with the exception of *protium* (see *Section 9.3*)) neutrons, and is positively charged; the nucleus of protium consists of a single proton. The electrons occupy a region of space around the nucleus. Nearly all the mass of an atom is concentrated in the nucleus, but the volume of the nucleus is only a tiny fraction of that of the atom; the radius of the nucleus is about 10^{-15} m while the atom itself is about 10^5 times larger than this. It follows that the density of the nucleus is enormous, more than 10^{12} times than of the metal Pb.

Although chemists tend to consider the electron, proton and neutron as the fundamental (or elementary) particles of an atom, particle physicists would disagree, since their research shows the presence of yet smaller particles.

1.3 Atomic number, mass number and isotopes

Nuclides, atomic number and mass number

A *nuclide* is a particular type of atom and possesses a characteristic *atomic number*, Z, which is equal to the number of protons in the nucleus; because the atom is electrically neutral, Z also equals the number of electrons. The *mass number*, A, of a nuclide is the number of protons *and* neutrons in the nucleus. A shorthand method of showing the atomic number and mass number of a nuclide along with its symbol, E, is:

Mass number $\longrightarrow {}^{A}_{Z}\text{E} \longleftarrow$ Element symbol
Atomic number $\longrightarrow$

e.g. ${}^{20}_{10}\text{Ne}$

Atomic number = Z = number of protons in the nucleus = number of electrons

Mass number = A = number of protons + number of neutrons

Number of neutrons = A − Z

Relative atomic mass

Since the electrons are of minute mass, the mass of an atom essentially depends upon the number of protons and neutrons in the nucleus. As Table 1.1 shows, the mass of a single atom is a very small, non-integral number, and for convenience we adopt a system of *relative atomic masses*. The key point of this formalism is that by defining the *atomic mass unit* (u) such that it has a value of $\approx 1.660 \times 10^{-27}$ kg, we can effectively consider the mass of each proton and neutron to be ≈ 1u. This leads to a set of *relative atomic masses* (A_r). The scale is based upon measurements taken for carbon and all atomic masses are thus stated *relative to* ${}^{12}_{6}\text{C} = 12.0000$.

Isotopes

Nuclides of the same element possess the same number of protons and electrons but may have different mass numbers; the number of protons and electrons defines the element but the number of neutrons may vary. Nuclides of a particular element that differ in the number of neutrons and, therefore, their mass number, are called *isotopes* (see *Appendix 5*). Isotopes of some elements occur naturally while others may be produced artificially.

Elements that occur naturally with only one nuclide are *monotopic* and include phosphorus, ${}^{31}_{15}\text{P}$ and fluorine, ${}^{19}_{9}\text{F}$. Elements that exist as mixtures of isotopes include C (${}^{12}_{6}\text{C}$ and ${}^{13}_{6}\text{C}$) and O (${}^{16}_{8}\text{O}$, ${}^{17}_{8}\text{O}$ and ${}^{18}_{8}\text{O}$). Since the atomic number is constant for a given element, isotopes are often distinguished only by stating the atomic masses, e.g. ${}^{12}\text{C}$ and ${}^{13}\text{C}$.

Worked example 1.1 Relative atomic mass

Calculate the value of A_r for naturally occurring chlorine if the distribution of isotopes is 75.77% ${}^{35}_{17}\text{Cl}$ and 24.23% ${}^{37}_{17}\text{Cl}$.

The relative atomic mass of chlorine is the weighted mean of the mass numbers of the two isotopes:

$$\text{Relative atomic mass, } A_r = \left(\frac{75.77}{100} \times 35\right) + \left(\frac{24.23}{100} \times 37\right)$$

$$= 35.48$$

BOX 1.1   **CHEMICAL AND THEORETICAL BACKGROUND**

Isotopes and allotropes

Do not confuse *isotope* and *allotrope!* Sulfur exhibits both isotopes and allotropes. Isotopes of sulfur (with percentage naturally occurring abundances) are $^{32}_{16}S$ (95.02%), $^{33}_{16}S$ (0.75%), $^{34}_{16}S$ (4.21%), $^{36}_{16}S$ (0.02%).

Allotropes of an element are different structural modifications of that element. Allotropes of sulfur include cyclic structures, e.g. S_6 (see below) and S_8 (Figure 1.1c), and S_x-chains of various lengths (poly*catena*sulfur).

Further examples of isotopes and allotropes appear throughout the book.

S_6 Part of the helical chain of S_∞

Self-study exercises

1. If A_r for Cl is 35.48, what is the ratio of $^{35}Cl:^{37}Cl$ present in a sample of Cl atoms containing naturally occurring Cl?

                   [*Ans.* 3.127:1]

2. Calculate the value of A_r for naturally occurring Cu if the distribution of isotopes is 69.2% ^{63}Cu and 30.8% ^{65}Cu.

                   [*Ans.* 63.6]

3. Why in question 2 is it adequate to write ^{63}Cu rather than $^{63}_{29}Cu$?
                   [*Ans.* see Section 1.3]

Figure 1.1c; five peaks are observed due to combinations of the isotopes of sulfur. (See problem 1.3 at the end of thic chapter.)

> Isotopes of an element have the same atomic number, Z, but different atomic masses.

1.4 Successes in early quantum theory

We saw in Section 1.2 that electrons in an atom occupy a region of space around the nucleus. The importance of electrons in determining the properties of atoms, ions and molecules, including the bonding between or within them, means that we must have an understanding of the electronic structures of each species. No adequate discussion of

Isotopes can be separated by *mass spectrometry* and Figure 1.1a shows the isotopic distribution in naturally occurring Ru. Compare this plot (in which the most abundant isotope is set to 100) with the values listed in Appendix 5. Figure 1.1b shows a mass spectrometric trace for molecular S_8, the structure of which is shown in

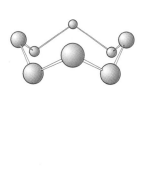

               (a)   **(b)**   **(c)**

Fig. 1.1 Mass spectrometric traces for (a) atomic Ru and (b) molecular S_8; the mass:charge ratio is m/z and in these traces $z = 1$. (c) The molecular structure of S_8.

electronic structure is possible without reference to *quantum theory* and *wave mechanics*. In this and the next few sections, we review some of the crucial concepts. The treatment is mainly qualitative, and for greater detail and more rigorous derivations of mathematical relationships, the references at the end of Chapter 1 should be consulted.

The development of quantum theory took place in two stages. In the older theories (1900–1925), the electron was treated as a particle, and the achievements of greatest significance to inorganic chemistry were the interpretation of atomic spectra and assignment of electronic configurations. In more recent models, the electron is treated as a wave (hence the name *wave mechanics*) and the main successes in chemistry are the elucidation of the basis of stereochemistry and methods for calculating the properties of molecules (exact *only* for species involving light atoms).

Since all the results obtained by using the older quantum theory may also be obtained from wave mechanics, it may seem unnecessary to refer to the former; indeed, sophisticated treatments of theoretical chemistry seldom do. However, most chemists often find it easier and more convenient to consider the electron as a particle rather than a wave.

Some important successes of classical quantum theory

Historical discussions of the developments of quantum theory are dealt with adequately elsewhere, and so we focus only on some key points of *classical* quantum theory (in which the electron is considered to be a particle).

At low temperatures, the radiation emitted by a hot body is mainly of low energy and occurs in the infrared, but as the temperature increases, the radiation becomes successively dull red, bright red and white. Attempts to account for this observation failed until, in 1901, Planck suggested that energy could be absorbed or emitted only in *quanta* of magnitude ΔE related to the frequency of the radiation, ν, by equation 1.1. The proportionality constant is h, the Planck constant ($h = 6.626 \times 10^{-34}$ J s).

$$\Delta E = h\nu \qquad \text{Units: } E \text{ in J; } \nu \text{ in s}^{-1} \text{ or Hz} \qquad (1.1)$$

$$c = \lambda\nu \qquad \text{Units: } \lambda \text{ in m; } \nu \text{ in s}^{-1} \text{ or Hz} \qquad (1.2)$$

The hertz, Hz, is the SI unit of frequency.

Since the frequency of radiation is related to the wavelength, λ, by equation 1.2, in which c is the speed of light in a vacuum ($c = 2.998 \times 10^8$ m s^{-1}), we can rewrite equation 1.1 in the form of equation 1.3 and relate the energy of radiation to its wavelength.

$$\Delta E = \frac{hc}{\lambda} \qquad (1.3)$$

On the basis of this relationship, Planck derived a relative intensity/wavelength/temperature relationship which was in good agreement with experimental data. This derivation is not straightforward and we shall not reproduce it here.

One of the most important applications of early quantum theory was the interpretation of the atomic spectrum of hydrogen on the basis of the Rutherford–Bohr model of the atom. When an electric discharge is passed through a sample of dihydrogen, the H_2 molecules dissociate into atoms, and the electron in a particular *excited* H atom may be *promoted* to one of many high energy levels. These states are transient and the electron falls back to a lower energy state, emitting energy as it does so. The consequence is the observation of *spectral lines* in the emission spectrum of hydrogen; the spectrum (a small part of which is shown in Figure 1.2) consists of groups of discrete lines corresponding to electronic transitions, each of *discrete energy*. As long ago as 1885, Balmer pointed out that the wavelengths of the spectral lines observed in the visible region of the atomic spectrum of hydrogen obeyed equation 1.4, in which R is the Rydberg constant for hydrogen, $\bar{\nu}$ is the wavenumber in cm^{-1}, and n is an integer 3, 4, 5 ... This series of spectral lines is known as the *Balmer series*.

Wavenumber = reciprocal of wavelength; convenient (non-SI) units are 'reciprocal centimetres', cm^{-1}

$$\bar{\nu} = \frac{1}{\lambda} = R\left(\frac{1}{2^2} - \frac{1}{n^2}\right) \qquad (1.4)$$

$$R = \text{Rydberg constant for hydrogen}$$
$$= 1.097 \times 10^7 \text{ m}^{-1} = 1.097 \times 10^5 \text{ cm}^{-1}$$

Other series of spectral lines occur in the ultraviolet (Lyman series) and infrared (Paschen, Brackett and Pfund series). All

Fig. 1.2 Part of the emission spectrum of atomic hydrogen. Groups of lines have particular names, e.g. Balmer and Lyman series.

Fig. 1.3 Some of the transitions that make up the Lyman and Balmer series in the emission spectrum of atomic hydrogen.

lines in all the series obey the general expression given in equation 1.5 where $n' > n$. For the Lyman series, $n = 1$, for the Balmer series, $n = 2$, and for the Paschen, Brackett and Pfund series, $n = 3$, 4 and 5 respectively. Figure 1.3 shows some of the allowed transitions of the Lyman and Balmer series in the emission spectrum of atomic H. Note the use of the word *allowed*; the transitions must obey *selection rules* to which we return in Section 20.6.

$$\bar{\nu} = \frac{1}{\lambda} = R\left(\frac{1}{n^2} - \frac{1}{n'^2}\right) \tag{1.5}$$

Bohr's theory of the atomic spectrum of hydrogen

In 1913, Niels Bohr combined elements of quantum theory and classical physics in a treatment of the hydrogen atom. He stated two postulates for an electron in an atom:

- *Stationary states* exist in which the energy of the electron is constant; such states are characterized by *circular orbits* about the nucleus in which the electron has an angular momentum mvr given by equation 1.6. The integer, n, is the *principal quantum number*.

$$mvr = n\left(\frac{h}{2\pi}\right) \tag{1.6}$$

where m = mass of electron; v = velocity of electron; r = radius of the orbit; h = the Planck constant; $h/2\pi$ may be written as $\hbar$.

- Energy is absorbed or emitted only when an electron moves from one stationary state to another and the energy change is given by equation 1.7 where n_1 and n_2 are the principal quantum numbers referring to the energy levels E_{n_1} and E_{n_2} respectively.

$$\Delta E = E_{n_2} - E_{n_1} = h\nu \tag{1.7}$$

If we apply the Bohr model to the H atom, the radius of each allowed circular orbit can be determined from equation 1.8. The origin of this expression lies in the centrifugal force acting on the electron as it moves in its circular orbit; for the orbit to be maintained, the centrifugal force must equal the force of attraction between the negatively charged electron and the positively charged nucleus.

$$r_n = \frac{\varepsilon_0 h^2 n^2}{\pi m_e e^2} \tag{1.8}$$

where ε_0 = permittivity of a vacuum

$\qquad = 8.854 \times 10^{-12}\,\mathrm{F\,m^{-1}}$

$\qquad h$ = Planck constant = $6.626 \times 10^{-34}\,\mathrm{J\,s}$

$\qquad n = 1, 2, 3 \ldots$ describing a given orbit

$\qquad m_e$ = electron rest mass = $9.109 \times 10^{-31}\,\mathrm{kg}$

$\qquad e$ = charge on an electron (elementary charge)

$\qquad = 1.602 \times 10^{-19}\,\mathrm{C}$

From equation 1.8, substitution of $n = 1$ gives a radius for the first orbit of the H atom of $5.292 \times 10^{-11}\,\mathrm{m}$, or

52.92 pm. This value is called the *Bohr radius* of the H atom and is given the symbol a_0.

An increase in the principal quantum number from $n = 1$ to $n = \infty$ has a special significance; it corresponds to the ionization of the atom (equation 1.9) and the ionization energy, *IE*, can be determined by combining equations 1.5 and 1.7, as shown in equation 1.10. Values of *IE*s are quoted *per mole of atoms*:

One mole of a substance contains the Avogadro number, *L*, of particles:

$$L = 6.022 \times 10^{23}\,\text{mol}^{-1}$$

$$\text{H(g)} \longrightarrow \text{H}^+(\text{g}) + \text{e}^- \tag{1.9}$$

$$IE = E_\infty - E_1 = \frac{hc}{\lambda} = hcR\left(\frac{1}{1^2} - \frac{1}{\infty^2}\right) \tag{1.10}$$

$$= 2.179 \times 10^{-18}\,\text{J}$$

$$= 2.179 \times 10^{-18} \times 6.022 \times 10^{23}\,\text{J mol}^{-1}$$

$$= 1.312 \times 10^6\,\text{J mol}^{-1}$$

$$= 1312\,\text{kJ mol}^{-1}$$

Although the SI unit of energy is the joule, ionization energies are often expressed in electron volts (eV) (1 eV = $96.4853 \approx 96.5\,\text{kJ mol}^{-1}$).

Impressive as the success of the Bohr model was when applied to the H atom, extensive modifications were required to cope with species containing more than one electron; we shall not pursue this further here.

1.5 An introduction to wave mechanics

The wave-nature of electrons

The quantum theory of radiation introduced by Max Planck and Albert Einstein implies a particle theory of light, in addition to the wave theory of light required by the phenomena of interference and diffraction. In 1924, Louis de Broglie argued that if light were composed of particles and yet showed wave-like properties, the same should be true of electrons and other particles. This phenomenon is referred to as *wave–particle duality*. The de Broglie relationship (equation 1.11) combines the concepts of classical mechanics with the idea of wave-like properties by showing that a particle with momentum mv (m = mass and v = velocity of the particle) possesses an associated wave of wavelength λ.

$$\lambda = \frac{h}{mv} \quad \text{where } h \text{ is the Planck constant.} \tag{1.11}$$

An important physical observation which is a consequence of the de Broglie relationship is that electrons accelerated to a velocity of $6 \times 10^6\,\text{m s}^{-1}$ (by a potential of 100 V) have an associated wavelength of $\approx 120\,\text{pm}$ and such electrons are diffracted as they pass through a crystal. This phenomenon is the basis of electron diffraction techniques used to determine structures of chemical compounds.

The uncertainty principle

If an electron has wave-like properties, there is an important and difficult consequence: it becomes impossible to know exactly both the momentum and position of the electron *at the same instant in time*. This is a statement of Heisenberg's *uncertainty principle*. In order to get around this problem, rather than trying to define its exact position and momentum, we use the *probability of finding the electron* in a given volume of space. The probability of finding an electron at a given point in space is determined from the function ψ^2 where ψ is a mathematical function which describes the behaviour of an electron-wave; ψ is the *wavefunction*.

The probability of finding an electron at a given point in space is determined from the function ψ^2 where ψ is the *wavefunction*.

The Schrödinger wave equation

Information about the wavefunction is obtained from the Schrödinger wave equation which can be set up and solved either exactly or approximately; the Schrödinger equation can be solved *exactly only* for a species containing a nucleus and *only one* electron (e.g. ^1H, $^4_2\text{He}^+$), i.e. a *hydrogen-like* system.

A *hydrogen-like atom or ion* contains a nucleus and only one electron.

The Schrödinger wave equation may be represented in several forms and in Box 1.2 we examine its application to the motion of a particle in a one-dimensional box; equation 1.12 gives the form of the Schrödinger wave equation that is appropriate for motion in the *x*-direction:

$$\frac{d^2\psi}{dx^2} + \frac{8\pi^2 m}{h^2}(E - V)\psi = 0 \tag{1.12}$$

where m = mass, E = total energy and V = potential energy of the particle.

Of course, in reality, electrons move in three-dimensional space and an appropriate form of the Schrödinger wave equation is given in equation 1.13.

$$\frac{\partial^2\psi}{\partial x^2} + \frac{\partial^2\psi}{\partial y^2} + \frac{\partial^2\psi}{\partial z^2} + \frac{8\pi^2 m}{h^2}(E - V)\psi = 0 \tag{1.13}$$

Solving this equation will not concern us, although it is useful to note that it is advantageous to work in spherical polar coordinates (Figure 1.4). When we look at the results obtained from the Schrödinger wave equation, we talk in terms of the *radial and angular parts of the wavefunction*,

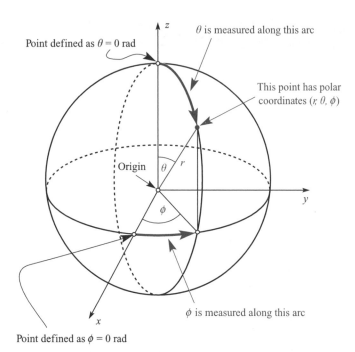

Fig. 1.4 Definition of the polar coordinates (r, θ, ϕ) for a point shown here in blue; r is the radial coordinate and θ and ϕ are angular coordinates. Cartesian axes (x, y and z) are also shown.

and this is represented in equation 1.14 where $R(r)$ and $A(\theta, \phi)$ are radial and angular wavefunctions respectively.[†]

$$\psi_{\text{Cartesian}}(x, y, z) \equiv \psi_{\text{radial}}(r)\psi_{\text{angular}}(\theta, \phi) = R(r)A(\theta, \phi)$$
(1.14)

Having solved the wave equation, what are the results?

- The wavefunction ψ is a solution of the Schrödinger equation and describes the behaviour of an electron in a region of space called the *atomic orbital.*
- We can find energy values that are associated with particular wavefunctions.
- The quantization of energy levels arises naturally from the Schrödinger equation (see *Box 1.2*).

A wavefunction ψ is a mathematical function that contains detailed information about the behaviour of an electron. An atomic wavefunction ψ consists of a radial component, $R(r)$, and an angular component, $A(\theta, \phi)$. The region of space defined by a wavefunction is called an atomic orbital.

1.6 Atomic orbitals

The quantum numbers n, l and m_l

An atomic orbital is usually described in terms of three integral *quantum numbers*. We have already encountered the

principal quantum number, n, in the Bohr model of the hydrogen atom. The principal quantum number is a positive integer with values lying between the limits $1 \leq n \leq \infty$; allowed values arise when the radial part of the wavefunction is solved.

Two more quantum numbers, l and m_l, appear when the angular part of the wavefunction is solved. The quantum number l is called the *orbital quantum number* and has allowed values of $0, 1, 2 \ldots (n - 1)$. The value of l determines the shape of the atomic orbital, and the *orbital angular momentum* of the electron. The value of the *magnetic quantum number*, m_l, gives information about the directionality of an atomic orbital and has integral values between $+l$ and $-l$.

Each atomic orbital may be uniquely labelled by a set of three quantum numbers: n, l and m_l.

Worked example 1.2   Quantum numbers: atomic orbitals

Given that the principal quantum number, n, is 2, write down the allowed values of l and m_l, and determine the number of atomic orbitals possible for $n = 3$.

For a given value of n, the allowed values of l are $0, 1, 2 \ldots (n - 1)$, and those of m_l are $-l \ldots 0 \ldots +l$.

For $n = 2$, allowed values of $l = 0$ or 1.

For $l = 0$, the allowed value of $m_l = 0$.

For $l = 1$, allowed values of $m_l = -1, 0, +1$

Each set of three quantum numbers defines a particular atomic orbital, and, therefore, for $n = 2$, there are four

[†] The radial component in equation 1.14 depends on the quantum numbers n and l, whereas the angular component depends on l and m_l, and the components should really be written as $R_{n,l}(r)$ and $A_{l,m_l}(\theta, \phi)$.

Particle in a box

The following discussion illustrates the so-called *particle in a one-dimensional box* and illustrates quantization arising from the Schrödinger wave equation.

The *Schrödinger wave equation* for the motion of a particle in one dimension is given by:

$$\frac{d^2\psi}{dx^2} + \frac{8\pi^2 m}{h^2}(E - V)\psi = 0$$

where m is the mass, E is the total energy and V is the potential energy of the particle. The derivation of this equation is considered in the set of exercises at the end of Box 1.2. For a given system for which V and m are known, we can use the Schrödinger equation to obtain values of E (the *allowed energies of the particle*) and ψ (the *wavefunction*). The wavefunction itself has no physical meaning, but ψ^2 is a probability (see main text) and for this to be the case, ψ must have certain properties:

- ψ must be finite for all values of x;
- ψ can only have one value for any value of x;
- ψ and $\frac{d\psi}{dx}$ must vary continuously as x varies.

Now, consider a particle that is undergoing simple-harmonic wave-like motion in one dimension, i.e. we can fix the direction of wave propagation to be along the x axis (the choice of x is arbitrary). Let the motion be further constrained such that the particle cannot go outside the fixed, vertical walls of a box of width a. There is no force acting on the particle *within* the box and so the potential energy, V, is zero; if we take $V = 0$, we are placing limits on x such that $0 \leq x \leq a$, i.e. the particle cannot move outside the box. The only restriction that we place on the total energy E is that it must be positive and cannot be infinite. There is one further restriction that we shall simply state: the *boundary condition* for the particle in the box is that ψ must be zero when $x = 0$ and $x = a$.

Now let us rewrite the Schrödinger equation for the specific case of the particle in the one-dimensional box where $V = 0$:

$$\frac{d^2\psi}{dx^2} = -\frac{8\pi^2 mE}{h^2}\psi$$

which may be written in the simpler form:

$$\frac{d^2\psi}{dx^2} = -k^2\psi \quad \text{where} \quad k^2 = \frac{8\pi^2 mE}{h^2}$$

The solution to this (a well-known general equation) is:

$$\psi = A\sin kx + B\cos kx$$

where A and B are integration constants. When $x = 0$, $\sin kx = 0$ and $\cos kx = 1$; hence, $\psi = B$ when $x = 0$. However, the boundary condition above stated that $\psi = 0$ when $x = 0$, and this is only true if $B = 0$. Also from the boundary condition, we see that $\psi = 0$ when $x = a$, and hence we can rewrite the above equation in the form:

$$\psi = A\sin ka = 0$$

Since the probability, ψ^2, that the particle will be at points between $x = 0$ and $x = a$ cannot be zero (i.e. the particle must be somewhere inside the box), A cannot be zero and the last equation is only valid if:

$$ka = n\pi$$

where $n = 1, 2, 3 \ldots$; n cannot be zero as this would make the probability, ψ^2, zero meaning that the particle would no longer be in the box.

Combining the last two equations gives:

$$\psi = A\sin\frac{n\pi x}{a}$$

and, from earlier:

$$E = \frac{k^2 h^2}{8\pi^2 m} = \frac{n^2 h^2}{8ma^2}$$

where $n = 1, 2, 3 \ldots$; n is the *quantum number* determining the energy of a particle of mass m confined within a one-dimensional box of width a. So, the limitations placed on the value of ψ have led to *quantized energy levels*, the spacing of which is determined by m and a.

The resultant motion of the particle is described by a series of standing sine waves, three of which are illustrated below. The wavefunction ψ_2 has a wavelength of a, while wavefunctions ψ_1 and ψ_3 possess wavelengths of $\frac{a}{2}$ and $\frac{2a}{3}$ respectively.

Each of the waves in the diagram has an amplitude of zero at the origin (i.e. at the point $a = 0$); points at which $\psi = 0$ are called *nodes*. For a given particle of mass m, the separations of the energy levels vary according to n^2, i.e. the spacings are not equal.

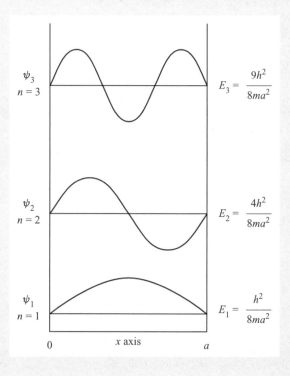

ψ_3
$n = 3$
$$E_3 = \frac{9h^2}{8ma^2}$$

ψ_2
$n = 2$
$$E_2 = \frac{4h^2}{8ma^2}$$

ψ_1
$n = 1$
$$E_1 = \frac{h^2}{8ma^2}$$

0   x axis a

Self-study exercises

Consider a particle that is undergoing simple-harmonic wave-like motion in one dimension, with the wave propagation along the x axis. The general equation for the wave is:

$$\psi = A \sin \frac{2\pi x}{\lambda}$$

where A is the amplitude of the wave.

1. If $\psi = A \sin \frac{2\pi x}{\lambda}$, find $\frac{d\psi}{dx}$ and hence show that

$$\frac{d^2\psi}{dx^2} = -\frac{4\pi^2}{\lambda^2}\psi.$$

2. If the particle in the box is of mass m and moves with velocity v, what is its kinetic energy, KE? Using the de Broglie equation (1.11), write an expression for KE in terms of m, h and λ.

3. The equation you derived in part (2) applies only to a particle moving in a space in which the potential energy, V, is constant, and the particle can be regarded as possessing only kinetic energy, KE. If the potential energy of the particle does vary, the total energy, $E = KE + V$. Using this information and your answers to parts (1) and (2), derive the Schrödinger equation (stated on p. 8) for a particle in a one-dimensional box.

atomic orbitals with the sets of quantum numbers:

$n = 2, \quad l = 0, \quad m_l = 0$

$n = 2, \quad l = 1, \quad m_l = -1$

$n = 2, \quad l = 1, \quad m_l = 0$

$n = 2, \quad l = 1, \quad m_l = +1$

Self-study exercises

1. If m_l has values of $-1, 0, +1$, write down the corresponding value of l.   [*Ans. l = 1*]

2. If l has values 0, 1, 2 and 3, deduce the corresponding value of n.   [*Ans. n = 4*]

3. For $n = 1$, what are the allowed values of l and m_l?   [*Ans. l = 0; m_l = 0*]

The distinction among the *types* of atomic orbital arises from their *shapes* and *symmetries*. The four types of atomic orbital most commonly encountered are the *s*, *p*, *d* and *f* orbitals, and the corresponding values of l are 0, 1, 2 and 3 respectively. Each atomic orbital is labelled with values of n and l, and hence we speak of 1*s*, 2*s*, 2*p*, 3*s*, 3*p*, 3*d*, 4*s*, 4*p*, 4*d*, 4*f* etc. orbitals.

| For an *s* orbital, $l = 0$. | For a *p* orbital, $l = 1$. |
| For a *d* orbital, $l = 2$. | For an *f* orbital, $l = 3$. |

Worked example 1.3   Quantum numbers: types of orbital

Using the rules that govern the values of the quantum numbers n and l, write down the possible types of atomic orbital for $n = 1$, 2 and 3.

The allowed values of l are integers between 0 and $(n-1)$.
For $n = 1$, $l = 0$.
The only atomic orbital for $n = 1$ is the 1*s* orbital.

For $n = 2$, $l = 0$ or 1.
The allowed atomic orbitals for $n = 2$ are the 2*s* and 2*p* orbitals.

For $n = 3$, $l = 0$, 1 or 2.
The allowed atomic orbitals for $n = 3$ are the 3*s*, 3*p* and 3*d* orbitals.

Self-study exercises

1. Write down the possible types of atomic orbital for $n = 4$.   [*Ans. 4s, 4p, 4d, 4f*]

2. Which atomic orbital has values of $n = 4$ and $l = 2$?   [*Ans. 4d*]

3. Give the three quantum numbers that describe a 2*s* atomic orbital.   [*Ans. n = 2, l = 0, m_l = 0*]

Degenerate orbitals possess the same energy.

Now consider the consequence on these orbital types of the quantum number m_l. For an *s* orbital, $l = 0$ and m_l can only equal 0. This means that for any value of n, there is only one *s* orbital; it is said to be singly degenerate. For a *p* orbital, $l = 1$, and there are three possible m_l values of: $+1$, 0, -1. This means that there are three *p* orbitals for a given value of n when $n \geq 2$; the set of *p* orbitals is said to be triply or three-fold degenerate. For a *d* orbital, $l = 2$, and there are five possible values of m_l: $+2$, $+1$, 0, -1, -2, meaning that for a given value of n ($n \geq 3$), there are five *d* orbitals; the set is said to be five-fold degenerate. As an exercise, you should show that there are seven *f* orbitals in a degenerate set for a given value of n ($n \geq 4$).

For a given value of n ($n \geq 1$) there is one *s* atomic orbital.
For a given value of n ($n \geq 2$) there are three *p* atomic orbitals.
For a given value of n ($n \geq 3$) there are five *d* atomic orbitals.
For a given value of n ($n \geq 4$) there are seven *f* atomic orbitals.

Table 1.2 Solutions of the Schrödinger equation for the hydrogen atom which define the $1s$, $2s$ and $2p$ atomic orbitals. For these forms of the solutions, the distance r from the nucleus is measured in atomic units.

Atomic orbital	n	l	m_l	Radial part of the wavefunction, $R(r)^{\ddagger}$	Angular part of wavefunction, $A(\theta, \phi)$
$1s$	1	0	0	$2e^{-r}$	$\dfrac{1}{2\sqrt{\pi}}$
$2s$	2	0	0	$\dfrac{1}{2\sqrt{2}}(2 - r)\,e^{-\left(\frac{r}{2}\right)}$	$\dfrac{1}{2\sqrt{\pi}}$
$2p_x$	2	1	+1	$\dfrac{1}{2\sqrt{6}}r\,e^{-\left(\frac{r}{2}\right)}$	$\dfrac{\sqrt{3}(\sin\theta\cos\phi)}{2\sqrt{\pi}}$
$2p_z$	2	1	0	$\dfrac{1}{2\sqrt{6}}r\,e^{-\left(\frac{r}{2}\right)}$	$\dfrac{\sqrt{3}(\cos\theta)}{2\sqrt{\pi}}$
$2p_y$	2	1	−1	$\dfrac{1}{2\sqrt{6}}r\,e^{-\left(\frac{r}{2}\right)}$	$\dfrac{\sqrt{3}(\sin\theta\sin\phi)}{2\sqrt{\pi}}$

$\ddagger$ For the $1s$ atomic orbital, the formula for $R(r)$ is actually: $2\left(\dfrac{Z}{a_0}\right)^{\frac{3}{2}}e^{-\left(\frac{Zr}{a_0}\right)}$

but for the hydrogen atom, $Z = 1$ and $a_0 = 1$ atomic unit. Other functions are similarly simplified.

The radial part of the wavefunction, $R(r)$

The mathematical forms of some of the wave functions for the H atom are listed in Table 1.2. Figure 1.5 shows plots of the radial parts of the wavefunction, $R(r)$ against distance, r, from the nucleus for the $1s$ and $2s$ atomic orbitals of the hydrogen atom, and Figure 1.6 shows plots of $R(r)$ against r for the $2p$, $3p$, $4p$ and $3d$ atomic orbitals; the nucleus is at $r = 0$.

From Table 1.2, we see that the radial parts of the wavefunctions decay exponentially as r increases, but the decay is slower for $n = 2$ than for $n = 1$. This means that the likelihood of the electron being further from the nucleus increases as n increases. This pattern continues for higher values of n. The exponential decay can be seen clearly

in Figure 1.5a. Several points should be noted from the plots of the radial parts of wavefunctions in Figures 1.5 and 1.6:

- s atomic orbitals have a finite value of $R(r)$ at the nucleus;
- for all orbitals other than s, $R(r) = 0$ at the nucleus;
- for the $1s$ orbital, $R(r)$ is always positive; for the first orbital of other types (i.e. $2p$, $3d$, $4f$), $R(r)$ is positive everywhere except at the origin;
- for the second orbital of a given type (i.e. $2s$, $3p$, $4d$, $5f$), $R(r)$ may be positive or negative but the wavefunction has only one sign change; the point at which $R(r) = 0$ (not including the origin) is called a radial node;
- for the third orbital of a given type (i.e. $4p$, $5d$, $6f$), $R(r)$ has two sign changes, i.e. it possesses two radial nodes.

(a)

Distance from nucleus/atomic units

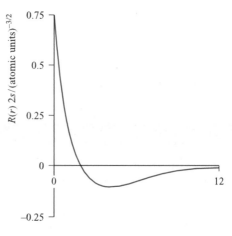

(b)

Distance from nucleus/atomic units

Fig. 1.5 Plots of the radial parts of the wavefunction, $R(r)$ against distance, r, from the nucleus for (a) the $1s$ and (b) the $2s$ atomic orbitals of the hydrogen atom; the nucleus is at $r = 0$. The vertical scales for the two plots are different but the horizontal scales are the same.

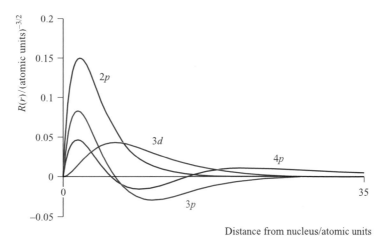

Fig. 1.6 Plots of radial parts of the wavefunction $R(r)$ against r for the 2p, 3p, 4p and 3d atomic orbitals; the nucleus is at $r = 0$.

ns orbitals have $(n - 1)$ radial nodes.
np orbitals have $(n - 2)$ radial nodes.
nd orbitals have $(n - 3)$ radial nodes.
nf orbitals have $(n - 4)$ radial nodes.

The radial distribution function, $4\pi r^2 R(r)^2$

Let us now consider how we might represent atomic orbitals in three-dimensional space. We said earlier that a useful description of an electron in an atom is the *probability of finding the electron* in a given volume of space. The function ψ^2 (see *Box 1.3*) is proportional to the *probability density* of the electron at a point in space. By considering values of ψ^2 at points around the nucleus, we can define a *surface boundary* which encloses the volume of space in which the electron will spend, say, 95% of its time. This effectively gives us a physical representation of the atomic orbital, since ψ^2 may be described in terms of the radial and angular components $R(r)^2$ and $A(\theta, \phi)^2$.

First consider the radial components. A useful way of depicting the probability density is to plot a *radial distribution function* (equation 1.15) and this allows us to envisage the region in space in which the electron is found.

$$\text{Radial distribution function} = 4\pi r^2 R(r)^2 \qquad (1.15)$$

The radial distribution functions for the 1s, 2s and 3s atomic orbitals of hydrogen are shown in Figure 1.7, and Figure 1.8 shows those of the 3s, 3p and 3d orbitals. *Each* function is zero at the nucleus, following from the r^2 term and the fact that at the nucleus $r = 0$. Since the function depends on $R(r)^2$, it is always positive in contrast to $R(r)$, plots for which are shown in Figures 1.5 and 1.6. Each plot of $4\pi r^2 R(r)^2$ shows at least one maximum value for the function, corresponding to a distance from the nucleus at which the electron has the highest probability of being found. Points at which $4\pi r^2 R(r)^2 = 0$ (ignoring $r = 0$) correspond to radial nodes where $R(r) = 0$.

The angular part of the wavefunction, $A(\theta, \phi)$

Now let us consider the angular parts of the wavefunctions, $A(\theta, \phi)$, for different types of atomic orbitals. These are *independent* of the principal quantum number as Table 1.2 illustrates for $n = 1$ and 2. Moreover, for s orbitals, $A(\theta, \phi)$ is independent of the angles θ and ϕ and is of a constant value. Thus, an s orbital is spherically symmetric about the nucleus. We noted above that a set of p orbitals is triply degenerate; by convention they are given the labels p_x, p_y and p_z. From Table 1.2, we see that the angular part of the p_z wavefunction is independent of ϕ; the orbital can be

---

BOX 1.3   **CHEMICAL AND THEORETICAL BACKGROUND**

Notation for ψ^2 and its normalization

Although we use ψ^2 in the text, it should strictly be written as $\psi\psi^*$ where ψ^* is the complex conjugate of ψ. In the x-direction, the probability of finding the electron between the limits x and $(x + dx)$ is proportional to $\psi(x)\psi^*(x)\,dx$. In three-dimensional space this is expressed as $\psi\psi^*\,d\tau$ in which we are considering the probability of finding the electron in a volume element $d\tau$. For just the radial part of the wavefunction, the function is $R(r)R^*(r)$.

In all of our mathematical manipulations, we must ensure that the result shows that the electron is *somewhere*

(i.e. it has not vanished!) and this is done by *normalizing* the wavefunction to unity. This means that the probability of finding the electron somewhere in space is taken to be 1. Mathematically, the normalization is represented as follows:

$$\int \psi^2\,d\tau = 1 \qquad \text{or more correctly} \qquad \int \psi\psi^*\,d\tau = 1$$

and this effectively states that the integral ($\int$) is over all space ($d\tau$) and that the total integral of ψ^2 (or $\psi\psi^*$) must be unity.

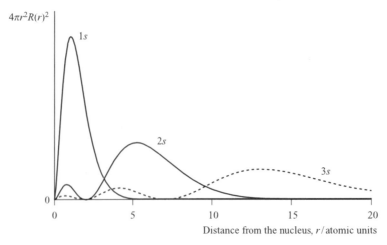

Fig. 1.7 Radial distribution functions, $4\pi r^2 R(r)^2$, for the $1s$, $2s$ and $3s$ atomic orbitals of the hydrogen atom.

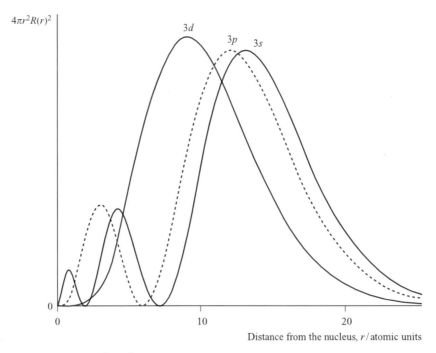

Fig. 1.8 Radial distribution functions, $4\pi r^2 R(r)^2$, for the $3s$, $3p$ and $3d$ atomic orbitals of the hydrogen atom.

represented as two spheres (touching at the origin)[†], the centres of which lie on the z axis. For the p_x and p_y orbitals, $A(\theta, \phi)$ depends on both the angles θ and ϕ; these orbitals are similar to p_z but are oriented along the x and y axes.

Although we must not lose sight of the fact that wavefunctions are mathematical in origin, most chemists find such functions hard to visualize and prefer pictorial representations of orbitals. The boundary surfaces of the s and three p atomic orbitals are shown in Figure 1.9. The different colours of the *lobes* are significant. The boundary

surface of an s orbital has a constant *phase*, i.e. the amplitude of the wavefunction associated with the boundary surface of the s orbital has a constant sign. For a p orbital, there is *one* phase change with respect to the boundary surface and this occurs at a *nodal plane* as is shown for the p_z orbital in Figure 1.9. The amplitude of a wavefunction may be positive or negative; this is shown using + and − signs, or by shading the lobes as in Figure 1.9.

Just as the function $4\pi r^2 R(r)^2$ represents the probability of finding an electron at a distance r from the nucleus, we use a function dependent upon $A(\theta, \phi)^2$ to represent the probability in terms of θ and ϕ. For an s orbital, squaring $A(\theta, \phi)$ causes no change in the spherical symmetry, and the surface boundary for the s atomic orbital shown in Figure 1.10 looks similar to that in Figure 1.9. For the p orbitals

---

[†] In order to emphasize that ϕ is a continuous function we have extended boundary surfaces in representations of orbitals to the nucleus, but for p orbitals, this is strictly not true if we are considering $\approx 95\%$ of the electronic charge.

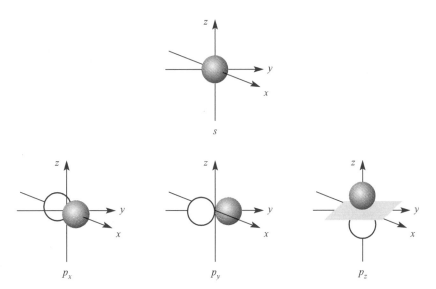

Fig. 1.9 Boundary surfaces for the angular parts of the $1s$ and $2p$ atomic orbitals of the hydrogen atom. The nodal plane shown in grey for the $2p_z$ atomic orbital lies in the xy plane.

however, going from $A(\theta, \phi)$ to $A(\theta, \phi)^2$ has the effect of elongating the lobes as illustrated in Figure 1.10. Squaring $A(\theta, \phi)$ necessarily means that the signs (+ or −) disappear, but in practice chemists often indicate the amplitude by a sign or by shading (as in Figure 1.10) because of the importance of the signs of the wavefunctions with respect to their overlap during bond formation (see *Section 1.13*).

Finally, Figure 1.11 shows the boundary surfaces for five hydrogen-like d orbitals. We shall not consider the mathematical forms of these wavefunctions, but merely represent the orbitals in the conventional manner. Each d orbital possesses *two* nodal planes and as an exercise you should recognize where these planes lie for each orbital. We consider d orbitals in more detail in Chapters 19 and 20, and f orbitals in Chapter 24.

Orbital energies in a hydrogen-like species

Besides providing information about the wavefunctions, solutions of the Schrödinger equation give orbital energies, E (energy levels), and equation 1.16 shows the dependence of E on the principal quantum number for *hydrogen-like species*.

$$E = -\frac{k}{n^2} \qquad k = \text{a constant} = 1.312 \times 10^3 \,\text{kJ mol}^{-1}$$

$$(1.16)$$

For each value of n there is only one energy solution and for *hydrogen-like species*, all atomic orbitals with the same principal quantum number (e.g. $3s$, $3p$ and $3d$) are degenerate.

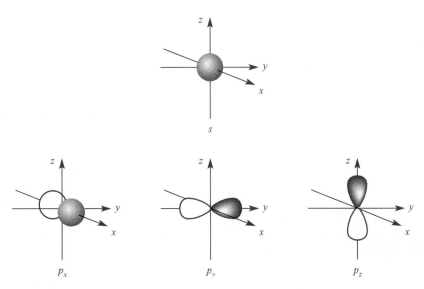

Fig. 1.10 Representations of an s and a set of three degenerate p atomic orbitals. The lobes of the p_x orbital are elongated like those of the p_y and p_z but are directed along the axis that passes through the plane of the paper.

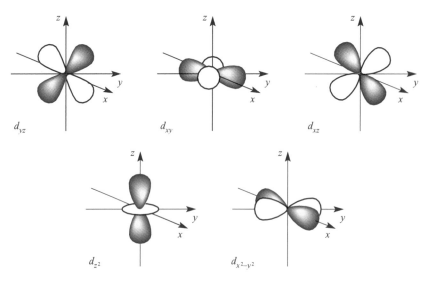

Fig. 1.11 Representations of a set of five degenerate d atomic orbitals.

Size of orbitals

For a given atom, a series of orbitals with different values of n but the same values of l and m_l (e.g. 1s, 2s, 3s, 4s, ...) differ in their relative size (spatial extent). The larger the value of n, the larger the orbital, although this relationship is not linear. An increase in size also corresponds to an orbital being more *diffuse*.

The spin quantum number and the magnetic spin quantum number

Before we place electrons into atomic orbitals, we must define two more quantum numbers. In a classical model, an electron is considered to spin about an axis passing through it and to have *spin angular momentum* in addition to orbital angular momentum (see *Box 1.4*). The *spin quantum number*, s, determines the magnitude of the spin angular momentum of an electron and has a value of $\frac{1}{2}$. Since angular momentum is a vector quantity, it must have direction, and this is determined by the *magnetic spin quantum number*, m_s, which has a value of $+\frac{1}{2}$ or $-\frac{1}{2}$.

Whereas an atomic orbital is defined by a unique set of *three* quantum numbers, an electron in an atomic orbital is defined by a unique set of *four* quantum numbers: n, l, m_l and m_s. As there are only two values of m_s, an orbital can only accommodate two electrons.

> An orbital is fully occupied when it contains two electrons which are *spin-paired*; one electron has a value of $m_s = +\frac{1}{2}$ and the other, $m_s = -\frac{1}{2}$.

Worked example 1.4   Quantum numbers: an electron in an atomic orbital

Write down two possible sets of quantum numbers that describe an electron in a 2s atomic orbital. What is the physical significance of these unique sets?

The 2s atomic orbital is defined by the set of quantum numbers $n = 2$, $l = 0$, $m_l = 0$.

An electron in a 2s atomic orbital may have one of two sets of four quantum numbers:

$$n = 2, \quad l = 0, \quad m_l = 0, \quad m_s = +\tfrac{1}{2}$$

or

$$n = 2, \quad l = 0, \quad m_l = 0, \quad m_s = -\tfrac{1}{2}$$

If the orbital were fully occupied with two electrons, one electron would have $m_s = +\frac{1}{2}$, and the other electron would have $m_s = -\frac{1}{2}$, i.e. the two electrons would be spin-paired.

Self-study exercises

1. Write down two possible sets of quantum numbers to describe an electron in a 3s atomic orbital.
   [*Ans.* $n = 3$, $l = 0$, $m_l = 0$, $m_s = +\frac{1}{2}$; $n = 3$, $l = 0$, $m_l = 0$, $m_s = -\frac{1}{2}$]

2. If an electron has the quantum numbers $n = 2$, $l = 1$, $m_l = -1$ and $m_s = +\frac{1}{2}$ which type of atomic orbital is it occupying?   [*Ans.* 2p]

The ground state of the hydrogen atom

So far we have focused on the atomic orbitals of hydrogen and have talked about the probability of finding an electron in different atomic orbitals. The most energetically favourable (stable) state of the H atom is its *ground state* in which the single electron occupies the 1s (lowest energy) atomic orbital. The electron can be promoted to higher energy orbitals (see *Section 1.4*) to give *excited states*.

> The notation for the ground state electronic configuration of an H atom is $1s^1$ signifying that one electron occupies the 1s atomic orbital.

BOX 1.4

Angular momentum, the inner quantum number, *j*, and spin–orbit coupling

The value of *l* determines not only the shape of an orbital but also the amount of orbital angular momentum associated with an electron in it:

$$\text{Orbital angular momentum} = \left[\sqrt{l(l+1)}\right]\frac{h}{2\pi}$$

The axis through the nucleus about which the electron (considered classically) can be thought to rotate defines the direction of the orbital angular momentum. The latter gives rise to a magnetic moment whose direction is in the same sense as the angular vector and whose magnitude is proportional to the magnitude of the vector. An electron in an *s* orbital (*l* = 0) has no orbital angular momentum, an electron in a *p* orbital (*l* = 1) has angular momentum $\sqrt{2}(h/2\pi)$, and so on. The orbital angular momentum vector has $(2l+1)$ possible directions in space corresponding to the $(2l+1)$ possible values of m_l for a given value of *l*.

We are particularly interested in the component of the angular momentum vector along the *z* axis; this has a different value for each of the possible orientations that this vector can take up. The actual magnitude of the *z* component is given by $m_l(h/2\pi)$. Thus, for an electron in a *d* orbital (*l* = 2), the orbital angular momentum is $\sqrt{6}(h/2\pi)$, and the *z* component of this may have values of $+2(h/2\pi)$, $+(h/2\pi)$, 0, $-(h/2\pi)$ or $-2(h/2\pi)$. The orbitals in a sub-shell of given *n* and *l*, are, as we have seen, degenerate. If, however, the atom is placed in a magnetic field, this degeneracy is removed. And, if we arbitrarily define the direction of the magnetic field as the *z* axis, electrons in the various *d* orbitals will interact to different extents with the magnetic field as a consequence of their different values of the *z* components of their angular momentum vectors (and, hence, orbital magnetic moment vectors).

An electron also has spin angular momentum which can be regarded as originating in the rotation of the electron about its own axis. The magnitude of this is given by:

$$\text{Spin angular momentum} = \left[\sqrt{s(s+1)}\right]\frac{h}{2\pi}$$

where *s* = spin quantum number (see text). The axis defines the direction of the spin angular momentum vector, but

again it is the orientation of this vector with respect to the *z* direction in which we are particularly interested. The *z* component is given by $m_s(h/2\pi)$; since m_s can only equal $+\frac{1}{2}$ or $-\frac{1}{2}$, there are only two possible orientations of the spin angular momentum vector, and these give rise to *z* components of magnitude $+\frac{1}{2}(h/2\pi)$ and $-\frac{1}{2}(h/2\pi)$.

For an electron having both orbital and spin angular momentum, the total angular momentum vector is given by:

$$\text{Total angular momentum} = \left[\sqrt{j(j+1)}\right]\frac{h}{2\pi}$$

where *j* is the so-called inner quantum number; *j* may have values of $(l+s)$ or $(l-s)$, i.e. $l+\frac{1}{2}$ or $l-\frac{1}{2}$. (When *l* = 0 and the electron has no orbital angular momentum, the total angular momentum is $\left[\sqrt{s(s+1)}\right]\frac{h}{2\pi}$ because *j* = *s*.)

The *z* component of the total angular momentum vector is now $j(h/2\pi)$ and there are $(2j+1)$ possible orientations in space.

For an electron in an *ns* orbital (*l* = 0), *j* can only be $\frac{1}{2}$. When the electron is promoted to an *np* orbital, *j* may be $\frac{3}{2}$ or $\frac{1}{2}$, and the energies corresponding to the different *j* values are not quite equal. In the emission spectrum of sodium, for example, transitions from the $3p_{3/2}$ and $3p_{1/2}$ levels to the $3s_{1/2}$ level therefore correspond to slightly different amounts of energy, and this *spin–orbit coupling* is the origin of the doublet structure of the strong yellow line in the spectrum of atomic sodium. The fine structure of many other spectral lines arises in analogous ways, though the number actually observed depends on the difference in energy between states differing only in *j* value and on the resolving power of the spectrometer. The difference in energy between levels for which $\Delta j = 1$ (the spin–orbit coupling constant, λ) increases with the atomic number of the element involved; e.g. that between the $np_{3/2}$ and $np_{1/2}$ levels for Li, Na and Cs is 0.23, 11.4 and 370 cm^{-1} respectively.

For further information: see Box 20.5.

1.7 Many-electron atoms

The helium atom: two electrons

The preceding sections have been devoted to hydrogen-like species containing one electron, the energy of which depends only on *n* (equation 1.16); the atomic spectra of such species contain only a few lines associated with changes in the value of *n* (Figure 1.3). It is *only* for such species that the Schrödinger equation has been solved exactly.

The next simplest atom is He (*Z* = 2), and for its two electrons, three electrostatic interactions must be considered:

- attraction between electron (1) and the nucleus;
- attraction between electron (2) and the nucleus;
- repulsion between electrons (1) and (2).

The *net* interaction will determine the energy of the system.

In the ground state of the He atom, two electrons with $m_s = +\frac{1}{2}$ and $-\frac{1}{2}$ occupy the 1*s* atomic orbital, i.e. the electronic configuration is $1s^2$. For all atoms except hydrogen-like species, orbitals of the same principal quantum number but differing *l* are *not* degenerate. If one of the $1s^2$ electrons is promoted to an orbital with *n* = 2, the energy of the system depends upon whether the electron goes into a 2*s* or 2*p* atomic orbital, because each situation gives rise to different electrostatic interactions. However, there is no energy distinction among the three different $2p$ atomic orbitals. If promotion is to an orbital with $n = 3$, different amounts of energy are needed depending upon whether $3s$, $3p$ or $3d$ orbitals are involved, although there is no energy difference among the three $3p$ atomic orbitals, or among the five $3d$ atomic orbitals. The emission spectrum of He arises as the electrons fall back to lower energy states or to the ground state and it follows that the spectrum contains many more lines than that of atomic H.

In terms of obtaining wavefunctions and energies for the atomic orbitals of He, it has not been possible to solve the Schrödinger equation exactly and only approximate solutions are available. For atoms containing more than two electrons, it is even more difficult to obtain accurate solutions to the wave equation.

> In a *multi-electron atom*, orbitals with the same value of n but different values of l are *not* degenerate.

Ground state electronic configurations: experimental data

Now consider the ground state electronic configurations of isolated atoms of all the elements (Table 1.3). These are experimental data, and are nearly always obtained by analyzing atomic spectra. Most atomic spectra are too complex for discussion here and we take their interpretation on trust.

We have already seen that the ground state electronic configurations of H and He are $1s^1$ and $1s^2$ respectively. The $1s$ atomic orbital is fully occupied in He and its configuration is often written as [He]. In the next two elements, Li and Be, the electrons go into the $2s$ orbital, and then from B to Ne, the $2p$ orbitals are occupied to give the electronic configurations [He]$2s^2 2p^m$ ($m = 1–6$). When $m = 6$, the energy level (or *shell*) with $n = 2$ is fully occupied, and the configuration for Ne can be written as [Ne]. The filling of the $3s$ and $3p$ atomic orbitals takes place in an analogous sequence from Na to Ar, the last element in the series having the electronic configuration [Ne]$3s^2 2p^6$ or [Ar].

With K and Ca, successive electrons go into the $4s$ orbital, and Ca has the electronic configuration [Ar]$4s^2$. At this point, the pattern changes. To a first approximation, the 10 electrons for the next 10 elements (Sc to Zn) enter the $3d$ orbitals, giving Zn the electronic configuration $4s^2 3d^{10}$. There are some irregularities (see *Table 1.3*) to which we return later. From Ga to Kr, the $4p$ orbitals are filled, and the electronic configuration for Kr is [Ar]$4s^2 3d^{10} 4p^6$ or [Kr].

From Rb to Xe, the general sequence of filling orbitals is the same as that from K to Kr although there are once again irregularities in the distribution of electrons between s and d atomic orbitals (see *Table 1.3*).

From Cs to Rn, electrons enter f orbitals for the first time; Cs, Ba and La have configurations analogous to those of Rb, Sr and Y, but after that the configurations change as we begin the sequence of the *lanthanoid*

elements (see *Chapter 24*).[†] Cerium has the configuration [Xe]$4f^1 6s^2 5d^1$ and the filling of the seven $4f$ orbitals follows until an electronic configuration of [Xe]$4f^{14} 6s^2 5d^1$ is reached for Lu. Table 1.3 shows that the $5d$ orbital is not usually occupied for a lanthanoid element. After Lu, successive electrons occupy the remaining $5d$ orbitals (Hf to Hg) and then the $6p$ orbitals to Rn which has the configuration [Xe]$4f^{14} 6s^2 5d^{10} 6p^6$ or [Rn]. Table 1.3 shows some irregularities along the series of d-block elements.

For the remaining elements in Table 1.3 beginning at francium (Fr), filling of the orbitals follows a similar sequence as that from Cs but the sequence is incomplete and some of the heaviest elements are too unstable for detailed investigations to be possible. The metals from Th to Lr are the *actinoid* elements, and in discussing their chemistry, Ac is generally considered with the actinoids (see *Chapter 24*).

A detailed inspection of Table 1.3 makes it obvious that there is no one sequence that represents accurately the occupation of different sets of orbitals with increasing atomic number. The following sequence is *approximately* true for the relative energies (lowest energy first) of orbitals in *neutral atoms*:

$$1s < 2s < 2p < 3s < 3p < 4s < 3d < 4p < 5s < 4d < 5p$$
$$< 6s < 5d \approx 4f < 6p < 7s < 6d \approx 5f$$

The energies of different orbitals are close together for high values of n and their relative energies can change significantly on forming an ion (see *Section 19.2*).

Penetration and shielding

Although it is not possible to calculate the dependence of the energies of orbitals on atomic number with the degree of accuracy that is required to obtain agreement with all the electronic configurations listed in Table 1.3, some useful information can be gained by considering the different *screening effects* that electrons in different atomic orbitals have on one another. Figure 1.12 shows the radial distribution functions for the $1s$, $2s$ and $2p$ atomic orbitals of the H atom. (It is a common approximation to assume hydrogen-like wavefunctions for multi-electron atoms.) Although values of $4\pi r^2 R(r)^2$ for the $1s$ orbital are much greater than those of the $2s$ and $2p$ orbitals at distances relatively close to the nucleus, the values for the $2s$ and $2p$ orbitals are still significant. We say that the $2s$ and $2p$ atomic orbitals *penetrate* the $1s$ atomic orbital; calculations show that the $2s$ atomic orbital is more penetrating than the $2p$ orbital.

Now let us consider the arrangement of the electrons in Li ($Z = 3$). In the ground state, the $1s$ atomic orbital is fully occupied and the third electron could occupy either a $2s$ or $2p$ orbital. Which arrangement will possess the lower

---

[†] The IUPAC recommends the names lanthanoid and actinoid in preference to lanthanide and actinide; the ending '-ide' usually implies a negatively charged ion.

Table 1.3 Ground state electronic configurations of the elements up to $Z = 103$.

Atomic number	Element	Ground state electronic configuration	Atomic number	Element	Ground state electronic configuration
1	H	$1s^1$	53	I	$[\text{Kr}]5s^24d^{10}5p^5$
2	He	$1s^2 = [\text{He}]$	54	Xe	$[\text{Kr}]5s^24d^{10}5p^6 = [\text{Xe}]$
3	Li	$[\text{He}]2s^1$	55	Cs	$[\text{Xe}]6s^1$
4	Be	$[\text{He}]2s^2$	56	Ba	$[\text{Xe}]6s^2$
5	B	$[\text{He}]2s^22p^1$	57	La	$[\text{Xe}]6s^25d^1$
6	C	$[\text{He}]2s^22p^2$	58	Ce	$[\text{Xe}]4f^16s^25d^1$
7	N	$[\text{He}]2s^22p^3$	59	Pr	$[\text{Xe}]4f^36s^2$
8	O	$[\text{He}]2s^22p^4$	60	Nd	$[\text{Xe}]4f^46s^2$
9	F	$[\text{He}]2s^22p^5$	61	Pm	$[\text{Xe}]4f^56s^2$
10	Ne	$[\text{He}]2s^22p^6 = [\text{Ne}]$	62	Sm	$[\text{Xe}]4f^66s^2$
11	Na	$[\text{Ne}]3s^1$	63	Eu	$[\text{Xe}]4f^76s^2$
12	Mg	$[\text{Ne}]3s^2$	64	Gd	$[\text{Xe}]4f^76s^25d^1$
13	Al	$[\text{Ne}]3s^23p^1$	65	Tb	$[\text{Xe}]4f^96s^2$
14	Si	$[\text{Ne}]3s^23p^2$	66	Dy	$[\text{Xe}]4f^{10}6s^2$
15	P	$[\text{Ne}]3s^23p^3$	67	Ho	$[\text{Xe}]4f^{11}6s^2$
16	S	$[\text{Ne}]3s^23p^4$	68	Er	$[\text{Xe}]4f^{12}6s^2$
17	Cl	$[\text{Ne}]3s^23p^5$	69	Tm	$[\text{Xe}]4f^{13}6s^2$
18	Ar	$[\text{Ne}]3s^23p^6 = [\text{Ar}]$	70	Yb	$[\text{Xe}]4f^{14}6s^2$
19	K	$[\text{Ar}]4s^1$	71	Lu	$[\text{Xe}]4f^{14}6s^25d^1$
20	Ca	$[\text{Ar}]4s^2$	72	Hf	$[\text{Xe}]4f^{14}6s^25d^2$
21	Sc	$[\text{Ar}]4s^23d^1$	73	Ta	$[\text{Xe}]4f^{14}6s^25d^3$
22	Ti	$[\text{Ar}]4s^23d^2$	74	W	$[\text{Xe}]4f^{14}6s^25d^4$
23	V	$[\text{Ar}]4s^23d^3$	75	Re	$[\text{Xe}]4f^{14}6s^25d^5$
24	Cr	$[\text{Ar}]4s^13d^5$	76	Os	$[\text{Xe}]4f^{14}6s^25d^6$
25	Mn	$[\text{Ar}]4s^23d^5$	77	Ir	$[\text{Xe}]4f^{14}6s^25d^7$
26	Fe	$[\text{Ar}]4s^23d^6$	78	Pt	$[\text{Xe}]4f^{14}6s^15d^9$
27	Co	$[\text{Ar}]4s^23d^7$	79	Au	$[\text{Xe}]4f^{14}6s^15d^{10}$
28	Ni	$[\text{Ar}]4s^23d^8$	80	Hg	$[\text{Xe}]4f^{14}6s^25d^{10}$
29	Cu	$[\text{Ar}]4s^13d^{10}$	81	Tl	$[\text{Xe}]4f^{14}6s^25d^{10}6p^1$
30	Zn	$[\text{Ar}]4s^23d^{10}$	82	Pb	$[\text{Xe}]4f^{14}6s^25d^{10}6p^2$
31	Ga	$[\text{Ar}]4s^23d^{10}4p^1$	83	Bi	$[\text{Xe}]4f^{14}6s^25d^{10}6p^3$
32	Ge	$[\text{Ar}]4s^23d^{10}4p^2$	84	Po	$[\text{Xe}]4f^{14}6s^25d^{10}6p^4$
33	As	$[\text{Ar}]4s^23d^{10}4p^3$	85	At	$[\text{Xe}]4f^{14}6s^25d^{10}6p^5$
34	Se	$[\text{Ar}]4s^23d^{10}4p^4$	86	Rn	$[\text{Xe}]4f^{14}6s^25d^{10}6p^6 = [\text{Rn}]$
35	Br	$[\text{Ar}]4s^23d^{10}4p^5$	87	Fr	$[\text{Rn}]7s^1$
36	Kr	$[\text{Ar}]4s^23d^{10}4p^6 = [\text{Kr}]$	88	Ra	$[\text{Rn}]7s^2$
37	Rb	$[\text{Kr}]5s^1$	89	Ac	$[\text{Rn}]6d^17s^2$
38	Sr	$[\text{Kr}]5s^2$	90	Th	$[\text{Rn}]6d^27s^2$
39	Y	$[\text{Kr}]5s^24d^1$	91	Pa	$[\text{Rn}]5f^27s^26d^1$
40	Zr	$[\text{Kr}]5s^24d^2$	92	U	$[\text{Rn}]5f^37s^26d^1$
41	Nb	$[\text{Kr}]5s^14d^4$	93	Np	$[\text{Rn}]5f^47s^26d^1$
42	Mo	$[\text{Kr}]5s^14d^5$	94	Pu	$[\text{Rn}]5f^67s^2$
43	Tc	$[\text{Kr}]5s^24d^5$	95	Am	$[\text{Rn}]5f^77s^2$
44	Ru	$[\text{Kr}]5s^14d^7$	96	Cm	$[\text{Rn}]5f^77s^26d^1$
45	Rh	$[\text{Kr}]5s^14d^8$	97	Bk	$[\text{Rn}]5f^97s^2$
46	Pd	$[\text{Kr}]5s^04d^{10}$	98	Cf	$[\text{Rn}]5f^{10}7s^2$
47	Ag	$[\text{Kr}]5s^14d^{10}$	99	Es	$[\text{Rn}]5f^{11}7s^2$
48	Cd	$[\text{Kr}]5s^24d^{10}$	100	Fm	$[\text{Rn}]5f^{12}7s^2$
49	In	$[\text{Kr}]5s^24d^{10}5p^1$	101	Md	$[\text{Rn}]5f^{13}7s^2$
50	Sn	$[\text{Kr}]5s^24d^{10}5p^2$	102	No	$[\text{Rn}]5f^{14}7s^2$
51	Sb	$[\text{Kr}]5s^24d^{10}5p^3$	103	Lr	$[\text{Rn}]5f^{14}7s^26d^1$
52	Te	$[\text{Kr}]5s^24d^{10}5p^4$			

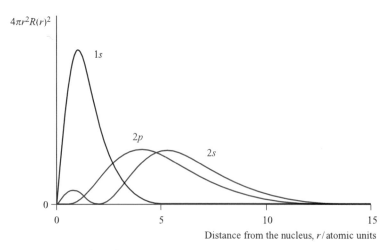

Fig. 1.12 Radial distribution functions, $4\pi r^2 R(r)^2$, for the $1s$, $2s$ and $2p$ atomic orbitals of the hydrogen atom.

energy? An electron in a $2s$ or $2p$ atomic orbital experiences the *effective charge*, Z_{eff}, of a nucleus partly *shielded* by the $1s$ electrons. Since the $2p$ orbital penetrates the $1s$ orbital less than a $2s$ orbital does, a $2p$ electron is shielded more than a $2s$ electron. Thus, occupation of the $2s$ rather than the $2p$ atomic orbital gives a lower energy system. Although we should consider the energies of the *electrons* in atomic orbitals, it is common practice to think in terms of the orbital energies themselves: $E(2s) < E(2p)$. Similar arguments lead to the sequence $E(3s) < E(3p) < E(3d)$ and $E(4s) < E(4p) < E(4d) < E(4f)$. As we move to atoms of elements of higher atomic number, the energy differences between orbitals with the same value of n become smaller, the validity of assuming hydrogen-like wavefunctions becomes more doubtful, and predictions of ground states become less reliable. The treatment above also ignores electron–electron interactions within the same principal quantum shell.

A set of empirical rules for estimating the effective nuclear charges experienced by electrons in different atomic orbitals is described in Box 1.5.

1.8 The periodic table

In 1869 and 1870 respectively, Dmitri Mendeléev and Lothar Meyer stated that the *properties of the elements can be represented as periodic functions of their atomic weights*, and set out their ideas in the form of a *periodic table*. As new elements have been discovered, the original form of the periodic table has been extensively modified, and it is now recognized that *periodicity* is a consequence of the variation in ground state electronic configurations. A modern periodic table (Figure 1.13) emphasizes the blocks of 2, 6, 10 and 14 elements which result from the filling of the s, p, d and f atomic orbitals respectively. An exception is He that, for reasons of its chemistry, is placed in a *group* with Ne, Ar, Kr, Xe and Rn. A more

detailed periodic table is given on the inside front-cover of the book.

The IUPAC (International Union of Pure and Applied Chemistry) has produced guidelines[†] for naming blocks and groups of elements in the periodic table. In summary,

- blocks of elements may be designated by use of the letters s, p, d or f (Figure 1.13);
- elements (except H) in groups 1, 2 and 13–18 are designated *main group elements*;
- with the exception of group 18, the first two elements of each main group are called *typical elements*;
- elements in groups 3–11 (i.e. those with partially filled d orbitals) are designated *transition elements*;

Note the distinction between a transition and d-block element. Elements in groups 3–12 inclusive are collectively called d-block elements, but by the IUPAC rulings, a transition metal is an element, an atom of which possesses an incomplete d-shell or which gives rise to a cation with an incomplete d-shell. Thus elements in group 12 are not classed as transition elements. Collective names for some of the groups of elements in the periodic table are given in Table 1.4.

1.9 The *aufbau* principle

Ground state electronic configurations

In the previous two sections, we have considered experimental electronic configurations and have seen that the organization of the elements in the periodic table depends on the number, and arrangement, of electrons that each element

---

[†] IUPAC: *Nomenclature of Inorganic Chemistry* (*Recommendations 1990*), ed. G.J. Leigh, Blackwell Scientific Publications, Oxford.

BOX 1.5 **CHEMICAL AND THEORETICAL BACKGROUND**

Effective nuclear charge and Slater's rules

Slater's rules

Effective nuclear charges, Z_{eff}, experienced by electrons in different atomic orbitals may be estimated using *Slater's rules*. These rules are based on experimental data for electron promotion and ionization energies, and Z_{eff} is determined from the equation:

$$Z_{eff} = Z - S$$

where Z = nuclear charge, Z_{eff} = effective nuclear charge, S = screening (or shielding) constant.

Values of S may be estimated as follows:

1. Write out the electronic configuration of the element in the following order and groupings: $(1s)$, $(2s, 2p)$, $(3s, 3p)$, $(3d)$, $(4s, 4p)$, $(4d)$, $(4f)$, $(5s, 5p)$ etc.
2. Electrons in any group higher in this sequence than the electron under consideration contribute nothing to S.
3. Consider a particular electron in an ns or np orbital:
   (i) Each of the other electrons in the (ns, np) group contributes $S = 0.35$.
   (ii) Each of the electrons in the $(n-1)$ shell contributes $S = 0.85$.
   (iii) Each of the electrons in the $(n-2)$ or lower shells contributes $S = 1.00$.
4. Consider a particular electron in an nd or nf orbital:
   (i) Each of the other electrons in the (nd, nf) group contributes $S = 0.35$.
   (ii) Each of the electrons in a lower group than the one being considered contributes $S = 1.00$.

An example of how to apply Slater's rules

Question: Confirm that the experimentally observed electronic configuration of K, $1s^2 2s^2 2p^6 3s^2 3p^6 4s^1$, is energetically more stable than the configuration $1s^2 2s^2 2p^6 3s^2 3p^6 3d^1$.

For K, $Z = 19$.

Applying Slater's rules, the effective nuclear charge experienced by the $4s$ electron for the configuration $1s^2 2s^2 2p^6 3s^2 3p^6 4s^1$ is:

$$Z_{eff} = Z - S$$
$$= 19 - [(8 \times 0.85) + (10 \times 1.00)]$$
$$= 2.20$$

The effective nuclear charge experienced by the $3d$ electron for the configuration $1s^2 2s^2 2p^6 3s^2 3p^6 3d^1$ is:

$$Z_{eff} = Z - S$$
$$= 19 - (18 \times 1.00)$$
$$= 1.00$$

Thus, an electron in the $4s$ (rather than the $3d$) atomic orbital is under the influence of a greater effective nuclear charge and in the ground state of potassium, it is the $4s$ atomic orbital that is occupied.

Slater versus Clementi and Raimondi values of Z_{eff}

Slater's rules have been used to estimate ionization energies, ionic radii and electronegativities. More accurate effective nuclear charges (see table below) have been calculated by Clementi and Raimondi by using *self-consistent field* (SCF) methods, and indicate much higher Z_{eff} values for the d electrons. However, the simplicity of Slater's approach makes this an attractive method for 'back-of-the-envelope' estimations of Z_{eff}.

Values of Z_{eff}, calculated by Clementi and Raimondi, for elements with $Z = 1$–10; e.g. Z_{eff} of a $2s$ electron of a Be atom is 1.91.

	H		He
Z	1		2
$1s$	1.00		1.69

	Li	Be	B	C	N	O	F	Ne
Z	3	4	5	6	7	8	9	10
$1s$	2.69	3.68	4.68	5.67	6.66	7.66	8.65	9.64
$2s$	1.28	1.91	2.58	3.22	3.85	4.49	5.13	5.76
$2p$			2.42	3.14	3.83	4.45	5.10	5.76

Self-study exercises

1. Show that Slater's rules give a value of $Z_{eff} = 1.95$ for a $2s$ electron in a Be atom. Compare this value with that calculated by Clementi and Raimondi.

2. Show that Slater's rules give a value of $Z_{eff} = 5.20$ for a $2p$ electron of F. Compare this value with that calculated by Clementi and Raimondi.

3. Use Slater's rules to estimate values of Z_{eff} for (a) a $4s$ and (b) a $3d$ electron in a V atom.

[*Ans.* (a) 3.30; (b) 4.30]

4. Using your answer to question 3, explain why the valence configuration of the ground state of a V^+ ion is likely to be $3d^3 4s^1$ rather than $3d^2 4s^2$.

possesses. Establishing the ground state electronic configuration of an atom is the key to understanding its chemistry, and we now discuss the *aufbau* principle (*aufbau* means 'building up' in German) which is used in conjunction with Hund's rules and the Pauli exclusion principle to determine electronic ground state configurations:

- Orbitals are filled in order of energy, the lowest energy orbitals being filled first.

- Hund's first rule (often referred to simply as Hund's rule): in a set of degenerate orbitals, electrons may not be spin-paired in an orbital until *each* orbital in the set contains one electron; electrons singly occupying orbitals in a degenerate set have parallel spins, i.e. they have the same values of m_s.

- Pauli exclusion principle: no two electrons in the same atom may have the same set of n, l, m_l and m_s quantum numbers; it follows that each orbital can accommodate

	s-block elements		d-block elements											p-block elements					
Group 1	Group 2	Group 3	Group 4	Group 5	Group 6	Group 7	Group 8	Group 9	Group 10	Group 11	Group 12	Group 13	Group 14	Group 15	Group 16	Group 17	Group 18
1 H																	2 He
3 Li	4 Be											5 B	6 C	7 N	8 O	9 F	10 Ne
11 Na	12 Mg											13 Al	14 Si	15 P	16 S	17 Cl	18 Ar
19 K	20 Ca	21 Sc	22 Ti	23 V	24 Cr	25 Mn	26 Fe	27 Co	28 Ni	29 Cu	30 Zn	31 Ga	32 Ge	33 As	34 Se	35 Br	36 Kr
37 Rb	38 Sr	39 Y	40 Zr	41 Nb	42 Mo	43 Tc	44 Ru	45 Rh	46 Pd	47 Ag	48 Cd	49 In	50 Sn	51 Sb	52 Te	53 I	54 Xe
55 Cs	56 Ba	57–71 La–Lu	72 Hf	73 Ta	74 W	75 Re	76 Os	77 Ir	78 Pt	79 Au	80 Hg	81 Tl	82 Pb	83 Bi	84 Po	85 At	86 Rn
87 Fr	88 Ra	89–103 Ac–Lr	104 Rf	105 Db	106 Sg	107 Bh	108 Hs	109 Mt									

f-block elements

Lanthanoids	58 Ce	59 Pr	60 Nd	61 Pm	62 Sm	63 Eu	64 Gd	65 Tb	66 Dy	67 Ho	68 Er	69 Tm	70 Yb	71 Lu
Actinoids	90 Th	91 Pa	92 U	93 Np	94 Pu	95 Am	96 Cm	97 Bk	98 Cf	99 Es	100 Fm	101 Md	102 No	103 Lr

Fig. 1.13 The modern periodic table in which the elements are arranged in numerical order according to the number of protons (and electrons) they possess. The division into *groups* places elements with the same number of valence electrons into vertical columns within the table. Under IUPAC recommendations, the groups are labelled from 1 to 18 (Arabic numbers). The vertical groups of three *d*-block elements are called *triads*. Rows in the periodic table are called *periods*. The first period contains H and He, but the row from Li to Ne is sometimes referred to as the first period. Strictly, the lanthanoids include the 14 elements Ce–Lu, and the actinoids include Th–Lr; however, common usage places La with the lanthanoids, and Ac with the actinoids (see Chapter 24).

Table 1.4 IUPAC recommended names for groups of elements in the periodic table.

Group number	Recommended name
1	Alkali metals
2	Alkaline earth metals
15	Pnicogens‡
16	Chalcogens
17	Halogens
18	Noble gases

‡ The name pnicogen is proposed but is not (as yet) IUPAC-approved.

a maximum of two electrons with different m_s values (different spins = spin-paired).

Worked example 1.5   Using the *aufbau* principle

Determine (with reasoning) the ground state electronic configurations of (a) Be ($Z = 4$) and (b) P ($Z = 15$).

The value of Z gives the number of electrons to be accommodated in atomic orbitals in the ground state of the atom.

Assume an order of atomic orbitals (lowest energy first) as follows: $1s < 2s < 2p < 3s < 3p$

(a) Be $Z = 4$

Two electrons (spin-paired) are accommodated in the lowest energy $1s$ atomic orbital.

The next two electrons (spin-paired) are accommodated in the $2s$ atomic orbital.

The ground state electronic configuration of Be is therefore $1s^2 2s^2$.

(b) P $Z = 15$

Two electrons (spin-paired) are accommodated in the lowest energy $1s$ atomic orbital.

The next two electrons (spin-paired) are accommodated in the $2s$ atomic orbital.

The next six electrons are accommodated in the three degenerate $2p$ atomic orbitals, two spin-paired electrons per orbital.

The next two electrons (spin-paired) are accommodated in the $3s$ atomic orbital.

Three electrons remain and applying Hund's rule, these singly occupy each of the three degenerate $3p$ atomic orbitals.

The ground state electronic configuration of P is therefore $1s^2 2s^2 2p^6 3s^2 3p^3$.

Self-study exercises

1. Where, in the above argument, is the Pauli exclusion principle applied?

2. Will the three electrons in the P $3p$ atomic orbitals possess the same or different values of the spin quantum number?
[*Ans.* Same; parallel spins]

3. Show, with reasoning, that the ground state electronic configuration of O ($Z = 8$) is $1s^2 2s^2 2p^4$.

Worked example 1.6   The ground state electronic configurations of the noble gases

The atomic numbers of He, Ne, Ar and Kr are 2, 10, 18 and 36 respectively. Write down the ground state electronic configurations of these elements and comment upon their similarities or differences.

Apply the *aufbau* principle using the atomic orbital energy sequence:

$1s < 2s < 2p < 3s < 3p < 4s < 3d < 4p$

The ground state electronic configurations are:

He	$Z = 2$	$1s^2$
Ne	$Z = 10$	$1s^2 2s^2 2p^6$
Ar	$Z = 18$	$1s^2 2s^2 2p^6 3s^2 3p^6$
Kr	$Z = 36$	$1s^2 2s^2 2p^6 3s^2 3p^6 4s^2 3d^{10} 4p^6$

Each element Ne, Ar and Kr has a ground state electronic configuration $\ldots ns^2 np^6$. Helium is the odd one out, but still possesses a filled quantum level; this is a characteristic property of a noble gas.

Self-study exercises

1. Values of Z for Li, Na, K and Rb are 3, 11, 19 and 37 respectively. Write down their ground state configurations and comment on the result.
[*Ans.* All are of the form [X]ns^1 where X is a noble gas.]

2. How are the ground state electronic configurations of O, S and Se ($Z = 8$, 16, 34 respectively) alike? Give another element related in the same way.
[*Ans.* All are of the form [X]$ns^2 np^4$ where X is a noble gas; Te or Po]

Valence and core electrons

The configuration of the outer or *valence electrons* is of particular significance. These electrons determine the chemical properties of an element. Electrons that occupy lower energy quantum levels are called *core electrons*. For an element of low atomic number, the core and valence electrons are readily recognized by looking at the ground state electronic configuration. That of oxygen is $1s^2 2s^2 2p^4$. The core electrons of oxygen are those in the $1s$ atomic orbital; the six electrons with $n = 2$ are the valence electrons.

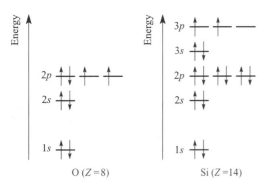

Fig. 1.14 Diagrammatic representations of the ground state electronic configurations of O and Si. The complete configurations are shown here, but it is common to simply indicate the valence electrons. For O, this consists of the $2s$ and $2p$ levels, and for Si, the $3s$ and $3p$ levels.

Diagrammatic representations of electronic configurations

The notation we have used to represent electronic configurations is convenient and is commonly adopted, but sometimes it is also useful to indicate the relative energies of the electrons. Figure 1.14 gives qualitative energy level diagrams which describe the ground state electronic configurations of O and Si.

Worked example 1.7 Quantum numbers for electrons

Confirm that the configuration shown for oxygen in Figure 1.14 is consistent with each electron possessing a unique set of four quantum numbers.

Each atomic orbital is designated by a unique set of three quantum numbers:

$$
\begin{array}{llll}
1s & n=1 & l=0 & m_l=0 \\
2s & n=2 & l=0 & m_l=0 \\
2p & n=2 & l=1 & m_l=-1 \\
   & n=2 & l=1 & m_l=0 \\
   & n=2 & l=1 & m_l=+1
\end{array}
$$

If an atomic orbital contains two electrons, they must have opposite spins so that the sets of quantum numbers for the two electrons are different: e.g. in the $1s$ atomic orbital:

one electron has $\quad n=1 \quad l=0 \quad m_l=0 \quad m_s=+\frac{1}{2}$

the other electron has $\quad n=1 \quad l=0 \quad m_l=0 \quad m_s=-\frac{1}{2}$

[This discussion is extended in Box 20.5]

Self-study exercises

1. Show that the electronic configuration $1s^2 2s^2 2p^1$ for B corresponds to each electron having a unique set of four quantum numbers.

2. The ground state of N is $1s^2 2s^2 2p^3$. Show that each electron in the $2p$ level possesses a unique set of four quantum numbers.

1.10 Ionization energies and electron affinities

Ionization energies

The ionization energy of hydrogen (equations 1.9 and 1.10) was discussed in Section 1.4; since the H atom only has one electron, no additional ionization processes can occur. For multi-electron atoms, successive ionizations are possible.

The first ionization energy, IE_1, of an atom is the internal energy change at 0 K, $\Delta U(0\,\mathrm{K})$, associated with the removal of the first valence electron (equation 1.17); the energy change is defined for a *gas phase* process. The units are kJ mol^{-1} or electron volts (eV).[†]

$$X(g) \longrightarrow X^+(g) + e^- \tag{1.17}$$

It is often necessary to incorporate ionization energies into thermochemical calculations (e.g. Born–Haber or Hess cycles) and it is convenient to define an associated *enthalpy change*, $\Delta H(298\,\mathrm{K})$. Since the difference between $\Delta H(298\,\mathrm{K})$ and $\Delta U(0\,\mathrm{K})$ is very small (see *Box 1.6*), values of IE can be used in thermochemical cycles so long as extremely accurate answers are not required.

The first ionization energy (IE_1) of a gaseous atom is the internal energy change at 0 K associated with the removal of the first valence electron:

$$X(g) \longrightarrow X^+(g) + e^-$$

For thermochemical cycles, an associated *change in enthalpy* is used:

$$\Delta H(298\,\mathrm{K}) \approx \Delta U(0\,\mathrm{K})$$

The second ionization energy, IE_2, of an atom refers to step 1.18; note that this is equivalent to the first ionization of the ion X^+. Equation 1.19 describes the step corresponding to the third ionization energy, IE_3, of X, and successive ionizations are similarly defined:

$$X^+(g) \longrightarrow X^{2+}(g) + e^- \tag{1.18}$$

$$X^{2+}(g) \longrightarrow X^{3+}(g) + e^- \tag{1.19}$$

[†] An electron volt is a non-SI unit with a value of $\approx 1.60218 \times 10^{-19}$ J; to compare eV and kJ mol^{-1} units, it is necessary to multiply by the Avogadro number. $1\,\mathrm{eV} = 96.4853 \approx 96.5\,\mathrm{kJ\,mol}^{-1}$.

BOX 1.6

The relationship between ΔU and ΔH

The relationship between the change in internal energy and change in enthalpy of the system for a reaction at a given temperature is given by the equation:

$$\Delta U = \Delta H - P\Delta V$$

where P is the pressure and ΔV is the change in volume. The $P\Delta V$ term corresponds to the work done, e.g. in expanding the system against the surroundings as a gas is liberated during a reaction. Often in a chemical reaction, the pressure P corresponds to atmospheric pressure (1 atm = 101 300 Pa, or 1 bar = 10^5 Pa).

In general, the work done by or on the system is much smaller than the enthalpy change, making the $P\Delta V$ term negligible with respect to the values of ΔU and ΔH. Thus:

$$\Delta U(T\,\mathrm{K}) \approx \Delta H(T\,\mathrm{K})$$

However, in Section 1.10, we are considering two different temperatures and state that:

$$\Delta U(0\,\mathrm{K}) \approx \Delta H(298\,\mathrm{K})$$

In order to assess the variation in ΔH with temperature, we apply Kirchhoff's equation:

$$\Delta C_P = \left(\frac{\partial \Delta H}{\partial T}\right)_P$$

the integrated form of which (integrating between the limits of the temperatures 0 and 298 K) is:

$$\int_0^{298} \mathrm{d}(\Delta H) = \int_0^{298} \Delta C_P\, \mathrm{d}T$$

Integrating the left-hand side gives:

$$\Delta H(298\,\mathrm{K}) - \Delta H(0\,\mathrm{K}) = \int_0^{298} \Delta C_P\, \mathrm{d}T$$

Consider the ionization of an atom X:

$$X(g) \longrightarrow X^+(g) + e^-(g).$$

If X, X^+ and e^- are all ideal monatomic gases, then the value of C_P for each is $\frac{5}{2}R$ (where R is the molar gas constant = $8.314 \times 10^{-3}\,\mathrm{kJ\,K^{-1}\,mol^{-1}}$), giving for the reaction a value of ΔC_P of $\frac{5}{2}R$. Therefore:

$$\Delta H(298\,\mathrm{K}) - \Delta H(0\,\mathrm{K}) = \int_0^{298} \tfrac{5}{2}R\, \mathrm{d}T$$

$$= \left(\frac{5 \times 8.314 \times 10^{-3}}{2}\right)[T]_0^{298}$$

$$= 6.2\,\mathrm{kJ\,mol^{-1}}$$

Inspection of typical values of ionization energies in Appendix 8 shows that a correction of this magnitude is relatively insignificant.

Values of ionization energies for the elements are listed in Appendix 8. Figure 1.15 shows the variation in the values of IE_1 as a function of Z. Several repeating patterns are apparent and some features to note are:

- the high values of IE_1 associated with the noble gases;
- the very low values of IE_1 associated with the group 1 elements;

- the *general* increase in values of IE_1 as a given period is crossed;
- the discontinuity in values of IE_1 on going from an element in group 15 to its neighbour in group 16;
- the decrease in values of IE_1 on going from an element in group 2 or 12 to its neighbour in group 13.
- the rather similar values of IE_1 for a given row of *d*-block elements.

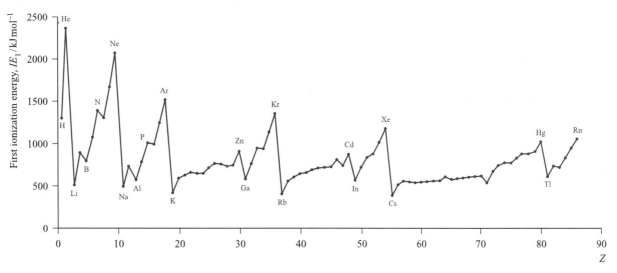

Fig. 1.15 The values of the first ionization energies of the elements up to Rn.

Exchange energies

Filled and half-filled shells are often referred to as possessing a 'special stability'. However, this is misleading, and we should really consider the *exchange energy* of a given configuration. This can only be justified by an advanced quantum mechanical treatment but we can summarize the idea as follows. Consider two electrons in *different* orbitals. The repulsion between the electrons if they have anti-parallel spins is greater than if they have parallel spins, e.g. for a p^2 configuration:

$$\uparrow \quad \downarrow \quad \underline{\quad} \qquad \text{versus} \qquad \uparrow \quad \uparrow \quad \underline{\quad}$$

The difference in energy between these two configurations is the *exchange energy*, K, i.e. this is the extra stability that the right-hand configuration has with respect to the left-hand one. The total exchange energy is expressed in terms of K (the actual value of K depends on the atom or ion):

$$\text{Exchange energy} = \sum \frac{N(N-1)}{2} K$$

where N = number of electrons with parallel spins.

For further discussion, see:
A.B. Blake (1981) *Journal of Chemical Education*, vol. 58, p. 393.
B.J. Duke (1978) *Education in Chemistry*, vol. 15, p.186.
D.M.P. Mingos (1998) *Essential Trends in Inorganic Chemistry*, Oxford University Press, Oxford, p. 14.

Each of these trends can be rationalized in terms of ground state electronic configurations. The noble gases (except for He) possess ns^2np^6 configurations which are particularly stable (see *Box 1.7*) and removal of an electron requires a great deal of energy. The ionization of a group 1 element involves loss of an electron from a singly occupied ns orbital with the resultant X^+ ion possessing a noble gas configuration. The *general* increase in IE_1 across a given period is a consequence of an increase in Z_{eff}. A group 15 element has a ground state electronic configuration ns^2np^3 and the np level is *half-occupied*. A certain stability (see *Box 1.7*) is associated with such configurations and it is more difficult to ionize a group 15 element than its group 16 neighbour. In going from Be (group 2) to B (group 13), there is a marked decrease in IE_1 and this may be attributed to the relative stability of the filled shell $2s^2$ configuration compared with the $2s^22p^1$ arrangement; similarly, in going from Zn (group 12) to Ga (group 13), we need to consider the difference between $4s^23d^{10}$ and $4s^23d^{10}4p^1$ configurations. Trends among IE values for d-block metals are discussed in Section 19.3.

Electron affinities

The first electron affinity (EA_1) is *minus* the internal energy change (equation 1.20) for the gain of an electron by a *gaseous* atom (equation 1.21). The second electron affinity of atom Y is defined for process 1.22. Each reaction occurs in the gas phase.

$$EA = -\Delta U(0\,\text{K}) \tag{1.20}$$

$$Y(g) + e^- \longrightarrow Y^-(g) \tag{1.21}$$

$$Y^-(g) + e^- \longrightarrow Y^{2-}(g) \tag{1.22}$$

As we saw for ionization energies, it is convenient to define an enthalpy change associated with each of the reactions 1.21 and 1.22 and we approximate $\Delta_{\text{EA}}H(298\,\text{K})$ to $\Delta_{\text{EA}}U(0\,\text{K})$. Selected values of these enthalpy changes are given in Table 1.5.

> The first electron affinity, EA_1, of an atom is *minus* the internal energy change at 0 K associated with the gain of one electron by a gaseous atom:
>
> $$Y(g) + e^- \longrightarrow Y^-(g)$$
>
> For thermochemical cycles, an associated *enthalpy* is used:
>
> $$\Delta_{\text{EA}}H(298\,\text{K}) \approx \Delta_{\text{EA}}U(0\,\text{K}) = -EA$$

Table 1.5 Approximate *enthalpy changes* $\Delta_{\text{EA}}H(298\,\text{K})$ associated with the attachment of an electron to an atom or anion.[‡]

Process	$\approx \Delta_{\text{EA}}H / \text{kJ mol}^{-1}$
$H(g) + e^- \longrightarrow H^-(g)$	-73
$Li(g) + e^- \longrightarrow Li^-(g)$	-60
$Na(g) + e^- \longrightarrow Na^-(g)$	-53
$K(g) + e^- \longrightarrow K^-(g)$	-48
$N(g) + e^- \longrightarrow N^-(g)$	≈ 0
$P(g) + e^- \longrightarrow P^-(g)$	-72
$O(g) + e^- \longrightarrow O^-(g)$	-141
$O^-(g) + e^- \longrightarrow O^{2-}(g)$	$+798$
$S(g) + e^- \longrightarrow S^-(g)$	-201
$S^-(g) + e^- \longrightarrow S^{2-}(g)$	$+640$
$F(g) + e^- \longrightarrow F^-(g)$	-328
$Cl(g) + e^- \longrightarrow Cl^-(g)$	-349
$Br(g) + e^- \longrightarrow Br^-(g)$	-325
$I(g) + e^- \longrightarrow I^-(g)$	-295

[‡] Tables of data differ in whether they list values of EA or $\Delta_{\text{EA}}H$ and it is essential to take note of the sign convention being used.

The attachment of an electron to an *atom* is usually exothermic. Two electrostatic forces oppose one another: the repulsion between the valence shell electrons and the additional electron, and the attraction between the nucleus and the incoming electron. In contrast, *repulsive* interactions are dominant when an electron is added to an *anion* and the process is endothermic (Table 1.5).

1.11 Bonding models: an introduction

In Sections 1.11–1.13 we summarize valence bond (VB) and molecular orbital (MO) theories of homonuclear bond formation (see *Section 1.12*), and include practice in generating *Lewis structures*. For further details, readers are guided to the first-year texts listed at the end of the chapter.

A historical overview

The foundations of modern chemical bonding theory were laid in 1916–1920 by G.N. Lewis and I. Langmuir who suggested that ionic species were formed by electron transfer, while electron sharing was important in covalent molecules. In some cases, it was suggested that the shared electrons in a bond were provided by one of the atoms but that once the bond (sometimes called a *coordinate bond*) is formed, it *is indistinguishable from a 'normal' covalent bond*.

> In a *covalent* species, electrons are shared between atoms. In an *ionic* species, one or more electrons are transferred between atoms to form ions.

Modern views of atomic structure are, as seen earlier, based largely on the applications of wave mechanics to atomic systems. Modern views of *molecular structure* are based on applying wave mechanics to molecules; such studies provide answers as to how and why atoms combine. The Schrödinger equation can be written to describe the behaviour of electrons in molecules, but it can be solved only approximately. Two such methods are the valence bond approach, developed by Heitler and Pauling, and the molecular orbital approach associated with Hund and Mulliken:

- *Valence bond theory* treats the formation of a molecule as arising from the bringing together of complete atoms which, when they interact, to a large extent retain their original character.
- *Molecular orbital theory* allocates electrons to molecular orbitals formed by the overlap (interaction) of atomic orbitals.

Although familiarity with both VB and MO concepts is necessary, it is often the case that a given situation is more conveniently approached by using one or other of these models. We begin with the conceptually simple approach of Lewis for representing the bonding in covalent molecules.

Lewis structures

Lewis presented a simple, but useful, method of describing the arrangement of valence electrons in molecules. The approach uses dots (or dots and crosses) to represent the number of *valence electrons*, and the nuclei are indicated by appropriate elemental symbols. A basic premise of the theory is that electrons in a molecule should be paired; the presence of a single (odd) electron indicates that the species is a *radical*.

Diagram **1.1** shows the Lewis structure for H_2O with the O−H bonds designated by pairs of dots (electrons); an alternative representation is given in structure **1.2** where each line stands for *one pair* of electrons, i.e. a *single covalent bond*. Pairs of valence electrons which are not involved in bonding are *lone pairs*.

$$
\begin{array}{cc}
\text{H} & \text{H} \\
\bullet\bullet & | \\
\text{:O: H} & \text{:O—H} \\
\bullet\bullet & \bullet\bullet \\
(1.1) & (1.2)
\end{array}
$$

The Lewis structure for N_2 shows that the N−N bond is composed of three pairs of electrons and is a *triple bond* (structures **1.3** and **1.4**). Each N atom has one lone pair of electrons. The Lewis structures **1.5** and **1.6** for O_2 indicate the presence of a double bond, with each O atom bearing two lone pairs of electrons.

$$
\begin{array}{cc}
\text{:N:N:} & \text{:N} \equiv \text{N:} \\
(1.3) & (1.4) \\
\text{:O:O:} & \text{:O} = \text{O:} \\
(1.5) & (1.6)
\end{array}
$$

Lewis structures give the connectivity of an atom in a molecule, the bond order and the number of lone pairs and these may be used to derive structures using the valence-shell electron-pair repulsion model (see *Section 1.19*).

1.12 Homonuclear diatomic molecules: valence bond (VB) theory

Uses of the term *homonuclear*

The word *homonuclear* is used in two ways:

- A *homonuclear covalent bond* is one formed between two atoms of the same element, e.g. the H—H bond in H_2, the O=O bond in O_2, and the O—O bond in H_2O_2 (Figure 1.16).
- A *homonuclear molecule* contains one type of element. Homonuclear diatomic molecules include H_2, N_2 and F_2, homonuclear triatomics include O_3 (ozone) and examples of larger homonuclear molecules are P_4, S_8 and C_{60}.

Covalent bond distance, covalent radius and van der Waals radius

Three important definitions are needed before we discuss covalent bonding.

The length of a covalent bond (*bond distance*), d, is the *internuclear separation* and may be determined experimentally by microwave spectroscopy or diffraction methods (X-ray, neutron or electron diffraction). It is convenient to define the covalent radius, r_{cov}, of an atom: for an atom X, r_{cov} is half of the covalent bond length of a homonuclear X—X *single* bond. Thus, $r_{cov}(S)$ can be determined from the solid state structure of S_8 (Figure 1.1c) determined by X-ray diffraction methods or, better still, by averaging the values of the bond distances of S—S single bonds found for all the allotropes of sulfur.

> For an atom X, the value of the single bond *covalent radius*, r_{cov}, is half of the internuclear separation in a homonuclear X—X single bond.

The α- and β-forms of sulfur (orthorhombic and monoclinic sulfur, respectively) both crystallize with the S_8 molecules stacked in a regular arrangement; the packing in the α-form (density = $2.07\,g\,cm^{-3}$) is more efficient than that in the β-form (density = $1.94\,g\,cm^{-3}$). Van der Waals forces operate between the molecules, and half of the distance of closest approach of two sulfur atoms belonging to *different* S_8 rings is defined as the van der Waals radius, r_v, of sulfur. The weakness of the bonding is evidenced by the fact that S_8 vaporizes, retaining the ring structure, without absorbing much energy. The van der Waals radius of an element is necessarily larger than its covalent radius, e.g. r_v and r_{cov} for S are 185 and 103 pm respectively. Van der

Waals forces encompass dispersion and dipole–dipole interactions; dispersion forces are discussed in the latter part of Section 5.13 and dipole moments in Section 1.16. Values of r_v and r_{cov} are listed in *Appendix 6*.

> The van der Waals radius, r_v, of an atom X is half of the distance of closest approach of two non-bonded atoms of X.

The valence bond (VB) model of bonding in H_2

Valence bond theory considers the interactions between separate atoms as they are brought together to form molecules. We begin by considering the formation of H_2 from two H atoms, the nuclei of which are labelled H_A and H_B, and the electrons of which are 1 and 2, respectively. When the atoms are so far apart that there is no interaction between them, electron 1 is exclusively associated with H_A, while electron 2 resides with nucleus H_B. Let this state be described by a wavefunction ψ_1.

When the H atoms are close together, we cannot tell which electron is associated with which nucleus since, although *we* gave them labels, the two nuclei are actually indistinguishable, as are the two electrons. Thus, electron 2 could be with H_A and electron 1 with H_B. Let this be described by the wavefunction ψ_2.

Equation 1.23 describes an overall description of the covalently bonded H_2 molecule; $\psi_{covalent}$ is a *linear combination* of wavefunctions ψ_1 and ψ_2. The equation contains a *normalization factor*, N (see *Box 1.3*).

$$\psi_{covalent} = \psi_+ = N(\psi_1 + \psi_2) \tag{1.23}$$

Another linear combination of ψ_1 and ψ_2 can be written as shown in equation 1.24:

$$\psi_- = N(\psi_1 - \psi_2) \tag{1.24}$$

In terms of the spins of electrons 1 and 2, ψ_+ corresponds to spin-pairing, and ψ_- corresponds to parallel spins (non-spin-paired). Calculations of the energies associated with these states as a function of the internuclear separation of H_A and H_B show that, while ψ_- represents a repulsive state (high energy), the energy curve for ψ_+ reaches a minimum value when the internuclear separation, d, is 87 pm and this corresponds to an H—H bond dissociation energy, ΔU, of $303\,kJ\,mol^{-1}$. While these are near enough to the experimental values of $d = 74\,pm$ and $\Delta U = 458\,kJ\,mol^{-1}$ to suggest that the model has some validity, they are far enough from them to indicate that the expression for ψ_+ needs refining.

> The bond dissociation energy (ΔU) and enthalpy (ΔH) values for H_2 are defined for the process:
>
> $$H_2(g) \longrightarrow 2H(g)$$

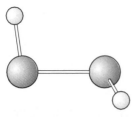

Fig. 1.16 The structure of hydrogen peroxide, H_2O_2; O atoms are shown in blue.

Improvements to equation 1.23 can be made by:

- allowing for the fact that each electron screens the other from the nuclei to some extent;
- taking into account the possibility that *both* electrons 1 and 2 may be associated with either H_A *or* H_B, i.e. allowing for the transfer of one electron from one nuclear centre to the other to form a pair of ions $H_A^+H_B^-$ or $H_A^-H_B^+$.

The latter modification is dealt with by writing two additional wavefunctions, ψ_3 and ψ_4 (one for each ionic form) and so equation 1.23 can be rewritten in the form of equation 1.25. The coefficient c indicates the relative contributions made by two sets of wavefunctions; remember that for a *homonuclear diatomic* such as H_2, the situations described by ψ_1 and ψ_2 are equally probable, as are those described by ψ_3 and ψ_4.

$$\psi_+ = N[(\psi_1 + \psi_2) + c(\psi_3 + \psi_4)] \qquad (1.25)$$

Since the wavefunctions ψ_1 and ψ_2 arise from an internuclear interaction involving the *sharing* of electrons among nuclei, and ψ_3 and ψ_4 arise from electron *transfer*, we can simplify equation 1.25 to 1.26 in which the overall wavefunction, $\psi_{molecule}$, is composed of covalent and ionic terms:

$$\psi_{molecule} = N[\psi_{covalent} + (c \times \psi_{ionic})] \qquad (1.26)$$

Based on this model of H_2, calculations with $c \approx 0.25$ give values of 75 pm for $d(H-H)$ and 398 kJ mol^{-1} for the bond dissociation energy. Modifying equation 1.26 still further leads to a value of ΔU very close to the experimental value, but details of the procedure are beyond the scope of this book.[†]

Now consider the physical significance of equations 1.25 and 1.26. The wavefunctions ψ_1 and ψ_2 represent the structures shown in **1.7** and **1.8**, while ψ_3 and ψ_4 represent the ionic forms **1.9** and **1.10**.

$H_A(1)H_B(2)$	$H_A(2)H_B(1)$	$[H_A(1)(2)]^-$ H_B^+	H_A^+ $[H_B(1)(2)]^-$
(1.7)	(1.8)	(1.9)	(1.10)

Dihydrogen is described as a *resonance hybrid* of these contributing *resonance* or *canonical structures*. In the case of H_2, an example of a homonuclear diatomic molecule which is necessarily symmetrical, we simplify the picture to **1.11**. Each of structures **1.11a**, **1.11b** and **1.11c** is a resonance structure and the double headed arrows indicate the *resonance* between them. The contributions made by **1.11b** and **1.11c** are equal. The term 'resonance hybrid' is

somewhat unfortunate but is too firmly established to be eradicated.

$$H—H \longleftrightarrow H^+ \ H^- \longleftrightarrow H^- \ H^+$$

$$(1.11a) \qquad\qquad (1.11b) \qquad\qquad (1.11c)$$

A crucial point about resonance structures is that they *do not exist as separate species*. Rather, they indicate extreme bonding pictures, the combination of which gives a description of the molecule overall. In the case of H_2, the contribution made by resonance structure **1.11a** is significantly greater than that of **1.11b** or **1.11c**.

Notice that **1.11a** describes the bonding in H_2 in terms of a *localized two-centre two-electron*, 2c-2e, covalent bond. A particular resonance structure will always indicate a localized bonding picture, although the combination of several resonance structures may result in the description of the bonding in the species as a whole as being delocalized (see *Section 4.4*).

The valence bond (VB) model applied to F_2, O_2 and N_2

The valence bond model describes bonds in terms of the spin-pairing of electrons. The number of unpaired electrons in an atom, whether in its ground state or an accessible excited state, indicates how many bonds it will form.

Consider the formation of F_2. The ground state electronic configuration of F is $[He]2s^22p^5$, and the presence of the single unpaired electron indicates the formation of an F–F single bond. We can write down resonance structures **1.12** to describe the bonding in F_2, with the expectation that the covalent contribution will predominate.

$$F—F \longleftrightarrow F^+ \ F^- \longleftrightarrow F^- \ F^+$$

$$(1.12)$$

The formation of O_2 involves the combination of two O atoms with ground state electronic configurations of $1s^22s^22p^4$. Each O atom has two unpaired electrons and so VB theory predicts the formation of an O=O double bond. Since VB theory works on the premise that electrons are paired wherever possible, the model predicts that O_2 is diamagnetic. One of the notable failures of VB theory is its inability to predict the observed *paramagnetism* of O_2. As we shall see, molecular orbital theory is fully consistent with O_2 being a diradical. When two N atoms ($[He]2s^22p^3$) combine to give N_2, an N≡N triple bond results. Of the possible resonance structures, the predominant form is covalent and this gives a satisfactory picture of the bonding in N_2.

> In a *diamagnetic* species, all electrons are spin-paired; a diamagnetic substance is repelled by a magnetic field. A *paramagnetic* species contains one or more unpaired electrons; a paramagnetic substance is attracted by a magnetic field.

[†] For detailed discussion, see: R. McWeeny (1979) *Coulson's Valence*, 3rd edn, Oxford University Press, Oxford.

1.13 Homonuclear diatomic molecules: molecular orbital (MO) theory

An overview of the MO model

In molecular orbital (MO) theory, we begin by placing the nuclei of a given molecule in their equilibrium positions and then calculate the *molecular orbitals* (i.e. regions of space spread over the entire molecule) that a single electron might occupy. Each MO arises from interactions between orbitals of atomic centres in the molecule, and such interactions are:

- allowed if the *symmetries* of the atomic orbitals are compatible with one another;
- efficient if the region of *overlap* between the two atomic orbitals is significant;
- efficient if the atomic orbitals are relatively close in energy.

An important ground-rule of MO theory is that *the number of MOs that can be formed must equal the number of atomic orbitals of the constituent atoms.*

Each MO has an associated energy and, to derive the electronic ground state of a molecule, the available electrons are placed, according to the *aufbau* principle, in MOs beginning with that of lowest energy. The sum of the individual energies of the electrons in the orbitals (after correction for electron–electron interactions) gives the total energy of the molecule.

Molecular orbital theory applied to the bonding in H_2

An approximate description of the MOs in H_2 can be obtained by considering them as *linear combinations of atomic orbitals* (LCAOs). Each of the H atoms has one $1s$ atomic orbital; let the two associated wavefunctions be ψ_1 and ψ_2. In Section 1.6, we mentioned the importance of the *signs of the wavefunctions* with respect to their overlap during bond formation. The sign of the wavefunction associated with the $1s$ atomic orbital may be either $+$ or $-$. Just as transverse waves interfere in a constructive (in-phase) or destructive (out-of-phase) manner, so too do orbitals. Mathematically, we represent the possible combinations of the two $1s$ atomic orbitals by equations 1.27 and 1.28, where N and N^* are the normalization factors. Whereas ψ_{MO} is an in-phase (*bonding*) interaction, ψ^*_{MO} is an out-of-phase (*antibonding*) interaction.

$$\psi_{\text{MO(in-phase)}} = \psi_{\text{MO}} = N[\psi_1 + \psi_2] \tag{1.27}$$

$$\psi_{\text{MO(out-of-phase)}} = \psi^*_{\text{MO}} = N^*[\psi_1 - \psi_2] \tag{1.28}$$

The values of N and N^* are determined using equations 1.29 and 1.30 where S is the *overlap integral*. This is a measure of the extent to which the regions of space described by the two

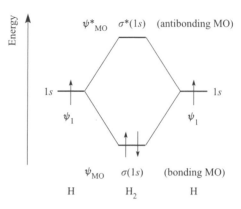

Fig. 1.17 An orbital interaction diagram for the formation of H_2 from two hydrogen atoms. By the *aufbau* principle, the two electrons occupy the lowest energy (bonding) molecular orbital.

wavefunctions ψ_1 and ψ_2 coincide. Although we mentioned earlier that orbital interaction is efficient if the region of overlap between the two atomic orbitals is significant, the numerical value of S is still much less than unity and is often neglected giving the approximate results shown in equations 1.29 and 1.30:

$$N = \frac{1}{\sqrt{2(1+S)}} \approx \frac{1}{\sqrt{2}} \tag{1.29}$$

$$N^* = \frac{1}{\sqrt{2(1-S)}} \approx \frac{1}{\sqrt{2}} \tag{1.30}$$

The interaction between the H $1s$ atomic orbitals on forming H_2 may be represented by the energy level diagram in Figure 1.17. The bonding MO, ψ_{MO}, is stabilized with respect to the $1s$ atomic orbitals while the antibonding MO, ψ^*_{MO}, is destabilized.[†] Each H atom contributes one electron and, by the *aufbau* principle, the two electrons occupy the lower of the two MOs in the H_2 molecule and are spin-paired (Figure 1.17). It is important to remember that in MO theory *we construct the orbital interaction diagram first and then put in the electrons according to the aufbau principle.*

The bonding and antibonding MOs in H_2 are given the symmetry labels σ and σ^* ('*sigma*' and '*sigma-star*') or, more fully, $\sigma(1s)$ and $\sigma^*(1s)$ to indicate their atomic orbital origins. In order to define these labels, consider the pictorial representations of the two MOs. Figure 1.18a shows that when the $1s$ atomic orbitals interact in-phase, the two wavefunctions reinforce each other, especially in the region of space between the nuclei. The two electrons occupying this MO will be found predominantly between the two nuclei, and the build-up of electron density reduces

---

[†] The difference between the energies of the $1s$ atomic orbitals and ψ^*_{MO} is slightly greater than between those of the $1s$ atomic orbitals and ψ_{MO}, i.e. an antibonding MO is slightly more antibonding than the corresponding bonding MO is bonding; the origin of this effect is beyond the scope of this book.

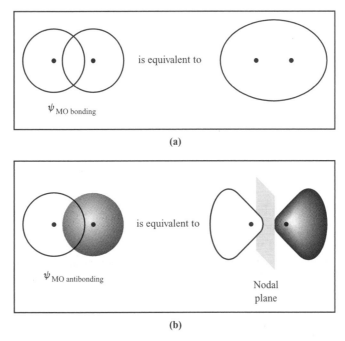

Fig. 1.18 Schematic representations of (a) the bonding and (b) the antibonding molecular orbitals in the H_2 molecule. The H nuclei are represented by black dots. The shaded orbital lobes could equally well be marked with a + sign, and the unshaded lobes with a − sign to indicate the sign of the wavefunction.

internuclear repulsion. Figure 1.18b illustrates that the out-of-phase interaction results in a *nodal plane between the two H nuclei*. If the antibonding orbital were to be occupied, there would be a zero probability of finding the electrons at any point on the nodal plane. This lack of

electron density raises the internuclear repulsion and, as a result, destabilizes the MO.

Now let us return to the σ and σ^* labels. An MO has σ-symmetry if it is symmetrical with respect to a line joining the two nuclei; i.e. if you rotate the orbital about the

---

CHEMICAL AND THEORETICAL BACKGROUND

The parity of MOs for a molecule that possesses a centre of inversion

We consider symmetry in Chapter 3, but it is useful at this point to consider the labels that are commonly used to describe the *parity of a molecular orbital*. A *homonuclear* diatomic molecule (e.g. H_2, Cl_2) possesses a centre of inversion (centre of symmetry), and the parity of an MO describes the way in which the orbital behaves with respect to this centre of inversion.

First find the centre of inversion in the molecule; this is the point through which you can draw an infinite number of straight lines such that each line passes through a pair or similar points, one on each side of the centre of symmetry and at equal distances from it:

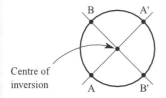

Centre of inversion

Point A is related to A' by passing through the centre of inversion. Similarly, B is related to B'

Now ask the question: 'Does the wavefunction have the same *sign* at the same distance but in opposite directions from the centre of symmetry?'

If the answer if 'yes', then the orbital is labelled *g* (from the word *gerade*, German for 'even'). If the answer if 'no', then the orbital is labelled *u* (from the word *ungerade*, German for 'odd'). For example, the σ-bonding MO in H_2 is labelled σ_g, while the antibonding MO is σ_u.

Parity labels *only* apply to MOs in molecules that possess a centre of inversion, e.g. homonuclear X_2, octahedral EX_6 and square planar EX_4 molecules. Heteronuclear XY, or tetrahedral EX_4 molecules, for example, do not possess a centre of inversion.

Self-study exercise

Identify which of the MOs in Figure 1.20 should be labelled *g* and which labelled *u*.

internuclear axis (the axis joining the two nuclear centres marked in Figure 1.18), there is no phase change. A σ^*-orbital must exhibit two properties:

- the σ label means that rotation of the orbital about the internuclear axis generates no phase change, *and*
- the * label means that there is a nodal plane *between* the nuclei, and this plane is orthogonal to the internuclear axis.

The ground state electronic configuration of H_2 may be written using the notation $\sigma(1s)^2$ indicating that two electrons occupy the $\sigma(1s)$ MO.

The orbital interaction diagram shown in Figure 1.17 can be used to predict several properties of the H_2 molecule. Firstly, the electrons are paired and so we expect H_2 to be diamagnetic as is found experimentally. Secondly, the formal bond order can be found using equation 1.31; for H_2 this gives a bond order of one.

Bond order $= \frac{1}{2}$[(Number of bonding electrons)

$\quad$ − (Number of antibonding electrons)] (1.31)

We cannot measure the bond order experimentally but we can make some useful correlations between bond order and the experimentally measurable bond distances and bond dissociation energies or enthalpies. Along a series of species related by electron gain (reduction) or loss (oxidation), inspection of the corresponding MO diagram shows how the bond order may change (assuming that there are no gross changes to the energy levels of the orbitals). For example, the oxidation of H_2 to $[H_2]^+$ (a change brought about by the action of an electric discharge on H_2 at low pressures) can be considered in terms of the removal of one electron from Figure 1.17. The bond order of $[H_2]^+$ is thus (equation 1.31) 0.5, and we would expect the H−H bond to be weakened. Experimentally,

the bond dissociation energy, ΔU, for H_2 is $458\,kJ\,mol^{-1}$ and for $[H_2]^+$ is $269\,kJ\,mol^{-1}$. Similar correlations can be made between bond order and bond length: the lower the bond order, the larger the internuclear separation; the experimentally determined bond lengths of H_2 and $[H_2]^+$ are 74 and 105 pm. While such correlations are useful, they must be treated with caution and *only* used in series of closely related species.

The bonding in He₂, Li₂ and Be₂

Molecular orbital theory can be applied to any homonuclear diatomic molecule, but as more valence atomic orbitals become available, the MO diagram becomes more complex. Treatments of the bonding in He_2, Li_2 and Be_2 are similar to that for H_2. In practice, He does not form He_2, and the construction of an MO diagram for He_2 is a useful exercise because it rationalizes this observation. Figure 1.19a shows that when the two $1s$ atomic orbitals of two He atoms interact, σ and σ^* MOs are formed as in H_2. However, each He atom contributes two electrons, meaning that in He_2, both the bonding *and* antibonding MOs are fully occupied. The bond order (equation 1.31) is zero and so the MO picture of He_2 is consistent with its non-existence. Using the same notation as for H_2, the ground state electronic configuration of He_2 is $\sigma(1s)^2\sigma^*(1s)^2$.

The ground state electronic configuration of Li ($Z = 3$) is $1s^2 2s^1$ and when two Li atoms combine, orbital overlap occurs efficiently between the $1s$ atomic orbitals and between the $2s$ atomic orbitals. To a first approximation we can ignore $1s$–$2s$ overlap since the $1s$ and $2s$ orbital energies are poorly matched. An approximate orbital interaction diagram for the formation of Li_2 is given in Figure 1.19b. Each Li atom provides three electrons, and the six electrons in Li_2 occupy the lowest energy MOs to give a ground state

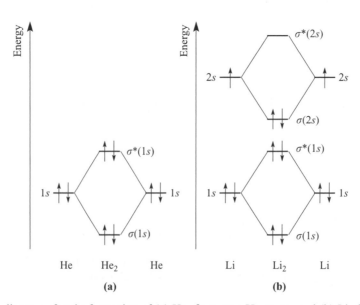

Fig. 1.19 Orbital interaction diagrams for the formation of (a) He_2 from two He atoms and (b) Li_2 from two Li atoms.

electronic configuration of $\sigma(1s)^2 \sigma^*(1s)^2 \sigma(2s)^2$. Effectively, we could ignore the interaction between the core $1s$ atomic orbitals since the net bonding is determined by the interaction between the valence atomic orbitals, and a simpler, but informative, electronic ground state is $\sigma(2s)^2$. Figure 1.19b also shows that Li_2 is predicted to be diamagnetic in keeping with experimental data. By applying equation 1.31, we see that MO theory gives a bond order in Li_2 of one. Note that the terminology 'core and valence orbitals' is equivalent to that for 'core and valence electrons' (see *Section 1.9*).

Like Li, Be has available $1s$ and $2s$ atomic orbitals for bonding; these atomic orbitals constitute the *basis set of orbitals*. An orbital interaction diagram similar to that for Li_2 (Figure 1.19b) is appropriate. The difference between Li_2 and Be_2 is that Be_2 has two more electrons than Li_2 and these occupy the $\sigma^*(2s)$ MO. The predicted bond order in Be_2 is thus zero. In practice, this prediction is essentially fulfilled, but there is evidence for an extremely unstable Be_2 species with bond length 245 pm and bond energy 10 kJ mol^{-1}.

> A *basis set of orbitals* is composed of those which are available for orbital interactions.

In each of Li_2 and Be_2, it is unnecessary to include the core ($1s$) atomic orbitals in order to obtain a useful bonding picture. This is true more generally, and throughout this book, MO treatments of bonding focus only on the interactions between the valence orbitals of the atoms concerned.

The bonding in F_2 and O_2

The valence shell of an F atom contains $2s$ and $2p$ atomic orbitals, and the formation of an F_2 molecule involves $2s$–$2s$ and $2p$–$2p$ orbital interactions. Before we can construct an MO diagram for the formation of F_2, we must consider what types of interactions are possible between p atomic orbitals.

By convention, each p atomic orbital is directed along one of the three Cartesian axes (Figure 1.10), and, in considering the formation of a diatomic X_2, it is convenient to fix the positions of the X nuclei on one of the axes. In diagram **1.13**, the nuclei are placed on the z axis, but this choice of axis is arbitrary. Defining these positions also defines the relative orientations of the two sets of p orbitals (Figure 1.20).

(1.13)

Figures 1.20a and 1.20b show the in-phase and out-of-phase combinations of two $2p_z$ atomic orbitals. In terms of the region between the nuclei, the p_z–p_z interaction is not dissimilar from that of two s atomic orbitals (Figure 1.18) and the symmetries of the resultant MOs are consistent

with the σ and σ^* labels. Thus, the direct interaction of two p atomic orbitals (i.e. when the orbitals lie along a common axis) leads to $\sigma(2p)$ and $\sigma^*(2p)$ MOs. The p_x orbitals of the two atoms X can overlap only in a sideways manner, an interaction which has a smaller overlap integral than the direct overlap of the p_z atomic orbitals. The in-phase and out-of-phase combinations of two $2p_x$ atomic orbitals are shown in Figures 1.20c and 1.20d. The bonding MO is called a π-orbital ('*pi-orbital*'), and its antibonding counterpart is a π^*-orbital ('*pi-star-orbital*'). Note the positions of the nodal planes in each MO. An MO has π-symmetry if it is asymmetrical with respect to a line joining the two nuclei: i.e. if you rotate the orbital about the internuclear axis (the z axis in Figure 1.20), there is a phase change. A π^*-orbital must exhibit two properties:

- the π label means that rotation of the orbital about the internuclear axis generates a phase change, *and*
- the * label means that there must be a nodal plane *between* the nuclei.

The overlap between two p_y atomic orbitals generates an MO which has the same symmetry properties as that derived from the combination of the two p_x atomic orbitals, but the $\pi(p_y)$ MO lies in a plane perpendicular to that of the $\pi(p_x)$. The $\pi(p_x)$ and $\pi(p_y)$ MOs lie at the same energy: they are *degenerate*. The $\pi^*(p_y)$ and $\pi^*(p_x)$ MOs are similarly related.

Now let us return to the formation of F_2. The valence orbitals of F are the $2s$ and $2p$, and Figure 1.21 shows a general orbital interaction diagram for the overlap of these orbitals. We may assume to a first approximation that the energy separation of the fluorine $2s$ and $2p$ atomic orbitals (the s–p separation) is sufficiently great that only $2s$–$2s$ and $2p$–$2p$ orbital interactions occur. Notice that the stabilization of the $\pi(2p_x)$ and $\pi(2p_y)$ MOs relative to the $2p$ atomic orbitals is less than that of the $\sigma(2p)$, consistent with the relative efficiencies of orbital overlap discussed above. In F_2 there are 14 electrons to be accommodated and, according to the *aufbau* principle, this gives a ground state electronic configuration of $\sigma(2s)^2 \sigma^*(2s)^2 \sigma(2p_z)^2 \pi(2p_x)^2 \pi(2p_y)^2 \pi^*(2p_x)^2 \pi^*(2p_y)^2$. The MO picture for F_2 is consistent with its observed diamagnetism. The predicted bond order is one, in keeping with the result of the VB treatment (see *Section 1.12*).

Figure 1.21 can also be used to describe the bonding in O_2. Each O atom has six valence electrons ($2s^2 2p^4$) and the total of 12 electrons in O_2 give an electronic ground state of $\sigma(2s)^2 \sigma^*(2s)^2 \sigma(2p_z)^2 \pi(2p_x)^2 \pi(2p_y)^2 \pi^*(2p_x)^1 \pi^*(2p_y)^1$. This result is one of the triumphs of early MO theory: the model correctly predicts that O_2 possesses two unpaired electrons and is paramagnetic. From equation 1.31, the bond order in O_2 is two.

What happens if the s–p separation is small?

A comparison of theoretical with experimental data for F_2 and O_2 indicates that the approximations we have made

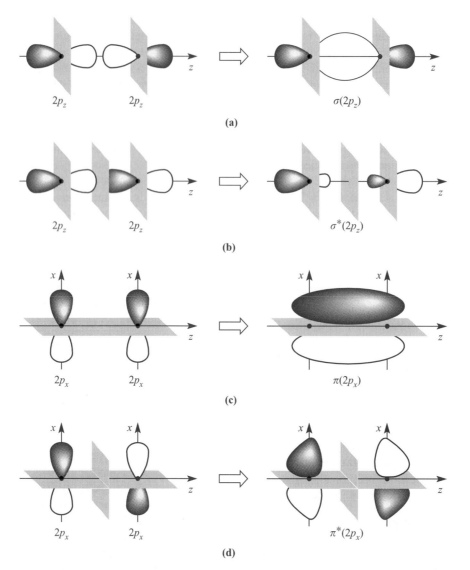

Fig. 1.20 The overlap of two $2p$ atomic orbitals for which the atomic nuclei are defined to lie on the z axis: (a) direct overlap along the z axis gives a $\sigma(2p_z)$ MO (bonding); (b) the formation of the $\sigma^*(2p_z)$ MO (antibonding); (c) sideways overlap of two $2p_x$ atomic orbitals gives a $\pi(2p_x)$ MO (bonding); (d) the formation of $\pi^*(2p_x)$ MO (antibonding). Atomic nuclei are marked in black and nodal planes in grey.

above are appropriate. However, this is not the case if the s–p energy difference is relatively small. In going from Li to F, the effective nuclear charge experienced by an electron in a $2s$ or $2p$ atomic orbital increases and the orbital energy decreases. This is shown in Figure 1.22; the trend is non-linear and the s–p separation increases significantly from B to F. The relatively small s–p separation observed for B and C means that the approximation made when construct-ing the orbital interaction diagram in Figure 1.21 is no longer valid when we construct similar diagrams for the formation of B_2 and C_2. Here, *orbital mixing* may occur[†] between

orbitals of similar symmetry and energy with the result that the ordering of the MOs in B_2, C_2 and N_2 differs from that in F_2 and O_2. Figure 1.23 compares the energy levels of the MOs and the ground state electronic configurations of the diatomics X_2 for X = B, C, N, O and F. Notice the so-called σ–π *crossover* that occurs between N_2 and O_2.

Since the MO approach is a theoretical model, what experimental evidence is there for this σ–π crossover? The actual electronic configurations of molecules are nearly always determined spectroscopically, particularly by *photo-electron spectroscopy*, a technique in which electrons in different orbitals are distinguished by their ionization energies (see *Box 4.2*). Experimental data support the orbital orderings shown in Figure 1.23. Table 1.6 lists experimental bond distances and bond dissociation enthalpies for

[†] This effect is dealt with in detail but at a relatively simple level in Chapter 3 of C.E. Housecroft and E.C. Constable (1997) *Chemistry: An Integrated Approach*, Addison Wesley Longman, Harlow.

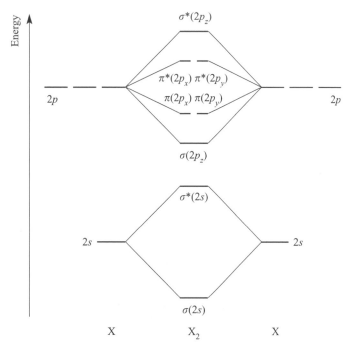

Fig. 1.21 A general orbital interaction diagram for the formation of X_2 in which the valence orbitals of atom X are the $2s$ and $2p$. In constructing this diagram we assume that the s–p separation is sufficiently large that no orbital mixing occurs.

diatomics of the second period including Li_2 and Be_2, and also gives their bond orders calculated from MO theory. Since the nuclear charges change along the series, we should not expect all bonds of order one to have the same bond dissociation enthalpy. However, the general relationship between the bond order, dissociation enthalpy and distance is unmistakable. Table 1.6 also states whether a given molecule is diamagnetic or paramagnetic. We have already seen that MO theory correctly predicts (as does VB theory) that Li_2 is diamagnetic. Similarly, both the MO and VB models are consistent with the diamagnetism of C_2, N_2 and F_2. The paramagnetism of O_2 is predicted by MO theory as we have already seen, and this result is independent of whether the crossover of the $\sigma(2p)$ and $\pi(2p)$ occurs or not

(Figure 1.23). However, the MO model is only consistent with B_2 being paramagnetic *if* the $\pi(2p)$ level is at a lower energy than the $\sigma(2p)$; consider in Figure 1.23 what would happen if the relative orbital energies were reversed.

Worked example 1.8   Molecular orbital theory: properties of diatomics

The bond dissociation enthalpies for the nitrogen–nitrogen bond in N_2 and $[N_2]^-$ are 945 and 765 kJ mol^{-1} respectively. Account for this difference in terms of MO theory, and state whether $[N_2]^-$ is expected to be diamagnetic or paramagnetic.

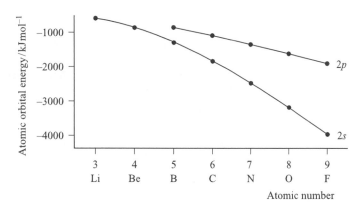

Fig. 1.22 In crossing the period from Li to F, the energies of the $2s$ and $2p$ atomic orbitals decrease owing to the increased effective nuclear charge.

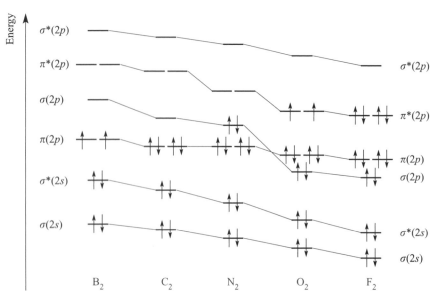

Fig. 1.23 Changes in the energy levels of the MOs and the ground state electronic configurations of homonuclear diatomic molecules involving first row *p*-block elements.

Each N atom has the ground state configuration of $[He]2s^22p^3$.

An MO diagram for N_2, assuming only $2s$–$2s$ and $2p$–$2p$ orbital interactions, can be constructed, the result being as shown in Figure 1.23. From this diagram, the bond order in N_2 is 3.0.

The change from N_2 to $[N_2]^-$ is a one-electron reduction and, assuming that Figure 1.23 is still applicable, an electron is added to a $\pi^*(2p)$ orbital. The calculated bond order in $[N_2]^-$ is therefore 2.5.

The lower bond order of $[N_2]^-$ than N_2 is consistent with a lower bond dissociation enthalpy.

The electron in the $\pi^*(2p)$ orbital is unpaired and $[N_2]^-$ is expected to be paramagnetic.

Self-study exercises

1. Using Figure 1.23 as a basis, account for the fact that $[N_2]^+$ is paramagnetic.

2. Using MO theory, rationalize why the N−N bond distance in $[N_2]^+$ is longer (112 pm) than in N_2 (109 pm).

    [*Ans.* Loss of electron from $\sigma(2p)$ MO]

3. Use Figure 1.23 to rationalize why the bond orders in $[N_2]^+$ and $[N_2]^-$ are *both* 2.5.

1.14 The octet rule

The ground state electronic configurations in Table 1.3 map out a pattern illustrating that filled quantum levels provide 'building blocks' within the electronic configurations of the heavier elements. Worked example 1.6 emphasized that each noble gas is characterized by having a filled quantum level; with the exception of He, this configuration is of the

Table 1.6 Experimental data and bond orders for homonuclear diatomic molecules X_2 in which X is an atom in the period Li to F.

Diatomic	Bond distance / pm	Bond dissociation enthalpy / kJ mol^{-1}	Bond order	Magnetic data
Li$_2$	267	110	1	Diamagnetic
Be$_2$[‡]	–	–	0	–
B$_2$	159	297	1	Paramagnetic
C$_2$	124	607	2	Diamagnetic
N$_2$	110	945	3	Diamagnetic
O$_2$	121	498	2	Paramagnetic
F$_2$	141	159	1	Diamagnetic

[‡] See text on p. 31.

form ns^2np^6. In the early development of bonding models (see *Section 1.11*), the *octet rule* was commonly cited as a means of rationalizing the formation of a particular compound (or ion) which involved an *s*- or *p*-block element. However, the concept of the octet rule is rather limited since it is, strictly, only valid for $n = 2$. Further, many molecules, especially neutral compounds of boron, simply do not contain enough valence electrons for each atom to be associated with eight electrons. Ions (e.g. Mg^{2+} and O^{2-}) often exist only in environments in which electrostatic interaction energies compensate the energies needed to form the ions from atoms.

> An atom obeys the *octet rule* when it gains, loses or shares electrons to give an *outer* shell containing eight electrons with the configuration ns^2np^6.

We have already noted the exception of He, but for $n \geq 3$, the presence of *nd* atomic orbitals leads to the possibility of *expanding the octet*. Although the octet rule is still useful at an elementary level, we must bear in mind that it is restricted to a relatively small number of elements. Its extension to the *18-electron rule* which takes into account the filling of *ns*, *np* and *nd* sub-levels is discussed in Sections 20.4 and 23.3.

Worked example 1.9   The octet rule and the expansion of the octet

In which of the following covalent compounds is the central atom obeying the octet rule: (a) CH_4; (b) H_2S; (c) ClF_3? Rationalize the apparent violation of the rule where it occurs.

(a) CH_4: A C atom has 4 valence electrons and forms 4 covalent bonds by sharing electrons with 4 H atoms to give an octet. This can be represented by the Lewis structure:

$$
\begin{array}{c}
\text{H} \\
\bullet\bullet \\
\text{H} \bullet\!\bullet \text{C} \bullet\!\bullet \text{H} \\
\bullet\bullet \\
\text{H}
\end{array}
$$

(b) H_2S: An S atom has 6 valence electrons and forms 2 covalent bonds by sharing electrons with 2 H atoms to give an octet. The appropriate Lewis structure which shows that S obeys the octet rule in H_2S is:

$$
\begin{array}{c}
\bullet\bullet \\
\text{H} \bullet\!\bullet \text{S} \bullet\bullet \\
\bullet\bullet \\
\text{H}
\end{array}
$$

(c) ClF_3: A Cl atom has 7 valence electrons and should form only one single covalent bond to obtain an octet (e.g.

as in Cl_2). In ClF_3, the octet is expanded:

$$
\begin{array}{c}
\text{F} \\
\bullet\bullet \\
\text{F} \bullet\!\bullet \text{Cl} \bullet\!\bullet \text{F} \\
\bullet\bullet\; \bullet\bullet
\end{array}
$$

The expansion of the octet is possible because the valence electronic configuration of Cl is $3s^23p^5$ and there is an empty set of $3d$ orbitals (close in energy to the $3p$ atomic orbitals) that may be involved in bonding.

Self-study exercises

1. Show that N in NF_3 obeys the octet rule.

2. Show that Se in H_2Se obeys the octet rule, but in $SeCl_4$, it does not.

3. In which of the following molecules is the octet rule violated by the central atom: (a) H_2S; (b) HCN; (c) SO_2; (d) CO_2; (e) SO_3?   [*Ans.* (c); (e)]

1.15 Electronegativity values

In a homonuclear diatomic molecule X_2, the electron density in the region between the nuclei is symmetrical; each X nucleus has the same effective nuclear charge. On the other hand, the disposition of electron density in the region between the two nuclei of a *heteronuclear* diatomic molecule $X-Y$ may be asymmetrical. If the effective nuclear charge of Y is greater than that of X, the pair of electrons in the $X-Y$ covalent bond will be drawn towards Y and away from X.

Pauling electronegativity values, χ^P

In the early 1930s, Linus Pauling established the concept of *electronegativity* which he defined as 'the power of an atom in a molecule to attract electrons to itself' (the electron withdrawing power of an atom). The symbol for electronegativity is χ but we distinguish between different electronegativity scales by use of a superscript; here χ^P for Pauling. Pauling first developed the idea in response to the observation that experimentally determined bond dissociation enthalpy values for heteronuclear bonds were often at variance with those obtained by simple additivity rules. Equation 1.32 shows the relationship between the bond dissociation enthalpy, D, of the homonuclear diatomic X_2 and the enthalpy change of atomization, Δ_aH°, of X. Effectively, this partitions bond enthalpy into a contribution made by each atom, and in this case, the contributions are equal.

$$\Delta_aH^\circ(X) = \tfrac{1}{2} \times D(X-X) \tag{1.32}$$

In equation 1.33, we apply the same type of additivity to the bond in the heteronuclear diatomic XY. Estimates obtained for $D(X-Y)$ using this method sometimes agree quite well with experimental data (e.g. ClBr and ClI), but often differ significantly (e.g. HF and HCl) as worked example 1.10 shows.

$$D(X-Y) = \tfrac{1}{2} \times [D(X-X) + D(Y-Y)] \tag{1.33}$$

Worked example 1.10   Bond enthalpy additivity

Given that $D(H-H)$ and $D(F-F)$ in H_2 and F_2 are 436 and 158 kJ mol^{-1}, estimate the bond dissociation enthalpy of HF using a simple additivity rule. Compare the answer with the experimental value of 570 kJ mol^{-1}.

Assume that we may transfer the contribution made to $D(H-H)$ by an H atom to $D(H-F)$, and similarly for F.

$$
\begin{aligned}
D(H-F) &= \tfrac{1}{2} \times [D(H-H) + D(F-F)] \\
&= \tfrac{1}{2} \times [436 + 158] \\
&= 297 \text{ kJ mol}^{-1}
\end{aligned}
$$

Clearly, this model is unsatisfactory since it grossly underestimates the value of $D(H-F)$ which, experimentally, is found to be 570 kJ mol^{-1}.

Self-study exercises

1. Given that $D(H-H)$, $D(Cl-Cl)$, $D(Br-Br)$ and $D(I-I)$ in H_2, Cl_2, Br_2 and I_2 are 436, 242, 193 and 151 kJ mol^{-1} respectively, estimate (by the above method) values of $D(H-X)$ in HCl, HBr and HI.   [*Ans.* 339; 315; 294 kJ mol^{-1}]

2. Compare your answers to question 1 with experimental values of 432, 366 and 298 kJ mol^{-1} for $D(H-X)$ in HCl, HBr and HI.

Within the framework of the VB approach, Pauling suggested that the difference, ΔD, between an experimental value of $D(X-Y)$ and that obtained using equation 1.33 could be attributed to the ionic contribution to the bond (equation 1.26). The greater the *difference* in electron attracting powers (the *electronegativities*) of atoms X and Y, the greater the contribution made by X^+Y^- (or X^-Y^+), and the greater the value of ΔD. Pauling determined an approximately self-consistent scale of electronegativities, χ^P, as follows. He first converted ΔD values (obtained from $D_{\text{experimental}} - D_{\text{calculated}}$, the calculated value coming from equation 1.33) from units of kJ mol^{-1} to eV in order to obtain a numerically small value of ΔD. He then arbitrarily related ΔD to the difference in electronegativity values between atoms X and Y (equation 1.34).

$$\Delta \chi = \chi^P(Y) - \chi^P(X) = \sqrt{\Delta D} \qquad \text{units of } \Delta D = \text{eV} \tag{1.34}$$

Electronegativity, χ^P, was defined by Pauling as 'the power of an atom in a molecule to attract electrons to itself'.

Over the years, the availability of more accurate thermochemical data has allowed Pauling's initial values of χ^P to be more finely tuned. Values listed in Table 1.7 are those in current use. An element of intuition is required in deciding whether X or Y has the higher electronegativity value and in order to avoid giving an element a negative value of χ^P, $\chi^P(H)$ has been taken as 2.2. Although equation 1.34 implies that the units of χ^P are eV$^{\frac{1}{2}}$, it is not customary to give units to electronegativity values; by virtue of their different definitions, values of χ on different electronegativity scales (see below) possess different units.

In Table 1.7, more than one value of χ^P is listed for some elements. This follows from the fact that the electron withdrawing power of an element varies with its oxidation state (see *Section 7.1*); remember that the Pauling definition of χ^P refers to an atom *in a compound*. Electronegativity values also vary with bond order. Thus for C, χ^P has the values of 2.5 for a C−C bond, 2.75 for a C=C bond and 3.3 for a C≡C bond. For most purposes, the value of $\chi^P(C) = 2.6$ suffices, although the variation underlines the fact that such values must be used with caution.

Following from the original concept of electronegativity, various scales based upon different ground rules have been devised. We focus on two of the more commonly used scales, those of Mulliken and of Allred and Rochow; χ values from these scales are *not directly comparable* with Pauling values, although *trends* in the values should be similar (Figure 1.24). Scales may be adjusted so as to be comparable with the Pauling scale.

Mulliken electronegativity values, χ^M

In one of the simplest approaches to electronegativity, Mulliken took the value of χ^M for an atom to be the mean of the values of the first ionization energy, IE_1, and the first electron affinity, EA_1 (equation 1.35).

$$\chi^M = \frac{IE_1 + EA_1}{2} \qquad \text{where } IE_1 \text{ and } EA_1 \text{ are in eV} \tag{1.35}$$

Allred–Rochow electronegativity values, χ^{AR}

Allred and Rochow chose as a measure of electronegativity of an atom the electrostatic force exerted by the effective nuclear charge Z_{eff} (estimated from Slater's rules, see *Box 1.5*) on the valence electrons. The latter are assumed to reside at a distance from the nucleus equal to the covalent radius, r_{cov}, of the atom. Equation 1.36 gives the method of calculating values of the Allred–Rochow electronegativity, χ^{AR}.

$$\chi^{AR} = \left(3590 \times \frac{Z_{\text{eff}}}{r_{\text{cov}}^2}\right) + 0.744 \qquad \text{where } r_{\text{cov}} \text{ is in pm} \tag{1.36}$$

Table 1.7 Pauling electronegativity (χ^P) values for the s- and p-block elements.

Group 1	Group 2		Group 13	Group 14	Group 15	Group 16	Group 17
H 2.2							
Li 1.0	Be 1.6		B 2.0	C 2.6	N 3.0	O 3.4	F 4.0
Na 0.9	Mg 1.3		Al(III) 1.6	Si 1.9	P 2.2	S 2.6	Cl 3.2
K 0.8	Ca 1.0	(d-block elements)	Ga(III) 1.8	Ge(IV) 2.0	As(III) 2.2	Se 2.6	Br 3.0
Rb 0.8	Sr 0.9		In(III) 1.8	Sn(II) 1.8 Sn(IV) 2.0	Sb 2.1	Te 2.1	I 2.7
Cs 0.8	Ba 0.9		Tl(I) 1.6 Tl(III) 2.0	Pb(II) 1.9 Pb(IV) 2.3	Bi 2.0	Po 2.0	At 2.2

Since, however, Slater's rules are partly empirical and covalent radii are unavailable for some elements, the Allred–Rochow scale is no more rigid or complete than the Pauling one.

Electronegativity: final remarks

Despite the somewhat dubious scientific basis of all of the three methods described above, the trends in electronegativities obtained by them are roughly in agreement as Figure 1.24 exemplifies. The most useful of the scales for application in inorganic chemistry is probably the Pauling scale, which, being based empirically on thermochemical data, can reasonably be used to predict similar data. For example, if the electronegativities of two elements X and Y have been derived from the single covalent bond enthalpies of HX, HY, X_2, Y_2 and H_2, we can estimate the bond dissociation enthalpy of the bond in XY with a fair degree of reliability.

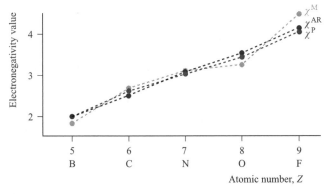

Fig. 1.24 Although electronegativity values for a given element from different scales cannot be expected to be the same, *trends* in values along a series of elements are comparable. This is illustrated with values of χ^P (Pauling; blue), χ^M (Mulliken; grey) and χ^{AR} (Allred–Rochow; black) for first row elements from the p-block.

Worked example 1.11   Estimation of a bond dissociation enthalpy from χ^P values

Using the following data, estimate a value for $D(Br-F)$:

$D(F-F) = 158 \text{ kJ mol}^{-1}$ $D(Br-Br) = 224 \text{ kJ mol}^{-1}$

$\chi^P(F) = 4.0$ $\chi^P(Br) = 3.0$

First, use the values of χ^P to find ΔD:

$$\sqrt{\Delta D} = \chi^P(F) - \chi^P(Br) = 1.0$$

$$\Delta D = 1.0^2 = 1.0$$

This gives the value in eV; convert to kJ mol^{-1}:

$$1.0 \text{ eV} \approx 96.5 \text{ kJ mol}^{-1}$$

ΔD is defined as follows:

$$\Delta D = [D(\text{Br}-\text{F})_{\text{experimental}}] - \{\tfrac{1}{2} \times [D(\text{Br}-\text{Br}) + D(\text{F}-\text{F})]\}$$

So an estimate of $D(\text{Br}-\text{F})$ is given by:

$$\begin{aligned} D(\text{Br}-\text{F}) &= \Delta D + \{\tfrac{1}{2} \times [D(\text{Br}-\text{Br}) + D(\text{F}-\text{F})]\} \\ &= 96.5 + \{\tfrac{1}{2} \times [224 + 158]\} \\ &= 287.5 \text{ kJ mol}^{-1} \end{aligned}$$

[This compares with an experimental value of $250.2 \text{ kJ mol}^{-1}$.]

Self-study exercises

1. Use the following data to estimate the bond dissociation enthalpy of BrCl: $D(\text{Br}-\text{Br}) = 224 \text{ kJ mol}^{-1}$; $D(\text{Cl}-\text{Cl}) = 242 \text{ kJ mol}^{-1}$; $\chi^P(\text{Br}) = 3.0$; $\chi^P(\text{Cl}) = 3.2$.
   [*Ans.* $\approx 237 \text{ kJ mol}^{-1}$; actual experimental value $= 218 \text{ kJ mol}^{-1}$]

2. Use the following data to estimate the bond dissociation enthalpy of HF: $D(\text{H}-\text{H}) = 436 \text{ kJ mol}^{-1}$; $D(\text{F}-\text{F}) = 158 \text{ kJ mol}^{-1}$; $\chi^P(\text{H}) = 2.2$; $\chi^P(\text{F}) = 4.0$.
   [*Ans.* $\approx 610 \text{ kJ mol}^{-1}$; actual experimental value $= 570 \text{ kJ mol}^{-1}$]

In this book we avoid the use of the concept of electronegativity as far as possible and base the systemization of descriptive inorganic chemistry on rigidly defined and independently measured thermochemical quantities such as ionization energies, electron affinities, bond dissociation enthalpies, lattice energies and hydration enthalpies. However, some mention of electronegativity values is unavoidable.

1.16 Dipole moments

Polar diatomic molecules

The symmetrical electron distribution in the bond of a homonuclear diatomic renders the bond *non-polar*. In a heteronuclear diatomic, the electron withdrawing powers of the two atoms may be different, and the bonding electrons are drawn closer towards the more electronegative atom. The bond is *polar* and possesses an *electric dipole moment* (μ). Be careful to distinguish between *electric* and *magnetic* dipole moments (see *Section 20.8*).

The dipole moment of a diatomic XY is given by equation 1.37 where d is the distance between the point electronic charges (i.e. the internuclear separation), e is the charge on the electron (1.602×10^{-19} C) and q is point charge. The SI unit of μ is the coulomb metre (C m) but for convenience, μ

tends to be given in units of debyes (D) where $1 \text{ D} = 3.336 \times 10^{-30} \text{ C m}$.

$$\mu = q \times e \times d \tag{1.37}$$

Worked example 1.12 Dipole moments

The dipole moment of a gas phase HBr molecule is 0.827 D. Determine the charge distribution in this diatomic if the bond distance is 141.5 pm. ($1 \text{ D} = 3.336 \times 10^{-30} \text{ C m}$)

To find the charge distribution we need to find q using the expression:

$$\mu = qed$$

Units must be consistent:

$$d = 141.5 \times 10^{-12} \text{ m}$$

$$\mu = 2.779 \times 10^{-30} \text{ C m}$$

$$\begin{aligned} q &= \frac{\mu}{ed} \\ &= \frac{2.779 \times 10^{-30}}{1.602 \times 10^{-19} \times 141.5 \times 10^{-12}} \\ &= 0.123 \text{ (no units)} \end{aligned}$$

The charge distribution can be written as: $\overset{+0.123}{\text{H}} \longrightarrow \overset{-0.123}{\text{Br}}$ since we know that Br is more electronegative than H.

Self-study exercises

1. The bond length in HF is 92 pm, and the dipole moment is 1.83 D. Determine the charge distribution in the molecule.
   [*Ans.* $\overset{+0.41}{\text{H}} \longrightarrow \overset{-0.41}{\text{F}}$]

2. The bond length in ClF is 163 pm. If the charge distribution is: $\overset{+0.11}{\text{Cl}} \longrightarrow \overset{-0.11}{\text{F}}$, show that the molecular dipole moment is 0.86 D.

In worked example 1.12, the result indicates that the electron distribution in HBr is such that effectively 0.123 electrons have been transferred from H to Br. The partial charge separation in a polar diatomic molecule can be represented by use of the symbols δ^+ and δ^- assigned to the appropriate nuclear centres, and a barred arrow represents the direction in which the dipole moment acts. This is shown for HF in structure **1.14**. Keep in mind that a dipole moment is a vector quantity.

$$\overset{\delta^+}{\text{H}} \underline{\hspace{1.5em}} \overset{\delta^-}{\text{F}}$$
$$\longmapsto$$
$$(1.14)$$

A word of caution: attempts to calculate the degree of ionic character of the bonds in heteronuclear diatomics from their observed dipole moments and the moments calculated on the basis of charge separation neglect the effects of any lone pairs of electrons and are therefore of doubtful validity. The significant effects of lone pairs are illustrated below in Example 3.

Molecular dipole moments

Polarity is a *molecular property*. For polyatomic species, the net molecular dipole moment depends upon the magnitudes and relative directions of all the bond dipole moments in the molecule. In addition, lone pairs of electrons may contribute significantly to the overall value of μ. We consider three examples below, using the Pauling electronegativity values of the atoms involved to give a indication of individual bond polarities. This practice is useful but must be treated with caution as it can lead to spurious results, e.g. when the bond multiplicity is not taken into account when assigning a value of χ^P. Experimental values of molecular electric dipole moments are determined by microwave spectroscopy or other spectroscopic methods.

Example 1: CF_4

(1.15)

The values of $\chi^P(C)$ and $\chi^P(F)$ are 2.6 and 4.0, respectively, indicating that each $C-F$ bond is polar in the sense $C^{\delta+}-F^{\delta-}$. The CF_4 molecule (**1.15**) is tetrahedral and the four bond moments (each a vector of equivalent magnitude) oppose and cancel one another. The effects of the F lone pairs also cancel out, and the net result is that CF_4 is non-polar.

Example 2: H_2O

(1.16)

For O and H, $\chi^P = 3.4$ and 2.2, respectively, showing that each $O-H$ bond is polar in the sense $O^{\delta-}-H^{\delta+}$. Since the H_2O molecule is non-linear, resolution of the two bond vectors gives a resultant dipole moment which acts in the direction shown in structure **1.16**. In addition, the O atom has two lone pairs of electrons which will reinforce the overall moment. The experimental value of μ for H_2O in the gas phase is 1.85 D.

Example 3: NH_3 and NF_3

The molecules NH_3 and NF_3 have trigonal pyramidal structures (**1.17**), and have dipole moments of 1.47 and 0.24 D respectively. This significant difference may be rationalised

X = H or F

(1.17)

by considering the bond dipole moments and the effects of the N lone pair. The values of $\chi^P(N)$ and $\chi^P(H)$ are 3.0 and 2.2, so each bond is polar in the sense $N^{\delta-}-H^{\delta+}$. The resultant dipole moment acts in a direction which is reinforced by the lone pair. Ammonia is a polar molecule with N carrying a partial negative charge. In NF_3, each $N-F$ bond is polar in the sense $N^{\delta+}-F^{\delta-}$ since F is more electronegative ($\chi^P(F) = 4.0$) than N. The resultant dipole moment *opposes* the effects of the lone pair, rendering the NF_3 molecule far less polar than NH_3.

Clearly, molecular shape is an important factor in determining whether a molecule is polar or not and question 1.31 at the end of the chapter considers this further.

1.17 MO theory: heteronuclear diatomic molecules

In this section, we return to MO theory and apply it to heteronuclear diatomic molecules. In each of the orbital interaction diagrams constructed in Section 1.13 for *homonuclear* diatomics, the resultant MOs contained *equal* contributions from each atomic orbital involved. This is represented in equation 1.27 for the bonding MO in H_2 by the fact that each of the wavefunctions ψ_1 and ψ_2 contributes equally to ψ_{MO}, and the representations of the MOs in H_2 (Figure 1.18) depict *symmetrical* orbitals. Now we look at representative examples of diatomics in which the MOs may contain *different* atomic orbital contributions, a scenario that is typical for heteronuclear diatomics.

First, we must consider likely restrictions when we are faced with the possibility of combining different types of atomic orbitals.

Which orbital interactions should be considered?

At the beginning of Section 1.13 we stated some general requirements that should be met for orbital interactions to take place efficiently. We stated that orbital interactions are allowed if the *symmetries* of the atomic orbitals are compatible with one another. In our approach to the bonding in a diatomic, we made the assumption that only the interactions between *like* atomic orbitals, e.g. $2s-2s$, $2p_z-2p_z$, need be considered. Such interactions are *symmetry-allowed*, and in addition, in a *homonuclear* diatomic the energies of like atomic orbitals on the two atoms are exactly matched.

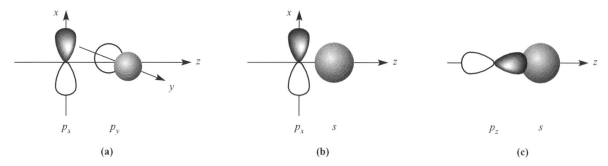

Fig. 1.25 Overlap between atomic orbitals is not always allowed by symmetry. Combinations (a) and (b) lead to non-bonding situations but (c) is symmetry-allowed and gives rise to a bonding interaction.

In a heteronuclear diatomic, we often encounter two atoms that have different basis sets of atomic orbitals, or have sets of similar atomic orbitals lying at different energies. For example, in CO, although both C and O possess valence $2s$ and $2p$ atomic orbitals, the greater effective nuclear charge of O means that its atomic orbitals lie at a lower energy than those of C. Before we look more closely at some examples of heteronuclear diatomics, let us briefly consider some symmetry-allowed and disallowed orbital interactions. It is important to remember that we are looking at these symmetry properties *with respect to the internuclear axis*. In our earlier discussion of homonuclear diatomics (e.g. Figure 1.21), we ignored the possibility of overlap between the p_x and p_y orbitals. Such an interaction between orthogonal p atomic orbitals (Figure 1.25a) would give a zero overlap integral. Similarly, for nuclei lying on the z axis, interaction between p_x and p_z, or p_y and p_z, orbitals gives zero overlap. An interaction between an s and a p atomic orbital *may* occur depending upon the orientation of the p orbital. In Figure 1.25b, overlap would be partly bonding and partly antibonding and the net effect is a *non-bonding* interaction. On the other hand, Figure 1.25c shows an s–p interaction that *is* allowed by symmetry. Whether or not this leads to effective overlap depends upon the relative energies of the two atomic orbitals. This is illustrated in Figure 1.26 for a diatomic XY. Let the interaction between ψ_X and ψ_Y be symmetry-allowed; the orbital energies are not the same but are close enough that overlap between the orbitals is efficient. The orbital interaction diagram shows that the energy of the bonding MO is closer to $E(\psi_Y)$ than to $E(\psi_X)$ and the consequence of this is that the bonding orbital possesses *greater Y than X character*. This is expressed in equation 1.38 in which $c_2 > c_1$. For the antibonding MO, the situation is reversed, and ψ_X contributes more than ψ_Y; in equation 1.39, $c_3 > c_4$.

$$\psi_{MO} = N[(c_1 \times \psi_X) + (c_2 \times \psi_Y)] \qquad (1.38)$$

$$\psi_{MO}^* = N^*[(c_3 \times \psi_X) + (c_4 \times \psi_Y)] \qquad (1.39)$$

The energy separation ΔE in Figure 1.26 is critical. If it is large, interaction between ψ_X and ψ_Y will be inefficient (the overlap integral is very small). In the extreme case, there is no interaction at all and both ψ_X and ψ_Y appear in the XY

molecule as unperturbed *non-bonding* atomic orbitals. We illustrate this below.

Hydrogen fluoride

The ground state configurations of H and F are $1s^1$ and $[He]2s^2 2p^5$ respectively. Since $Z_{eff}(F) > Z_{eff}(H)$, the F $2s$ and $2p$ atomic orbital energies are significantly lowered with respect to the H $1s$ atomic orbital (Figure 1.27).

We now have to consider which atomic orbital interactions are symmetry-allowed and then ask whether the atomic orbitals are sufficiently well energy-matched. First, define the axis set for the orbitals; let the nuclei lie on the z axis. Overlap between the H $1s$ and F $2s$ orbitals is allowed by symmetry, but the energy separation is very large (note the break on the energy axis in Figure 1.27). Overlap between the H $1s$ and F $2p_z$ atomic orbitals is also symmetry-allowed and there is a reasonable orbital energy match. As Figure 1.27 shows, an interaction occurs leading to σ and σ^* MOs; the σ-orbital has greater F than H character. The two F $2p_x$ and $2p_y$ atomic orbitals become non-bonding orbitals in HF since no net bonding interaction with the H $1s$ atomic orbital is possible. Once the orbital interaction diagram has been constructed, the

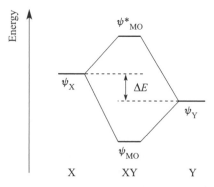

Fig. 1.26 The relative energies of atomic orbitals of X and Y will dictate whether an interaction (formally allowed by symmetry) will lead to efficient overlap or not. Here, an interaction occurs but the contribution made by ψ_Y to ψ_{MO} is greater than that made by ψ_X, while ψ_X contributes more than ψ_Y to the antibonding MO.

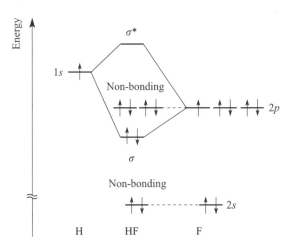

Fig. 1.27 An orbital interaction diagram for the formation of HF. Only the valence atomic orbitals and electrons are shown. The break in the vertical (energy) axis indicates that the energy of the F $2s$ atomic orbital is much lower than is actually shown.

eight valence electrons are accommodated as shown in Figure 1.27, giving a bond order of one in HF. The MO picture of HF indicates that the electron density is greater around the F than H nucleus; the model is consistent with a polar H—F bond in the sense $H^{\delta^+}-F^{\delta^-}$. [**Exercise**: Sketch pictorial representations of the σ and σ^* MOs in HF]

Carbon monoxide

In Chapter 23 we discuss the chemistry of compounds containing metal–carbon bonds (*organometallic compounds*) of which *metal carbonyls* of the type $M_x(CO)_y$ are one group. In order to investigate the way in which CO bonds to metals, we must appreciate the electronic structure of the carbon monoxide molecule.

Before constructing an orbital interaction diagram for CO, we must take note of the following:

- $Z_{eff}(O) > Z_{eff}(C)$;
- the energy of the O $2s$ atomic orbital is lower than that of the C $2s$ atomic orbital;
- the $2p$ level in O is at lower energy than that in C;
- the $2s$–$2p$ energy separation in O is greater than that in C (Figure 1.22).

We could generate an approximate orbital interaction diagram by assuming that only $2s$–$2s$ and $2p$–$2p$ overlap occurs, but, as a consequence of the relative atomic orbital energies, such a picture is too simplistic. Figure 1.28a gives a more accurate MO picture of the electronic structure of CO obtained computationally, although even this is over-simplified. Figure 1.28b illustrates more fully the extent of orbital mixing, but for our discussion, the simplified picture presented in Figure 1.28a suffices. Two of the more important features to notice are:

- The highest occupied MO (HOMO) is σ-bonding and possesses predominantly *carbon* character; occupation

of this MO effectively creates an outward pointing lone pair centred on C.
- A degenerate pair of $\pi^*(2p)$ MOs make up the lowest unoccupied MOs (LUMOs); each MO possesses more C than O character.

Pictorial representations of the HOMO and one of the LUMOs are given in Figure 1.28.

HOMO = highest occupied molecular orbital.
LUMO = lowest unoccupied molecular orbital.

1.18 Isoelectronic molecules

Two species are *isoelectronic* if they possess the same *total* number of electrons.

If two species contain the same number of electrons they are *isoelectronic*; CH_4, $[BH_4]^-$ and $[NH_4]^+$ are isoelectronic, since each contains a total of 10 electrons. Two other series of isoelectronic species are N_2, CO and $[NO]^+$, and $[SiF_6]^{2-}$, $[PF_6]^-$ and SF_6.

The word *isoelectronic* is often used in the context of meaning 'same number of *valence electrons*', although strictly such usage should always be qualified; e.g. HF, HCl and HBr are isoelectronic *with respect to their valence electrons*.

The isoelectronic principle is simple but important. Often, species that are isoelectronic possess the same structure and are *isostructural*. However, if the term is used loosely and only the valence electrons are considered, this expectation may not hold. With respect to their valence electrons, CO_2 with SiO_2 are isoelectronic, but whereas CO_2 is a linear molecule, SiO_2 possesses an infinite lattice in which Si is 4-coordinate (see *Section 13.9*).

1.19 Molecular shape and the VSEPR model

Valence-shell electron-pair repulsion theory

The shapes of molecules containing a central p-block atom tend to be controlled by the number of electrons in the valence shell of the central atom. The *valence-shell electron-pair repulsion* (VSEPR) theory provides a simple model for predicting the shapes of such species. The model combines original ideas of Sidgwick and Powell with extensions developed by Nyholm and Gillespie, and may be summarized as follows:

- Each valence shell electron pair of the central atom E in a molecule EX_n containing E–X single bonds is stereochemically significant, and repulsions between them determine the molecular shape.

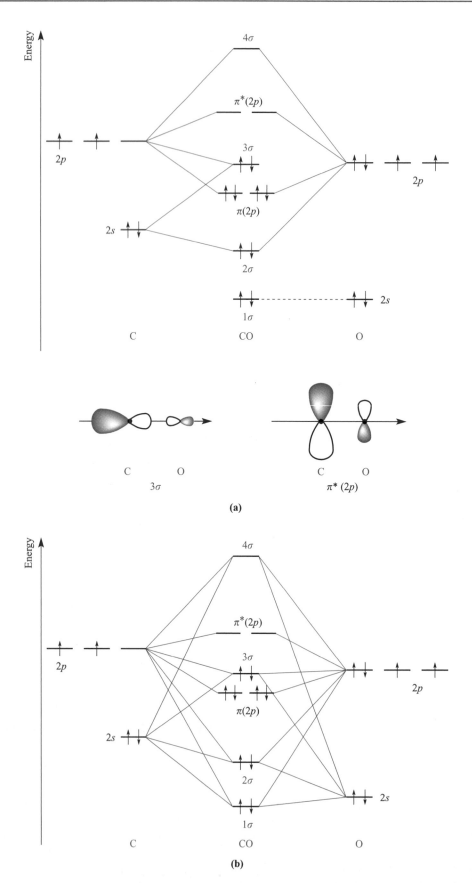

Fig. 1.28 (a) A simplified orbital interaction diagram for CO which allows for the effects of some orbital mixing. The labels 1σ, $2\sigma \ldots$ rather than $\sigma(2s) \ldots$ are used because some orbitals contain both s and p character. (b) A more rigorous (but still qualitative) orbital interaction diagram for CO.

- Electron–electron repulsions decrease in the sequence:

  lone pair–lone pair > lone pair–bonding pair > bonding pair–bonding pair.

- Where the central atom E is involved in multiple bond formation to atoms X, electron–electron repulsions decrease in the order:

  triple bond–single bond > double bond–single bond > single bond–single bond.

- Repulsions between the bonding pairs in EX_n depend on the difference between the electronegativities of E and X; electron–electron repulsions are less the more the E–X bonding electron density is drawn away from the central atom E.

The VSEPR theory works best for simple halides of the *p*-block elements, but may also be applied to species with other substituents. However, the model does *not* take *steric factors* (i.e. the relative sizes of substituents) into account.

In a molecule EX_n, there is a minimum energy arrangement for a given number of electron pairs. In $BeCl_2$ (Be, group 2), repulsions between the two pairs of electrons in the valence shell of Be are minimized if the Cl–Be–Cl unit is linear. In BCl_3 (B, group 13), electron–electron repulsions are minimized if a trigonal planar arrangement of electron pairs (and thus Cl atoms) is adopted. The structures in the left-hand column of Figure 1.29 represent the minimum energy structures for EX_n molecules for $n = 2$–8 and in which there are no lone pairs of electrons associated with E. Table 1.8 gives further representations of these structures, along with their *ideal* bond angles. Ideal bond angles may be expected when all the X substituents are identical, but in, for example, BF_2Cl (**1.18**) some distortion occurs because Cl is larger than F, and the shape is only *approximately* trigonal planar.

∠ F–B–F = 118°

(1.18)   (1.19)

The presence of lone pairs is taken into account using the guidelines above and the 'parent structures' in Figure 1.29. In H_2O (**1.19**), repulsions between the two bonding pairs and two lone pairs of electrons lead to a tetrahedral arrangement but owing to the inequalities between the lone pair–lone pair, lone pair–bonding pair and bonding pair–bonding pair interactions, distortion from an ideal arrangement arises and this is consistent with the observed H–O–H bond angle of 104.5°.

Worked example 1.13 VSEPR theory

Predict the structures of (a) XeF_2 and (b) $[XeF_5]^-$.

Xe is in group 18 and possesses eight electrons in its valence shell. F is in group 17, has seven valence electrons and forms one covalent single bond. Before applying the VSEPR model, decide which is the central atom in the molecule. In each of XeF_2 and $[XeF_5]^-$, Xe is the central atom.

(a) XeF_2. Two of the eight valence electrons of the Xe atom are used for bonding (two Xe–F single bonds), and so around the Xe centre there are two bonding pairs of electrons and three lone pairs.

The parent shape is a trigonal bipyramid (Figure 1.29) with the three lone pairs in the equatorial plane to minimize lone pair-lone pair repulsions. The XeF_2 molecule is therefore linear:

(b) $[XeF_5]^-$. The electron from the negative charge is conveniently included within the valence shell of the central atom. Five of the nine valence electrons are used for bonding and around the Xe centre there are five bonding pairs and two lone pairs of electrons.

The parent shape is a pentagonal bipyramid (Figure 1.29) with the two lone pairs opposite to each other to minimize lone pair–lone pair repulsions. The $[XeF_5]^-$ anion is therefore pentagonal planar:

When structures are determined by diffraction methods, *atom* positions are located. Thus, in terms of a structural descriptor XeF_2 is linear and $[XeF_5]^-$ is pentagonal planar. In the diagrams above we show two representations of each species, one with the lone pairs to emphasize the origin of the prediction from the VSEPR model.

Self-study exercise

Show that VSEPR theory is in agreement with the following molecular shapes:

BCl_3	trigonal planar
$[IF_5]^{2-}$	pentagonal planar
$[NH_4]^+$	tetrahedral
SF_6	octahedral

Limitations of VSEPR theory

The generalizations of the VSEPR model are useful, but the theory has limitations when it comes to the *detailed* structures of 'difficult' species such as ClF_3. Figure 1.30a

2-Coordinate

3-Coordinate

4-Coordinate

5-Coordinate

6-Coordinate

7-Coordinate

8-Coordinate

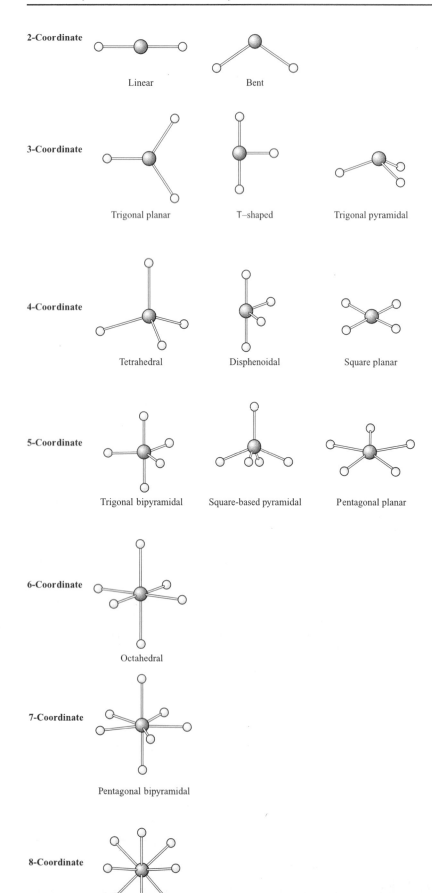

Linear

Bent

Trigonal planar

T–shaped

Trigonal pyramidal

Tetrahedral

Disphenoidal

Square planar

Trigonal bipyramidal

Square-based pyramidal

Pentagonal planar

Octahedral

Pentagonal bipyramidal

Square antiprism

Fig. 1.29 Common shapes for molecules of type EX_n or ions of type $[EX_n]^{m+/-}$. The structures in the left-hand column are 'parent' shapes used in VSEPR theory.

Table 1.8 'Parent' shapes for EX_n molecules ($n = 2$–8).

Formula EX_n	Coordination number of atom E	Shape	Spatial representation	Ideal bond angles ($\angle X$–E–X) / degrees
EX_2	2	Linear	X—E—X	180
EX_3	3	Trigonal planar		120
EX_4	4	Tetrahedral		109.5
EX_5	5	Trigonal bipyramidal		$\angle X_{ax}$–E–$X_{eq} = 90$ $\angle X_{eq}$–E–$X_{eq} = 120$
EX_6	6	Octahedral		$\angle X_1$ – E – $X_2 = 90$
EX_7	7	Pentagonal bipyramidal		$\angle X_{ax}$–E–$X_{eq} = 90$ $\angle X_{eq}$–E–$X_{eq} = 72$
EX_8	8	Square antiprismatic		$\angle X_1$–E–$X_2 = 78$ $\angle X_1$–E–$X_3 = 73$

shows the observed structure of ClF_3, and VSEPR theory may be used to *rationalize* this **T**-shaped arrangement (Figure 1.30b). The small departure of the F–Cl–F bond angle from 90° (Table 1.8) may be attributed to lone pair–bonding pair repulsion. The feature of much greater significance is the large difference in Cl–F bond distances which implies a substantial difference in bond order between the axial and equatorial bonds. It is impossible to predict accurately the structure of ClF_3, or of the related BrF_3 and IF_3. [*Exercise*: What structural possibilities, other than that in Figure 1.30, does VSEPR theory give for ClF_3?]

The difference in axial (ax) and equatorial (eq) bond distances for structures based on the trigonal bipyramid is not unique to ClF_3. For example, in PF_5, they are 158 and 153 pm respectively, and in SF_4 (**1.20**) they are 165 and

155 pm. Bond distance differences are also observed in BrF_5 (**1.21**), the structure of which is derived from an octahedral arrangement with a lone pair occupying one site. The Br atom lies a little below the plane containing the equatorial F atoms ($\angle F_{ax}$–Br–$F_{eq} = 84.5°$) and the Br–F_{ax} and Br–F_{eq} bond distances are 168 and 178 pm respectively.

(1.20)   (1.21)

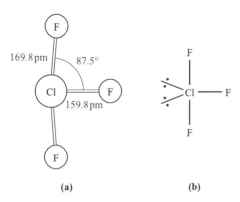

Fig. 1.30 (a) The experimentally determined structure of ClF_3 and (b) the rationalization of this structure using VSEPR theory.

The isoelectronic species IF_7 and $[TeF_7]^-$ are predicted by VSEPR theory to be pentagonal bipyramidal and this is observed. However, electron diffraction data for IF_7 and X-ray diffraction data for $[Me_4N][TeF_7]$ reveal that the equatorial F atoms are not coplanar, a result that cannot be predicted by the VSEPR model. Moreover, in IF_7, the $I-F_{ax}$ and $I-F_{eq}$ distances are 179 and 186 pm respectively, and in $[TeF_7]^-$, the $Te-F_{ax}$ bond distance is 179 pm and the $Te-F_{eq}$ distances lie in the range 183 to 190 pm.

Among species in which VSEPR theory appears to fail are $[SeCl_6]^{2-}$, $[TeCl_6]^{2-}$ and $[BrF_6]^-$. When characterized as alkali metal salts, these anions are found to possess *regular octahedral* structures *in the solid state*, whereas VSEPR theory suggests shapes based on there being seven electron pairs around the central atom. Although these structures cannot readily be predicted, we can rationalize them in terms of having a *stereochemically inactive* pair of electrons. Stereochemically inactive lone pairs are usually observed for the heaviest members of a periodic group, and the tendency for valence shell s electrons to adopt a non-bonding role in a molecule is called the *stereochemical inert pair effect*. Similarly, $[SbCl_6]^-$ and $[SbCl_6]^{3-}$ *both* possess regular octahedral structures. Finally, consider $[XeF_8]^{2-}$, $[IF_8]^-$ and $[TeF_8]^{2-}$. As expected from VSEPR theory, $[IF_8]^-$ and $[TeF_8]^{2-}$ are square antiprismatic; this structure is related to the cube but with one face of the cube rotated through $45°$. However, $[XeF_8]^{2-}$ also adopts this structure, indicating that the lone pair of electrons is stereochemically inactive.

It is important to note that whereas VSEPR theory may be applicable to p-block species, it is *not* appropriate for those of the d-block (see *Chapters 19–23*).

If the presence of a lone pair of electrons influences the shape of a molecule or ion, the lone pair is *stereochemically active*. If it has no effect, the lone pair is *stereochemically inactive*. The tendency for the pair of *valence s* electrons to adopt a non-bonding role in a molecule or ion is termed the *stereochemical inert pair effect*.

1.20 Molecular shape: geometrical isomerism

In this section we discuss *geometrical isomerism*. Examples are taken from both p- and d-block chemistry. Other types of isomerism are described in Section 19.8.

If two species have the same molecular formulae and the same structural framework, but differ in the spatial arrangement of different atoms or groups about a central atom or a double bond, then the compounds are *geometrical isomers*.

Square planar species

In a square planar species such as $[ICl_4]^-$ or $[PtCl_4]^{2-}$ (**1.22**), the four Cl atoms are equivalent. Similarly, in $[PtCl_3(PMe_3)]^-$ (**1.23**), there is only one possible arrangement of the groups around the square planar Pt(II) centre. (The use of arrows or lines to depict bonds in coordination compounds is discussed in Section 6.11.)

$$
\begin{bmatrix}
& Cl & \\
& | & \\
Cl & \!\!-\!Pt\!-\!\! & Cl \\
& | & \\
& Cl &
\end{bmatrix}^{2-}
\qquad
\begin{bmatrix}
& PMe_3 & \\
& | & \\
Cl & \!\!-\!Pt\!-\!\! & Cl \\
& | & \\
& Cl &
\end{bmatrix}^{-}
$$

(1.22) (1.23)

The introduction of two PMe_3 groups to give $[PtCl_2(PMe_3)_2]$ leads to the possibility of two *geometrical isomers*, i.e. two possible spatial arrangements of the groups around the square planar Pt(II) centre. These are shown in structures **1.24** and **1.25** and the names *cis* and *trans* refer to the positioning of the Cl (or PMe_3) groups, adjacent to or opposite one another.

cis-isomer   *trans*-isomer

(1.24) (1.25)

Square planar species of the general form EX_2Y_2 or EX_2YZ may possess *cis*- and *trans*-isomers.

Octahedral species

There are two types of geometrical isomerism associated with octahedral species. In EX_2Y_4, the X groups may be

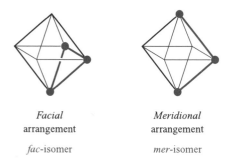

Facial
arrangement

fac-isomer

Meridional
arrangement

mer-isomer

Fig. 1.31 The origin of the names *fac*- and *mer*-isomers.

mutually *cis* or *trans* as shown for $[SnF_4Me_2]^{2-}$ (**1.26** and **1.27**). In the solid state structure of $[NH_4]_2[SnF_4Me_2]$, the anion is present as the *trans*-isomer.

cis-isomer

(1.26)

trans-isomer

(1.27)

If an octahedral species has the general formula EX_3Y_3, then the X groups (and also the Y groups) may be arranged so as to define one face of the octahedron or may lie in a plane that also contains the central atom E (Figure 1.31). These geometrical isomers are labelled *fac* (facial) and *mer* (meridional) respectively. In $[PCl_4][PCl_3F_3]$, the $[PCl_3F_3]^-$ anion exists as both *mer*- and *fac*-isomers (**1.28** and **1.29**).

fac-isomer

(1.28)

mer-isomer

(1.29)

An *octahedral* species containing two identical groups (e.g. of type EX_2Y_4) may possess *cis*- and *trans*-arrangements of these groups. An octahedral species containing three identical groups (e.g. of type EX_3Y_3) may possess *fac*- and *mer*-isomers.

Trigonal bipyramidal species

In trigonal bipyramidal EX_5, there are two types of X atom: axial and equatorial. This leads to the possibility of geometrical isomerism when more than one type of substituent is attached to the central atom. Iron pentacarbonyl, $Fe(CO)_5$, is trigonal bipyramidal and when one CO is exchanged for PPh_3, two geometrical isomers are possible depending on

whether the PPh_3 ligand is axially (**1.30**) or equatorially (**1.31**) sited.

(1.30)

(1.31)

For trigonal bipyramidal EX_2Y_3, three geometrical isomers (**1.32** to **1.34**) are possible depending on the relative positions of the X atoms. Steric factors may dictate which isomer is preferred for a given species; e.g. in the static structure of PCl_3F_2, the F atoms occupy the two axial sites, and the larger Cl atoms reside in the equatorial plane.

(1.32)

(1.33)

(1.34)

In a *trigonal bipyramidal* species, geometrical isomerism arises because of the presence of *axial* and *equatorial* sites.

High coordination numbers

The presence of axial and equatorial sites in a pentagonal bipyramidal molecule leads to geometrical isomerism in a similar manner to that in a trigonal bipyramidal species. In a square antiprismatic molecule EX_8, each X atom is identical (Figure 1.29). Once two or more different atoms or groups are present, e.g. EX_6Y_2, geometrical isomers are possible. As an exercise, draw out the four possibilities for square antiprismatic EX_6Y_2.

Double bonds

In contrast to a single (σ) bond where free rotation is generally assumed, rotation about a double bond is not a low energy process. The presence of a double bond may therefore lead to geometrical isomerism as is observed for N_2F_2. Each N atom carries a lone pair as well as forming one N–F single bond and a N=N double bond. Structures **1.35** and **1.36** show the *trans*- and *cis*-isomers[†] of N_2F_2 respectively.

(1.35)

(1.36)

[†] In organic chemistry, IUPAC nomenclature uses the prefix (*E*)- for a *trans*-arrangement of groups and (*Z*)- for a *cis*-arrangement, but for inorganic compounds, the terms *trans*- and *cis*- remain in use.

Summary

This chapter gives a 'review base' for the remainder of the book. Readers will be familiar with most of the topics, and in the reading list below, more basic texts are suggested for more in-depth coverage.

Suggested further reading

First-year chemistry: basic principles

C.E. Housecroft and E.C. Constable (1997) *Chemistry: An Integrated Approach*, Addison Wesley Longman, Harlow – A readable text that covers first-year general chemistry and provides detailed background of all the material that is taken as assumed knowledge in this book.

C.E. Housecroft and E.C. Constable (1997) *Self-Study Workbook* to accompany *Chemistry: An Integrated Approach*, Addison Wesley Longman, Harlow – This workbook provides adequate practice in problem-solving in topics covered in many first-year general chemistry courses.

P.W. Atkins and L.L. Jones (1997) *Chemistry: Molecules, Matter, and Change*, 3rd edn, Freeman, New York – This first-year text is suitable for reviewing basic topics.

S.S. Zumdahl (1995) *Chemical Principles*, 2nd edn, Heath, Lexington – A useful first-year text for an overview of basic concepts.

Physical methods

E.A.V. Ebsworth, D.W.H. Rankin and S. Cradock (1991) *Structural Methods in Inorganic Chemistry*, 2nd edn, Blackwell Scientific Publications, Oxford – A readable text which gives details of the important methods by which chemists determine structural details of compounds.

B.K. Hunter and J.K.M. Sanders (1993) *Modern NMR Spectroscopy: A Guide for Chemists*, 2nd edn, Oxford University Press, Oxford – An excellent text that provides the theory behind most of the NMR spectroscopic techniques that you are likely to need in conjunction with this book.

Quantum mechanics and bonding

P.W. Atkins (1998) *Physical Chemistry*, 6th edn, Oxford University Press, Oxford – This text provides a solid and well-tested background in physical chemistry.

R.J. Gillespie and I. Hargittai (1991) *The VSEPR Model of Molecular Geometry*, Allyn and Bacon, Boston – A text with numerous examples that takes the VSEPR model from basic principles to a quantum mechanical basis.

R.J. Gillespie (1992) *Chemical Society Reviews*, vol. 21, p. 59 – 'The VSEPR model revisited'; an up-to-date account of the VSEPR model with new applications.

R. McWeeny (1979) *Coulson's Valence*, 3rd edn, Oxford University Press, Oxford – A general treatment of chemical bonding with a detailed mathematical approach.

M.J. Winter (1994) *Chemical Bonding*, Oxford University Press, Oxford – This text approaches chemical bonding non-mathematically and is aimed at first-year undergraduate students.

Structure and Bonding (1987) vol. 66 – A volume containing articles dealing with different aspects of electronegativity.

Problems

1.1 Using the list of naturally occurring isotopes in Appendix 5, determine the number of electrons, protons and neutrons present in an atom of each isotope of (a) Al, (b) Br and (c) Fe, and give appropriate notation to show these data for each isotope.

1.2 Hydrogen possesses three isotopes, but tritium (^{3}H), which is radioactive, occurs as less than 1 in 10^{17} atoms in a sample of natural hydrogen. If the value of A_r for hydrogen is 1.008, estimate the percentage abundance of protium, ^{1}H, and deuterium, ^{2}H (or D) present in a sample of natural hydrogen. Point out any assumptions that you make. Explain why your answers are not the same as those quoted in Appendix 5.

1.3 (a) By using the data in Appendix 5, account for the isotopic distribution shown in Figure 1.1b. (b) The mass spectrum of S_8 shows other peaks at lower values of m/z. By considering the structure of S_8 shown in Figure 1.1c, suggest the origin of these lower-mass peaks.

1.4 Four of the lines in the Balmer series are at 656.28, 486.13, 434.05 and 410.17 nm. Show that these wavelengths are consistent with equation 1.4.

1.5 Using the Bohr model, determine the values of the radii of the second and third orbits of the hydrogen atom.

1.6 Write down the sets of quantum numbers that defines the (a) 1s, (b) 4s, (c) 5s atomic orbitals.

1.7 Write down the three sets of quantum numbers that define the three 3p atomic orbitals.

1.8 How many atomic orbitals make up the set with $n = 4$ and $l = 3$? What label is given to this set of orbitals? Write down a set of quantum numbers that defines each orbital in the set.

1.9 Which of the following species are hydrogen-like: (a) H^+; (b) He^+; (c) He^-; (d) Li^+; (e) Li^{2+}?

1.10 (a) Will a plot of $R(r)$ for the 1s atomic orbital of He^+ be identical to that of the H atom (Figure 1.5a)? [Hint: Look at Table 1.2.] (b) On the *same axis set*, sketch approximate representations of the function $4\pi r^2 R(r)^2$ for H and He^+.

1.11 Using equation 1.16, determine the energies of atomic orbitals of hydrogen with $n = 1, 2, 3, 4$ and 5. What can you say about the relative spacings of the energy levels?

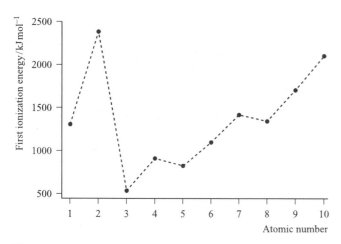

Fig. 1.32 Graph for problem 1.17.

1.12 Write down (with reasoning) the ground state electronic configurations of (a) Li, (b) F, (c) S, (d) Ca, (e) Ti, (f) Al.

1.13 Draw energy level diagrams to show the ground state electronic configurations of only the *valence* electrons in an atom of (a) F, (b) Al, and (c) Mg.

1.14 (a) Write down an equation that defines the process to which the value of IE_4 of Sn refers. Is this process exothermic or endothermic? (b) To what overall process does a value of $(IE_1 + IE_2 + IE_3)$ for Al refer?

1.15 The first four ionization energies of an atom X are 403, 2633, 3900 and 5080 kJ mol^{-1}. Suggest to what periodic group X belongs and give reasons for your choice.

1.16 In Figure 1.15, identify the trends in the first ionization energies of the elements in (a) descending group 1, (b) descending group 13, (c) crossing the first row of the *d*-block; (d) crossing the row of elements from B to Ne, (e) going from Xe to Cs, and (f) going from P to S. Rationalize each of the trends you have described.

1.17 Figure 1.32 shows the values of IE_1 for the first ten elements. (a) Label each point with the symbol of the appropriate element. (b) Give detailed reasons for the observed trend in values.

1.18 (a) Using the data in Table 1.5, determine a value for ΔH for the process:

$$O(g) + 2e^- \longrightarrow O^{2-}(g)$$

(b) Comment on the relevance of the sign and magnitude of your answer to part (a) in the light of the fact that many metal oxides with ionic lattices are thermodynamically stable.

1.19 Draw Lewis structures to describe the bonding in the following molecules: (a) F_2; (b) BF_3; (c) NH_3; (d) H_2Se; (e) H_2O_2; (f) $BeCl_2$; (g) SiH_4; (h) PF_5.

1.20 Use the Lewis structure model to deduce the type of nitrogen–nitrogen bond present in (a) N_2H_4, (b) N_2F_4, (c) N_2F_2 and (d) $[N_2H_5]^+$.

1.21 Draw out the resonance structures for the O_3 molecule. What can you conclude about the net bonding picture?

1.22 (a) Use VB theory to describe the bonding in the diatomic molecules Li_2, B_2 and C_2. (b) Experimental data show that Li_2 and C_2 are diamagnetic whereas B_2 is paramagnetic. Is the VB model consistent with these facts?

1.23 Using VB theory and the Lewis structure model, determine the bond order in (a) H_2, (b) Na_2, (c) S_2, (d) N_2 and (e) Cl_2. Is there any ambiguity with finding the bond orders by this method?

1.24 Does VB theory indicate that the diatomic molecule He_2 is a viable species? Rationalize your answer.

1.25 (a) Use MO theory to determine the bond order in each of $[He_2]^+$ and $[He_2]^{2+}$. (b) Does the MO picture of the bonding in these ions suggest that they are viable species?

1.26 (a) Construct an MO diagram for the formation of O_2; show only the participation of the valence orbitals of the oxygen atoms. (b) Use the diagram to rationalize the following trend in O–O bond distances: O_2, 121 pm; $[O_2]^+$, 112 pm; $[O_2]^-$, 134 pm; $[O_2]^{2-}$, 149 pm. (c) Which of these species are paramagnetic?

1.27 Using the data in Table 1.7, determine which of the following covalent single bonds is polar and (if appropriate) in which direction the dipole moment will act. (a) N–H; (b) F–Br; (c) C–H; (d) P–Cl; (e) N–Br.

1.28 Pick out *pairs* of isoelectronic species from the following list; not all species have a 'partner': HF; CO_2; SO_2; NH_3; PF_3; SF_4; SiF_4; $SiCl_4$; $[H_3O]^+$; $[NO_2]^+$; $[OH]^-$; $[AlCl_4]^-$.

1.29 Use the VSEPR model to predict the structures of (a) H_2Se, (b) $[BH_4]^-$, (c) NF_3, (d) SbF_5, (e) $[H_3O]^+$, (f) IF_7, (g) $[I_3]^-$, (h) $[I_3]^+$, (i) SO_3.

1.30 Use VSEPR theory to rationalize the structure of SOF_4 shown in Figure 1.33. What are the bond orders of (a) each S–F bond and (b) the S–O bond?

1.31 Determine the shapes of each of the following molecules and then, using the data in Table 1.7, state whether each is expected to be polar or not: (a) H_2S; (b) CO_2; (c) SO_2; (d) BF_3; (e) PF_5; (f) *cis*-N_2F_2; (g) *trans*-N_2F_2; (h) HCN.

1.32 State whether you expect the following species to possess geometrical isomers and, if so, draw their structures and give them distinguishing labels: (a) BF_2Cl; (b) $POCl_3$; (c) $MePF_4$; (d) $[PF_2Cl_4]^-$.

Fig. 1.33 The structure of SOF_4.

Chapter 2 Nuclear properties

2.1 Introduction

In this chapter we are concerned with *nuclear* properties and reactions involving the nucleus. When they occur naturally, such transformations of the nucleus lead to it being radioactive; transformations may also be brought about artificially and the energy released in nuclear fission reactions is harnessed in the nuclear fuels industry. The techniques of nuclear magnetic resonance (NMR) and Mössbauer spectroscopies owe their existence to properties of particular nuclei.

2.2 Nuclear binding energy

Mass defect and binding energy

The mass of an atom of ^1H is exactly equal to the sum of the masses of one proton and one electron. However, the atomic mass of any other atom is *less than* the sum of the masses of the protons, neutrons and electrons present. This *mass defect* is a measure of the *binding energy* of the protons and neutrons in the nucleus, and the loss in mass and liberation of energy are related by Einstein's equation 2.1. Mass defects also apply to ordinary chemical reactions, but in these the loss of mass is extremely small and is generally ignored.

$$\Delta E = \Delta m c^2 \qquad (2.1)$$

where ΔE = energy liberated, Δm = loss of mass, and c = speed of light in a vacuum = $2.998 \times 10^8 \, m \, s^{-1}$.

Although nuclear binding energies are derived in terms of *atomic* mass, it would be more logical to derive them from *nuclear* masses since the mass defect is a phenomenon arising from the combination of the particles in the nucleus.

However, accurate values of nuclear, as distinct from atomic, masses are known only for elements of low atomic number where it is possible to remove all the electrons in a mass spectrometer.

Worked example 2.1 Nuclear binding energy

Assuming that the mass defect originates solely from the interaction of protons and neutrons in the nucleus, estimate the nuclear binding energy of ^7_3Li given the following data:

Observed atomic mass of ^7_3Li = 7.016 00 u
$1 \, u \approx 1.660 \, 54 \times 10^{-27} \, kg$
Electron rest mass = $9.109 \, 39 \times 10^{-31} \, kg$
Proton rest mass = $1.672 \, 62 \times 10^{-27} \, kg$
Neutron rest mass = $1.674 \, 93 \times 10^{-27} \, kg$
$c = 2.998 \times 10^8 \, m \, s^{-1}$

The actual mass of a ^7_3Li atom

$$= 7.016 \, 00 \times 1.660 \, 54 \times 10^{-27}$$

$$= 1.165 \, 03 \times 10^{-26} \, kg$$

The sum of the masses of the protons, neutrons and electrons in a ^7_3Li atom

$$= (3 \times 9.109 \, 39 \times 10^{-31}) + (3 \times 1.672 \, 62 \times 10^{-27})$$
$$+ (4 \times 1.674 \, 93 \times 10^{-27})$$

$$= 1.172 \, 03 \times 10^{-26} \, kg$$

The difference in mass = Δm

$$= (1.172 \, 03 \times 10^{-26}) - (1.165 \, 03 \times 10^{-26})$$

$$= 0.007 \, 00 \times 10^{-26} \, kg$$

Fig. 2.1 Variation in average binding energy per nucleon as a function of mass number. Note that the energy scale is positive, meaning that the nuclei with the highest values of the binding energies release the greatest amount of energy upon formation.

Nuclear binding energy = $\Delta E = \Delta mc^2$

$= (0.007\,00 \times 10^{-26}) \times (2.998 \times 10^8)^2 \text{ kg m}^2 \text{ s}^{-2}$

$= 6.291\,60 \times 10^{-12} \text{ J per atom}$

$\approx 6.29 \times 10^{-12} \text{ J per atom} \qquad (\text{J} = \text{kg m}^2\text{s}^{-2})$

Self-study exercises

1. Estimate the nuclear binding energy of $^4_2$He given that its observed atomic mass is 4.002 60 u; other necessary data are given above. [*Ans.* $\approx 4.53 \times 10^{-12}$ J per atom]

2. If the nuclear binding energy of an atom of $^9_4$Be is 9.3182×10^{-12} J per atom, calculate the atomic mass of $^9_4$Be; other necessary data are given above. [*Ans.* 9.012 18 u]

The binding energy of 6.29×10^{-12} J calculated in worked example 2.1 for $^7_3$Li is for a single nucleus. This corresponds to 3.79×10^{12} J or 3.79×10^9 kJ per mole of nuclei, i.e. a huge amount of energy is liberated when the fundamental particles combine to form a mole of atoms. Its magnitude can readily be appreciated if the value 3.79×10^9 kJ mol^{-1} is compared with the heat liberated when one mole of *n*-butane is burnt $(\Delta_c H^o(298\,\text{K}) = -2857\,\text{kJ per mole of butane})$.

The average binding energy per nucleon

In comparing the binding energies of different nuclei, it is more useful to consider the average binding energy per *nucleon*, i.e. per particle in the nucleus. For $^7_3$Li, this is

given in equation 2.2, assuming that the only particles of significance in the nucleus are protons and neutrons.[†]

For $^7_3$Li, binding energy per nucleon

$$= \frac{6.29 \times 10^{-12}}{7} = 8.98 \times 10^{-13} \text{ J} \qquad (2.2)$$

Since values of nuclear binding energies are always very large, it is often convenient to quote them in mega electron volts (MeV) as in Figure 2.1 which shows the variation in binding energy per nucleon as a function of mass number. These values represent the energy released per nucleon upon the formation of the nucleus from its fundamental particles, and so the plot in Figure 2.1 can be used to give a measure of the relative stabilities of nuclei with respect to decomposition into those particles. The nucleus with the greatest binding energy is $^{56}_{26}$Fe and this is therefore the most stable nucleus. In general, nuclei with mass numbers around 60 have the highest average binding energies per nucleon, and it is these elements (e.g. Fe, Ni) that are believed to constitute the bulk of the Earth's core.

Figure 2.1 also shows that nuclei with mass numbers of 4, 12 and 16 have relatively high binding energies per nucleon, implying particular stabilities associated with $^4_2$He, $^{12}_6$C and $^{16}_8$O. These nuclei tend to be those used as projectiles in the synthesis of the heaviest nuclei (see *Section 2.6*). Finally in Figure 2.1, note that the binding energy per nucleon decreases appreciably for mass numbers >100.

[†] This assumption is valid for this exercise, but other particles (within the realm of the particle physicist) do exist.

The data in Figure 2.1 are of crucial significance for the application of nuclear reactions as energy sources. A reaction involving nuclei will be exothermic if:

- a heavy nucleus is divided into two nuclei of medium mass (so-called *nuclear fission*, see *Section 2.5*), or
- two light nuclei are combined to give one nucleus of medium mass (so-called *nuclear fusion*, see *Section 2.8*).

2.3 Radioactivity

Nuclear emissions

Nuclear transformations generally possess very high activation barriers and are usually very slow, but even so spontaneous changes of many heavy nuclides (e.g. $^{238}_{92}$U and $^{232}_{90}$Th) have been known since the 19th century. When one nuclide decomposes to form a different nuclide, it is said to be *radioactive*. In such nuclear changes, three types of emission were initially recognized by Rutherford:

- α-particles (now known to be helium *nuclei*, $[^4_2\text{He}]^{2+}$);
- β-particles (electrons emitted *from the nucleus* and having high kinetic energies);
- γ-radiation (high energy X-rays).

An example of spontaneous radioactive decay is that of carbon-14, which takes place by loss of a β-particle to give nitrogen-14 (equation 2.3) and this decay is the basis of radiocarbon dating (see *Section 2.9*). The emission of a β-particle results in an increase in the atomic number by one and leaves the mass number unchanged.

$$^{14}_{6}\text{C} \longrightarrow {}^{14}_{7}\text{N} + \beta^- \qquad (2.3)$$

More recent work has shown that the decay of some nuclei involves the emission of three other types of particle:

- the positron (β^+);
- the neutrino (ν_e);
- the antineutrino.

A positron is of equal mass but opposite charge to an electron. A neutrino and antineutrino possess near zero masses, are uncharged and accompany the emission from the nucleus of a positron and an electron respectively. The symbol used for a positron is β^+, and in this book we denote a β-particle in equations by β^- (as in equation 2.3) for clarity.

The energies associated with the emissions of α- and β-particles and γ-radiation are significantly different. An α-particle is emitted with an energy in the range $\approx (6-16) \times 10^{-13}$ J, and this means that an α-particle penetrates a few centimetres of air, causing ionization of some molecules. A barrier of a few sheets of paper or a very thin metal foil is sufficient to stop a stream of α-particles (Figure 2.2). The health risk associated with α-particles arises from their ingestion. A β-particle is emitted with an

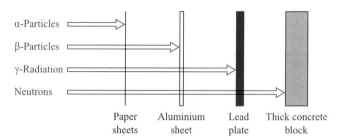

Fig. 2.2 A comparison of the penetrating powers of α-particles, β-particles, γ-radiation and neutrons. Neutrons are especially penetrating and their use in a nuclear reactor calls for concrete-wall shields of ≥ 2 m in thickness.

energy $\approx (0.03-5.0) \times 10^{-13}$ J, but since they are much lighter than α-particles, β-particles travel much faster and have a greater range. The penetrating power of β-particles exceeds that of α-particles and an aluminium barrier is required to stop them (Figure 2.2). Whereas α-particles emitted by a particular nucleus usually have the same energy, the energies of β-particles from a particular nuclide show a continuous distribution up to a maximum value. This observation was initially surprising since nuclei have discrete energy levels, and it led to the postulate that another particle of variable energy (the antineutrino) was emitted simultaneously.

γ-Radiation has a very short wavelength and very high energy (see *Appendix 4*). Its emission often accompanies the loss of α- or β-particles. This phenomenon arises because the *daughter nuclide* (the product of α- or β-particle loss) is often in an excited state, and energy in the form of γ-radiation is emitted as the transition from excited to ground state occurs. The energies of γ-radiations are in the same range as those of β-particles, but their penetrating power is far greater; a Pb shield (several centimetres thick) is required to absorb γ-radiation (Figure 2.2).

Nuclear transformations

Equation 2.3 gave an example of a spontaneous nuclear transformation. Since the loss of a β-particle is accompanied by a one-unit increase in atomic number and a retention in mass number, it effectively converts a neutron into a proton.

Since an α-particle is a helium nucleus (i.e. $[^4_2\text{He}]^{2+}$), its emission lowers the atomic number by two and the mass number by four. Equation 2.4 illustrates the radioactive decay of uranium-238 to thorium-234. The loss of the α-particle is accompanied by emission of γ-radiation, but the latter affects neither the atomic nor mass number. The α-particle in equation 2.4 is shown as *neutral* helium gas; as they are emitted, α-particles readily pick up electrons from the environment.

$$^{238}_{92}\text{U} \longrightarrow {}^{234}_{90}\text{Th} + {}^4_2\text{He} + \gamma \qquad (2.4)$$

Many *nuclear reactions*, as opposed to ordinary chemical reactions, change the identity of (*transmute*) the starting element. Steps involving the loss of an α- or β-particle may

Fig. 2.3 The decay series from $^{238}_{92}$U to $^{206}_{82}$Pb. Only the last nuclide in the series, $^{206}_{82}$Pb, is stable with respect to further decay. [*Exercise:* Three of the nuclides are not labelled. What are their identities?]

be part of a *decay series* (Figure 2.3). The initial nuclide is $^{238}_{92}$U and this spontaneously decays with the loss of an α-particle to $^{234}_{90}$Th. Once formed, $^{234}_{90}$Th decays by loss of a β-particle to $^{234}_{91}$Pa, which itself loses a β-particle. The decay series continues with successive nuclides losing either an α- or β-particle until ultimately the stable isotope $^{206}_{82}$Pb is produced. Not every step in the series takes place at the same rate.

The kinetics of radioactive decay

Radioactive decay of *any* nuclide follows *first order kinetics*. However, the observed kinetics of the decay may be complicated by the decay of the daughter nuclide. In the discussion below, we consider only a *single* decay step.

In a first order process, the rate of the reaction:

A ⟶ products

at a particular time, t, depends upon the concentration of the reactant A present at time, t. Radioactive decay processes are often conveniently considered in terms of the number of nuclei, N, present and equation 2.5 gives the appropriate rate equation.

$$\text{Rate of decay} = -\frac{dN}{dt} = kN \qquad (2.5)$$

where t = time and k = first order rate constant.

The integrated form of this rate equation may be written as in equation 2.6, or in the form of equation 2.7 which emphasizes that the decay is exponential.

$$\ln N - \ln N_0 = -kt \qquad (2.6)$$

where $\ln = \log_e$, N = number of nuclides at time t and N_0 = number of nuclides present at $t = 0$

$$\frac{N}{N_0} = e^{-kt} \qquad (2.7)$$

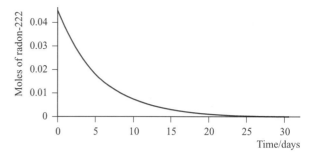

Fig. 2.4 Radioactive decay follows first order kinetics and a plot of the number of nuclides against time is an exponential decay curve. The graph shows a decay curve for radon-222 which has a half-life of 3.82 days.

From equation 2.6, it follows that a plot of $\ln N$ against t is linear, and the rate constant k is found from the gradient of the line (see *problem 2.5* at the end of the chapter).

Figure 2.4 shows the first order decay of $^{222}_{86}$Rn, and the exponential curve is typical of any radioactive decay process. A characteristic feature is that the time taken for the number of nuclides present at time t, N_t, to decrease to half their number, $\frac{N_t}{2}$, is constant. This time period is called the *half-life*, $t_{\frac{1}{2}}$, of the nuclide.

> The *half-life* of a radioactive nuclide is the time taken for the number of nuclides present at time t, N_t, to fall to half of its value, $\frac{N_t}{2}$.

Worked example 2.2   Radioactive decay

In Figure 2.4, there are initially 0.045 moles of radon-222 present. Estimate a value for the half-life of $^{222}_{86}$Rn.

Firstly determine the time taken for the number of moles of $^{222}_{86}$Rn to decrease from 0.045 to half this value (0.0225); this is the *first half-life*. From the graph, $(t_{\frac{1}{2}})_1 \approx 3.8$ days.

For greater accuracy, you should read off from the graph at least three half-lives and take an average value.

[The actual value of $t_{\frac{1}{2}}$ for $^{222}_{86}$Rn is 3.82 days.]

Self-study exercises

1. Read off *three* half-lives from Figure 2.4 and show that each is 3.8 days.

2. If $t_{\frac{1}{2}}$ for $^{222}_{86}$Rn is 3.8 days, how long does it take for 0.050 mmol to decay to 0.0062 mmol?   [*Ans.* 11.4 days]

The half-life is related to the rate constant and equation 2.8 is derived from equation 2.6 by substituting values of $N = \frac{N_0}{2}$ and $t = t_{\frac{1}{2}}$.

Table 2.1 The natural radioactive decay series from $^{238}_{92}U$ to $^{206}_{82}Pb$ (see Figure 2.3); (yr = year; d = day; min = minute; s = second).

Nuclide	Symbol	Particle emitted	Half-life
Uranium-238	$^{238}_{92}U$	α	4.5×10^9 yr
Thorium-234	$^{234}_{90}Th$	β^-	24.1 d
Protactinium-234	$^{234}_{91}Pa$	β^-	1.18 min
Uranium-234	$^{234}_{92}U$	α	2.48×10^5 yr
Thorium-230	$^{230}_{90}Th$	α	8.0×10^4 yr
Radium-226	$^{226}_{88}Ra$	α	1.62×10^3 yr
Radon-222	$^{222}_{86}Rn$	α	3.82 d
Polonium-218	$^{218}_{84}Po$	α	3.05 min
Lead-214	$^{214}_{82}Pb$	β^-	26.8 min
Bismuth-214	$^{214}_{83}Bi$	β^-	19.7 min
Polonium-214	$^{214}_{84}Po$	α	1.6×10^{-4} s
Lead-210	$^{210}_{82}Pb$	β^-	19.4 yr
Bismuth-210	$^{210}_{83}Bi$	β^-	5.0 d
Polonium-210	$^{210}_{84}Po$	α	138 d
Lead-206	$^{206}_{82}Pb$	none	Non-radioactive

$$\left.\begin{aligned} \ln\left(\frac{N_0}{2}\right) - \ln N_0 &= -\ln 2 = -k t_{\frac{1}{2}} \\ t_{\frac{1}{2}} &= \frac{\ln 2}{k} \end{aligned}\right\} \qquad (2.8)$$

The values of half-lives of naturally occurring radioactive nuclides vary enormously, e.g. 4.5×10^9 yr for $^{238}_{92}U$ and 1.6×10^{-4} s for $^{214}_{84}Po$. Table 2.1 lists half-life data for nuclides involved in the decay series in Figure 2.3.

The rate of an ordinary chemical reaction depends on temperature (the Arrhenius equation relates the rate constant, k, to the temperature, T, in kelvin), but radioactive decay differs and is *temperature-independent*.

Units of radioactivity

The SI unit of radioactivity is the becquerel (Bq) and is equal to one nuclear disintegration per second. The unit is named after Henri Becquerel who discovered the phenomenon of radioactivity in 1896. A non-SI unit also in use is the curie (Ci), where $1\,\text{Ci} = 3.7 \times 10^{10}\,\text{Bq}$; the curie is named after Marie Curie who discovered the elements radium and polonium.

2.4 Artificial isotopes

Bombardment of nuclei by high-energy α-particles and neutrons

The last section described *naturally occurring* radioactive processes. Similar transformations occur when nuclei are bombarded with high-energy neutrons or positively charged particles; the former are particularly effective since, being uncharged, they are not subject to electrostatic repulsion by nuclei. Such nuclear reactions take place with *conservation of atomic number and mass number* and provide a means of generating artificial isotopes. Equation 2.9 shows the reaction that occurs when an Al foil is bombarded with α-particles which have been given high energies in a *cyclotron* (an accelerating machine). The nuclear transformation may also be written using the notation $^{27}_{13}Al(\alpha,n)^{30}_{15}P$ which has the general form shown in equation 2.10.[†]

$$^{27}_{13}Al + {}^4_2He \longrightarrow {}^{30}_{15}P + {}^1_0n \qquad (2.9)$$

$$\text{Initial nuclide}\begin{pmatrix} \text{incoming} & \text{outgoing} \\ \text{particles} & , & \text{particles} \\ \text{or quanta} & \text{or quanta} \end{pmatrix}\text{final nuclide} \quad (2.10)$$

The product of reaction 2.9 rapidly decays ($t_{\frac{1}{2}} = 3.2\,\text{min}$) according to equation 2.11. The loss of a positron from the nucleus effectively converts a proton into a neutron.

$$^{30}_{15}P \longrightarrow {}^{30}_{14}Si + \beta^+ \qquad (2.11)$$

High-energy (or 'fast') neutrons are produced by the *nuclear fission* of $^{235}_{92}U$ and have energies of $\approx 1\,\text{MeV}$ (see *Section 2.5*). The bombardment of sulfur-32 with fast neutrons (equation 2.12) gives an artificial isotope of phosphorus, but $^{32}_{15}P$ has a half-life of 14.3 days and decays by β-particle emission (equation 2.13).

$$^{32}_{16}S + {}^1_0n \underset{\text{fast}}{\longrightarrow} {}^{32}_{15}P + {}^1_1H \qquad (2.12)$$

$$^{32}_{15}P \longrightarrow {}^{32}_{16}S + \beta^- \qquad (2.13)$$

Bombardment of nuclei by 'slow' neutrons

An important process for the production of artificial radioactive isotopes is the (n,γ) reaction which is brought about by the bombardment of nuclei with '*slow*' or *thermal neutrons*. The neutrons are formed by fission of $^{235}_{92}U$ nuclei and their kinetic energy is reduced by elastic collisions with low atomic number nuclei (e.g. $^{12}_6C$ or ^2_1H) during passage through graphite or deuterium oxide (heavy water). A thermal neutron has an energy of $\approx 0.05\,\text{eV}$. In reaction 2.14, naturally occurring phosphorus-31 (the *target* nucleus) is converted into artificial phosphorus-32.

$$^{31}_{15}P + {}^1_0n \underset{\text{slow}}{\longrightarrow} {}^{32}_{15}P + \gamma \qquad (2.14)$$

The production of artificial nuclides has two important consequences:

- the production of artificial isotopes of elements that do not possess naturally occurring radioisotopes;

[†] In *nuclear* equations, we do not keep track of electrons unless they are of a nuclear origin.

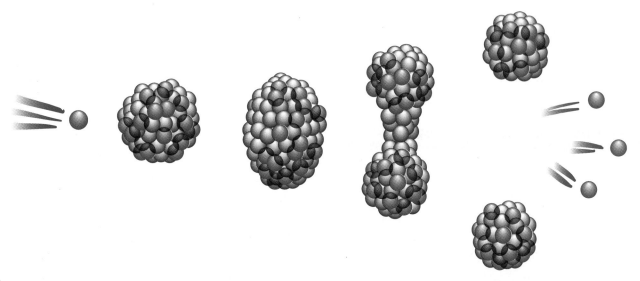

Fig. 2.5 A schematic representation of the collision of a thermal neutron with a heavy nuclide leading to fission into two nuclides of lower mass numbers and the release of (in this case) three neutrons. The fission is accompanied by the release of large amounts of energy. [Redrawn with permission from P. Fenwick (1990) *Reprocessing and the Environment*, Hobsons, Cambridge.]

● the synthesis of the *transuranium elements*, nearly all of which are exclusively man-made.

The *transuranium elements* ($Z \geq 93$) are almost exclusively all man-made. Other man-made elements include technetium (Tc), promethium (Pm), astatine (At) and francium (Fr).

Different nuclei show wide variations in their ability to absorb neutrons, and also in their probabilities of undergoing other nuclear reactions; such probabilities are often expressed as the *cross-section* of a nucleus for a particular nuclear reaction. For example, the nuclides $^{12}_{6}C$, $^{2}_{1}H$ and $^{1}_{1}H$ have very low cross-sections with respect to the capture of thermal neutrons, but $^{10}_{5}B$ and $^{113}_{48}Cd$ possess very high cross-sections.

2.5 Nuclear fission

The fission of uranium-235

From the energy scale in Figure 2.1 it is clear that large amounts of energy are released upon the fission of very heavy nuclei. The action of thermal neutrons on $^{235}_{92}U$ results in a reaction of the general type shown in equation 2.15 where the fission process is variable; Figure 2.5 shows a schematic representation of the process. Reaction 2.16 gives a typical example; once formed, yttrium-95 and iodine-138 decay by β-particle emission with half-lives of 10.3 min and 6.5 s respectively.

$$\underset{\text{slow}}{^{235}_{92}U} + {}^{1}_{0}n \longrightarrow \text{fission products} + \underset{\text{fast}}{x{}^{1}_{0}n} + \text{energy} \quad (2.15)$$

$$^{235}_{92}U + {}^{1}_{0}n \longrightarrow {}^{95}_{39}Y + {}^{138}_{53}I + 3{}^{1}_{0}n \quad (2.16)$$

A particular reaction path during nuclear fission is called a *reaction channel*, and the yields of different nuclei in the fission of $^{235}_{92}U$ indicate that it is more favourable to form two isotopes lying in the approximate mass ranges 100 to 90 and 134 to 144, than two nuclides with masses <90 and >144, or >100 and <134. Equation 2.16 illustrates the general point that *the sum of the mass numbers of the two fission products plus the neutrons must equal 236*. The average number of neutrons released per nucleus undergoing fission is ≈2.5 and the energy liberated (2×10^{10} kJ mol^{-1} of $^{235}_{92}U$) is about two million times that obtained by burning an equal mass of coal. Since each neutron can initiate another nuclear reaction, a *branching chain reaction* (Figure 2.6) is

Fig. 2.6 A representation of a branched chain reaction in which each step of the reaction produces two neutrons, each of which can initiate the fission of a $^{235}_{92}U$ nuclide. If left uncontrolled, such a chain reaction would lead to a violently explosive situation.

possible. If this involves a quantity of $^{235}_{92}U$ larger than a certain *critical mass*, a violent explosion occurs liberating enormous amounts of energy. This is the principle behind fission-type nuclear bombs and illustrates that extreme precautions are required when handling $^{235}_{92}U$ on an industrial scale.

Worked example 2.3   Balancing nuclear equations

Identify the second nuclide formed in the fission reaction:

$$^{235}_{92}U + {}^{1}_{0}n \longrightarrow {}^{103}_{42}Mo + ? + 2{}^{1}_{0}n$$

The reaction must proceed with conservation of mass number and of charge. The mass numbers are denoted by the superscripts, and the charges by the subscripts (i.e. the number of protons).

Let the unknown product be $^{A}_{Z}E$.

$$Z = 92 - 42 = 50$$

$$A = 235 + 1 - 103 - 2 = 131$$

The value of Z identifies the element as Sn (see the periodic table on the inside front cover of the book).

The nuclide is $^{131}_{50}Sn$.

Self-study exercises

1. Identify the second nuclide formed in the reaction:

$$^{235}_{92}U + {}^{1}_{0}n \longrightarrow {}^{92}_{36}Kr + ? + 2{}^{1}_{0}n \qquad [Ans.\ {}^{142}_{56}Ba]$$

2. Identify the second nuclide formed in the reaction:

$$^{235}_{92}U + {}^{1}_{0}n \longrightarrow {}^{141}_{55}Cs + ? + 2{}^{1}_{0}n \qquad [Ans.\ {}^{93}_{37}Rb]$$

BOX 2.1   APPLICATIONS

Electricity from nuclear power

Nuclear power is now used in a number of countries as a source of electrical power. The fuel in all commercial nuclear reactors is uranium, but of naturally occurring uranium only 0.7% is $^{235}_{92}U$, the radionuclide required for the fission process. Enrichment of the uranium is usually carried out but, even then, $^{235}_{92}U$ constitutes only a few percent of the uranium used as the fuel source. Using nuclear power on a commercial basis is a controversial issue; the public is made very aware of the problems involved in disposing of nuclear waste. The chart below shows the variation in the amount of electricity supplied by nuclear power electricity in selected countries around the world.

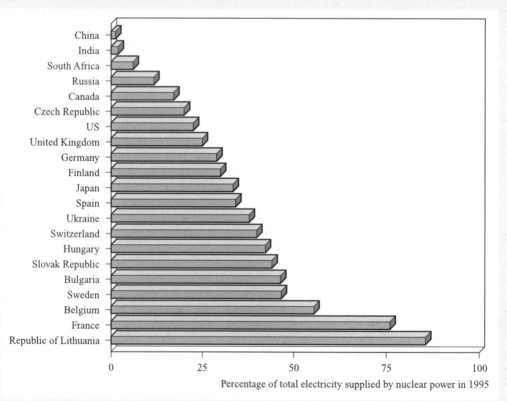

Percentage of total electricity supplied by nuclear power in 1995

[Data: BNFL: Data booklet, 1996 edition.]

The production of energy by nuclear fission

Nuclear fission can be successfully harnessed for the production of nuclear energy (see *Box 2.1*). This source of energy does not contribute to atmospheric pollution in the sense that there are no emissions of the gaseous oxides of carbon, nitrogen and sulfur associated with fossil fuels. Disadvantages of nuclear power include the problems of disposing of radioactive isotopes generated as fission products, and the risks involved if a nuclear reactor 'goes critical'.

The production of energy by nuclear fission in a nuclear reactor must be a controlled process. Neutrons released from the fission of $^{235}_{92}U$ lose most of their kinetic energy by passage through a *moderator* (graphite or D_2O). They then undergo one of two nuclear reactions. The first is capture by $^{235}_{92}U$ leading to further fission; the second is

capture by $^{238}_{92}U$ (scheme 2.17). Such isotope production is called *breeding*.

$$\left. \begin{array}{l} ^{238}_{92}U + ^{1}_{0}n \longrightarrow {}^{239}_{92}U + \gamma \\[1mm] ^{239}_{92}U \xrightarrow{-\beta^-} {}^{239}_{93}Np \xrightarrow{-\beta^-} {}^{239}_{94}Pu \end{array} \right\} \qquad (2.17)$$

The occurrence of a potentially catastrophic branching chain reaction is prevented by controlling the neutron concentration in the nuclear reactor by inserting boron-containing steel, boron carbide or cadmium control rods. The choice of material follows from the high cross-section for neutron capture exhibited by $^{10}_{5}B$ and $^{113}_{48}Cd$.

Nuclear reprocessing

Eventually, the $^{235}_{92}U$ fuel in a nuclear reactor becomes spent, and, rather than being disposed of, it is *reprocessed*. This

---

The disaster at Chernobyl

The name of Chernobyl (near Kiev, Ukraine) became known throughout the world on 26 April 1986 when reactor number 4 exploded. The power in the nuclear reactor is estimated to have increased from $\approx$200 MW to 3800 MW (MW = megawatt) in 2.5 s, and it took only another 1.5 s for the power to reach 120$\times$ its normal value. Energy well in excess of that required to melt the fuel in the reactor was generated within a mere 20 s. In the ensuing explosion, the reactor lid weighing $\approx$10^6 kg was blown off, allowing radioactive material to escape into the atmosphere where prevailing winds carried it to Scandinavia within a couple of days, and eastwards towards Japan over the following week. The release of radioactive material was exacerbated by graphite fires that started in the reactor and continued to burn for several days. It was about two weeks before the radiation levels from the reactor had been reduced to less dangerous levels.

Estimates of the total radiation released from the Chernobyl disaster vary but it may have been as great as

178 MCi; 1 Ci is roughly equal to the activity of 1 g of radium. Thirty-one people died on the night of the explosion from radiation or burns, and there were 200 known casualties from radiation sickness. In the longer term, Chernobyl has left the world with a number of long-lived radioisotopes distributed in the atmosphere. The main health risks come from $^{131}_{53}I$ ($t_{\frac{1}{2}} = 8.02$ days), $^{134}_{55}Cs$ ($t_{\frac{1}{2}} = 2.06$ yr) and $^{137}_{55}Cs$ ($t_{\frac{1}{2}} = 30.2$ yr). While the half-life of $^{131}_{53}I$ is much shorter than those of $^{134}_{55}Cs$ or $^{137}_{55}Cs$, it is easily taken up by the thyroid gland and may cause cancer. Exposure to $^{131}_{53}I$ by people and animals in the few days after the disaster was unavoidable, and the graph below indicates how incidences of thyroid cancer in children in the Ukraine increased since the Chernobyl accident. The final death toll from Chernobyl remains an unknown statistic, and in 1995 the World Health Organization (WHO) called for further research into the radiation effects to be carried out.

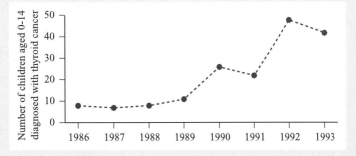

[Data from: I.A. Likhtarev *et al.* (1995), *Nature*, vol. 375, p. 365.]

Further reading

C.H. Atwood (1988) *Journal of Chemical Education*, vol. 65, p. 1037 – 'Chernobyl: What happened?'.

I.A. Likhtarev *et al.* (1995) *Nature*, vol. 375, p. 365 – 'Thyroid cancer in the Ukraine'.

both recovers uranium and separates $^{235}_{92}U$ from the fission products. Short-lived radioactive products are initially allowed to decay while the spent fuel is retained in *pond storage*; after this period, the uranium is converted into the soluble salt $[UO_2][NO_3]_2$ (see *Box 6.3*). In the series of reactions 2.18–2.21, the nitrate is converted into UF_6.

$$[UO_2][NO_3]_2 \xrightarrow{570\,K} UO_3 + NO + NO_2 + O_2 \quad (2.18)$$
$$\text{Hydrated salt}$$

$$UO_3 + H_2 \xrightarrow{970\,K} UO_2 + H_2O \quad (2.19)$$

$$UO_2 + 4HF \longrightarrow UF_4 + 2H_2O \quad (2.20)$$

$$UF_4 + F_2 \xrightarrow{720\,K} UF_6 \quad (2.21)$$

At this stage, the UF_6 contains both $^{235}_{92}U$ and $^{238}_{92}U$. Application of Graham's law of effusion:

$$\text{Rate of effusion} \propto \frac{1}{\sqrt{\text{Molecular mass}}}$$

shows that $^{235}_{92}UF_6$ can be separated from $^{238}_{92}UF_6$ by subjecting them to a centrifugal force; molecules of the two isotopically labelled compounds move to the outer wall of their container at different rates. The result is the isolation of $^{235}_{92}U$-*enriched UF*$_6$. After this process, the hexafluoride is converted back to uranium-235 metal, thereby regenerating fuel for reuse in the nuclear reactor.

2.6 Syntheses of transuranium elements

The transuranium elements are shown in Table 2.2 and have all been discovered since 1940. By 1955, the table extended to mendelevium and, by 1997, to meitnerium ($Z = 109$). All of these 'new' elements have been produced synthetically (see

Table 2.2 The transuranium elements. The names are those recently agreed by the IUPAC.

Z	Name of element	Symbol
93	Neptunium	Np
94	Plutonium	Pu
95	Americium	Am
96	Curium	Cm
97	Berkelium	Bk
98	Californium	Cf
99	Einsteinium	Es
100	Fermium	Fm
101	Mendelevium	Md
102	Nobelium	No
103	Lawrencium	Lr
104	Rutherfordium	Rf
105	Dubnium	Db
106	Seaborgium	Sg
107	Bohrium	Bh
108	Hassium	Hs
109	Meitnerium	Mt

also *Section 24.5*) by the bombardment of particular heavy nuclides with particles such as neutrons (e.g. equation 2.17) and $^{12}_{6}C^{n+}$ or $^{18}_{8}O^{n+}$ ions (equations 2.22 and 2.23).

$$^{249}_{97}Bk + ^{18}_{8}O \longrightarrow ^{260}_{103}Lr + ^{4}_{2}He + 3^{1}_{0}n \quad (2.22)$$

$$^{248}_{96}Cm + ^{18}_{8}O \longrightarrow ^{261}_{104}Rf + 5^{1}_{0}n \quad (2.23)$$

The scale on which these transmutations is carried out is *extremely* small, and in some cases has been described as 'atom-at-a-time' chemistry. The target materials in equations 2.22 and 2.23 are actinoid elements (see *Chapter 24*), which, although artificially prepared, have relatively long half-lives ($^{249}_{97}Bk$, $t_{\frac{1}{2}} = 300$ days; $^{248}_{96}Cm$, $t_{\frac{1}{2}} = 3.5 \times 10^5$ yr). Studying the product nuclides is extremely difficult, not only because of the tiny quantities of materials involved but also because of their short half-lives ($^{260}_{103}Lr$, $t_{\frac{1}{2}} = 3$ min; $^{261}_{104}Rf$, $t_{\frac{1}{2}} = 65$ s).

2.7 The separation of radioactive isotopes

In forming artificial radioactive isotopes, problems of isolation are often encountered. For example, a product may decay quickly with the result that the initial product is contaminated with the daughter nuclide.

Chemical separation

The methods used to separate a desired isotope depend on whether or not the starting material and the product are isotopes of the same element (e.g. equation 2.14). If they are not, the problem is essentially one of chemical separation of a small amount of one element from large amounts of one or more others. Methods of separation include volatilization, electrodeposition, solvent extraction, ion-exchange or precipitation on a 'carrier'. For example, in the process $^{64}_{30}Zn(n,p)^{64}_{29}Cu$, the target (after bombardment with fast neutrons) is dissolved in dilute HNO_3 and the Cu is deposited electrolytically. This method is successful because of the significant difference between the reduction potentials $E^o(Cu^{2+}/Cu) = +0.34$ V and $E^o(Zn^{2+}/Zn) = -0.76$ V (see *Chapter 7*).

The Szilard–Chalmers effect

In an (n,γ) reaction, the product (unless it decays rapidly) is an isotope of the target element. Since isotopes of an element have identical *chemical* properties, chemical separation methods cannot be applied. Instead, use is made of the *Szilard–Chalmers effect*: if the nuclear reaction is accompanied by homolytic bond cleavage (brought about by the γ-radiation emitted in the reaction), radicals of the product isotope are scavenged and thereby separated from the target isotope. An example is the formation of $^{128}_{53}I$ from naturally occurring $^{127}_{53}I$. The target isotope is used in the form of ethyl iodide and is subjected to thermal neutron bombardment. A significant amount of the $^{128}_{53}I$ formed is liberated as atomic iodine-128 and these atoms (radicals) combine either with each other to form $^{128}_{53}I_2$ or react with added $^{127}_{53}I_2$ in an exchange reaction to give $^{127}_{53}I^{128}_{53}I$. Molecular iodine (present in aqueous solution in the presence of iodide ion as $[I_3]^-$, see *Section 16.7*) can be separated from ethyl iodide by reduction with aqueous sodium sulfite (equation 2.24).

$$I_2 + [SO_3]^{2-} + H_2O \longrightarrow 2I^- + [SO_4]^{2-} + 2H^+ \qquad (2.24)$$

For this method to be useful, there must be no rapid exchange reaction between target and product (equation 2.25) and hence an alkyl halide rather than an alkali metal halide is chosen for the irradiation.

$$C_2H_5(^{127}_{53}I) + ^{128}_{53}I \not\longrightarrow C_2H_5(^{128}_{53}I) + ^{127}_{53}I \qquad (2.25)$$

2.8 Nuclear fusion

Figure 2.1 showed that the fusion of two nuclei of low mass liberates immense amounts of energy. An example is the formation of helium-4 from deuterium and tritium (equation 2.26).

$$^2_1H \quad + \quad ^3_1H \quad \longrightarrow ^4_2He + ^1_0n \qquad (2.26)$$
$$\text{Deuterium} \quad \text{Tritium}$$

Compared with fission reactions, nuclear fusion has the advantage that large quantities of radioactive products are

not formed. However, the activation energies for fusion reactions are very high and, up to the present time, it has been possible to overcome the barrier only by supplying the energy from a *fission* reaction to drive a *fusion* reaction. This is the principle behind the hydrogen or thermonuclear bomb; tritium is expensive and inconvenient ($t_{\frac{1}{2}}$ is only 12 yr), but can be prepared from lithium deuteride enriched in ^6Li. A fusion explosion generated by compression of a few kilograms of plutonium brings about reactions such as 2.26–2.29.

$$^2_1H + ^2_1H \longrightarrow ^3_1H + ^1_1H \qquad (2.27)$$
$$^2_1H + ^2_1H \longrightarrow ^3_2He + ^1_0n \qquad (2.28)$$
$$^6_3Li + ^1_0n \longrightarrow ^4_2He + ^3_1H \qquad (2.29)$$

Fusion reactions are believed to take place in the Sun and start at temperatures above 10^7 K; reactions 2.30–2.32 have been suggested as the chief source of the Sun's energy.

$$^1_1H + ^1_1H \longrightarrow ^2_1H + \beta^+ + \nu_e \qquad (2.30)$$
$$^1_1H + ^2_1H \longrightarrow ^3_2He + \gamma \qquad (2.31)$$
$$^3_2He + ^3_2He \longrightarrow ^4_2He + 2^1_1H \qquad (2.32)$$

2.9 Applications of isotopes

The applications of isotopes, both radioactive and stable, are now so numerous that the examples in this section are

---

BOX 2.3   **APPLICATIONS**

Radioisotopes in medicine

The uses of radioisotopes in medicine are extremely important. Certain elements are readily absorbed by particular organs in a human body, and this is capitalized upon in the use of *radiotracers* (introduced by food- or drug-intake) to probe the function of human organs. An advantage of the technique is that it is *non-invasive*.

Although the uptake of ^{131}I by the thyroid gland was a health risk after the Chernobyl disaster (Box 2.2), *controlled* uptake has medical applications. If a patient ingests ^{131}I (e.g. as a solution of ^{131}I-labelled NaI), the isotope is quickly absorbed by the thyroid gland and the size and state of the gland can be detected by monitoring the radioactivity emitted. For ^{131}I, $t_{\frac{1}{2}} \approx 8$ days, and the dose administered soon decays.

Molybdenum-99 ($t_{\frac{1}{2}} = 2.8$ days) decays by β-emission to give the metastable radioisotope of technetium, ^{99m}Tc ($t1_{\frac{1}{2}} = 6.0$ h); ^{99m}Tc is usually generated in the form of $[^{99m}TcO_4]^-$, produced by decay of $[^{99}MoO_4]^{2-}$. Complexes of ^{99m}Tc (see *Section 22.8* and *Box 22.7*) are used as diagnostic imaging agents in the brain, heart and kidneys.

In addition to using radioisotopes to examine patients, the γ-radiation emitted may be used in cancer treatment. Cobalt

occurs naturally as the non-radioactive isotope ^{59}Co (100% abundance) but can be converted to ^{60}Co; ^{60}Co is a β-emitter with $t_{\frac{1}{2}} = 5.27$ yr:

$$^{59}_{27}Co + ^1_0n \longrightarrow ^{60}_{27}Co + \gamma$$

$$^{60}_{27}Co \longrightarrow ^{60}_{28}Ni + \beta^- + \gamma$$

Further reading

M.F. Hawthorne (1993) *Angewandte Chemie, International Edition in English*, vol. 32, p. 950 – 'The role of chemistry in the development of boron neutron capture therapy'.

R.C. Elder and K. Tepperman (1994) 'Metal-based drugs & imaging agents' in *Encyclopedia of Inorganic Chemistry*, ed. R.B. King, Wiley, Chichester, vol. 4, p. 2165.

See also Box 2.5: Magnetic resonance imaging (MRI); Section 12.9: Figures 12.17b and 12.17c and accompanying discussion; Section 22.8: footnote references for radiopharmaceuticals.

necessarily selective. Many applications involve the use of isotopes as 'tracers' in which all isotopes of an element are regarded as being chemically equivalent. Some uses (such as the observation of the kinetic isotope effect or shifts in infrared spectroscopic absorptions) depend on the small, but significant, differences in properties between isotopes of a given element.

Infrared (IR) spectroscopy

When the hydrogen atom in an $X-H$ bond is exchanged for deuterium (see *Section 9.3*), the *reduced mass* of the pair of bonded atoms changes and shifts the position of the absorption in the IR spectrum due to the $X-H$ stretching mode. Shifts of this kind can be used to confirm assignments in IR spectra. For example, $N-H$, $O-H$ and $C-H$ bonds all absorb around $3000-3600 \, cm^{-1}$, but if a compound is shaken with D_2O,[†] usually only the OH and NH groups undergo rapid *deuterium exchange reactions* (equation 2.33); H attached directly to C exchanges extremely slowly except in cases where it is acidic (e.g. a terminal alkyne).

$$R-OH + D_2O \rightleftharpoons R-OD + HOD \qquad (2.33)$$

By observing which IR spectroscopic bands shift (and by how much), it is possible to confirm the assignment of an $N-H$, $O-H$ or $C-H$ absorption.

> **Worked example 2.4   The effects of deuteration on $\bar{\nu}_{O-H}$ in an IR spectrum**

An absorption at $3650 \, cm^{-1}$ in the IR spectrum of a compound X has been assigned to an $O-H$ stretching mode. To what wavenumber is this band expected to shift upon deuteration? What assumption have you made in the calculation?

The $O-H$ vibrational wavenumber, $\bar{\nu}$, is related to the reduced mass (μ) by the equation:

$$\bar{\nu}_{O-H} \propto \frac{1}{\sqrt{\mu_{O-H}}}$$

where the reduced mass is given by the equation:

$$\frac{1}{\mu} = \frac{1}{m_1} + \frac{1}{m_2}$$

in which m_1 and m_2 are the masses of the O and H atoms in kg.

For the comparison of the $O-H$ and $O-D$ vibrational wavenumbers, we can write:

$$\frac{\bar{\nu}_{O-D}}{\bar{\nu}_{O-H}} = \sqrt{\frac{\mu_{O-H}}{\mu_{O-D}}}$$

and since we are now dealing with a *ratio*, it is not necessary to convert the atomic masses to kg. The relative atomic masses of O, H and D are, approximately, 16, 1 and 2, respectively. The reduced masses of $O-H$ and $O-D$ bonds are found as follows:

$$\frac{1}{\mu_{O-H}} = \frac{1}{m_1} + \frac{1}{m_2} = \frac{1}{16} + 1 = 1.0625 \qquad \mu_{O-H} = 0.9412$$

$$\frac{1}{\mu_{O-D}} = \frac{1}{m_1} + \frac{1}{m_2} = \frac{1}{16} + \frac{1}{2} = 0.5625 \qquad \mu_{O-D} = 1.7778$$

The vibrational wavenumber of the $O-D$ bond is therefore:

$$\bar{\nu}_{O-D} = \bar{\nu}_{O-H} \times \sqrt{\frac{\mu_{O-H}}{\mu_{O-D}}} = 3650 \times \sqrt{\frac{0.9412}{1.7778}} = 2656 \, cm^{-1}$$

The calculation makes the assumption that the force constants of $O-H$ and $O-D$ bonds are the same. The full equation relating $\bar{\nu}$ to μ is:

$$\bar{\nu} = \frac{1}{2\pi c} \sqrt{\frac{k}{\mu}}$$

where c is the speed of light in a vacuum, and k is the force constant of the bond.

Self-study exercises

1. An absorption at $3337 \, cm^{-1}$ in the vibrational spectrum of NH_3 shifts to $x \, cm^{-1}$ in ND_3. Determine x. [*Ans.* $2437 \, cm^{-1}$]

2. An absorption at $3161 \, cm^{-1}$ in an IR spectrum is assigned to a $C-H$ stretching mode. At what wavenumber will this band appear upon deuteration? [*Ans.* $2320 \, cm^{-1}$]

Spectroscopic studies of isotopically substituted molecules do not always involve special synthetic chemistry. For many elements, natural isotopic abundances ensure that ordinary compounds contain several species. For example, in GeH_3Cl, naturally occurring isotopes of Cl (^{35}Cl and ^{37}Cl) and Ge (^{70}Ge, ^{72}Ge, ^{74}Ge and ^{76}Ge) are all present in proportions sufficient to give rise to observable pure rotational spectra of $^{70}GeH_3{}^{35}Cl$, $^{70}GeH_3{}^{37}Cl$ etc. Where special syntheses are required, they must be designed so as to make the best possible use of the isotope to be incorporated, e.g. deuterated ammonia, ND_3, would not be prepared by exchange between NH_3 and D_2O since a large proportion of the deuterium would be wasted by conversion to HOD. A better method is the reaction between D_2O and Mg_3N_2.

Kinetic isotope effects

Isotopic labelling may be used to probe the mechanism of a reaction. Consider the case where the rate-determining step

---

[†] Up until this point in this chapter, we have used the full notation for isotopes, e.g. ^2_1H, but for the most part in this book, we shall adopt the less rigorous, but the nonetheless unambiguous, notation showing only the mass number, e.g. ^2H. In addition, we introduce the label D for deuterium.

Fig. 2.7 The zero point energy (corresponding to the lowest vibrational state) of a C−D bond is lower than that of a C−H bond and this results in the bond dissociation enthalpy, D, of the C−D bond being greater than that of the C−H bond.

of a reaction involves breaking a particular C−H bond. Labelling the compound with deuterium (not always a trivial matter experimentally!) at that site will mean that a C−D rather than a C−H bond is broken. The bond dissociation energy of a C−D bond is higher than that of a C−H bond because the zero point energy is lowered when the reduced mass, μ, of a bond is increased, i.e. $\mu(C−D) > \mu(C−H)$ (Figure 2.7). Since it requires more energy to break a C−D than a C−H bond, the rate-determining step should proceed more slowly for the deuterated compound. This observation is known as the *kinetic isotope effect* and is quantified by comparing the rate constants, k_H and k_D, for the reactions involving the non-deuterated and deuterated compounds respectively. If the value of the ratio $k_H:k_D > 1$, then a kinetic isotope effect has been observed.

The *zero point energy* of a molecule corresponds to the energy of its lowest vibrational level (vibrational ground state).

Radiocarbon dating

Radiocarbon dating is a technique used widely by archaeologists to date articles composed of organic material (e.g. wood), and the importance of the method was recognized in 1960 by the award of the Nobel Prize in Chemistry to its developer, W.F. Libby. The method relies on the fact that one isotope of carbon, $^{14}_{6}C$, is radioactive ($t_{\frac{1}{2}} = 5730$ yr) and decays according to equation 2.34.

$$^{14}_{6}C \longrightarrow {}^{14}_{7}N + \beta^-$$ (2.34)

In a *living* plant, the ratio of $^{14}_{6}C:^{12}_{6}C$ is constant. Although carbon-14 decays, it is reformed at the same rate by collisions

between high-energy neutrons and atmospheric nitrogen-14 (equation 2.35).

$$^{14}_{7}N + {}^{1}_{0}n \longrightarrow {}^{14}_{6}C + {}^{1}_{1}H$$ (2.35)

The process of photosynthesis in living plants ensures that the uptake of carbon-14 (and carbon-12 and carbon-13) in the form of CO_2 is continuous. Once a plant dies, no further $^{14}_{6}C$ enters the system and the carbon-14 present decays, with the result that the $^{14}_{6}C:^{12}_{6}C$ ratio gradually changes with time. Provided that we assume that the $^{14}_{6}C:^{12}_{6}C$ ratio in living species has not altered over an archaeological timescale, then it is possible to date an artifact by measuring the $^{14}_{6}C:^{12}_{6}C$ ratio. Unfortunately, this ratio *has* altered, but corrections may be made by using information gained from extremely old, but still living, trees such as the American bristlecone which grows in the mountains of eastern California.[†]

Worked example 2.5   Radiocarbon dating

The β-activity of 1 g of carbon from the wood of a recently felled tree is 0.26 Bq. If the activity of 1 g of carbon isolated from the wood of an Egyptian mummy case is 0.16 Bq under the same conditions, estimate the age of the mummy case. (^{14}C: $t_{\frac{1}{2}} = 5730$ yr.)

First, use the half-life to determine the rate constant for the decay of ^{14}C. From equation 2.8:

$$k = \frac{\ln 2}{t_{\frac{1}{2}}} = \frac{\ln 2}{5730} = 1.210 \times 10^{-4}\,\text{yr}^{-1}$$

The integrated rate equation (equation 2.6) for radioactive decay is:

$$\ln N - \ln N_0 = -kt$$

or

$$\ln\left(\frac{N}{N_0}\right) = -kt$$

in which N is the activity at time t and N_0 is the activity at $t = 0$. The activity of the recently felled tree corresponds to $t = 0$. It is not necessary to convert the units of k to s^{-1} (to be consistent with Bq) because units of Bq cancel in the ratio of $\frac{N}{N_0}$.

$$\ln\left(\frac{0.16}{0.26}\right) = -1.210 \times 10^{-4} \times t$$

$$t = 4010\,\text{yr}$$

...continued

[†] For further details, see: I. Robertson and J. Waterhouse (1998) *Chemistry in Britain*, vol. 34, January issue, p. 27 – 'Trees of knowledge'.

Self-study exercises

1. The β-activity of 0.9 g of C from the wood of a present-day tree is 0.25 Bq. If the activity of 0.9 g of carbon isolated from the wood of an ancient artifact is 0.19 Bq under the same conditions, estimate the age of the artifact.

   [*Ans.* 2268 yr]

2. The β-activity of 1 g of C from recently felled timber is 0.26 Bq. An ancient artifact is thought to be 3500 years old. What β-activity from a 1 g sample would confirm this age?

   [*Ans.* 0.17 Bq]

Analytical applications

The use of radioisotopes in analysis (see also *Section 16.3*) includes determinations of solubilities of sparingly soluble salts and vapour pressures of rather involatile substances, and investigations of solid solution formation and adsorption of precipitates.

As an example, we consider the measurement of the solubility of strontium sulfate which, at 298 K, is 0.11 g dm^{-3}. Naturally occurring strontium contains four isotopes, none of which is radioactive. The radioisotope ^{90}Sr ($t_{\frac{1}{2}} = 28.1$ yr) is produced from the fission of ^{235}U and is commercially available. A uniform mixture of ^{90}SrSO$_4$ and the inactive salt SrSO$_4$ is prepared and the radioactivity of the combined sample is measured; this gives a standard value for the activity per gram of sample. A saturated aqueous solution is then prepared using the same uniform mixture, and is evaporated to dryness. The activity of the residue is measured and the amount of solid material can be accurately determined using this and the standard data. This method is called *isotope dilution analysis* (see *problem 2.16*)

2.10 Sources of ^{2}H and ^{13}C

In the laboratory, ^{2}H (D) and ^{13}C are commonly encountered even though both occur naturally only in low abundance (0.015 and 1.1% respectively).

Deuterium: electrolytic separation of isotopes

Solvents for NMR spectroscopy, enriched in deuterium to an extent of ≥99%, are commercially available. The separation of deuterium from naturally occurring hydrogen is achieved electrolytically with the isotope in the form of D$_2$O. When an aqueous solution of NaOH (natural isotopic abundances) is electrolyzed (equation 2.36) using an Ni electrode, the separation factor defined in equation 2.37 is ≈6. The choice of electrode is critical to the optimization of this value.

$$\left.\begin{array}{ll} \textit{At the anode}: & 2H_2O \rightleftharpoons O_2 + 4H^+ + 4e^- \\ \textit{At the cathode}: & 4H_2O + 4e^- \rightleftharpoons 2H_2 + 4[OH]^- \end{array}\right\} \quad (2.36)$$

$$\text{Separation factor} = \frac{\left(\dfrac{H}{D}\right)_{gas}}{\left(\dfrac{H}{D}\right)_{solution}} \quad (2.37)$$

The electrolysis is continued until ≈90% of the liquid has been converted into O$_2$ and H$_2$; most of the residual liquid is then neutralized with CO$_2$, and the water distilled and added to the remaining electrolyte. This process is repeated to give ≤99.9% D$_2$O. In the later stages of the separation, the gas evolved at the cathode is burned to yield partially enriched deuterium oxide that can be electrolyzed further. Cheap electrical power is, of course, essential for the economic concentration of D$_2$O by this method.

Carbon-13: chemical enrichment

Carbon-13 enriched compounds such as ^{13}CO, H^{13}CN, [^{13}CN]$^-$ or ^{13}CO$_2$ are prepared by various methods and we focus upon methods involving chemical equilibria in which the label is transferred from one species to another.

$$H^{12}CN(g) + [^{13}CN]^-(aq) \rightleftharpoons H^{13}CN(g) + [^{12}CN]^-(aq) \quad (2.38)$$

For the isotope exchange reaction 2.38, the equilibrium constant, K, is 1.026 (298 K). The fact that K is not unity arises from a small difference in the standard free energy between reactants and products, which follows from differences in zero point energies (Figure 2.7). For equilibrium 2.38, products are favoured (albeit slightly) over reactants. As the system involves two phases, it is particularly suitable for isotopic enrichment, with the ^{13}C label moving from one phase to the other, and can readily be made the basis of a multistage process. Equilibrium 2.39 shows a further example although here, a catalyst is required.

$$^{13}CO_2(g) + [H^{12}CO_3]^-(aq) \rightleftharpoons {}^{12}CO_2(g) + [H^{13}CO_3]^-(aq)$$
$$K = 1.012 \quad (2.39)$$

2.11 Multinuclear NMR spectroscopy in inorganic chemistry

In this section, we introduce the applications of nuclear magnetic resonance (NMR) spectroscopy to inorganic systems, not only to determine the numbers and environments of particular nuclei, but also to investigate (usually in solution) the dynamic behaviour of molecular species. A detailed description of the technique of NMR spectroscopy is beyond the scope of this book, but appropriate references are listed at the end of the chapter. In the discussion that follows, we assume that readers are already familiar with the concepts of ^{1}H and ^{13}C NMR spectroscopies, including homonuclear ^{1}H–^{1}H and heteronuclear ^{13}C–^{1}H spin–spin coupling. A factual summary is given in Box 2.4.

Which nuclei are suitable for NMR spectroscopic studies?

A wide range of nuclei may be observed by NMR spectroscopy, but the inherent properties of some nuclei (e.g. a large quadrupole moment) may make their observation difficult. The main criterion is that the nucleus possesses a value of the nuclear spin quantum number $I \geq \frac{1}{2}$ (Table 2.3). Secondly, it is advantageous (but not essential) for the nucleus to occur in significant abundance. Carbon-13 is an example of a low abundant isotope which is, nevertheless, extensively used for NMR spectroscopy; isotopic enrichment may be used to improve signal:noise ratios. A third requirement is that the nucleus possesses a relatively short *spin-relaxation time* (τ_1); this property depends not only on the nucleus itself but also on its molecular environment. Some elements exhibit more than one NMR active nucleus and the choice for experimental observation may depend upon the relative inherent values of τ_1. For example, ^{6}Li and ^{7}Li are NMR active, but whereas τ_1 values for ^{7}Li are typically <3 s, those for ^{6}Li lie in the range ≈ 10–80 s; ^{7}Li is thus more appropriate for NMR spectroscopic observation and this choice is also favoured by the fact that ^{7}Li is more abundant (92.5%) than ^{6}Li. Another nuclear property that may mitigate against easy observation is the *quadrupole moment* arising from a non-spherical charge distribution of the nucleus and which is associated with values of $I > \frac{1}{2}$. Although the possession of a quadrupole moment leads to short values of τ_1, it generally causes the signals in the NMR spectrum to be broad (e.g. ^{11}B). Signal broadening is also seen in the spectra of nuclei *attached* to nuclei with quadrupole moments, e.g. the ^{1}H NMR spectrum of protons attached to ^{11}B.

Chemical shift ranges

The range of chemical shifts over which NMR spectroscopic signals appear is dependent on the nucleus. The most commonly observed nucleus is ^{1}H and, in organic compounds, a *spectral window* from $\delta + 15$ to 0 usually encompasses most signals. In inorganic compounds, the window may have to be widened if, for example, ^{1}H nuclei attached to metal centres are to be observed, or if signals are *paramagnetically shifted*. The chemical shift range for ^{13}C NMR spectra is typically $\delta + 250$ to -50, for ^{31}P NMR spectra, $\approx \delta + 300$ to -300, and for ^{77}Se NMR spectra $\approx \delta + 2000$ to -1000. Figure 2.8 illustrates the change in chemical shift for the ^{31}P nucleus on going from triphenylphosphine to the corresponding oxide. Such a shift to higher frequency is typical when a tertiary phosphine (R_3P) is oxidized, and also tends to occur when a phosphine ligand coordinates to a *d*-block metal centre.

Spin–spin coupling

The number and spins of the *attached nuclei* determine the *multiplicity* (number of lines) and pattern of the NMR spectroscopic signal of the observed nucleus. The coupling constant between nuclei X and Y is denoted as J_{XY} and is measured in Hz.

In general the multiplicity of an NMR spectroscopic signal can be determined using equation 2.40 where the nucleus being observed is coupling (see *Box 2.4*) to *n* equivalent nuclei with quantum number *I*.

Multiplicity (number of lines) $= 2nI + 1$ (2.40)

Case study 1: ^{31}P NMR spectrum of $[PF_6]^-$

The ^{31}P NMR spectrum of a salt containing the octahedral $[PF_6]^-$ ion exhibits a binomial septet (Figure 2.9) consistent with six equivalent ^{19}F nuclei ($I = \frac{1}{2}$) attached to the central ^{31}P centre. The large value of J_{PF} 708 Hz is typical of ^{31}P–^{19}F coupling constants for *directly attached* nuclei; the magnitudes of coupling constants usually diminish with nuclear separation, but a consequence of large values for directly attached nuclei is that *long range couplings* may be observed (see below).

Case study 2: ^{31}P NMR spectrum of $Ph_2PCH_2CH_2P(Ph)CH_2CH_2PPh_2$

(2.1)

Structure **2.1** shows that $Ph_2PCH_2CH_2P(Ph)CH_2CH_2PPh_2$ contains two phosphorus environments, labelled a and b; the ^{31}P NMR spectrum exhibits two signals with an integral ratio of 1:2. For directly attached inequivalent phosphorus atoms, values of J_{PP} are typically 450–600 Hz; in compound **2.1**, *long range coupling* between non-equivalent ^{31}P nuclei is observed. The signals due to atoms P_b and P_a are a triplet and doublet respectively; values of J_{PP} (29 Hz) measured from the two signals are necessarily equal. Additionally, coupling between the ^{31}P and closest ^{1}H nuclei may be observed. Two types of heteronuclear NMR spectra are routinely recorded: one in which coupling to protons is observed and one in which protons are instrumentally *decoupled* from the observed nucleus.

The notation ^{31}P$\{^1$H$\}$ means proton-decoupled ^{31}P; corresponding notations are used for other proton-decoupling.

Table 2.3   Properties of selected NMR active nuclei. A complete list is available from *WebElements* on the World Wide Web at: http://www.webelements.com/

Nucleus	Natural abundance / %	I	Frequency of observation / MHz (referred to ^{1}H at 100 MHz)‡	Chemical shift reference $(\delta = 0)^*$
^{1}H	>99.9	$\frac{1}{2}$	100	SiMe$_4$
^{2}H	0.015	1	15.35	SiMe$_4$
^{7}Li	92.5	$\frac{3}{2}$	38.9	LiCl (1 M in H$_2$O)
^{11}B	80.1	$\frac{3}{2}$	32.1	F$_3$B·OEt$_2$
^{13}C	1.1	$\frac{1}{2}$	25.1	SiMe$_4$
^{17}O	0.04	$\frac{5}{2}$	13.5	H$_2$O
^{19}F	100	$\frac{1}{2}$	94.0	CFCl$_3$
^{23}Na	100	$\frac{3}{2}$	26.45	NaCl (1 M in H$_2$O)
^{27}Al	100	$\frac{5}{2}$	26.1	[Al(H$_2$O)$_6$]$^{3+}$
^{29}Si	4.7	$\frac{1}{2}$	19.9	SiMe$_4$
^{31}P	100	$\frac{1}{2}$	40.5	H$_3$PO$_4$ (85%, aq)
^{77}Se	7.6	$\frac{1}{2}$	19.1	SeMe$_2$
^{103}Rh	100	$\frac{1}{2}$	3.2	Rh (metal)
^{117}Sn	7.7	$\frac{1}{2}$	35.6	SnMe$_4$
^{119}Sn	8.6	$\frac{1}{2}$	37.3	SnMe$_4$
^{129}Xe	26.4	$\frac{1}{2}$	27.7	XeOF$_4$
^{183}W	14.3	$\frac{1}{2}$	4.2	Na$_2$WO$_4$ (in D$_2$O)
^{195}Pt	33.8	$\frac{1}{2}$	21.5	Na$_2$[PtCl$_6$]
^{199}Hg	16.8	$\frac{1}{2}$	17.9	HgMe$_2$

‡ The operating frequency of an instrument is defined by the field of the magnet and is designated by the frequency at which the ^{1}H nuclei of SiMe$_4$ resonate.

* It is important to quote the reference when reporting NMR spectra since alternative references may be used.

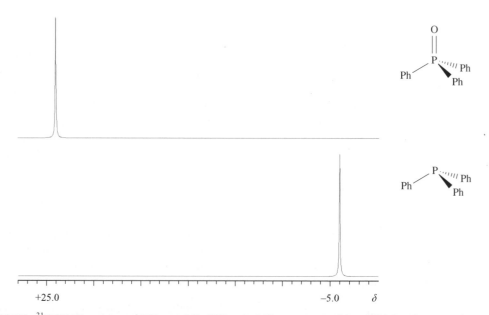

Fig. 2.8   The 162 MHz ^{31}P NMR spectra of PPh$_3$ and O=PPh$_3$. A shift to more positive δ (higher frequency) generally accompanies the oxidation of a tertiary phosphine and recording the ^{13}P NMR spectrum of a phosphine before use in the laboratory is an easy way of checking the purity of phosphines which are readily oxidized in air.

Fig. 2.9 The 162 MHz ^{31}P NMR spectrum of a salt of $[PF_6]^-$ consists of a binomial septet. The value of J_{PF} can be measured between any pair of adjacent lines in the signal.

Case study 3: ^{11}B NMR spectrum of $[BH_4]^-$

The ^{11}B NMR spectrum of Na$[BH_4]$ is shown in Figure 2.10. The 1:4:6:4:1 pattern of signal integrals corresponds to the binomial quintet expected for four equivalent ^{1}H nuclei coupling to ^{11}B. Although $I = \frac{3}{2}$ for ^{11}B, it is the $I = \frac{1}{2}$ of the attached protons that determines the nature of the signal in the ^{11}B NMR spectrum of $[BH_4]^-$.

Case study 4: ^{31}P{^{1}H} NMR spectrum of PhMe$_2$P·BH$_3$

Figure 2.11 shows the structure of the adduct PhMe$_2$P·BH$_3$ and its ^{31}P{^{1}H} NMR spectrum. The signal is a four-line multiplet (but *not* a binomial quartet) and arises primarily from coupling between ^{31}P and ^{11}B nuclei. For ^{11}B, $I = \frac{3}{2}$;

this means there are four spin states with values $+\frac{3}{2}$, $+\frac{1}{2}$, $-\frac{1}{2}$ and $-\frac{3}{2}$. There is an *equal probability* that the ^{31}P nucleus will 'see' the ^{11}B nucleus in each of the four spin states, and this gives rise to the ^{31}P signal being split into four equal intensity lines: a 1:1:1:1 multiplet. The observed signal is complicated by the fact the ^{11}B has an 80% abundance and the second isotope, ^{10}B, is also NMR active ($I = 3$). It too couples to the ^{31}P nucleus giving a 7-line multiplet (1:1:1:1:1:1:1) but the value of $J_{^{31}P^{10}B}$ is smaller than $J_{^{31}P^{11}B}$. The result is two overlapping signals, but the dominant feature is the 1:1:1:1 multiplet, the signal shape of which is affected by both the underlying 7-line multiplet and relaxation effects.

Fig. 2.10 The 128 MHz ^{11}B NMR spectrum of a solution of NaBH$_4$ in CD$_3$C(O)CD$_3$. The value of J_{BH} can be measured between any pair of adjacent lines in the signal.

BOX 2.4   **CHEMICAL AND THEORETICAL BACKGROUND**

NMR spectroscopy: a factual résumé

NMR active nuclei and isotope abundance

Many nuclei possess a property described as spin. The nuclear spin (nuclear angular momentum) is quantized and is described by the spin quantum number I which can have values of 0, $\frac{1}{2}$, 1, $\frac{3}{2}$, 2, $\frac{5}{2}$ etc. If the value of I for a nucleus is zero, the nucleus is *NMR inactive*, e.g. ^{12}C. For both ^1H and ^{13}C, $I = \frac{1}{2}$ and these nuclei are *NMR active*. In this book, we encounter other NMR active nuclei with different (non-zero) values of I. In the absence of an applied magnetic field, the different nuclear spin states of a nucleus are degenerate. However, when a magnetic field is applied, they are split (become non-degenerate) and this allows nuclear spin transitions to occur when radiofrequency (RF) radiation is absorbed.

When a ^1H NMR spectrum of a hydrogen-containing compound is recorded, virtually all the H atoms in the sample contribute to the observed spectrum; in a naturally occurring hydrogen sample, the abundance of ^1H is 99.985%. The fact that only 1% of naturally occurring carbon is ^{13}C means that if a ^{13}C NMR spectrum of a carbon-containing compound is recorded, only 1% of the carbon atoms present are observed. This has important ramifications in regard of $^1H-^{13}C$ coupling as we see below.

Resonance frequencies and chemical shifts

A particular nucleus (e.g. ^1H, ^{13}C, ^{31}P) absorbs characteristic radiofrequencies, i.e. it *resonates* at a characteristic frequency. If an NMR spectrometer is tuned to a particular resonance frequency, *only* a selected NMR active nucleus is observed. For example, only ^1H nuclei are observed if a 400 MHz spectrometer is tuned to 400 MHz, but if the same spectrometer is retuned to 162 MHz, only ^{31}P nuclei are observed. This is analogous to tuning a radio and receiving only one station at a time.

Each signal in an NMR spectrum is denoted by a *chemical shift value*, δ, a value that is given relative to the signal observed for a specified reference compound (see below). The δ value is given by the equation:

$$\delta = \frac{(\nu - \nu_0) \times 10^6}{\nu_0} = \frac{\Delta\nu \times 10^6}{\nu_0}$$

where $\Delta\nu$ is the frequency difference, in Hz, between the signal of interest and some defined reference frequency (ν_0). The multiplier of 10^6 is included so that values of δ are of convenient magnitudes. The chemical shift refers to a frequency difference in *parts per million* (ppm) and chemical shifts may be reported as ppm *or* δ values, this being the same *numerical* value for a given signal.

The standard reference (for which δ is defined as 0) for both ^1H and ^{13}C NMR spectroscopies is tetramethylsilane, $SiMe_4$ (TMS); see also Table 2.3. When the NMR spectrum of a compound is recorded, signals due to particular nuclei are said to be *shifted* with respect to the standard reference signal. A shift to positive δ is 'shifted to higher frequency' and a shift to negative δ is 'shifted to lower frequency'.

Older terminology which may still be encountered relates a positive δ value to a 'downfield shift' and a negative δ value to an 'upfield shift'.

Solvents for solution studies

Samples for solution NMR spectroscopy are generally prepared using *deuterated solvents*. One reason for this is that, were non-deuterated solvents to be used (e.g. CH_3Cl in place of CD_3Cl) for a ^1H NMR spectroscopic experiment, the signals due to the solvent would 'swamp' those due to the sample. Deuterated solvents are commercially available, typically with >99.5% ^2H label incorporated. The remaining unlabelled compound provides a useful *internal reference* signal in the ^1H NMR spectrum of the sample under study.

Integration of signals and signal broadening

Under normal conditions of measuring ^1H NMR spectra, the ratio of the peak areas (*integrals*) of the signals in the spectrum is proportional to the number of nuclei giving rise to the signals. For example, in a ^1H NMR spectrum of $HC\equiv CCH_3$, two signals with relative integrals 1:3 are observed. However, the integration of signals must be treated with caution since the peak integral is dependent upon the *relaxation time* of the nucleus in question, i.e. the time taken for the nucleus to relax from an excited to ground state during the NMR spectroscopic experiment. (Further details of this phenomenon may be found in references cited at the end of this chapter.) One particular problem is the relative integrals of signals in a ^{13}C NMR spectrum.

In some cases, signals may be broadened and this can affect the measurement of the relative integrals of signals. For example, signals arising from protons attached to N are broadened due to *quadrupolar relaxation* by ^{14}N ($I = 1$). Exchange with solvent protons also causes broadening, e.g.:

$$CH_3CH_2OH + HOH \rightleftharpoons CH_3CH_2OH + HOH$$

Homonuclear spin–spin coupling: $^1H-^1H$

A ^1H nucleus ($I = \frac{1}{2}$) may be in one of two spin states ($m_I = +\frac{1}{2}$, $m_I = -\frac{1}{2}$) and the energy difference between the spin states depends on the applied magnetic field of the NMR spectrometer. Consider a system in which there are two magnetically non-equivalent ^1H nuclei, H_A and H_B. There are two possible situations:

- The local magnetic field generated by the spin of H_A is *not* detected by H_B; the ^1H NMR spectrum consists of two resonances, each a *singlet* because there is *no coupling* between the two ^1H nuclei.
- H_A *is* affected by the magnetic fields associated with H_B; the ^1H NMR signal for H_A is *split into two equal lines* depending on which of the two spin states of H_B (equal

probabilities) it 'sees'. Similarly, the signal for H_B is composed of two equal lines. Protons H_A and H_B *couple* with each other and the spectrum consists of two *doublets*.

The separation between the two lines in each of the doublets described above must be equal, and this splitting is called the *coupling constant, J*, and is measured in hertz (Hz). In general, coupling to one proton gives a doublet, to two equivalent protons gives a triplet, to three equivalent protons gives a quartet, and so on. The relative intensities of the lines in the *multiplet* are given by a binomial distribution, readily determined using a Pascal's triangle:

$$
\begin{array}{ccccccccc}
 & & & & 1 & & & & \leftarrow \text{singlet} \\
 & & & 1 & & 1 & & & \leftarrow \text{doublet} \\
 & & 1 & & 2 & & 1 & & \leftarrow \text{triplet} \\
 & 1 & & 3 & & 3 & & 1 & \leftarrow \text{quartet} \\
1 & & 4 & & 6 & & 4 & & 1 \quad \leftarrow \text{quintet}
\end{array}
$$

Self-study exercise

The 100 MHz ^1H spectrum of butanone is shown below, and consists of a quartet, a singlet and a triplet. The coupling constants J for the triplet and quartet are equal. Account for the observed spectrum.

Heteronuclear spin–spin coupling: $^{13}C-^1H$

Each of the nuclei ^1H and ^{13}C has a magnetic spin quantum number $I = \frac{1}{2}$, and when ^{13}C and ^1H nuclei are in close proximity, they can couple. However, in molecules containing a natural isotopic distribution of carbon atoms, only 1% are ^{13}C nuclei. From a statistical consideration, it follows that in a ^1H NMR spectrum of, for example, acetone, $^{13}C-^1H$ coupling is *not* observed, although it *is* observed in the ^{13}C NMR spectrum of the *same* sample. The ^{13}C NMR spectrum of acetone exhibits a singlet due to the C=O carbon atom, and a quartet due to the two equivalent methyl ^{13}C nuclei.

Self-study exercise

Why do you not observe $^{13}C-^{13}C$ coupling in the ^{13}C NMR spectrum of acteone?

Case study 5: ^{19}F NMR spectrum of $[XeF_5]^-$

The planar $[XeF_5]^-$ ion contains five equivalent F atoms (see *worked example 1.13*). Both the ^{19}F and ^{129}Xe nuclei are NMR active: ^{19}F, $I = \frac{1}{2}$, 100% abundance; ^{129}Xe, $I = \frac{1}{2}$, 26.4%. The ^{19}F NMR spectrum of $[XeF_5]^-$ is shown in Figure 2.12. The equivalence of the ^{19}F nuclei gives rise to one signal; 26.4% of the F centres are attached to ^{129}Xe while the remainder are bonded to other Xe nuclei. The spectrum can be interpreted in terms of a singlet (the central line) due to 73.6 % of the ^{19}F nuclei plus an overlapping doublet due to the 26.4% of the ^{19}F nuclei that couple to ^{129}Xe. The centre of the doublet coincides with the position of the singlet because *all* the ^{19}F nuclei resonate at the same frequency. The two side peaks in Figure 2.12 are called *satellite peaks*.

Stereochemically non-rigid species

The NMR spectroscopic examples discussed so far have assumed that, with the exception of free rotation about single bonds, the molecule or ion is static in solution. For the majority of organic and inorganic species, this assumption is valid, but the possibility of *stereochemical non-rigidity* (*fluxionality*) *on the NMR spectroscopic timescale* must be

Fig. 2.11 The 162 MHz $^{31}P\{^1H\}$ NMR spectrum of the adduct $PhMe_2P \cdot BH_3$. The four line pattern is *not* a binomial quartet but an approximate 1:1:1:1 multiplet.

considered. Five-coordinate species such as $Fe(CO)_5$, **2.2**, PF_5, **2.3**, and BrF_5, **2.4**, constitute one group of compounds for which the activation barrier for dynamic behaviour in solution is relatively low, and exchange of substituents is facile.

The inclusion of the qualifier 'on the NMR spectroscopic timescale' is important. The timescale of the NMR spectroscopic technique (10^{-1} to 10^{-5} s, depending on the observed nucleus) is relatively long, and is significantly longer than that of IR spectroscopy; $Fe(CO)_5$ appears static on the IR spectroscopic timescale, but dynamic within the timescale of a ^{13}C NMR spectroscopic experiment. Lowering the temperature slows down the dynamic behaviour, and *may* make it slower than the spectroscopic timescale. However, some fluxional processes have very low energy barriers; even at 103 K, the axial and equatorial CO groups in $Fe(CO)_5$ exchange positions and the ^{13}C NMR spectrum consists of one signal corresponding to the average ^{13}C environment. On the other hand, the room temperature solution ^{19}F NMR spectrum of BrF_5 exhibits a doublet and a binomial quintet (due to ^{19}F–^{19}F coupling) with relative integrals of 4 : 1, and this is consistent with structure **2.4**. Above 450 K, one signal is observed, indicating that the five F atoms are equivalent on the NMR timescale, i.e. BrF_5 molecule is fluxional. On going from the low to high temperature limit, the two signals *coalesce* to give a single resonance.

The usual dynamic process in which 5-coordinate species are involved in solution is *Berry pseudo-rotation*. Although ligand–ligand repulsions are minimized in a trigonal bipyramidal arrangement, only a small amount of energy is needed to convert it into a square-based pyramid. The interconversion involves small perturbations of the bond angles subtended at the central atom, and continued repetition of the process results in each substituent 'visiting' both equatorial and axial sites in the trigonal bipyramidal structure (Figure 2.13).

Exchange processes in solution

A number of hydrated cations in aqueous solution undergo exchange with the solvent at rates slow enough to be

Fig. 2.12 The 376 MHz ^{19}F NMR spectrum of $[XeF_5]^-$, simulated using literature parameters. The isotopic abundance of ^{129}Xe is 26.4%; the centre of the doublet coincides with the position of the singlet.

Trigonal bipyramid
(Atoms 2 and 3 are
in axial sites.)

Square-based pyramid

Trigonal bipyramid
(Atoms 4 and 5 are
in axial sites.)

Fig. 2.13 Berry pseudo-rotation interconverts one trigonal bipyramidal structure into another via a square-based pyramidal intermediate. The numbering scheme illustrates that axial and equatorial sites in the trigonal bipyramid are interchanged.

observed on the NMR spectroscopic timescale by using ^{17}O isotopic labelling; ^{17}O has $I = \frac{5}{2}$, while both ^{16}O and ^{18}O are NMR inactive. Different chemical shifts are observed for the

^{17}O nuclei in bulk and coordinated water, and from the signal intensity ratios, hydration numbers can be obtained. For example, Al^{3+} has been shown to be present as $[Al(H_2O)_6]^{3+}$.

Reactions such as that in equation 2.41 are known as *redistribution reactions*.

$$PCl_3 + P(OEt)_3 \rightleftharpoons PCl_2(OEt) + PCl(OEt)_2 \qquad (2.41)$$

A *redistribution reaction* is one in which substituents exchange between species but the types and numbers of each type of bond remain the same.

The position of equilibrium can be followed by using ^{31}P NMR spectroscopy since each of the four species has a characteristic chemical shift. Rate data are obtained by following the variation in relative signal integrals with time, and equilibrium constants (and hence values of ΔG° since $\Delta G^{\circ} = -RT \ln K$) can be found from the relative signal integrals when no further change takes place (i.e. equilibrium has been established); by determining ΔG° at

---

BOX 2.5

APPLICATIONS

Magnetic resonance imaging (MRI)

Magnetic resonance imaging (MRI) is a clinical technique to obtain an image of, for example, a human organ or tumour. The image is generated from information obtained from the ^{1}H NMR spectroscopic signals of water. The signal intensity depends upon the proton relaxation times and the concentration of water. The relaxation times can be altered, and the image enhanced, by using *MRI contrast agents*. Coordination complexes containing paramagnetic Gd^{3+}, Fe^{3+} or Mn^{2+} are potentially suitable as contrast agents, and of these, complexes containing the Gd^{3+} ion have so far proved to be especially useful. To minimize toxic side-effects in patients, Gd^{3+} must be introduced in the form of a complex that will not dissociate in the body, and chelating ligands are particularly suitable (see *Chapter 6* for a discussion of stability constants). Excretion is also an important consideration; complexes must not remain in the body any longer than is necessary. One of the successful ligands in use is derived from H_5DTPA; after the intravenous injection to introduce $[Gd(DPTA)]^{2-}$, clearance through the kidneys takes about 30 min.

H_5DTPA

If an image of a certain organ is required, it is important to find a contrast agent that *targets* that organ, e.g. gadolinium(III) complexes are used to target the liver.

Dependence upon the observation of proton signals in some organs (e.g. lungs) presents problems with respect to MRI. The use of ^{129}Xe magnetic imaging has been tested as a means of overcoming some of the difficulties associated with proton observation. Under the right conditions, gaseous ^{129}Xe taken into mouse lungs allows excellent images to be observed.

Further reading

M.S. Albert, G.D. Cates, B. Driehuys, W. Happer, B. Saam, C.S. Springer and A. Wishnia (1994) *Nature*, vol. 370, p. 199 – 'Biological magnetic resonance imaging using laser-polarized ^{129}Xe'.

J.F. Desreux and V. Jacques (1995) in *Handbook of Metal–Ligand Interactions in Biological Fluids*, ed. G. Berthon, vol. 2, p. 1109, Dekker, New York – 'Role of metal–ligand interactions in the design of MRI contrast agents'.

S.H. Koenig and R.D. Brown (1995) in *Handbook of Metal–Ligand Interactions in Biological Fluids*, ed. G. Berthon, vol. 2, p. 1093, Dekker, New York – 'Relaxivity of MRI magnetic contrast agents. Concepts and principles'.

R.A. Moats, S.E. Fraser and T.J. Meade (1997) *Angewandte Chemie, International Edition in English*, vol. 36, p. 726 – 'A "smart" magnetic resonance imaging agent that reports on specific enzymic activity'.

P.G. Morris (1986) *Nuclear Magnetic Resonance Imaging in Medicine and Biology*, Clarendon Press, Oxford.

Table 2.4 Properties of selected nuclei observed by Mössbauer spectroscopy. The radioisotope source provides the γ-radiation required for the Mössbauer effect.

Nucleus observed	Natural abundance / %	Ground spin state	Excited spin state	Radioisotope source[‡]
^{57}Fe	2.2	$\frac{1}{2}$	$\frac{3}{2}$	^{57}Co
^{119}Sn	8.6	$\frac{1}{2}$	$\frac{3}{2}$	^{119m}Sn
^{99}Ru	12.7	$\frac{3}{2}$	$\frac{5}{2}$	^{99}Rh
^{197}Au	100	$\frac{3}{2}$	$\frac{1}{2}$	^{197m}Pt

[‡] m = metastable

different temperatures, values of ΔH° and ΔS° can be found using equations 2.42 and 2.43.

$$\Delta G^\circ = \Delta H^\circ - T\Delta S^\circ \tag{2.42}$$

$$\frac{d \ln K}{dT} = \frac{\Delta H^\circ}{RT^2} \tag{2.43}$$

Values of ΔH° for these types of reactions are almost zero, the redistribution of the groups being driven by an increase in the entropy of the system.

2.12 Mössbauer spectroscopy in inorganic chemistry

Mössbauer spectroscopy is by no means as widely used as NMR spectroscopy, and its brief coverage here reflects this.

The technique of Mössbauer spectroscopy

The *Mössbauer effect* is the emission and resonant absorption of nuclear γ-rays studied under conditions such that the nuclei have negligible recoil velocities when γ-rays are emitted or absorbed. This is only achieved by working with *solid samples* in which the nuclei are held rigidly in a crystal lattice. The energy, and thus the frequency of the γ-radiation involved, corresponds to the transition between the ground state and the short-lived excited state of the nuclide concerned. Table 2.4 lists properties of several nuclei which can be observed using Mössbauer spectroscopy.

We illustrate the study of the Mössbauer effect by reference to ^{57}Fe spectroscopy. The basic apparatus includes a radioactive source, a solid absorber with the ^{57}Fe-containing sample and a γ-ray detector. For ^{57}Fe samples, the radioactive source is ^{57}Co and is incorporated into stainless steel; the ^{57}Co source decays by capture of an extra-nuclear electron to give the excited state of ^{57}Fe which emits γ-radiation as it decays to its ground state. If ^{57}Fe is present in the same form in both source and absorber, resonant absorption occurs and no radiation is transmitted. However, if the ^{57}Fe in the source and absorber is present in two different forms, absorption does *not* occur and γ-radiation reaches the detector. Moving the source at different velocities towards or

away from the ^{57}Fe absorber has the effect of varying the energy of the γ-radiation (i.e. by the Doppler effect). The velocity of movement required to bring about maximum absorption relative to stainless steel (defined as an arbitrary zero for iron) is called the *isomer shift* of ^{57}Fe in the sample, with units of mm s^{-1}.

What can isomer shift data tell us?

The isomer shift gives a measure of the electron density on the ^{57}Fe centre, and isomer shift values can be used to determine the oxidation state of the Fe atom. Similarly, in ^{197}Au Mössbauer spectroscopy, isomer shifts can be used to distinguish between Au(I) and Au(III). Three specific examples are chosen here from iron chemistry.

The cation [Fe(NH$_3$)$_5$(NO)]$^{2+}$ has presented chemists with an ambiguity in terms of the description of the bonding which has, in some instances, been described in terms of an [NO]$^+$ unit bound to an Fe(I) centre. Results of ^{57}Fe Mössbauer spectroscopy have revealed that the correct description is that of an [NO]$^-$ ligand bound to an Fe(III) centre.

The formal oxidation states of the iron centres in [Fe(CN)$_6$]$^{4-}$ and [Fe(CN)$_6$]$^{3-}$ are +2 and +3; however, the

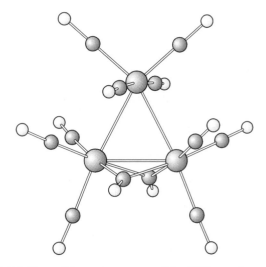

Fig. 2.14 The solid state structure of Fe$_3$(CO)$_{12}$ as determined by X-ray diffraction methods. The molecule contains two Fe environments by virtue of the arrangement of the CO groups.

closeness of the isomer shifts for these species suggests that the actual oxidation states are similar and this may be interpreted in terms of the extra electron in $[Fe(CN)_6]^{4-}$ being delocalized on the cyano ligands rather than the iron centre.

Differences in isomer shifts can be used to distinguish between different iron environments in the same molecule: the existence of two signals in the Mössbauer spectrum of $Fe_3(CO)_{12}$ provided the first evidence for the presence of two types of iron atom in the solid state structure (Figure 2.14), a fact that has been confirmed by X-ray diffraction methods.

Summary

This chapter has been concerned with the properties of the nucleus, aspects of nuclear energy, radioactivity, and two spectroscopic techniques (NMR and Mössbauer) that owe their existence to nuclear properties.

The following terms were introduced in this chapter. Do you know what they mean?

- ❑ neutron
- ❑ proton
- ❑ nucleon
- ❑ nuclide
- ❑ mass number
- ❑ mass defect
- ❑ binding energy
- ❑ radioactive decay
- ❑ first order rate equation
- ❑ first order rate constant
- ❑ half-life
- ❑ α-particle
- ❑ β-particle (β^-)
- ❑ γ-radiation
- ❑ positron (β^+)
- ❑ neutrino (ν_e)
- ❑ antineutrino
- ❑ transmutation of an element
- ❑ nuclear fission
- ❑ nuclear fusion
- ❑ slow (thermal) neutron
- ❑ fast neutron
- ❑ transuranium element
- ❑ isotopic enrichment
- ❑ zero point energy
- ❑ isotope exchange reaction
- ❑ kinetic isotope effect
- ❑ spectroscopic timescale
- ❑ nuclear spin quantum number, I
- ❑ chemical shift (in NMR spectroscopy)
- ❑ spin–spin coupling (in NMR spectroscopy)
- ❑ proton-decoupled NMR spectrum
- ❑ multiplicity of an NMR spectroscopic signal
- ❑ satellite peaks (in an NMR spectrum)
- ❑ stereochemically non-rigid
- ❑ fluxionality
- ❑ Berry pseudo-rotation
- ❑ redistribution reaction
- ❑ Mössbauer effect
- ❑ isomer shift (in Mössbauer spectroscopy)

Suggested further reading

Basic reaction kinetics

C.E. Housecroft and E.C. Constable (1997) *Chemistry: An Integrated Approach*, Addison Wesley Longman, Harlow – Chapter 10 covers first order reaction kinetics with worked examples, and includes mathematical background for the integration of rate equations.

Nuclear chemistry

G.R. Choppin, J.-O. Liljenzin and J. Rydberg (1995) *Radiochemistry and Nuclear Chemistry*, 2nd edn, Butterworth-Heinemann, Oxford – An excellent general account of both the subjects and their chemical and practical applications.

G. Friedlander, J.W. Kennedy, E.S. Macias and J.M. Miller (1981) *Nuclear and Radiochemistry*, 3rd edn, Wiley, New York – A general textbook of radiochemistry and its applications.

J. Godfrey, R. McLachlan and C.H. Atwood (1991) *Journal of Chemical Education*, vol. 68, p. 819 – An article entitled 'Nuclear reactions versus inorganic reactions' provides a useful comparative survey and includes a résumé of the kinetics of radioactive decay.

N.N. Greenwood and A. Earnshaw (1997) *Chemistry of the Elements,* 2nd edn, Butterworth-Heinemann, Oxford – Chapter 1 gives an account of the origins of the elements and of nuclear processes.

D.C. Hoffmann (1994) *Chemical & Engineering News,* 2 May issue, p. 24 – An article entitled 'The heaviest elements' which gives a good feeling for the problems and fascination involved in working with the transuranium elements.

D.C. Hoffmann and G.R. Choppin (1986) *Journal of Chemical Education*, vol. 63, p. 1059 – A discussion of high level nuclear waste.

D.C. Hoffmann and D.M. Lee (1999) *Journal of Chemical Education*, vol. 76, p. 331 – An excellent article that covers the development and future prospects of 'atom-at-a-time' chemistry.

NMR and Mössbauer spectroscopies

C. Brevard and P. Granger (1981) *Handbook of High Resolution Multinuclear NMR*, Wiley-Interscience, New York – A reference book listing nuclear properties, standard references, typical chemical shift ranges and coupling constants.

C.E. Housecroft (1994) *Boranes and Metallaboranes: Structure, Bonding and Reactivity*, 2nd edn, Ellis Horwood, Hemel Hempstead – Chapter 2 includes an account of the interpretation of ^{11}B and ^1H NMR spectra of boranes and their derivatives.

B.K. Hunter and J.K.M. Sanders (1993) *Modern NMR Spectroscopy: A Guide for Chemists*, 2nd edn, Oxford University Press, Oxford – An excellent, detailed and readable text.

A.G. Maddock (1997) *Mössbauer Spectroscopy: Principles and Applications*, Horwood Publishing, Chichester – A comprehensive account of the technique and its applications.

R.V. Parish (1990), *NMR, NQR, EPR and Mössbauer Spectroscopy in Inorganic Chemistry*, Ellis Horwood, Chichester – A text dealing with the theory, applications and interpretation of spectra; includes end-of-chapter problems.

J.K.M. Sanders, E.C. Constable, B.K. Hunter and C.M. Pearce (1993) *Modern NMR Spectroscopy: A Workbook of Chemical Problems*, 2nd edn, Oxford University Press, Oxford – An invaluable collection of NMR spectroscopic problem-solving exercises.

Problems

2.1 For each of the following isotopes, state the number of neutrons, protons and electrons present: (a) $^{19}_{9}F$; (b) $^{59}_{27}Co$; (c) $^{235}_{92}U$.

2.2 What do you understand by the terms: (a) atomic number; (b) mass number; (c) mass defect; (d) binding energy per nucleon?

2.3 Using the data in Appendix 5, plot a representation of the mass spectrum of naturally occurring atomic Ba.

2.4 Radium-224 is radioactive and decays by emitting an α-particle. (a) Write an equation for this process. (b) The decay of radium-224 produces helium gas. Rutherford and Geiger determined that α-particles were emitted from $^{224}_{88}Ra$ at a rate of $7.65 \times 10^{12}\,s^{-1}\,mol^{-1}$, and that this corresponded to a rate of helium production of $2.90 \times 10^{-10}\,dm^3\,s^{-1}$ at 273 K, 1 bar. If 1 mole of helium occupies $22.7\,dm^3$ (273 K, 1 bar), estimate a value for the Avogadro constant.

2.5 Use the following data to determine the half-life of $^{218}_{84}Po$ and the rate constant for the decay of $^{218}_{84}Po$.

Time / s	0	200	400	600	800	1000
Moles $^{218}_{84}Po$	0.250	0.110	0.057	0.025	0.012	0.005

2.6 The half-life of strontium-90 is 29.1 years. Determine the rate constant for the decay of strontium-90 in units of s^{-1}. [The SI unit of time is the second.]

2.7 Complete the following table which refers to possible nuclear reactions of a nuclide:

Reaction type	Change in number of protons	Change in number of neutrons	Change in mass number	Is a new element formed?
α-particle loss				
β-particle loss				
Positron loss				
(n,γ) reaction				

2.8 For each step in Figure 2.3, identify the particle emitted.

2.9 Interpret the following notational forms of nuclear reactions: (a) $^{58}_{26}Fe(2n,\beta)^{60}_{27}Co$; (b) $^{55}_{25}Mn(n,\gamma)^{56}_{25}Mn$; (c) $^{32}_{16}S(n,p)^{32}_{15}P$; (d) $^{23}_{11}Na(\gamma,3n)^{20}_{11}Na$.

2.10 Identify the second fission product in the following reactions:
(a) $^{235}_{92}U + ^{1}_{0}n \longrightarrow ^{142}_{56}Ba + ? + 2^{1}_{0}n$
(b) $^{235}_{92}U + ^{1}_{0}n \longrightarrow ^{137}_{52}Te + ? + 2^{1}_{0}n$

2.11 In each of the following reactions, are the incoming neutrons 'fast' or 'slow' ? Give reasons for your choices.
(a) $^{14}_{7}N + ^{1}_{0}n \longrightarrow ^{14}_{6}C + ^{1}_{1}H$
(b) $^{238}_{92}U + ^{1}_{0}n \longrightarrow ^{239}_{92}U + \gamma$
(c) $^{235}_{92}U + ^{1}_{0}n \longrightarrow ^{85}_{34}Se + ^{148}_{58}Ce + 3^{1}_{0}n$

2.12 Determine the half-life of Bk given that a plot of $\ln N$ against t is linear with a gradient of $-0.0023\,day^{-1}$ where N is the number of nuclides present at time t.

2.13 The IR spectrum of naturally occurring CO shows an absorption at $2170\,cm^{-1}$ assigned to the vibrational mode of the molecule. If the sample is enriched in ^{13}C, what change do you expect to see when the IR spectrum is re-recorded?

2.14 If the oxide P_4O_6 is dissolved in an aqueous solution of sodium carbonate, compound **A** of formula Na_2HPO_3 may be crystallized from solution. The IR spectrum of **A** contains a band at $2300\,cm^{-1}$. The corresponding band in the IR spectrum of **B** (obtained by an analogous method from P_4O_6 and Na_2CO_3 dissolved in D_2O) is at $1630\,cm^{-1}$. On recrystallization of **A** from D_2O, however, its IR spectrum is not affected. Discuss the interpretation of these observations.

2.15 Why is the method of isotope dilution analysis used to determine the solubility of sparingly soluble salts rather than a method depending upon mass determination?

2.16 A small amount of the radioactive isotope $^{212}_{82}Pb$ was mixed with a quantity of a non-radioactive lead salt containing $0.0100\,g$ lead ($A_r = 207$). The whole sample was dissolved in aqueous solution and lead(II) chromate ($PbCrO_4$) was precipitated by the addition of a soluble chromate salt. Evaporation of $10\,cm^3$ of the supernatant liquid gave a residue having a radioactivity of 4.17×10^{-5} that of the original quantity of $^{212}_{82}Pb$. Calculate the solubility of lead(II) chromate in $mol\,dm^{-3}$.

In questions 2.17 to 2.27, refer to Table 2.3 for isotopic abundances where needed.

2.17 Why is a coupling constant measured in Hz and is not recorded as a chemical shift difference?

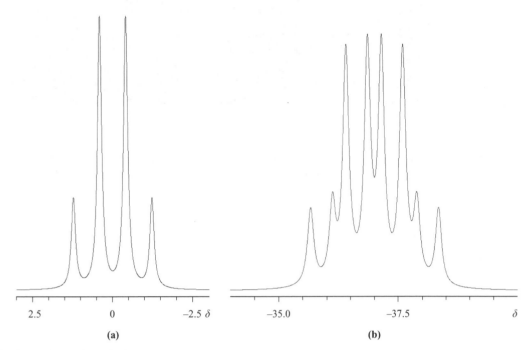

Fig. 2.15 Figure for problem 2.23.

2.18 Long range couplings are often observed between ^{31}P and ^{19}F nuclei, between ^{31}P and ^{1}H nuclei, but not between remote non-equivalent ^{1}H nuclei. What does this tell you about the relative magnitudes of values of J_{PF}, J_{PH} and J_{HH} for the respective pairs of nuclei when they are directly attached?

2.19 Rationalize the fact that the ^{13}C NMR spectrum of CF_3CO_2H consists of two binomial quartets with coupling constants of 44 and 284 Hz respectively.

2.20 How might you use ^{31}P NMR spectroscopy to distinguish between Ph_2PH and Ph_3P?

2.21 The ^{31}P NMR spectrum of PMe_3 consists of a binomial decet (J 2.7 Hz). (a) Account for this observation. (b) Predict the nature of the ^{1}H NMR spectrum of PMe_3.

2.22 The ^{29}Si NMR spectrum of compound **2.5** shows a triplet with a coupling constant of 194 Hz. (a) Rationalize these data and (b) predict the nature of the signal in the ^{1}H NMR spectrum of **2.5** that is assigned to the silicon-bound protons. [^{29}Si: 4.7% abundant; $I = \frac{1}{2}$]

(2.5)

2.23 Figure 2.15 shows the ^{11}B NMR spectra of (a) THF·BH$_3$ (**2.6**) and (b) PhMe$_2$P·BH$_3$. Interpret the observed coupling patterns and mark on the figure where you would measure relevant coupling constants.

(2.6)

2.24 (a) Predict the structure of SF_4 using the VSEPR model. (b) Account for the fact that at 298 K and in solution the ^{19}F NMR spectrum of SF_4 exhibits a singlet but that at 175 K, two equal intensity triplets are observed.

2.25 The ^{19}F NMR spectrum of each of the following molecules exhibits one signal. For which species is this observation consistent with a static molecular structure as predicted by VSEPR theory: (a) SiF$_4$; (b) PF$_5$; (c) SF$_6$; (d) SOF$_2$; (e) CF$_4$?

2.26 Outline the mechanism of Berry pseudo-rotation, giving two examples of molecules that undergo this process.

2.27 Is it correct to interpret the phrase 'static solution structure' as meaning necessarily rigid? Use the following molecules to exemplify your answer: PMe$_3$; OPMe$_3$; PPh$_3$; SiMe$_4$.

TOPICS

- Symmetry operators and symmetry elements
- Point groups
- An introduction to character tables
- Infrared spectroscopy
- Chiral molecules

3.1 Introduction

Within chemistry, symmetry is important both at a molecular level and within crystalline systems, and an understanding of symmetry is essential in discussions of molecular spectroscopy and calculations of molecular properties. A discussion of *crystal symmetry* is not appropriate in this book, and we introduce only *molecular symmetry*. For qualitative purposes, it is sufficient to refer to the shape of a molecule using terms such as tetrahedral, octahedral or square planar. However, the common use of these descriptors is not always precise, e.g. consider the structures of BF_3, **3.1**, and BF_2H, **3.2**, both of which are planar. A molecule of BF_3 is correctly described as being trigonal planar, since its symmetry properties are fully consistent with this description; all the F−B−F bond angles are 120° and the B−F bond distances are all identical (131 pm). It is correct to say that the boron centre in BF_2H, **3.2**, is in a *pseudo-trigonal planar* environment but the molecular symmetry properties are not the same as those of BF_3. The F−B−F bond angle in BF_2H is smaller than the two H−B−F angles, and the B−H bond is shorter (119 pm) than the B−F bonds (131 pm).

(3.1)   (3.2)

The descriptor *symmetrical* implies that a species possesses a number of indistinguishable configurations. When structure **3.1** is rotated in the plane of the paper through 120°, the resulting structure is indistinguishable from the first; another 120° rotation results in a third indistinguishable molecular orientation (Figure 3.1). This is *not* true if we carry out the same rotational operations on BF_2H.

Group theory is the mathematical treatment of symmetry. In this chapter, we introduce the fundamental language of group theory (*symmetry operator*, *symmetry element*, *point group* and *character table*) and exemplify its applications in inorganic chemistry. The chapter does not set out to give a comprehensive survey of molecular symmetry, but rather to introduce some common terminology and its meaning.

3.2 Symmetry operations and symmetry elements

In Figure 3.1, we applied 120° rotations to BF_3 and saw that each rotation generated a representation of the molecule that was indistinguishable from the first. Each rotation is an example of a *symmetry operation*.

A *symmetry operation* is an operation performed on an object which leaves it in a configuration that is indistinguishable from, and superimposable on, the original configuration.

The rotations described in Figure 3.1 were performed about an axis perpendicular to the plane of the paper and passing through the boron atom; the axis is an example of a *symmetry element*.

F | 120° rotation ⟹ F | 120° rotation ⟹ F |

F–B–F   F–B–F   F–B–F

Fig. 3.1 Rotation of the trigonal planar BF_3 molecule through 120° generates a representation of the structure that is indistinguishable from the first; one F atom is marked in blue simply as a label. A second 120° rotation gives another indistinguishable structural representation.

A symmetry operation is carried out with respect to points, lines or planes, the latter being the *symmetry elements*.

Rotation about an *n*-fold axis of symmetry

The symmetry operation of rotation about an *n*-fold axis (the symmetry element) is denoted by the symbol C_n, in which the angle of rotation is $\dfrac{360°}{n}$; *n* is an integer, e.g. 2, 3 or 4. Applying this notation to the BF_3 molecule in Figure 3.1 gives a value of $n = 3$ (equation 3.1), and therefore we say that the BF_3 molecule contains a C_3 *rotation axis*; in this case, the axis lies perpendicular to the plane containing the molecule.

$$\text{Angle of rotation} = 120° = \frac{360°}{n} \tag{3.1}$$

In addition, BF_3 also contains three 2-fold (C_2) rotation axes, each coincident with a B−F bond as shown in Figure 3.2.

If a molecule possesses more than one type of *n*-axis, the axis of highest value of *n* is called the *principal axis*; it is the axis of *highest molecular symmetry*. For example, in BF_3, the C_3 axis is the principal axis.

Reflection through a plane of symmetry (mirror plane)

If reflection of all parts of a molecule through a plane produces an indistinguishable configuration, the plane is a

plane of symmetry; the symmetry operation is one of reflection and the symmetry element is the mirror plane (denoted by σ). For BF_3, the plane containing the molecular framework (the shaded plane shown in Figure 3.2) is a *mirror plane*. In this case, the plane lies perpendicular to the vertical principal axis and is denoted by the symbol σ_h.

The framework of atoms in a linear, bent or planar molecule can always be drawn in a plane, but this plane can be labelled σ_h *only* if the molecule possesses a C_n axis *perpendicular* to the plane. If the plane *contains* the principal axis, it is labelled σ_v. Consider the H_2O molecule. This possesses a C_2 axis (Figure 3.3) but it also contains *two* mirror planes, one containing the H_2O framework, and one perpendicular to it. Each plane contains the principal axis of rotation and so may be denoted as σ_v but in order to distinguish between them, we use the notations σ_v and σ_v'. The σ_v label refers to the plane that bisects the H−O−H bond angle and the σ_v' label refers to the plane in which the molecule lies.

A special type of σ plane which contains the principal rotation axis, but which bisects the angle between two adjacent 2-fold axes is labelled σ_d.

In the notation for planes of symmetry, σ, the subscripts h, v and d stand for horizontal, vertical and dihedral respectively.

Reflection through a centre of symmetry (inversion centre)

If reflection of *all* parts of a molecule through the centre of the molecule produces an indistinguishable configuration, the centre is a *centre of symmetry*, also called a *centre of inversion* (see also *Box 1.8*); it is designated by the symbol *i*.

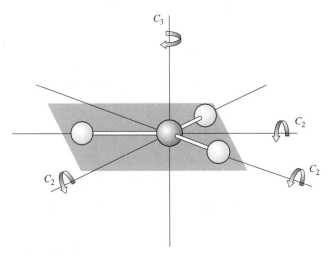

Fig. 3.2 The 3-fold (C_3) and three 2-fold (C_2) axes of symmetry possessed by the trigonal planar BF_3 molecule.

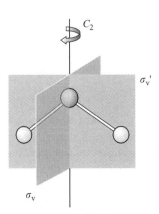

Fig. 3.3 The H_2O molecule possesses one C_2 axis and two mirror planes.

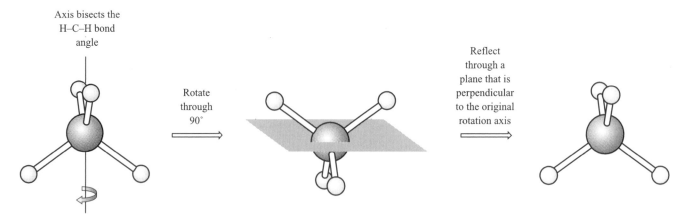

Fig. 3.4 An improper rotation (or rotation–reflection), S_n, involves rotation about $\dfrac{360°}{n}$ followed by reflection through a plane that is perpendicular to the rotation axis. The diagram illustrates the operation about one of the S_4 axes in CH_4; three S_4 operations are possible for the CH_4 molecule. [*Exercise*: where are the three rotation axes for the three S_4 operations in CH_4?]

Each of the molecules CO_2 (**3.3**), *trans*-N_2F_2 (see *worked example 3.1*), SF_6 (**3.4**), and benzene (**3.5**) possesses a centre of symmetry, but H_2S (**3.6**), *cis*-N_2F_2 and SiH_4 (**3.7**) do not.

O=C=O

(3.3)

(3.4)

(3.5)

(3.6)

(3.7)

Rotation about an axis, followed by reflection through a plane perpendicular to this axis

If rotation through $\dfrac{360°}{n}$ about an axis, followed by reflection through a plane perpendicular to that axis, yields an indistinguishable configuration, the axis is an *n*-fold rotation–reflection axis, also called an *n-fold improper rotation axis*. It is denoted by the symbol S_n. Tetrahedral species of the type XY_4 (all Y groups must be equivalent) possess three S_4 axes, and the operation of one S_4 rotation–reflection in the CH_4 molecule is illustrated in Figure 3.4.

Identity operator

All objects can be operated upon by the identity operator E. This is the simplest operator (although it may not be easy to appreciate why we identify such an operator!) and effectively

identifies the molecular configuration. The operator E leaves the molecule unchanged.

Worked example 3.1   Symmetry properties of *cis*- and *trans*-N_2F_2

How do the rotation axes and planes of symmetry in *cis*- and *trans*-N_2F_2 differ?

First draw the structures of *cis*- and *trans*-N_2F_2; both are planar molecules.

cis   *trans*

1. The identity operator E applies to each isomer.
2. Each isomer possesses a plane of symmetry which contains the molecular framework. However, their labels differ (see point 5 below).
3. The *cis*-isomer contains a C_2 axis which lies in the plane of the molecule, but the *trans*-isomer contains a C_2 axis which bisects the N–N bond and is perpendicular to the plane of the molecule.

4. The *cis*- (but not the *trans*-) isomer contains a mirror plane, σ_v, lying perpendicular to the plane of the molecule

and bisecting the N−N bond:

5. The consequence of the different types of C_2 axes, and the presence of the σ_v plane in the *cis*-isomer, is that the symmetry planes containing the *cis*- and *trans*-N_2F_2 molecular frameworks are labelled σ_v' and σ_h respectively.

Self-study exercises

1. How do the rotation axes and planes of symmetry in Z- and E-CFH=CFH differ?

2. How many planes of symmetry do (a) $F_2C=O$, (b) ClFC=O and (c) $[HCO_2]^-$ possess? [*Ans.* (a) 2; (b) 1; (c) 2]

Worked example 3.2 Symmetry operations in NH₃

The symmetry operators for NH_3 are E, C_3 and $3\sigma_v$. (a) Draw the structure of NH_3. (b) What is the meaning of the E operator? (c) Draw a diagram to show the rotation and reflection symmetry operations.

(a) The molecule is trigonal pyramidal.

(b) The E operator is the identity operator and it leaves the molecule unchanged.
(c) The C_3 axis passes through the N atom, perpendicular to a plane containing the three H atoms. Each σ_v plane contains one N−H bond and bisects the opposite H−N−H bond angle.

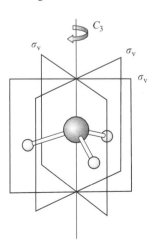

Self-study exercises

1. What symmetry operators are lost in going from NH_3 to NH_2Cl? [*Ans.* C_3; two σ_v]

2. Compare the symmetry operators possessed by NH_3, NH_2Cl, $NHCl_2$ and NCl_3.

3. Draw a diagram to show the symmetry operators of $NClF_2$. [*Ans.* Show one σ_v; only other operator is E]

Worked example 3.3 Trigonal planar BCl₃ *versus* trigonal pyramidal PCl₃

What symmetry elements do BCl_3 and PCl_3 (a) have in common and (b) not have in common?

PCl_3 is trigonal pyramidal (use VSEPR theory) and so possesses the same symmetry elements as NH_3 in worked example 3.2. These are E, C_3 and $3\sigma_v$.

BCl_3 is trigonal planar (use VSEPR) and possesses all the above symmetry elements:

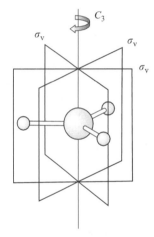

In addition, BCl_3 contains a σ_h plane and three C_2 axes (see *Figure 3.2*). Further, rotation through 120° about the C_3 axis, followed by reflection through the plane perpendicular to this axis (the σ_h plane), generates a molecular configuration indistinguishable from the first – this is an improper rotation S_3.

Conclusion:
The symmetry elements that BCl_3 and PCl_3 have in common are E, C_3 and $3\sigma_v$.

The symmetry elements possessed by BCl_3 but not by PCl_3 are σ_h, $3C_2$ and S_3.

Self-study exercises

1. Show that BF_3 and $F_2C=O$ have the following symmetry elements in common: E, two mirror planes, one C_2.

2. How do the symmetry elements of ClF_3 and BF_3 differ? [*Ans*: BF_3, as for BCl_3 above; ClF_3, E, σ_v', σ_v, C_2]

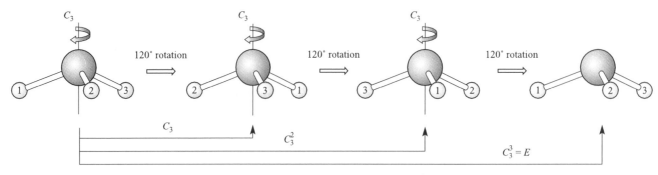

Fig. 3.5 Successive C_3 rotations in NH_3 are distinguished using the notation C_3, C_3^2 and C_3^3. The effect of the last operation is the same as that of the identity operator acting on NH_3 in the initial configuration.

3.3 Successive operations

As we have seen in Section 3.2, a particular symbol is used to denote a specific symmetry operation. To say that NH_3 possesses a C_3 axis tells us that we can rotate the molecule through 120° and end up with a molecular configuration that is indistinguishable from the first. However, it takes three such operations to give a configuration of the NH_3 molecule that *exactly* coincides with the first. The three separate 120° rotations are identified by using the notation in Figure 3.5. We cannot *actually* distinguish between the three H atoms, but for clarity they are labelled H(1), H(2) and H(3) in the figure. Since the third rotation, C_3^3, returns the NH_3 molecule to its initial configuration, we can write equation 3.2, or, in general, equation 3.3.

$$C_3^3 = E \qquad (3.2)$$

$$C_n^n = E \qquad (3.3)$$

Similar statements can be written to show the combined effects of successive operations. For example, an improper axis of rotation, S_n, can be written in the form of equation 3.4.

$$S_n = C_n \times \sigma_h \qquad (3.4)$$

Self-study exercise

$[PtCl_4]^{2-}$ is square planar; to what rotational operation is C_4^2 equivalent?

3.4 Point groups

The number and nature of the symmetry elements of a given molecule are conveniently denoted by its *point group*, and give rise to labels such as C_2, C_{3v}, D_{3h}, D_{2d}, T_d, O_h or I_h. These point groups belong to the classes of C groups, D groups and special groups, the latter containing groups that possess special symmetries, i.e. tetrahedral, octahedral and icosahedral.

To describe the symmetry of a molecule in terms of one symmetry element (e.g. a rotation axis) provides information only about this property. Each of BF_3 and NH_3 possesses a 3-fold axis of symmetry, but their structures and overall symmetries are different; BF_3 is trigonal planar and NH_3 is trigonal pyramidal. On the other hand, if we describe the symmetries of these molecules in terms of their respective point groups (D_{3h} and C_{3v}), we are providing information about *all* their symmetry elements.

Before we look at some representative point groups, we emphasize that it is not essential to memorize the symmetry elements of a particular point group. These are listed in *character tables* (see *Section 3.5* and *Box 4.1*) which are widely available.

Table 3.1 summarizes the most important classes of point group and gives their characteristic types of symmetry elements; E is, of course, common to every group. Some particular features of significance are given below.

C_1 point group

Molecules that appear to have no symmetry at all, e.g. **3.8**, must possess the symmetry element E and effectively possess at least one C_1 axis of rotation. They therefore belong to the C_1 point group, although since $C_1 = E$, the rotational symmetry operation is ignored when we list the symmetry elements of this point group.

$$\begin{array}{c} H \\ | \\ F \diagdown C \diagup\!\!\!\!\diagup Cl \\ | \\ Br \end{array}$$

(3.8)

$C_{\infty v}$ point group

C_∞ signifies the presence of an ∞-fold axis of rotation, i.e. that possessed by a linear molecule (Figure 3.6); for the molecular

Table 3.1 Characteristic symmetry elements of some important classes of point groups. The characteristic symmetry elements of the T_d, O_h and I_h are omitted because the point groups are readily identified (see Figure 3.7). No distinction is made in this table between σ_v and σ_d planes of symmetry. For *complete* lists of symmetry elements, character tables should be consulted.

Point group	Characteristic symmetry elements	Comments
C_s	E, one σ plane	
C_i	E, inversion centre	
C_n	E, one (principal) n-fold axis	
C_{nv}	E, one (principal) n-fold axis, n σ_v planes	
C_{nh}	E, one (principal) n-fold axis, one σ_h plane, one S_n-fold axis which is coincident with the C_n axis	The S_n axis necessarily follows from the C_n axis and σ_h plane. For $n = 2$, 4 or 6, there is also an inversion centre.
D_{nh}	E, one (principal) n-fold axis, n C_2 axes, one σ_h plane, n σ_v planes, one S_n-fold axis	The S_n axis necessarily follows from the C_n axis and σ_h plane. For $n = 2$, 4, or 6, there is also an inversion centre.
D_{nd}	E, one (principal) n-fold axis, n C_2 axes, n σ_v planes, one S_{2n}-fold axis	For $n = 3$ or 5, there is also an inversion centre.
T_d		Tetrahedral
O_h		Octahedral
I_h		Icosahedral

species to belong to the $C_{\infty v}$ point group, it must also possess an infinite number of σ_v planes but *no* σ_h plane or inversion centre. These criteria are met by asymmetrical diatomics such as HF, CO and [CN]$^-$ (Figure 3.6a), and linear polyatomics (throughout this book, polyatomic is used to mean a species containing three or more atoms) that do not possess a centre of symmetry, e.g. OCS and HCN.

$D_{\infty h}$ point group

Symmetrical diatomics (e.g. H_2, [O_2]$^{2-}$) and linear polyatomics that contain a centre of symmetry (e.g. [N_3]$^-$, CO_2, HC≡CH) possess a σ_h plane in addition to a C_∞ axis and an infinite number of σ_v planes (Figure 3.6b). These species belong to the $D_{\infty h}$ point group.

T_d, O_h or I_h point groups

Molecular species that belong to the T_d, O_h or I_h point groups (Figure 3.7) possess many symmetry elements, although it is seldom necessary to identify them all before the appropriate point group can be assigned. Species with tetrahedral symmetry include SiF_4, [ClO_4]$^-$, [$CoCl_4$]$^{2-}$, [NH_4]$^+$, P_4 (Figure 3.8a) and B_4Cl_4 (Figure 3.8b), and those with octahedral symmetry include SF_6, [PF_6]$^-$, W(CO)$_6$ (Figure 3.8c) and [Fe(CN)$_6$]$^{3-}$. There is no centre of symmetry in a tetrahedron but there is one in an octahedron, and this distinction has consequences with regard to the observed electronic spectra of tetrahedral and octahedral metal complexes (see *Section 20.6*). Members of the icosahedral point group are uncommon, e.g. [$B_{12}H_{12}$]$^{2-}$ (Figure 3.8d).

Determining the point group of a molecule or molecular ion

The application of a *systematic* approach to the assignment of a point group is essential, otherwise there is the risk that

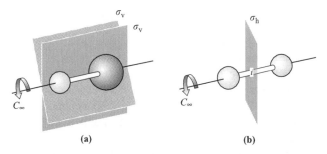

Fig. 3.6 Linear molecular species can be classified according to whether they possess a centre of symmetry (inversion centre) or not. All linear species possess a C_∞ axis of rotation and an infinite number of σ_v planes; in (a), two such planes are shown and these planes are omitted from (b) for clarity. Diagram (a) shows an asymmetrical diatomic belonging to the point group $C_{\infty v}$, and (b) shows a symmetrical diatomic belonging to the point group $D_{\infty h}$.

Tetrahedron   Octahedron   Icosahedron

Fig. 3.7 The tetrahedron (T_d symmetry), octahedron (O_h symmetry) and icosahedron (I_h symmetry) possess four, six and twelve vertices respectively, and four, eight and twenty equilateral-triangular faces respectively.

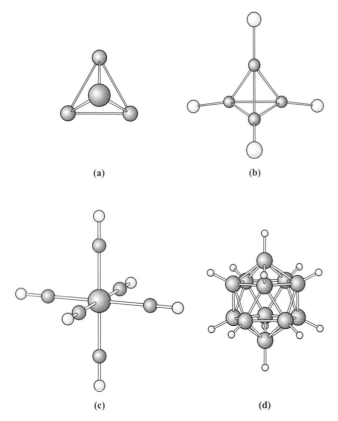

Fig. 3.8 The molecular structures of (a) P_4, (b) B_4Cl_4 (the B atoms are shown in blue), (c) $[W(CO)_6]$ (the W atom is shown in blue and the C atoms in grey) and (d) $[B_{12}H_{12}]^{2-}$ (the B atoms are shown in blue).

symmetry elements will be missed with the consequence that an incorrect assignment is made. Figure 3.9 shows a procedure that may be adopted; some of the less common point groups (e.g. S_n, T, O) are omitted from the scheme. Notice that it is *not* necessary to find all the symmetry elements (e.g. improper axes) in order to determine the point group.

We illustrate the application of Figure 3.9 with reference to four worked examples, with an additional example in Section 3.7. Before assigning a point group to a molecule, its structure must be determined by, for example, microwave spectroscopy, or X-ray, electron or neutron diffraction methods.

Worked example 3.4   Point group assignments: 1

Determine the point group of *trans*-N_2F_2.

First draw the structure.

$$\text{F} \diagdown \atop {\text{N}=\text{N}} \diagdown \atop \text{F}$$

Apply the strategy shown in Figure 3.9:

START $\Longrightarrow$

Is the molecule linear?	No
Does *trans*-N_2F_2 have T_d, O_h or I_h symmetry?	No
Is there a C_n axis?	Yes; a C_2 axis perpendicular to the plane of the paper and passing through the midpoint of the N–N bond
Are there two C_2 axes perpendicular to the principal axis?	No
Is there a σ_h plane (perpendicular to the principal axis)?	Yes

$\Longrightarrow$ **STOP**

The point group is C_{2h}.

Self-study exercises

1. Show that the point group of *cis*-N_2F_2 is C_{2v}.

2. Show that the point group of *E*-CHCl=CHCl is C_{2h}.

Worked example 3.5   Point group assignments: 2

Determine the point group of PF_5.

First, draw the structure.

$$\begin{array}{c} \text{F} \\ | \\ \text{F}-\text{P}^{\text{....}\text{F}}_{\diagdown \text{F}} \\ | \\ \text{F} \end{array}$$

In the trigonal bipyramidal arrangement, the three equatorial F atoms are equivalent, and the two axial F atoms are equivalent.

Apply the strategy shown in Figure 3.9:

START $\Longrightarrow$

Is the molecule linear?	No
Does PF_5 have T_d, O_h or I_h symmetry?	No
Is there a C_n axis?	Yes; a C_3 axis containing the P and two axial F atoms
Are there three C_2 axes perpendicular to the principal axis?	Yes; each lies along a $P–F_{eq}$ bond
Is there a σ_h plane (perpendicular to the principal axis)?	Yes; it contains the P and three F_{eq} atoms.

$\Longrightarrow$ **STOP**

The point group is D_{3h}.

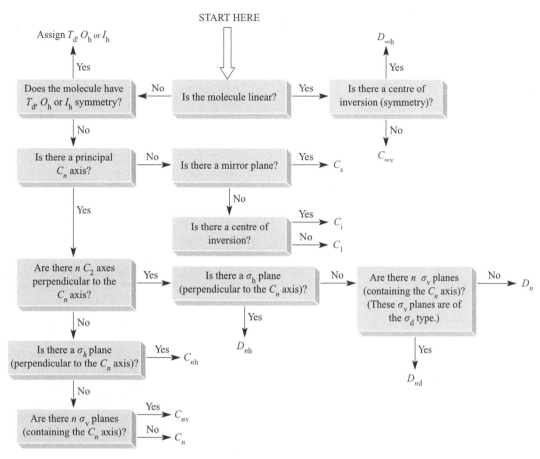

Fig. 3.9 Scheme for assigning point groups of molecules and molecular ions. Apart from the cases of $n = 1$ or ∞, n most commonly has values of 2, 3, 4, 5 or 6.

Self-study exercises

1. Show that BF_3 belongs to the D_{3h} point group.

2. Show that OF_2 belongs to the C_{2v} point group.

Worked example 3.6 ■ Point group assignments: 3

To what point group does $POCl_3$ belong ?

The structure of $POCl_3$ is:

$$
\begin{array}{c}
\mathrm{O} \\
\|\| \\
\mathrm{Cl}\text{----}\underset{|}{\overset{}{\mathrm{P}}}\text{''''}\mathrm{Cl} \\
\mathrm{Cl}
\end{array}
$$

Apply the strategy shown in Figure 3.9:

START $\Longrightarrow$

Is the molecule linear?	No
Does $POCl_3$ have T_d, O_h or I_h symmetry?	No (remember that although this molecule is loosely considered as being tetrahedral in shape, it does *not* possess tetrahedral symmetry)
Is there a C_n axis?	Yes; a C_3 axis running along the O–P bond
Are there 3 C_2 axes perpendicular to the principal axis?	No
Is there a σ_h plane (perpendicular to the principal axis)?	No
Are there n σ_v planes (containing the principal axis)?	Yes; each contains the one Cl and the O and P atoms

$\Longrightarrow$ **STOP**

The point group is C_{3v}.

Self-study exercises

1. Show that $CHCl_3$ possesses C_{3v} symmetry, but that CCl_4 belongs to the T_d point group.

2. Assign point groups to (a) $[NH_4]^+$ and (b) NH_3.

[*Ans.* (a) T_d; (b) C_{3v}]

Three projections of the cyclic structure of S_8 are shown below; all S–S bond distances are equivalent, as are all S–S–S bond angles. To what point group does S_8 belong?

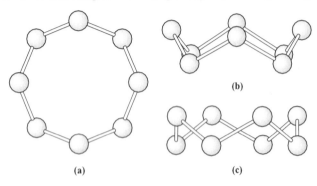

(a)   (b)   (c)

Follow the scheme in Figure 3.9:

START $\Longrightarrow$

Is the molecule linear?	No
Does S_8 have T_d, O_h or I_h symmetry?	No
Is there a C_n axis?	Yes; a C_4 axis running through the centre of the ring; perpendicular to the plane of the paper in diagram (a)
Are there 4 C_2 axes perpendicular to the principal axis?	Yes; these are most easily seen from diagram (c)
Is there a σ_h plane (perpendicular to the principal axis)?	No
Are there n σ_d planes (containing the principal axis)?	Yes; these are most easily seen from diagrams (a) and (c)

$\Longrightarrow$ **STOP**

The point group is D_{4d}.

Self-study exercises

1. Copy diagram (a) above. Show on the figure where the C_4 axis and the four C_2 axes lie.

2. S_6 has the chair conformation shown in Box 1.1. Confirm that this molecule contains a centre of inversion.

3.5 Character tables: an introduction

While Figure 3.9 provides a point group assignment using certain diagnostic symmetry elements, it may be necessary to establish whether any additional symmetry elements are exhibited by a molecule in a given point group.

Table 3.2 The character table for the C_{2v} point group. Explanations of the numbers in the table are given in Box 4.1. For more character tables, see Appendix 3.

C_{2v}	E	C_2	$\sigma_v(xz)$	$\sigma_v'(yz)$		
A_1	1	1	1	1	z	z, x^2, y^2, z^2
A_2	1	1	-1	-1	R_z	xy
B_1	1	-1	1	-1	x, R_y	x, xz
B_2	1	-1	-1	1	y, R_x	y, yz

Each point group has an associated *character table*, and that for the C_{2v} point group is shown in Table 3.2. The point group is indicated at the top left-hand corner and the symmetry elements possessed by a member of the point group are given across the top row of the character table. The H_2O molecule has C_{2v} symmetry and when we looked at the symmetry elements of H_2O in Figure 3.3, we labelled the two perpendicular planes. In the character table, taking the z axis as coincident with the principal axis, the σ_v and σ_v' planes are defined as lying in the xz and yz planes, respectively. Placing the molecular framework in a convenient orientation with respect to a Cartesian set of axes has many advantages, one of which is that the atomic orbitals on the central atom point in convenient directions. We return to this in Section 4.4.

The left-hand column in a character table gives a list of *symmetry labels*. These are used to label the symmetry properties of, for example, molecular orbitals or modes of molecular vibrations. We shall not delve further into this topic at this stage although we shall use the symmetry labels (see *Box 4.1*) at various points later in the book.

Appendix 3 gives character tables for the most commonly encountered point groups, and each table has the same format as that in Table 3.2.

3.6 Why do we need to recognize symmetry elements?

So far in this chapter, we have described the possible symmetry elements that a molecule might possess and, on the basis of these symmetry properties, we have illustrated how a molecular species can be assigned to a particular point group. Now we address some of the reasons why the recognition of symmetry elements in a molecule is important to the inorganic chemist.

Most of the applications of symmetry fall into one of the following categories:

- constructing molecular and hybrid orbitals;
- interpreting spectroscopic (e.g. vibrational and electronic) properties;
- determining whether a molecular species is chiral.

We return to the applications of symmetry within MO theory in Chapter 4, and to spectroscopic properties in Chapter 20. The next two sections deal briefly with the consequences of symmetry on observed bands in infrared spectra and with the relationship between molecular symmetry and chirality.

3.7 Infrared spectroscopy

The discussion that follows is necessarily selective and is pitched at a simplistic level. Although in this section we derive the number of vibrational modes for some simple molecules, for more complicated species it is necessary to use character tables. The reading list at the end of the chapter gives sources of detailed discussions of the relationship between group theory and normal modes of vibration.

Infrared (IR) and Raman (see *Box 3.1*) spectroscopies are branches of *vibrational spectroscopy* and the former technique is the more widely available of the two in student teaching laboratories.

How many vibrational modes are there for a given molecular species?

Vibrational spectroscopy is concerned with the observation of the *degrees of vibrational freedom*, the number of which can be determined as follows. The motion of a molecule containing n atoms can conveniently be described in terms

of the three Cartesian axes; the molecule has $3n$ *degrees of freedom* which together describe the *translational*, *vibrational* and *rotational* motions of the molecule.

The translational motion of a molecule (i.e. movement through space) can be described in terms of 3 degrees of freedom relating to the three Cartesian axes. If there are $3n$ degrees of freedom in total and 3 degrees of freedom for translational motion, it follows that there must be $(3n - 3)$ degrees of freedom for rotational and vibrational motion. For a *non-linear molecule* there are 3 degrees of rotational freedom, but for a *linear molecule*, there are 2 degrees of rotational freedom. This difference arises because there is no rotation about the molecular axis in a linear molecule. Having taken account of translational and rotational motion, the number of degrees of vibrational freedom can be determined (equations 3.5 and 3.6).

Number of degrees of vibrational freedom for a
$$\textit{non-linear} \text{ molecule} = 3n - 6 \qquad (3.5)$$

Number of degrees of vibrational freedom for a
$$\textit{linear} \text{ molecule} = 3n - 5 \qquad (3.6)$$

For example, from equation 3.6, the linear CO_2 molecule has four *normal modes of vibration* and these are shown in Figure 3.10. Two of the modes are *degenerate*; i.e. they possess the same energy and could be represented in a single diagram with the understanding that one vibration occurs in the plane of the paper and another, identical in energy, takes place in a plane perpendicular to the first.

BOX 3.1   **CHEMICAL AND THEORETICAL BACKGROUND**

Raman spectroscopy

Infrared and Raman spectroscopies are both concerned with the study of molecular vibrations, and while IR spectroscopy is used routinely in the practical laboratory, Raman spectroscopy is a more specialized technique. When radiation of a particular frequency, ν, (usually from a laser source) falls on a molecule, some radiation is scattered. The scattered radiation is of two types:

- Rayleigh scattering involves radiation of frequency, ν_0, equal to that of the incident radiation, and
- Raman scattering involves radiation of frequencies $\nu_0 \pm \nu$ where ν is a fundamental frequency of a vibrational mode of the molecule.

The selection rules for Raman and IR active vibrations are different. *A vibrational mode is Raman active if the polarizability of the molecule changes during the vibration.* Changes in polarizability (for Raman spectra) are not as easy to visualize as changes in electric dipole moments (for IR spectra) and in most cases it is necessary to use group theory to determine whether or not a mode will be Raman active.

A combination of IR and Raman spectroscopic data is often of great use. Molecules with a centre of symmetry are

subject to the *rule of mutual exclusion* which states that in such a molecule, a vibrational mode which is IR active is Raman inactive, and *vice versa*. Thus, for a molecule with an inversion centre, a 'missing' absorption in the IR spectrum *may* be observed in the Raman spectrum; note, however, the presence of symmetry elements other than the inversion centre does result in some exceptions to the rule of mutual exclusion and it must be applied with caution. We exemplify the rule with reference to CO_2 (Figure 3.10). The two vibrational modes which are asymmetric *with respect to the inversion centre* (i.e. the carbon atom) are IR active and Raman inactive, while the symmetric stretch is IR inactive but Raman active. Thus, the value of $1333 \, cm^{-1}$ for this latter vibration can be confirmed from a Raman spectrum.

For more detailed accounts of the Raman effect, see:

E.A.V. Ebsworth, D.W.H. Rankin and S. Cradock (1991) *Structural Methods in Inorganic Chemistry*, 2nd edn, Blackwell Scientific Publications, Oxford, Chapter 5.

K. Nakamoto (1986) *Infrared and Raman Spectra of Inorganic and Coordination Compounds*, 3rd edn, Wiley, New York.

(a) Symmetric stretch
IR inactive

(b) Asymmetric stretch
IR active ($2349\,cm^{-1}$)

(c) Bend (deformation)
IR active ($667\,cm^{-1}$)

(d) Bend (deformation)
IR active ($667\,cm^{-1}$)

Fig. 3.10 The vibrational modes of CO_2 ($D_{\infty h}$); in each mode of vibration, the carbon atom remains stationary. Vibrations (a) and (b) are stretching modes. Bending mode (c) occurs in the plane of the paper, while bend (d) occurs in a plane perpendicular to that of the paper; the + signs designate motion towards the reader. The two bending modes require the same amount of energy and are therefore *degenerate*.

Selection rule for an infrared active mode of vibration

One of the important consequences of precisely denoting molecular symmetry is seen in infrared spectroscopy. An IR spectrum records the frequency of a molecular vibration, but not all modes of vibration of a particular molecule give rise to observable absorption bands in the IR spectrum. This is because the following *selection rule* must be obeyed: *for a vibrational mode to be IR active, it must give rise to a change in the molecular dipole moment* (see *Section 1.16*).

> For a mode of vibration to be infrared (IR) active, it must give rise to a change in the molecular electric dipole moment.

In the discussions of IR spectroscopy that follow, we are concerned only with *fundamental* absorptions, these being the dominant features of IR spectra.

> The transition from the vibrational ground state to the first excited state is the *fundamental* transition.

Linear and bent triatomic molecules

We can readily illustrate the effect of molecular symmetry on molecular dipole moments, and thus on infrared active modes of vibration, by considering the linear molecule CO_2. The two C–O bond distances are equal (116 pm) and the molecule is readily identified as being 'symmetrical'; strictly, CO_2 possesses $D_{\infty h}$ symmetry. As a consequence of its symmetry, CO_2 is non-polar. Although both the asymmetric stretch and the bend (Figure 3.10) give rise to a change in dipole moment (generated transiently as the vibration occurs), the symmetric stretch does not. Thus, only two fundamental absorptions are observed in the IR spectrum of CO_2.

Now consider SO_2 which is a bent molecule (C_{2v}). Figure 3.11 shows the three normal modes of vibration; *all* give rise to a change in molecular dipole moment and are therefore IR active. A comparison of these results for CO_2 and SO_2 illustrates that vibrational spectroscopy can be used to determine whether an X_3 or XY_2 species is linear or bent.

Linear molecules of the general type XYZ (e.g. OCS or HCN) possess $C_{\infty v}$ symmetry and their IR spectra are expected to show three absorptions; the symmetric stretching, asymmetric stretching and bending modes are all IR active.

> A stretching mode is designated by the symbol ν, while a deformation (bending) is denoted by δ.
> For example: ν_{CO} stands for the stretch of a C–O bond.

Trigonal planar and trigonal pyramidal molecules

A molecule of the type XY_3 with D_{3h} symmetry (i.e. trigonal planar) undergoes the normal modes of vibration shown in Figure 3.12. The symmetric stretch is not accompanied by a change in molecular dipole moment and is *not* IR active. The remaining three normal modes are IR active and so molecules such as SO_3, BF_3 and BCl_3 exhibit three absorptions in their IR spectra. The IR spectra of anions such as $[NO_3]^-$ and $[CO_3]^{2-}$ may also be recorded, but the counterion may also give rise to IR spectroscopic bands. Therefore, simple salts such as those of the alkali metals are chosen because they give spectra in which the bands can be assigned to the anion.

A set of normal modes of vibration similar to those shown in Figure 3.12 can also be drawn for a trigonal pyramidal (C_{3v}) XY_3 species such as NH_3, PCl_3, AsF_3, $[H_3O]^+$ and $[SO_3]^{2-}$. Each mode is IR active and so we expect to observe four absorptions in the IR spectrum of such a species.

Symmetric stretch
IR active ($1151\,cm^{-1}$)

Asymmetric stretch
IR active ($1362\,cm^{-1}$)

Scissoring (symmetric bend)
IR active ($518\,cm^{-1}$)

Fig. 3.11 The vibrational modes of SO_2 (C_{2v}).

Fig. 3.12 The vibrational modes of SO_3 (D_{3h}); only three are IR active. The + and − notation is used to show the 'up' and 'down' motion of the atoms during the mode of vibration. [*Exercise*: two of the modes are labelled as being degenerate: why is this?]

Differences in the IR spectra of XY_3 molecules possessing C_{3v} or D_{3h} symmetry is a method of distinguishing between these structures. Further, XY_3 molecules with **T**-shaped structures (e.g. ClF_3) belong to the C_{2v} point group, and vibrational spectroscopy may be used to distinguish their structures from those of C_{3v} or D_{3h} XY_3 species.

Tetrahedral and square planar species

An XY_4 molecule with T_d symmetry has nine normal modes of vibration (Figure 3.13) but only six of them are IR active. The IR spectra of species such as CCl_4, $TiCl_4$, OsO_4, $[ClO_4]^-$ and $[SO_4]^{2-}$ exhibit *two* absorptions because of degeneracies.

There are nine normal modes of vibration for a square planar (D_{4h}) XY_4 molecule but, as can be seen from Figure 3.14, there are only three IR absorptions. Among compounds of the *p*-block elements, D_{4h} XY_4 structures are rare; the observation of absorptions at 586, 291 and 161 cm^{-1} in the IR spectrum of XeF_4 is consistent with the structure predicted by VSEPR theory.

Worked example 3.8   Infrared spectra

The IR spectrum of $SnCl_2$ exhibits absorptions at 352, 334 and 120 cm^{-1}. What shape do these data suggest for the molecule, and is this result consistent with VSEPR theory?

For linear $SnCl_2$ ($D_{\infty h}$), the asymmetric stretch and the bend are IR active, but the symmetric stretch is IR inactive (no change in molecular dipole moment).

For bent $SnCl_2$, the symmetric stretching, asymmetric stretching and scissoring modes are all IR active.

The data therefore suggest that $SnCl_2$ is bent, and this is consistent with the VSEPR model since there is a lone pair in addition to two bonding pairs of electrons:

$$\overset{\displaystyle ..}{\underset{Cl \diagdown \ \diagup Cl}{Sn}}$$

Self-study exercises

1. The vibrational modes of XeF_2 are at 555, 515 and 213 cm^{-1} but only two are IR active. Explain why this is consistent with XeF_2 having a linear structure.

2. How many IR active vibrational modes does CS_2 possess, and why? Hint: CS_2 is isostructural with CO_2.

Fig. 3.13 The vibrational modes of CH_4 (T_d), only two of which are IR active.

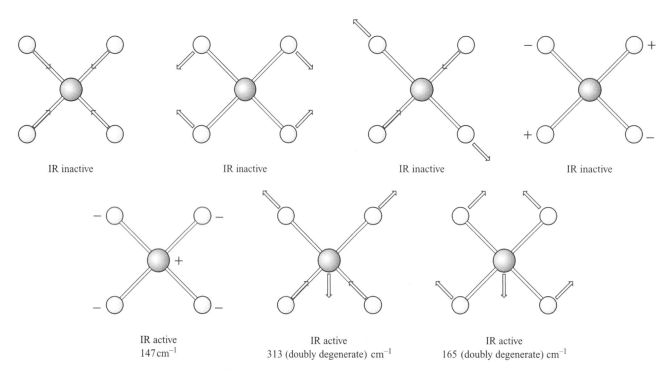

IR inactive   IR inactive   IR inactive   IR inactive

IR active
147 cm^{-1}

IR active
313 (doubly degenerate) cm^{-1}

IR active
165 (doubly degenerate) cm^{-1}

Fig. 3.14 The vibrational modes of $[PtCl_4]^{2-}$ (D_{4h}); only the three modes (two of which are degenerate) shown in the lower row are IR active. The + and − notation is used to show the 'up' and 'down' motion of the atoms during the mode of vibration.

Observing IR spectroscopic absorptions: practical problems

We have just described how to establish the number of vibrational degrees of freedom for a simple molecule with n atoms, how to deduce the total number of normal modes of vibration, and so determine the number of absorptions expected in its IR spectrum. Our premise for using IR spectroscopy to distinguish between, for example, an XY_3 molecule having C_{3v} or D_{3h} symmetry, depends upon being able to observe *all* the expected absorptions. However, a 'normal' laboratory IR spectrometer only spans the range between 4000 and 200 cm^{-1} and so if the vibration in question absorbs outside this range, the corresponding band will remain unobserved. An example is $[PtCl_4]^{2-}$ (Figure 3.14) where two of the three IR active vibrational modes are below 200 cm^{-1}; a specialized far-infrared spectrometer may be used to observe such absorptions.

Samples for IR spectroscopy are often prepared in cells with optical windows which themselves absorb within the 4000 and 200 cm^{-1} range; common materials are NaCl and KBr and these materials 'cut off' at 650 and 385 cm^{-1} respectively with the effect that absorptions (due to the sample) below these values are masked by the absorption due to the optical window. 'Solution cells' are used, not only for neat liquid samples but for solutions of the sample in a suitable solvent. This adds a further problem, since absorptions due to the solvent may mask those of the sample; in regions of strong solvent absorption, the transmitted radiation is essentially zero and so no absorptions at frequencies due to the sample may be detected.

3.8 Chiral molecules

A molecule is chiral if it possesses a *non-superimposable mirror image*. Helical chains such as Se_∞ (Figure 3.15a) may be right- or left-handed and are chiral. Six-coordinate complexes such as $[Cr(acac)_3]$ ([acac]$^-$, see *Table 6.7*) in which there are three didentate chelating ligands also possess non-superimposable mirror images (Figure 3.15b). Chiral molecules can rotate the plane of plane-polarized light (Figure 3.16). This property is known as *optical activity* and the two mirror images are known as *optical isomers* or *enantiomers*. Enantiomers rotate the light to equal extents, but in opposite directions, the dextrorotatory (*d*) enantiomer to the right and the laevorotatory (*l*) to the left (see *Box 3.2*). The amount of rotation *and* its sign depend upon the wavelength of the incident light. At this point, we note that the observation of optical activity depends upon *chemical* properties of the chiral molecule; if the two enantiomers interconvert rapidly to give an equilibrium mixture containing equal amounts of the two forms, no overall rotation occurs. A mixture of equal amounts of two enantiomers is called a *racemate* or *racemic mixture*. Chiral complexes and the separation of enantiomers are discussed further in Section 19.8.

The rotation, α, may be measured in an instrument called a *polarimeter*. In practice, the amount of rotation depends upon the wavelength of the light, temperature and the concentration of compound present in solution. The *specific rotation*, $[\alpha]$, for a chiral compound in solution is given by equation 3.7. Light of a single frequency is used for specific

Fig. 3.15 Pairs of enantiomers are non-superimposable mirror images of each other. (a) Helical Se_∞ has either a right- or left-handedness. (b) The six-coordinate complex $[Cr(acac)_3]$ contains three identical didentate, chelating ligands; the labels Λ and Δ describe the absolute configuration of the molecule (see *Box 3.2*).

rotation measurements and a common choice is the *sodium D-line* in the emission spectrum of atomic sodium; the specific rotation at this wavelength is denoted as $[\alpha]_D$.

$$[\alpha] = \frac{\alpha}{c \times \ell} \tag{3.7}$$

in which α = observed rotation, ℓ = path length of solution in the polarimeter (in dm) and c = concentration (in $g\,cm^{-3}$).

The importance of chirality is clearly seen in, for example, dramatic differences in the activities of different enantiomers of chiral drugs.[†]

[†] A relevant article is: E. Thall (1996) *Journal of Chemical Education*, vol. 73, p. 481 – 'When drug molecules look in the mirror'.

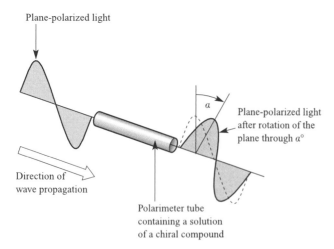

Plane-polarized light

Direction of wave propagation

Plane-polarized light after rotation of the plane through $\alpha°$

α

Polarimeter tube containing a solution of a chiral compound

Fig. 3.16 One enantiomer of a chiral compound rotates the plane of polarized light through a characteristic angle, $\alpha°$; the instrument used to measure this rotation is called a polarimeter. The direction indicated (a clockwise rotation as we view the light as it emerges from the polarimeter) is designated as $+\alpha°$. The other enantiomer of the same compound would rotate the plane of polarized light through an angle $-\alpha°$.

A helical chain such as Se_∞ is easy to recognize, but it is not always such a facile task to identify a chiral compound by attempting to convince oneself that there is, or is not, a non-superimposable mirror image. Symmetry considerations come to our aid: a chiral molecular species must lack an improper (S_n) axis of symmetry.

> A chiral molecule lacks an improper (S_n) axis of symmetry.

Another commonly used criterion for identifying a chiral species is the lack of an inversion centre, i, and plane of symmetry, σ. However, both of these properties are compatible with the criterion given above, since we can rewrite the symmetry operations i and σ in terms of the improper rotations S_2 and S_1 respectively. (See problem 3.25 at the end of the chapter.) However, a word of caution: there are a few species that are non-chiral (achiral) despite lacking an inversion centre, i, and plane of symmetry, σ.

Worked example 3.9   Chiral species

The oxalate ligand, $[C_2O_4]^{2-}$, is a didentate ligand and the structure of the complex ion $[Fe(ox)_3]^{3-}$ is shown below. The view in the right-hand diagram is along one O−Fe−O axis. Confirm that the point group to which the ion belongs is D_3 and that members of this point group are chiral.

BOX 3.2

CHEMICAL AND THEORETICAL BACKGROUND

Nomenclature of chiral compounds

The nomenclature of chiral compounds is complicated. Historically, compounds were described in terms of the sign of the rotation of plane-polarized light; the rotation was denoted ($+$) or d for dextrorotatory, and ($-$) or l for laevorotatory. The sign and magnitude of rotation are often dependent on the wavelength of light and this was incorporated in the descriptor: $(-)_{589}$ or $(-)_D$ (where D stands for the sodium D-line at a wavelength of 589 nm). Whilst this system is useful provided that the wavelength is specified, it is purely defined in terms of an observable (the rotation); there is *no direct relationship* with the absolute configuration of the molecule.

This problem was first addressed in organic chemistry where a chosen reference compound, glyceraldehyde, was arbitrarily assigned, one absolute configuration to the ($+$) and the other to the ($-$) enantiomer. The ($+$) form was assigned a D absolute configuration and the ($-$) form, an L configuration. Chemical transformations between organic molecules then allowed the assignment of D or L absolute configurations to be related to the arbitrarily assigned glyceraldehyde configuration. A consequence is that, for many organic molecules, the ($-$) enantiomer may possess a D (not an L) configuration! Additionally, it is not always easy to relate a D or L configuration of a highly complicated organic molecule back to the configuration of glyceraldehyde. As a matter of interest, the original arbitrarily assigned configuration to ($+$)-(D) glyceraldehyde has been shown to be correct by anomalous dispersion X-ray experiments.

In order to describe the absolute configuration of an organic molecule, the Cahn−Ingold−Prelog system was introduced. The descriptors R and S refer to the *absolute* arrangement of the groups about a centre. A complete description of a molecule will include both the sign of the rotation and the absolute configuration, e.g. $(+)_{589}$-(R).

Unfortunately, the Cahn−Ingold−Prelog rules are not directly applicable to most inorganic systems. For example, the three chelating ligands in $[Cr(en)_3]^{3+}$ (en $=$ $H_2NCH_2CH_2NH_2$) are identical and 'priorities' (an integral part of the Cahn−Ingold−Prelog rules) cannot be assigned to individual nitrogen-donor atoms. Descriptions based upon the observable rotation are, of course, useful, for example, $(+)_{589}$-$[Cr(en)_3]^{3+}$ and $(-)_{589}$-$[Cr(en)_3]^{3+}$. However, these convey no information about the absolute configurations of the complexes.

A number of schemes have been introduced to describe the configurations of such compounds, the most useful of which is the IUPAC recommended Λ and Δ system. This is exemplified in Figure 3.15b with the structures of the enantiomers of $[Cr(acac)_3]$.

For further discussion: see *Box 19.2* and *Section 19.8*.

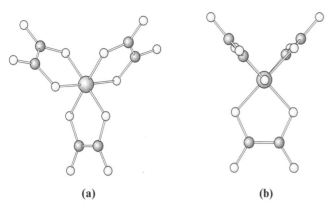

(a)   **(b)**

Using the scheme in Figure 3.9:

START $\Longrightarrow$

Is the molecular ion linear?	No
Does it have T_d, O_h or I_h symmetry?	No
Is there a C_n axis?	Yes; a C_3 axis; perpendicular to the plane of the paper in diagram (a)
Are there 3 C_2 axes perpendicular to the principal axis?	Yes; one runs vertically through the Fe centre in diagram (b)
Is there a σ_h plane (perpendicular to the principal axis)?	No
Are there n σ_d planes (containing the principal axis)?	No

$\Longrightarrow$ **STOP**

The point group is D_3.

No centre of symmetry or planes of symmetry have been identified and this alone is sufficient to confirm that molecular species in the D_3 point group are chiral.

Self-study exercise

By referring to the character table (Appendix 3) for the D_3 point group, confirm that the symmetry elements of the D_3 point group do not include i, σ or S_n axis.

Summary

This chapter has introduced the fundamental ideas of symmetry and group theory. You should be able to determine the symmetry elements of simple molecular species, and (by using Figure 3.9) assign a simple molecule to an appropriate point group. You should understand how IR spectroscopy can provide evidence for the structures of simple XY_n ($n = 2, 3, 4$) species, and appreciate the symmetry properties associated with a molecule being chiral.

The following terms have been introduced in this chapter. Do you know what they mean?

- ❑ symmetry element
- ❑ symmetry operator
- ❑ identity operator (E)
- ❑ rotation axis (C_n)
- ❑ plane of reflection (σ_h, σ_v or σ_d)
- ❑ centre of symmetry or inversion centre (i)
- ❑ improper rotation axis (S_n)
- ❑ point group
- ❑ translational degrees of freedom
- ❑ rotational degrees of freedom
- ❑ vibrational degrees of freedom
- ❑ normal mode of vibration
- ❑ degenerate modes of vibration
- ❑ selection rule (for an IR active mode)
- ❑ fundamental absorption
- ❑ chiral species
- ❑ enantiomer (optical isomer)
- ❑ racemic mixture
- ❑ specific rotation

Suggested further reading

Symmetry and group theory

P.W. Atkins, M.S. Child and C.S.G. Phillips (1970) *Tables for Group Theory*, Oxford University Press, Oxford – A set of character tables with useful additional notes and symmetry diagrams.

R.L. Carter (1998) *Molecular Symmetry and Group Theory*, Wiley, New York – An introduction to molecular symmetry and group theory as applied to chemical problems including vibrational spectroscopy.

F.A. Cotton (1990) *Chemical Applications of Group Theory*, 3rd edn, Wiley, New York – A more mathematical treatment of symmetry and its importance in chemistry.

G. Davidson (1991) *Group Theory for Chemists*, Macmillan, London – An excellent introduction to group theory with examples and exercises.

J.E. Huheey, E.A. Keiter and R.L. Keiter (1993) *Inorganic Chemistry: Principles of Structure and Reactivity*, 4th edn, Harper Collins, New York – Chapter 3 provides a useful, and readable, introduction to symmetry and group theory.

S.F.A. Kettle (1985) *Symmetry and Structure*, Wiley, Chichester – A detailed, but readable, account of symmetry and group theory.

A. Rodger and P.M. Rodger (1995) *Molecular Geometry*, Butterworth-Heinemann, Oxford – A useful, clear text for student use.

D.F. Shriver and P.W. Atkins (1999) *Inorganic Chemistry*, 3rd edn, Oxford University Press, Oxford – Contains a clear and concise introduction to symmetry and symmetry-related topics.

A.F. Wells (1984) *Structural Inorganic Chemistry*, 5th edn, Oxford University Press, Oxford – A definitive work on structural inorganic chemistry; chapter 2 gives a concise introduction to crystal symmetry.

Infrared spectroscopy

E.A.V. Ebsworth, D.W.H. Rankin and S. Cradock (1991) *Structural Methods in Inorganic Chemistry*, 2nd edn, Blackwell Scientific Publications, Oxford – Chapter 5 deals with vibrational spectroscopy in detail.

S.F.A. Kettle (1985) *Symmetry and Structure*, Wiley, Chichester – Chapter 9 deals with the relationship between molecular symmetry and molecular vibrations.

K. Nakamoto (1997) *Infrared and Raman Spectra of Inorganic and Coordination Compounds*, 5th edn, Wiley, New York – Part A: *Theory and Applications in Inorganic Chemistry* – An invaluable reference book for all practising experimental inorganic chemists, and including details of normal coordinate analysis.

Problems

Some of these questions require the use of Figure 3.9

3.1 Give the structures of the following molecules: (a) BCl_3; (b) SO_2; (c) PBr_3; (d) CS_2; (e) CHF_3. Which molecules are polar?

3.2 In group theory, what is meant by the symbols (a) E, (b) σ, (c) C_n and (d) S_n? What is the distinction between planes labelled σ_h, σ_v, σ_v' and σ_d?

3.3 For each of the following two-dimensional shapes, determine the highest order rotation axis of symmetry.

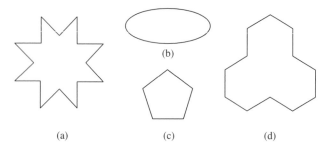

(a)   (b)   (c)   (d)

3.4 Draw the structure of SO_2 and identify its symmetry properties.

3.5 The structure of H_2O_2 was shown in Figure 1.16. Apart from the operator E, H_2O_2 possesses only one other symmetry operator. What is it?

3.6 By drawing appropriate diagrams, illustrate the fact that BF_3 possesses a 3-fold axis, three 2-fold axes, and four planes of symmetry. Give appropriate labels to these symmetry elements.

3.7 Using the answer to problem 3.6 to help you, deduce which symmetry elements are lost on going from (a) BF_3 to $BClF_2$ and (b) $BClF_2$ to $BBrClF$. (c) Which symmetry element (apart from E) is common to all three molecules?

3.8 Which of the following molecules or ions contain (a) a C_3 axis but no σ_h plane, and (b) a C_3 axis *and* a σ_h plane: NH_3; SO_3; PBr_3; $AlCl_3$; $[SO_4]^{2-}$; $[NO_3]^-$?

3.9 Which of the following molecules contains a C_4 axis *and* a σ_h plane: CCl_4; $[ICl_4]^-$; $[SO_4]^{2-}$; SiF_4; XeF_4?

3.10 How many mirror planes do each of the following molecules contain: (a) SF_4; (b) H_2S; (c) SF_6; (d) SOF_4; (e) SO_2; (f) SO_3?

3.11 (a) What structure would you expect Si_2H_6 to possess? (b) Draw the structure of the conformer most favoured in terms of steric energy. (c) Does this conformer possess an inversion centre? (d) Draw the structure of the conformer least favoured in terms of steric energy. (e) Does this conformer possess an inversion centre?

3.12 Which of the following species contain inversion centres? (a) BF_3; (b) SiF_4; (c) XeF_4; (d) PF_5; (e) $[XeF_5]^-$; (f) SF_6; (g) C_2F_4; (h) $H_2C=C=CH_2$.

3.13 Explain what is meant by an ∞-fold axis of rotation.

3.14 To which point group does NF_3 belong?

3.15 The point group of $[AuCl_2]^-$ is $D_{\infty h}$. What shape is this ion?

3.16 Determine the point group of SF_5Cl.

3.17 The point group of BrF_3 is C_{2v}. Draw the structure of BrF_3 and compare your answer with the predictions of VSEPR theory.

3.18 In worked example 1.13, we predicted the structure of the $[XeF_5]^-$ ion. Confirm that this structure is consistent with D_{5h} symmetry.

3.19 Assign a point group to each member in the series (a) CCl_4, (b) CCl_3F, (c) CCl_2F_2, (d) $CClF_3$ and (e) CF_4.

3.20 (a) Deduce the point group of SF_4. (b) Is SOF_4 in the same point group?

3.21 Which of the following point groups possesses the highest number of symmetry elements: (a) O_h; (b) T_d; (c) I_h?

3.22 Determine the number of degrees of vibrational freedom for each of the following: (a) SO_2; (b) SiH_4; (c) HCN; (d) H_2O; (e) BF_3.

3.23 How many normal modes of vibration are IR active for (a) H_2O, (b) SiF_4, (c) PCl_3, (d) $AlCl_3$, (e) CS_2 and (f) HCN?

3.24 Explain what is meant by the terms (a) chiral; (b) enantiomer; (c) helical chain.

3.25 Confirm that the symmetry operation of (a) inversion is equivalent to an S_2 improper rotation, and (b) reflection through a plane is equivalent to an S_1 improper rotation.

Chapter 4   Bonding in polyatomic molecules

TOPICS

■ Hybridization of atomic orbitals

■ Molecular orbital theory: ligand group orbitals

■ Delocalized bonding

■ Partial molecular orbital treatments

4.1 Introduction

In Chapter 1, we considered three approaches to the bonding in diatomic molecules:

● Lewis structures;
● valence bond (VB) theory;
● molecular orbital (MO) theory.

In this chapter we extend the discussion to polyatomic molecules (i.e. those containing three or more atoms). Within the valence bond model, treatment of a molecule XY_n ($n \geq 2$) raises the question of compatibility (or not) between the positions of the Y atoms and the directionalities of the atomic orbitals on the central atom X. Although an s atomic orbital is spherically symmetric, other atomic orbitals possess directional properties (see *Section 1.6*). Consider H_2O: Figure 4.1 illustrates that, if the atoms of the H_2O molecule lie in (for example) the yz plane, the directionalities of the $2p_y$ and $2p_z$ atomic orbital of oxygen are not compatible with the directionalities of the two O−H bonds. Although we could define the z axis to coincide with one O−H bond, the y axis could not (at the same time) coincide with the other O−H bond. Hence, there is a problem in trying to derive a localized bonding scheme in terms of an atomic orbital basis set (see *Section 1.13*). In the next section we describe a bonding model within valence bond (VB) theory that overcomes this problem. After we have considered how VB theory views the bonding in a range of XY_n species, we move on to the problems of applying molecular orbital theory to polyatomic species.

A *polyatomic species* contains three or more atoms.

4.2 Valence bond theory: hybridization of atomic orbitals

What is orbital hybridization?

The word 'hybridization' means 'mixing' and when used in the context of atomic orbitals, it describes a way of deriving *spatially directed orbitals* which may be used within VB theory. Like all bonding theories, *orbital hybridization is a model*, and should *not* be taken to be a real phenomenon.

Hybrid orbitals may be formed by mixing the characters of atomic orbitals that are close in energy. The character of a hybrid orbital depends on the atomic orbitals involved and their percentage contributions. The labels given to hybrid orbitals reflect the contributing atomic orbitals, e.g. an sp hybrid possesses equal amounts of s and p orbital character.

Hybrid orbitals are generated by mixing the characters of atomic orbitals.

The reason for creating a set of hybrid orbitals is to produce a convenient bonding scheme for a particular molecular species. An individual hybrid orbital points along a given internuclear axis within the framework of the molecule under consideration, and use of a set of hybrid orbitals provides a bonding picture in terms of *localized σ-bonds*. In working through the rest of this section, notice that each hybridization scheme for an atom X in a molecule XY_n is appropriate only for a particular shape, the shape being defined by the number of attached groups and any lone pairs.

A set of *hybrid orbitals* provides a bonding picture for a molecule in terms of *localized σ-bonds*.

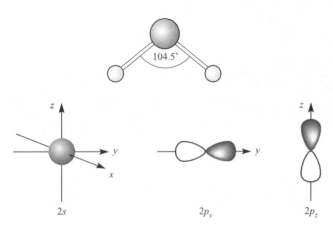

Fig. 4.1 A comparison of the shape of the H_2O molecule (the framework of which is taken as lying in the yz plane) with the spatial properties of the $2s$, $2p_y$ and $2p_z$ atomic orbitals of oxygen.

sp Hybridization: a scheme for linear species

The notation sp means that one s atomic orbital and one p atomic orbital mix to form a set of two hybrid orbitals with different directional properties.

One possible combination of a $2s$ atomic orbital and $2p_x$ atomic orbital is shown in Figure 4.2a. In the figure, the colour of the orbital lobe corresponds to a particular phase (see *Section 1.6*) and the addition of the $2s$ component reinforces one lobe of the $2p_x$ atomic orbital but diminishes the other. Equation 4.1 represents the combination mathematically. The wavefunction $\psi_{sp\,\text{hybrid}}$ describes a normalized (see *Section 1.12*) sp hybrid orbital which possesses 50% s and 50% p character. Although equation 4.1 and Figure 4.2a refer to the combination of $2s$ and $2p_x$ atomic orbitals, this could just as well be $2s$ with $2p_y$ or $2p_z$, or $3s$ with $3p_x$, and so on.

$$\psi_{sp\,\text{hybrid}} = \frac{1}{\sqrt{2}}(\psi_{2s} + \psi_{2p_x}) \tag{4.1}$$

Now comes an important general rule: *if we begin with n atomic orbitals, we must end up with n orbitals after hybridization.* Figure 4.2b and equation 4.2 show the second possibility for the combination of a $2s$ and a $2p_x$ atomic orbital. The sign change for the combination changes the phase of the $2p_x$

Fig. 4.2 The formation of two sp hybrid orbitals from one $2s$ atomic orbital and one $2p$ atomic orbital.

orbital and so the resultant hybrid points in the opposite direction to the one shown in Figure 4.2a. (Remember that p atomic orbitals have vector properties.)

$$\psi_{sp\,\text{hybrid}} = \frac{1}{\sqrt{2}}(\psi_{2s} - \psi_{2p_x}) \tag{4.2}$$

Equations 4.1 and 4.2 represent two wavefunctions which are equivalent in every respect *except for their directionalities* with respect to the x axis. Although the orbital energies of the initial $2s$ and $2p_x$ atomic orbitals were different, mixing leads to two hybrid orbitals of equal energy.

The model of sp hybridization can be used to describe the σ-bonding in a linear molecule such as $BeCl_2$ in which the Be−Cl bonds are of equal length. The ground state electronic configuration of Be is $[He]2s^2$ and the valence shell contains the $2s$ atomic orbital and three $2p$ atomic orbitals (Figure 4.3). If we use two of these atomic orbitals, treating them separately, to form two localized Be−Cl bonds, we cannot rationalize the bond equivalence. However, if we take the $2s$ atomic orbital and one $2p$ atomic orbital, mix their characters to form sp hybrids, and use one hybrid orbital to form one Be−Cl interaction and the other hybrid orbital for the second interaction, then the equivalence of the Be−Cl interactions is a natural consequence of the bonding picture. Effectively, we are representing the valence state of Be in a linear molecule as consisting of two degenerate sp hybrids, each containing one electron; this is represented by the notation $(sp)^2$. Figure 4.3 represents the change from the ground state electronic configuration of Be to an sp valence state. This is a *theoretical state* which can be used to describe σ-bonding in a linear molecule.

sp^2 Hybridization: a scheme for trigonal planar species

The notation sp^2 means that one s and two p atomic orbitals mix to form a set of three hybrid orbitals with different directional properties.

Let us consider the combination of $2s$, $2p_x$ and $2p_y$ atomic orbitals. The final hybrid orbitals must be equivalent in every way except for their directional properties; sp^2 hybrids must contain the same amount of s character as each other and the same amount of p character as one another. We begin by giving one-third of the $2s$ character to each sp^2 hybrid orbital. The remaining two-thirds of each hybrid orbital consists of $2p$ character, and the normalized wavefunctions are given in equations 4.3 to 4.5.

$$\psi_{sp^2\,\text{hybrid}} = \frac{1}{\sqrt{3}}\psi_{2s} + \sqrt{\frac{2}{3}}\psi_{2p_x} \tag{4.3}$$

$$\psi_{sp^2\,\text{hybrid}} = \frac{1}{\sqrt{3}}\psi_{2s} - \frac{1}{\sqrt{6}}\psi_{2p_x} + \frac{1}{\sqrt{2}}\psi_{2p_y} \tag{4.4}$$

$$\psi_{sp^2\,\text{hybrid}} = \frac{1}{\sqrt{3}}\psi_{2s} - \frac{1}{\sqrt{6}}\psi_{2p_x} - \frac{1}{\sqrt{2}}\psi_{2p_y} \tag{4.5}$$

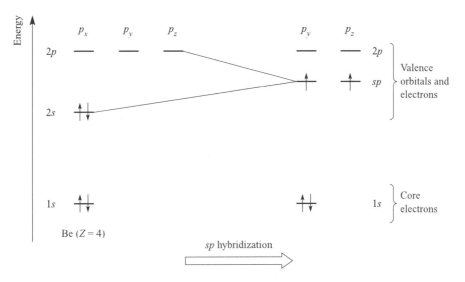

Fig. 4.3 Scheme to show the formation of the *sp* hybridized valence state of a beryllium atom from its ground state. This is a formalism and is not a 'real' observation, e.g. the valence state *cannot* be observed by spectroscopic techniques. The choice of using the $2p_x$ orbital for hybridization is arbitrary.

Figure 4.4 gives a pictorial representation of the way in which the three sp^2 hybrid orbitals are constructed. Remember that a change in sign for the atomic wavefunction means a change in phase. The resultant directions of the lower two hybrid orbitals in Figure 4.4 are determined by resolving the vectors associated with the $2p_x$ and $2p_y$ atomic orbitals.

The model of sp^2 hybridization can be used to describe the σ-bonding in trigonal planar molecules such as BH_3. The valence state of the B atom is $(sp^2)^3$ (i.e. three sp^2 hybrid orbitals, each with one electron) and the equivalence of the B–H interactions follows by considering that each interaction is formed by the overlap of one B sp^2 hybrid orbital with the $1s$ atomic orbital of an H atom (Figure 4.5). Each

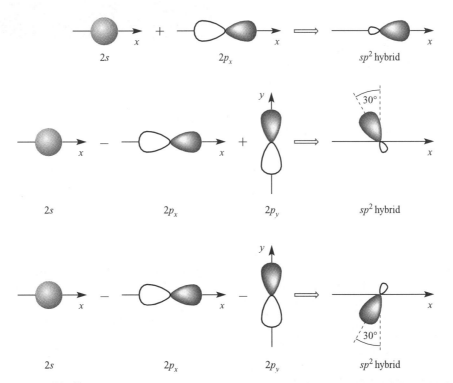

Fig. 4.4 The formation of three sp^2 hybrid orbitals from one $2s$ atomic orbital and two $2p$ atomic orbitals. The choice of p_x and p_y is arbitrary. (If we started with $2p_x$ and $2p_z$ atomic orbitals, the hybrids would lie in the xz plane; using the $2p_y$ and $2p_z$ atomic orbitals gives hybrid orbitals in the yz plane.) The directionalities of the hybrid orbitals follow from the relative contributions of the atomic orbitals (see equations 4.3–4.5).

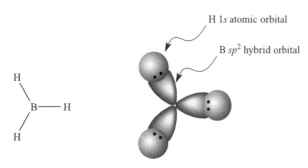

H 1s atomic orbital

B sp^2 hybrid orbital

Fig. 4.5 The bonding in trigonal planar BH_3 can be conveniently described in terms of the interactions between a set of sp^2 hybrid orbitals centred on the B atom and three H 1s atomic orbitals. Three pairs of electrons are available (three electrons from B and one from each H) to give three 2c-2e σ-bonds.

H atom contributes one electron to the bonding scheme and, so, each B–H σ-bond is a localized 2c-2e interaction (see *Section 1.12*). A diagram similar to that shown in Figure 4.3 can be constructed to show the formation of a valence state for the trigonal planar B atom.

sp^3 Hybridization: a scheme for tetrahedral species

The notation sp^3 means that one *s* and three *p* atomic orbitals mix to form a set of four hybrid orbitals with different directional properties.

A similar scheme to those described above can be derived to generate four sp^3 hybrid orbitals from one 2s and three 2p atomic orbitals. The sp^3 hybrid orbitals are described by the normalized wavefunctions in equations 4.6–4.9 and are shown pictorially in Figure 4.6a. Each sp^3 hybrid orbital possesses 25% *s* character and 75% *p* character, and the set of four equivalent orbitals define a tetrahedral framework.

$$\psi_{sp^3 \text{ hybrid}} = \tfrac{1}{2}(\psi_{2s} + \psi_{2p_x} + \psi_{2p_y} + \psi_{2p_z}) \tag{4.6}$$

$$\psi_{sp^3 \text{ hybrid}} = \tfrac{1}{2}(\psi_{2s} + \psi_{2p_x} - \psi_{2p_y} - \psi_{2p_z}) \tag{4.7}$$

$$\psi_{sp^3 \text{ hybrid}} = \tfrac{1}{2}(\psi_{2s} - \psi_{2p_x} + \psi_{2p_y} - \psi_{2p_z}) \tag{4.8}$$

$$\psi_{sp^3 \text{ hybrid}} = \tfrac{1}{2}(\psi_{2s} - \psi_{2p_x} - \psi_{2p_y} + \psi_{2p_z}) \tag{4.9}$$

In Figure 4.6b we illustrate how the tetrahedral structure of CH_4 relates to a cubic-framework. This relationship is important because it allows us to describe a tetrahedron in terms of a Cartesian axis set. Within valence bond theory, the bonding in CH_4 can conveniently be described in terms of an sp^3 valence state for C, i.e. four degenerate orbitals, each containing one electron. Each hybrid orbital overlaps with the 1s atomic orbital of one H atom to generate one of four equivalent, localized 2c-2e C–H σ-interactions.

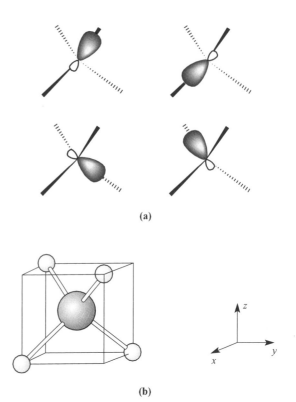

(a)

(b)

Fig. 4.6 (a) The directions of the orbitals that make up a set of four sp^3 hybrid orbitals correspond to a tetrahedral array. (b) The relationship between a tetrahedron and a cube; in CH_4, the four H atoms occupy alternate corners of a cube, and the cube is easily related to a Cartesian axis set.

Worked example 4.1   Hybridization scheme for the nitrogen atom in NH_3

Use VSEPR theory to account for the structure of NH_3, and suggest an appropriate hybridization scheme for the N atom.

The ground state electronic configuration of N is $[\text{He}]2s^2 2p^3$.

Three of the five valence electrons are used to form three N–H single bonds, leaving one lone pair.

The structure is trigonal pyramidal, derived from a tetrahedral arrangement of electron pairs:

The N atom has four valence atomic orbitals: $2s$, $2p_x$, $2p_y$ and $2p_z$. An sp^3 hybridization scheme gives a tetrahedral arrangement of hybrid orbitals, appropriate for accommodating the four pairs of electrons:

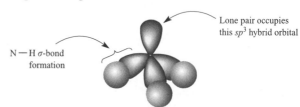

N–H σ-bond formation

Lone pair occupies this sp^3 hybrid orbital

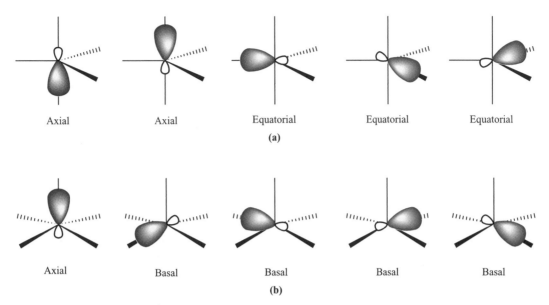

Fig. 4.7 A schematic representation of sp^3d hybridization. (a) A combination of s, p_x, p_y, p_z and d_{z^2} atomic orbitals gives a set of five sp^3d hybrid orbitals corresponding to a trigonal bipyramidal arrangement; the axial sp^3d hybrid orbitals are directed along the z axis. (b) A combination of s, p_x, p_y, p_z and $d_{x^2-y^2}$ atomic orbitals gives a set of five sp^3d hybrid orbitals corresponding to a square-based pyramidal arrangement; the axial sp^3d hybrid orbital is directed along the z axis.

Self-study exercises

1. Use VSEPR theory to account for the tetrahedral structure of $[NH_4]^+$.

2. Rationalize why H_2O is bent but XeF_2 is linear.

3. Give a suitable hybridization scheme for the central atom in each of the following: (a) $[NH_4]^+$; (b) H_2S; (c) BBr_3; (d) NF_3; (e) $[H_3O]^+$.

[*Ans.* (a) sp^3; (b) sp^3; (c) sp^2; (d) sp^3; (e) sp^3]

Other hybridization schemes

For molecular species with other than linear, trigonal planar or tetrahedral-based structures, it is usual to involve d orbitals within valence bond theory. We shall see later that this is not necessarily the case within molecular orbital theory, and, even within the VB model, the *extent* of d orbital participation is a controversial subject. We can often do without the involvement of d orbitals if we accept that the covalent bonds in the species concerned are relatively weak (e.g. XeF_4 and polyhalide ions). *Real* molecules do not have to conform to simple theories of valence, nor must they conform to the $sp^n d^m$ schemes that we consider in this book. Nevertheless, it is convenient to visualize the bonding in molecules in terms of a range of simple hybridization schemes.

The mixing of s, p_x, p_y, p_z and d_{z^2} atomic orbitals gives a set of five sp^3d hybrid orbitals, the mutual orientations of which correspond to a trigonal bipyramidal arrangement (Figure 4.7a). The five sp^3d hybrid orbitals are *not* equivalent and divide into sets of two axial and three equatorial orbitals; the axial orbital lobes lie along the z axis.[†] The model of sp^3d hybridization can be used to describe the σ-bonding in trigonal bipyramidal molecules such as PF_5.

The σ-bonding framework in a square-pyramidal species may also be described in terms of an sp^3d hybridization scheme. The change in spatial disposition of the five hybrid orbitals from trigonal bipyramidal to square-based pyramidal is a consequence of the participation of a different d orbital. Hybridization of s, p_x, p_y, p_z and $d_{x^2-y^2}$ atomic orbitals generates a set of five sp^3d hybrid orbitals (Figure 4.7b).

Hybridization of s, p_x, p_y, p_z, d_{z^2} and $d_{x^2-y^2}$ atomic orbitals gives six sp^3d^2 hybrid orbitals corresponding to an octahedral arrangement. The bonding in SF_6 or $[PF_6]^-$ can be described in terms of sp^3d^2 hybridization of the central atom. If we remove the z-components from this set (i.e. p_z and d_{z^2}) and hybridize only the s, p_x, p_y and $d_{x^2-y^2}$ atomic orbitals, the resultant set of four sp^2d hybrid orbitals corresponds to a square planar arrangement e.g. $[PtCl_4]^{2-}$.

Each set of hybrid orbitals is associated with a particular shape, although this may not coincide with the molecular shape if lone pairs also have to be accommodated:

- sp linear
- sp^2 trigonal planar
- sp^3 tetrahedral
- sp^3d (d_{z^2})   trigonal bipyramidal
- sp^3d ($d_{x^2-y^2}$)   square-based pyramidal
- sp^3d^2 octahedral
- sp^2d square planar

[†]Choice of coincidence between the z axis and the axial lobes is convenient and tends to be conventional.

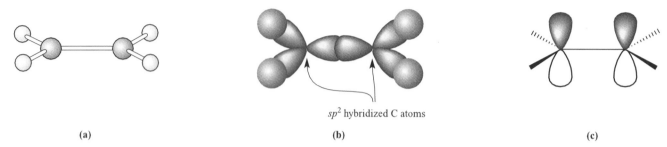

Fig. 4.8 (a) Ethene is a planar molecule with H−C−H and C−C−H bond angles close to 120°. (b) An sp^2 hybridization scheme is appropriate to describe the σ-bonding framework. (c) This leaves a $2p$ atomic orbital on each C atom; overlap between them gives a C−C π-interaction.

4.3 Valence bond theory: multiple bonding in polyatomic molecules

In the previous section, we emphasized that hybridization of some or all of the valence atomic orbitals of the central atom in an XY_n species provided a scheme for describing the X−Y σ-bonding. In, for example, the formation of sp, sp^2 and sp^3d hybrid orbitals, some p or d atomic orbitals remain unhybridized and, if appropriate, may be participate in the formation of π-bonds. In this section we use the examples of C_2H_4, HCN and BF_3 to illustrate how multiple bonds in polyatomic molecules are treated within VB theory. Before considering the bonding in any molecule, the ground state electronic configurations of the atoms involved should be noted.

C_2H_4

C $[He]2s^2 2p^2$
H $1s^1$

Ethene, C_2H_4, is a planar molecule (Figure 4.8a) with C−C−H and H−C−H bond angles of 121.3° and 117.4° respectively. Thus, each C centre is approximately trigonal planar and the σ-bonding framework within C_2H_4 can be described in terms of an sp^2 hybridization scheme (Figure 4.8b). The three σ-interactions per C atom use three of the

four valence electrons, leaving one electron occupying the unhybridized $2p$ atomic orbital. The interaction between the two $2p$ atomic orbitals (Figure 4.8c) and the pairing of the two electrons in these atomic orbitals, generates a C−C π-interaction. The bond order of the C−C bond in C_2H_4 is therefore two, in keeping with Lewis structure **4.1**. The π-component of the overall carbon–carbon bond is weaker than the σ-component and hence a C=C double bond, though stronger than a C−C single bond, is not twice as strong; the C−C bond enthalpy terms in C_2H_4 and C_2H_6 are 598 and 346 kJ mol^{-1} respectively.

$$\begin{array}{c} H \qquad\qquad H \\ \diagdown \qquad\quad \diagup \\ C = C \\ \diagup \qquad\quad \diagdown \\ H \qquad\qquad H \end{array}$$

(4.1)

HCN

C $[He]2s^2 2p^2$
N $[He]2s^2 2p^3$
H $1s^1$

Figure 4.9a shows the linear HCN molecule, a Lewis structure (**4.2**) for which indicates the presence of an H−C single bond, a C≡N triple bond, and a lone pair of electrons on N. An sp hybridization scheme is appropriate for both C

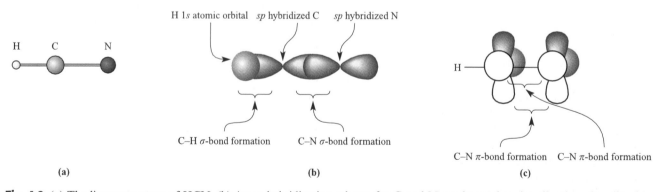

Fig. 4.9 (a) The linear structure of HCN. (b) An sp hybridization scheme for C and N can be used to describe the σ-bonding in HCN. (c) The π-character in the C−N bond arises from $2p$–$2p$ overlap.

$$H — C \equiv N :$$

$$(4.2)$$

and N; it is consistent with the linear arrangement of atoms around C and with the placement of the lone pair on N as far away as possible from the bonding electrons. Figure 4.9b shows the σ-bonding framework in HCN (each region of orbital overlap is occupied by a pair of electrons) and the outward pointing sp hybrid on N that accommodates the lone pair. If we arbitrarily define the HCN axis as the z axis, then after the formation of the σ-interactions, a $2p_x$ and a $2p_y$ atomic orbital remain on each of the C and N atoms; each atomic orbital contains one electron. Overlap between the two $2p_x$ and between the two $2p_y$ orbitals leads to two π-interactions (Figure 4.9c). The overall C−N bond order is three, consistent with Lewis structure **4.2**.

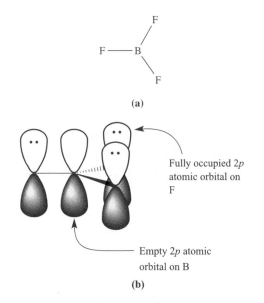

(a)

Fully occupied $2p$ atomic orbital on F

Empty $2p$ atomic orbital on B

(b)

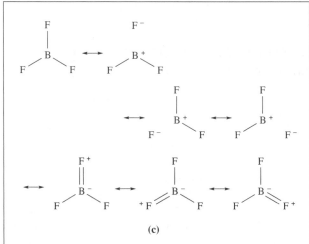

(c)

Fig. 4.10 (a) BF_3 possesses a trigonal planar structure. (b) $2p$–$2p$ overlap between B and F leads to the formation of a π-interaction. (c) Boron–fluorine double bond character is also deduced by considering the resonance structures for BF_3; only those forms that contribute significantly are shown.

BF_3

B $[He]2s^2 2p^1$
F $[He]2s^2 2p^5$

Boron trifluoride (Figure 4.10a) is trigonal planar (D_{3h}); sp^2 hybridization is appropriate for the B atom. Each of the three B−F σ-interactions arises by overlap of an sp^2 hybrid on the B atom with, for example, an sp^2 orbital on the F atom. After the formation of the σ-bonding framework, the B atom is left with an *unoccupied* $2p$ atomic orbital lying perpendicular to the plane containing the BF_3 molecule. As Figure 4.10b shows, this is ideally set up for interaction with a *filled* $2p$ atomic orbital on one of the F atoms to give a localized B−F π-interaction. Notice that the two electrons occupying this π-bonding orbital both originate from the F atom. This picture of the bonding in BF_3 is analogous to one of the resonance forms shown in blue in Figure 4.10c; all three resonance forms (see *Section 1.12*) are needed to account for the experimental observation that all three B−F bonds are of equal length (131 pm).

Worked example 4.2   Valence bond treatment of the bonding in $[NO_3]^-$

(a) The $[NO_3]^-$ ion has D_{3h} symmetry. What does this tell you about its structure? (b) Draw a set of resonance structures (focusing only on those that contribute significantly) for the nitrate ion. (c) Use an appropriate hybridization scheme to describe the bonding in $[NO_3]^-$.

(a) If $[NO_3]^-$ has D_{3h} symmetry, it must be planar, possess O−N−O bond angles of 120°, and have equal N−O bond distances.
(b) First, write down the electronic configurations for N ($Z = 7$) and O ($Z = 8$).

N $[He]2s^2 2p^3$ O $[He]2s^2 2p^4$

There is an additional electron from the negative charge giving a total of 24 valence electrons.

Both N and O are expected to obey the octet rule and so the most important resonance forms are expected to be:

(c) Using a hybridization scheme, we should end up with a bonding picture that corresponds to that depicted by the resonance structures.

An sp^2 hybridized nitrogen centre is consistent with the trigonal planar shape of $[NO_3]^-$. Allow the hybrid orbitals to overlap with suitable orbitals from oxygen; a choice of sp^2 hybridization on the O atom provides suitable orbitals to accommodate the oxygen lone pairs. Occupation of each bonding orbital by a pair of electrons gives three equivalent N—O σ-bonds:

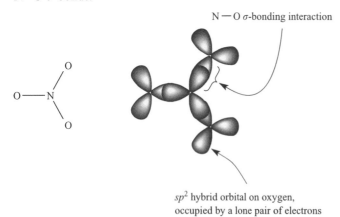

N — O σ-bonding interaction

sp^2 hybrid orbital on oxygen, occupied by a lone pair of electrons

Of the 24 valence electrons, 18 are accommodated either in σ-bonds or as oxygen lone pairs.

The next step is to consider multiple bonding character. Each N and O atom has an unused $2p$ atomic orbital lying perpendicular to the plane of the molecule; overlap between the $2p$ atomic orbital on nitrogen with one of those on an oxygen atom gives rise to *one* localized π-bond. The six remaining valence electrons are allocated as follows:

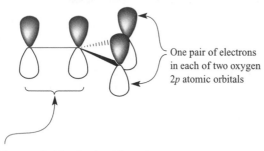

One pair of electrons in each of two oxygen $2p$ atomic orbitals

One pair of electrons for N — O π-bond formation

The combination of the σ- and π-bonding pictures gives one nitrogen–oxygen double bond and two single bonds. Three such schemes must be drawn (with the π-character in one of each of the N–O bonds) in order that the overall scheme is in keeping with the observed D_{3h} symmetry of $[NO_3]^-$.

Self-study exercises

1. Why are resonance structures containing two N=O double bonds not included in the set shown above for $[NO_3]^-$?

2. Use an appropriate hybridization scheme to describe the bonding in $[BO_3]^{3-}$.

π-Bond formation between different types of orbitals

The examples above illustrate $(p–p)$π-bond formation, but $(p–d)$π-interactions are also possible. However, the role of the d atomic orbitals in main group chemistry is controversial

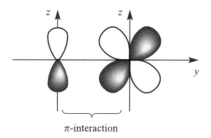

π-interaction

Fig. 4.11 The sideways overlap of a p_z orbital and a d_{yz} orbital leads to a π-interaction.

as examples later in the book illustrate. Figure 4.11 shows the interaction between a p_z atomic orbital and a d_{yz} atomic orbital. Such overlap probably contributes to the bonding in $[SO_4]^{2-}$, SOF_4 (see *problem 4.7* at the end of the chapter) and $[ClO_4]^-$. In general, however, strong π-bonds are formed only between atoms in which the overlap is between two $2p$ atomic orbitals, e.g. C–C, C–O, C–N, O–N and N–N. Additionally, π-overlap between metal d-orbitals and main group atom p orbitals is important in organometallic complexes; we return to this in Chapter 23.

4.4 Molecular orbital theory: the ligand group orbital approach and application to triatomic molecules

Despite its successes, the application of valence bond theory to the bonding in polyatomic molecules leads to conceptual difficulties. The method dictates that bonds are localized and, as a consequence, sets of resonance structures and bonding pictures involving hybridization schemes become rather tedious to establish, even for relatively small molecules (e.g. see *Figure 4.10c*). We therefore turn our attention to molecular orbital (MO) theory.

Molecular orbital diagrams: moving from a diatomic to polyatomic species

As part of our treatment of the bonding in diatomics in Section 1.13, we constructed MO diagrams such as Figures 1.21, 1.27 and 1.28. In each diagram, the atomic orbitals of the two atoms were represented on the right- and left-hand sides of the diagram with the MOs in the middle. Correlation lines connecting the atomic and molecular orbitals were constructed to produce a readily interpretable diagram.

Now consider the situation for a triatomic molecule such as CO_2. The molecular orbitals contain contributions from the atomic orbitals of three atoms, and we are presented with a problem of trying to draw an MO diagram involving four sets of orbitals (three sets of atomic orbitals and one of molecular orbitals). A description of the bonding in CF_4 involves five sets of atomic orbitals and one set of molecular orbitals, i.e. a six-component problem. Similarly, SF_6 is an

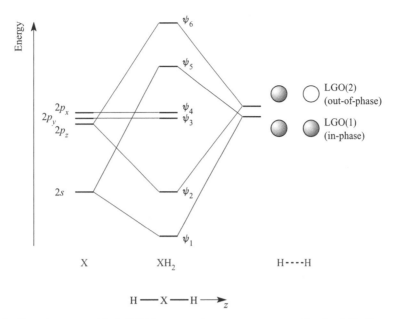

Fig. 4.12 Application of the ligand group orbital (LGO) approach to construct a qualitative MO diagram for the formation of a linear XH_2 molecule from the interactions of the valence orbitals of X ($2s$ and $2p$ atomic orbitals) and an H- - -H fragment. For clarity, the lines marking the $2p$ orbital energies are drawn apart, although these atomic orbitals are actually degenerate.

eight-component problem. It is obvious that such MO diagrams are complicated and, probably, difficult both to construct and interpret. In order to overcome this difficulty, it is common to resolve the MO description of a polyatomic molecule into a three-component problem, a method known as the *ligand group orbital (LGO) approach*.

Ligand group orbitals: MO approach to the bonding in linear XH_2

Initially, we illustrate the ligand group orbital approach by considering the bonding in a linear triatomic XH_2 in which the valence orbitals of X are the $2s$ and $2p$ atomic orbitals. Let us orient the H−X−H framework so that it coincides with the z axis as shown in Figure 4.12. Consider the two $1s$ atomic orbitals of the two H atoms. Each $1s$ atomic orbital has two possible phases and, when the *two 1s orbitals are taken as a group*, there are two possible phase combinations. These are called *ligand group orbitals* (LGOs) and are shown at the right-hand side of Figure 4.12.[†] Effectively, we are *transforming* the description of the bonding in XH_2 from one in which the basis sets are the atomic orbitals of atoms X and H, into one in which the basis sets are the atomic orbitals of atom X and the ligand group orbitals of an H- - -H fragment. This is a valuable approach for polyatomic molecules.

> The number of ligand group orbitals formed = the number of atomic orbitals used.

In constructing an MO diagram for XH_2 (Figure 4.12), we consider the interactions of the valence atomic orbitals of X with the ligand group orbitals of the H- - -H fragment. Ligand group orbital LGO(1) has the correct symmetry to interact with the $2s$ atomic orbital of X, giving an MO with H−X−H σ-bonding character. The symmetry of LGO(2) is matched to that of the $2p_z$ atomic orbital of X. The resultant bonding MOs and their antibonding counterparts are shown in Figure 4.13, and the MO diagram in Figure 4.12 shows the corresponding orbital interactions. The $2p_x$ and $2p_y$ atomic orbitals of X become non-bonding orbitals in XH_2. The final step in the construction of the MO diagram is to place the available electrons in the MOs according to the *aufbau* principle (see *Section 1.9*). An important result of the MO treatment of the bonding in XH_2 is that the σ-bonding character in orbitals ψ_1 and ψ_2 is spread over all three atoms, indicating that the bonding character is *delocalized* over the H−X−H framework. Delocalized bonding is a general result within MO theory.

A bent triatomic: H_2O

Now let us follow a similar procedure to that outlined above and apply MO theory to the bonding in H_2O in terms of the interactions between the valence atomic orbitals of the O atom and the ligand group orbitals of the H- - -H fragment. First we must define a convenient axis set. In Figure 4.14, the H_2O framework lies in the yz plane with the z axis coincident with the C_2 axis. The valence orbitals of O are the $2s$, $2p_x$, $2p_y$ and $2p_z$ atomic orbitals. The ligand group orbitals of the

[†] In Figure 4.12, the energies of the two ligand group orbitals are close together because the H nuclei are far apart; compare this to the situation in the H_2 molecule (Figure 1.18). Similarly, in Figure 4.15, the LGOs for the H_3 fragment form two sets (all in-phase, and the degenerate pair of orbitals) but their respective energies are close because of the large H- - -H separations.

ψ_1 ψ_2 ψ_3 ψ_4 ψ_5 ψ_6

X—H bonding character   Centred on atom X   X—H antibonding character

Fig. 4.13 Schematic representations of the MOs in linear XH_2. The wavefunction labels correspond to those in Figure 4.12.

H---H unit are shown on the right-hand side of Figure 4.14 and are the same as those of the H---H fragment in linear XH_2 (Figure 4.12). The significant difference in the treatment of the bonding for linear XH_2 and bent H_2O is the *relative* positions of the atoms in the molecules.

The MO diagram shown in Figure 4.14 is constructed as follows. The $2s$ and $2p_z$ atomic orbitals of the O atom possess the correct symmetry to interact with LGO(1) of the H---H fragment and their interactions must lead to *three* MOs, ψ_1, ψ_3 and ψ_6. On symmetry grounds, ψ_1 could also include $2p_z$ character, but $2s$ character dominates because of the energy separation of the $2s$ and $2p_z$ atomic orbitals. The interaction between the $2p_y$ atomic orbital and LGO(2) leads to MOs ψ_2 and ψ_5 which possess H—O—H bonding and antibonding character respectively. The oxygen $2p_x$ atomic orbital becomes non-bonding.

The eight valence electrons in H_2O occupy the MOs according to the *aufbau* principle, and this gives rise to two occupied H—O—H bonding MOs and two occupied MOs with mainly oxygen character. (To appreciate this fully, see end of chapter *problem 4.12*.) Although this bonding model for H_2O is approximate, it is *qualitatively* adequate for most descriptive purposes.

4.5 Molecular orbital theory applied to the polyatomic molecules BH_3, NH_3 and CH_4

In this section, we consider the bonding in BH_3 and NH_3 in order to investigate the differences between trigonal planar and pyramidal species with σ-bonding interactions. We follow this with a look at the way in which MO theory views the bonding in CH_4.

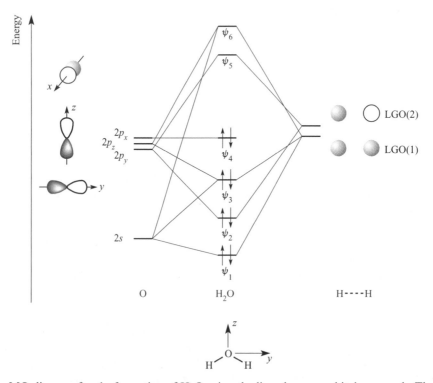

Fig. 4.14 A qualitative MO diagram for the formation of H_2O using the ligand group orbital approach. The two H atoms in the H_2 fragment are out of bonding range with each other, their positions being analogous to those in H_2O. For clarity, the lines marking the oxygen $2p$ orbital energies are drawn apart, despite their being degenerate.

BOX 4.1   **CHEMICAL AND THEORETICAL BACKGROUND**

Assigning symmetry labels to orbitals

Throughout this book, we try to avoid excessive use of symmetry labels such as a_1, b_2, e_g and t_{2g} to identify particular types of orbitals. Their use, however, in Chapter 20 is inevitable and we give some explanation of the labels in Box 20.2.

In the present discussion, we introduce the relationships between group theory and orbital symmetry by illustrating how symmetry labels for the MOs in H$_2$O arise. As you work through this Box, you should also refer to Figure 4.14; the ligand group orbitals for the H---H fragment are those discussed in the main text.

The H$_2$O molecule belongs to the C_{2v} point group and part of the character table for this group is shown below:

C_{2v}	E	C_2	$\sigma_v(xz)$	$\sigma_v'(yz)$
A_1	1	1	1	1
A_2	1	1	−1	−1
B_1	1	−1	1	−1
B_2	1	−1	−1	1

The inclusion of the xz and yz terms in the last two columns of the character table specifies that the H$_2$O molecule is taken to lie in the yz plane, i.e. the z axis coincides with the principal axis (see *Section 3.2*).

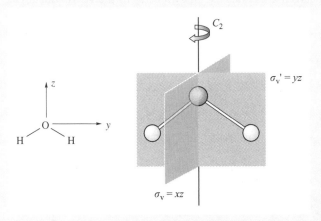

The character table has several important features.

- The labels in the first column (under the point group symbol) tell us the symmetry types of orbitals that are permitted within the specified point group.
- The numbers in the column headed E (the identity operator) indicate the degeneracy of each type of orbital; in the C_{2v} point group, all orbitals have a degeneracy of one, i.e. they are non-degenerate.
- Each row of numbers following a given symmetry label indicates how a particular orbital behaves when operated upon by each symmetry operation. A number 1 means that the orbital is unchanged by the operation, a −1 means the orbital changes sign, and a 0 means that the orbital changes in some other way.

To illustrate its use, let us consider the $2s$ atomic orbital of the O atom *in water* (i.e. the O atom as represented on the left-hand side of Figure 4.14).

2s atomic orbital of oxygen

Apply each symmetry operation of the C_{2v} point group in turn. Applying the E operator leaves the $2s$ atomic orbital unchanged; rotation about the C_2 axis leaves the atomic orbital unchanged; reflections through the σ_v and σ_v' planes leave the $2s$ atomic orbital unchanged. These results correspond to the following row of characters:

E	C_2	$\sigma_v(xz)$	$\sigma_v'(yz)$
1	1	1	1

and this matches those for the symmetry type A_1 in the C_{2v} character table. We therefore label the $2s$ atomic orbital on the oxygen atom *in water* as an a_1 orbital. (Lower case letters are used for the orbital label, but upper case for the symmetry type in the character table.) The same test is now carried out on each atomic orbital of the O atom and on each LGO for the H---H fragment.

LGO(1) of the H-----H fragment   LGO(2) of the H-----H fragment

For example, LGO(1) should also be given the symmetry label a_1, but LGO(2) is given the label b_2 because applying each symmetry operation to this orbital corresponds to the following row of characters:

E	C_2	$\sigma_v(xz)$	$\sigma_v'(yz)$
1	−1	−1	1

In the construction of an orbital correlation diagram, only orbitals with the *same* symmetry labels can interact (see the exercise below).

The example above is a relatively simple one, and while demonstrating the application of group theory to the symmetry of orbitals, it hides the fact that labelling orbitals in other molecules and fragments may be difficult. For example, if we are dealing with a degenerate set of orbitals (e.g. symmetry type E), then the entry in the character table is the *sum* of the characters that describe the results of a given symmetry operation on each of the orbitals in the degenerate set.

An important ramification of the relationship between group theory and orbitals is that ligand group orbitals for a fragment of a molecule *must* possess one of the symmetry types appropriate to the point group of the molecule in question.

Self-study exercises

With reference to Figure 4.14, answer the following questions about the application of MO theory to the bonding in H_2O.

1. Assign an appropriate symmetry label to each atomic orbital of oxygen, and each molecular orbital in water.

2. Confirm that the symmetry label of an MO is the same as that of *each* orbital that contributes to it; first, you may need to do question 4.12 at the end of the chapter!

3. Why does the oxygen $2p_x$ atomic orbital become non-bonding in H_2O?

4. Why may the characters of the oxygen $2s$ and $2p_z$ mix upon the formation of H_2O?

Further reading

J. Barrett (1991) *Understanding Inorganic Chemistry: The Underlying Physical Principles*, Ellis Horwood (Simon & Schuster), New York – Chapters 2 and 4.

F.A. Cotton (1990) *Chemical Applications of Group Theory*, 3rd edn., Wiley, New York – Chapter 8.

G. Davidson (1991) *Group Theory for Chemists*, Macmillan, London – Chapters 6, 7 and 10.

S.F.A. Kettle (1985) *Symmetry and Structure*, Wiley, Chichester – Chapters 2–6 (and more).

BH_3

The existence of BH_3 in the gas phase has been established even though the molecule readily dimerizes; the bonding in B_2H_6 is described in Section 4.7. The BH_3 molecule is trigonal planar (D_{3h}) and by considering the orbital interactions between the atomic orbitals of the B atom and the LGOs of an appropriate H_3 fragment, we can establish a molecular bonding scheme. We begin by choosing a convenient axis set as shown at the bottom of Figure 4.15; the z axis coincides

with the principal rotation axis of BH_3 and all of the atoms lie in the xy plane. The ligand group orbitals of the H_3 fragment are derived by taking linear combinations of the three H $1s$ atomic orbitals. The in-phase combination, LGO(1), is unique, but LGO(2) and LGO(3), each of which contains one nodal plane, are degenerate. In LGO(3), the nodal plane passes through one of the H nuclear positions and there is *no* contribution from this H atom to the LGO. The permitted combinations of the $1s$ atomic orbitals are dictated by the symmetry of the H_3 fragment (D_{3h});

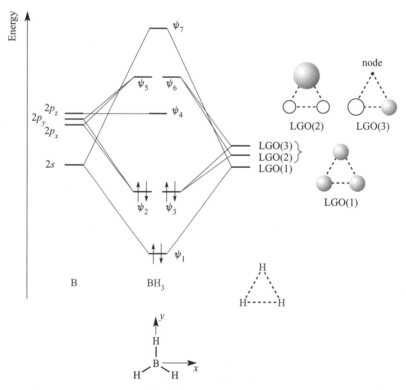

Fig. 4.15 A qualitative MO diagram for the formation of BH_3 using the ligand group orbital approach. The three H atoms in the H_3 fragment are out of bonding range with each other, their positions being analogous to those in the BH_3 molecule. Orbitals LGO(2) and LGO(3) form a degenerate pair, although for clarity, the lines marking their orbital energies are drawn apart; similarly for the three $2p$ atomic orbitals of boron. [*Exercise*: where do the nodal planes lie in LGO(2) and LGO(3)?]

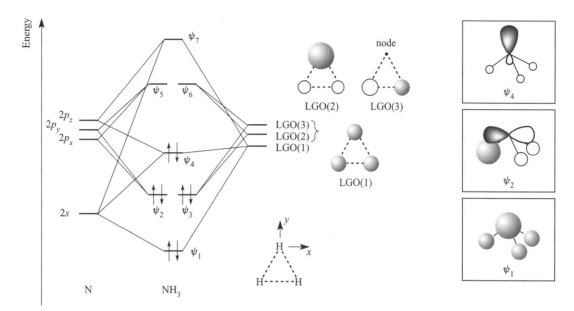

Fig. 4.16 A qualitative MO diagram for the formation of NH_3 using the ligand group orbital approach. Orbitals LGO(2) and LGO(3) are degenerate, but for clarity the lines marking their orbital energies are drawn apart; similarly for the three $2p$ atomic orbitals of nitrogen. The diagrams on the right-hand side show schematic representations of MOs ψ_1, ψ_2 and ψ_4. The relative sizes of the orbital lobes indicate their approximate relative contributions to each MO.

we shall not detail this further but some explanation of the relationship between orbital symmetry and group theory is given in Box 4.1. Figure 4.15 gives an approximate indication of the relative contributions made by each $1s$ atomic orbital to each ligand group orbital, i.e. look at the relative sizes of the orbital lobes in LGO(1), LGO(2) and LGO(3).

The MO diagram is constructed by matching the symmetries of the ligand group orbitals to those of the atomic orbitals of the B atom. Interaction is allowed between the $2s$ atomic orbital and LGO(1) giving a pair of MOs, ψ_1 and ψ_7, which have bonding and antibonding character respectively. The $2p_y$ and $2p_x$ atomic orbitals interact with LGO(2) and LGO(3) respectively to give the degenerate pairs of MOs ψ_2 and ψ_3 (bonding) and ψ_5 and ψ_6 (antibonding). The boron $2p_z$ atomic orbital becomes a non-bonding MO.

The MO approach describes the bonding in BH_3 in terms of three MOs (ψ_1, ψ_2 and ψ_3 in Figure 4.15). Orbital ψ_1 possesses σ-bonding character which is *delocalized over all four atoms*. Orbitals ψ_2 and ψ_3 also exhibit delocalized character, and the bonding in BH_3 is described by considering a *combination of all three bonding MOs*.

NH₃

The NH_3 molecule is trigonal pyramidal (C_{3v}) and a bonding scheme can be derived by considering the interaction between the atomic orbitals of the N atom and the ligand group orbitals of an appropriate H_3 fragment. A convenient axis set has the z axis coincident with the C_3 axis of NH_3 (see *worked example 3.2*); the x and y axes are directed as shown in Figure 4.16. The ligand group orbitals shown in Figure 4.16

are the same as those for the H_3 fragment in BH_3. However, a comparison of Figures 4.15 and 4.16 shows that the interactions between the ligand group orbitals and the atomic orbitals of the central atom are *not* the same in BH_3 and NH_3. This is readily understood in terms of the position of the B or N nucleus with respect to the three H nuclei: the B atom is in the plane of the H atoms, but the N atom lies above it. In BH_3, the $2p_z$ atomic orbital becomes a non-bonding MO, but in NH_3, it can participate in bonding. By matching the symmetries of the atomic orbitals of the N atom with those of the ligand group orbitals of the H_3 fragment, the qualitative MO diagram in Figure 4.16 can be constructed. On symmetry grounds, MO ψ_1 could also contain N $2p_z$ character, but the energy separation of the $2s$ and $2p$ atomic orbitals is such that $2s$ character predominates; this is analogous to the case for H_2O described above. Schematic representations of ψ_1, ψ_2 and ψ_4 are shown in Figure 4.16.

After constructing the MO diagram, the eight valence electrons are placed in the MOs according to the *aufbau* principle. The HOMO (highest occupied molecular orbital) is ψ_4 and its character is essentially that of a lone pair of electrons on the N atom. Again, we see that delocalization of the σ-bonding character in each of the bonding MOs results from the molecular orbital treatment.

CH₄

In using the MO approach to describe the bonding in CH_4, it is convenient to relate the tetrahedron to the cube (Figure 4.6). The valence orbitals of the C atom are the $2s$, $2p_x$, $2p_y$ and $2p_z$ atomic orbitals (Figure 4.17a) and four ligand group orbitals can be constructed using the H $1s$ atomic

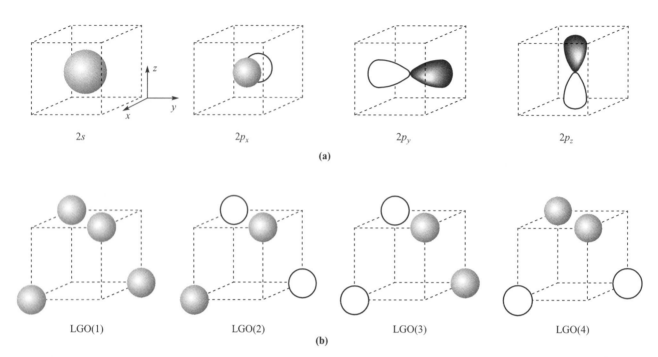

Fig. 4.17 The ligand group orbital approach to the bonding in CH_4. (a) The $2s$, $2p_x$, $2p_y$ and $2p_z$ atomic orbitals of carbon. (b) The four hydrogen $1s$ atomic orbitals combine to generate four ligand group orbitals (LGOs).

orbitals as shown in Figure 4.17b: LGO(1) is unique, while LGO(2), LGO(3) and LGO(4) are related to one another by rotation. By comparing Figures 4.17a and 4.17b, the symmetries of the four ligand group orbitals can be readily matched to those of the $2s$, $2p_x$, $2p_y$ and $2p_z$ atomic orbitals of the C atom. This allows us to construct a qualitative MO diagram (Figure 4.18) in which the interactions between the carbon atomic orbitals and the ligand group orbitals of the H_4 fragment lead to four MOs with delocalized σ-bonding character and four antibonding MOs.

Fig. 4.18 A qualitative MO diagram for the formation of CH_4 from the orbital basis set shown in Figure 4.17. Six electrons occupy MOs $\psi_2 - \psi_4$.

A comparison of the MO and VB bonding models

When we considered how valence bond theory can be used to describe the bonding in BH_3, CH_4 and NH_3, we used appropriate hybridization schemes such that bonds known to be structurally equivalent would be equivalent in the bonding scheme. One hybrid orbital contributed to each *localized* X–H (X = B, C or N) bond. On the other hand, the results of MO theory indicate that the bonding character is *delocalized*. Moreover, in each of BH_3, NH_3 and CH_4, there are two different *types* of bonding MO: a unique MO involving the $2s$ atomic orbital of the central atom, and a degenerate set of two (in BH_3 and NH_3) or three (in CH_4) MOs involving the $2p$ atomic orbitals of the central atom. Evidence for these orderings of MOs comes from photoelectron spectroscopy (see *Box 4.2*). How can the results of MO theory account for the experimentally observed equivalence of the X–H bonds in a given molecule?

As we have already mentioned, it is essential to understand that, in MO theory, the bonding in a molecule is described by combining the characters of *all* the occupied MOs with bonding character. Take CH_4 as an example. Orbital ψ_1 (Figure 4.18) is spherically symmetric and provides equal bonding character in all four C–H interactions. Orbitals ψ_2, ψ_3 and ψ_4 are related to one another by rotations through 90°; since they are degenerate, they must be considered as a set and not as individual orbitals. Taken together, orbitals ψ_2, ψ_3 and ψ_4 provide a picture of four equivalent C–H bonding interactions and, therefore, the overall picture is one of C–H bond equivalence.

BOX 4.2

CHEMICAL AND THEORETICAL BACKGROUND

Photoelectron spectroscopy (PES)

The energies of *occupied* atomic orbitals and molecular orbitals can be studied by *photoelectron spectroscopy* (PES). In a PES experiment, an atom or molecule is irradiated with electromagnetic radiation of energy E, causing electrons to be ejected from the system. Each electron possesses a characteristic *binding energy* and must absorb an amount of energy equal to, or in excess of, this binding energy if it is to be ejected. The energy of an ejected electron is that in excess of the binding energy assuming that E is greater than the binding energy.

Excess energy of electron = E − (binding energy of electron)

Since the excess energy can be measured and E is known, the binding energy can be determined. *Koopmans' theorem*

relates the binding energy of the electron to the energy of the atomic or molecular orbital in which it resides. This relationship allows photoelectron spectroscopy to be used to estimate the energies of occupied orbitals, and, thus, obtain information about the ordering of orbitals in a particular atomic or molecular species.

Further reading

R.L. DeKock and H.B. Gray (1980) *Chemical Structure and Bonding*, Benjamin/Cummings, Menlo Park – This includes a more detailed discussion of the application of PES.

4.6 Molecular orbital theory: bonding analyses soon become complicated

In this section, we consider the bonding in BF_3 using the ligand group orbital approach. Although BF_3 is a fairly simple molecule, the following discussion demonstrates the complexity of the treatment when the atomic orbital basis set of each atom contains both s and p orbitals. For the D_{3h} BF_3 molecule, it is convenient to choose the z axis to coincide with the principal rotation axis; this also permits a direct comparison between the results for BF_3 and BH_3 (Figure 4.15). The valence orbitals of the B atom are the $2s$, $2p_x$, $2p_y$ and $2p_z$ atomic orbitals, and of the F atom are the $2s$, $2p_x$, $2p_y$ and $2p_z$ atomic orbitals. A ligand group orbital approach treats the bonding analysis in BF_3 in terms of the interactions between the four atomic orbitals of the B atom and the twelve ligand group orbitals of a D_{3h} F_3 fragment. Figure 4.19 shows the characters of the ligand group orbitals, with LGO(1) being that of lowest energy.[†]

We now take each ligand group orbital in turn and attempt to match its symmetry properties with those of the atomic orbitals of a B atom placed at the point in the F_3 triangle indicated in the top left-hand diagram in Figure 4.19. The results of this exercise are listed in Table 4.1 and we can see that an MO diagram constructed on the grounds of symmetry matching is going to be complicated. We have, however, ignored several facts, including:

- the relative energies of the orbitals in each basis set;

- the spatial properties of the orbitals, e.g. LGO(1) and LGO(4) may both interact with the B $2s$ atomic orbital, but the net overlap integral (see *Section 1.13*) will not be the same;

- the fact that an MO in BF_3 may contain character from more than one LGO, provided their symmetries are compatible, e.g. LGO(1) and LGO(4).

A simplified picture of the bonding in BF_3 indicates that the MOs with bonding character arise from interactions between:

- LGO(1) and the B $2s$ atomic orbital;
- LGO(2) and the B $2p_y$ atomic orbital;
- LGO(3) and the B $2p_x$ atomic orbital;
- LGO(5) and the B $2p_z$ atomic orbital.

The picture can be improved by including contributions from LGO(4), LGO(11) and LGO(12) which interact with the B $2s$, $2p_y$ and $2p_x$ atomic orbitals respectively, and whose characters may appear in the same or different MOs as those of LGO(1), LGO(2) and LGO(3) respectively. Orbitals LGO(6), LGO(7), LGO(8), LGO(9) and LGO(10) effectively become non-bonding MOs in BF_3; the extent of overlap between LGO(6) and boron $2p_y$ and between LGO(7) and B $2p_x$ turns out to be small despite the fact that, on symmetry grounds alone, interactions were allowed.

Figure 4.20 shows a *simplified* MO diagram for the formation of BF_3. Molecular orbital ψ_1 contains mainly $2s$ and LGO(1) character; there is also a small contribution from LGO(4). Orbitals ψ_2 and ψ_3 are degenerate; ψ_2 contains $2p_y$, LGO(2) and LGO(11) character; ψ_3 contains $2p_x$, LGO(3) and LGO(12) character. Each of these three MOs is therefore a B−F σ-bonding orbital with the character *delocalized* over all three B−F interactions. Orbital ψ_4 in Figure 4.20 arises from the interaction between the $2p_z$ atomic orbital and LGO(5); it possesses B−F π-bonding character, delocalized over all four atoms. From this simplified MO

[†]The construction of these LGOs allows mixing between the $2p_x$, $2p_y$ and $2p_z$ atomic orbitals and follows from symmetry arguments. In this book we shall not pursue this further; for relevant discussions see, for example: S.F.A. Kettle (1985) *Symmetry and Structure*, Wiley, Chichester.

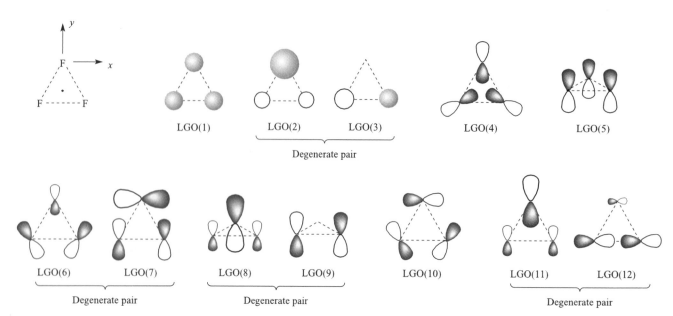

Fig. 4.19 Schematic representations of the ligand group orbitals (LGOs) for a D_{3h} F_3 fragment, the geometry of which is analogous to that in BF_3 (the position of the B atom is marked by the dot in the top left-hand diagram); the F_3 triangle lies in the xy plane. Orbitals LGO(5), LGO(8) and LGO(9) contain contributions from the $2p_z$ atomic orbitals, directed perpendicular to the F_3 triangle. The relative sizes of the lobes in each diagram *approximately* represent the relative contributions made by the fluorine atomic orbitals to each ligand group orbital.

picture of BF_3, and using the same arguments about bond order that we introduced in Section 1.13, we estimate that the π-bond order *per B−F bond* is $\frac{1}{3}$. We must reiterate that this is a simplified bonding picture, but one which is consistent with there being partial π-character in each of the B−F bonds. The observed bond distance in BF_3 is 131 pm compared to the sum of r_{cov}(B) and r_{cov}(F) of 159 pm.

We can now appreciate that attempts to construct MO diagrams without the aid of computer programs (of which many PC versions are now available) is non-trivial. However, one can often use the MO approach to gain insight

into particular aspects of the bonding in a molecule rather than to elucidate a full bonding picture.

4.7 Molecular orbital theory: learning to use the theory objectively

The aim of this section is not to establish complete bonding pictures for molecules using MO theory, but rather to

Table 4.1 Symmetry matches between the ligand group orbitals (LGOs) shown in Figure 4.19 and the atomic orbitals of a B atom placed in the F_3 triangle at the position shown by the dot in the top left-hand diagram in Figure 4.19; the F_3 fragment lies in the xy plane.

Ligand group orbital	Boron atomic orbital	Comment
LGO(1)	$2s$	Leads to B−F σ-bonding character
LGO(2)	$2p_y$	Leads to B−F σ-bonding character
LGO(3)	$2p_x$	Leads to B−F σ-bonding character
LGO(4)	$2s$	Leads to B−F σ-bonding character
LGO(5)	$2p_z$	Leads to B−F π-bonding character
LGO(6)	$2p_y$	Leads to B−F σ-bonding character
LGO(7)	$2p_x$	Leads to B−F σ-bonding character
LGO(8)	No match[‡]	Non-bonding MO in BF_3
LGO(9)	No match[‡]	Non-bonding MO in BF_3
LGO(10)	No match[‡]	Non-bonding MO in BF_3
LGO(11)	$2p_y$	Leads to B−F σ-bonding character
LGO(12)	$2p_x$	Leads to B−F σ-bonding character

[‡] On symmetry grounds, each LGO may be matched to an atomic orbital of the boron atom, but the net interaction is a non-bonding one.

Fig. 4.20 A qualitative MO diagram for the formation of BF_3; the ligand group orbitals (LGOs) are shown in Figure 4.19. The light grey rectangle in the stack of MOs in BF_3 represents a group of eight non-bonding MOs. The diagram is an over-simplification of the bonding in BF_3, but is sufficiently detailed to account for the B–F bonds possessing partial π-character.

develop an objective way of using the MO model to rationalize particular features about a molecule. This often involves drawing a *partial MO diagram* for the molecule in question. In each example below, the reader should consider the implications of this partial treatment: it can be dangerous because bonding features, other than those upon which one is focusing, are ignored. However, with care and practice, the use of partial MO treatments is extremely valuable as a method of understanding structural and chemical properties in terms of bonding and we shall make use of it later in the book.

CO_2

A bonding scheme for CO_2 using *sp* hybridized orbitals leaves the $2p$ atomic orbitals on each atom available to form two, localized CO π-bonds. If the z axis is defined as in structure **4.3**, then one π-bond is formed by the overlap of C and O $2p_x$ atomic orbitals, and the other by the overlap of $2p_y$ atomic orbitals.

(4.3)

Let us now consider how molecular orbital theory handles the formation of π-bonds in CO_2. We can use the ligand group orbital approach, and consider the interaction between the atomic orbitals of C and the LGOs of an O---O fragment (with O atoms on the z axis). Since we are only concerned with the π-system, we can restrict our discussion to interactions involving the $2p_x$ and $2p_y$ atomic orbitals. Further, in drawing a partial MO diagram, we can focus on interactions involving the $2p_x$ orbitals, so long as we remember that related and equivalent orbital interactions involving the $2p_y$ atomic orbitals also exist.

Figure 4.21 illustrates the two ligand group orbitals of the O---O fragment, formed by taking in-phase and out-of-phase combinations of the oxygen $2p_x$ atomic orbitals. Only LGO(1) has the correct symmetry to overlap with the carbon $2p_x$ atomic orbital, giving rise to two MOs: ψ_1 possesses π-bonding character delocalized across all three atoms, and ψ_3 is the corresponding antibonding MO. The fragment orbital LGO(2) becomes non-bonding in CO_2. Figure 4.21 shows that four valence electrons are available to occupy MOs ψ_1 and ψ_2; this number of electrons is determined after considering the number of valence electrons occupying the lower energy CO σ-bonding MOs.

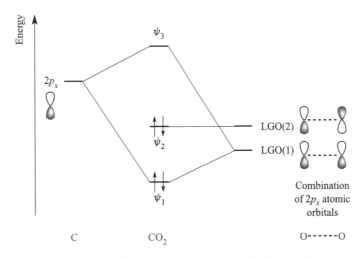

Fig. 4.21 A partial MO diagram that illustrates the formation of delocalized C–O π-bonds using the ligand group orbital approach. The CO_2 molecule is defined as lying on the z axis. A similar diagram can be drawn to show the overlap of $2p_y$ orbitals to give another π-interaction.

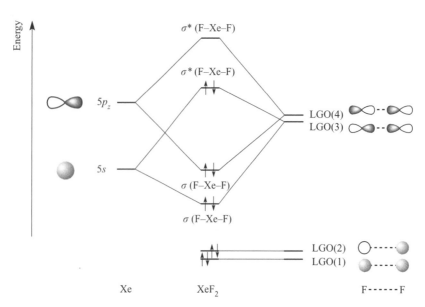

Fig. 4.22 A qualitative, partial MO diagram for the formation of XeF$_2$ using the ligand group orbital approach.

It may be concluded from the molecular orbital treatment that there are two orthogonal π-bonding interactions in CO$_2$, each of which is delocalized over the O$-$C$-$O framework. Two pairs of electrons are associated with these interactions and so the net π-bond order *per C$-$O interaction* is one.

Self-study exercise

Work out a qualitative MO description for the σ-bonding in CO$_2$ and show that this picture is consistent with leaving four electrons to occupy the π-type MOs shown in Figure 4.21.

XeF$_2$

Xenon difluoride is a linear molecule and a valence bond approach to the bonding provides a localized picture with two 2c-2e Xe$-$F bonds. By using a molecular orbital approach a rather different interpretation of the bonding in XeF$_2$ is obtained.

The valence orbitals of F are the $2s$, $2p_x$, $2p_y$ and $2p_z$ atomic orbitals, and those of Xe are the $5s$, $5p_x$, $5p_y$ and $5p_z$ atomic orbitals.[†] Initially, let us consider only orbital interactions which can lead to σ-bonding interactions. We define the z axis to coincide with the molecular axis (as in **4.3**) and construct ligand group orbitals for the F$---$F fragment using the $2s$ and $2p_z$ atomic orbitals. These are shown on the right-hand side of Figure 4.22. Although the energies of the ligand group orbitals containing F $2p$ character are compatible with those of the Xe $5s$ and $5p$ atomic orbitals,

the ligand group orbitals with F $2s$ character are too low in energy for these orbitals to interact to any great extent with the Xe atomic orbitals. Thus, we can make the approximation that fluorine-based orbitals LGO(1) and LGO(2) become essentially non-bonding MOs in XeF$_2$. With this assumption in mind, the MO diagram shown in Figure 4.22 is constructed. There are sufficient valence electrons (four from Xe and three from each F) to occupy all but the highest lying MO. Figure 4.22 shows that two MOs with F$-$Xe$-$F bonding character and one MO with F$-$Xe$-$F antibonding character are occupied. The *net bond order per Xe$-$F interaction* is therefore $\frac{1}{2}$ because the net bonding character arises from one MO which is described as a three-centre two-electron (3c-2e) bonding interaction.

> In a 3c-2e bonding interaction, two electrons occupy a bonding MO which is delocalized over three atomic centres.

Let us now consider why we are justified in omitting π-type interactions from our qualitative bonding model for XeF$_2$. The Xe $5p_x$ and $5p_y$ atomic orbitals may interact with ligand group orbitals of the F$---$F fragment which contain contributions from the F $2p_x$ and $2p_y$ atomic orbitals. This gives rise to a doubly degenerate set of π-bonding MOs, a degenerate set of π-type non-bonding MOs centred on the fluorine atoms, and a degenerate set of π^*-MOs: i.e. a scheme similar to that in Figure 4.21 can also be drawn for XeF$_2$. There are enough valence electrons to fill *all* these π-type MOs and consequently there is no net π-bonding in XeF$_2$. (See problem 4.17 at the end of the chapter.)

[NO$_3$]$^-$

In worked example 4.2, we considered the bonding in [NO$_3$]$^-$ using a VB approach. Three resonance structures (one of

---

[†] We assume that there is no participation by the Xe $4d$ atomic orbitals; any participation by the $5d$ atomic orbitals remains a matter for debate.

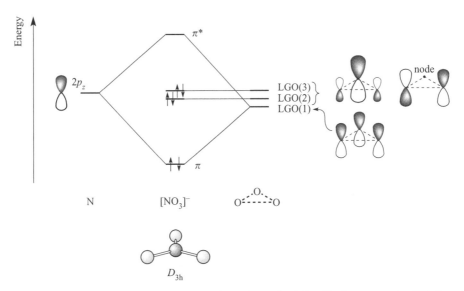

Fig. 4.23 A qualitative, partial MO diagram to illustrate the formation of a delocalized π-system in $[NO_3]^-$; a ligand group orbital approach is used.

which is **4.4**) are needed to account for the equivalence of the N–O bonds, in which the net bond order per N–O bond is 1.33. Molecular orbital theory allows us to represent the N–O π-system in terms of delocalized interactions.

(4.4)

Let us define an axis set such that the $[NO_3]^-$ ion (D_{3h}) lies in the xy plane. We can conveniently consider the π-interactions in $[NO_3]^-$ in terms of the interactions between the N $2p_z$ atomic orbital and the three ligand group orbitals formed by combining the $2p_z$ atomic orbitals of the O atoms. The latter are analogous to those we drew for the F_3 fragment in BF_3 (i.e. LGO(5), LGO(8) and LGO(9) in Figure 4.19). A partial MO diagram for $[NO_3]^-$ which focuses only on π-type orbitals is shown in Figure 4.23. The interactions between the N $2p_z$ atomic orbital and LGO(1) give rise to a π-bonding MO and its antibonding counterpart. Orbitals LGO(2) and LGO(3) become non-bonding MOs. Six electrons are available to occupy the three lowest lying MOs in Figure 4.23; the number of electrons can be deduced by considering that of the 24 valence electrons in $[NO_3]^-$, six occupy σ-bonding MOs, 12 occupy MOs with essentially non-bonding character derived from oxygen-centred LGOs, leaving six for the π-type MOs (see end of chapter *problem 4.18*).

Molecular orbital theory therefore gives a picture of $[NO_3]^-$ in which there is *one* occupied MO with π-character and this is delocalized over all four atoms giving an N–O π-*bond order* of $\frac{1}{3}$. This is in agreement with the valence bond picture, but it is perhaps easier to visualize the delocalized bonding scheme than the resonance between three contributing forms of the

type of structure **4.4**. The bonding in the isoelectronic species $[CO_3]^{2-}$ and $[BO_3]^{3-}$ (both D_{3h}) can be treated in a similar manner.

SF$_6$

The valence bond model of octahedral SF_6, **4.5**, provides a picture of the bonding in which there are six equivalent S–F single bonds; an sp^3d^2 hybridization scheme can be used and each of the six valence electrons of the S atom is involved in bond formation. Within this model, the expansion of the octet of electrons around S assumes the participation of the 3d atomic orbitals in bonding. However, the involvement of the 3d orbitals is not a clear-cut matter; they are energetically less accessible for bonding than the 3s and 3p valence orbitals. By using MO theory, we can generate a bonding scheme for SF_6 *with or without* the participation of the 3d atomic orbitals; in forming six S–F bonds with only the 3s and 3p valence orbitals of sulfur, SF_6 is classed as a *hypervalent molecule*.

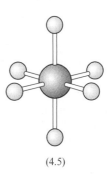

(4.5)

Consider the construction of a partial MO diagram for the formation of SF_6 using *only* the 3s and 3p atomic orbitals as the basis set of the S atom. It is convenient to orient the molecule so that the F atoms lie on the x, y and z axes.

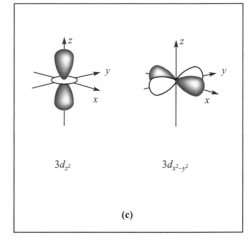

Fig. 4.24 (a) Schematic representations of the four ligand group orbitals (LGOs) for an octahedral (O_h) F_6 fragment which are the main contributors to the bonding in SF_6 in a model in which the basis set of S consists only of the $3s$ and $3p$ atomic orbitals. The representations of LGO(2), LGO(3) and LGO(4) are oversimplified. (b) Representations of the two ligand group orbitals for an octahedral F_6 fragment which have the correct symmetries to overlap with the $3d$ atomic orbitals shown in diagram (c).

Ligand group orbitals can now be constructed for the F_6 fragment using $2s$, $2p_x$, $2p_y$ and $2p_z$ atomic orbitals. However, for a qualitative picture of the bonding, we can assume that the energies of the LGOs containing fluorine $2s$ character are too low for these orbitals to interact significantly with the sulfur atomic orbitals and so we focus only on those ligand group orbitals that possess fluorine $2p$ character. From the eighteen $2p$ atomic orbitals (three from each F atom), we can construct eighteen ligand group orbitals, but only four of these (Figure 4.24a) have significant σ-interactions with the $3s$ and $3p$ atomic orbitals of the S atom; the remaining fourteen ligand group orbitals become essentially non-bonding MOs in SF_6. Figure 4.25a shows a partial MO diagram for the formation of SF_6. On symmetry grounds, interactions can occur between the $3s$ atomic orbital and LGO(1), the $3p_z$ atomic orbital and LGO(2), the $3p_y$ atomic orbital and LGO(3), and the $3p_x$ atomic orbital and LGO(4). This leads to four bonding MOs consisting of a unique MO and a triply degenerate set of MOs. Four corresponding antibonding MOs are also formed. There are 48 valence electrons in

SF_6 (6 from S and 7 from each F), and so by considering the fact that only those MOs (in the valence band) with F $2s$ character will be lower in energy than those shown in Figure 4.25a, we can see that MOs ψ_1 to ψ_4 will be fully occupied. The σ^*-orbitals remain empty because the remaining valence electrons are accommodated in fluorine-centred non-bonding MOs, not shown in the partial MO diagram in Figure 4.25a. Although this picture is an oversimplified state of affairs, Figure 4.25a can be used to show that, using a basis set with only the $3s$ and $3p$ atomic orbitals of sulfur gives four bonding MOs in SF_6. Orbital ψ_1 possesses bonding character delocalized over the six S−F interactions. Each of MOs ψ_2, ψ_3 and ψ_4 contains bonding character delocalized over two S−F bonds but, *taken as a set*, ψ_2, ψ_3 and ψ_4 treat the six bonds in an equivalent manner. Thus, the model is consistent with the observed equivalence of the S−F bonds, but the bond order of each bond is only $\frac{2}{3}$ (i.e. four pairs of electrons for six S−F bonding interactions).

Now let us expand the basis set of sulfur to include the $3d$ atomic orbitals. Of the fourteen ligand group orbitals of the

(a)

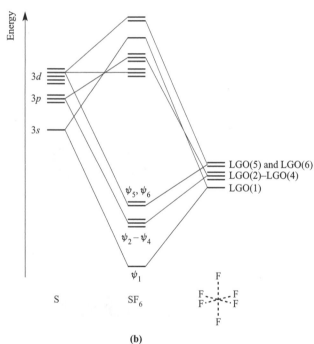

(b)

Fig. 4.25 Qualitative, partial MO diagrams for the formation of SF_6 using the ligand group orbital approach with a basis set for sulfur that is composed of (a) the $3s$ and $3p$ atomic orbitals and (b) the $3s$, $3p$ and $3d$ atomic orbitals.

F_6 fragment that we ignored in our first bonding model, two (shown in Figure 4.24b) have symmetries that are compatible with those of two of the $3d$ atomic orbitals (Figure 4.24c). Interaction can occur between the $3d_{z^2}$ atomic orbital and LGO(5), and between the $3d_{x^2-y^2}$ atomic orbital and LGO(6). The inclusion of the $3d$ orbitals means that the MO diagram for the formation of SF_6 can be approximately represented as in Figure 4.25b. The interactions involving the

S $3s$ and $3p$ atomic orbitals and orbitals LGO(1) to LGO(4) are effectively the same as in Figure 4.25a, but now there are additional interactions leading to the formation of two more bonding and two more antibonding MOs. Three of the five $3d$ atomic orbitals (d_{xy}, d_{xz} and d_{yz}) become non-bonding MOs in SF_6. By considering the number of valence electrons as we did in the first bonding model for SF_6, we see that all the bonding MOs in Figure 4.25b are occupied. The consequence of allowing the $3d$ atomic orbitals to participate in the bonding in SF_6 is that the S–F bond order is one, since there are six bonding MOs with bonding character delocalized over the six S–F bonds.

We stress that, although our MO treatments of SF_6 are able to describe the bonding on the basis of either the participation or non-participation of the sulfur d orbitals, it is a weakness of the treatment that we cannot unambiguously describe the 'real' bonding picture. The participation of $3d$ atomic orbitals in SF_6 and other main group molecules remains a controversial issue (see end of chapter reading list).

A more advanced problem: B_2H_6

Two common features of boron hydrides (see *Sections 12.5 and 12.11*) are that the B atoms are usually attached to more than three atoms and that *bridging* H atoms are often present. Although a valence bond model has been developed by Lipscomb to deal with the problems of generating localized bonding schemes in boron hydrides,[†] the bonding in these compounds is not readily described in terms of VB theory. The structure of B_2H_6 (D_{2h} symmetry) is shown in Figure 4.26. Features of particular interest are that:

● despite having only one valence electron, each *bridging* H atom is attached to *two* B atoms;

● despite having only three valence electrons, each B atom is attached to four H atoms;

Fig. 4.26 The structure of B_2H_6 determined by electron diffraction.

[†]For detailed discussion of the VB model (called *styx* rules) see: W.N. Lipscomb (1963) *Boron Hydrides*, Benjamin, New York; a summary of *styx* rules and further discussion of the use of MO theory for boron hydrides are given in: C.E. Housecroft (1994) *Boranes and Metallaboranes: Structure, Bonding and Reactivity*, 2nd edn, Ellis Horwood, Chichester.

- the B−H bond distances are not all the same and suggest two types of B−H bonding interactions.

Often, B_2H_6 is described as being *electron deficient*; it is a dimer of BH_3 and possesses 12 valence electrons. The formation of the B−H−B bridges can be envisaged as in structure **4.6**. Whereas each terminal B−H interaction is taken to be a localized 2c-2e bond, each bridging unit is considered as a 3c-2e bonding interaction. Each *half* of the 3c-2e interaction is expected to be weaker than a terminal 2c-2e bond and this is consistent with the observed bond distances in Figure 4.26. Bonding pictures for B_2H_6 which assume either sp^3 or sp^2 hybridized B centres are frequently adopted, but this approach is not entirely satisfactory.

(4.6)

Although the molecular orbital treatment given below is an oversimplification, it still provides valuable insight into the distribution of electron density in B_2H_6. Using the ligand group orbital approach, we can consider the interactions between the *pair* of bridging H atoms and the residual B_2H_4 fragment (Figure 4.27a). First we construct

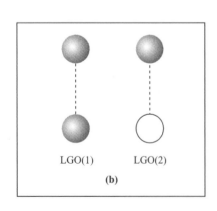

Fig. 4.27 (a) The structure of B_2H_6 can be broken down into H_2B---BH_2 and H---H fragments. (b) The ligand group orbitals (LGOs) for the H---H fragment. (b) The six lowest energy LGOs for the B_2H_4 unit; the nodal plane in LGO(3) is shown.

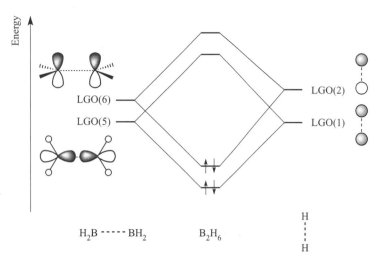

Fig. 4.28 A qualitative, partial MO diagram showing the formation of the B−H−B bridging interactions.

ligand group orbitals for the H---H fragment. The $1s$ atomic orbitals of the two H atoms can combine in- or out-of-phase (Figure 4.27b). Now we construct ligand group orbitals for the B_2H_4 fragment. The valence orbitals of the B atom are the $2s$ and $2p$ atomic orbitals, and 12 LGOs can be constructed; the six lowest energy ligand group orbitals are represented in Figure 4.27c. The higher energy ligand group orbitals possess antibonding B−H or B---B character and can be derived by considering the orbitals shown in Figure 4.27c.

Self-study exercise

Before proceeding further in this section, work out how the ligand group orbitals in Figure 4.27c have been constructed.

The next step in developing the bonding model for B_2H_6 is to attempt to match the symmetries of the ligand group orbitals of the H---H fragment with the ligand group orbitals of the B_2H_4 fragment. When comparing the ligand group orbitals of the two fragments, it is crucial to keep in mind the relative positions of the atoms; as defined in Figure 4.27a, the bridging H atoms lie in the xz plane while the B_2H_4 unit is in the yz plane. On symmetry grounds, LGO(1) of the H---H unit can interact with LGO(1) *and* LGO(5) of the B_2H_4 fragment, while LGO(2) of the H---H fragment can interact with LGO(6) of the B_2H_4 unit. It is clear that LGO(2) and LGO(4) will become non-bonding MOs in B_2H_6. At first glance, LGO(3) might be thought to be compatible with LGO(2) of the H---H fragment, but the two H atoms lie in the nodal plane of LGO(3), and so it also becomes non-bonding in B_2H_6. Although virtually impossible to assess in a qualitative picture, the combination of the boron-based LGO(5) with the H---H LGO(1) is better matched in terms of energy than the LGO(1)–LGO(1) combination.

We are now in a position to construct an *approximate*, partial MO diagram for B_2H_6. The orbital interactions shown in Figure 4.28 lead to two bonding and two antibonding MOs; note that the terms 'bonding' and 'antibonding' are used *with respect to the bridging interactions* that are the focus of this particular ligand group analysis. Consideration of the number of valence electrons available leads us to deduce that both the bonding MOs will be occupied. An important conclusion of the MO model is that the boron–hydrogen bridge character is delocalized over all *four* atoms of the bridging unit in B_2H_6. Since there are two such bonding MOs containing four electrons, this result is consistent with the 3c-2e B−H−B model that we described earlier.

Summary

In this chapter we have considered ways of approaching the bonding in small polyatomic molecules. Within VB theory, the construction of sets of resonance structures soon becomes tedious (and difficult) for polyatomic species. Hybridization schemes, each appropriate for a given spatial arrangement of atoms and lone pairs of electrons, are very useful and provide a means of developing descriptions of σ-bonding frameworks and associated π-bonding. There are limitations however, since the atoms themselves are not restricted to the hybridization schemes that we commonly adopt. The use of MO theory is widespread. However, for all but a few simple molecules, it is difficult to produce a qualitative MO diagram without the aid of an appropriate computer program; such programs are readily available for use on personal computers. The last section in this chapter illustrated some of the approximations that one can make to simplify molecular orbital problems, but it must be emphasized that great care is needed in developing such bonding models since many assumptions are made. The partial treatments given here introduce some of the descriptions that appear later in this book.

The following terms were introduced in this chapter. Do you know what they mean?

❑ orbital hybridization
❑ sp, sp^2, sp^3, sp^3d, sp^2d and sp^3d^2 hybridization
❑ ligand group orbital (LGO) approach
❑ basis set of orbitals
❑ delocalized bonding interaction
❑ symmetry matching of orbitals
❑ energy matching of orbitals
❑ 3c-2e bonding interaction

Suggested further reading

J.K. Burdett (1997) *Chemical Bonds, A Dialog*, Wiley, New York – An original résumé of modern valence theory presented in the form of a 19th century style dialogue between teacher and pupil.

G. Davidson (1991) *Group Theory for Chemists*, Macmillan, London – Chapter 10 provides a useful discussion and also illustrates the use of group theory.

R.L. DeKock and H.B. Gray (1980) *Chemical Structure and Bonding*, Benjamin/Cummings, Menlo Park – A readable text, treating VB and MO theories and giving examples of the relationship between photoelectron spectra and MO energy levels.

H.B. Gray (1994) *Chemical Bonds*, University Science Books, California – An introduction to atomic and molecular structure with numerous illustrations.

S.F.A. Kettle (1985) *Symmetry and Structure*, Wiley, Chichester – An advanced discussion which includes carefully explained applications of group theory.

L. Pauling (1960) *The Nature of the Chemical Bond*, 3rd edn, Cornell University Press, Ithaca – A classic book dealing with covalent, metallic and hydrogen bonding from the viewpoint of VB theory.

M.J. Winter (1994) *Chemical Bonding*, Oxford University Press, Oxford – Chapters 5 and 6 give a basic introduction to hybridization and MO theory in polyatomics.

Role of *d*-orbitals in main group compounds

O.J. Curnow (1998) *Journal of Chemical Education*, vol. 75, p. 910 – 'A simple qualitative molecular orbital/valence bond description of the bonding in main group "hypervalent" molecules'.

Problems

4.1 (a) State what is meant by the *hybridization of atomic orbitals*. (b) Why does VB theory sometimes use hybrid orbital rather than atomic orbital basis sets? (c) Show that equations 4.1 and 4.2 correspond to normalized wavefunctions.

4.2 Figure 4.4 shows the formation of three sp^2 hybrid orbitals (see *equations 4.3–4.5*). (a) Confirm that the directionalities of the three hybrids are as specified in the figure. (b) Show that equations 4.3 and 4.5 correspond to normalized wavefunctions.

4.3 Use the information given in Figure 4.6b and equations 4.6 to 4.9 to reproduce the directionalities of the four sp^3 hybrid orbitals shown in Figure 4.6a.

4.4 (a) Derive a set of diagrams similar to those in Figures 4.2 and 4.4 to describe the formation of sp^2d hybrid orbitals. (b) What is the percentage character of each sp^2d hybrid orbital in terms of the constituent atomic orbitals?

4.5 Use VSEPR theory to predict the shapes of the following species and suggest an appropriate hybridization scheme for each central atom: (a) SiF_4; (b) $[ICl_4]^-$; (c) NF_3; (d) F_2O; (e) SbF_5; (f) $[AsF_6]^-$; (g) ClF_3; (h) BF_3.

4.6 (a) The structures of *cis*- and *trans*-N_2F_2 were shown in worked example 3.1. Give an appropriate hybridization scheme for the N atoms in each isomer. (b) What hybridization scheme is appropriate for the O atoms in H_2O_2 (Figure 1.16)?

4.7 (a) SOF_4 has C_{2v} symmetry. What is its structure? (b) Suggest an appropriate bonding scheme for SOF_4 within VB theory.

4.8 (a) Draw the structure of $[CO_3]^{2-}$. (b) If all the C−O bond distances are equal, write a set of resonance structures to describe the bonding in $[CO_3]^{2-}$. (c) Describe the bonding in $[CO_3]^{2-}$ in terms of a hybridization scheme and compare the result with that obtained in part (b).

4.9 (a) Is CO_2 linear or bent? (b) What hybridization is appropriate for the C atom? (c) Outline a bonding scheme for CO_2 using the hybridization scheme you have suggested. (d) What C−O bond order does your scheme imply? (e) Draw a Lewis structure for CO_2. Is this structure consistent with the results you obtained in parts (c) and (d)?

4.10 What is meant by a *ligand group orbital*?

4.11 VB and MO approaches to the bonding in linear XH_2 (X has 2*s* and 2*p* valence atomic orbitals) give pictures in which the X−H bonding is localized and delocalized respectively. Explain how this difference arises.

4.12 Table 4.2 gives the results of a Fenske−Hall self-consistent field (SCF) quantum chemical calculation for H_2O using an orbital basis set of the atomic orbitals of O and the LGOs of an H---H fragment. The axis set is as defined in Figure 4.14. (a) Use the data to construct pictorial representations of the MOs of H_2O and confirm that

Table 4.2 Results of a self-consistent field quantum chemical calculation for H_2O using an orbital basis set of the atomic orbitals of the O atom and the ligand group orbitals of an H---H fragment. The axis set is defined in Figure 4.14.

Atomic orbital or ligand group orbital	Percentage character of MOs with the sign of the eigenvector given in parentheses					
	ψ_1	ψ_2	ψ_3	ψ_4	ψ_5	ψ_6
O $2s$	71 (+)	0	7 (−)	0	0	22 (−)
O $2p_x$	0	0	0	100 (+)	0	0
O $2p_y$	0	59 (+)	0	0	41 (−)	0
O $2p_z$	0	0	85 (−)	0	0	15 (+)
H---H LGO(1)	29 (+)	0	8 (+)	0	0	63 (+)
H---H LGO(2)	0	41 (−)	0	0	59 (−)	0

Figure 4.14 is consistent with the results of the calculation. (b) How does MO theory account for the presence of lone pairs in H_2O?

4.13 Refer to Figure 4.15 and the accompanying discussion. (a) Why does the B $2p_z$ atomic orbital become a non-bonding MO in BH_3? (b) Draw schematic representations of each bonding and antibonding MO in BH_3.

4.14 (a) The inserts at the right-hand side of Figure 4.16 show three of the MOs in NH_3. Sketch representations of the other four MOs. (b) In the text, ψ_4 is described as possessing 'essentially' lone pair character. What additional character does ψ_4 possess?

4.15 Use a ligand group orbital approach to describe the bonding in $[NH_4]^+$. Draw schematic representations of each of the bonding MOs.

4.16 The I−I bond distance in I_2 (gas phase) is 267 pm, in the $[I_3]^+$ ion is 268 pm, and in $[I_3]^-$ is 290 pm (for the $[AsPh_4]^+$ salt). (a) Draw Lewis structures for these species. Do these representations account for the variation in bond distance? (b) Use MO theory to describe the bonding and deduce the I−I bond order in each species. Are your results consistent with the structural data?

4.17 (a) On the basis that the $4d$ electrons can be ignored, how many valence electrons does XeF_2 possess? (b) Combine the results of the π-interactions discussed in the text with the information given in Figure 4.22 to construct an approximate MO diagram for XeF_2 showing both σ-, π- and non-bonding MOs. (c) Which of the MOs in your diagram are occupied? (d) Determine the bond order of each Xe−F bond in XeF_2, and confirm that the presence of the π-type interactions does not affect the bond order.

4.18 Using Figures 4.19, 4.20 and 4.23 to help you, compare the MO pictures of the bonding in BF_3 and $[NO_3]^-$. What approximations have you made in your bonding analyses?

4.19 By considering the structures of the following molecules, confirm that the point group assignments are correct: (a) BH_3, D_{3h}; (b) NH_3, C_{3v}; (c) B_2H_6, D_{2h}. [Hint: use Figure 3.9.]

4.20 In the description of the bonding of B_2H_6, we draw the conclusion that the two bonding MOs in Figure 4.28 have B−H bonding character delocalized over the four bridge atoms. (a) What other character do these MOs possess? (b) Does your answer to (a) alter the conclusion that this approximate MO description is consistent with the valence bond idea of there being two 3c-2e bridge bonds?

4.21 In $[B_2H_7]^-$ (**4.7**), each B atom is *approximately* tetrahedral. (a) How many valence electrons are present in the anion? (b) Assume that each B atom is sp^3 hybridized. After localization of the three terminal B−H bonds per B, what B-centred orbital remains for use in the bridging interaction? (c) Following from your answer to part (b), construct an approximate orbital diagram to show the formation of $[B_2H_7]^-$ from two BH_3 units and H^-. What does this approach tell you about the nature of the B−H−B bridge?

(4.7)

Chapter 5 Structures and energetics of metallic and ionic solids

5.1 Introduction

In the solid state, both metallic and ionic compounds possess ordered arrays of atoms or ions and form crystalline materials with *lattice* structures. Studies of their structures may conveniently be considered as related topics because both are concerned with the packing of *spherical* atoms or ions. However, differences in *bonding* result in quite distinct properties for metallic and ionic solids. Metallic bonding is characterized by strong, non-directional cohesive forces and high electrical conductivity. Such conductivity implies the presence of bonding electrons delocalized over the whole crystal. Ionic bonding in the solid state arises from non-directional electrostatic interactions between charged species (ions), e.g. Na^+ and Cl^- in rock salt. Ionic solids are *insulators*.

> An *anion* is a negatively charged ion and a *cation* is a positively charged ion.

Although metallic and ionic solids have three-dimensional structures, it does *not* follow that three-dimensional structures are necessarily metallic or ionic. Diamond, for example, is a non-metal (see *Sections 5.11* and *5.12*). In Sections 1.12 and 1.15, we considered the inclusion of ionic contributions to 'covalent' bonding pictures. Later in this chapter we shall discuss how including some covalent character in a predominantly ionic model comes closer to reality for some so-called 'ionic' compounds.

5.2 Packing of spheres

Many readers will be familiar with descriptions of metal lattices based upon the packing of spherical atoms, and in this section we provide a brief résumé of common types of packing, and introduce the terms *unit cell* and *interstitial hole*.

Cubic and hexagonal close-packing

Let us place a number of equal sized spheres in a rectangular box, with the restriction that there must be a *regular arrangement* of spheres. Figure 5.1 shows the most efficient way in which to cover the floor of the box. Such an arrangement is *close-packed*, and spheres that are not on the edges of the assembly are in contact with six other spheres within the layer. A motif of hexagons is produced within the assembly. Figure 5.2 shows part of the same close-packed arrangement of spheres; hollows lie between the spheres and we can build a second layer of spheres upon the first by placing spheres in these hollows. However, if we arrange the spheres in the second layer so that close-packing is again achieved, it is possible to occupy only every other hollow. This is shown on going from Figure 5.2a to 5.2b, in which spheres are placed so that they rest in the hollows indicated.

Now consider the hollows that are visible in the top layer of spheres in Figure 5.2b. There are *two distinct types of hollows*: hollow **A** lies directly over a *sphere* in the first layer, while hollows labelled **C** lie over *hollows* in the first layer. The consequence of this is that when a third layer of spheres is constructed, two different close-packed arrangements are possible. Figure 5.2c shows spheres in the third

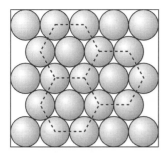

Fig. 5.1 If the floor of a rectangular box is covered with spheres of equal size, with the restriction that the arrangement must be ordered, then the most efficient arrangement is a *close-packed* one. There is a repeating hexagonal motif which is shown in blue.

layer lying in hollows of type **C**, and Figure 5.2d shows a sphere placed in hollow **A**. The arrangements shown can, of course, be extended sideways, and the sequences of layers can be repeated such that the fourth layer of spheres is equivalent to the first, and so on. The two close-packed arrangements are distinguished in that one contains *two repeating layers*, ABABAB..., while the second contains *three repeating layers*, ABCABC... (Figures 5.2d and 5.2c).

> *Close-packing of spheres* results in the most efficient use of the space available; 74% of the space is occupied by the spheres.

The ABABAB... and ABCABC... packing arrangements are called *hexagonal close-packing* (hcp) and *cubic close-packing* (ccp), respectively. In each structure, any given sphere is surrounded by (and touches) 12 other spheres and is said to have 12 *nearest neighbours*, to have a *coordination number* of 12, or to be *12-coordinate*. Figure 5.3 shows representations of the ABABAB... and ABCABC... arrangements which illustrate how this coordination number arises; in these diagrams, 'ball-and-stick' representations of the lattice are used to allow the connectivities to be seen. This type of representation is commonly used *but does not imply* that the spheres do not touch one another.

The unit cell: hexagonal and cubic close-packing

A *unit cell* is a fundamental concept in solid state chemistry, and is the smallest repeating unit of the structure which carries *all* the information necessary to construct *unambiguously* an infinite lattice.

> The smallest repeating unit in a solid state lattice is a *unit cell*.

The unit cells in Figure 5.4 characterize cubic (ccp) and hexagonal close-packing (hcp). Whereas these respective descriptors are not obviously associated with the packing sequences shown in Figures 5.2 and 5.3, their origins are clear in the unit cell diagrams. Cubic close-packing is also called *face-centred cubic* (fcc) packing, and this name clearly reflects the nature of the unit cell shown in Figure 5.4a. The relationship between the ABABAB... sequence and the hcp unit cell is easily recognized; the latter consists of parts of three ABA layers. However, it is harder to see the

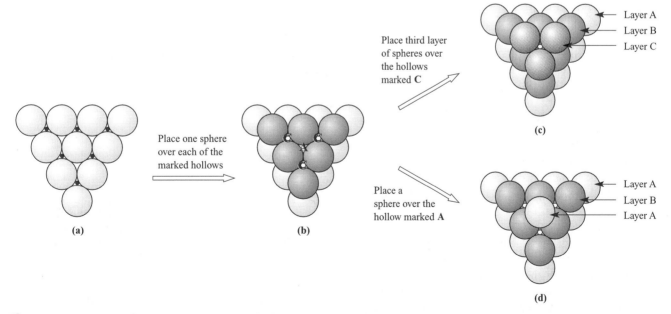

Fig. 5.2 (a) One layer of close-packed spheres exhibits hollows which form a regular pattern. (b) When a second layer of spheres is placed on top of the first so that the new spheres lie in the hollows of the first layer, it is only possible for every other hollow to be occupied, i.e. spheres lie in the hollows marked by blue dots in diagram (a). In the second layer, there are two types of hollow: hollow **A** lies over a sphere in the first layer, but hollows of type **C** lie over a hollow in the first layer. (c) A third layer of spheres can be constructed by resting spheres in hollows **C** and this gives an ABC arrangement. (d) A third layer of spheres can be constructed by resting a sphere in hollow **A** and this gives an ABA arrangement.

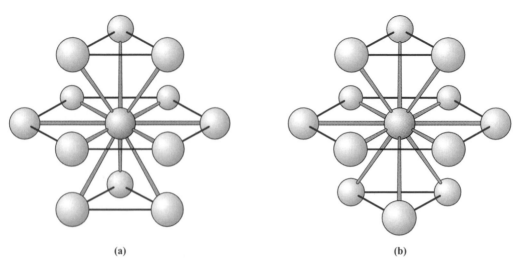

(a)   **(b)**

Fig. 5.3 In both the (a) ABA and (c) ABC close-packed arrangements, the coordination number of each atom is twelve.

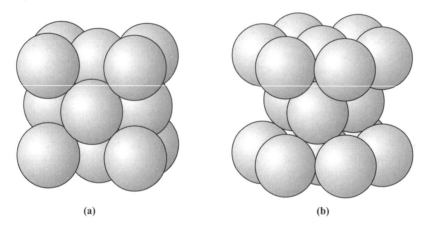

(a)   **(b)**

Fig. 5.4 Unit cells of (a) a cubic close-packed (face-centred cubic) lattice and (b) a hexagonal close-packed lattice.

ABCABC... sequence within the ccp unit cell since the close-packed layers are not parallel to the base of the unit cell but instead lie along the body-diagonal of the cube.

Interstitial holes: hexagonal and cubic close-packing

Close-packed structures contain *octahedral* and *tetrahedral holes* (or *sites*). Figure 5.5 shows two representations of two layers of close-packed spheres: Figure 5.5a is a 'space-filling' representation, while in Figure 5.5b, the sizes of the spheres have been reduced so that connectivity lines can be shown (a 'ball-and-stick' diagram). This illustrates that the spheres lie at the corners of either tetrahedra or octahedra; conversely, the spheres pack such that there are octahedral and tetrahedral holes between them. There is one octahedral hole per sphere, and there are twice as many tetrahedral as octahedral holes in a close-packed array; the octahedral holes are larger than the tetrahedral sites. Whereas a tetrahedral hole can accommodate a sphere of radius ≤ 0.23 times that of the close-packed spheres, a sphere of radius 0.41 times that of the close-packed spheres fits into an octahedral hole.

Non-close-packing: simple cubic and body-centred cubic arrays

Spheres are not always packed as efficiently as in close-packed arrangements; ordered arrays can be constructed in

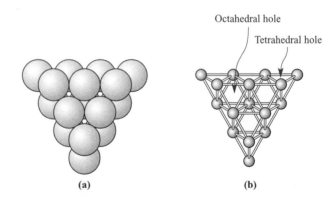

Octahedral hole

Tetrahedral hole

(a)   **(b)**

Fig. 5.5 Two layers of close-packed atoms shown (a) with the spheres touching, and (b) with the sizes of the spheres reduced so that connectivity lines are visible. In (b), the tetrahedral and octahedral holes are indicated.

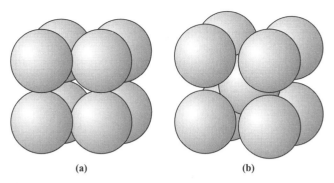

Fig. 5.6 Unit cells of (a) a simple cubic lattice and (b) a body-centred cubic lattice.

which the space occupied by the spheres is less than the 74% found for a close-packed arrangement.

If spheres are placed so as to define a network of cubic frameworks, the unit cell is a simple cube (Figure 5.6a). In the extended lattice, each sphere has a coordination number of six. The hole within each cubic unit is not large enough to accommodate a sphere equal in size to those in the array, but if the eight spheres in the cubic cell are pulled apart slightly, another sphere is able to fit inside the hole. The result is the *body-centred cubic* (bcc) arrangement (Figure 5.6b). The coordination number of each sphere in a bcc lattice is eight.

5.3 The packing-of-spheres model applied to the structures of elements

In Section 5.2, we considered some of the ways in which *hard spheres* may pack together to give ordered arrays. Although the idea of hard, spherical atoms is at odds with modern quantum theory, the packing-of-spheres model is extremely useful for depicting many solid state structures. The model is applicable to the group 18 elements because they are monatomic, to metals, and to H_2 and F_2 because these diatomic molecules are freely rotating and so can be regarded as spherical entities.

Group 18 elements in the solid state

The group 18 elements are the 'noble gases' (see *Chapter 17*), and Table 5.1 lists selected physical data for these elements. Each element (with the exception of helium, see footnote in Table 5.1) solidifies only at low temperatures. The enthalpy changes accompanying the fusion processes are very small, consistent with the fact that weak van der Waals forces operate between the atoms in the solid state. In the crystal-line solid, ccp structures are adopted by each of solid Ne, Ar, Kr and Xe.

H_2 and F_2 in the solid state

The liquefaction of gaseous H_2 occurs at 20.4 K[†] and solidi-fication at 14.0 K. However, even in the solid state, H_2 molecules have sufficient energy to rotate about a fixed lattice point and consequently the space occupied by each diatomic can be represented by a sphere. In the solid state, these spheres adopt an hcp arrangement.

Difluorine solidifies at 53 K, and on cooling to 45 K, a phase change occurs to give a distorted close-packed structure. This description is applicable because, like H_2, each F_2 molecule rotates freely about a fixed lattice-point. (The second phase above 45 K has a more complicated structure.)

The application of the packing-of-spheres model to the crystalline structures of H_2 and F_2 is *only* valid because they contain freely rotating molecules. Other diatomics such as the heavier halogens do not behave in this manner (see *Section 16.4*).

Metallic elements in the solid state

With the exception of Hg, all metals are solid at 298 K; the statement 'solid at room temperature' is ambiguous because the low melting points of Cs (301 K) and Ga (303 K) mean that in some countries, these metals are liquids. Table 5.2 shows that most metals crystallize with ccp, hcp or bcc lattices. However, many metals are *polymorphic* and exhibit more than one structure depending upon the conditions of temperature and/or pressure; we return to this later.

On the basis of the hard sphere model, close-packing represents the most efficient use of space with a common packing efficiency of 74%. The bcc structure is not much less efficient in packing terms, for although there are only eight nearest neighbours, each at a distance x (compared with twelve in the close-packed lattices), there are six more neighbours at distances of $1.15x$, leading to a packing efficiency of 68%.

Among the few metals that adopt structures other than ccp, hcp or bcc lattices are those in group 12. The structures of Zn and Cd are based upon hcp lattices but distortion leads to each atom having only six nearest neighbours (within the same layer of atoms) and six others at a greater distance. Mercury adopts a distorted simple cubic lattice, with the distortion leading to a coordination number of six. Manganese stands out among the *d*-block metals as having an unusual structure; the atoms are arranged in a complex cubic lattice such that there are four environments with coordination numbers of 12, 13 or 16. Atypical structures are also exhibited by most of the *p*-block metals. In group 13, Al and Tl adopt ccp and hcp lattices respectively, but Ga and In adopt quite different structures. Atoms of Ga are organized so that there is only one nearest neighbour (at 249 pm), with six next-nearest neighbours lying at distances within the

---

[†] All phase changes mentioned in this chapter are at atmospheric pressure, unless otherwise stated.

Table 5.1 Selected physical data for the group 18 elements.

Element	Melting point / K	$\Delta_{fus}H$(mp) / kJ mol^{-1}	Boiling point / K	$\Delta_{vap}H$(bp) / kJ mol^{-1}	Van der Waals radius (r_v) / pm
Helium	‡	–	4.2	0.08	99
Neon	24.5	0.34	27	1.71	160
Argon	84	1.12	87	6.43	191
Krypton	116	1.37	120	9.08	197
Xenon	161	1.81	165	12.62	214
Radon	202	–	211	18	–

‡ Helium cannot be solidified under atmospheric pressure, the pressure condition for which all other phase changes in the table are considered.

range 270 and 279 pm, i.e. there is a tendency for the atoms to pair together. Indium forms a distorted ccp lattice, and the twelve near neighbours separate into two groups, four at 325 pm and eight at 338 pm. In group 14, Pb adopts a ccp structure, but in white Sn (the stable allotrope at 298 K), each atom possesses a coordination number of only six (grey Sn, see *Section 5.4*). Metals with coordination numbers of less than eight are among those that are the most volatile.

Table 5.2 Structures (at 298 K), melting points (K) and values of the standard enthalpies of atomization of the metallic elements. ◆ = hcp; ⊕ = ccp (fcc); ● = bcc

Be
◆ ← Metal lattice type
1560 ← Melting point (K)
324 ← Standard enthalpy of atomization (kJ mol^{-1})
112 ← Metallic radius for 12-coordinate atom (pm)

1	2	3	4	5	6	7	8	9	10	11	12	13	14	15
Li ● 454 161 157	Be ◆ 1560 324 112													
Na ● 371 108 191	Mg ◆ 923 146 160											Al ⊕ 933 330 143		
K ● 337 90 235	Ca ◆ 1115 178 197	Sc ◆ 1814 378 164	Ti ◆ 1941 470 147	V ● 2183 514 135	Cr ● 2180 397 129	Mn see text 1519 283 137	Fe ● 1811 418 126	Co ◆ 1768 428 125	Ni ⊕ 1728 430 125	Cu ⊕ 1358 338 128	Zn see text 693 130 137	Ga see text 303 277 153		
Rb ● 312 82 250	Sr ◆ 1050 164 215	Y ◆ 1799 423 182	Zr ◆ 2128 609 160	Nb ● 2750 721 147	Mo ● 2896 658 140	Tc ◆ 2430 677 135	Ru ◆ 2607 651 134	Rh ⊕ 2237 556 134	Pd ⊕ 1828 377 137	Ag ⊕ 1235 285 144	Cd see text 594 112 152	In see text 430 243 167	Sn see text 505 302 158	
Cs ● 301 78 272	Ba ● 1000 178 224	La ◆ 1193 423 188	Hf ◆ 2506 619 159	Ta ● 3290 782 147	W ● 3695 850 141	Re ◆ 3459 774 137	Os ◆ 3306 787 135	Ir ⊕ 2719 669 136	Pt ⊕ 2041 566 139	Au ⊕ 1337 368 144	Hg see text 234 61 155	Tl ◆ 577 182 171	Pb ⊕ 600 195 175	Bi † 544 210 182

† See *Figure 14.2c* and associated text.

5.4 Polymorphism in metals

Polymorphism: phase changes in the solid state

It is generally convenient to consider the structures of metals in terms of the observed lattice type at 298 K and atmospheric pressure,[†] but these data do not tell the whole story. When subject to changes in temperature and/or pressure, the structure of a metal may change; each form of the metal is a particular *polymorph*. For example, scandium undergoes a reversible transition from an hcp lattice (α-Sc) to bcc lattice (β-Sc) at 1610 K. Some metals undergo more than one change: at atmospheric pressure, Mn undergoes transitions from the α- to β-form at 983 K, from the β- to γ-form at 1352 K, and from γ- to σ-Mn at 1416 K. Although α-Mn adopts a complex lattice (see above), the β-polymorph has a somewhat simpler structure containing two 12-coordinate Mn environments, the γ-form possesses a distorted ccp structure, and the σ-polymorph adopts a bcc lattice. Phases that form at high temperatures may be *quenched* to lower temperatures (i.e. rapidly cooled with retention of structure), allowing the structure to be determined at ambient temperatures. Thermochemical data show that there is usually very little difference in energy between different polymorphs of an element.

> If a substance exists in more than one crystalline form, it is *polymorphic*.

An interesting example of polymorphism is observed for tin. At 298 K and 1 bar pressure, β-Sn (white tin) is the thermodynamically stable polymorph but lowering the temperature to 286 K results in a slow transition to α-Sn (grey tin). The β ⟶ α transition is accompanied by a change in coordination number from 6 to 4, and α-Sn adopts a diamond-type lattice (see *Figure 5.19*). The density of Sn *decreases* from 7.31 to 5.75 g cm^{-3} during the β ⟶ α transition, whereas it is more usual for there to be a increase in density in going from a higher to lower temperature polymorph.

Phase diagrams

In order to appreciate the effects on an element of changing the temperature and pressure, a *phase diagram* must be consulted. Figure 5.7 shows the phase diagram for Fe; each line on the diagram is a *phase boundary* and crossing a boundary (i.e. changing the phase of the metal) requires a change of temperature and/or pressure. For example, at 298 K and

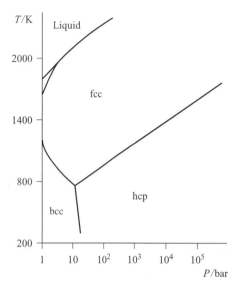

Fig. 5.7 A pressure–temperature phase diagram for iron.

1 bar pressure, Fe has a bcc structure (α-Fe). Raising the temperature to 1185 K (still at 1 bar) results in a transition to γ-Fe with an fcc structure. A transition from α- to γ-Fe also occurs by increasing the pressure on Fe maintained at, e.g., 800 K.

5.5 Metallic radii

The *metallic radius*, r_{metal}, is defined as half of the distance between the nearest-neighbour atoms in a solid state metallic lattice. However, structural data for different polymorphs of the same metal indicate that r_{metal} varies with the coordination number. For example, the ratio of the interatomic distances (and, therefore, of r_{metal}) in a bcc polymorph to those in close-packed forms of the same metal is 0.97:1.00, corresponding to a change in coordination number from 8 to 12. If the coordination number decreases further, r_{metal} also decreases:

Coordination number	12	8	6	4
Relative radius	1.00	0.97	0.96	0.88

> The *metallic radius* is half of the distance between the *nearest-neighbour* atoms in a solid state metal lattice, and is dependent upon coordination number.

The values of r_{metal} listed in Table 5.2 refer to 12-coordinate metal centres; since not all metals actually adopt structures with 12-coordinate atoms, some values of r_{metal} have been estimated. The need for a *consistent* set of data is obvious if one is to make meaningful comparisons within a periodic sequence of elements. Values of r_{metal} (Table 5.2) increase down each of groups 1, 2, 13 and 14. In each of the triads of the *d*-block elements, r_{metal} generally increases

---

[†] Although we often refer to 'atmospheric pressure', a pressure of 1 bar (1.00×10^5 Pa) has been defined by the IUPAC as the *standard pressure*. Until 1982, the standard pressure was 1 atmosphere (1 atm = 101 300 Pa) and this pressure remains in use in some tables of physical data.

on going from the first to second row element, but there is little change on going from the second to third row metal. This latter observation is due to the presence of a filled $4f$ level, and the so-called *lanthanoid contraction* (see *Sections 22.3 and 24.3*).

Worked example 5.1   Metallic radii

Use values of r_{metal} in Table 5.2 to deduce an appropriate value for the metallic radius (a) r_K in metallic K at 298 K and 1 bar pressure, and (b) r_{Sn} in α-Sn. Is the answer for part (b) consistent with the observed interatomic distance in α-Sn of 280 pm?

The values of r_{metal} in Table 5.2 refer to 12-coordinate metal atoms, and values of K and Sn are 235 and 158 pm respectively.

(a) The structure of K at 298 K and 1 bar pressure is bcc, and the coordination number of each K atom is eight. From the relative radii listed in the text:

$$\frac{r_{12\text{-coordinate}}}{r_{8\text{-coordinate}}} = \frac{1}{0.97}$$

The appropriate radius for a K atom in a bcc lattice is:

$$r_{8\text{-coordinate}} = 0.97 \times (r_{12\text{-coordinate}}) = 0.97 \times 235 = 228 \text{ pm}$$

(b) In α-Sn, each Sn atom is four-coordinate. From the relative radii listed in the text:

$$\frac{r_{12\text{-coordinate}}}{r_{4\text{-coordinate}}} = \frac{1}{0.88}$$

The radius for a Sn atom in α-Sn is estimated from:

$$r_{4\text{-coordinate}} = 0.88 \times (r_{12\text{-coordinate}}) = 0.88 \times 158 = 139 \text{ pm}$$

The interatomic distance is twice the value of r_{metal}, and so the calculated value of the Sn–Sn distance of 278 pm is in good agreement with the observed value of 280 pm.

Self-study exercises

Use data in Table 5.2.
1. Estimate a value for the metallic radius, r_{Na}, in metallic Na (298 K, 1 bar).   [*Ans.* 185 pm]
2. The internuclear separation of two Na atoms in the metal (298 K, 1 bar) is 372 pm. Estimate a value of r_{metal} appropriate for 12-coordination.   [*Ans.* 192 pm]

5.6 Melting points and standard enthalpies of atomization of metals

The melting points of the metallic elements are given in Table 5.2 and periodic trends are easily observed. The metals with the lowest melting points are in groups 1, 12, 13 (with the exception of Al), 14 and 15. These metals are, in general, those that do *not* adopt close-packed structures in the solid state. The particularly low melting points of the alkali metals (and correspondingly low values of the standard enthalpies of fusion which range from 3.0 kJ mol^{-1} for Li to 2.1 kJ mol^{-1} for Cs) often give rise to interesting practical observations. For example, when a piece of potassium is dropped on to water, exothermic reaction 5.1 occurs, providing enough heat energy to melt the unreacted metal; the molten potassium continues to react vigorously.

$$2K + 2H_2O \longrightarrow 2KOH + H_2 \qquad (5.1)$$

Values of the standard enthalpies of atomization, $\Delta_a H^\circ(298 \text{ K})$, (or sublimation) in Table 5.2 refer to the processes defined in equation 5.2, and correspond to the destruction of the metallic lattice. Mercury is an exception, since at 298 K it is a liquid.

$$\frac{1}{n} M_n(\text{standard state}) \longrightarrow M(g) \qquad (5.2)$$

Those metals with the lowest values of $\Delta_a H^\circ(298 \text{ K})$ are again those with other than close-packed structures. Since $\Delta_a H^\circ$ appears in thermochemical cycles such as the Born–Haber cycle (see *Section 5.14*), it is clear that $\Delta_a H^\circ$ is important in accounting for the reactivity patterns of these metals.

In general, there appears to be a rough correlation between values of $\Delta_a H^\circ(298 \text{ K})$ and the number of unpaired electrons. In any long period (K to Ga, Rb to Sn, and Cs to Bi in Table 5.2), the maximum values are reached in the middle of the *d*-block (with the exception of Mn which has the atypical structure described in *Section 5.3*).

5.7 Alloys and intermetallic compounds

The physical properties of many metals render them unsuitable for fabrication and engineering purposes. By combining two or more metals, or metals with non-metals, one can form *alloys* with enhanced properties such as strength, malleability, ductility, hardness or resistance to corrosion. For example, adding Sn to Pb gives Pb-based alloys with applications as solders; by varying the Pb:Sn ratio, the melting point of the solder can be modified and tailored to the needs of particular applications.

> An *alloy* is an intimate mixture or, in some cases, a compound of two or more metals, or metals and non-metals; alloying improves the physical properties and resistance to corrosion, heat etc. of the material.

Alloys are manufactured by combining the component elements in the molten state followed by cooling. If the melt is quenched (cooled rapidly), the distribution of the two types of metal atoms in the *solid solution* will be

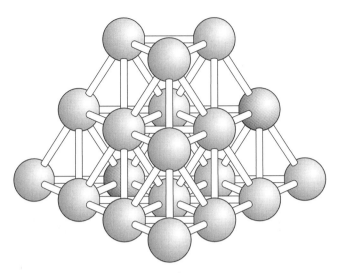

Fig. 5.8 In a substitutional alloy, some of the atom sites in the host lattice (shown in grey) are occupied by solute atoms (shown in blue).

random; the element in excess is termed the *solvent*, and the minor component is the *solute*. Slow cooling may result in a more ordered distribution of the solute atoms. The subject of alloys is not simple, and we shall introduce it only by considering the classes of *substitutional* and *interstitial* alloys, and *intermetallic compounds*.

Substitutional alloys

In a substitutional alloy, atoms of the solute occupy sites in the lattice of the solvent metal (Figure 5.8). To maintain the original lattice structure of the host metal, atoms of both components should be of a similar size. The solute atoms must also tolerate the same coordination environment as atoms in the host lattice. An example of a substitutional alloy is sterling silver (used for silver cutlery and jewellery) which contains 92.5% Ag and 7.5% Cu; elemental Ag and Cu both adopt ccp lattices and $r_{metal}(Ag) \approx r_{metal}(Cu)$ (Table 5.2).

Interstitial alloys

A close-packed lattice contains tetrahedral *and* octahedral interstitial holes (see *Figure 5.5*). Assuming a hard-sphere model for the atomic lattice,[†] one can calculate that an atom of radius 0.41 times that of the atoms in the close-packed array can occupy an octahedral hole, while significantly smaller atoms may be accommodated in tetrahedral sites.

We illustrate interstitial alloys by discussing *carbon steels* in which C atoms occupy a small proportion of the octahedral holes in an Fe lattice. α-Iron possesses a bcc structure at 298 K (1 bar pressure), and a transition to γ-Fe (ccp)

occurs at 1185 K; over the range 1674 to 1803 K, α-Fe is again observed (Figure 5.7). Carbon steels are extremely important industrially (see *Box 5.1*), and there are three basic types designated by their carbon content. *Low carbon steel* contains between 0.03 and 0.25% carbon and is used for steel sheeting, e.g. in the motor vehicle industry and in the manufacture of steel containers. *Medium carbon steel* contains 0.25–0.70% C, and is suited for uses such as bolts, screws, machine parts, connecting rods and railings. The strongest of the carbon steels, *high carbon steel*, contains 0.8–1.5% C and finds applications in a variety of cutting and drilling tools. The corrosion of carbon steels is a disadvantage of the material, but coatings can be applied to inhibit such action. *Galvanized steel* possesses a Zn coating; Zn has a low mechanical strength but a high resistance to corrosion and combined with the high mechanical strength of the steel, galvanized steel meets the demands of many industrial applications. If the Zn coating is scratched revealing the Fe beneath, it is the Zn that oxidizes in preference to the Fe; the scratched Zn coating behaves as a *sacrificial anode* (see *Box 7.3*).

An alternative method of enhancing the properties of steel is to alloy it with another metal, M; this combines both interstitial and substitutional alloy structures, with C occupying holes in the Fe lattice, and M occupying lattice sites. *Stainless steel* is an example of an *alloy steel*; it contains 13–18% Cr and 8–12% Ni, and the C content is very low. For high-wear resistance (e.g. in rail and tram tracks), Mn is alloyed with steel. Other alloy steels contain Ti, V, Co or W, and each solute metal confers specific properties on the finished product. Specific steels are described in *Sections 21.2* and *22.2*.

Intermetallic compounds

When melts of some metal mixtures solidify, the alloy formed may possess a definite lattice type that is different from those of the pure metals. Such systems are classified as *intermetallic compounds*, e.g. β-brass, CuZn. At 298 K, Cu has a ccp lattice and Zn has a structure related to an hcp array, but β-brass adopts a bcc structure. The relative proportions of the two metals are crucial to the alloy being described as an intermetallic compound. Alloys labelled 'brass' may have variable compositions, and the α-phase is a substitutional alloy possessing the ccp structure of Cu with Zn functioning as the solute. β-Brass exists with Cu:Zn stoichiometries around 1:1, but increasing the percentage of Zn leads to a phase transition to γ-brass (sometimes written as Cu_5Zn_8, although the composition is not fixed), followed by a transition to ε-brass which has an approximate stoichiometry of 1:3.[‡]

[†] It is important not to lose sight of the fact that the hard-sphere model is approximate and conflicts with the wave-mechanical view of the atom.

[‡] The variation of phases with temperature and Cu:Zn stoichiometry is more complex than this description implies; see N.N. Greenwood and A. Earnshaw (1997) *Chemistry of the Elements*, 2nd edn, Butterworth-Heinemann, Oxford, p. 1178.

Iron and steel production

The major raw materials for the commercial production of Fe are haematite (Fe_2O_3), magnetite (Fe_3O_4) and siderite ($FeCO_3$) (see also *Section 21.2*). The extraction of iron is carried out on an enormous scale to meet the consumer demands for both iron and steel. In 1995, Russia, China, the US and Japan led the world in the production of steel:

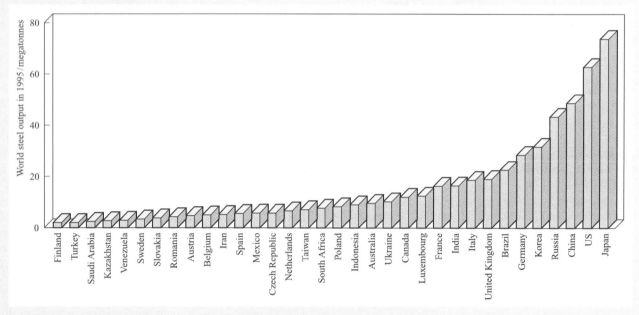

[*Source of data*: Metal Bulletin Feb-16-1996 © Metal Bulletin plc; with permission]

The industrial manufacturing processes for iron and steel can be summarized as follows. Iron ore is mixed with limestone ($CaCO_3$) and coke in a blast furnace in which temperatures vary from ≈750 to 2250 K. Carbon is converted to CO in the highest temperature zone, but both C and CO may reduce the iron ore:

$$2C + O_2 \longrightarrow 2CO$$

$$Fe_2O_3 + 3C \longrightarrow 2Fe + 3CO$$

$$Fe_2O_3 + 3CO \longrightarrow 2Fe + 3CO_2$$

The function of the limestone is to remove impurities and the product of these reactions is *slag* which contains, for example, calcium silicate. Molten Fe from the furnace is collected and cooled in salt-moulds as *pig iron* which contains 2–4% C plus small amounts of P, Si, S and Mn. After remelting and moulding, the product is *cast iron*; this is brittle and its exact nature depends upon the relative amounts of secondary elements. A high Si content results in the C being in the form of graphite, and the cast iron so-formed is called *grey cast iron*. On the other hand, *white cast iron* forms when the Si content is low and carbon is present within the iron–carbon phase *cementite*, Fe_3C.

The *puddling process* is used to convert cast iron to wrought iron; oxidation of C, S, and other impurities leaves wrought iron with <0.2% C content. Unlike cast iron, wrought iron is tough and malleable and is readily worked; its applications, in wrought iron railings and window and door grills, are widespread.

Iron can be converted into steel by the Bessemer, Siemens electric arc or basic oxygen processes. The Bessemer process was the first to be patented, but the Siemens electric arc and basic oxygen processes are used in modern steel production. In the basic oxygen process, O_2 oxidizes the carbon in pig iron, reducing its content to the levels required for commercial steel (see main text).

Recycling

Considering the period over which steel has been commercially produced, steel recycling has become important only relatively recently. Since the late 1980s, Corus (formerly British Steel) has broadened its efforts in regard of recycling steel cans and in 1992, one-eighth of the 12 billion cans in the UK market were recovered for recycling. In the US, the Steel Recycling Institute encourages steel-can recycling, a process that has grown from 15% of cans being reclaimed in 1988, to 60% in 1997.

Further reading

N.N. Greenwood and A. Earnshaw (1997), *Chemistry of the Elements*, 2nd edn, Butterworth-Heinemann, Oxford, p. 1072.

F.J. Berry (1993) 'Industrial chemistry of iron and its compounds' in *Chemistry of Iron*, ed. J. Silver, Blackie, Glasgow.

5.8 Bonding in metals and semiconductors

If we consider the various structure types adopted by metals and then try to provide a model for localized metal–metal bonding, we run into a problem: there are not enough valence shell orbitals or electrons for each metal atom to form 2-centre 2-electron bonds with all its neighbours. For example, an alkali metal has eight near-neighbours (Table 5.2), but only one valence electron. We must therefore use a bonding model with multi-centre orbitals (see *Sections 4.4–4.7*). Further, the fact that metals are good electrical conductors means that the multi-centre orbitals must spread over the whole metal crystal so that we can account for the electron mobility. Several bonding theories have been described, and *band theory* is the most general. Before discussing band theory, we review *electrical conductivity* and *resistivity*.

Electrical conductivity and resistivity

> An *electrical conductor* offers a low resistance (measured in ohms, Ω) to the flow of an electrical current (measured in amperes, A).

The electrical resistivity of a substance measures its resistance to an electrical current (equation 5.3); for a wire of uniform cross-section, the resistivity (ρ) is given units of ohm metre ($\Omega\,m$).

$$\text{Resistance (in } \Omega) = \frac{\text{resistivity (in } \Omega\,m)}{\text{cross-sectional area of wire (in m}^2)} \times \text{length of wire (in m)}$$

$$R = \frac{\rho \times l}{a}$$

(5.3)

Figure 5.9 shows the variation in resistivity of three metals with temperature. In each case, ρ increases with temperature, and the electrical conductivity (which is the inverse of the

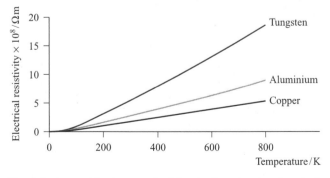

Fig. 5.9 A metal is characterized by the fact that its *electrical resistivity increases* as the temperature increases, i.e. its *electrical conductivity decreases* as the temperature increases.

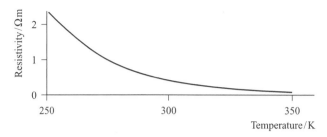

Fig. 5.10 A semiconductor, such as germanium, is characterized by the fact that its electrical resistivity *decreases* as the temperature increases; thus, its electrical conductivity *increases* as the temperature increases.

resistance) decreases as the temperature is raised. This property distinguishes a metal from a *semiconductor* which is a material in which the electrical conductivity increases as the temperature increases (Figure 5.10).

> The *electrical conductivity* of a metal decreases with temperature; that of a semiconductor increases with temperature.

Band theory of metals and insulators

The fundamental concept of band theory is to consider the energies of the molecular orbitals in an assembly of metal atoms. An MO diagram describing the bonding in a metallic solid is characterized by having groups of MOs (i.e. *bands*) which are very close in energy. We can readily see how bands arise by constructing an approximate MO diagram for lithium metal, Li_n.

The valence orbital of an Li atom is the $2s$ atomic orbital, and Figure 5.11 shows schematic MO diagrams for the

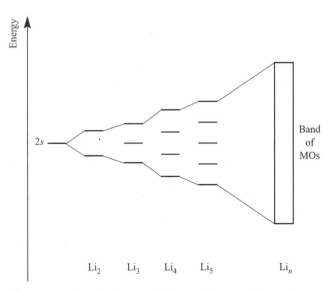

Fig. 5.11 The interaction of two $2s$ atomic orbitals in Li_2 leads to the formation of two MOs. With three Li atoms, three MOs are formed, and so on. For Li_n, there are n molecular orbitals, but because the $2s$ atomic orbitals are all of the same energy, the energies of the MOs are very close together and comprise a *band* of orbitals.

formation of species incorporating different numbers of Li atoms (see *Section 1.13*). If two Li atoms combine, the overlap of the two 2*s* atomic orbitals leads to the formation of two MOs; if three Li atoms combine, three MOs are formed, and so on. For *n* Li atoms, there are *n* MOs, but because the 2*s* atomic orbitals possess the same energy, the energies of the resultant MOs are very close together and so are termed a *band* of orbitals. Now, let us apply the *aufbau* principle and consider the occupation of the MOs in Figure 5.11. Each Li atom contributes one electron; in Li₂, this leads to the lowest MO being filled, and in Li₃, the lowest MO is fully occupied and the next MO is half-filled. In Li*ₙ*, the band must be half-occupied. Since the band of MOs in Li*ₙ* contains contributions from all the Li atoms, the model provides a delocalized picture of the bonding in the metal, and the metal–metal bonding is *non-directional*. Moreover, because the energies of the MOs within the band are very close together and not all the MOs are populated in the ground state, electrons can move into vacant MOs *within the band* under the influence of an electric field. Because of the delocalization, we can readily rationalize the movement of electrons from one Li atom to another, and understand why electrical conductivity results. This model indicates that electrical conductivity is a characteristic property of *partially filled bands* of MOs. In theory, no resistance should oppose the flow of a current if the nuclei are arranged at the points of a perfectly ordered lattice, and the increased *thermal population* of higher energy levels within the band at higher temperatures might be expected to lead to an increase in the electrical conductivity. In practice, however, thermal vibrations of the nuclei produce electrical resistance and this effect is sufficiently enhanced at higher temperatures so as to result in a *decrease* in the conductivity of the metal as the temperature increases.

> A *band* is a group of MOs, the energy differences between which are so small that the system behaves as if a continuous, non-quantized variation of energy within the band is possible.

The model just described for Li is oversimplified; bands are also formed by the overlap of higher energy (unoccupied) atomic orbitals and the 2*p* band actually overlaps with the 2*s* band to some extent since the *s*–*p* separation in atomic Li is relatively small (see *Figure 1.21*). This is also true for Be and, of course, this is of great significance since the ground state electronic configuration of Be is [He]2*s*²; were the energy separation of the 2*s* and 2*p* bands in Be large, the 2*s* band would be fully occupied and Be would be an insulator. In reality, the 2*s* and 2*p* bands overlap, and generate, in effect, a single, partially occupied band, thereby giving Be its metallic character. Figure 5.12a–c illustrates that:

- a fully occupied band separated from the next (empty) band by a large energy separation (the *band gap*) leads to the material being an insulator;

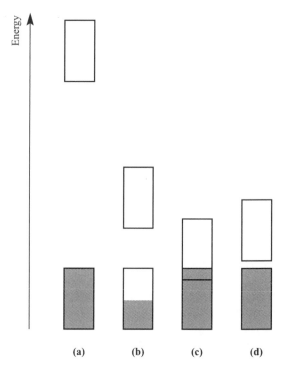

Fig. 5.12 The relative energies of occupied and empty bands in (a) an insulator, (b) a metal in which the lower band is only *partially* occupied, (c) a metal in which the occupied and empty bands overlap, and (d) a semiconductor.

- a partially occupied band leads to the material being metallic;
- metallic character is also consistent with the overlap of an occupied and a vacant band.

> A *band gap* occurs when there is a significant energy difference between two bands.

The Fermi level

The energy level of the highest occupied orbital in a metal at absolute zero is called the *Fermi level*. At this temperature, the electronic configuration predicted by the *aufbau* principle appertains and so, in Li for example, the Fermi level lies exactly at the centre of the half-filled band. For other metals, the Fermi level lies at or near the centre of the band. At temperatures above 0 K, electrons thermally populate MOs just above the Fermi level, and some energy levels just below it remain unoccupied. In the case of a metal, the thermal populations of different energy states cannot be described in terms of a Boltzmann distribution, but are instead given by the *Fermi–Dirac distribution*.[†]

[†] For a mathematical treatment of Fermi–Dirac statistics, see Appendix 17 in M. Ladd (1994) *Chemical Bonding in Solids and Fluids*, Ellis Horwood, Chichester.

Band theory of semiconductors

Figure 5.12d illustrates a situation in which a fully occupied band is separated from an unoccupied band by a *small band gap*. This property characterizes a *semiconductor*. In this case, electrical conductivity depends upon there being sufficient energy available for thermal population of the upper band, and it follows that the conductivity increases as the temperature is raised. In the next section, we look more closely at the types and properties of semiconductors.

5.9 Semiconductors

Intrinsic semiconductors

In the macromolecular structures of diamond, silicon, germanium and α-tin, each atom is tetrahedrally sited (see *Figure 5.19*). An atom of each element provides four valence orbitals and four valence electrons, and, in the bulk element, this leads to the formation of a fully occupied band and an unoccupied band lying at higher energy. The corresponding band gap can be measured spectroscopically since it is equal to the energy needed to promote an electron across the energy gap. For C, Si, Ge and α-Sn, the band gaps are 520, 106, 64 and $8 \, \text{kJ mol}^{-1}$ respectively. The variation down

group 14 leads to C being an insulator, while for α-Sn, the band structure approaches that of a single, partially occupied band and this allotrope of Sn tends towards being metallic.

Each of Si, Ge and α-Sn is classed as an *intrinsic semiconductor*, the extent of occupation of the upper band increasing with increasing temperature. Electrons present in the upper *conduction* band act as charge carriers and result in the semiconductor being able to conduct electricity. Additionally, the removal of electrons from the lower *valence* band creates *positive holes* into which electrons can move, again leading to the ability to conduct charge.

A *charge carrier* in a *semiconductor* is either a positive hole or an electron that is able to conduct electricity.

Extrinsic (n- and p-type) semiconductors

The semiconducting properties of Si and Ge can be enhanced by *doping* these elements with atoms of a group 13 or group 15 element. Doping involves the introduction of only a minutely small proportion of dopant atoms, less than 1 in 10^6, and extremely pure Si or Ge must first be produced. The reduction of SiO_2 in an electric furnace gives Si, and the Czochralski process (see *Box 5.2*) is used to draw single crystals of Si from the melt. We describe how dopants are introduced into semiconductors in *Section 27.6*.

BOX 5.2   **APPLICATIONS**

The production of pure silicon for semiconductors

Semiconductors demand the use of silicon of extreme purity. The native element does not occur naturally and silica (SiO_2) and silicate minerals are its principal sources. Silicon can be extracted from silica by reduction with carbon in an electric furnace, but the product is far too impure for the semiconductor industry. A number of purification methods are used, but of these, two are important for producing single crystals of Si.

Zone melting

Beginning with a polycrystalline Si rod, a small zone (which lies perpendicular to the direction of the rod) is melted. The focus-point of the zone is gradually moved along the length of the rod; under carefully controlled conditions, cooling, which takes place behind the melt-zone, produces single crystals while impurities migrate along the rod with the molten material. Since the first experiments in the 1950s to develop this technique, the method has been adapted commercially and involves many passes of the melt-zone along the silicon rod before crystals suitable for use in semiconductors are obtained.

The Czochralski process

The principle of the Czochralski process is to draw single crystals of Si from the molten element. The thermal

decomposition of ultra-pure $SiHCl_3$ is first used to obtain Si of high purity, and the polycrystalline or powdered element is then placed in a crucible, surrounded by a heating device. Controlled drawing conditions permit single crystals ($\approx 2\text{–}3 \, \text{cm}$ in diameter) to be drawn from the Si melt; the drawing-wire attached to the crystal being grown is rotated in a direction countering the rotation of the crucible; the conditions aim to provide a uniform distribution within the crystal of any remaining impurities. The crucible material is obviously critical; for example, if quartz is used, O atoms may be introduced into the Si crystals.

Further reading

Gmelin Handbook of Inorganic Chemistry (1984), *Silicon* Part A1 'History', System number 15, Springer-Verlag, Berlin, p. 51.

T.J. Trentler, K.M. Hickman, S.C. Goel, A.M. Viano, P.C. Gibbons and W.E. Buhro (1995) *Science*, vol. 270, p. 1791.

See also: Section 13.6 (hydrides of group 14 elements) and Section 27.6 (chemical vapour deposition).

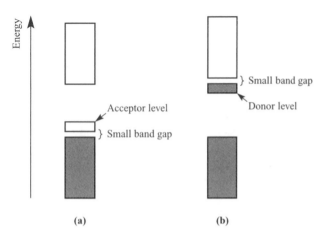

Fig. 5.13 (a) In a p-type semiconductor (e.g. Ga-doped Si), electrical conductivity arises from thermal population of an acceptor level which leaves vacancies (positive holes) in the lower band. (b) In an n-type semiconductor (e.g. As-doped Si), a donor level is close in energy to the conduction band.

Extrinsic semiconductors contain dopants; a *dopant* is an impurity introduced into a semiconductor in minute amounts to enhance its electrical conductivity.

In Ga-doped Si, the substitution of a Ga (group 13) for a Si (group 14) atom in the bulk solid produces an electron-deficient site. This introduces a discrete, unoccupied level into the band structure (Figure 5.13a). The band gap that separates this level from the lower lying occupied band is small ($\approx 10\,\text{kJ}\,\text{mol}^{-1}$) and thermal population of the *acceptor level* is possible. The acceptor levels remain discrete if the concentration of Ga atoms is low, and in these circumstances the electrons in them do not contribute directly to the electrical conductance of the semiconductor. However, the positive holes left behind in the valence band act as charge carriers; one can think either in terms of an electron moving into the hole, thereby leaving another hole into which another electron can move and so on, or in terms of the movement of positive holes (in the opposite direction to the electron migration). This gives rise to a *p-type* (*p* stands for positive) semiconductor. Other group 13 dopants for Si are B and Al.

In As-doped Si, replacing an Si (group 14) by an As (group 15) atom introduces an electron-rich site. The extra electrons occupy a discrete level below the conduction band (Figure 5.13b), and, because of the small band gap ($\approx 10\,\text{kJ}\,\text{mol}^{-1}$), electrons from the *donor level* can thermally populate the conduction band where they are free to move. Electrical conduction can be described in terms of the movement of negatively charged electrons and this generates an *n-type* (*n* stands for negative) semiconductor. Phosphorus atoms can similarly be used as dopants in silicon.

The n- and p-type semiconductors are *extrinsic semiconductors*, and their precise properties are controlled by the choice and concentration of dopant. Semiconductors are discussed further in *Section 27.6*.

5.10 Sizes of ions

Before we embark upon a discussion of the structures of ionic solids, we must say something about the sizes of ions, and define the term *ionic radius*. The process of ionization (e.g. equation 5.4) results in a contraction of the species owing to an increase in the effective nuclear charge. Similarly, when an atom gains an electron (e.g. equation 5.5), the imbalance between the number of protons and electrons causes the anion to be larger than the original atom.

$$Na(g) \longrightarrow Na^+(g) + e^- \qquad (5.4)$$

$$F(g) + e^- \longrightarrow F^-(g) \qquad (5.5)$$

Ionic radii

Although from a wave-mechanical viewpoint, the radius of an individual ion has no precise physical significance, for purposes of descriptive crystallography, it is convenient to have a compilation of values obtained by partitioning measured interatomic distances in 'ionic' compounds. Values of the *ionic radius* (r_{ion}) may be derived from X-ray diffraction data. However, experimental data only give the *internuclear distance* and we generally take this to be the sum of the ionic radii of the cation and anion (equation 5.6).

Internuclear distance between
a cation and the closest anion
in a lattice $\qquad = r_{cation} + r_{anion} \qquad (5.6)$

Equation 5.6 assumes a hard sphere model for the ions, with ions of opposite charge touching one another in the crystal lattice. Use of such an approximation means that the assignment of individual radii is somewhat arbitrary. Among many approaches to this problem we mention three.

Landé assumed that in the solid state structures of the lithium halides, LiX, the anions were in contact with one another (see *Figure 5.15a* and the corresponding discussion). Landé took half of each anion–anion distance to be the radius of that anion, and then obtained r_{Li^+} by substituting into equation 5.6 values of r_{X^-} and the measured internuclear Li–X distances.

Pauling considered a series of alkali metal halides, each member of which contained isoelectronic ions (NaF, KCl, RbBr, CsI). In order to partition the ionic radii, he assumed that the radius of each ion was inversely proportional to its actual nuclear charge less an amount due to screening effects. The latter were estimated using Slater's rules (see *Box 1.5*).

Goldschmidt and, more recently, Shannon and Prewitt, concentrated on the analysis of experimental data (mostly fluorides and oxides) with the aim of obtaining a set of ionic radii which, when combined in pairs (equation 5.6), reproduced the observed internuclear distances. In view of the approximate nature of the concept of the ionic radius, no great importance should be attached to small differences in quoted values so long as self-consistency is maintained in any one set of data. Further, some dependence of ionic size on coordination number is expected if we consider the different electrostatic interactions that a particular ion experiences in differing environments in an ionic crystal; r_{ion} for a given ion increases slightly with an increase in coordination number.

Values of ionic radii for selected ions are listed in Appendix 6. Ionic radii are sometimes quoted for species such as Si^{4+} and Cl^{7+}, but such data are highly artificial. The sums of the appropriate ionization energies of Si and Cl make it inconceivable that such ions exist in stable species. Nonetheless, a value for the radius of 'Cl^{7+}' can be calculated by subtracting $r_{O^{2-}}$ from the Cl–O internuclear distance on $[ClO_4]^-$.

We should mention that in the few cases in which the variation in electron density in a crystal has been accurately determined (e.g. NaCl), the minimum electron density does not in fact occur at distances from the nuclei indicated by the ionic radii in general use; e.g. in LiF and NaCl, the minima are found at 92 and 118 pm from the nucleus of the cation, whereas tabulated values of r_{Li^+} and r_{Na^+} are 76 and 102 pm, respectively. Such data make it clear that discussing lattice structures in terms of the ratio of the ionic radii is, at best, only a rough guide. For this reason, we restrict our discussion of *radius ratio rules* to that in Box 5.3.

Periodic trends in ionic radii

Figure 5.14 illustrates trends in ionic radii on descending representative groups and on crossing the first row of the *d*-block. In each case, r_{ion} corresponds to that of a 6-coordinate ion. The cation size increases on descending groups 1 and 2, as does the anion size on descending group 17. Figure 5.14 also allows comparisons of the relative sizes of cations and anions in alkali metal and alkaline earth metal halide salts (see *Section 5.11*).

The right-hand side of Figure 5.14 illustrates the small variation in size for M^{3+} and M^{2+} ions of the *d*-block metals. As expected, the decrease in nuclear charge in going from Fe^{3+} to Fe^{2+}, and from Mn^{3+} to Mn^{2+}, causes an increase in r_{ion}.

5.11 Ionic lattices

In this section we describe some common structure types adopted by ionic compounds of general formulae MX, MX_2 or M_2X, as well as that of the mineral *perovskite*, $CaTiO_3$. Such structures are usually determined by X-ray diffraction methods. Different ions scatter X-rays to differing extents depending on the total number of electrons in the ion and, consequently, different types of ions can generally be

---

BOX 5.3   **CHEMICAL AND THEORETICAL BACKGROUND**

Radius ratio rules

The structures of many ionic crystals can be rationalized *to a first approximation* by considering the relative sizes and relative numbers of the ions present. For monatomic ions, cations are *usually* smaller than anions (see Appendix 6), although examples such as KF and CsF show that this is not always true. The *radius ratio* $\dfrac{r_+}{r_-}$ can be used to make a first guess at the likely coordination number and geometry around the cation using a set of simple rules:

Value of $\dfrac{r_+}{r_-}$	Predicted coordination number of cation	Predicted coordination geometry of cation
<0.15	2	Linear
0.15–0.22	3	Trigonal planar
0.22–0.41	4	Tetrahedral
0.41–0.73	6	Octahedral
>0.73	8	Cubic

For a given compound stoichiometry, predictions about the coordination type of the cation necessarily make predictions about the coordination type of the anion. Use of radius ratios meets with some success, but there are *many* limitations. We can exemplify this by looking at the group 1 halides. The

ionic radii are as follows:

Cation	Li⁺	Na⁺	K⁺	Rb⁺	Cs⁺
r_+ / pm	76	102	138	149	170

Anion	F⁻	Cl⁻	Br⁻	I⁻
r_- / pm	133	181	196	220

For LiF, the radius ratio is 0.57 and so an octahedral coordination around the Li⁺ cation is predicted; for LiF, this corresponds to an NaCl type lattice, in agreement with the observed structure. In fact each of the group 1 halides (except CsCl, CsBr and CsI) at 298 K and 1 bar pressure adopts the NaCl type structure; CsCl, CsBr and CsI adopt the CsCl type lattice. Radius ratio rules predict the correct structures in only some cases; they predict tetrahedral coordination for the cations in LiBr and LiI, and cubic coordination in NaF, KF, KCl, RbF, RbCl, RbBr and CsF (in addition to CsCl, CsBr and CsI). Radius ratio rules give only one prediction for any one ionic crystal, and some compounds undergo phase changes under the influence of temperature and pressure, e.g. when CsCl is sublimed onto an amorphous surface, it crystallizes with the NaCl structure and, under high pressure conditions, RbCl adopts a CsCl type lattice.

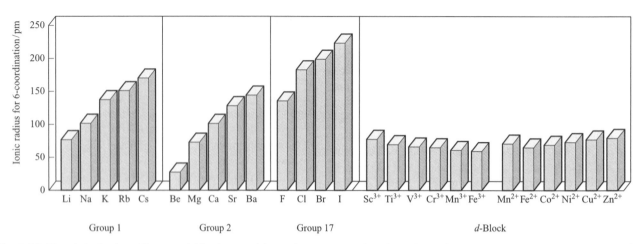

Fig. 5.14 Trends in ionic radii, r_{ion}, within the metal ions of groups 1 and 2, the anions of group 17, and metal ions from the first row of the *d*-block.

distinguished from one another. Use of X-ray diffraction methods does have some limitations. Firstly, the location of light atoms (e.g. H) in the presence of much heavier atoms is difficult and, sometimes, impossible. Neutron diffraction (in which neutrons are diffracted by *nuclei*) may be used as a complementary technique. Secondly, X-ray diffraction is seldom able to identify the state of ionization of the species present; only for a few substances (e.g. NaCl) has the electron density distribution been determined with sufficient accuracy for this purpose.

Throughout our discussion, we refer to 'ionic' lattices, suggesting the presence of discrete ions. Although a *spherical ion model* is used to describe the structures, we shall see in Section 5.13 that this picture is unsatisfactory for some compounds in which covalent contributions to the bonding are significant. Useful as the hard sphere model is in acquiring a basic grasp of common crystal structure types, it must be clearly understood that it is at odds with modern quantum theory. As we saw in Chapter 1, the wavefunction of an electron does not suddenly drop to zero with increasing distance from the nucleus, and in a close-packed or any other crystal, there is a finite electron density everywhere. Thus *all treatments of the solid state based upon the hard sphere model are approximations.*

Each structure type is designated by the name of one of the compounds crystallizing with that structure, and phrases such as 'CaO adopts an NaCl structure' are commonly found in the chemical literature.

The rock salt (NaCl) lattice

In salts of formula MX, the coordination numbers of M and X must be *equal*.

Rock salt (or halite, NaCl) occurs naturally as cubic crystals, which, when pure, are colourless or white. Figure 5.15 shows two representations of the unit cell (see *Section 5.2*) of NaCl. Figure 5.15a illustrates the way in which the ions occupy the space available; the larger Cl^- ions ($r_{Cl^-} = 181$ pm) define an

fcc arrangement with the Na^+ ions ($r_{Na^+} = 102$ pm) occupying the octahedral holes. This description relates the structure of the ionic lattice to the close-packing-of-spheres model. Such a description is often employed, but is not satisfactory for salts such as KF; while this adopts an NaCl lattice, the K^+ and F^- ions are almost the same size ($r_{K^+} = 138$, $r_{F^-} = 133$ pm) (see *Box 5.3*). Although Figure 5.15a is relatively realistic, it hides most of the structural details of the unit cell and is difficult to reproduce. The more open representation shown in Figure 5.15b tends to be more useful.

The complete NaCl lattice is built up by placing unit cells next to one another so that ions residing in the corner, edge or face sites (Figure 5.15b) are *shared* between adjacent unit cells. Bearing this in mind, Figure 5.15b shows that *each* Na^+ and Cl^- ion is 6-coordinate in the crystal lattice, while within a single unit cell, the octahedral environment is defined completely only for the central Na^+ ion.

Figure 5.15b is not a unique representation of a unit cell of the NaCl lattice. It is equally valid to draw a unit cell with Na^+ ions in the corner sites; such a cell has a Cl^- ion in the unique central site. This shows that the Na^+ ions are also in an fcc arrangement, and the NaCl lattice could therefore be described in terms of two interpenetrating fcc lattices, one consisting of Na^+ ions and one of Cl^- ions.

Among the many compounds that crystallize with the NaCl lattice are NaF, NaBr, NaI, NaH, halides of Li, K and Rb, AgF, AgCl, AgBr, MgO, CaO, SrO, BaO, MnO, CoO, NiO, MgS, CaS, SrS and BaS.

Worked example 5.2 Compound stoichiometry from a unit cell

Show that the structure of the unit cell for sodium chloride (Figure 5.15b) is consistent with the formula NaCl.

In Figure 5.15b, 14 Cl^- ions and 13 Na^+ ions are shown. However, all but one of the ions is shared between two or more unit cells.

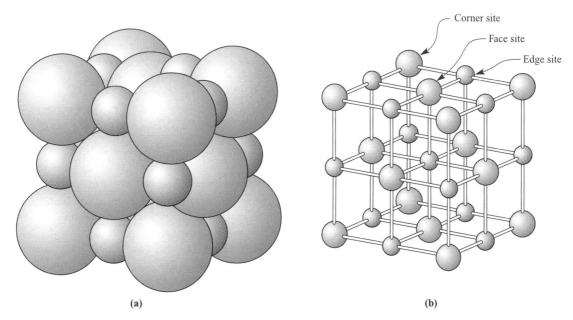

(a) **(b)**

Fig. 5.15 Two representations of the unit cell of NaCl: (a) shows a space-filling representation, and (b) shows a 'ball-and-stick' representation which reveals the coordination environments of the ions. The Cl^- ions are shown in grey and the Na^+ ions in blue; since both types of ions are in equivalent environments, a unit cell with Na^+ ions in the corner sites is also valid. There are four types of site in the unit cell: central (not labelled), face, edge and corner positions.

There are four types of site:

- unique central position (the ion belongs entirely to the unit cell shown);
- face site (the ion is shared between two unit cells);
- edge sites (the ion is shared between four unit cells);
- corner site (the ion is shared between eight unit cells).

The total number of Na^+ and Cl^- ions belonging to the unit cell is calculated as follows:

Site	Number of Na^+ ions	Number of Cl^-
Central	1	0
Face	0	$(6 \times \frac{1}{2}) = 3$
Edge	$(12 \times \frac{1}{4}) = 3$	0
Corner	0	$(8 \times \frac{1}{8}) = 1$
TOTAL	4	4

The ratio of Na^+:Cl^- ions is $4:4 = 1:1$
This ratio is consistent with the formula NaCl.

Self-study exercises

1. Show that the structure of the unit cell for caesium chloride (Figure 5.16) is consistent with the formula CsCl.

2. MgO adopts an NaCl lattice. How many Mg^{2+} and O^{2-} ions are present per unit cell? [*Ans.* 4 of each]

3. The unit cell of AgCl (NaCl type lattice) can be drawn with Ag^+ ions at the corners of the cell, or Cl^- at the corners. Confirm that the number of Ag^+ and Cl^- ions per unit cell remains the same whichever arrangement is considered.

The caesium chloride (CsCl) lattice

In the CsCl lattice, each ion is surrounded by eight others of opposite charge. A single unit cell (Figure 5.16a) makes the connectivity obvious only for the central ion. However, by extending the lattice, one sees that it is constructed of interpenetrating cubes (Figure 5.16b), and the coordination number of each ion is seen. Because the Cs^+ and Cl^- ions are in the same environments, it is valid to draw a unit cell either with Cs^+ or Cl^- at the corners of the cube. Note the relationship between the structure of the unit cell and bcc packing.

The CsCl structure is relatively uncommon but is also adopted by CsBr, CsI, TlCl and TlBr. At 298 K, NH_4Cl and NH_4Br possess CsCl lattices; $[NH_4]^+$ is treated as a spherical ion (Figure 5.17), an approximation that can

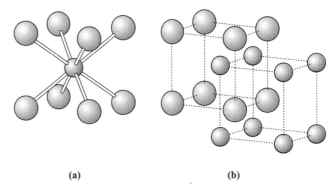

(a) **(b)**

Fig. 5.16 (a) The unit cell of CsCl; Cs^+ ions are shown in grey and Cl^- in blue, but the unit cell could also be drawn with the Cs^+ ion in the central site. (b) One way to describe the CsCl lattice is in terms of interpenetrating cubic units of Cs^+ and Cl^- ions.

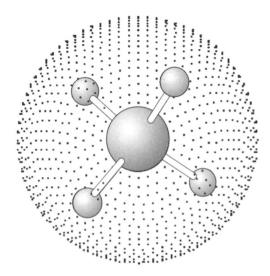

Fig. 5.17 The $[NH_4]^+$ ion can be treated as a sphere in descriptions of solid state lattices; some other ions (e.g. $[BF_4]^-$, $[PF_6]^-$) can be treated similarly.

made for a number of simple ions in the solid state due to their rotating or lying in random orientations about a fixed point. Above 457 and 411 K respectively, NH_4Cl and NH_4Br adopt NaCl lattices.

The fluorite (CaF_2) lattice

In salts of formula MX_2, the coordination number of X must be *half* that of M.

Calcium fluoride occurs naturally as the mineral *fluorite* (fluor-spar). Figure 5.18a shows a unit cell of CaF_2. Each cation is 8-coordinate and each anion 4-coordinate; six of the Ca^{2+} ions are shared between two unit cells and the 8-coordinate environment can be appreciated by envisaging two adjacent unit cells. (*Exercise*: How does the coordination number of 8 for the remaining Ca^{2+} ions arise?) Other compounds that adopt this lattice type include group 2 metal fluorides, $BaCl_2$, and the dioxides of the *f*-block metals.

The antifluorite lattice

If the cation and anion sites in Figure 5.18a are exchanged, the coordination number of the anion becomes *twice* that of the anion, and it follows that the compound formula is M_2X. This arrangement corresponds to the antifluorite structure, and is adopted by the group 1 metal oxides and sulfides of type M_2O and M_2S; Cs_2O is an exception.

The zinc blende (ZnS) lattice: a diamond-type network

Figure 5.18b shows the structure of zinc blende (ZnS). A comparison of this with Figure 5.18a reveals a relationship between the structures of zinc blende and CaF_2; in going from Figure 5.18a to 5.18b, half of the anions are removed and the ratio of cation:anion changes from 1:2 to 1:1.

An alternative description is that of a *diamond-type network*. Figure 5.19a gives a representation of the structure of diamond; each C atom is tetrahedrally sited and the structure is very rigid. This structure type is also adopted by Si, Ge and α-Sn (grey tin). Figure 5.19b (with atom labels that relate it to Figure 5.19a) shows a view of the diamond network that is comparable with the unit cell of zinc blende in Figure 5.18b. Every other site in the diamond-type array is occupied by either a zinc or a sulfur centre. The fact that we are comparing the structure of an apparently ionic compound (ZnS) with that of a covalently bonded species should not cause concern. As we have already mentioned, the hard sphere ionic model is a convenient approximation but does not allow for the fact that the bonding in many compounds such as ZnS is neither wholly ionic nor wholly covalent.

At 1296 K, zinc blende undergoes a transition to wurtzite, the structure of which we consider later; zinc blende and wurtzite are *polymorphs*. Zinc(II) sulfide occurs naturally both as zinc blende (also called *sphalerite*) and wurtzite, although the former is more abundant and is the major ore for Zn production. Although zinc blende is thermodynamically favoured at 298 K by 13 kJ mol^{-1}, the transition from

(a)

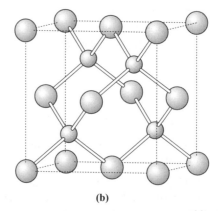

(b)

Fig. 5.18 (a) The unit cell of CaF_2; the Ca^{2+} ions are shown in grey and the F^- ions in blue. (b) The unit cell of zinc blende (ZnS); the zinc centres are shown in grey and the sulfur centres in blue. Both sites are equivalent, and the unit cell could be drawn with the S^{2-} ions in the grey sites.

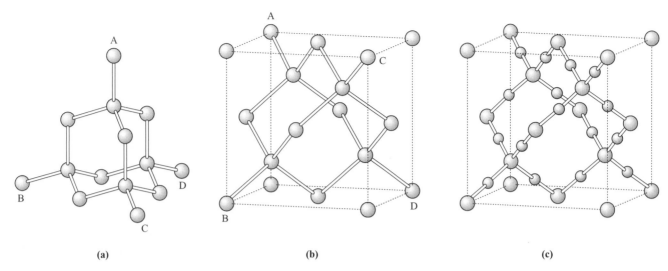

(a)   (b)   (c)

Fig. 5.19 (a) A typical representation of the diamond lattice. (b) Reorientation of the network shown in (a) provides a representation that can be compared with the unit cell of zinc blende (Figure 5.18b); the atom labels correspond to those in diagram (a). This lattice is also adopted by Si, Ge and α-Sn. (c) The unit cell of β-cristobalite, SiO_2, with the oxygen atoms shown in blue.

wurtzite to zinc blende is *extremely* slow allowing both minerals to exist in nature. This scenario resembles that of the diamond ⟶ graphite transition (see *Chapter 13*), graphite being thermodynamically favoured at 298 K. If the latter transition were *not* infinitesimally slow, diamonds would lose their place in the world gemstone market!

The β-cristobalite (SiO_2) lattice

Before discussing the structure of wurtzite, we consider β-cristobalite, the structure of which is related to that of the diamond-type network. β-Cristobalite is one of several forms of SiO_2 (see *Figure 13.13*). Figure 5.19c shows the unit cell of the β-cristobalite lattice; comparison with Figure 5.19b shows that it is related to the structure of Si by placing an O atom between adjacent Si atoms. The idealized structure shown in Figure 5.19c has an Si−O−Si bond angle of 180° whereas in practice this angle is 147° (almost the same as in

$(SiH_3)_2O$, ∠Si−O−Si = 144°) indicating that the bonding in SiO_2 is *not* purely electrostatic.

The wurtzite (ZnS) structure

Wurtzite is a second polymorph of ZnS; in contrast to the cubic symmetry of zinc blende, wurtzite has hexagonal symmetry. In the three unit cells shown in Figure 5.20, the 12 ions in corner sites define a hexagonal prism. Each of the zinc and sulfur centres is tetrahedrally sited, and a unit cell in which Zn^{2+} and S^{2-} are interchanged with respect to Figure 5.20 is equally valid.

The rutile (TiO_2) structure

The mineral rutile occurs in granite rocks and is an important industrial source of TiO_2 (see *Box 21.3*). Figure 5.21 shows the unit cell of rutile. The coordination numbers of titanium and

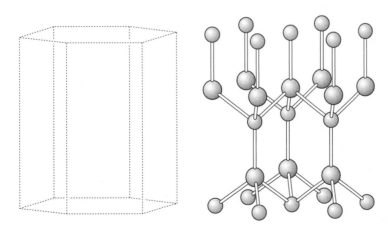

Fig. 5.20 Three unit cells of wurtzite (a second polymorph of ZnS) defines a hexagonal prism; the Zn^{2+} ions are shown in grey and the S^{2-} ions in blue. Both ions are tetrahedrally sited and an alternative unit cell could be drawn by interchanging the ion positions.

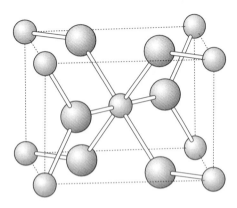

Fig. 5.21 The unit cell of rutile (one polymorph of TiO_2); the titanium centres are shown in grey and the oxygen centres in blue.

oxygen are six (octahedral) and three (trigonal planar) respectively, consistent with the 1:2 stoichiometry of rutile. Two of the O^{2-} ions shown in Figure 5.21 reside fully within the unit cell, while the other four are in face-sharing positions.

The rutile lattice is adopted by SnO_2 (*cassiterite*, the most important tin-bearing mineral), MnO_2 (*pyrolusite*) and PbO_2.

The CdI_2 and $CdCl_2$ lattices: layer structures

Many compounds of formula MX_2 crystallize in so-called *layer structures*, a typical one being CdI_2 which has hexagonal symmetry. This lattice can be described in terms of I^- ions arranged in an hcp array with Cd^{2+} ions occupying the octahedral holes in every other layer (Figure 5.22 in which the hcp array is denoted by the ABAB layers). Extending the lattice infinitely gives a structure which can be

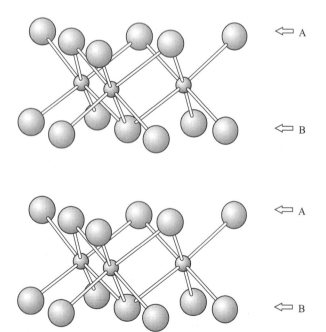

Fig. 5.22 Parts of two layers of the CdI_2 lattice; Cd^{2+} ions are shown in blue and I^- ions in grey. The I^- ions are arranged in an hcp array.

described in terms of 'stacked sandwiches', each 'sandwich' consisting of a layer of I^- ions, a parallel layer of Cd^{2+} ions, and another parallel layer of I^- ions; each 'sandwich' is electrically neutral. Only weak van der Waals forces operate between the 'sandwiches' (the central gap between the layers in Figure 5.22) and this leads to CdI_2 crystals exhibiting pronounced cleavage planes parallel to the layers.

> If a crystal breaks along a plane related to the lattice structure, the plane is called a *cleavage plane*.

Other compounds crystallizing with a CdI_2 lattice include $MgBr_2$, MgI_2, CaI_2, iodides of many *d*-block metals, and many metal hydroxides including $Mg(OH)_3$ (the mineral *brucite*) in which the $[OH]^-$ ions are treated as spheres for the purposes of structural description.

The $CdCl_2$ lattice is related to the CdI_2 layer-structure but with the Cl^- ions in a *cubic* close-packed arrangement. Examples of compounds adopting this structure are $FeCl_2$ and $CoCl_2$. Other layer structures include *talc* and *mica* (see *Section 13.9*).

The perovskite ($CaTiO_3$) lattice: a double oxide

Perovskite is an example of a *double oxide*; it does not, as the formula might imply, contain $[TiO_3]^{2-}$, but is a mixed Ca(II) and Ti(IV) oxide. Figure 5.23a shows one representation of a unit cell of perovskite (see *problem 5.13* at the end of the chapter). The cell is cubic, with Ti(IV) centres at the corners of the cube, and O^{2-} ions in the 12 edge sites. The 12-coordinate Ca^{2+} ion lies at the centre of the unit cell. Each Ti(IV) centre is six-coordinate, and this can be appreciated by considering the assembly of adjacent unit cells in the crystal lattice.

Many double oxides or fluorides such as $BaTiO_3$, $SrFeO_3$, $NaNbO_3$, $KMgF_3$ and $KZnF_3$ crystallize with a perovskite lattice. Deformations of the lattice may be caused as a consequence of the relative sizes of the ions, e.g. in $BaTiO_3$, the Ba^{2+} ion is relatively large ($r_{Ba^{2+}} = 142\,pm$ compared with $r_{Ca^{2+}} = 100\,pm$) and causes a displacement of each Ti(IV) centre such that there is one short Ti−O contact. This leads to $BaTiO_3$ possessing *ferroelectric* properties (see *Section 27.6*).

The structures of some high temperature superconductors are also related to that of perovskite. Another mixed oxide lattice is that of *spinel*, $MgAl_2O_4$ (see *Box 12.6*).

5.12 Crystal structures of semiconductors

This section draws attention to some of the common lattice types adopted by semiconductors. The diamond-type network (often referred to an *adamantine solid* structure) is adopted by Si and Ge; the addition of dopants occurs without structural change. Related to this network is the zinc

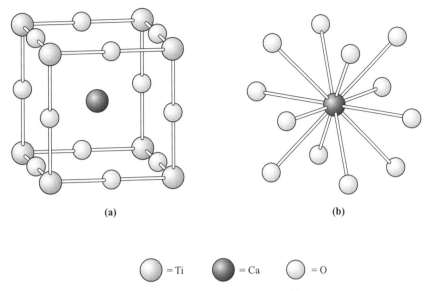

Fig. 5.23 (a) One representation of a unit cell of perovskite ($CaTiO_3$); (b) the Ca^{2+} ion is 12-coordinate with respect to the O^{2-} ions.

blende lattice and among compounds adopting this structure are GaAs, InAs, GaP, ZnSe, ZnTe, CdS, CdSe, CdTe, HgS, HgSe and HgTe. Each binary compound (including zinc blende) is an intrinsic semiconductor. The wurtzite lattice is also important in semiconducting materials; ZnO, CdSe and InN are examples of compounds adopting this structure.

5.13 Lattice energy: estimates from an electrostatic model

The *lattice energy*, $\Delta U(0\,K)$, of an ionic compound is the change in internal energy that accompanies the formation of one mole of the solid from its constituent gas-phase ions at $0\,K$.

For a salt MX_n, equation 5.7 defines the reaction, the energy change for which corresponds to the lattice energy.

$$M^{n+}(g) + nX^-(g) \longrightarrow MX_n(s) \qquad (5.7)$$

The lattice energy can be *estimated* by assuming an electrostatic model for the solid state ionic lattice; the ions are considered to be point charges. Later in this chapter, we consider to what extent this approximation is true.

Coulombic attraction within an isolated ion-pair

Before we consider an ionic lattice, let us review the appropriate equation for the change in internal energy when two oppositely charged ions M^{z+} and X^{z-} are brought together from infinite separation to form the *isolated ion-pair*, MX (equation 5.8).

$$M^{z+}(g) + X^{z-}(g) \longrightarrow MX(g) \qquad (5.8)$$

Let the ions carry charges of z_+e and z_-e where e is the electronic charge and z_+ and z_- are integers. The ions attract each other, and energy is released as the ion-pair is formed. The change in internal energy can be estimated from equation 5.9 by considering the Coulombic attraction between the ions. For an isolated ion-pair:

$$\Delta U = -\left(\frac{|z_+|\,|z_-|e^2}{4\pi\varepsilon_0 r}\right) \qquad (5.9)$$

where: ΔU = change in internal energy (unit = joules); $|z_+|$ = modulus[†] of the positive charge (for K^+, $|z_+| = 1$; for Mg^{2+}, $|z_+| = 2$); $|z_-|$ = modulus[†] of the negative charge (for F^-, $|z_-| = 1$; for O^{2-}, $|z_-| = 2$); e = charge on the electron = 1.602×10^{-19} C; ε_0 = permittivity of a vacuum = 8.854×10^{-12} F m^{-1}; r = internuclear distance between the ions (units = m).

Coulombic interactions in an ionic lattice

Now consider a salt MX which has an NaCl lattice. A study of the coordination geometry in Figure 5.15 (remembering that the lattice extends indefinitely) shows that each M^{z+} ion is surrounded by:

- 6 X^{z-} ions, each at a distance r
- 12 M^{z+} ions, each at a distance $\sqrt{2}r$
- 8 X^{z-} ions, each at a distance $\sqrt{3}r$
- 6 M^{z+} ions, each at a distance $\sqrt{4}r = 2r$

and so on.

The change in Coulombic energy when an M^{z+} ion is brought from infinity to its position in the lattice is given by equation 5.10.

[†] The modulus of a real number is its *positive* value, e.g. $|z_+|$ and $|z_-|$ are both positive.

$$\Delta U = -\frac{e^2}{4\pi\varepsilon_0}\left[\left(\frac{6}{r}|z_+||z_-|\right) - \left(\frac{12}{\sqrt{2}r}|z_+|^2\right)\right.$$
$$\left. + \left(\frac{8}{\sqrt{3}r}|z_+||z_-|\right) - \left(\frac{6}{\sqrt{4}r}|z_+|^2\right)\cdots\right]$$
$$= -\frac{|z_+||z_-|e^2}{4\pi\varepsilon_0 r}\left[6 - \left(\frac{12|z_+|}{\sqrt{2}|z_-|}\right) + \left(\frac{8}{\sqrt{3}}\right)\right.$$
$$\left. - \left(3\frac{|z_+|}{|z_-|}\right)\cdots\right] \tag{5.10}$$

The ratio of the charges on the ions, $\frac{|z_+|}{|z_-|}$, is constant for a given type of structure (e.g. 1 for NaCl) and so the series in square brackets in equation 5.10 (which slowly converges and may be summed algebraically) is a function only of the crystal geometry. Similar series can be written for other crystal lattices, but for a particular structure type, the series is independent of $|z_+|$, $|z_-|$ and r. Erwin Madelung first evaluated such series in 1918, and the values appropriate for various lattice types are *Madelung constants, A* (see *Table 5.4*). Equation 5.10 can therefore be written in the more simple form of equation 5.11, in which the lattice energy is estimated in joules *per mole* of compound.

$$\Delta U = -\frac{LA|z_+||z_-|e^2}{4\pi\varepsilon_0 r} \tag{5.11}$$

where L = Avogadro number = $6.022 \times 10^{23}\,\mathrm{mol}^{-1}$, and A = Madelung constant (no units)

Although we have derived this expression by considering the ions that surround M^{z+}, the same equation results by starting from a central X^{z-} ion.

Born forces

Coulombic interactions are not the only forces operating in a real ionic lattice. The ions have finite size, and electron–electron and nucleus–nucleus repulsions also arise; these are *Born forces*. Equation 5.12 gives the simplest expression for the increase in repulsive energy upon assembling the lattice from gaseous ions.

$$\Delta U = \frac{LB}{r^n} \tag{5.12}$$

where B = repulsion coefficient, and n = Born exponent.

Values of the Born exponent (Table 5.3) can be evaluated from compressibility data and depend on the electronic configurations of the ions involved; effectively, this says that n shows a dependence on the sizes of the ions.

> **Worked example 5.3 Born exponents**

Using the values given in Table 5.3, determine an appropriate Born exponent for BaO.

Table 5.3 Values of the Born exponent, n, given for an ionic compound MX in terms of the electronic configuration of the ions [M$^+$][X$^-$]. The value of n for an ionic compound is determined by averaging the component values, e.g. for MgO, $n = 7$; for LiCl, $n = \frac{5+9}{2} = 7$.

Electronic configuration of the ions in an ionic compound MX	Examples of ions	n (no units)
[He][He]	H$^-$, Li$^+$	5
[Ne][Ne]	F$^-$, O^{2-}, Na$^+$, Mg^{2+}	7
[Ar][Ar], or [3d^{10}][Ar]	Cl$^-$, S^{2-}, K$^+$, Ca^{2+}, Cu$^+$	9
[Kr][Kr] or [4d^{10}][Kr]	Br$^-$, Rb$^+$, Sr^{2+}, Ag$^+$	10
[Xe][Xe] or [5d^{10}][Xe]	I$^-$, Cs$^+$, Ba^{2+}, Au$^+$	12

Ba^{2+} is isoelectronic with Xe, and so $n = 12$
O^{2-} is isoelectronic with Ne, and $n = 7$
The value of n for BaO $= \dfrac{12+7}{2} = 9.5$

Self-study exercises:

Use data in Table 5.3.

1. Calculate an appropriate Born exponent for MgO. [*Ans.* 7]

2. Calculate an appropriate Born exponent for AgF.
   [*Ans.* 8.5]

3. What is the change in the Born exponent in going from BaO to SrO?
   [*Ans.* −1]

The Born–Landé equation

In order to write an expression for the lattice energy that takes into account both the Coulombic and Born interactions in an ionic lattice, we combine equations 5.11 and 5.12 to give equation 5.13.

$$\Delta U(0\,\mathrm{K}) = -\frac{LA|z_+||z_-|e^2}{4\pi\varepsilon_0 r} + \frac{LB}{r^n} \tag{5.13}$$

We evaluate B in terms of the other components of the equation by making use of the fact that at the equilibrium separation where $r = r_0$, the differential $\dfrac{\mathrm{d}\Delta U}{\mathrm{d}r} = 0$. Differentiating with respect to r gives equation 5.14, and rearrangement gives an expression for B (equation 5.15).

$$0 = \frac{LA|z_+||z_-|e^2}{4\pi\varepsilon_0 r_0^2} - \frac{nLB}{r_0^{n+1}} \tag{5.14}$$

$$B = \frac{A|z_+||z_-|e^2 r_0^{n-1}}{4\pi\varepsilon_0 n} \tag{5.15}$$

Combining equations 5.13 and 5.15 gives an expression for the lattice energy that is based on an electrostatic model

Table 5.4 Madelung constants, A, for selected lattice types. Values of A are numerical and have no units.

Lattice type	A
Sodium chloride (NaCl)	1.7476
Caesium chloride (CsCl)	1.7627
Wurtzite (α-ZnS)	1.6413
Zinc blende (β-ZnS)	1.6381
Fluorite (CaF_2)	2.5194
Rutile (TiO_2)	2.408[a]
Cadmium iodide (CdI_2)	2.355[a]

[a] For these structures, the value depends slightly on the lattice parameters for the unit cell.

and takes into account Coulombic attractions, Coulombic repulsions and Born repulsions between ions in the crystal lattice. Equation 5.16 is the *Born–Landé equation*.

$$\Delta U(0\,\mathrm{K}) = -\frac{LA|z_+||z_-|e^2}{4\pi\varepsilon_0 r_0}\left(1 - \frac{1}{n}\right) \quad (5.16)$$

Because of its simplicity, the Born–Landé expression is the one that chemists tend to use; many chemical problems involve the use of estimated lattice energies, e.g. for hypothetical compounds. Often lattice energies are incorporated into thermochemical cycles, and so an associated *enthalpy* change is needed (see *Section 5.14*).

Madelung constants

Values of Madelung constants for selected lattices are given in Table 5.4. Remembering that these values are derived by considering the coordination environments (near and far neighbours) of ions in the crystal lattice, it may seem surprising that, for example, the values for the NaCl and CsCl lattices (Figures 5.15 and 5.16) are similar. This is simply a consequence of the infinite nature of the structures: although the first (attractive) term in the algebraic series for A is greater by a factor of $\frac{8}{6}$ for the CsCl lattice, the second (repulsive) term is also greater, and so on.

Table 5.4 shows that Madelung constants for MX_2 structures are $\approx 50\%$ higher than those for MX lattices. We return to this difference in Section 5.16.

Worked example 5.4 Use of the Born–Landé equation

Sodium fluoride adopts the NaCl type lattice. Estimate the lattice energy of NaF using an electrostatic model.

Data required:

$L = 6.022 \times 10^{23}\,\mathrm{mol}^{-1}$ $A = 1.7476$ $e = 1.602 \times 10^{-19}\,\mathrm{C}$
$\varepsilon_0 = 8.854 \times 10^{-12}\,\mathrm{F\,m}^{-1}$ Born exponent for NaF = 7
Internuclear Na–F distance = 231 pm

The change in internal energy (the lattice energy) is given by the Born–Landé equation:

$$\Delta U(0\,\mathrm{K}) = -\frac{LA|z_+||z_-|e^2}{4\pi\varepsilon_0 r_0}\left(1 - \frac{1}{n}\right)$$

r must be in m: 231 pm $= 2.31 \times 10^{-10}\,\mathrm{m}$

$$\Delta U_0 = -\left(\frac{\begin{array}{c}6.022 \times 10^{23} \times 1.7476 \times 1 \\ \times 1 \times (1.602 \times 10^{-19})^2\end{array}}{4 \times 3.142 \times 8.854 \times 10^{-12} \times 2.31 \times 10^{-10}}\right) \times \left(1 - \frac{1}{7}\right)$$

$$= -900\,624\,\mathrm{J\,mol}^{-1}$$

$$\approx -901\,\mathrm{kJ\,mol}^{-1}$$

Self-study exercises

1. Estimate the lattice energy of KF (NaCl lattice) using an electrostatic model; the K–F internuclear separation is 266 pm.
   [*Ans.* $-798\,\mathrm{kJ\,mol}^{-1}$]

2. By assuming an electrostatic model, estimate the lattice energy of MgO (NaCl lattice); values of r_{ion} are listed in Appendix 6.
   [*Ans.* $-3926\,\mathrm{kJ\,mol}^{-1}$]

Refinements to the Born–Landé equation

Lattice energies obtained from the Born–Landé equation are *approximate*, and for more accurate evaluations of their values, several improvements to the equation can be made. The most important of these arises by replacing the $\frac{1}{r^n}$ term in equation 5.12 by $e^{-\frac{r}{\rho}}$, a change reflecting the fact that wavefunctions show an exponential dependence on r; ρ is a constant that can be expressed in terms of the compressibility of the crystal. This refinement results in the lattice energy being given by the *Born–Mayer equation* (equation 5.17).

$$\Delta U(0\,\mathrm{K}) = -\frac{LA|z_+||z_-|e^2}{4\pi\varepsilon_0 r_0}\left(1 - \frac{\rho}{r_0}\right) \quad (5.17)$$

The constant ρ has a value of 35 pm for all alkali metal halides. Note that r_0 appears in the Born repulsive term (compare equations 5.16 and 5.17).

Further refinements in lattice energy calculations include the introduction of terms for the *dispersion energy* and the *zero-point energy* (see *Section 2.9*). Dispersion forces[†] arise from momentary fluctuations in electron density which produce temporary dipole moments that, in turn, induce dipole moments in neighbouring species. Dispersion forces are also referred to as *induced-dipole–induced-dipole interactions*. They are non-directional and give rise to a dispersion

---

[†] Dispersion forces are also known as London dispersion forces.

energy that is related to the internuclear separation, r, and the *polarizability*, α, of the atom (or molecule) according to equation 5.18.

$$\text{Dispersion energy} \propto \frac{\alpha}{r^6} \tag{5.18}$$

The polarizability of a species is a measure of the degree to which it may be distorted, e.g. by the electric field due to an adjacent atom or ion. In the hard sphere model of ions in lattices, we assume that there is no polarization of the ions. This is a gross approximation. The polarizability increases rapidly with an increase in atomic size, and large ions (or atoms or molecules) give rise to relatively large induced dipoles and, thus, significant dispersion forces. Values of α can be obtained from measurements of the relative permittivity (*dielectric constant*, see *Section 8.2*) or the refractive index of the substance in question.

In NaCl, the contributions to the total lattice energy ($-766\,\text{kJ mol}^{-1}$) made by electrostatic attractions, electrostatic and Born repulsions, dispersion energy and zero-point energy are -860, $+99$, -12 and $+7\,\text{kJ mol}^{-1}$ respectively. In fact, the error introduced by neglecting the last two terms (which always tend to compensate each other) is very small.

Overview

Lattice energies derived using the electrostatic model are often referred to as 'calculated' values to distinguish them from values obtained using thermochemical cycles. It should, however, be appreciated that values of r_0 obtained from X-ray diffraction studies are *experimental* quantities and may conceal departures from ideal ionic behaviour. In addition, the actual charges on ions may well be less than their formal charges. Nevertheless, the concept of lattice energy is of immense importance in inorganic chemistry.

5.14 Lattice energy: the Born–Haber cycle

By considering the *definition* of lattice energy, it is easy to see why these quantities are not measured directly. However, an associated *lattice enthalpy* of a salt can be related to several other quantities by a thermochemical cycle called the *Born–Haber cycle*. If the anion in the salt is a halide, then all the other quantities in the cycle have been determined independently; the reason for this statement will become clearer when we look at applications of lattice energies in Section 5.16.

Let us consider a general metal halide MX_n. Figure 5.24 shows a thermochemical cycle describing the formation of crystalline MX_n from its constituent elements in their standard states. The quantity $\Delta_{\text{lattice}}H^\circ(298\,\text{K})$ is the enthalpy change that accompanies the formation of the crystalline salt from the gaseous ions under standard conditions. The same approximation is made as for ionization energies and electron affinities (see *Section 1.10*), i.e. $\Delta U(0\,\text{K}) \approx \Delta H(298\,\text{K})$; relatively little error is introduced by using this approximation. A value of $\Delta_{\text{lattice}}H^\circ$ can be determined using equation 5.19 (by application of Hess's Law of constant heat summation) and represents an *experimental value* since it is derived from experimentally determined data.

$$\Delta_f H^\circ(MX_n,s) = \Delta_a H^\circ(M,s) + \frac{n}{2}D(X_2,g)$$
$$+ \Sigma IE(M,g) + n\Delta_{EA}H(X,g)$$
$$+ \Delta_{\text{lattice}}H^\circ(MX_n,s) \tag{5.19}$$

Rearranging this expression and introducing the approximation that the lattice energy $\Delta U(0\,\text{K}) \approx \Delta_{\text{lattice}}H(298\,\text{K})$ gives equation 5.20. All the quantities on the right-hand side of the equation are obtained from standard tables of data. (Enthalpies of atomization: see Appendix 10;

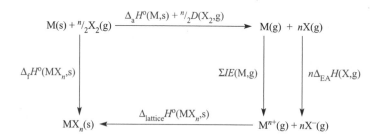

$\Delta_a H^\circ(M,s)$	= Enthalpy of atomization of metal M
$D(X_2,g)$	= Dissociation enthalpy of X_2 = 2 × Enthalpy of atomization of X
$\Sigma IE(M,g)$	= Sum of the ionization energies for the processes $M(g) \rightarrow M^+(g) \rightarrow M^{2+}(g)... \rightarrow M^{n+}(g)$
$\Delta_{EA}H(X,g)$	= Enthalpy change associated with the attachment of an electron
$\Delta_f H^\circ(MX_n,s)$	= Standard enthalpy of formation
$\Delta_{\text{lattice}}H^\circ(MX_n,s)$	= Lattice enthalpy change (see text)

Fig. 5.24 A Born–Haber thermochemical cycle for the formation of a salt MX_n. This gives an *enthalpy change* associated with the formation of the ionic lattice MX_n.

ionization energies: see Appendix 8; electron affinities: see Appendix 9.)

$$\Delta U(0\,K) \approx \Delta_f H^\circ(MX_n,s) - \Delta_a H^\circ(M,s)$$

$$- \frac{n}{2} D(X_2,g) - \Sigma IE(M,g) - n\Delta_{EA}H(X,g) \tag{5.20}$$

Worked example 5.5 Application of the Born–Haber cycle

Given that the standard enthalpy of formation at 298 K of CaF_2 is $-1228\,kJ\,mol^{-1}$, determine the lattice energy for CaF_2 using appropriate data from the Appendices.

First, construct an appropriate thermochemical cycle:

Values that need to be found in the Appendices are:

Appendix 10: $\Delta_a H^\circ(Ca,s) = 178\,kJ\,mol^{-1}$
$D(F_2,g) = 2\Delta_a H^\circ(F,g) = 158\,kJ\,mol^{-1}$
Appendix 8: $IE_1(Ca,g) = 590$; $IE_2(Ca,g) = 1145\,kJ\,mol^{-1}$
Appendix 9: $\Delta_{EA}H(F,g) = -328\,kJ\,mol^{-1}$

Use of Hess's Law gives:

$$\Delta U(0\,K) \approx \Delta_f H^\circ(CaF_2,s) - \Delta_a H^\circ(Ca,s)$$

$$- D(F_2,g) - \Sigma IE(Ca,g) - 2\Delta_{EA}H(F,g)$$

$$\approx -1228 - 178 - 158 - 590 - 1145 + 2(328)$$

$$\Delta U(0\,K) \approx -2643\,kJ\,mol^{-1}$$

Self-study exercises

Use data from the Appendices.

1. If $\Delta_f H^\circ(298\,K)$ for $CaCl_2 = -795\,kJ\,mol^{-1}$, determine its lattice energy. [*Ans.* $-2252\,kJ\,mol^{-1}$]

2. If the lattice energy of $CsF = -744\,kJ\,mol^{-1}$, determine $\Delta_f H^\circ(298\,K)$ for the compound. [*Ans.* $-539\,kJ\,mol^{-1}$]

5.15 Lattice energy: 'calculated' versus 'experimental' values

If we take NaCl as a typical example, $\Delta U(0\,K)$ determined by using a Born–Haber cycle is $-783\,kJ\,mol^{-1}$. The value calculated (using an experimental value of r_0 from X-ray diffraction data) from the Born–Mayer equation is $-761\,kJ\,mol^{-1}$; a more refined calculation, the basis of which was outlined in Section 5.13, gives $-768\,kJ\,mol^{-1}$. This level of agreement is observed for all the alkali metal halides (including those of Li), and for the group 2 metal fluorides. While this is not rigid proof that all these compounds are wholly ionic, the close agreement does support our use of the electrostatic model as a basis for discussing the thermochemistry of these compounds.

For compounds with layer structures, the situation is different. There is a significant difference between the calculated ($-1986\,kJ\,mol^{-1}$) and experimental ($-2435\,kJ\,mol^{-1}$) values of $\Delta U(0\,K)$ for CdI_2, indicating that the electrostatic model is unsatisfactory; we noted earlier that in the CdI_2 lattice (Figure 5.22), van der Waals forces operate between layers of adjacent I^- centres. The electrostatic model is similarly found to be unsatisfactory for Cu(I) halides (zinc blende lattice) and for AgI (wurtzite lattice). For the Ag(I) halides, the discrepancy between $\Delta U(0\,K)_{calculated}$ and $\Delta U(0\,K)_{experimental}$ follows the sequence AgF < AgCl < AgBr < AgI. Contributions due to covalent character in the lattice are significant for the larger halides, and are the origin of the decreasing solubility of the Ag(I) halides in water on going from AgF to AgI (see *Section 6.9*).

5.16 Applications of lattice energies

We now consider some typical applications of lattice energies; further examples are given in later chapters.

Estimation of electron affinities

The availability of laser photodetachment techniques has permitted more accurate experimental determinations of electron affinities. Even so, tables of electron affinities list some calculated values, in particular for the formation of multiply charged ions. One method of estimation uses the Born–Haber cycle, with a value for the lattice energy derived using an electrostatic model. Compounds for which this is valid are limited (see *Section 5.15*).

Consider the estimation of $\Delta_{EA}H^\circ(298\,K)$ for the process 5.21.

$$O(g) + 2e^- \longrightarrow O^{2-}(g) \tag{5.21}$$

We can apply the Born–Haber cycle to a metal oxide having a lattice type of known Madelung constant, and for which an electrostatic model is a reasonably valid approximation. Magnesium(II) oxide fits these criteria: it has an NaCl lattice, r_0 has been accurately determined by X-ray diffraction methods, and compressibility data are available; an electrostatic model gives $\Delta U(0\,K) = -3975\,kJ\,mol^{-1}$. All other quantities in the appropriate Born–Haber cycle are independently measurable and a value for $\Delta_{EA}H^\circ(298\,K)$ for reaction 5.21 can be evaluated. A series of similar values for $\Delta_{EA}H^{o}(298\,K)$ for reaction 5.21 can be obtained using different group 2 metal oxides.

The attachment of two electrons to an O atom can be considered in terms of the consecutive processes in scheme 5.22, and accepted values for the associated enthalpy changes for the two steps are -141 and $+798\,kJ\,mol^{-1}$.

$$\left.\begin{array}{l} O(g) + e^{-} \longrightarrow O^{-}(g) \\ O^{-}(g) + e^{-} \longrightarrow O^{2-}(g) \end{array}\right\} \qquad (5.22)$$

The second step is highly *endothermic*. It appears that the only reason the O^{2-} ion exists is the high lattice energies of oxide salts, e.g. $\Delta U(0\,K)$ for Na_2O, K_2O, MgO and CaO are -2481, -2238, -3795 and $-3414\,kJ\,mol^{-1}$.

Fluoride affinities

Fluoride acceptors such as BF_3, AsF_5 and SbF_5 readily form the anions $[BF_4]^{-}$, $[AsF_6]^{-}$ and $[SbF_6]^{-}$ respectively, and the F^{-} affinity for each acceptor can be determined using a thermochemical cycle such as that in scheme 5.23.

$$\begin{array}{ccc} KBF_4(s) & \xrightarrow{\Delta H^{o}_1} & KF(s) + BF_3(g) \\ \Big\uparrow {\scriptstyle \Delta_{lattice}H^{o}(KBF_4,\,s)} & & \Big\uparrow {\scriptstyle \Delta_{lattice}H^{o}(KF,\,s)} \\ K^{+}(g) + [BF_4]^{-}(g) & \xleftarrow{\Delta H^{o}_2} & K^{+}(g) + F^{-}(g) + BF_3(g) \end{array}$$

$$(5.23)$$

The high-temperature form of KBF_4 crystallizes with a CsCl lattice and we can estimate the lattice energy using an electrostatic model, assuming that the $[BF_4]^{-}$ ion can be treated as a sphere (see *Figure 5.17*). The lattice energy of KF is known, and ΔH^{o}_1 can be determined from the temperature variation of the dissociation pressure of solid KBF_4. Use of Hess's Law allows ΔH^{o}_2 to be determined; this value $(-360\,kJ\,mol^{-1})$ corresponds to the enthalpy change associated with the attachment of F^{-} to BF_3.

Estimation of standard enthalpies of formation and disproportionation

For well-established ionic compounds, it is seldom the case that the lattice energy is known while the standard enthalpy of formation is not. However, in theoretical studies of hypothetical compounds, one may wish to estimate a value of $\Delta_f H^{o}(298\,K)$ using a Born–Haber cycle incorporating a calculated value of the lattice energy. The earliest example of this method addressed the question of whether it was conceivable that neon might form a salt $Ne^{+}Cl^{-}$. On the basis that the size of the Ne^{+} ion would be similar to that of Na^{+}, and that NeCl would possess an NaCl lattice, the lattice energy of NeCl was estimated to be $\approx -840\,kJ\,mol^{-1}$. This leads to a value of $\Delta_f H^{o}(NeCl, s) \approx +1010\,kJ\,mol^{-1}$, the very high ionization energy of Ne $(2081\,kJ\,mol^{-1})$ being

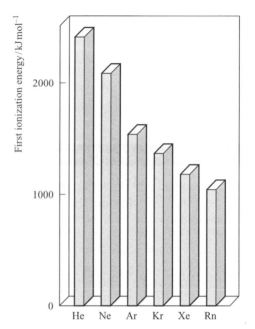

Fig. 5.25 The trend in the values of the first ionization energies of the noble gases (group 18).

responsible for making the process so highly endothermic and unlikely to occur in practice.

Much later, lattice energy considerations pointed towards the feasibility of preparing the first compound of a noble gas; the first ionization energies of Xe and O_2 are similar, and the discovery that O_2 reacted with PtF_6 to give $[O_2]^{+}[PtF_6]^{-}$ led to the suggestion that Xe (see *Chapter 17*) might also react with PtF_6. The trend in first ionization energies on descending group 18 is shown in Figure 5.25; although radon is the easiest to ionize, it is highly radioactive and xenon is more readily handled in the laboratory. The reaction between Xe and PtF_6 was successful, although the exact nature of the product 'Xe[PtF$_6$]' remains uncertain, even though it is ≈ 40 years since the reaction was first studied by Neil Bartlett.

A further example considers the possible formation of CaF (in contrast to the more usual CaF_2). Here, a simple Born–Haber cycle is not helpful since CaF is not thermodynamically unstable with respect to decomposition into its *constituent elements*, but is unstable with respect to *disproportionation* (equation 5.24).

$$2CaF(s) \longrightarrow Ca(s) + CaF_2(s) \qquad (5.24)$$

A species *disproportionates* if it undergoes simultaneous oxidation and reduction.

The thermochemical cycle to be considered is given in equation 5.25, in which the values of $\Delta_a H^{o}(Ca, s)$ $(178\,kJ\,mol^{-1})$ and the difference between IE_1 and IE_2 for Ca $(-555\,kJ\,mol^{-1})$ are significantly smaller in magnitude than the lattice energy of CaF_2 $(-2610\,kJ\,mol^{-1})$.

$$2CaF(s) \xrightarrow{\Delta H^\circ} Ca(s) + CaF_2(s)$$

$$\Bigg\uparrow 2\Delta_{lattice}H^\circ(CaF,s) \qquad \Bigg\downarrow \Delta_{lattice}H^\circ(CaF_2,s) - \Delta_a H^\circ(Ca,s)$$

$$2Ca^+(g) + 2F^-(g) \xleftarrow{\hspace{2cm}} Ca(g) + Ca^{2+}(g) + 2F^-(g)$$

$$IE_1(Ca,g) - IE_2(Ca,g)$$

$$(5.25)$$

The magnitude and sign of the enthalpy change, ΔH°, for the disproportionation reaction therefore depend largely on the balance between the lattice energy of CaF_2 and twice the lattice energy of CaF. The value of $\Delta U(0\,K)$ for CaF_2 will significantly exceed that of CaF because:

- $|z_+|$ for Ca^{2+} is twice that of Ca^+;
- r_0 for Ca^{2+} is smaller than that of Ca^+;
- Madelung constants for MX_2 structures are ≈ 1.5 times those of MX lattices (see *Table 5.4*).

The net result is that ΔH° for the disproportionation reaction shown in equation 5.25 is negative.

The Kapustinskii equation

A problem in estimating the lattice energy of a hypothetical compound is deciding what lattice type to assume. Attempts have been made to make use of the fact that Madelung constants for MX and MX_2 lattice types (Table 5.4) are in an approximate ratio 2:3. In 1956, Kapustinskii derived what has become the best known *general* expression for estimating lattice energies, and one form of this is given in equation 5.26.

$$\Delta U(0\,K) = \frac{(1.07 \times 10^5)v|z_+||z_-|}{r_+ + r_-} \qquad (5.26)$$

where v = number of ions in the formula of the salt (e.g. 2 for NaCl, 3 for CaF_2); r_+ and r_- = radius for 6-coordinate cation and anion, respectively, in pm.

This expression has its origins in the Born–Landé equation, with a value of 9 for the Born exponent (the value for NaCl) and half the value of the Madelung constant for NaCl; the inclusion of the factor v shows why *half* of A is included. Although the Kapustinskii equation is useful, it is a gross *approximation* and values obtained in this way must be treated with caution.

5.17 Defects in solid state lattices: an introduction

So far in this chapter, we have assumed implicitly that all the pure substances considered have ideal lattices in which every site is occupied by the correct type of atom or ion. This state only appertains at $0\,K$, and above this temperature, *lattice defects* are always present; the energy required to create a

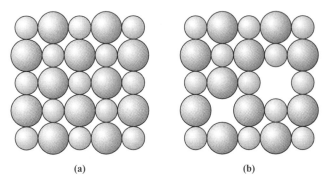

Fig. 5.26 (a) Part of one face of an ideal NaCl lattice; compare this with Figure 5.15. (b) A Schottky defect involves vacant cation and anion sites; equal numbers of cations and anions must be absent to maintain electrical neutrality.

defect is more than compensated for by the resulting increase in entropy of the structure. There are various types of lattice defects, but we shall introduce only the *Schottky* and *Frenkel defects*. Solid state defects are discussed further in *Chapter 27*. Spinels and defect spinels are introduced in *Box 12.1*.

Schottky defect

A Schottky defect consists of an atom or ion vacancy in a crystal lattice, but the stoichiometry of a compound (and thus electrical neutrality) must be retained. In a metal lattice, a vacant atom site may be present. Examples of Schottky defects in ionic lattices are a vacant cation *and* a vacant anion site in an MX salt, or a vacant cation *and* two vacant anion sites in an MX_2 salt. Figure 5.26 illustrates a Schottky defect in an NaCl lattice; holes are present (Figure 5.26b) where ions are expected on the basis of the ideal lattice (Figure 5.26a).

Frenkel defect

In a Frenkel defect, an atom or ion occupies a normally vacant site, leaving its 'own' lattice site vacant. Figure 5.27

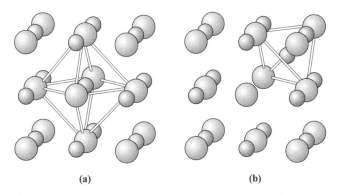

Fig. 5.27 Silver bromide adopts an NaCl lattice. (a) An ideal lattice can be described in terms of Ag^+ ions occupying octahedral holes in a cubic close-packed array of bromide ions. (b) A Frenkel defect in AgBr involves the migration of Ag^+ ions into tetrahedral holes; in the diagram, one Ag^+ ion occupies a tetrahedral hole which was originally vacant in (a), leaving the central octahedral hole empty.

illustrates this for AgBr, which adopts an NaCl lattice. In Figure 5.27a, the central Ag^+ ion is in an octahedral hole with respect to the fcc arrangement of Br^- ions. Migration of the Ag^+ ion to one of the previously unoccupied tetrahedral holes (Figure 5.27b) generates a Frenkel defect in the lattice. This type of defect is possible if there is a relatively large difference in size between cation and anion; in AgBr, the cation must be accommodated in a tetrahedral hole which is significantly smaller than the octahedral site. More generally, Frenkel defects are observed in lattices which are relatively open and in which the coordination number is low.

Experimental observation of Schottky and Frenkel defects

There are several methods that may be used to study the occurrence of Schottky and Frenkel defects in stoichiometric crystals, but the simplest, in principle, is to measure the density of the crystal extremely accurately. Low concentrations of Schottky defects lead to the observed density of a crystal being lower than that calculated from X-ray diffraction and data based on the size and structure of the unit cell. On the other hand, since the Frenkel defect does not involve a change in the number of atoms or ions present, no such density differences will be observed.

Summary

The following terms were introduced in this chapter. Do you know what they mean?
- ❑ close-packing (of spheres or atoms)
- ❑ cubic close-packed (ccp) lattice
- ❑ hexagonal close-packed (hcp) lattice
- ❑ face-centred cubic (fcc) lattice
- ❑ simple cubic lattice
- ❑ body-centred cubic (bcc) lattice
- ❑ coordination number (in a lattice)
- ❑ unit cell
- ❑ interstitial hole
- ❑ polymorph
- ❑ phase diagram
- ❑ metallic radius
- ❑ alloy
- ❑ electrical resistivity
- ❑ band theory
- ❑ band gap
- ❑ insulator
- ❑ semiconductor
- ❑ intrinsic and extrinsic semiconductors
- ❑ n- and p-type semiconductors
- ❑ doping (a semiconductor)
- ❑ ionic radius
- ❑ NaCl lattice

- ❑ CsCl lattice
- ❑ CaF_2 (fluorite) lattice
- ❑ Antifluorite lattice
- ❑ Zinc blende lattice
- ❑ Diamond network
- ❑ Wurtzite lattice
- ❑ β-Cristobalite lattice
- ❑ TiO_2 (rutile) lattice
- ❑ CdI_2 and $CdCl_2$ (layer) lattices
- ❑ Perovskite lattice
- ❑ Lattice energy
- ❑ Born–Landé equation
- ❑ Madelung constant
- ❑ Born exponent
- ❑ Born–Haber cycle
- ❑ Disproportionation
- ❑ Kapustinskii equation
- ❑ Schottky defect
- ❑ Frenkel defect

Suggested further reading

Packing of spheres and structures of ionic lattices

C.E. Housecroft and E.C. Constable (1997) *Chemistry: An Integrated Approach*, Addison Wesley Longman, Harlow – Chapters 6 and 7 give detailed accounts at an introductory level.

A.F. Wells (1984) *Structural Inorganic Chemistry*, 5th edn, Clarendon Press, Oxford – Chapters 4 and 6 present careful descriptions, ranging from basic to more advanced material.

Dictionary of Inorganic Compounds (1992), Chapman and Hall, London – The introduction to vol. 4 gives a useful summary of structure types.

Structure determination

W. Clegg (1998) *Crystal Structure Determination*, Oxford University Press, Oxford – An introduction to the subject with examples.

Alloys

B.C. Giessen (1994) in *Encyclopedia of Inorganic Chemistry*, ed. R.B. King, Wiley, Chichester, vol. 1, p. 90 – A detailed overview of alloys with further references.

A.F. Wells (1984) *Structural Inorganic Chemistry*, 5th edn, Clarendon Press, Oxford – Chapter 29 provides excellent coverage of metal and alloy lattice types.

Semiconductors

M. Hammonds (1998) *Chemistry & Industry*, p. 219 – 'Getting power from the sun' illustrates the application of the semiconducting properties of Si.

C.E. Stanton, S.T. Nguyen, J.M. Kesselman, P.E. Laaibinis and N.S. Lewis (1994) in *Encyclopedia of Inorganic Chemistry*, ed. R.B. King, Wiley, Chichester, vol. 7, p. 3725 – An up-to-date general survey of semiconductors which defines pertinent terminology and gives pointers for further reading.

J. Wolfe (1998) *Chemistry & Industry*, p. 224 – 'Capitalising on the sun' describes the applications of Si and other materials in solar cells.

Solid State: for more general information

A.K. Cheetham and P. Day (1992) *Solid State Chemistry*, Clarendon Press, Oxford.

M. Ladd (1994) *Chemical Bonding in Solids and Fluids*, Ellis Horwood, Chichester.

L. Smart and E. Moore (1992) *Solid State Chemistry: An Introduction*, Chapman and Hall, London.

A.R. West (1999) *Basic Solid State Chemistry*, 2nd edn, Wiley-VCH, Weinheim.

Problems

5.1 Outline the similarities and differences between cubic and hexagonal close-packed arrangements of spheres, paying particular attention to (a) coordination numbers, (b) interstitial holes and (c) unit cells.

5.2 State the coordination number of a sphere in each of the following arrangements: (a) ccp; (b) hcp; (c) bcc; (d) fcc; (e) simple cubic.

5.3 (a) Lithium metal undergoes a phase change at 80 K (1 bar pressure) from the α- to β-form; one form is bcc and the other is a close-packed lattice. Suggest, with reasons, which form is which. What name is given to this type of structural change? (b) Suggest why tin buttons on nineteenth century military uniforms crumbled in exceptionally cold winters.

5.4 Refer to Table 5.2. (a) Write an equation that defines the process for which the standard enthalpy of atomization of cobalt is defined. (b) Suggest reasons for the trend in standard enthalpies of atomization on descending group 1. (c) Outline possible reasons for the trend in values of $\Delta_a H^\circ$ on going from Cs to Bi.

5.5 'Titanium dissolves nitrogen to give a solid solution of composition $TiN_{0.2}$; the metal lattice defines an hcp arrangement.' Explain what is meant by this statement, and suggest whether, on the basis of this evidence, $TiN_{0.2}$ is likely to be an interstitial or substitutional alloy. Relevant data may be found in Appendix 6 and Table 5.2.

5.6 What do you understand by the 'band theory of metals'?

5.7 (a) Draw a representation of the structure of diamond and give a description of the bonding. (b) Is the same picture of the bonding appropriate for silicon which is isostructural with diamond? If not, suggest an alternative picture of the bonding.

5.8 (a) Give a definition of electrical resistivity and state how it is related to electrical conductivity. (b) At 273–290 K, the electrical resistivities of diamond, Si, Ge and α-Sn are approximately 1×10^{11}, 1×10^{-3}, 0.46 and 11×10^{-8} Ω m. Rationalize this trend in values. (c) How does the change in electrical resistivity with temperature vary for a typical metal and for a semiconductor?

5.9 Distinguish between an intrinsic and extrinsic semiconductor, giving examples of materials that fall into these classes, and further classifying the types of extrinsic semiconductors.

5.10 The metallic, covalent and ionic radii of Al are 143, 130 and 54 pm respectively; the value of r_{ion} is for a 6-coordinate ion. (a) How is each of these quantities defined? (b) Suggest reasons for the trend in values.

5.11 With reference to the NaCl, CsCl and TiO_2 lattice types, explain what is meant by (a) coordination number, (b) unit cell, (c) ion sharing between unit cells, and (d) determination of the formula of an ionic salt from the unit cell.

5.12 Determine the number of formula units of (a) CaF_2 in a unit cell of fluorite, and (b) TiO_2 in a unit cell of rutile.

5.13 (a) Confirm that the unit cell for perovskite shown in Figure 5.23a is consistent with the stoichiometry $CaTiO_3$. (b) A second unit cell can be drawn for perovskite; this has Ti(IV) at the centre of a cubic cell; Ti(IV) is in an octahedral environment with respect to the O^{2-} ions. In what sites must the Ca^{2+} lie in order that the unit cell depicts the correct compound stoichiometry? Draw a diagram to illustrate this unit cell.

5.14 (a) Give a definition of lattice energy. Does your definition mean that the associated enthalpy of reaction will be positive or negative? (b) Use the Born–Landé equation to calculate a value for the lattice energy of KBr, for which $r_0 = 328$ pm. KBr adopts an NaCl lattice; other data may be found in Tables 5.3 and 5.4.

5.15 Using data from the Appendices and the fact that $\Delta_f H^\circ(298\,\text{K}) = -859\,\text{kJ mol}^{-1}$, calculate a value for the lattice energy of $BaCl_2$. Outline any assumptions that you have made.

5.16 (a) Given that $\Delta U(0\,\text{K})$ and $\Delta_f H^\circ(298\,\text{K})$ for MgO are -3795 and $-602\,\text{kJ mol}^{-1}$ respectively, derive a value for $\Delta_{EA} H^\circ(298\,\text{K})$ for the reaction:

$$O(g) + 2e^- \longrightarrow O^{2-}(g)$$

Other data may be found in the Appendices. (b) Compare the calculated value with that obtained using electron affinity data from Appendix 9, and suggest reasons for any differences.

5.17 Discuss the interpretation of the following:

(a) $\Delta_f H^\circ(298\,\text{K})$ becomes less negative along the series LiF, NaF, KF, RbF, CsF, but more negative along the series LiI, NaI, KI, RbI, CsI.

(b) The thermal stability of the isomorphous sulfates of Ca, Sr and Ba with respect to decomposition into the metal oxide (MO) and SO_3 increases in the sequence $CaSO_4 < SrSO_4 < BaSO_4$.

5.18 Data from Tables 5.3 and 5.4 are needed for this question. (a) Estimate the lattice energy of CsCl if the Cs–Cl internuclear distance is 356.6 pm. (b) Now consider a polymorph of CsCl that crystallizes with an NaCl lattice; estimate its lattice energy given that the Cs–Cl distance is 347.4 pm. (c) What conclusions can you draw from your answers to parts (a) and (b)?

Chapter 6   Acids, bases and ions in aqueous solution

6.1 Introduction

The importance of water as a medium for inorganic reactions stems not only from the fact that it is far more readily available than any other solvent, but also because of the abundance of accurate physicochemical data for aqueous solutions compared with the relative scarcity of such data for solutions in non-aqueous solvents. This chapter is concerned mainly with *equilibria* and we review calculations involving acid–base equilibrium constants.

Liquid water is approximately 55 molar H_2O, a fact commonly overlooked in the study of classical physical chemistry where, by convention, we take the *activity* (see *Section 6.3*) (and hence, the approximate concentration) of water to be unity.[†]

Worked example 6.1   Molarity of water

Show that pure water is approximately 55 molar.

Density of water $= 1\,\mathrm{g\,cm^{-3}}$
Thus, $1000\,\mathrm{cm^3}$ (or $1\,\mathrm{dm^3}$) has a mass of $1000\,\mathrm{g}$
For H_2O, $M_r = 18$

Number of moles in $1000\,\mathrm{g} = \dfrac{1000}{18} = 55.5 =$ number of moles per $\mathrm{dm^3}$

Therefore, the concentration of pure water $\approx 55\,\mathrm{mol\,dm^{-3}}$.

[†]The use of [ ] for concentration should not be confused with the use of [ ] to show the presence of an ion. For example, [OH]⁻ means 'hydroxide ion', but [OH⁻] means 'the concentration of hydroxide ions'.

Self-study exercises

1. How many moles of H_2O are there per $100\,\mathrm{g}$ of pure water?
   [*Ans.* 5.55 moles]

2. Show that 99.9% deuterated water is ≈ 50 molar.

6.2 Properties of water

Structure and hydrogen bonding

At atmospheric pressure, solid H_2O can adopt one of two polymorphs, depending upon the conditions of crystallization. At higher pressures, five polymorphs exist which differ in their arrangement of the oxygen atoms in the crystal lattice. We shall be concerned here only with the normal form of ice.

The structure of ice has been accurately determined using neutron diffraction techniques; X-ray diffraction is not suitable for accurately locating the H atoms (see the beginning of *Section 5.11*). Ice possesses an infinite lattice (Figure 6.1). The key to making the structure rigid is *intermolecular hydrogen bonding* (see also *Section 9.6*). The hydrogen-bonded network may be described in terms of a wurtzite lattice (see *Figure 5.20*) in which the O atoms occupy the sites of *both* the Zn and S centres; this places each O atom in a tetrahedral environment with respect to other O atoms. Each O atom is involved in four hydrogen bonds, through the use of two lone pairs and two H atoms (Figure 6.1). The hydrogen bonds are asymmetrical (O–H distances $= 101\,\mathrm{pm}$ and $175\,\mathrm{pm}$) and non-linear; each H atom lies slightly off the O····O line, so that the intramolecular H–O–H bond angle is $105°$. The wurtzite lattice is very open, and as a result, ice has a relatively low density ($0.92\,\mathrm{g\,cm^{-3}}$). On

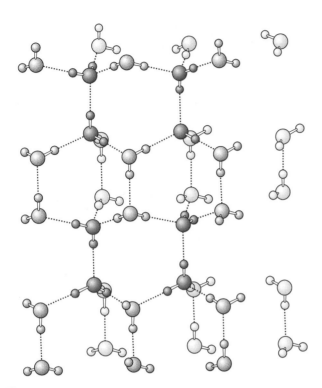

Fig. 6.1 Part of the structure of ice; it consists of a three-dimensional network of hydrogen-bonded H_2O molecules. (Based on: L. Pauling (1960) *The Nature of the Chemical Bond*, Cornell University Press, Ithaca.)

melting (273 K), the lattice partly collapses, allowing some of the lattice cavities to be occupied by H_2O molecules. Consequently, the density increases, reaching a maximum at 277 K; between 277 and 373 K, thermal expansion is the dominant effect, causing the density to decrease (Figure 6.2). Even at the boiling point (373 K), much of the hydrogen bonding remains and is responsible for water having high values of the enthalpy and entropy of vaporization (Table 6.1 and see *Section 9.6*). The strength of a hydrogen bond in ice or water is $\approx 25\ kJ\ mol^{-1}$, and within the bulk liquid, intermolecular bonds are continually being formed and broken (thus transferring a proton between species) and the lifetime of a given H_2O molecule is only $\approx 10^{-12}$ s.

When water acts as a solvent, hydrogen bonds between water molecules are destroyed as water–solute interactions

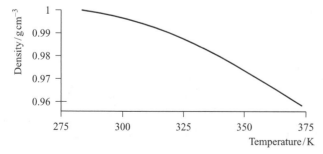

Fig. 6.2 The variation in the value of the density of water between 283 and 373 K.

Table 6.1 Selected physical properties of water

Property	Value
Melting point / K	273.00
Boiling point / K	373.00
Enthalpy of fusion, $\Delta_{fus}H^{\circ}(273\ K)\ /\ kJ\ mol^{-1}$	6.01
Enthalpy of vaporization, $\Delta_{vap}H^{\circ}(373\ K)\ /\ kJ\ mol^{-1}$	40.65
Entropy of vaporization, $\Delta_{vap}S^{\circ}(373\ K)\ /\ J\ mol^{-1}\ K^{-1}$	109
Relative permittivity (at 298 K)	78.39
Dipole moment, μ / debye	1.84

form; the latter may be ion–dipole interactions (e.g. when NaCl dissolves) or new hydrogen bonds (e.g. when H_2O and MeOH mix).

The self-ionization of water

Water itself is ionized to a very small extent (equation 6.1) and the value of the self-ionization constant, K_w (equation 6.2), shows that the equilibrium lies well to the left-hand side. The self-ionization in equation 6.1 is also called autoprotolysis.

$$2H_2O(l) \rightleftharpoons [H_3O]^+(aq) + [OH]^-(aq) \qquad (6.1)$$
$$\text{Water} \qquad \text{Oxonium ion} \qquad \text{Hydroxide ion}$$

$$K_w = [H_3O^+][OH^-] = 1 \times 10^{-14} \qquad \text{(at 298 K)} \qquad (6.2)$$

Although we use concentrations in equation 6.2, this is an approximation, and we return to this equilibrium in Section 6.3.

In aqueous solution, protons are solvated and so it is more correct to write $[H_3O]^+(aq)$ than $H^+(aq)$. Even this is oversimplified because the oxonium ion is further hydrated and species such as $[H_5O_2]^+$ (see *Figure 9.1*), $[H_7O_3]^+$ and $[H_9O_4]^+$ are also present.

If a pure liquid partially dissociates into ions, it is *self-ionizing*.

Water as a Brønsted acid or base

A *Brønsted acid* can act as a proton donor, and a *Brønsted base* can function as a proton acceptor.

Equilibrium 6.1 illustrates that water can function both as a Brønsted acid and a Brønsted base. In the presence of other Brønsted acids or bases, the role of water depends on the relative strengths of the various species in solution. When HCl is bubbled into water, the gas dissolves and equilibrium 6.3 is established.

$$HCl(aq) + H_2O(l) \rightleftharpoons [H_3O]^+(aq) + Cl^-(aq) \qquad (6.3)$$

Hydrogen chloride is a much stronger acid than water, and equilibrium 6.3 lies well over to the right-hand side, so much so that hydrochloric acid is regarded as being fully

BOX 6.1   **CHEMICAL AND THEORETICAL BACKGROUND**

The equilibrium constants K_a, K_b and K_w

In dealing with acid–base equilibria in aqueous solution, three equilibrium constants are of special significance:

- K_a is the acid dissociation constant.
- K_b is the base dissociation constant.
- K_w is the self-ionization constant of water.

Essential equations relating to acid–base equilibria are listed below. Expressions involving concentrations are approximations, since we should strictly be using activities (see main text). Moreover, for a weak acid, HA, we *assume* that the concentration in aqueous solution of the dissociated acid *at equilibrium* is negligible with respect to the concentration of acid presently initially; similarly for a weak base.

For a general weak acid HA in aqueous solution:

$$K_a = \frac{[H_3O^+][A^-]}{[HA]}$$

For a general weak base B in aqueous solution:

$$K_b = \frac{[BH^+][OH^-]}{[B]}$$

$$pK_a = -\log K_a \qquad K_a = 10^{-pK_a}$$

$$pK_b = -\log K_b \qquad K_b = 10^{-pK_b}$$

$$K_w = [H_3O^+][OH^-] = 1.00 \times 10^{-14}$$

$$pK_w = -\log K_w = 14.00$$

$$K_w = K_a \times K_b$$

$$pH = -\log [H_3O^+]$$

Review example 1: Calculate the pH of aqueous 0.020 M acetic acid ($K_a = 1.7 \times 10^{-5}$)

The equilibrium in aqueous solution is:

$$MeCO_2H(aq) + H_2O(l) \rightleftharpoons [MeCO_2]^-(aq) + [H_3O]^+(aq)$$

and K_a is given by:

$$K_a = \frac{[MeCO_2^-][H_3O^+]}{[MeCO_2H][H_2O]} = \frac{[MeCO_2^-][H_3O^+]}{[MeCO_2H]}$$

since $[H_2O]$ is taken to be unity where we are dealing with *equilibrium concentrations*.

Since $[MeCO_2^-] = [H_3O^+]$

$$K_a = \frac{[H_3O^+]^2}{[MeCO_2H]}$$

$$[H_3O^+] = \sqrt{K_a \times [MeCO_2H]}$$

The *initial* concentration of $MeCO_2H$ is 0.020 mol dm^{-3}, and since the degree of dissociation is very small, the *equilibrium* concentration of $MeCO_2H \approx 0.020$ mol dm^{-3}.

$$[H_3O^+] = \sqrt{1.7 \times 10^{-5} \times 0.020}$$

$$[H_3O]^+ = 5.8 \times 10^{-4} \, \text{mol dm}^{-3}$$

The pH value can now be determined:

$$pH = -\log [H_3O^+]$$

$$= -\log (5.8 \times 10^{-4})$$

$$= 3.2$$

Review example 2: Find the concentration of $[OH]^-$ present in a 5.0×10^{-5} mol dm^{-3} solution of Ca(OH)$_2$

At a concentration of 5.0×10^{-5} mol dm^{-3}, Ca(OH)$_2$ is *fully* ionized, with two moles of $[OH]^-$ provided by each mole of Ca(OH)$_2$.

$$[OH^-] = 2 \times 5.0 \times 10^{-5} = 1.00 \times 10^{-4} \, \text{mol dm}^{-3}$$

To find the pH, we need to find $[H_3O^+]$:

$$K_w = [H_3O^+][OH^-] = 1.00 \times 10^{-14} \text{ (at 298 K)}$$

$$[H_3O^+] = \frac{1 \times 10^{-14}}{1 \times 10^{-4}} = 1.00 \times 10^{-10} \, \text{mol dm}^{-3}$$

$$pH = -\log [H_3O^+] = 10.0$$

Review example 3: The value of K_a for HCN is 4.0×10^{-10}. What is the value of pK_b for $[CN]^-$?

K_a for HCN and K_b for $[CN]^-$ are related by the expression:

$$K_a \times K_b = K_w = 1.00 \times 10^{-14} \text{ (at 298 K)}$$

$$K_b = \frac{K_w}{K_a} = \frac{1.00 \times 10^{-14}}{4.0 \times 10^{-10}} = 2.5 \times 10^{-5}$$

$$pK_b = -\log K_b = 4.6$$

---

dissociated, i.e. it is a *strong acid*. Water accepts a proton to form $[H_3O]^+$, and thus behaves as a Brønsted base. In the reverse direction, $[H_3O]^+$ acts as a *weak* acid and Cl$^-$ as a *weak* base; they are, respectively, the *conjugate acid* and *conjugate base* of H$_2$O and HCl.

In an aqueous solution of NH$_3$, water behaves as a Brønsted acid, donating H$^+$ (equation 6.4); $[NH_4]^+$ is the conjugate acid of NH$_3$, while $[H_3O]^+$ is the conjugate acid of H$_2$O. Conversely, NH$_3$ is the conjugate base of $[NH_4]^+$, and $[OH]^-$ is the conjugate base of H$_2$O.

$$NH_3(aq) + H_2O(l) \rightleftharpoons [NH_4]^+(aq) + [OH]^-(aq) \qquad (6.4)$$

The difference in acid strengths between H$_2$O and $[NH_4]^+$ (or the difference between the base strengths of NH$_3$ and $[OH]^-$)

is relatively small, but is sufficient to enable water to act as a stronger acid than in the bulk state. This is seen by comparing the values of the equilibrium constants in equations 6.2 and 6.5.

$$K = \frac{[NH_4^+][OH^-]}{[NH_3]} = 1.8 \times 10^{-5} \qquad \text{(at 298 K)} \qquad (6.5)$$

Conjugate acids and bases are related as follows:

HA(aq) + H$_2$O(l) $\rightleftharpoons$ A$^-$(aq) + [H$_3$O]$^+$(aq)

conjugate   conjugate   conjugate   conjugate
acid 1   base 2   base 1   acid 2

conjugate | acid–base pair

conjugate acid–base pair

6.3 Definitions and units in aqueous solution

In this section, we discuss the conventions and units generally used in the study of aqueous solutions. In some respects, these are *not* the same as those used in many other branches of chemistry. At the level of working within this text and, often, in the practical laboratory, certain approximations can be made, but it is crucial to understand their limitations.

Molarity and molality

A one molar aqueous solution (1 M or 1 mol dm^{-3}) contains one mole of solute dissolved in a sufficient volume of water to give 1 dm^3 *of solution*. In contrast, if one mole of solute is dissolved in 1 kg *of water*, the solution is said to be one molal (1 mol kg^{-1}).

Standard state

We are already used to the concept of standard state in respect of pure solids, liquids and gases. The standard state of a liquid or solid substance, whether pure or in a mixture, or for a solvent is taken as the state of the pure substance at 298 K and 1 bar pressure (1 bar = 1.00×10^5 Pa); the standard state of a gas is that of the pure gas at 298 K, 1 bar pressure and exhibiting ideal gas behaviour.

For a *solute in a solution*, the definition of its standard state is referred to a situation of *infinite dilution*: it is the state (a hypothetical one) at standard molality (m^o), 1 bar pressure, and exhibiting infinitely diluted solution behaviour. In the standard state, interactions between solute molecules or ions are negligible.

Activity

When the concentration of a solute is greater than about 0.1 mol dm^{-3}, interactions between the solute molecules or ions are significant, and the *effective* and real concentrations are no longer equal. It becomes necessary to define a new quantity called the *activity* which is a measure of concentration but takes into account the interactions between the solution species. The *relative activity*, a_i, of a component i is dimensionless and is defined by equation 6.6 where μ_i is the chemical potential of component i, μ_i^o is the standard chemical potential of i, R is the molar gas constant, and T is the temperature in kelvin.[†]

$$\mu_i = \mu_i^o + RT \ln a_i \qquad (6.6)$$

The *activity* of any pure substance in its standard state is defined to be unity.

The relative activity of a solute is related to its *molality* by equation 6.7 where γ_i is the activity coefficient of the solute, and m_i and m_i^o are the molality and standard state molality, respectively. Since the latter is defined as being unity, equation 6.7 reduces to equation 6.8.

$$a_i = \frac{\gamma_i m_i}{m_i^o} \qquad (6.7)$$

$$a_i = \gamma_i m_i \qquad (6.8)$$

Although all thermodynamic expressions dealing with aqueous solutions should strictly be expressed in terms of activities, inorganic chemists, and students in particular, may be dealing with situations in which two criteria are true:

● problem solving involves very dilute solutions ($\leq 1 \times 10^{-3}$ mol dm^{-3});

● very accurate answers are not required.

If these criteria hold, then we can approximate the activity of the solute to its concentration, the latter being measured, most often, in *molarity*. We use this approximation throughout the book, but it is crucial to keep in mind the limitations of this approach.

6.4 Brønsted acids and bases

Carboxylic acids: examples of mono-, di- and polybasic acids

In organic compounds, acidity is quite often associated with the presence of a carboxylic acid group (CO$_2$H) and it is relatively easy to determine the number of *ionizable hydrogen atoms* in the system. Acetic acid,[‡] **6.1**, is a *monobasic acid* since it can donate only one proton. Ethanedioic acid (oxalic acid) **6.2** can donate two protons and so is a *dibasic*

[†] For further discussion, see: P.W. Atkins (1998) *Physical Chemistry*, 6th edn, Oxford University Press, Oxford, p. 182.
[‡] The systematic name for MeCO$_2$H is ethanoic acid, but acetic acid is the IUPAC-accepted trivial name.

acid. The *tetrabasic acid* **6.3** and the anions derived from it are commonly encountered in coordination chemistry; the trivial name of N,N,N',N'-ethylenediaminetetraacetic acid (see *Table 6.7*) is generally abbreviated to H_4EDTA.

(6.1)   (6.2)

(6.3)

Equilibrium 6.9 describes the dissociation of $MeCO_2H$ in aqueous solution; it is a weak acid with $K_a = 1.75 \times 10^{-5}$ at 298 K.

$$MeCO_2H(aq) + H_2O(l) \rightleftharpoons [H_3O]^+(aq) + [MeCO_2]^-(aq)$$

Acetic acid   Acetate ion   (6.9)

(Ethanoic acid)   (Ethanoate ion)

Acids **6.2** and **6.3** undergo *stepwise dissociation* in aqueous solution, and equations 6.10 and 6.11 describe the steps for oxalic acid.

$$K_a(1) = 5.90 \times 10^{-2} \quad (298\,K) \qquad (6.10)$$

$$K_a(2) = 6.40 \times 10^{-5} \quad (298\,K) \qquad (6.11)$$

Each dissociation step has an associated equilibrium constant (acid dissociation constant), and it is general for polybasic acids that $K_a(1) > K_a(2)$, and so on; it is more difficult to remove H^+ from an anion than from a neutral species. Values of equilibrium constants may be temperature-dependent, and the inclusion of the temperature to which the stated value applies is important. In general, quoted values usually refer to 293 or 298 K. In this book, unless otherwise stated, values of K_a refer to 298 K.

Inorganic acids

In inorganic chemistry, *hydrogen halides* and *oxoacids* are of particular significance in terms of acidic behaviour in aqueous solution. Each of the hydrogen halides is monobasic (equation 6.12) and for X = Cl, Br and I, the equilibrium lies far to the right-hand side, making these strong acids.

In each case, $K_a > 1$; note that this means that the pK_a values are negative (pK_a HCl ≈ -7; HBr ≈ -9; HI ≈ -11) since $pK_a = -\log K_a$. In many instances, equation 6.12 for X = Cl, Br or I is written showing only the forward reaction, thereby emphasizing strong acid behaviour. Hydrogen fluoride on the other hand is a weak acid ($pK_a = 3.45$).

$$HX(aq) + H_2O(l) \rightleftharpoons [H_3O]^+(aq) + X^-(aq) \qquad (6.12)$$

> The IUPAC definition of an *oxoacid* is 'a compound which contains oxygen, at least one other element, at least one hydrogen bound to oxygen, and which produces a conjugate base by proton loss.'

Examples of oxoacids include hypochlorous acid (HOCl), perchloric acid ($HClO_4$), nitric acid (HNO_3), sulfuric acid (H_2SO_4) and phosphoric acid (H_3PO_4). Many well-recognized common names exist for oxoacids, and the IUPAC has recommended that such names be retained. In this book, we follow this recommendation, although in Box 6.2, we introduce systematic nomenclature.

A wide variety of oxoacids exists and later chapters introduce many of them. Note that:

- oxoacids may be mono-, di- or polybasic;
- not all the hydrogen atoms in an oxoacid are necessarily ionizable.

Nitric acid, nitrous acid and hypochlorous acid are examples of monobasic acids; HNO_3 is essentially fully ionized in aqueous solution (equation 6.13), but HNO_2 and HOCl behave as weak acids (equations 6.14 and 6.15).

$$HNO_3(aq) + H_2O(l) \rightleftharpoons [H_3O]^+(aq) + [NO_3]^-(aq)$$

Nitrate ion

$$pK_a = -1.64 \qquad (6.13)$$

$$HNO_2(aq) + H_2O(l) \rightleftharpoons [H_3O]^+(aq) + [NO_2]^-(aq)$$

Nitrite ion

$$pK_a = 3.37\ (285\,K) \qquad (6.14)$$

$$HOCl(aq) + H_2O(l) \rightleftharpoons [H_3O]^+(aq) + [OCl]^-(aq)$$

Hypochlorite ion

$$pK_a = 4.53 \qquad (6.15)$$

Sulfuric acid is dibasic; in aqueous solution, the first dissociation step lies well over to the right-hand side (equation 6.16), but $[HSO_4]^-$ is a weaker acid (equation 6.17). Two series of salts can be isolated, e.g. sodium hydrogensulfate(1−) and sodium sulfate.

$$H_2SO_4(aq) + H_2O(l) \rightleftharpoons [H_3O]^+ + [HSO_4]^-$$

Hydrogensulfate(1−) ion

$$pK_a \approx -2.0 \qquad (6.16)$$

$$[HSO_4]^-(aq) + H_2O(l) \rightleftharpoons [H_3O]^+ + [SO_4]^{2-}$$

Sulfate ion

$$pK_a = 1.92 \qquad (6.17)$$

BOX 6.2

Oxoacid nomenclature recommended by the IUPAC

Although a systematic method exists for naming oxoacids, many such laboratory reagents are well recognized by their common names (e.g. sulfuric acid, nitric acid, nitrous acid). A problem with many common names is that they often hide structural information, and, most importantly, do not indicate the number of ionizable hydrogen atoms. Where large families of acids exist, e.g. the sulfur oxoacids, systematic names are usually more helpful than traditional ones.

Hydrogen nomenclature (for p-block oxoacids)

Systematic *hydrogen nomenclature* considers the oxoacid as a hydrogen salt of the corresponding conjugate base. In this way, the number of ionizable hydrogen atoms is immediately obvious; furthermore, structural information about the

conjugate base(s) is built into the name. As an example, consider nitric acid (HNO_3), the conjugate base of which is the nitrate ion $[NO_3]^-$. The systematic name for $[NO_3]^-$ is trioxonitrate(1−). This name carries sufficient information for us to draw the structure of the anion; within the systematic nomenclature, *every* anion derived from an oxoacid has the suffix '-ate', in contrast to traditional usage of '-ate' or '-ite'. The charge of the anion is included in parentheses at the end of the name. The name of the parent oxoacid now follows by adding the word 'hydrogen' before the anion name: HNO_3 is hydrogen trioxonitrate(1−). In simple cases, the charge can be omitted since it adds no additional information. It is also acceptable to replace the charge by the oxidation state (as an upper case Roman numeral) of the central atom, e.g. trioxonitrate(V). Further examples are:

Formula	Traditional name	Hydrogen nomenclature
H_2SO_4	Sulfuric acid	Dihydrogen tetraoxosulfate(2−)
		Dihydrogen tetraoxosulfate(VI)
H_3PO_4	Phosphoric acid	Trihydrogen tetraoxophosphate(3−)
	(Orthophosphoric acid)	Trihydrogen tetraoxophosphate(V)
H_3PO_2	Phosphinic acid	Hydrogen dihydridodioxophosphate(1−)
	(Hypophosphorous acid)	Hydrogen dihydridodioxophosphate(I)

Acid nomenclature (p-block and d-block oxoacids)

This is an alternative method of nomenclature, but one that is less easily interpretable in terms of structure than the hydrogen nomenclature. The name consists of two words, the second always being 'acid'. The name gives the number of oxygen atoms attached to the central atom, the identity of the central atom and (usually) the oxidation state of the central atom. The first word in the name always ends in '-ic', in contrast to traditional usage of '-ic' or '-ous'. Selected examples are listed in the table:

Formula	Traditional name	Hydrogen nomenclature
HNO_3	Nitric acid	Trioxonitric(V) acid
HNO_2	Nitrous acid	Dioxonitric(III) acid
H_2SO_4	Sulfuric acid	Tetraoxosulfuric(VI) acid
$HClO_4$	Perchloric acid	Tetraoxochloric(VII) acid

For detailed rules, see: *IUPAC: Nomenclature of Inorganic Chemistry (Recommendations 1990)*, ed. G.J. Leigh, Blackwell Scientific Publications, Oxford, p. 122 and p. 248.

Tables of data and the existence of crystalline salts can sometimes be misleading, as is the case for 'sulfurous acid'. It is *not* possible to isolate pure H_2SO_3, even though we often refer to 'sulfurous acid' and values of acid dissociation constants are available (equations 6.18 and 6.19).

$$H_2SO_3(aq) + H_2O(l) \rightleftharpoons [H_3O]^+(aq) + [HSO_3]^-(aq)$$
$$\text{Hydrogensulfite(1−) ion}$$

$$pK_a = 1.82 \quad (6.18)$$

$$[HSO_3]^-(aq) + H_2O(l) \rightleftharpoons [H_3O]^+(aq) + [SO_3]^{2-}(aq)$$
$$\text{Sulfite ion}$$

$$pK_a = 6.92 \quad (6.19)$$

An aqueous solution of 'sulfurous acid' can be prepared by dissolving SO_2 in water (equation 6.20), but the equilibrium constant indicates that such solutions contain mainly dissolved SO_2. A similar situation arises for 'carbonic acid', H_2CO_3 (see *Section 13.9*).

$$SO_2(aq) + H_2O(l) \rightleftharpoons H_2SO_3(aq) \quad K < 10^{-9} \quad (6.20)$$

In the oxoacids above, *each* hydrogen atom is attached to oxygen in the free acid, and the number of H atoms corresponds to the basicity of the acid. However, this is not always the case: e.g. although phosphinic acid has the formula H_3PO_2, there is only one O−H bond (structure

6.4) and H_3PO_2 is *monobasic* (equation 6.21). Further examples of this type are given in *Section 14.11*.

$$\begin{array}{c} O \\ \| \\ H \text{---} P\text{''''}OH \\ | \\ H \end{array}$$

(6.4)

$$H_3PO_2(aq) + H_2O(l) \rightleftharpoons [H_3O]^+(aq) + [H_2PO_2]^-(aq)$$

(6.21)

Inorganic bases: hydroxides

Many inorganic bases are hydroxides, and the term *alkali* is commonly used. The group 1 hydroxides NaOH, KOH, RbOH and CsOH are strong bases, being essentially fully ionized in aqueous solution; LiOH is weaker ($pK_b = 0.2$).

Inorganic bases: nitrogen bases

The term 'nitrogen bases' tends to suggest ammonia and organic amines (RNH_2), but there are a number of important inorganic nitrogen bases related to NH_3. Ammonia dissolves in water, and functions as a weak base, accepting H^+ to form the ammonium ion (equation 6.4). Although solutions of NH_3 in water are often referred to as ammonium hydroxide, it is not possible to isolate solid samples of 'NH_4OH'. Confusion may arise from tables of data for the dissociation constants for bases; some tables quote K_b or pK_b, while others list values of K_a or pK_a. For the relationship between K_a and K_b, see *Box 6.1*. Thus, a value of pK_a for 'ammonia' of 9.25 is really that of the ammonium ion and refers to equilibrium 6.22, while a value of pK_b of 4.75 refers to equilibrium 6.4.

$$[NH_4]^+(aq) + H_2O(l) \rightleftharpoons [H_3O]^+(aq) + NH_3(aq)$$

$$pK_a = 9.25 \quad (6.22)$$

Worked example 6.2   Relationship between pK_a and pK_b for a weak base

The degree of dissociation of NH_3 in aqueous solution can be described in terms of either a value of K_a or K_b. Deduce a relationship between the values of pK_a and pK_b.

K_b refers to the equilibrium:

$$NH_3(aq) + H_2O(l) \rightleftharpoons [NH_4]^+(aq) + [OH]^-(aq)$$

$$K_b = \frac{[NH_4^+][OH^-]}{[NH_3]}$$

K_a refers to the equilibrium:

$$[NH_4]^+(aq) + H_2O(l) \rightleftharpoons [H_3O]^+(aq) + NH_3(aq)$$

$$K_a = \frac{[NH_3][H_3O^+]}{[NH_4^+]}$$

Combining the two expressions gives:

$$\frac{[NH_4^+]}{[NH_3]} = \frac{K_b}{[OH^-]} = \frac{[H_3O^+]}{K_a}$$

$$K_b \times K_a = [H_3O^+][OH^-]$$

The right-hand side product is equal to the self-dissociation constant for water, K_w

$$K_b \times K_a = K_w = 1.00 \times 10^{-14}$$

and so:

$$pK_b + pK_a = pK_w = 14.00$$

Self-study exercises

1. If pK_a for the conjugate acid of $PhNH_2$ is 4.63, what is pK_b for $PhNH_2$? To what equilibria do K_a and K_b refer?
   [*Ans.* 9.37; work out by analogy to those for NH_3 above]

2. For N_2H_4, $pK_b = 6.05$. What is K_b?   [*Ans.* 8.91×10^{-7}]

Hydrazine, N_2H_4, **6.5**, is a weak Brønsted base ($pK_b = 6.05$), weaker than NH_3; it reacts with strong acids to give hydrazinium salts (equation 6.23).

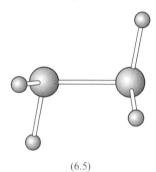

(6.5)

$$N_2H_4(aq) + HCl(aq) \longrightarrow [N_2H_5]Cl(aq) \quad (6.23)$$

The value of pK_b for hydroxylamine, NH_2OH, is 8.04, showing it to be a weaker base than either NH_3 or N_2H_4.

6.5 The energetics of acid dissociation in aqueous solution

Hydrogen halides

The strengths of different acids in aqueous solutions tend to be discussed in elementary textbooks on a qualitative basis. In the case of the hydrogen halides, an exact treatment in terms of independently measurable thermodynamic quantities is *almost* possible. Consider the dissociation of HX (X is F, Cl, Br or I) in aqueous solution (equilibrium 6.24 or 6.25):

$$HX(aq) + H_2O(l) \rightleftharpoons [H_3O]^+(aq) + X^-(aq) \quad (6.24)$$

Fig. 6.3 The energetics of the dissociation of a hydrogen halide, HX (X = F, Cl, Br or I), in aqueous solution can be considered in terms of a cycle of steps. The significance of each step is discussed in the text.

$$HX(aq) \rightleftharpoons H^+(aq) + X^-(aq) \qquad (6.25)$$

The factors that influence the degree of dissociation are summarized in Figure 6.3. Equation 6.26 relates K_a for the dissociation of HX in aqueous solution to ΔG°, and the latter depends on both changes in enthalpy and entropy (equation 6.27).

$$\Delta G^\circ = -RT \ln K \qquad (6.26)$$

$$\Delta G^\circ = \Delta H^\circ - T\Delta S^\circ \qquad (6.27)$$

A Hess cycle relates ΔH° for each of steps (1) to (6) in Figure 6.3 to that of the solution dissociation step. In Figure 6.3, step (2) is the cleavage of the H−X bond for the gas phase molecule. Steps (3) and (5) are the ionization of the gaseous H atom and the hydration of the gaseous H^+ ion, respectively, and are common to all four hydrogen halides. Step (4) is the attachment of an electron to the gaseous X atom, and the associated enthalpy change is $\Delta_{EA}H$ (see *Table 1.5*). Step (6) is the hydration of gaseous X^-.

Step (1) causes some experimental difficulty. It is the reverse of the dissolution of gaseous HX in water to form solvated *undissociated* HX. Since HCl, HBr and HI are essentially fully dissociated in aqueous solution, measurement of enthalpy or entropy changes for step (1) must be estimated from somewhat unsatisfactory comparisons with noble gases and methyl halides. For HF, which is a weak acid in dilute aqueous solution, it might appear that values of ΔH° and ΔS° for step (1) could be obtained directly. However, IR spectroscopic data indicate that the species present in solution is the strongly hydrogen-bonded ion-pair $F^- \cdots HOH_2^+$.

We shall focus mainly on the conclusions drawn from calculations using the cycle in Figure 6.3.[†] Firstly, consider the change in enthalpy for the dissociation of HX(aq). Since values of ΔH° for each of steps (3) and (5) are independent of the halide, it is the sum of the values of ΔH° for steps (1), (2), (4) and (6) that determines the trend in the values of ΔH° for reaction 6.25. Figure 6.4 summarizes the data and

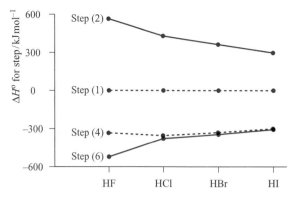

Fig. 6.4 Trends in the values of ΔH° for steps (1), (2), (4) and (6) defined in Figure 6.3. [Data: W.E. Dasent (1984) *Inorganic Energetics*, 2nd edn, Cambridge University Press, and references cited therein.]

illustrates why there is, in fact, rather little difference between the values of the overall enthalpy change for each of the hydrogen halides. Each reaction is exothermic, with ΔH° values in the order HF < HCl < HBr ≈ HI. If we now consider the $T\Delta S^\circ$ term for reaction 6.25 for each halide, the effect of its inclusion is rather dramatic, and leads to ΔG° for reaction 6.25 for X = F being positive while values of ΔG° for HCl, HBr and HI are negative (Table 6.2). From equation 6.26, calculated values of pK_a can be obtained, and these are listed in Table 6.2; for comparison, the *experimental* value of pK_a for HF is 3.45. Of great significance is that pK_a for HF is positive compared with negative values for HCl, HBr and HI. The enthalpy of dissolution of HF ($-\Delta H^\circ$ for step(1)) is significantly larger than those for the other hydrogen halides, and this, along with the much stronger bond in HF, outweighs the more negative enthalpy of hydration of F^-, making ΔH° for the dissociation process much less negative for HF than any of the other halides (Table 6.2). Entropy effects, although smaller, contribute in the same direction. It is easy to see that an explanation of the relative acid strengths of the hydrogen halides is not a trivial exercise. Moreover, electronegativity does *not* enter into the discussion: one must exercise care because it is all too easy to conclude from electronegativity (see *Table 1.7*) that HF is expected to be the strongest acid in the series.

Table 6.2 Thermodynamic data and calculated values of pK_a for the dissociation of the hydrogen halides in aqueous solution. The values of ΔH°, $T\Delta S^\circ$, ΔG° and pK_a refer to the dissociation process shown in Figure 6.3. For steps (3) and (5) in Figure 6.3, the values of ΔH° are 1312 and $-1091\,kJ\,mol^{-1}$ respectively.

	HF	**HCl**	**HBr**	**HI**
$\Delta H^\circ / kJ\,mol^{-1}$	−22	−63	−71	−68
$T\Delta S^\circ / kJ\,mol^{-1}$	−30	−10	−4	+3
$\Delta G^\circ / kJ\,mol^{-1}$	+8	−53	−67	−71
Calculated pK_a	1.4	−9.3	−11.7	−12.4

[†] For a fuller discussion, see: W.E. Dasent (1984) *Inorganic Energetics*, 2nd edn, Cambridge University Press, Chapter 5.

Table 6.3   Examples of series of oxoacids $EO_n(OH)_m$ for an element E; not all experimentally determined values of pK_a are known to the same degree of accuracy.

Formula of acid	$EO_n(OH)_m$ notation	Oxidation state of E	$pK_a(1)$	$pK_a(1)$ estimated by using Bell's rule
HNO_2	$N(O)(OH)$	$+3$	3.37	3
HNO_3	$N(O)_2(OH)$	$+5$	-1.64	-2
H_2SO_3	$S(O)(OH)_2$	$+4$	1.82	3
H_2SO_4	$S(O)_2(OH)_2$	$+6$	≈-3	-2
$HOCl$	$Cl(OH)$	$+1$	7.53	8
$HClO_2$	$Cl(O)(OH)$	$+3$	2.0	3
$HClO_3$	$Cl(O)_2(OH)$	$+5$	-1.0	-2
$HClO_4$	$Cl(O)_3(OH)$	$+7$	≈-8	-7

H_2S, H_2Se and H_2Te

Similar cycles to that in Figure 6.3 can be constructed for H_2S, H_2Se and H_2Te, allowing values of K_a to be estimated. Equations 6.28 to 6.30 give the first dissociation steps.

$$H_2S(aq) + H_2O(l) \rightleftharpoons [H_3O]^+(aq) + [HS]^-(aq)$$

$$pK_a(1) = 7.04 \qquad (6.28)$$

$$H_2Se(aq) + H_2O(l) \rightleftharpoons [H_3O]^+(aq) + [HSe]^-(aq)$$

$$pK_a = 4.0 \qquad (6.29)$$

$$H_2Te(aq) + H_2O(l) \rightleftharpoons [H_3O]^+(aq) + [HTe]^-(aq)$$

$$pK_a = 3.0 \qquad (6.30)$$

Although the explanation of the trend in values is not simple, and some data must be estimated (rather than being experimentally determined), it is apparent that the decrease in the $X-H$ bond strength with the increasing atomic number of X plays an important role in accounting for what is often thought to be a puzzling observation: as group 16 is descended and X becomes more metallic, its hydride becomes more acidic.

6.6 Trends within a series of oxoacids $EO_n(OH)_m$

For some elements with varying oxidation states, series of oxoacids with different numbers of oxygen atoms may exist (Table 6.3). There is no adequate thermodynamic treatment for rationalizing the observed trends within a series, but there are certain empirical methods for estimating K_a. The best known of these is Bell's rule (equation 6.31) which relates the first acid dissociation constant to the number of 'hydrogen-free' O atoms in an acid of formula $EO_n(OH)_m$.

$$pK_a \approx 8 - 5n \qquad (6.31)$$

Table 6.3 illustrates some comparisons between experimentally determined values of pK_a and those estimated

from Bell's rule. Of course, this empirical approach does not take into account the effects of changing element E.

It is often the case (experimentally) that successive values of pK_a for members of a series $EO_n(OH)_m$ differ by about 4 or 5. The increase in acid strength with increase in the number of O atoms attached to atom E is generally attributed to the greater possibility in the conjugate base of delocalization of negative charge on to the O atoms.

6.7 Aquated cations: formation and acidic properties

Water as a Lewis base

Although in this chapter we are mainly concerned with *Brønsted* acids and bases, it is important not to lose sight of the definition of *Lewis* acids and bases, and relevant to this chapter is the fact that water functions as a Lewis base when it acts as a solvent.

A *Lewis acid* is an electron acceptor, and a *Lewis base* is an electron donor.

When a metal salt dissolves in water, the cation and anion are hydrated; we discuss the energetics of this process in Section 6.9, but for now, we consider the interactions between the individual ions (freed from their ionic lattice on dissolution) and the solvent molecules. Consider the dissolution of NaCl; Figure 6.5a shows a schematic representation of the formation of the inner hydration shell around Na^+. The $O \cdots Na$ interaction can be described in terms of an *ion–dipole interaction*, while the solvation of the anion can be described in terms of the formation of hydrogen bonds between Cl^- and H atoms of surrounding H_2O molecules.

Hydration is the specific case of solvation when the solvent is water.

Fig. 6.5 (a) The first hydration shell of an Na^+ ion; ion–dipole interactions operate between the Na^+ ion and the H_2O molecules. (b) If the metal–oxygen bond possesses significant covalent character, the first hydration shell can be reasonably represented showing oxygen-to-metal ion coordinate bonds; however, there is also an ionic contribution to the bonding interaction.

Figure 6.5b shows another representation of a hexaaqua ion. Each O atom donates a pair of electrons to the metal M^{n+} ion, and each H_2O molecule acts as a Lewis base while the metal ion functions as a Lewis acid. We are implying that the M−O interaction is essentially covalent, in contrast to the case for Na^+ in Figure 6.5a. In practice, the character of the metal $\cdots$ oxygen interaction varies with the nature of the metal ion and relevant to this is the electro-neutrality principle (see *Section 19.6*).

The configurations **6.6** and **6.7** have been established in the first hydration shell for *dilute* solutions of LiCl and NaCl by detailed neutron diffraction studies. In concentrated solutions, the plane of the water molecule in **6.6** makes an angle of up to $50°$ with the $M^{n+} \cdots O$ axis (Figure 6.6) implying interaction of the cation with a lone pair of electrons rather than an ion–dipole interaction.

For both the cations and anion in NaCl and LiCl, there are six H_2O molecules in the primary hydration shell (Figure 6.5). Spectroscopic studies suggest that the hydration of other halide ions is similar to that of Cl^-, but for more complex anions, very few data are available. For a limited

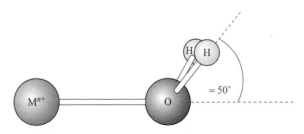

Fig. 6.6 If the plane of each water molecule in $[M(H_2O)_6]^{n+}$ makes an angle of $\approx 50°$ with the $M^{n+} \cdots O$ axis, it suggests that the metal–oxygen interaction involves the use of an oxygen lone pair.

number of hydrated cations, tracer methods and electronic and NMR spectroscopies provide reliable information about coordination number and stoichiometry.

Aquated cations as Brønsted acids

In the aqueous chemistry of cations, *hydrolysis* refers to the reversible loss of H^+ from an aqua species. The term hydrolysis is, however, used in a wider context, e.g. the reaction:

$$PCl_3 + 3H_2O \longrightarrow H_3PO_3 + 3HCl$$

is a hydrolysis process.

Aquated cations can act as Brønsted acids by loss of H^+ from a coordinated water molecule (equation 6.32).

$$[M(H_2O)_6]^{n+}(aq) + H_2O(l)$$
$$\rightleftharpoons [H_3O]^+(aq) + [M(H_2O)_5(OH)]^{(n-1)+}(aq) \quad (6.32)$$

The position of the equilibrium (and thus, the strength of the acid) depends on the degree to which the O−H bonds are *polarized*, and this is affected by the charge density of the cation (equation 6.33).

$$\text{Charge density of an ion} = \frac{\text{charge on the ion}}{\text{surface area of the ion}} \quad (6.33)$$

Surface area of sphere $= 4\pi r^2$

When H_2O coordinates to M^{n+}, charge is withdrawn towards the metal centre, leaving the H atoms more δ^+ (structure **6.8**) than in bulk water. Small cations such as Li^+, Mg^{2+}, Al^{3+}, Fe^{3+} and Ti^{3+} possess high charge densities, and in the corresponding hydrated ions, the H atoms carry significant positive charge. The pK_a values for $[Al(H_2O)_6]^{3+}$ and $[Ti(H_2O)_6]^{3+}$ (equations 6.34 and 6.35) illustrate the effect when the charge on the ion is high.

$$[Al(H_2O)_6]^{3+}(aq) + H_2O(l)$$
$$\rightleftharpoons [Al(H_2O)_5(OH)]^{2+}(aq) + [H_3O]^+(aq) \quad pK_a = 5.0$$
$$(6.34)$$

$$[Ti(H_2O)_6]^{3+}(aq) + H_2O(l)$$
$$\rightleftharpoons [Ti(H_2O)_5(OH)]^{2+}(aq) + [H_3O]^+(aq) \quad pK_a = 3.9$$
$$(6.35)$$

It is instructive to compare acid strengths of hexaaqua ions with other acids. The pK_a values of $MeCO_2H$ (equation 6.9) and HOCl (equation 6.15) are similar to that of $[Al(H_2O)_6]^{3+}$, while pK_a for $[Ti(H_2O)_6]^{3+}$ is close to that of HNO_2 (equation 6.14).

(a)

(b)

Fig. 6.7 (a) A schematic representation of the structure of the dinuclear cation $[Cr_2(\mu\text{-}OH)_2(H_2O)_8]^{4+}$. (b) The structure (X-ray diffraction) of this cation as determined for the salt $[Cr_2(\mu\text{-}OH)_2(H_2O)_8][2,4,6\text{-}Me_3C_6H_2SO_3]_4\cdot4H_2O$ [L. Spiccia *et al.* (1987) *Inorg. Chem.*, vol. 26, p. 474].

The characteristic colour of the $[Fe(H_2O)_6]^{3+}$ ion is purple, but aqueous solutions appear yellow due to the formation of the hydroxo species $[Fe(H_2O)_5(OH)]^{2+}$ and $[Fe(H_2O)_4(OH)_2]^+$ (equations 6.36 and 6.37).

$$[Fe(H_2O)_6]^{3+}(aq) + H_2O(l)$$
$$\rightleftharpoons [Fe(H_2O)_5(OH)]^{2+}(aq) + [H_3O]^+(aq) \qquad pK_a = 2.0$$
$$(6.36)$$

$$[Fe(H_2O)_5(OH)]^{2+}(aq) + H_2O(l)$$
$$\rightleftharpoons [Fe(H_2O)_4(OH)_2]^+(aq) + [H_3O]^+(aq) \qquad pK_a = 3.3$$
$$(6.37)$$

The facile dissociation of $[Fe(H_2O)_6]^{3+}$ means that its aqueous solutions must be stabilized by the addition of acid, which (by Le Chatelier's principle) drives equilibrium 6.36 to the left-hand side.

Proton loss is, in some cases, accompanied by the formation of dinuclear or polynuclear species in aqueous solution. For example, after the dissociation of H^+ from $[Cr(H_2O)_6]^{3+}$, the product undergoes an intermolecular condensation (equation 6.38). The resulting dichromium species (Figure 6.7) contains *bridging*† hydroxy groups.

$$2[Cr(H_2O)_5(OH)]^{2+}(aq)$$
$$\rightleftharpoons [(H_2O)_4Cr(\mu\text{-}OH)_2Cr(H_2O)_4]^{4+}(aq) + 2H_2O(l) \quad (6.38)$$

---

† The prefix μ means that the specified group is in a *bridging* position; μ$_3$ means a bridge between three atoms etc.

A similar reaction occurs in the corresponding V(III) system. On going from V(III) to V(IV), the charge density on the vanadium centre is sufficient to permit the dissociation of two protons from *one* coordinated H_2O, and the blue oxo-vanadium(IV) or vanadyl ion, **6.9**, is formed. It is common for this cation to be written simply as $[VO]^{2+}$, even though this is not a 'naked' vanadium oxo species.

(6.9)

6.8 Amphoteric oxides and hydroxides

Amphoteric behaviour

If an oxide or hydroxide is able to act as either an acid or a base, it is said to be *amphoteric*.

Some oxides and hydroxides are able to react with both acids and bases, thereby functioning as both bases and acids, respectively. Water is probably the most common example, but in this section we consider the *amphoteric* nature of metal oxides and hydroxides. Aluminium oxide, Al_2O_3, reacts with acids (equation 6.39) and with hydroxide ion (equation 6.40).

$$\gamma\text{-}Al_2O_3(s) + 3H_2O(l) + 6[H_3O]^+(aq)$$
$$\longrightarrow 2[Al(H_2O)_6]^{3+}(aq) \qquad (6.39)$$

$$Al_2O_3(s) + 3H_2O(l) + 2[OH]^-(aq) \longrightarrow 2[Al(OH)_4]^-(aq)$$
$$(6.40)$$

The hexaaqua ion, **6.10**, may be isolated as, for example, the sulfate salt after reaction with H_2SO_4. The ion $[Al(OH)_4]^-$, **6.11**, can be isolated as, for example, the Na^+ salt if the source of hydroxide is NaOH.

(6.10)

(6.11)

Similarly, aluminium hydroxide is amphoteric (equations 6.41 and 6.42).

$$Al(OH)_3(s) + KOH(aq) \longrightarrow K[Al(OH)_4](aq) \qquad (6.41)$$

$$Al(OH)_3(s) + 3HNO_3(aq) \longrightarrow Al(NO_3)_3(aq) + 3H_2O(l)$$
$$(6.42)$$

□ = Non-metallic elements   ▨ = Metallic elements

Fig. 6.8 The so-called 'diagonal line' divides metals from non-metals, although some elements that lie next to the line (e.g. Si) are semi-metals.

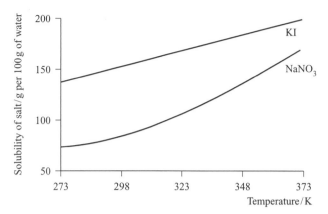

Fig. 6.9 The temperature-dependence of the solubilities in water of potassium iodide and sodium nitrate.

Periodic trends in amphoteric properties

As we discuss in later chapters, the character of the oxides of the elements across a row of the periodic table (*s*- and *p*-blocks) changes from basic to acidic, consistent with a change from metallic to non-metallic character of the element. Elements that lie close to the so-called 'diagonal line' (Figure 6.8) possess amphoteric oxides and hydroxides; in group 2, $Be(OH)_2$ and BeO are amphoteric, but $M(OH)_2$ and MO (M = Mg, Ca, Sr or Ba) are basic. Among the oxides of the *p*-block, Al_2O_3, Ga_2O_3, In_2O_3, GeO, GeO_2, SnO, SnO_2, PbO, PbO_2, As_2O_3, Sb_2O_3 and Bi_2O_3 are amphoteric. Within group 13, Ga_2O_3 is more acidic than Al_2O_3, whereas In_2O_3 is more *basic* than either Al_2O_3 or Ga_2O_3; for most of its chemistry, In_2O_3 can be regarded as having a basic rather than amphoteric nature. In group 14, both the metal(II) and metal(IV) oxides of Ge, Sn and Pb are amphoteric; in group 15, only the lower oxidation state oxides exhibit amphoteric behaviour, with the M_2O_5 oxides being acidic. For the oxides M_2O_3, basic character predominates as the group is descended: $As_2O_3 < Sb_2O_3 < Bi_2O_3$.

6.9 Solubilities of ionic salts

Solubility and saturated solutions

When an ionic solid, MX, is added to water, equilibrium 6.43 is established (if the ions formed are singly charged). When equilibrium is reached, the solution is *saturated*.

$$MX(s) \rightleftharpoons M^+(aq) + X^-(aq) \qquad (6.43)$$

The *solubility* of the solid *at a specified temperature* is the mass of solid that dissolves when equilibrium is reached in the presence of an excess of solid, divided by the mass of the solvent. The solubility of a salt is a dimensionless quantity, but is regularly expressed in terms of mass of solid (in g) per 100 g of water (or a similar quantity). The inclusion of temperature is vital, since solubility may depend significantly on temperature as is illustrated for KI and $NaNO_3$ in Figure 6.9.

> Tabulated values of *solubilities of ionic salts* refer to the maximum amount of solid that will dissolve in a given mass of water to give a saturated solution.

Sparingly soluble salts and solubility products

If the solubility of an ionic salt is extremely small (i.e. a saturated solution contains very few ions), the salt is said to be *sparingly soluble*. Such salts may include some that we might loosely refer to as being 'insoluble', for example AgCl and $BaSO_4$. Equation 6.44 shows the equilibrium that is established in aqueous solution when CaF_2 dissolves.

$$CaF_2(s) \rightleftharpoons Ca^{2+}(aq) + 2F^-(aq) \qquad (6.44)$$

An expression for the equilibrium constant should strictly be given in terms of the activities (see *Section 6.3*) of the species involved, but since we are dealing with very dilute solutions, we may express K in terms of concentrations (equation 6.45).

$$K = \frac{[Ca^{2+}][F^-]^2}{[CaF_2]} \qquad (6.45)$$

The activity of any solid is, *by convention*, unity. The equilibrium constant is thereby given in terms of the *equilibrium concentrations* of the dissolved ions and is referred to as the *solubility product*, or *solubility constant*, K_{sp} (equation 6.46).

$$K_{sp} = [Ca^{2+}][F^-]^2 \qquad (6.46)$$

Values of K_{sp} for a range of sparingly soluble salts are listed in Table 6.4.

Table 6.4 Values of K_{sp}(298 K) for selected sparingly soluble salts.

Compound	Formula	K_{sp}(298 K)
Barium sulfate	$BaSO_4$	1.07×10^{-10}
Calcium carbonate	$CaCO_3$	4.96×10^{-9}
Calcium hydroxide	$Ca(OH)_2$	4.68×10^{-6}
Calcium phosphate	$Ca_3(PO_4)_2$	2.07×10^{-33}
Iron(II) hydroxide	$Fe(OH)_2$	4.87×10^{-17}
Iron(II) sulfide	FeS	6.00×10^{-19}
Iron(III) hydroxide	$Fe(OH)_3$	2.64×10^{-39}
Lead(II) iodide	PbI_2	8.49×10^{-9}
Lead(II) sulfide	PbS	3.00×10^{-28}
Magnesium carbonate	$MgCO_3$	6.82×10^{-6}
Magnesium hydroxide	$Mg(OH)_2$	5.61×10^{-12}
Silver(I) chloride	$AgCl$	1.77×10^{-10}
Silver(I) bromide	$AgBr$	5.35×10^{-13}
Silver(I) iodide	AgI	8.51×10^{-17}
Silver(I) chromate	Ag_2CrO_4	1.12×10^{-12}
Silver(I) sulfate	Ag_2SO_4	1.20×10^{-5}

Worked example 6.3 Solubility product

The solubility product for PbI_2 is 8.49×10^{-9} (298 K). Calculate the solubility of PbI_2 in g per 100 g of water.

The equilibrium for the dissolution of lead(II) iodide is:

$$PbI_2(s) \rightleftharpoons Pb^{2+}(aq) + 2I^-(aq)$$

$$K_{sp} = [Pb^{2+}][I^-]^2$$

One mole of PbI_2 dissolves to give one mole of Pb^{2+} and two moles of I^-, and the solubility of PbI_2 (in $mol\,dm^{-3}$) equals the concentration of aqueous Pb^{2+}. Since $[I^-] = 2[Pb^{2+}]$, we can rewrite the expression for K_{sp}, and thus find $[Pb^{2+}]$:

$$K_{sp} = 4[Pb^{2+}]^3$$

$$8.49 \times 10^{-9} = 4[Pb^{2+}]^3$$

$$[Pb^{2+}] = \sqrt[3]{2.12 \times 10^{-9}} = 1.28 \times 10^{-3}\,mol\,dm^{-3}$$

The solubility of PbI_2 is thus $1.28 \times 10^{-3}\,mol\,dm^{-3}$ at 298 K.

Converting to g per 100 g of water:

$$M_r\,PbI_2 = 461$$

$$\text{Solubility of } PbI_2 = \frac{1.28 \times 10^{-3} \times 461}{10}$$

$$= 0.059\,\text{g per 100 g of water at 298 K}$$

Self-study exercises

1. The solubility product for Ag_2SO_4 is 1.20×10^{-5} (298 K). What is the solubility of Ag_2SO_4 in (a) $mol\,dm^{-3}$, and (b) g per 100 g of water?
   [*Ans.* (a) $1.44 \times 10^{-2}\,mol\,dm^{-3}$; (b) 0.45 g per 100 g]

2. If the solubility of AgI is $2.17 \times 10^{-6}\,g\,dm^{-3}$ calculate K_{sp}.
   [*Ans.* 8.50×10^{-17}]

The energetics of the dissolution of an ionic salt: $\Delta_{sol}G^\circ$

We can consider the equilibrium between a solid salt MX and its ions in saturated aqueous solution in terms of the thermodynamic cycle in equation 6.47.

$$(6.47)$$

where $\Delta_{lattice}G^\circ$ = standard free energy change accompanying the formation of the ionic lattice from gaseous ions; $\Delta_{hyd}G^\circ$ = standard free energy change accompanying the hydration of the gaseous ions; and $\Delta_{sol}G^\circ$ = standard free energy change accompanying the dissolution of the ionic salt.

In this cycle, $\Delta_{sol}G^\circ$ is related by equation 6.48 to the equilibrium constant, K, for the dissolution process; for a sparingly soluble salt, the equilibrium constant is K_{sp}.

$$\Delta_{sol}G^\circ = -RT \ln K \quad (6.48)$$

In principle, it is possible to use free energy data to calculate values of K and this is particularly valuable for accessing values of K_{sp}; however, there are two difficulties with determining values of $\Delta_{sol}G^\circ$ using cycle 6.47. First, $\Delta_{sol}G^\circ$ is a small difference between two much larger quantities (equation 6.49), neither of which is usually accurately known. The situation is made worse by the exponential relationship between $\Delta_{sol}G^\circ$ and K. Second, hydration energies are not very accessible quantities, as we shall discuss later on.

$$\Delta_{sol}G^\circ = \Delta_{hyd}G^\circ - \Delta_{lattice}G^\circ \quad (6.49)$$

An alternative method of accessing values of $\Delta_{sol}G^\circ$ is by using equation 6.50, which relates the free energies of formation for the species involved to the free energy change for reaction 6.43.

$$\Delta_{sol}G^\circ = \Delta_f G^\circ(M^+,aq) + \Delta_f G^\circ(X^-,aq) - \Delta_f G^\circ(MX,s) \quad (6.50)$$

Values of $\Delta_f G^\circ(M^+,aq)$ and $\Delta_f G^\circ(X^-,aq)$ can often be determined from standard reduction potentials (see *Appendix 11*) using equation 6.51, and tables giving values of $\Delta_f G^\circ(MX,s)$ for a wide range of salts are readily available. Equation 6.51 and its uses are discussed in detail in *Chapter 7*, and worked example 7.7 is especially relevant.

$$\Delta G^\circ = -zFE^\circ \quad (6.51)$$

where F = Faraday constant = $96\,485\,C\,mol^{-1}$.

The magnitude of $\Delta_{sol}G^\circ$ depends upon the balance between the corresponding $T\Delta_{sol}S^\circ$ and $\Delta_{sol}H^\circ$ terms (equation 6.52).

$$\Delta_{sol}G^\circ = \Delta_{sol}H^\circ - T\Delta_{sol}S^\circ \quad (6.52)$$

Table 6.5 Solubilities and values of the changes in free energy, enthalpy and entropy of solution for sodium fluoride, chloride and bromide; the entropy change is given in the form of a $T\Delta_{sol}S^{\circ}$ term ($T = 298$ K).

Compound	Solubility / g per 100 g of water at 298 K	Solubility / mol per 100 g of water at 298 K	$\Delta_{sol}G^{\circ}$ / kJ mol^{-1}	$\Delta_{sol}H^{\circ}$ / kJ mol^{-1}	$T\Delta_{sol}S^{\circ}$ / kJ mol^{-1}
NaF	4.2	0.10	+7.9	+0.9	−7.0
NaCl	36	0.62	−8.6	+3.9	+12.5
NaBr	91	0.88	−17.7	−0.6	+17.1

Thermochemical experiments (i.e. measuring the heat evolved or taken in during dissolution of an ionic salt) provide a method of determining values of the enthalpy change, $\Delta_{sol}H^{\circ}$. If $\Delta_{sol}G^{\circ}$ has been determined, then $\Delta_{sol}S^{\circ}$ can be derived using equation 6.52. Observed trends in the values of these thermodynamic parameters are not easily discussed, since a wide variety of factors contribute to the signs and magnitudes of $\Delta_{sol}S^{\circ}$ and $\Delta_{sol}H^{\circ}$, and hence to $\Delta_{sol}G^{\circ}$ and the actual solubility of a given salt. Table 6.5 lists relevant data for sodium halides. The increase in solubility on going from NaF to NaBr corresponds to a progressively more negative value for $\Delta_{sol}G^{\circ}$, and the $\Delta_{sol}H^{\circ}$ and $T\Delta_{sol}S^{\circ}$ terms *both* contribute to this trend. In contrast, the silver halides show the opposite behaviour, with the solubility in aqueous solution following the sequence AgF > AgCl > AgBr > AgI. While the values of the $T\Delta_{sol}S^{\circ}$ term become more positive on going from AgF to AgI (i.e. the same trend as in Table 6.5 for the sodium halides), the $\Delta_{hyd}H^{\circ}$ term also becomes more positive (i.e. the opposite trend to that for NaX). Combined in equation 6.52, these lead to values of $\Delta_{sol}G^{\circ}$ for AgF, AgCl, AgBr and AgI of −12.7, +55.8, +68.7 and +91.3 kJ mol^{-1} respectively. The origin of this result lies in the non-electrostatic contribution to the lattice energy which progressively stabilizes the solid with respect to aqueous ions on going from AgF to AgI (see *Section 5.15*). Even from a consideration of only two sets of metal halides, it is clear that providing general explanations for the observed trends in the solubilities of ionic salts is not possible.

The energetics of the dissolution of an ionic salt: hydration ·of ions

We have already seen (equation 6.47) that the free energy change accompanying the hydration of an ionic salt contributes towards the solubility of the salt, and we have also mentioned that values of $\Delta_{hyd}G^{\circ}$ and the corresponding enthalpy and entropy changes are not readily accessible quantities. In this section, we look more closely at $\Delta_{hyd}G^{\circ}$, $\Delta_{hyd}H^{\circ}$ and $\Delta_{hyd}S^{\circ}$; equation 6.53 gives the general hydration processes to which these quantities refer.

$$\left.\begin{array}{l} M^{+}(g) \longrightarrow M^{+}(aq) \\ X^{-}(g) \longrightarrow X^{-}(aq) \end{array}\right\} \tag{6.53}$$

The primary problem is that individual ions cannot be studied in isolation, and experimental measurements of

$\Delta_{hyd}H^{\circ}$ are restricted to those involving pairs of ions that do not interact. Even then, the problem is non-trivial.

In principle, the value of $\Delta_{hyd}G^{\circ}$ (in J mol^{-1}) for an ion of charge ze and radius r_{ion} (in m) can be calculated on the basis of electrostatics using equation 6.54.

$$\Delta_{hyd}G^{\circ} = -\frac{Lz^{2}e^{2}}{8\pi\varepsilon_{0}r_{ion}}\left(1 - \frac{1}{\varepsilon_{r}}\right) \tag{6.54}$$

where L = Avogadro number = 6.022×10^{23} mol^{-1}; e = charge on the electron = 1.602×10^{-19} C; ε_{0} = permittivity of a vacuum = 8.854×10^{-12} F m^{-1}; and ε_{r} = relative permittivity of the water (dielectric constant) = 78.7.

In practice, this expression gives unsatisfactory results since the relative permittivity (see *Section 8.2*) of bulk water is not valid within the region of the ion, and available values of r_{ion} refer to ionic lattices rather than hydrated ions.

The simplest way of obtaining thermodynamic functions of hydration for individual ions rests on the assumption that very large ions such as $[Ph_4As]^{+}$ and $[BPh_4]^{-}$ have the same values of $\Delta_{hyd}G^{\circ}$ etc. From data for salts containing appropriate cation–anion pairs (e.g. $[Ph_4As][BPh_4]$, $[Ph_4As]Cl$ and $K[BPh_4]$), data for the individual ions can be derived (e.g. K^{+} and Cl^{-}). However, direct *experimental* measurements involving $[Ph_4As][BPh_4]$ are not feasible because of the low solubility of this salt in water. Hence, data for this compound come from theory.

An alternative method for obtaining thermodynamic functions of hydration, is based upon an arbitrary assignment of a value of $\Delta_{hyd}H^{\circ}(H^{+},g) = 0$. From this starting point, and using values of $\Delta_{hyd}H^{\circ}$ for a range of ionic *salts* and the hydrogen halides, a self-consistent set of *relative* hydration enthalpies can be obtained. More sophisticated methods are based upon the estimation of $\Delta_{hyd}H^{\circ}(H^{+},g) = -1019$ kJ mol^{-1}, and Table 6.6 lists corresponding absolute values of $\Delta_{hyd}H^{\circ}$ for a range of ions.

Values of hydration entropies, $\Delta_{hyd}S^{\circ}$, can be derived by assigning (by convention) a value of zero for the absolute entropy, S°, of gaseous H^{+}. Table 6.6 lists values of $\Delta_{hyd}S^{\circ}$ for selected ions, and the corresponding values of $\Delta_{hyd}G^{\circ}$ are obtained by substitution of $\Delta_{hyd}S^{\circ}$ and $\Delta_{hyd}H^{\circ}$ into equation 6.52 ($T = 298$ K). Inspection of Table 6.6 reveals several points of interest:

- Highly charged ions have more negative values of $\Delta_{hyd}H^{\circ}$ and $\Delta_{hyd}S^{\circ}$ than singly charged ions; the more negative enthalpy term is rationalized in terms of simple

Table 6.6 Absolute values of $\Delta_{hyd}H^\circ$, $\Delta_{hyd}S^\circ$, $\Delta_{hyd}G^\circ$ (at 298 K), and ionic radii for selected ions.

Ion	$\Delta_{hyd}H^\circ / kJ\,mol^{-1}$	$\Delta_{hyd}S^\circ / J\,K^{-1}\,mol^{-1}$	$T\Delta_{hyd}S^\circ / kJ\,mol^{-1}$ (for $T = 298$ K)	$\Delta_{hyd}G^\circ / kJ\,mol^{-1}$	$r_{ion} / pm^{\ddagger}$
H^+	−1091	−130	−39	−1052	–
Li^+	−519	−140	−42	−477	76
Na^+	−404	−110	−33	−371	102
K^+	−321	−70	−21	−300	138
Rb^+	−296	−70	−21	−275	149
Cs^+	−271	−60	−18	−253	170
Mg^{2+}	−1931	−320	−95	−1836	72
Ca^{2+}	−1586	−230	−69	−1517	100
Sr^{2+}	−1456	−220	−66	−1390	126
Ba^{2+}	−1316	−200	−60	−1256	142
Al^{3+}	−4691	−530	−158	−4533	54
La^{3+}	−3291	−430	−128	−3163	105
F^-	−504	−150	−45	−459	133
Cl^-	−361	−90	−27	−334	181
Br^-	−330	−70	−21	−309	196
I^-	−285	−50	−15	−270	220

‡Values of r_{ion} refer to a coordination number of 6 in the solid state.

electrostatic attraction, and the more negative $\Delta_{hyd}S^\circ$ values can be considered in terms of highly charged ions imposing more order on H_2O molecules in the environment of the ion.

- For ions of a given charge, $\Delta_{hyd}H^\circ$ and $\Delta_{hyd}S^\circ$ show some dependence on ion size (i.e. r_{ion}); smaller ions possess more negative values of both $\Delta_{hyd}H^\circ$ and $\Delta_{hyd}S^\circ$.
- The variation in $\Delta_{hyd}H^\circ$ outweighs that in $T\Delta_{hyd}S^\circ$, and as a result, the most negative values of $\Delta_{hyd}G^\circ$ arise for small ions (comparing those with a constant charge), and for highly charged ions (comparing those of similar size).
- For monatomic ions of about the same size, (e.g. K^+ and F^-), anions are more strongly hydrated than cations (more negative $\Delta_{hyd}G^\circ$).

Solubilities: some concluding remarks

Let us now return to equation 6.47, and relate the observed solubility of a salt to the magnitude of the difference between $\Delta_{lattice}G^\circ$ and $\Delta_{hyd}G^\circ$, and in particular to the sizes of the ions involved.

First, we reiterate that $\Delta_{sol}G^\circ$ is generally a *relatively* small value, being the difference between two much larger values ($\Delta_{lattice}G^\circ$ and $\Delta_{hyd}G^\circ$). Moreover, as Table 6.5 illustrates, $\Delta_{sol}G^\circ$ can be either positive or negative, whereas $\Delta_{lattice}G^\circ$ and $\Delta_{hyd}G^\circ$ are always negative values (provided they are defined as in equation 6.47).

As we saw in Table 6.6, of the two terms $\Delta_{hyd}H^\circ$ and $T\Delta_{hyd}S^\circ$, the dominant factor in determining the magnitude of $\Delta_{hyd}G^\circ$ is $\Delta_{hyd}H^\circ$; similarly, for $\Delta_{lattice}G^\circ$, the dominant factor is $\Delta_{lattice}H^\circ$. Thus, in considering the relationship between the solubility of a salt and the sizes of the component ions, we can turn our attention to the relationships between r_{ion}, $\Delta_{hyd}H^\circ$ and $\Delta_{lattice}H^\circ$ given in equations 6.55

and 6.56. The actual *values* of $\Delta_{hyd}H^\circ$ and $\Delta_{lattice}H^\circ$ are always negative.

$$\Delta_{lattice}H^\circ \propto \frac{1}{r_+ + r_-} \tag{6.55}$$

$$\Delta_{hyd}H^\circ \propto \frac{1}{r_+} + \frac{1}{r_-} \tag{6.56}$$

r_+ = radius of cation; r_- = radius of anion

Now consider the application of these two expressions to a series of salts of similar lattice type. If $r_- \gg r_+$, equation 6.55 shows that there will be little variation in $\Delta_{lattice}H^\circ$. However, upon dissolution, if $r_- \gg r_+$, $\Delta_{hyd}H^\circ$(cation) will be much more negative than $\Delta_{hyd}H^\circ$(anion) for all values of r_+. Thus, $\Delta_{hyd}H^\circ$ will be roughly proportional to $\frac{1}{r_+}$. Thus, along a series of related salts with increasing r_+, but with $r_- \gg r_+$, $\Delta_{lattice}H^\circ$ will remain nearly constant while $\Delta_{hyd}H^\circ$ becomes *less* negative. Hence, $\Delta_{sol}H^\circ$ (and thus $\Delta_{sol}G^\circ$) will become less negative (equation 6.57) and solubility will decrease.

$$\Delta_{sol}H^\circ = \Delta_{hyd}H^\circ - \Delta_{lattice}H^\circ \tag{6.57}$$

Such a series is exemplified by the alkali metal hexachloroplatinates; the hydrated sodium salt has a very high solubility, while the solubilities of $K_2[PtCl_6]$, $Rb_2[PtCl_6]$ and $Cs_2[PtCl_6]$ are 2.30×10^{-3}, 2.44×10^{-4} and 1.04×10^{-4} mol per 100 g of water (at 293 K). A similar trend is observed for alkali metal hexafluorophosphates (MPF_6).

Although the above, and similar, arguments are qualitative, they provide a helpful means of assessing the pattern in solubilities for series of *ionic* salts; we stress 'ionic' because equations 6.55 and 6.56 assume an electrostatic model. Our discussions in *Section 5.15* indicated how partial covalent character in silver halides affects solubility trends.

6.10 Common-ion effect

So far, we have discussed aqueous solutions containing a single, dissolved ionic salt, MX. Now we consider the effect of adding a second salt which has one of its ions in common with the first salt.

> If a salt MX is added to an aqueous solution containing the solute MY (the ion M^{n+} is common to both salts), the presence of the dissolved M^{n+} ions suppresses the dissolution of MX compared with that in pure water; this is the *common-ion effect*.

The origin of the common-ion effect is seen by applying Le Chatelier's principle. In equation 6.58, the presence of Cl^- in solution (from a soluble salt such as KCl) will suppress the dissolution of AgCl.

$$AgCl(s) \rightleftharpoons Ag^+(aq) + Cl^-(aq) \qquad (6.58)$$

The effect is analogous to that of mixing a weak acid with the salt of that acid (e.g. acetic acid and sodium acetate) to form a buffer solution.

Worked example 6.4   The common-ion effect

The value of K_{sp} for AgCl is 1.77×10^{-10} (at 298 K). Compare the solubility of AgCl in water and in $0.01 \, mol \, dm^{-3}$ hydrochloric acid.

First, determine the solubility of AgCl in water.

$$AgCl(s) \rightleftharpoons Ag^+(aq) + Cl^-(aq)$$

$$K_{sp} = [Ag^+][Cl^-] = 1.77 \times 10^{-10}$$

Since the concentrations of $[Ag^+]$ and $[Cl^-]$ in aqueous solution are equal, we can write:

$$[Ag^+]^2 = 1.77 \times 10^{-10}$$

$$[Ag^+] = 1.33 \times 10^{-5} \, mol \, dm^{-3}$$

The solubility of AgCl is therefore $1.33 \times 10^{-5} \, mol \, dm^{-3}$.

Now consider the solubility of AgCl in $0.01 \, mol \, dm^{-3}$ HCl aqueous solution.

HCl is essentially fully dissociated and thus, $[Cl^-] = 0.01 \, mol \, dm^{-3}$.

	$AgCl(s) \rightleftharpoons$	$Ag^+(aq) +$	$Cl^-(aq)$
Initial aqueous ion concentrations / $mol \, dm^{-3}$:		0	0.01
Equilibrium concentrations / $mol \, dm^{-3}$:		x	$(0.01 + x)$

$$K_{sp} = 1.77 \times 10^{-10} = [Ag^+][Cl^-]$$

$$1.77 \times 10^{-10} = x(0.01 + x)$$

Since x is obviously much less than 0.01, we can make the approximation that $0.01 + x \approx 0.01$.

$$1.77 \times 10^{-10} \approx 0.01x$$

$$x \approx 1.77 \times 10^{-8} \, mol \, dm^{-3}$$

The solubility of AgCl is therefore $1.77 \times 10^{-8} \, mol \, dm^{-3}$.

Conclusion: the solubility of AgCl is ≈ 1000 times less in $0.01 \, mol \, dm^{-3}$ aqueous HCl solution than in water.

Self-study exercises

Use K_{sp} data above.

1. How much more soluble is AgCl in water than in $5 \times 10^{-3} \, mol \, dm^{-3}$ aqueous HCl at 298 K?

   [*Ans.* ≈ 375 times]

2. What is the solubility of AgCl in $0.02 \, mol \, dm^{-3}$ aqueous KCl? [*Ans.* $8.85 \times 10^{-9} \, mol \, dm^{-3}$]

Worked example 6.4 illustrates the use of the common-ion effect in gravimetric analysis; AgCl is always precipitated from a solution containing a slight excess of a common ion, Cl^- or Ag^+, in the determination of silver or chloride respectively.

> *Gravimetric analysis* is a quantitative technique in which the material under study is isolated as a precipitate.

6.11 Coordination complexes: an introduction

Definitions and terminology

Although we discuss complex formation in detail later in the book, in this section we introduce some general principles concerning the coordination of *ligands* to ions in aqueous solution. The word *ligand* is derived from the Latin verb '*ligare*' meaning 'to bind'.

> In a *coordination complex*, a central atom or ion is coordinated by one or more molecules or ions (*ligands*) which act as Lewis bases, forming *coordinate bonds* with the central atom or ion; the latter acts as a Lewis acid. Atoms in the ligands that are directly bonded to the central atom or ion are *donor atoms*.

Examples of coordination complexes include those involving *d*-block metal ions (e.g. $[Co(NH_3)_6]^{2+}$, **6.12**) and species with a central *p*-block element (e.g. $[BF_4]^-$, **6.13**, and $H_3B\cdot THF$, **6.14**) (THF = tetrahydrofuran) although **6.14** is unstable with respect to hydrolysis in aqueous solution. Equations 6.59–6.61 show the formation of these coordination complexes.

> In a complex:
> - a *line* is used to denote the interaction between an *anionic* ligand and the acceptor;
> - an *arrow* is used to show the donation of an electron pair from a *neutral* ligand to an acceptor.

(6.12)

(6.13)

(6.14)

$$[Co(H_2O)_6]^{2+} + 6NH_3 \rightleftharpoons [Co(NH_3)_6]^{2+} + 6H_2O \quad (6.59)$$

$$BF_3 + F^- \rightleftharpoons [BF_4]^- \quad (6.60)$$

$$BH_3 + THF \rightleftharpoons H_3B\cdot THF \quad (6.61)$$

> When a Lewis base donates a pair of electrons to a Lewis acid, a *coordinate bond* is formed and the resulting species is an *adduct*. The centred-dot in, for example, $H_3B\cdot THF$ indicates the formation of an adduct.

In $[BF_4]^-$, the B−F bond formed in reaction 6.60 *is identical* to the other three B−F bonds; all are 2c-2e covalent bonds. In structures **6.12**–**6.14**, the coordinate bond between the central atom or ion and a *neutral ligand* is denoted by an *arrow*, but if the ligand is *anionic*, the coordinate bond is indicated by a line. This convention is sometimes ignored, for example, when the stereochemistry of the coordination complex is illustrated; compare **6.12** with **6.15** which shows the octahedral environment of the Co(II) centre.

(6.15)

Investigating coordination complex formation

The formation of complexes in aqueous solution may be studied by a number of methods, of which testing the modifications of chemical properties is only one, and a somewhat unreliable one at that. *All* reactions are equilibria, and chemical tests are often only investigations of *relative* values of equilibrium constants. For example, in an aqueous solution of an Ag^+ salt saturated with NH_3, nearly all the Ag^+ is present as the complex $[Ag(NH_3)_2]^+$ (equation 6.62).

$$Ag^+(aq) + 2NH_3(aq) \rightleftharpoons [Ag(NH_3)_2]^+(aq) \quad (6.62)$$

On adding a chloride-containing solution, *no* AgCl precipitate is observed; however, the addition of an iodide-containing solution results in the precipitation of silver iodide. These observations can be rationalized as follows: AgI ($K_{sp} = 8.51 \times 10^{-17}$) is much less soluble in aqueous solution than AgCl ($K_{sp} = 1.77 \times 10^{-10}$). The fact that no AgCl is precipitated means that the equilibrium constant for reaction 6.62 is sufficiently large that the AgCl formed is soluble in the solution (i.e. very little uncomplexed Ag^+ is available for combination with Cl^-). On the other hand, the solubility of AgI is so low, that even the formation of a small amount produces a precipitate.

Physical methods (e.g. electronic and vibrational spectroscopic, solubility or conductivity measurements) provide more reliable information and, in some cases, allow the determination of equilibrium constants for complex formation.

Neutral complexes are usually only sparingly soluble in water, but are often readily soluble in organic solvents. For example, the red complex $[Fe(acac)_3]$ (Figure 6.10) (Hacac is the abbreviation for acetylacetone, the systematic name for which is pentane-2,4-dione) can be extracted from aqueous solution into benzene or chloroform, and the formation of $[Fe(acac)_3]$ is used as a means of extracting Fe(III) from solution. Pentane-2,4-dione is a *β-diketone* and deprotonation gives $[acac]^-$, a *β-diketonate* (equation 6.63). The formation of $[Fe(acac)_3]$ in aqueous solution involves equilibria 6.63 and 6.64.

$$K_a = 1 \times 10^{-9} \quad (6.63)$$

$$Fe^{3+}(aq) + 3[acac]^-(aq) \rightleftharpoons [Fe(acac)_3](aq)$$

$$K = 1 \times 10^{26} \quad (6.64)$$

The amount of complex formed depends on the pH of the solution. If the pH is too low, H^+ ions compete with Fe^{3+} ions for the ligand (i.e. the back reaction 6.63 competes with the forward reaction 6.64). If the pH is too high,

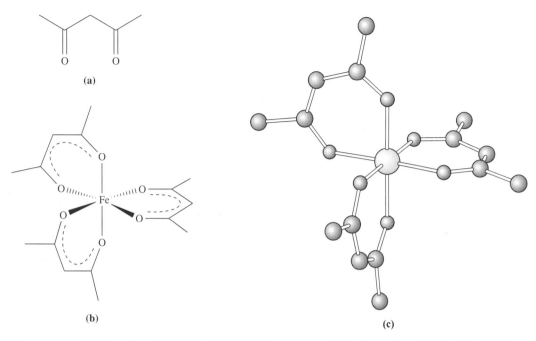

Fig. 6.10 (a) The structure of Hacac (see *Table 6.7*); (b) Fe(III) forms an octahedral complex with [acac]⁻; (c) the structure of the coordination complex [Fe(acac)₃], determined by X-ray diffraction. [J. Iball *et al.* (1967) *Acta Crystallogr.*, vol. 23, p. 239.]

BOX 6.3 APPLICATIONS

The use of solvent extraction in nuclear reprocessing

In Section 2.5, we discussed the production of energy by nuclear fission, and the reprocessing of nuclear fuels. We described how short-lived radioactive products decay during pond storage, and how uranium is converted into $[UO_2][NO_3]_2$ and, finally, UF_6. One of the complicating factors in this process is that the fuel to be reprocessed contains plutonium and fission products in addition to uranium. Two different solvent extraction processes are needed to effect separation.

Tributyl phosphate (TBP)

Stage 1: separation of the fission products from plutonium and uranium nitrates

The mixture to be separated contains $[UO_2]^{2+}$ and Pu(IV) nitrates, as well as metal ions such as $^{90}_{38}Sr^{2+}$. Kerosene is added to the aqueous solution of metal salts, giving a *two-phase* system (i.e. these solvents are immiscible). Tributyl phosphate (TBP, a phosphate ester) is added to form complexes with the uranium-containing and plutonium ions, extracting them into the kerosene layer. The fission products remain in the aqueous solution, and separation of the solvent layers thus achieves separation of the fission products from Pu- and U-containing species. Repeated extractions from the aqueous layer by the same process increases the efficiency of the separation.

Stage 2: separation of plutonium and uranium nitrates

The kerosene fraction is now subjected to a second solvent extraction. Addition of iron(II) sulfamate, $Fe(NH_2SO_3)_2$, and shaking of the kerosene fraction with water, results in the formation of plutonium(III) nitrate which is partitioned into the aqueous layer. $[UO_2][NO_3]_2$ resists reduction, is complexed by TBP and extracted into the organic layer. Separation of the two solvent fractions thus separates the uranium and plutonium salts; repeated extractions result in a highly efficient separation. The extraction of $[UO_2][NO_3]_2$ from kerosene back into an aqueous phase can be achieved by adding nitric acid; under these conditions, the uranium-TBP complex dissociates and $[UO_2][NO_3]_2$ returns to the aqueous layer.

Fe(III) is precipitated as $Fe(OH)_3$ for which $K_{sp} = 2.64 \times 10^{-39}$. Thus, there is an optimum pH for the extraction of Fe(III) from aqueous media using Hacac and a given organic solvent (e.g. $CHCl_3$). Although we have defined ligands as being *Lewis* bases, most are also Brønsted bases, and accurate pH control is of great importance in studies of complex formation. Solvent extraction is important in the analytical and industrial separation of many metals (see *Box 6.3*).

Solvent extraction involves the extraction of a substance using a suitable solvent; in a two-phase solvent system, the solute is extracted from one solvent into another, the extracting solvent being chosen so that impurities remain in the original solvent.

6.12 Stability constants of coordination complexes

As we saw earlier, metal ions in aqueous solution are hydrated; the aqua species may be denoted as $M^{z+}(aq)$ where this often represents the hexaaqua ion $[M(H_2O)_6]^{z+}$. Now consider the addition of a neutral ligand L to the solution, and the formation of a series of complexes $[M(H_2O)_5L]^{z+}$, $[M(H_2O)_4L_2]^{z+} \ldots [ML_6]^{z+}$. Equilibria 6.65–6.70 show the stepwise displacements of coordinated H_2O by L.

$$[M(H_2O)_6]^{z+}(aq) + L(aq)$$
$$\rightleftharpoons [M(H_2O)_5L]^{z+}(aq) + H_2O(l) \quad (6.65)$$

$$[M(H_2O)_5L]^{z+}(aq) + L(aq)$$
$$\rightleftharpoons [M(H_2O)_4L_2]^{z+}(aq) + H_2O(l) \quad (6.66)$$

$$[M(H_2O)_4L_2]^{z+}(aq) + L(aq)$$
$$\rightleftharpoons [M(H_2O)_3L_3]^{z+}(aq) + H_2O(l) \quad (6.67)$$

$$[M(H_2O)_3L_3]^{z+}(aq) + L(aq)$$
$$\rightleftharpoons [M(H_2O)_2L_4]^{z+}(aq) + H_2O(l) \quad (6.68)$$

$$[M(H_2O)_2L_4]^{z+}(aq) + L(aq)$$
$$\rightleftharpoons [M(H_2O)L_5]^{z+}(aq) + H_2O(l) \quad (6.69)$$

$$[M(H_2O)L_5]^{z+}(aq) + L(aq)$$
$$\rightleftharpoons [ML_6]^{z+}(aq) + H_2O(l) \quad (6.70)$$

The equilibrium constant, K_1, for reaction 6.65 is given by equation 6.71; $[H_2O]$ (strictly, the *activity* of H_2O) is unity (see *Section 6.3*) and does not appear in the expression for K.

$$K_1 = \frac{[M(H_2O)_5L^{z+}]}{[M(H_2O)_6^{z+}][L]} \quad (6.71)$$

In the formation of a complex $[ML_6]^{z+}$ from $[M(H_2O)_6]^{z+}$, each displacement of a coordinated water molecule by ligand L has a characteristic *stepwise stability constant*, K_1, K_2, K_3, K_4, K_5 or K_6.

Alternatively, we may consider the overall formation of $[ML_6]^{z+}$ (equation 6.72); in order to distinguish stepwise and overall stability constants, the symbol β is generally used for the latter. Equation 6.73 gives an expression for β_6 for $[ML_6]^{z+}$. We must refer to β_6 and not just β, because overall stability constants for the products of each of reactions 6.65–6.70 can also be defined.

$$[M(H_2O)_6]^{z+}(aq) + 6L(aq) \rightleftharpoons [ML_6]^{z+}(aq) + 6H_2O(l) \quad (6.72)$$

$$\beta_6 = \frac{[ML_6^{z+}]}{[M(H_2O)_6^{z+}][L]^6} \quad (6.73)$$

Values of K and β can be related; e.g. for equilibrium 6.72, β_6 can be expressed in terms of the six stepwise stability constants according to equations 6.74.

$$\left. \begin{aligned} \beta_6 &= K_1 \times K_2 \times K_3 \times K_4 \times K_5 \times K_6 \\ &\text{or} \\ \log \beta_6 &= \log K_1 + \log K_2 + \log K_3 \\ &\quad + \log K_4 + \log K_5 + \log K_6 \end{aligned} \right\} \quad (6.74)$$

For the formation of a complex $[ML_n]^{z+}$ from $[M(H_2O)_m]^{z+}$ and ligand L, the overall stability constant β_n is given by the expression:

$$\beta_n = \frac{[ML_n^{z+}]}{[M(H_2O)_m^{z+}][L]^n}$$

Worked example 6.5   Formation of $[Ni(H_2O)_{6-x}(NH_3)_x]^{2+}$

Results of a pH study using a glass electrode (in 2 M NH_4NO_3 aqueous solution) give values of the stepwise stability constants (at 303 K) of $[Ni(H_2O)_{6-x}(NH_3)_x]^{2+}$ ($x = 1$–6) as: $\log K_1 = 2.79$; $\log K_2 = 2.26$; $\log K_3 = 1.69$; $\log K_4 = 1.25$; $\log K_5 = 0.74$; $\log K_6 = 0.03$. Calculate (a) β_6 for $[Ni(NH_3)_6]^{2+}$, and (b) $\Delta G^\circ_1(303 \text{ K})$. (c) If the value of $\Delta H^\circ_1(303 \text{ K}) = -16.8 \text{ kJ mol}^{-1}$, calculate $\Delta S^\circ_1(303 \text{ K})$. ($R = 8.314 \text{ J K}^{-1} \text{ mol}^{-1}$)

(a) $\beta_6 = K_1 \times K_2 \times K_3 \times K_4 \times K_5 \times K_6$

$\log \beta_6 = \log K_1 + \log K_2 + \log K_3$

$\qquad + \log K_4 + \log K_5 + \log K_6$

$\log \beta_6 = 2.79 + 2.26 + 1.69 + 1.25 + 0.74 + 0.03$

$\qquad = 8.76$

$\beta_6 = 5.75 \times 10^8$

(b) $\Delta G^{\circ}{}_1(303\,\text{K})$ refers to the stepwise formation of $[\text{Ni}(\text{H}_2\text{O})_5(\text{NH}_3)]^{2+}$.

$$\Delta G^{\circ}{}_1(303\,\text{K}) = -RT \ln K_1$$

$$= -(8.314 \times 10^{-3} \times 303)\ln 10^{2.79}$$

$$= -16.2\,\text{kJ mol}^{-1}$$

(c) $\Delta G^{\circ}{}_1 = \Delta H^{\circ}{}_1 - T\Delta S^{\circ}{}_1$

$$\Delta S^{\circ}{}_1 = \frac{\Delta H^{\circ}{}_1 - \Delta G^{\circ}{}_1}{T}$$

$$\Delta S^{\circ}{}_1(303\,\text{K}) = \frac{-16.8 - (-16.2)}{303}$$

$$= -1.98 \times 10^{-3}\,\text{kJ K}^{-1}\,\text{mol}^{-1}$$

$$= -1.98\,\text{J K}^{-1}\,\text{mol}^{-1}$$

Self-study exercises

These questions refer to $[\text{Ni}(\text{H}_2\text{O})_{6-x}(\text{NH}_3)_x]^{2+}$ ($x = 1\text{--}6$), with data quoted at 303 K.

1. Determine $\Delta G^{\circ}{}_2(303\,\text{K})$ if $\log K_2 = 2.26$.
   [*Ans.* $-13.1\,\text{kJ mol}^{-1}$]

2. If $\Delta S^{\circ}{}_1(303\,\text{K}) = -1.98\,\text{J K}^{-1}\,\text{mol}^{-1}$, confirm that $\Delta H^{\circ}{}_1(303\,\text{K}) = -16.8\,\text{kJ mol}^{-1}$, given that $\log K_1 = 2.79$.

3. Given the values $\log K_1 = 2.79$, $\log K_2 = 2.26$ and $\log K_3 = 1.69$, use the appropriate value to determine $\Delta G^{\circ}(303\,\text{K})$ for the equilibrium:

$$[\text{Ni}(\text{H}_2\text{O})_4(\text{NH}_3)_2]^{2+} + \text{NH}_3 \rightleftharpoons [\text{Ni}(\text{H}_2\text{O})_3(\text{NH}_3)_3]^{2+} + \text{H}_2\text{O}$$

[*Ans.* $-9.8\,\text{kJ mol}^{-1}$]

Determination of stability constants

For a given aqueous solution containing known concentrations of a metal ion M^{z+} and ligand L, it may have been found that only *one* coordination complex of known formula is present in solution. If this is the case, then the stability constant for this complex can be obtained directly from a determination of the concentration of uncomplexed M^{z+}, L or complexed M^{z+} in that solution. Such determinations can be made by polarographic or potentiometric measurements (if a suitable reversible electrode exists), by pH measurements (if the ligand is the conjugate base of a weak acid), or by ion-exchange, spectrophotometric (i.e. observation of electronic spectra and use of the Beer–Lambert Law), NMR spectroscopic or distribution methods.

Trends in stepwise stability constants

Figure 6.11 shows that for the formation of the complex ions $[\text{Al}(\text{H}_2\text{O})_{6-x}\text{F}_x]^{(3-x)+}$ ($x = 1\text{--}6$), the stepwise stability constants become smaller as more F^- ligands are introduced;

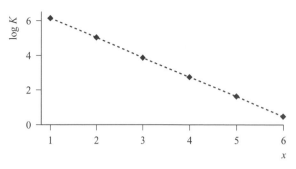

Fig. 6.11 Stepwise stability constants for the formation of $[\text{Al}(\text{H}_2\text{O})_{6-x}\text{F}_x]^{(3-x)+}$ ($x = 1\text{--}6$).

a similar trend is also observed in the formation of $[\text{Ni}(\text{H}_2\text{O})_{6-x}(\text{NH}_3)_x]^{2+}$ ($x = 1\text{--}6$) in worked example 6.5. This decrease in values of K is typical of many systems; however, the trend is not always so smooth as in Figure 6.11. (Stability constants are discussed further in *Section 20.10*.)

Thermodynamic considerations of complex formation: an introduction

A detailed discussion of the thermodynamics of complex formation in aqueous solution lies beyond the scope of this book, but we discuss briefly entropy changes that accompany the formation of coordination compounds in solution, and the so-called *chelate effect*. In *Chapter 20*, we look further at the thermodynamics of complex formation.

We saw in Section 6.9 that highly charged ions have more negative values of $\Delta_{\text{hyd}}S^{\circ}$ than singly charged ions, and this can be viewed in terms of the highly charged ions imposing more order on H_2O molecules in the environment of the ion. When complex formation occurs between highly charged cations and anions, with a resulting partial or total cancellation of charges, the changes in enthalpy for these processes are significantly *negative*. However, the accompanying changes in entropy are significantly *positive* because less order is imposed on the H_2O molecules around the complex ion than around the uncomplexed, metal cations and anionic ligands. The corresponding values of ΔG° are, therefore, substantially negative indicating that very stable complexes are formed. For example, $\Delta S^{\circ}(293\,\text{K})$ for reaction 6.75 is $+113\,\text{J K}^{-1}\,\text{mol}^{-1}$; the ligand in equation 6.75 is $[\text{EDTA}]^{4-}$.[†]

[†] In the *solid state*, the complex formed between Ca^{2+} and $[\text{EDTA}]^{4-}$ is cation-dependent and is 7- or 8-coordinate; the additional coordination sites are occupied by H_2O, and similarly in $[\text{Mg}(\text{EDTA})(\text{H}_2\text{O})]^{2-}$.

$$(6.75)$$

Another source of increase in entropy is important: when we are dealing with *comparable* uncharged ligands (e.g. $MeNH_2$ and $H_2NCH_2CH_2NH_2$), *polydentate* ligands form more stable complexes than *monodentate* ones.

The number of donor atoms through which a ligand coordinates to a metal ion is defined as the *denticity* of the ligand; a monodentate ligand possesses one donor atom (e.g. NH_3), a didentate ligand two (e.g. $[acac]^-$) and so on. In general, a ligand with more than one donor atom is termed polydentate.

Coordination of a polydentate ligand to an ion leads to the formation of a *chelate ring*, and five such rings can be seen in $[Ca(EDTA)]^{2-}$ in equation 6.75. The word *chelate* is derived from the Greek for a crab's claw. Table 6.7 lists some common ligands; en, $[ox]^{2-}$ and bpy form five-membered chelate rings on coordination to a metal ion, whereas coordination of $[acac]^-$ gives a six-membered ring (Figure 6.10). Both ring-sizes are particularly favourable; the stability of a chelate ring diminishes as its size increases beyond six atoms, and ring-strain causes the formation of three- and four-membered rings to be relatively unfavourable.

Let us now look quantitatively at the displacement of four water molecules by either four NH_3, four $MeNH_2$ or two en ligands in relatively comparable systems (equations 6.76–6.78). Care is essential in making comparisons; NH_3 and $MeNH_2$ are approximate (but not perfect) models for half of the ligand en.

$$Cd^{2+}(aq) + 4NH_3(aq) \rightleftharpoons [Cd(NH_3)_4]^{2+}(aq)$$

$$\log \beta_4 = 7.02 \qquad (6.76)$$

$$Cd^{2+}(aq) + 4MeNH_2(aq) \rightleftharpoons [Cd(MeNH_2)_4]^{2+}(aq)$$

$$\log \beta_4 = 6.55 \qquad (6.77)$$

$$Cd^{2+}(aq) + 2en(aq) \rightleftharpoons [Cd(en)_2]^{2+}(aq)$$

$$\log \beta_2 = 10.30 \qquad (6.78)$$

The values of overall stability constants for these reactions correspond to values of $\Delta G^o(298\,K)$ of -40.0, -37.4 and $-58.8\,kJ\,mol^{-1}$. The values of $\log \beta$ and ΔG^o illustrate that the formation of the chelated complex is more thermo-

dynamically favourable than the formations of either $[Cd(NH_3)_4]^{2+}$ or $[Cd(MeNH_2)_4]^{2+}$; this phenomenon is often referred to as the *chelate effect*. The role of entropy can be assessed as follows. Measured changes in enthalpy accompanying reactions 6.76–6.78 are -52.5, -57.3 and $-55.7\,kJ\,mol^{-1}$, respectively, and combining free energy and enthalpy data gives values of $\Delta S^o(298\,K)$ of -41.9, -66.8 and $+10.4\,J\,K^{-1}\,mol^{-1}$, respectively. The significant feature here is that ΔS^o for reaction 6.78 is more positive than for either of reactions 6.76 or 6.77, and this indicates a more favourable situation in terms of entropy factors. We can also see from these data that $\Delta S^o(298\,K)$ for reaction 6.79 is substantially positive ($+52.3\,J\,K^{-1}\,mol^{-1}$), again underlining the fact that the formation of the chelated complex is favoured.

$$[Cd(NH_3)_4]^{2+}(aq) + 2en(aq)$$

$$\rightleftharpoons [Cd(en)_2]^{2+}(aq) + 4NH_3(aq) \qquad (6.79)$$

Most of these observations can be accounted for in terms of the number of monodentate molecules displaced per polydentate ligand introduced; however, entropy effects associated with the desolvation of the ligands prior to coordination also play a role. In this example, the relative thermodynamic stability of $[Cd(en)_2]^{2+}$ compared with $[Cd(NH_3)_4]^{2+}$ or $[Cd(MeNH_2)_4]^{2+}$ appears to be purely an entropy effect, but a word of caution is needed. In other cases, *both* enthalpy and entropy factors are important, making general discussions of the chelate effect difficult.[†]

So far, we have considered only the coordination of monodentate or acyclic polydentate ligands. A wealth of coordination chemistry involves *macrocyclic ligands* (see *Section 10.8*), which include the family of crown ethers (for example, 18-crown-6, **6.16**, and benzo-12-crown-4, **6.17**), and the encapsulating *cryptand ligands* (see *Figure 10.7*).

(6.16)

(6.17)

Complex stability is enhanced when a macrocyclic ligand replaces a comparable acyclic (open-chain) ligand. For

---

[†] For more in-depth discussions of the chelate and macrocyclic effects, see: M. Gerloch and E.C. Constable (1994) *Transition Metal Chemistry: The Valence Shell in d-Block Chemistry*, VCH, Weinheim (Chapter 8); J. Burgess (1988) *Ions in Solution: Basic Principles of Chemical Interactions*, Ellis Horwood, Chichester (Chapter 6); L.F. Lindoy (1989) *The Chemistry of Macrocyclic Ligand Complexes*, Cambridge University Press, Cambridge (Chapter 6).

Table 6.7 Names and structures of selected ligands.

Name of ligand	Abbreviation (if any)	Denticity	Structure with donor atoms marked in blue
Water		Monodentate	
Ammonia		Monodentate	
Tetrahydrofuran	THF	Monodentate	
Pyridine	py	Monodentate	
Ethane-1,2-diamine[‡]	en	Didentate	
Dimethylsulfoxide	DMSO	Monodentate	
Acetylacetonate ion	[acac]⁻	Didentate	
Oxalate or ethanedioate ion	[ox]²⁻	Didentate	
2,2'-Bipyridine	bpy or bipy	Didentate	
1,10-phenanthroline	phen	Didentate	
1,4,7-triazaheptane[‡]	dien	Tridentate	
1,4,7,10-tetraazadecane[‡]	trien	Tetradentate	
N,N,N',N'-ethylenediaminetetraacetate ion[*]	[EDTA]⁴⁻	Hexadentate	See equation 6.75

[‡]The older names (still in use) for ethane-1,2-diamine, 1,4,7-triazaheptane and 1,4,7,10-tetraazadecane are ethylenediamine, diethylenetriamine and triethylenetetramine.
[*]Although not systematic by the IUPAC rules, this is the commonly accepted name for this anion.

example, values of $\log K_1$ for complexes **6.18** and **6.19** are 23.9 and 28.0 respectively, revealing the far greater thermodynamic stability of the macrocyclic complex.

(6.18)   (6.19)

It is not easy to generalize about the origins of the macrocyclic effect. In considering comparable open- and closed-chain complexes such as **6.18** and **6.19**, entropy factors tend, in most cases, to favour the formation of the macrocyclic complex. However, the enthalpy term does not always favour the macrocyclic complex, although the value of ΔG° (i.e. the ultimate arbiter) always favours the formation of the macrocycle. We shall consider the formation of macrocyclic compounds further in *Chapter 10*.

6.13 Factors affecting the stabilities of complexes containing only monodentate ligands

Although there is no single generalization relating values of stability constants of complexes of *different cations* with the *same ligand*, a number of useful correlations exist, and in this section we explore some of the most important of them.

Ionic size and charge

The stabilities of complexes of the non-d-block metal ions of a given charge normally decrease with increasing cation size (the 'size' of the ion is in a crystallographic sense). Thus, for a complex with a given ligand, L, the order of stability is $Ca^{2+} > Sr^{2+} > Ba^{2+}$. Similar behaviour is found for the lanthanoid M^{3+} ions.

> For ions of similar size, the stability of a complex with a specified ligand increases substantially as the ionic charge increases, e.g. $Li^+ < Mg^{2+} < Al^{3+}$.

For a metal with two (or more) oxidation states, the more highly charged ion is the smaller; the effects of size and charge reinforce each other, leading to greater stability for complexes involving the higher oxidation state metal ion.

Table 6.8 Stability constants for the formation of Fe(III) and Hg(II) halides $[FeX]^{2+}(aq)$ and $[HgX]^+(aq)$; see equations 6.80 and 6.81.

Metal ion	$\log K_1$			
	X = F	**X = Cl**	**X = Br**	**X = I**
$Fe^{3+}(aq)$	6.0	1.4	0.5	–
$Hg^{2+}(aq)$	1.0	6.7	8.9	12.9

Hard and soft metal centres and ligands

When we consider the acceptor properties of metal ions towards ligands (i.e. Lewis acid–Lewis base interactions), two classes of metal ion can be identified, although the distinction between them is not clear-cut. Consider equilibria 6.80 and 6.81.

$$Fe^{3+}(aq) + X^-(aq) \rightleftharpoons [FeX]^{2+}(aq) \qquad (6.80)$$

$$Hg^{2+}(aq) + X^-(aq) \rightleftharpoons [HgX]^+(aq) \qquad (6.81)$$

Table 6.8 gives stability constants for the complexes $[FeX]^{2+}$ and $[HgX]^+$ for different halide ions; while the stabilities of the Fe^{3+} complexes *decrease* in the order $F^- > Cl^- > Br^-$, those of the Hg^{2+} complexes *increase* in the order $F^- < Cl^- < Br^- < I^-$. More generally, it is observed that for the lighter s- and p-block cations, early d-block metals, lanthanoids and actinoids, the stabilities of halide complexes follow the sequence $F^- > Cl^- > Br^- > I^-$; this order is reversed for halide complexes of later d-block metal ions, tellurium, polonium and thallium. Similar patterns are found for other donor atoms: ligands with O- or N-donors form more stable complexes with lighter s- and p-block cations, early d-block metal and f-block cations, whereas ligands containing S- or P-donors show a preference for later d-block metal ions and heavier p-block cations.

Cations (Lewis acids) and ligands (Lewis bases) are classified as being 'hard' (less polarizable) or 'soft' (more polarizable) (see end of *Section 5.13*); earlier names for these are class (a) or class (b) cations and ligands, respectively. The principle of *hard and soft acids and bases* has been developed from ideas of Ahrland, Chatt and Davies, and later Pearson, and can be used to rationalize observed patterns in the stabilities of complexes; effectively, hard cations are better matched with hard ligands, than with soft ones, and vice versa.

> *Hard cations* form more stable complexes with *hard ligands*, while *soft cations* form more stable complexes with *soft ligands*.

Representative cations and ligands are classified in Table 6.9, but a significant number of species lie in the middle. 'Intermediate' cations may form coordination compounds with either hard or soft ligands; similarly, 'intermediate'

Table 6.9 Selected hard and soft metal centres (Lewis acids) and ligands (Lewis bases) and those that exhibit intermediate behaviour. Ligand abbreviations are defined in Table 6.7; R = alkyl and Ar = aryl.

	Ligands (Lewis bases)	Metal centres (Lewis acids)
Hard; class (a)	F^-, Cl^-, H_2O, ROH, R_2O, $[OH]^-$, $[RO]^-$, $[RCO_2]^-$, $[CO_3]^{2-}$, $[NO_3]^-$, $[PO_4]^{3-}$, $[SO_4]^{2-}$, $[ClO_4]^-$, $[ox]^{2-}$, NH_3, RNH_2	Li^+, Na^+, K^+, Rb^+, Cs^+, Mn^{2+}, Zn^{2+}, Al^{3+}, Ga^{3+}, In^{3+}, Sc^{3+}, Cr^{3+}, Fe^{3+}, Co^{3+}, Y^{3+}, $[VO]^{2+}$, $[VO_2]^+$
Soft; class (b)	I^-, H^-, R^-, $[CN]^-$ (*C*-bound), CO (*C*-bound), RNC, RSH, R_2S, $[RS]^-$, $[SCN]^-$ (*S*-bound), R_3P, R_3As, R_3Sb, alkenes, arenes	Zero oxidation state metal centres, Tl^+, Cu^+, Ag^+, Au^+, $[Hg_2]^{2+}$, Pb^{2+}, Hg^{2+}, Cd^{2+}, Pd^{2+}, Pt^{2+}, Tl^{3+}
Intermediate	Br^-, $[N_3]^-$, py, $[SCN]^-$ (*N*-bound), $ArNH_2$, $[NO_2]^-$, $[SO_3]^{2-}$	Sn^{2+}, Fe^{2+}, Co^{2+}, Ni^{2+}, Cu^{2+}, Zn^{2+}, Ru^{3+}, Rh^{3+}, Ir^{3+}

ligands show no preference for hard or soft cations. Soft ligands such as the *C*-donors CO, RNC, alkenes and arenes, and *P*-donor phosphine ligands, are important in stabilizing low oxidation state metal centres in organometallic complexes (see *Chapter 23*).

The principle of hard and soft acids and bases is qualitatively useful, but lacks a satisfactory quantitative basis; in particular, its value in the study of equilibria in aqueous solution is doubtful, since complex formation under these conditions involves displacement of H_2O by ligands and is thus competitive rather than a simple combination reaction.

Summary

In this chapter, we have discussed the role of water as a solvent, and its ability to function both as a Brønsted acid and base, and as a Lewis base. The degree of dissociation of another Brønsted acid in aqueous solution depends upon the difference in acid strengths between H_2O and the acid in question; similarly for Brønsted bases. We have also considered the factors that influence the solubility of ionic salts in water, and have introduced the concept of the displacement of water molecules from the coordination sphere of a solvated ion by other ligands.

The following terms were introduced in this chapter.
Do you know what they mean?
❑ self-ionization
❑ self-ionization constant of water, K_w
❑ Brønsted acid
❑ Brønsted base
❑ conjugate acid and base pair
❑ molality (as distinct from molarity)
❑ standard state of a solute in solution
❑ activity
❑ acid dissociation constant, K_a
❑ base dissociation constant, K_b
❑ mono-, di- and polybasic acids

❑ stepwise dissociation (of an acid or base)
❑ Bell's rule
❑ Lewis base
❑ Lewis acid
❑ ion–dipole interaction
❑ hydration shell (of an ion)
❑ hexaaqua ion
❑ hydrolysis (of a hydrated cation)
❑ use of the prefix μ, μ_3 ...
❑ polarization of a bond
❑ charge density of an ion
❑ amphoteric
❑ 'diagonal line' in the periodic table
❑ saturated solution
❑ solubility (of an ionic solid)
❑ sparingly soluble
❑ solubility product
❑ standard enthalpy (or free energy, or entropy) of hydration
❑ standard enthalpy (or free energy, or entropy) of solution
❑ common-ion effect
❑ gravimetric analysis
❑ solvent extraction
❑ stepwise stability constant (of a complex)
❑ overall stability constant (of a complex)
❑ ligand
❑ denticity (of a ligand)
❑ chelate
❑ chelate effect
❑ macrocyclic effect
❑ hard and soft cations (acids) and ligands (bases)

You should be able to give equations to relate the following quantities:
❑ pH and $[H_3O^+]$
❑ K_a and pK_a
❑ pK_a and pK_b
❑ K_a and K_b
❑ ΔG° and K
❑ ΔG°, ΔH° and ΔS°

Suggested reading list

H₂O: structure

A.F. Goncharov, V.V. Struzhkin, M.S. Somayazulu, R.J. Hemley and H.K. Mao (1996) *Science*, vol. 273, p. 218 – An article entitled 'Compression of ice at 210 gigapascals: Infrared evidence for a symmetric hydrogen-bonded phase'.

A.F. Wells (1984) *Structural Inorganic Chemistry*, 5th edn, Clarendon Press, Oxford – Chapter 15 includes a description of the various polymorphs of ice and illustrates the phase diagram of H_2O.

Acid–base equilibria: review material

C.E. Housecroft and E.C. Constable (1997) *Chemistry: An Integrated Approach*, Addison Wesley Longman, Harlow – Chapter 11 includes acid–base equilibria in aqueous solutions, and reviews calculations involving pH, pK_a and pK_b.

Ions in aqueous solution

J. Burgess (1978) *Metal Ions in Solution*, Ellis Horwood, Chichester – A thorough treatment of most aspects of metal ions in both aqueous and non-aqueous media.

J. Burgess (1988) *Ions in Solution: Basic Principles of Chemical Interaction*, Ellis Horwood, Chichester – A very readable introduction to the chemistry of ions in aqueous solution.

W.E. Dasent (1984) *Inorganic Energetics*, 2nd edn, Cambridge University Press, Cambridge – Chapter 5 discusses in detail the energetics of salt dissolution in aqueous solution.

D.A. Johnson (1982) *Some Thermodynamic Aspects of Inorganic Chemistry*, 2nd edn, Cambridge University Press, Cambridge

– Contains a useful discussion of the solubility of ionic salts in aqueous solution.

Y. Marcus (1985) *Ion Solvation*, Wiley, New York – A detailed and thorough account of this subject.

A.G. Sharpe (1990) *Journal of Chemical Education*, vol. 67, p. 309 – A short review of the solvation of halide ions and its chemical significance.

E.B. Smith (1982) *Basic Chemical Thermodynamics*, 3rd edn, Clarendon Press, Oxford – Chapter 7 introduces the concept of activity in a very understandable fashion.

Stability constants

H.M.N.H. Irving (1994) 'Stability constants and their determination' in *Encyclopedia of Inorganic Chemistry*, ed. R.B. King, Wiley, Chichester, vol. 7, p. 3907 – A good account of stability constants at a level somewhat above the coverage in this book.

A.E. Martell and R.J. Motekaitis (1988) *Determination and Use of Stability Constants*, VCH, New York – A detailed account of the experimental methods for the determination of stability constants, and an overview of their applications.

Chemical Society Special Publication no. 17 (1964) *Stability Constants*, and Special Publication no. 25 (1971) *Stability Constants Supplement No. 1* – A useful source of data, but the uncritical inclusion of data from many different sources means the user must be cautious.

Hardness and softness

R.G. Pearson (1997) *Chemical Hardness*, Wiley-VCH, Weinheim – By the originator of the theory of chemical hardness, this book provides an account of its applications in chemistry.

Problems

6.1 The values of $pK_a(1)$ and $pK_a(2)$ for chromic acid (H_2CrO_4) are 0.74 and 6.49 respectively. (a) Determine values of K_a for each dissociation step. (b) Write equations to represent the dissociation steps of chromic acid in aqueous solution.

6.2 Four pK_a values (1.0, 2.0, 7.0, 9.0) are tabulated for the acid $H_4P_2O_7$. Write equations to show the dissociation steps in aqueous solution and assign, with reasoning, a pK_a value to each step.

6.3 The values of pK_a for CH_3CO_2H and CF_3CO_2H are 4.75 and 0.23, both of which are very nearly independent of temperature. Suggest reasons for this difference.

6.4 (a) To what equilibria do the values of $pK_a(1) = 10.71$ and $pK_a(2) = 7.56$ for the conjugate acid of $H_2NCH_2CH_2NH_2$ refer? (b) Calculate the corresponding values of pK_b and write equations to show the equilibria to which these values refer.

6.5 (a) Write equations to show how you expect compounds **6.20** to **6.24** to dissociate in aqueous solution. (b) Suggest how compound **6.25** will react with NaOH in aqueous solution. What salts would it be possible to isolate?

(6.20) (6.21) (6.22)

(6.23) (6.24)

(6.25)

6.6 In aqueous solution, boric acid behaves as a weak acid ($pK_a = 9.1$) and the following equilibrium is established:

$$B(OH)_3(aq) + 2H_2O(l) \rightleftharpoons [B(OH)_4]^-(aq) + [H_3O]^+(aq)$$

(a) Draw the structures of $B(OH)_3$ and $[B(OH)_4]^-$. (b) How would you classify the acidic behaviour of $B(OH)_3$? (c) The formula of boric acid may also be written as H_3BO_3; compare the acidic behaviour of this acid with that of H_3PO_3.

6.7 When NaCN dissolves in water, the resulting solution is basic. Account for this observation given that pK_a for HCN is 9.31.

6.8 Write equations to illustrate the amphoteric behaviour of $[HCO_3]^-$ in aqueous solution.

6.9 Which of the following oxides are likely to be acidic, basic, or amphoteric in aqueous solution: (a) MgO; (b) SnO; (c) CO_2; (d) P_2O_5; (e) Sb_2O_3; (f) SO_2; (g) Al_2O_3; (h) BeO?

6.10 Explain what is meant by the terms (a) saturated solution; (b) solubility; (c) sparingly soluble salt; (d) solubility product (solubility constant).

6.11 Write down expressions for K_{sp} for the following ionic salts: (a) AgCl; (b) $CaCO_3$; (c) CaF_2.

6.12 Using your answers to question 6.11, write down expressions for the solubility (in $mol\,dm^{-3}$) of (a) AgCl, (b) $CaCO_3$ and (c) CaF_2 in terms of K_{sp}.

6.13 Calculate the solubility of $BaSO_4$ at 298 K in g per 100 g of water given that $K_{sp} = 1.07 \times 10^{-10}$.

6.14 Outline the changes that occur (a) to the salt, and (b) to the water molecules, when solid NaF dissolves in water. How do these changes affect (qualitatively) the entropy of the system.

6.15 (a) Using standard reduction potentials from Appendix 11, determine values of $\Delta_f G^\circ(K^+, aq)$ and $\Delta_f G^\circ(F^-, aq)$. (b) Hence, find $\Delta_{sol} G^\circ(KF, s)$ at 298 K, if $\Delta_f G^\circ(KF, s) = -537.8\,kJ\,mol^{-1}$. (c) What does the value for $\Delta_{sol} G^\circ(KF, s)$ imply about the solubility of KF in water?

6.16 Using data from Appendix 11, and the value for the standard free energy of formation for PbS of $-99\,kJ\,mol^{-1}$, determine a value for K_{sp} for this salt.

6.17 (a) Discuss the factors that contribute towards KCl being a readily soluble salt (35 g per 100 g H_2O at 298 K). (b) Develop your answer to part (a) by using the following data: $\Delta_{hyd} H^\circ(K^+, g) = -330\,kJ\,mol^{-1}$; $\Delta_{hyd} H^\circ(Cl^-, g) = -370\,kJ\,mol^{-1}$; $\Delta_{lattice} H^\circ(KCl, s) = -715\,kJ\,mol^{-1}$.

6.18 Potassium chromate is used as an indicator in titrations for the determination of chloride ion. At the end point of a titration of an aqueous solution of a metal chloride salt (e.g. NaCl) against silver nitrate solution in the presence of potassium chromate, red Ag_2CrO_4 precipitates. Give equations for the pertinent reactions occurring during the titration, and, using relevant data from Table 6.4, explain how the indicator works.

6.19 The formation of a buffer solution is an example of the common-ion effect. Explain how a buffer works with reference to a solution containing acetic acid and sodium acetate.

6.20 Calculate the solubility (in g per 100 g H_2O) of AgBr ($K_{sp} = 5.35 \times 10^{-13}$) (a) in aqueous solution, and (b) in 0.5 M KBr solution.

6.21 Discuss the interpretation of the observation that magnesium oxide is more soluble in aqueous magnesium chloride than in pure water.

6.22 Soda-water is made by saturating H_2O with CO_2. If one titrates soda-water with alkali using phenolphthalein as indicator, one obtains a fading end-point. What does this suggest?

6.23 What explanation can you offer for the decrease in solubility of the alkaline earth metal sulfates in the sequence $CaSO_4 > SrSO_4 > BaSO_4$?

6.24 Construct a thermochemical cycle for the decomposition of the phosphonium halides according to the equation:

$$PH_4X(s) \rightleftharpoons PH_3(g) + HX(g)$$

and use it to account for the fact that the most stable phosphonium halide is the iodide.

6.25 (a) Give expressions to define the stepwise stability constants for equilibria 6.66 and 6.68. (b) For each of the complex ions formed in steps 6.66 and 6.68, gives expressions to define the overall stability constants, β_2 and β_4.

6.26 A pH study using a glass electrode at 303 K for complex formation between Al^{3+} ions and $[acac]^-$ (Table 6.7) in aqueous solution gives values of $\log K_1$, $\log K_2$ and $\log K_3$ as 8.6, 7.9 and 5.8. (a) To what equilibria do these values refer? (b) Determine values for $\Delta G_1^\circ(303\,K)$, $\Delta G_2^\circ(303\,K)$ and $\Delta G_3^\circ(303\,K)$ and comment on the relative ease with which successive ligand displacement reactions occur.

6.27 How many chelate rings are present in each of the following complexes? Assume that all the donor atoms are involved in coordination. (a) $[Cu(trien)]^{2+}$; (b) $[Fe(ox)_3]^{3-}$; (c) $[Ru(bpy)_3]^{2+}$; (d) $[Co(dien)_2]^{3+}$; (e) $[K(18\text{-crown-}6)]^+$.

Chapter 7 Reduction and oxidation

7.1 Introduction

This chapter is concerned with equilibria involving oxidation and reduction processes. Firstly, we review concepts that will be familiar to most readers: definitions of oxidation and reduction, and the use of oxidation states (oxidation numbers).

Oxidation and reduction

The terms oxidation and reduction are applied in a number of different ways, and one must remain versatile in their uses.

Oxidation refers to gaining oxygen, losing hydrogen or losing one or more electrons. *Reduction* refers to losing oxygen, gaining hydrogen or gaining one or more electrons.

Oxidation and reduction steps complement one another, e.g. in reaction 7.1, magnesium is oxidized, while oxygen is reduced.

$$2Mg + O_2 \longrightarrow 2MgO \qquad (7.1)$$

oxidation

reduction

This reaction could be written in terms of the two half-reactions 7.2 and 7.3, but it is important to remember that neither reaction occurs in isolation.

$$Mg \longrightarrow Mg^{2+} + 2e^- \qquad oxidation \qquad (7.2)$$

$$O_2 + 4e^- \longrightarrow 2O^{2-} \qquad reduction \qquad (7.3)$$

Redox is an abbreviation for reduction–oxidation.

In an *electrolytic cell*, the passage of an electrical current initiates a redox reaction, e.g. in the Downs process (see *Section 8.12 and Figure 10.1*) for the manufacture of Na and Cl_2 (equation 7.4).

$$\left. \begin{array}{l} Na^+ + e^- \longrightarrow Na \\ Cl^- \longrightarrow \frac{1}{2}Cl_2 + e^- \end{array} \right\} \qquad (7.4)$$

In a *galvanic cell*, a spontaneous redox reaction occurs and generates an electrical current (see *Section 7.2*).

Many reactions are more complicated than those shown above, and interpreting them in terms of oxidation and reduction steps requires care. The assignment of oxidation states (or oxidation numbers) facilitates this process.

Oxidation states

Oxidation states can be assigned to each atom of an element in a compound but *are a formalism*. We assume that readers of this book are already familiar with this concept, but practice is given in problems 7.1 and 7.2 at the end of the chapter. The oxidation state of an *element* is taken to be zero, irrespective of whether the element exists as atoms (e.g. Ne), molecules (e.g. O_2, P_4) or an infinite lattice (e.g. Si). In addition, in the assignment of oxidation states to elements in a compound, any *homonuclear bond* is ignored. For example, in H_2O_2, **7.1**, the oxidation state of each O atom is -1.

(7.1)

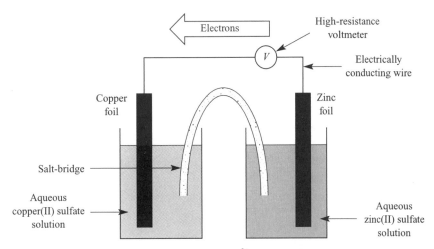

Fig. 7.1 A representation of the Daniell cell. In the left-hand cell, Cu^{2+} ions are reduced to copper metal, and in the right-hand cell, zinc metal is oxidized to Zn^{2+} ions. The cell diagram is written as: $Zn(s)\,|\,Zn^{2+}(aq)\vdots Cu^{2+}(aq)\,|\,Cu(s)$.

An *oxidation* process is accompanied by an increase in the oxidation state of the element involved; conversely, a decrease in the oxidation state corresponds to a *reduction* step.

In reaction 7.5, the oxidation state of Cl in HCl is -1, and in Cl_2 is 0; the change indicates an *oxidation* step. In $KMnO_4$, the oxidation state of Mn is $+7$, while in $MnCl_2$ it is $+2$, i.e. $[MnO_4]^-$ is *reduced* to Mn^{2+}.

$$16HCl(conc) + 2KMnO_4(s)$$

$$\longrightarrow 5Cl_2(g) + 2KCl(aq) + 2MnCl_2(aq) + 8H_2O(l) \quad (7.5)$$

reduction

oxidation

The *net change* in oxidation states involved in the oxidation and reduction steps in a given reaction *must balance*. In reaction 7.5:

- the net change in oxidation state for $Mn = 2 \times (-5) = -10$;
- the net change in oxidation state for $Cl = 10 \times (+1) = +10$.

Although in some formulae, fractional oxidation states might be suggested, the IUPAC[†] recommends that such usage be avoided; e.g. in $[O_2]^-$, it is preferable to consider the group as a whole than to assign an oxidation state of $-\frac{1}{2}$ to each O atom.

The *net change in oxidation states* for the oxidation and reduction steps in a given reaction must balance.

[†] *IUPAC: Nomenclature of Inorganic Chemistry (Recommendations 1990)*, ed. G.J. Leigh, Blackwell Scientific Publications, Oxford, p. 47.

Stock nomenclature

Although we write the oxidation state of Mn in $[MnO_4]^-$ as $+7$, this must not be taken to imply the presence of an Mn^{7+} ion (which, on electrostatic grounds, would be extremely unlikely). *Stock nomenclature* uses Roman numerals to indicate oxidation state: e.g. manganate(VII), hexaaquacobalt(II) and tetraoxoiodate(VII). This gives the oxidation state of the central atom without implying the presence of discrete, highly charged ions.

7.2 Standard reduction potentials, E°, and relationships between E°, ΔG° and K

Half-cells and galvanic cells

One type of simple electrochemical *half-cell* consists of a metal strip dipping into a solution of its ions, e.g. a Cu strip immersed in an aqueous solution of a Cu(II) salt. No chemical reaction occurs in such a half-cell, although an equation describing the half-cell refers (by convention) to the appropriate *reduction* process (equation 7.6); the reaction is written as an *equilibrium*.

$$Cu^{2+}(aq) + 2e^- \rightleftharpoons Cu(s) \quad (7.6)$$

When two such half-cells are combined in an electrical circuit, a redox reaction occurs *if* there is a potential difference between the half-cells. This is illustrated in Figure 7.1 by the Daniell cell in which a Cu^{2+}/Cu half-cell (equation 7.6) is combined with Zn^{2+}/Zn half-cell (equation 7.7).

$$Zn^{2+}(aq) + 2e^- \rightleftharpoons Zn(s) \quad (7.7)$$

The two solutions in the Daniell cell are connected by a *salt-bridge* (e.g. gelatine containing aqueous KCl or KNO_3), which allows the passage of ions between the half-cells without allowing the Cu(II) and Zn(II) solutions to mix too quickly. When the Daniell cell is assembled, redox reaction 7.8 occurs *spontaneously*.

$$Zn(s) + Cu^{2+}(aq) \longrightarrow Zn^{2+}(aq) + Cu(s) \qquad (7.8)$$

The Daniell cell is an example of a *galvanic cell*; in this type of electrochemical cell, *electrical work* is done *by the system*. The potential difference, E_{cell}, between the two half-cells can be measured (in volts, V) on a voltmeter in the circuit (Figure 7.1) and the value of E_{cell} is related to the change in free energy for the cell reaction. Equation 7.9 gives this relationship under standard conditions, where E^{o}_{cell} is the *standard cell potential*.

$$\Delta G^{o} = -zFE^{o}_{cell} \qquad (7.9)$$

where F = Faraday constant = $96\,485\,C\,mol^{-1}$; z = number of moles of electrons transferred *per mole of reaction*; ΔG^{o} is in $J\,mol^{-1}$; and E^{o}_{cell} is in volts.

Standard conditions for an electrochemical cell are defined as follows:

- unit activity for *each* component in the cell (for *dilute* solutions, activity is approximated to concentration, see *Section 6.3*);
- the pressure of any gaseous component is 1 bar (10^{5} Pa);[†]
- a solid component is in its standard state;
- the temperature is 298 K.

The equilibrium constant, K, for the cell reaction is related to ΔG^{o} by equation 7.10, and to E^{o}_{cell} by equation 7.11.

$$\Delta G^{o} = -RT \ln K \qquad (7.10)$$

$$\ln K = -\frac{zFE^{o}_{cell}}{RT} \qquad (7.11)$$

where $R = 8.314\,J\,K^{-1}\,mol^{-1}$.

For a cell reaction that is thermodynamically favourable:

- E^{o}_{cell} is positive;
- ΔG^{o} is negative;
- $K > 1$.

For $z = 1$, a value of $E^{o}_{cell} = 0.6\,V$ corresponds to a value of $\Delta G^{o} \approx -60\,kJ\,mol^{-1}$ and $K \approx 10^{10}$ at 298 K, i.e. this indicates a thermodynamically favourable cell reaction, one that will tend towards completion.

Worked example 7.1 The Daniell cell

The standard cell potential (at 298 K) for the Daniell cell is 1.10 V. Calculate the corresponding values of ΔG^{o} and K and comment on the thermodynamic viability of the cell reaction:

$$Zn(s) + Cu^{2+}(aq) \longrightarrow Zn^{2+}(aq) + Cu(s)$$

$$(F = 96\,485\,C\,mol^{-1}; R = 8.314 \times 10^{-3}\,kJ\,K^{-1}\,mol^{-1})$$

The equation needed is:

$$\Delta G^{o} = -zFE^{o}_{cell}$$

and z is 2 for the cell reaction:

$$Zn(s) + Cu^{2+}(aq) \longrightarrow Zn^{2+}(aq) + Cu(s)$$

$$\Delta G^{o} = -zFE^{o}_{cell}$$
$$= -2 \times 96\,485 \times 1.10$$
$$= -212\,267\,J \text{ per mole of reaction}$$
$$\approx -212\,kJ \text{ per mole of reaction}$$

$$\ln K = -\frac{\Delta G^{o}}{RT} = -\frac{212}{298 \times 8.314 \times 10^{-3}}$$

$$\ln K = 85.6$$

$$K = 1.45 \times 10^{37}$$

The large, negative value of ΔG^{o} and a value of K which is $\gg 1$ correspond to a thermodynamically favourable reaction, one which virtually goes to completion.

Self-study exercises

1. For the Daniel cell, $\log K = 37.2$. Calculate ΔG^{o} for the cell.
   [*Ans:* $-212\,kJ\,mol^{-1}$]

2. The value of ΔG^{o} for the Daniel cell is $-212\,kJ\,mol^{-1}$. Calculate E^{o}_{cell}.
   [*Ans:* 1.10 V]

It is possible to obtain values for E^{o}_{cell} *experimentally*, although it is usual in the laboratory to work with solutions of concentrations $<1\,mol\,dm^{-3}$, and thus measure values of E_{cell} (rather than *standard* cell potentials). Such values are dependent on solution concentration (strictly, activity), and E_{cell} and E^{o}_{cell} are related by the Nernst equation (see *equation 7.21*).[‡]

It is also possible to *calculate* E^{o}_{cell} (and the corresponding value of ΔG^{o}) using values of *standard reduction potentials* for half-cells, and this is the more routine method of evaluating the thermodynamic viability of redox reactions.

Defining and using standard reduction potentials, E^{o}

Tabulated values of *standard reduction potentials*, E^{o}, refer to single electrodes. For example, for the half-cell reaction 7.6, the value of $E^{o}_{Cu^{2+}/Cu} = +0.34\,V$. However, it is impossible to measure the potential of an individual electrode and the universal practice is to express all such potentials relative to that of the *standard hydrogen electrode*. The latter consists

---

[†] The standard pressure is given in some tables of data as 1 atm (101 300 Pa), but at the level of accuracy of most tables, this makes no difference to the values of E^{o}.

[‡] For an introduction to galvanic cells and the Nernst equation see: C.E. Housecroft and E.C. Constable (1997) *Chemistry: An Integrated Approach*, Addison Wesley Longman, Harlow, Chapter 12; for a more detailed discussion, see: P.W. Atkins (1998) *Physical Chemistry*, 6th edn, Oxford University Press, Oxford, Chapter 10.

of H^+ ions at concentration of $1\,mol\,dm^{-3}$ (strictly, unit activity) in equilibrium with H_2 at 1 bar pressure (equation 7.12). This electrode is taken to have a standard reduction potential $E^\circ = 0\,V$ at all temperatures.

$$2H^+(aq, 1\,mol\,dm^{-3}) + 2e^- \rightleftharpoons H_2(g, 1\,bar) \qquad (7.12)$$

Having defined this half-cell, it is now possible to combine it with another half-cell, measure E°_{cell}, and, thus, to find E° for the second half-cell. In order to obtain the correct sign (by convention) for the half-cell, equation 7.13 must be applied.

$$E^\circ_{cell} = [E^\circ_{reduction\,process}] - [E^\circ_{oxidation\,process}] \qquad (7.13)$$

For example, if Zn metal is placed into dilute acid, H_2 is evolved. Thus, when the standard hydrogen electrode is connected in a galvanic cell with a Zn^{2+}/Zn electrode, reaction 7.14 is the spontaneous cell process.

$$Zn(s) + 2H^+(aq) \longrightarrow Zn^{2+}(aq) + H_2(g) \qquad (7.14)$$

The oxidation process is Zn going to Zn^{2+}, and the reduction process involves H^+ ions being converted to H_2. For this cell, the *measured* value of E°_{cell} is 0.76 V, and, thus, $E^\circ_{Zn^{2+}/Zn} = -0.76\,V$ (equation 7.15). Note that no sign need be included with E_{cell} because it is always positive for the spontaneous reaction.

$$\left.\begin{array}{l} 0.76 = 0 - E^\circ_{Zn^{2+}/Zn} \\[4pt] E^\circ_{Zn^{2+}/Zn} = -0.76\,V \end{array}\right\} \qquad (7.15)$$

Selected values of standard reduction potentials are listed in Table 7.1 (see also *Appendix 11*). Most of these values have been obtained directly from potential difference measurements, but a few values have been calculated from data obtained by calorimetric methods. This latter technique is

Table 7.1 Selected standard reduction potentials (at 298 K); further data are listed in Appendix 11. The concentration of each substance in aqueous solution is $1\,mol\,dm^{-3}$ and the pressure of a gaseous component is 1 bar ($10^5\,Pa$). Note that where the half-cell contains $[OH]^-$, the value of E° refers to $[OH^-] = 1\,mol\,dm^{-3}$, and the notation $E^\circ_{[OH^-]=1}$ should be used (see *Box 7.1*).

Reduction half-equation	$E^\circ\,/\,V$
$Li^+(aq) + e^- \rightleftharpoons Li(s)$	-3.04
$K^+(aq) + e^- \rightleftharpoons K(s)$	-2.93
$Ca^{2+}(aq) + 2e^- \rightleftharpoons Ca(s)$	-2.87
$Na^+(aq) + e^- \rightleftharpoons Na(s)$	-2.71
$Mg^{2+}(aq) + 2e^- \rightleftharpoons Mg(s)$	-2.37
$Al^{3+}(aq) + 3e^- \rightleftharpoons Al(s)$	-1.66
$Mn^{2+}(aq) + 2e^- \rightleftharpoons Mn(s)$	-1.19
$Zn^{2+}(aq) + 2e^- \rightleftharpoons Zn(s)$	-0.76
$Fe^{2+}(aq) + 2e^- \rightleftharpoons Fe(s)$	-0.44
$Cr^{3+}(aq) + e^- \rightleftharpoons Cr^{2+}(aq)$	-0.41
$Fe^{3+}(aq) + 3e^- \rightleftharpoons Fe(s)$	-0.04
$2H^+(aq, 1\,mol\,dm^{-3}) + 2e^- \rightleftharpoons H_2(g, 1\,bar)$	0
$Cu^{2+}(aq) + e^- \rightleftharpoons Cu^+(aq)$	$+0.15$
$AgCl(s) + e^- \rightleftharpoons Ag(s) + Cl^-(aq)$	$+0.22$
$Cu^{2+}(aq) + 2e^- \rightleftharpoons Cu(s)$	$+0.34$
$[Fe(CN)_6]^{3-}(aq) + e^- \rightleftharpoons [Fe(CN)_6]^{4-}(aq)$	$+0.36$
$O_2(g) + 2H_2O(l) + 4e^- \rightleftharpoons 4[OH]^-(aq)$	$+0.40$
$I_2(aq) + 2e^- \rightleftharpoons 2I^-(aq)$	$+0.54$
$Fe^{3+}(aq) + e^- \rightleftharpoons Fe^{2+}(aq)$	$+0.77$
$Ag^+(aq) + e^- \rightleftharpoons Ag(s)$	$+0.80$
$[Fe(bpy)_3]^{3+}(aq) + e^- \rightleftharpoons [Fe(bpy)_3]^{2+}(aq)^\ddagger$	$+1.03$
$Br_2(aq) + 2e^- \rightleftharpoons 2Br^-(aq)$	$+1.09$
$[Fe(phen)_3]^{3+}(aq) + e^- \rightleftharpoons [Fe(phen)_3]^{2+}(aq)^\ddagger$	$+1.12$
$O_2(g) + 4H^+(aq) + 4e^- \rightleftharpoons 2H_2O(l)$	$+1.23$
$[Cr_2O_7]^{2-}(aq) + 14H^+(aq) + 6e^- \rightleftharpoons 2Cr^{3+}(aq) + 7H_2O(l)$	$+1.33$
$Cl_2(aq) + 2e^- \rightleftharpoons 2Cl^-(aq)$	$+1.36$
$[MnO_4]^-(aq) + 8H^+(aq) + 5e^- \rightleftharpoons Mn^{2+}(aq) + 4H_2O(l)$	$+1.51$
$Co^{3+}(aq) + e^- \rightleftharpoons Co^{2+}(aq)$	$+1.92$
$[S_2O_8]^{2-}(aq) + 2e^- \rightleftharpoons 2[SO_4]^{2-}(aq)$	$+2.01$
$F_2(aq) + 2e^- \rightleftharpoons 2F^-(aq)$	$+2.87$

‡ bpy = 2,2'-bipyridine; phen = 1,10-phenanthroline (see *Table 6.7*)

for systems that cannot be investigated in aqueous media because of solvent decomposition (e.g. $F_2/2F^-$) or for which equilibrium is established only very slowly, such that the electrode is non-reversible (e.g. O_2, $4H^+/2H_2O$). Table 7.1 is organized such that the half-cell with the most positive E^o is at the bottom of the table. The most powerful *oxidizing agent* among the oxidized species in Table 7.1 is F_2, i.e. F_2 is readily reduced to F^- ions. Conversely, at the top of the table, Li is the most powerful *reducing agent*, i.e. Li is readily oxidized to Li^+.

Worked example 7.2 Using standard reduction potentials to calculate E^o_{cell}

The following two half-reactions correspond to two half-cells that are combined to form an electrochemical cell:

$$[MnO_4]^-(aq) + 8H^+(aq) + 5e^- \rightleftharpoons Mn^{2+}(aq) + 4H_2O(l)$$

$$Fe^{3+}(aq) + e^- \rightleftharpoons Fe^{2+}(aq)$$

(a) What is the spontaneous cell reaction? (b) Calculate E^o_{cell}.

(a) First, look up values of E^o for the half-reactions.

$$Fe^{3+}(aq) + e^- \rightleftharpoons Fe^{2+}(aq) \qquad E^o = +0.77\,V$$

$$[MnO_4]^-(aq) + 8H^+(aq) + 5e^- \rightleftharpoons Mn^{2+}(aq) + 4H_2O(l)$$
$$E^o = +1.51\,V$$

The relative values show that, in aqueous solution under standard conditions, $[MnO_4]^-$ is a more powerful oxidizing agent than Fe^{3+}. The spontaneous cell reaction is therefore:

$$[MnO_4]^-(aq) + 8H^+(aq) + 5Fe^{2+}(aq)$$
$$\rightarrow Mn^{2+}(aq) + 4H_2O(l) + 5Fe^{3+}(aq)$$

(b) The cell potential difference is the difference between the standard reduction potentials of the two half-cells:

$$E^o_{cell} = [E^o_{reduction\ process}] - [E^o_{oxidation\ process}]$$
$$= (+1.51) - (+0.77)$$
$$= 0.74\,V$$

Self-study exercises

For these exercises, refer to Appendix 11 for data.

1. The following two half-cells are combined:

$$Zn^{2+}(aq) + 2e^- \rightleftharpoons Zn(s)$$

$$Ag^+(aq) + e^- \rightleftharpoons Ag(s)$$

Calculate E^o_{cell}, and state whether the spontaneous reaction reduces Ag^+, or oxidizes Ag. [*Ans.* 1.56 V]

2. For the following cell reaction,

$$2[S_2O_3]^{2-} + I_2 \rightarrow [S_4O_6]^{2-} + 2I^-$$

write down the two half-cells, and hence determine E^o_{cell}. [*Ans.* 0.46 V]

Although a positive value of E^o_{cell} indicates a spontaneous process, it is more revealing to consider the corresponding value of ΔG^o (equation 7.9). The latter takes into account the number of electrons transferred during the reaction, as well as the magnitude and sign of the cell potential. For example, to investigate the reaction between Fe and aqueous Cl_2, we consider redox couples 7.16–7.18.

$$Fe^{2+}(aq) + 2e^- \rightleftharpoons Fe(s) \qquad E^o = -0.44\,V \qquad (7.16)$$
$$Fe^{3+}(aq) + 3e^- \rightleftharpoons Fe(s) \qquad E^o = -0.04\,V \qquad (7.17)$$
$$Cl_2(aq) + 2e^- \rightleftharpoons 2Cl^-(aq) \qquad E^o = +1.36\,V \qquad (7.18)$$

These data indicate that either reaction 7.19 or 7.20 may occur.

$$Fe(s) + Cl_2(aq) \rightleftharpoons Fe^{2+}(aq) + 2Cl^-(aq)$$
$$E^o_{cell} = 1.80\,V \qquad (7.19)$$

$$2Fe(s) + 3Cl_2(aq) \rightleftharpoons 2Fe^{3+}(aq) + 6Cl^-(aq)$$
$$E^o_{cell} = 1.40\,V \qquad (7.20)$$

The value of E^o_{cell} is positive for both reactions, and from their relative magnitudes, it might be thought that reaction 7.19 is favoured over reaction 7.20. *Caution is needed:* the true state of affairs is evident only by comparing values of ΔG^o. For reaction 7.19 (where $z = 2$), $\Delta G^o = -347\,kJ$ per mole of reaction, while for reaction 7.20 ($z = 6$), $\Delta G^o = -810\,kJ$ per mole of reaction. *Per mole of Fe*, the values of ΔG^o are -347 and $-405\,kJ$, revealing that reaction 7.20 is thermodynamically favoured over reaction 7.19. This example shows how important it is to consider changes in free energy, rather than simply the cell potentials.

Dependence of reduction potentials on cell conditions

The discussion above centred on *standard* reduction potentials (see *Box 7.1*). However, laboratory experiments seldom occur under standard cell conditions, and a change in conditions can cause a significant change in the ability of a reagent to act as a reducing or oxidizing agent.

Consider a Zn^{2+}/Zn half-cell (at 298 K) in which $[Zn^{2+}] = 0.1\,mol\,dm^{-3}$, i.e. *non-standard* conditions. The Nernst equation (equation 7.21) shows how the reduction potential varies with the concentrations of the species present.

$$E = E^o - \left\{ \frac{RT}{zF} \times \left(\ln \frac{[reduced\ form]}{[oxidized\ form]} \right) \right\} \qquad (7.21)^\dagger$$

---

† The Nernst equation is sometimes written in the form:

$$E = E^o - \left\{ \frac{RT}{zF} \times \ln Q \right\}$$

where Q (the quotient in equation 7.21) is the *reaction quotient*.

BOX 7.1　　　　　　　　　　　　　　　　　　　　　　　

Notation for standard reduction potentials

In an electrochemical cell under standard conditions, the concentration of each substance in aqueous solution is $1 \, \text{mol dm}^{-3}$. Thus, in Table 7.1, each half-cell listed contains the specified solution species at a concentration of $1 \, \text{mol dm}^{-3}$. This leads to the reduction of O_2 being represented by two half-reactions, depending upon the cell conditions:

$$O_2(g) + 4H^+(aq) + 4e^- \rightleftharpoons 2H_2O(l)$$

$$E^o = +1.23 \, \text{V when } [H^+] = 1 \, \text{mol dm}^{-3}, \text{ i.e. pH} = 0$$

$$O_2(g) + 2H_2O(l) + 4e^- \rightleftharpoons 4[OH]^-(aq)$$

$$E^o = +0.40 \, \text{V when } [OH^-] = 1 \, \text{mol dm}^{-3}, \text{ i.e. pH} = 14$$

Similar situations arise for other species in which the value of the electrode potential is pH-dependent. For clarity, therefore, we have adopted the following notation: for half-cells for which the electrode potential is pH-dependent, E^o refers to $[H^+] = 1 \, \text{mol dm}^{-3}$ (pH = 0); for other pH values, the concentration of $[H]^+$ or $[OH]^-$ is specifically stated, for example, $E_{[H^+]=0.1}$ or $E_{[OH^-]=0.05}$. For the case of $[OH^-] = 1 \, \text{mol dm}^{-3}$, this refers to standard conditions, and the notation used is $E^o_{[OH^-]=1}$.

Application of the Nernst equation to the Zn^{2+}/Zn half-cell ($E^o = -0.76 \, \text{V}$) gives $E = -0.79 \, \text{V}$ for $[Zn^{2+}] = 0.1 \, \text{mol dm}^{-3}$ (equation 7.22); the concentration (strictly, activity) of Zn metal is taken to be unity. The more negative value of E, corresponding to a more positive value of ΔG, signifies that it is more difficult to reduce Zn^{2+} at the lower concentration.

$$E = -0.76 - \left\{ \frac{8.314 \times 298}{2 \times 96\,485} \times \left(\ln \frac{1}{[Zn^{2+}]} \right) \right\} \quad (7.22)$$

Now consider the effect of pH (pH $= -\log[H^+]$) on the oxidizing ability of $[MnO_4]^-$ in aqueous solution at 298 K. The critical point is that half-reaction 7.23 contains H^+ ions.

$$[MnO_4]^-(aq) + 8H^+(aq) + 5e^- \rightleftharpoons Mn^{2+}(aq) + 4H_2O(l)$$

$$E^o = +1.51 \, \text{V} \quad (7.23)$$

By applying the Nernst equation, we write equation 7.24, remembering that the concentration (strictly, activity) of H_2O is, by convention, unity.

$$E = 1.51 - \left\{ \frac{8.314 \times 298}{5 \times 96\,485} \times \left(\ln \frac{[Mn^{2+}]}{[MnO_4^-][H^+]^8} \right) \right\} \quad (7.24)$$

Thus, as $[H^+]$ increases (i.e. the pH of the solution is lowered), the value of E becomes more positive. The fact that the oxidizing power of $[MnO_4]^-$ is lower in dilute acid than in concentrated acid, explains why, for example, $[MnO_4]^-$ will not oxidize Cl^- in neutral solution, but liberates Cl_2 from concentrated HCl.

The potentials for the reduction of water ($[H^+] = 10^{-7} \, \text{mol dm}^{-3}$) to H_2, and for the reduction of O_2 to H_2O (the reverse of the oxidation of H_2O to O_2) are of particular significance in aqueous solution chemistry. They provide general guidance (subject to the limitations of thermodynamic versus kinetic control) concerning the nature of chemical species that can exist under aqueous conditions. For reduction process 7.25, $E^o = 0 \, \text{V}$ (by definition).

$$2H^+(aq, 1 \, \text{mol dm}^{-3}) + 2e^- \rightleftharpoons H_2(g, 1 \, \text{bar}) \quad (7.25)$$

If the pressure of H_2 is maintained at 1 bar, application of the Nernst equation (equation 7.21) allows us to calculate E over a range of values of $[H^+]$. For neutral water (pH 7), $E_{[H^+]=10^{-7}} = -0.41 \, \text{V}$, and at pH 14, $E^o_{[OH^-]=1} = -0.83 \, \text{V}$. Whether or not the water (pH 7) or molar aqueous alkali (pH 14) is reduced by a species present in solution depends upon the reduction potential of that species relative to that of the $2H^+/H_2$ couple. Bear in mind that we might be considering the reduction of H_2O to H_2 as a *competitive* process which could occur in preference to the desired reduction. The potential of $-0.83 \, \text{V}$ for the $2H^+/H_2$ electrode in molar alkali is of limited importance in isolation. Many M^{z+}/M systems that should reduce water under these conditions are prevented from doing so by the formation of a coating of hydroxide or hydrated oxide. Others, which are less powerfully reducing, bring about reduction because they are modified by complex formation; an example is the formation of $[Zn(OH)_4]^{2-}$ in alkaline solution (equation 7.26). The value of $E^o = -0.76 \, \text{V}$ for the Zn^{2+}/Zn half-cell (Table 7.1) applies *only* to *hydrated* Zn^{2+} ions; when they are in the form of the stable hydroxo complex $[Zn(OH)_4]^{2-}$, $E^o_{[OH^-]=1} = -1.20 \, \text{V}$ (equation 7.27).

$$Zn^{2+}(aq) + 4[OH]^-(aq) \longrightarrow [Zn(OH)_4]^{2-}(aq) \quad (7.26)$$

$$[Zn(OH)_4]^{2-}(aq) + 2e^- \rightleftharpoons Zn(s) + 4[OH]^-(aq)$$

$$E^o_{[OH^-]=1} = -1.20 \, \text{V} \quad (7.27)$$

Now consider the reduction of O_2 to H_2O, or the oxidation of H_2O to O_2, by a species present in the cell. Equation 7.28 gives the relevant half-reaction.

$$O_2(g) + 4H^+(aq) + 4e^- \rightleftharpoons 2H_2O(l) \quad E^o = +1.23 \, \text{V} \quad (7.28)$$

For a 1 bar pressure of O_2 at 298 K, applying the Nernst equation shows that the half-cell potential becomes $+0.82 \, \text{V}$ in neutral water, and $+0.40 \, \text{V}$ in molar aqueous alkali. So, from a thermodynamic standpoint, O_2 in the presence of water should oxidize any system with a reduction potential less positive than $+1.23 \, \text{V}$ at pH 0 (i.e.

$[H]^+ = 1\,mol\,dm^{-3}$), +0.82 V at pH 7 and +0.40 V at pH 14. Conversely, any system with a half-cell potential more positive than +1.23 V should, at pH 0, oxidize water to O_2, and so on.

We cannot emphasize enough that care has to be taken when considering such processes. Just as the half-cell potentials of the reduction processes considered above vary with experimental conditions, so too do the reduction potentials of other electrodes. It is essential to bear this in mind when using tables of E^o values which are *only* appropriate under *standard conditions*.

Worked example 7.3 Oxidation of Cr^{2+} ions in O_2-free, acidic, aqueous solution

Explain why an acidic, aqueous solution of Cr^{2+} ions liberates H_2 from solution (assume standard conditions). What will be the effect of raising the pH of the solution?

Firstly, write down the half-reactions that are relevant to the question:

$$Cr^{3+}(aq) + e^- \rightleftharpoons Cr^{2+}(aq) \qquad E^o = -0.41\,V$$
$$2H^+(aq) + 2e^- \rightleftharpoons H_2(g) \qquad E^o = 0\,V$$

The following redox reaction will occur:

$$2Cr^{2+}(aq) + 2H^+(aq) \longrightarrow 2Cr^{3+}(aq) + H_2(g)$$

In order to check its thermodynamic feasibility, calculate ΔG^o.

$E^o{}_{cell} = 0.41\,V$

At 298 K:

$$\Delta G^o = -zFE^o{}_{cell}$$
$$= -(2 \times 96\,485 \times 0.41)$$
$$= -79.1 \times 10^3\,J\,per\,mole\,of\,reaction$$
$$= -79.1\,kJ\,per\,mole\,of\,reaction$$

Thus, the reaction is thermodynamically favourable, indicating that aqueous Cr^{2+} ions are not stable in acidic (1 M), aqueous solution. [Note: In fact, this reaction is quite slow.]

Raising the pH of the solution lowers the concentration of H^+ ions. Let us (arbitrarily) consider a value of pH 3.0. The $2H^+/H_2$ electrode now has a new reduction potential.

$$E = E^o - \left\{ \frac{RT}{zF} \times \left(\ln\frac{1}{[H^+]^2} \right) \right\}$$
$$= 0 - \left\{ \frac{8.314 \times 298}{2 \times 96\,485} \times \left(\ln\frac{1}{(1 \times 10^{-3})^2} \right) \right\}$$
$$= -0.18\,V$$

Now we must consider the following combination of half-cells, taking Cr^{3+}/Cr^{2+} still to be under standard conditions:

$$Cr^{3+}(aq) + e^- \rightleftharpoons Cr^{2+}(aq) \qquad E^o = -0.41\,V$$
$$2H^+(aq) + 2e^- \rightleftharpoons H_2(g) \qquad E = -0.18\,V$$

$E_{cell} = 0.23\,V$

At 298 K:

$$\Delta G = -zFE^o{}_{cell}$$
$$= -(2 \times 96\,485 \times 0.23)$$
$$= -44.4 \times 10^3\,J\,per\,mole\,of\,reaction$$
$$= -44.4\,kJ\,per\,mole\,of\,reaction$$

Thus, although the reaction still has a negative value of ΔG, the increase in pH has made the oxidation of Cr^{2+} less thermodynamically favourable.

[Note: pH plays another important role: at pH values only a few units above zero, precipitation of hydroxides (particularly of Cr^{3+}) will occur.]

Self-study exercises

1. Calculate E_{cell} for the reduction of H^+ to H_2 at pH 2.0 (Why is this not $E^o{}_{cell}$?) [*Ans.* −0.12 V]

2. For the half-cell: $O_2 + 4H^+ + 4e^- \rightleftharpoons 2H_2O$
   $E^o = +1.23\,V$. Derive a relationship to show how E depends on pH at 298 K and $P(O_2) = 1$ bar. Hence show that at pH 14, $E = +0.41\,V$.

7.3 The effect of complex formation or precipitation on M^{z+}/M reduction potentials

In the previous section, we saw that, in the presence of $[OH]^-$, the potential for the reduction of Zn^{2+} to Zn is significantly different from that of hydrated Zn^{2+}. In this section, we extend this discussion, and discuss how metal ions can be stabilized with respect to reduction by the formation of a precipitate or coordination complex.

Half-cells involving silver halides

Under standard conditions, Ag^+ ions are reduced to Ag (equation 7.29), but if the concentration of Ag^+ is lowered, application of the Nernst equation shows that the reduction potential becomes less positive (i.e. ΔG is less negative). Consequently, reduction of Ag^+ to Ag becomes less easy; in other words, Ag^+ has been stabilized with respect to reduction.

$$Ag^+(aq) + e^- \rightleftharpoons Ag(s) \qquad E^o = +0.80\,V \qquad (7.29)$$
$$\Delta G^o = -77.2\,kJ\,per\,mole\,of\,Ag$$

In practice, a lower concentration of Ag^+ ions can be achieved by dilution of the aqueous solution, but it can

Reference electrodes

Equation 7.31 shows the reduction reaction that occurs in the *silver chloride/silver electrode*, which is written in the form $Cl^-(aq)\,|\,AgCl\,|\,Ag$ (each vertical bar denotes a phase boundary). This is an example of a half-cell which is constructed by coating a wire of metal M with a solid salt (MX) and immersing this electrode in an aqueous solution containing X^- ions; $[X^-]$ at unit activity $\approx 1\,mol\,dm^{-3}$ for the standard electrode.

- Ag wire coated in AgCl(s)

- Aqueous solution containing $1\,mol\,dm^{-3}$ Cl^-

- Porous plug

This electrode ($E^\circ = +0.222\,V$) is used as a reference electrode, being much more convenient to handle in the laboratory than the standard hydrogen electrode; an electrode that requires a cylinder of H_2 at 1 bar pressure is not ideal for routine experimental work! Other reduction potentials may be quoted *with respect to the silver chloride/silver electrode*, and this effectively gives a scale of relative values on which the standard reduction potential for the reference electrode is set to 0 V.

Another reference electrode which is constructed in a similar manner is the *calomel electrode*, $2Cl^-(aq)\,|\,Hg_2Cl_2\,|\,2Hg$. The half-cell reaction is:

$$Hg_2Cl_2(s) + 2e^- \rightleftharpoons 2Hg(l) + 2Cl^-(aq) \quad E^\circ = +0.268\,V$$

The *standard* calomel electrode contains $1\,M$ KCl solution (at 298 K). Clearly, this design of this electrode is not as straightforward as that of the $Cl^-(aq)\,|\,AgCl\,|\,Ag$ electrode; mercury is a liquid at 298 K, and contact into an electrical circuit is made by means of a Pt wire which dips into the liquid Hg, itself surrounded by a coating of Hg(I) chloride (calomel).

---

also be brought about by removal of Ag^+ ions from solution by the formation of a stable complex or by precipitation of a sparingly soluble salt (see *Section 6.9*). Consider the formation of AgCl (equation 7.30) for which $K_{sp} = 1.77 \times 10^{-10}$ and ΔG° can be found using equation 7.10.

$$Ag^+(aq) + Cl^-(aq) \rightleftharpoons AgCl(s)$$
$$\Delta G^\circ = +55.6\,kJ \text{ per mole of AgCl} \quad (7.30)$$

Reduction of Ag(I) when it is in the form of solid AgCl occurs according to reaction 7.31, and the relationship between equilibria 7.29–7.31 allows us to find, by difference, ΔG° for reaction 7.31. This leads to a value of $E^\circ = +0.22\,V$ for this half-cell (see *Box 7.2*).

$$AgCl(s) + e^- \rightleftharpoons Ag(s) + Cl^-(aq)$$
$$\Delta G^\circ = -21.6\,kJ \text{ per mole of AgCl} \quad (7.31)$$

The difference in values of E° for half-reactions 7.29 and 7.31 indicates that it is more difficult to reduce Ag(I) in the form of solid AgCl than as hydrated Ag^+.

Silver iodide ($K_{sp} = 8.51 \times 10^{-17}$) is less soluble than AgCl in aqueous solution, and so reduction of Ag(I) in the form of solid AgI is thermodynamically less favourable than reduction of AgCl. However, AgI is much more soluble in aqueous KI than AgCl is in aqueous KCl solution. The species present in the iodide solution is the complex $[AgI_3]^{2-}$, the overall stability constant (see *Section 6.12*) for which is $\approx 10^{14}$ (equation 7.32). Following a similar procedure to

that detailed above, we can use this value to determine that the half-cell corresponding to reduction process 7.33 has a value of $E^\circ = -0.03\,V$.

$$Ag^+(aq) + 3I^-(aq) \rightleftharpoons [AgI_3]^{2-}(aq) \quad \beta_3 \approx 10^{14} \quad (7.32)$$
$$[AgI_3]^{2-}(aq) + e^- \rightleftharpoons Ag(s) + 3I^-(aq)$$
$$E^\circ = -0.03\,V \quad (7.33)$$

Again, Ag(I) has been stabilized with respect to reduction, but this time to a greater extent: the value of E° indicates that Ag in the presence of $[AgI_3]^{2-}$ and I^- (both $1\,mol\,dm^{-3}$) is as powerful a reducing agent as H_2 in the presence of H^+ (under standard conditions).

Modifying the relative stabilities of different oxidation states of a metal

Just as we can 'tune' the reducing power of Ag by manipulation of the solution species or precipitates present, we can also alter the relative stabilities of two oxidation states of a metal, both of which are subject to removal by precipitation or complexation. As an example, consider the Mn^{3+}/Mn^{2+} couple, for which equation 7.34 is appropriate for aqua species.

$$Mn^{3+}(aq) + e^- \rightleftharpoons Mn^{2+}(aq) \quad E^\circ = +1.54\,V \quad (7.34)$$

In alkaline solution, both metal ions are precipitated, but Mn(III) much more completely than Mn(II) since values of K_{sp} for $Mn(OH)_3$ and $Mn(OH)_2$ are $\approx 10^{-36}$ and

$\approx 2 \times 10^{-13}$, respectively. Precipitation has the effect of significantly changing the half-cell potential for the reduction of Mn(III); in solutions in which $[OH^-] = 1 \, mol \, dm^{-3}$, Mn(III) is stabilized with respect to reduction to Mn(II) as the value of $E^\circ_{[OH^-]=1}$ for equation 7.35 illustrates.

$$Mn(OH)_3(s) + e^- \rightleftharpoons Mn(OH)_2(s) + [OH]^-(aq)$$

$$E^\circ_{[OH^-]=1} = +0.15 \, V \qquad (7.35)$$

Worked example 7.4   Oxidation of Mn(II) to Mn(III)

Using data from equations 7.34 and 7.35, and from Table 7.1, explain why Mn(II) is not oxidized by O_2 in solutions at pH 0, but is oxidized by O_2 in solutions in which $[OH^-]$ is $1 \, mol \, dm^{-3}$.

First, find the half-equations that are relevant to the question; note that pH 0 corresponds to standard conditions in which $[H^+] = 1 \, mol \, dm^{-3}$.

$$Mn(OH)_3(s) + e^- \rightleftharpoons Mn(OH)_2(s) + [OH]^-(aq)$$

$$E^\circ_{[OH^-]=1} = +0.18 \, V$$

$$O_2(g) + 2H_2O(l) + 4e^- \rightleftharpoons 4[OH]^-(aq)$$

$$E^\circ_{[OH^-]=1} = +0.40 \, V$$

$$O_2(g) + 4H^+(aq) + 4e^- \rightleftharpoons 2H_2O(l) \qquad E^\circ = +1.23 \, V$$

$$Mn^{3+}(aq) + e^- \rightleftharpoons Mn^{2+}(aq) \qquad E^\circ = +1.54 \, V$$

From this table of reduction potentials (arranged with the most positive value at the bottom of the table), it is seen that $Mn^{3+}(aq)$ is the most powerful oxidizing agent of the species listed. Thus, under acidic conditions (pH 0), O_2 cannot oxidize $Mn^{2+}(aq)$.

In alkaline medium with $[OH^-] = 1 \, mol \, dm^{-3}$, O_2 is able to oxidize $Mn(OH)_2$:

$$O_2(g) + 2H_2O(l) + 4Mn(OH)_2(s) \rightleftharpoons 4Mn(OH)_3(s)$$

$$E^\circ_{cell} = 0.40 - 0.18$$

$$= 0.22 \, V$$

$$\Delta G^\circ = -nFE^\circ_{cell}$$

$$= -(4 \times 96\,485 \times 0.22)$$

$$= -84\,907 \, J \text{ per mole of reaction}$$

$$\approx -85 \, kJ \text{ per mole of reaction}$$

or

$$\approx -21 \, kJ \text{ per mole of } Mn(OH)_2$$

The large negative value of ΔG° indicates that the oxidation of $Mn(OH)_2$ is thermodynamically favoured.

Self-study exercises

1. Why is the notation $E^\circ_{[OH^-]=1}$ used rather than E° for the first two equilibria in the list above?   [*Ans.* See Box 7.1]

BOX 7.3   **APPLICATIONS**

Undersea steel structures: sacrificial anodes and cathodic protection

In Chapter 5, we discussed structural and manufacturing aspects of steel, and the fact that galvanized steel possesses a protective Zn coating. Uses of Zn-coated steel include ships' hulls, undersea pipelines and oil-rigs, i.e. structures that are in contact with seawater. In the presence of H_2O, O_2 *and* an electrolyte (e.g. seawater), steel is subject to corrosion. There is always the possibility that coated steel will be scratched, and that this surface imperfection will permit rusting of the iron beneath it to occur. The Zn coating, however, behaves as a *sacrificial anode*. The actual process of corrosion are not simple, but can be summarized as follows:

$$Zn^{2+}(aq) + 2e^- \rightleftharpoons Zn(s) \qquad E^\circ = -0.76 \, V$$

$$Fe^{2+}(aq) + 2e^- \rightleftharpoons Fe(s) \qquad E^\circ = -0.44 \, V$$

$$O_2(g) + 2H_2O(l) + 4e^- \rightleftharpoons 4[OH]^-(aq))$$

$$E^\circ_{[OH^-]=10^{-7}} = +0.82 \, V$$

In the absence of Zn, Fe is oxidized and is precipitated in the form of $Fe(OH)_2$; if sufficient O_2 is available, further oxidation results in the formation of the familiar red-brown ('rust-coloured') $Fe_2O_3 \cdot H_2O$ (see *Section 21.9*). For Zn-coated

steel, a scratch in the Zn surface means that the oxidation of Zn or Fe is a competitive process; determination of ΔG° for the possible redox processes shows that oxidation of Zn is thermodynamically more favourable than that of Fe, and so the corrosion (rusting) of the steel is inhibited. Furthermore, the Zn^{2+} ions precipitate as $Zn(OH)_2$ ($K_{sp} = 7 \times 10^{-17}$), forming a deposit around the scratched area providing further protection for the steel.

While the anodic oxidation of the zinc coating gives some protection to steel structures, the problems arising from the rusting of steel which is in prolonged contact with seawater are serious. A successful protective measure is to attach metal blocks to, for example, undersea pipelines, the metal being chosen so as to function as an anode in an electrochemical cell in which the seawater is the electrolyte and the Fe of the pipeline is forced to be the cathode. This method of protection (known as *cathodic protection*) is somewhat different from a zinc coating acting as a sacrificial anode. The metal blocks are typically of Mg or Zn, and gradually corrode as anodic oxidation occurs; the iron is never able to function as an anode (and so will not corrode), provided that the metal blocks are regularly renewed.

2. For the reaction:

$O_2(g) + 2H_2O(l) + 4Mn(OH)_2(s) \rightleftharpoons 4Mn(OH)_3(s)$

with $[OH^-] = 1 \, mol \, dm^{-3}$, $\Delta G^\circ = -21.2 \, kJ$ per mole of $Mn(OH)_2$. Find E°_{cell} for the reaction shown in the equation.
[*Ans*. 0.22 V]

Most *d*-block metals resemble Mn in that higher oxidation states are more stable (with respect to reduction) in alkaline rather than acidic solutions. This follows from the fact that the hydroxide of the metal in its higher oxidation state is much less soluble than the hydroxide of the metal in its lower oxidation state.

Analogous principles apply when metal ions in different oxidation states form complexes with the same ligand; usually, the metal ion in the higher oxidation state is stabilized to a greater extent than that in the lower oxidation state. Equations 7.36 and 7.37 show the reduction of hexaaqua and hexaammine complexes of Co(III); remember that $M^{z+}(aq)$ represents $[M(H_2O)_n]^{z+}(aq)$ (see *Section 6.12*).

$Co^{3+}(aq) + e^- \rightleftharpoons Co^{2+}(aq)$ $E^\circ = +1.92 \, V$ (7.36)

$[Co(NH_3)_6]^{3+}(aq) + e^- \rightleftharpoons [Co(NH_3)_6]^{2+}(aq)$

$E^\circ = +0.11 \, V$ (7.37)

It follows from these data that the overall formation constant for $[Co(NH_3)_6]^{3+}$ is $\approx 10^{31}$ times greater than that for $[Co(NH_3)_6]^{2+}$.

A similar comparison can be made for the reduction of the hexaaqua ion of Fe^{3+} and the cyano complex (equations 7.38 and 7.39), and leads to the conclusion that the overall formation constant for $[Fe(CN)_6]^{3-}$ is $\approx 10^7$ times greater than that of $[Fe(CN)_6]^{4-}$.

$Fe^{3+}(aq) + e^- \rightleftharpoons Fe^{2+}(aq)$ $E^\circ = +0.77 \, V$ (7.38)

$[Fe(CN)_6]^{3-}(aq) + e^- \rightleftharpoons [Fe(CN)_6]^{4-}(aq)$

$E^\circ = +0.36 \, V$ (7.39)

Some organic ligands, notably 1,10-phenanthroline and 2,2'-bipyridine (Table 6.7), stabilize the *lower* of two oxidation states of a metal. This is apparent from the values of E° for the appropriate half-reactions in Table 7.1.

7.4 Disproportionation reactions

Disproportionation

Some redox reactions involve *disproportionation* (see *Section 5.16*), e.g. reactions 7.40 and 7.41.

$2Cu^+(aq) \rightleftharpoons Cu^{2+}(aq) + Cu(s)$ (7.40)

$3[MnO_4]^{2-}(aq) + 4H^+(aq)$

$\rightleftharpoons 2[MnO_4]^-(aq) + MnO_2(s) + 2H_2O(l)$ (7.41)

Reaction 7.40 takes place when Cu_2SO_4 (prepared by reacting Cu_2O and dimethyl sulfate) is added to water, while reaction 7.41 occurs when acid is added to a solution of K_2MnO_4. Equilibrium constants for such disproportionation reactions can be calculated from reduction potentials as in worked example 7.5.

Worked example 7.5   Disproportionation of copper(I)

Using appropriate data from Table 7.1, determine K (at 298 K) for the equilibrium:

$2Cu^+(aq) \rightleftharpoons Cu^{2+}(aq) + Cu(s)$

Three redox couples in Table 7.1 involve Cu(I), Cu(II) and Cu metal:

(1) $Cu^{2+}(aq) + e^- \rightleftharpoons Cu^+(aq)$ $E^\circ = +0.15 \, V$

(2) $Cu^{2+}(aq) + 2e^- \rightleftharpoons Cu(s)$ $E^\circ = +0.34 \, V$

(3) $Cu^+(aq) + e^- \rightleftharpoons Cu(s)$ $E^\circ = +0.52 \, V$

The disproportionation of Cu(I) is the result of combining half-reactions (1) and (3). Thus:

$E^\circ_{cell} = 0.52 - 0.15$

$= 0.37 \, V$

$\Delta G^\circ = -zFE^\circ_{cell}$

$= -(1 \times 96\,485 \times 0.37 \times 10^{-3})$

$= -35.7 \, kJ$ per mole of reaction

$\ln K = -\dfrac{\Delta G^\circ}{RT}$

$= \dfrac{35.7}{8.314 \times 10^{-3} \times 298}$

$K = 1.81 \times 10^6$

The value indicates that disproportionation is thermodynamically favourable.

Self-study exercises

1. For the disproportionation of Cu(I) to Cu and Cu(II), $K(298 \, K) = 1.81 \times 10^6$. Calculate ΔG° for the reaction, per mole of Cu(I).   [*Ans*. $-17.85 \, kJ \, mol^{-1}$]

2. By considering redox couples in Appendix 11 which contain Cr^{2+}, Cr^{3+} and Cr metal, confirm that Cr^{2+} will *not* disproportionate into Cr and Cr^{3+}.

Stabilizing species against disproportionation

Species that are unstable with respect to disproportionation, such as Cu^+ in aqueous solution, may be stabilized under appropriate conditions. For example, Cu^+ can be stabilized by precipitation as a sparingly soluble salt such as $CuCl$ ($K_{sp} = 1.72 \times 10^{-7}$) or by the formation in solution of a complex ion such as $[Cu(CN)_4]^{3-}$. In the case of $[MnO_4]^{2-}$ (equation 7.41), all that is necessary is to make the solution alkaline so as to remove the H^+ ions involved in bringing about the disproportionation.

7.5 Potential diagrams

For an element exhibiting several different oxidation states in aqueous solution, we must consider a number of different half-reactions in order to obtain a clear picture of its solution chemistry. Consider manganese as an example; aqueous solution species may contain manganese in oxidation states ranging from Mn(II) to Mn(VII), and equations 7.42–7.46 give half-reactions for which standard reduction potentials can be determined experimentally.

$$Mn^{2+}(aq) + 2e^- \rightleftharpoons Mn(s) \qquad E^o = -1.19\,V \qquad (7.42)$$

$$[MnO_4]^-(aq) + e^- \rightleftharpoons [MnO_4]^{2-}(aq)$$
$$E^o = +0.56\,V \qquad (7.43)$$

$$MnO_2(s) + 4H^+(aq) + 2e^- \rightleftharpoons Mn^{2+}(aq) + 2H_2O$$
$$E^o = +1.23\,V \qquad (7.44)$$

$$[MnO_4]^-(aq) + 8H^+(aq) + 5e^- \rightleftharpoons Mn^{2+}(aq) + 4H_2O(l)$$
$$E^o = +1.51\,V \qquad (7.45)$$

$$Mn^{3+}(aq) + e^- \rightleftharpoons Mn^{2+}(aq) \qquad E^o = +1.54\,V \qquad (7.46)$$

These potentials may be used to derive values of E^o for other half-reactions such as 7.47, care being taken to remember that different numbers of electrons are involved in different reduction steps and, thus, to calculate E^o by first finding the corresponding value of ΔG^o.

$$[MnO_4]^-(aq) + 4H^+(aq) + 3e^- \rightleftharpoons MnO_2(s) + 2H_2O(l)$$
$$E^o = +1.69\,V \qquad (7.47)$$

Self-study exercise

Confirm that the value of E^o for half-equation 7.47 can be obtained from E^o values for half-reactions 7.44 and 7.45, but that the method of working must involve determination of ΔG^o values for the reactions.

Standard reduction potentials are often tabulated as in Appendix 11, but it is also useful to present data in the form of a *potential diagram* (also known as Latimer diagrams).

Figure 7.2 gives potential diagrams for Mn under conditions of $[H^+] = 1\,mol\,dm^{-3}$ (pH 0) and $[OH^-] = 1\,mol\,dm^{-3}$ (pH 14); reading from left to right, species are arranged in order of decreasing oxidation state of Mn. The $[MnO_4]^-$ ion (usually in the form of $KMnO_4$) is a common oxidizing agent, and equations 7.45 or 7.47 are the half-reactions that one would usually consider appropriate in acidic solution. The potential diagram (acidic solution) shows an intermediate Mn(VI) species between $[MnO_4]^-$ and MnO_2. However, values of E^o show that the $[HMnO_4]^-/MnO_2$ couple is a more powerful oxidant (more negative ΔG^o) than the $[MnO_4]^-/[HMnO_4]^-$ couple. This means that $[HMnO_4]^-$ will not accumulate during the reduction of $[MnO_4]^-$ to MnO_2. An alternative way of considering the instability of $[HMnO_4]^-$ in aqueous solution at pH 0 is to note from the potential diagram that $[HMnO_4]^-$ is unstable

Fig. 7.2 Potential diagrams (Latimer diagrams) for manganese in aqueous solution at pH 0 (i.e. $[H^+] = 1\,mol\,dm^{-3}$), and in aqueous solution at pH 14. For such diagrams, it is essential to specify the pH, and the reason is obvious by comparing the two diagrams.

with respect to disproportionation (equation 7.48). Similarly, at pH 0, Mn^{3+} is unstable with respect to disproportionation to MnO_2 and Mn^{2+} (equation 7.49).

$$3[HMnO_4]^-(aq) + H^+(aq)$$
$$\rightleftharpoons MnO_2(s) + 2[MnO_4]^-(aq) + 2H_2O(l) \qquad (7.48)$$
$$2Mn^{3+}(aq) + 2H_2O(l) \rightleftharpoons Mn^{2+}(aq) + MnO_2(s) + 4H^+(aq) \qquad (7.49)$$

We saw in Section 7.2 that the value of the reduction potential for a half-reaction depends on cell conditions, and where the half-reaction involves H^+ or $[OH]^-$ ions, the reduction potential varies with pH. Moreover, the extent of variation depends on the number of moles of H^+ or $[OH]^-$ per mole of reaction. It follows that the potential diagrams in Figure 7.2 are appropriate *only* at the stated pH values; a new potential diagram is needed for every value of pH, and, therefore, *caution is needed* when using these diagrams.

In using potential diagrams, it is essential to remember that the reduction potential for one step may *not* be derived simply by summation of reduction potentials for steps which contribute to the desired redox half-reaction. For example, in Figure 7.2, for the reduction of $[MnO_4]^{2-}$ in alkaline solution to MnO_2, $E^\circ = +0.60\,V$, and this is *not* the sum of the standard reduction potentials for the reduction of $[MnO_4]^{2-}$ to $[MnO_4]^{3-}$ followed by reduction of $[MnO_4]^{3-}$ to MnO_2. Account must be taken of the number of electrons transferred in each step, and the most foolproof way of doing this is to determine the corresponding values of ΔG° for each step.

Worked example 7.6   Potential diagrams

The following potential diagram summarizes some of the redox chemistry of iron in aqueous solution. Calculate the value of E° for the reduction of $Fe^{3+}(aq)$ to iron metal.

$$Fe^{3+}(aq) \xrightarrow{+0.77} Fe^{2+}(aq) \xrightarrow{-0.44} Fe(s)$$
$$\underline{\hspace{6cm}}$$
$$E^\circ$$

Although there are short cuts to this problem, the most rigorous method is to determine $\Delta G^\circ(298\,K)$ for each step.

Fe^{3+} to Fe^{2+} is a one-electron reduction.

$$\Delta G^\circ_1 = -zFE^\circ$$
$$= -[1 \times 96\,485 \times 10^{-3} \times 0.77]$$
$$= -74.3\,kJ \text{ per mole of } Fe^{3+}$$

Fe^{2+} to Fe is a two-electron reduction.

$$\Delta G^\circ_2 = -zFE^\circ$$
$$= -[2 \times 96\,485 \times 10^{-3} \times (-0.44)]$$
$$= +84.9\,kJ \text{ per mole of } Fe^{2+}$$

Next, find ΔG° for the reduction of Fe^{3+} to Fe:

$$\Delta G^\circ = \Delta G^\circ_1 + \Delta G^\circ_2$$
$$= -74.3 + 84.9$$
$$= +10.6\,kJ \text{ per mole of } Fe^{3+}$$

Fe^{3+} to Fe is a three-electron reduction; the standard reduction potential for the process is found from the corresponding value of ΔG°:

$$E^\circ = -\frac{\Delta G^\circ}{zF}$$
$$= -\frac{10.6}{3 \times 96\,485 \times 10^{-3}}$$
$$= -0.04\,V$$

Self-study exercises

1. Although the method given here is probably the 'safest' way to perform the calculation, substitution of a value for the Faraday constant may in fact be excluded. Why?

2. Construct a potential diagram for the reduction of aqueous Cr^{3+} to Cr^{2+}, followed by reduction to Cr. Values of E° for the Cr^{3+}/Cr^{2+} and Cr^{2+}/Cr couples are -0.41 and $-0.91\,V$, respectively. Calculate E° for the Cr^{3+}/Cr couple.
   [*Ans.* $-0.74\,V$]

7.6 The relationships between standard reduction potentials and some other quantities

Factors influencing the magnitudes of standard reduction potentials

In this section, we first consider factors that influence the magnitude of E° for the Na^+/Na and Ag^+/Ag couples, by correlating these values with those of other, independently determined thermodynamic quantities. This comparison allows us to investigate the reasons why, in aqueous media, Na is so much more reactive than Ag, and gives an example that can be extended to other pairs or families of species.

Whereas the standard reduction potential for half-reaction 7.51 is readily measurable in aqueous solution (see *Section 7.2*), that for half-reaction 7.50 must be determined by a rather elaborate set of experiments involving Na amalgam electrodes (amalgams, see *Box 22.3*).

$$Na^+(aq) + e^- \rightleftharpoons Na(s) \qquad E^\circ = -2.71\,V \qquad (7.50)$$
$$Ag^+(aq) + e^- \rightleftharpoons Ag(s) \qquad E^\circ = +0.80\,V \qquad (7.51)$$

We can represent the general half-equation for M^+ reduction as taking place in steps as shown in Figure 7.3. Since all standard reduction potentials are measured with respect to

Fig. 7.3 The half-reaction for the reduction of M^{z+} ions to M, or H^+ to H_2, can be considered in terms of three contributing steps for which thermodynamic data may be determined independently.

Table 7.2 Factors determining the magnitude of the standard reduction potentials for the Na^+/Na and Ag^+/Ag couples in aqueous solution (pH 0); steps (1), (2) and (3) are defined in Figure 7.3.

Redox couple	ΔH^o for step (1) / kJ mol^{-1}	ΔH^o for step (2) / kJ mol^{-1}	ΔH^o for step (3) / kJ mol^{-1}	Overall ΔH^o / kJ mol^{-1}	Calculated E^o / V [‡]
Na^+/Na	404	−492	−108	−196	−2.52
$H^+/\frac{1}{2}H_2$	1091	−1312	−218	−439	0
Ag^+/Ag	480	−733	−285	−538	+1.03

[‡]Values of E^o are found by dividing by $-zF$ ($z = 1$), and scaling to give $E^o(H^+/\frac{1}{2}H_2) = 0\,V$.

the standard hydrogen electrode (for which, by convention, ΔH^o, ΔG^o and ΔS^o are all zero), we must also consider the second thermodynamic cycle (involving absolute values) in Figure 7.3. Table 7.2 lists values of ΔH^o for steps in the cycles defined in Figure 7.3; in an exact treatment, we ought to consider values of ΔG^o, but to a first approximation, we can ignore entropy changes (which largely cancel one another out in this case). From the thermodynamic data, we derive calculated values of E^o and these are given in the right-hand column of Table 7.2. There is good agreement between these values and the experimental ones for half-reactions 7.50 and 7.51. The enthalpy changes for steps (2) and (3) are both negative, and this is a general result for all elements. The *sign* of E^o is determined by the extent to which $\Delta H^o(1)$ offsets $[\Delta H^o(2) + \Delta H^o(3)]$.

Similar analyses for other metals can be carried out. For example, Cu and Zn are adjacent *d*-block metals, and it is interesting to investigate factors that contribute to the difference between E^o values for the Cu^{2+}/Cu and Zn^{2+}/Zn redox couples, and thus reveal how a balance of thermodynamic

factors governs the spontaneous reaction that occurs in the Daniell cell (reaction 7.8). Table 7.3 lists relevant thermodynamic data; it is apparent that a crucial factor in making $E^o_{Cu^{2+}/Cu}$ significantly more positive than $E^o_{Zn^{2+}/Zn}$ is the greater enthalpy of atomization of Cu compared to that of Zn. Thus, what is often regarded as a purely 'physical' property plays a very important role in influencing chemical behaviour. Finally, if we were to consider factors influencing values of E^o for half-reaction 7.52, we would find that the variation in hydration enthalpies plays an important part (oxidizing power of halogens, see *Section 16.4*).

$$X_2 + 2e^- \rightleftharpoons 2X^- \qquad (X = F, Cl, Br, I) \qquad (7.52)$$

Values of $\Delta_f G^o$ for aqueous ions

In Section 6.9, we saw that the standard free energies of formation of aqueous ions can often be determined from E^o values. Worked example 7.7 provides an illustration of the use of reduction potential data in a calculation of a standard free energy of solution of an ionic salt.

Table 7.3 Factors determining the magnitude of the standard reduction potentials for the Cu^{2+}/Cu and Zn^{2+}/Zn couples in aqueous solution (pH 0); steps (1), (2) and (3) are defined in Figure 7.3.

Redox couple	ΔH^o for step (1) / kJ mol^{-1}	ΔH^o for step (2) / kJ mol^{-1}	ΔH^o for step (3) / kJ mol^{-1}	Overall ΔH^o / kJ mol^{-1}	Calculated E^o / V [‡]
Zn^{2+}/Zn	2047	−2644	−130	−727	−0.78
$H^+/\frac{1}{2}H_2$	1091	−1312	−218	−439	0
Cu^{2+}/Cu	2099	−2702	−338	−941	+0.33

[‡]Values of E^o are found by dividing by $-zF$, and scaling to give $E^o(H^+/\frac{1}{2}H_2) = 0\,V$.

Worked example 7.7   Determination of $\Delta_{sol}G^\circ$ for an ionic salt

Calculate the value of $\Delta_{sol}G^\circ(298\,K)$ for NaBr given that $\Delta_fG^\circ(NaBr,s)$ is $-349.0\,kJ\,mol^{-1}$. ($F = 96\,485\,C\,mol^{-1}$).

The process to be considered is:

$$NaBr(s) \rightleftharpoons Na^+(aq) + Br^-(aq)$$

and the equation needed is:

$$\Delta_{sol}G^\circ = \Delta_fG^\circ(Na^+,aq) + \Delta_fG^\circ(Br^-,aq) - \Delta_fG^\circ(NaBr,s)$$

To find $\Delta_fG^\circ(Na^+,aq)$ and $\Delta_fG^\circ(Br^-,aq)$, we need (from Appendix 11) the standard reduction potentials for the processes:

$$Na^+(aq) + e^- \rightleftharpoons Na(s) \qquad E^\circ = -2.71\,V$$

$$\tfrac{1}{2}Br_2(l) + e^- \rightleftharpoons Br^-(aq) \qquad E^\circ = +1.09\,V$$

Now determine Δ_fG° for each aqueous ion, remembering that the standard reduction potential refers to the reverse of the formation of $Na^+(aq)$:

$$\Delta G^\circ = -nFE^\circ$$

$$-\Delta_fG^\circ(Na^+,aq) = -\frac{96\,485 \times (-2.71)}{1000} = 261.5\,kJ\,mol^{-1}$$

$$\Delta_fG^\circ(Br^-,aq) = -\frac{96\,485 \times 1.09}{1000} = -105.2\,kJ\,mol^{-1}$$

$$\Delta_{sol}G^\circ = \Delta_fG^\circ(Na^+,aq) + \Delta_fG^\circ(Br^-,aq) - \Delta_fG^\circ(NaBr,s)$$

$$= -261.5 + (-105.2) - (-349.0)$$

$$= -17.7\,kJ\,mol^{-1}$$

Self-study exercises

1. Calculate the value of $\Delta_{sol}G^\circ(298\,K)$ for NaCl given that $\Delta_fG^\circ(NaCl,s)$ is $-384.0\,kJ\,mol^{-1}$.   [*Ans.* $-8.7\,kJ\,mol^{-1}$]

2. $\Delta_{sol}G^\circ(298\,K)$ for $NaF = +7.9\,kJ\,mol^{-1}$. Determine $\Delta_fG^\circ(NaF,s)$ at 298 K.   [*Ans.* $-546.3\,kJ\,mol^{-1}$]

7.7 Applications of redox reactions to the extraction of elements from their ores

The Earth's environment is an oxidizing one and, in nature, many elements occur as oxides, sulfides or other compounds in which the element is in an oxidized form, e.g. tin occurs as *cassiterite* (SnO_2), and lead as *galena* (PbS). The extraction of these elements from their ores depends on redox chemistry. Heating cassiterite with carbon reduces Sn(IV) to Sn(0) (equation 7.53), and Pb is extracted from galena by reaction sequence 7.54.

Fig. 7.4 An Ellingham diagram showing how the standard free energies of formation, Δ_fG°, of several metal oxides and carbon monoxide (the blue line) vary with temperature. Values of Δ_fG° refer to formation reactions involving a half-mole of O_2: $M + \tfrac{1}{2}O_2 \rightarrow MO$, $\tfrac{1}{2}M + \tfrac{1}{2}O_2 \rightarrow \tfrac{1}{2}MO_2$, or $\tfrac{2}{3}M + \tfrac{1}{2}O_2 \rightarrow \tfrac{1}{3}M_2O_3$. The points marked ♦ or ◇ are the melting or boiling points, respectively, of the elemental metal.

$$SnO_2 + C \xrightarrow{\Delta} Sn + CO_2 \tag{7.53}$$

$$PbS \xrightarrow{O_2, \Delta} PbO \xrightarrow{C\ or\ CO, \Delta} Pb \tag{7.54}$$

Examples of this type are numerous, and similar extraction processes are described in *Box 5.1* and *Chapters 21* and *22*.

Ellingham diagrams

The choice of a reducing agent and the conditions for a particular extraction process can be assessed by using an *Ellingham diagram* such as that in Figure 7.4. This illustrates how Δ_fG° for a range of metal oxides and CO varies with temperature. In order that values are mutually comparable, Δ_fG° refers to the free energy of formation *per half-mole of* O_2.[†] Thus for SrO, Δ_fG° refers to reaction 7.55, and for Al_2O_3 it corresponds to reaction 7.56.

$$Sr + \tfrac{1}{2}O_2 \rightarrow SrO \tag{7.55}$$

$$\tfrac{2}{3}Al + \tfrac{1}{2}O_2 \rightarrow \tfrac{1}{3}Al_2O_3 \tag{7.56}$$

In Figure 7.4, each plot is either linear (e.g. NiO) or has two linear sections (e.g. ZnO); for the latter, there is a change in gradient at the melting point of the metal.

[†] Other data could have been plotted, e.g. values of Δ_fG° per mole of O_2. *Consistency* is the keyword!

Three general results arise from Figure 7.4:

- as the temperature increases, each *metal* oxide becomes *less* thermodynamically stable (less negative $\Delta_f G^\circ$);
- CO becomes *more* thermodynamically stable at higher temperatures (more negative $\Delta_f G^\circ$);
- the *relative* stabilities of the oxides at any given temperature can be seen directly from an Ellingham diagram.

The third point indicates how an Ellingham diagram can be applied. For example, at 1000 K, CO is more thermodynamically stable than SnO_2, and carbon can be used at 1000 K to reduce SnO_2 (equation 7.53). On the other hand, reduction of FeO by carbon occurs at $T > 1000$ K.

The second point has a very important consequence: among the metal oxides in Figure 7.4, the extraction of *any* of the metals from their respective oxides could involve carbon as the reducing agent. In fact at $T > 1800$ K, a greater range of metal oxides than in Figure 7.4 may be reduced by carbon. However, on an industrial scale, this method of obtaining a metal from its oxide is often not commercially viable. Alternative methods of extracting metals from their ores are described in later chapters in the book.

Summary

In this chapter, we began by reviewing some principles concerning redox equilibria; we considered thermodynamic relationships between ΔG°, K and E°_{cell} and illustrated the way in which the reduction potential for a half-cell depends on cell conditions. Reduction potential data may be presented in different ways: tables of E° values, potential (Latimer) diagrams. Trends in E° values are not necessarily easy to understand; hydration enthalpies of ions in solution, ionization energies for the formation of ions, and enthalpies of atomization of elements are factors that contribute to the value of E°. Finally, we considered how redox reactions may be utilized for the extraction of elements from their ores.

The following terms were introduced in this chapter.
Do you know what they mean?
- ☐ oxidation
- ☐ reduction
- ☐ oxidation state (oxidation number)
- ☐ half-reaction (half-equation)
- ☐ electrolytic cell
- ☐ galvanic cell
- ☐ standard conditions for a half-cell
- ☐ standard hydrogen electrode
- ☐ standard reduction potential, E°
- ☐ standard cell potential, E°_{cell}
- ☐ potential diagram (Latimer diagram)
- ☐ Ellingham diagram

Important thermodynamic equations

$$E^\circ_{cell} = [E^\circ_{reduction\ process}] - [E^\circ_{oxidation\ process}]$$

$$\Delta G^\circ = -zFE^\circ_{cell}$$

$$\Delta G^\circ = -RT \ln K$$

$$E = E^\circ - \left\{ \frac{RT}{zF} \times \left(\ln \frac{[\text{reduced form}]}{[\text{oxidized form}]} \right) \right\}$$

(Nernst equation)

Suggested further reading

A.J. Bard, R. Parsons and J. Jordan (1985) *Standard Potentials in Aqueous Solution*, Marcel Dekker, New York – A critical compilation of values, the successor to Latimer's famous treatment of this subject.

J. Burgess (1978) *Metal Ions in Solution*, Ellis Horwood, Chichester & Halsted Press, New York – A thorough treatment of most aspects of metal ions in both aqueous and non-aqueous solutions.

R.G. Compton and G.H.W. Sanders (1996) *Electrode Potentials*, Oxford University Press, Oxford – A useful introduction to electrochemical equilibria and electrochemical principles.

D.A. Johnson (1982) *Some Thermodynamic Aspects of Inorganic Chemistry*, 2nd edn, Cambridge University Press, Cambridge – Contains a very useful discussion of solubility and redox potentials.

W.L. Jolly (1991) *Modern Inorganic Chemistry*, 2nd edn, McGraw-Hill, New York – Contains a treatment of redox potentials which complements that given in this chapter by discussing some systems involving non-metals.

Problems

7.1 Give the oxidation state of each element in the following compounds and ions; Pauling electronegativity values in Appendix 7 may be useful: (a) CaO; (b) H_2O; (c) HF; (d) $FeCl_2$; (e) XeF_6; (f) OsO_4; (g) Na_2SO_4; (h) $[PO_4]^{3-}$; (i) $[PdCl_4]^{2-}$; (j) $[ClO_4]^-$; (k) $[Cr(H_2O)_6]^{3+}$.

7.2 What oxidation state change does each *metal* undergo in the following reactions or half-reactions?

(a) $[Cr_2O_7]^{2-} + 14H^+ + 6e^- \longrightarrow 2Cr^{3+} + 7H_2O$
(b) $2K + 2H_2O \longrightarrow 2KOH + H_2$

(c) $Fe_2O_3 + 2Al \xrightarrow{\Delta} 2Fe + Al_2O_3$
(d) $[MnO_4]^- + 2H_2O + 3e^- \longrightarrow MnO_2 + 4[OH]^-$

7.3 Which of the following reactions are redox reactions? In those that are, identify the oxidation and reduction processes.

(a) $N_2 + 3Mg \xrightarrow{\Delta} Mg_3N_2$
(b) $N_2 + O_2 \longrightarrow 2NO$
(c) $2NO_2 \longrightarrow N_2O_4$
(d) $SbF_3 + F_2 \longrightarrow SbF_5$

(e) $6HCl + As_2O_3 \longrightarrow 2AsCl_3 + 3H_2O$

(f) $2CO + O_2 \longrightarrow 2CO_2$

(g) $MnO_2 + 4HCl \longrightarrow MnCl_2 + Cl_2 + 2H_2O$

(h) $[Cr_2O_7]^{2-} + 2[OH]^- \rightleftharpoons 2[CrO_4]^{2-} + H_2O$

7.4 In each redox reaction in question 7.3, confirm that the net increases and decreases in oxidation states balance each other.

7.5 Using data from Table 7.1, write down the spontaneous cell process, and calculate E°_{cell} and ΔG° for the following combinations of half-cells:

(a) $Ag^+(aq) + e^- \rightleftharpoons Ag(s)$
    with $Zn^{2+}(aq) + 2e^- \rightleftharpoons Zn(s)$

(b) $Br_2(aq) + 2e^- \rightleftharpoons 2Br^-(aq)$
    with $Cl_2(aq) + 2e^- \rightleftharpoons 2Cl^-(aq)$

(c) $[Cr_2O_7]^{2-}(aq) + 14H^+(aq) + 6e^-$
    $\rightleftharpoons 2Cr^{3+}(aq) + 7H_2O(l)$
    with $Fe^{3+}(aq) + e^- \rightleftharpoons Fe^{2+}(aq)$

7.6 Use the data in Appendix 11 to rationalize *quantitatively* why:

(a) Mg liberates H_2 from dilute HCl, but Cu does not;

(b) Br_2 liberates I_2 from aqueous KI solution, but does not liberate Cl_2 from aqueous KCl solution;

(c) the role of Fe^{3+} ions as an oxidizing agent is influenced by the presence of certain ligands in solution;

(d) a method of growing Ag crystals is to immerse a zinc foil in an aqueous solution of $AgNO_3$.

7.7 Consider the half-reaction:

$$[MnO_4]^-(aq) + 8H^+(aq) + 5e^- \rightleftharpoons Mn^{2+}(aq) + 4H_2O(l)$$
$$E^\circ = +1.51\,V$$

If the ratio of concentrations of $[MnO_4]^- : Mn^{2+}$ is 100:1, determine E at pH values of (a) 0.5; (b) 2.0; and (c) 3.5 ($T = 298\,K$). Over this pH range, how does the ability of permanganate(VII) (when being reduced to Mn^{2+}) to oxidize aqueous chloride, bromide or iodide ions change?

7.8 (a) Using appropriate data from Appendix 11, determine E°_{cell} for the disproportionation of H_2O_2. (b) Calculate ΔG° for this process. (c) Comment on the fact that H_2O_2 can be stored without significant decomposition, unless, for example, traces of MnO_2, $[OH]^-$ or iron metal are added.

7.9 Use the following experimental data to determine $E^\circ_{Cu^{2+}/Cu}$, and comment on the need (or not) to make use of *all* the data given.

$[Cu^{2+}]/mol\,dm^{-3}$	0.001	0.005	0.010	0.050
E/V	0.252	0.272	0.281	0.302

7.10 (a) Calculate $E_{Ag^+/Ag}$ for a half-cell in which the concentration of silver(I) ions is $0.1\,mol\,dm^{-3}$ ($T = 298\,K$). (b) Are silver(I) ions more or less easily reduced by zinc in

this solution than under standard conditions? Quantify your answer in thermodynamic terms.

7.11 Given that K_{sp} for AgI is 8.51×10^{-17}, and $E^\circ_{Ag^+/Ag} = +0.80\,V$, calculate E° for the reduction step:

$$AgI(s) + e^- \rightleftharpoons Ag(s) + I^-(aq)$$

and hence confirm the statement in Section 7.3 that reduction of silver(I) when in the form of solid AgI is thermodynamically less favourable than reduction of AgCl.

7.12 Using data from Table 7.1 and from Section 7.3, explain why H_2 is evolved when powdered Ag is heated with a concentrated solution of HI.

7.13 Calculate the overall formation constant for $[Fe(CN)_6]^{3-}$, given that the overall formation constant for $[Fe(CN)_6]^{4-}$ is $\approx 10^{35}$, and that:

$$Fe^{3+}(aq) + e^- \rightleftharpoons Fe^{2+}(aq) \qquad E^\circ = +0.77\,V$$
$$[Fe(CN)_6]^{3-}(aq) + e^- \rightleftharpoons [Fe(CN)_6]^{4-}(aq) \quad E^\circ = +0.36\,V$$

7.14 Using data in Appendix 11, determine which of the following species is thermodynamically unstable with respect to disproportionation (and under what conditions) in aqueous solution: (a) Fe^{2+}; (b) Sn^{2+}; (c) $[ClO_3]^-$.

7.15 Determine $\Delta G^\circ(298\,K)$ for the reaction:

$$2CuCl(s) \rightleftharpoons Cu^{2+}(aq) + 2Cl^-(aq) + Cu(s)$$

given the following data:

$$2Cu^+(aq) \rightleftharpoons Cu^{2+}(aq) + Cu(s) \quad K = 1.81 \times 10^6$$
$$CuCl(s) \rightleftharpoons Cu^+(aq) + Cl^-(aq) \quad K_{sp} = 1.72 \times 10^{-7}$$

What does the value of ΔG° tell you about the tendency of precipitated CuCl to disproportionate?

7.16 Using appropriate data from equations 7.42 to 7.46, confirm the value of E° given for equation 7.47.

7.17 Write balanced half-equations corresponding to the steps shown in the potential diagrams in Figure 7.2.

7.18 (a) Use data from Appendix 11 to construct a potential diagram showing the redox chemistry of vanadium in aqueous solution at pH 0. (b) Use your diagram to establish whether any vanadium species is unstable with respect to disproportionation.

7.19 The following potential diagram summarizes the results of electrochemical studies of the aqueous solution (pH 0) chemistry of uranium:

$$[UO_2]^{2+} \xrightarrow{+0.06} [UO_2]^{2+} \xrightarrow{+0.61} U^{4+} \xrightarrow{-0.61} U^{3+} \xrightarrow{-1.80} U$$

$$+0.33$$

Use the information to deduce as much as possible about the chemistry of uranium under these conditions.

7.20 The following potential diagram is part of that illustrating the redox chemistry of chlorine in aqueous solution at pH 0. (a) Calculate the value of E° for the reduction of $[ClO_3]^-$ to $HClO_2$. (b) Justify why, *in this case*, the value of

E° can simply be taken to be the mean of $+1.15$ and $+1.28$ V.

$$[ClO_3]^- \xrightarrow{+1.15} ClO_2 \xrightarrow{+1.28} HClO_2$$

E°

7.21 By constructing thermodynamic cycles analogous to those shown in Figure 7.3, discuss the factors that contribute to the trend in values of E° for the group 1 metals Li to Cs. [$\Delta_{hyd}H^{\circ}$: see *Table 6.6*. *IE* and $\Delta_{atom}H^{\circ}$: see Appendices 8 and 10.]

Chapter 8 Non-aqueous media

TOPICS

- Relative permittivity
- Acid–base behaviour in non-aqueous solvents
- Ammonia
- Hydrogen fluoride
- Sulfuric acid

- Fluorosulfonic acid
- Bromine trifluoride
- Dinitrogen tetraoxide
- Molten (fused) salts

8.1 Introduction

Although many inorganic reactions take place in aqueous solution, water is not always a suitable solvent; some reagents may react with H_2O (e.g. the alkali metals) and non-polar molecules are insoluble in water. This chapter discusses *non-aqueous solvents*. The use of solvents other than water is commonplace for the organic chemist, and such compounds include dichloromethane, hexane, toluene and ethers such as diethyl ether, **8.1**, tetrahydrofuran, **8.2**, and diglyme, **8.3**.

(8.1)   (8.2)

(8.3)

These solvents are of significant use to the inorganic chemist, but also available are more exotic solvents such as liquid NH_3, liquid SO_2, H_2SO_4, BrF_3 and liquid salts such as $[pyBu][AlCl_4]$, **8.4**.

(8.4)

We can conveniently place non-aqueous solvents into the following categories:

- protic solvents (e.g. HF, H_2SO_4, MeOH);
- aprotic solvents (e.g. N_2O_4, BrF_3);
- coordinating solvents (e.g. MeCN, Et_2O, Me_2CO).

A *protic solvent* undergoes *self-ionization* (see *Section 6.2*) to provide protons which are solvated. *If* it undergoes self-ionization, an *aprotic solvent* does so without the formation of protons.

As we discuss the properties and uses of some non-aqueous solvents, we must keep in mind that the extent to which non-aqueous solvents can be used is limited by the fact that many are highly reactive.

Quantitative data are scarce for non-aqueous media, and, in solvents of relative permittivity lower than that of water, data are difficult to interpret because of ion-association. Although we shall make some general observations in this chapter, no integrated treatment of inorganic chemistry in non-aqueous solvents is yet possible, and much of our discussion centres on the properties and uses of selected, individual solvents.

8.2 Relative permittivity

Before beginning a discussion of non-aqueous solvents, we must define the *relative permittivity*, also referred to as the *dielectric constant*, of a substance. In a vacuum, the Coulombic potential energy of a system of two unit electronic charges is given by equation 8.1 where ε_0 is the (absolute)

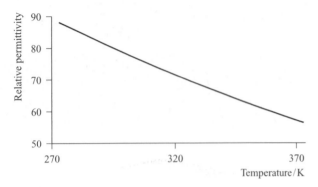

Fig. 8.1 Variation in the relative permittivity (dielectric constant) of water as a function of temperature.

permittivity of a vacuum $(8.854 \times 10^{-12}\,F\,m^{-1})$, e is the charge on the electron $(1.602 \times 10^{-19}\,C)$ and r is the separation (in metres) between the point charges.

$$\text{Coulombic potential energy} = \frac{e^2}{4\pi\varepsilon_0 r} \qquad (8.1)$$

If a material is placed between the charges, the force is reduced by an amount that depends upon the *relative permittivity* of the material; the new Coulombic potential energy is given by equation 8.2 where ε_r is the relative permittivity of the material. Since it is a *relative* quantity, ε_r is dimensionless.

$$\text{Coulombic potential energy} = \frac{e^2}{4\pi\varepsilon_0\varepsilon_r r} \qquad (8.2)$$

For example, at 298 K, ε_r of water (the dielectric constant) is 78.7, but as Figure 8.1 shows, ε_r varies with temperature. A value of 78.7 can be considered to be a 'high' value and from equation 8.2, we see that in aqueous solution, the force between two point charges (or two ions) is considerably reduced compared with that in a vacuum. Thus we can consider a dilute aqueous solution of a salt to contain well-separated, non-interacting ions.

Table 8.1 lists dielectric constants for water and a range of common organic solvents. The *absolute* permittivity

of a solvent is found using equation 8.3, but it is usual to discuss solvent properties in terms of the relative values.

$$\text{Absolute permittivity of a material} = \varepsilon_0\varepsilon_r \qquad (8.3)$$

Table 8.1 also gives the dipole moment of each solvent; in general, the trend in values of dipole moments (μ) follows that in values of the relative permittivities for solvents having related structures. Ion–solvent interactions are favoured (i.e. to facilitate the dissolution of an ionic salt) by using a solvent with a large dipole moment, but for maximum effect, the solvent molecule should also be small, and both ends of it should be able to interact with the ions in the same way that water interacts with cations through the oxygen atoms (see *Figure 6.5*) and with anions through the hydrogen atoms. Thus, ammonia ($\varepsilon_r = 25.0$, $\mu = 1.47\,D$) is a better solvent (see *Section 8.6*) for ionic salts than dimethylsulfoxide or nitromethane, even though these have ε_r values of 46.7 and 35.9, and dipole moments of 3.96 and 3.46 D, respectively.

8.3 Energetics of ionic salt transfer from water to an organic solvent

In this section, we consider the changes in enthalpy and free energy that accompany the transfer of simple ions from water to some organic solvents of high relative permittivity. These data provide us with an idea of the *relative* abilities of water and these organic liquids to act as solvents with regard to the ions considered. Since most organic liquids are soluble in water to some extent, or are completely miscible with water, thermodynamic data for the dissolution of salts are usually obtained by considering the two solvents separately; data for the transfer of ions ($\Delta_{transfer}G^\circ$ and $\Delta_{transfer}H^\circ$) can be derived from the differences between the values corresponding to the dissolution processes in the two solvents. Our discussion centres on the four organic solvents: methanol (**8.5**), formamide (**8.6**), *N,N*-dimethylformamide

Table 8.1 Relative permittivity (dielectric constant) values at 298 K (unless otherwise stated) for selected organic solvents.

Solvent	Formula[‡]	Relative permittivity, ε_r	Dipole moment, μ / debye
Formamide	$HC(O)NH_2$	109 (293 K)	3.73
Water	H_2O	78.7	1.85
Acetonitrile	MeCN	37.5 (293 K)	3.92
N,N-Dimethylformamide (DMF)	$HC(O)NMe_2$	36.7	3.86
Nitromethane	$MeNO_2$	35.9 (303 K)	3.46
Methanol	MeOH	32.7	1.70
Ethanol	EtOH	24.3	1.69
Dichloromethane	CH_2Cl_2	9.1 (293 K)	1.60
Tetrahydrofuran	C_4H_8O (structure **8.2**)	7.6	1.75
Diethyl ether	Et_2O	4.3 (293 K)	1.15
Benzene	C_6H_6	2.3	0

[‡] Me = methyl; Et = ethyl.

(DMF, **8.7**) and acetonitrile (**8.8**), relative permittivities and dipole moments for which are listed in Table 8.1.

$$\text{Me}-\overset{\text{O}}{}-\text{H} \qquad \text{HC}\overset{\text{NH}_2}{\underset{\text{O}}{\diagdown}} \qquad \text{HC}\overset{\text{NMe}_2}{\underset{\text{O}}{\diagdown}} \qquad \text{MeC}\equiv\text{N}$$

(8.5)   (8.6)   (8.7)   (8.8)

In an analogous approach to that discussed in Section 6.9, we can make the assumption that very large ions such as $[Ph_4As]^+$ and $[BPh_4]^-$ have the same values of $\Delta_{transfer}G^o$ and $\Delta_{transfer}H^o$. By considering a series of $[Ph_4As]X$ and $M[BPh_4]$ salts (in conjunction with $[Ph_4As][BPh_4]$), it is possible to obtain the thermodynamic data given in Table 8.2, where $\Delta_{transfer}H^o$ and $\Delta_{transfer}G^o$ refer to the transfer of the specified ion from water to the organic solvent. A positive value of $\Delta_{transfer}G^o$ indicates an unfavourable transfer, while a negative value corresponds to a favourable process.

The data in Table 8.2 show that the large, non-polar $[Ph_4As]^+$ and $[BPh_4]^-$ ions are more solvated in each organic solvent than in water; enthalpy and entropy effects both contribute in the same direction. Alkali metal ions exhibit no simple pattern of behaviour, although in each solvent, values of $\Delta_{transfer}H^o$ and $\Delta_{transfer}G^o$ are less positive for the alkali metal ions than for the halide ions. For the halide ions, transfer from water to the organic media is thermodynamically unfavourable, but we can go further than this generalization. Methanol and formamide are capable of forming hydrogen bonds between the H atoms of the OH or NH_2 groups and the halide ions in solution; MeCN and DMF do not possess this capability. Not only are the values of $\Delta_{transfer}G^o$ for the halide ion significantly more positive for MeCN and DMF than for MeOH and formamide, but the variation in values among the halide ions is much greater. We may conclude that halide ions (and F^- and Cl^- in particular) are much less strongly solvated in solvents in which hydrogen bonding is not possible than in those in which hydrogen-bonded interactions can form

(this, of course, includes water). This difference is the origin of the solvent dependence of reactions involving halide ions. A well-known example is the bimolecular reaction 8.4 for which the rate increases from X = F to I in aqueous solution, but decreases in N,N-dimethylformamide.

$$CH_3Br + X^- \longrightarrow CH_3X + Br^- \qquad (X = F, Cl \text{ or } I) \qquad (8.4)$$

Fluoride ion in solvents with which it is not able to form hydrogen bonds is sometimes described as 'naked', but this term is misleading; in DMF, it still has a free energy of solvation of about $-400\,kJ\,mol^{-1}$ ($\approx 60\,kJ\,mol^{-1}$ less negative than in water) and so is still very much less reactive than in the gas phase.

8.4 Acid–base behaviour in non-aqueous solvents

Strengths of acids and bases

When we dealt with acid–base behaviour in aqueous solution in Chapter 6, we saw that the strength of an acid HX (equation 8.5) depended upon the relative proton donor abilities of HX and $[H_3O]^+$.

$$HX(aq) + H_2O(l) \rightleftharpoons [H_3O]^+(aq) + X^-(aq) \qquad (8.5)$$

Similarly, the strength of a base, B, in aqueous solution depends upon the relative proton accepting abilities of B and $[OH]^-$ (equation 8.6).

$$B(aq) + H_2O(l) \rightleftharpoons [BH]^+(aq) + [OH]^-(aq) \qquad (8.6)$$

Tabulated values of K_a (or K_b) generally refer to the ionizations of acids in *aqueous solution*, and in stating that 'HCl is a strong acid', we assume an aqueous medium. However, if HCl is dissolved in acetic acid, the extent of ionization is far less than in water and HCl behaves as a weak acid.

Table 8.2 Values of $\Delta_{transfer}H^o$ and $\Delta_{transfer}G^o$ for the transfer of ions from water to an organic solvent.

Ion	Methanol		Formamide		N,N-Dimethylformamide		Acetonitrile	
	$\Delta_{transfer}H^o$ /kJ mol^{-1}	$\Delta_{transfer}G^o$ /kJ mol^{-1}	$\Delta_{transfer}H^o$ /kJ mol^{-1}	$\Delta_{transfer}G^o$ /kJ mol^{-1}	$\Delta_{transfer}H^o$ /kJ mol^{-1}	$\Delta_{transfer}G^o$ /kJ mol^{-1}	$\Delta_{transfer}H^o$ /kJ mol^{-1}	$\Delta_{transfer}G^o$ /kJ mol^{-1}
F^-	12	20	20	25	–	≈ 60	–	71
Cl^-	8	13	4	14	18	48	19	42
Br^-	4	11	–1	11	1	36	8	31
I^-	–2	7	–7	7	–15	20	–8	17
Li^+	–22	4	–6	–10	–25	–10	–	25
Na^+	–20	8	–16	–8	–32	–10	–13	15
K^+	–19	10	–18	–4	–36	–10	–23	8
$[Ph_4As]^+$ and $[BPh_4]^-$	–2	–23	–1	–24	–17	–38	–10	–33

Levelling and differentiating effects

Non-aqueous solvents that are good proton acceptors (e.g. NH_3) encourage acids to ionize in them; thus, in a *basic solvent*, all acids are strong. The solvent is said to exhibit a *levelling effect* on the acid, since the strength of the dissolved acid cannot exceed that of the protonated solvent. For example, in aqueous solution, no acidic species can exist that is a stronger acid than $[H_3O]^+$. In an acidic solvent (e.g. $MeCO_2H$, H_2SO_4), ionization of bases is facilitated; most acids are relatively weak under these conditions, and some even ionize as bases.

We noted above that HCl, when dissolved in acetic acid, behaves as a weak acid. Hydrogen bromide and iodide behave similarly but the *extent of ionization* of the three hydrogen halides varies along the series: HI > HBr > HCl. This contrasts with the fact that all three compounds are classed as strong acids (i.e. fully ionized) in aqueous solution. Thus, acetic acid exerts a *differentiating effect* on the acidic behaviour of HCl, HBr and HI, whereas water does not.

'Acids' in acidic solvents

The effects of dissolving 'acids' in acidic non-aqueous solvents can be dramatic. When dissolved in H_2SO_4, $HClO_4$ (for which pK_a in aqueous solution is −8) is practically non-ionized and HNO_3 ionizes according to equation 8.7.

$$HNO_3 + 2H_2SO_4 \rightleftharpoons [NO_2]^+ + [H_3O]^+ + 2[HSO_4]^- \quad (8.7)$$

Reaction 8.7 can be regarded as the summation of equilibria 8.8–8.10, and it is the presence of the nitryl ion,[†] $[NO_2]^+$, that is responsible for the use of an HNO_3/H_2SO_4 mixture in the nitration of aromatic compounds.

$$HNO_3 + H_2SO_4 \rightleftharpoons [H_2NO_3]^+ + [HSO_4]^- \quad (8.8)$$

$$[H_2NO_3]^+ \rightleftharpoons [NO_2]^+ + H_2O \quad (8.9)$$

$$H_2O + H_2SO_4 \rightleftharpoons [H_3O]^+ + [HSO_4]^- \quad (8.10)$$

These examples signify caution: *just because we name a compound an 'acid', it may not behave as one in non-aqueous media.* Later we consider superacid media in which even hydrocarbons may be protonated (see *Section 8.9*).

Acids and bases: a solvent-oriented definition

A Brønsted acid is a proton donor, and a Brønsted base accepts protons. In aqueous solution, $[H_3O]^+$ is formed and in bulk water, self-ionization corresponds to the transfer of a proton from one solvent molecule to another (equation 8.11) illustrating amphoteric behaviour (see *Section 6.8*).

$$2H_2O \rightleftharpoons [H_3O]^+ + [OH]^- \quad (8.11)$$

In liquid NH_3 (see *Section 8.6*), proton transfer leads to the formation of $[NH_4]^+$ (equation 8.12), and, in a liquid ammonia solution, an acid may be described as a substance

that produces $[NH_4]^+$ ions, while a base produces $[NH_2]^-$ ions.

$$2NH_3 \rightleftharpoons \underset{\text{ammonium ion}}{[NH_4]^+} + \underset{\text{amide ion}}{[NH_2]^-} \quad (8.12)$$

This solvent-oriented definition can be widened to include behaviour in any solvent which undergoes self-ionization.

> In a *self-ionizing solvent*, an acid is a substance that produces the cation characteristic of the solvent, and a base is a substance that produces the anion characteristic of the solvent.

Liquid dinitrogen tetraoxide, N_2O_4, undergoes the self-ionization shown in equation 8.13. In this medium, nitrosyl salts such as $[NO][ClO_4]$ behave as acids, and metal nitrates (e.g. $NaNO_3$) behave as bases.

$$N_2O_4 \rightleftharpoons [NO]^+ + [NO_3]^- \quad (8.13)$$

In some ways, this acid–base terminology is unfortunate, since there are other, more common descriptors (e.g. Brønsted, Lewis, hard and soft). However, the terminology has been helpful in suggesting lines of research for the study of non-aqueous systems, and its use will probably continue.

8.5 Self-ionizing and non-ionizing non-aqueous solvents

In the sections that follow, we shall consider selected inorganic non-aqueous solvents in some detail. The solvents chosen for discussion are all self-ionizing and can be divided into two categories:

- proton containing (NH_3, HF, H_2SO_4, $HOSO_2F$);
- aprotic (BrF_3, N_2O_4).

One notable exception to the solvents we shall study is liquid SO_2. The solvent-based definition of acids and bases described above was first put forward for SO_2 for which the self-ionization process 8.14 was proposed.

$$2SO_2 \rightleftharpoons [SO]^{2+} + [SO_3]^{2-} \quad (8.14)$$

Unlike other self-ionization equilibria that we shall discuss, reaction 8.14 requires the separation of doubly charged ions, and on these grounds alone, the establishment of this equilibrium must be considered improbable. Its viability is also questioned by the fact that thionyl chloride, $SOCl_2$ (the only reported acid in the solvent), does *not* exchange ^{35}S nor ^{18}O with the liquid SO_2 solvent. Selected properties of SO_2 are given in Table 8.3, and its liquid range is compared with those of other solvents in Figure 8.2. Liquid SO_2 is an effective, inert solvent for both organic compounds (e.g. amines, alcohols, carboxylic acids, esters) and covalent inorganic compounds (e.g. Br_2, CS_2, PCl_3, $SOCl_2$, $POCl_3$)

[†] The nitryl ion is also called the nitronium ion.

Table 8.3   Selected physical properties of sulfur dioxide, SO_2.

Property / units	Value
Melting point / K	200.3
Boiling point / K	263.0
Density of liquid / $g\,cm^{-3}$	1.43
Dipole moment / D	1.63
Relative permittivity	17.6 (at boiling point)

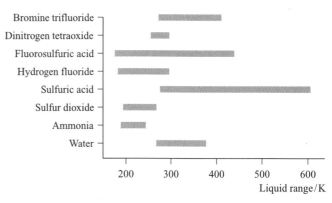

Fig. 8.2 Liquid ranges for water and selected non-aqueous solvents.

and is quite a good ionizing medium for such compounds as Ph_3CCl (giving $[Ph_3C]^+$). It is also used for the syntheses of some group 16 and 17 cationic species. For example, $[I_3]^+$ and $[I_5]^+$ (equation 8.15) have been isolated as the $[AsF_6]^-$ salts from the reactions of AsF_5 and I_2 in liquid SO_2, the

---

BOX 8.1   **APPLICATIONS**

Supercritical CO_2

The critical point on a phase diagram of a substance marks the limiting (highest) temperature and pressure for a liquid $\rightleftharpoons$ vapour equilibrium. For CO_2, the critical temperature (T_c) is 304.1 K and the critical pressure (P_c) is 7.37 MPa (73.7 bar). At higher temperatures, a *supercritical fluid* may exist, which has properties that make it of great interest as a solvent. In addition, it is non-inflammable and relatively non-toxic. The uses of supercritical CO_2 as a solvent are rapidly increasing as literature publications between 1995 and 2000 illustrate. New applications include a medium for radical polymerizations in which it is crucial to use a solvent that is inert with respect to the free radicals present in the reaction mixture. An example is the polymerization of $CH_2=CHC(O)NH_2$, which under these conditions yields an extremely high molecular weight homopolymer:

$$n\ \ \begin{array}{c} H \quad\quad H \\ \ \ \diagdown \quad \diagup \\ C = C \\ \diagup \quad\quad \diagdown \\ H \quad\quad\ \ C = O \\ \quad\quad\quad | \\ \quad\quad H_2N \end{array} \xrightarrow[\text{CO}_2\ (330\,\text{K},\ 345\,\text{bar})]{\text{Radical initiator}} \left(\begin{array}{c} H \quad\quad H \\ | \quad\quad\ | \\ -C - C - \\ | \quad\quad\ | \\ H \quad\quad\ C = O \\ \quad\quad\quad | \\ \quad\quad H_2N \end{array}\right)_n$$

Applications lie in all facets of chemistry, e.g. supercritical CO_2 as a solvent for the extraction of the pungent compounds in Australian-grown ginger (*Zingiber officinale*). The advantage of using supercritical CO_2 over other extraction methods such as steam-distillation is that a lower degree of natural compound degradation during extraction is observed.

The range of compounds soluble in supercritical CO_2 limits its uses; in many cases, this can be overcome by the use of surfactants that possess a water-soluble head and CO_2-compatible tail. It has been demonstrated that water 'pockets' can be dispersed within the supercritical CO_2 solvent, allowing aqueous chemistry to be carried out in what is essentially a non-aqueous environment. An advantage of this scenario is that reagents not normally soluble in water, but soluble in supercritical CO_2, can be brought into intimate contact with water-soluble reagents.

Further reading

F.A. Adamsky and E.J. Beckman (1994) *Macromolecules*, vol. 27, p. 312 – 'Inverse emulsion polymerization of acrylamide in supercritical carbon dioxide'.

J.P. Bartley (1995) *Journal of Science and Food Agriculture*, vol. 68, p. 215 – 'A new method for the determination of pungent compounds in ginger *(Zingiber officinale)*'.

J.F. Brennecke (1996) *Chemistry & Industry*, p. 831 – 'New applications of supercritical fluids'.

M.J. Clarke, K.L. Harrison, K.P. Johnston and S.M. Howdle (1997) *Journal of the American Chemical Society*, vol. 119, p. 6399 – 'Water in supercritical carbon dioxide microemulsions: Spectroscopic investigation of a new environment for aqueous inorganic chemistry'.

product depending on the molar ratio of the reactants. Reactions of selenium with AsF_5 (at 350 K) or SbF_5 (at 250 K) in liquid SO_2 have yielded the salts $[Se_4][AsF_6]_2$ and $[Se_8][SbF_6]_2$ respectively.

$$3AsF_5 + 5I_2 \xrightarrow{\text{liquid } SO_2} 2[I_5][AsF_6] + AsF_3 \qquad (8.15)$$

In addition to the examples given in this chapter, important applications of non-aqueous solvents include the separation of uranium and plutonium in nuclear technology (see *Box 6.3*), and the analytical separation of many metals. Supercritical CO_2 is a non-aqueous solvent for which applications are rapidly increasing in number (see *Box 8.1*).

8.6 Liquid ammonia

Liquid ammonia has been widely studied, and in this section we discuss its properties and the types of reactions that occur in it, making comparisons between liquid ammonia and water.

Physical properties

Selected properties of NH_3 are listed in Table 8.4 and are compared with those of water; it has a liquid range of 44.3 K (Figure 8.2). The lower boiling point than water suggests that hydrogen bonding in liquid NH_3 is less extensive than in liquid H_2O, and this is further illustrated by the values of $\Delta_{vap}H^o$ (23.3 and 40.7 kJ mol^{-1} for NH_3 and H_2O respectively). This is consistent with the presence of one lone pair on the nitrogen atom in NH_3 compared with two on the oxygen atom in H_2O.

The relative permittivity of NH_3 is considerably less than that of H_2O and, as a consequence, the ability of liquid NH_3 to dissolve ionic compounds is generally significantly less than that of water. Exceptions include $[NH_4]^+$ salts, iodides and nitrates which are usually readily soluble. For example, AgI, which is sparingly soluble in water, dissolves easily in liquid NH_3 (solubility = 206.8 g per 100 g of NH_3), a fact that indicates that both the Ag^+ and I^- ions interact strongly with the solvent; Ag^+ forms an ammine complex (see *Section 22.12*). Changes in solubility patterns in going from water to

Table 8.4 Selected physical properties of ammonia, NH_3, and water, H_2O.

Property / units	NH_3	H_2O
Melting point / K	195.3	273.0
Boiling point / K	239.6	373.0
Density of liquid / g cm^{-3}	0.77	1.00
Dipole moment / D	1.47	1.85
Relative permittivity	25.0	78.7
	(at melting point)	(at 298 K)
Self-ionization constant	5.1×10^{-27}	1.0×10^{-14}

liquid NH_3 lead to some interesting precipitation reactions in NH_3. Whereas in aqueous solution, $BaCl_2$ reacts with $AgNO_3$ to precipitate AgCl, in liquid NH_3, AgCl and $Ba(NO_3)_2$ react to precipitate $BaCl_2$. Most chlorides (and almost all fluorides) are practically insoluble in liquid NH_3. Molecular organic compounds are generally more soluble in NH_3 than in H_2O.

Self-ionization

As we have already mentioned, liquid NH_3 undergoes self-ionization (equation 8.12), and the small value of K_{self} (Table 8.4) indicates that the equilibrium lies far over to the left-hand side. The $[NH_4]^+$ and $[NH_2]^-$ ions have ionic mobilities approximately equal to those of alkali metal and halide ions; this contrasts with the situation in water in which $[H_3O]^+$ and $[OH]^-$ are much more mobile than other singly charged ions.

Reactions in liquid NH_3

We described above some precipitations that differ in liquid NH_3 and H_2O. Equation 8.16 shows a further example; the solubility of KCl is 0.04 g per 100 g NH_3, compared with 34.4 g per 100 g H_2O.

$$KNO_3 + AgCl \longrightarrow \underset{ppt}{KCl} + AgNO_3 \qquad (8.16)$$

In water, neutralization reactions follow the general reaction 8.17; the solvent-oriented definition of acids and bases allows us write an analogous reaction (equation 8.18) for a neutralization process in liquid NH_3.

$$\text{Acid} + \text{Base} \longrightarrow \text{Salt} + \text{Water} \quad \textit{in aqueous solution} \qquad (8.17)$$

$$\text{Acid} + \text{Base} \longrightarrow \text{Salt} + \text{Ammonia} \quad \textit{in liquid ammonia} \qquad (8.18)$$

Thus, in liquid NH_3, reaction 8.19 is a neutralization process which may be followed by conductivity or potentiometry, or by the use of an indicator such as phenolphthalein **8.9**; this indicator is colourless but is deprotonated by a strong base such as $[NH_2]^-$ to a red anion just as it is by $[OH]^-$ in aqueous solution.

(8.9)

$$NH_4Br + KNH_2 \longrightarrow KBr + 2NH_3 \qquad (8.19)$$

Liquid NH_3 is an ideal solvent for reactions requiring a strong base, since the amide ion is strongly basic.

As we discussed in Section 8.4, the behaviour of 'acids' is solvent-dependent. In aqueous solution, sulfamic acid, H_2NSO_2OH **8.10**, behaves as a monobasic acid according to equation 8.20, but in liquid NH_3 it can function as a dibasic acid (equation 8.21).

(8.10)

$$H_2NSO_2OH(aq) + H_2O(l)$$
$$\rightleftharpoons [H_3O]^+(aq) + [H_2NSO_2O]^-(aq) \quad K_a = 1.01 \times 10^{-1}$$
(8.20)

$$H_2NSO_2OH + 2KNH_2 \longrightarrow K_2[HNSO_2O] + 2NH_3 \quad (8.21)$$

The levelling effect of liquid NH_3 means that the strongest acid possible in this medium is $[NH_4]^+$. Solutions of ammonium halides in NH_3 may be used as acids, for example in the preparation of silane or arsane (equations 8.22 and 8.23). Germane, GeH_4, can be prepared from Mg_2Ge in a reaction analogous to the preparation of SiH_4.

$$Mg_2Si + 4NH_4Br \longrightarrow SiH_4 + 2MgBr_2 + 4NH_3 \quad (8.22)$$

$$Na_3As + 3NH_4Br \longrightarrow AsH_3 + 3NaBr + 3NH_3 \quad (8.23)$$

A saturated solution of NH_4NO_3 in liquid NH_3 (which has a vapour pressure of less than 1 bar even at 298 K) dissolves many metal oxides and even some metals; nitrate to nitrite reduction often accompanies the dissolution of metals. Metals that form insoluble hydroxides under aqueous conditions, form insoluble amides in liquid NH_3, e.g. $Zn(NH_2)_2$. Just as $Zn(OH)_2$ dissolves in the presence of excess hydroxide ion (equation 8.24), $Zn(NH_2)_2$ reacts with amide ion to form soluble salts containing anion **8.11** (equation 8.25).

$$Zn^{2+} + 2[OH]^- \longrightarrow Zn(OH)_2 \xrightarrow{\text{excess } [OH]^-} [Zn(OH)_4]^{2-}$$
(8.24)

$$Zn^{2+} + 2[NH_2]^- \longrightarrow Zn(NH_2)_2 \xrightarrow{\text{excess } [NH_2]^-} [Zn(NH_2)_4]^{2-}$$
(8.25)

(8.11)

Parallels can be drawn between the behaviour of metal nitrides in liquid NH_3 and that of metal oxides in aqueous media. Many similar analogies can be drawn.

Solutions of s-block metals in liquid NH_3

All of the group 1 metals and the group 2 metals Ca, Sr and Ba dissolve in liquid NH_3 to give metastable solutions from which the group 1 metals can be recovered unchanged; the group 2 metals are recoverable as solids of composition $[M(NH_3)_6]$. Yellow $[Li(NH_3)_4]$ and blue $[Na(NH_3)_4]$ may also be isolated at low temperatures.

Dilute solutions of the metals are bright blue, the colour arising from the short wavelength tail of a broad and intense absorption band in the infrared region of the spectrum. The electronic spectra in the visible region of solutions of all the s-block metals are the same, indicating the presence of a species common to all the solutions: this is the solvated electron (equation 8.26).

$$M \xrightarrow{\text{dissolve in liquid } NH_3} M^+(solv) + e^-(solv) \quad (8.26)$$

Each dilute solution of metal in liquid NH_3 occupies a volume greater than the sum of the volumes of the metal plus solvent. These data suggest that the electrons occupy cavities of radius 300–400 pm. Very dilute solutions of the metals are paramagnetic, and the magnetic susceptibility corresponds to that calculated for the presence of one free electron per metal atom.

As the concentration of a solution of an s-block metal in liquid NH_3 increases, the molar conductivity initially decreases, reaching a minimum at $\approx 0.05\, \text{mol dm}^{-3}$. Thereafter, the molar conductivity increases, and in saturated solutions is comparable to that of the metal itself. Such saturated solutions are no longer blue and paramagnetic, but are bronze and diamagnetic; they are essentially 'metal-like' and have been described as *expanded metals*. The conductivity data can be described in terms of:

- process 8.26 at low concentrations;
- association of $M^+(solv)$ and $e^-(solv)$ at concentrations around $0.05\, \text{mol dm}^{-3}$;
- metal-like behaviour at higher concentrations.

However, in order to rationalize the fact that the magnetic susceptibilities of solutions *decrease* as the concentration increases, it is necessary to invoke equilibria 8.27 at higher concentrations.

$$\left.\begin{array}{l} 2M^+(solv) + 2e^-(solv) \rightleftharpoons M_2(solv) \\ M(solv) + e^-(solv) \rightleftharpoons M^-(solv) \end{array}\right\} \quad (8.27)$$

The blue solutions of alkali metals in liquid NH_3 decompose very slowly, liberating H_2 (equation 8.28) as the solvent is reduced.

$$2NH_3 + 2e^- \longrightarrow 2[NH_2]^- + H_2 \quad (8.28)$$

Although reaction 8.28 is thermodynamically favoured, there is a significant kinetic barrier. Decomposition is catalyzed by many d-block metal compounds, e.g. by stirring the solution with a rusty Fe wire. Ammonium salts (which

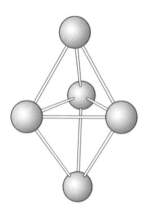

Fig. 8.3 The Zintl ion $[Sn_5]^{2-}$ has a trigonal bipyramidal cluster structure.

Table 8.5 Selected standard reduction potentials (298 K) in aqueous and liquid ammonia media; the concentration of each solution is $1\,mol\,dm^{-3}$.

Reduction half-equation	E° / V in aqueous solution	E° / V in liquid ammonia
$Li^{+} + e^{-} \rightleftharpoons Li$	−3.04	−2.24
$K^{+} + e^{-} \rightleftharpoons K$	−2.93	−1.98
$Na^{+} + e^{-} \rightleftharpoons Na$	−2.71	−1.85
$Zn^{2+} + 2e^{-} \rightleftharpoons Zn$	−0.76	−0.53
$2H^{+} + 2e^{-} \rightleftharpoons H_2$ (g, 1 bar)	0.00	0.00
$Cu^{2+} + 2e^{-} \rightleftharpoons Cu$	+0.34	+0.43
$Ag^{+} + e^{-} \rightleftharpoons Ag$	+0.80	+0.83

are strong acids in liquid NH_3) decompose immediately (equation 8.29).

$$2[NH_4]^{+} + 2e^{-} \longrightarrow 2NH_3 + H_2 \qquad (8.29)$$

Dilute solutions of alkali metals in liquid NH_3 have many applications as reducing agents; reactions 8.30 to 8.34 (in which e^{-} represents the electron generated in reaction 8.26) provide examples and others are mentioned later in the book. In each of reactions 8.30–8.34, the anion shown is isolated as an alkali metal salt, the cation being provided from the alkali metal dissolved in the liquid NH_3.

$$2GeH_4 + 2e^{-} \longrightarrow 2[GeH_3]^{-} + H_2 \qquad (8.30)$$

$$O_2 + e^{-} \longrightarrow \underset{\text{superoxide ion}}{[O_2]^{-}} \qquad (8.31)$$

$$O_2 + 2e^{-} \longrightarrow \underset{\text{peroxide ion}}{[O_2]^{2-}} \qquad (8.32)$$

$$[MnO_4]^{-} + e^{-} \longrightarrow [MnO_4]^{2-} \qquad (8.33)$$

$$[Fe(CO)_5] + 2e^{-} \longrightarrow [Fe(CO)_4]^{2-} + CO \qquad (8.34)$$

Early synthetic routes to *Zintl ions* (see *Section 13.7*) involved reduction of Ge, Sn or Pb in solutions of Na in liquid NH_3. The method has been developed with the addition of the macrocyclic ligand cryptand-222 (crypt-222) (see *Section 10.8*) which encapsulates the Na^{+} ion and allows the isolation of salts of the type [Na(crypt-222)]$_2$[Sn$_5$] (equation 8.35). Zintl ions produced in this way include $[Sn_5]^{2-}$ (Figure 8.3), $[Pb_5]^{2-}$, $[Pb_2Sb_2]^{2-}$, $[Bi_2Sn_2]^{2-}$, $[Ge_9]^{2-}$, $[Ge_9]^{4-}$ and $[Sn_9Tl]^{3-}$.

$$Sn \xrightarrow[\text{Na in liquid } NH_3]{} \underset{\text{Zintl phase}}{NaSn_{1.0-1.7}}$$

$$\xrightarrow[\text{in 1,2-diaminoethane}]{\text{2,2,2-crypt}} [Na(crypt-222)]_2[Sn_5] \qquad (8.35)$$

Redox reactions in liquid NH_3

Reduction potentials for the reversible reduction of metal ions to the corresponding metal in aqueous solution and in liquid NH_3 are listed in Table 8.5. Note that the values

follow the same general trend, but that the oxidizing ability of each metal ion is solvent-dependent. Reduction potentials for oxidizing systems cannot be obtained in liquid NH_3 owing to the ease with which the solvent is oxidized.

Information deduced from reduction potentials, and from lattice energies and solubilities, indicates that H^{+} and *d*-block M^{n+} ions have more negative absolute standard free energies of solvation in NH_3 than in H_2O; for alkali metal ions, values of $\Delta_{solv}G^{\circ}$ are about the same in the two solvents. These data are consistent with the observation that the addition of NH_3 to aqueous solutions of *d*-block M^{n+} ions results in the formation of ammine complexes such as $[M(NH_3)_6]^{n+}$ whereas alkali metal ions are not complexed by NH_3.

8.7 Liquid hydrogen fluoride

Physical properties

Hydrogen fluoride attacks silica glass (equation 8.36) thereby corroding glass reaction vessels, and it is only relatively recently that HF has found applications as a non-aqueous solvent. It can be handled in polytetrafluoroethene (PTFE) containers, or, if absolutely free of water, in Cu or Monel steel (a nickel alloy) equipment.

$$4HF + SiO_2 \longrightarrow SiF_4 + 2H_2O \qquad (8.36)$$

Hydrogen fluoride has a liquid range from 190 to 292.5 K (Figure 8.2); the relative permittivity is 84 at 273 K, rising to 175 at 200 K. Liquid HF undergoes self-ionization (equilibrium 8.37), for which $K_{self} \approx 2 \times 10^{-12}$ at 273 K.

$$3HF \rightleftharpoons \underset{\text{fluoronium ion}}{[H_2F]^{+}} + \underset{\substack{\text{difluorohydrogenate(1-)} \\ \text{ion or hydrogendifluoride ion}}}{[HF_2]^{-}} \qquad (8.37)$$

The difference in electronegativities of H ($\chi^{P} = 2.2$) and F ($\chi^{P} = 4.0$) results in the presence of extensive intermolecular hydrogen bonding in the liquid. Chains and rings of various

BOX 8.2   **CHEMICAL AND THEORETICAL BACKGROUND**

The structure of the [HF₂]⁻ anion

The single crystal structures of a number of salts containing $[HF_2]^-$ or $[DF_2]^-$ (i.e. deuterated species) have been determined by X-ray or neutron diffraction techniques; these include $[NH_4][HF_2]$, $Na[HF_2]$, $K[HF_2]$, $Rb[HF_2]$, $Cs[HF_2]$ and $Tl[HF_2]$.

The anion is linear, and its formation is a consequence of the H and F atoms being involved in strong hydrogen bonding:

$$\left[\text{F} \longrightarrow \text{H} \longrightarrow \text{F} \right]^-$$

In the solid state structures reported, the F---F distance is ≈227 pm. This value is greater than twice the H−F bond length in HF (2×92 pm), but an H····F hydrogen-bond will always be weaker and longer than a two-centre covalent H−F bond. However, comparison of the values gives some indication of the strength of the hydrogen bonding in $[HF_2]^-$. (See also Figure 8.4.)

---

sizes are formed, and some of these, e.g. cyclic $(HF)_6$, persist in the vapour.

Acid–base behaviour in liquid HF

Using the solvent-oriented definition that we introduced in Section 8.4, a species that produces $[H_2F]^+$ ions in liquid HF is an acid, and one that produces $[HF_2]^-$ is a base.

Many organic compounds are soluble in liquid HF, and in the cases of, for example, amines and carboxylic acids, protonation of the organic species accompanies dissolution (equation 8.38). Proteins react immediately with liquid HF, and it produces very serious skin-burns.

$$MeCO_2H + 2HF \longrightarrow [MeC(OH)_2]^+ + [HF_2]^- \qquad (8.38)$$

Most inorganic salts are converted to the corresponding fluorides when dissolved in liquid HF, but only a few of these are soluble. Fluorides of the *s*-block metals, silver and thallium(I) dissolve to give salts such as $K[HF_2]$ and $K[H_2F_3]$, and thus exhibit basic character. Similarly, NH_4F is basic in liquid HF. Studies of the Me_4NF–HF system over a range of compositions and temperatures reveal the formation of the compounds of composition $Me_4NF \cdot nHF$ ($n = 2, 3, 5$ and 7). X-Ray diffraction studies for compounds

with $n = 2, 3$ and 5 have confirmed the structures of $[H_2F_3]^-$ (Figure 8.4a), $[H_3F_4]^-$ (Figure 8.4b) and $[H_5F_6]^-$, in which strong hydrogen bonding is an important feature (see *Section 9.6*).

Among molecular fluorides, CF_4 and SiF_4 are insoluble in liquid HF, but F^- acceptors such as AsF_5 and SbF_5 dissolve according to equation 8.39 to give very strongly acidic solutions. Less potent fluoride acceptors such as BF_3 function as weak acids in liquid HF (equation 8.40); PF_5 behaves as a *very* weak acid (equation 8.41). On the other hand, ClF_3 and BrF_3 act as F^- donors (equation 8.42) and behave as bases.

$$EF_5 + 2HF \rightleftharpoons [H_2F]^+ + [EF_6]^- \qquad E = As\ or\ Sb \qquad (8.39)$$

$$BF_3 + 2HF \rightleftharpoons [H_2F]^+ + [BF_4]^- \qquad (8.40)$$

$$PF_5 + 2HF \rightleftharpoons [H_2F]^+ + [PF_6]^- \qquad (8.41)$$

$$BrF_3 + HF \rightleftharpoons [BrF_2]^+ + [HF_2]^- \qquad (8.42)$$

Few protic acids are able to exhibit acidic behaviour in liquid HF, on account of the competition between HF and the solute as H^+ donors; perchloric acid and fluorosulfonic acid (equations 8.43) do act as acids.

$$HOSO_2F + HF \rightleftharpoons [H_2F]^+ + [SO_3F]^- \qquad (8.43)$$

(a)

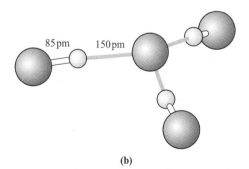

(b)

Fig. 8.4 The structures of the anions (a) $[H_2F_3]^-$ and (b) $[H_3F_4]^-$, determined by low temperature X-ray diffraction for the $[Me_4N]^+$ salts. The distances given are the average values for like internuclear separations; the experimental error on each distance is between ±3−6 pm [D. Mootz *et al.* (1987) *Z. Anorg. Allg. Chem.*, vol. 544, p. 159]. Colour code: F, blue; H, pale grey.

With SbF_5, HF forms a *superacid* (equation 8.44) which is capable of protonating *very* weak bases including hydrocarbons (see *Section 8.9*).

$$2HF + SbF_5 \rightleftharpoons [H_2F]^+ + [SbF_6]^- \qquad (8.44)$$

Electrolysis in liquid HF

Electrolysis in liquid HF is an important preparative route to both inorganic and organic fluorine-containing compounds, many of which are difficult to access by other routes. Anodic oxidation in liquid HF involves half-reaction 8.45 and with NH_4F as substrate, the products of the subsequent fluorination are NFH_2, NF_2H and NF_3.

$$2F^- \rightleftharpoons F_2 + 2e^- \qquad (8.45)$$

Anodic oxidation of water gives OF_2, of SCl_2 yields SF_6, of acetic acid yields CF_3CO_2H and of trimethylamine produces $(CF_3)_3N$.

8.8 Sulfuric acid

Physical properties

Selected physical properties of H_2SO_4 are given in Table 8.6; it is a liquid at 298 K, and the long liquid range (Figure 8.2) contributes towards making this a widely used non-aqueous solvent. Disadvantages of liquid H_2SO_4 are its high viscosity (27 times that of water at 298 K) and high value of $\Delta_{vap}H^\circ$. Both these properties arise from extensive intermolecular hydrogen bonding, and make it difficult to remove the solvent by evaporation from reaction mixtures. Dissolution of a solute in H_2SO_4 is only favourable if new interactions can be established to compensate for the loss of the extensive hydrogen bonding. Generally, this is only possible if the solute is ionic.

The value of the equilibrium constant for the self-ionization process 8.46 is notably large; in addition, other equilibria such as 8.47 are involved to a lesser extent.

$$2H_2SO_4 \rightleftharpoons [H_3SO_4]^+ + \underset{\substack{\text{hydrogensulfate}(1-) \\ \text{ion}}}{[HSO_4]^-}$$

$$K_{self} = 2.7 \times 10^{-4} \quad (8.46)$$

$$2H_2SO_4 \rightleftharpoons [H_3O]^+ + [HS_2O_7]^- \quad K_{self} = 5.1 \times 10^{-5} \quad (8.47)$$

8.12

Table 8.6 Selected physical properties of sulfuric acid, H_2SO_4.

Property / units	Value
Melting point / K	283.4
Boiling point / K	≈ 603
Density of liquid / $g\,cm^{-3}$	1.84
Relative permittivity	110 (at 292 K)
Self-ionization constant	2.7×10^{-4} (at 298 K)

(8.12)

Acid–base behaviour in liquid H_2SO_4

Sulfuric acid is a highly acidic solvent and most other 'acids' are neutral or behave as bases in it; we have already noted the basic behaviour of HNO_3. Initial proton transfer (equation 8.8) leads to the formation of the 'protonated acid' $[H_2NO_3]^+$, and in such cases, the resulting species often eliminates water (equation 8.9). Protonation of H_2O follows (equation 8.10).

The nature of such reactions can be examined by an ingenious combination of cryoscopic and conductivity measurements. Cryoscopy gives ν, the *total* number of particles produced per molecule of solute. The ionic mobilities[†] of $[H_3SO_4]^+$ and $[HSO_4]^-$ are very high, and the conductivity in H_2SO_4 is almost entirely due to the presence of $[H_3SO_4]^+$ and/or $[HSO_4]^-$. These ions carry the electrical current by proton-switching mechanisms, thus avoiding the need for migration through the viscous solvent. Conductivity measurements tell us γ, the number of $[H_3SO_4]^+$ or $[HSO_4]^-$ ions produced per molecule of solute. For a solution of acetic acid in H_2SO_4, experiment shows that $\nu = 2$ and $\gamma = 1$, consistent with reaction 8.48.

$$MeCO_2H + H_2SO_4 \longrightarrow [MeC(OH)_2]^+ + [HSO_4]^- \quad (8.48)$$

For nitric acid, $\nu = 4$ and $\gamma = 2$ corresponding to reaction 8.49, and for boric acid, $\nu = 6$ and $\gamma = 2$, consistent with reaction 8.50.

$$HNO_3 + 2H_2SO_4 \longrightarrow [NO_2]^+ + [H_3O]^+ + 2[HSO_4]^- \quad (8.49)$$

$$H_3BO_3 + 6H_2SO_4 \longrightarrow [B(HSO_4)_4]^- + 3[H_3O]^+ + 2[HSO_4]^- \quad (8.50)$$

For the $[B(HSO_4)_4]^-$ ion (**8.13**) to be formed, $H[B(HSO_4)_4]$ must act as a strong acid in H_2SO_4 solution; $H[B(HSO_4)_4]$ is a stronger acid even than HSO_3F (see *Section 8.9*). The ionization constants (in H_2SO_4) for HSO_3F and $H[B(HSO_4)_4]$ are 3×10^{-3} and 0.4, respectively.

(8.13)

[†] For discussions of ion transport see: P.W. Atkins (1998) *Physical Chemistry*, 6th edn, Oxford University Press, Oxford, Chapter 24; J. Burgess (1988) *Ions in Solution: Basic Principles of Chemical Interactions*, Ellis Horwood, Chichester, Chapter 2.

The species '$H[B(HSO_4)_4]$' has not been isolated as a pure compound, but a solution of this acid can be prepared by dissolving boric acid in *oleum* (equation 8.51) (see *Section 15.9*) and can be titrated conductiometrically against a solution of a strong base such as $KHSO_4$ (equation 8.52).

$$H_3BO_3 + 2H_2SO_4 + 3SO_3 \longrightarrow [H_3SO_4]^+ + [B(HSO_4)_4]^-$$
$$(8.51)$$

$$H[B(HSO_4)_4] + KHSO_4 \longrightarrow K[B(HSO_4)_4] + H_2SO_4 \quad (8.52)$$

> In a *conductiometric titration*, the end point is found by monitoring changes in the electrical conductivity of the solution; at the end point, the conductivity reaches a minimum value.

Few species function as strong acids in H_2SO_4 medium; perchloric acid (a potent acid in aqueous solution) is essentially non-ionized in H_2SO_4 and behaves only as a very weak acid.

In some cases (in contrast to equation 8.48), the cations formed from carboxylic acids are unstable, e.g. HCO_2H and $H_2C_2O_4$ (equation 8.53) decompose with loss of CO.

$$\begin{array}{c} CO_2H \\ | \\ CO_2H \end{array} + H_2SO_4 \longrightarrow CO + CO_2 + [H_3O]^+ + [HSO_4]^-$$

$$(8.53)$$

8.9 Fluorosulfonic acid

Physical properties

Table 8.7 lists some of the physical properties of fluorosulfonic acid,[†] HSO_3F, **8.14**; it has a relatively long liquid range (Figure 8.2) and a high dielectric constant. It is far less viscous than H_2SO_4 (by a factor of ≈ 16) and, like H_2SO_4 but unlike HF, can be handled in glass apparatus.

$$\underset{O}{\overset{O}{\underset{\|}{\underset{F}{\overset{\|}{O}}}}} \ S \cdots OH$$

(8.14)

Equation 8.54 shows the self-ionization of HSO_3F.

$$2HSO_3F \rightleftharpoons [H_2SO_3F]^+ + [SO_3F]^- \quad (8.54)$$

Superacids

Extremely potent acids, capable of protonating even hydrocarbons, are termed *superacids* and include mixtures of HF

Table 8.7 Selected physical properties of fluorosulfonic acid, HSO_3F.

Property / units	Value
Melting point / K	185.7
Boiling point / K	438.5
Density of liquid / $g\,cm^{-3}$	1.74
Relative permittivity	120 (at 298 K)
Self-ionization constant	4.0×10^{-8} (at 298 K)

and SbF_5 (equation 8.44) and HSO_3F and SbF_5 (equation 8.55). The latter mixture is called *magic acid* (one of the strongest acids known) and is available commercially under this name. Antimony(V) fluoride is a strong Lewis acid and forms an adduct with F^- (from HF) or $[SO_3F]^-$ (from HSO_3F). Figure 8.5 shows the crystallographically determined structure of the related adduct $SbF_5OSO(OH)CF_3$.

$$2HSO_3F + SbF_5 \rightleftharpoons [H_2SO_3F]^+ + [F_5SbOSO_2F]^- \quad (8.55)$$
$$\mathbf{8.15}$$

(8.15)

Equilibrium 8.55 is an over-simplification of the SbF_5–HSO_3F system, but represents the system sufficiently for most purposes. The species present depend on the ratio of $SbF_5:HSO_3F$, and at higher concentrations of SbF_5, species including $[SbF_6]^-$, $[Sb_2F_{11}]^{2-}$, HS_2O_6F and HS_3O_9F may exist.

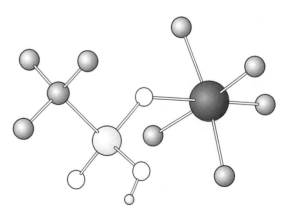

Fig. 8.5 The solid state structure (X-ray diffraction) of $SbF_5OSO(OH)CF_3$ [D. Mootz *et al.* (1991) *Z. Naturforsch., Teil B*, vol. 46, p. 1659]. Colour code: Sb, black; F, dark blue; S, pale blue; O, white; C, grey; H, pale grey.

[†] Fluorosulfonic acid is also called fluorosulfuric acid, and the IUPAC name is hydrogen fluorotrioxosulfate.

In superacidic media, hydrocarbons act as bases, and this is an important route to the formation of carbenium ions;[†] e.g. deprotonation of 2-methylpropane yields the trimethyl-carbenium ion (equation 8.56). Phosphorus(III) halides can be converted to phosphonium cations $[HPX_3]^+$, carbonic acid to the unstable cation $[C(OH)_3]^+$, and $Fe(CO)_5$ to $[HFe(CO)_5]^+$.

$$Me_3CH + [H_2SO_3F]^+ \longrightarrow [Me_3C]^+ + H_2 + HSO_3F \quad (8.56)$$
$$\underset{\substack{\text{generated in} \\ \text{magic acid}}}{}$$

8.10 Bromine trifluoride

In this and the next section, we consider two *aprotic* non-aqueous solvents.

Physical properties

Bromine trifluoride is a pale yellow liquid at 298 K; selected physical properties are given in Table 8.8 and the compound is discussed again in *Section 16.7*. Bromine trifluoride is an extremely powerful fluorinating agent and fluorinates essentially every species that dissolves in it. However, massive quartz is kinetically stable towards BrF_3 and the solvent can be handled in quartz vessels. Alternatively, metal (e.g. Ni) containers can be used; the metal surface becomes protected by a thin layer of metal fluoride.

The proposed self-ionization of BrF_3 (equation 8.57) has been substantiated by the isolation and characterization of acids and bases, and by conductiometric titrations of them (see below). Using the solvent-based acid–base definitions, an acid in BrF_3 is a species that produces $[BrF_2]^+$ (**8.16**), and a base is one that gives $[BrF_4]^-$ (**8.17**).

$$2BrF_3 \rightleftharpoons [BrF_2]^+ + [BrF_4]^- \quad (8.57)$$

Behaviour of fluoride salts and molecular fluorides in BrF₃

Bromine trifluoride acts as a Lewis acid, readily accepting F^-. When dissolved in BrF_3, alkali metal fluorides, BaF_2 and AgF combine with the solvent to give salts containing the $[BrF_4]^-$ anion, e.g. $K[BrF_4]$ (equation 8.58), $Ba[BrF_4]_2$ and $Ag[BrF_4]$. On the other hand, if the fluoride solute is a more powerful F^- acceptor than BrF_3, salts containing $[BrF_2]^+$ may be formed, e.g. equations 8.59–8.61.

[†] A *carbenium* ion is also called a *carbocation*; the older name of *carbonium* ion is also in use.

Table 8.8 Selected physical properties of bromine trifluoride, BrF_3.

Property / units	Value
Melting point / K	281.8
Boiling point / K	408
Density of liquid / $g\,cm^{-3}$	2.49
Relative permittivity	107
Self-ionization constant	8.0×10^{-3} (at 281.8 K)

$$KF + BrF_3 \longrightarrow K^+ + [BrF_4]^- \quad (8.58)$$

$$SbF_5 + BrF_3 \longrightarrow [BrF_2]^+ + [SbF_6]^- \quad (8.59)$$

$$SnF_4 + 2BrF_3 \longrightarrow 2[BrF_2]^+ + [SnF_6]^{2-} \quad (8.60)$$

$$AuF_3 + BrF_3 \longrightarrow [BrF_2]^+ + [AuF_4]^- \quad (8.61)$$

Conductiometric measurements on solutions containing $[BrF_2][SbF_6]$ and $Ag[BrF_4]$, or $[BrF_2]_2[SnF_6]$ and $K[BrF_4]$ exhibit minima at 1:1 and 1:2 molar ratios of reactants respectively. These data support the formulation of neutralization reactions 8.62 and 8.63.

$$\underset{\text{acid}}{[BrF_2][SbF_6]} + \underset{\text{base}}{Ag[BrF_4]} \longrightarrow Ag[SbF_6] + 2BrF_3 \quad (8.62)$$

$$\underset{\text{acid}}{[BrF_2]_2[SnF_6]} + \underset{\text{base}}{2K[BrF_4]} \longrightarrow K_2[SnF_6] + 4BrF_3 \quad (8.63)$$

Reactions in BrF₃

Much of the chemistry studied in BrF_3 media involves fluorination reactions, and the preparation of highly fluorinated species. For example, the salt $Ag[SbF_6]$ can be prepared in liquid BrF_3 from elemental Ag and Sb in a 1:1 molar ratio (equation 8.64), while $K_2[SnF_6]$ is produced when KCl and Sn are combined in a 2:1 molar ratio in liquid BrF_3 (equation 8.65).

$$Ag + Sb \xrightarrow{\text{in } BrF_3} Ag[SbF_6] \quad (8.64)$$

$$2KCl + Sn \xrightarrow{\text{in } BrF_3} K_2[SnF_6] \quad (8.65)$$

In contrast to the situation for H_2SO_4 where we noted that it is difficult to separate reaction products from the solvent by evaporation, BrF_3 can be removed *in vacuo* ($\Delta_{vap}H^\circ = 47.8\,kJ\,mol^{-1}$). The syntheses of many other inorganic fluoro-derivatives can be carried out in a similar manner to reactions 8.64 or 8.65, and equations 8.66–8.69 give further examples.

$$Ag + Au \xrightarrow{\text{in } BrF_3} Ag[AuF_4] \quad (8.66)$$

$$KCl + VCl_4 \xrightarrow{\text{in } BrF_3} K[VF_6] \quad (8.67)$$

$$2ClNO + SnCl_4 \xrightarrow{\text{in } BrF_3} [NO]_2[SnF_6] \quad (8.68)$$

$$Ru + KCl \xrightarrow{\text{in } BrF_3} K[RuF_6] \quad (8.69)$$

Some of the compounds prepared by this method can also be made using F_2 as the fluorinating agent, but use of F_2 generally requires higher reaction temperatures and the reactions are not always as product-specific.

Non-aqueous solvents that behave similarly to BrF_3 in that they are good oxidizing *and* fluorinating agents include ClF_3, BrF_5 and IF_5.

8.11 Dinitrogen tetraoxide

Physical properties

The data in Table 8.9 and Figure 8.2 emphasize the very short liquid range of N_2O_4. Despite this and the low relative permittivity (which makes it a poor solvent from most inorganic compounds), the preparative uses of N_2O_4 justify its inclusion in this chapter.

$$N_2O_4 \rightleftharpoons [NO]^+ + [NO_3]^- \qquad (8.70)$$

The proposed self-ionization process for N_2O_4 is given in equation 8.70, but conductivity data indicate that this can occur only to an extremely small extent; physical evidence for this equilibrium is lacking. However, the presence of $[NO_3]^-$ in the solvent is indicated by the rapid exchange of nitrate ion between liquid N_2O_4 and $[Et_4N][NO_3]$ (which is soluble owing to its very low lattice energy). In terms of the solvent-oriented acid–base definition, acidic behaviour in N_2O_4 is characterized by the production of $[NO]^+$, and basic behaviour by the formation of $[NO_3]^-$. This terminology assumes the operation of equilibrium 8.70. A few reactions in liquid N_2O_4 can be rationalized in terms of equilibrium 8.71, but there is no physical evidence to confirm this proposal.

$$N_2O_4 \rightleftharpoons [NO_2]^+ + [NO_2]^- \qquad (8.71)$$

Reactions in N_2O_4

Reactions carried out in liquid N_2O_4 generally utilize the fact that N_2O_4 is a good oxidizing (see *Box 8.3*) and nitrating

Table 8.9 Selected physical properties of dinitrogen tetraoxide, N_2O_4.

Property / units	Value
Melting point / K	261.8
Boiling point / K	294.2
Density of liquid / $g\,cm^{-3}$	1.49 (at 273 K)
Relative permittivity	2.42 (at 291 K)

agent. Electropositive metals such as Li and Na react in liquid N_2O_4 liberating NO (equation 8.72).

$$Li + N_2O_4 \longrightarrow LiNO_3 + NO \qquad (8.72)$$

Less reactive metals may react rapidly if ClNO, $[Et_4N][NO_3]$ or an organic donor such as MeCN is present. These observations can be explained as follows.

- ClNO can be considered to be a very weak acid in liquid N_2O_4, and hence encourages reaction with metals (equation 8.73).

$$Sn + 2ClNO \xrightarrow{\text{in liquid } N_2O_4} SnCl_2 + 2NO \qquad (8.73)$$

- $[Et_4N][NO_3]$ functions as a base in liquid N_2O_4 and its action on metals such as Zn and Al arises from the formation of nitrato complexes (equation 8.74) analogous to hydroxo complexes in an aqueous system; Figure 8.6 shows the structure of $[Zn(NO_3)_4]^{2-}$.

$$Zn + 2[Et_4N][NO_3] + 2N_2O_4$$
$$\longrightarrow [Et_4N]_2[Zn(NO_3)_4] + 2NO \qquad (8.74)$$

- Organic donor molecules appear to facilitate reactions with metals by increasing the degree of self-ionization of the solvent as a result of adduct formation with the $[NO]^+$ cation; e.g. Cu dissolves in liquid N_2O_4/MeCN according to equation 8.75, and Fe behaves similarly, dissolving to give $[NO][Fe(NO_3)_4]$.

$$Cu + 3N_2O_4 \xrightarrow{\text{in presence of MeCN}} [NO][Cu(NO_3)_3] + 2NO \qquad (8.75)$$

BOX 8.3   **APPLICATIONS**

Liquid N_2O_4 as a fuel in the Apollo missions

During the Apollo Moon missions, a fuel was needed that was suitable for landing on, and taking off from, the Moon's surface. The fuel chosen was a mixture of liquid N_2O_4 and derivatives of hydrazine (N_2H_4). Dinitrogen tetraoxide is a powerful oxidizing agent and contact with, for example, $MeNHNH_2$, leads to immediate oxidation of the latter:

$$5N_2O_4 + 4MeNHNH_2 \longrightarrow 9N_2 + 12H_2O + 4CO_2$$

The reaction is highly exothermic, and at the operating temperatures, all products are gases.

Safety is of utmost importance; the fuels clearly must not contact each other before the required moment of landing or lift-off. Further, $MeNHNH_2$ is extremely toxic.

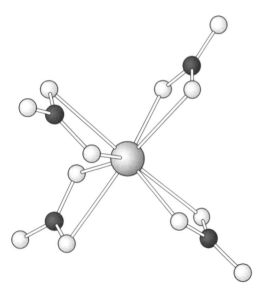

Fig. 8.6 The solid state structure (X-ray diffraction) of the $[Zn(NO_3)_4]^{2-}$ anion in the salt $[Ph_4As]_2[Zn(NO_3)_4]$. Each $[NO_3]^-$ ligand is coordinated to the Zn(II) centre through two O atoms, with one short (average 206 pm) and one long (average 258 pm) Zn–O interaction [C. Bellitto *et al.* (1976) *J. Chem. Soc., Dalton Trans.*, p. 989]. Colour code: Zn, blue; N, black; O, pale grey.

The presence of $[NO]^+$ cations in compounds such as $[NO][Cu(NO_3)_3]$, $[NO][Fe(NO_3)_4]$, $[NO]_2[Zn(NO_3)_4]$ and $[NO]_2[Mn(NO_3)_4]$ is confirmed by the appearance of a characteristic absorption (ν_{NO}) at $\approx 2300\ cm^{-1}$ in the infrared spectra of the complexes.

Just as hydrolysis of a compound may occur in water (see *Section 6.7*), solvolysis such as reaction 8.76 can take place in liquid N_2O_4. Such reactions are of synthetic importance as routes to *anhydrous* metal nitrates.

$$ZnCl_2 + 2N_2O_4 \longrightarrow Zn(NO_3)_2 + 2ClNO \qquad (8.76)$$

In many of the reactions carried out in liquid N_2O_4, the products are solvates, for example $[Fe(NO_3)_3]\cdot 1.5N_2O_4$, $[Cu(NO_3)_2]\cdot N_2O_4$, $[Sc(NO_3)_2]\cdot 2N_2O_4$ and $[Y(NO_3)_3]\cdot 2N_2O_4$. Such formulations may, in some cases, be correct, with molecules of N_2O_4 present, analogous to water molecules of crystallization in crystals isolated from an aqueous system. However, the results of X-ray diffraction studies on some solvated compounds illustrate the presence, not of N_2O_4 molecules, but of $[NO]^+$ and $[NO_3]^-$ ions. Two early examples to be crystallographically characterized were $[Sc(NO_3)_3]\cdot 2N_2O_4$ and $[Y(NO_3)_3]\cdot 2N_2O_4$, for which the formulations $[NO]_2[Sc(NO_3)_5]$ and $[NO]_2[Y(NO_3)_5]$ were confirmed. In the $[Y(NO_3)_5]^{2-}$ anion, the Y(III) centre is 10-coordinate with didentate nitrato ligands, while in $[Sc(NO_3)_5]^{2-}$, the Sc(III) centre is 9-coordinate with one $[NO_3]^-$ ligand being monodentate (see also *Section 24.7*).

8.12 Ionic liquids

The use of *ionic liquids* (also called *molten* or *fused salts*) as reaction media is a relatively new area, although molten conditions have been well established in industrial processes (e.g. the Downs process, *Figure 10.1*) for many years. While some 'molten salts' are hot as the term suggests, others operate at ambient temperatures and the term 'ionic liquids' is more appropriate. This section provides only a brief introduction to the area.

The term *eutectic* is commonly encountered in this field. The reason for forming a eutectic mixture is to provide a molten system at a convenient working temperature. For example, the melting point of NaCl is 1073 K, but is lowered if $CaCl_2$ is added as in the Downs process. An

---

BOX 8.4　　　　　　　　　　　　　　　　　　　　　　　　　　　**APPLICATIONS**

Towards electrically powered vehicles

Electrically powered vehicles are powered by rechargeable battery packs containing several hundred separate batteries. Although the ultimate destruction of batteries is a source of pollution, the fact that electric cars do not emit exhaust gases allows them to be termed 'environmentally friendly'. Currently, electric vehicles are still a novelty, but legislation in, for example, California, which will come into operation in the early part of the 21st century, is forcing car manufacturers to 'think electric'. The present status of battery technology means that the range between recharging-stops is rather small (typically 80–150 km) and the process of recharging typically takes several hours. Electric cars are also considerably more expensive than conventional vehicles. Design of suitable and improved batteries is topical, but research competes with that into ways of reducing polluting emissions from conventional motor vehicles.

Some of the batteries being considered and developed for use in electrically powered vehicles utilize molten salts, and operate at high temperatures. Two examples containing a molten LiCl–KCl eutectic as the electrolyte are the lithium silicon/iron disulfide and lithium/iron sulfide batteries.

Further reading

A.F. Sammells (1983) *Journal of Chemical Education*, vol. 60, p. 320 – 'Fuel cells and electrochemical energy storage'.
C.A. Vincent (1996) *Chemistry & Industry*, p. 688 – 'Recent developments in battery technology'.
B. Scrosati and C.A. Vincent (1997) *Modern Batteries*, 2nd edn, Arnold, London.

environmentally 'friendly' application of a LiCl–KCl eutectic is highlighted in *Box 8.4*.

> A *eutectic* is a mixture of two substances and is characterized by a sharp melting point lower than that of either of the components; a eutectic behaves as though it were a single substance.

Molten salt solvent systems

When an ionic salt such as NaCl melts, the ionic lattice (see *Figure 5.15*) collapses, but some order is still retained. Evidence for this comes from X-ray diffraction patterns, from which *radial distribution functions* reveal that the average coordination number (with respect to cation–anion interactions) of each ion in liquid NaCl is ≈four, compared to six in the crystalline lattice. For cation–cation or anion–anion interactions, the coordination number is higher, although, as in the solid state, the internuclear distances are larger than for cation–anion separations. The solid-to-liquid transition is accompanied by an increase in volume of ≈10–15%. The number of ions in the melt can be determined in a similar way to that described in Section 8.8 for H_2SO_4 systems; in molten NaCl, $\nu = 2$.

Other alkali metal halides behave in a similar manner to NaCl, but metal halides in which the bonding has a significant covalent contribution (e.g. Hg(II) halides) form melts in which equilibria such as 8.77 are established. In the solid state, $HgCl_2$ forms a molecular lattice, and layer structures are adopted by $HgBr_2$ (distorted CdI_2 lattice) and HgI_2.

$$2HgBr_2 \rightleftharpoons [HgBr]^+ + [HgBr_3]^- \qquad (8.77)$$

In terms of the solvent-oriented description of acid–base chemistry in a non-aqueous solvent, equation 8.77 illustrates that, in molten $HgBr_2$, species producing $[HgBr]^+$ ions may be considered to act as acids, and those providing $[HgBr_3]^-$ ions function as bases. In most molten salts, however, the application of this type of acid–base definition is not appropriate.

An important group of molten salts with more convenient operating temperatures contain the tetrachloroaluminate ion, $[AlCl_4]^-$; an example is an $NaCl$–Al_2Cl_6 mixture. The melting point of Al_2Cl_6 is 463 K (at 2.5 bar), and its addition to NaCl (melting point, 1073 K) results in a 1:1 medium with a melting point of 446 K. In this and other Al_2Cl_6–alkali metal chloride melts, equilibria 8.78 and 8.79 are established, with the additional formation of $[Al_3Cl_{10}]^-$ (see *Section 12.6*).

$$Al_2Cl_6 + 2Cl^- \rightleftharpoons 2[AlCl_4]^- \qquad (8.78)$$

$$2[AlCl_4]^- \rightleftharpoons [Al_2Cl_7]^- + Cl^- \qquad (8.79)$$

Ionic salts at ambient temperatures

Another well-established and useful system consists of Al_2Cl_6 with an organic salt such as butylpyridinium chloride, [pyBu]Cl; reaction 8.80 occurs to give [pyBu][AlCl₄], **8.4**, and

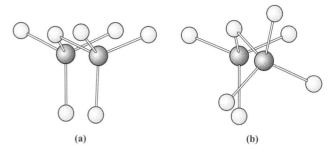

Fig. 8.7 The crystallographically determined structure of the $[Al_2Cl_7]^-$ ion. In the compound $[(C_6Me_6)_3Zr_3Cl_6][Al_2Cl_7]_2$, the anions adopt one of two different conformations: (a) an eclipsed conformation and (b) a staggered conformation [F. Stollmaier *et al.* (1981) *J. Organomet. Chem.*, vol. 208, p. 327]. Colour code: Al, blue; Cl, grey.

in the molten state, the $[Al_2Cl_7]^-$ ion, **8.18**, is formed according to equilibrium 8.79. In the solid state, X-ray diffraction data for several salts illustrate that $[Al_2Cl_7]^-$ can adopt either a staggered or an eclipsed conformation (Figure 8.7). Raman spectroscopic data (see *Box 3.1*) have shown that $[Al_2Cl_7]^-$ is a more dominant species in molten Al_2Cl_6–[pyBu]Cl than in the Al_2Cl_6–alkali metal chloride systems.

$$Al_2Cl_6 + 2[pyBu]Cl \rightleftharpoons 2[pyBu][AlCl_4] \qquad (8.80)$$

$$\left[\begin{array}{c} Cl \quad\quad Cl \quad\quad Cl \\ Cl^{\prime\prime\prime\prime}\!-\!Al \quad\quad Al\!-\!^{\prime\prime\prime\prime}Cl \\ Cl \quad\quad\quad Cl \end{array} \right]^-$$

(8.18)

The beauty of [pyBu][AlCl₄] and similar systems (see below) is that they are conducting liquids at 298 K and are therefore extremely valuable as ionic solvents, dissolving a wide range of inorganic and organic compounds. A further advantage of these solvents is their long liquid ranges. Their potential is great, and applications are currently being researched.

Reactions in and applications of molten salt media

Manufacturing processes in which metals are extracted from molten metal salts are important examples of the uses of molten salts and include the Downs process, and the production of Li by electrolysis of molten LiCl, and of Be and Ca from $BeCl_2$ and $CaCl_2$, respectively.

It is not possible here to survey the many types of reactions that have been carried out in molten salt media, and we have chosen examples to illustrate the range of possibilities. Some unusual cations have been isolated as products from reactions in molten salt media. For example, the reaction of Bi and $BiCl_3$ in KCl–$BiCl_3$ solvent at ≈570 K yields $[Bi_9]_2[BiCl_5]_4[Bi_2Cl_8]$ which contains $[Bi_9]^{5+}$, $[BiCl_5]^{2-}$ and $[Bi_2Cl_8]^{2-}$. In a melt containing $AlCl_3$ and MCl (M = Na or K) at ≈530 K, Bi and $BiCl_3$ react to form $[Bi_5]^{2+}$ (a trigonal bipyramidal species like $[Sn_5]^{2-}$, Figure 8.3) and $[Bi_8]^{2+}$, which are isolated as the $[AlCl_4]^-$ salts.

(8.19)   (8.20)

Electrochemical and spectroscopic studies of anionic d-block metal chloro complexes and organometallic compounds (which may be unstable in some solvents) may be performed in Al_2Cl_6–ethylpyridinium chloride, Al_2Cl_6–butylpyridinium chloride and Al_2Cl_6–[1-methyl-3-ethylimidazolium chloride] (the 1-methyl-3-ethylimidazolium cation is shown in structure **8.19**) systems, all of which are ionic liquids at room temperature. An example of such a study is the observation of the electronic absorption spectrum of $[RuO_2Cl_4]^{2-}$, **8.20**, a species that decomposes in aqueous solution. Problems of oxide contaminants in these melts can be overcome by the addition of the highly poisonous gas, $COCl_2$. This has been illustrated in a study of the electrochemistry of $TiCl_4$ in an Al_2Cl_6–[**8.19**]Cl melt; if the system contains the contaminant $[TiOCl_4]^{2-}$ in addition to the desired $[TiCl_6]^{2-}$, the addition of $COCl_2$ (see *Section 13.8*) successfully removes the contaminant (reaction 8.81).

$$[TiOCl_4]^{2-} + COCl_2 \longrightarrow [TiCl_6]^{2-} + CO_2 \qquad (8.81)$$

Protonated contaminants may also be a problem, e.g. the formation of $[HMo_2Cl_8]^{3-}$ when $[Mo_2Cl_8]^{4-}$ salts are studied in molten salt media. Such contaminants can be scavenged using $EtAlCl_2$.

Summary

This chapter has been concerned with selected examples of non-aqueous solvents. Because this is an inorganic chemistry textbook, the emphasis has been on inorganic solvents. However, it is important to recognize the role that non-aqueous solvents such as MeCN, CH_2Cl_2 and ethers play both in inorganic and organic synthesis, see for example the review by R.E. Morris and S.J. Weigel (1997) *Chemical Society Reviews*, vol. 26, p. 309 ('The synthesis of molecular sieves from non-aqueous solvents').

The following terms were introduced in this chapter.
Do you know what they mean?
- [ ] non-aqueous solvent
- [ ] relative permittivity
- [ ] coordinating solvent
- [ ] protic solvent
- [ ] aprotic solvent
- [ ] solvent-oriented acid and base
- [ ] levelling effect
- [ ] differentiating effect
- [ ] conductiometric titration
- [ ] superacid
- [ ] ionic liquids (molten or fused) salts
- [ ] eutectic

Suggested further reading

General: non-aqueous solvents

C.C. Addison (1980) *Chemical Reviews*, vol. 80, p. 21 – An article focusing on the uses of N_2O_4 and HNO_3 in non-aqueous systems.

J.R. Chipperfield (1999) *Non-Aqueous Solvents*, Oxford University Press, Oxford – A book in the OCP ('Primer') series which gives a good introduction to the topic.

R.J. Gillespie and J. Passmore (1971) *Accounts of Chemical Research*, vol. 4, p. 413 – An article that highlights the uses of non-aqueous solvents (HF, SO_2 and HSO_3F) in the preparation of polycations.

Gmelin Handbook of Inorganic Chemistry (1982): *Fluorine*, Supplement volume 3, System number 5, Springer-Verlag, Berlin – This includes a detailed account of solubilities and reactions in liquid HF.

K.M. Mackay, R.A. Mackay and W. Henderson (1996) *Modern Inorganic Chemistry*, 5th edn, Blackie, London – Chapter 6 gives a general introduction to non-aqueous solvents.

G. Mamantov and A.I. Popov, eds (1994) *Chemistry of Non-aqueous Solutions: Recent Advances*, VCH, New York – A collection of reviews covering up-to-date topics in the field of non-aqueous solvents.

Y. Marcus (1985) *Ion Solvation*, Wiley, New York – An in-depth discussion of the solvation by organic liquids.

A.G. Massey (1990) *Main Group Chemistry*, Ellis Horwood, Chichester – Chapter 12 includes an introduction to non-aqueous solvents.

T.C. Waddington (1969) *Non-aqueous Solvents*, Nelson, London – A slightly fuller and more advanced treatment than is given in this chapter.

Metals in liquid NH_3

J.L. Dye (1984) *Progress in Inorganic Chemistry*, vol. 32, p. 327.

P.P. Edwards (1982) *Advances in Inorganic Chemistry and Radiochemistry*, vol. 25, p. 135.

Superacids

R.J. Gillespie (1968) *Accounts of Chemical Research*, vol. 1, p. 202.

G.A. Olah, G.K.S. Prakash and J. Sommer (1985) *Superacids*, Wiley, New York.

G.A. Olah, G.K.S. Prakash and J. Sommer (1979) *Science*, vol. 206, p. 13.

Ionic liquids

M. Freemantle (1998) *Chemical & Engineering News*, vol. 76, issue 13, p. 32.

C.L. Hussey (1983) *Advances in Molten Salt Chemistry*, vol. 5, p. 185.

K.R. Seddon (1997) *Journal of Chemical Technology and Biotechnology*, vol. 68, p. 351.

Problems

8.1 (a) Give four examples of non-aqueous solvents commonly used in *organic* chemistry, and give one example of a reaction that is carried out in each solvent. (b) Assess the relative importance of the use of aqueous and non-aqueous media in organic and inorganic *general* synthesis.

8.2 Explain what is meant by the relative permittivity of a solvent. What information does this property provide in terms of assisting you to choose a solvent for a given reaction?

8.3 Which of the following solvents are polar: (a) acetonitrile; (b) water; (c) acetic acid; (d) fluorosulfonic acid; (e) dichloromethane; (f) bromine trifluoride; (g) hexane; (h) THF; (i) DMF; (j) liquid sulfur dioxide; (k) benzene?

8.4 Suggest likely products for the following reactions (which are balanced on the left-hand sides) in liquid NH_3.

(a) $ZnI_2 + 2KNH_2 \longrightarrow$
(b) Zinc-containing product of (a) with an excess of KNH_2
(c) $Mg_2Ge + 4NH_4Br \longrightarrow$
(d) $MeCO_2H + NH_3 \longrightarrow$
(e) $O_2 \xrightarrow{\text{Na in liquid } NH_3}$
(f) $HC{\equiv}CH + KNH_2 \longrightarrow$

How does reaction (d) differ from the behaviour of $MeCO_2H$ in aqueous solution?

8.5 Discuss the following observations:

(a) Zinc dissolves in a solution of sodium amide in liquid NH_3 with liberation of H_2; careful addition of ammonium iodide to the resulting solution produces a white precipitate which dissolves if an excess of ammonium iodide is added.
(b) Addition of K to H_2O results in a vigorous reaction; addition of K to liquid NH_3 gives a bright blue solution, which over a period of time liberates H_2.

8.6 Early in the study of chemical reactions in liquid NH_3, it was noted that nitrogen compounds behave in liquid NH_3 in a manner similar to analogous oxygen-containing species in water. For example, $K[NH_2]$ has an analogue in $K[OH]$, and $[NH_4]Cl$ is analogous to $[H_3O]Cl$. What would be the corresponding compounds in the nitrogen-system to the following from the oxygen-system: (a) H_2O_2; (b) HgO; (c) HNO_3; (d) MeOH; (e) H_2CO_3; (f) $[Cr(H_2O)_6]Cl_3$?

8.7 Give an explanation for the following observations: AlF_3 has only a low solubility in liquid HF, but a combination of NaF and AlF_3 leads to dissolution of the reagents; when BF_3 is added to the solution, a precipitate forms.

8.8 Write equations to show what happens when each of the following dissolves in liquid HF: (a) ClF_3; (b) MeOH; (c) Et_2O; (d) CsF; (e) SrF_2; (f) $HClO_4$.

8.9 $H_2S_2O_7$ functions as a monobasic acid in H_2SO_4. (a) Write an equation to show what happens when $H_2S_2O_7$ dissolves in H_2SO_4. (b) Assess the strength of $H_2S_2O_7$ as an acid given that the ionization constant is 1.4×10^{-2}.

8.10 Suggest (giving equations) how the following species behave in H_2SO_4: (a) H_2O; (b) NH_3; (c) HCO_2H (given that it decomposes); (d) H_3PO_4 (if $\nu = 2$; $\gamma = 1$); (e) HCl (if $\nu = 3$; $\gamma = 1$).

8.11 Compare the behaviour of nitric acid in aqueous and sulfuric acid solutions, giving examples from both inorganic and organic chemistries of the uses of HNO_3 in these two media.

8.12 Discuss the following observations:

(a) The alkene $Ph_2C{=}CH_2$ forms a conducting solution in liquid HCl; when such a solution is titrated conductiometrically with a solution of BCl_3 in liquid HCl, a sharp end point is reached when the molar ratio of $Ph_2C{=}CH_2{:}BCl_3$ is 1:1.
(b) For a solution of N_2O_4 in H_2SO_4, values of $\nu = 6$ and $\gamma = 3$ have been determined.

8.13 Confirm that the structures of $[BrF_2]^+$ and $[BrF_4]^-$ (**8.16** and **8.17**) are consistent with VSEPR theory.

8.14 How would you attempt to demonstrate that $AsCl_3$ ionizes slightly according to the equation:

$$2AsCl_3 \rightleftharpoons [AsCl_2]^+ + [AsCl_4]^-$$

and that there exist acids and bases in the $AsCl_3$ system?

8.15 (a) Describe the bonding in the $[Al_2Cl_7]^-$ anion (**8.18**).
(b) Equilibria 8.78 and 8.79 describe part of the NaCl–Al_2Cl_6 system; additionally $[Al_3Cl_{10}]^-$ is present. Write an equation to show how $[Al_3Cl_{10}]^-$ may be formed, and suggest a structure for this anion.

8.16 Suggest structures for the $[BiCl_5]^{2-}$ and $[Bi_2Cl_8]^{2-}$ anions, the formation of which was described in Section 8.12.

Chapter 9   Hydrogen

1	2		13	14	15	16	17	18
H								He
Li	Be		B	C	N	O	F	Ne
Na	Mg		Al	Si	P	S	Cl	Ar
K	Ca		Ga	Ge	As	Se	Br	Kr
Rb	Sr	d-block	In	Sn	Sb	Te	I	Xe
Cs	Ba		Tl	Pb	Bi	Po	At	Rn
Fr	Ra							

9.1 Hydrogen: the simplest atom

An atom of hydrogen consists of one proton (constituting the nucleus) and one electron. This simplicity of atomic structure means that H is of great importance in theoretical chemistry, and has been central in the development of atomic and bonding theories (see *Chapter 1*). The nuclear properties of the hydrogen atom are essential to the technique of ^1H NMR spectroscopy (see *Section 2.11*).

In this chapter, we extend our discussions of hydrogen, looking at the properties of the H^+ and H^- ions, properties and reactivity of H_2, and aspects of binary hydrides.

A *binary compound* is one composed of only two different elements.

9.2 The H^+ and H^- ions

The hydrogen ion (proton)

The ionization energy of hydrogen (defined for reaction 9.1) is $1312 \, kJ \, mol^{-1}$, a value that is high enough to preclude the existence of H^+ ions under ordinary conditions.

$$H(g) \longrightarrow H^+(g) + e^- \tag{9.1}$$

However, as we discussed in Chapter 6, the *hydrated* proton or *oxonium ion*, $[H_3O]^+$, is an important species in aqueous solution; $\Delta_{hyd}H^\circ(H^+,g) = -1091 \, kJ \, mol^{-1}$ (see *Section 6.9*). The $[H_3O]^+$ ion (**9.1**) is a well-defined species which has been crystallographically characterized in various salts. The ions $[H_5O_2]^+$ (Figure 9.1) and $[H_9O_4]^+$ have also been isolated in crystalline acid hydrates.

$$\left[\begin{array}{c} H \overset{O\cdots\cdots H}{\diagdown} \\ H \end{array} \right]^+$$

(9.1)

When crystals of a compound are grown from a solvent, they may contain *solvent of crystallization*; if the solvent is water, the compound is a *hydrate*. The formula of the solvated compound shows the molar ratio in which the solvent of crystallization is present, e.g. $CuSO_4 \cdot 5H_2O$, copper(II) sulfate pentahydrate or copper(II) sulfate–water (1/5).

The hydride ion

The enthalpy change $\Delta_{EA}H(298 \, K)$ (see *Section 1.10*) associated with the attachment of an electron to an H atom (reaction 9.2) is $-73 \, kJ \, mol^{-1}$.

$$H(g) + e^- \longrightarrow H^-(g) \tag{9.2}$$

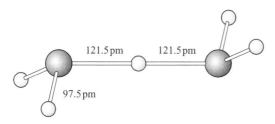

Fig. 9.1 The structure of $[H_5O_2]^+$ determined by neutron diffraction in the compound $[V(H_2O)_6][H_5O_2][CF_3SO_3]_4$. [F.A. Cotton *et al.* (1984) *J. Am. Chem. Soc.*, vol. 106, p. 5319.]

All alkali metal hydrides (see *Sections 9.7* and *10.4*) crystallize with the NaCl lattice. From diffraction data and the ionic radii of the metal ions (*Appendix 6*) the radius of H^- can be estimated using equation 9.3; it varies from 130 pm (in LiH) to 154 pm (in CsH) and can be considered similar to that of F^- (133 pm).

$$\text{Internuclear distance} = r_{\text{cation}} + r_{\text{anion}} \quad (9.3)$$

The large increase in size on going from the H atom ($r_{\text{cov}} = 37$ pm) to the H^- ion arises from interelectronic repulsion when a second electron enters the $1s$ atomic orbital. The smaller r_{H^-} in LiH may suggest some degree of covalent bonding, but calculated and experimental values of lattice energies (see *Sections 5.13* to *5.16*) for *each* of the group 1 metal hydrides are in good agreement, suggesting that an electrostatic model is appropriate for each compound.

Hydrides of the *s*-block metals (excluding Be) can be made by heating the metal with H_2.

$$\tfrac{1}{2}H_2(g) + e^- \longrightarrow H^-(g) \quad \Delta_r H = \tfrac{1}{2}D(H-H) + \Delta_{EA}H \quad (9.4)$$
$$= \Delta_a H^\circ + \Delta_{EA}H$$
$$= +145 \text{ kJ mol}^{-1}$$

When we compare $\Delta_r H$ for reaction 9.4 with those for the formations of F^- and Cl^- from F_2 and Cl_2 (-249 and -228 kJ mol^{-1}, respectively), we understand why, since H^- is about the same size as F^-, ionic hydrides are relatively unstable species with respect to dissociation into their constituent elements. Salt-like hydrides of metals in high oxidation states are most unlikely to exist. (More about binary hydrides in *Section 9.7*.)

9.3 Isotopes of hydrogen

Protium and deuterium

Hydrogen possesses three isotopes, *protium*, *deuterium* and *tritium*, selected properties of which are given in Table 9.1; the isotopes of hydrogen exhibit greater differences in physical and chemical properties than isotopes of any other element. The origin of the differences between H and D, or between

Table 9.1 Selected properties of the isotopes of hydrogen

	Protium	Deuterium	Tritium
Symbols[‡]	^{1}H or H	^{2}H or D	^{3}H or T
Natural abundance	99.985%	0.0156%	<1 in 10^{17} atoms
Isotopic mass / u	1.0078	2.0141	3.0160
Nuclear spin	$\tfrac{1}{2}$	1	$\tfrac{1}{2}$

[‡] Strictly, ^{1}H should be written as $^1_1$H, ^{2}H as $^2_1$H and ^{3}H as $^3_1$H, but the less rigorous symbols are generally used.

pairs of compounds such as H_2O and D_2O, lies in the difference in mass, which in turn affects their fundamental vibrational wavenumbers and zero point energies (see *Figure 2.7* and *worked example 2.4*). The fundamental vibrations for H_2, HD and D_2 are at 4159, 3630 and 2990 cm^{-1}, respectively, and from these data, the zero-point energies of H_2 and D_2 are calculated to be 26.0 and 18.4 kJ mol^{-1}, respectively. The total electronic binding energies for these molecules (represented by the overlap of their atomic wavefunctions) are the *same*, and so it follows that their dissociation energies (e.g. Figure 2.7) differ by 7.6 kJ mol^{-1}, with the D–D bond being stronger than the H–H bond. Similarly, an X–D bond is stronger than the corresponding X–H bond (where X is any element), and this difference is the basis of the kinetic isotope effect (see *Section 2.9*).

Deuterated compounds

A deuterium label in heavy water is indicated by writing $[^2H_2]$water or water-d_2, and similarly for other labelled compounds.

Compounds in which H atoms have been replaced by D are used for a variety of purposes, e.g. as solvents in ^{1}H NMR spectroscopy (see *Box 2.4*). In a fully deuterated material, the D-for-H exchange can have significant effects on the properties of the compound as is shown in Table 9.2 for H_2O and D_2O; the difference in boiling points indicates that intermolecular hydrogen bonding (see *Sections 6.2* and *9.6*) is stronger in D_2O than in H_2O. The major industrial use of D_2O is as a moderator in nuclear reactors; D has a much lower cross-section for neutron capture than H and D_2O is a suitable material for reducing the energies of fast neutrons (see *Section 2.4*) produced in fission without appreciably diminishing the neutron flux.

Many fully or partially deuterated compounds are available commercially, and the extent of *deuterium labelling* (see *Section 2.9*) can be determined by mass spectrometry, density measurements (after conversion into water) or IR spectroscopy.

Tritium

Tritium (Table 9.1) occurs in the upper atmosphere and is formed naturally by reaction 9.5, involving neutrons arriving from outer space. Tritium (see *Section 2.8*) was first obtained synthetically by the bombardment of deuterium compounds

Table 9.2 Selected properties of H_2O and D_2O ('heavy water').

Property	H_2O	D_2O
Melting point / K	273.00	276.83
Boiling point / K	373.00	374.42
Temperature of maximum density / K[‡]	277.0	284.2
Maximum density / g cm^{-3}	0.999 95	1.105 3
Relative permittivity (at 298 K)	78.39	78.06
K_w (at 298 K)	1×10^{-14}	2×10^{-15}
Symmetric stretch,[*] ν_1 (gaseous molecule) / cm^{-1}	3657	2671

[‡] See Figure 6.2.
[*] The symmetric stretching mode is illustrated (for SO_2) in Figure 3.11.

such as $[ND_4]_2SO_4$ with fast neutrons, but is now prepared from lithium deuteride enriched in ^6_3Li.

$$^{14}_{7}N + ^1_0n \longrightarrow ^{12}_{6}C + ^3_1H \qquad (9.5)$$

Tritium is radioactive, a weak β-emitter with $t_{\frac{1}{2}} = 12.3$ yr. It is used extensively as a tracer, in both chemical and biochemical studies; its weak radioactivity, rapid excretion and failure to concentrate in vulnerable organs make it one of the least toxic radioisotopes.

9.4 Dihydrogen

Occurrence

Hydrogen is the most abundant element in the universe, and, after oxygen and silicon, is the third most abundant element on Earth, where it occurs mainly in the form of water or combined with carbon in organic molecules (hydrocarbons, plant and animal material). In the Earth's atmosphere, H_2 occurs to an extent of less than 1 ppm by volume, but those of Jupiter, Neptune, Saturn and Uranus contain large amounts of H_2 (see *Box 9.1*).

Physical properties

Dihydrogen is a colourless, odourless gas, sparingly soluble in all solvents, and at 298 K and 1 bar pressure, it conforms closely to the ideal gas laws. The solid state structure of H_2 can be described in terms of an hcp lattice (see *Section 5.3*), but values of the melting point, enthalpy of fusion, boiling point and enthalpy of vaporization are all very low (Table 9.3), consistent with there being only weak van der Waals forces between the H_2 molecules. The covalent bond in H_2 is unusually strong for a single bond in a diatomic molecule.

Synthesis and uses

In the laboratory, H_2 may be prepared by electrolysis of water (H_2 is liberated at the cathode), but small quantities of H_2 are most conveniently prepared by reactions between dilute acids and suitable metals (e.g. Fe, Zn, equation 9.6), by treating metals that form amphoteric hydroxides (e.g. Zn, Al) with aqueous alkali (equation 9.7) or by reacting metal hydrides with water (equation 9.8).

$$Zn(s) + 2HCl(aq) \longrightarrow ZnCl_2(aq) + H_2(g) \qquad (9.6)$$

BOX 9.1   CHEMICAL AND THEORETICAL BACKGROUND

Metallic character of hydrogen?

The atmospheres of Saturn, Uranus, Jupiter and Neptune are rich in H_2. The *cores* of Saturn and Jupiter are composed of hydrogen subjected to extreme conditions, and it is probable this possesses metallic character. However, establishing the metallic character of hydrogen on Earth is proving to be an extremely difficult task. A report in 1996 from the Livermore Laboratory (US) described how, when a thin layer of *liquid* H_2 is subjected to enormous pressure, changes in conductivity are observed that are consistent with the formation of metallic hydrogen. However, attempts to impart metallic character to *solid* H_2 have so far been unsuccessful. Recently, it has been discovered that, under extremely high pressures,

H_2 (normally a non-polar molecule) undergoes a redistribution of electronic charge such that the ionic contribution to the bonding (represented by the resonance form H^+-H^-) becomes important. This remarkable finding may go some way to helping to explain why attempts to form metallic hydrogen in the solid state have not met with success.

Further reading

P.P. Edwards and F. Hensel (1997) *Nature*, vol. 388, p. 621 – 'Will solid hydrogen ever be a metal?'

Table 9.3 Selected physical properties of H_2.

Physical property	Value
Melting point / K	13.66
Boiling point / K	20.13
Enthalpy of vaporization / kJ mol^{-1}	0.904
Enthalpy of fusion / kJ mol^{-1}	0.117
Density / g dm^{-3}	0.090
Bond dissociation enthalpy / kJ mol^{-1}	435.99
Interatomic distance / pm	74.14
Standard entropy / J K^{-1} mol^{-1}	130.7

$$2Al(s) + 2NaOH(aq) + 6H_2O(l)$$
$$\longrightarrow 2Na[Al(OH)_4](aq) + 3H_2(g) \quad (9.7)$$

$$CaH_2(s) + 2H_2O(l) \longrightarrow Ca(OH)_2(aq) + 2H_2(g) \quad (9.8)$$

Group 1 metals liberate H_2 from water (equation 9.9), but such reactions are not suitable for preparative use because of their extreme vigour. Many other metals that, on thermodynamic grounds, would be expected to react in this way are made *kinetically inert* (see *Section 25.2*) by the presence of a thin film of insoluble metal oxide. Such metals are *passivated*. Although Be is passivated and does not react with water even on heating, the other group 2 metals react with H_2O to give H_2, reactivity increasing down the group; Mg does not react with *cold* water.

$$2K + 2H_2O \longrightarrow 2KOH + H_2 \quad (9.9)$$

A metal is *passivated* if it possesses a surface-coating of the metal oxide which protects it from reaction with, for example, water.

Dihydrogen has industrial applications, the most important being in the Haber process (see *Sections 14.5* and *26.6*), the hydrogenation of unsaturated fats (to produce, for example, margarine), and the production of organic compounds such as methanol (equation 9.10).

$$CO + 2H_2 \xrightarrow[\approx 550K, \ 50 \ bar]{Cu/ZnO \ catalyst} CH_3OH \quad (9.10)$$

For these industrial uses, H_2 is produced *in situ* (because the very low density and boiling point make transport costs unacceptably high). The reagents in reaction 9.10 are collectively called *synthesis gas;* the mixture is manufactured by the *water-gas shift reaction* – reaction of carbon or a hydrocarbon (e.g. CH_4) with steam followed by partial treatment of the CO produced with water vapour (equation 9.11).

$$\left.\begin{array}{l} CH_4 + H_2O \xrightarrow{1200 \ K, \ Ni \ catalyst} CO + 3H_2 \\[2mm] CO + H_2O \xrightarrow{700 \ K, \ iron \ oxide \ catalyst} CO_2 + H_2 \end{array}\right\} \quad (9.11)$$

The CO_2 is absorbed in, for example, K_2CO_3 solution from which it may be recovered by heating. The ratio of H_2:CO in the product mix can be altered, making this reaction both a source of synthesis gas and of H_2. Although equation 9.11 shows *heterogeneous catalysts*, the use of *homogeneous catalysts* is also viable (see *Chapter 26*). Equation 9.11 illustrates the use of CH_4 as the precursor; this represents an oil-based feedstock, and is one of several suitable, low molecular weight hydrocarbons produced in the cracking of crude petroleum. The alternative use of carbon (i.e. coal) means that the water-gas shift reaction can be adapted to meet commercial feedstocks.

In the future, depletion of specific fuel resources may make H_2 the major alternative source of energy, and an alternative to nuclear power; such a change would lead to the so-called *hydrogen economy*. Energy may be produced directly by combustion (H_2 and O_2 combine explosively, and this reaction is used to power the space shuttle's lift-off) or electrochemically in fuel cells (an expensive method that is used in spacecraft where water produced is reconsumed). The ready availability of H_2O makes it an attractive raw material, but production of H_2 from H_2O inevitably requires a large net *input* of energy for which solar sources are environmentally acceptable, e.g. energy collected using photovoltaic cells could be used to electrolyze water. The *photolytic* production of H_2 from H_2O is also possible, although a catalyst is required since water is transparent to light. Equation 9.12 represents such a process: the catalyst, A, exists in two oxidation states, the oxidized form is A(ox) and the reduced form, A(red). The search for suitable photocatalysts is being actively researched; one example is the complex $[Ru(bpy)_3]^{3+}$ (Figure 9.2) which undergoes the reversible redox process 9.13 (see *Figure 22.19* and discussion).

$$\left.\begin{array}{l} H_2O + 2A(ox) \xrightarrow{h\nu} \tfrac{1}{2}O_2 + 2H^+ + 2A(red) \\[2mm] 2H^+ + 2A(red) \longrightarrow H_2 + 2A(ox) \end{array}\right\} \quad (9.12)$$

$$[Ru(bpy)_3]^{3+} + e^- \rightleftharpoons [Ru(bpy)_3]^{2+} \quad (9.13)$$

A *photolytic process* (*photolysis*) is initiated by light; in an equation, this is indicated by $h\nu$ over the arrow; the reactants are said to be *photolyzed*.

Photosynthesis uses sunlight as the energy source; conversion of CO_2 and H_2O into carbohydrates and O_2 by chlorophyll-containing plants is tantamount to photolysis of H_2O followed by reduction of CO_2 by H_2. This natural process can be modified so that some H_2 is liberated, and certain blue-green algae are effective for this purpose.

Interstitial metal hydrides (see *Section 9.7*) are of current interest as hydrogen storage vessels.

These methods of H_2 production remain at the experimental stage, but are of great potential importance.

Reactivity

Dihydrogen is not very reactive under ambient conditions, but the lack of reactivity is kinetic (rather than thermodynamic) in origin, and arises from the strength of the H−H bond (Table 9.3). The branching-chain reaction of

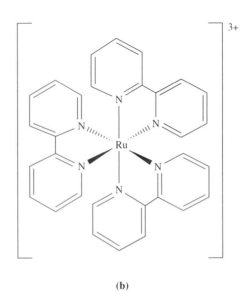

Fig. 9.2 (a) The structure of $[Ru(bpy)_3]^{3+}$ (bpy = 2,2′-bipyridine) determined by X-ray diffraction for the compound $[Ru(bpy)_3][PF_6]_3$ [M. Biner *et al.* (1992) *J. Am. Chem. Soc.*, vol. 114, p. 5197], and (b) a schematic representation of $[Ru(bpy)_3]^{3+}$.

H_2 and O_2 is initiated by sparking and the resulting explosion (or 'pop' on a small scale) is well known in the qualitative test for H_2. Part of the reaction scheme is given (in a simplified form) in equations 9.14–9.18; efficient branching results in a rapid, explosive reaction, and is the reason why it is effective in rocket fuels.

$$H_2 \longrightarrow 2H^{\bullet} \qquad\qquad \textit{initiation} \qquad (9.14)$$

$$H_2 + O_2 \longrightarrow 2OH^{\bullet} \qquad \textit{initiation} \qquad (9.15)$$

$$H^{\bullet} + O_2 \longrightarrow OH^{\bullet} + {}^{\bullet}O^{\bullet} \qquad \textit{branching} \qquad (9.16)$$

$${}^{\bullet}O^{\bullet} + H_2 \longrightarrow OH^{\bullet} + H^{\bullet} \qquad \textit{branching} \qquad (9.17)$$

$$OH^{\bullet} + H_2 \longrightarrow H_2O + H^{\bullet} \qquad \textit{propagation} \qquad (9.18)$$

Halogens react with H_2 (equation 9.19) with the ease of reaction decreasing down group 17. Even at low temperatures, F_2 reacts explosively with H_2 in a radical chain reaction. In the light-induced reaction of Cl_2 and H_2, the initiation step is the homolytic cleavage of the Cl–Cl bond to give $Cl^{\bullet}$ radicals (equation 9.20) which react with H_2 to give $H^{\bullet}$ and HCl in one of a series of steps in the radical chain; HCl can be formed either in a propagation or termination step.

$$H_2 + X_2 \longrightarrow 2HX \qquad X = F, Cl, Br, I \qquad (9.19)$$

$$Cl_2 \xrightarrow{h\nu} 2Cl^{\bullet} \qquad\qquad\qquad\qquad (9.20)$$

BOX 9.2	**CHEMICAL AND THEORETICAL BACKGROUND**

The $[H_3]^+$ ion

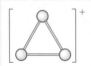

The equilateral triangular $[H_3]^+$ ion may appear to be a theoretical novelty, and indeed, it has been the subject of many theoretical studies. However, just as Jupiter has provided challenges in regard to metallic hydrogen (see *Box 9.1*), it has also proved to be the source of exciting spectroscopic data, analysis of which has confirmed the existence of $[H_3]^+$. The atmosphere of Jupiter consists mainly of dihydrogen, and the formation of $[H_3]^+$ has been explained in terms of the ionization of H_2, brought about by collisions between H_2 molecules and charged particles (with *extremely*

high kinetic energies) which originate from Jupiter's magnetosphere:

$$H_2 \longrightarrow [H_2]^+ + e^-$$

It is proposed that further collisions between H_2 and $[H_2]^+$ lead to the formation of $[H_3]^+$.

$$H_2 + [H_2]^+ \longrightarrow [H_3]^+ + H$$

The chemistry of this cation in the atmosphere of Jupiter is a subject for future research.

Further reading

S. Miller and J. Tennyson (1992) *Chemical Society Reviews*, vol. 22, p. 281 – '$[H_3]^+$ in space'.

Reactions of H_2 with Br_2 or I_2 occur only at higher temperatures and also involve the initial fission of the X_2 molecule. For Br_2 (but not for I_2) the mechanism is a radical chain (equation sequence 9.21).

$$\left.\begin{aligned} Br_2 &\longrightarrow 2Br^{\cdot} \\ Br^{\cdot} + H_2 &\longrightarrow HBr + H^{\cdot} \\ H^{\cdot} + Br_2 &\longrightarrow HBr + Br^{\cdot} \\ HBr + H^{\cdot} &\longrightarrow Br^{\cdot} + H_2 \\ Br^{\cdot} &\longrightarrow Br_2 \end{aligned}\right\} \quad (9.21)$$

Dihydrogen reacts with many metals when heated to give metal hydrides, MH_n, although these are not necessarily stoichiometric (e.g. $TiH_{1.7}$, see *Section 5.7*). By the action of an electric discharge, H_2 is partially dissociated into atoms, particularly at low pressures. This provides a reactive source of the element, and facilitates combination with elements (e.g. Sn and As) that do not react directly with H_2.

The reaction between N_2 and H_2 (equation 9.22) is of major industrial importance. However, the reaction is extremely slow and mixtures of N_2 and H_2 remain indefinitely unchanged; manipulation of the temperature and pressure and the use of a catalyst are essential. (More about catalysts and their industrial applications in *Chapter 26*.)

$$3H_2(g) + N_2(g) \rightleftharpoons 2NH_3(g) \quad (9.22)$$

Interaction between a catalytic surface and H_2 weakens and aids cleavage of the H—H bond (*Figure 9.3*). On an industrial scale, the hydrogenation of enormous numbers of unsaturated organic compounds is carried out on surfaces of metals such as Ni, Pd and Pt. The use of homogeneous catalysts is becoming increasingly important, e.g. reaction 9.23 (the *hydroformylation process*).

$$RHC=CH_2 + H_2 + CO \xrightarrow{\text{Co}_2(\text{CO})_8 \text{ catalyst}} RCH_2CH_2CHO \quad (9.23)$$

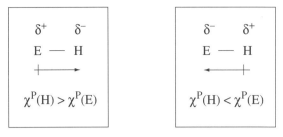

Fig. 9.4 The direction of the dipole moment in a polar E—H bond depends upon the relative electronegativity values; Pauling electronegativity values, χ^P, are given in *Appendix 7*.

9.5 Polar and non-polar E—H bonds

Although we refer to compounds of the type EH_n (E = any element) as *hydrides*, and this tends to suggest the presence of H^- (or at least, $H^{\delta-}$), the difference in electronegativity values between E and H means that the E—H bond may be non-polar, or polar in either of the senses shown in Figure 9.4. For H, $\chi^P = 2.2$ and a number of E—H bonds in which E is a *p*-block element (e.g. B—H, C—H, Si—H, P—H) are essentially non-polar. Since metals are electropositive, the H atom in an M—H bond carries a δ^- partial charge. In contrast, N, O and F are more electronegative than H, and in N—H, O—H and F—H bonds, the H atom carries a δ^+ partial charge.

The molecular environment of an X—H bond also influences the magnitude of the bond dipole and properties associated with the bond; this is demonstrated by a comparison of the pK_a values for CH_3CO_2H (4.75) and CF_3CO_2H (0.23).

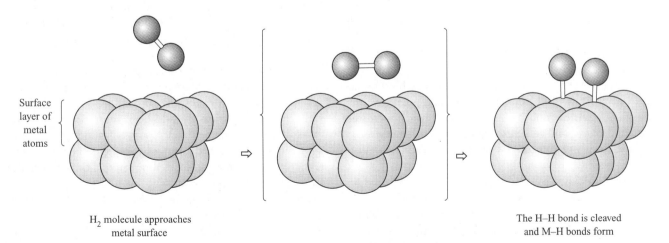

Fig. 9.3 A schematic representation of the interaction of an H_2 molecule with a metal surface to give *adsorbed* hydrogen atoms. The scheme does not imply anything about the detailed mechanism of the process. Further details about heterogeneous catalysis are given in Chapter 26.

Table 9.4 Typical values for the enthalpies of dissociation of different types of hydrogen bonds.

Category of hydrogen bond	Hydrogen bond ($\cdots$)	Approximate dissociation enthalpy / $kJ\,mol^{-1}$
Asymmetrical	$F-H\cdots FH$	29
Asymmetrical	$HO-H\cdots OH_2$	25
Asymmetrical	$H_2N-H\cdots NH_3$	17
Asymmetrical	$HS-H\cdots SH_2$	7
Symmetrical	$F\cdots H\cdots F$ in $[HF_2]^-$	165

9.6 Hydrogen bonding

The hydrogen bond

A *hydrogen bond* is formed between an H atom attached to an electronegative atom, and an electronegative atom that possesses a lone pair of electrons.

Physical and solid state structural data for many compounds provide evidence for the formation of intermolecular hydrogen bonds. Such interactions arise between an H atom attached to an electronegative atom, and an electronegative atom bearing a lone pair of electrons, i.e. $X-H\cdots Y$ where atom Y may or may not be the same as X. The strengths of hydrogen bonds are greater than those of normal van der Waals interactions (i.e. dipole–dipole or dispersion forces; see the latter part of *Section 5.13*) and Table 9.4 lists representative examples.

We have already described the hydrogen-bonded network in ice (see *Section 6.2*). Here, as in most hydrogen-bonded interactions, the H atom is *asymmetrically* positioned with respect to the two atoms with which it interacts. Association in carboxylic acids (see *Box 9.3*) is a consequence of hydrogen bonding. In a typical $X-H\cdots Y$ interaction, the $X-H$ covalent bond is *slightly* longer and weaker than a comparable bond in the absence of hydrogen bonding. In such cases, the interaction may be considered in terms of a weak electrostatic interaction between a covalently bonded H with a δ^+ charge, and a lone pair of electrons on the adjacent atom. Some experimental observations cannot be rationalized within a purely electrostatic model, and point towards a covalent contribution, the importance of which increases as the hydrogen bond becomes stronger: it is now recognized that 'hydrogen bonding' covers more than one type of interaction.

Table 9.4 shows typical values of bond dissociation enthalpies of some hydrogen bonds. An example of how such data can be obtained comes from the dissociation of a carboxylic acid dimer in the vapour state: the position of equilibrium 9.24 is temperature-dependent, and ΔH° for the reaction can be obtained from the variation of K_p with temperature.

$$R-C\begin{array}{c}O-H\cdots\cdots O\\ \\O\cdots\cdots H-O\end{array}C-R \;\rightleftharpoons\; 2RCO_2H \qquad (9.24)$$

For formic acid (methanoic acid) ΔH° for the dissociation in equation 9.24 is found to be $+60\,kJ\,mol^{-1}$, or the value can be expressed as $+30\,kJ$ per mole of hydrogen bonds. This quantity is often referred to as the hydrogen-bond energy, but this is not strictly correct since other bonds change *slightly* when hydrogen bonds are broken (Figures 9.5a and 9.5b).

In some hydrogen-bonded interactions, the H atom is *symmetrically* positioned, e.g. in $[HF_2]^-$ (see Figure 9.8) or $[H_5O_2]^+$ (Figure 9.1). In the formation of $[HF_2]^-$, (equation 9.25), appreciable stretching of the original covalent $H-F$ bond takes place, to give two equivalent $H\cdots F$ interactions.

$$HF + F^- \longrightarrow [HF_2]^- \qquad (9.25)$$

The bonding in symmetrical $X\cdots H\cdots X$ interactions is best considered in terms of a 3c-2e interaction, i.e. as a delocalized interaction such as was described for B_2H_6 in Section 4.7. Each $H\cdots F$ bond is relatively strong (Table 9.4), with the bond dissociation enthalpy being of a similar magnitude to that of the $F-F$ bond in F_2 ($158\,kJ\,mol^{-1}$); compare this to the bond dissociation enthalpy of HF ($570\,kJ\,mol^{-1}$).

The use of the qualitative descriptors 'strong' and 'weak' for hydrogen bonds is common, e.g. strong $O-H\cdots O$ interactions are typified by $O\cdots O$ separations of 250–265 pm, while $O\cdots O$ distances of >280 pm would be considered weak. Accurate neutron and X-ray diffraction data[†] have shown that, in so-called 'strong hydrogen bonds' involving $O-H\cdots O$ interactions, shortening of the $O\cdots O$ distance from 280 to 240 pm is accompanied by a change from asymmetrical, electrostatic hydrogen bonds to symmetrical, covalent interactions.

[†] P. Gilli *et al.* (1994) *Journal of the American Chemical Society*, vol. 116, p. 909.

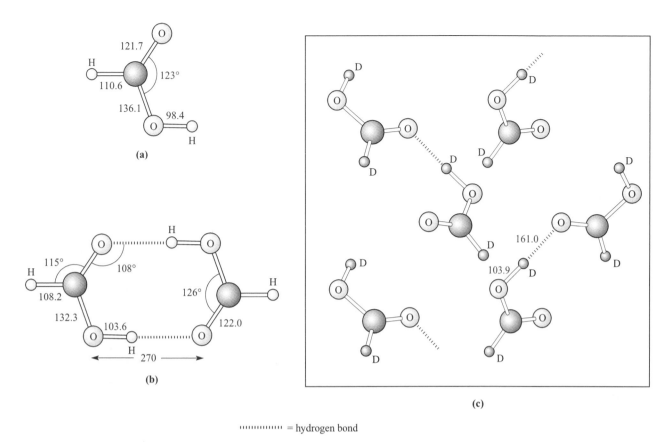

============================ = hydrogen bond

Fig. 9.5 In the vapour state, formic acid exists both as a (a) monomer and (b) dimer, the structures of which have been determined by electron diffraction. (c) In the solid state, a more complex assembly is formed as revealed in a neutron diffraction study of deuterated formic acid, DCO_2D; the figure shows part of the packing diagram for the unit cell. [A. Albinati *et al.* (1978) *Acta Crystallogr., Sect. B*, vol. 34, p. 2188.] Distances are in pm.

Trends in boiling points, melting points and enthalpies of vaporization for *p*-block binary hydrides

It is generally expected that the melting and boiling points of members of a series of related molecular compounds increase with increasing molecular size, owing to an increase in inter-molecular dispersion forces. This is seen, for example, along a homologous series of alkanes. However, a comparison of the melting and boiling points of *p*-block hydrides, EH_n, provides evidence for hydrogen bonding. Figure 9.6 shows that, for E = group 14 element, melting and boiling points follow the expected trends, but for E = group 15, 16 or 17 element, the first member of the group shows anomalous behaviour, i.e. the melting and boiling points of NH_3, H_2O and HF are higher than expected when compared to their heavier congeners. Figure 9.7 illustrates that values of $\Delta_{vap}H$ show a similar pattern. It is tempting to think that Figures 9.6 and 9.7 indicate that the hydrogen bonding in H_2O is stronger than in HF; certainly, the values for H_2O appear to be particularly high. However, this is not a sound conclusion; boiling points and values of $\Delta_{vap}H$ relate to differences between the liquid and gaseous states, and there is independent evidence that while H_2O is hydrogen-bonded in the liquid but not in the vapour state, HF is strongly hydrogen-bonded in both.

Deviations from Trouton's empirical rule (equation 9.26) are another way of expressing the data in Figures 9.6 and 9.7. For HF, H_2O and NH_3, $\Delta_{vap}S = 116$, 109 and 97 J mol^{-1} K^{-1} respectively. Hydrogen bonding in each liquid lowers its entropy, and make the *change* in the entropy on going from liquid to vapour larger than it would have been had hydrogen bonding not played an important role.

For *liquid* $\rightleftharpoons$ *vapour*: $\quad \Delta_{vap}S = \dfrac{\Delta_{vap}H}{bp}$

$$\approx 88 \, \text{J K}^{-1} \, \text{mol}^{-1} \qquad (9.26)$$

Infrared spectroscopy

The IR spectrum of a hydrate, alcohol or carboxylic acid exhibits a characteristic absorption around 3500 cm^{-1} assigned to the ν(OH) mode. The typical broadness of this band can be explained in terms of the involvement of the O$-$H hydrogen atom in hydrogen bonding. In cases where we can compare the stretching frequencies of the same molecule with and without hydrogen-bonded association (e.g. liquid water and water vapour), a shift is observed to higher wavenumber as hydrogen bonding is lost.

BOX 9.3   **CHEMICAL AND THEORETICAL BACKGROUND**

Intermolecular hydrogen bonding in the solid state: carboxylic acids

Figure 6.1 illustrated how hydrogen bonding between H_2O molecules in the solid state leads to the formation of a rigid network. Hydrogen bonding between carboxylic acid molecules leads to aggregation in the solid state (see *Figure 9.5*), and for a *difunctional* carboxylic acid such as fumaric acid or benzene-1,4-dicarboxylic acid, this has the potential to lead to the formation of ribbon-like arrays.

Fumaric acid

Benzene-1,4-dicarboxylic acid
(terephthalic acid)

This potential for constructing network structures can be increased by adding further carboxylic acid groups, although the relative orientations of the functionalities are important if a two- or three-dimensional framework is to be established in the solid state. Benzene-1,3,5-tricarboxylic acid (trimesic acid) possesses three CO_2H groups, symmetrically disposed about the C_6-ring core:

Benzene-1,3,5-tricarboxylic acid

In the solid state, the molecules are held together by hydrogen-bonded interactions to form sheets. These sheets interpenetrate to form a complex array containing channels; part of the solid state lattice (determined by X-ray diffraction) is shown below:

Molecules containing carboxylic acid functionalities are not confined to organic systems; for example, the C=C double bond in fumaric acid can interact with a low oxidation state metal centre (see *Chapter 23*) to form organometallic compounds such as $Fe(CO)_4(\eta^2\text{-}HO_2CCHCHCO_2H)$; the η^2-prefix (see *Box 18.1*) indicates that the two carbon atoms of the C=C bond of the fumaric acid residue are linked to the Fe centre. Hydrogen bonding can occur between adjacent pairs of molecules as is depicted below, and such interactions extend through the solid state lattice to produce an extensive, three-dimensional array.

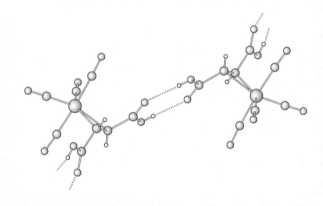

Further reading

J.S. Moore and S. Lee (1994) *Chemistry & Industry*, p. 556 – 'Crafting molecular based solids'.

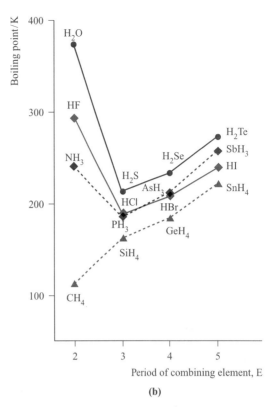

Fig. 9.6 Trends in (a) melting and (b) boiling points for some *p*-block hydrides, EH$_n$.

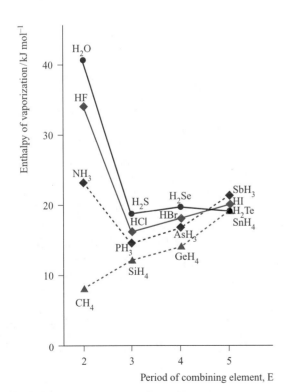

Fig. 9.7 Trends in values of $\Delta_{vap}H$ (measured at the boiling point of the liquid) for some *p*-block hydrides, EH$_n$.

Similar observations are noted for other hydrogen-bonded systems.

Solid state structures

The presence of hydrogen bonding has important effects on the solid state structures of many compounds as we have already discussed for ice (*Section 6.2*) and carboxylic acids (*Box 9.3*). The *solid* state structures of some simple carboxylic acids are more complex than one might at first imagine. Figure 9.5c shows part of the solid state packing diagram for deuterated formic acid; the orientation of the DCO$_2$D molecules allows the assembly of a more extensive hydrogen-bonded network than simple dimers. The solid state structure of acetic acid is similarly complex.

The structure of solid HF consists of zig-zag chains (Figure 9.8a) although the positions of the H atoms are not accurately known. Structural parameters are available for a number of salts containing [HF$_2$]$^-$, and include neutron diffraction data for the deuterated species. The anion is linear with the H atom positioned symmetrically between the two F atoms (Figure 9.8b); the H−F distance is relatively short, consistent with strong hydrogen bonding (see *Table 9.4* and earlier discussion).

Although hydrogen bonds commonly involve F, O or N, this is not an exclusive picture. Examples include the solid state of HCN which exhibits a linear chain with C−H····N interactions, the 1:1 complex formed between acetone and

Fig. 9.8 (a) The solid state structure of HF consists of zig-zag chains. (b) The structure of the $[HF_2]^-$ ion, determined by X-ray and neutron diffraction for the K^+ salt.

chloroform, and the existence of salts containing the $[HCl_2]^-$ anion.

Hydrogen bonding in biological systems

We cannot leave the topic of hydrogen bonding without mentioning its important role in biological systems, one of the best known being the formation of the double helical structure of DNA (deoxyribonucleic acid). The structures of adenine and thymine are exactly matched to permit hydrogen bonding between them, and they are referred to as complementary bases; guanine and cytosine form the second base-pair (Figure 9.9). The hydrogen bonding between these base pairs in the strands of DNA leads to the assembly of the double helix.[†]

9.7 Binary hydrides: classification and general properties

Detailed chemistries of most of the hydrides are considered in later chapters.

Classification

The four major classes into which it is convenient to place binary hydrides are:

- metallic
- saline (salt-like)
- molecular
- polymeric

with a number of hydrides falling into intermediate or borderline categories.

Interstitial metal hydrides

Hydrogen atoms are small enough to occupy the interstitial holes in a metal lattice and the absorption of H_2 by a variety of metals (and also alloys) leads to the formation of metal hydrides in which hydrogen atoms reside in interstitial cavities, *interstitial metal hydrides*. For example, non-stoichiometric hydrides $TiH_{1.7}$, $HfH_{1.98}$ and $HfH_{2.10}$ are formed when titanium and hafnium react with H_2. Niobium

forms a series of non-stoichiometric hydrides of formula NbH_x ($0 < x \leq 1$) and at low hydrogen content, the bcc structure of Nb metal is retained. An interesting property of these metal hydrides is their ability to release H_2 upon heating, and this leads to their use as 'hydrogen storage vessels'. The nickel–metal hydride battery has recently been developed for use as a power source in electric cars; although expensive, its performance statistics are promising, with a life-span is about eight years, and a range between recharging stops of $\approx 160\,km$ (see *Boxes 8.4* and *10.3*).

Saline hydrides

Saline hydrides are formed when the group 1 or 2 metals (except Be) are heated with H_2. All are white, high melting solids (e.g. LiH, mp = 953 K; NaH, mp = 1073 K with decomposition); the group 1 hydrides crystallize with the NaCl lattice, and the presence of the H^- ion (see *Section 9.2*) is indicated by the good agreement between lattice energies obtained from Born–Haber cycles and from X-ray and compressibility data. Additional evidence comes from the fact that the electrolysis of molten LiH liberates H_2 at the *anode* (equation 9.27).

$$\left.\begin{array}{ll} 2H^- \longrightarrow H_2 + 2e^- & \textit{at the anode} \\ Li^+ + e^- \longrightarrow Li & \textit{at the cathode} \end{array}\right\} \quad (9.27)$$

The reactivity of the group 1 hydrides increases with an increase in atomic number and ionic size of the metal; in keeping with this, values of $\Delta_f H^\circ$ become less negative, with that of LiH being significantly more negative than those of the other alkali metal hydrides. Table 9.5 lists factors that contribute towards this trend; since the hydride ion is a common factor in the series, we need to look at the extent to which the value of $\Delta_{lattice} H^\circ$ offsets the sum of $\Delta_a H^\circ$ and IE_1 in order to reconcile the trend in values of $\Delta_f H^\circ$ (equation 9.28). The H^- ion is similar in size to F^-, and thus, the trend parallels that observed for alkali metal fluorides.

$$(9.28)$$

[†] For a discussion of DNA, see: C.K. Mathews and K.E. van Holde (1996) *Biochemistry*, 2nd edn, Benjamin/Cummings, New York, Chapter 4.

Adenine - - - - - thymine base-pair

Guanine - - - - - cytosine base-pair

Fig. 9.9 The complementary base-pairs in DNA interact through hydrogen bonds. The backbones of each strand in DNA consists of sugar units (to which the bases are attached) connected by phosphate groups.

Table 9.5 Values of the $\Delta_f H^{\circ}(298\,K)$ of the alkali metal hydrides, MH, depend upon the relatives magnitudes of $\Delta_a H^{\circ}(298\,K)$ and IE_1 of the metals, and the lattice energies, $\approx \Delta_{lattice} H^{\circ}(298\,K)$, of MH.

Metal	$\Delta_a H^{\circ}(M)$ / kJ mol^{-1}	$IE_1(M)$ / kJ mol^{-1}	$\Delta_{lattice} H^{\circ}$ / kJ mol^{-1}	$\Delta_f H^{\circ}(MH)$ / kJ mol^{-1}
Li	161	521	−920	−90.5
Na	108	492	−808	−56.3
K	90	415	−714	−57.7
Rb	82	405	−685	−52.3
Cs	78	376	−644	−54.2

Saline hydrides react immediately with protic solvents such as H_2O (equation 9.29), NH_3 or EtOH, showing that the H^- ion is an extremely strong base. Widespread use is made of NaH and KH as deprotonating agents (e.g. reaction 9.30).

$$NaH + H_2O \longrightarrow NaOH + H_2 \qquad (9.29)$$

$$Ph_2PH + NaH \longrightarrow Na[PPh_2] + H_2 \qquad (9.30)$$

Of the saline hydrides, LiH, NaH and KH are the most commonly used, but their moisture-sensitivity means that reaction conditions must be water-free. Of particular significance are the reactions between LiH and Al_2Cl_6 to give

Remarkable optical properties of yttrium hydrides

In 1996, a report appeared in *Nature* of experiments in which a 500 nm thick film of yttrium (coated with a 5–20 nm layer of palladium to prevent aerial oxidation) was subjected to 10^5 Pa pressure of H_2 gas at room temperature. As H_2 diffused through the Pd layer, the latter catalyzed the dissociation of H_2 into H atoms which then entered the yttrium lattice. A series of observations followed:

- initially the yttrium film was a reflecting surface, i.e. a mirror;
- a few minutes after H atoms entered the lattice, a partially reflecting surface was observed and this was attributed to the formation of YH_2;

- after more hydrogen had been taken up and a composition of $YH_{2.86}$ had been reached, the surface became yellow and transparent.

These remarkable changes were shown to be reversible. The accommodation of the H atoms within the metal lattice is not simple, because the lattice of yttrium atoms undergoes a phase transition from an initially fcc to hcp structure; the fcc lattice is present in the β-YH_2 phase.

For details of these observations and photographs depicting the mirror to non-reflector transitions, see: J.N. Huiberts, R. Griessen, J.H. Rector, R.J. Wijngaarden, J.P. Dekker, D.G. de Groot and N.J. Koeman (1996) *Nature*, vol. 380, p. 231.

---

lithium tetrahydridoaluminate(1−), Li[AlH$_4$], (also called lithium aluminium hydride or *lithal*) and between NaH and B(OMe)$_3$ or BCl$_3$ (equations 9.31 and 9.32) to give sodium tetrahydroborate(1−), commonly known as sodium borohydride (see *Section 12.5*).[†] The compounds Li[AlH$_4$], Na[BH$_4$] and NaH find wide applications as reducing agents, e.g. reactions 9.33 and 9.34.

$$4NaH + B(OMe)_3 \xrightarrow{520\,K} Na[BH_4] + 3NaOMe \qquad (9.31)$$

$$4NaH + BCl_3 \longrightarrow Na[BH_4] + 3NaCl \qquad (9.32)$$

$$ECl_4 \xrightarrow{Li[AlH_4]} EH_4 \qquad E = Si, Ge\ or\ Sn \qquad (9.33)$$

$$[ZnMe_4]^{2-} \xrightarrow{Li[AlH_4]} [ZnH_4]^{2-} \qquad (9.34)$$

Molecular hydrides

Covalent hydrides with molecular structures are formed by the *p*-block elements in groups 13 to 17 with the exception of Al (see *Section 12.5*) and Bi; BiH$_3$ is thermally unstable, decomposing above 198 K, and little is known about PoH$_2$. Hydrides of the halogens, sulfur and nitrogen are prepared by reacting these elements with H_2 under appropriate conditions (e.g. reaction 9.22); the remaining hydrides are formed by treating suitable metal salts with water, aqueous acid or NH$_4$Br in liquid NH$_3$, or by use of [BH$_4$]$^-$ or [AlH$_4$]$^-$, e.g. reaction 9.33. Specific syntheses are given in later chapters.

Most molecular hydrides are volatile and have simple structures which comply with VSEPR theory (see *Section 1.19*). However, BH$_3$, (**9.2**), although known in the gas phase, dimerizes to give B$_2$H$_6$, (**9.3**), and GaH$_3$ behaves similarly; B$_2$H$_6$ and Ga$_2$H$_6$ are described in *Section 12.5*.

(9.2)   (9.3)

A number of molecular hydrido complex anions of the *d*-block metals are known, and their structures can be rationalized in terms of the Kepert model (see *Section 19.7*). The anion [NiH$_4$]$^{4-}$ is tetrahedral. X-ray diffraction data have confirmed a square-based pyramidal structure for [CoH$_5$]$^{4-}$ (Figure 9.10a), and [IrH$_5$]$^{4-}$ adopts a similar structure; these anions have been isolated as the salts Mg$_2$CoH$_5$ and M$_2$IrH$_5$ (M = Mg, Ca or Sr). Alkaline earth metal ions have also been used to stabilize salts containing [FeH$_6$]$^{4-}$, [RuH$_6$]$^{4-}$ and [OsH$_6$]$^{4-}$ (Figure 9.10b). The reaction of Na[ReO$_4$] with Na in EtOH yields Na$_2$[ReH$_9$], and the K$^+$ and [Et$_4$N]$^+$ salts have been prepared by *metathesis* from the Na$^+$ salt; K$_2$[TcH$_9$] has been prepared from [TcO$_4$]$^-$ and K in EtOH in the presence of 1,2-diaminoethane.

> A *metathesis reaction* involves an exchange, for example:
>
> $$AgNO_3 + NaCl \longrightarrow AgCl + NaNO_3$$

Neutron diffraction data for K$_2$[ReH$_9$] have confirmed a 9-coordinate Re atom in a tricapped trigonal prismatic environment (Figure 9.10c); [TcH$_9$]$^{2-}$ is assumed to be similar to [ReH$_9$]$^{2-}$. Despite there being two H environments in [ReH$_9$]$^{2-}$, only one signal is observed in the solution ^{1}H NMR spectrum indicating that the dianion is stereochemically non-rigid on the NMR spectroscopic timescale (see *Section 2.11*). Palladium(II) and platinum(II) form the square planar [PdH$_4$]$^{2-}$ and [PtH$_4$]$^{2-}$. The salt K$_2$[PtH$_4$] is made by reacting Pt with KH under H$_2$ (1–10 bar, 580–700 K); 'K$_3$PtH$_5$' also forms but structural data show this to contain [PtH$_4$]$^{2-}$ and H$^-$ ions. A high pressure of H$_2$ is

---

[†] A system of coordination nomenclature is used for anions containing H$^-$, and, thus, [AlH$_4$]$^-$ is called the tetrahydridoaluminate(1−) ion; anions containing boron are exceptions, and [BH$_4$]$^-$ is the tetrahydroborate(1−) ion.

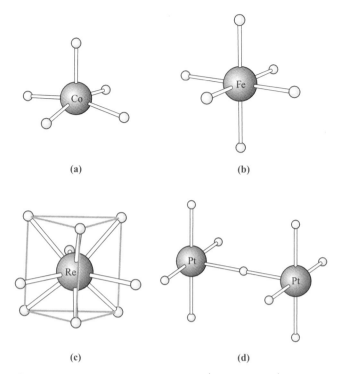

Fig. 9.10 The structures of (a) $[CoH_5]^{4-}$, (b) $[FeH_6]^{4-}$, (c) $[ReH_9]^{2-}$ and (d) $[Pt_2H_9]^{5-}$.

also needed to form $Li_5[Pt_2H_9]$ but, once formed, it is stable with respect to H_2 loss; the structure of $[Pt_2H_9]^{5-}$ is shown in Figure 9.10d. The Pt(IV) complex $K_2[PtH_6]$ results if KH and Pt sponge are heated (775 K) under 1500–1800 bar H_2; neutron diffraction confirms that deuterated $[PtD_6]^{2-}$ is octahedral.

Polymeric hydrides

Polymeric hydrides (white solids) are formed by Be and Al. In BeH_2 (Figure 9.11), each Be centre is tetrahedral, giving a chain structure in which multi-centre bonding of the type described for B_2H_6 is present. The structure of AlH_3 consists of an infinite lattice, in which each Al(III) centre is in an AlH_6-octahedral site; H atoms bridge pairs of Al centres.

Intermediate hydrides

Not all hydrides can be placed in the above categories, e.g. those of Pd, Cu, lanthanoids and actinoids. Palladium reversibly absorbs large amounts of H_2 or D_2 (but no other gases, a fact that is of great importance in the separation of H_2 from gaseous mixtures). The absorbed hydrogen has a high

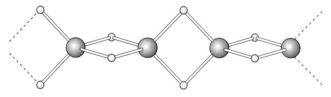

Fig. 9.11 Part of the polymeric chain structure of BeH_2; Be atoms are shown in blue.

mobility, but the form in which it is present has not been established, although the limiting composition is known to be $\approx PdH_{0.7}$.

Summary

In Chapter 9, we have considered some of the properties of the hydrogen atom, and have discussed the physical and chemical properties of dihydrogen. The polarity of an X–H bond depends upon the identity of X, and on the molecular environment of the X–H bond. For X = N, O or F (and to a lesser extent, for some other X atoms), the formation of hydrogen bonds is important, and has a profound influence on physical and structural properties. Binary hydrides are conveniently classified as being either metallic, saline, molecular or polymeric, although some hydrides have intermediate character; we have looked briefly at the methods of synthesis and structures of some of these hydrides, and details of molecular hydrides X_mH_n will be considered in later chapters under sections dealing with the relevant element X.

The following terms were introduced in this chapter. Do you know what they mean?

- ❑ hydrogen ion (proton)
- ❑ oxonium ion
- ❑ hydrate
- ❑ solvent of crystallization
- ❑ hydride ion
- ❑ protium
- ❑ deuterium
- ❑ tritium
- ❑ deuterium labelling
- ❑ passivate
- ❑ synthesis gas
- ❑ water-gas shift reaction
- ❑ heterogeneous catalyst
- ❑ homogeneous catalyst
- ❑ hydrogen economy
- ❑ hydrogen bonding
- ❑ asymmetrical hydrogen bond
- ❑ symmetrical hydrogen bond
- ❑ anomalous properties of HF, H_2O and NH_3
- ❑ Trouton's rule
- ❑ binary compound
- ❑ metallic hydride
- ❑ saline (salt-like) hydride
- ❑ molecular hydride
- ❑ polymeric hydride
- ❑ metathesis

Suggested further reading

Hydrogen/dihydrogen

M. Kakiuchi (1994) 'Hydrogen: Inorganic chemistry' in *Encyclopedia of Inorganic Chemistry*, ed. R.B. King, Wiley, Chichester, vol. 3, p. 1444 – An up-to-date account including detailed discussion of isotopes and isotope effects.

C.A. McAuliffe (1980) *Hydrogen and Energy*, Macmillan, London – A short and readable account of energy sources.

Hydrogen bonding

J. Emsley, O.P.A. Hoyte and R.E. Overill (1977) *Journal of the Chemical Society, Chemical Communications*, p. 225 – 'The strongest hydrogen bond? *Ab initio*-LCAO-MO-SCF calculations' looks at the strength of the hydrogen bond between formic acid and fluoride ion.

J. Emsley (1980) *Chemical Society Reviews*, vol. 9, p. 91 – A review of strong hydrogen bonds.

P. Gilli, V. Bertolasi, V. Ferretti and G. Gilli (1994) *Journal of the American Chemical Society*, vol. 116, p. 909 – 'Covalent nature of the strong homonuclear hydrogen bond. Study of the O–H····O system by crystal structure correlation methods'.

A.F. Goncharov, V.V. Struzhkin, M.S. Somayazulu, R.J. Hemley and H.K. Mao (1996) *Science*, vol. 273, p. 218 – 'Compression of ice at 210 gigapascals: Infrared evidence for a symmetric hydrogen-bonded phase'.

G.A. Jeffery (1997) *An Introduction to Hydrogen Bonding*, Oxford University Press, Oxford – A text that introduces modern ideas on hydrogen bonding.

K. Manchester (1997) *Chemistry & Industry*, p. 835 – 'Masson Gulland: hydrogen bonding in DNA' gives a historical perspective on the importance of hydrogen bonding in DNA.

C.K. Mathews and K.E. van Holde (1996) *Biochemistry*, 2nd edn, Benjamin/Cummings, New York, Chapter 4 – An introduction to the structure and properties of DNA.

Metal hydrides

K. Yvon (1994) 'Hydrides: solid state transition metal complexes' in *Encyclopedia of Inorganic Chemistry*, ed. R.B. King, Wiley, Chichester, vol. 3, p. 1401 – A well-illustrated review covering hydrido ions and interstitial hydrides.

Problems

9.1 Confirm that the difference in values of $\nu(\text{O–H})$ and $\nu(\text{O–D})$ given in Table 9.2 is consistent with the isotopic masses of H and D.

9.2 (a) Outline the reasons why it is necessary to use deuterated solvents in ^1H NMR spectroscopy. (b) Draw the structures of THF-d_8 and DMF-d_7.

9.3 For deuterium, $I = 1$. In a fully labelled sample of CDCl_3, what is observed in the ^{13}C NMR spectrum?

9.4 In ^1H NMR spectra in which the solvent is acetonitrile-d_3, labelled to an extent of 99.6%, a multiplet is observed at δ 1.94. How does this multiplet arise, and what is its appearance? [D, $I = 1$; H, $I = \frac{1}{2}$]

9.5 How would you attempt to prepare a sample of pure HD and to establish the purity of the product?

9.6 The IR spectrum of a $0.01 \, \text{mol dm}^{-3}$ solution of *tert*-butanol in CCl_4 shows a sharp peak at $3610 \, \text{cm}^{-1}$; in the IR spectrum of a similar $1.0 \, \text{mol dm}^{-3}$ solution, this absorption is much diminished in intensity, but a very strong, broad peak at $3330 \, \text{cm}^{-1}$ is observed. Rationalize these observations.

9.7 Suggest an explanation for the fact that solid CsCl, but not LiCl, absorbs HCl at low temperatures.

9.8 Suggest a structure for the $[\text{H}_9\text{O}_4]^+$ ion.

9.9 Write a brief, but critical, account of 'the hydrogen bond'.

9.10 (a) Write equations for the reactions of KH with NH_3 and with ethanol. (b) Identify the conjugate acid–base pairs in each reaction.

9.11 Write equations for the following processes, noting appropriate conditions:

(a) electrolysis of water;
(b) electrolysis of molten LiH;
(c) CaH_2 reacting with water;
(d) Mg treated with dilute nitric acid;
(e) combustion of H_2;
(f) reaction of H_2 with CuO.

9.12 Solutions of H_2O_2 are used as bleaching agents. For the decomposition of H_2O_2 to H_2O and O_2, $\Delta G^\circ = -116.7 \, \text{kJ mol}^{-1}$. Why can H_2O_2 be stored for periods of time without significant decomposition?

9.13 Magnesium hydride possesses a rutile lattice. (a) Sketch a unit cell of rutile. (b) What are the coordination numbers and geometries of the Mg and H centres in this structure?

9.14 Confirm the stoichiometry of aluminium hydride as 1:3 from the text description of the infinite structure.

9.15 Discuss the bonding in BeH_2 in terms of a suitable hybridization scheme. Relate this to a bonding description for Ga_2H_6.

9.16 Suggest explanations for the following trends in data.

(a) In gas phase CH_4, NH_3 and H_2O, $\angle \text{H–C–H} = 109.5°$, $\angle \text{H–N–H} = 106.7°$ and $\angle \text{H–O–H} = 104.5°$.
(b) The dipole moments (in the gas phase) of NH_3 and NH_2OH are 1.47 and 0.59 D.
(c) The ratios of $\Delta_{\text{vap}}H$:bp for NH_3, N_2H_4, PH_3, P_2H_4, SiH_4 and Si_2H_6 are, respectively 97.3, 108.2, 78.7, 85.6, 75.2 and 81.9 $\text{J K}^{-1} \text{mol}^{-1}$. However, for HCO_2H, the ratio is 60.7 $\text{J K}^{-1} \text{mol}^{-1}$.

Group 1: the alkali metals

1	2		13	14	15	16	17	18
H								He
Li	Be		B	C	N	O	F	Ne
Na	Mg		Al	Si	P	S	Cl	Ar
K	Ca		Ga	Ge	As	Se	Br	Kr
Rb	Sr	*d*-block	In	Sn	Sb	Te	I	Xe
Cs	Ba		Tl	Pb	Bi	Po	At	Rn
Fr	Ra							

10.1 Introduction

The alkali metals – lithium, sodium, potassium, rubidium, caesium and francium – are members of group 1 of the periodic table, and each has a ground state valence electronic configuration ns^1. Discussions of these metals usually neglect the heaviest member of the group; only artificial isotopes of francium are known, the longest lived, $^{223}_{87}Fr$, having $t_{\frac{1}{2}} = 21.8\,min$.

We have already covered several aspects of the chemistry of the alkali metals as follows:

- ionization energies of metals (*Section 1.10*);
- structures of metal lattices (*Section 5.3*);
- metallic radii, r_{metal} (*Section 5.5*);
- melting points and standard enthalpies of atomization of metals (*Section 5.6*);
- ionic radii, r_{ion} (*Section 5.10*);
- NaCl and CsCl ionic lattices (*Section 5.11*);
- energetics of the dissolution of MX (*Section 6.9*);

- standard reduction potentials, $E^o{}_{M^+/M}$ (*Section 7.6*);
- energetics of MX transfer from water to organic solvents (*Section 8.3*);
- alkali metals in liquid NH_3 (Section 8.6);
- saline hydrides, MH (*Section 9.7*).

10.2 Occurrence, extraction and uses

Occurrence

Sodium and potassium are abundant in the Earth's biosphere (2.6% and 2.4% respectively) but do not occur naturally in the elemental state. The main sources of Na and K (see *Box 10.1*) are *rock salt* (almost pure NaCl), natural brines and seawater, *sylvite* (KCl), *sylvinite* (KCl/NaCl) and *carnallite* ($KCl\cdot MgCl_2\cdot 6H_2O$). Other Na- and K-containing minerals such as borax ($Na_2[B_4O_5(OH)_4]\cdot 8H_2O$; see *Sections 12.2* and *12.7*) and saltpetre (KNO_3, see *Section 14.2*) are commercially important sources of other elements (e.g. B and N respectively). Unlike many inorganic chemicals, NaCl need not be manufactured since large natural deposits are available. Evaporation of seawater yields a mixture of salts, but since NaCl represents the major component of the mixture, its production in this manner is a viable operation. In contrast to Na and K, natural abundances of Li, Rb and Cs are small (% abundance Rb > Li > Cs); these metals occur as various silicate minerals, e.g. *spodumene* ($LiAlSi_2O_6$).

Extraction

Sodium, economically much the most important of the alkali metals, is manufactured by the Downs process in which molten NaCl (see *Section 8.12*) is electrolyzed (scheme

BOX 10.1 RESOURCES, ENVIRONMENTAL AND BIOLOGICAL

Potassium salts: resources and commercial demand

In statistical tables of mineral production, 'potash' and 'K$_2$O equivalents' are listed, but in fact refer to soluble potassium salts. (Strictly, potash refers only to KOH.) World production of 'potash' has risen from 0.32 Mt in 1900 to >30 Mt at the end of the 20th century, with the major producers being Canada and the former Soviet Union, followed by Germany. Major industrial countries such as the US must import large amounts of 'potash' to meet commercial demands, and the graph below shows the balance of imports and home produced 'K$_2$O equivalents' of potassium salts from 1988 to 1997. About 95% of the 'potash' produced is destined to be used in the form of fertilizers.

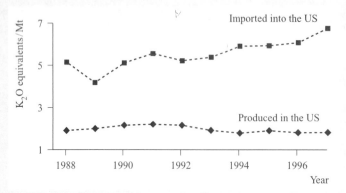

[Data: *Chemical & Engineering News* (1999), June 28 issue, p. 34 (converted to SI units)]

10.1); CaCl$_2$ is added to reduce the operating temperature to about 870 K, since pure NaCl melts at 1073 K. The design of the electrolysis cell (Figure 10.1) is critical to prevent reformation of NaCl by recombination of Na and Cl$_2$. Use of the Downs process for Cl$_2$ production is described in *Section 16.2*.

At the cathode:	$Na^+(l) + e^- \longrightarrow Na(l)$
At the anode:	$2Cl^-(l) \longrightarrow Cl_2(g) + 2e^-$
Overall reaction:	$2Na^+(l) + 2Cl^-(l) \longrightarrow 2Na(l) + Cl_2(g)$

$$(10.1)$$

Lithium is extracted from LiCl in a similar electrolytic process; LiCl is first obtained from spodumene by heating with CaO to give LiOH, which is then converted to the chloride. Potassium can be obtained electrolytically from KCl, but a more efficient method of extraction is the action of Na vapour on molten KCl in a counter-current fractionating tower. This yields a Na–K alloy which can be separated into its components by distillation. In similar fashion, Rb and Cs can be obtained from RbCl and CsCl, small quantities of which are produced as by-products from extraction of Li from spodumene.

Fig. 10.1 A schematic representation of the electrolysis cell used in the Downs process to produce sodium commercially from NaCl. The products (Na and Cl$_2$) must be kept separate from each other to prevent recombination to form NaCl.

Small amounts of Na, K, Rb and Cs can be obtained by thermal decomposition of their azides (equation 10.2); an application of NaN$_3$ is in car airbags (see *equation 14.4*). Lithium cannot be obtained from an analogous reaction because the products recombine, yielding the nitride, Li$_3$N (see *equation 10.6*).

$$2NaN_3 \xrightarrow{\text{570 K, } in \, vacuo} 2Na + 3N_2 \qquad (10.2)$$

Major uses of the alkali metals and their compounds

Lithium has the lowest density (0.53 g cm^{-3}) of all known metals. It is used in the manufacture of alloys, and in certain glasses and ceramics. Lithium carbonate has recently found application in the treatment of manic-depressive disorders, although large amounts of lithium salts damage the central nervous system. A LiCl–KCl eutectic is used in some batteries (see *Box 8.4*).

Sodium, potassium and their compounds have many uses of which selected examples are given here. Sodium–potassium alloy is used as a heat-exchange coolant in nuclear reactors. A major use of Na–Pb alloy was in the production of the anti-knock agent PbEt$_4$, but the increasing demand for unleaded fuels renders this of decreasing importance. The varied applications of compounds of Na include those in the paper, glass, detergent, chemical and metal industries; Figure 10.2 summarizes uses of NaCl and NaOH. In 1997, the total demand in the US for NaCl was ≈49.5 Mt, the

major consumption of NaCl being in the manufacture of NaOH, Cl$_2$ (see *Box 10.4*) and Na$_2$CO$_3$ (see *Section 10.7*). A large fraction of salt is used for winter road-deicing (Figure 10.2a and *Box 11.5*); however, in addition to the corrosive effects of NaCl, recent environmental concerns have focused on the side-effects on roadside vegetation and run-off into water sources, and Switzerland, for example, operates reduced-salt schemes.

Both Na and K are involved in various electrophysiological functions in higher animals. The [Na$^+$]:[K$^+$] ratio is different in intra- and extra-cellular fluids, and the concentration gradients of these ions across cell membranes are the origin of the trans-membrane potential difference that, in nerve and muscle cells, is responsible for the transmission of nerve impulses. A balanced diet therefore includes both Na$^+$ and K$^+$ salts. Potassium is also an essential plant nutrient, and K$^+$ salts are widely used as fertilizers. Uses of Li and Na in batteries are highlighted in *Box 10.3*, and the use of KO$_2$ in breathing masks is described in *Section 10.6*.

Many organic syntheses involve Li, Na or their compounds, and uses of the reagents Na[BH$_4$] and Li[AlH$_4$] are widespread. Alkali metals and some of their compounds also have uses in catalysts, e.g. the formation of MeOH from H$_2$ and CO (equation 9.10) where doping the catalyst with Cs makes it more effective.

10.3 Physical properties

General properties

The alkali metals illustrate, more clearly than any other group of elements, the influence of increase in atomic and ionic size on physical and chemical properties. Thus, the group 1 metals are often chosen to illustrate general principles. Some important physical properties of the group 1 metals are given in Table 10.1. Some important points arising from these data are listed below; see *Section 6.9* for detailed discussion of the energetics of ion hydration.

- With increasing atomic number, the atoms become larger and the strength of metallic bonding (see *Section 5.8*) decreases.

- The effect of increasing size evidently outweighs that of increasing nuclear charge, since the ionization energies decrease from Li to Cs (see *Figure 1.15*). The values of IE_2 for all the alkali metals are so high that the formation of M^{2+} ions under chemically reasonable conditions is not viable.

- Values of $E^{\circ}_{\text{M}^+/\text{M}}$ are related to energy changes accompanying the processes:

$$M(s) \longrightarrow M(g) \qquad \textit{atomization}$$
$$M(g) \longrightarrow M^+(g) \qquad \textit{ionization}$$
$$M^+(g) \longrightarrow M^+(aq) \qquad \textit{hydration}$$

and down group 1, differences in these energy changes

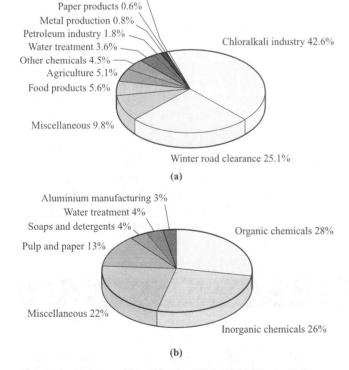

Textiles and dyeing 0.5%
Paper products 0.6%
Metal production 0.8%
Petroleum industry 1.8%
Water treatment 3.6%
Other chemicals 4.5%
Agriculture 5.1%
Food products 5.6%
Miscellaneous 9.8%
Chloralkali industry 42.6%
Winter road clearance 25.1%

(a)

Aluminium manufacturing 3%
Water treatment 4%
Soaps and detergents 4%
Pulp and paper 13%
Miscellaneous 22%
Organic chemicals 28%
Inorganic chemicals 26%

(b)

Fig. 10.2 (a) Uses of NaCl in the US in 1990 [Data: U.S. Geological Survey]; (b) industrial uses of NaOH in Western Europe in 1994. [Data: *Chemistry & Industry*, (1995) p. 832.]

Table 10.1 Some physical properties of the alkali metals, M, and their ions, M^+.

Property	Li	Na	K	Rb	Cs
Atomic number, Z	3	11	19	37	55
Ground state electronic configuration	$[He]2s^1$	$[Ne]3s^1$	$[Ar]4s^1$	$[Kr]5s^1$	$[Xe]6s^1$
Enthalpy of atomization, $\Delta_a H^o(298\,K)/kJ\,mol^{-1}$	161	108	90	82	78
Dissociation enthalpy of M−M bond in M_2 $(298\,K)/kJ\,mol^{-1}$	110	74	55	49	44
Melting point, mp / K	453.5	371	336	312	301.5
Boiling point, bp / K	1615	1156	1032	959	942
Standard enthalpy of fusion, $\Delta_{fus} H^o(mp)/kJ\,mol^{-1}$	3.0	2.6	2.3	2.2	2.1
First ionization energy, IE_1 / kJ mol^{-1}	520.2	495.8	418.8	403.0	375.7
Second ionization energy, IE_2 / kJ mol^{-1}	7298	4562	3052	2633	2234
Metallic radius, r_{metal} / pm‡	152	186	227	248	265
Ionic radius, r_{ion} / pm*	76	102	138	149	170
Standard enthalpy of hydration of M^+, $\Delta_{hyd} H^o(298\,K)/kJ\,mol^{-1}$	−519	−404	−321	−296	−271
Standard entropy of hydration of M^+, $\Delta_{hyd} S^o(298\,K)/J\,K^{-1}\,mol^{-1}$	−140	−110	−70	−70	−60
Standard free energy of hydration of M^+, $\Delta_{hyd} G^o(298\,K)/kJ\,mol^{-1}$	−477	−371	−300	−275	−253
Standard reduction potential, $E^o_{M^+/M}$ / V	−3.04	−2.71	−2.93	−2.98	−3.03
NMR active nuclei (% abundance, nuclear spin)	^6Li (7.5, $I=1$): ^7Li (92.5, $I=\frac{3}{2}$)	^{23}Na (100, $I=\frac{3}{2}$)	^{39}K (93.3, $I=\frac{3}{2}$): ^{41}K (6.7, $I=\frac{3}{2}$)	^{85}Rb (72.2, $I=\frac{5}{2}$): ^{87}Rb (27.8, $I=\frac{3}{2}$)	^{133}Cs (100, $I=\frac{7}{2}$)

‡ For 8-coordinate atom in body-centred cubic metal; compare values for 12-coordinate atoms in Appendix 6.

* For 6-coordination.

almost cancel out resulting in similar $E^o_{M^+/M}$ values. The lower reactivity of Li towards H_2O is *kinetic* rather than thermodynamic in origin; Li is a harder and higher melting metal, is less rapidly dispersed, and reacts more slowly than its heavier congeners.

In general, the chemistry of the group 1 metals is dominated by compounds containing M^+ ions. However, a small number of compounds containing the M^- ion (M = Na, K, Rb or Cs) are known (see *Section 10.8*), and the organometallic chemistry of the group 1 metals is a growing area and is described further in *Chapter 18*.

Considerations of lattice energies calculated using an electrostatic model provide a satisfactory understanding for the fact that ionic compounds are central to the chemistry of Na, K, Rb and Cs. That Li shows a so-called 'anomalous' behaviour and exhibits a *diagonal relationship* to Mg can be explained in terms of similar energetic considerations.

Atomic spectra and flame tests

In the vapour state, the alkali metals exist as atoms or M_2 molecules; the strength of the M−M covalent bond decreases down the group (Table 10.1). Excitation of the outer ns^1 electron of the M atom occurs easily and emission spectra are readily observed. We have already described the use of the *sodium D-line* in the emission spectrum of atomic Na for specific rotation measurements (see *Section 3.8*). When the salt of an alkali metal is treated with concentrated HCl (giving a volatile metal chloride) and is heated strongly in the non-luminous Bunsen flame, a characteristic flame colour is observed (Li, crimson; Na, yellow; K, lilac; Rb, red-violet; Cs, blue) and this *flame* test is used in *qualitative* analysis to identify the M^+ ion. In *quantitative* analysis, use is made of the characteristic atomic spectrum in *flame photometry* or *atomic absorption spectroscopy*.

BOX 10.2   **APPLICATIONS**

Keeping time with caesium

In 1993, the National Institute of Standards and Technology brought into use a caesium-based atomic clock which keeps international standard time to within one second in 10^6 years; the system depends upon repeated transitions from the ground to a specific excited state of atomic Cs, and the monitoring of the frequency of the electromagnetic radiation emitted.

Radioactive isotopes

In addition to the radioactivity of Fr, 0.02% of naturally occurring K consists of ^{40}K, a β-particle and positron emitter with $t_{\frac{1}{2}} = 1.26 \times 10^9$ yr. This provides the human body with a natural source of radioactivity, albeit at very low levels. Radioactive Cs isotopes from Chernobyl were described in *Box 2.2*.

NMR active nuclei

Each of the alkali metals has at least one NMR active nucleus (Table 10.1), although not all nuclei are of sufficient sensitivity to permit their routine use. For examples of NMR spectroscopy utilizing *s*-block metals, see *Section 2.11* and *worked example 18.1*.

10.4 The metals

Appearance

The metals Li, Na, K and Rb are silvery-white, but Cs has a golden-yellow cast. All are soft, Li the least so, and the trend is consistent with their melting points (Table 10.1). The particularly low melting point of Cs means that it may be a liquid at ambient temperatures in some countries.

Reactivity

We have already described the behaviour of the metals in liquid NH_3 (see *Section 8.6*).

Although Li, Na and K are stored under a hydrocarbon solvent to prevent reaction with atmospheric O_2 and water vapour, they can be handled in air, provided undue exposure is avoided; Rb and Cs should be handled in an inert atmosphere. Lithium reacts quickly with water (equation 10.3); Na reacts vigorously, and K, Rb and Cs react violently with the ignition of H_2 produced.

$$2Li + 2H_2O \longrightarrow 2LiOH + H_2 \qquad (10.3)$$

All the metals react with the halogens (equation 10.4) and H_2 when heated (equation 10.5). The energetics of metal hydride formation are essentially like those of metal halide formation, being expressed in terms of a Born–Haber cycle (see *Section 5.14*). Contrast this with the treatment for the oxides discussed in Section 10.6.

$$2M + X_2 \longrightarrow 2MX \qquad X = halogen \qquad (10.4)$$

$$2M + H_2 \longrightarrow 2MH \qquad (10.5)$$

Only Li reacts with N_2 and reaction 10.6 occurs at 298 K to give red-brown, moisture-sensitive lithium nitride. Solid Li_3N has an interesting lattice structure (Figure 10.3) and a high ionic conductivity (see *Section 27.3*). Reactions with O_2 are discussed in Section 10.6.

$$6Li + N_2 \longrightarrow 2Li_3N \qquad (10.6)$$

Acetylides, M_2C_2, are formed when Li or Na is heated with carbon; these compounds can also be prepared by treating the metal with C_2H_2 in liquid NH_3. Reactions between graphite and K, Rb or Cs lead to the formation of

---

BOX 10.3   **APPLICATIONS**

Alkali metal ion batteries

The sodium/sulfur battery operates around 570–620 K and consists of a molten sodium anode and a liquid sulfur cathode separated by a solid β-alumina electrolyte (see *Section 27.3*). The cell reaction is:

$$2Na(l) + nS(l) \longrightarrow Na_2S_n(l) \qquad E_{cell} = 2.0\,V$$

and this is reversed when the battery is recharged by changing the polarity of the cell. Trials with the sodium/sulfur battery (or *Beta battery*) in battery-powered cars are promising, but further development for the commercial market is still needed (see *Box 8.4*).

Several properties of lithium, including its highly negative reduction potential, make it suitable for battery use. For example, the lithium/iron sulfide battery contains a lithium anode and FeS_2 cathode ($E_{cell} = 1.5\,V$) and finds use in cameras. A recent advancement has been the development of solid-state lithium-containing batteries; the lithium-ion battery has a cell potential of 3.6 V and consists of a positive $LiCoO_2$ electrode separated from a graphite electrode by a solid electrolyte across which Li^+ ions can migrate when the cell is charging. The Li^+ ions are *intercalated* by the graphite (see *Section 13.4*) and return to the $LiCoO_2$

electrode when the cell is discharged. The cell reaction can be represented as follows:

$$LiCoO_2 + 6C(graphite) \underset{discharge}{\overset{charge}{\rightleftharpoons}} LiC_6 + CoO_2$$

The crucial factor in this battery is that both electrodes are able to as act as *hosts* for the Li^+ ions, and the system has been termed a 'rocking-chair' cell to reflect the fact that the Li^+ ions 'rock' back and forth between the two host materials during charging and discharging. Lithium-ion batteries have application in, for example, laptop and notebook computers and have potential use in electric cars.

Further reading

P.G. Bruce (1997) *Chemical Communications*, p. 1817 – 'Solid-state chemistry of lithium power sources'.

J.R. Owen (1997) *Chemical Society Reviews*, vol. 26, p. 259 – 'Rechargeable lithium batteries'.

C.A. Vincent (1996) *Chemistry & Industry*, p. 688 – 'Recent developments in battery technology'.

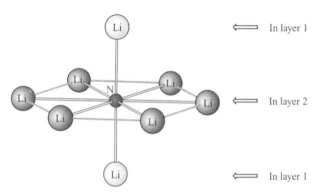

⇐ In layer 1

⇐ In layer 2

⇐ In layer 1

Fig. 10.3 The solid state structure of Li_3N consists of layers of N^{3-} and Li^+ (blue) ions (ratio 1 : 2) alternating with layers of Li^+ (pale grey) ions; the latter are arranged such that they lie over the N^{3-} ions. Each N centre is in a hexagonal bipyramidal (8-coordinate) environment; there are two types of Li^+ ion, those in layer 1 are 2-coordinate, and those in layer 2 are 3-coordinate with respect to the N centres (see *problem 10.12* at the end of the chapter).

intercalation compounds, C_nM ($n = 8, 24, 36, 48, 60$) which are discussed further in *Section 13.4*.

The alkali metals dissolve in Hg to give amalgams (see *Box 22.3*). Sodium amalgam (which is a liquid only when the percentage of Na is low) is a useful reducing agent in inorganic and organic chemistry; it can be used in aqueous media because there is a large overpotential for the discharge of H_2.

10.5 Halides

The MX halides (see *Chapter 5* for structures) are prepared by direct combination of the elements (equation 10.4) and all the halides have large negative $\Delta_f H^\circ$ values. However, Table 10.2 shows that for X = F, values of $\Delta_f H^\circ(MX)$ become *less negative* down the group, while the reverse

trend is true for X = Cl, Br and I. For a given metal, $\Delta_f H^\circ(MX)$ always becomes less negative on going from MF to MI. These generalizations can be explained in terms of a Born–Haber cycle. Consider the formation of MX (equation 10.7) and refer to Figure 5.24.

$$\Delta_f H^\circ(MX)$$
$$= \underbrace{\{\Delta_a H^\circ(M) + IE_1(M)\}}_{\text{metal-dependent term}} + \underbrace{\{\tfrac{1}{2} D(X_2) + \Delta_{EA} H(X)\}}_{\text{halide-dependent term}}$$
$$+ \Delta_{\text{lattice}} H^\circ(MX) \tag{10.7}$$

For MF, the variable quantities are $\Delta_a H^\circ(M)$, $IE_1(M)$ and $\Delta_{\text{lattice}} H^\circ(MF)$, and similarly for each of MCl, MBr and MI. The sum of $\Delta_a H^\circ(M)$ and $IE_1(M)$ gives for the formation of Li^+ 681, of Na^+ 604, of K^+ 509, of Rb^+ 485 and of Cs^+ 454 kJ mol^{-1}. For the fluorides, the trend in the values of $\Delta_f H^\circ(MF)$ depends on the relative values of $\{\Delta_a H^\circ(M) + IE_1(M)\}$ and $\Delta_{\text{lattice}} H^\circ(MF)$ (Table 10.2), and similarly for chlorides, bromides and iodides. Inspection of the data shows that the variation in $\{\Delta_a H^\circ(M) + IE_1(M)\}$ is *less* than the variation in $\Delta_{\text{lattice}} H^\circ(MF)$, but *greater* than the variation in $\Delta_{\text{lattice}} H^\circ(MX)$ for X = Cl, Br and I. This is because lattice energy is proportional to $1/(r_+ + r_-)$ (see *Section 5.13*) and so variation in $\Delta_{\text{lattice}} H^\circ(MX)$ for a given halide is greatest when r_- is smallest (for F^-) and least when r_- is largest (for I^-). Considering the halides of a given metal (equation 10.7), the small change in the term $\{\tfrac{1}{2} D(X_2) + \Delta_{EA} H(X)\}$ ($-249, -228, -213, -188$ kJ mol^{-1} for F, Cl, Br, I respectively) is outweighed by the decrease in $\Delta_{\text{lattice}} H^\circ(MX)$. In Table 10.2, note that the *difference* between the values of $\Delta_f H^\circ(MF)$ and $\Delta_f H^\circ(MI)$ *decreases* significantly as the size of the M^+ ion *increases*.

The solubilities of the alkali metal halides in water are determined by a delicate balance between lattice energies and free energies of hydration ($\Delta_{\text{sol}} G^\circ$ and $\Delta_{\text{hyd}} G^\circ$, see

Table 10.2 Standard enthalpies of formation ($\Delta_f H^\circ$) and lattice energies ($\Delta_{\text{lattice}} H^\circ$) of alkali metal halides, MX.

M	$\Delta_f H^\circ(MX)$ / kJ mol^{-1}				$\Delta_{\text{lattice}} H^\circ(MX)$ / kJ mol^{-1}			
	Halide ion size increases →				Halide ion size increases →			
	F	Cl	Br	I	F	Cl	Br	I
Li	−616	−409	−351	−270	−1030	−834	−788	−730
Na	−577	−411	−361	−288	−910	−769	−732	−682
K	−567	−436	−394	−328	−808	−701	−671	−632
Rb	−558	−435	−395	−334	−774	−680	−651	−617
Cs	−553	−443	−406	−347	−744	−657	−632	−600

Metal ion size increases ↓

Section 6.9); LiF, with a very high lattice energy, is only sparingly soluble, but solubility relationships among the other halides call for detailed discussion beyond the scope of this book.[†] The salts LiCl, LiBr, LiI and NaI are soluble in some oxygen-containing organic solvents, e.g. LiCl dissolves in THF and MeOH; complexation of the Li^+ or Na^+ ion by the *O*-donor solvents is likely in all cases (see *Section 10.8*). Both LiI and NaI are very soluble in liquid NH_3, forming complexes; the unstable complex $[Na(NH_3)_4]I$ has been isolated and contains a tetrahedrally coordinated Na^+ ion.

In the vapour state, alkali metal halides are present mainly as ion-pairs, but measurements of $M-X$ bond distances and electric dipole moments suggest that covalent contributions to the bonding, particularly in the lithium halides, are important.

10.6 Oxides and hydroxides

Oxides, peroxides, superoxides, suboxides and ozonides

When the group 1 metals are heated in an excess of air or in O_2, the principal products obtained depend on the metal: lithium *oxide*, Li_2O (equation 10.8), sodium *peroxide*, Na_2O_2 (equation 10.9), and the *superoxides* KO_2, RbO_2 and CsO_2 (equation 10.10).

$$4Li + O_2 \longrightarrow 2Li_2O \qquad \textit{oxide formation} \qquad (10.8)$$

$$2Na + O_2 \longrightarrow Na_2O_2 \qquad \textit{peroxide formation} \qquad (10.9)$$

$$K + O_2 \longrightarrow KO_2 \qquad \textit{superoxide formation} \qquad (10.10)$$

The oxides Na_2O, K_2O, Rb_2O and Cs_2O can be obtained impure by using a limited air supply, but are better prepared by thermal decomposition of the peroxides or superoxides. The colours of the oxides vary from white to orange; Li_2O and Na_2O form white crystals while K_2O is pale yellow, Rb_2O yellow and Cs_2O orange. A peroxide of lithium can be obtained by the action of H_2O_2 on an ethanolic solution of LiOH, but it decomposes on heating. Sodium peroxide (widely used as an oxidizing agent) is manufactured by heating Na metal on Al trays in air; when pure, Na_2O_2 is colourless and the faint yellow colour usually observed is due to the presence of small amounts of NaO_2. The superoxides and peroxides contain the paramagnetic $[O_2]^-$ and diamagnetic $[O_2]^{2-}$ ions respectively (see *problem 10.13* at the end of the chapter). Superoxides have magnetic moments of $\approx 1.73\mu_B$ consistent with one unpaired electron.

Partial oxidation of Rb and Cs at low temperatures yields *suboxides* such as Rb_6O, Rb_9O_2, Cs_7O and $Cs_{11}O_3$. Their structures consist of octahedral units of metal ions with the oxygen residing at the centre; the octahedra are fused

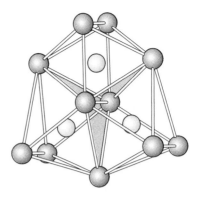

Fig. 10.4 The structure of the suboxide $Cs_{11}O_3$ consists of three oxygen-centred, face-sharing octahedral units; the shared faces are shaded in blue. Colour code: Cs, blue; O, pale grey.

together by sharing faces (Figure 10.4). The formulae of the suboxides are misleading in terms of the oxidation states. Each contains M^+ and O^{2-} ions, and, for example, the formula of Rb_6O is better written as $(Rb^+)_6(O^{2-})\cdot 4e^-$, indicating the presence of free electrons.

The alkali metal oxides, peroxides and superoxides react with water according to equations 10.11–10.13. One use of KO_2 is in breathing masks where it absorbs H_2O producing O_2 for respiration, and KOH which absorbs exhaled CO_2 (reaction 10.14).

$$M_2O + H_2O \longrightarrow 2MOH \qquad (10.11)$$

$$M_2O_2 + 2H_2O \longrightarrow 2MOH + H_2O_2 \qquad (10.12)$$

$$2MO_2 + 2H_2O \longrightarrow 2MOH + H_2O_2 + O_2 \qquad (10.13)$$

$$KOH + CO_2 \longrightarrow KHCO_3 \qquad (10.14)$$

Sodium peroxide reacts with CO_2 to give Na_2CO_3, rendering it suitable for use in air purification in confined spaces (e.g. in submarines); KO_2 acts similarly but more effectively.

The thermal stabilities of the group 1 peroxides with respect to decomposition to the oxide (equation 10.15) depend on cation size; Li_2O_2 is the least stable peroxide, while Cs_2O_2 is the most stable. The stabilities of the superoxides (with respect to decomposition to M_2O_2 and O_2) follow a similar trend.

$$M_2O_2(s) \longrightarrow M_2O(s) + \tfrac{1}{2}O_2(g) \qquad (10.15)$$

Ozonides, MO_3, containing the paramagnetic, bent $[O_3]^-$ ion (see *Section 15.4*), are known for all the alkali metals. The salts KO_3, RbO_3 and CsO_3 can be prepared from the peroxides or superoxides by reaction with ozone, but this method fails, or gives low yields, for LiO_3 and NaO_3. These ozonides have recently been prepared in liquid ammonia by the interaction of CsO_3 with an ion-exchange resin loaded with either Li^+ or Na^+ ions. The ozonides are violently explosive.

An *ion-exchange resin* consists of a solid phase (e.g. a zeolite) which contains acidic or basic groups which may exchange with cations or anions, respectively, from solutions washed through the resin; an important application is in water purification (see *Box 15.3*).

[†] For further discussion, see: W. E. Dasent (1984) *Inorganic Energetics*, 2nd edn, Cambridge University Press, Cambridge, Chapter 5.

BOX 10.4   **APPLICATIONS**

The chloralkali industry

The *chloralkali industry* produces huge quantities of NaOH and Cl_2 by the electrolysis of *aqueous* NaCl (brine).

At the anode: $2Cl^-(aq) \longrightarrow Cl_2(g) + 2e^-$

At the cathode: $2H_2O(l) + 2e^- \longrightarrow 2[OH]^-(aq) + H_2(g)$

Three types of electrolysis cell are available:

- the mercury cell, which employs a mercury cathode;
- the diaphragm cell, which uses an asbestos diaphragm separating the steel cathode and graphite or platinum-coated titanium anode;
- the membrane cell in which a cation-exchange membrane, with high permeability to Na^+ ions and low permeability to Cl^- and $[OH]^-$ ions, is placed between the anode and cathode.

On environmental grounds, the chloralkali industry is being pressured to replace mercury and diaphragm cells by the membrane cell. This is not the only environmental concern facing the industry; demand for Cl_2 has fallen in the pulp and paper industry and in the production of chlorofluorocarbons, the latter being phased out as a result of the *Montreal Protocol for the Protection of the Ozone Layer*. Nevertheless, overall demand for Cl_2 remains high, much being used in the production of chloroethene (polyvinylchloride, PVC). Uses of Cl_2 are summarized in *Figure 16.1*.

Aqueous NaOH from the electrolytic process is evaporated to give solid NaOH (caustic soda) as a white, translucent solid which is fused and cast into sticks, or made into flakes or pellets. Uses of NaOH are summarized in Figure 10.2b.

The chloralkali industry illustrates an interesting market problem. While the electrolysis of brine produces NaOH and Cl_2 in a *fixed molar ratio*, the markets for the two chemicals are different and unrelated. Interestingly, prices of the two chemicals follow opposite trends; in times of recession, demand for Cl_2 falls more sharply than that of NaOH with the result that the price of Cl_2 falls as stocks build up. Conversely, industrial demand for Cl_2 increases faster than that of NaOH when the economy is strong; consequently, the price of the alkali falls as stocks increase. The net result is clearly important to the long term stability of the chloralkali industry as a whole.

Further reading

N. Botha (1995) *Chemistry & Industry*, p. 832 – 'The outlook for the world chloralkali industry'.

Hydroxides

In 1995, NaOH ranked eighth among chemicals manufactured in the US (see *Box 10.4*); NaOH is used throughout organic and inorganic chemistry wherever a cheap alkali is needed, and industrial uses are summarized in Figure 10.2b. Solid NaOH (mp 591 K) is often handled as flakes or pellets, and dissolves in water with considerable evolution of heat. Potassium hydroxide (mp 633 K) closely resembles NaOH in preparation and properties. It is more soluble in EtOH, in which it produces a low concentration of ethoxide ions (equation 10.16); this gives rise to the use of *ethanolic* KOH in organic synthesis.

$$C_2H_5OH + [OH]^- \rightleftharpoons [C_2H_5O]^- + H_2O \qquad (10.16)$$

The crystal structures of the group 1 hydroxides are usually complicated, but the high temperature form of KOH has the NaCl lattice, with the $[OH]^-$ ions undergoing rotation rendering them pseudo-spherical.

The reactions of alkali metal hydroxides (see *Section 6.4*) with acids and acidic oxides call for no special mention. However, reactions with CO are of interest since they give metal formates (methanoates), e.g. reaction 10.17.

$$NaOH + CO \xrightarrow{450\,K} HCO_2Na \qquad (10.17)$$

Many non-metals disproportionate when treated with aqueous alkali: P_4 gives PH_3 and $[H_3PO_2]^-$, S_8 gives S^{2-} and a mixture of oxoanions, and Cl_2 reacts to give Cl^- and $[OCl]^-$ or $[ClO_3]^-$ (see also *Section 16.8*). Non-metals that do not form stable hydrides, and amphoteric metals, react with aqueous MOH to yield H_2 and oxoanions, e.g. reaction 10.18.

$$2Al + 2NaOH + 6H_2O \longrightarrow 2Na[Al(OH)_4] + 3H_2 \qquad (10.18)$$

10.7 Salts of oxoacids: carbonates and hydrogencarbonates

The properties of alkali metal salts of most oxoacids depend on the oxoanion present and not on the cation; thus we tend to discuss oxoacids under the appropriate acid. However, we single out the carbonates and hydrogencarbonates because of their importance.

In many countries, sodium carbonate (soda ash) and sodium hydrogencarbonate (commonly called sodium bicarbonate) are manufactured by the Solvay process (Figure 10.5), but this is being superseded where natural sources of the mineral *trona*, $Na_2CO_3 \cdot NaHCO_3 \cdot 2H_2O$, are available (e.g. in the US). Figure 10.5 shows that NH_3 can be recycled, but most waste $CaCl_2$ is dumped (e.g. into the sea) or used in winter road-clearance (see *Box 11.5*). Sodium carbonate is used in the chemical, glass, pulp and paper industries, in the manufacture of detergents and for water-treatment; in 1997, the total consumption in the US was ≈6.5 Mt. Sodium hydrogencarbo-

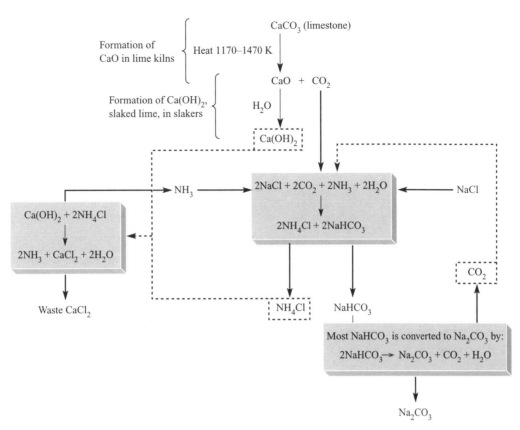

Fig. 10.5 Schematic representation of the Solvay process for the manufacture of Na_2CO_3 and $NaHCO_3$, from $CaCO_3$, NH_3 and NaCl. The recycling parts of the process are shown with blue, broken lines.

nate, although a direct product in the Solvay process, is also manufactured by passing CO_2 through aqueous Na_2CO_3 or by dissolving trona in H_2O saturated with CO_2. Its uses include those as a foaming agent, a food additive (e.g. baking powder) and an effervescent in pharmaceutical products. The Solvay company has now developed a process for using $NaHCO_3$ in pollution control, e.g. by neutralizing SO_2 or HCl in industrial and other waste emissions.

There are some notable differences between Na^+ and other alkali metal $[CO_3]^{2-}$ and $[HCO_3]^-$ salts. Whereas $NaHCO_3$ can be separated in the Solvay process by *precipitation*, the same is not true of $KHCO_3$. Hence, K_2CO_3 is produced, not via $KHCO_3$, but by the reaction of KOH with CO_2; K_2CO_3 has uses in the manufacture of certain glasses and ceramics. Among its applications, $KHCO_3$ is used as a buffering agent in water treatment and wine production. Lithium carbonate (see also *Section 10.2*) is only sparingly soluble in water; '$LiHCO_3$' has not been isolated. The thermal stabilities of the group 1 metal carbonates with respect to reaction 10.19, increase down the group as r_{M^+} increases, lattice energy being a crucial factor. Such a trend in stability is common to all series of oxo-salts of the alkali metals.

$$M_2CO_3 \xrightarrow{\Delta} M_2O + CO_2 \qquad (10.19)$$

The solid state structures of $NaHCO_3$ and $KHCO_3$ exhibit hydrogen bonding (see *Section 9.6*). In $KHCO_3$, the anions

associate in pairs (Figure 10.6a) whereas in $NaHCO_3$, infinite chains are present (Figure 10.6b). In each case, the hydrogen bonds are asymmetrical. Sodium silicates are also of great commercial importance: see *Sections 13.2* and *13.9*.

10.8 Aqueous solution chemistry

Hydrated ions

We introduced hydrated alkali metal cations in *Sections 6.7* and *6.9*. Many Li^+ salts containing small anions (e.g. LiF, Li_2CO_3) are sparingly soluble in water, but for large anions, the Li^+ salts are soluble while the K^+, Rb^+ and Cs^+ salts are sparingly soluble (e.g. $MClO_4$, $M_2[PtCl_6]$ for M = K, Rb or Cs). In *dilute* solutions, alkali metal ions rarely form complexes, but where these are formed, e.g. with $[P_2O_7]^{4-}$ and $[EDTA]^{4-}$ (see *Table 6.7*), the normal order of stability constants is $Li^+ > Na^+ > K^+ > Rb^+ > Cs^+$. In contrast, when the aqueous ions are adsorbed on an *ion-exchange resin*, the order of the strength of adsorption is usually $Li^+ < Na^+ < K^+ < Rb^+ < Cs^+$, suggesting that the *hydrated ions* are adsorbed, since hydration energies decrease along this series and the total

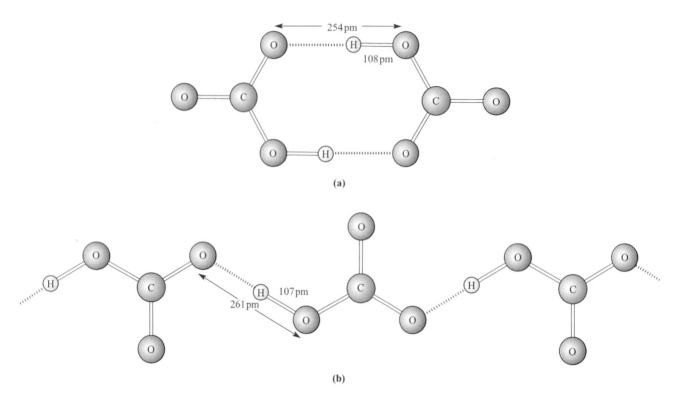

Fig. 10.6 In the solid state, hydrogen bonding results in anion association in $NaHCO_3$ and $KHCO_3$, and the formation of (a) dimers in $NaHCO_3$ and (b) infinite chains in $KHCO_3$.

interaction (i.e. primary hydration plus secondary interaction with more water molecules) is greatest for Li.

Complex ions

Unlike simple inorganic ligands, *polyethers* and, in particular, *cyclic polyethers* complex alkali metal ions quite strongly. The *crown ethers* are cyclic ethers which include 1,4,7,10,13,16-hexaoxacyclooctadecane (Figure 10.7a), the common name for which is 18-crown-6; this nomenclature gives the total number (C + O) and number of O atoms in the ring. Figure 10.7b shows the structure of the [K(18-crown-6)]⁺ cation; the K⁺ ion is coordinated by the six *O*-donors. The radius of the cavity† inside the 18-crown-6 ring is 140 pm, and this compares with values of r_{ion} for the alkali metal ions ranging from 76 pm for Li⁺ to 170 pm for Cs⁺ (Table 10.1). The radius of the K⁺ ion (138 pm) is well-matched to that of the macrocycle, and stability constants for the formation of [M(18-crown-6)]⁺ (equation 10.20) in acetone follow the sequence K⁺ > Rb⁺ > Cs⁺ ≈ Na⁺ > Li⁺.

$$M^+ + \text{18-crown-6} \rightleftharpoons [M(\text{18-crown-6})]^+ \qquad (10.20)$$

Different crown ethers have different cavity sizes (e.g. the radii of 15-crown-5 and 21-crown-7 cavities are 90 and

170 pm respectively), and selective complexation of M⁺ ions is possible by cavity-ion radius matching. Complexes formed by such macrocyclic ligands are appreciably more stable than those formed by closely related open chain ligands (see *Section 6.12*).

The crown ether-complexed alkali metal ions are large and hydrophobic, and their salts tend to be soluble in *organic* solvents. For example, whereas $KMnO_4$ is water-soluble but insoluble in benzene, [K(18-crown-6)][MnO_4] is soluble in benzene; mixing benzene with aqueous $KMnO_4$ leads to the purple colour being transferred from the aqueous to benzene layer. This phenomenon is very useful in preparative organic chemistry, the *anions* being little solvated and, therefore, highly reactive.

> A *cryptand* is a polycyclic ligand containing a cavity; when the ligand coordinates to a metal ion, the complex ion is called a *cryptate*.

Figure 10.7c shows the structure of the *cryptand* ligand 4,7,13,16,21,24-hexaoxa-1,10-diazabicyclo[8.8.8]hexacosane, commonly called cryptand-222 or crypt-222, where the 222 notation gives the number of *O*-donor atoms in each of the three chains. Cryptand-222 is an example of a *bicyclic* ligand which can *encapsulate* an alkali metal ion. Cryptands protect the complexed metal cation even more effectively than do crown ethers. They show selective coordination behaviour; cryptands-211, -221 and -222 with cavity radii of 80, 110 and 140 pm respectively, form their most stable

---

† The concept of 'cavity size' is not as simple as it may appear; for further discussion, see the suggested reading list under 'macrocyclic effect' at the end of the chapter.

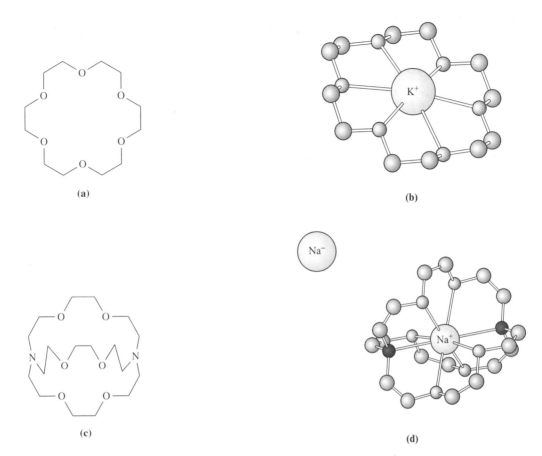

Fig. 10.7 The structures of (a) the macrocyclic polyether 18-crown-6, (b) the [K(18-crown-6)]$^+$ cation for the [Ph$_3$Sn]$^-$ salt (X-ray diffraction) [T. Birchall *et al.* (1988) *J. Chem. Soc., Chem. Commun.*, p. 877], (c) the cryptand ligand crypt-[222], and (d) [Na(crypt-[222])]$^+$Na$^-$ (X-ray diffraction) [F.J. Tehan *et al.* (1974) *J. Am. Chem. Soc.*, vol. 96, p. 7203].

alkali metal complexes with Li$^+$, Na$^+$ and K$^+$ respectively (see *Table 10.1* for r_{ion}).

$$2Na \rightleftharpoons Na^+ + Na^- \qquad (10.21)$$

The ability of crypt-222 to shift equilibrium 10.21 to the right-hand side is striking. This is observed when crypt-222 is added to Na dissolved in ethylamine, and the isolated product is the diamagnetic, golden-yellow [Na(crypt-222)]$^+$Na$^-$ (Figure 10.7d). The solid state structure indicates that the effective radius of the *sodide* ion is $\approx$230 pm, i.e. Na$^-$ is similar in size to I$^-$. Similar *alkalides* have been prepared containing K$^-$, Rb$^-$ and Cs$^-$. In these reactions, the cryptand : metal molar ratio is 1 : 2. If the reaction is carried out using a greater proportion of ligand, paramagnetic black *electrides* can be isolated, e.g. [Cs(crypt-222)$_2$]$^+$e$^-$ in which the electron is trapped in a cavity of radius $\approx$240 pm. Electrides can also be prepared using crown ethers, and examples of crystallographically confirmed complexes are [Cs(15-crown-5)$_2$]$^+$e$^-$, [Cs(18-crown-6)$_2$]$^+$e$^-$ and [Cs(18-crown-6)(15-crown-5)]$^+$e$^-$·18-crown-6. The arrangement of the electron-containing cavities in the solid state has a profound effect on the electrical conductivities of these materials; the conductivity of [Cs(18-crown-6)(15-crown-5)]$^+$e$^-$·18-crown-6 (in which the electron-cavities

form rings) is $\approx$10^6 times greater than that of either [Cs(15-crown-5)$_2$]$^+$e$^-$ or [Cs(18-crown-6)$_2$]$^+$e$^-$ (in which the free electron-cavities are organized in chains).

Cryptands have also been used to isolate crystalline LiO$_3$ and NaO$_3$ as [Li(crypt-211)][O$_3$] and [Na(crypt-222)][O$_3$] respectively, and further applications of these encapsulating ligands are in the isolation of alkali metal salts of *Zintl ions* (see *Sections 8.6* and *13.7*). Sodium and potassium cryptates are interesting models for biologically occurring materials (such as the polypeptide valinomycin, Figure 10.8a) involved in the transfer of Na$^+$ and K$^+$ across cell membranes. Figure 10.8b shows the structure of [K(valinomycin)]$^+$ and illustrates the way in which the ligand can adopt a conformation so as to wrap itself around the K$^+$ ion.

10.9 Non-aqueous coordination chemistry

A growing number of complexes (generally air- and moisture-sensitive) involving alkali metal ions with *O*- or *N*-donor ligands and formed in non-aqueous media are now

BOX 10.5 APPLICATIONS

Large cations for large anions 1

Alkali metal ions encapsulated within crown ether or cryptand ligands are often used as a source of 'large cations' to aid the crystallization of salts containing large anions. An example is the compound $[K(crypt-222)]_2[C_{60}] \cdot 4C_6H_5Me$ which contains the fulleride $[C_{60}]^{2-}$. The space filling diagram shows part of the packing diagram of $[K(crypt-222)]_2[C_{60}] \cdot 4C_6H_5Me$; solvent molecules have been removed for clarity. The $[K(crypt-222)]^+$ cations have similar overall dimensions to the fulleride dianions, allowing the ions to pack efficiently in the crystal lattice.

[Data from: T.F. Fassler *et al.* (1997) *Angew. Chem., Int. Ed. Engl.*, vol. 36, p. 486.]

See also: *Box 23.2 – Large cations for large anions: 2.*

known, although the chemistry of the later group 1 metals is not so widely developed as that of Li. A general method of synthesis is to prepare an alkali metal salt in the presence of a coordinating ligand. For example, in $[\{LiCl(HMPA)\}_4]$, use of the bulky ligand HMPA (hexamethylphosphoramide)

10.1 results in the isolation of a discrete complex rather than an extended LiCl lattice; the complex contains the cubic Li_4Cl_4 core shown in **10.2**. Increasing the size of the halogen tends to reduce the nuclearity of the product, e.g. $[Li_2Br_2(HMPA)_3]$, **10.3**. The bonding in these complexes is

R = CHMe₂

(a)

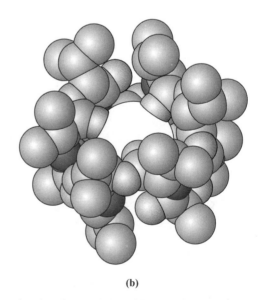

(b)

Fig. 10.8 (a) The structure of valinomycin and (b) a space-filling diagram showing the structure of $[K(valinomycin)]^+$, determined by X-ray diffraction of the salt $[K(valinomycin)]_2[I_3][I_5]$; H atoms are omitted for clarity [K. Neupert-Laves *et al.* (1975) *Helv. Chim. Acta*, vol. 58, p. 432]. Colour code: O, blue; N black; C, grey; K^+ ion, white.

of interest; **10.2** can be viewed in terms of a central aggregate of Li$^+$ and Cl$^-$ ions, and in general, the bonding should be considered to be predominantly ionic.

(10.1)

(10.2)

R = P(NMe$_2$)$_3$

(10.3)

R = P(NMe$_2$)$_3$

Amidolithium complexes of type RR′NLi (e.g. R and R′ = alkyl, aryl, silyl) exhibit a fascinating structural diversity; as above, *bulky* amido ligands are essential for complex stabilization. Planar Li$_2$N$_2$-rings are common structural units, and these appear in a variety of laddered structures which may be polymeric or discrete molecular as in [{^{t}BuHNLi}$_8$] (Figure 10.9).

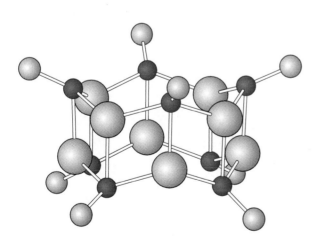

Fig. 10.9 The structure of [{LiNH^{t}Bu}$_8$] determined by X-ray diffraction; hydrogen and methyl-carbon atoms have been omitted for clarity [N.D.R. Barnett *et al.* (1996) *J. Chem. Soc., Chem. Commun.*, p. 2321]. Colour code: Li, blue; N, black; C, grey.

Summary

This chapter has described some descriptive chemistry of the group 1 (alkali) metals, and has introduced examples of complexes containing crown ether and cryptand ligands.

The following terms were introduced in this chapter.
Do you know what they mean?

☐ amalgam
☐ peroxide ion
☐ superoxide ion
☐ ozonide ion
☐ ion-exchange (ion-exchange resin)
☐ crown ether
☐ cryptand
☐ alkalide
☐ electride

Suggested further reading

W. Hesse, M. Jansen and W. Schnick (1989) *Progress in Solid State Chemistry*, vol. 19, p. 47 – A review of alkali metal oxides, peroxides, superoxides and ozonides.

F.S. Mair and R. Snaith (1994) 'Alkali metals: Inorganic chemistry' in *Encyclopedia of Inorganic Chemistry*, ed. R.B. King, Wiley, Chichester, vol. 1, p. 35 – A recent survey with a large number of references into the primary literature.

A.F. Wells (1984) *Structural Inorganic Chemistry*, 5th edn, Clarendon Press, Oxford – A well-illustrated and detailed account of the structures of alkali metal compounds.

Macrocyclic effect
These four references give excellent accounts of the macrocyclic effect:

J. Burgess (1988) *Ions in Solution: Basic Principles of Chemical Interactions*, Ellis Horwood, Chichester, Chapter 6.

E.C. Constable (1996) *Metals and Ligand Reactivity*, revised edn, VCH, Weinheim, Chapter 6.

E.C. Constable (1999) *Coordination Chemistry of Macrocyclic Compounds*, Oxford University Press, Oxford, Chapter 5.

L.F. Lindoy (1989) *The Chemistry of Macrocyclic Ligand Complexes*, Cambridge University Press, Cambridge, Chapter 6.

Alkalides and electrides
J.L. Dye (1977) *Journal of Chemical Education*, vol. 54, p. 333 – 'The formation of Na$^-$'.

J.L. Dye (1979) *Angewandte Chemie International Edition in English*, vol. 18, p. 587 – 'Compounds of alkali metal anions'.

M.J. Wagner, R.H. Huang, J.L. Eglin and J.L. Dye (1994) *Nature*, vol. 368, p. 726 – Report of the electride [Cs(18-crown-6)(15-crown-5)]$^+$e$^-$·18-crown-6, its structure and electrical conductivity, with references to other work in the area.

Problems

10.1 (a) Write down, in order, the names and symbols of the metals in group 1; check your answer by reference to the first page of this chapter. (b) Give a *general* notation that shows the ground state electronic configuration of each metal.

10.2 Explain why, for a given alkali metal, the second ionization energy is very much higher than the second.

10.3 Describe the solid state structures of (a) the alkali metals and (b) the alkali metal chlorides, and comment on trends down the group.

10.4 Discuss trends in (a) melting points, and (b) ionic radii, r_+, for the metals on descending group 1.

10.5 (a) Describe the bonding in the M_2 diatomics (M = Li, Na, K, Rb, Cs) in terms of valence bond and molecular orbital theories. (b) Account for the trend in metal–metal bond dissociation energies given in Table 10.1.

10.6 (a) Write an equation for the decay of ^{40}K by loss of a positron. (b) Determine the volume of gas produced when 1 g of ^{40}K decays according to this equation. (c) The decay of ^{40}K is the basis of a method for dating rock samples. Suggest how this method works.

10.7 Comment on the following observations:

(a) Li is the only alkali metal to form a stable nitride.
(b) The mobilities of the alkali metal ions in aqueous solution follow the sequence
$Li^+ < Na^+ < K^+ < Rb^+ < Cs^+$.
(c) E° for $M^+(aq) + e^- \rightleftharpoons M(s)$ is nearly constant (see *Table 10.1*) for the alkali metals.

10.8 Suggest what will happen when a mixture of LiI and NaF is heated.

10.9 Very often, samples for IR spectroscopy are prepared as solid state discs by grinding the compound for analysis with an alkali metal halide. Suggest why the IR spectra of $K_2[PtCl_4]$ in KBr and KI discs might be different.

10.10 Suggest why KF is a better reagent than NaF for replacement of chlorine in organic compounds by fluorine by the autoclave reaction:

10.11 Suggest why the solubility of sodium sulfate in water increases to 305 K and then decreases.

10.12 By considering Figure 10.3 and the packing of the units shown into an infinite lattice, show that (a) the ratio of $Li^+:N^{3-}$ ions in layer 2 is 2:1, and (b) the stoichiometry of the compound is Li_3N.

10.13 Construct approximate MO diagrams for $[O_2]^-$ and $[O_2]^{2-}$ and confirm that $[O_2]^-$ is paramagnetic, while $[O_2]^{2-}$ is diamagnetic.

10.14 What general type of reaction is equilibrium 10.21? Confirm your answer by considering the oxidation state changes involved. Give two other examples of this general type of reaction.

10.15 Write down the formulae of the following ions: (a) superoxide; (b) peroxide; (c) ozonide; (d) azide; (e) nitride; (f) sodide.

10.16 Write a brief account of the uses of the alkali metals and their compounds, with reference to relevant industrial processes.

10.17 Alkali metal cyanides, MCN, are described as *pseudohalides*. (a) Draw the structure of the cyanide ion, and give a description of its bonding. (b) Interpret the structure of NaCN if it possesses an NaCl lattice.

10.18 Give an account of what happens when Na dissolves in liquid NH_3.

10.19 Write balanced equations for the following reactions:

(a) sodium hydride with water;
(b) potassium hydroxide with acetic acid;
(c) thermal decomposition of sodium azide;
(d) potassium peroxide with water;
(e) sodium fluoride with boron trifluoride;
(f) electrolysis of molten KBr;
(g) electrolysis of aqueous NaCl.

Chapter 11

The group 2 metals

TOPICS

- Occurrence, extraction and uses
- Physical properties
- The metals
- Halides

- Oxides and hydroxides
- Salts of oxoacids
- Complex ions in aqueous solution
- Complexes with amido or alkoxy ligands

1	2			13	14	15	16	17	18
H									He
Li	Be			B	C	N	O	F	Ne
Na	Mg			Al	Si	P	S	Cl	Ar
K	Ca			Ga	Ge	As	Se	Br	Kr
Rb	Sr	*d*-block		In	Sn	Sb	Te	I	Xe
Cs	Ba			Tl	Pb	Bi	Po	At	Rn
Fr	Ra								

11.1 Introduction

The relationships among the elements in group 2 – beryllium, magnesium, calcium, strontium, barium and radium – are very like those among the alkali metals. However, Be stands apart from the other group 2 metals to a *greater* extent than does Li from its homologues. For example, whereas Li^+ and Na^+ salts (with a common counterion) usually crystallize with the same lattice type, this is not true for Be(II) and Mg(II) compounds. Beryllium compounds tend either to be covalent or to contain the hydrated $[Be(H_2O)_4]^{2+}$ ion. The high values of the enthalpy of atomization (Appendix 10) and ionization energies (Appendix 8) of the Be atom, and the small size and consequent high charge density of a naked Be^{2+} ion, mitigate against the formation of naked Be^{2+}. Further, the restriction of the valence shell of Be to an octet of electrons excludes the formation of more than four localized 2c-2e bonds by a Be atom. It is noteworthy that Be is the only Group 2 metal not to form a stable complex with $[EDTA]^{4-}$ (see *Table 6.7*).

The elements Ca, Sr, Ba and Ra are collectively known as the *alkaline earth metals*. We shall have little to say about radium; it is radioactive and is formed as $^{226}_{88}Ra$ (α-emitter, $t_{\frac{1}{2}} = 1622\,yr$) in the $^{238}_{92}U$ decay series; uses of radium-226 in cancer treatment have generally been superseded by other radioisotopes. The properties of radium and its compounds can be inferred by extrapolation from those of corresponding Ca, Sr and Ba compounds.

We have already described some aspects of the chemistry of the group 2 elements as follows:

- ionization energies of metals (*Section 1.10*);
- bonding in diatomic Be_2 (*Section 1.13*);
- bonding schemes for $BeCl_2$ (*Sections 1.19* and *4.2*);
- structures of metals (*Table 5.2*);
- structures of halides and oxides, see CaF_2, CdI_2 and NaCl lattices (*Section 5.11*);
- lattice energy treatment of disproportionation of CaF into Ca and CaF_2 (*Section 5.16*);
- solubility products, e.g. for CaF_2 (*Section 6.9*);
- hydration of metal ions (*Section 6.9*);
- saline hydrides, MH_2 (*Section 9.7*).

11.2 Occurrence, extraction and uses

Occurrence

Beryllium occurs principally as the silicate mineral *beryl*, $Be_3Al_2[Si_6O_{18}]$ (silicates, see *Section 13.9*); it is also found in many natural minerals, and precious forms include *emerald* and *aquamarine*. Magnesium and calcium are the eighth and fifth most abundant elements, respectively, in the Earth's crust, and Mg, the third most abundant in the sea. The

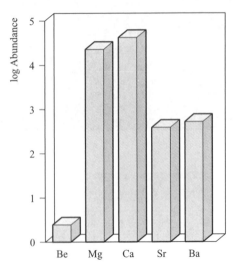

Fig. 11.1 Relative abundances in the Earth's crust of the alkaline earth metals (excluding Ra); the data are plotted on a logarithmic scale. The units of abundance are ppm.

elements Mg, Ca, Sr and Ba are widely distributed in minerals and as dissolved salts in seawater; some important minerals are *dolomite* ($CaCO_3 \cdot MgCO_3$), *magnesite* ($MgCO_3$), *olivine* (($Mg,Fe)_2SiO_4$), *carnallite* ($KCl \cdot MgCl_2 \cdot 6H_2O$), $CaCO_3$ (in the forms of *chalk*, *limestone* and *marble*), *gypsum* ($CaSO_4 \cdot 2H_2O$), *strontianite* ($SrCO_3$) and *barytes* ($BaSO_4$). The natural abundances of Be, Sr and Ba are far less than those of Mg and Ca (Figure 11.1).

Extraction

Of the group 2 metals, only Mg is manufactured on a large scale (see *Box 11.1*). Dolomite is thermally decomposed to a mixture of MgO and CaO, and MgO is reduced by ferrosilicon in Ni vessels (equation 11.1); Mg is removed by distillation *in vacuo*.

$$2MgO + 2CaO + FeSi \xrightarrow{1450\,K} 2Mg + Ca_2SiO_4 + Fe$$
(11.1)

Extraction of Mg by electrolysis of fused $MgCl_2$ is also important and is applied to the extraction of the metal from seawater. The first step is precipitation (see *Table 6.4*) of $Mg(OH)_2$ by addition of $Ca(OH)_2$ (*slaked lime*), produced from $CaCO_3$ (available as various calcareous deposits, see *Figure 10.5*). Neutralization with hydrochloric acid (equation 11.2) and evaporation of water gives $MgCl_2 \cdot xH_2O$, which, after heating at 990 K, yields the anhydrous chloride. Electrolysis of molten $MgCl_2$ and solidification of Mg completes the process (equation 11.3).

$$2HCl + Mg(OH)_2 \longrightarrow MgCl_2 + 2H_2O$$
(11.2)

At the cathode: $Mg^{2+}(l) + 2e^- \longrightarrow Mg(l)$ ⎱
At the anode: $2Cl^-(l) \longrightarrow Cl_2(g) + 2e^-$ ⎰
(11.3)

Beryllium is obtained from *beryl* by first heating with Na_2SiF_6, extracting the water-soluble BeF_2 formed, and precipitating $Be(OH)_2$. Beryllium is then produced either by reduction of BeF_2 (equation 11.4), or by electrolysis of $BeCl_2$ fused with NaCl.

$$BeF_2 + Mg \xrightarrow{1550\,K} Be + MgF_2$$
(11.4)

The production of Ca is by electrolysis of fused $CaCl_2$ and CaF_2; Sr and Ba are extracted by reduction of the corresponding oxides by Al, or by electrolysis of MCl_2 (M = Sr, Ba).

BOX 11.1   **RESOURCES, ENVIRONMENTAL AND BIOLOGICAL**

Recycling of materials: magnesium

Recycling of materials became increasingly important during the last decades of the 20th century, and will continue to have a significant influence on chemical industries. A large fraction of the total Mg consumed is in the form of Al/Mg alloys (see *Figure 11.2*), and recycling of Al cans necessarily means recovery of Mg. The graph below shows the variation in total consumption of primary Mg in the US from 1960 to 1990, and the increasing trend towards recovering the metal.

[Data from U.S. Geological Survey]

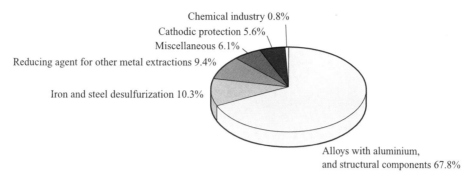

Fig. 11.2 Uses of Mg in the US in 1990 [data from U.S. Geological Survey]; for a discussion of *cathodic protection*, see *Box 7.3*.

Major uses of the group 2 metals and their compounds

Caution! Beryllium and soluble barium compounds are extremely toxic.

Beryllium is one of the lightest metals known, is non-magnetic, and has a high thermal conductivity and a very high melting point (1560 K); these properties, combined with inertness towards aerial oxidation, render it of industrial importance. It is used in the manufacture of body-parts in high-speed aircraft and missiles, and in communication satellites. Because of its low electron density, Be is a poor absorber of electromagnetic radiation and, as a result, is used in X-ray tube windows. Its high melting point and low cross-section for neutron capture (see *Section 2.4*) make Be useful in the nuclear energy industry.

Figure 11.2 summarizes the major uses of Mg. The presence of Mg in Mg/Al alloys imparts greater mechanical strength and resistance to corrosion, and improves fabrication properties; Mg/Al alloys are used in aircraft and automobile body-parts and lightweight tools. Miscellaneous uses (Figure 11.2)

include flares, fireworks and photographic flashlights, and medical applications such as indigestion powders (*milk of magnesia*, $Mg(OH)_2$) and a purgative (*Epsom salts*, $MgSO_4 \cdot 7H_2O$). Both Mg^{2+} and Ca^{2+} ions are catalysts for diphosphate–triphosphate (see *Box 14.10*) transformations in biological systems; Mg^{2+} is an essential constituent of chlorophylls in green plants (see *Section 11.8*).

Uses of compounds of calcium far outnumber those of the metal, and CaO ranked fifth among chemicals manufactured in the US in 1995; production in the US has continued to increase, reaching 22.5 Mt in 1998. Calcium oxide (quicklime or lime) is produced by calcining limestone (see *Figure 10.5*) and a major use is as a component in building mortar. Dry sand and CaO mixtures can be stored and transported; on adding water, and as CO_2 is absorbed, the mortar sets as solid $CaCO_3$ (scheme 11.5). The sand in the mortar is a binding agent.

$$\left.\begin{array}{l} CaO(s) + H_2O(l) \longrightarrow Ca(OH)_2(s) \quad \Delta_rH^\circ = -65\,kJ\,mol^{-1} \\ \text{Quicklime} \qquad\qquad\qquad\quad \text{Slaked lime} \\ Ca(OH)_2(s) + CO_2(g) \longrightarrow CaCO_3(s) + H_2O(l) \end{array}\right\}$$

$$(11.5)$$

BOX 11.2   **RESOURCES, ENVIRONMENTAL AND BIOLOGICAL**

Desulfurization processes to limit SO₂ emissions

Current awareness of the effects of environmental pollution has been instrumental in the development of *desulfurization processes*; this includes desulfurization of fossil fuels and flue gases from a variety of sources. The aim in a flue gas desulfurization process, for example, is to optimize the removal of SO_2 from emissions into the atmosphere. One important method of desulfurization in commercial operation throughout the world, is based upon the neutralization reactions between $Ca(OH)_2$ or $CaCO_3$ and sulfuric acid. Flue gases containing SO_2 are passed through absorbers containing slaked lime or limestone. The reactions occurring are:

$$SO_2 + H_2O \rightleftharpoons H^+ + [HSO_3]^-$$

$$H^+ + [HSO_3]^- + \tfrac{1}{2}O_2 \longrightarrow 2H^+ + [SO_4]^{2-}$$

$$2H^+ + [SO_4]^{2-} + Ca(OH)_2 \longrightarrow CaSO_4 \cdot 2H_2O$$

or

$$2H^+ + [SO_4]^{2-} + H_2O + CaCO_3 \longrightarrow CaSO_4 \cdot 2H_2O + CO_2$$

An advantage of the system is that $CaSO_4 \cdot 2H_2O$, *gypsum*, is non-toxic and is not a waste product; it has a number of commercial applications, for example in the production of plaster of Paris (see *Section 11.7*) and cement.

In an alternative desulfurization process, NH_3 replaces $Ca(OH)_2$ or $CaCO_3$ and the final sulfur-containing product is $[NH_4]_2[SO_4]$. Again, the sulfur is removed in the form of a commercially desirable chemical, since $[NH_4]_2[SO_4]$ has applications as a fertilizer.

For related information: see *Box 15.5*, and *Box 22.6*.

Table 11.1   Some physical properties of the alkaline earth metals, M, and their ions, M^{2+}.

Property	Be	Mg	Ca	Sr	Ba	Ra
Atomic number, Z	4	12	20	38	56	88
Ground state electronic configuration	$[He]2s^2$	$[Ne]3s^2$	$[Ar]4s^2$	$[Kr]5s^2$	$[Xe]6s^2$	$[Rn]7s^2$
Enthalpy of atomization, $\Delta_a H^{\circ}(298\,K)/kJ\,mol^{-1}$	324	146	178	164	178	130
Melting point, mp / K	1560	923	1115	1040	1000	973
Boiling point, bp / K	≈ 3040	1380	1757	1657	1913	1413
Standard enthalpy of fusion, $\Delta_{fus} H^{\circ}(mp)/kJ\,mol^{-1}$	7.9	8.5	8.5	7.4	7.1	–
First ionization energy, $IE_1/kJ\,mol^{-1}$	899.5	737.7	589.8	549.5	502.8	509.3
Second ionization energy, $IE_2/kJ\,mol^{-1}$	1757	1451	1145	1064	965.2	979.0
Third ionization energy, $IE_3/kJ\,mol^{-1}$	14850	7733	4912	4138	3619	3300
Metallic radius, $r_{metal}/pm^{\ddagger}$	112	160	197	215	224	–
Ionic radius, r_{ion}/pm^{*}	27	72	100	126	142	148
Standard enthalpy of hydration of M^{2+}, $\Delta_{hyd} H^{\circ}(298\,K)/kJ\,mol^{-1}$	−2500	−1931	−1586	−1456	−1316	–
Standard entropy of hydration of M^{2+}, $\Delta_{hyd} S^{\circ}(298\,K)/J\,K^{-1}\,mol^{-1}$	−300	−320	−230	−220	−200	–
Standard free energy of hydration of M^{2+}, $\Delta_{hyd} G^{\circ}(298\,K)/kJ\,mol^{-1}$	−2410	−1836	−1517	−1390	−1256	–
Standard reduction potential, $E^{\circ}_{M^{2+}/M}/V$	−1.85	−2.37	−2.87	−2.89	−2.90	−2.92

‡ For 12-coordinate atoms.

* For 4-coordination for Be^{2+}, and 6-coordination for other M^{2+} ions.

Other important uses of lime are in the steel industry (see *Box 5.1*), pulp and paper manufacturing, and extraction of Mg. Calcium carbonate is in huge demand in, for example, steel, glass, cement and concrete manufacturing, and the Solvay process (*Figure 10.5*). Recent applications of $CaCO_3$ and $Ca(OH)_2$ with environmental significance are in desulfurization processes (see *Box 11.2*). Large quantities of $Ca(OH)_2$ are used to manufacture bleaching powder, $Ca(OCl)_2 \cdot Ca(OH)_2 \cdot CaCl_2 \cdot 2H_2O$ (see *Sections 16.2* and *16.9*) and in water treatment (see *equation 11.28*).

Calcium fluoride occurs naturally as the mineral fluorspar, and is commercially important as the raw material for the manufacture of HF (equation 11.6) and F_2 (see *Section 16.2*). Smaller amounts of CaF_2 are used as a flux in the steel industry, for welding electrode coatings, and in glass manufacture; prisms and cell windows made from CaF_2 are used in spectrophotometers.

$$CaF_2 + 2H_2SO_4 \xrightarrow[\text{conc}]{} 2HF + Ca(HSO_4)_2 \qquad (11.6)$$

Uses of Ba as a 'getter' in vacuum tubes arise from its high reactivity with gases including O_2 and N_2.

11.3 Physical properties

General properties

Selected physical properties of the group 2 elements are listed in Table 11.1. The intense radioactivity of Ra makes it impossible to obtain all the data for this element. Some general points to note from Table 11.1 are as follows.

- The general trend in decreasing values of IE_1 and IE_2 down the group (see *Section 1.10*) is broken by the increase in going from Ba to Ra, attributed to the *thermodynamic 6s inert pair effect* (see *Box 12.3*).
- High values of IE_3 preclude the formation of M^{3+} ions.
- Quoting a value of r_{ion} for beryllium assumes that the Be^{2+} ion is present in BeF_2 and BeO, a questionable assumption.
- There are no simple explanations for the irregular group variations in properties such as melting points and $\Delta_a H^{\circ}$.
- Values of E° for the M^{2+}/M couple are fairly constant (with the exception of Be), and can be explained in a similar way as for the group 1 metals (see *Sections 7.6* and *10.3*).

Flame tests

As for the alkali metals, emission spectra for the group 2 metals are readily observed and flame tests (see *Section 10.3*) can be used to distinguish between Ca-, Sr- and Ba-containing compounds: Ca (orange-red, but pale green when viewed through blue glass), Sr (crimson, but violet through blue glass), Ba (apple-green).

Radioactive isotopes

The isotope ^{90}Sr is a β-emitter ($t_{\frac{1}{2}} = 29.1\,yr$) and a fission product of uranium. In the event of a nuclear energy plant disaster or through the dumping of nuclear waste, there is a danger that grass, and then milk, may be contaminated with ^{90}Sr and that it may be incorporated with calcium phosphate into bone.† For discussion of ^{226}Ra, see *Section 11.1*.

† For further details, see: D.C. Hoffman and G.R. Choppin (1986) *Journal of Chemical Education*, vol. 63, p. 1059 – 'Chemistry related to isolation of high-level nuclear waste'.

11.4 The metals

Appearance

Beryllium and magnesium are greyish metals, while the remaining group 2 metals are soft and silver-coloured. The metals are malleable, ductile and quite brittle; in air, the shiny surface of each metal quickly tarnishes.

Reactivity

Beryllium and magnesium are passivated (equation 11.7) and are kinetically inert to O_2 and H_2O at ambient temperatures. However, Mg *amalgam* liberates H_2 from water, since no coating of oxide forms on its surface; Mg metal reacts with steam or hot water (equation 11.8).

$$2Be + O_2 \longrightarrow \underset{\substack{\text{protective oxide} \\ \text{coating on metal}}}{2BeO} \tag{11.7}$$

$$\underset{\text{steam}}{Mg + 2H_2O} \longrightarrow Mg(OH)_2 + H_2 \tag{11.8}$$

Beryllium and magnesium dissolve readily in non-oxidizing acids; magnesium is attacked by nitric acid, whereas beryllium reacts with dilute HNO_3 but is passivated by concentrated nitric acid. Magnesium does not react with aqueous alkali, whereas Be forms an *amphoteric* hydroxide (see *Section 11.6*).

The metals Ca, Sr and Ba exhibit similar chemical behaviours, generally resembling, but being slightly less reactive than, Na. They react with water and acids liberating H_2, and the similarity with Na extends to dissolution in liquid NH_3 to give blue solutions containing solvated electrons. From these solutions, it is possible to isolate hexaammines, $[M(NH_3)_6]$ (M = Ca, Sr, Ba), but these slowly decompose to amides (equation 11.9).

$$[M(NH_3)_6] \longrightarrow M(NH_2)_2 + 4NH_3 + H_2 \qquad M = Ca, Sr, Ba \tag{11.9}$$

When heated, all the group 2 metals combine with O_2, N_2, sulfur or halogens (equations 11.10–11.13).

$$2M + O_2 \xrightarrow{\Delta} 2MO \tag{11.10}$$

$$3M + N_2 \xrightarrow{\Delta} M_3N_2 \tag{11.11}$$

$$8M + S_8 \xrightarrow{\Delta} 8MS \tag{11.12}$$

$$M + X_2 \xrightarrow{\Delta} MX_2 \qquad X = F, Cl, Br, I \tag{11.13}$$

BOX 11.3   **APPLICATIONS**

CaC$_2$: worldwide production

The overall trend in worldwide production of CaC_2 is in a downward direction. Analysis of the market explains this, in part, in terms of a switch from ethyne (which is manufactured from CaC_2) to ethene as a precursor in the organic chemical industry. However, trends in different regions of the world (shown below between 1962 and 1982) reflect differing strategies. For example, in South Africa where coal (rather than oil) reserves constitute the available raw materials, production of CaC_2 has increased. The increase in Eastern Europe seen in the 1980s is now declining, in line with Western nations.

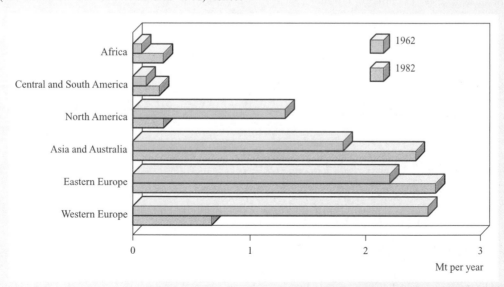

[Data from: *Ullman's Encyclopedia of Industrial Inorganic Chemicals and Products* (1998) Wiley-VCH, Weinheim.]

Differences between the first and later members of group 2 are illustrated by the formation of hydrides and carbides. When heated with H_2, Ca, Sr and Ba form saline hydrides, MH_2, but Mg reacts only under high pressure. In contrast, BeH_2 (which is polymeric, *Figure 9.11*) is prepared from beryllium alkyls (see *Section 18.3*). Beryllium combines with carbon at high temperatures to give Be_2C which possesses an antifluorite lattice (see *Section 5.11*). The other group 2 metals form carbides MC_2 which contain the $[C{\equiv}C]^{2-}$ ion, and adopt NaCl lattices that are elongated along one axis. Whereas Be_2C reacts with water according to equation 11.14, the carbides of the later metals hydrolyze to yield C_2H_2 (equation 11.15 and *Box 11.3*); CaH_2 is used as a drying agent (see *Box 11.4*) but its reaction with water is highly exothermic.

$$Be_2C + 4H_2O \longrightarrow 2Be(OH)_2 + CH_4 \qquad (11.14)$$

$$MC_2 + 2H_2O \longrightarrow M(OH)_2 + C_2H_2 \qquad M = Mg, Ca, Sr, Ba \qquad (11.15)$$

The carbide Mg_2C_3 (which contains the linear $[C_3]^{4-}$ ion, **11.1**, isoelectronic with CO_2) is formed by heating MgC_2, or by reaction of Mg dust with pentane vapour at 950 K. Reaction of Mg_2C_3 with water produces $MeC{\equiv}CH$.

$$\left[\, C{=}C{=}C \,\right]^{4-}$$

(11.1)

11.5 Halides

Beryllium halides

Anhydrous beryllium halides are covalent. The fluoride, BeF_2, is obtained as a glass (sublimation point 1073 K) from the thermal decomposition of $[NH_4]_2[BeF_4]$, itself prepared from BeO and NH_3 in an excess of aqueous HF. Molten BeF_2 is virtually a non-conductor of electricity, and the fact that solid BeF_2 adopts a β-cristobalite lattice (see *Section 5.11*) is consistent with its being a covalent solid. Beryllium difluoride is very soluble in water, the formation of $[Be(H_2O)_4]^{2+}$ (see *Section 11.8*) being thermodynamically favourable (Table 11.1).

Anhydrous $BeCl_2$ (mp 688 K, bp 793 K) can be prepared by reaction 11.16. This is a standard method of preparing a metal chloride that cannot be made by dehydration of hydrates obtained from aqueous media. In the case of Be, $[Be(H_2O)_4]^{2+}$ is formed and attempted dehydration of $[Be(H_2O)_4]Cl_2$ yields the hydroxide, not the chloride (equation 11.17).

$$2BeO + CCl_4 \xrightarrow{1070\,K} 2BeCl_2 + CO_2 \qquad (11.16)$$

$$[Be(H_2O)_4]Cl_2 \xrightarrow{\Delta} Be(OH)_2 + 2H_2O + 2HCl \qquad (11.17)$$

A *deliquescent* substance absorbs water from the surrounding air and eventually forms a liquid.

In the vapour state above 1020 K, $BeCl_2$ is monomeric and has a linear structure; at lower temperatures, the vapour also contains planar dimers. It forms colourless, deliquescent crystals containing infinite chains; the coordination environment of each Be centre is tetrahedral and the Be–Cl distances are longer than in the monomer (Figure 11.3). In Section 4.2, we described the bonding in monomeric $BeCl_2$ in terms of *sp* hybridization. In the polymer, each Be atom can be considered to be sp^3 hybridized and a localized σ-bonding scheme is appropriate in which each Cl donates a lone pair of electrons into an empty hybrid orbital on an adjacent Be atom (Figure 11.3c). The formation of this

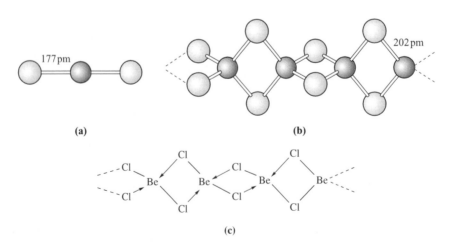

Fig. 11.3 (a) The linear structure of $BeCl_2$ in the gas phase; (b) the solid state polymeric structure of $BeCl_2$ is similar to that of BeH_2 (Figure 9.11), although the bonding in these two compounds is *not* the same; (c) in $BeCl_2$, there are sufficient valence electrons to invoke 2c-2e Be–Cl bonds. Colour code: Be, blue; Cl, pale grey.

chain demonstrates the Lewis acidity of beryllium dihalides; $BeCl_2$ acts as a Friedel–Crafts catalyst (i.e. like $AlCl_3$), and the formation of adducts is illustrated by $[BeF_4]^{2-}$, $[BeCl_4]^{2-}$ and $BeCl_2 \cdot 2L$ (L = ether, aldehyde, ketone).

Halides of Mg, Ca, Sr and Ba

The fluorides of Mg(II), Ca(II), Sr(II) and Ba(II) are ionic, have high melting points, and are sparingly soluble in water, the solubility increasing slightly with increasing cation size (K_{sp} for MgF_2, CaF_2, SrF_2 and BaF_2 = 7.42×10^{-11}, 1.46×10^{-10}, 4.33×10^{-9} and 1.84×10^{-7} respectively). Whereas MgF_2 adopts a rutile lattice (see *Figure 5.21*), CaF_2, SrF_2 and BaF_2 crystallize with the fluorite structure (*Figure 5.18*). In the vapour state, linear MgF_2 molecules are present, but CaF_2, SrF_2 and BaF_2 are bent. In contrast to the behaviour of BeF_2, none of the later metal fluorides behaves as a Lewis acid.

Magnesium chloride, bromide and iodide crystallize from aqueous solution as hydrates which undergo partial hydrolysis when heated. The anhydrous salts are, therefore, prepared by reaction 11.18.

$$Mg + X_2 \longrightarrow MgX_2 \qquad X = Cl, Br, I \qquad (11.18)$$

A *hygroscopic* solid absorbs water from the surrounding air but does not become a liquid.

Anhydrous MCl_2, MBr_2 and MI_2 (M = Ca, Sr and Ba) can be prepared by dehydration of the hydrated salts. These anhydrous halides are *hygroscopic* and $CaCl_2$ (manufactured as a by-product in the Solvay process, see *Figure 10.5*) is used as a laboratory drying agent (see *Box 11.4*) and for road deicing (see *Box 11.5*). In the solid state, many of the anhydrous halides possess complicated layer structures such as the CdI_2 lattice (*Figure 5.22*). Most of these halides are somewhat soluble in polar solvents such as ethers or pyridine, and a number of crystalline complexes have been isolated. Octahedral coordination has been confirmed by X-ray diffraction studies of complexes including *trans*-$[MgBr_2(py)_4]$, *trans*-$[MgBr_2(THF)_4]$, *cis*-$[MgBr_2(diglyme)(THF)]$ (Figure 11.4a) and *trans*-$[CaI_2(THF)_4]$. In $[MgBr_2(THF)_2]$, octahedral coordination in the solid state is achieved by the formation of a chain structure (Figure 11.4b); py = pyridine, THF = tetrahydrofuran (see *Table 6.7*). The larger sizes of the heavier metals permit higher coordination numbers, e.g. pentagonal bipyramidal *trans*-$[SrBr_2(py)_5]$, **11.2**. In organic chemistry, $MgBr_2$ is used as a catalyst for esterification reactions, and $MgBr_2 \cdot 2Et_2O$ is

Inorganic elements and compounds as drying agents

It is useful to distinguish between different classes of *drying agent* as being reagents that react with water either *reversibly* or *irreversibly*; the former can be regenerated, usually by heating, while the latter (sometimes classed as *dehydrating* agents) cannot. Caution is always needed when choosing a drying agent for the following reasons:

- the substance from which water is being removed may react with the drying agent;
- dehydrating agents often react vigorously with water and should not be used to dry very wet solvents, for which a predrying stage is appropriate;
- magnesium perchlorate, $Mg(ClO_4)_2$, although an extremely efficient drying agent, is best avoided because of the risk of explosions.

Many drying or dehydrating agents are compounds of group 1 or 2 metals; concentrated H_2SO_4, molecular sieves and silica gel (see *Section 13.2*) are also commonly used to absorb water, while P_4O_{10} (often written as 'P_2O_5') is a highly effective dehydrating agent.

Agents for drying or predrying solvents

Typically, anhydrous salts that absorb water as solvate are suitable for removing water from solvents. Anhydrous $MgSO_4$, $CaCl_2$, $CaSO_4$, Na_2SO_4 and K_2CO_3 are hygroscopic and of these, $CaSO_4$ and $MgSO_4$ are particularly efficient and inert drying agents.

Drying agents that react irreversibly with H_2O

Drying agents in this category include Ca and Mg (for alcohols), CaH_2 (for a range of solvents, but not lower alcohols nor aldehydes), $LiAlH_4$ (for hydrocarbons and ethers) and sodium. The latter, generally extruded as wire, is extremely efficient for removing water from hydrocarbons or ethers, but reacts with, for example, alcohols, and is not suitable for drying halogenated solvents.

Drying agents for use in desiccators and drying tubes

Suitable agents for drying samples in desiccators are anhydrous $CaCl_2$, $CaSO_4$, KOH and P_4O_{10}. Gases may be dried by passage through drying tubes packed with a suitable agent, but possible reaction of the gas with the drying agent must be considered. Although P_4O_{10} is a common choice for use in desiccators, reaction with water results in the formation of a brown, viscous layer on the surface of the anhydrous powder, thereby curtailing its dehydrating ability (see *Section 14.10*).

(a)

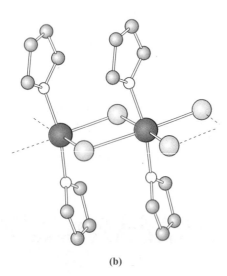

(b)

Fig. 11.4 The structures (X-ray diffraction) of (a) [MgBr$_2$(diglyme)(THF)] (diglyme = MeOCH$_2$CH$_2$OCH$_2$CH$_2$OMe) [N. Metzler *et al.* (1994) *Z. Naturforsch., Teil B*, vol. 49, p. 1448] and (b) [MgBr$_2$(THF)$_2$] [R. Sarma *et al.* (1977) *J. Am. Chem. Soc.*, vol. 99, p. 5289]; H atoms have been omitted. Colour code: Mg, dark blue; Br, pale blue; O, white; C, grey.

commercially available, being a catalyst for the conversion of aliphatic epoxides to the corresponding ketones.

(11.2)

11.6 Oxides and hydroxides

Oxides and peroxides

Beryllium oxide, BeO, is formed by ignition of Be or its compounds in O$_2$. It is an insoluble white solid which adopts a wurtzite lattice (see *Figure 5.20*). The oxides of the other group 2 metals are usually prepared by thermal decomposition of the corresponding carbonate (equation 11.19, for which temperature T refers to $P(CO)_2 = 1$ bar).

$$MCO_3 \xrightarrow{T\,K} MO + CO_2 \qquad \begin{cases} M = Mg & T = 813\,K \\ Ca & 1173\,K \\ Sr & 1563\,K \\ Ba & 1633\,K \end{cases}$$

(11.19)

Figure 11.5 shows the trend in melting points of the oxides; MgO, CaO, SrO and BaO crystallize with an NaCl lattice and the decrease in melting point reflects the decrease in lattice energy as the cation size increases (Table 11.1). The high melting point of MgO makes it suitable as a refractory material (see *Box 11.6*).

Refractory materials are suitable for use in furnace linings; such a material has a high melting point, low electrical conductivity and high thermal conductivity, and is chemically inert at the high operating temperatures of the furnace.

BOX 11.5 **APPLICATIONS**

CaCl$_2$: action against the cold and effective dust control

In 1998, production of CaCl$_2$ in the US was 0.7 Mt, significantly lower than CaO (22.5 Mt) but nonetheless indicating its great commercial importance. Approximately 30% of the CaCl$_2$ manufactured is used for deicing of roads and public walkways, and is sometimes applied in combination with NaCl. This application of CaCl$_2$ arises because the anhydrous salt takes up water in an exothermic reaction; the heat evolved melts surrounding ice and is sufficient to maintain effective deicing action at temperatures as low as 222 K. The second major use of CaCl$_2$ is in dust control. Again this application relies on the ability of anhydrous CaCl$_2$ to absorb water, this time from the atmosphere. Addition of anhydrous CaCl$_2$ to dry road surfaces and hard shoulders provides a means of trapping water, helping to aggregate the dust particles.

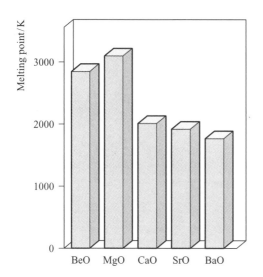

Fig. 11.5 The melting points of the group 2 metal oxides.

The action of water on MgO slowly converts it to $Mg(OH)_2$ which is sparingly soluble. Oxides of Ca, Sr and Ba react rapidly and exothermically with water, and absorb CO_2 from the atmosphere (equation 11.5). The conversion of CaO to calcium carbide and its subsequent hydrolysis (reaction 11.20) is industrially important (see *Box 11.3*), although, as an organic precursor, ethyne is being superseded by ethene.

$$\left.\begin{array}{l} CaO + 3C \xrightarrow{2300\,K} CaC_2 + CO \\[4pt] CaC_2 + 2H_2O \longrightarrow Ca(OH)_2 + C_2H_2 \end{array}\right\} \qquad (11.20)$$

Group 2 metal peroxides, MO_2, are known for M = Mg, Ca, Sr and Ba, and (as for the alkali metal peroxides), the stability with respect to the decomposition reaction 11.21 increases with the size of the M^{2+} ion. This trend arises from the difference between the lattice energies of MO and MO_2 (for a given M) which becomes smaller as r_+ increases; $\Delta_{lattice}H^\circ(MO,s)$ is always more negative than $\Delta_{lattice}H^\circ(MO_2,s)$.

$$MO_2 \longrightarrow MO + \tfrac{1}{2}O_2 \qquad (11.21)$$

All the peroxides are strong oxidizing agents. Magnesium peroxide (used in toothpastes) is manufactured by reacting

$MgCO_3$ or MgO with H_2O_2. Calcium peroxide is prepared by cautious dehydration of $CaO_2 \cdot 8H_2O$, itself made by reaction 11.22.

$$Ca(OH)_2 + H_2O_2 + 6H_2O \longrightarrow CaO_2 \cdot 8H_2O \qquad (11.22)$$

The reactions of SrO and BaO with O_2 (600 K, 200 bar pressure, and 850 K, respectively) yield SrO_2 and BaO_2. Pure BaO_2 has not been isolated and the commercially available material contains BaO and $Ba(OH)_2$. Reactions of the peroxides with acids (equation 11.23) generate H_2O_2.

$$SrO_2 + 2HCl \longrightarrow SrCl_2 + H_2O_2 \qquad (11.23)$$

Hydroxides

Beryllium hydroxide is amphoteric and this sets it apart from the hydroxides of the other group 2 metals which are basic. In the presence of excess $[OH]^-$, $Be(OH)_2$ behaves as a Lewis acid (equation 11.24), forming the tetrahedral complex ion **11.3**, but $Be(OH)_2$ also reacts with acids, e.g. reaction 11.25.

(11.3)

$$Be(OH)_2 + 2[OH]^- \longrightarrow [Be(OH)_4]^{2-} \qquad (11.24)$$

$$Be(OH)_2 + H_2SO_4 \longrightarrow BeSO_4 + 2H_2O \qquad (11.25)$$

The water solubilities of $M(OH)_2$ (M = Mg, Ca, Sr, Ba) increase down the group, as do their thermal stabilities with respect to decomposition into MO and H_2O. Magnesium hydroxide acts as a weak base, whereas $Ca(OH)_2$, $Sr(OH)_2$ and $Ba(OH)_2$ are strong bases. *Soda lime* is a mixture of NaOH and $Ca(OH)_2$ and is manufactured from CaO and aqueous NaOH; it is easier to handle than NaOH and is commercially available, being used, for example, as an absorbent for CO_2, and in qualitative tests for $[NH_4]^+$ salts, amides, imides and related compounds which evolve NH_3 when heated with soda lime.

BOX 11.6   APPLICATIONS

MgO: refractory material

When one looks for a commercially viable refractory oxide, MgO (*magnesia*) is high on the list: it has a very high melting point (3073 K), can withstand heating above 2300 K for long periods, and is relatively inexpensive. Magnesia is fabricated into bricks for lining furnaces in steelmaking. Incorporating chromium ore into the refractory bricks increases their resistance to thermal shock. Magnesia bricks are also

widely used in night-storage radiators: MgO conducts heat extremely well, but also has the ability to store it. In a radiator, the bricks absorb heat which is generated by electrically heated filaments during periods of 'off-peak' consumer rates, and then radiate the thermal energy over relatively long periods.

11.7 Salts of oxoacids

In this section, we give selected coverage of group 2 metal salts of oxoacids, paying attention only to compounds of special interest or importance.

Most beryllium salts of strong oxoacids crystallize as soluble hydrates. Beryllium carbonate tends to hydrolyze, giving a salt containing $[Be(H_2O)_4]^{2+}$ (see *Section 11.8*). $BeCO_3$ can be isolated only by precipitation under an atmosphere of CO_2. This tendency towards hydrolysis is also illustrated by the formation of *basic beryllium acetate* $[Be_4(\mu_4\text{-}O)(\mu\text{-}O_2CMe)_6]$ (rather than $Be(MeCO_2)_2$) by the action of $MeCO_2H$ on $Be(OH)_2$. Figure 11.6 shows the structure of $[Be_4(\mu_4\text{-}O)(\mu\text{-}O_2CMe)_6]$; the central oxygen atom is bonded to four Be centres, each of which is tetrahedrally sited. A similar structure is observed in the basic nitrate $[Be_4(\mu_4\text{-}O)(\mu\text{-}O_2NO)_6]$ which is formed in reaction sequence 11.26.

$$BeCl_2 \xrightarrow{N_2O_4} [NO]_2[Be(NO_3)_4] \xrightarrow{323\,K} Be(NO_3)_2$$

$$\downarrow {\scriptstyle 398\,K} \qquad (11.26)$$

$$[Be_4(\mu_4\text{-}O)(\mu\text{-}O_2NO)_6]$$

The carbonates of Mg and the later metals are sparingly soluble in water; their thermal stabilities (equation 11.19) increase with cation size, and this trend can be rationalized in terms of lattice energies. The metal carbonates are much more soluble in a solution of CO_2 than in water due to the formation of $[HCO_3]^-$. However, salts of the type '$M(HCO_3)_2$' have not been isolated. *Hard water* contains Mg^{2+} and Ca^{2+} ions which complex with the stearate ions in soaps, producing insoluble 'scum' in household baths and basins. *Temporary hardness* is due to the presence of hydrogencarbonate salts and can be overcome by boiling (which shifts equilibrium 11.27 to the right-hand side causing

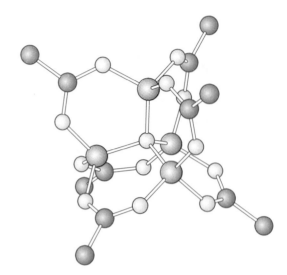

Fig. 11.6 The structure of basic beryllium acetate, $[Be_4(\mu_4\text{-}O)(\mu\text{-}O_2CMe)_6]$ (X-ray diffraction) [A. Tulinsky *et al.* (1959) *Acta Crystallogr.*, vol. 12, p. 623]; hydrogen atoms have been omitted. Colour code: Be, blue; C, grey; O, pale grey.

$CaCO_3$, or similarly $MgCO_3$, to precipitate) or by adding an appropriate amount of $Ca(OH)_2$ (again causing precipitation, equation 11.28).

$$Ca(HCO_3)_2(aq) \rightleftharpoons CaCO_3(s) + CO_2(g) + H_2O(l) \quad (11.27)$$

$$Ca(HCO_3)_2(aq) + Ca(OH)_2(aq) \longrightarrow 2CaCO_3(s) + 2H_2O(l) \quad (11.28)$$

Permanent hardness is caused by other Mg^{2+} and Ca^{2+} salts (e.g. sulfates). The process of *water softening* involves passing the hard water through a cation-exchange resin (see *Section 10.6*). Washing-machine detergents contain 'builders' that remove Mg^{2+} and Ca^{2+} ions from washing-water; polyphosphates have been used for this purpose, but because phosphates are damaging to the environment (see *Box 14.10*), zeolites (see *Section 13.9*) are used in preference.

BOX 11.7   **APPLICATIONS**

Gypsum plasters

The building industry is the major consumer of gypsum plasters. Gypsum, $CaSO_4 \cdot 2H_2O$, is mined on a large scale worldwide, and is calcined to form the β-hemihydrate, $CaSO_4 \cdot \frac{1}{2}H_2O$. The hemihydrate is referred to as *plaster of Paris*, the name being derived from Montmartre in Paris where gypsum was quarried. Hydration of the hemihydrate with a carefully controlled amount of H_2O initially gives a slurry which hardens as $CaSO_4 \cdot 2H_2O$ crystallizes. Crystals are needle-like and it is their intergrowth that provides gypsum with its strength and suitability for the building trade. Calcined gypsum which is stored for long periods may age by absorbing water, and this affects the rehydration process. The setting process of gypsum plasters may be accelerated or slowed down by suitable additives, e.g. <0.1% of citric acid is sufficient to retard the crystallization process. Gypsum plasters suitable for applying to walls have been developed so that additives are already present with the hemihydrate. Building contractors commonly use prefabricated gypsum plasterboards and tiles. Plasterboards are fabricated by pouring a hemihydrate–water–additive slurry onto cardboard sheets ≈0.5 mm thick. After completing the lamination by applying a second sheet of cardboard, the plasterboard is dried. The incorporation of fibreglass (see *Box 12.5*) into plasterboards is also possible, giving fibreboard products. An advantage of gypsum plasterboards as partition walls is their degree of fire resistance.

Calcium carbonate occurs naturally in two crystalline forms, *calcite* and the metastable *aragonite*. The energy difference between them is $<5\,kJ\,mol^{-1}$ with calcite being the thermodynamically favoured form. However, aragonite is kinetically stable with respect to conversion to calcite. Aragonite can be prepared in the laboratory by precipitation of $CaCO_3$ from hot aqueous solution.

Sulfates of Mg and Ca have important applications and those of $CaSO_4$ are described in *Section 15.2*. Hydrated calcium sulfate ($CaSO_4 \cdot 2H_2O$, *gypsum*) occurs naturally and is also a product of desulfurization processes involving $Ca(OH)_2$ or $CaCO_3$ (see *Box 11.2*). Gypsum crystals cleave easily owing to the presence of layers which are held together by hydrogen bonding. When gypsum is heated at $\approx 400\,K$, it forms the hemihydrate $CaSO_4 \cdot \frac{1}{2}H_2O$ (*plaster of Paris*), and if this is mixed with water, the material expands slightly as the dihydrate is regenerated (see *Box 11.7*).

Calcium phosphate is described in *Section 14.2*.

A hydrate $X \cdot nH_2O$ in which $n = \frac{1}{2}$ is called a *hemihydrate*; if $n = 1\frac{1}{2}$, it is a *sesquihydrate*.

11.8 Complex ions in aqueous solution

Aqua species of beryllium

We have already noted that there is a high tendency to form $[Be(H_2O)_4]^{2+}$ in aqueous media. In ^{17}O-enriched water, exchange between coordinated water and solvent is slow on the NMR spectroscopic timescale, permitting the nature of the hydrated ion to be established. The tetrahedral coordination sphere ($Be\text{--}O = 162.0\,pm$) has been established in the solid state structure of $[Be(H_2O)_4][O_2CC\equiv CCO_2]$ (Figure 11.7). The charge density of Be^{2+} is high and solutions of beryllium salts are acidic (see *Section 6.7*). Reaction 11.29 is an over-simplistic representation of the dissociation, since various condensation processes occur, e.g. reaction 11.30, and hydroxo-bridged species are also present.

$$[Be(H_2O)_4]^{2+} + H_2O \rightleftharpoons [Be(H_2O)_3(OH)]^+ + [H_3O]^+ \tag{11.29}$$

$$4[Be(H_2O)_4]^{2+} + 2H_2O$$
$$\rightleftharpoons 2[(H_2O)_3Be\text{--}O\text{--}Be(H_2O)_3]^{2+} + 4[H_3O]^+ \tag{11.30}$$

Aqua species of Mg^{2+}, Ca^{2+}, Sr^{2+} and Ba^{2+}

Using ^{17}O-labelled water, it is possible to confirm the hydration number in $[Mg(H_2O)_6]^{2+}$ by use of ^{17}O NMR spectroscopy, but for the aqua ions of Ca^{2+}, Sr^{2+} and Ba^{2+}, exchange between coordinated water and solvent is too fast on the NMR timescale for such studies. An octahedral coordination sphere has been established for $[Mg(H_2O)_6]^{2+}$ in a

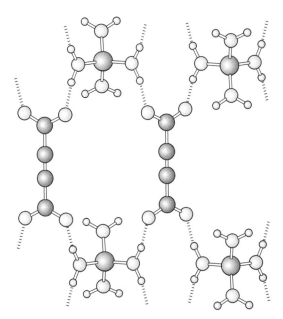

Fig. 11.7 Part of the packing diagram of $[Be(H_2O)_4][O_2CC\equiv CCO_2]$ showing hydrogen bonding between $[Be(H_2O)_4]^{2+}$ cations and $[O_2CC\equiv CCO_2]^{2-}$ anions; the structure was determined by neutron diffraction [C. Robl *et al.* (1992) *J. Solid State Chem.*, vol. 96, p. 318]. Colour code: Be, blue; C, grey; O, pale blue; H, small, pale grey.

number of crystalline salts, and also for $[Ca(H_2O)_6]^{2+}$. The $[Mg(H_2O)_6]^{2+}$ ion dissociates to some extent in aqueous solution, but hydrated cations of the later metals are not appreciably ionized and solutions of their salts derived from strong acids are neutral.

Complexes with ligands other than water

In this section we consider complexes formed in aqueous solution in which the metal centre is coordinated by O- and N-donor ligands to give cationic species. Two important ligands are $[EDTA]^{4-}$ (see *equation 6.75*) and $[P_3O_{10}]^{5-}$ (see *Figure 14.13*). Both form water-soluble complexes with Mg^{2+} and the heavier metal ions, and are *sequestering agents* used in water-softening to remove Mg^{2+} and Ca^{2+} ions.

Macrocyclic ligands, including crown ethers and cryptands (see *Sections 6.12* and *10.8*), form stable complexes with Mg^{2+}, Ca^{2+}, Sr^{2+} and Ba^{2+}. In an analogous manner to that noted for group 1 cations, selectivity corresponding to matching of cation (Table 11.1) and ligand-cavity sizes is observed. Thus, values of the stability constants for complexation with cryptand-222 (cavity radius 140 pm) in water follow the sequence $Ba^{2+} > Sr^{2+} \gg Ca^{2+} > Mg^{2+}$. An important class of macrocyclic ligands are the porphyrins and the parent compound is shown in Figure 11.8a; deprotonation of the two NH groups of a porphyrin gives a dianionic porphyrinato ligand. Chlorophylls, the pigments in green plants involved in photosynthesis, are porphyrinato derivatives containing Mg^{2+} coordinated within a square planar array of the four N donor atoms. The structure of chlorophyll a is shown in Figure 11.8b; the extensive conjugation

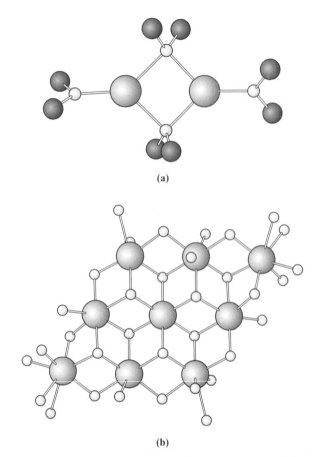

Fig. 11.8 The structures of (a) porphyrin and (b) chlorophyll a.

in the ring system means that the molecule absorbs light in the visible region (λ_{max} 660 nm) and this initiates a complicated series of reactions involving other systems containing Mn or Fe. Note that it is the *ligand* (not Mg^{2+}) that is involved in these redox reactions.

11.9 Complexes with amido or alkoxy ligands

In Section 11.5, we described group 2 metal halide complexes such as $[CaI_2(THF)_4]$ and *trans*-$[SrBr_2(py)_5]$. The number of complexes of the group 2 metals with *N*- or *O*-donor ligands continues to grow, notably those incorporating sterically demanding amido or alkoxy ligands.

With the bulky bis(trimethylsilyl)amido ligand, each of the M^{2+} ions forms at least one type of complex. In the gas phase, monomeric $[Be\{N(SiMe_3)_2\}_2]$ contains a linear N−Be−N unit; in the solid state structure of $[Mg\{N(SiMePh_2)_2\}_2]$, $\angle N-Mg-N = 162.8°$, the deviation from linearity being attributed to weak dipolar interactions between the electropositive metal centre and the electron density of the aromatic rings. Coordination numbers of three and four for Mg(II), Ca(II), Sr(II) and Ba(II) are seen in dimers $[M\{N(SiMe_3)_2\}_2]_2$ or solvated monomers, e.g. tetrahedral $[Ba\{N(SiMe_3)_2\}_2(THF)_2]$. The structure of $[Ca\{N(SiMe_2)_2\}_2]_2$ is shown in Figure 11.9a, and similar structures have been crystallographically confirmed for the analogous Mg, Sr and Ba compounds as well as for $[Mg\{N(CH_2Ph)_2\}_2]_2$.

While alkoxy derivatives of the alkaline earth metals have been known for many years, the area has undergone significant expansion since 1990. Much of this interest stems from the fact that calcium, strontium and barium alkoxides are potential precursors for high temperature superconductors

Fig. 11.9 The structures (determined by X-ray diffraction) of (a) $[Ca_2\{N(SiMe_3)_2\}_2\{\mu\text{-}N(SiMe_3)_2\}_2]$ in which the methyl groups have been omitted [M. Westerhausen *et al.* (1991) *Z. Anorg. Allg. Chem.*, vol. 604, p. 127] and (b) $[Ca_9(OCH_2CH_2OMe)_{18}(HOCH_2CH_2OMe)_2]$ for which only the $Ca_9(\mu_3\text{-}O)_8(\mu\text{-}O)_8O_{20}$ core is shown (four of the ligands in $[Ca_9(OCH_2CH_2OMe)_{18}(HOCH_2CH_2OMe)_2]$ are terminally attached, leaving four oxygen atoms non-coordinated to Ca^{2+} centres) [S.C. Goel *et al.* (1991) *J. Am. Chem. Soc.*, vol. 113, p. 1844]. Colour code: Ca, blue; O and N, pale grey; Si, black.

(see *Chapter 27*) and volatile compounds suitable for *chemical vapour deposition* (CVD) studies are being sought. Mononuclear complexes include several of the type $[M(OR)_2(THF)_3]$, e.g. $[Ca(OC_6H_2\text{-}2,6\text{-}^tBu_2\text{-}4\text{-}Me)_2(THF)_3]$. Some interesting high nuclearity species have also been isolated, including $[Ba_4(\mu_4\text{-}O)(\mu\text{-}OC_6H_2(CH_2NMe_2)_3\text{-}2,4,6)_6]$, formed by treating BaI_2 with $K[OC_6H_2(CH_2NMe_2)_3\text{-}2,4,6]$ in THF, and $[Ca_9(OCH_2CH_2OMe)_{18}(HOCH_2CH_2OMe)_2]$ (Figure 11.9b), produced by reacting Ca metal with 2-methoxyethanol in hexane.

Summary

This chapter has described some descriptive chemistry of the group 2 metals; magnesium and the later metals are classed as the alkaline earth metals. Distinct differences have been noted between the chemistry of beryllium and its heavier congeners.

The following terms were introduced in this chapter. Do you know what they mean?

- ❑ deliquescent
- ❑ hygroscopic
- ❑ refractory material
- ❑ permanent and temporary hardness of water
- ❑ water-softening agent (sequestering agent)
- ❑ hemihydrate
- ❑ sesquihydrate
- ❑ porphyrin
- ❑ amido ligand
- ❑ alkoxy ligand

Suggested further reading

A.F. Wells (1984) *Structural Inorganic Chemistry*, 5th edn, Clarendon Press, Oxford – A full account of the structural chemistry of the group 2 metals and their compounds.

Special topics

D.L. Kepert, A.F. Waters and A.H. White (1996) *Australian Journal of Chemistry*, vol. 49, p. 117 – 'Synthesis and structural systematics of nitrogen base adducts of group 2 salts' (Part VIII in a series of papers covering this subject).

S. Mann (1995) *Chemistry & Industry*, p. 93 – 'Biomineral and biomimetics: Smart solutions to living in the material world'.

N.C.J. Strynadka and M.N.G. James (1994) – 'Calcium-binding proteins' in *Encyclopedia of Inorganic Chemistry*, ed. R.B. King, Wiley, Chichester, vol. 1, p. 477.

Problems

11.1 (a) Write down, in order, the names and symbols of the metals in group 2; check your answer by reference to the first page of this chapter. Which metals are classed as alkaline earth metals? (b) Give a *general* notation that shows the ground state electronic configuration of each metal.

11.2 Using data in Table 6.4, determine the relative solubilities of $Ca(OH)_2$ and $Mg(OH)_2$ and explain the relevance of your answer to the extraction of magnesium from seawater.

11.3 (a) Write an equation to show how Mg reacts with N_2 when heated. (b) Suggest how the product reacts with water.

11.4 The structure of magnesium carbide, MgC_2, is of the NaCl type, elongated along one axis. (a) Explain how this elongation arises. (b) What do you infer from the fact that there is no similar elongation in NaCN which also crystallizes with a NaCl lattice?

11.5 Write balanced equations for the following reactions:
(a) the thermal decomposition of $[NH_4]_2[BeF_4]$;
(b) the reaction between NaCl and $BeCl_2$;
(c) the dissolution of BeF_2 in water;

11.6 (a) Suggest a likely structure for the dimer of $BeCl_2$, present in the vapour phase below 1020 K. What hybridization scheme is appropriate for the Be centres? (b) $BeCl_2$ dissolves in diethyl ether to form monomeric $BeCl_2 \cdot 2Et_2O$; suggest a structure for this compound and give a description of the bonding.

11.7 MgF_2 has a TiO_2 lattice. (a) Sketch a unit cell of MgF_2, and (b) confirm the stoichiometry of MgF_2 using the solid state structure.

11.8 Discuss the trends in data in Table 11.2.

Table 11.2 Data for problem 11.8.

Metal, M	$\Delta_f H^\circ$ / kJ mol^{-1}			
	MF_2	MCl_2	MBr_2	MI_2
Mg	−1113	−642	−517	−360
Ca	−1214	−795	−674	−535
Sr	−1213	−828	−715	−567
Ba	−1200	−860	−754	−602

11.9 (a) How do anhydrous $CaCl_2$ and CaH_2 function as drying agents? (b) Compare the solid state structures and properties of $BeCl_2$ and $CaCl_2$.

11.10 How would you attempt to estimate the following?

(a) $\Delta_r H^\circ$ for the solid state reaction:
$$MgCl_2 + Mg \longrightarrow 2MgCl$$
(b) $\Delta_r H^\circ$ for the reaction:
$$CaCO_3(\text{calcite}) \longrightarrow CaCO_3(\text{aragonite})$$

11.11 (a) Identify the conjugate acid–base pairs in reaction 11.23. (b) Suggest how BaO_2 will react with water.

11.12 (a) Determine $\Delta_r H^\circ$ for the reactions of SrO and BaO with water, given that values of $\Delta_f H^\circ$(298 K) for SrO(s), BaO(s), $Sr(OH)_2$(s), $Ba(OH)_2$(s) and H_2O(l) are −592.0, −553.5, −959.0, −944.7 and −285.5 kJ mol^{-1} respectively. (b) Compare the values of $\Delta_r H^\circ$ with that for the reaction of CaO with water (equation 11.5), and comment on factors contributing to the trend in values.

11.13 (a) What qualitative test is used for CO_2? (b) What reaction takes place, and (c) what is observed in a positive test?

Table 11.3   Data for problem 11.14: log K for the formation of the complexes $[M(\text{crypt-222})]^{n+}$.

M^{n+}	Na^+	K^+	Rb^+	Mg^{2+}	Ca^{2+}	Sr^{2+}	Ba^{2+}
log K	4.2	5.9	4.9	2.0	4.1	13.0	>15

11.14 Discuss the data presented in Table 11.3; other relevant data are available in this book.

11.15 Write a short account that justifies the so-called *diagonal-relationship* between Li and Mg.

11.16 Suggest why MgO is more soluble in aqueous $MgCl_2$ solution than in pure water.

Chapter 12

The group 13 elements

TOPICS

- Occurrence, extraction and uses
- Physical properties
- The elements
- Simple hydrides
- Halides and complex halides
- Oxides, oxoacids, oxoanions and hydroxides

- Compounds containing nitrogen
- Aluminium to thallium: salts of oxoacids and aqueous solution chemistry
- Metal borides
- Electron deficient cluster compounds: an introduction

1	2		13	14	15	16	17	18
H								He
Li	Be		B	C	N	O	F	Ne
Na	Mg		Al	Si	P	S	Cl	Ar
K	Ca	d-block	Ga	Ge	As	Se	Br	Kr
Rb	Sr		In	Sn	Sb	Te	I	Xe
Cs	Ba		Tl	Pb	Bi	Po	At	Rn
Fr	Ra							

12.1 Introduction

The elements in group 13 – boron, aluminium, gallium, indium and thallium – show a wide variation in properties: B is a non-metal, Al is a metal but exhibits many chemical similarities to B, and the later elements essentially behave as metals. Although the M(III) oxidation state is characteristic for elements in the group, the M(I) state occurs for all elements except B, and for Tl this is the more stable oxidation state. Thallium shows similarities to elements outside those in group 13, and can be compared to the alkali metals, Ag, Hg and Pb, an observation that led Dumas to describe it as the 'duckbill platypus among elements'.

In contrast to the later elements, B forms a large number of so-called *electron-deficient* cluster compounds, the bonding in which poses problems within valence bond theory; we introduce these compounds in Section 12.11.

12.2 Occurrence, extraction and uses

Occurrence

The main sources of boron are *borax*, $Na_2[B_4O_5(OH)_4] \cdot 8H_2O$, and *kernite*, $Na_2[B_4O_5(OH)_4] \cdot 2H_2O$, with extensive deposits being worked commercially in the Mojave Desert, California. Aluminium is the most abundant metal in the Earth's crust, and occurs in aluminosilicates (see *Section 13.9*) such as *clays*, *micas* and *feldspars*, in *bauxite* (hydrated oxides) and, to a lesser extent, in *cryolite*, $Na_3[AlF_6]$. Gallium, indium and thallium occur in trace amounts as sulfides in various minerals.

Extraction

Of the group 13 elements, Al is of the greatest commercial importance with uses exceeding those of all metals except Fe; Figure 12.1 shows the dramatic rise in the production of Al in the US (the world's largest producer) since 1920. Its isolation from the widely available aluminosilicate minerals is prohibitively difficult; bauxite and cryolite are the chief ores, and both are consumed in the extraction process. Crude bauxite is a mixture of oxides (impurities include Fe_2O_3, SiO_2 and TiO_2) and is purified using the Bayer process. After addition of the crude ore to hot aqueous NaOH under pressure (which causes Fe_2O_3 to separate), the solution is seeded with $Al_2O_3 \cdot 3H_2O$ and cooled, or is treated with a stream of CO_2 to precipitate crystalline α-$Al(OH)_3$. Anhydrous Al_2O_3 (*alumina*) is produced by the action of heat. Electrolysis of molten Al_2O_3 gives Al at the cathode, but the melting point (2345 K) is high, and it is more practical and economical to use a mixture of cryolite and alumina as the electrolyte with an operating temperature for the melt

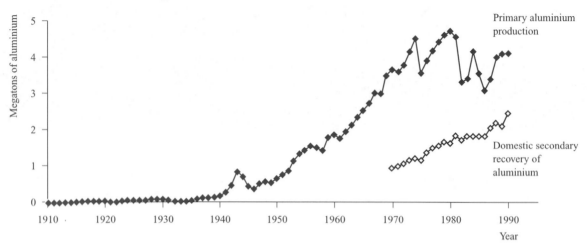

Fig. 12.1 Production of primary aluminium in the US between 1920 and 1990 has risen dramatically; the contribution that recycled aluminium has made to the market in the latter part of the 20th century has been considerable. The term 'secondary aluminium' refers to metal recovered from new and old scrap, and the total production is the sum of primary and secondary sources. [Data: U.S. Geological Survey.]

of 1220 K. The extraction is expensive in terms of the electrical power required, and Al production is often associated with hydroelectric schemes.

The first steps in the extraction of boron from borax are its conversion to boric acid (equation 12.1) and then to the oxide (equation 12.2); boron of low purity is obtained by reduction of the oxide by Mg, followed by washing the product with alkali, hydrochloric acid and then hydrofluoric acid.

$$Na_2[B_4O_5(OH)_4]\cdot 8H_2O + H_2SO_4$$
$$\longrightarrow 4B(OH)_3 + Na_2SO_4 + 5H_2O \quad (12.1)$$

$$2B(OH)_3 \xrightarrow{\Delta} B_2O_3 + 3H_2O \quad (12.2)$$

The product is a very hard, black solid of low electrical conductivity which is inert towards most acids, but is slowly attacked by concentrated HNO_3 or fused alkali. Pure boron is made by the vapour phase reduction of BBr_3 with H_2, or by pyrolysis of B_2H_6 or BI_3. At least four allotropes can be obtained under different conditions but transitions between them are extremely slow. For a discussion of the production of boron fibres, see *Section 27.7*.

Trace quantities of Ga can be extracted from crude bauxite, and Ga, In and Tl can be extracted electrolytically from enriched aqueous solutions of their salts, obtained as by-products of other element extraction processes. Demand for these elements is relatively small.

Major uses of the group 13 elements and their compounds

The widespread applications of Al are summarized in Figure 12.2; its strength can be increased by alloying with Cu or Mg. Aluminium oxide (see *Section 12.7*) has many important uses. *Corundum* (α-alumina) and *emery* (corundum mixed with the iron oxides *magnetite* and *haematite*) are extremely

hard and are used as abrasives; diamond is the only naturally occurring mineral harder than corundum. Gemstones including ruby, sapphire, oriental topaz, oriental amethyst and oriental emerald, result from the presence of trace metal salts in Al_2O_3; e.g. Cr(III) produces the red colour of ruby. Artificial crystals can be manufactured from bauxite in furnaces, and artificial rubies are important as components in lasers. The γ-form of Al_2O_3 is used as a catalyst and as a stationary phase in chromatography. Al_2O_3 fibres are described in *Section 27.7*.

The two commercially important borates are $Na_2[B_4O_5(OH)_4]\cdot 8H_2O$ (*borax*) and $Na_2[B_4O_5(OH)_4]\cdot 2H_2O$ (*kernite*); they are used in the manufacture of borosilicate glass (which, having a high refractive index, is suitable for optical lenses), borate-based detergents, flame retardants (see *Box 16.1*), agrochemicals and wood preservatives. Borax has been used in pottery glazes for many centuries and remains in use in the ceramics industry. The reaction between fused borax and metal oxides is the basis for using borax as a

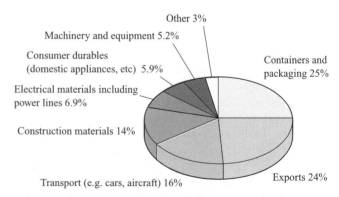

Fig. 12.2 Uses of aluminium in the US in 1990; the US is the world's largest producer of the metal and in 1990, a quarter of that manufactured was exported. [Data: U.S. Geological Survey.]

BOX 12.1   **RESOURCES, ENVIRONMENTAL AND BIOLOGICAL**

Borax and boric acid: toxicity and essentiality

Boron is an essential trace plant nutrient. Although its exact function has not yet been established, deprivation of boron affects plant growth and in boron-poor soils, crop yields are diminished. An important application of borax is in borate fertilizers. In contrast, the toxicities of boric acid and borax to animal life are sufficient for them to be used as insecticides, e.g. in ant and cockroach control. The level of toxicity is relatively low, but does cause some concern; e.g. borax and honey was, at one time, used to relieve the pain of teething in children, but this use is no longer recommended.

flux in brazing; when metals are being fused together, coatings of metal oxides must be removed to ensure good metal–metal contact at the point of fusion. Boric acid, $B(OH)_3$, is used on a large scale in the glass industry, as a flame retardant, as a component in buffer solutions and is also an antibacterial agent. The use of B_2O_3 in the glass industry is described in *Box 12.5*. Elemental boron is used in the production of impact-resistant steels and (because ^{10}B has a high cross-section for neutron capture, *Section 2.5*) in control rods for nuclear reactors. Amorphous B is used in pyrotechnics, giving a characteristic green colour when it burns.

Gallium and indium phosphides, arsenides and antimonides have important applications in the semiconductor industry (see *Sections 5.9* and *27.6*; *Boxes 13.3* and *18.4*). They are used as transistor materials and in light-emitting devices (LEDs) in, for example, pocket calculators; the colour of the light emitted depends on the band gap.

Indium is used in lead-free solders, for producing seals between glass, ceramics and metals (because In has the ability to bond to non-wettable materials), and for fabricating special mirrors which reduce headlight glare; uses of indium–tin oxide (ITO) are highlighted in *Box 12.7*.

Thallium sulfate used to be used to kill ants and rats, but the extremely high toxicity levels of Tl compounds are now well recognized and all Tl-containing species must be treated with caution.

12.3 Physical properties

Table 12.1 lists selected physical properties of the group 13 elements. Despite the discussion of ionization energies that follows, there is no evidence for the formation of free M^{3+}

Table 12.1 Some physical properties of the Group 13 elements, M, and their ions.

Property	B	Al	Ga	In	Tl
Atomic number, Z	5	13	31	49	81
Ground state electronic configuration	$[He]2s^2 2p^1$	$[Ne]3s^2 3p^1$	$[Ar]3d^{10}4s^2 4p^1$	$[Kr]4d^{10}5s^2 5p^1$	$[Xe]4f^{14}5d^{10}6s^2 6p^1$
Enthalpy of atomization, $\Delta_a H^{\circ}(298\,K)\,/\,kJ\,mol^{-1}$	582	330	277	243	182
Melting point, mp / K	2453‡	933	303	430	576.5
Boiling point, bp / K	4273	2792	2477	2355	1730
Standard enthalpy of fusion, $\Delta_{fus}H^{\circ}(mp)\,/\,kJ\,mol^{-1}$	50.2	10.7	5.6	3.3	4.1
First ionization energy, $IE_1\,/\,kJ\,mol^{-1}$	800.6	577.5	578.8	558.3	589.4
Second ionization energy, $IE_2\,/\,kJ\,mol^{-1}$	2427	1817	1979	1821	1971
Third ionization energy, $IE_3\,/\,kJ\,mol^{-1}$	3660	2745	2963	2704	2878
Fourth ionization energy, $IE_4\,/\,kJ\,mol^{-1}$	25 030	11 580	6200	5200	4900
Metallic radius, $r_{metal}\,/\,pm^*$	–	143	153	167	171
Covalent radius, $r_{cov}\,/\,pm$	88	130	122	150	155
Ionic radius, $r_{ion}\,/\,pm^{**}$	–	54 (Al^{3+})	62 (Ga^{3+})	80 (In^{3+})	89 (Tl^{3+})
					159 (Tl^+)
Standard reduction potential, $E^{\circ}\,(M^{3+}/M)\,/\,V$	–	−1.66	−0.55	−0.34	+0.72
Standard reduction potential, $E^{\circ}\,(M^+/M)\,/\,V$	–	–	−0.2	−0.14	−0.34
NMR active nuclei (% abundance, nuclear spin)	^{10}B (19.6, $I = 3$) ^{11}B (80.4, $I = \frac{3}{2}$)	^{27}Al (100, $I = \frac{5}{2}$)	^{69}Ga (60.4, $I = \frac{3}{2}$) ^{70}Ga (39.6, $I = \frac{3}{2}$)	^{113}In (4.3, $I = \frac{9}{2}$)	^{203}Tl (29.5, $I = \frac{1}{2}$) ^{205}Tl (70.5, $I = \frac{1}{2}$)

‡ For β-rhombohedral boron.

* Only the values for Al, In and Tl (the structures of which are close-packed) are strictly comparable; see text (Section 5.3) for Ga.

** There is no evidence for the existence of simple cationic boron under chemical conditions; values of r_{ion} for M^{3+} refer to 6-coordination; for Tl^+, r_{ion} refers to 8-coordination.

Relativistic effects

Among many generalizations about heavier elements are two that depend on quantum theory for explanation:

- the ionization energies of the 6s electrons are anomalously high, leading to the marked stabilization of Hg(0), Tl(I), Pb(II) and Bi(III) compared with Cd(0), In(I), Sn(II) and Sb(III);
- whereas bond energies usually decrease down a group of *p*-block elements, they often increase down a group of *d*-block metals, both in the elements themselves and their compounds.

These observations can be accounted for (though often far from simply) if Einstein's theory of relativity is combined with quantum mechanics, in which case, they are attributed to *relativistic effects*. We focus here on *chemical* generalizations.

According to the theory of relativity, the mass m of a particle increases from its rest mass m_0 when its velocity v approaches the speed of light, c, and m is then given by the equation:

$$m = \frac{m_0}{\sqrt{1 - \left(\dfrac{v}{c}\right)^2}}$$

For a one-electron system, the Bohr model of the atom (which, despite its shortcomings, gives the correct value for the ionization energy) leads to the velocity of the electron being expressed by the equation:

$$v = \frac{Ze^2}{2\varepsilon_0 nh}$$

where Z = atomic number, e = charge on the electron, ε_0 = permittivity of a vacuum, h = Planck constant.

For $n = 1$ and $Z = 1$, v is only $\approx \left(\dfrac{1}{137}\right)c$, but for $Z = 80$, $\dfrac{v}{c}$ becomes ≈ 0.58, leading to $m \approx 1.2m_0$. Since the radius of the Bohr orbit is given by the equation:

$$r = \frac{Ze^2}{4\pi\varepsilon_0 mv^2}$$

the increase in m results in an approximately 20% contraction of the radius of the 1s ($n = 1$) orbital; this is called *relativistic contraction*. Other s orbitals are affected in a similar way and as a consequence, when Z is high, s orbitals have diminished overlap with orbitals of other atoms. A detailed treatment shows that p orbitals (which have a low electron density near to the nucleus) are less affected. On the other hand, d orbitals (which are more effectively screened from the nuclear charge by the contracted s and p orbitals) undergo a *relativistic expansion*; a similar argument applies to f orbitals. The relativistic contraction of the s orbitals means that for an atom of high atomic number, there is an extra energy of attraction between s electrons and the nucleus. This is manifested in higher ionization energies for the 6s electrons, contributing to the *thermodynamic 6s inert pair effect* which is discussed further in Box 12.3.

---

ions in compounds of the group 13 elements under normal conditions, other than, perhaps, some trifluorides.

Electronic configurations and oxidation states

While the elements have an outer electronic configuration ns^2np^1 and a larger difference between IE_1 and IE_2 than between IE_2 and IE_3 (i.e. comparing the removal of a *p* with that of an *s* electron), the relationships between the electronic structures of the group 13 elements and those of the preceding noble gases are more complex than for the group 1 and 2 elements discussed in Chapters 10 and 11. For Ga and In, the electronic structures of the species formed after the removal of three valence electrons are $[Ar]3d^{10}$ and $[Kr]4d^{10}$ respectively, while for Tl, the corresponding species has the configuration $[Xe]4f^{14}5d^{10}$. Thus, whereas for B and Al, the value of IE_4 (Table 12.1) refers to the removal of an electron from a noble gas configuration, this is not the case for the three later elements; the difference between IE_3 and IE_4 is not nearly so large for Ga, In and Tl as for B and Al. On going down group 13, the observed discontinuities in values of IE_2 and IE_3, and the differences between them (Table 12.1), originate in the failure of the *d* and *f*-electrons

(which have a low screening power, see *Section 1.7*) to compensate for the increase in nuclear charge. This failure is also reflected in the relatively small difference between values of r_{ion} for Al^{3+} and Ga^{3+}. For Tl, *relativistic effects* (see *Box 12.2*) are also involved.

On descending group 13, the trend in IE_2 and IE_3 shows *increases* at Ga and Tl (Table 12.1), and this leads to a marked increase in stability of the +1 oxidation state for these elements. In the case of Tl (the only salt-like trihalide of which is TlF_3), this is termed the *thermodynamic 6s inert pair effect* (see *Box 12.3*) so called to distinguish it from the *stereochemical inert pair effect* mentioned in *Section 1.19*. Similar effects are seen for Pb (group 14) and Bi (group 15), for which the most stable oxidation states are +2 and +3 respectively, rather than +4 and +5. The inclusion in Table 12.1 of E° values for the M^{3+}/M and M^+/M redox couples for the later group 13 elements reflects the variable accessibility of the M^+ state within the group.

Although an oxidation state of +3 (and for Ga and Tl, +1) is characteristic of a group 13 element, most of the group 13 elements also form compounds in which a *formal* oxidation state of +2 is suggested, e.g. B_2Cl_4 and $GaCl_2$. However,

BOX 12.3 **CHEMICAL AND THEORETICAL BACKGROUND**

The thermodynamic 6s inert pair effect

We confine attention here to the conversion of a metal halide MX_n into MX_{n+2}:

$$MX_n + X_2 \longrightarrow MX_{n+2}$$

In the simplest possible case, both halides are ionic solids and the energy changes involved are:

- absorption of the lattice energy of MX_n;
- absorption of $IE_{(n+1)} + IE_{(n+2)}$ to convert $M^{n+}(g)$ into $M^{(n+2)+}(g)$;
- liberation of the enthalpy of formation of $2X^-(g)$ (which is nearly constant for X = F, Cl, Br and I, see *Appendices 9* and *10*);
- liberation of the lattice energy of MX_{n+2}.

For a given M, the difference between the lattice energies of MX_n and MX_{n+2} is greatest for X = F, so if any saline halide MX_{n+2} is formed, it will be the fluoride. This treatment is probably a good representation of the conversion of TlF into TlF_3, and PbF_2 into PbF_4.

If, however, the halides are covalent compounds, the energy changes in the conversion are quite different. In this case, n times the M−X bond energy in MX_n and $2\Delta_f H^\circ(X,g)$ have to be absorbed, while $(n+2)$ times the M−X bond energy in MX_{n+2} is liberated; $IE_{(n+1)}$ and $IE_{(n+2)}$ are not involved. The most important quantities in determining whether the conversion is possible are now the M−X bond energies in the two halides. The limited experimental data available indicate that both sets of M−X bond

energies decrease along the series F > Cl > Br > I, and that the M−X bond energy is always greater in MX_n than in MX_{n+2}. The overall result is that formation of MX_{n+2} is most likely for X = F. (The use of bond energies relative to ground-state atoms is unfortunate, but is inevitable since data are seldom available for valence state atoms; in principle, it would be better to consider the promotion energy for the change from one valence state of M to another, followed by a term representing the energy liberated when each valence state of M forms M−X bonds. However, this is beyond our present capabilities.)

The third possibility for the MX_n to MX_{n+2} conversion, and the one most likely in practice, is that MX_n is an ionic solid and MX_{n+2} is a covalent molecule. The problem now involves many more quantities and is too complicated for discussion here. Representative changes are the conversions of $TlCl$ to $TlCl_3$, and of $PbCl_2$ to $PbCl_4$.

Finally, we must consider the effect of varying M down a group. In general, ionization energies (see *Appendix 8*) and lattice energies of compounds *decrease* as atomic and ionic radii (see *Appendix 6*) *increase*. It is where there is actually an *increase* in ionization energies, as is observed for the valence s electrons of Tl, Pb and Bi, that we get the clearest manifestations of the *thermodynamic 6s inert pair effect*. Where covalent bond formation is involved, a really satisfactory discussion of this inert pair effect is not yet possible, but the attempt at formulation of the problem can nevertheless be a rewarding exercise.

---

caution is needed. In B_2Cl_4, the +2 oxidation state arises because of the presence of a B−B bond, and $GaCl_2$ is the mixed oxidation state species $Ga[GaCl_4]$.

NMR active nuclei

All the group 13 elements possess at least one isotope that is NMR active (Table 12.1). In particular, routine use is made of ^{11}B NMR spectroscopy in the characterization of B-containing compounds (e.g. *Figure 2.10*). The ^{205}Tl nucleus is readily observed, and, since Tl^+ behaves similarly to Na^+ and K^+, replacement of these group 1 metal ions by Tl^+ allows ^{205}Tl NMR spectroscopy to be used to investigate Na- or K-containing biological systems.

12.4 The elements

Appearance

Impure (amorphous) boron is a brown powder, but the pure element forms shiny, silver-grey crystals; properties including its high melting point and low electrical conductivity make B an important refractory material (see *Section*

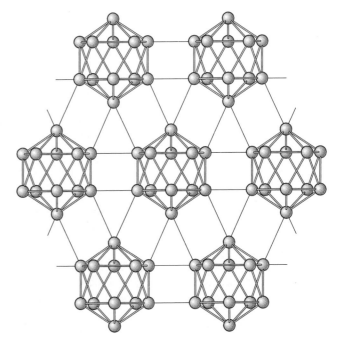

Fig. 12.3 Part of one layer of the infinite lattice of α-rhombohedral boron, showing the B_{12}-icosahedral building blocks which are covalently linked to give a rigid, infinite lattice.

11.6). Aluminium is a hard, white metal. Thermodynamically, it should react with air and water but it is resistant owing to the formation of an oxide layer, 10^{-6} to 10^{-4} mm thick. A thicker layer of Al_2O_3 can be obtained by making Al the anode in the electrolysis of H_2SO_4; the result is *anodized aluminium* which will take up dyes and pigments to produce a strong and decorative finish. Gallium is a silver-coloured metal with a particularly long liquid range (303–2477 K). Indium and thallium are soft metals, and In has the unusual property of emitting a high-pitched 'cry' when the metal is bent.

Structures of the elements

The structures of the group 13 *metals* were described in *Section 5.3* and *Table 5.2*. The first 'allotrope' of boron to

be documented was the α-tetragonal form, but this has been reformulated as a carbide or nitride, $B_{50}C_2$ or $B_{50}N_2$, the presence of C or N arising as a result of synthetic conditions. This carbidic phase is *not* the same as the boron carbide B_4C (more correctly formulated as $B_{13}C_2$) which has a structure related to that of β-rhombohedral B. The standard state of B is the β-rhombohedral form, but the structure of α-rhombohedral B makes an easier starting point in our discussion. Both the α- and β-rhombohedral allotropes contain icosahedral B_{12}-units (Figures 12.3 and 12.4a); the bonding in elemental B is covalent, and within each B_{12}-unit, it is delocalized. We return to bonding descriptions in boron cluster compounds in Section 12.11, but for now note that the connectivity of each B atom in Figures 12.3 and 12.4 exceeds the number of valence electrons available per B.

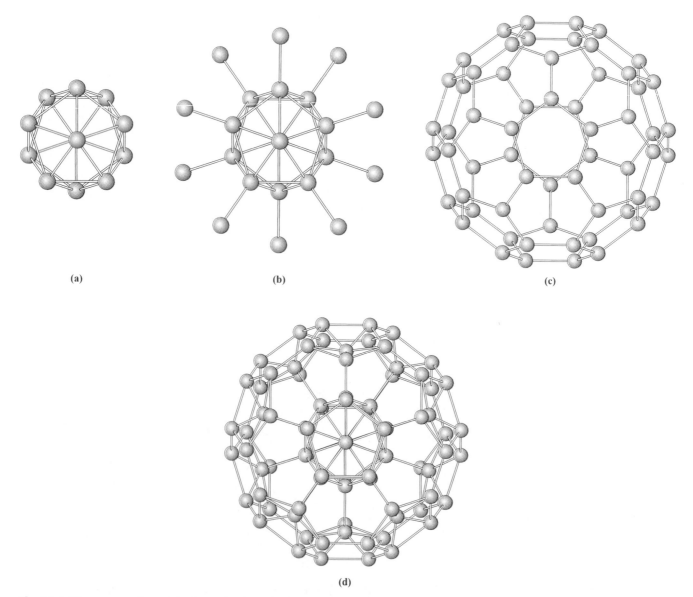

(a)   **(b)**   **(c)**

(d)

Fig. 12.4 The construction of the B_{84}-unit, the main building block of the infinite lattice of β-rhombohedral boron. (a) In the centre of the unit is a B_{12}-icosahedron, and (b) to each of these twelve, another boron atom is covalently bonded. (c) A B_{60}-cage is the outer 'skin' of the B_{84}-unit. (d) The final B_{84}-unit can be described in terms of covalently bonded sub-units $(B_{12})(B_{12})(B_{60})$.

α-Rhombohedral boron consists of B_{12}-icosahedra covalently linked by B–B bonds to form an infinite lattice. A readily interpretable picture of the lattice is to consider each icosahedron as an approximate sphere, and the overall structure as a ccp array of B_{12}-icosahedra, one layer of which is shown in Figure 12.3. However, note that this is an infinite covalent lattice, as distinct from the close-packed metal lattices described in Chapter 5.

The structure of β-rhombohedral B consists of B_{84}-units, connected through B_{10}-units. Each B_{84}-unit is conveniently viewed in terms of the subunits shown in Figure 12.4; their interrelationship is described in the figure caption, but an interesting point to note is the structural relationship between the B_{60}-subunit shown in Figure 12.4c and the fullerene C_{60} (*Figure 13.3*). The covalent lattices of both α- and β-rhombohedral B are extremely rigid, making crystalline B very hard, with a high melting point (2453 K for β-rhombohedral B).

Reactivity

Boron is inert under normal conditions except for attack by F_2. At high temperatures, it reacts with most non-metals (exceptions include H_2), most metals and with NH_3; the formations of metal borides (see *Section 12.10*) and boron nitride (see *Section 12.8*) are of particular importance.

The reactivities of the heavier group 13 elements contrast with that of the first member of the group. Aluminium readily oxidizes in air (see above); it dissolves in dilute mineral acids (e.g. reaction 12.3) but is passivated by concentrated HNO_3. Aluminium reacts with aqueous NaOH or KOH liberating H_2 (equation 12.4).

$$Al + 3H_2SO_4 \longrightarrow Al(SO_4)_3 + 3H_2 \qquad (12.3)$$
$$\text{dilute, aq}$$

$$2Al + 2MOH + 6H_2O \longrightarrow 2M[Al(OH)_4] + 3H_2$$
$$(M = Na, K) \qquad (12.4)$$

Reactions of Al with halogens at room temperature or with N_2 on heating give the Al(III) halides or nitride. Aluminium is often used to reduce metal oxides, e.g. in the *thermite process* (equation 12.5) which is highly exothermic.

$$2Al + Fe_2O_3 \longrightarrow Al_2O_3 + 2Fe \qquad (12.5)$$

Gallium, indium and thallium dissolve in most acids to give salts of Ga(III), In(III) or Tl(I), but only Ga liberates H_2 from aqueous alkali. All three metals react with halogens at, or just above, 298 K; the products are of the type MX_3 with the exceptions of reactions 12.6 and 12.7.

$$2Tl + 2Br_2 \longrightarrow Tl[TlBr_4] \qquad (12.6)$$

$$3Tl + 2I_2 \longrightarrow Tl_3I_4 \qquad (12.7)$$

12.5 Simple hydrides

Neutral hydrides

With three valence electrons, each group 13 element might be expected to form a hydride MH_3. Although the existence of BH_3 has been established in the gas phase, its propensity to dimerize means that B_2H_6 (diborane(6), **12.1**) is, in practice, the simplest hydride of boron.

(12.1)

We have already discussed the structure of and bonding in B_2H_6 (*Sections 9.7* and *4.7*); the reader is reminded of the presence of 3c-2e (delocalized, 3-centre 2-electron) B–H–B interactions. In worked example 12.1, the ^{11}B and ^1H NMR spectra of B_2H_6 are analyzed.

> **Worked example 12.1 Multinuclear NMR spectroscopy: B_2H_6**
>
> Predict the (a) ^{11}B and (b) ^1H NMR spectra of B_2H_6. (c) What would you observe in the proton decoupled ^{11}B NMR spectrum of B_2H_6? [^1H, 100%, $I = \frac{1}{2}$; ^{11}B, 80.4%, $I = \frac{3}{2}$]. Information needed:
>
> - In the ^1H NMR spectrum, coupling to ^{10}B (see *Table 12.1*) can, to a first approximation, be ignored.[†]
> - A general point in the NMR spectra of boranes is that:
>
> $$J(^{11}B-^1H_{terminal}) > J(^{11}B-^1H_{bridge})$$
>
> - Ignore couplings between boron nuclei.
>
> (a) First, draw the structure of B_2H_6; there is one B environment, and two H environments:
>
>
> Consider the ^{11}B NMR spectrum. There is one signal, but each ^{11}B nuclei couples to two terminal ^1H nuclei and two bridging ^1H nuclei. The signal therefore appears as a triplet

[†]For further details, see: C.E. Houscroft (1994) *Boranes and Metallaboranes: Structure, Bonding and Reactivity*, 2nd edn, Ellis Horwood, Hemel Hempstead, Chapter 3, and references cited therein.

of triplets:

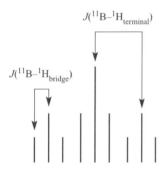

The exact nature of the observed spectrum depends upon the actual values of $J(^{11}B-^1H_{terminal})$ and $J(^{11}B-^1H_{bridge})$

(b) In the ^1H NMR spectrum, there will be two signals, with relative integrals 2:4 (bridge H:terminal H).

Consider first the signal due to the terminal protons. For ^{11}B, $I = \frac{3}{2}$, meaning that there are four spin states with values $+\frac{3}{2}$, $+\frac{1}{2}$, $-\frac{1}{2}$ and $-\frac{3}{2}$. There is an *equal probability* that each terminal ^1H will 'see' the ^{11}B nucleus in each of the four spin states, and this gives rise to the ^1H signal being split into four equal intensity lines: a 1:1:1:1 multiplet.

Now consider the bridging protons. Each ^1H nucleus couples to *two* ^{11}B nuclei, and the signal will be a 1:2:3:4:3:2:1 multiplet since the combined nuclear spins of the two ^{11}B nuclei can adopt seven orientations, but not with equal probabilities:

(c) The proton-decoupled ^{11}B NMR spectrum (written as the $^{11}B\{^1H\}$ NMR spectrum) will exhibit a singlet, since all $^{11}B-^1H$ coupling has been removed (see *Section 2.11, case study 2*).

Self-study exercises

1. Refer to the spectral diagram in part (a) above. (i) Which part of the signal is the triplet due to $^{11}B-^1H_{terminal}$ spin–spin coupling? (ii) Indicate where else on the above diagram you could measure values of $J(^{11}B-^1H_{terminal})$ and $J(^{11}B-^1H_{bridge})$.

2. Refer to the spectral diagram in part (b) above. (i) Confirm the 1:2:3:4:3:2:1 intensities by considering the coupling to one ^{11}B nucleus and then adding in the effects of coupling to the second ^{11}B nucleus. (ii) Where else in the spectrum could you measure values of $J(^1H-^{11}B)$?

3. The $[BH_4]^-$ ion has a tetrahedral structure. Explain why the ^1H NMR spectrum exhibits a 1:1:1:1 multiplet, while the ^{11}B NMR spectrum shows a binomial quintet.

[*Ans.* Refer to case study 3 in Section 2.11]

No Al analogue of B_2H_6 exists, although monomeric AlH_3 has been isolated at low temperature in a matrix. In the solid state, X-ray and neutron diffraction data have shown that aluminium hydride consists of a three-dimensional network in which each Al centre is octahedral, involved in six Al–H–Al 3c-2e interactions. Digallane, Ga_2H_6, was fully characterized in the early 1990s, and electron diffraction data show it to be structurally similar to B_2H_6 (Ga–H$_{term}$ = 152 pm, Ga–H$_{bridge}$ = 171 pm, Ga–H–Ga = 98°). The existence of neutral, binary hydrides of In and Tl has not been confirmed. The hydrides of the group 13 elements are extremely air and moisture sensitive, and handling them requires the use of high vacuum techniques with all-glass apparatus.

$$\text{(12.2)}$$

Diborane(6) is an important reagent in synthetic organic chemistry, and reactions 12.8 and 12.9 give two convenient laboratory preparations; the structure of *diglyme*, used as solvent in reactions 12.8 and 12.9, is shown in diagram **12.2**. Reaction 12.10 is the basis for an industrial synthesis of B_2H_6.

$$3Na[BH_4] + 4Et_2O\cdot BF_3$$
$$\xrightarrow{\text{diglyme, 298 K}} 2B_2H_6 + 3Na[BF_4] + 4Et_2O \quad \text{(12.8)}$$

$$2Na[BH_4] + I_2 \xrightarrow{\text{diglyme, 298 K}} B_2H_6 + 2NaI + H_2 \quad \text{(12.9)}$$

$$2BF_3 + 6NaH \xrightarrow{\text{450 K}} B_2H_6 + 6NaF \quad \text{(12.10)}$$

Diborane(6) is a colourless gas (bp 180.5 K) which is rapidly decomposed by water (equation 12.11). Like other boron hydrides (see *Section 12.11*), B_2H_6 has a small positive value of $\Delta_f H^\circ$ (+36 kJ mol^{-1}); mixtures with air or O_2 are liable to inflame or explode (reaction 12.12).

$$B_2H_6 + 6H_2O \longrightarrow 2B(OH)_3 + 6H_2 \quad \text{(12.11)}$$

$$B_2H_6 + 3O_2 \longrightarrow B_2O_3 + 3H_2O$$
$$\Delta_r H^\circ = -2138 \text{ kJ per mole of } B_2H_6 \quad \text{(12.12)}$$

Digallane, Ga_2H_6, is prepared by reaction 12.13; the product condenses at low temperature as a white solid (mp 223 K) but decomposes above 253 K.

$$\text{(12.13)}$$

Figure 12.5 summarizes some reactions of B_2H_6 and Ga_2H_6. Compared with the much studied B_2H_6, Ga_2H_6 has received only recent attention, and not all reaction types can be

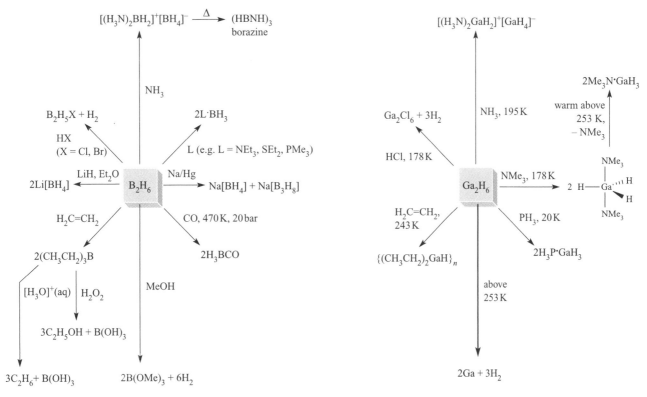

Fig. 12.5 Selected reactions of B_2H_6 and Ga_2H_6; all reactions of Ga_2H_6 must be carried out at low temperature since it decomposes above 253 K to gallium and dihydrogen. Borazine (top left-hand diagram) is discussed further in Section 12.8.

compared. However, three points should be noted:

- Ga_2H_6 is *unlike* B_2H_6 in that Ga_2H_6 rapidly decomposes to its constituent elements;
- Ga_2H_6 and B_2H_6 both react with HCl, but in the case of the borane, substitution of a terminal H by Cl is observed, whereas both terminal and bridging H atoms can be replaced in Ga_2H_6.
- Ga_2H_6 is *like* B_2H_6 in that it reacts with Lewis bases.

This last class of reaction is well documented and the examples in Figure 12.5 illustrate two reaction types with the steric demands of the Lewis base being an important factor in determining which pathway predominates. For example, two NH_3 molecules can attack the *same* B or Ga centre, resulting in *asymmetric cleavage* of the E_2H_6 molecule; in contrast, reactions with more sterically demanding Lewis bases tend to cause *symmetric cleavage* (equation 12.14).

Many of the reactions of B_2H_6 involve the non-isolable BH_3, and a value of $150 \, kJ \, mol^{-1}$ has been estimated for the dissociation enthalpy of B_2H_6 into $2BH_3$. Using this value, we can compare the Lewis acid strengths of BH_3, boron trihalides (BX_3) and boron trialkyls, and find that BH_3 lies between BX_3 and BMe_3 in behaviour towards simple Lewis bases such as NMe_3. However, only BH_3 forms adducts with CO and PF_3. Both CO and PF_3 are capable of acting as both electron donors (each using a lone pair of electrons centred on C or P respectively) and electron acceptors (using empty π^* or *d* orbitals in CO or PF_3 respectively). Formation of $OC \cdot BH_3$ and $F_3P \cdot BH_3$ suggests that BH_3 can also act in both capacities. Its electron acceptance is readily understood in terms of an empty atomic orbital, i.e. B has four valence atomic orbitals, but only three are used for bonding in BH_3. Electron donation by BH_3 is ascribed to *hyperconjugation* analogous to that proposed for a methyl group in organic compounds.[†]

Aluminium hydride can be prepared by reaction 12.15; the solvent can be Et_2O, but the formation of etherate complexes $(Et_2O)_nAlH_3$ complicates the synthesis.

$$3Li[AlH_4] + AlCl_3 \longrightarrow \frac{4}{n}[AlH_3]_n + 3LiCl \quad (12.15)$$

[†] For a discussion of hyperconjugation, see: J. March (1992) *Advanced Organic Chemistry: Reactions, Mechanisms and Structure*, 4th edn, Wiley, New York.

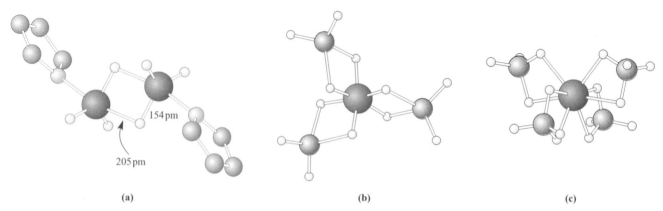

Fig. 12.6 (a) The structure of $[Al_2H_6(THF)_2]$ (X-ray diffraction at 173 K); hydrogen atoms have been omitted from the THF ligands [I.B. Gorrell *et al.* (1993) *J. Chem. Soc., Chem. Commun.*, p. 189]. (b) The structure of $[Al(BH_4)_3]$ deduced from spectroscopic studies. (c) The structure of $[Al(BH_4)_4]^-$ (X-ray diffraction) in the salt $[Ph_3MeP][Al(BH_4)_4]$ [D. Dou *et al.* (1994) *Inorg. Chem.*, vol. 33, p. 5443]. Colour code: B, dark blue; Al, black; H, pale grey; O, pale blue; C, grey.

Above 423 K, $[AlH_3]_n$ is unstable with respect to decomposition to the elements, and this thermal instability has potential for generating thin films of Al. Aluminium hydride reacts with Lewis bases, e.g. to give $Me_3N \cdot AlH_3$ (see *reaction 12.22*), in which the Al centre is tetrahedrally coordinated. As is general among the *p*-block elements, later elements in a group may exhibit higher coordination numbers than earlier congeners, and one example is $THF \cdot AlH_3$, the solid state structure of which is dimeric, albeit with asymmetrical Al−H−Al bridges (Figure 12.6a).

The $[MH_4]^-$ ions

We have already described (*Section 9.7*) the syntheses and reducing properties of $[BH_4]^-$ and $[AlH_4]^-$, and reactions 12.8 and 12.9 showed the use of $Na[BH_4]$ (the most important salt containing the $[BH_4]^-$ ion) as a precursor to B_2H_6. Sodium tetrahydroborate(1−) is a white non-volatile crystalline solid, a typical ionic salt with an NaCl lattice. It is stable in dry air and soluble in water, being kinetically, rather than thermodynamically, stable in water. Although insoluble in Et_2O, it dissolves in THF and polyethers. Despite the salt-like properties of $Na[BH_4]$, derivatives with some other metals are covalent, involving M−H−B 3c-2e interactions. An example is $[Al(BH_4)_3]$ (Figure 12.6b) in which the $[BH_4]^-$ ion behaves as a *didentate ligand* as in structure **12.3**. In *trans*-$[V(BH_4)_2(Me_2PCH_2CH_2PMe_2)_2]$, each $[BH_4]^-$ ligand is *monodentate* (**12.4**), forming one B−H−V bridge, and in $[Zr(BH_4)_4]$, the 12-coordinate Zr(IV) centre is surrounded by four *tridentate* ligands (**12.5**). Complex formation may (equation 12.16) or may not (equation 12.17) be accompanied by reduction of the central metal.

(12.5)

$$2[VCl_4(THF)_2] + 10[BH_4]^-$$
$$\longrightarrow 2[V(BH_4)_4]^- + 8Cl^- + B_2H_6 + H_2 + 4THF \quad (12.16)$$

$$HfCl_4 + 4[BH_4]^- \longrightarrow [Hf(BH_4)_4] + 4Cl^- \quad (12.17)$$

Although $[Al(BH_4)_3]$ is a widely cited example of a tetrahydroborate(1−) complex of Al(III), the first complex to be characterized by X-ray diffraction, $[Ph_3MeP][Al(BH_4)_4]$ (Figure 12.6c), was not reported until 1994. It is prepared by reaction 12.18 and provides the first example of a molecular species containing an 8-coordinate Al(III) centre; the coordination sphere is approximately dodecahedral (see *Figure 19.8*).

$$[Al(BH_4)_3] + [BH_4]^- \longrightarrow [Al(BH_4)_4]^- \quad (12.18)$$

In solution, many covalent complexes containing the $[BH_4]^-$ ligand exhibit dynamic behaviour which may be observed on the NMR spectroscopic timescale. For example, the room temperature ^1H NMR spectrum of $[Al(BH_4)_3]$ shows only one signal.

The salt $Li[AlH_4]$ is a widely used reducing and hydrogenating agent; it is obtained as a white solid by reaction 12.19 or 12.20, and is stable in dry air but is decomposed by water (equation 12.21).

$$4LiH + AlCl_3 \longrightarrow 3LiCl + Li[AlH_4] \quad (12.19)$$

$$Li + Al + 2H_2 \xrightarrow{250\,bar,\,400\,K,\,ether} Li[AlH_4] \quad (12.20)$$

$$Li[AlH_4] + 4H_2O \longrightarrow LiOH + Al(OH)_3 + 4H_2 \quad (12.21)$$

Adducts of aluminium hydride can be obtained from $[AlH_4]^-$ (e.g. reaction 12.22) and some of these compounds

(12.3)   (12.4)

are important reducing agents and polymerization catalysts in organic chemistry.

$$3Li[AlH_4] + AlCl_3 + 4Me_3N \longrightarrow 4Me_3N \cdot AlH_3 + 3LiCl$$
(12.22)

The compounds $Li[EH_4]$ for E = Ga, In and Tl have been prepared at low temperatures, (e.g. reaction 12.23) but are thermally unstable.

$$4LiH + GaCl_3 \longrightarrow Li[GaH_4] + 3LiCl$$
(12.23)

12.6 Halides and complex halides

Boron halides: BX_3 and B_2X_4

Boron trihalides are monomeric under ordinary conditions, possess trigonal planar structures (**12.6**), and are much more volatile than the corresponding compounds of Al. Boron trifluoride is a colourless gas (bp 172 K), BCl_3 and BBr_3 are colourless liquids (BCl_3, mp 166 K, bp 285 K; BBr_3, mp 227 K, bp 364 K), while BI_3 is a white solid (mp 316 K).

B–X distance:
X = F   131 pm
X = Cl 174 pm
X = Br 189 pm
X = I   210 pm

(12.6)

Equation 12.24 shows the usual synthesis of BF_3; excess H_2SO_4 removes the H_2O formed. Boron trifluoride fumes strongly in moist air and is partially hydrolyzed by excess H_2O (equation 12.25). With small amounts of H_2O at low temperatures, the adducts $BF_3 \cdot H_2O$ and $BF_3 \cdot 2H_2O$ are obtained.

$$B_2O_3 + 3CaF_2 + 3H_2SO_4 \longrightarrow 2BF_3 + 3CaSO_4 + 3H_2O$$
(12.24)

$$4BF_3 + 6H_2O \longrightarrow 3[H_3O]^+ + 3[BF_4]^- + B(OH)_3$$
(12.25)

Pure tetrafluoroboric acid, HBF_4, is *not* isolable but is commercially available in Et_2O solution, or as solutions formulated as $[H_3O][BF_4] \cdot 4H_2O$. It can also be formed by reaction 12.26.

$$B(OH)_3 + 4HF \longrightarrow [H_3O]^+ + [BF_4]^- + 2H_2O$$
(12.26)

Tetrafluoroboric acid is a very strong acid, and mixtures of HF and BF_3 are extremely strong proton donors, although not quite as strong as those of HF and SbF_5 (see *Section 8.7*). Salts containing the $[BF_4]^-$ ion are frequently encountered in synthetic chemistry; the $[BF_4]^-$ ion (like $[PF_6]^-$, structure **14.24**) coordinates very weakly, if at all, to metal centres and is often used as an 'innocent' anion to precipitate

cations. For a discussion of the stability of KBF_4 with respect to $KF + BF_3$, see *Section 5.16*.

(12.7)

Boron trifluoride forms a range of complexes with ethers, nitriles and amines. It is commercially available as the adduct $Et_2O \cdot BF_3$ (**12.7**). Being a liquid at 298 K, it is a convenient means of handling BF_3 which has many applications as a catalyst in organic reactions, e.g. in Friedel–Crafts alkylations and acylations.

The reactions between B and Cl_2 or Br_2 yield BCl_3 or BBr_3 respectively, while BI_3 is prepared by reaction 12.27. All three trihalides are decomposed by water (equation 12.28), and react with inorganic or organic compounds containing labile protons to eliminate HX (X = Cl, Br, I). Thus, while BF_3 forms an adduct with NH_3, BCl_3 reacts in liquid NH_3 to form $B(NH_2)_3$.

$$BCl_3 + 3HI \xrightarrow{\Delta} BI_3 + 3HCl$$
(12.27)

$$BX_3 + 3H_2O \longrightarrow B(OH)_3 + 3HX \quad X = Cl, Br, I$$
(12.28)

Unlike $[BF_4]^-$, the ions $[BCl_4]^-$, $[BBr_4]^-$ and $[BI_4]^-$ are stabilized only in the presence of large cations such as $[^nBu_4N]^+$.

In mixtures containing two or three of BF_3, BCl_3 and BBr_3, exchange of the halogen atoms occurs to yield BF_2Cl, $BFBr_2$, BFClBr etc. and their formation can be monitored by using ^{11}B or ^{19}F NMR spectroscopy.

The thermodynamics of adduct formation by BF_3, BCl_3 and BBr_3 have been much discussed, and reactions with NMe_3 (Lewis base L) in the gas phase show that the order of adduct stabilities is $L \cdot BF_3 < L \cdot BCl_3 < L \cdot BBr_3$. Determinations of $\Delta_r H^\circ$ for reaction 12.29 in nitrobenzene solution reveal the same sequence.

$$py(soln) + BX_3(g) \longrightarrow py \cdot BX_3(soln) \qquad py =$$
(12.29)

This sequence is the opposite of that predicted on the basis of the electronegativities of the halogens, but by considering changes in bonding during adduct formation, we may rationalize the experimental observations. In BX_3, the B–X bonds contain partial π-character (Figure 12.7a) (see *Section 4.3*). Reaction with a Lewis base, L, leads to a change in stereochemistry at the B centre from trigonal planar to tetrahedral and, as a result, the π-contributions to the B–X bonds are lost (Figure 12.7b). This is demonstrated by the observation that the B–F bond length increases from 130 pm in BF_3 to 145 pm in $[BF_4]^-$. We can formally consider adduct formation to occur

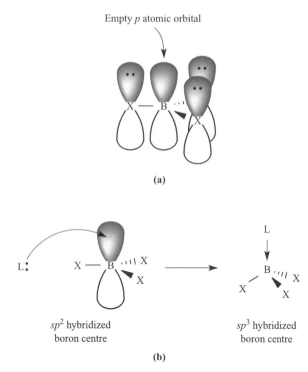

(a)

(b)

Fig. 12.7 (a) The formation of partial π-bonds in a trigonal planar BX_3 molecule can be considered in terms of the donation of electron density from filled p atomic orbitals on the X atoms into the empty $2p$ atomic orbital on boron. (b) Reaction of BX_3 with a Lewis base, L, results in a change from a trigonal planar (sp^2 boron centre) to tetrahedral (sp^3 boron centre) molecule.

in two steps: (i) the reorganization of trigonal planar to pyramidal B, and (ii) the formation of an L ⟶ B coordinate bond. The first step is endothermic, while the second is exothermic; the pyramidal BX_3 *intermediate* cannot be isolated and is only a *model* state. The observed ordering of adduct stabilities can now be understood in terms of the energy difference between that associated with loss of π-character (which is greatest for BF_3) and that associated with formation of the L ⟶ B bond. Evidence for the amount of π-character in BX_3 following the sequence $BF_3 > BCl_3 > BBr_3$ comes from the fact that the increase in the B–X bond distances in BX_3 (130, 176 and 187 pm for BF_3, BCl_3 and BBr_3) is greater than the increase in the values of r_{cov} for X (71, 99 and 114 pm for F, Cl and Br). It has been suggested that the presence of the π-bonding in boron trihalides is the reason why these molecules are monomeric, while the corresponding halides of the heavier group 13 elements are oligomeric (e.g. Al_2Cl_6); π-bonding is always stronger in compounds involving first-row elements (e.g. compare the chemistries of C and Si, or N and P, in Chapters 13 and 14).

(12.8)   (12.9)

Among the group 13 elements, B alone forms halides of the type $X_2B–BX_2$, although adducts of the type $LX_2M–MX_2L$ (M = Al, Ga; L = Lewis base) are closely related compounds, e.g. see structure **12.13**. At 298 K, B_2Cl_4 is a colourless, unstable liquid, and is prepared by co-condensing BCl_3 and Cu vapours on a surface cooled with liquid N_2; B_2Cl_4 is converted to B_2F_4 (a colourless gas at 298 K) by reaction with SbF_3. The compounds B_2Br_4 and B_2I_4 are, respectively, an easily hydrolyzed liquid and a pale yellow solid. In the solid state, B_2F_4 and B_2Cl_4 are planar (**12.8**), but in the vapour phase, B_2F_4 remains planar while B_2Cl_4 has a staggered structure (D_{2d}, **12.9**); B_2Br_4 adopts a staggered conformation in the vapour, liquid and solid phases. These preferences are not readily explained.

The thermal decomposition of B_2X_4 (X = Cl, Br, I) gives BX_3 and cluster molecules of type B_nX_n (X = Cl, n = 8–12; X = Br, n = 7–10; X = I, n = 8 or 9). Some degree of selectiveness can be achieved by fine tuning the reaction conditions (e.g. equation 12.30).

$$B_9Cl_9 \xleftarrow[\text{a few minutes}]{720\,K,} B_2Cl_4 \xrightarrow[\text{of } CCl_4,\text{ several days}]{373\,K,\text{ in presence}} B_8Cl_8 \qquad (12.30)$$

The cluster B_4Cl_4 can be obtained by passing an electrical discharge through BCl_3 in the presence of Hg. Figure 12.8 shows the structures of B_4Cl_4 and B_8Cl_8. Reactions of B_4Cl_4 may occur with retention of the cluster core (e.g. reaction 12.31) or its fragmentation (e.g. reaction 12.32), and reactions of B_8Cl_8 are often accompanied by cage expansion (e.g. reaction 12.33), an exception being Friedel–Crafts bromination which gives B_8Br_8.

$$B_4Cl_4 + 4Li^tBu \longrightarrow B_4{}^tBu_4 + 4LiCl \qquad (12.31)$$

$$B_4Cl_4 \xrightarrow{480\,K,\,CFCl_3} BF_3 + B_2F_4 \qquad (12.32)$$

$$B_8Cl_8 \xrightarrow{AlMe_3} B_9Cl_{9-n}Me_n \qquad n = 0–4 \qquad (12.33)$$

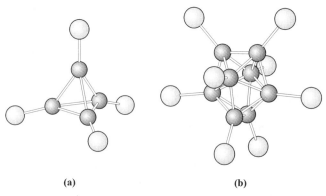

(a)   **(b)**

Fig. 12.8 The family of B_nX_n (X = Cl, Br, I) molecules possess cluster structures. (a) B_4Cl_4 has a tetrahedral core and (b) B_8Cl_8 possesses a dodecahedral cluster core. Colour code: B, blue; Cl, pale grey.

BOX 12.4   **APPLICATIONS**

Lewis acid pigment solubilization

Applications of pigments for coatings, printing and information storage are widespread, but the fabrication of thin films of pigments is difficult because of their insoluble nature. Dyes, on the other hand, are easier to manipulate. Research at the Xerox Corporation has shown that Lewis acid complexes can be utilized to solubilize and lay down thin films of certain pigments. For example, the photosensitive perylene derivative shown below forms an adduct with $AlCl_3$:

Complex formation occurs in $MeNO_2$ solution and the solution is then applied to the surface to be coated. Washing with water removes the Lewis acid leaving a thin film of the photosensitive pigment. The Lewis acid pigment solubilization (LAPS) technique has been used to fabricate multilayer photoconductors and appears to have a promising technological future.

Further reading

B.R. Hsieh and A.R. Melnyk (1998) *Chemistry of Materials*, vol. 10, p. 2313 – 'Organic pigment nanoparticle thin film devices via Lewis acid pigment solubilization'.

Analysis of the bonding in any of these clusters poses problems; if the terminal B–X bonds are considered to be localized 2c-2e interactions, then there are insufficient valence electrons remaining for a localized treatment of the B–B interactions in the B_n core. We return to this problem at the end of Section 12.11.

Al(III), Ga(III), In(III) and Tl(III) halides and their complexes

The trifluorides of Al, Ga, In and Tl are non-volatile solids, best prepared by fluorination of the metal (or one of its simple compounds) with F_2; AlF_3 is also prepared by reaction 12.34.

$$Al_2O_3 + 6HF \xrightarrow{970\,K} 2AlF_3 + 3H_2O \qquad (12.34)$$

Each trifluoride is high melting and has an infinite lattice structure. In AlF_3, each Al centre is octahedral, surrounded by six F atoms each of which links two Al centres. The octahedral AlF_6-unit is encountered in other Al fluorides: Tl_2AlF_5 contains polymeric chains composed of AlF_6-octahedra linked through opposite vertices (represented by either **12.10** or **12.11**), and in $TlAlF_4$ and $KAlF_4$, AlF_6 octahedra are linked through four vertices to form sheets.

(12.11)

Cryolite, $Na_3[AlF_6]$ (see *Section 12.2*) occurs naturally but is also synthesized (reaction 12.35) to meet commercial needs. The solid state structure of cryolite is related to the perovskite lattice (see *Figure 5.23*).

$$Al(OH)_3 + 6HF + 3NaOH \longrightarrow Na_3[AlF_6] + 6H_2O \quad (12.35)$$

Compounds MX_3 (M = Al, Ga or In; X = Cl, Br or I) are obtained by direct combination of the elements. They are relatively volatile and in the solid state possess layer lattices or lattices containing dimers M_2X_6. The vapours consist of dimeric molecules and these are also present in solutions of the compounds in inorganic solvents. Only at high temperatures does dissociation to monomeric MX_3 occur. In the monomer, the group 13 metal is trigonal planar, but in the dimer, a tetrahedral environment results from X $\longrightarrow$ M coordinate bond formation involving a halogen lone pair of electrons (Figure 12.9). Solid $AlCl_3$ adopts a layer lattice with octahedrally sited Al.

When water is dripped on to solid $AlCl_3$, vigorous hydrolysis occurs, but in *dilute* aqueous solution, $[Al(H_2O)_6]^{3+}$ (see *equation 6.34*) and Cl^- ions are present. In coordinating solvents such as Et_2O, $AlCl_3$ forms adducts such as $Et_2O\cdot AlCl_3$, structurally analogous to **12.6**. (A commercial application of $AlCl_3$ adducts is highlighted in *Box 12.4*.)

F F   F F   F F
\|/   \|/   \|/
—Al—F—Al—F—Al—F
/|\   /|\   /|\
F F   F F   F F

(12.10)

(a)

Cl   Cl

      Cl

Al   Al

      Cl

Cl   Cl

(b)

Fig. 12.9 (a) The structure of Al_2Cl_6 with bond distances determined in the vapour phase; the terminal M−X bond distances are similarly shorter than the bridging distances in Al_2Br_6, Al_2I_6, Ga_2Cl_6, Ga_2Br_6, Ga_2I_6 and In_2I_6. In $AlCl_3$ monomer, the Al−Cl distances are 206 pm. Colour code: Al, grey; Cl, pale grey. (b) A representation of the bonding in Al_2Cl_6 showing the Cl lone pair donation to Al.

Addition of Cl^- to $AlCl_3$ yields the tetrahedral $[AlCl_4]^-$ and this reaction is important in Friedel–Crafts acylations and alkylations, the initial steps in which are summarized in equation 12.36.

$$RC\overset{+}{\equiv}O + [AlCl_4]^- \xleftarrow{RC(O)Cl} AlCl_3 \xrightarrow{RCl} R^+ + [AlCl_4]^-$$

$$(12.36)$$

Gallium and indium trichlorides and tribromides also form adducts, but with coordination numbers of 4, 5 or 6: $[MCl_6]^{3-}$, $[MBr_6]^{3-}$, $[MCl_5]^{2-}$, $[MCl_4]^-$ and $[MBr_4]^-$ (M = Ga or In) and $L\cdot GaX_3$ or $L_3\cdot InX_3$ (L = neutral Lewis base). The square-based pyramidal structure of $[InCl_5]^{2-}$ has been confirmed by X-ray diffraction for the $[Et_4N]^+$ salt; this is not expected by VSEPR arguments, but one must bear in mind that energy differences between 5-coordinate geometries are often small and preferences can be tipped by, for example, crystal packing forces.

The Tl(III) halides are less stable than those of the earlier group 13 elements; $TlCl_3$ and $TlBr_3$ are very unstable with respect to conversion to the Tl(I) halides (equation 12.37).

$$TlBr_3 \longrightarrow TlBr + Br_2 \qquad (12.37)$$

The compound TlI_3 is isomorphous with the alkali metal triiodides and is really Tl(I) triiodide, **12.12**. However, when treated with excess I^-, an interesting redox reaction occurs with the formation of $[TlI_4]^-$ (see *Section 12.9*). The decrease in stability of the higher oxidation state on going from the binary fluoride to iodide is a general feature of all metals that exhibit more than one oxidation state. For ionic compounds, this is easily explained in terms of lattice energies, the increase in lattice energy accompanying an

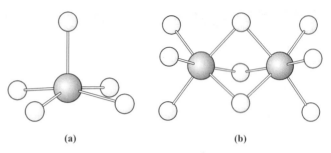

(a)   **(b)**

Fig. 12.10 (a) The structure of $[TlCl_5]^{2-}$ determined by X-ray diffraction for the salt $[H_3N(CH_2)_5NH_3][TlCl_5]$. [M.A. James *et al.* (1996) *Can. J. Chem.*, vol. 74, p. 1490.] (b) The crystallographically determined structure of $[Tl_2Cl_9]^{3-}$ in $Cs_3[Tl_2Cl_9]$. Colour code: Tl, blue; Cl, pale grey.

increase in oxidation state being greatest for the smallest anions.

$$Tl^+ \quad \left[I\!\!-\!\!-\!\!I\!\!-\!\!-\!\!I \right]^-$$

$$(12.12)$$

Thallium(III) exhibits coordination numbers higher than four in complex chlorides, prepared by addition of chloride salts to $TlCl_3$. In $[H_3N(CH_2)_5NH_3][TlCl_5]$, a square-based pyramidal structure for the anion has been confirmed (Figure 12.10a). In $K_3[TlCl_6]$, the anion has the expected octahedral structure, and in $Cs_3[Tl_2Cl_9]^{3-}$, the Tl(III) centres in the anion are also octahedral (Figure 12.10b).

Lower oxidation state Al, Ga, In and Tl halides

Aluminium(I) halides are formed in reactions of Al(III) halides with Al at 1270 K followed by rapid cooling; red AlCl is also formed by treating the metal with HCl at 1170 K. The monohalides are unstable with respect to disproportionation (equation 12.38).

$$3AlX \longrightarrow 2Al + AlX_3 \qquad (12.38)$$

The reaction of AlBr with PhOMe at 77 K followed by warming to 243 K yields $[Al_2Br_4(OMePh)_2]$, **12.13**; this is air- and moisture-sensitive and decomposes at 298 K, but represents a close relation of the X_2B-BX_2 compounds described earlier. The compound $Ga_2Br_4py_2$ (py = pyridine) is structurally similar to **12.13** and the Ga−Ga bond length of 242 pm corresponds to a single bond (r_{cov} = 122 pm).

Me

Br   O —— Ph

Br

        Al —— Al

                  Br

Ph —— O

            Br

Me

$$(12.13)$$

Gallium(I) chloride forms when $GaCl_3$ is heated at 1370 K, but has not been isolated as a pure compound. When $GaCl_3$ is heated with Ga, a compound of stoichiometry '$GaCl_2$' is formed, but crystallographic and magnetic data show this is $Ga^+[GaCl_4]^-$. The mixed In(I)/In(III) compound $In[InCl_4]$ is prepared in a similar way to its Ga analogue; InCl can also be isolated from the $InCl_3$/In reaction mixture and has a deformed NaCl lattice.

Thallium(I) halides, TlX, are stable compounds which in some ways resemble Ag(I) halides. Thallium(I) fluoride is very soluble in water, but TlCl, TlBr and TlI are sparingly soluble; the trend in solubilities can be traced to increased covalent contributions in the 'ionic' lattices for the larger halides, a situation that parallels the trend for the Ag(I) halides (see *Section 5.15*). In the solid state, TlF has a distorted NaCl lattice, while TlCl and TlBr adopt CsCl structures. Thallium(I) iodide is dimorphic; below 443 K, the yellow form adopts a lattice derived from an NaCl structure in which neighbouring layers are slipped with respect to each other and, above 443 K, the red form crystallizes with a CsCl lattice. Under high pressures, TlCl, TlBr and TlI become metallic in character.

12.7 Oxides, oxoacids, oxoanions and hydroxides

It is a general observation that, within the *p*-block, basic character increases down a group. Thus:

- boron oxides are exclusively acidic;
- aluminium and gallium oxides are amphoteric;
- indium and thallium oxides are exclusively basic.

Thallium(I) oxide is soluble in water and the resulting hydroxide is as strong a base a KOH.

Boron oxides, oxoacids and oxoanions

The principal oxide of boron, B_2O_3, is obtained as a vitreous solid by dehydration of boric acid at red heat (equation 12.2), or in a crystalline form by controlled dehydration. The latter possesses a three-dimensional, covalent structure comprising planar BO_3 units (B–O = 138 pm) which share O atoms, but which are mutually twisted with respect to each other to give a rigid lattice. Under high pressure and at 803 K, a transition to a more dense form occurs, the change in density being 2.56 to 3.11 $g\,cm^{-3}$. This second polymorph contains tetrahedral BO_4 units, which are irregular because three O atoms are shared among three BO_4 units, while one atom connects two BO_4 units. Heating B_2O_3 with B at 1273 K gives BO; its structure has not been established, but the fact that reaction with water yields $(HO)_2BB(OH)_2$ suggests it contains B–B bonds. Trigonal planar and tetrahedral B exemplified in the polymorphs of B_2O_3 occur frequently in boron–oxygen chemistry.

The commercial importance of B_2O_3 is in its use in the borosilicate glass industry (see *Box 12.5*). As a Lewis acid, B_2O_3 is a valuable catalyst; BPO_4 (formed by reacting B_2O_3 with P_4O_{10}) catalyzes the hydration of alkenes and dehydration of amides to nitriles. The structure of BPO_4 can be considered in terms of SiO_2 (see *Section 13.9*) in which alternate Si atoms have been replaced by B or P atoms.

Water is taken up slowly by B_2O_3 giving $B(OH)_3$ (ortho-boric or boric acid), but above 1270 K, molten B_2O_3 reacts rapidly with steam to give $B_3O_3(OH)_3$ (metaboric acid, Figure 12.11a). Industrially, boric acid is obtained from borax (reaction 12.1), and heating $B(OH)_3$ converts it to

---

BOX 12.5   **APPLICATIONS**

B_2O_3 in the glass industry

The glass industry in Western Europe and the US accounts for about half the B_2O_3 consumed. Fused B_2O_3 dissolves metal oxides to give metal borates. Fusion with Na_2O or K_2O results in a viscous molten phase, rapid cooling of which produces a glass; fusion with appropriate metal oxides leads to coloured metal borate glasses. Borosilicate glass is of particular commercial importance. It is formed by fusing B_2O_3 and SiO_2 together; a metal oxide component sometimes may be added. Borosilicate glasses include *Pyrex* which is used to manufacture most laboratory glassware as well as kitchenware. It contains a high proportion of SiO_2 and exhibits a low linear coefficient of expansion. Pyrex glass can be heated and cooled rapidly without breaking, and is resistant to attack by alkalis or acids. The refractive index of Pyrex is 1.47, and if a piece of clean Pyrex glassware is immersed in a mixture of $MeOH/C_6H_6$, 16/84 by weight, it

seems to 'disappear'; this gives a quick way of testing if a piece of glassware is made from Pyrex. Although the linear coefficient of expansion of silica glass is lower than that of Pyrex glass (0.8 versus 3.3), the major advantage of borosilicate over silica glass is its workability. The softening point (i.e. the temperature at which the glass can be worked and blown) of fused silica glass is 1983 K, while that of Pyrex is 1093 K.

Fibreglass falls into two categories: textile fibres and insulation fibreglass. Of the textile fibres, alumino-borosilicate glass has the most widespread applications. The fibres possess high tensile strength and low thermal expansion, and are used in reinforced plastics. Insulation fibreglass includes glass wool which contains ≈55–60% SiO_2, ≈3% Al_2O_3, ≈10–14% Na_2O, 3–6% B_2O_3 plus other components such as CaO, MgO and ZrO_2.

(a)

(b)

Fig. 12.11 (a) The structure of metaboric acid, $B_3O_3(OH)_3$. (b) Schematic representation of part of one layer of the solid state lattice of boric acid (orthoboric acid), $B(OH)_3$; covalent bonds within each molecule are highlighted in bold, and intermolecular hydrogen bonds are shown by blue hashed lines. The hydrogen bonds are asymmetrical, with $O-H = 100\,pm$, and $O\cdots O = 270\,pm$.

$B_3O_3(OH)_3$. Both boric acids have layer structures in which molecules are linked by hydrogen bonds; the slippery feel of $B(OH)_3$ and its use as a lubricant are consequences of the layers (Figure 12.11b). In aqueous solution, $B(OH)_3$ behaves as a weak acid, but is a *Lewis* rather than Brønsted acid (equation 12.39). Complex formation with 1,2-diols leads to an increase in acid strength (equation 12.40).

$$B(OH)_3(aq) + 2H_2O(l) \rightleftharpoons [B(OH)_4]^-(aq) + [H_3O]^+(aq)$$

$$pK_a = 9.1 \qquad (12.39)$$

$$(12.40)$$

Many borate anions exist and metal borates such as *colemanite* $(Ca[B_3O_4(OH)_3]\cdot H_2O)$, *borax* $(Na_2[B_4O_5(OH)_4]\cdot 8H_2O)$, *kernite* $(Na_2[B_4O_5(OH)_4]\cdot 2H_2O)$ and *ulexite* $(NaCa[B_5O_6(OH)_6]\cdot 5H_2O)$ occur naturally. The solid state structures of borates are well established, and Figure 12.12 shows selected anions. In planar BO_3 groups, $B-O \approx 136\,pm$, but in tetrahedral BO_4 units, $B-O \approx 148\,pm$. This increase is similar to that observed on going from BF_3 to $[BF_4]^-$ (see *Section 12.6*) and suggests that $B-O$ π-bonding involving O lone pairs is present in *planar* BO_3 units. This is lost on going to a tetrahedral BO_4 unit. While solid state data abound, less is known about the nature of borate anions in aqueous solution. It is possible to distinguish between trigonal planar and tetrahedral B using ^{11}B NMR spectroscopy and data show that species containing only 3-coordinate B are unstable in solution and rapidly convert to species with 4-coordinate B. The species present in solution are also pH- and temperature-dependent.

The reactions of $B(OH)_3$ with Na_2O_2, or borates with H_2O_2 yield sodium peroxoborate (commonly known as sodium perborate). This is an important constituent of washing powders because it hydrolyzes in water to give H_2O_2 and so is a bleaching agent. On an industrial scale, sodium peroxoborate is manufactured from $Na_2B_4O_7$ by electrolysis. The solid state structure of sodium peroxoborate has been

Fig. 12.12 The structures of selected borate anions; trigonal planar and tetrahedral B atoms are present, and each *tetrahedral* B carries a negative charge. The $[B_4O_5(OH)_4]^{2-}$ anion occurs in the minerals *borax* and *kernite*. In the pyroborate ion, $[B_2O_5]^{4-}$, the B—O—B bond angle depends on the cation present, e.g. $\angle$B—O—B = 153° in $Co_2B_2O_5$, and 131.5° in $Mg_2B_2O_5$.

determined by X-ray diffraction and contains anion **12.14**; the compound is formulated as $Na_2[B_2(O_2)_2(OH)_4]\cdot6H_2O$.

(12.14)

Aluminium oxides, oxoacids, oxoanions and hydroxides

Aluminium oxide occurs in two main forms: α-alumina (*corundum*) and γ-Al_2O_3 (*activated alumina*). The solid state structure of α-Al_2O_3 consists of an hcp array of O^{2-} ions with cations occupying two-thirds of the octahedral interstitial sites. α-Alumina is extremely hard and is relatively unreactive; its density (4.0 g cm^{-3}) exceeds that of γ-Al_2O_3 (3.5 g cm^{-3}) which has a defect spinel structure (see *Box 12.6* and *Section 20.9*). The α-form is made by dehydrating $Al(OH)_3$ or AlO(OH) at ≈1300 K, while dehydration of γ-AlO(OH) below 720 K gives γ-Al_2O_3. Both $Al(OH)_3$ or AlO(OH) occur as minerals: *diaspore*, α-AlO(OH), *boehmite*, γ-AlO(OH), and *gibbsite*, γ-$Al(OH)_3$; α-$Al(OH)_3$ (*bayerite*)

does not occur naturally but can be prepared by reaction 12.41. Precipitates of γ-AlO(OH) are formed when NH_3 is added to solutions of Al salts.

$$2Na[Al(OH)_4](aq) + CO_2(g)$$
$$\longrightarrow 2Al(OH)_3(s) + Na_2CO_3(aq) + H_2O(l) \quad (12.41)$$

The catalytic and adsorbing properties of γ-Al_2O_3, AlO(OH) and $Al(OH)_3$ make this group of compounds invaluable commercially. One use of $Al(OH)_3$ is as a *mordant*, i.e. it absorbs dyes and is used to fix them to fabrics. The amphoteric nature of γ-Al_2O_3 and $Al(OH)_3$ is illustrated in reactions 12.42–12.45; equation 12.44 shows the formation of an *aluminate* when $Al(OH)_3$ dissolves in excess alkali.

$$\gamma\text{-}Al_2O_3 + 3H_2O + 2[OH]^- \longrightarrow 2[Al(OH)_4]^- \quad (12.42)$$

$$\gamma\text{-}Al_2O_3 + 3H_2O + 6[H_3O]^+ \longrightarrow 2[Al(H_2O)_6]^{3+} \quad (12.43)$$

$$Al(OH)_3 + [OH]^- \longrightarrow [Al(OH)_4]^- \quad (12.44)$$

$$Al(OH)_3 + 3[H_3O]^+ \longrightarrow [Al(H_2O)_6]^{3+} \quad (12.45)$$

For use as the stationary phases in chromatography, acidic, neutral or basic forms of alumina (made by controlled addition of water to γ-Al_2O_3) are commercially available.

The electrical and/or magnetic properties of a number of mixed oxides of Al and other metals including members of the spinel family (see *Box 12.6*) and sodium β-alumina (see

BOX 12.6   **CHEMICAL AND THEORETICAL BACKGROUND**

'Normal' spinel and 'inverse' spinel lattices

A large group of minerals called *spinels* have the general formula AB_2X_4 in which X is most commonly oxygen and the oxidation states of metals A and B are +2 and +3 respectively; examples include $MgAl_2O_4$ (*spinel*, after which this structural group is named), $FeCr_2O_4$ (*chromite*) and Fe_3O_4 (*magnetite*, a mixed Fe(II), Fe(III) oxide). The spinel family also includes sulfides, selenides and tellurides, and may contain metal ions in the +4 and +2 oxidation states, e.g. $TiMg_2O_4$, usually written as Mg_2TiO_4. Our discussion below focuses on spinel-type compounds containing A^{2+} and B^{3+} ions.

The spinel lattice is not geometrically simple but can be considered in terms of a cubic close-packed array of O^{2-} ions with one-eighth of the tetrahedral holes occupied by A^{2+} ions and half of the octahedral holes occupied by B^{3+} ions. The unit cell contains eight formula units, i.e. $[AB_2X_4]_8$.

Some mixed metal oxides AB_2X_4 in which at least one of the metals is a *d*-block element (e.g. $CoFe_2O_4$) possess an *inverse spinel* structure which is derived from the spinel lattice by exchanging the sites of the A^{2+} ions with half of the B^{3+} ions.

The occupation of octahedral sites may be ordered or random, and structure types cannot be simply partitioned into 'normal' or 'inverse'. A parameter λ is used to provide information about the distribution of cations in the interstitial sites of the close-packed array of X^{2-} ions; λ indicates the proportion of B^{3+} ions occupying *tetrahedral* holes. For a normal spinel, $\lambda = 0$; for an inverse spinel, $\lambda = 0.5$. Thus, for $MgAl_2O_4$, $\lambda = 0$, and for $CoFe_2O_4$, $\lambda = 0.5$. Other spinel-type compounds have values of λ between 0 and 0.5; for example, for $MgFe_2O_4$, $\lambda = 0.45$ and for $NiAl_2O_4$, $\lambda = 0.38$. We discuss factors governing the preference for a normal or inverse spinel structure in Section 20.9.

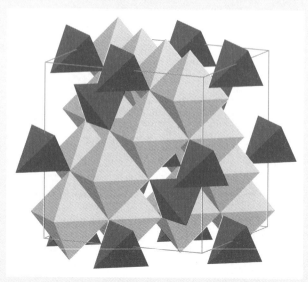

The inverse spinel structure of Fe_3O_4 showing the unit cell and the tetrahedral and octahedral environments of the Fe centres. The vertex of each tetrahedron is occupied by an O atom.

Section 27.3) have extremely important industrial applications. In this section, we single out $Ca_3Al_2O_6$ because of its role in cement manufacture, and because it contains a discrete aluminate ion. Calcium aluminates are prepared from CaO and Al_2O_3, the product depending on the stoichiometry of the reactants; $Ca_3Al_2O_6$ comprises Ca^{2+} and $[Al_6O_{18}]^{18-}$ ions and is a major component in Portland cement. The cyclic $[Al_6O_{18}]^{18-}$ ion, **12.15**, is isostructural with $[Si_6O_{18}]^{12-}$ (see *Section 13.9*) and the presence of these units in the solid state lattice imparts a very open structure which facilitates the formation of hydrates, a property crucial to the setting of cement.

(12.15)

BOX 12.7   **APPLICATIONS**

The unusual properties of indium–tin oxide (ITO)

Thin films of indium–tin oxide (ITO) have commercially valuable properties: it is transparent, electrically conducting and reflects IR radiation. Applications of ITO are varied. It is used as a coating material for flat-panel computer displays, for coating architectural glass panels, and in electrochromic devices. Coating motor vehicle and aircraft windscreens and motor vehicle rear windows allows them to be electrically heated for deicing purposes. A thin film of ITO (or related material) on the cockpit canopy of an aircraft such as the stealth plane renders this part of the plane radar-silent, contributing to the sophisticated design that allows the stealth plane to go undetected by radar.

Related information: see *Box 22.4 – Electrochromic 'smart' windows*

Oxides of Ga, In and Tl

The oxides and related compounds of the heavier group 13 metals call for less attention than those of Al. Gallium, like Al, forms more than one polymorph of Ga_2O_3, $GaO(OH)$ and $Ga(OH)_3$, and the compounds are amphoteric. This contrasts with the basic nature of In_2O_3, $InO(OH)$ and $In(OH)_3$. Thallium is unique among the group in exhibiting an oxide for the M(I) state: Tl_2O forms when Tl_2CO_3 is heated in N_2, and reacts with water (equation 12.46). Thallium(III) forms an oxide, but no hydroxide.

$$Tl_2O + H_2O \longrightarrow 2TlOH \qquad (12.46)$$

12.8 Compounds containing nitrogen

The BN unit is isoelectronic with C_2 and many boron–nitrogen analogues of carbon systems exist. However useful this analogy is *structurally*, a BN group does *not* mimic a CC unit *chemically*, and reasons for this difference can be understood by considering the electronegativity values $\chi^P(B) = 2.0$, $\chi^P(C) = 2.6$ and $\chi^P(N) = 3.0$.

Nitrides

Boron nitride, BN, is a robust (sublimation point = 2603 K), chemically rather inert compound which is used as a ceramic material (e.g. in crucible manufacture). Preparative routes include the high temperature reactions of borax with $[NH_4]Cl$, B_2O_3 with NH_3, and $B(OH)_3$ with $[NH_4]Cl$. High purity boron nitride can be made by reacting NH_3 with BF_3 or BCl_3. The fabrication of thin films of BN is described in *Section 27.6*. The common form of boron nitride has an ordered layer structure containing hexagonal rings (Figure 12.13, compare with *Figure 13.2*). The layers are arranged so that a B atom in one layer lies directly over an N atom in the next, and so on. The B–N distances within a layer

Fig. 12.13 Part of the layer structure of the common polymorph of boron nitride, BN. Hexagonal rings in adjacent layers lie over one another so that B and N atoms are eclipsed.

are much shorter than those between layers (Figure 12.13) and, in Table 12.2, it is compared with those in other B–N species. The B–N bonds are shorter than in adducts such as $Me_3N \cdot BBr_3$ in which a single boron–nitrogen bond can be assigned, and implies the presence of π-bonding in BN resulting from overlap between orthogonal N $2p$ (occupied) and B $2p$ (vacant) orbitals. The *interlayer* distance of 330 pm is consistent with van der Waals interactions, and boron nitride acts as a good lubricant, thus resembling graphite. Unlike graphite, BN is white and an insulator. This difference can be interpreted in terms of band theory (see *Section 5.8*), with the band gap in boron nitride being considerably greater than that in graphite because of the polarity of the B–N bond.

Table 12.2 Boron–nitrogen bond distances in selected species; all data are from X-ray diffraction studies (≤ 298 K).

Species	B–N distance / pm	Comment
$Me_3N \cdot BBr_3$	160.2	Single bond
$Me_3N \cdot BCl_3$	157.5	Single bond
Cubic-$(BN)_n$	157	Single bond
Hexagonal-$(BN)_n$	144.6	Intralayer distance, see *Figure 12.13*; some π-contribution
$B(NMe_2)_3$	143.9	Some π-contribution
$Mes_2\bar{B}=\overset{+}{N}H_2$	137.5	Double bond
$Mes_2\bar{B}=\overset{+}{N}=\bar{B}Mes_2{}^\ddagger$	134.5	Double bond
$^tBu\bar{B}\equiv\overset{+}{N}{}^tBu$	125.8	Triple bond

‡ $Mes = 2,4,6\text{-}Me_3C_6H_2$

Heating the layered form of BN at $\approx 2000\,K$ and $>50\,kbar$ pressure in the presence of catalytic amounts of Li_3N or Mg_3N_2 converts it to a more dense polymorph with the zinc blende structure (see *Section 5.11*). Table 12.2 shows that the $B-N$ bond distance in cubic-BN is similar to those in $R_3N\cdot BR_3$ adducts and longer than in the layered form of boron nitride; this further supports the existence of π-bonding within the layers of the latter. Structurally, the cubic form of BN resembles diamond (*Figure 5.19*) and the two materials are almost equally hard; crystalline cubic BN is called *borazon* and is used as an abrasive. A third polymorph of boron nitride with a wurtzite lattice is formed by compression of the layered form at $\approx 12\,kbar$.

Of the group 13 metals, only Al reacts directly with N_2 (at $1020\,K$) to form a nitride; AlN has a wurtzite lattice and is hydrolyzed to NH_3 by hot dilute alkali. Gallium and indium nitrides also crystallize with the wurtzite structure, and are more reactive than their B or Al counterparts. The importance of the group 13 metal nitrides, and of the related MP, MAs and MSb (M = Al, Ga, In) compounds lies in their applications in the semiconductor industry (see also *Section 18.4*).

Molecular species containing boron–nitrogen bonds

We have already described the formation of $B-N$ single bonds in adducts $R_3N\cdot BH_3$, and now we extend the discussion to include compounds with boron–nitrogen multiple bonds.

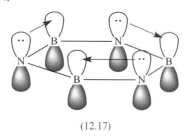

(12.16)

The hexagonal B_3N_3-motif in the layered form of boron nitride appears in a group of compounds called *borazines*. The parent compound $(HBNH)_3$, **12.16**, is isoelectronic and isostructural with benzene. It is prepared by reaction 12.47, from B_2H_6 (Figure 12.5) or from the *B*-chloro-derivative, itself prepared from BCl_3 (equation 12.48).

$$NH_4Cl + Na[BH_4] \xrightarrow{-NaCl,\, -H_2} H_3N\cdot BH_3 \xrightarrow{\Delta} (HBNH)_3 \quad (12.47)$$

$$BCl_3 + 3NH_4Cl \xrightarrow{420\,K,\, C_6H_5Cl} (ClBNH)_3$$
$$\xrightarrow{Na[BH_4]} (HBNH)_3 \quad (12.48)$$

The use of an alkylammonium chloride in place of NH_4Cl in reaction 12.48 leads to the formation of an *N*-alkyl derivative

$(ClBNR)_3$ which can be converted to $(HBNR)_3$ by treatment with $Na[BH_4]$.

(12.17)

Borazine is a colourless liquid (mp 215 K, bp 328 K) with an aromatic odour and *physical* properties that resemble those of benzene. The $B-N$ distances in the planar B_3N_3 ring are equal (144 pm) and close to those in the layered form of BN (Table 12.2). This is consistent with substantial, but not complete, delocalization of the N lone pairs around the ring as represented in **12.17**. Structure **12.16** gives one resonance form of borazine, analogous to a Kekulé structure for benzene. Despite the formal charge distribution, a consideration of the relative electronegativities of B and N indicates that B is susceptible to attack by nucleophiles while N attracts electrophiles (Figure 12.14). Thus, the reactivity of borazine contrasts sharply with that of benzene, although it must be remembered that C_6H_6 is *kinetically* inert towards the addition of, for example, HCl and H_2O. Equations 12.49 and 12.50 give representative reactions of borazine; the formula notation indicates the nature of the *B*- or *N*-substituents, e.g. $(ClHBNH_2)_3$ contains Cl attached to B.

$$(HBNH)_3 + 3HCl \longrightarrow (ClHBNH_2)_3 \quad \text{addition reaction} \quad (12.49)$$

$$(HBNH)_3 + 3H_2O \longrightarrow \{H(HO)BNH_2\}_3 \quad \text{addition reaction} \quad (12.50)$$

Each of the products of these reactions possesses a chair conformation (compare cyclohexane), and treatment of

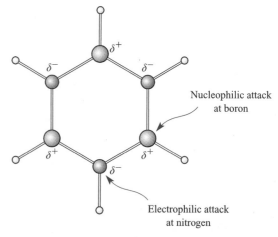

Fig. 12.14 The difference in electronegativities of boron and nitrogen leads to a charge distribution which makes the B atoms (shown in blue) and N atoms (shown in grey), respectively, susceptible to nucleophilic and electrophilic attack.

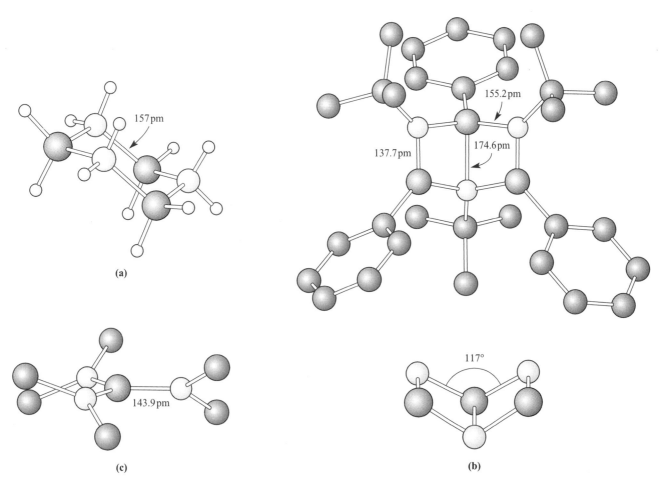

Fig. 12.15 The structures (determined by X-ray diffraction) of (a) $B_3N_3H_{12}$ [P.W.R. Corfield *et al.* (1973) *J. Am. Chem. Soc.*, vol. 95, p. 1480], (b) the Dewar borazine derivative N,N',N''-$^tBu_3$-B,B',B''-$Ph_3B_3N_3$ [P. Paetzold *et al.* (1991) *Z. Naturforsch., Teil B*, vol. 46, p. 853], (c) $B(NMe_2)_3$ [G. Schmid *et al.* (1982) *Z. Naturforsch., Teil B*, vol. 37, p. 1230, structure determined at 157 K]; H atoms in (b) and (c) have been omitted. Colour code: B, blue; N, pale grey; C, grey; H, pale grey.

$(ClHBNH_2)_3$ with $Na[BH_4]$ leads to the formation of $(H_2BNH_2)_3$ (Figure 12.15a).

(12.18)

Dewar borazine derivatives **12.18** can be stabilized by the introduction of sterically demanding substituents. Figure 12.15b shows the structure of N,N',N''-$^tBu_3$-B,B',B''-$Ph_3B_3N_3$; the 'open-book' conformation of the B_3N_3 framework mimics that of the C_6-unit in Dewar benzene. By comparing the bond distances in Figure 12.15b with those in Table 12.2, we see that the central B–N bond in **12.18** is longer than a typical single bond, the four distances of 155 pm (Figure 12.15b) are close to those expected for single bonds, and the two remaining B–N bond lengths correspond to double bonds. Dewar borazines are prepared by cyclotrimerization of iminoboranes $RBNR'$ (**12.19**),

(12.19)   (12.20)

although cyclooligomerization processes are not simple.[†] A family of $RBNR'$ compounds is now known, and can be rendered kinetically stable with respect to oligomerization by the introduction of bulky substituents and/or maintaining low temperatures. They can be made by elimination of a suitable species from compounds of type **12.20** (e.g. reaction 12.51) and possess very short B–N bonds (Table 12.2) consistent with triple bond character.

$$\xrightarrow{\Delta, \approx 10^{-3}\,bar} Me_3SiCl + \bar{R}B{\equiv}\overset{+}{N}R'$$

(12.51)

[†] For a detailed account, see: P. Paetzold (1987) *Advances in Inorganic Chemistry*, vol. 31, p. 123.

Compounds **12.20** can be made by reactions such as 12.52 or 12.53, and reaction 12.54 has been used to prepare Mes_2BNH_2 which has been structurally characterized. The B−N distance in Mes_2BNH_2 (Table 12.2) implies a double bond, and the planes containing the C_2B and NH_2 units are close to being coplanar as required for efficient overlap of the B and N $2p$ atomic orbitals in π-bond formation.

$$M[BH_4] + [R_2NH_2]Cl \longrightarrow H_2BNR_2 + MCl + 2H_2 \quad (12.52)$$

$$R_2BCl + R'_2NH + Et_3N \longrightarrow R_2BNR'_2 + [Et_3NH]Cl \quad (12.53)$$

$$Mes_2BF \xrightarrow[-NH_4F]{\text{Liquid } NH_3, \, Et_2O} Mes_2BNH_2 \quad (12.54)$$

Mes = mesityl

While considering the formation of B−N π-bonds, it is instructive to consider the structure of $B(NMe_2)_3$. As Figure 12.15c shows, each B and N atom is in a trigonal planar environment, and the B−N bond distances indicate partial π-character (Table 12.2) (see *Section 4.6*) as expected. Further, in the solid state structure, the twisting of the NMe_2 units, which is clearly apparent in Figure 12.15c, will mitigate against efficient $2p$–$2p$ atomic orbital overlap. Presumably, such twisting results from steric interactions and the observed structure of $B(NMe_2)_3$ provides an interesting example of a subtle balance of steric and electronic effects.

$$
\begin{array}{c}
\text{H}_2 \\
\text{N} \\
\diamond \\
\text{Me}_2\text{B} \qquad \text{BMe}_2 \\
\text{N} \\
\text{H}_2
\end{array}
$$

(12.21)

With less bulky substituents, compounds **12.20** readily dimerize. For example, Me_2BNH_2 forms the cyclodimer **12.21**. Whereas Me_2BNH_2 is a gas at room temperature (bp 274 K) and reacts rapidly with H_2O, dimer **12.21** has a melting point of 282 K and is kinetically stable towards hydrolysis by water.

$$
\begin{array}{c}
\text{R} \qquad\qquad\qquad \text{R} \\
\diagdown\;\; \overset{-}{\text{B}} = \overset{+}{\text{N}} = \overset{-}{\text{B}} \;\;\diagup \\
\diagup \qquad\qquad\qquad \diagdown\!\!\!\text{R} \\
\text{R} \qquad\qquad\qquad \text{R}
\end{array}
$$

(12.22)

Compounds **12.19** and **12.20** are analogues of alkynes and alkenes respectively, and allene analogues, **12.22**, can also be prepared, e.g. reaction 12.55. Crystallographic data for $[Mes_2BNBMes_2]^-$ reveal B−N bond lengths consistent with double bond character (Table 12.2) and the presence of B−N π-bonding is further supported by the fact that the planes containing the C_2B units are mutually orthogonal as shown in structure **12.22**.

$$Mes_2BNH_2 \xrightarrow[-2^nBuH]{2^nBuLi \text{ in } Et_2O} \{Li(OEt_2)NHBMes_2\}_2$$

$$\Big\downarrow \begin{array}{c} 2Mes_2BF \text{ in } Et_2O \\ -2LiF \end{array}$$

$$[Li(OEt_2)_3][Mes_2BNBMes_2] \xleftarrow[-^nBuH]{^nBuLi \text{ in } Et_2O} (Mes_2B)_2NH$$

(12.55)

Compounds containing B−P bonds and structurally analogous to the families of B−N compounds described above are also known.

Molecular species containing group 13 metal–nitrogen bonds

Coordinate M−N bond (M = heavier group 13 element) formation is exemplified in a number of complexes such as $trans$-$[GaCl_2(py)_4]^+$, and in $(Me_2AlNMe_2)_2$, which has a cyclic structure analogous to **12.21**. Coordinate bond formation also gives a series of Al_xN_y cluster-compounds by reactions such as 12.56 and 12.57; the structures of selected groups of clusters are shown in Figure 12.16, and the

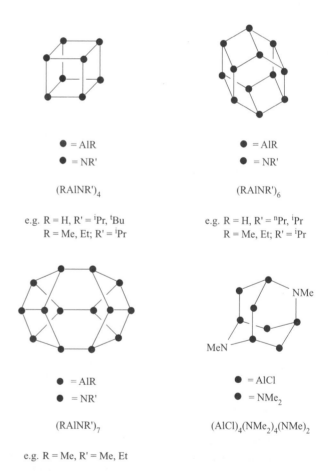

● = AlR
● = NR'

$(RAlNR')_4$

e.g. R = H, R' = ^iPr, ^tBu
R = Me, Et; R' = ^iPr

● = AlR
● = NR'

$(RAlNR')_6$

e.g. R = H, R' = ^nPr, ^iPr
R = Me, Et; R' = ^iPr

● = AlR
● = NR'

$(RAlNR')_7$

e.g. R = Me, R' = Me, Et

● = AlCl
● = NMe_2

$(AlCl)_4(NMe_2)_4(NMe)_2$

Fig. 12.16 The structures of some representative aluminium–nitrogen cluster compounds. Localized bonding schemes are appropriate for each cage (see *problem 12.17* at the end of the chapter).

bonding in the Al_xN_y cages is rationalized in terms of localized schemes.

$$nM[AlH_4] + nR'NH_2 \longrightarrow (HAlNR')_n + nMH + 2nH_2$$
$$M = Li, Na \qquad (12.56)$$

$$nAlR_3 + nR'NH_2 \longrightarrow (RAlNR')_n + 2nRH \qquad (12.57)$$

A number of related Ga-containing cages are known, as well a few Tl−N clusters, e.g. $Tl_2(MeSi)_2(N^tBu)_4$, **12.23**. However, in the latter and related compounds, the Tl atoms do not carry terminal substituents, another manifestation of the thermodynamic 6s inert pair effect (see *Box 12.3*).

● = Tl
● = N^{t}Bu

(12.23)

Multiple bonded compounds of the type observed for boron are not a feature of the later group 13 elements.

12.9 Aluminium to thallium: salts of oxoacids, aqueous solution chemistry and complexes

Aluminium sulfate and alums

The most important soluble oxosalts of Al are undoubtedly $Al_2(SO_4)_3 \cdot 16H_2O$ and the double sulfates $MAl(SO_4)_2 \cdot 12H_2O$ (*alums*). In alums, M^+ is usually K^+, Rb^+, Cs^+ or $[NH_4]^+$, but Li^+, Na^+ and Tl^+ compounds also exist; Al^{3+} may be replaced by another M^{3+} ion, but its size must be comparable and possible metals are Ga, In (but not Tl), Ti, V, Cr, Mn, Fe and Co. The sulfate ion in an alum can be replaced by $[SeO_4]^{2-}$. Alums occur naturally in *alum shales*, but are well known in crystal growth experiments; beautiful octahedral crystals are characteristic, e.g. in colourless $KAl(SO_4)_2 \cdot 12H_2O$ or purple $KFe(SO_4)_2 \cdot 12H_2O$. The purple colour of the latter arises from the presence of the $[Fe(H_2O)_6]^{3+}$ ion and, in all alums, the M^{3+} ion is octahedrally coordinated by six water ligands. The remaining water molecules are held in the crystal lattice by hydrogen bonds and connect the hydrated cations to the anions. Aluminium sulfate is used in water purification (see *Box 15.3*) for the removal of phosphate and of colloidal matter, the coagulation of which is facilitated by the high charge on the Al^{3+} cation. Intake of Al salts by humans, however, is suspected of causing Alzheimer's disease.

An *alum* has the general formula $M^IM^{III}(SO_4)_2 \cdot 12H_2O$.

Aqua ions

The M^{3+} aqua ions (M = Al, Ga, In, Tl) are acidic (see *equation 6.34*) and the acidity increases down the group. Solutions of their salts are appreciably hydrolyzed and salts of weak acids (e.g. carbonates and cyanides) cannot exist in aqueous solution. Solution NMR spectroscopic studies show that in acidic media, Al(III) is present as octahedral $[Al(H_2O)_6]^{3+}$, but raising the pH leads to the formation of polymeric species such as hydrated $[Al_2(OH)_2]^{4+}$ and $[Al_7(OH)_{16}]^{5+}$. Further increase in pH causes $Al(OH)_3$ to precipitate, and in alkaline solution, the aluminate anions $[Al(OH)_4]^-$ (tetrahedral) and $[Al(OH)_6]^{3-}$ (octahedral) and polymeric species such as $[(HO)_3Al(\mu\text{-}O)Al(OH)_3]^{2-}$ are present. The aqueous solution chemistry of Ga(III) resembles that of Al(III), but the later metals are not amphoteric (see *Section 12.7*).

Redox reactions in aqueous solution

The standard reduction potentials for the M^{3+}/M couples (Table 12.1) show that $Al^{3+}(aq)$ is much less readily reduced in aqueous solution than are the later M^{3+} ions. This can be attributed, in part, to the more negative free energy of hydration of the smaller Al^{3+} ion, but an important contributing factor (equation 12.58) in differentiating between the values of E^o for the Al^{3+}/Al and Ga^{3+}/Ga couples is the significant increase in the sum of the first three ionization energies (Table 12.1).

$$M^{3+}(aq) \xrightarrow{-\Delta_{hyd}H^o} M^{3+}(g) \xrightarrow{-\Sigma IE_{1-3}} M(g) \xrightarrow{-\Delta_a H^o} M(s)$$
$$(12.58)$$

Although In(I) can be obtained in low concentration by oxidation of an In anode in dilute $HClO_4$ (equation 12.59), the solution rapidly evolves H_2 and forms In(III).

$$In^{3+}(aq) + 2e^- \longrightarrow In^+(aq) \qquad E^o = -0.42\,V \qquad (12.59)$$

In contrast, E^o for the reduction of Tl(III) to Tl(I) in molar $HClO_4$ is +1.26 V, and under these conditions, Tl(III) is a powerful oxidizing agent. The value of E^o is, however, dependent on the anion present and complex formed (see *Section 7.3*); Tl(I) (like the alkali metal ions) forms few stable complexes in aqueous solution, whereas Tl(III) is strongly complexed by a variety of anions. For example, consider the presence of Cl^- in solution. Whereas TlCl is fairly insoluble, Tl(III) forms the soluble complex $[TlCl_4]^-$ and, at $[Cl^-] = 1\,mol\,dm^{-3}$, $E^o(Tl^{3+}/Tl^+) = +0.9\,V$. Thallium(III) forms a more stable complex with I^- than Cl^-, and at high $[I^-]$, $[TlI_4]^-$ is produced in solution even though $E^o(Tl^{3+}/Tl^+)$ is more positive than $E^o(I_2/2I^-)$ (+0.54 V) and TlI is sparingly soluble. Thus, while tabulated reduction potentials for the Tl^{3+}/Tl^+ and $I_2/2I^-$ couples might suggest that aqueous I^- will reduce Tl(III) to Tl(I) (see *Appendix 11*), in the presence of high concentrations of I^-, Tl(III) is stabilized. Indeed, the addition of I^- to solutions

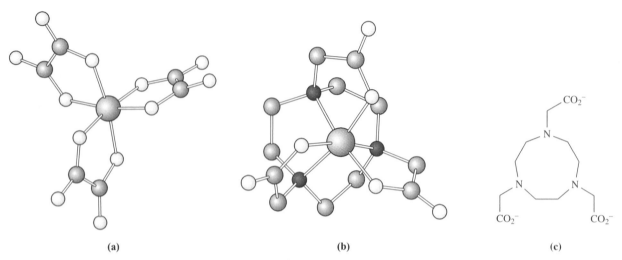

Fig. 12.17 The structures (X-ray diffraction) of (a) $[Al(ox)_3]^{3-}$ in the ammonium salt [N. Bulc *et al.* (1984) *Acta Crystallogr., Sect. C*, vol. 40, p. 1829], and (b) [GaL] [C.J. Broan *et al.* (1991) *J. Chem. Soc., Perkin Trans. 2*, p. 87] where ligand L^{3-} is shown in diagram (c). Hydrogen atoms have been omitted from (a) and (b); colour code: Al and Ga, blue; O, pale grey; C, grey; N, black.

of TlI_3 (see structure **12.12**), which contain $[I_3]^-$ (i.e. $I_2 + I^-$), brings about reaction 12.60 oxidizing Tl(I) to Tl(III).

$$TlI_3 + I^- \longrightarrow [TlI_4]^- \qquad (12.60)$$

In alkaline media, Tl(I) is also easily oxidized, since TlOH is soluble in water and hydrated Tl_2O_3 (which is in equilibrium with Tl^{3+} and $[OH]^-$ ions in solution) is very sparingly soluble in water ($K_{sp} \approx 10^{-45}$).

Coordination complexes of the M^{3+} ions

Increasing numbers of coordination complexes of the group 13 metal ions are now known. Octahedral coordination is common, e.g. in $[M(acac)_3]$ (M = Al, Ga, In), $[M(ox)_3]^{3-}$ (M = Al, Ga, In) and *mer*-$[Ga(N_3)_3(py)_3]$ (see *Table 6.7* for ligand abbreviations and structures). Figure 12.17a shows the structure of $[Al(ox)_3]^{3-}$. The complexes $[M(acac)_3]$ are structurally related to $[Fe(acac)_3]$ (see *Figure 6.10*); in Section 6.11, we discussed the influence of $[H^+]$ on the formation of $[Fe(acac)_3]$ and similar arguments apply to the group 13 metal ion complexes.

(12.24)

Deprotonation of 8-hydroxyquinoline gives the didentate ligand **12.24** which has a number of applications. For example, Al^{3+} may be extracted into organic solvents as the octahedral complex $[Al(\mathbf{12.24})_3]$ providing a weighable form of Al^{3+} for the gravimetric analysis of aluminium.

Complexes involving macrocyclic ligands with pendant carboxylate or phosphate groups have received attention in the development of highly stable metal complexes suitable for *in vivo* applications, e.g. tumour-seeking complexes containing

radioisotopes (see *Box 2.3*). The incorporation of ^{67}Ga (γ-emitter, $t_{\frac{1}{2}} = 3.2$ days), ^{68}Ga (β^+-emitter, $t_{\frac{1}{2}} = 68$ min) or ^{111}In (γ-emitter, $t_{\frac{1}{2}} = 2.8$ days) into such complexes yields potential radiopharmaceuticals. Figure 12.17c shows an example of a well-studied ligand which forms very stable complexes with Ga(III) and In(III) ($\log K \geq 20$). The way in which this ligand encapsulates the M^{3+} ion with the three *N*-donor atoms forced into a *fac*-arrangement is evident in Figure 12.17b.

12.10 Metal borides

Solid state metal borides are characteristically extremely hard, involatile, high melting and chemically inert materials which are industrially important with uses as refractory materials and in rocket cones and turbine blades, i.e. components that must withstand extreme stress, shock and high temperatures. Preparative routes to metal borides are varied, as are their structures. Some may be made by direct combination of the elements at high temperatures, and others from metal oxides (e.g. reactions 12.61 and 12.62).

$$Eu_2O_3 \xrightarrow{\text{boron carbide/carbon, } \Delta} EuB_6 \qquad (12.61)$$

$$TiO_2 + B_2O_3 \xrightarrow{\text{Na, } \Delta} TiB_2 \qquad (12.62)$$

Metal borides may be boron- or metal-rich, and general families include MB_3, MB_4, MB_6, MB_{10}, MB_{12}, M_2B_5 and M_3B_4 (B-rich), and M_3B, M_4B, M_5B, M_3B_2 and M_7B_3 (M-rich). The formulae bear no relation to those expected on the basis of the formal oxidation states of boron and metal.

The structural diversity of these materials is so great as to preclude a full discussion here, but we can conveniently consider them in terms of the categories shown in Table 12.3 which are identified in terms of the arrangement of the B

Table 12.3 Classification of the structures of solid state metal borides.

Description of the boron atom organization	Pictorial representation of the boron association	Examples of metal borides adopting each structure type
Isolated B atoms		Ni_3B, Mn_4B, Pd_5B_2, Ru_7B_3
Pairs of B atoms	B——B	Cr_5B_3
Chains		V_3B_4, Cr_3B_4, HfB, CrB, FeB
Linked double chains		Ta_3B_4
Sheets		MgB_2, TiB_2, CrB_2, Ti_2B_5, W_2B_5
Linked B_6 octahedra (see text)		CaB_6, LaB_6
Linked B_{12} icosahedra (see text; see also Figure 12.3)		ZrB_{12}, UB_{12}

(B–B links to adjacent icosahedra are not shown)

atoms within a host metal lattice. The structure-type of the MB_6 borides (e.g. CaB_6) can be envisaged by likening it to that of a CsCl lattice with B_6-units (Table 12.3) replacing Cl^- ions; however, the B–B distances *between* adjacent B_6-octahedra are similar to those *within* each unit and so a 'discrete ion' model is not actually appropriate.

The structure type of MB_{12} (e.g. UB_{12}) can similarly be described in terms of an NaCl lattice in which the Cl^- ions are replaced by B_{12}-icosahedra (Table 12.3).

Although this summary of metal borides is brief, it illustrates the complexity of structures frequently encountered in the chemistry of boron.

12.11 Electron-deficient borane clusters: an introduction

In this section, we introduce *electron-deficient clusters* containing boron, focusing on the small clusters $[B_6H_6]^{2-}$, B_5H_9 and B_4H_{10}. A comprehensive treatment of borane and carbaborane clusters is beyond the scope of this book, but more detailed accounts can be found in the references cited at the end of the chapter.

> An *electron-deficient* species possesses fewer valence electrons than are required for a localized bonding scheme; in a *cluster*, the atoms form a cage-like structure.

Boron hydrides

The pioneering work of Alfred Stock between 1912 and 1936 revealed that boron formed a range of hydrides of varying nuclearities. Since these early studies, the number of neutral and anionic boron hydrides has increased greatly, and the structures of three of the smaller boranes are shown in

Figure 12.18. The following classes of cluster are now recognized, along with others.

- In a *closo*-cluster, the atoms form a closed, deltahedral cage and have the general formula $[B_nH_n]^{2-}$ (e.g. $[B_6H_6]^{2-}$).
- In a *nido*-cluster, the atoms form an open cage which is derived from a closed deltahedron with one vertex unoccupied; general formulae are B_nH_{n+4}, $[B_nH_{n+3}]^-$, etc. (e.g. B_5H_9, $[B_5H_8]^-$).
- In an *arachno*-cluster, the atoms form an open cage which is derived from a closed deltahedron with two vertices unoccupied; general formulae are B_nH_{n+6}, $[B_nH_{n+5}]^-$ etc. (e.g. B_4H_{10}, $[B_4H_9]^-$).
- In a *hypho*-cluster, the atoms form an open cage which is derived from a closed deltahedron with three vertices unoccupied; this is a poorly exemplified group of compounds with general formulae B_nH_{n+8}, $[B_nH_{n+7}]^-$ etc.
- A *conjuncto*-cluster consists of two or more cages connected together through a shared atom, an external bond, a shared edge or a shared face (e.g. $\{B_5H_8\}_2$).

> A *deltahedron* is a polyhedron that possesses only *triangular* faces, e.g. an octahedron.

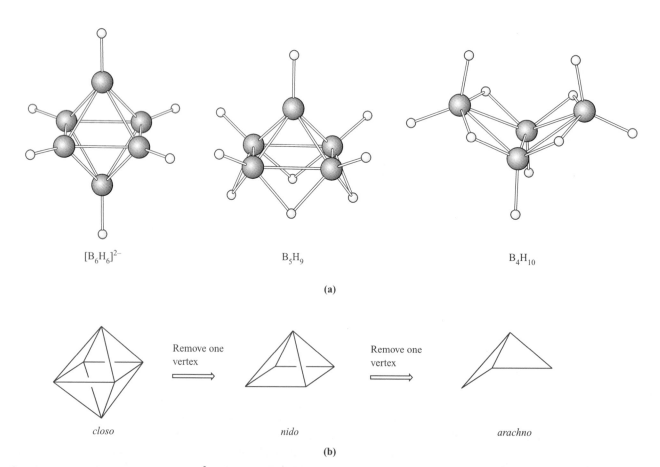

Fig. 12.18 (a) The structures of $[B_6H_6]^{2-}$, B_5H_9 and B_4H_{10}; colour code: B, blue; H, pale grey. (b) Schematic representation of the derivation of *nido* (with $n = 5$) and *arachno* (with $n = 4$) cages from a parent *closo* deltahedral cage with $n = 6$.

BOX 12.8   **CHEMICAL AND THEORETICAL BACKGROUND**

Nomenclature of boranes

The name of a borane denotes the number of boron atoms, the number of hydrogen atoms, and the overall charge. The number of boron atoms is given by a Greek prefix (di-, tri-, tetra-, penta-, hexa- etc.), the exception being for nine and eleven where the Latin nona- and undeca- are used. The number of hydrogen atoms is shown as an Arabic numeral in parenthesis at the end of the name (see below). The charge for an ion is shown at the end of the name; the nomenclature

for anions is also distinguished from that of neutral boranes (see examples below). As a prefix, the class of cluster (*closo-, nido-, arachno-, conjuncto-* etc.) should be stated.

- $[B_6H_6]^{2-}$ *closo*-hexahydrohexaborate(2−)
- B_4H_{10} *arachno*-tetraborane(10)
- B_5H_9 *nido*-pentaborane(9)
- B_6H_{10} *nido*-hexaborane(10)

At one time, there was considerable interest in the possibility of using boron hydrides as high energy fuels, but in practice, it is difficult to ensure complete combustion to B_2O_3, and involatile polymers tend to block exhaust ducts. Although interest in fuel applications has faded, boranes remain a fascination to structural and theoretical chemists.

(12.25)

The higher boranes can be prepared by controlled pyrolysis of B_2H_6 in the vapour phase. The pyrolysis of B_2H_6 in a hot–cold reactor (i.e. a reactor having an interface between two regions of extreme temperatures) gives, for example, B_4H_{10}, B_5H_{11} or B_5H_9 depending upon the temperature interface. Decaborane(14), $B_{10}H_{14}$, is produced by heating B_2H_6 at 453–490 K under static conditions. Such methods are complicated by the interconversion of one borane to another, and it has been desirable to seek selective syntheses. Salts of $[B_3H_8]^-$, **12.25** (obtained by the action of B_2H_6 on $Na[BH_4]$), are commercially available and are easier to handle than B_2H_6. Reactions 12.63 to 12.65 give preparations of B_4H_{10}, B_5H_9 and $[B_6H_6]^{2-}$.

$$4Na[B_3H_8] + 4HCl \longrightarrow 3B_4H_{10} + 3H_2 + 4NaCl \quad (12.63)$$

$$5[B_3H_8]^- + 5HBr \xrightarrow{-H_2} 5[B_3H_7Br]^-$$

$$\xrightarrow{373\,K} 3B_5H_9 + 4H_2 + 5Br^- \quad (12.64)$$

$$2Na[B_3H_8] \xrightarrow{433\,K\ in\ diglyme} Na_2[B_6H_6] + 5H_2 \quad (12.65)$$

(Diglyme: see structure **12.2**)

The dianion $[B_6H_6]^{2-}$ has a closed octahedral B_6 cage (Figure 12.18a) and is a *closo*-cluster. Each B atom is connected to four other B atoms within the cage, and to one terminal H. The structure of B_5H_9 (Figure 12.18a) consists

of a square-based pyramidal cage of B atoms, each of which carries one terminal H. The remaining four H atoms occupy B−H−B bridging sites around the square face of the cage. Figure 12.18a shows the structure of B_4H_{10} which has an open framework of two edge-sharing B_3 triangles. The inner B atoms carry one terminal H each, and two terminal H atoms are bonded to each of the outer B atoms; the remaining four H atoms are involved in B−H−B bridges. X-Ray diffraction data have shown that the B−B bond distances in $[B_6H_6]^{2-}$ are equal (169 pm), but in B_5H_9, the unbridged B−B edges (apical–basal, 166 pm) are shorter than the H-bridged edges (basal–basal, 172 pm); the apical and basal atoms in B_5H_9 are defined in structure **12.26**. A similar situation is observed in B_4H_{10} (H-bridged edges = 187 pm, unique B−B edge = 174 pm from electron diffraction data). The *range* of B−B distances in these three cages is significant and, in the light of the discussion of bonding that follows, it is instructive to compare these distances with twice the covalent radius of B (r_{cov} = 88 pm). Longer B−B edges are observed in other clusters (e.g. 197 pm in $B_{10}H_{14}$) but are still regarded as bonding interactions.

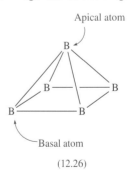

Apical atom

Basal atom

(12.26)

In a *formal sense*, we can consider the structure of B_5H_9 as being related to that of $[B_6H_6]^{2-}$ by removing one vertex from the B_6 octahedral cage (Figure 12.18b). Similarly, the B_4 cage in B_4H_{10} is related to that of B_5H_9 by the removal of another vertex. The removal of a vertex is accompanied by the addition of bridging H atoms. These observations lead us to a discussion of the bonding. The first point is that boron-containing and related clusters exhibit structures in which the bonding is *not* readily represented in terms of localized bonding models. This is in contrast to the situation

BOX 12.9 **CHEMICAL AND THEORETICAL BACKGROUND**

Bonding in $[B_6H_6]^{2-}$

In Section 23.5, we discuss the *isolobal principle*, and the relationship between the bonding properties of different cluster *fragments*. The bonding in boron-containing clusters and, more generally, in organometallic clusters, is conveniently dealt with in terms of molecular orbital theory. In this Box, we show how the *frontier orbitals* (i.e. the highest occupied and lowest unoccupied MOs) of six BH units combine to give the seven cluster bonding MOs in $[B_6H_6]^{2-}$. This *closo*-anion possesses an octahedral structure:

After accounting for the localized B–H bonding orbital (σ_{BH}) and its antibonding counterpart, a BH fragment has three

orbitals remaining which are classed as its frontier orbitals:

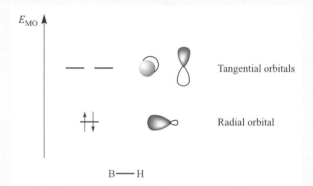

If we consider the BH fragments as being placed in the orientations shown in the structural diagram on the left, then the three frontier orbitals can be classified as one *radial* orbital (pointing into the B_6 cage) and two *tangential* orbitals (lying over the cluster surface). When the six BH-units come together, a total of (6×3) orbitals combine to give 18 MOs, 7 of which possess cluster-bonding character. The interactions that give rise to these bonding MOs are shown below. The 11 non-bonding and antibonding MOs are omitted from the diagram.

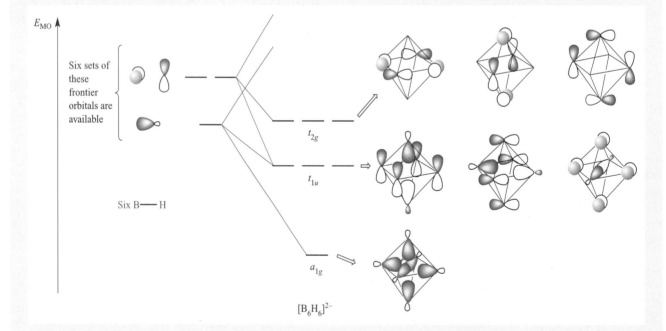

Once the molecular orbital interaction diagram has been constructed, the electrons that are available in $[B_6H_6]^{2-}$ can be accommodated in the lowest lying MOs. Each BH unit provides two electrons, and in addition the 2– charge provides two electrons. There is, therefore, a total of 7

electron pairs available which will completely occupy the seven bonding MOs shown in the diagram above. Relating this to Wade's rules, the MO approach shows that there are 7 electron-pairs for a *closo*-cage possessing 6 cluster vertices.

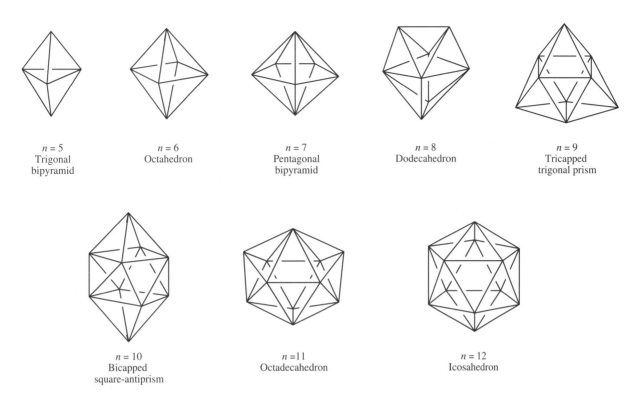

| $n = 5$ | $n = 6$ | $n = 7$ | $n = 8$ | $n = 9$ |
| Trigonal bipyramid | Octahedron | Pentagonal bipyramid | Dodecahedron | Tricapped trigonal prism |

| $n = 10$ | $n = 11$ | $n = 12$ |
| Bicapped square-antiprism | Octadecahedron | Icosahedron |

Fig. 12.19 The deltahedral cages with 5 to 12 vertices which are the parent cages used in conjunction with Wade's rules to rationalize borane cluster structures. As a general (but not foolproof) scheme, when removing vertices from these cages to generate *nido*-frameworks, remove a vertex of connectivity three from the trigonal bipyramid, any vertex from the octahedron or icosahedron, a 'cap' from the tricapped trigonal prism or bicapped square-antiprism, and a vertex of highest connectivity from the remaining deltahedra.

in B_2H_6, $[BH_4]^-$ and $[B_3H_8]^-$ where 2c-2e and 3c-2e interactions can adequately represent the distributions of valence electrons.[†] A satisfactory solution to this problem is to consider a delocalized approach and invoke MO theory (see *Box 12.9*). The situation has been greatly helped by an empirical set of rules developed by Wade, Williams and Mingos. The initial *Wade's rules* can be summarized as follows, and 'parent' deltahedra are shown in Figure 12.19:

- a *closo*-deltahedral cluster cage with n vertices requires $(n + 1)$ pairs of electrons which occupy $(n + 1)$ cluster bonding MOs;
- from a 'parent' *closo*-cage with n vertices, a set of more open cages (*nido*, *arachno* and *hypho*) can be derived, each of which possesses $(n + 1)$ pairs of electrons occupying $(n + 1)$ cluster bonding MOs;
- for a parent *closo*-deltahedron with n vertices, the related *nido*-cluster has $(n - 1)$ vertices and $(n + 1)$ pairs of electrons;
- for a parent *closo*-deltahedron with n vertices, the related *arachno*-cluster has $(n - 2)$ vertices and $(n + 1)$ pairs of electrons;

- for a parent *closo*-deltahedron with n vertices, the related *hypho*-cluster has $(n - 3)$ vertices and $(n + 1)$ pairs of electrons.

In counting the number of cluster-bonding electrons available in a borane, we first formally break down the cluster into fragments and determine the number of valence electrons that each fragment can contribute for cluster bonding. A procedure is as follows.

- Determine how many {BH}-units are present (i.e. assume each B atom carries a terminal hydrogen atom); each {BH}-unit provides two electrons for cage-bonding (of the three valence electrons of B, one is used to form a localized terminal B−H bond, leaving two for cluster bonding).
- Count how many additional H atoms there are; each provides one electron.
- Add up the number of electrons available from the cluster fragments and take account of any overall charge.
- The total number of electrons corresponds to $(n + 1)$ pairs of electrons, and thus, the number of vertices, n, of the parent deltahedron can be established.
- Each {BH}-unit occupies one vertex in the parent deltahedron, and from the number of vertices left vacant, the class of cluster can be determined; if vertices are non-equivalent, the first to be left vacant *tends* to be either one of highest connectivity or a 'cap' in 'capped' structures (e.g. $n = 9$ and 10 in Figure 12.19).

[†]A valence bond method called *styx* rules, devised by W.N. Lipscomb, provides a means of constructing bonding networks for boranes in terms of 3c-2e B−H−B interactions, 3c-2e B−B−B interactions, 2c-2e B−B bonds, and BH_2 units, but the method is only easily applied to a limited number of clusters.

● Additional H atoms are placed in bridging sites along B−B edges of an *open* face of the cluster, or in extra terminal sites, usually available if there are any B atoms of especially low connectivity.

Worked example 12.2   Using Wade's rules to rationalize a structure

Rationalize why $[B_6H_6]^{2-}$ adopts an octahedral cage.

There are 6 {BH}-units and no additional H atoms.
Each {BH}-unit provides 2 valence electrons.
There are two electrons from the 2− charge.
Total number of cage-bonding electrons available
$$= (6 \times 2) + 2 = 14 \text{ electrons}$$
$$= 7 \text{ pairs}$$
Thus, $[B_6H_6]^{2-}$ has 7 pairs of electrons with which to bond 6 {BH}-units.

This means that there are $(n + 1)$ pairs of electrons for n vertices, and so $[B_6H_6]^{2-}$ is a *closo*-cage, a 6-vertex deltahedron, i.e. the octahedron is adopted (see *Figure 12.19*).

Self-study exercises

Refer to Figure 12.19.

1. Rationalize why $[B_{12}H_{12}]^{2-}$ adopts an icosahedral structure for the boron cage.

2. Show that the observed bicapped square-antiprismatic structure of the boron cage in $[B_{10}H_{10}]^{2-}$ is consistent with Wade's rules.

3. In each of the following, rationalize the observed boron cage structure in terms of Wade's rules: (a) B_5H_9 (a square-based pyramid); (b) B_4H_{10} (two edge-fused triangles, Figure 12.18); (c) $[B_6H_9]^-$ (a pentagonal pyramid); (d) B_5H_{11} (an open network of three, edge-fused triangles).

Worked example 12.3   Using Wade's rules to predict a structure

Suggest a likely structure for $[B_5H_8]^-$.

There are 5 {BH}-units and 3 additional H atoms.
Each {BH}-unit provides 2 valence electrons.
There is one electron from the 1− charge.
Total number of cage-bonding electrons available
$$= (5 \times 2) + 3 + 1 = 14 \text{ electrons}$$
$$= 7 \text{ pairs}$$
7 pairs of electrons are consistent with the parent deltahedron having 6 vertices, i.e. $(n + 1) = 7$, and so $n = 6$.

The parent deltahedron is an octahedron and the B_5-core of $[B_5H_8]^-$ will be derived from an octahedron with one vertex left vacant:

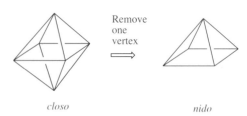

closo   *nido*

The three extra H atoms form B−H−B bridges along three of the four B−B edges of the open (square) face of the B_5-cage. The predicted structure of $[B_5H_8]^-$ is:

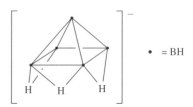

● = BH

Self-study exercises

Refer to Figure 12.19.

1. Confirm the following classifications within Wade's rules:
   (a) $[B_9H_9]^{2-}$, *closo*; (b) B_6H_{10}, *nido*; (c) B_4H_{10}, *arachno*; (d) $[B_8H_8]^{2-}$, *closo*; (e) $[B_{11}H_{13}]^{2-}$, *nido*.

2. Suggest likely structures for the following: (a) $[B_9H_9]^{2-}$; (b) B_6H_{10}; (c) B_4H_{10}; (d) $[B_8H_8]^{2-}$.
   [*Ans.* (a) tricapped trigonal prism; (b) pentagonal pyramid; (c) see *Figure 12.18*; (d) dodecahedron]

The types of reaction undergone by borane clusters depend upon the class and size of the cage; $[B_6H_6]^{2-}$, B_5H_9 and B_4H_{10} provide examples of a *closo*-hydroborate anion, a small *nido*-borane, and a small *arachno*-borane, respectively. In some ways, the reactivity patterns of the *closo*-hydroborate anions resemble those of aromatic organic molecules, a consequence of electron delocalization within the closed cage. The reactivities of B_5H_9 and B_4H_{10} have been well explored and typical reactions are given in Figures 12.20 and 12.21. The *nido*-B_5H_9 cluster is more reactive than *closo*-$[B_6H_6]^{2-}$, and *arachno*-B_4H_9 is more susceptible still to reactions involving cage degradation or cleavage. For example, B_4H_{10} is hydrolyzed by H_2O, while B_5H_9 is hydrolyzed only slowly by water but completely by alcohols. Many reactions involving *arachno*-B_4H_{10} with Lewis bases are known and Figure 12.21 illustrates cleavage with NH_3 (a small base) to give an ionic salt and by a more sterically demanding base to give neutral adducts; compare these reactions with those of B_2H_6 (equation 12.14). Carbon monoxide and PF_3, on the other hand, react with B_4H_{10} with elimination of H_2 and retention of the B_4 cage. Deprotonation of both B_4H_{10} and B_5H_9 can be achieved using NaH or KH and in each case H^+ is removed from a *bridging* site. This preference is quite general among boranes and can be rationalized in terms of redistribution of the two electrons

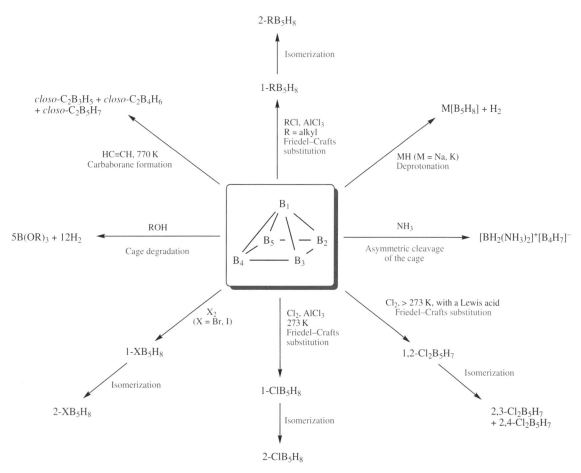

Fig. 12.20 Selected reactions of the *nido*-borane B_5H_9; the numbering scheme in the central structure is used to indicate positions of substitution in products that retain the B_5-core.

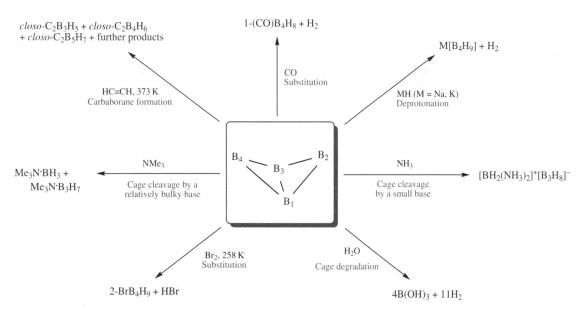

Fig. 12.21 Selected reactions of the *arachno*-borane B_4H_{10}; the numbering scheme in the central structure is used to denote positions of substitution in products that retain the B_4-core.

from the B−H−B bridge into a B−B interaction upon H^+ removal. Electrophiles react with B_5H_9 (Figure 12.20) with initial attack being at the apical B atom. Isomerizations to give the basally-substituted derivatives occur but have been shown by ^{10}B labelling studies to involve B_5 cage rearrangement rather than migration of the substituent. Both B_4H_{10} and B_5H_9 react with ethyne to generate a new family of cluster compounds, the *carbaboranes*. Structurally, carbaboranes resemble boranes, with structures rationalized in terms of Wade's rules (a CH unit provides one more electron for bonding than a BH-unit). The structures of the carbaborane products in Figures 12.20 and 12.21 are shown in **12.27–12.29**, although in each case only one cage-isomer is illustrated; an example of the application of Wade's rules to them is given in worked example 12.4.

(12.27)   (12.28)

(12.29)

Worked example 12.4   Applying Wade's rules to carbaborane structures

(a) Rationalize why the cage structure of $C_2B_4H_6$ is an octahedron. (b) How many cage isomers are possible?

(a) There are 4 {BH}-units, 2 {CH}-units and no additional H atoms.

Each {BH}-unit provides 2 valence electrons.
Each {CH}-unit provides 3 valence electrons.
Total number of cage-bonding electrons available
$$= (4 \times 2) + (2 \times 3) = 14 \text{ electrons}$$
$$= 7 \text{ pairs}$$

Thus, $C_2B_4H_6$ has 7 pairs of electrons with which to bond 6 cluster units.

There are $(n + 1)$ pairs of electrons for n vertices, and so $C_2B_4H_6$ is a *closo*-cage, a 6-vertex deltahedron, i.e. the octahedron is adopted (see *Figure 12.19*).

(b) In an octahedron, all vertices are equivalent. It follows that there are two possible arrangements of the two carbon and four boron atoms, leading to two cage isomers:

It is *not* possible to say anything about isomer preference using Wade's rules.

Self-study exercises

1. Rationalize the structures of carbaboranes (a) **12.27** and (b) **12.29**, and determine how many isomers of each are possible.
   [*Ans.* (a) 3; (b) 4]

2. The carbaborane $C_2B_{10}H_{12}$ has the same cage structure as $[B_{12}H_{12}]^{2-}$ (Figure 12.19, the icosahedron). (a) Rationalize this observation using Wade's rules. (b) How many isomers are possible for $C_2B_{10}H_{12}$?   [(b) *Ans.* 3]

Before leaving this introduction to boron clusters, we return briefly to the boron halides of type B_nX_n (X = halogen). Although these have deltahedral structures, they do not 'obey' Wade's rules. Formally, by Wade's rules, we may consider that each {BX}-unit in B_8X_8 provides two electrons for cage-bonding; but this approach gives an electron count (eight pairs) which is inconsistent with the observed closed, dodecahedral cage (Figure 12.8b). Similarly, B_4Cl_4 has a tetrahedral structure (Figure 12.8a) although a simple electron count gives only four electron pairs for cluster bonding. The apparent violation of Wade's rules arises because the symmetry of the B_n-cluster bonding MOs is appropriate to allow interaction with filled p atomic orbitals of the terminal halogens; donation of electrons from the terminal halogen atoms to boron can occur. One must therefore be aware that, while Wade's rules are extremely useful in many instances, apparent exceptions do exist and require more in-depth bonding analyses.

Summary

This chapter has been concerned with descriptive chemistry of the group 13 elements. The chemistry of the first member, B, and of its compounds is distinct from that of the later, metallic elements. In particular, the wide range of electron-deficient cluster compounds formed by B is not replicated by the later members of the group; an approach to the bonding in such clusters (Wade's rules) has been introduced.

The following terms were introduced in this chapter.
Do you know what they mean?

☐ thermodynamic 6s inert pair effect
☐ relativistic effect
☐ mordant
☐ cyclodimer
☐ alum
☐ electron-deficient cluster
☐ deltahedron
☐ Wade's rules

Suggested further reading

A.J. Downs, ed. (1993) *The Chemistry of Aluminium, Gallium, Indium and Thallium*, Kluwer, Dordrecht – Covers the chemistry and commercial aspects of these elements including applications to materials.

D.F. Shriver and M.A. Drezdzon (1986) *The Manipulation of Air-sensitive Compounds*, 2nd edn, Wiley-Interscience, New York – Many compounds of the group 13 elements are extremely sensitive to air and moisture; this book gives a detailed account of methods of handling such compounds.

A.F. Wells (1984) *Structural Inorganic Chemistry*, 5th edn, Clarendon Press, Oxford – Includes a full account of the structural chemistry of the elements and compounds in group 13.

Borane clusters

N.N. Greenwood and A. Earnshaw (1997) *Chemistry of the Elements*, 2nd edn, Butterworth-Heinemann, Oxford – Chapter 6 covers boron clusters in some detail.

N.N. Greenwood (1992) *Chemical Society Reviews*, vol. 21, p. 49 – 'Taking stock: The astonishing development of boron hydride cluster chemistry'.

C.E. Housecroft (1994) *Boranes and Metallaboranes: Structure, Bonding and Reactivity*, 2nd edn, Ellis Horwood, Hemel Hempstead – A clear and well-illustrated introduction to borane clusters and their derivatives.

C.E. Housecroft (1994) *Clusters of the p-Block Elements*, Oxford University Press, Oxford – An introductory survey of clusters containing p-block elements including boron.

K. Wade (1971) *Electron Deficient Compounds*, Nelson, London – A classic account of the boron hydrides and related electron-deficient compounds.

Other specialized topics

A.J. Downs and C.R. Pulham (1994) *Chemical Society Reviews*, vol. 23, p. 175 – 'The hydrides of aluminium, gallium, indium and thallium: A re-evaluation'.

P. Paetzold (1987) *Advances in Inorganic Chemistry*, vol. 31, p. 123 – 'Iminoboranes'

Problems

12.1 (a) Write down, in order, the names and symbols of the elements in group 13; check your answer by reference to the first page of this chapter. (b) Classify the elements in terms of metallic and non-metallic behaviour. (c) Give a *general* notation showing the ground state electronic configuration of each element.

12.2 Using the data in Table 12.1, draw a potential diagram for Tl and determine the value of $E^{\circ}(Tl^{3+}/Tl^{+})$.

12.3 Plot a graph to show the variation in values of IE_1, IE_2 and IE_3 for the group 13 elements (Table 12.1), and plot a similar graph to show the variation in values of IE_1 and IE_2 for the group 2 metals (Table 11.1). Account for differences in trends of IE_2 for the group 2 and 13 elements.

12.4 Write equations for the following processes, involved in the extraction of the elements from their ores:

(a) the reduction of boron oxide by Mg;
(b) the result of the addition of hot aqueous NaOH to a mixture of solid Al_2O_3 and Fe_2O_3;
(c) the reaction of CO_2 with aqueous $Na[Al(OH)_4]$.

12.5 Predict the following NMR spectra: (a) the ^{11}B NMR spectrum of $[BH_4]^-$; (b) the ^1H NMR spectrum of $[BH_4]^-$; (c) the ^{11}B NMR spectrum of the adduct $BH_3 \cdot PMe_3$; (d) the $^{11}B\{^1H\}$ NMR spectrum of $THF \cdot BH_3$. [^1H, 100%, $I = \frac{1}{2}$; ^{31}P, 100%, $I = \frac{1}{2}$; ^{11}B, 80.4%, $I = \frac{3}{2}$; ignore ^{10}B].

12.6 The thermite process is shown in equation 12.5. Determine $\Delta_r H^{\circ}$ for this reaction if $\Delta_f H^{\circ}(Al_2O_3,s,298\,K)$ and $\Delta_f H^{\circ}(Fe_2O_3,s,298\,K)$ $= -1675.7$ and $-824.2\,kJ\,mol^{-1}$, and comment on the relevance of this value to that of $\Delta_{fus} H(Fe,s) = 13.8\,kJ\,mol^{-1}$.

12.7 Explain how, during dimerization, each BH_3 molecule acts both as a Lewis base and a Lewis acid.

12.8 Describe the bonding in Ga_2H_6 and Ga_2Cl_6, both of which have structures of the type shown in **12.30**.

X = H or Cl

(12.30)

12.9 The ordering of the relative stabilities of adducts $L \cdot BH_3$ for some common adducts is, according to L: $Me_2O < THF < Me_2S < Me_3N < Me_3P < H^-$. In addition to answering each of the following, indicate how you could use NMR spectroscopy to confirm your proposals.

(a) What happens when Me_3N is added to a THF solution of $THF \cdot BH_3$?
(b) Will Me_2O displace Me_3P from $Me_3P \cdot BH_3$?

(c) Is $[BH_4]^-$ stable in THF solution with respect to a displacement reaction?

(d) Suggest what may be formed when $Ph_2PCH_2CH_2PPh_2$ is added to a THF solution of $THF \cdot BH_3$, the latter remaining in excess.

12.10 Suggest explanations for the following facts.

(a) $Na[BH_4]$ is very much less rapidly hydrolyzed by H_2O than is $Na[AlH_4]$.

(b) The rate of hydrolysis of B_2H_6 by water vapour is given by the equation:

$$\text{Rate} \propto (P_{B_2H_6})^{\frac{1}{2}}(P_{H_2O})$$

(c) A saturated aqueous solution of boric acid is neutral to the indicator bromocresol green (pH range 3.8–5.4), and a solution of $K[HF_2]$ is acidic to this indicator; when, however, excess boric acid is added to a solution of $K[HF_2]$, the solution becomes alkaline to bromocresol green.

12.11 Suggest likely products for the following reactions:

(a) $BCl_3 + EtOH \longrightarrow$

(b) $BF_3 + EtOH \longrightarrow$

(c) $BCl_3 + PhNH_2 \longrightarrow$

(d) $BF_3 + KF \longrightarrow$

12.12 (a) Write down the formula of cryolite. (b) Write down the formula of perovskite. (c) Cryolite is described as possessing a lattice structure closely related to that of perovskite. Suggest how this is possible when the stoichiometries of the two compounds do not appear to be compatible.

12.13 (a) Suggest structures for $[MBr_6]^{3-}$, $[MCl_5]^{2-}$ and $[MBr_4]^-$ (M = Ga or In).

(b) In the salt $[Et_4N]_2[InCl_5]$, the anion has a square-based pyramidal structure, as does $[TlCl_5]^{2-}$ in the salt $[H_3N(CH_2)_5NH_3][TlCl_5]$. Comment on these observations in the light of your answer to part (a).

(c) Suggest methods of preparing $[H_3N(CH_2)_5NH_3][TlCl_5]$ and $Cs_3[Tl_2Cl_9]$.

(d) Explain how magnetic data enable one to distinguish between the formulations $GaCl_2$ and $Ga[GaCl_4]$ for gallium dichloride.

12.14 Comment on each of the following observations.

(a) AlF_3 is almost insoluble in anhydrous HF, but dissolves if KF is present. Passage of BF_3 through the resulting solution causes AlF_3 to reprecipitate.

(b) The Raman spectra of germanium tetrachloride, a solution of gallium trichloride in concentrated hydrochloric acid, and fused gallium dichloride contain the following lines:

	Absorption / cm^{-1}			
$GeCl_4$	134	172	396	453
$GaCl_3$/HCl	114	149	346	386
$GaCl_2$	115	153	346	380

(c) When TlI_3, which is isomorphous with the alkali metal triiodides, is treated with aqueous NaOH, hydrated Tl_2O_3 is quantitatively precipitated.

12.15 Figure 12.6c shows the solid state structure of the $[Al(BH_4)_4]^-$ ion, present in $[Ph_3MeP][Al(BH_4)_4]$. In the light of these structural data, account for the following observations, recorded for the compound in solution. (a) At 298 K, the ^1H NMR spectrum of $[Ph_3MeP][Al(BH_4)_4]$ shows one, broad signal in addition to signals assigned to the cation; this pattern of signals is retained at 203 K. (b) In the ^{11}B NMR spectrum (298 K) of the same compound, a quintet is observed. (c) In the IR spectrum of $[Ph_3MeP][Al(BH_4)_4]$, absorptions due to bridging Al–H–B and terminal B–H interactions are both observed.

12.16 Write a brief account of the bonding and reactivity of borazine, which emphasizes the ways in which this compound is similar or dissimilar to benzene.

12.17 Give appropriate bonding descriptions for the aluminium–nitrogen compounds depicted in Figure 12.16.

12.18 Use Wade's rules to suggest likely structures for B_5H_9, $[B_8H_8]^{2-}$, $C_2B_{10}H_{12}$ and $[B_6H_9]^-$. Indicate, where appropriate, the possible occurrences of cage-isomers.

12.19 (a) Two-electron reduction of B_5H_9 followed by protonation is a convenient route to B_5H_{11}. What structural change (and why) do you expect the B_5 cage to undergo during this reaction?

(b) Account for the fact that the solution ^{11}B NMR spectrum of $[B_3H_8]^-$ (**12.25**) exhibits one signal which is a binomial nonet.

(c) The photolysis of B_5H_9 leads to the formation of a mixture of three isomers of $B_{10}H_{16}$. The products arise from the intermolecular elimination of H_2. Suggest the nature of the product, and the reason that three isomers are formed.

12.20 Suggest likely products for the following reactions, with the stoichiometries stated:

(a) $B_5H_9 + Br_2 \xrightarrow{298\,K}$

(b) $B_4H_{10} + PF_3 \longrightarrow$

(c) $1\text{-}BrB_5H_8 \xrightarrow{KH,\,195\,K}$

(d) $2\text{-}MeB_5H_8 \xrightarrow{ROH}$

Chapter 13

The group 14 elements

- Occurrence, extraction and uses
- Physical properties
- The elements
- Hydrides
- Carbides, silicides, germides, stannides and plumbides
- Halides and complex halides

- Oxides and oxoacids and hydroxides, including silicates
- Silicones
- Sulfides
- Cyanogen and silicon nitride
- Aqueous solution chemistry of tin and lead

1	2			13	**14**	15	16	17	18
H									He
Li	Be			B	**C**	N	O	F	Ne
Na	Mg			Al	**Si**	P	S	Cl	Ar
K	Ca	*d*-block		Ga	**Ge**	As	Se	Br	Kr
Rb	Sr			In	**Sn**	Sb	Te	I	Xe
Cs	Ba			Tl	**Pb**	Bi	Po	At	Rn
Fr	Ra								

13.1 Introduction

The elements in group 14 – carbon, silicon, germanium, tin and lead – show a gradation from C, which is non-metallic, to Pb, which, though its oxides are amphoteric, is mainly metallic in nature. The so-called '*diagonal line*' which is often drawn through the *p*-block to separate metallic from non-metallic elements, passes between Si and Ge, indicating that Si is non-metallic and Ge is metallic. However, this distinction is not definitive. In the solid state, Si and Ge possess a covalent diamond-type lattice (see *Figure 5.19a*), but their electrical resistivities (see *Section 5.8*) are significantly lower than that of diamond, indicating metallic behaviour. Silicon and germanium are classed as *semi-metals*[†] and we have already discussed their semiconducting properties (see *Section 5.9*).

[†] Under IUPAC recommendations, the term 'semi-metal' is preferred over 'metalloid'.

All members of group 14 exhibit an oxidation state of +4, but the +2 oxidation state increases in stability as the group is descended; carbenes exemplify the C(II) state but exist only as reaction intermediates, silicon dihalides are stable only at high temperatures, the Ge(II) and Sn(II) states are well established, and Pb(II) is more stable than the Pb(IV) state. In this respect, Pb resembles its periodic neighbours, Tl and Bi, with the inertness of the 6*s* electrons being a general feature of the last member of each of groups 13, 14 and 15 (see *Box 12.3*).

Carbon is essential to life on Earth, and most of its compounds lie within the remit of organic chemistry. Nonetheless, compounds of C that are formally classified as 'inorganic' abound and extend to *organometallic species*. (see *Chapters 18* and *23*).

13.2 Occurrence, extraction and uses

Occurrence

The two long-established crystalline allotropes of carbon, diamond and graphite occur naturally, as does amorphous carbon (e.g. in coal); diamonds occur in igneous rocks (e.g. in the Kimberley volcanic pipes, South Africa). Carbon dioxide constitutes only 0.04% of the atmosphere, and, although vital for photosynthesis, CO_2 is not a major source of carbon. During the 1990s, it was discovered that molecular allotropes of carbon, the *fullerenes* (see *Section 13.4*), occur naturally in a number of deposits in Australia, New Zealand and North America; however, laboratory synthesis remains the chief means of accessing these new allotropes.

Elemental Si does not occur naturally, but it constitutes 25.7% of the Earth's crust (Si is the second most abundant

BOX 13.1

RESOURCES, ENVIRONMENTAL AND BIOLOGICAL

Recycling: tin and lead

Recycling of tin and lead, particularly the latter, takes place on a huge scale. In Box 5.1, we described steel-can recycling operations. The tin used to coat steel cans is recovered using specialized detinning processes. In 1997 in the US, 12 000 t of tin was recovered, 66% of which originated in tin-plated cans; the recycled metal represented ≈22% of the tin consumed in the US in the following year.

Lead–acid storage batteries represent a major source of metal that is recovered. In 1998, ≈75% of refined Pb manufactured in the US originated from recycled metal, much of it (≈1 Mt) coming from spent batteries from vehicle and industrial sources.

element after O) in the form of sand, quartz, rock crystal, flint, agate and silicate minerals (see *Section 13.9*). In contrast, Ge makes up only 1.5 ppm of the Earth's crust, being present in trace amounts in a range of minerals (e.g. zinc ores) and in coal. The principal tin-bearing ore is *cassiterite* (SnO_2); important ores of lead are *galena* (PbS), *anglesite* ($PbSO_4$) and *cerussite* ($PbCO_3$).

Extraction and manufacture

Sources of natural graphite are supplemented by manufactured material formed by heating powdered coke (high temperature carbonized coal) with silica at ≈2800 K. Approximately 30% of diamonds for industrial use in the US are synthetic, usually produced by the metal (Ni or Pt) catalyzed conversion of graphite to diamond at ≈2000 K and 100 kbar pressure. Under less extreme conditions, diamond films may be grown using a chemical vapour deposition method, and hydrothermal processes are currently being investigated.[†] The manufacture of amorphous carbon (carbon black, used in synthetic rubbers) involves burning oil in a limited supply of air.

Silicon (not of high purity) is extracted from silica, SiO_2, by heating with C or CaC_2 in an electric furnace. Impure Ge can be obtained from flue dusts collected during the extraction of zinc from its ores, or by reducing GeO_2 with H_2 or C. For use in the electronic and semiconductor industries, ultrapure Si and Ge are required, and both can be obtained by zone-melting techniques (see *Box 5.2* and *Section 27.6*).

Tin is obtained from *cassiterite* (SnO_2) by reduction with C in a furnace (see *Section 7.7*), but a similar process cannot be applied to extract Pb from its sulfide ore since $\Delta_f G^\circ(CS_2, g)$ is +67 kJ mol^{-1}; thermodynamically viable processes involve reactions 13.1 or 13.2 at high temperatures. Both Sn and Pb are refined electrolytically. Recycling of Sn and Pb is highlighted in *Box 13.1*.

$$\left.\begin{array}{l} 2PbS + 3O_2 \longrightarrow 2PbO + 2SO_2 \\ PbO + C \longrightarrow Pb + CO \\ \text{or} \\ PbO + CO \longrightarrow Pb + CO_2 \end{array}\right\} \quad (13.1)$$

$$PbS + 2PbO \longrightarrow 3Pb + SO_2 \quad (13.2)$$

Uses

Diamond is one of the hardest substances known, and apart from its commercial value as a gemstone, it has applications in cutting tools and abrasives. The structural differences between diamond and graphite lead to remarkable differences in physical properties (see *Section 13.3*) and uses. Graphite is used in lubricants, pencils and electrodes, and as a refractory material. Charcoal (made by heating wood) and animal charcoal (produced by charring treated bones) are microcrystalline forms of graphite, supported, in the case of animal charcoal, on calcium phosphate. The adsorption properties of *activated charcoal* render it commercially important (see *Box 13.2*). Carbon fibres of great tensile strength (formed by heating oriented organic polymer fibres at ≥1750 K) contain graphite crystallites oriented parallel to the fibre axis, and are used to strengthen materials such as plastics. Carbon-composites are fibre-reinforced, chemically inert materials which possess high strength, rigidity, thermal stability, high resistance to thermal shock and retain their mechanical properties at high temperature. Such properties have led to their use in external body parts of the space shuttle (see *Section 27.7*).

Silicon has major applications in the electronic and semiconductor industries (see *Sections 5.8, 5.9* and *27.6*, and *Box 13.3*); applications of Ge as a semiconductor are gradually becoming fewer as new and more efficient semiconducting materials are developed. The optical properties of GeO_2 make it suitable for specialized use in certain lenses and in infrared detectors and spectrometers. Silica is an extremely important commercial material; it is the main component of glass, and large quantities of sand are consumed worldwide by the building industry. Quartz glass (formed on cooling fused SiO_2) can withstand sudden

[†] See for example: X.-Z. Zhao, R. Roy, K.A. Cherian and A. Badzian (1997) *Nature*, vol. 385, p. 513 – 'Hydrothermal growth of diamond in metal–C–H$_2$O systems'; R.C. DeVries (1997) *Nature*, vol. 385, p. 485 – 'Diamonds from warm water'.

BOX 13.2 **APPLICATIONS**

Activated charcoal: utilizing a porous structure

Activated charcoal is a finely divided form of amorphous carbon and is manufactured from organic materials (e.g. peat, wood) by heating in the presence of reagents that promote both oxidation and dehydration. Activated charcoal possesses a pore structure with a large internal surface area: *microporous* materials exhibit pores <2 nm wide, *macroporous* refers to activated charcoals with a pore size >50 nm, and *mesoporous* materials fall in between these extremes. The largest internal surface areas are found for microporous materials ($>700 \, m^2 \, g^{-1}$). The ability of the hydrophobic surface to adsorb small molecules is the key to the widespread applications of activated charcoal. (Comparisons should be made with the porous structures and applications of zeolites: see *Sections 13.9* and *26.6*.)

Early large-scale applications of activated charcoal were in gas masks in World War I. Various gas-filters including those in cooker extractors and mobile or bench-top laboratory fume-hoods contain activated charcoal filters. About 20%

of the activated charcoal that is produced is consumed in the sugar industry where it is used as a decolouring agent. Water purification uses large amounts of activated charcoal.

The porous structure means that activated charcoal is an excellent heterogeneous catalyst, especially when impregnated with a *d*-block metal such as palladium. On an industrial scale, it is used, for example, in the manufacture of phosgene (equation 13.40), and in laboratory syntheses, it has many uses, e.g.:

$$4CoCl_2 + O_2 + 4[NH_4]Cl + 20NH_3$$
$$\xrightarrow{\text{activated charcoal}} 4[Co(NH_3)_6]Cl_3 + 2H_2O$$

Further reading

A.J. Evans (1999) *Chemistry & Industry*, p. 702 – 'Cleaning air with carbon'.

temperature changes and has specialist uses; we discuss different types of glasses in Section 13.9. Silica gel (an amorphous form of silica, produced by treating aqueous sodium silicate with acid) is used as a drying agent, a stationary

phase in chromatography, and a heterogeneous catalyst. *Caution!* Inhalation of silica dusts may lead to the lung disease *silicosis*. Hydrated silica forms the exoskeletons of marine diatoms, but the role of Si in other biological systems

BOX 13.3 **APPLICATIONS**

Solar power: thermal and electrical

Harnessing energy from the Sun is, of course, an environmentally acceptable method of producing power. Conversion via heat exchange units (often referred to as solar panels) provides thermal energy to raise the temperature of swimming pools or to provide domestic hot water. Conversion via photovoltaic systems (often termed solar cells) produces electricity and involves the use of semiconductors. Initially, NASA's space programme was the driving force behind the development of solar cells, and applications in satellites and other space vessels remain at the cutting edge of design technology. However, we all now feel the benefits of solar cells which are used in items such as solar-powered calculators. Silicon has been the workhorse of this commercial operation. The thickness of a typical cell is 200–350 μm, and is constructed of an n-doped layer (which faces the sun), a p-doped layer and a metal-contact grid on the top and bottom surfaces. The latter are connected by a conducting wire. At the n–p junction, electrons move from the p-type to the n-type silicon, and 'holes' (see *Section 5.9*) move in the opposite direction; this leads to a flow of electricity around the circuit. Power output per cell is small, and a large number of cells must operate together to produce a viable

voltage supply. Weather conditions and the number of daylight hours are key factors that have to be accommodated if adequate solar power is to be generated for domestic or similar uses.

Other semiconductors in use in solar cells include GaAs (e.g. in space satellites), CdTe (a promising newcomer to solar cell development) and TiO_2 (used in the Grätzel cell which involves a novel design in which a TiO_2 film is coated with an organic dye).

Further reading

M. Hammonds (1998) *Chemistry & Industry*, p. 219 – 'Getting power from the sun'.

K. Kalyanasundaram and M. Grätzel (1999) in *Optoelectronic Properties of Inorganic Compounds*, ed. D.M. Roundhill and J.P. Fackler, Plenum Press, New York, p. 169 – 'Efficient photovoltaic solar cells based on dye sensitization of nanocrystalline oxide films'.

J. Wolfe (1998) *Chemistry & Industry*, p. 224 – 'Capitalising on the sun'.

Fig. 13.1 The declining use of leaded fuels in motor vehicles is illustrated by these statistics from the US. [Data: U.S. Geological Survey.]

is less well defined.[†] The applications of silicates and aluminosilicates are discussed in *Section 13.9*.

Tin-plating of steel cans improves corrosion resistance and is a major use of Sn. The metal is, however, soft and tin-alloys such as pewter, soldering metal, bronze and die-casting alloy have greater commercial value than pure Sn. High quality window glass is usually manufactured by the Pilkington process which involves floating molten glass on molten tin to produce a flat surface. Tin dioxide is an opacifier used in enamels and paints (also see *Section 27.4*), and its applications in gas sensors are the topic of *Box 13.9*. The use of tin-based chemicals as flame retardants (see *Box 16.1*) is increasing in importance.

Lead is a soft metal and has been widely used in the plumbing industry; this use has diminished as awareness of the toxicity of the metal has grown (see *Box 13.4*). Similarly, uses of Pb in paints have been reduced, and 'environmentally friendly' lead-free fuels are replacing leaded counterparts (Figure 13.1). Lead oxides are of great commercial importance, e.g. in the manufacture of 'lead crystal' glass. *Red lead*, Pb_3O_4, is used as a pigment and a corrosion-resistant coating for steel and iron. Large amounts of Pb are consumed in the automobile industry in lead–acid batteries in which the cell reaction is a combination of half-reactions 13.3 and 13.4; a normal automobile 12 V battery contains six cells connected in series.

$$PbSO_4(s) + 2e^- \rightleftharpoons Pb(s) + [SO_4]^{2-}(aq) \qquad E^o = -0.36\,V$$
$$(13.3)$$

$$PbO_2(s) + 4H^+(aq) + [SO_4]^{2-}(aq) + 2e^-$$
$$\rightleftharpoons PbSO_4(s) + 2H_2O(l) \qquad E^o = +1.69\,V \qquad (13.4)$$

13.3 Physical properties

Table 13.1 lists selected physical properties of the group 14 elements. A comparison with Table 12.1 shows there to be some similarities in *trends* down groups 13 and 14.

[†] For a thought-provoking account, see: J.D. Birchall (1995) *Chemical Society Reviews*, vol. 24, p. 351 – 'The essentiality of silicon in biology'.

Ionization energies and cation formation

On descending group 14, the trends in ionization energies reveal two particular points:

- the relatively large increases between values of IE_2 and IE_3 for each element;
- the discontinuities (i.e. *increases*) in the trends of values of IE_3 and IE_4 at Ge and Pb.

The sums of the first four ionization energies for any element suggest that it is unlikely that M^{4+} ions are formed. For example, although both SnF_4 and PbF_4 are non-volatile solids, neither has a symmetrical lattice structure in the solid state. Both SnO_2 and PbO_2 adopt the rutile lattice, but the fact that PbO_2 is brown argues against a formulation of $Pb^{4+}(O^{2-})_2$; agreement between values of lattice energies determined using a Born–Haber cycle and calculated from an electrostatic model is good for SnO_2, but is poor for PbO_2. Thus, values of the M^{4+} ionic radii (Table 13.1) should be treated with some caution.

Aqueous solution chemistry involving cations of the group 14 elements is restricted to Sn and Pb (see *Section 13.13*), and so Table 13.1 gives E^o values only for these metals.

Some energetic and bonding considerations

Table 13.2 lists some experimentally determined values for covalent bond enthalpy terms. When we try to interpret the chemistry of the group 14 elements on the basis of such bond energies, caution is necessary for two reasons:

- many thermodynamically favourable reactions are kinetically controlled;
- in order to use bond enthalpy terms successfully, *complete* reactions must be considered.

The first point is illustrated by considering that although the combustions of CH_4 and SiH_4 are both thermodynamically favourable, SiH_4 is spontaneously inflammable in air, whereas CH_4 explodes in air only when a spark provides the energy to overcome the activation barrier. In respect of the second point, consider reaction 13.5.

$$(13.5)$$

Table 13.1 Some physical properties of the Group 14 elements, M, and their ions.

Property	C	Si	Ge	Sn	Pb
Atomic number, Z	6	14	32	50	82
Ground state electronic configuration	$[He]2s^2 2p^2$	$[Ne]3s^2 3p^2$	$[Ar]3d^{10}4s^2 4p^2$	$[Kr]4d^{10}5s^2 5p^2$	$[Xe]4f^{14}5d^{10}6s^2 6p^2$
Enthalpy of atomization, $\Delta_a H^o(298\,K)\,/\,kJ\,mol^{-1}$	717	456	375	302	195
Melting point, mp / K	>3823‡	1687	1211	505	600
Boiling point, bp / K	5100	2628	3106	2533	2022
Standard enthalpy of fusion, $\Delta_{fus} H^o(mp)\,/\,kJ\,mol^{-1}$	104.6	50.2	36.9	7.0	4.8
First ionization energy, $IE_1\,/\,kJ\,mol^{-1}$	1086	786.5	762.2	708.6	715.6
Second ionization energy, $IE_2\,/\,kJ\,mol^{-1}$	2353	1577	1537	1412	1450
Third ionization energy, $IE_3\,/\,kJ\,mol^{-1}$	4620	3232	3302	2943	3081
Fourth ionization energy, $IE_4\,/\,kJ\,mol^{-1}$	6223	4356	4411	3930	4083
Metallic radius, $r_{metal}\,/\,pm$	–	–	–	158	175
Covalent radius, $r_{cov}\,/\,pm^*$	77	118	122	140	154
Ionic radius, $r_{ion}\,/\,pm^{**}$	–	–	53 (Ge^{4+})	74 (Sn^{4+}) 93 (Sn^{2+})	78 (Pb^{4+}) 119 (Pb^{2+})
Standard reduction potential, $E^o(M^{2+}/M)\,/\,V$	–	–	–	−0.14	−0.13
Standard reduction potential, $E^o(M^{4+}/M^{2+})\,/\,V$	–	–	–	+0.15	+1.69***
NMR active nuclei (% abundance, nuclear spin)	^{13}C (1.1, $I = \frac{1}{2}$)	^{29}Si (4.7, $I = \frac{1}{2}$)	^{73}Ge (7.8, $I = \frac{9}{2}$)	^{117}Sn (7.6, $I = \frac{1}{2}$); ^{119}Sn (8.6, $I = \frac{1}{2}$)	^{207}Pb (22.6, $I = \frac{1}{2}$)

‡ For diamond.

* Values for C, Si, Ge and Sn refer to diamond-type structures and thus refer to 4-coordination; the value for Pb also applies to a 4-coordinate centre.

** Values are for 6-coordination.

*** This value is for the half-reaction: $PbO_2(s) + 4H^+(aq) + [SO_4]^{2-}(aq) + 2e^- \rightleftharpoons PbSO_4(s) + 2H_2O(l)$.

Inspection of Table 13.2 shows that $E(C–H) > E(C–Cl)$, but the fact that the H–Cl bond ($431\,kJ\,mol^{-1}$) is significantly stronger than the Cl–Cl bond ($242\,kJ\,mol^{-1}$) results in reaction 13.5 being energetically favourable.

> *Catenation* is the tendency for covalent bond formation between atoms of a given element, e.g. C–C bonds in hydrocarbons or S–S bonds in polysulfides.

The particular strength of the C–C bond contributes towards the fact that catenation in carbon compounds is common. However, it must be stressed that *kinetic* as well as thermodynamic factors may be involved, and any detailed discussion of kinetic factors is subject to complications:

- Even when C–C bond breaking is the rate-determining step, it is the bond dissociation *energy* (zero point energy: see *Section 2.9*) rather than the enthalpy term that is important.
- Reactions are often bimolecular processes in which bond-making and bond-breaking occur simultaneously, and in such cases, the rate of reaction may bear no relationship to the difference between bond enthalpy terms of the reactants and products.

In contrast to the later elements in group 14, C tends not to expand its valence octet of electrons, and, while complexes such as $[SiF_6]^{2-}$ and $[Sn(OH)_6]^{2-}$ are known, carbon analogues are not. The fact that CCl_4 is kinetically inert towards hydrolysis but $SiCl_4$ is readily hydrolyzed by water has traditionally been ascribed to the availability of $3d$ orbitals on Si which can stabilize an associative transition state. This view has been challenged with the suggestion that the phenomenon is steric in origin associated purely with the lower accessibility of the C centre arising from the shorter C–Cl bonds with respect to the Si–Cl bonds.

The possible role of $(p–d)\pi$-bonding for Si and the later elements in group 14 remains a controversial issue (see *Section 4.7*) and we return to this in *Section 13.6*. On the other hand, $(p–p)\pi$-bonding leading to double to triple homonuclear bonds, which is so common in carbon chemistry, is relatively unimportant later in the group. A similar situation is observed in groups 15 and 16. The mesityl derivative **13.1** was the first compound containing an Si=Si bond to be characterized; in the Raman spectrum, an absorption at $529\,cm^{-1}$ is assigned to the ν(Si=Si) mode, and in the solid state structure, the Si–Si bond distance of 216 pm is less than twice the value of

Table 13.2 Some experimental covalent bond enthalpy terms ($kJ\,mol^{-1}$); the values for single bonds refer to the group 14 elements in tetrahedral environments.

C–C	C=C	C≡C	C–H	C–F	C–Cl	C–O	C=O
346	598	813	416	485	327	359	806
Si–Si			Si–H	Si–F	Si–Cl	Si–O	Si=O
226			326	582	391	466	642
Ge–Ge			Ge–H	Ge–F	Ge–Cl	Ge–O	
186			289	465	342	350	
Sn–Sn			Sn–H		Sn–Cl		
151			251		320		
					Pb–Cl		
					244		

BOX 13.4   RESOURCES, ENVIRONMENTAL AND BIOLOGICAL

Toxicity of lead

Lead salts are extremely toxic. The ingestion of a soluble lead salt can cause acute poisoning, and long-term exposure to a source of the metal (e.g. old water pipes, Pb-based paints) may result in chronic poisoning. Organolead(IV) compounds such as Et_4Pb, used as an anti-knock additive to leaded motor fuels, attack the nervous system. In a relevant piece of research, analysis of wines produced between 1962 and 1991 from grapes grown in roadside vineyards has shown some correlation between a decrease in Pb content and the introduction of unleaded fuels. Sequestering agents such as $[EDTA]^{4-}$ (see equation 6.75 and accompanying text) are

used to complex Pb^{2+} ions in the body, and their removal follows by natural excretion.

Further reading

R.A. Goyer (1988) in *Handbook on toxicity of inorganic compounds*, eds H.G. Seiler, H. Sigel and A. Sigel, Marcel Dekker, New York, p. 359 – 'Lead'.

R. Lobinski *et al.* (1994) *Nature*, vol. 370, p. 24 – 'Organolead in wine'.

r_{cov} (2×118 pm). Such species are stabilized with respect to polymerization by the presence of bulky substituents such as mesityl (in **13.1**), CMe_3 or $CH(SiMe_3)_2$. The central Si_2C_4-unit in **13.1** is planar, allowing overlap of orthogonal $3p$ orbitals for π-bond formation; the bulky mesityl substituents adopt a 'paddle-wheel' conformation minimizing steric interactions.[†]

Mesityl = Mes = 1,3,5-trimethylphenyl

(13.1)

The formation of $(p-p)\pi$-bonds between C and Si is also rare; an example is shown in equation 13.6.

$$Mes_2FSi \overset{H}{\underset{H}{\diagdown}}C=C\overset{H}{\underset{H}{\diagup}} + {}^tBuLi$$

$$\longrightarrow \quad \underset{Mes}{\overset{Mes}{\diagdown}}Si=C\overset{H}{\underset{{}^tBu}{\diagup}} + LiF \quad (13.6)$$

The first Ge=C double bond was reported in 1987, since when a number of examples have been reported, including $Mes_2Ge=CHCH_2{}^tBu$ which is stable at 298 K.

NMR active nuclei

Table 13.1 lists NMR active nuclei for the group 14 elements. Although the isotopic abundance of ^{13}C is only 1.1%, use of ^{13}C NMR spectroscopy is very important. The low abundance means that, unless a sample is isotopically enriched, satellite peaks in, for example, a ^1H NMR spectrum, will not be observed and application of ^{13}C as an NMR active nucleus lies in its *direct observation*. The appearance of satellite peaks due to coupling of an observed nucleus such as ^1H to ^{29}Si or ^{119}Sn is diagnostic (see case study 5 in *Section 2.11*). Direct observation of ^{29}Si nuclei is a routine means of characterizing Si-containing compounds. Tin-119 NMR spectroscopy (^{119}Sn being generally favoured over ^{117}Sn for direct observation) is also valuable; the chemical shift range is large and, as with many heteronuclei, δ values may provide an indication of coordination environments.

Mössbauer spectroscopy

The ^{119}Sn nucleus is suitable for Mössbauer spectroscopy (see *Section 2.12*) and isomer shift values can be used to distinguish between Sn(II) and Sn(IV) environments. The spectroscopic data may also provide information about the coordination number of the Sn centre.

13.4 Allotropes of carbon

Graphite and diamond: structure and properties

We have already described the rigid structure of diamond (*Figure 5.19a*). Diamond is not the thermodynamically most stable form of the element but is *metastable*. At room

---

[†] In a second structurally characterized polymorph, the orientations of the mesityl groups differ, see: R. Okazaki and R. West (1996) *Advances in Organometallic Chemistry*, vol. 39, p. 231.

temperature, the conversion of diamond into graphite is thermodynamically favoured (equation 13.7), making graphite the standard state of C at 298 K. However, reaction 13.7 is infinitely slow.

$$C(diamond) \longrightarrow C(graphite)$$

$$\Delta_r G^\circ (298\,K) = -2.9\,kJ\,mol^{-1} \qquad (13.7)$$

> A state is *metastable* if it exists without observable change even though it is thermodynamically unstable with respect to another state.

Diamond has a higher density than graphite ($\rho_{graphite} = 2.25$; $\rho_{diamond} = 3.51\,g\,cm^{-3}$), and this allows artificial diamonds to be made from graphite at high pressures. There are two structural modifications of graphite. The 'normal' form is α-graphite and can be converted to the β-form by grinding; a β ⟶ α-transition occurs above 1298 K. Both forms possess layered structures and Figure 13.2a shows 'normal' graphite. (Compare the structure of graphite with that of boron nitride in *Figure 12.13*.) The *intra*layer C−C bond distances are equal (142 pm) while the *inter*layer distances are 335 pm; a comparison of these distances with the values for C of $r_{cov} = 77$ pm and $r_v = 185$ pm indicates that while covalent bonding is present within each layer, only weak van der Waals interactions operate between adjacent layers. Graphite cleaves readily and is used as a lubricant; these facts follow directly from the weak interlayer interactions. The electrical conductivity (see *Section 5.8*) of α-graphite is direction-dependent; in a direction parallel to the layers, the electrical resistivity is $1.3 \times 10^{-5}\,\Omega\,m$ (at 293 K) but is $\approx 1\,\Omega\,m$ in a direction perpendicular to the layers. Each C atom has four valence electrons and forms three σ-bonds, leaving one electron to participate in delocalized π-bonding. The molecular π-orbitals extend over each layer, and while the bonding MOs are fully occupied, the energy gap between them and the vacant antibonding MOs is very small, allowing the electrical conductivity in the direction *parallel* to the layers to approach that of a metal. In contrast, the electrical resistivity of diamond is $1 \times 10^{11}\,\Omega\,m$, making diamond an excellent insulator.

Graphite is more reactive than diamond; it is oxidized by atmospheric O_2 above 970 K whereas diamond burns at >1170 K. Graphite reacts with hot, concentrated HNO_3 to give the aromatic compound $C_6(CO_2H)_6$. We consider some specific types of reactions below.

Graphite: intercalation compounds

Graphite possesses the remarkable property of forming many *intercalation* (*lamellar* or *graphitic*) compounds, the formation of which involves movement apart of the carbon layers and the penetration of atoms or ions between them. There are two general types of compound:

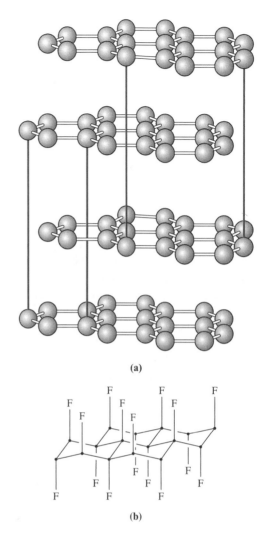

(a)

(b)

Fig. 13.2 (a) Part of the infinite layered-lattice of α-graphite ('normal' graphite); the layers are co-parallel, and atoms in *alternate* layers lie over each other. (b) Part of one layer of the structure of $(CF)_n$ for $n = 1$.

- colourless, non-conductors of electricity in which the carbon layers become buckled owing to saturation of the C atoms and loss of the π-system;
- coloured, electrical conductors in which the planarity and π-delocalization of the layers are retained.

Polymeric carbon monofluoride, $(CF)_n$ ($n \leq 1$), is a widely studied example of the first type of compound. It is formed when F_2 reacts with graphite at 720 K (or at lower temperatures in the presence of HF), although at 970 K, the product is monomeric CF_4. The fluorine content in materials formulated as $(CF)_n$ is variable and their colour varies, being white when $n \approx 1.0$. Carbon monofluoride possesses a layer structure, and is used as a lubricant, being more resistant to atmospheric oxidation at high temperatures than graphite. Part of one layer is shown in Figure 13.2b; in the idealized compound CF, each C atom is tetrahedral; each C−C bond distance within a layer is 154 pm, and between layers is 820 pm, i.e. more than double that in α-graphite.

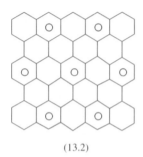

(13.2)

The second class of intercalation compound includes the blue graphite salts formed with strong acids in the presence of oxidizing agents, and the metallic-looking red or blue compounds formed when graphite reacts with group 1 metals. For example, when graphite is treated with an excess of K (and unreacted metal is washed out with Hg), a paramagnetic copper-coloured material formulated as $K^+[C_8]^-$ results. The penetration of K^+ ions between the layers causes structural changes in the graphite framework: the initially staggered layers (Figure 13.2a) become eclipsed, and the interlayer spacing increases from 335 to 540 pm. The K^+ ions lie above (or below) the centres of alternate C_6-rings, as indicated in structure **13.2**, forming layers of centred-hexagonal motifs. The electrical conductivity of KC_8 is greater than that of α-graphite, consistent with the addition of electrons to the delocalized π-system. Heating KC_8 leads to the formation of a series of decomposition products as the metal is eliminated (equation 13.8). The structures of these materials are related, there being one, two, three, four or five carbon layers respectively between layers of K^+ ions.

$$KC_8 \xrightarrow{\Delta} KC_{24} \xrightarrow{\Delta} KC_{36} \xrightarrow{\Delta} KC_{48} \xrightarrow{\Delta} KC_{60}$$
copper-coloured   blue

(13.8)

Such alkali metal intercalates are extremely reactive, igniting in air and exploding on contact with water. Potassium can be replaced by a *d*-block metal by reaction of KC_8 with metal chloride, but the choice of solvent for the reactions is critical, as is the nature of the *d*-block metal salt (e.g. $CuCl_2 \cdot 2H_2O$, $MnCl_2 \cdot 4H_2O$ for sources of Cu^{2+} and Mn^{2+}). Examples include MnC_{16}, FeC_{24} and CuC_{16} which contain Mn(II), Fe(III) and Cu(II) respectively.

In the metal-containing intercalation compounds, the carbon layers are reduced and become negatively charged. In contrast, in intercalation compounds formed with strong acids in the presence of oxidizing agents, the carbon layers lose electrons and become positively charged, e.g. graphite hydrogensulfate, $[C_{24}]^+[HSO_4]^- \cdot 24H_2O$, which is produced when graphite is treated with concentrated H_2SO_4 and a little HNO_3 or CrO_3. A related product forms when the acid is $HClO_4$; in this intercalate, the planar layers of carbon atoms are 794 pm apart and are separated by $[ClO_4]^-$ ions and acid molecules. Cathodic reduction of this material, or treatment with graphite, gives a series of compounds as $HClO_4$ is sequentially eliminated. These materials are better electrical conductors than

graphite, and this can be explained in terms of a positive-hole mechanism (see *Section 5.9*).

Other intercalation compounds include those formed with Cl_2, Br_2, ICl and halides such as KrF_2, UF_6 and $FeCl_3$. Reaction of graphite with $[O_2]^+[AsF_6]^-$ results in the formation of the salt $[C_8]^+[AsF_6]^-$. The catalytic properties of some graphite intercalation compounds render them of practical importance; e.g. KC_8 is a hydrogenation catalyst.

Fullerenes: synthesis and structure

In 1985, Kroto, Smalley and coworkers discovered that, by subjecting graphite to laser-radiation at >10 000 K, new allotropes of carbon were formed. The *fullerenes* are named after architect Buckminster Fuller, known for designing geodesic domes. Each fullerene is *molecular* and the family includes C_{60}, C_{70}, C_{76}, C_{78} and C_{84}. Several synthetic routes to fullerenes have been developed; C_{60} and C_{70} are the major components of the mixture formed when graphitic soot is produced as graphite rods are evaporated (by applying an electrical arc between them) in a helium atmosphere at ≈130 bar and the vapour condensed. Extraction of the soot into benzene yields a red solution from which C_{60} and C_{70} can be separated by chromatography. Hexane or benzene solutions of C_{60} are magenta while those of C_{70} are red. Both C_{60} and C_{70} are now available commercially, and this has encouraged rapid exploration of their chemical properties.

Figure 13.3a shows the structure of C_{60}. Although a number of X-ray diffraction studies of C_{60} have been carried out, the near-spherical shape of the molecule has led to frustrating orientational disorder (see *Section 18.3*) problems. The C_{60} molecule consists of an approximately spherical network of atoms which are connected in 5- and 6-membered rings; all the C atoms are equivalent, as indicated by the fact that the ^{13}C NMR spectrum of C_{60} exhibits one signal (δ +143). The rings are arranged such that no 5-membered rings are adjacent to each other; this is more easily seen in the schematic representation of C_{60} shown in Figure 13.3b which also gives a bonding scheme. Each C atom is covalently bonded to three others in an approximately trigonal planar arrangement; the relatively large surface of the 'sphere' means that there is only slight deviation from planarity at each C centre. There are two types of C−C bond: those at the junctions of two hexagonal rings (6,6-edges) are of length 139 pm, while those between a hexagonal and a pentagonal ring (5,6-edges) are longer, 145.5 pm. These differences indicate the presence of localized double and single bonds, and similar bonding descriptions are appropriate for other fullerene cages. We consider chemical evidence for the presence of C=C double bonds below. The C_{70} molecule has D_{5h} symmetry and is approximately ellipsoidal (Figure 13.4); it comprises 6- and 5-membered rings organized so that, as in C_{60}, 5-membered rings are never adjacent. The ^{13}C NMR spectrum of C_{70} confirms that there are five C environments in solution, consistent with the solid state structure (Figure 13.4a).

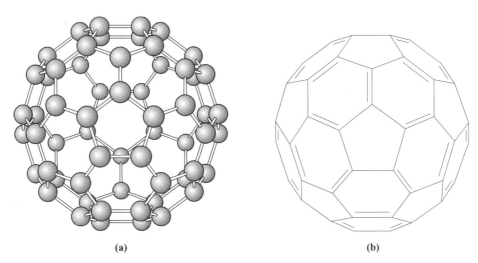

(a)   **(b)**

Fig. 13.3 (a) The structure of the fullerene C_{60}; the approximately spherical molecule is composed of fused 5- and 6-membered rings of carbon atoms. [X-ray diffraction at 173 K of the benzene solvate $C_{60}\cdot4C_6H_6$, M.F. Meidine *et al.* (1992) *J. Chem. Soc., Chem. Commun.*, p. 1534.] (b) A representation of C_{60}, in the same orientation as is shown in (a), but showing only the upper surface and illustrating the localized single and double carbon–carbon bonds.

Fullerenes: reactivity

Since efficient syntheses have been available, fullerenes (in particular C_{60}) have been the focus of an explosion of research. We provide a brief introduction to the chemical properties of C_{60}; organometallic derivatives are covered in *Section 23.10*, and the reading list at the end of the chapter gives more in-depth coverage.

The structural representation in Figure 13.3b suggests connected benzene rings, but the chemistry of C_{60} is *not* reminiscent of benzene. Its behaviour reflects the presence of *localized* double and single C–C bonds, e.g. C_{60} undergoes *addition* reactions. Birch reduction gives a mixture of polyhydrofullerenes (equation 13.9) with $C_{60}H_{32}$ being the dominant product; reoxidation occurs with the quinone

shown. Reaction 13.10 shows a selective route to $C_{60}H_{36}$; the hydrogen-transfer agent is 9,10-dihydroanthracene (DHA). In addition to being a selective method of hydrogenation, use of 9,9',10,10'-[D_4]dihydroanthracene provides a method of selective deuteration.

$$C_{60} \xrightleftharpoons[\text{2. }^t\text{BuOH}]{\text{1. Li / liquid NH}_3} C_{60}H_{18} + \ldots\ldots + C_{60}H_{36} \qquad (13.9)$$

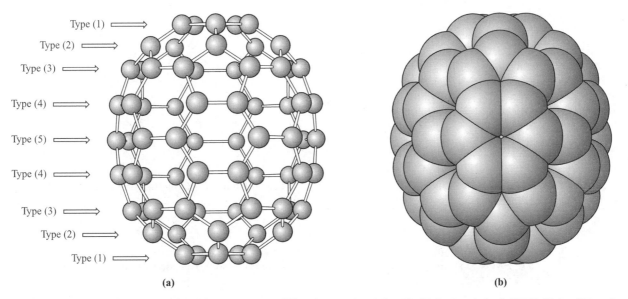

Type (1) $\Longrightarrow$
Type (2) $\Longrightarrow$
Type (3) $\Longrightarrow$
Type (4) $\Longrightarrow$
Type (5) $\Longrightarrow$
Type (4) $\Longrightarrow$
Type (3) $\Longrightarrow$
Type (2) $\Longrightarrow$
Type (1) $\Longrightarrow$

(a)   **(b)**

Fig. 13.4 The structure of C_{70} determined from an X-ray diffraction study of $C_{70}\cdot6S_8$ [H.B. Bürgi *et al.* (1993) *Helv. Chim. Acta*, vol. 76, p. 2155]: (a) a ball-and-stick representation showing the five carbon atom types, and (b) a space-filling diagram illustrating the ellipsoidal shape of the molecule.

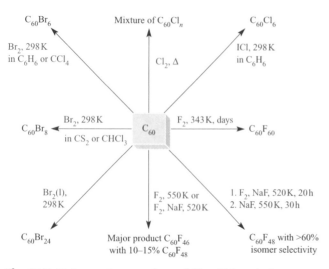

Fig. 13.5 Halogenation reactions of C_{60}. Although the number of possible isomers for products $C_{60}X_n$ where $2 \leq n \leq 58$ is, at the very least, large, some of the reactions (such as fluorination using NaF and F_2) are surprisingly selective.

$$C_{60}H_{18} \xleftarrow[\text{120-fold excess of}]{\substack{\text{623 K, sealed tube,} \\ \text{DHA, 24 h}}} C_{60} \xrightarrow[\text{120-fold excess of}]{\substack{\text{623 K, sealed tube,} \\ \text{DHA, several min}}} C_{60}H_{36}$$

DHA =

(13.10)

Additions of F_2, Cl_2 and Br_2 also occur, the degree and selectivity of halogenation depending on conditions (Figure 13.5). Just as going from benzene to cyclohexane causes a change from a planar to boat- or chair-shaped ring, addition of substituents to C_{60} causes deformation of the near-spherical surface. This is illustrated in Figure 13.6 which shows the structure of $C_{60}Br_{24}$, the surface of which includes both boat and chair C_6-rings; two of the 24 sp^3 C centres are highlighted in the figure.

The ene-like nature of C_{60} is reflected in a range of reactions such as the formation of the epoxide $C_{60}O$ (reaction 13.11); addition occurs at the junction of two 6-membered rings (a 6,6-bond), i.e. at a C=C bond (Figure 13.3b).

$$C_{60} \xrightarrow[]{O_2, \text{ benzene}} \text{(structure)}$$

(13.11)

Other reactions typical of double-bond character include the formation of cycloaddition products (exemplified schematically in equation 13.12), and some have been developed to prepare a range of rather exotic derivatives.

$$\text{(structure)}$$

(13.12)

Reactions of C_{60} with free radicals readily occur, e.g. photolysis of RSSR produces RS$^{\bullet}$ which reacts with C_{60} to give $C_{60}SR^{\bullet}$, although this is unstable with respect to regeneration of C_{60}. The stabilities of radical species $C_{60}Y^{\bullet}$ are highly dependent on the steric demands of Y. When the reaction of ^{t}Bu$^{\bullet}$ (produced by photolysis of a *tert*-butyl halide) with C_{60} is monitored by ESR spectroscopy (which detects the presence of unpaired electrons), the intensity of the signal due to the radical $C_{60}{}^t$Bu$^{\bullet}$ increases over the temperature range 300–400 K. These data are consistent with equilibrium 13.13, with reversible formation and cleavage of an inter-cage C–C bond.

(13.13)

The formation of methanofullerenes, $C_{60}CR_2$, occurs by reaction at either 5,6- or 6,6-edges in C_{60}. For the 6,6-addition products, the product of the reaction of C_{60} with diphenylazomethane is $C_{61}Ph_2$ (equation 13.14) and, initially, structural data suggested that the reaction was an example of 'cage-expansion' with the addition of the CPh_2 unit being concomitant with the cleavage of the C–C bond marked *a* in equation 13.14. This conclusion was at odds with NMR spectroscopic data and theoretical calculations, and a low temperature X-ray diffraction study of

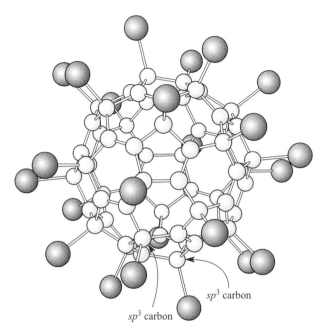

Fig. 13.6 The structure of $C_{60}Br_{24}$ determined by X-ray diffraction at 143 K [F.N. Tebbe *et al.* (1992) *Science*, vol. 256, p. 822]. The introduction of substituents results in deformation of the C_{60} surface; compare the structure of $C_{60}Br_{24}$ with that of C_{60} in Figure 13.3a which shows the C_{60} cage in a similar orientation.

compound **13.3** has now confirmed that 6,6-edge-bridged methanofullerenes should be described in terms of the C_{60} cage sharing a common C–C bond with a cyclopropane ring.

(13.14)

(13.3)

Theoretical studies on C_{60} show that the LUMO is triply degenerate and the HOMO–LUMO (see *Section 1.17*) separation is relatively small. It follows that reduction of C_{60} should be readily achieved. A number of charge transfer complexes have been prepared in which a suitable donor molecule transfers an electron to C_{60} as in equation 13.15.

(Liquid at 298 K)

(13.15)

This particular product is of importance because, on cooling to 16 K, it becomes *ferromagnetic* (see *Figure 20.25*). The electrochemical reduction of C_{60} results in the formation of a series of *fulleride* ions, $[C_{60}]^{n-}$ where $n = 1-6$. The midpoint potentials (obtained using cyclic voltammetry and measured with respect to the ferrocenium/ferrocene couple, $Fc^+/Fc = 0$ V, ferrocene, see *Section 23.13*) for the reversible one-electron steps at 213 K are given in scheme 13.16.

$$C_{60} \xrightleftharpoons{-0.81\ V} [C_{60}]^- \xrightleftharpoons{-1.24\ V} [C_{60}]^{2-} \xrightleftharpoons{-1.77\ V} [C_{60}]^{3-}$$

$$\xrightleftharpoons{-2.22\ V} [C_{60}]^{4-} \xrightleftharpoons{-2.71\ V} [C_{60}]^{5-} \xrightleftharpoons{-3.12\ V} [C_{60}]^{6-} \quad (13.16)$$

By titrating C_{60} in liquid NH_3 against a Rb/NH_3 solution (see *Section 8.6*) at 213 K, five successive reduction steps are observed and the $[C_{60}]^{n-}$ anions have been studied by vibrational and electronic spectroscopies. At low temperatures, some alkali metal fulleride salts of type $[M^+]_3[C_{60}]^{3-}$ become *superconducting* (see *Section 27.4*). The structures of the M_3C_{60} fullerides can be described in terms of M^+ ions occupying the interstitial holes in a lattice composed of close-packed, near-spherical C_{60} cages. In K_3C_{60} and Rb_3C_{60}, the $[C_{60}]^{3-}$ cages are arranged in a ccp lattice, and the cations fully occupy the octahedral and tetrahedral holes (Figure 13.7). The temperature at which a material becomes superconducting is its *critical temperature*, T_c. Values of T_c for K_3C_{60} and Rb_3C_{60} are 18 K and 28 K respectively, and for Cs_3C_{60} (in which the C_{60} cages adopt a bcc lattice), $T_c = 40$ K. Although Na_3C_{60} is structurally related to K_3C_{60} and Rb_3C_{60}, it is not superconducting. The paramagnetic $[C_{60}]^{2-}$ anion has been isolated as the $[K(crypt-222)]^+$ salt (reaction 13.17 and *Section 10.8*).

$$C_{60} \xrightarrow[\text{toluene/crypt-222}]{\text{DMF/K}} [K(crypt-222)]_2[C_{60}] \quad (13.17)$$

In the solid state, the $[C_{60}]^{2-}$ cages are arranged in layers with hexagonal packing, although the cages are well separated; $[K(crypt-222)]^+$ cations reside between the layers of fulleride anions.

Endohedral metallofullerenes are a remarkable series of compounds in which metal atoms are encapsulated within a fullerene cage; the general family is denoted as $M_x@C_n$.

Fig. 13.7 A representation of the structures of K_3C_{60} and Rb_3C_{60} in which the $[C_{60}]^{3-}$ cages are arranged in an fcc lattice with the M^+ ions occupying all the octahedral and tetrahedral holes. Some of the cations in the unit cell shown are hidden by $[C_{60}]^{3-}$ anions.

Examples of these compounds include $Sc_2@C_{84}$, $Y@C_{82}$ and $La_2@C_{80}$, with the larger fullerenes producing more stable compounds than C_{60}. The compounds are prepared by vaporizing graphite rods impregnated with an appropriate metal oxide or metal carbide. By use of ^{13}C and ^{139}La NMR spectroscopies, it has been shown that the two lanthanum atoms in $La_2@C_{80}$ undergo circular motion within the fullerene cage.

Carbon nanotubes

Carbon *nanotubes* were discovered in 1991 and consist of elongated cages, best thought of as rolled graphite-like sheets, i.e. in contrast to the fullerenes, nanotubes consist of networks of fused 6-membered rings. Nanotubes are very flexible and have great potential in materials science. As a result, research in this area is a 'hot topic' but is beyond the scope of this book; the end of the chapter reading list provides an entry into the area.

13.5 Structural and chemical properties of silicon, germanium, tin and lead

Structures

The solid state structures of Si, Ge, Sn and Pb and the trends from semiconductor to metal on descending the group have already been discussed:

- diamond-type lattice of Si, Ge and α-Sn (*Section 5.11* and *Figure 5.19*);
- polymorphism of Sn (*Section 5.4*);
- structure of Pb (*Section 5.3*);
- semiconducting properties (*Section 5.9*).

Chemical properties

Silicon is much more reactive than carbon; at high temperatures, Si combines with O_2, F_2, Cl_2, Br_2, I_2, S_8, N_2, P_4, C and B to give binary compounds. Silicon liberates H_2 from aqueous alkali (equation 13.18), but is insoluble in acids other than a mixture of concentrated HNO_3 and HF.

$$Si + 4[OH]^- \longrightarrow [SiO_4]^{4-} + 2H_2 \qquad (13.18)$$

On descending group 14, the electropositivity and reactivity of the elements increase. In general, Ge behaves in a similar manner to Si, but, being more electropositive, reacts with concentrated HNO_3 (forming GeO_2), and does not react with aqueous alkali. Reactions between Ge and HCl or H_2S yield $GeCl_4$ or GeS_2 respectively. Although high temperatures are needed for reactions between Sn and O_2 (to give SnO_2) or sulfur (giving SnS_2), the metal reacts readily with halogens to yield SnX_4. Tin is little affected by dilute HCl or H_2SO_4, but reacts with dilute HNO_3 (to give $Sn(NO_3)_2$ and NH_4NO_3) and with concentrated acids yielding $SnCl_2$ (from HCl) and $SnSO_4$ and SO_2 (from H_2SO_4). Hot aqueous alkali oxidizes the metal to Sn(IV) according to equation 13.19.

$$\left[\begin{array}{c} OH \\ HO_{\prime\prime\prime\prime\prime\cdots}\underset{|}{\overset{|}{Sn}}\cdots^{\prime\prime\prime\prime}OH \\ HO \blacktriangleright \quad \blacktriangleleft OH \\ OH \end{array} \right]^{2-}$$

(13.4)

$$Sn + 2[OH]^- + 4H_2O \longrightarrow [Sn(OH)_6]^{2-} + 2H_2 \qquad (13.19)$$

13.4

A *pyrophoric* material is spontaneously inflammable.

When finely divided, Pb is pyrophoric, but bulk pieces are passivated by coatings of, for example, PbO, and reaction with O_2 in air occurs only above $\approx 900\,K$. Lead reacts very slowly with dilute mineral acids, slowly evolves H_2 from hot concentrated HCl, and reacts with concentrated HNO_3 to give $Pb(NO_3)_2$ and oxides of nitrogen. For reactions of Pb with halogens, see *Section 13.8*.

13.6 Hydrides

Although the extensive chemistry of hydrocarbons (i.e. carbon hydrides) lies outside this book, we note several

Methane hydrates

A gas hydrate is an example of a *clathrate*, a crystalline solid comprising a *host* (a three-dimensional assembly of H_2O molecules which form cage-like arrays) and *guest* molecules (small molecules such as CH_4 which occupy the cavities in the host lattice). Gas hydrates occur naturally in the Arctic and in deep-sea sediments, and their importance lies in their ability to trap gases within crystalline masses, thereby acting rather like natural gas 'storage tanks'. It is possible that such deposits could be tapped for fuel sources, but on the other hand, any uncontrolled release of the huge amounts of CH_4 that is presently trapped inside these clathrates could add to the 'greenhouse' effect (see *Box 13.7*).

points for comparisons with later group 14 hydrides:

- Table 13.2 illustrated the relative strength of a C–H bond compared with C–Cl and C–O bonds, and this trend is *not* mirrored by later elements;
- CH_4 is chlorinated with some difficulty, whereas SiH_4 reacts violently with Cl_2;
- CH_4 is stable with respect to hydrolysis, but SiH_4 is readily attacked by water;
- SiH_4 is spontaneously inflammable in air and, although it is the *kinetic* stability of CH_4 with respect to reaction with O_2 at 298 K that is crucial, values of $\Delta_c H^\circ$ show that combustion of SiH_4 is more exothermic than that of CH_4;
- catenation is more common for C than the later group 14 elements, and hydrocarbon families are much more diverse than their Si, Ge, Sn and Pb analogues.

Binary hydrides

Silane, SiH_4, is formed when $SiCl_4$ or SiF_4 reacts with $Li[AlH_4]$ and is a source of pure Si (equation 13.20) for semi-conductors (see *Section 5.9*, *Box 5.2* and *Section 27.6*). Silanes Si_nH_{2n+2} with straight or branched chains are known for $1 \le n \le 10$, and Figure 13.8 compares the boiling points of the first five straight-chain silanes with their hydrocarbon analogues. Silanes are explosively inflammable in air (equation 13.21).

$$SiH_4 \xrightarrow{\Delta} Si + 2H_2 \tag{13.20}$$

$$SiH_4 + 2O_2 \longrightarrow SiO_2 + 2H_2O \tag{13.21}$$

A mixture of SiH_4, Si_2H_6, Si_3H_8 and Si_4H_{10} along with traces of higher silanes is obtained when Mg_2Si reacts with aqueous acid, but the non-specificity of this synthesis renders it of little practical value. By irradiating SiH_4 with a CO_2 laser, SiH_4 can be converted selectively into Si_2H_6. Silane is a colourless gas which is insoluble in water, reacts rapidly with alkalis (equation 13.22) and forms compounds of the type $M[SiH_3]$ with Na, K (equation 13.23), Rb and Cs. The crystalline salt $K[SiH_3]$ possesses an NaCl lattice and is a valuable synthetic reagent, e.g. equation 13.24.

$$SiH_4 + 2KOH + H_2O \longrightarrow K_2SiO_3 + 4H_2 \tag{13.22}$$

$$2SiH_4 + 2K \xrightarrow{\text{in } MeOCH_2CH_2OMe} 2K[SiH_3] + H_2 \tag{13.23}$$

$$Me_3ESiH_3 + KCl \xleftarrow{Me_3ECl}[\text{E} = \text{Si, Ge, Sn}] K[SiH_3] \xrightarrow{MeI} MeSiH_3 + KI \tag{13.24}$$

Germanes Ge_nH_{2n+2} (straight and branched chain isomers) are known for $1 \le n \le 9$. GeH_4 is less reactive than SiH_4; it is a colourless gas (bp 184 K, dec 488 K), insoluble in water, and prepared by treating GeO_2 with $Na[BH_4]$ although higher germanes are also formed. Discharges of various frequencies are finding increased use for this type of synthesis and have been used to convert GeH_4 into higher germanes, or mixtures of SiH_4 and GeH_4 into Ge_2H_6, $GeSiH_6$ and Si_2H_6. Mixed hydrides of Si and Ge, e.g. $GeSiH_6$ and $GeSi_2H_8$, are also formed when an intimate mixture of Mg_2Ge and Mg_2Si is treated with acid. Reactions between GeH_4 and alkali metals, M, in liquid NH_3 produce $M[GeH_3]$, and, like $[SiH_3]^-$, the $[GeH_3]^-$ ion is synthetically useful. The reaction of $SnCl_4$ with $Li[AlH_4]$ gives SnH_4 (bp 221 K) but this decomposes at 298 K into Sn and H_2; note the variation in reactivities: $SiH_4 > GeH_4 < SnH_4$. Plumbane, PbH_4, is poorly characterized and may not actually have been isolated. Significantly, however, replacement of H atoms by alkyl or aryl substituents is accompanied by increased stability (see *Section 18.5*).

Fig. 13.8 Boiling points of the straight-chain silanes, Si_nH_{2n+2}, and hydrocarbons C_nH_{2n+2}.

Fig. 13.9 Representative reactions of SiH_3Cl. The structures of $N(SiH_3)_3$ (determined by X-ray diffraction at 115 K) and $(H_3Si)_2O$ (determined by electron diffraction).

Halohydrides of silicon and germanium

Of compounds of the type SiH_nX_{4-n} (X = halogen, n = 1–3), $SiHCl_3$ is of particular importance in the purification of Si in the semiconductor industry (equation 13.25); both $SiHCl_3$ (mp 145 K, bp 306 K) and SiH_4 (see above) are suitable because they are volatile sources of Si.

$$Si(\text{impure}) + 3HCl \xrightarrow[-H_2]{670\,K} SiHCl_3$$

$$\xrightarrow[\substack{2.\,\text{CVD (chemical}\\ \text{vapour deposition)}}]{\substack{1.\,\text{Purification}\\ \text{by distillation}}} Si(\text{pure, polycrystalline}) \quad (13.25)$$

Another application of $SiHCl_3$ is the *hydrosilation* reaction (equation 13.26), a method of introducing an $SiCl_3$ group and an entry to organosilicon chemistry.

$$RCH=CH_2 + SiHCl_3 \longrightarrow RCH_2CH_2SiCl_3 \quad (13.26)$$

$$SiH_4 + nHX \xrightarrow{\Delta,\,AlCl_3} SiH_{4-n}X_n + nH_2$$
$$n = 1 \text{ or } 2 \quad (13.27)$$

The halo-derivatives SiH_2X_2 and SiH_3X (X = Cl, Br, I) can be prepared from SiH_4 (equation 13.27) and some reactions of SiH_3Cl (bp 243 K) are shown in Figure 13.9. The ease with which SiH_nX_{4-n} compounds hydrolyze releasing HX means that they must be handled in moisture-free conditions. The preparation and reactivity of GeH_3Cl resemble those of SiH_3Cl.

The structures of trisilylamine, $N(SiH_3)_3$, and disilyl ether, $(H_3Si)_2O$ are shown in Figure 13.9. The NSi_3 skeleton in $N(SiH_3)_3$ is planar and the N–Si bond distance of 173 pm is shorter than the sum of the covalent radii, Σr_{cov} (see *Appendix 6*); similarly, in $(H_3Si)_2O$, the Si–O–Si bond

angle of 144° is large (compare 111° in Me_2O) and the Si–O bonds of 163 pm are shorter than Σr_{cov}. Trigermyl-amine is isostructural with $N(SiH_3)_3$, but $P(SiH_3)_3$ is pyramidal with P–Si bonds of length 225 pm. In $(H_3Si)_2S$, the Si–S–Si bond angle is 97° and the Si–S bond distances (214 pm) are consistent with a bond order of one. For many years, these data have been taken as an indication that N and O take part in $(p-d)\pi$-bonding with Si, there being no corresponding interactions in Si–P or Si–S bonds. However, the issue is by no means clear-cut, and recent arguments centre around the planarity of $N(SiH_3)_3$ (and related strengthening of Si–N and Si–O bonds) being due to $n(N) \longrightarrow \sigma^*(Si–H)$ electron donation, where $n(N)$ represents the non-bonding (lone pair) electrons of the N atom. This is so-called *negative hyperconjugation*,[†] and is analogous to the donation of electrons from a *d*-block metal centre to a σ^*-orbital of a PR_3 ligand that we describe in Section 20.4. The possibility of $(p-d)\pi$-bonding should not be confused with the $(p-p)\pi$-bonding which occurs in, for example, Si=N bonds (with a formal bond order of two) in compounds such as $^tBu_2Si=NSi^tBu_3$, **13.5**. Notice that in **13.5** the nitrogen atom is in a *linear* environment and can be considered to have a stereochemically inactive lone pair, possibly involved in π-interactions.

$$\underset{169\,\text{pm}}{^tBu_2Si}\overset{157\,\text{pm}}{=\!=\!=}N—Si^tBu_3$$

(13.5)

[†] Negative hyperconjugation: see Y. Mo, Y. Zhang and J. Gao (1999) *Journal of the American Chemical Society*, vol. 121, p. 5737 and references cited in this paper.

13.7 Carbides, silicides, germides, stannides and plumbides

Carbides

Classifying carbides is not simple, but some useful categories are:

- saline (salt-like) carbides which produce mainly CH_4 when hydrolyzed;
- those containing the $[C{\equiv}C]^{2-}$ ion;
- those containing the $[C{=}C{=}C]^{4-}$ ion;
- interstitial carbides;
- solid state carbides with other lattice structures;
- fulleride salts (see *Section 13.4*);
- endohedral metallofullerenes (see *Section 13.4*).

Examples of saline carbides are Be_2C (see *Section 11.4* and *equation 11.14*) and Al_4C_3, both made by heating the constituent elements at high temperatures. Although their solid state structures contain isolated C centres which are converted to CH_4 on reaction with H_2O, it is unlikely that the 'C^{4-}' ion is present since the interelectronic repulsion energy would be enormous.

Carbides containing the $[C{\equiv}C]^{2-}$ (acetylide) ion include Na_2C_2, K_2C_2, MC_2 (M = Mg, Ca, Sr, Ba), Ag_2C_2 and Cu_2C_2; they evolve C_2H_2 when treated with water (see *equation 11.15*). Calcium carbide is manufactured (see *Box 11.3*) as a grey solid by heating CaO with coke at $\approx 2300\,K$, and when pure, it is colourless. It adopts a distorted NaCl lattice, the axis along which the $[C{\equiv}C]^{2-}$ are aligned being lengthened; the C–C bond distance is 119 pm, compared with 120 pm in C_2H_2. The reaction between CaC_2 and N_2 (equation 13.28) is used commercially for the production of calcium cyanamide, a nitrogenous fertilizer (equation 13.29); the cyanamide ion, **13.6**, is isoelectronic with CO_2.

$$^-N{=\!=}C{=\!=}N^-$$

$$(13.6)$$

$$CaC_2 + N_2 \xrightarrow{1300\,K} CaNCN + C \qquad (13.28)$$

$$CaNCN + 3H_2O \longrightarrow CaCO_3 + 2NH_3 \qquad (13.29)$$

Equations 13.30 and 13.31 show syntheses of Na_2C_2, Ag_2C_2 and Cu_2C_2; the group 11 carbides are heat- and shock-sensitive, and explosive when dry.

$$2NaNH_2 + C_2H_2 \xrightarrow{\text{in liquid } NH_3} Na_2C_2 + 2NH_3 \qquad (13.30)$$

$$2[M(NH_3)_2]^+ + C_2H_2 \longrightarrow M_2C_2 + 2[NH_4]^+ + 2NH_3$$
$$M = Ag, Cu \qquad (13.31)$$

Carbides of formula MC_2 do not necessarily contain the acetylide ion. The room temperature form of ThC_2 (Th is an actinoid metal, see *Chapter 24*) adopts an NaCl lattice but is not isostructural with CaC_2; in ThC_2, the C_2-units ($d_{CC} = 133$ pm) in alternating layers lie in different orientations. The solid state structure of LaC_2 contains C_2-units with $d_{CC} = 129$ pm. Unlike CaC_2 which is an insulator, ThC_2 and

LaC_2 have metallic appearances and are electrical conductors. The C–C bond lengths can be rationalized in terms of structures approximating to $Th^{4+}[C_2]^{4-}$ and $La^{3+}[C_2]^{3-}$; compared with $[C_2]^{2-}$, the extra electrons in $[C_2]^{4-}$ and $[C_2]^{3-}$ reside in antibonding MOs, thus weakening the C–C interaction. However, the conducting properties and diamagnetism of ThC_2 and LaC_2 show that this is an oversimplified description since electron delocalization into a conduction band (see *Section 5.8*) must occur. Hydrolysis of these carbides is also atypical of a $[C_2]^{2-}$-containing species, e.g. the reaction of ThC_2 and H_2O yields mainly C_2H_2, C_2H_6 and H_2.

Carbides containing $[C{=}C{=}C]^{4-}$ are rare; they include Mg_2C_3 (see end of *Section 11.4*) which liberates propyne upon hydrolysis.

The structures of the so-called *interstitial carbides* (formed by heating C with d-block metals having $r_{metal} > 130$ pm, e.g. Ti, Zr, V, Mo, W) may be described in terms of a close-packed metal lattice with C atoms occupying octahedral holes (see *Figure 5.5*). In carbides of type M_2C (e.g. V_2C, Nb_2C) the metal atoms are in an hcp lattice and half of the octahedral sites are occupied; in the MC type (e.g. TiC and WC), the metal atoms adopt a ccp structure and all the octahedral holes are occupied. These interstitial carbides are important refractory materials; characteristically they are very hard and infusible, have melting points $>2800\,K$ and, in contrast to the acetylide derivatives, do not react with water. Tungsten carbide, WC, is one of the hardest substances known and is widely used in cutting tools and dies.

Transition metals with $r_{metal} < 130$ pm (e.g. Cr, Fe, Co, Ni) form carbides with a range of stoichiometries (e.g. Cr_3C_2, Fe_3C) which possess complicated structures involving C–C bonding. In Cr_3C_2 (formed by reaction 13.32), the Cr atoms form a lattice of edge-sharing trigonal prisms each occupied by a C atom such that carbon chains run through the structure with C–C distances comparable to single bonds.

$$3Cr_2O_3 + 13C \xrightarrow[\text{presence of } H_2]{1870\,K,\text{ in}} 2Cr_3C_2 + 9CO \qquad (13.32)$$

Carbides of this type are hydrolyzed by water or dilute acid to give mixtures of hydrocarbons and H_2.

Silicides

The structures of the metal silicides (prepared by direct combination of the elements at high temperatures) are diverse, and a full discussion of the structures is beyond the scope of this book.[†] Some examples of their solid state structural types are:

- isolated Si atoms (e.g. Mg_2Si, Ca_2Si);
- Si_2-units (e.g. U_3Si_2);
- Si_4-units (e.g. NaSi, KSi, CsSi)
- Si_n-chains (e.g. CaSi);
- planar or puckered hexagonal networks of Si atoms (e.g. β-USi_2, $CaSi_2$);

[†] For further details, see: A.F. Wells (1984) *Structural Inorganic Chemistry*, 5th edn, Clarendon Press, Oxford, p. 987.

• three-dimensional network of Si atoms (e.g. $SrSi_2$, α-USi_2).

The Si_4-units present in the alkali metal silicides are noteworthy. The $[Si_4]^{4-}$ anion is isoelectronic with P_4 and the solid state structures of several group 1 metal silicides contain tetrahedral Si_4-units, but these are not isolated anions. The structure of Cs_4Si_4 comes close to featuring discrete, tetrahedral $[Si_4]^{4-}$ ions, but significant cation–anion interactions exist. The silicide K_3LiSi_4 possesses tetrahedral Si_4-units linked by Li^+ ions to give infinite chains, and in K_7LiSi_8, pairs of Si_4-units are connected as shown in structure **13.7** with additional interactions involving K^+ ions.

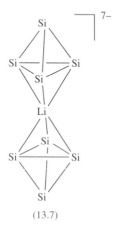

(13.7)

Silicides are hard materials, but their melting points are generally lower than those of the metal carbides. Treatment of Mg_2Si with dilute acids gives mixtures of silanes (see *Section 13.6*). The properties of some silicides make them useful as refractory materials (e.g. Fe_3Si and $CrSi_2$); Fe_3Si is used in magnetic tapes and disks to increase their thermal stability.

Germides, stannides and plumbides

Germanium, tin and lead do not form solid state binary compounds with metals. In contrast, the formation of *Zintl ions* (see *Section 8.6*) for Ge, Sn and Pb differentiates them from the earlier group 14 elements. The synthesis of $[Sn_5]^{2-}$ (equation 8.35) typifies the preparations of other Zintl ions. Diamagnetic Zintl ions include $[M_4]^{4-}$ (M = Ge, Sn or Pb), $[Sn_5]^{2-}$, $[Pb_5]^{2-}$, $[Sn_9]^{4-}$, $[Ge_9]^{2-}$, $[Ge_9]^{4-}$, $[Ge_{10}]^{2-}$, $[Sn_8Tl]^{3-}$, $[Sn_9Tl]^{3-}$ and $[Pb_2Sb_2]^{2-}$. Paramagnetic ions are exemplified by $[Sn_9]^{3-}$ and $[Ge_9]^{3-}$. The structure of $[Sn_5]^{2-}$ was shown in Figure 8.3, and Figure 13.10 depicts $[Sn_9]^{4-}$ and $[Ge_9]^{3-}$ and illustrates the deltahedral families of group 14 Zintl ions. Bonding in these ions is delocalized, and for the diamagnetic species, Wade's rules can be used to rationalize the observed structures. Wade's rules (see *Section 12.11*) were developed for boranes and, if a lone pair is localized outside the cage, a group 14 element can contribute two electrons to cluster bonding thereby mimicking a

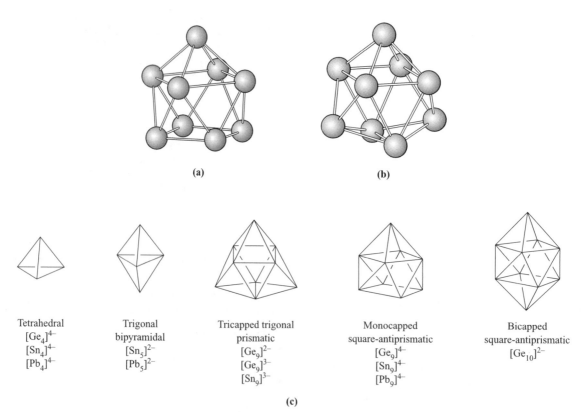

(a)

(b)

| Tetrahedral
$[Ge_4]^{4-}$
$[Sn_4]^{4-}$
$[Pb_4]^{4-}$ | Trigonal
bipyramidal
$[Sn_5]^{2-}$
$[Pb_5]^{2-}$ | Tricapped trigonal
prismatic
$[Ge_9]^{2-}$
$[Ge_9]^{3-}$
$[Sn_9]^{3-}$ | Monocapped
square-antiprismatic
$[Ge_9]^{4-}$
$[Sn_9]^{4-}$
$[Pb_9]^{4-}$ | Bicapped
square-antiprismatic
$[Ge_{10}]^{2-}$ |

(c)

Fig. 13.10 The structures, established by X-ray diffraction, of (a) $[Sn_9]^{4-}$, determined for the salt $[Na(crypt-222)]_4[Sn_9]$ [J.D. Corbett *et al.* (1977) *J. Am. Chem. Soc.*, vol. 99, p. 3313], and (b) $[Ge_9]^{3-}$, determined for the compound [K(crypt-222)]$_3$[Ge$_9$]·PPh$_3$ [C. Belin *et al.* (1991) *New J. Chem.*, vol. 15, p. 931]; for discussion of cryptand ligands including crypt-222, see *Section 10.8*. (c) Schematic representations of structure types for selected Zintl ions.

BH unit. More strictly, each group 14 element is *isolobal* with a BH unit (see *Section 23.5*).

Worked example 13.1 Structures of Zintl ions

Rationalize the structure of $[Sn_9]^{4-}$ shown in Figure 13.10a.

There are 9 Sn atoms and each provides 2 valence electrons, assuming that each atom carries a lone pair of electrons.

There are four electrons from the 4− charge.

Total number of cage-bonding electrons available

$$= (9 \times 2) + 4 = 22 \text{ electrons}$$

$$= 11 \text{ pairs}$$

Thus, $[Sn_9]^{4-}$ has 11 pairs of electrons with which to bond 9 Sn atoms.

This means that there are $(n + 2)$ pairs of electrons for n vertices, and so $[Sn_9]^{4-}$ is a *nido*-cage, based on an 10-vertex deltahedron (see Figure 12.19) with one vertex vacant. This corresponds to the observed structure of a monocapped square-antiprism.

Self-study exercises

1. By referring to Figures 12.19 and 13.10c, rationalize the structures of:
   (a) $[Ge_4]^{4-}$; (b) $[Sn_5]^{2-}$; (c) $[Ge_9]^{2-}$; (d) $[Ge_{10}]^{2-}$.

2. Rationalize why $[Sn_5]^{2-}$ and $[Pb_5]^{2-}$ are isostructural.

3. Rationalize why $[Pb_5]^{2-}$ adopts the same cluster structure as $C_2B_3H_5$. [Hint: Look back to worked example 12.4]

Reaction conditions are critical to the selective formation of a Zintl ion. The alloy KSn_2 reacts with crypt-222 (see *Section 10.8*) in 1,2-diaminoethane to give $[K(crypt-222)]_3[Sn_9]$ containing the paramagnetic $[Sn_9]^{3-}$ ion. However, reaction times must be less than two days, since longer periods favour the formation of $[K(crypt-222)]_4[Sn_9]$ containing the diamagnetic $[Sn_9]^{4-}$ ion. The paramagnetic clusters $[Sn_9]^{3-}$ and $[Ge_9]^{3-}$ both adopt *distorted* tricapped trigonal prismatic structures (Figure 13.10b).

13.8 Halides and complex halides

Carbon halides

Selected physical properties of the tetrahalides of C and Si are listed in Table 13.3. The carbon tetrahalides differ markedly from those of the later group 14 elements: they are inert towards water and dilute alkali and do not form complexes with metal halides. The distinction has been attributed to the absence of *d* orbitals in the valence shell of a C atom; look back at the electronic versus steric debate, outlined in Section 13.3. However, one must be cautious. In the case of CX_4 being inert towards attack by water, the 'lack of C *d* orbitals' presupposes that the reaction would proceed through a 5-coordinate intermediate (i.e. as is proposed for hydrolysis of silicon halides). Of course, it is impossible to establish the mechanism of a reaction that does not occur! Certainly, CF_4 and CCl_4 are *thermodynamically* unstable with respect to hydrolysis; compare the value of $\Delta_r G^\circ$ for equation 13.33 with that of -290 kJ mol^{-1} for the hydrolysis of $SiCl_4$.

$$CCl_4(l) + 2H_2O(l) \longrightarrow CO_2(g) + 4HCl(aq)$$

$$\Delta_r G^\circ = -380 \text{ kJ mol}^{-1} \qquad (13.33)$$

Carbon tetrafluoride is extremely inert and may be prepared by the reaction of SiC and F_2, with the second product, SiF_4, being removed by passage through aqueous NaOH. Equation 13.34 shows a convenient laboratory-scale synthesis of CF_4 from graphite-free calcium cyanamide (see *structure* **13.6**); trace amounts of CsF are added to prevent the formation of NF_3.

$$CaNCN + 3F_2 \xrightarrow{\text{CsF, 298 K, 12 h}} CF_4 + CaF_2 + N_2 \qquad (13.34)$$

Uncontrolled fluorination of an organic compound usually leads to decomposition because large amounts of heat are evolved (equation 13.35).

$$\Delta_r H^\circ = -480 \text{ kJ mol}^{-1} \qquad (13.35)$$

Table 13.3 Selected physical properties of the carbon and silicon tetrahalides.

Property	CF$_4$	CCl$_4$	CBr$_4$	CI$_4$	SiF$_4$	SiCl$_4$	SiBr$_4$	SiI$_4$
Melting point / K	89	250	363	444 (dec)	183	203	278.5	393.5
Boiling point / K	145	350	462.5	–	187	331	427	560.5
Appearance at 298 K	Colourless gas	Colourless liquid	Colourless solid	Dark red solid	Colourless gas, fumes in air	Colourless, fuming liquid	Colourless, fuming liquid	Colourless solid

The preparation of a fully fluorinated organic compound tends therefore to be carried out in an inert solvent (the vaporization of which consumes the heat liberated) in a reactor packed with gold- or silver-plated copper turnings (which similarly absorb heat but may also play a catalytic role). Other methods include use of CoF_3 or AgF_2 as fluorinating agents, or electrolysis in liquid HF (see *Section 8.7*).

Fluorocarbons (see also *Section 16.3*) have boiling points close to those of the corresponding hydrocarbons but have higher viscosities. They are inert towards concentrated alkalis and acids, and dissolve only in non-polar organic solvents. Their main applications are as high temperature lubricants. *Freons* are chlorofluorocarbons (CFCs) or chlorofluorohydrocarbons, made by partial replacement of chlorine as in, for example, the first step of scheme 13.36. Although CFCs have been used extensively in aerosol propellants, air-conditioners, foams for furnishings, refrigerants and solvents, concern over their role in the depletion of the ozone-layer is now resulting in rapid phasing out of their use as is described in *Box 13.6*.

$$CHCl_3 \xrightarrow[SbCl_5, SbF_3]{HF} CHF_2Cl \xrightarrow{970\,K} C_2F_4 + HCl \qquad (13.36)$$

Two important plastics are manufactured from chloro-fluoro-compounds. The monomer for the commercially named *Teflon* or PTFE is C_2F_4 (tetrafluoroethene) which is prepared by reaction 13.36; polymerization occurs in the presence of water with an organic peroxide catalyst. Teflon is an inert white solid, stable up to 570 K; it has widespread domestic applications, e.g. non-stick coatings for kitchenware. The monomer CF_2=CFCl is used to manufacture the commercial polymer *Kel-F*. Both Teflon and Kel-F are used in laboratory equipment such as sealing tape and washers, parts in gas cylinder valves and regulators, coatings for stirrer bars, and sleeves for glass joints operating under vacuum.

Carbon tetrachloride (Table 13.3) is produced by chlorination of CH_4 at 520–670 K or by the reaction sequence 13.37 in which the CS_2 is recycled.

$$\left.\begin{array}{l} CS_2 + 3Cl_2 \xrightarrow{Fe\ catalyst} CCl_4 + S_2Cl_2 \\ CS_2 + 2S_2Cl_2 \longrightarrow CCl_4 + 6S \\ 6S + 3C \longrightarrow 3CS_2 \end{array}\right\} \qquad (13.37)$$

In the past, CCl_4 has been widely used as a solvent and for the chlorination of inorganic compounds. However, its high toxicity and the fact that photochemical or thermal decomposition results in the formation of $CCl_3^{\bullet}$ and $Cl^{\bullet}$ radicals has led to its manufacture and use being controlled by environmental legislation. Reactions 13.38 and 13.39 give preparations of CBr_4 and CI_4 (Table 13.3). Both compounds are toxic and are easily decomposed to their elements; CI_4 decomposes slowly in the presence of H_2O, giving CHI_3 and I_2.

$$3CCl_4 + 4AlBr_3 \longrightarrow 3CBr_4 + 4AlCl_3 \qquad (13.38)$$

$$CCl_4 + 4C_2H_5I \xrightarrow{AlCl_3} CI_4 + 4C_2H_5Cl \qquad (13.39)$$

Carbonyl chloride (*phosgene*), **13.8**, is a highly toxic, colourless gas (bp 281 K) with a choking smell, and was used in World War I chemical warfare. It is manufactured by reaction 13.40, and is used industrially in the production of diisocyanates (for polyurethane polymers), polycarbonates and 1-naphthyl-*N*-methylcarbamate, **13.9** (for insecticides).

$$CO + Cl_2 \xrightarrow{activated\ carbon\ catalyst} COCl_2 \qquad (13.40)$$

BOX 13.6 **RESOURCES, ENVIRONMENTAL AND BIOLOGICAL**

CFCs and the Montreal Protocol

The *ozone layer* is a stratum in the atmosphere 15–30 km above the Earth's surface, and it protects life on the Earth from UV radiation originating from the Sun because O_3 absorbs strongly in the ultraviolet region of the spectrum. A possible effect of UV radiation on humans is skin cancer. Chlorofluorocarbons (CFCs) are atmospheric pollutants which contribute towards the depletion of the ozone layer. In 1987, the 'Montreal Protocol for the Protection of the Ozone Layer' was established and legislation was implemented to phase out the use of CFCs: an almost complete phase-out of CFCs was required by 1996 for industrial nations, with developing nations following this ban by 2010. Taking the 1986 European consumption of CFCs as a standard (100%), the graph below illustrates how the usage of these chemicals (e.g. aerosol propellants, refrigerants) was reduced between 1986 and 1993. A recent step has been to phase out asthma inhalers that use CFCs; in the UK, the turnover of CFCs in inhalers is 800 t per year. Replacements for CFCs are HFCs (hydrofluorocarbons); research has suggested HFCs are environmentally more acceptable than CFCs. The Montreal Protocol is not confined to CFCs: potential ozone depleters such as CH_3Br are also targetted for phasing out (see *Box 16.3*).

[Data from *Chemistry & Industry*, 1994, p. 323.]

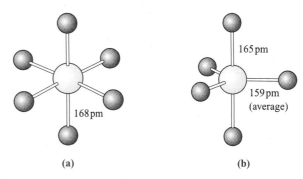

(13.8)   (13.9)   (13.10)

(a)   **(b)**

Fig. 13.11 Solid-state structures (X-ray diffraction) of (a) octahedral $[SiF_6]^{2-}$, determined for the salt $[C(NH_2)_3]_2[SiF_6]$ [A. Waskowska (1997) *Acta Crystallogr., Sect. C*, vol. 53, p. 128] and (b) trigonal bipyramidal $[SiF_5]^-$, determined for the compound $[Et_4N][SiF_5]$ [D. Schomburg *et al.* (1984) *Inorg. Chem.*, vol. 23, p. 1378]. Colour code: Si, pale grey; F, blue.

Fluorination of $COCl_2$ using SbF_3 yields $COClF$ and COF_2 which, like $COCl_2$, are unstable to water, and react with NH_3 (to give urea, **13.10**) and alcohols (to give esters). Reaction of $COCl_2$ with SbF_5 yields the linear cation $[ClCO]^+$. Its presence in the condensed phase has been established by vibrational spectroscopic studies. Reaction between COF_2 and SbF_5, however, gives an adduct $F_2CO \cdot SbF_5$ rather than $[FCO]^+[SbF_6]^-$.

Silicon halides

Many fluorides and chlorides of Si are known, but we confine our discussion to SiF_4 and $SiCl_4$ (Table 13.3) and some of their derivatives. Silicon and Cl_2 react to give $SiCl_4$, and SiF_4 can be obtained by fluorination of $SiCl_4$ using SbF_3, or by reaction 13.41; compare with equations 12.24 and 14.64.

$$SiO_2 + 2H_2SO_4 + 2CaF_2 \longrightarrow SiF_4 + 2CaSO_4 + 2H_2O \tag{13.41}$$

Both SiF_4 and $SiCl_4$ are molecular with tetrahedral structures. They react readily with water, but the former is only partially hydrolyzed (compare equations 13.42 and 13.43). Controlled hydrolysis of $SiCl_4$ results in the formation of $(Cl_3Si)_2O$, through the intermediate $SiCl_3OH$.

$$2SiF_4 + 4H_2O \longrightarrow SiO_2 + 2[H_3O]^+ + [SiF_6]^{2-} + 2HF \tag{13.42}$$

$$SiCl_4 + 2H_2O \longrightarrow SiO_2 + 4HCl \tag{13.43}$$

The formation of $[SiF_6]^{2-}$, the hexafluorosilicate ion (Figure 13.11a), illustrates the ability of Si to act as an F^- acceptor and increase its coordination number beyond four. Hexafluorosilicates are best prepared by reactions of SiF_4 with metal fluorides in aqueous HF; the K^+ and Ba^{2+} salts are sparingly soluble. In aqueous solution, fluorosilicic acid is a strong acid, but pure H_2SiF_6 has not been isolated. The $[SiF_5]^-$ ion (Figure 13.11b) is formed in the reaction of SiO_2 with aqueous HF, and may be isolated as a tetraalkylammonium ion. Silicon tetrachloride does not react with alkali metal chlorides, although lattice energy considerations suggest that it might be possible to stabilize the $[SiCl_6]^{2-}$ ion using a large quaternary ammonium cation.

Halides of germanium, tin and lead

There are many similarities between the tetrahalides of Ge and Si, and GeX_4 (X = F, Cl, Br or I) is prepared by direct combination of the elements. At 298 K, GeF_4 is a colourless gas, $GeCl_4$, a colourless liquid, and GeI_4 a red-orange solid (mp 417 K); $GeBr_4$ melts at 299 K. Each hydrolyzes, liberating HX. Unlike $SiCl_4$, $GeCl_4$ accepts Cl^- (e.g. reaction 13.44).

$$GeCl_4 + 2[Et_4N]Cl \xrightarrow{\text{in } SOCl_2} [Et_4N]_2[GeCl_6] \tag{13.44}$$

The Si(II) halides SiF_2 and $SiCl_2$ can be obtained only as unstable species (by action of SiF_4 or $SiCl_4$ on Si at ≈ 1500 K) which polymerize to cyclic products. In contrast, Ge forms stable dihalides; GeF_2, $GeCl_2$ and $GeBr_2$ are produced when Ge is heated with GeX_4, but the products disproportionate on heating (equation 13.45).

$$2GeX_2 \xrightarrow{\Delta} GeX_4 + Ge \tag{13.45}$$

Reaction between GeF_2 and F^- gives $[GeF_3]^-$, and several salts containing $[GeCl_3]^-$ are known (e.g. equation 13.46).

$$\left.\begin{array}{l} Ge + RbCl \xrightarrow{\text{in } 6\,M\,HCl} Rb[GeCl_3] \\ GeCl_2 + [Et_4N]Cl \longrightarrow [Et_4N][GeCl_3] \end{array}\right\} \tag{13.46}$$

The preference for the +2 over +4 oxidation state increases down the group, the change being due to the thermodynamic 6*s* inert pair effect (*Box 12.3*). Whereas members of the GeX_4 family are more stable than GeX_2, PbX_2 halides are more stable than PbX_4. Tin tetrafluoride (which forms hygroscopic crystals, see *Section 11.5*) is prepared from $SnCl_4$ and HF. At 298 K, SnF_4 is a white solid and has a sheet structure **13.11** with octahedral Sn atoms. At 978 K, SnF_4 sublimes to give a vapour containing tetrahedral molecules. SnF_4 is thermally stable, but PbF_4 (which has the same solid state structure as SnF_4) decomposes into PbF_2 and F_2 when heated, and must be prepared by the action of F_2 or halogen fluorides (see *Section 16.7*) on Pb compounds.

(13.11)

● = F

(13.12)

205 pm

218 pm

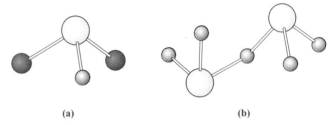

(a)

(b)

Fig. 13.12 The structures of (a) $[SnCl_2F]^-$ and (b) $[Sn_2F_5]^-$ from the solid state structure (X-ray diffraction) of $[Co(en)_3][SnCl_2F][Sn_2F_5]Cl$ (en, see *Table 6.7*); each Sn atom is in a trigonal pyramidal environment [I.E. Rakov *et al.* (1995) *Koord. Khim.*, vol. 21, p. 16]. Colour code: Sn, grey; F, blue; Cl, black.

Tin(II) fluoride is water-soluble and can be prepared in aqueous media. In contrast, PbF_2 is only sparingly soluble. One form of PbF_2 adopts a CaF_2 lattice (see *Figure 5.18a*), while the solid state structure of SnF_2 consists of puckered Sn_4F_8 rings, **13.12**, with each Sn being trigonal pyramidal consistent with the presence of a lone pair. In structures **13.11** and **13.12**, the Sn−F bridge bonds are longer than the terminal bonds, a feature that is common in this type of structure. Many tin fluoride compounds show a tendency to form F−Sn−F bridges in the solid state, as we illustrate later.

Tin(IV) chloride, bromide and iodide are made by combining the respective elements and resemble their Si and Ge analogues. The compounds hydrolyze, liberating HX, but hydrates such as $SnCl_4 \cdot 4H_2O$ can also be isolated. The reaction of Sn and HCl gives $SnCl_2$, a white solid which is partially hydrolyzed by water. The hydrate $SnCl_2 \cdot 2H_2O$ is commercially available and is used as a reducing agent. In the solid state, $SnCl_2$ has a puckered-layer structure, but discrete, bent molecules are present in the gas phase.

The Sn(IV) halides are Lewis acids, their ability to accept halide ions (e.g. reaction 13.47) following the order $SnF_4 > SnCl_4 > SnBr_4 > SnI_4$.

$$2KCl + SnCl_4 \xrightarrow[\text{in presence of HCl(aq)}]{} K_2[SnCl_6] \qquad (13.47)$$

Similarly, $SnCl_2$ accepts Cl^- to give trigonal pyramidal $[SnCl_3]^-$, but the existence of discrete anions in the solid state is cation-dependent, e.g. in $Cs[SnCl_3]$, there are long-range interactions between Sn and Cl atoms of adjacent anions. The $[SnF_5]^-$ ion can be formed from SnF_4, but in the solid state, it is polymeric with bridging F atoms and octahedral Sn centres. Bridge formation is similarly observed in Na^+ salts of $[Sn_2F_5]^-$ and $[Sn_3F_{10}]^{4-}$, formed by reacting NaF and SnF_2 in aqueous solution. Figure 13.12 shows the structures of the $[SnCl_2F]^-$ and $[Sn_2F_5]^-$ ions.

Lead tetrachloride is obtained as an oily liquid by the reaction of cold concentrated H_2SO_4 on $[NH_4]_2[PbCl_6]$; the latter is made by passing Cl_2 through a saturated solution of $PbCl_2$ in aqueous NH_4Cl. The ease with which $[PbCl_6]^{2-}$ is obtained is a striking example of stabilization of a higher oxidation state by complexation (see *Section 7.3*); in contrast, $PbCl_4$ is hydrolyzed by water and decomposes to $PbCl_2$ and Cl_2 when gently heated. The Pb(II) halides are considerably more stable than their Pb(IV) analogues and are crystalline solids at 298 K; they can be precipitated by mixing aqueous solutions of soluble halide and soluble Pb(II) salts (e.g. equation 13.48). Note that few Pb(II) salts are very soluble in water.

$$Pb(NO_3)_2(aq) + 2NaCl(aq) \longrightarrow PbCl_2(s) + 2NaNO_3(aq) \qquad (13.48)$$

Lead(II) chloride is much more soluble in hydrochloric acid than in water owing to the formation of $[PbCl_4]^{2-}$. In the solid state, $PbCl_2$ has a complicated structure with 9-coordinate Pb centres, but PbF_2 has the fluorite structure (*Figure 5.18a*). The yellow diiodide adopts the CdI_2 lattice (*Figure 5.22*).

13.9 Oxides, oxoacids and hydroxides

Oxides and oxoacids of carbon

Unlike the later elements in group 14, carbon forms stable, volatile monomeric oxides: CO and CO_2. A comment on the difference between CO_2 and SiO_2 can be made in the light of the thermochemical data in Table 13.2: the C=O bond enthalpy term is *more* than twice that for the C−O bond, while the Si=O bond enthalpy term is *less* than twice that of the Si−O bond. In rationalizing these differences, there is some justification for saying that the C=O bond is strengthened relative to Si=O by $(p–p)\pi$ contributions, and that the Si−O bond is strengthened relative to the C−O bond by $(p–d)\pi$-bonding (but see comments at the end of *Section 13.6*). Irrespective of the interpretation of

Table 13.4 Selected properties of CO and CO_2.

Property	CO	CO_2
Melting point / K	68	–
Boiling point / K	82	195 (sublimes)
$\Delta_f H^\circ(298\,K) / kJ\,mol^{-1}$	−110.5	−393.5
$\Delta_f G^\circ(298\,K) / kJ\,mol^{-1}$	−137	−394
Bond energy / $kJ\,mol^{-1}$	1075	806
C−O bond distance / pm	112.8	116.0
Dipole moment / D	0.11	0

Table 13.5 Selected low temperature baths involving dry ice.[‡]

Bath components	Temperature / K
Dry ice + ethane-1,2-diol	258
Dry ice + heptan-3-one	235
Dry ice + acetonitrile	231
Dry ice + ethanol	201
Dry ice + acetone	195
Dry ice + diethyl ether	173

[‡] To construct a bath, add *small* pieces of solid CO_2 to the solvent. Initial sublimation of the CO_2 ceases as the bath temperature decreases to the point where solid dry ice persists. The bath temperature is maintained by occasionally adding small pieces of dry ice. See also *Table 14.1*.

the enthalpy terms however, the data indicate that (ignoring enthalpy and entropy changes associated with vaporization) SiO_2 is stable with respect to conversion into molecular O=Si=O, while CO_2 is stable with respect to the formation of a macromolecular species containing 4-coordinate C and C−O single bonds. However, an extended solid phase of CO_2 has recently been prepared by laser-heating a molecular phase at 1800 K and under 40 gigaPa pressure; the vibrational spectrum of the new phase indicates that it is structurally similar to quartz (see below).[†]

Carbon monoxide is a colourless gas, formed when C burns in a restricted supply of O_2; small scale preparations involve the dehydration of methanoic acid (equation 13.49) and CO is manufactured by reduction of CO_2 using coke heated above 1070 K or by the water–gas shift reaction (see *Section 9.4*). Industrially, CO is very important and we consider some relevant catalytic processes in *Chapter 26*.

$$HCO_2H \xrightarrow{\text{conc } H_2SO_4} CO + H_2O \qquad (13.49)$$

Carbon monoxide is almost insoluble in water under normal conditions and does not react with aqueous NaOH, but at high pressures and temperatures, HCO_2H and $Na[HCO_2]$ are formed respectively. Carbon monoxide combines with F_2, Cl_2 and Br_2 (as in equation 13.40), sulfur and selenium. The high toxicity of CO arises from the formation of a stable complex with haemoglobin (see *Section 28.3*) with the consequent inhibition of O_2 transport in the body. The oxidation of CO to CO_2 is the basis of quantitative analysis for CO (equation 13.50) with the I_2 formed being removed and titrated against thiosulfate; CO is similarly oxidized by a mixture of MnO_2, CuO and Ag_2O at ambient temperatures and this reaction is used in respirators.

$$I_2O_5 + 5CO \longrightarrow I_2 + 5CO_2 \qquad (13.50)$$

The thermodynamics of the oxidation of carbon is of immense importance in metallurgy as we have already discussed in *Section 7.7*.

Selected physical properties of CO and CO_2 are given in Table 13.4; bonding models are described in *Sections 1.7* and *4.7*. The bond in CO is the strongest known in a stable

molecule and confirms the efficiency of $(p–p)\pi$ bonding between C and O. However, considerations of the bonding provide no simple explanation as to why the dipole moment of CO is so low.

In an excess of O_2, C burns to give CO_2. Solid CO_2 is called *dry ice* and readily sublimes (Table 13.4) but may be kept in insulated containers for laboratory use in, e.g. low temperature baths (Table 13.5). *Supercritical CO_2* has become a much studied and versatile solvent (see *Box 8.1*). Small scale laboratory syntheses of gaseous CO_2 usually involve reactions such as 13.51; for the industrial production of CO_2, see *Figure 10.5* and *Section 9.4*.

$$CaCO_3 + 2HCl \longrightarrow CaCl_2 + CO_2 + H_2O \qquad (13.51)$$

Carbon dioxide is the world's major environmental source of acid and its low solubility in water is of immense biochemical and geochemical significance. An aqueous solution of CO_2 is only weakly acidic, but since most of the CO_2 in solution is present in molecular form, it does not follow that carbonic acid, H_2CO_3 (see below), is a very weak acid; the true pK_a value of H_2CO_3 for the first dissociation step is 3.9. Moreover, something that is of great biological and industrial importance is the fact that combination of CO_2 with water is a relatively slow process; this can be shown by titrating a saturated solution of CO_2 against aqueous NaOH using phenolphthalein as indicator. Neutralization of CO_2 occurs by two routes. For pH < 8, the main pathway is by direct hydration (equation 13.52) which shows pseudo-first order kinetics. At pH > 10, the main pathway is by attack of hydroxide ion (equation 13.53). The overall rate of process 13.53 (which is first order in both CO_2 and $[OH]^-$) is greater than that of process 13.52.

$$\left.\begin{array}{l} CO_2 + H_2O \longrightarrow H_2CO_3 \\ H_2CO_3 + [OH]^- \longrightarrow [HCO_3]^- + H_2O \end{array}\right\} \begin{array}{l} \textit{slow} \\ \textit{very fast} \end{array} \qquad (13.52)$$

$$\left.\begin{array}{l} CO_2 + [OH]^- \longrightarrow [HCO_3]^- \\ [HCO_3]^- + [OH]^- \longrightarrow [CO_3]^{2-} + H_2O \end{array}\right\} \begin{array}{l} \textit{slow} \\ \textit{very fast} \end{array} \qquad (13.53)$$

Until 1993, there was no evidence that free carbonic acid had been isolated, although an unstable ether adduct is formed

[†] Quartz-like CO_2, see: V. Iota, C.S. Yoo and H. Cynn (1999) *Science*, vol. 283, p. 1510.

when dry HCl reacts with $NaHCO_3$ suspended in Me_2O at 243 K, and there is mass spectrometric evidence for H_2CO_3 being a product of the thermal decomposition of $[NH_4][HCO_3]$. However, IR spectroscopic data now indicate that H_2CO_3 can be isolated using a cryogenic method in which glassy MeOH solution layers of $KHCO_3$ (or Cs_2CO_3) and HCl are quenched on top of each other at 78 K and the reaction mixture warmed to 300 K. It remains a fact that, under ambient conditions, H_2CO_3 is not a readily studied species.

$$O = C \begin{array}{c} O^- \\ \\ O^- \end{array}$$

(13.13)

The carbonate ion is planar and possesses D_{3h} symmetry with all C–O bonds of length 129 pm. A delocalized bonding picture involving $(p–p)\pi$ interactions is appropriate, and VB theory describes the ion in terms of three resonance structures of which one is **13.13**. The C–O bond distance in $[CO_3]^{2-}$ is longer than in CO_2 (Table 13.4) and is consistent with a formal bond order of 1.33. Most metal carbonates, other than those of the group 1 metals (see *Section 10.7*), are sparingly soluble in water. A general method of preparing peroxo salts can be used to convert K_2CO_3 to $K_2C_2O_6$; the electrolysis of aqueous K_2CO_3 at 253 K using a high current density produces a salt believed to contain the peroxocarbonate ion **13.14**. An alternative route involves the reaction of CO_2 with KOH in 86% aqueous H_2O_2 at 263 K. The colour of the product is variable and probably depends upon the presence of impurities such as KO_3; the electrolytic method gives a blue material whereas the product from the second route is orange. Peroxocarbonates are also believed to be intermediates in the reactions of CO_2 with superoxides (see *Section 10.6*).

(13.14)   116 pm
$$O=C=C=C=O$$
               129 pm
(13.15)

A third oxide of carbon is the suboxide C_3O_2 which possesses the linear structure **13.15**. It is obtained by dehydrating malonic acid, $CH_2(CO_2H)_2$, using P_2O_5 at 420 K. At room temperature, it is a gas (bp 279 K), but polymerizes above 288 K to form a red-brown paramagnetic material.

Silica, silicates and aluminosilicates

Silica, SiO_2, is an involatile solid and occurs in many different forms, nearly all of which possess lattice structures constructed of tetrahedral SiO_4 building blocks, often represented as in structure **13.16**. Each unit is connected to

β-quartz $\underset{slow}{\overset{1143 \text{ K}}{\rightleftharpoons}}$ β-tridymite $\underset{slow}{\overset{1742 \text{ K}}{\rightleftharpoons}}$ β-cristobalite $\underset{slow}{\overset{1983 \text{ K}}{\rightleftharpoons}}$ liquid

846 K ‖ fast   393–433 K ‖ fast   473–548 K ‖ fast

α-quartz   α-tridymite   α-cristobalite

Fig. 13.13 Transition temperatures between polymorphs of SiO_2.

the next by sharing an oxygen atom to give Si–O–Si bridges. At atmospheric pressure, three polymorphs of silica exist; each is stable within a characteristic temperature range, but possesses a low-temperature (α) and high-temperature (β) modification (Figure 13.13). The structure of β-cristobalite and its relationship to that of diamond was shown in *Figure 5.19*. The different polymorphs of silica resemble β-cristobalite in having tetrahedral SiO_4-units, but each is made unique by exhibiting a different arrangement of these building blocks. α-Quartz has an interlinked helical chain structure and is optically active because the chain has a handedness. It is also *piezoelectric* and is therefore used in crystal oscillators and filters for frequency control and in electromechanical devices such as microphones and loudspeakers.

(13.16)

A *piezoelectric* crystal is one that generates an electric field (i.e. develops charges on opposite crystal faces when subjected to mechanical stress) or that undergoes some change to atomic positions when an electric field is applied to it; such crystals must lack a centre of symmetry (e.g. contain tetrahedral arrangements of atoms). Their ability to transform electrical oscillations into mechanical vibration, and vice versa, is the basis of their use in, e.g., crystal oscillators.

Transitions from one polymorph to another involve initial Si–O bond cleavage and require higher temperatures than the changes between α- and β-forms of one polymorph. When liquid silica cools, it forms a non-crystalline glass consisting of an infinite lattice assembled from SiO_4 tetrahedra connected in a random manner. Only a few oxides form glasses (e.g. B_2O_3, SiO_2, GeO_2, P_2O_5 and As_2O_5) since the criteria for a *random* assembly are:

- the coordination number of the non-oxygen element must be three or four (a coordination number of two gives a chain and greater than four gives too rigid a structure);
- only one O atom must be shared between any two non-oxygen atoms (greater sharing leads to too rigid an assembly).

When silica glass is cooled to $\approx 1750 \, K$, it becomes plastic and can be worked in an oxy-hydrogen flame. *Silica glass* apparatus is highly insensitive to thermal shock owing to the low coefficient of thermal expansion of silica. *Borosilicate glass* (Pyrex) contains 10–15% B_2O_3 and has a lower melting point than silica glass. *Soda glass* contains added alkali which converts some of the Si−O−Si bridges in the silica network into terminal Si=O groups, reducing the melting point below that of borosilicate glass.

In all forms of silica mentioned so far, the Si−O bond length is $\approx 160 \, pm$ and the Si−O−Si bond angle $\approx 144°$, values close to those in $(H_3Si)_2O$ (Figure 13.9). By heating silica under very high pressure, a rutile form (see *Figure 5.21*) containing 6-coordinate Si is formed in which the Si−O bond length is 179 pm (compare with the sum of $r_{cov}(Si) = 118 \, pm$ and $r_{cov}(O) = 73 \, pm$). This form of silica is more dense and less reactive than ordinary forms. Silica is not attacked by acids other than HF with which it form $[SiF_6]^{2-}$. Fusion of SiO_2 with alkali leads to the formation of silicates.

Although esters of type $Si(OR)_4$ (equation 13.54) are known, no well-defined 'silicic acid' (H_4SiO_4) has been established.

$$SiCl_4 + 4ROH \longrightarrow Si(OR)_4 + 4HCl \qquad (13.54)$$

Normal silica is only very slowly attacked by alkali, but *silicates* are readily formed by fusion of SiO_2 and metal

---

BOX 13.7 **RESOURCES, ENVIRONMENTAL AND BIOLOGICAL**

'Greenhouse' gases

Carbon dioxide normally comprises $\approx 0.04\%$ by volume of the Earth's atmosphere from which it is removed and returned according to the carbon cycle:

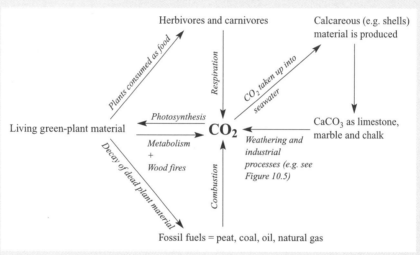

The balance is a delicate one, and the increase in combustion of fossil fuels and decomposition of limestone for cement manufacture in recent years have given rise to fears that a consequent increase in the CO_2 content of the atmosphere may lead to a 'greenhouse' effect, raising the temperature of the atmosphere. This arises because the sunlight that reaches the Earth's surface has its maximum energy in the visible region of the spectrum where the atmosphere is transparent. However, the energy maximum of the Earth's thermal radiation is in the infrared, where CO_2 absorbs strongly (see *Figure 3.10*). Even a small increase in the CO_2 component of the atmosphere might have serious effects because of its effects on the extent of the polar ice caps and glaciers, and because of the sensitivity of reaction rates to even small temperature changes. The danger is enhanced by the cutting down and burning of tropical rain forests which would otherwise reduce the CO_2 content of the atmosphere by photosynthesis.

The second major 'greenhouse' gas is CH_4 which is produced by the anaerobic decomposition of organic material; the old name of 'marsh gas' came about because bubbles of CH_4 escape from marshes. Flooded areas such as rice paddy fields produce large amounts of CH_4, and ruminants (e.g. cows, sheep and goats) also expel sizeable quantities of CH_4. Although the latter is a natural process, recent increases in the numbers of domestic animals around the world are naturally leading to increased release of CH_4 into the atmosphere.

Further reading

G.D. Farquhar (1997) *Science*, vol. 278, p. 1411 – 'Carbon dioxide and vegetation'.

J.G. Ferry (1997) *Science*, vol. 278, p. 1413 – 'Methane: Small molecule, big impact'.

A. Kendall, A. McDonald and A. Williams (1997) *Chemistry & Industry*, p. 342 – 'The power of biomass'.

J.D. Mahlman (1997) *Science*, vol. 278, p.1416 – 'Uncertainties in projections of human-caused climate warming'.

A. Moss (1992) *Chemistry & Industry*, p. 334 – 'Methane from ruminants in relation to global warming'.

See also *Box 15.6*: Volcanic emissions

hydroxides, oxides or carbonates. The range of known silicates is large and they, and the *aluminosilicates* (see later), are extremely important, both in nature and for commercial and industrial purposes.

Sodium silicates of variable composition are made by heating sand (which is impure quartz containing e.g. iron(III) oxide) with Na_2CO_3 at $\approx 1600\,K$. If the sodium content is high ($Na:Si \approx 3.2–4:1$), the silicates are water-soluble and the resulting alkaline solution (*water glass*) contains ions such as $[SiO(OH)_3]^-$ and $[SiO_2(OH)_2]^{2-}$; water glass is used commercially in detergents where it controls the pH and degrades fats by hydrolysis. If the Na-content is low, the silicate ions consist of large polymeric species and their Na^+ salts are insoluble in water. Equilibrium between the different species is attained rapidly at $pH > 10$, and more slowly in less alkaline solutions.

The Earth's crust is largely composed of silica and silicate minerals which form the principal constituents of all rocks and of the sands, clays and soils that result from degradation of rocks. Most inorganic building materials are based on silicate minerals and include natural silicates such as sandstone, granite and slate, as well as manufactured materials such as cement, concrete and ordinary glass. The latter is manufactured by fusing together limestone, sand and Na_2CO_3. Clays are used in the ceramics industry and mica as an electrical insulator. Fibrous asbestos once had extensive use in heat- and fire-resistant materials, but the health risks associated with the inhalation of asbestos fibres are now well-established and alternative heat- and fire-proofing materials are replacing asbestos. We discuss uses of zeolites later in the section.

It is universal practice to describe silicates in terms of a purely ionic model. However, although we might *write* Si^{4+}, the 4+ charge is unlikely on ionization energy grounds and is incompatible with the commonly observed Si—O—Si bond angle of $\approx 140°$. Figure 13.14 compares the ionic radii

of ions commonly present in silicates; the value for the 'Si^{4+}' ion is an estimate. Since the Al^{3+} and Si^{4+} ions are similar sizes, replacement is common and leads to the formation of aluminosilicates. If Al^{3+} replaces Si^{4+}, however, an extra singly charged cation must be present to maintain electrical neutrality. Thus, in the feldspar *orthoclase*, $KAlSi_3O_8$, the anion $[AlSi_3O_8]^-$ is readily recognized as being related to SiO_2 (i.e. $[AlSi_3O_8]^-$ is isoelectronic with Si_4O_8) and $[AlSi_3O_8]^-$ possesses the structure of quartz with one-quarter of the Si replaced by aluminium; the K^+ ions occupy cavities in the relatively open lattice. Double replacements are also common, e.g. $\{Na^+ + Si^{4+}\}$ replaced by $\{Ca^{2+} + Al^{3+}\}$ (look at the radii comparisons in Figure 13.14).

The overwhelming majority of silicates have structures based on SiO_4 tetrahedra (**13.16**) which, by sharing O atoms, assemble into small groups such as **13.17**, cyclic motifs, infinite chains, infinite layers or infinite three-dimensional networks. Sharing an atom only involves *corners* of tetrahedra; sharing an edge would bring two O^{2-} ions too close together.

(13.17)

Of the metal ions most commonly occurring in silicates, the coordination numbers with respect to O^{2-} ions are: 4 for Be^{2+}, 4 or 6 for Al^{3+}, 6 for Mg^{2+}, Fe^{3+} or Ti^{4+}, 6 or 8 for Na^+, and 8 for Ca^{2+}.

Figure 13.15 illustrates the structures of some silicate anions; $[Si_2O_7]^{6-}$ is shown in structure **13.17**. The simplest silicates contain the $[SiO_4]^{4-}$ ion and include Mg_2SiO_4 (*olivine*) and the synthetic $\beta\text{-}Ca_2SiO_4$ (which is an important constituent of cement, setting to a hard mass when finely ground and mixed with water). The mineral *thortveitite*, $Sc_2Si_2O_7$ (a major source of scandium), contains discrete $[Si_2O_7]^{6-}$ ions. The cyclic ions $[Si_3O_9]^{6-}$ and $[Si_6O_{18}]^{12-}$ occur in $Ca_3Si_3O_9$ ($\alpha\text{-}wollastonite$) and $Be_3Al_2Si_6O_{18}$ (*beryl*) respectively, while $[Si_4O_{12}]^{8-}$ is present in the synthetic salt $K_8Si_4O_{12}$. Short chain silicates are not common, although $[Si_3O_{10}]^{8-}$ occurs in a few rare minerals. Cage structures have been observed in some synthetic silicates and two examples are shown in Figure 13.16.

If the SiO_4 tetrahedra sharing two corners form an infinite chain, the $Si:O$ ratio is $1:3$ (Figure 13.15). Such chains are present in $CaSiO_3$ ($\beta\text{-}wollastonite$) and $CaMg(SiO_3)_2$ (*diopside*, a member of the *pyroxene* group of minerals which possess $[SiO_3]_n^{2n-}$ chains). Although infinite chains are present in these minerals, the relative orientations of the chains are different. *Asbestos* consists of a group of fibrous minerals some of which (e.g. $Ca_2Mg_5(Si_4O_{11})_2(OH)_2$,

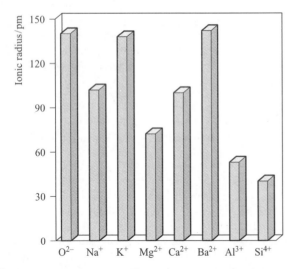

Fig. 13.14 Ionic radii of selected ions involved in silicates. These data can be used to rationalize cation replacements in silicates.

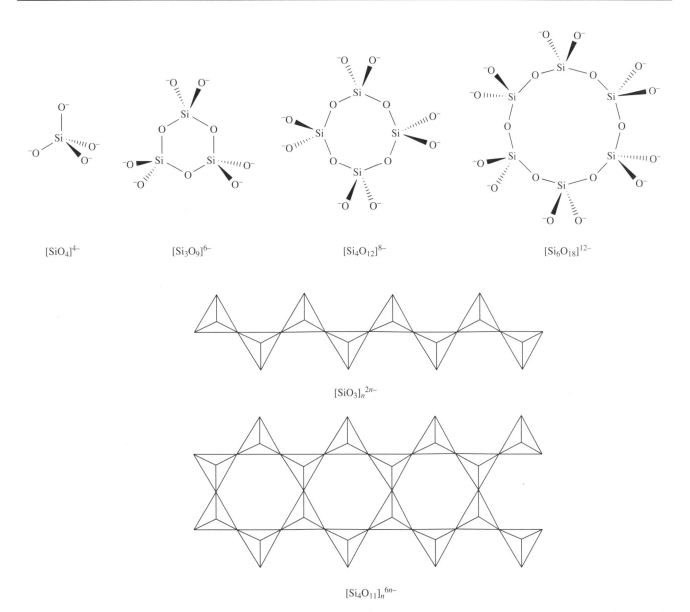

$[SiO_4]^{4-}$ $[Si_3O_9]^{6-}$ $[Si_4O_{12}]^{8-}$ $[Si_6O_{18}]^{12-}$

$[SiO_3]_n^{2n-}$

$[Si_4O_{11}]_n^{6n-}$

Fig. 13.15 Schematic representations of the structures of selected silicates. Conformational details of the rings are omitted. In the polymeric structures, each tetrahedron represents an SiO_4-unit as shown in structure **13.16**.

tremolite) contain the double chain silicate $[Si_4O_{11}]_n^{6n-}$, shown in Figure 13.15. More extended cross-linking of chains produces layer structures of composition $[Si_2O_5]^{2-}$; ring sizes within the layers may vary. Such sheets occur in *micas* and are responsible for the characteristic cleavage of these minerals into thin sheets. *Talc*, characterized by its softness, has the composition $Mg_3(Si_2O_5)_2(OH)_2$; Mg^{2+} ions are sandwiched between composite layers each containing $[Si_2O_5]^{2-}$ sheets and $[OH]^-$ ions, and the assembly can be represented by the sequence $\{Si_2O_5^{2-}\}\{OH^-\}\{Mg^{2+}\}_3\{OH^-\}\{Si_2O_5^{2-}\}$. This is electrically neutral, allowing talc to cleave readily in a direction parallel to the sandwich. A consequence of this cleavage is that talc is used as a dry lubricant, e.g. in personal care preparations.

Infinite sharing of all four oxygen atoms of the SiO_4 tetrahedra gives a composition SiO_2 (see earlier) but partial replacement of Si by Al leads to anions $[AlSi_nO_{2n+2}]^-$, $[Al_2Si_nO_{2n+2}]^{2-}$ etc. Minerals belonging to this group include *orthoclase* ($KAlSi_3O_8$), *albite* ($NaAlSi_3O_8$), *anorthite* ($CaAl_2Si_2O_8$) and *celsian* ($BaAl_2Si_2O_8$). Feldspars are aluminosilicate salts of K^+, Na^+, Ca^{2+} or Ba^{2+} and constitute an important class of rock-forming minerals; they include *orthoclase*, *celsian*, *albite* and *anorthite*. In feldspars, the holes in the structure that accommodate the cations are quite small. In *zeolites*, the cavities are much larger and can accommodate not only cations, but also molecules such as H_2O, CO_2, MeOH and hydrocarbons. Commercially and industrially, zeolites (both natural and synthetic) are extremely important. The Al:Si ratio varies widely among zeolites; Al-rich systems are hydrophilic and their ability to take up H_2O leads to their use as laboratory drying agents (molecular sieves). Different zeolites contain different sized

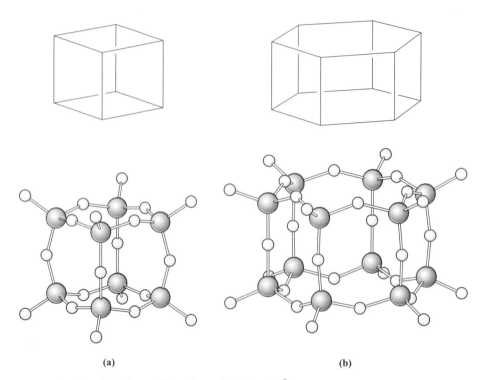

Fig. 13.16 The structures, elucidated by X-ray diffraction, of (a) $[Si_8O_{20}]^{8-}$, determined for the salt $[Me_4N]_8[Si_8O_{20}]\cdot65H_2O$ [M. Wiebcke *et al.* (1993) *Microporous Materials*, vol. 2, p. 55], and (b) $[Si_{12}O_{30}]^{12-}$, determined for the salt $K_2[\alpha$-cyclodextrin]$_2[Si_{12}O_{30}]\cdot36H_2O$ [K. Benner *et al.* (1997) *Angew. Chem., Int. Ed. Engl.*, vol. 36, p. 743]. The silicon atoms in (a) and (b) define a cube and hexagonal prism respectively. Colour code: Si, blue; O, pale grey.

cavities and channels, permitting a choice of zeolite to effect selective molecular adsorption. Silicon-rich zeolites are hydrophobic. Catalytic uses of zeolites (see *Sections 26.5* and *26.6*) are widespread, e.g. the synthetic zeolite ZSM-5 with composition $Na_n[Al_nSi_{96-n}O_{192}]\cdot\approx16H_2O$ ($n < 27$) catalyzes benzene alkylation, xylene isomerization and conversion of methanol to hydrocarbons (for motor fuels). Figure 13.17 illustrates the cavities present in zeolite H-ZSM-5.[†] Electrical neutrality upon Al-for-Si replacement can also be achieved by converting O^- to a terminal OH group. These groups are strongly acidic and means that such zeolites are excellent ion-exchange (see *Section 10.6*) materials and have applications in, for example, water purification and washing powders (see *Section 11.7*).

> *Zeolites* are crystalline, hydrated aluminosilicates that possess framework structures containing regular channels and/or cavities; the cavities contain H_2O molecules and cations (usually group 1 or 2 metal ions).

Oxides, hydroxides and oxoacids of germanium, tin and lead

The dioxides of Ge, Sn and Pb are involatile solids. Germanium dioxide closely resembles SiO_2, and exists in both

quartz and rutile forms. It dissolves in concentrated HCl forming $[GeCl_6]^{2-}$ and in alkalis to give *germanates*. While these are not as important as silicates, we should note that many silicates do possess germanate analogues, but there are germanates that have no silicate counterparts (e.g. the product of reaction 13.55).

$$5GeO_2 + 2Li_2O \xrightarrow{\text{molten}} Li_4[Ge_5O_{12}] \qquad (13.55)$$

Germanium monoxide is prepared by dehydration of the yellow hydrate, obtained by reaction of $GeCl_2$ with aqueous NH_3, or by heating $Ge(OH)_2$, obtained from $GeCl_2$ and water. The monoxide, which is amphoteric, is not as well characterized as GeO_2, and disproportionates at high temperature (equation 13.56).

$$2GeO \xrightarrow{970\,K} GeO_2 + Ge \qquad (13.56)$$

Solid SnO_2 and PbO_2 adopt the rutile lattice (*Figure 5.21*); SnO_2 occurs naturally as cassiterite but can easily be prepared by oxidation of Sn. In contrast, the formation of PbO_2 requires the action of powerful oxidizing agents such as alkaline hypochlorite on Pb(II) compounds. When freshly prepared, SnO_2 is soluble in many acids (equation 13.57) but it exhibits amphoteric behaviour and also reacts with alkalis; reaction 13.58 occurs in strongly alkaline media to give a stannate.

$$SnO_2 + 6HCl \longrightarrow 2[H_3O]^+ + [SnCl_6]^{2-} \qquad (13.57)$$

$$SnO_2 + 2KOH + 2H_2O \longrightarrow K_2[Sn(OH)_6] \qquad (13.58)$$

[†] Zeolites are generally known by acronyms that reflect the research or industrial companies of origin, e.g. ZSM stands for Zeolite Socony Mobil.

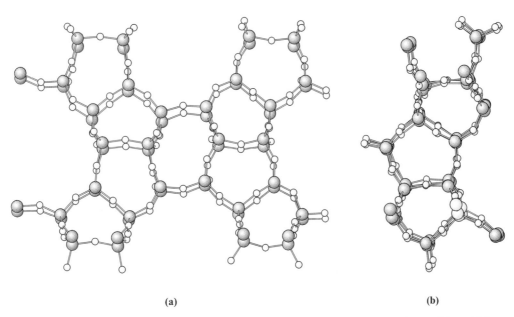

(a)

(b)

Fig. 13.17 The structure of H-ZSM-5 zeolite ($Al_{0.08}Si_{23.92}O_{48}$) is typical of a zeolite in possessing cavities which can accommodate guest molecules. (a) and (b) show two orthogonal views of the host lattice; the structure was determined by X-ray diffraction for the zeolite hosting 1,4-dichlorobenzene [H. van Koningsveld *et al.* (1996) *Acta Crystallogr.*, *Sect. B*, vol. 52, p. 140]. Colour code: (Si, Al), blue; O, pale grey.

In contrast, PbO_2 shows acidic (but no basic) properties, forming $[Pb(OH)_6]^{2-}$ when treated with alkali. Crystalline salts such $K_2[Sn(OH)_6]$ and $K_2[Pb(OH)_6]$ can be isolated.

The monoxides SnO and PbO (red form, *litharge*) possess layer structures in which each metal centre is at the apex of a square-based pyramidal array (Figure 13.18). It is generally accepted that each metal centre bears a lone pair of electrons occupying an orbital pointing towards the space between the layers; thus, electronic effects contribute to the preference for this asymmetric structure. Litharge is the more important form of PbO, but a yellow form also exists. While PbO can be prepared by heating the metal in

---

BOX 13.8 **APPLICATIONS**

Kaolin, smectite and hormite clays: from ceramics to natural absorbers

Crystalline clays (aluminosilicate minerals) are categorized according to structure. Clays in the *kaolin* or *china clay* group (e.g. *kaolinite*, $Al_2Si_2O_5(OH)_4$) possess sheet structures with alternating layers of linked SiO_4 tetrahedra and AlO_6 octahedra. *Smectite* clays (e.g. *sodium montmorillonite*, $Na[Al_5MgSi_{12}O_{30}(OH)_6]$) also have layer structures, with cations (e.g. Na^+, Ca^{2+}, Mg^{2+}) situated between the aluminosilicate layers. Interactions between the layers are weak, and water molecules readily penetrate the channels causing the lattice to expand; the volume of montmorillonite increases several times over as water is absorbed. *Hormite* clays (e.g. palygorskite) possess structures in which chains of SiO_4 tetrahedra are connected by octahedral AlO_6 or MgO_6 units; these clays exhibit outstanding adsorbent and absorbent properties.

Within industry and commerce, terms other than the mineral classifications are common. *Ball clay* is a type of kaolin particularly suited to the manufacture of ceramics: in 1998, 30% of the ball clay produced in the US was used for tile manufacture, 21% for sanitary ware, 10% for pottery and 39% for miscellaneous ceramics. Kaolinite (which is white and soft) is of great importance in the paper industry for coatings and as a filler; of the 9.8 Mt produced in the US in 1998, 56% was consumed in paper manufacture. Worldwide, the US, UK and former Soviet Union are the major producers of kaolin type clays.

Smectite clays tend to be referred to as *bentonite*, the name deriving from the rock in which the clays occur; 4 Mt of bentonite were mined in the US in 1998. *Fuller's earth* is a general term used commercially to describe hormite clays; 2.5 Mt were produced in 1998 in the US. Applications of smectite and hormite clays stem from their ability to absorb water, swelling as they do so. Drilling fluids rely on the outstanding, reversible behaviour of sodium montmorillonite as it takes in water: the property of *thixotropy*. When static, or at low drill speeds, an aqueous suspension of the clay is highly viscous owing to the absorption of water by lattice and the realignment of the charged aluminosilicate layers. At high drill speeds, electrostatic interactions between the layers are destroyed and the drill-fluid viscosity decreases. Fuller's earth clays are remarkably effective absorbents and two major applications are in pet litter, and in granules which can be applied to minor oil spillages (e.g. at fuel stations).

BOX 13.9   APPLICATIONS

Sensing gases

Detecting the presence of toxic gases can be carried out by IR spectroscopic means, but such techniques do not lend themselves to a domestic market. Capitalizing on the n-type semiconducting properties of SnO_2 has led to its use in gas sensors, and sensors that detect gases such as CO, hydrocarbons or solvent (alcohols, ketones, esters etc.) vapours, are commercially available and are now in common use in underground car parking garages, automatic ventilation systems, fire alarms and gas-leak detectors. The presence of even small amounts of the target gases results in a significant increase in the electrical conductivity of SnO_2, and this change is used to provide a measure of the gas concentration, triggering a signal or alarm if a pre-set threshold level is detected. The increase in electrical conductivity arises as follows. Adsorption of oxygen on to an SnO_2 surface draws electrons from the conduction band. The operating temperature of an SnO_2 sensor is 450–750 K and in the presence of a reducing gas such as CO or hydrocarbon, the SnO_2 surface loses oxygen and at the same time, electrons return to the conduction band of the bulk solid resulting in an increase in the electrical conductivity.

Doping the SnO_2 with Pd or Pt increases the sensitivity of a detector.

Tin(IV) oxide sensors play a major role in the commercial market and can be used to detect all the following gases, but other sensor materials include:

- ZnO, Ga_2O_3 and TiO_2/V_2O_5 for CH_4 detection;
- La_2CuO_4, Cr_2O_3/MgO and $Bi_2Fe_4O_9$ for C_2H_5OH vapour detection;
- ZnO, Ga_2O_3, ZrO_2 and WO_3 for H_2 detection;
- ZnO, TiO_2 (doped with Al and In) and WO_3 for NO_x;
- ZnO, Ga_2O_3, Co_3O_4 and TiO_2 (doped with Pt) for CO detection.

Further reading

W. Göpel and G. Reinhardt (1996) in *Sensors Update*, eds H. Baltes, W. Göpel and J. Hesse, VCH, Weinheim, vol. 1, p. 47 – 'Metal oxides sensors'.

For more information on semiconductors: see *Sections 5.8 and 5.9*.

---

air above 820 K, SnO is sensitive to oxidation and is best prepared by thermal decomposition of tin(II) oxalate; PbO can also be made by dehydrating $Pb(OH)_2$. Both SnO and PbO are amphoteric, but the oxo-anions formed from them, like those from GeO, are not well characterized. Of the group 14 elements, only lead forms a mixed oxidation state oxide; Pb_3O_4 (*red lead*) is obtained by heating PbO in air above 720 K, and is better formulated as $2PbO \cdot PbO_2$. In the solid state, two Pb environments are present. Nitric acid reacts with Pb_3O_4 (according to equation 13.59), while treatment with glacial acetic acid yields a mixture of $Pb(CH_3CO_2)_2$ and $Pb(CH_3CO_2)_4$, the

latter compound being an important reagent in organic chemistry; the two acetate salts can be separated by crystallization.

$$Pb_3O_4 + 4HNO_3 \longrightarrow PbO_2 + 2Pb(NO_3)_2 + 2H_2O \quad (13.59)$$

13.10 Silicones

Although silicones are organometallic compounds, they are conveniently described in this chapter because of their structural similarities to silicates. Hydrolysis of Me_nSiCl_{4-n} ($n = 1–3$) might be expected to give the derivatives $Me_nSi(OH)_{4-n}$ ($n = 1–3$). By analogy with carbon analogues, we might expect Me_3SiOH to be stable (except with respect to dehydration at higher temperatures), but $Me_2Si(OH)_2$ and $MeSi(OH)_3$ to undergo dehydration to $Me_2Si=O$ and $MeSiO_2H$ respectively. However, at the beginning of Section 13.9, we indicated that an Si=O bond is energetically less favourable than two Si–O bonds. As a consequence, hydrolysis of Me_nSiCl_{4-n} ($n = 1–3$) yields *silicones* which are oligomeric products (e.g. reaction 13.60) containing the tetrahedral groups **13.18–13.20** in which each O atom represents part of an Si–O–Si bridge. Diols can condense to give chains (**13.21**) or rings (e.g. **13.22**). Hydrolysis of $MeSiCl_3$ produces a cross-linked polymer.

Fig. 13.18 Two views (a) from the side and (b) from above of a part of one layer of the SnO and red PbO lattices; the metal atoms are shown in blue.

$$2Me_3SiOH \longrightarrow Me_3SiOSiMe_3 + H_2O \quad (13.60)$$

Diverse applications of silicones

Silicone products have many commercial roles. At one end of the market, they are crucial ingredients in personal care products: silicones are the components of shampoos and conditioners that improve the softness and silkiness of hair, and are also used in shaving foams, toothpastes, anti-perspirants, cosmetics, hair-styling gels and bath oils. At the other end of the spectrum, silicones find very different applications in silicone greases, sealants, varnishes, waterproofing materials, synthetic rubbers and hydraulic fluids. Silicones tend to be viscous oils which are immiscible with water, but for use in shampoos, silicones may be dispersed in water to give emulsions.

(13.18)

(13.19)

(13.20)

(13.21)

(13.22)

Silicone polymers have a range of structures and applications (see *Box 13.10*), and, in their manufacture, control of the polymerization is essential. The methylsilicon chlorides are co-hydrolyzed, or the initial products of hydrolysis are equilibrated by heating with H_2SO_4 which catalyzes the conversion of cyclic oligomers into chain polymers, bringing about redistribution of the terminal $OSiMe_3$ groups. For example, equilibration of $HOSiMe_2(OSiMe_2)_nOSiMe_2OH$ with $Me_3SiOSiMe_3$ leads to the polymer $Me_3Si(OSiMe_2)_nOSiMe_3$. Cross-linking, achieved by co-hydrolysis of Me_2SiCl_2 and $MeSiCl_3$, leads, after heating at 520 K, to silicone resins that are hard and inert; tailoring the product so that it possesses a smaller degree of cross-linking results in the formation of silicone rubbers.

13.11 Sulfides

The disulfides of C, Si, Ge and Sn show the gradation in properties that might be expected to accompany the increasingly metallic character of the elements. Pertinent properties of these sulfides are given in Table 13.6. Lead(IV) is too powerful an oxidizing agent to coexist with S^{2-}, and PbS_2 is not known.

Carbon disulfide is made by heating charcoal with sulfur at 1200 K, or by passing CH_4 and sulfur vapour over Al_2O_3 at 950 K. It is highly toxic (by inhalation and absorption through

Table 13.6   Selected properties of ES_2 (E = C, Si, Ge, Sn).

Property	CS_2	SiS_2	GeS_2	SnS_2
Melting point / K	162	1363 (sublimes)	870 (sublimes)	873 (dec.)
Boiling point / K	319	–	–	–
Appearance at 298 K	Volatile liquid, foul odour	White needle-like crystals	White powder or crystals	Golden-yellow crystals
Structure at 298 K	Linear molecule S=C=S	Solid state, chain[‡]	Three-dimensional lattice with Ge_3S_3 and larger rings with shared vertices.[‡]	CdI_2 type lattice (see *Figure 5.22*)

[‡] At high pressures and temperatures, SiS_2 and GeS_2 adopt a β-cristobalite lattice (see *Figure 5.19c*).

the skin) and extremely flammable, but is an excellent solvent which is used in the production of rayon and cellophane. Carbon disulfide is insoluble in water, but is, by a narrow margin, thermodynamically unstable with respect to hydrolysis to CO_2 and H_2S. However, this reaction has a high kinetic barrier and is very slow. Unlike CO_2, CS_2 polymerizes under high pressure to give a black solid with the chain structure **13.23**. When shaken with solutions of group 1 metal sulfides, CS_2 dissolves readily to give trithiocarbonates, M_2CS_3, which contain the $[CS_3]^{2-}$ ion **13.24**, a sulfur analogue of $[CO_3]^{2-}$. Salts are readily isolated, e.g. Na_2CS_3 forms yellow needles (mp 353 K). The free acid H_2CS_3 separates as an oil when salts are treated with hydrochloric acid (equation 13.61), and behaves as a weak acid in aqueous solution: $pK_a(1) = 2.68$, $pK_a(2) = 8.18$.

$$BaCS_3 + 2HCl \xrightarrow{273\ K} BaCl_2 + H_2CS_3 \quad (13.61)$$

(13.23)   (13.24)

(13.25)

The action of an electric discharge on CS_2 results in the formation of C_3S_2, **13.25** (compare with **13.15**), a red liquid which decomposes at room temperature, producing a black polymer $(C_3S_2)_x$. When heated, C_3S_2 explodes. In contrast to CO, CS is a short-lived radical species which decomposes at 113 K; it has, however, been observed in the upper atmosphere.

Several salts of the $[C_2S_4]^{2-}$ anion are known (made by, for example, reaction 13.62), although the free acid (an analogue of oxalic acid) has not been isolated.

$$[CH_3CS_2]^- + 2[S_x]^{2-}$$
$$\longrightarrow [C_2S_4]^{2-} + [HS]^- + H_2S + [S_{2x-4}]^{2-} \quad (13.62)$$

In $[Et_4N]_2[C_2S_4]$, the anion has D_{2d} symmetry, i.e. the dihedral angle between the planes containing the two CS_2-units is 90° (structure **13.26**), whereas in $[Ph_4P]_2[C_2S_4]\cdot 6H_2O$, this angle is 79.5°. Contrast these data with the planarity of $[C_2O_4]^{2-}$ in salts such as $K_2C_2O_4$ and $BaC_2O_4\cdot 2H_2O$.

(13.26)

Silicon disulfide is prepared by heating Si in sulfur vapour. Both the structure of this compound (Table 13.6) and the chemistry of SiS_2 show no parallels with SiO_2; SiS_2 is instantly hydrolyzed (equation 13.63).

$$SiS_2 + 2H_2O \longrightarrow SiO_2 + 2H_2S \quad (13.63)$$

The disulfides of Ge and Sn (Table 13.6) are precipitated when H_2S is passed into acidic solutions of Ge(IV) and Sn(IV) compounds. Some sulfides have cluster structures, e.g. $[Ge_4S_{10}]^{4-}$, prepared by reaction 13.64.

$$4GeS_2 + 2S^{2-} \xrightarrow{\text{Aqueous solution in presence of Cs}^+} [Ge_4S_{10}]^{4-} \quad (13.64)$$

(13.27)   (13.28)

Tin(IV) forms a number of thiostannates containing discrete anions, e.g. Na_4SnS_4 contains the tetrahedral $[SnS_4]^{4-}$ ion, and $Na_4Sn_2S_6$ and $Na_6Sn_2S_7$ contain anions **13.27** and **13.28** respectively.

The monosulfides of Ge, Sn and Pb are all obtained by precipitation from aqueous media. Both GeS and SnS crystallize with layer structures similar to that of black phosphorus (see *Section 14.4*). Lead(II) sulfide adopts the NaCl lattice; its formation as a black precipitate ($K_{sp} \approx 10^{-30}$) is observed in the qualitative test for H_2S (equation 13.65) and the very low solubility suggests that it is not a purely ionic compound.

$$Pb(NO_3)_2 + H_2S \longrightarrow \underset{\text{black ppt}}{PbS} + 2HNO_3 \quad (13.65)$$

Pure PbS is a p-type semiconductor when S-rich, and an n-type when Pb-rich. It exhibits *photoconductivity* and has applications in photoconductive cells, transistors, and in photographic exposure meters.

If a material is a *photoconductor*, it absorbs light with the result that electrons from the valence band are excited into the conducting band; thus, the electrical conductivity increases on exposure to light.

13.12 Cyanogen and silicon nitride

In discussing bonds formed between the group 14 elements and nitrogen, two compounds of importance emerge: cyanogen, C_2N_2, and silicon nitride.

Cyanogen and its derivatives

The CN˙ radical is a *pseudo-halogen*, i.e. its chemistry resembles that of a halogen atom, X; it forms C_2N_2, HCN and

$[CN]^-$, analogues of X_2, HX and X^-. Although C_2N_2 and HCN are thermodynamically unstable with respect to decomposition into their elements, hydrolysis by H_2O, and oxidation by O_2, they and $[CN]^-$ are *kinetically* stable enough for them to be well-established and much studied species.

Cyanogen, C_2N_2, is a toxic, extremely flammable gas (mp 245 K, bp 252 K) which is liable to react explosively with some powerful oxidants. Although $\Delta_f H^\circ(C_2N_2, 298 K) = +297 \text{ kJ mol}^{-1}$, pure C_2N_2 can be stored for long periods without decomposition. Reactions 13.66 and 13.67 give two syntheses of C_2N_2; reaction 13.67 illustrates the pseudo-halide like nature of $[CN]^-$ which is oxidized by Cu(II) in an analogous fashion to the oxidation of I^- to I_2. Cyanogen is manufactured by air-oxidation of HCN over a silver catalyst.

$$Hg(CN)_2 + HgCl_2 \xrightarrow{570 \text{ K}} C_2N_2 + Hg_2Cl_2 \qquad (13.66)$$

$$2CuSO_4 + 4NaCN \xrightarrow[\text{solution, } \Delta]{\text{aqueous}} C_2N_2 + 2CuCN + 2Na_2SO_4 \qquad (13.67)$$

$$\underset{137 \text{ pm}}{N \equiv C \overset{116 \text{ pm}}{\underline{}} C \equiv N}$$

(13.29)

Cyanogen has the linear structure **13.29** and the short C–C distance indicates considerable electron delocalization. It burns in air with a very hot, violet flame (equation 13.68), and resembles the halogens in that it is hydrolyzed by alkali (equation 13.69) and undergoes thermal dissociation to $CN^\bullet$ at high temperatures.

$$C_2N_2 + 2O_2 \longrightarrow 2CO_2 + N_2 \qquad (13.68)$$

$$C_2N_2 + 2[OH]^- \longrightarrow [OCN]^- + [CN]^- + H_2O \qquad (13.69)$$

$$\underset{106.5 \text{ pm}}{H \overset{115 \text{ pm}}{\underline{}} C \equiv N}$$

(13.30)

(13.31)

Hydrogen cyanide, HCN, **13.30**, is an extremely toxic and flammable, colourless volatile liquid (mp 260 K, bp 299 K) with a high dielectric constant due to strong hydrogen bonding; it has a characteristic smell of bitter almonds. The pure liquid polymerizes to $HC(NH_2)(CN)_2$ and $(H_2N)(NC)C=C(CN)(NH_2)$ mixed with higher molecular mass polymers, and in the absence of a stabilizer such as H_3PO_4, polymerization may be explosive. In the presence of traces of H_2O and NH_3, HCN forms adenine, **13.31**, and on reduction, gives $MeNH_2$. It is thought that HCN was one of the small molecules in the early atmosphere of the Earth, and played an important role in the formation of many biologically important compounds. Hydrogen cyanide is prepared on a small scale by adding acid to NaCN, and industrially by reactions 13.70 and 13.71.

$$2CH_4 + 2NH_3 + 3O_2 \xrightarrow[1250-1550 \text{ K, 2 bar}]{\text{Pt/Rh,}} 2HCN + 6H_2O \qquad (13.70)$$

$$CH_4 + NH_3 \xrightarrow{\text{Pt, } 1450-1550 \text{ K}} HCN + 3H_2 \qquad (13.71)$$

Many organic syntheses involve HCN, and it is of great industrial importance, a large fraction going into the production of 1,4-dicyanobutane (adiponitrile) for nylon manufacture, and cyanoethene (acrylonitrile) for production of acrylic fibres.

In aqueous solution, HCN behaves as a weak acid ($pK_a = 9.31$) and is slowly hydrolyzed (equation 13.72). An older name for hydrocyanic acid is prussic acid.

$$HCN + 2H_2O \longrightarrow [NH_4]^+ + [HCO_2]^- \qquad (13.72)$$

$$2HCN + Na_2CO_3 \xrightarrow{\text{aqueous solution}} 2NaCN + H_2O + CO_2 \qquad (13.73)$$

The neutralization of aqueous HCN by Na_2CO_3, $NaHCO_3$ or $Na[HCO_2]$ generates NaCN, the most important salt of the acid. It is manufactured by reaction 13.73, and has widespread uses in organic chemistry (e.g. for the formation of C–C bonds); it is also used in the extraction of Ag and Au. For discussion of the extraction of Ag and Au, and treat-

---

BOX 13.11 **RESOURCES, ENVIRONMENTAL AND BIOLOGICAL**

Hydrogen cyanide in plant material

A number of plants and their fruits, e.g. apricot and plum kernels, grape and apple seeds, are natural sources of HCN. The origin of the HCN is a cyanoglucoside, *amygdalin* (a sugar derivative) which is present in the fruit stones and seeds; hydrolysis of amygdalin releases HCN. Cassava is an important root crop grown in tropical regions as a source of starch, and, for example, it is used for the production of tapioca. Cassava plants may be either a sweet or bitter

variety; bitter cassava contains larger quantities of cyanoglucosides which liberate HCN when the roots are crushed or chewed. In order to render the root crop safe as a foodstuff, bitter cassava must be subjected to careful treatment of shredding, pressure and heat. A beneficial side effect is the natural defence that cassava has against, for example, insect pests.

ment of [CN]$^-$ waste, see *equation 22.4* and *Box 22.2*. At 298 K, NaCN and KCN adopt the NaCl lattice, each [CN]$^-$ ion freely rotating (or having random orientations) about a fixed point in the lattice and having an effective ionic radius of ≈ 190 pm. At lower temperatures, transitions to structures of lower symmetry occur, e.g. NaCN undergoes a cubic to hexagonal transition below 283 K. Crystals of NaCN and KCN are deliquescent, and both salts are soluble in water and are highly toxic. Fusion of KCN and sulfur gives potassium thiocyanate, KSCN.

Mild oxidizing agents convert [CN]$^-$ to cyanogen (equation 13.67) but with more powerful oxidants such as PbO or *neutral* [MnO$_4$]$^-$, cyanate ion, **13.32**, is formed (reaction 13.74). Potassium cyanate reverts to the cyanide on heating (equation 13.75).

$$PbO + KCN \longrightarrow Pb + K[OCN] \tag{13.74}$$

$$2K[OCN] \xrightarrow{\Delta} 2KCN + O_2 \tag{13.75}$$

$$O{=}C{=}N^- \longleftrightarrow {}^-O{-}C{\equiv}N$$
$$(13.32)$$

(13.33)

Two acids can be derived from **13.32**: HOCN (cyanic acid or hydrogen cyanate) and HNCO (isocyanic acid, **13.33**). It has now been established that HOCN and HNCO are not in equilibrium with each other. Isocyanic acid (pK_a = 3.66) is obtained by heating urea (equation 13.76) but rapidly trimerizes, although heating the trimer regenerates the monomer.

(13.76)

keto-Tautomer of cyanuric acid

The fulminate ion, [CNO]$^-$, is isomeric with the cyanate ion; fulminate salts can be reduced to cyanides but cannot be prepared by oxidation of them. The free acid readily polymerizes but is stable for short periods in Et$_2$O at low temperature. Metal fulminates are highly explosive; mercury(II) fulminate may be prepared by reaction 13.77 and is a dangerous initiatory explosive.

$$2Na[CH_2NO_2] + HgCl_2 \longrightarrow Hg(CNO)_2 + 2H_2O + 2NaCl \tag{13.77}$$

(13.34)   (13.35)

Cyanogen chloride, **13.34** (mp 266 K, bp 286 K), is prepared by the reaction of Cl$_2$ with NaCN or HCN, and readily trimerizes to **13.35** which has applications in the manufacture of dyestuffs and herbicides.

Silicon nitride

The wide applications of silicon nitride, Si$_3$N$_4$, as a ceramic and refractory material and in the form of whiskers justify its inclusion here; thin films of Si$_3$N$_4$: see *Section 27.6*. It is a white, chemically inert amorphous powder, which can be formed by reaction 13.78, or by combining Si and N$_2$ above 1650 K.

$$SiCl_4 + 4NH_3 \xrightarrow{-4HCl} Si(NH_2)_4 \xrightarrow{\Delta} Si(NH)_2 \xrightarrow{\Delta} Si_3N_4 \tag{13.78}$$

The two main polymorphs, α- and β-Si$_3$N$_4$, possess similar infinite lattice structures in which Si and N are in tetrahedral and approximately trigonal planar environments respectively. Another new refractory material is Si$_2$N$_2$O, made from Si and SiO$_2$ under N$_2$/Ar atmosphere at 1700 K; it possesses puckered hexagonal nets of alternating Si and N atoms, the sheets being linked by Si−O−Si bonds.

13.13 Aqueous solution chemistry and salts of oxoacids of tin and lead

Table 13.1 lists standard reduction potentials for the M^{4+}/M^{2+} and M^{2+}/M (M = Sn, Pb) couples. The value of $E^\circ(Sn^{4+}/Sn^{2+}) = +0.15$ V shows that Sn(II) salts in aqueous solution are readily oxidized by O$_2$; further, hydrolysis of Sn^{2+} to species such as [Sn$_2$O(OH)$_4$]$^{2-}$ and [Sn$_3$(OH)$_4$]$^{2+}$ is extensive. Aqueous solutions of Sn(II) salts are therefore usually acidified and complex ions are then likely to be present, e.g. if SnCl$_2$ is dissolved in dilute hydrochloric acid, [SnCl$_3$]$^-$ forms. In alkaline solutions, the dominant species is [Sn(OH)$_3$]$^-$. Extensive hydrolysis of Sn(IV) species in aqueous solution also occurs unless sufficient acid is present to complex the Sn(IV); thus, in aqueous HCl, Sn(IV) is present as [SnCl$_6$]$^{2-}$. In alkaline solution at high pH, [Sn(OH)$_6$]$^{2-}$ (**13.4**) is the main species and salts of this ion, e.g. K$_2$[Sn(OH)$_6$], can be isolated.

In comparison to their Sn(II) analogues, Pb(II) salts are much more stable in aqueous solution with respect to hydrolysis and oxidation. The most important *soluble* oxosalts are Pb(NO$_3$)$_2$ and Pb(CH$_3$CO$_2$)$_2$; the fact that many water-insoluble Pb(II) salts dissolve in a mixture of $[NH_4][CH_3CO_2]$ and CH_3CO_2H reveals that Pb(II) is strongly complexed by acetate. Most Pb(II) oxo-salts are, like the halides, sparingly soluble in water; $PbSO_4$ ($K_{sp} = 1.8 \times 10^{-8}$) dissolves in concentrated H_2SO_4.

The Pb^{4+} ion does not exist in aqueous solution, and the value of $E^\circ(Pb^{4+}/Pb^{2+})$ given in Table 13.1 is for the half-reaction 13.79 which forms part of the familiar lead–acid battery (see *equations 13.3* and *13.4*). For half-reaction 13.79, the fourth-power dependence of the half-cell potential upon $[H^+]$ immediately explains why the relative stabilities of Pb(II) and Pb(IV) depend upon the pH of the solution (see *Section 7.2*).

$$PbO_2(s) + 4H^+(aq) + 2e^- \rightleftharpoons Pb^{2+}(aq) + 2H_2O(l)$$

$$E^\circ = +1.45\,V \qquad (13.79)$$

Thus, for example, PbO_2 oxidizes concentrated HCl to Cl_2, but Cl_2 oxidizes Pb(II) in alkaline solution to PbO_2. It may be noted that thermodynamically, PbO_2 should oxidize water at pH = 0, and the usefulness of the lead–acid battery depends on there being a high overpotential for O_2 evolution.

Yellow crystals of $Pb(SO_4)_2$ may be obtained by electrolysis of fairly concentrated H_2SO_4 using a Pb anode; however, in cold water, it is hydrolyzed to PbO_2, as are Pb(IV) acetate and $[NH_4]_2[PbCl_6]$ (see *Section 13.8*). The complex ion $[Pb(OH)_6]^{2-}$ forms when PbO_2 dissolves in concentrated KOH solution, but on dilution of the solution, PbO_2 is reprecipitated.

Summary

This chapter has given an introduction to the descriptive chemistry of the group 14 elements. Following from what we saw for group 13, the chemistries of the first group 14 element, carbon, and its compounds are clearly different from those of the later elements. Moreover, the metallic character and stabilization of the +2 rather than +4 oxidation state is most prevalent for Sn and Pb.

The following terms were introduced in this chapter.
Do you know what they mean?
❑ catenation
❑ metastable
❑ Zintl ion
❑ pyrophoric
❑ piezoelectric
❑ photoconductor

Suggested further reading

Carbon: fullerenes and nanotubes

P.R. Birkett and M. Terrones (1999) *Chemistry in Britain*, May issue, p. 45 – 'Stretching the point'.

J. Cook, J. Sloan and M.L.H. Green (1996) *Chemistry & Industry*, p. 600 – 'Carbon nanotube chemistry'.

J.D. Crane and H.W. Kroto (1994) 'Carbon: Fullerenes' in *Encyclopedia of Inorganic Chemistry*, ed. R.B. King, Wiley, Chichester, vol. 2, p. 531.

Th. Henning and F. Salama (1998) *Science*, vol. 282, p. 2204 – 'Carbon in the universe'.

A. Hirsch (1994) *The Chemistry of the Fullerenes*, Thieme, Stuttgart.

H.W. Kroto (1992) *Angewandte Chemie, International Edition in English*, vol. 31, p. 111 – 'C_{60}: Buckminsterfullerene, the celestial sphere that fell to earth'.

H. Terrones, M. Terrones and W.K. Hsu (1995) *Chemical Society Reviews*, vol. 24, p. 341 – 'Beyond C_{60}: Graphite structures for the future'.

C. Thilgen, A. Herrmann and F. Diederich (1997) *Angewandte Chemie, International Edition in English*, vol. 36, p. 2268 – 'The covalent chemistry of higher fullerenes: C_{70} and beyond'.

Silicates and zeolites

J.M. Thomas (1990) *Philosopical Transactions of the Royal Society*, vol. A333, p. 173 – A Bakerian Lecture, well illustrated, that contains a general account of zeolites and their applications.

A.F. Wells (1984) *Structural Inorganic Chemistry*, 5th edn, Clarendon Press, Oxford – Chapter 23 contains a full account of silicate structures.

Other topics

P. Ettmayer and W. Lengauer (1994) 'Carbides: Transition metal solid state chemistry' in *Encyclopedia of Inorganic Chemistry*, ed. R.B. King, Wiley, Chichester, vol. 2, p. 519.

M.J. Hynes and B. Jonson (1997) *Chemical Society Reviews*, vol. 26, p. 133 – 'Lead, glass and the environment'.

S.M. Kauzlarich, ed. (1996) *Chemistry, Structure and Bonding of Zintl Phases and Ions: Selected Topics and Recent Advances*, Wiley, New York.

R. Okazaki and R. West (1996) *Advances in Organometallic Chemistry*, vol. 39, p. 231 – 'Chemistry of stable disilenes'.

S.T. Oyama (1996) *The Chemistry of Transition Metal Carbides and Nitrides*, Kluwer, Dordrecht.

A. Sekiguchi and H. Sakurai (1995) *Advances in Organometallic Chemistry*, vol. 37, p. 1 – 'Cage and cluster compounds of silicon, germanium and tin'.

P.J. Smith, ed. (1998) *Chemistry of Tin*, 2nd edn, Blackie, London.

See also *Chapter 5 reading list:* **semiconductors**.

Problems

13.1 (a) Write down, in order, the names and symbols of the elements in group 14; check your answer by reference to the first page of this chapter. (b) Classify the elements in terms of metallic, semi-metallic or non-metallic behaviour. (c) Give a *general* notation showing the ground state electronic configuration of each element.

13.2 Comment on the trends in values of (a) melting points, (b) $\Delta_{atom}H^\circ(298\,K)$ and $\Delta_{fus}H^\circ(mp)$ for the elements on descending group 14.

13.3 How does the structure of graphite account for (a) its use as a lubricant, (b) the design of graphite electrodes, and (c) the fact that diamond is the more stable allotrope at very high pressures.

13.4 Figure 13.7 shows a unit cell of K_3C_{60}. From the structural information given, confirm the stoichiometry of this fulleride.

13.5 Give four examples of reactions of C_{60} that are consistent with the presence of C=C bond character.

13.6 Comment on each of the following observations.

(a) The carbides Mg_2C_3 and CaC_2 liberate propyne and ethyne respectively when treated with water, reaction between ThC_2 and water produces mixtures composed mainly of C_2H_2, C_2H_6 and H_2, but no reaction occurs when water is added to TiC.

(b) Mg_2Si reacts with $[NH_4]Br$ in liquid NH_3 to give silane.

(c) Compound **13.36** is hydrolyzed by aqueous alkali at the same rate as the corresponding Si–D compound.

(13.36)

13.7 (a) Suggest why the NSi_3 skeleton in $N(SiMe_3)_3$ is planar. (b) Suggest reasons why, at 298 K, CO_2 and SiO_2 are not isostructural.

13.8 Predict the shapes of the following molecules or ions: (a) ClCN; (b) OCS; (c) $[SiH_3]^-$; (d) $[SnCl_5]^-$; (e) Si_2OCl_6; (f) $[Ge(C_2O_4)_3]^{2-}$; (g) $[PbCl_6]^{2-}$; (h) $[SnS_4]^{4-}$.

13.9 The observed structure of $[Sn_9Tl]^{3-}$ is a bicapped square-antiprism. (a) Confirm that this is consistent with Wade's rules. (b) How many isomers (retaining the bicapped square-antiprism core) of $[Sn_9Tl]^{3-}$ are possible?

13.10 Compare and contrast the structures and chemistries of the hydrides of the group 14 elements, and give pertinent examples to illustrate structural and chemical differences between BH_3 and CH_4, and between AlH_3 and SiH_4.

13.11 Write equations for: (a) the hydrolysis of $GeCl_4$; (b) the reaction of $SiCl_4$ with aqueous NaOH; (c) the 1:1 reaction of CsF with GeF_2; (d) the hydrolysis of SiH_3Cl; (e) the hydrolysis of SiF_4; (f) the 2:1 reaction of $[Bu_4P]Cl$ with $SnCl_4$. In each case suggest the structure of the product containing the group 14 element.

13.12 Rationalize the following signal multiplicities in the ^{119}Sn NMR spectra of some halo-anions and, where possible, use the data to distinguish between geometric isomers [^{19}F 100% $I = \frac{1}{2}$]: (a) $[SnCl_5F]^{2-}$ doublet; (b) $[SnCl_4F_2]^{2-}$ isomer A, triplet; isomer B, triplet; (c) $[SnCl_3F_3]^{2-}$ isomer A, doublet of triplets; isomer B, quartet; (d) $[SnCl_2F_4]^{2-}$ isomer A, quintet; isomer B, triplet of triplets; (e) $[SnClF_5]^{2-}$ doublet of quintets; (f) $[SnF_6]^{2-}$ septet.

13.13 What would you expect to form when:

(a) Sn is heated with concentrated aqueous NaOH;
(b) SO_2 is passed over PbO_2;
(c) CS_2 is shaken with aqueous NaOH;
(d) SiH_2Cl_2 is hydrolyzed by water;
(e) four molar equivalents of $ClCH_2SiCl_3$ react with three equivalents of $Li[AlH_4]$ in Et_2O solution?

13.14 Suggest one method for the estimation of each of the following quantities:

(a) Δ_rH° for the conversion:
$GeO_2(quartz) \longrightarrow GeO_2(rutile)$;
(b) the Pauling electronegativity value, χ^P, of Si;
(c) the purity of a sample of $Pb(MeCO_2)_4$ prepared in a laboratory experiment.

13.15 By referring to Figure 7.4, deduce whether carbon could be used to extract Sn from SnO_2 at (a) 500 K; (b) 750 K; (c) 1000 K. Justify your answer.

13.16 Comment on the following observations.

(a) the pyroxenes $CaMgSi_2O_6$ and $CaFeSi_2O_6$ are isomorphous;
(b) the feldspar $NaAlSi_3O_8$ may contain up to 10% of $CaAl_2Si_2O_8$;
(c) the mineral *spodumene*, $LiAlSi_2O_6$, is isostructural with *diopside*, $CaMgSi_2O_6$, but when it is heated it is transformed into a polymorph having the quartz structure with the Li^+ ions in the interstices.

13.17 Table 13.7 gives values of the symmetric and asymmetric stretches of the heteronuclear bonds in CO_2, CS_2 and $(CN)_2$, although the molecules are indicated only by the labels I, II and III. (a) Assign an identity to each of I, II and III. (b) State whether the stretching modes listed in Table 13.7 are IR active or inactive.

13.18 Account for the fact that when aqueous solution of KCN is added to a solution of aluminium sulfate, a precipitate of $Al(OH)_3$ forms.

13.19 What would you expect to be the hydrolysis products of (a) cyanic acid, (b) isocyanic acid and (c) thiocyanic acid?

Table 13.7 Data for problem 13.17.

Compound	ν_1(symmetric) / cm^{-1}	ν_3(asymmetric) / cm^{-1}
I	2330	2158
II	658	1535
III	1333	2349

Chapter 14

The group 15 elements

- Occurrence, extraction and uses
- Physical properties
- The elements
- Hydrides
- Nitrides, phosphides and arsenides
- Halides, oxohalides and complex halides
- Oxides of nitrogen

- Oxoacids of nitrogen
- Oxides of phosphorus, arsenic, antimony and bismuth
- Oxoacids of phosphorus
- Oxoacids of arsenic, antimony and bismuth
- Phosphazenes
- Sulfides and selenides
- Aqueous solution chemistry

1	2		13	14	**15**	16	17	18
H								He
Li	Be		B	C	**N**	O	F	Ne
Na	Mg		Al	Si	**P**	S	Cl	Ar
K	Ca	*d*-block	Ga	Ge	**As**	Se	Br	Kr
Rb	Sr		In	Sn	**Sb**	Te	I	Xe
Cs	Ba		Tl	Pb	**Bi**	Po	At	Rn
Fr	Ra							

14.1 Introduction

The rationalization of the properties of the group 15 elements (nitrogen, phosphorus, arsenic, antimony and bismuth) and their compounds is difficult, despite there being some general similarities in trends of the group 13, 14 and 15 elements, e.g. increase in metallic character and stabilities of lower oxidation states on descending the group. Although the 'diagonal' line (*Figure 6.8*) can be drawn between As and Sb, formally separating non-metallic and metallic elements, the distinction is not well defined and should be treated with caution.

Very little of the chemistry of the group 15 elements is that of simple ions. Although metal nitrides and phosphides that react with water are usually considered to contain N^{3-} and P^{3-} ions, electrostatic considerations make it doubtful whether these ionic formulations are correct. The only definite case of a simple cation in a *chemical* environment

is that of Bi^{3+}, and nearly all the chemistry of the group 15 elements involves covalently bonded compounds. The thermochemical basis of the chemistry of such species is much harder to establish than that of ionic compounds. In addition, they are much more likely to be *kinetically* inert, both to substitution reactions (e.g. NF_3 to hydrolysis, $[H_2PO_2]^-$ to deuteration), and to oxidation or reduction when these processes involve making or breaking covalent bonds, as well as the transfer of electrons. Nitrogen, for example, forms a range of oxoacids and oxoanions, and in aqueous media can exist in all oxidation states from +5 to −3, e.g. $[NO_3]^-$, N_2O_4, $[NO_2]^-$, NO, N_2O, N_2, NH_2OH, N_2H_4, NH_3. Tables of standard reduction potentials (usually calculated from thermodynamic data) or potential diagrams (see *Section 7.5*) are of limited use in summarizing the relationships between these species. Although they provide information about the thermodynamics of possible reactions, they say nothing about the kinetics. Much the same is true about the chemistry of phosphorus. The chemistry of the first two members of group 15 is far more extensive than that of As, Sb and Bi, and we can mention only a small fraction of the known inorganic compounds of N and P. In our discussions, we shall need to emphasize *kinetic* factors more than in earlier chapters.

14.2 Occurrence, extraction and uses

Occurrence

Naturally occurring N_2 makes up 78% (by volume) of the Earth's atmosphere and contains ≈0.36% ^{15}N. The latter is useful for isotopic labelling and can be obtained in concentrated form by chemical exchange processes similar to those exemplified for ^{13}C in *Section 2.10*. Because of the availability of N_2 in the atmosphere and its requirement by living organisms (in which N is present as proteins), the *fixing of nitrogen* in forms in which it may be assimilated by plants is of great importance. Attempts to devise synthetic nitrogen-fixation processes (see *Section 28.3*) that mimic the action of bacteria living in root nodules of leguminous plants have not yet been successful, although N_2 can be fixed by other processes, e.g. its industrial conversion to NH_3 (see *Section 14.5*). The only natural source of nitrogen suitably 'fixed' for uptake by plants is crude $NaNO_3$ (*Chile saltpetre* or *sodanitre*) which occurs in the deserts of South America.

Phosphorus is an essential constituent of plant and animal tissue; calcium phosphate occurs in bones and teeth, and phosphate esters of nucleotides (e.g. DNA, *Figure 9.9*) are of immense biological significance (see *Box 14.10*). Phosphorus occurs naturally in the form of *apatites*, $Ca_5X(PO_4)_3$, the important minerals being *fluorapatite* (X = F), *chloro-apatite* (X = Cl) and *hydroxyapatite* (X = OH); major deposits of the apatite-containing ore *phosphate rock* occur in North Africa, North America, Asia and the Middle East. Although native arsenic occurs, commercial sources of the element are *mispickel* (*arsenopyrite*, FeAsS), *realgar* (As_4S_4) and *orpiment* (As_2S_3). Native antimony is rare and the only commercial ore is *stibnite* (Sb_2S_3). Bismuth also occurs native, and as the ores *bismuthinite* (Bi_2S_3) and *bismite* (Bi_2O_3).

Extraction

The industrial separation of N_2 is discussed in Section 14.4. Mining of phosphate rock takes place on a vast scale (in 1997, 138 Mt were mined worldwide), with the majority destined for the production of fertilizers (see *Box 14.9*) and animal feed supplements. Elemental phosphorus is extracted from phosphate rock (which approximates in composition to $Ca_3(PO_4)_2$) by heating with sand and coke in an electric furnace (equation 14.1); phosphorus vapour distils out and is condensed under water to yield white phosphorus.

$$2Ca_3(PO_4)_2 + 6SiO_2 + 10C \xrightarrow{\approx 1700\,K} P_4 + 6CaSiO_3 + 10CO$$

$$(14.1)$$

The principal source of As is FeAsS, and the element is extracted by heating (equation 14.2) and condensing the As sublimate. An additional method is air-oxidation of arsenic sulfide ores to give As_2O_3 which is then reduced by C; As_2O_3 is also recovered on a large scale from flue dusts in Cu and Pb smelters.

$$FeAsS \xrightarrow{\Delta(\text{in absence of air})} FeS + As \qquad (14.2)$$

Antimony is obtained from stibnite by reduction using scrap iron (equation 14.3) or by conversion to Sb_2O_3 followed by reduction with C.

$$Sb_2S_3 + 3Fe \longrightarrow 2Sb + 3FeS \qquad (14.3)$$

The extraction of Bi from its sulfide or oxide ores involves reduction with carbon (via the oxide when the ore is Bi_2S_3), but the metal is also obtained as a byproduct of Pb, Cu, Sn, Ag and Au refining processes.

Uses

In the US, N_2 ranks second in industrial chemicals, and a large proportion of N_2 is converted to NH_3 (see *Box 14.2*). Gaseous N_2 is widely used to provide inert atmospheres, both industrially (e.g. in the electronics industry during the production of transistors etc.) and in laboratories. Liquid N_2 (bp 77 K) is an important coolant (Table 14.1) with applications in some freezing processes. Nitrogen-based chemicals are extremely important, and include nitrogenous fertilizers (see *Box 14.2*), nitric acid (see *Box 14.7*) and nitrate salts, explosives such as nitroglycerine (**14.1**) and trinitrotoluene (TNT, **14.2**), nitrite salts (e.g. in the curing of meat where they prevent discoloration by inhibiting oxidation of blood), cyanides and azides (e.g. in motor vehicle airbags where decomposition produces N_2 to inflate the airbag, see *equation 14.4*).

(14.1)

(14.2)

By far the most important application of phosphorus is in phosphate fertilizers and in *Box 14.9*, we highlight this use and possible associated environmental problems. Bone ash

Table 14.1 Selected low temperature baths involving liquid N_2.[‡]

Bath contents	Temperature / K
Liquid N_2 + cyclohexane	279
Liquid N_2 + acetonitrile	232
Liquid N_2 + octane	217
Liquid N_2 + heptane	182
Liquid N_2 + hexa-1,5-diene	132

[‡] To prepare a liquid N_2 *slush bath*, liquid N_2 is poured into an appropriate solvent which is constantly stirred. See also *Table 13.5*.

Phosphorus-containing nerve gases

Development of nerve gases during the latter half of the 20th century became coupled not just with their actual use, but with the threat of potential use during war. Two examples are Sarin and Soman which function by enzyme inhibition in the nervous system; inhalation of ≈ 1 mg is fatal.

Sarin   Soman

As the 21st century begins, policies of many countries are for chemical weapon disarmament, and programmes for the destruction of stock-piled nerve gases have been enforced.

A problem for those involved in developing destruction processes is to ensure that end-products are harmless. Sarin, for example, may be destroyed by hydrolysis:

$$(Me_2HCO)P(O)(Me)F + H_2O$$

$$\rightarrow (Me_2HCO)P(O)(Me)OH + HF$$

$$\downarrow H_2O$$

$$Me_2HCOH + MeP(O)(OH)_2$$

and the use of aqueous NaOH results in the formation of effectively harmless sodium salts.

Further reading

Y.-C. Yang (1995) *Chemistry & Industry*, p. 334 – 'Chemical reactions for neutralizing chemical warfare agents'.

(calcium phosphate) is used in the manufacture of bone china. Most white phosphorus is converted to H_3PO_4, or to compounds such as P_4O_{10}, P_4S_{10}, PCl_3 and $POCl_3$. Phosphoric acid is industrially very important and is used on a large scale in the production of fertilizers, detergents and food additives. It is responsible for the sharp taste of many soft drinks, and is used to remove oxide and scale from the surfaces of iron and steel. Phosphorus trichloride is also manufactured on a large scale; it is a precursor to many organophosphorus compounds, including nerve gases (see *Box 14.1*), flame retardants (see *Box 16.1*) and insecticides. Phosphorus is important in steel manufacture and phosphor bronzes. Red phosphorus (see *Section 14.4*) is used in safety matches and in the generation of smoke (e.g. fireworks, smoke bombs).

Arsenic salts and arsines are extremely toxic, and uses of arsenic compounds in weedkillers, wood preservatives, sheep- and cattle-dips, and poisons against vermin are less widespread than was once the case. Antimony compounds are less toxic, but large doses result in liver damage; potassium antimony tartrate (*tartar emetic*) was used medicinally as an emetic and expectorant but has now been replaced by less toxic reagents. Bismuth is one of the less toxic heavy metals and compounds, such as the subcarbonate $(BiO)_2CO_3$, find use in stomach remedies including treatments for peptic ulcers.

Arsenic is a doping agent in semiconductors (see *Section 5.9*) and GaAs has widespread uses in solid-state devices and semiconductors. Uses of As include those in the semiconductor industry, in alloys (e.g. it increases the strength of Pb) and in batteries; Sb_2O_3 is used in paints, adhesives and plastics, and as a flame retardant (see *Box 16.1*). Uses of Sb_2S_3 include those in photoelectric devices and electrophotographic recording materials, and as a flame retardant.

Bismuth is also combined into alloys (e.g. with Sn) and Bi-containing compounds such as BiOCl have applications in cosmetic products (e.g. creams, hair dyes and tints), as oxidation catalysts and in high-temperature superconductors; Bi_2O_3 has many uses in the glass and ceramics industry, and for catalysts and magnets.

14.3 Physical properties

Table 14.2 lists selected physical properties of the group 15 elements. Some observations regarding ionization energies are that:

- they increase rather sharply after removal of the *p* electrons;
- they decrease only slightly between P and As (similar behaviour to that between Al and Ga, and between Si and Ge);
- for removal of the *s* electrons, there is an increase between Sb and Bi, just as between In and Tl, and between Sn and Pb (see *Box 12.3*).

Values of $\Delta_a H^{\circ}$ decrease steadily from N to Bi, paralleling similar trends in groups 13 and 14.

Bonding considerations

Analogies between groups 14 and 15 are seen if we consider certain bonding aspects. Table 14.3 lists some covalent bond enthalpy terms for group 15 elements. Data for most single bonds follow trends reminiscent of those in group 14 (Table 13.2); e.g. N forms stronger bonds with H than

Table 14.2 Some physical properties of the group 15 elements and their ions.

Property	N	P	As	Sb	Bi
Atomic number, Z	7	15	33	51	83
Ground state electronic configuration	$[He]2s^2 2p^3$	$[Ne]3s^2 3p^3$	$[Ar]3d^{10}4s^2 4p^3$	$[Kr]4d^{10}5s^2 5p^3$	$[Xe]4f^{14}5d^{10}6s^2 6p^3$
Enthalpy of atomization, $\Delta_a H^\circ$(298 K) / kJ mol^{-1}	473[‡]	315	302	264	210
Melting point, mp / K	63	317	887 sublimes	904	544
Boiling point, bp / K	77	550	–	2023	1837
Standard enthalpy of fusion, $\Delta_{fus}H^\circ$(mp) / kJ mol^{-1}	0.71	0.66	24.44	19.87	11.30
First ionization energy, IE_1 / kJ mol^{-1}	1402	1012	947.0	830.6	703.3
Second ionization energy, IE_2 / kJ mol^{-1}	2856	1907	1798	1595	1610
Third ionization energy, IE_3 / kJ mol^{-1}	4578	2914	2735	2440	2466
Fourth ionization energy, IE_4 / kJ mol^{-1}	7475	4964	4837	4260	4370
Fifth ionization energy, IE_5 / kJ mol^{-1}	9445	6274	6043	5400	5400
Metallic radius, r_{metal} / pm	–	–	–	–	182
Covalent radius, r_{cov} / pm[*]	75	110	122	143	152
Ionic radius, r_{ion} / pm[**]	171 (N^{3-})	–	–	–	103 (Bi^{3+})
NMR active nuclei (% abundance, nuclear spin)	^{14}N (99.6, $I = 1$) ^{15}N (0.4, $I = \frac{1}{2}$)	^{31}P (100, $I = \frac{1}{2}$)	^{75}As (100, $I = \frac{3}{2}$)	^{121}Sb (57.3, $I = \frac{5}{2}$) ^{123}Sb (42.7, $I = \frac{7}{2}$)	^{209}Bi (100, $I = \frac{9}{2}$)

[‡] For nitrogen, $\Delta_a H^\circ = \frac{1}{2} \times$ dissociation energy of N_2.
[*] For 3-coordination.
[**] For 6-coordination.

does P, but weaker bonds with F, Cl or O. These observations, together with the absence of stable P-containing analogues of N_2, NO, HCN, $[N_3]^-$ and $[NO_2]^+$ (**14.3–14.7**) indicate that strong $(p–p)\pi$-bonding is important in N chemistry (but not for the heavier elements), while $(p–d)\pi$-bonding *may* play a role in the chemistry of P and the later elements. The non-existence of nitrogen pentahalides and the slowness of the reaction between NF_3 and H_2O, has been interpreted in terms of the lack of d-orbitals in the valence shell of nitrogen. Conversely, the reactivity of, for example, the phosphorus trihalides, and the existence of pentahalides of P, and pentafluorides and pentachlorides of As, Sb and Bi have been rationalized by the ability of these group 15 elements to expand their valence shells beyond eight by virtue of d-orbital participation. However, just as we showed in *Section 4.7* that it is possible to rationalize

the bonding in SF_6 *without* the use of the $3d$ atomic orbitals on sulfur, it is also possible to interpret the bonding in hypervalent compounds of P, As, Sb and Bi *without* invoking the use of the d-orbitals. Whether or not these atomic orbitals are used is not clear-cut, and discussions along the lines of those outlined in *Sections 4.7* and *13.6* are also appropriate for the group 15 elements. Double bond formation between the heavier group 15 elements is discussed in *Section 18.6*.

$$N \equiv N \qquad \overset{\bullet}{N} = O \qquad H - C \equiv N$$
$$\text{(14.3)} \qquad \text{(14.4)} \qquad \text{(14.5)}$$

$$\overset{-}{N} = \overset{+}{N} = \overset{-}{N} \qquad O = \overset{+}{N} = O$$
$$\text{(14.6)} \qquad \text{(14.7)}$$

The largest *difference* between groups 14 and 15 lies in the relative strengths of the $N \equiv N$ (in N_2) and $N - N$ (in N_2H_4) bonds compared with those of $C \equiv C$ and $C - C$ bonds (Tables 14.3 and 13.2). There is some uncertainty about a value for the $N = N$ bond enthalpy term because of difficulty in choosing a reference compound, but the approximate value given in Table 14.3 is seen to be *more* than twice that of the $N - N$ bond, whereas the $C = C$ bond is significantly *less* than twice as strong as the $C - C$ bond (Table 13.2). While N_2 is thermodynamically stable with respect to oligomerization to species containing $N - N$ bonds, $HC \equiv CH$ is thermodynamically unstable with respect to species with $C - C$ bonds. [See *problem 14.2* at the end of the chapter.] Similarly, the dimerization of P_2 to tetrahedral P_4 is thermodynamically favourable. The σ- and π-contributions that contribute to the very high strength of the $N \equiv N$ bond (which makes many nitrogen compounds endothermic and most of the others only *slightly* exothermic) were discussed

Table 14.3 Some covalent bond enthalpy terms (kJ mol^{-1}); the values for single bonds refer to the group 15 elements in 3-coordinate environments, and values for triple bonds are for dissociation of the appropriate diatomic molecule.

N–N 160	N=N ≈400[‡]	N≡N 946	N–H 391	N–F 272	N–Cl 193	N–O 201
P–P 209		P≡P 490	P–H 322	P–F 490	P–Cl 319	P–O 340
As–As 180			As–H 247	As–F 464	As–Cl 317	As–O 330
					Sb–Cl 312	
					Bi–Cl 280	

[‡] See text.

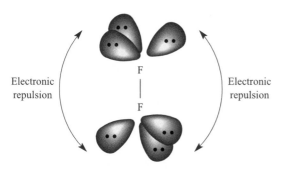

Electronic repulsion ⟵ ⟶ **Electronic repulsion**

F

F

Fig. 14.1 Schematic representation of the electronic repulsion, believed to weaken the F−F bond in F_2.

in *Section 1.13*. However, the particular weakness of the N−N single bond calls for comment. The O−O ($146 \, kJ \, mol^{-1}$ in H_2O_2) and F−F ($159 \, kJ \, mol^{-1}$ in F_2) bonds are also very weak, much weaker than S−S or Cl−Cl bonds. In N_2H_4, H_2O_2 and F_2, the N, O or F atoms carry lone pairs, and it is believed that the N−N, O−O and F−F bonds are weakened by repulsion between lone pairs on adjacent atoms (Figure 14.1). Lone pairs on larger atoms (e.g. in Cl_2) are further apart and experience less mutual repulsion. Each N atom in N_2 also has a non-bonding lone pair, but they are directed away from each other. Table 14.3 illustrates that N−O, N−F and N−Cl are also rather weak and, again, interactions between lone pairs of electrons can be used to rationalize these data. However, when N is singly bonded to an atom with no lone pairs (e.g. H), the bond is strong. In pursuing such arguments, we must remember that in a heteronuclear bond, extra energy contributions may be attributed to partial ionic character (see *Section 1.15*).

Another important difference between N and the later group 15 elements is the ability of N to take part in strong hydrogen bonding (see *Sections 9.6* and *14.5*). This arises from the much higher electronegativity of N ($\chi^P = 3.0$) compared with values for the later elements (χ^P values: P, 2.2; As, 2.2; Sb, 2.1; Bi, 2.0). This essentially has no counterpart within group 14,[†] although it is mirrored in the hydrogen bonding associated with O and F in groups 16 and 17.

NMR active nuclei

Nuclei that are NMR active are listed in Table 14.2. Routinely, ^{31}P NMR spectroscopy is used in characterizing P-containing species; see for example case studies 1, 2 and 4 in *Chapter 2*. Chemical shifts are usually reported with respect to $\delta = 0$ for 85% aqueous H_3PO_4, but other reference compounds are used, e.g. trimethylphosphite, $P(OMe)_3$. The chemical shift range for ^{31}P is large.

Radioactive isotopes

Although the only naturally occurring isotope of phosphorus is ^{31}P, sixteen radioactive isotopes are known. Of these, ^{32}P is the most important (see *equations 2.12* and *2.13*) with its half-life of 14.3 days making it suitable as a tracer.

14.4 The elements

Nitrogen

Dinitrogen is obtained industrially by fractional distillation of liquid air, and the product contains some Ar and traces of O_2. Dioxygen can be removed by addition of a small amount of H_2 and passage over a Pt catalyst, or by bubbling the gas through an aqueous solution of $CrCl_2$. Small amounts of N_2 can be prepared by thermal decomposition of sodium azide (equation 14.4) or by reactions 14.5 or 14.6. The latter should be carried out cautiously because of the risk of explosion; ammonium nitrite (NH_4NO_2) is potentially explosive, as is ammonium nitrate which is a powerful oxidant and a component of dynamite. In car air-bags, the decomposition of NaN_3 is initiated by an electrical impulse.[‡]

$$2NaN_3(s) \xrightarrow{\Delta} 2Na + 3N_2 \tag{14.4}$$

$$NH_4NO_2(aq) \xrightarrow{\Delta} N_2 + 2H_2O \tag{14.5}$$

$$2NH_4NO_3(s) \xrightarrow{>570\,K} 2N_2 + O_2 + 4H_2O \tag{14.6}$$

Dinitrogen is generally unreactive. It combines slowly with Li at ambient temperatures (*equation 10.6*), and, when heated, with the group 2 metals, Al (*Section 12.8*), Si, Ge (*Section 13.5*) and many *d*-block metals. The reaction between CaC_2 and N_2 is used industrially for manufacturing the nitrogenous fertilizer calcium cyanamide (*equations 13.28* and *13.29*). Many elements (e.g. Na, Hg, S) which are inert towards N_2 do react with atomic nitrogen, produced by passing N_2 through an electric discharge. At ambient temperatures, N_2 is reduced to hydrazine (N_2H_4) by vanadium(II) and magnesium hydroxides. We consider the reaction of N_2 with H_2 later in the chapter.

A large number of *d*-block metal complexes containing coordinated N_2 are known (see *Figure 14.6* and *equations 22.92* and *22.93* and *discussion*); N_2 is isoelectronic with CO and the bonding in complexes containing the N_2 ligand can be described in a similar manner to that in metal carbonyl complexes (see *Chapter 23*).

[†] Carbaborane (see *Section 12.11*) C−H bonds, for example, have been shown to participate in hydrogen-bonded C−H····O interactions in the solid state; see for example: M.G. Davidson, T.G. Hibbert, J.A.K. Howard, A. Mackinnon and K. Wade (1996) *Chemical Communications*, p. 2285.

[‡] A. Madlung (1996) *Journal of Chemical Education*, vol. 73, p. 347 – 'The chemistry behind the air bag'.

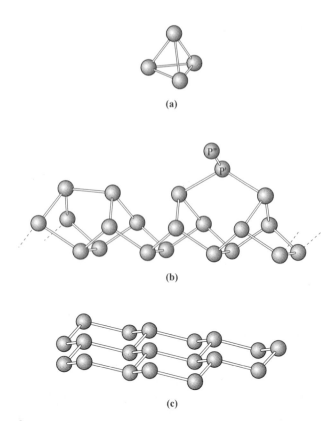

Fig. 14.2 (a) The tetrahedral P_4 molecule found in white phosphorus. (b) Part of one of the chain-like arrays of atoms present in the infinite lattice of Hittorf's phosphorus; the repeat unit contains 21 atoms, and atoms P' and P" are equivalent atoms in adjacent chains, with chains connected through P'–P" bonds. (c) Part of one layer of puckered six-membered rings present in black phosphorus and the rhombohedral allotropes of arsenic, antimony and bismuth.

Phosphorus

Phosphorus exhibits complicated allotropy; eleven forms have been reported, of which at least five are crystalline. Crystalline white phosphorus contains tetrahedral P_4 molecules (Figure 14.2a) in which the P–P distances (221 pm) are consistent with single bonds ($r_{cov} = 110$ pm). White phosphorus is *defined* as the standard state of the element, but is actually metastable (equation 14.7) (see *Section 13.4*). The lower stability of the white form probably originates from strain associated with the 60° bond angles.

$$\text{P} \xleftarrow{\Delta_f H^\circ = -39.3 \text{ kJ mol}^{-1}} \tfrac{1}{4}\text{P}_4 \xrightarrow{\Delta_f H^\circ = -17.6 \text{ kJ mol}^{-1}} \text{P}$$

Black — White — Red

$$(14.7)$$

White phosphorus is manufactured by reaction 14.1, and heating this allotrope in an inert atmosphere at ≈540 K produces red phosphorus. Several crystalline forms of red phosphorus exist, and all probably possess infinite lattices.[†]

[†] For recent details, see: H. Hartl (1995) *Angewante Chemie International Edition in English*, vol. 34, p. 2637 – 'New evidence concerning the structure of amorphous red phosphorus'.

Hittorf's phosphorus (also called violet phosphorus) is a well-characterized form of the red allotrope and its complicated structure is best described in terms of interlocking chains (Figure 14.2b). Non-bonded chains lie parallel to each other to give layers, and the chains in one layer lie at right-angles to the chains in the next layer, being connected by the P'–P" bonds shown in Figure 14.2b. All P–P bond distances are ≈222 pm, indicating covalent single bonds. Black phosphorus is the most stable allotrope and is obtained by heating white phosphorus under high pressure. Its appearance and electrical conductivity resemble those of graphite, and it possesses a double layer lattice of puckered six-membered rings (Figure 14.2c); P–P distances within a layer are 220 pm and the shortest interlayer P–P distance is 390 pm. On melting, all allotropes give a liquid containing P_4 molecules, and these are also present in the vapour; above 1070 K or at high pressures, P_4 is in equilibrium with P_2 (**14.8**).

$$\overset{\text{187 pm}}{\text{P}\equiv\equiv\text{P}}$$

$$(14.8)$$

Most of the chemical differences between the allotropes of phosphorus are due to differences in activation energies for reactions. Black phosphorus is kinetically inert and does not ignite in air even at 670 K. Red phosphorus is intermediate in reactivity between the white and black allotropes. It is not poisonous, is insoluble in organic solvents, does not react with aqueous alkali, and ignites in air above 520 K. It reacts with halogens, sulfur and metals, but less vigorously than does white phosphorus. The latter is a soft, waxy solid which becomes yellow on exposure to light; it is very poisonous, being readily absorbed into the blood and liver. White phosphorus is soluble in benzene, PCl_3 and CS_2 but is virtually insoluble in water, and is stored under water to prevent oxidation. In moist air, it undergoes *chemiluminescent* oxidation, emitting a green glow and slowly forming P_4O_8 (see *Section 14.10*) and some O_3; the chain reaction involved is extremely complicated.

A *chemiluminescent* reaction is one that is accompanied by the emission of light.

$$\text{P}_4 \xrightarrow{323 \text{ K, O}_2} \text{P}_4\text{O}_{10}$$

$$(14.8)$$

Above 323 K, white phosphorus inflames yielding phosphorus(V) oxide (equation 14.8); in a limited supply of air, P_4O_6 may form. White phosphorus combines violently with all of the halogens giving PX_3 (X = F, Cl, Br, I) or PX_5 (X = F, Cl, Br) depending on the relative amounts of P_4 and X_2. Concentrated HNO_3 oxidizes P_4 to H_3PO_4, and with hot aqueous NaOH, reaction 14.9 occurs with H_2, some P_2H_4 being formed.

$$P_4 + 3NaOH + 3H_2O \longrightarrow 3NaH_2PO_2 + PH_3 \qquad (14.9)$$

$$23P_4 + 12LiPH_2 \longrightarrow 6Li_2P_{16} + 8PH_3 \qquad (14.10)$$

Reaction 14.10 yields Li_2P_{16}, while Li_3P_{21} and Li_4P_{26} can be obtained by altering the ratio of $P_4 : LiPH_2$. The structures of the phosphide ions $[P_{16}]^{2-}$, **14.9**, $[P_{21}]^{3-}$, **14.10**, and $[P_{26}]^{4-}$ are related to one chain in Hittorf's phosphorus (Figure 14.2b).

(14.9)

(14.10)

Like N_2, P_4 can act as a ligand in *d*-block metal complexes.

Arsenic, antimony and bismuth

Arsenic vapour contains As_4 molecules, and the unstable yellow form of solid As probably also contains these units; at relatively low temperatures, Sb vapour contains molecular Sb_4. At room temperature and pressure, As, Sb and Bi are grey solids with lattice structures resembling that of black phosphorus (Figure 14.2c). On descending the group, although intralayer bond distances increase as expected,

similar increases in interlayer spacing do not occur, and the coordination number of each atom effectively changes from three (Figure 14.2c) to six (three atoms within a layer and three in the next layer).

Arsenic, antimony and bismuth burn in air (equation 14.11) and combine with halogens (see *Section 14.7*).

$$4M + 3O_2 \xrightarrow{\Delta} 2M_2O_3 \qquad M = As, Sb \text{ or } Bi \qquad (14.11)$$

They are not attacked by non-oxidizing acids but react with concentrated HNO_3 to give H_3AsO_4 (hydrated As_2O_5), hydrated Sb_2O_5 and $Bi(NO_3)_3$ respectively, and with concentrated H_2SO_4 to produce As_4O_6, $Sb_2(SO_4)_3$ and $Bi_2(SO_4)_3$ respectively. None of the elements reacts with aqueous alkali, but As is attacked by fused NaOH (equation 14.12).

$$2As + 6NaOH \longrightarrow 2Na_3AsO_3 + 3H_2 \qquad (14.12)$$
$$\text{sodium arsenite}$$

14.5 Hydrides

Trihydrides, EH_3 (E = N, P, As, Sb and Bi)

Each group 15 element forms a trihydride, selected properties of which are given in Table 14.4; the lack of data for BiH_3 stems from its instability. The variation in boiling points (*Figure 9.6b*, Table 14.4) is one of the strongest pieces of evidence for hydrogen bond formation by nitrogen; NH_3 also has a greater value of $\Delta_{vap}H^\circ$ and surface tension than the later trihydrides. Thermal stabilities of these compounds decrease down the group (BiH_3 decomposes above 228 K), and this trend is reflected in the bond enthalpy terms (Table 14.3). Ammonia is the only trihydride to possess a negative value of $\Delta_f H^\circ$ (Table 14.4).

Ammonia is obtained by the action of H_2O on the nitrides of Li or Mg (equation 14.13), by heating $[NH_4]^+$ salts with base (e.g. reaction 14.14), or by reducing a nitrate or nitrite in alkaline solution with Zn or Al (e.g. reaction 14.15).

Table 14.4 Selected data for the group 15 trihydrides, EH_3.

	NH_3	PH_3	AsH_3	SbH_3	BiH_3
Name (IUPAC recommended)[‡]	Ammonia (azane)	Phosphine (phosphane)	Arsine (arsane)	Stibine (stibane)	Bismuthane
Melting point / K	195.5	140	157	185	–
Boiling point / K	240	185.5	210.5	256	290*
$\Delta_{vap}H^\circ$(bp) / kJ mol^{-1}	23.3	14.6	16.7	21.3	–
$\Delta_f H^\circ$(298 K) / kJ mol^{-1}	−45.9	5.4	66.4	145.1	277*
Dipole moment / D	1.47	0.57	0.20	0.12	–
E−H bond distance / pm	101.2	142.0	151.1	170.4	–
∠H−E−H / deg	106.7	93.3	92.1	91.6	–

[‡] The common names for the first four trihydrides in the group are generally used; bismuthane is the IUPAC name and no trivial name is recommended.
* Estimated value.

$$Li_3N + 3H_2O \longrightarrow NH_3 + 3LiOH \qquad (14.13)$$

$$2NH_4Cl + Ca(OH)_2 \longrightarrow 2NH_3 + CaCl_2 + 2H_2O \qquad (14.14)$$

$$[NO_3]^- + 4Zn + 6H_2O + 7[OH]^- \longrightarrow NH_3 + 4[Zn(OH)_4]^{2-}$$
$$(14.15)$$

Trihydrides of the later elements are best made by method 14.16, or by acid hydrolysis of phosphides, arsenides, antimonides or bismuthides. Phosphine can be made by reaction 14.17, $[PH_4]I$ being prepared from P_2I_4 (see *Section 14.7*).

$$ECl_3 \xrightarrow{\text{Li[AlH}_4\text{] in Et}_2\text{O}} EH_3 \quad E = P, As, Sb, Bi \qquad (14.16)$$

$$[PH_4]I + KOH \longrightarrow PH_3 + KI + H_2O \qquad (14.17)$$

The industrial manufacture of NH_3 (see *Figure 26.11*) involves the Haber process (reaction 14.18), and the manufacture of H_2 (see *Section 9.4*) required contributes significantly to the overall cost of the process.

$$N_2 + 3H_2 \rightleftharpoons 2NH_3 \quad \begin{cases} \Delta_r H^o(298\,K) = -92\,kJ\,mol^{-1} \\ \Delta_r G^o(298\,K) = -33\,kJ\,mol^{-1} \end{cases}$$
$$(14.18)$$

The Haber process is a classic application of physicochemical principles to a system in equilibrium. The decrease in number of moles of gas means that $\Delta_r S^o(298\,K)$ is negative. For industrial viability, NH_3 must be formed in optimum yield *and* at a reasonable rate; increasing the temperature increases the rate of reaction, but decreases the yield since the forward reaction is exothermic. At a given temperature, both the equilibrium yield and the reaction rate are increased by working at high pressures; the presence of a suitable catalyst (see *Section 26.6*) also increases the rate; the rate-determining step is the dissociation of N_2 into N atoms chemisorbed onto the catalyst. The optimum reaction conditions are $T = 723\,K$, $P = 202\,600\,kPa$, and Fe_3O_4 mixed with K_2O, SiO_2 and Al_2O_3 as the heterogeneous catalyst; the Fe_3O_4 is reduced to give the catalytically active α-Fe. The NH_3 formed is either liquefied or dissolved in H_2O to form a saturated solution of specific gravity 0.880.

Ammonia is a colourless gas with a pungent odour; Table 14.4 lists selected properties and structural data for the trigonal pyramidal molecule **14.11**, the barrier to inversion for which is very low ($24\,kJ\,mol^{-1}$). Oxidation products of NH_3 depend on conditions. Reaction 14.19 occurs on combustion in O_2, but at $\approx 1200\,K$ in the presence of a Pt/Rh

---

Ammonia: an industrial giant

Ammonia is manufactured on a huge scale and, in 1996, ranked sixth among US industrial chemicals. The graph below illustrates the scale of production in the US and the steady growth in output towards the end of the 20th century.

[Data: *Chemical & Engineering News* (1999) July 28 issue, p. 34.]

Agriculture demands vast quantities of fertilizers to supplement soil nutrients; this is critical when the same land is used year after year for crop production. Essential nutrients are N, P, K (the three required in largest amounts), Ca, Mg and S plus trace elements. Direct use and its conversion into other nitrogenous fertilizers accounts for $\approx 80\%$ of all NH_3 produced. In addition to NH_3 itself, the nitrogen-rich compounds $[NH_4][NO_3]$ and $CO(NH_2)_2$ (urea) are of prime importance, along with $[NH_4]_2[HPO_4]$ which has the benefit of supplying both N and P nutrients; $[NH_4]_2[SO_4]$ accounts for a smaller portion of the market. Of the remaining 20% of NH_3 produced, about half is consumed in the synthetic fibre industry (e.g. nylon-6, nylon-6,6 and rayon). Manufacture of explosives (see structures **14.1** and **14.2**) accounts for $\approx 10\%$ of NH_3 output, and the remaining 5% goes into miscellaneous uses.

Phosphorus-containing fertilizers are highlighted in Box 14.9.

catalyst and a contact time of $\approx 1\,\text{ms}$, the less exothermic reaction 14.20 takes place. This reaction forms part of the manufacturing process for HNO_3 (see *Section 14.9*).

(14.11)

$$4NH_3 + 3O_2 \longrightarrow 2N_2 + 6H_2O \qquad (14.19)$$

$$4NH_3 + 5O_2 \xrightarrow{\text{Pt/Rh}} 4NO + 6H_2O \qquad (14.20)$$

The solubility of NH_3 in water is greater than that of any other gas, doubtless because of hydrogen bond formation between NH_3 and H_2O. The equilibrium constant (at 298 K) for reaction 14.21 shows that nearly all the dissolved NH_3 is *non-ionized*, consistent with the fact that even dilute solutions retain the characteristic smell of NH_3. Since $K_w = 10^{-14}$, it follows that the aqueous solutions of $[NH_4]^+$ salts of strong acids (e.g. NH_4Cl) are slightly acidic (equation 14.22). (See *worked example 6.2* for the relationship between pK_a and pK_b.)

$$NH_3(aq) + H_2O(l) \rightleftharpoons [NH_4]^+(aq) + [OH]^-(aq)$$
$$K_b = 1.8 \times 10^{-5} \qquad (14.21)$$

$$[NH_4]^+(aq) + H_2O(l) \rightleftharpoons [H_3O]^+(aq) + NH_3(aq)$$
$$K_a = 5.6 \times 10^{-10} \qquad (14.22)$$

Ammonium salts are easily prepared by neutralization reactions, e.g. equation 14.23. Industrial syntheses are carried out using the Solvay process (*Figure 10.5*), or reactions 14.24 and 14.25; both ammonium sulfate and nitrate are important fertilizers, and NH_4NO_3 is a component of some explosives (see *equation 14.6*).

$$NH_3 + HBr \longrightarrow NH_4Br \qquad (14.23)$$

$$CaSO_4 + 2NH_3 + CO_2 + H_2O \longrightarrow CaCO_3 + [NH_4]_2[SO_4] \qquad (14.24)$$

$$NH_3 + HNO_3 \longrightarrow NH_4NO_3 \qquad (14.25)$$

Detonation of NH_4NO_3 may be initiated by another explosion, and ammonium perchlorate is similarly metastable with respect to oxidation of the $[NH_4]^+$ cation by the anion; NH_4ClO_4 is used in solid rocket propellants, e.g. in the booster rockets of the space shuttle. 'Technical ammonium carbonate' (used in smelling salts) is actually a mixture of $[NH_4][HCO_3]$ and $[NH_4][NH_2CO_2]$ (ammonium carbamate); the latter is prepared by passing NH_3, CO_2 and steam into a lead chamber, and smells strongly of NH_3 because carbamic acid (**14.12**) is an extremely weak acid.

(14.12)

Ammonium salts often crystallize with lattices similar to those of the corresponding K^+, Rb^+ or Cs^+ salts. The $[NH_4]^+$ ion can be approximated to a sphere (see *Figure 5.17*) with $r_{ion} = 150\,\text{pm}$, a value similar to that of Cs^+. However, if, in the solid state, there is potential for hydrogen bonding involving the $[NH_4]^+$ ions, ammonium salts adopt structures unlike those of their alkali metal analogues, e.g. NH_4F possesses a wurtzite rather than an NaCl lattice. The majority of $[NH_4]^+$ salts are soluble in water, one exception being $[NH_4]_2[PtCl_6]$; solubilities of alkali metal $[PtCl_6]^{2-}$ salts were discussed in *Section 6.9*.

Phosphine (Table 14.4) is an extremely toxic, colourless gas which is much less soluble in water than is NH_3. The $P-H$ bond is not polar enough to form hydrogen bonds with H_2O. In contrast to NH_3, aqueous solutions of PH_3 are neutral, but in liquid NH_3, PH_3 acts as an acid (e.g. equation 14.26).

$$K + PH_3 \xrightarrow{\text{liquid } NH_3} K^+ + [PH_2]^- + \tfrac{1}{2}H_2 \qquad (14.26)$$

Phosphonium halides, PH_4X, are formed by treating PH_3 with HX but only the iodide is stable under ambient conditions; the chloride is unstable above 243 K and the bromide decomposes at 273 K. The $[PH_4]^+$ ion is decomposed by water (equation 14.27). Phosphine acts as a Lewis base and a range of adducts (including those with low oxidation state d-block metal centres) are known. Combustion of PH_3 yields H_3PO_4.

$$[PH_4]^+ + H_2O \longrightarrow PH_3 + [H_3O]^+ \qquad (14.27)$$

The hydrides AsH_3 and SbH_3 resemble those of PH_3 (Table 14.4), but they are less stable with respect to decomposition into their elements. Both AsH_3 and SbH_3 are extremely toxic gases, and SbH_3 is liable to explode. They are less basic than PH_3, but can be protonated with HF in the presence of AsF_5 or SbF_5 (equation 14.28). The salts $[AsH_4][AsF_6]$, $[AsH_4][SbF_6]$ and $[SbH_4][SbF_6]$ form air- and moisture-sensitive crystals which decompose well below 298 K.

$$AsH_3 + HF + AsF_5 \longrightarrow [AsH_4]^+ + [AsF_6]^- \qquad (14.28)$$

Hydrides E_2H_4 (E = N, P, As)

Hydrazine, N_2H_4, is a colourless liquid (mp 275 K, bp 386 K), miscible with water and with a range of organic solvents, and is corrosive and toxic; its vapour forms explosive mixtures with air. Although $\Delta_fH^\circ(N_2H_4, 298\,\text{K}) = +50.6\,\text{kJ\,mol}^{-1}$, N_2H_4 at ambient temperatures is *kinetically* stable with respect to N_2 and H_2. Alkyl derivatives of hydrazine (see *equation 14.35*) have been used as rocket fuels, e.g. combined with N_2O_4 in the Apollo missions. N_2H_4 has uses in the agricultural and plastics industries, and in the removal of O_2 from industrial water boilers to minimize corrosion (the reaction gives N_2 and H_2O). Hydrazine is obtained by the Raschig reaction (the basis for the industrial synthesis) which involves the partial

BOX 14.3

Thermal decompositions of arsine and stibine: the Marsh test

The thermal decomposition of AsH_3 and SbH_3 is the basis for the Marsh test, which provides a classic example of an analytical technique important in forensic science. The arsenic or antimony-containing material is first converted to AsH_3 or SbH_3 (e.g. by treatment with Zn and acid, liberating H_2 and the trihydride). Subsequent passage of the gaseous mixture through a heated tube causes the hydrides to decompose, forming brown-black deposits of the elements:

$$2EH_3(g) \xrightarrow{\Delta} 2E(s) + 3H_2(g) \qquad E = As \text{ or } Sb$$

If both As and Sb are present, the relative positions of the deposits establishes their identity; the lower thermal stability of SbH_3 means that it decomposes before AsH_3 in the tube. Treatment of the residues with aqueous NaOCl also distinguishes them since only arsenic dissolves.

oxidation of NH_3 (equation 14.29). Glue or gelatine is added to inhibit side-reaction 14.30 which otherwise consumes the N_2H_4 as it is formed; the additive removes traces of metal ions that catalyze reaction 14.30.

$$NH_3 + NaOCl \longrightarrow NH_2Cl + NaOH \qquad \textit{fast}$$

$$NH_3 + NH_2Cl + NaOH \longrightarrow N_2H_4 + NaCl + H_2O \quad \textit{slow}$$

$$(14.29)$$

$$2NH_2Cl + N_2H_4 \longrightarrow N_2 + 2NH_4Cl \qquad (14.30)$$

Hydrazine is obtained from the Raschig process as the monohydrate and is used in this form for many purposes. Dehydration is difficult, and direct methods to produce anhydrous N_2H_4 include reaction 14.31.

$$2NH_3 + [N_2H_5][HSO_4] \longrightarrow N_2H_4 + [NH_4]_2[SO_4] \qquad (14.31)$$

In aqueous solution, N_2H_4 usually forms $[N_2H_5]^+$ (hydrazinium) salts, but some salts of $[N_2H_6]^{2+}$ have been isolated, e.g. $[N_2H_6][SO_4]$. The pK_b values for hydrazine are given in equations 14.32 and 14.33, and the first step shows N_2H_4 to be a weaker base than NH_3 (equation 14.21).

$$N_2H_4(aq) + H_2O \rightleftharpoons [N_2H_5]^+ + [OH]^- \quad K_b(1) = 8.9 \times 10^{-7}$$

$$(14.32)$$

$$[N_2H_5]^+(aq) + H_2O \rightleftharpoons [N_2H_6]^{2+} + [OH]^- \quad K_b(2) \approx 10^{-14}$$

$$(14.33)$$

Both N_2H_4 and $[N_2H_5]^+$ are reducing agents, and reaction 14.34 is used for the determination of hydrazine.

$$N_2H_4 + KIO_3 + 2HCl \longrightarrow N_2 + KCl + ICl + 3H_2O \qquad (14.34)$$

Figure 14.3a shows the structure of N_2H_4, and the *gauche* conformation (Figure 14.3a and 14.3b) is also adopted by P_2H_4 in the gas phase. In the solid state, P_2H_4 has a staggered conformation (Figure 14.3c) while the related N_2F_4 exhibits both conformers. The eclipsed conformation (which would maximize lone pair–lone pair repulsions) is not observed.

Diphosphane, P_2H_4, is a colourless liquid (mp 174 K, bp 329 K), and is toxic and spontaneously inflammable; when

heated, it forms higher phosphanes. Diphosphane is formed as a minor product in several reactions in which PH_3 is prepared (e.g. reaction 14.9) and may be separated from PH_3 by condensation in a freezing mixture. It exhibits no basic properties.

Chloramine and hydroxylamine

$$(14.13)$$

The reactions of NH_3 and Cl_2 (diluted with N_2) or aqueous NaOCl (the first step in reaction 14.29) yield chloramine, **14.13**, the compound responsible for the odour of water containing nitrogenous matter that has been sterilized with Cl_2. Chloramine is unstable, and violently explosive, and is usually handled in dilute solutions (e.g. in H_2O or Et_2O).

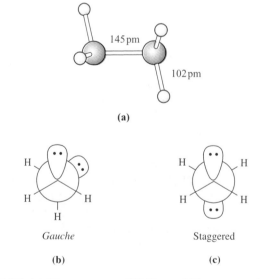

Fig. 14.3 (a) The structure of N_2H_4, and Newman projections showing (b) the observed *gauche* conformation, and (c) the possible staggered conformation. An eclipsed conformation is also possible.

$$[NO_3]^- \xrightarrow{+0.79} N_2O_4 \xrightarrow{+1.07} HNO_2 \xrightarrow{+1.00} NO \xrightarrow{+1.59} N_2O \xrightarrow{+1.77} N_2 \xrightarrow{-1.87} [NH_3OH]^+ \xrightarrow{+1.41} [N_2H_5]^+ \xrightarrow{+1.28} NH_3$$

(with -0.05 spanning from N_2O to $[NH_3OH]^+$, and $+0.27$ spanning from N_2 to NH_3)

Fig. 14.4. Potential diagram for nitrogen at pH $= 0$.

Its reaction with Me_2NH (equation 14.35) yields the rocket fuel 1,1-dimethylhydrazine.

$$NH_2Cl + 2Me_2NH \longrightarrow Me_2NNH_2 + [Me_2NH_2]Cl \quad (14.35)$$

Reaction 14.36 is one of several routes to hydroxylamine, NH_2OH, which is usually handled as a salt (e.g. the sulfate) or in aqueous solution. The free base can be obtained from its salts by treatment with NaOMe in MeOH.

$$2NO + 3H_2 + H_2SO_4$$
$$\xrightarrow{\text{platinized charcoal catalyst}} [NH_3OH]_2[SO_4] \quad (14.36)$$

Pure NH_2OH forms white, hygroscopic crystals (see *Section 11.5*), which melt at 306 K and explode at higher temperatures. It is a weaker base than NH_3 or N_2H_4. Many of its reactions arise from the great variety of redox reactions in which it takes part in aqueous solution, e.g. it reduces Fe(III) in acidic solution (equation 14.37) but oxidizes Fe(II) in the presence of alkali (equation 14.38).

$$2NH_2OH + 4Fe^{3+} \longrightarrow N_2O + 4Fe^{2+} + H_2O + 4H^+ \quad (14.37)$$

$$NH_2OH + 2Fe(OH)_2 + H_2O \longrightarrow NH_3 + 2Fe(OH)_3 \quad (14.38)$$

More powerful oxidizing agents (e.g. $[BrO_3]^-$) oxidize NH_2OH to HNO_3. The formation of N_2O in most oxidations of NH_2OH is an interesting example of the triumph of kinetic over thermodynamic factors. Consideration of the potential diagram (see *Section 7.5*) in Figure 14.4 shows that, on thermodynamic grounds, the expected product from the action of weak oxidizing agents on $[NH_3OH]^+$ (i.e. NH_2OH in acidic solution) would be N_2, but it seems that the reaction occurs by steps 14.39. A use of NH_2OH is as an antioxidant in photographic developers.

$$\left.\begin{array}{l} NH_2OH \longrightarrow NOH + 2H^+ + 2e^- \\ 2NOH \longrightarrow HON{=}NOH \\ HON{=}NOH \longrightarrow N_2O + H_2O \end{array}\right\} \quad (14.39)$$

Figure 14.4 also shows that, at pH $= 0$, NH_2OH is unstable with respect to disproportionation into N_2 and NH_3 or N_2H_4; in fact, hydroxylamine does slowly decompose to N_2 and NH_3.

Hydrogen azide and azide salts

Sodium azide, NaN_3, is obtained from molten sodium amide by reaction 14.40 (or by reacting $NaNH_2$ with $NaNO_3$ at

450 K), and treatment of NaN_3 with H_2SO_4 yields hydrogen azide, HN_3.

$$2NaNH_2 + N_2O \xrightarrow{460\,K} NaN_3 + NaOH + NH_3 \quad (14.40)$$

Hydrogen azide is a colourless liquid (mp 193 K, bp 309 K); it is dangerously explosive ($\Delta_f H^\circ(l, 298\,K) = +264\,kJ\,mol^{-1}$) and highly poisonous. Aqueous solutions of HN_3 are weakly acidic (equation 14.41).

$$HN_3 + H_2O \rightleftharpoons [H_3O]^+ + [N_3]^- \qquad pK_a = 4.75 \quad (14.41)$$

The structure of HN_3 is shown in Figure 14.5a, and a consideration of the resonance structures in Figure 14.5b provides an explanation for the asymmetry of the NNN-unit. The azide ion is isoelectronic with CO_2, and the symmetrical structure of $[N_3]^-$ (Figure 14.5c) is consistent with the bonding description in Figure 14.5d. A range of azide salts is known; Ag(I), Cu(II) and Pb(II) azides, which are insoluble in water, are explosive, and $Pb(N_3)_2$ is used as an initiator for less sensitive explosives. On the other hand, group 1 metal

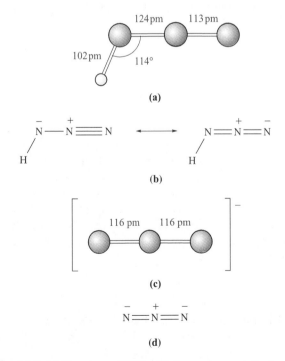

Fig. 14.5 (a) Structure of HN_3, (b) the major contributing resonance forms of HN_3, (c) the structure of the azide ion (the ion is symmetrical but bond distances vary slightly in different salts), and (d) the principal resonance structure of $[N_3]^-$. Colour code: N, blue; H, pale grey.

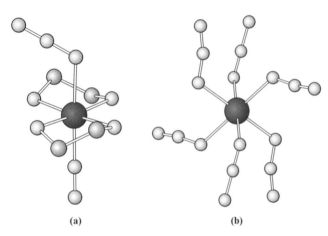

(a) **(b)**

Fig. 14.6 The structures (X-ray diffraction) of (a) *trans*-[Ru(en)$_2$(N$_2$)(N$_3$)]$^+$ in the [PF$_6$]$^-$ salt (H atoms omitted) [B.R. Davis *et al.* (1970) *Inorg. Chem.*, vol. 9, p. 2768] and (b) [Sn(N$_3$)$_6$]$^{2-}$ structurally characterized as the [Ph$_4$P]$^+$ salt [D. Fenske *et al.* (1983) *Z. Naturforsch., Teil B*, vol. 38, p. 1301]. Colour code: N, blue; Ru and Sn, black; C, grey.

azides decompose quietly when heated (equations 10.2 and 14.4). The azide group, like CN$^•$ (though to a lesser extent) shows similarities to a halogen and is another example of a pseudo-halogen (see *Section 13.12*); however, no N$_6$ molecule (i.e. a dimer of N$_3^•$ and so an analogue of an X$_2$ halogen) has yet been prepared. Like halide ions, the azide ion acts as a ligand in a wide variety of metal complexes, e.g. [Au(N$_3$)$_4$]$^-$, *trans*-[TiCl$_4$(N$_3$)$_2$]$^{2-}$, *cis*-[Co(en)$_2$(N$_3$)$_2$]$^+$, *trans*-[Ru(en)$_2$(N$_2$)(N$_3$)]$^+$ (which is also an example of a dinitrogen complex, Figure 14.6a) and [Sn(N$_3$)$_6$]$^{2-}$ (Figure 14.6b).

$$\left[N \!\equiv\! N \!-\! N \!=\! N \!=\! N \right]^+$$

(14.14)

The reaction of HN$_3$ with [N$_2$F][AsF$_6$] (prepared by reaction 14.52) in HF at 195 K results in the elimination of HF and formation of [N$_5$][AsF$_6$]. Designing the synthesis of [N$_5$]$^+$ (**14.14**) was not trivial. Precursors in which the N≡N and N=N bonds are preformed are critical, but should not involve gaseous N$_2$ since this is too inert. The HF solvent provides a heat sink for the exothermic reaction, the product being potentially explosive.

14.6 Nitrides, phosphides, arsenides, antimonides and bismuthides

Nitrides

Classifying nitrides is not simple, but nearly all nitrides fall into one of the following groups, although, as we have

seen for the borides and carbides, some care is needed in attempting to generalize:

- saline nitrides of the group 1 and 2 metals, and aluminium;
- covalently bonded nitrides of the non-metallic elements (see *Sections 12.8*, *13.12* and *15.10* for BN, C$_2$N$_2$, Si$_3$N$_4$ and S$_4$N$_4$);
- interstitial nitrides of *d*-block metals.

The classification of 'saline nitride' implies the presence of the N^{3-} ion, and as we discussed in Section 14.1, this is unlikely. However, it is usual to consider Li$_3$N, Be$_3$N$_2$, Mg$_3$N$_2$, Ca$_3$N$_2$, Sr$_3$N$_2$, Ba$_3$N$_2$ and AlN in terms of ionic formulations. Hydrolysis of saline nitrides liberates NH$_3$.

Nitrides of the *d*-block metals are hard, inert solids which resemble metals in appearance, and have high melting points and electrical conductivities (see *Box 14.4*). They can be prepared from the metal or metal hydride with N$_2$ or NH$_3$ at high temperatures Most possess structures in which the nitrogen atoms occupy octahedral holes in a close-packed metal lattice. Full occupancy of these holes leads to the stoichiometry MN (e.g. TiN, ZrN, HfN, VN, NbN); cubic close-packing of the metal atoms and a NaCl lattice for the nitride MN is favoured for metals in the earliest groups of the *d*-block.

Phosphides

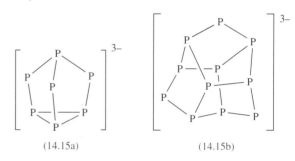

(14.15a) (14.15b)

Most elements combine with phosphorus to give binary phosphides; exceptions include Hg, Pb, Sb, Bi and Te. Types of solid state phosphides are very varied,† and simple classification is not possible. Phosphides of the *d*-block metals tend to be inert, metallic-looking compounds with high melting points and electrical conductivities. Their formulae are often deceptive in terms of the oxidation state of the metal and their structures may contain isolated P centres, P$_2$ groups, or rings, chains or layers of P atoms. The group 1 and 2 metals form compounds M$_3$P and M$_3$P$_2$ respectively which are hydrolyzed by water and can be considered to be ionic. The alkali metals also form phosphides which contain groups of P atoms forming chains or cages, the cages being either [P$_7$]$^{3-}$ (**14.15a**) or [P$_{11}$]$^{3-}$ (**14.15b**); e.g. LiP contains infinite helical chains, K$_4$P$_3$

---

† A detailed account is beyond the scope of this book; an excellent review is included in the further reading at the end of the chapter.

Industrial applications of metal nitrides

Nitrides of the *d*-block metals are hard, are resistant to wear and chemical attack including oxidation, and have very high melting points. These properties render nitrides such as TiN, ZrN and HfN invaluable for protecting high speed cutting tools. The applied coatings are extremely thin (typically $\leq 10 \, \mu m$), but nonetheless significantly prolong the lifetimes of tools that operate under the toughest of work conditions. Nitride coatings can be applied using the technique of chemical vapour deposition (see *Section 27.6*), or by forming

a surface layer of Fe_3N or Fe_4N by reacting the prefabricated steel tool with N_2.

Layers of TiN, ZrN, HfN or TaN are applied as diffusion barriers in semiconducting devices. The barrier layer ($\approx 100 \, nm$ thick) is fabricated between the semiconducting material (e.g. GaAs or Si) and the protective metallic (e.g. Au or Ni) coating, and prevents diffusion of metal atoms into the GaAs or Si device.

For related information: see the discussions of boron nitride, silicon nitride and ceramic coatings in *Section 27.6*.

---

contains $[P_3]^{4-}$ chains, Rb_4P_6 has planar $[P_6]^{4-}$ rings, Cs_3P_7 contains $[P_7]^{3-}$ cages, and Na_3P_{11} features $[P_{11}]^{3-}$ cages. The latter examples are phosphorus-rich species. Some other members of this class such as Ba_3P_{14} and Sr_3P_{14} contain $[P_7]^{3-}$ cages, while phosphides such as BaP_{10}, CuP_7, Ag_3P_{11}, MP_4 (e.g. M = Mn, Tc, Re, Fe, Ru, Os) and TlP_5 contain more extended arrays of P atoms, and two further examples (**14.9** and **14.10**) have already been mentioned.

Arsenides, antimonides and bismuthides

Metal arsenides, antimonides and bismuthides can be prepared by direct combination of the metal and group 15 element. Like the phosphides, classification is not simple, and structure types vary. We shall focus on two specific examples: GaAs and NiAs.

Gallium arsenide is an important semiconductor and crystallizes with a zinc blende lattice (see *Figure 5.18b*).

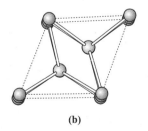

(a)	(b)

Fig. 14.7 Two views of the unit cell of the nickel arsenide (NiAs) lattice; As centres are shown in blue and Ni centres in grey. View (a) emphasizes the trigonal prismatic coordination environment of the As centres, while (b) (which views (a) from above) illustrates more clearly that the unit cell is not a cuboid.

Slow hydrolysis occurs in moist air and protection of semiconductor devices from the air is essential; N_2 is often used as a 'blanket gas'.

Nickel arsenide, NiAs, gives its name to a well-known structure type, being adopted by a number of *d*-block metal arsenides, antimonides, sulfides, selenides and tellurides. The lattice can be described as a hexagonal close-packed (hcp) array of As atoms with Ni atoms occupying octahedral holes. Although such a description might conjure up the concept of an *ionic* lattice, the bonding in NiAs is certainly *not* purely ionic. Figure 14.7 shows a unit cell of NiAs. The placement of the Ni atoms in *octahedral* holes in the hcp arrangement of As atoms means that the coordination environment of the As centres is *trigonal prismatic*. Although each Ni atom has six As neighbours at 243 pm, there are two Ni neighbours at a distance of only 252 pm (compare $r_{metal}(Ni) = 125 \, pm$) and there is almost certainly Ni–Ni bonding running through the structure. This is consistent with the observation that NiAs conducts electricity.

14.7 Halides, oxohalides and complex halides

Nitrogen halides

Since nitrogen is restricted to an octet of valence electrons, it does not form pentahalides. Thus, important nitrogen halides are NX_3 (X = F, Cl), N_2F_4 and N_2F_2, selected properties for which are listed in Table 14.5; NBr_3 and NI_3 exist but are less well characterized than NF_3 and NCl_3.

Nitrogen trifluoride is made either by reaction 14.42 which must be carried out in a controlled manner, or by electrolysis of anhydrous NH_4F/HF mixtures.

$$4NH_3 + 3F_2 \xrightarrow{\text{Cu catalyst}} NF_3 + 3NH_4F \qquad (14.42)$$

NF_3 is the most stable of the trihalides of nitrogen, being the only one to have a negative value of $\Delta_f H^\circ$ (Table 14.5). It is a colourless gas which is resistant to attack by acids and

Table 14.5 Selected data for nitrogen fluorides and trichloride.

	NF$_3$	NCl$_3$	N$_2$F$_4$	*cis*-N$_2$F$_2$	*trans*-N$_2$F$_2$
Melting point / K	66	<233	108.5	<78	101
Boiling point / K	144	<344; explodes at 368	199	167	162
$\Delta_f H^\circ$(298 K) / kJ mol^{-1}	−132.1	230.0	−8.4	69.5	82.0
Dipole moment / D	0.24	0.39	0.26‡	0.16	0
N−N bond distance / pm	–	–	149	121	122
N−X bond distance / pm	137	176	137	141	140
Bond angles / deg	∠F−N−F 102.5	∠Cl−N−Cl 107	∠F−N−F 103 ∠N−N−F 101	∠N−N−F 114	∠N−N−F 106

‡ *Gauche* conformation.

alkalis, but is decomposed by sparking with H$_2$ (equation 14.43). The resistance towards hydrolysis parallels that observed for the carbon tetrahalides (*Section 13.8*).

$$2NF_3 + 3H_2 \longrightarrow N_2 + 6HF \qquad (14.43)$$

(14.16)

The gas phase structure of NF$_3$ is trigonal pyramidal (**14.16**), and the molecular dipole moment is very small (Table 14.5). In contrast to NH$_3$ and PF$_3$, NF$_3$ shows no donor properties.

Nitrogen trichloride is an oily, yellow liquid at 289 K, but it is highly endothermic and dangerously explosive (Table 14.5). The difference in stabilities of NF$_3$ and NCl$_3$ lies in the relative bond strengths of N−F over N−Cl, and of Cl$_2$ over F$_2$. Nitrogen trichloride can be prepared by reaction 14.44, with the equilibrium being drawn to the right-hand side by extracting NCl$_3$ into a suitable organic solvent. Diluted with air, NCl$_3$ is used for bleaching flour since hydrolysis by moisture forms HOCl (see *Section 16.9*). Alkalis hydrolyze NCl$_3$ according to equation 14.45.

$$NH_4Cl + 3Cl_2 \rightleftharpoons NCl_3 + 4HCl \qquad (14.44)$$

$$2NCl_3 + 6[OH]^- \longrightarrow N_2 + 3[OCl]^- + 3Cl^- + 3H_2O \qquad (14.45)$$

Nitrogen tribromide is more reactive than NCl$_3$, and explodes at temperatures as low as 175 K. It can be prepared by reaction 14.46, attempts to make it by treating NCl$_3$ with Br$_2$ being unsuccessful.

$$(Me_3Si)_2NBr + 2BrCl \xrightarrow{\text{in pentane, 186 K}} NBr_3 + 2Me_3SiCl \qquad (14.46)$$

Although pure NI$_3$ has not been isolated, an explosive compound NH$_3$·NI$_3$ is formed when aqueous [I$_3$]$^-$ is treated with concentrated aqueous NH$_3$.

The nitrogen fluorides N$_2$F$_4$ and N$_2$F$_2$ can be obtained from reactions 14.47 and 14.48; properties of these fluorides are listed in Table 14.5, and both fluorides are explosive.

$$2NF_3 \xrightarrow{\text{Cu, 670 K}} N_2F_4 + CuF_2 \qquad (14.47)$$

$$2N_2F_4 + 2AlCl_3 \xrightarrow{203 \text{ K}} \textit{trans}\text{-}N_2F_2 + 3Cl_2 + N_2 + 2AlF_3$$

$$\downarrow 373 \text{ K}$$

$$\textit{cis}\text{-}N_2F_2 \qquad (14.48)$$

The structure of N$_2$F$_4$ resembles that of hydrazine, except that both the *gauche* and *trans* (staggered) conformers (Figure 14.3) are present in the liquid and gas phases. At temperatures above 298 K, N$_2$F$_4$ reversibly dissociates into blue NF$_2$· radicals which undergo many interesting reactions (e.g. equations 14.49–14.51).

$$2NF_2 + S_2F_{10} \longrightarrow 2F_2NSF_5 \qquad (14.49)$$

$$2NF_2 + Cl_2 \longrightarrow 2NClF_2 \qquad (14.50)$$

$$NF_2 + NO \longrightarrow F_2NNO \qquad (14.51)$$

(14.17)　　　(14.18)

Dinitrogen difluoride, N$_2$F$_2$, exists in both the *trans*- and *cis*-forms (**14.17** and **14.18**), with the *cis*-isomer being thermodynamically the more stable of the two but also the more reactive. Reaction 14.48 gives a selective method of preparing *trans*-N$_2$F$_2$; isomerization by heating gives a mixture of isomers from which *cis*-N$_2$F$_2$ can be isolated by treatment with AsF$_5$ (reaction 14.52).

Mixture of isomers:

$$\begin{cases} \textit{cis}\text{-}N_2F_2 \xrightarrow{AsF_5} [N_2F]^+[AsF_6]^- \xrightarrow{NaF/HF} \textit{cis}\text{-}N_2F_2 \\ \textit{trans}\text{-}N_2F_2 \xrightarrow{AsF_5} \text{No reaction} \end{cases}$$

$$(14.52)$$

Reaction 14.52 illustrates the ability of N$_2$F$_2$ to donate F$^-$ to *strong* acceptors such as AsF$_5$ and SbF$_5$, a reaction type shared by N$_2$F$_4$ (equations 14.53 and 14.54). The cation

$[NF_4]^+$ is formed in reaction 14.55. We return to the properties of AsF_5 and SbF_5 later.

$$N_2F_4 + AsF_5 \longrightarrow [N_2F_3]^+[AsF_6]^- \tag{14.53}$$

$$N_2F_4 + 2SbF_5 \longrightarrow [N_2F_3]^+[Sb_2F_{11}]^- \tag{14.54}$$

$$NF_3 + F_2 + SbF_5 \longrightarrow [NF_4]^+[SbF_6]^- \tag{14.55}$$

Oxofluorides and oxochlorides of nitrogen

(14.19)

Several oxofluorides and oxochlorides of nitrogen are known, but all are unstable gases or volatile liquids which are rapidly hydrolyzed. Nitrosyl halides FNO, ClNO and BrNO are formed in reactions of NO with F_2, Cl_2 and Br_2 respectively; structural details are shown in **14.19**. Nitryl fluoride, FNO_2, and nitryl chloride, $ClNO_2$, are prepared, respectively, by fluorination of N_2O_4 (reaction 14.56) and oxidation of ClNO (using e.g. Cl_2O or O_3). Both are planar molecules; FNO_2 is isoelectronic with $[NO_3]^-$.

$$N_2O_4 + 2CoF_3 \xrightarrow{570\,K} 2FNO_2 + 2CoF_2 \tag{14.56}$$

The oxohalides FNO, ClNO, FNO_2 and $ClNO_2$ combine with suitable fluorides or chlorides to give salts containing $[NO]^+$ or $[NO_2]^+$, e.g. reactions 14.57–14.59. The complex fluorides may also be conveniently prepared in liquid BrF_3 (see *Section 8.10*).

$$\underset{\substack{\text{halide}\\\text{acceptor}}}{FNO + AsF_5} \longrightarrow [NO]^+[AsF_6]^- \tag{14.57}$$

$$ClNO + SbCl_5 \longrightarrow [NO]^+[SbCl_6]^- \tag{14.58}$$

$$FNO_2 + BF_3 \longrightarrow [NO_2]^+[BF_4]^- \tag{14.59}$$

The main factor involved in the change from covalent to ionic halide is believed to be the enthalpy change accompanying the attachment of the halide ion to the halide acceptor.

Fluorination of FNO using iridium(VI) fluoride leads to the unstable F_3NO. Resonance structures **14.20** and **14.21** can be written to depict the bonding, and the short N–O bond (116 pm) and long N–F bonds (143 pm) suggest that contributions from **14.21** (and similar structures) are important.

(14.20)   (14.21)

Phosphorus halides

Phosphorus forms the halides PX_3 (X = F, Cl, Br and I) and PX_5 (X = F, Cl and Br); PI_5 is unknown. Most are made by direct combination of the elements with the product determined by which element is in excess; PF_3, however, must be made by reaction 14.60 and a convenient synthesis of PF_5 is from KPF_6 (see below). The halides are all hydrolyzed by water (e.g. equation 14.61), although PF_3 reacts only slowly.

$$PCl_3 + AsF_3 \longrightarrow PF_3 + AsCl_3 \tag{14.60}$$

$$PCl_3 + 3H_2O \longrightarrow H_3PO_3 + 3HCl \tag{14.61}$$

(14.22)

Each of the trihalides has a trigonal pyramidal structure, **14.22**. Phosphorus trifluoride is a very poisonous, colourless and odourless gas. It has the ability (like CO, see *Section 23.2*) to form complexes with metals and Lewis acids such as BH_3, and its toxicity arises from complex formation with haemoglobin. The formation of $[PF_4]^-$ had not been observed until recently; PF_3 reacts with Me_4NF in MeCN to give $[Me_4N][PF_4]$. Phosphorus trichloride is a colourless liquid (mp 179.5 K, bp 349 K) which fumes in moist air (equation 14.61) and is toxic. Its reactions include those in equations 14.62; $POCl_3$ is an important reagent for the preparation of phosphate esters.

$$PCl_3 \begin{cases} \xrightarrow{O_2} POCl_3 \\ \xrightarrow{X_2\ (X=halogen)} PCl_3X_2 \\ \xrightarrow{NH_3} P(NH_2)_3 \end{cases} \tag{14.62}$$

(14.23)

Single crystal X-ray diffraction (at 109 K) data show that PF_5 has a trigonal bipyramidal structure, **14.23**. In solution, the molecule is fluxional on the NMR spectroscopic timescale and a doublet is observed in the ^{19}F NMR spectrum, i.e. all ^{19}F environments are equivalent and couple with the ^{31}P nucleus. This stereochemical non-rigidity is an example of Berry pseudo-rotation (see *Figure 2.13*). Electron diffraction data show that in the gas phase, PCl_5 has a molecular, trigonal bipyramidal structure (P–Cl_{ax} = 214, P–Cl_{eq} = 202 pm), provided that thermal dissociation into PCl_3 and Cl_2 is prevented by the presence of an excess of Cl_2. In the solid state, however, tetrahedral $[PCl_4]^+$ (P–Cl = 197 pm) and octahedral $[PCl_6]^-$ (P–Cl = 208 pm) ions are present, and the compound crystallizes with a CsCl lattice (*Figure 5.16*). In contrast, PBr_5 (which dissociates in the gas phase to PBr_3 and Br_2) crystallizes in the form of $[PBr_4]^+Br^-$. The mixed halide PF_3Cl_2 is of particular interest; it is obtained from the reaction of PF_3 and Cl_2 as a gas (bp 280 K) which has a molecular structure with equatorial Cl atoms, but when PCl_5 reacts with AsF_3 in $AsCl_3$ solution, the solid product $[PCl_4]^+[PF_6]^-$ (mp ≈ 403 K) is isolated. Solid PI_5 has not been isolated,[†] but the isolation of the salt $[PI_4][AsF_6]$ by the reaction of PI_3 and $[I_3][AsF_6]$ has confirmed the existence of the tetrahedral $[PI_4]^+$ ion.

$$\left[\begin{array}{c} F \\ F_{\prime\prime\prime\prime\prime} \overset{|}{\underset{|}{P}} {}^{\prime\prime\prime\prime\prime} F \\ F \quad \quad F \\ F \end{array} \right]^{-}$$

(14.24)

Phosphorus pentafluoride is a strong Lewis acid and forms stable complexes with amines and ethers. The hexafluorophosphate ion, $[PF_6]^-$, **14.24**, is made in aqueous solution by reacting H_3PO_4 with concentrated HF; $[PF_6]^-$ is isoelectronic and isostructural with $[SiF_6]^{2-}$ (see *Figure 13.11a*). Salts such as $[NH_4][PF_6]$ are commercially available, and $[PF_6]^-$ is used to precipitate salts containing large organic or complex cations. Solid KPF_6 (prepared as in Figure 14.8) decomposes on heating to give PF_5 and this route is a useful means of preparing PF_5. Phosphorus pentachloride is an important reagent, and is made industrially by the reaction of PCl_3 and Cl_2; selected reactions are given in Figure 14.8.

Of the lower halides P_2X_4, the most important is the red, crystalline P_2I_4 (mp 398 K) which can be made by reacting white phosphorus with I_2 in CS_2. In the solid state, molecules of P_2I_4 adopt a *trans* (staggered) conformation (see *Figure 14.3*). In many of its reactions, P_2I_4 undergoes P–P bond fission, e.g. dropping H_2O on P_2I_4 in an inert atmosphere produces $[PH_4]I$.

Phosphoryl trichloride, $POCl_3$

$$\overset{O}{\underset{\substack{115° \\ Cl \\ 199 \text{ pm}}}{\overset{\parallel 145 \text{ pm}}{\underset{Cl}{\overset{}{P}}}} {}^{\prime\prime\prime\prime\prime} \overset{Cl}{\underset{103°}{}}$$

(14.25)

Of the phosphorus oxohalides, the most important is $POCl_3$, prepared by reaction of PCl_3 with O_2. Phosphoryl trichloride

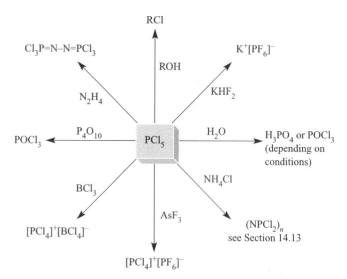

Fig. 14.8 Selected reactions of PCl_5.

is a colourless, fuming liquid (mp 275 K, bp 378 K), which is readily hydrolyzed by water liberating HCl. The vapour contains discrete molecules (**14.25**). Some of the many uses of $POCl_3$ are as a phosphorylating and chlorinating agent, and as a reagent in the preparation of phosphate esters.

Arsenic and antimony halides

Arsenic forms the halides AsX_3 (X = F, Cl, Br, I) and AsX_5 (X = F, Cl). The trihalides $AsCl_3$, $AsBr_3$ and AsI_3 can be made by direct combination of the elements, and reaction 14.63 is another route to $AsCl_3$. Reaction 14.64 is used to prepare AsF_3 (mp 267 K, bp 330 K) despite the fact that AsF_3 (like the other trihalides) is hydrolyzed by water; the H_2O formed in the reaction is removed with excess H_2SO_4. This reaction should be compared with reactions 12.24 and 13.41. Glass containers are not practical for AsF_3 as it reacts with silica in the presence of moisture.

$$As_2O_3 + 6HCl \longrightarrow 2AsCl_3 + 3H_2O \quad (14.63)$$

$$As_2O_3 + 3H_2SO_4 + 3CaF_2$$
$$\longrightarrow 3AsF_3 + 3CaSO_4 + 3H_2O \quad (14.64)$$

In the solid, liquid and gas states, AsF_3 and $AsCl_3$ have molecular, trigonal pyramidal structures. With an appropriate reagent, AsF_3 may act as either an F^- donor or acceptor (equations 14.65 and 14.66); compare this with the behaviours of BrF_3 (*Section 8.10*) and $AsCl_3$ (equation 14.67) which finds some use as a non-aqueous solvent.

$$AsF_3 + KF \longrightarrow K^+[AsF_4]^- \quad (14.65)$$

$$AsF_3 + SbF_5 \longrightarrow [AsF_2]^+[SbF_6]^- \quad (14.66)$$

$$2AsCl_3 \rightleftharpoons [AsCl_2]^+ + [AsCl_4]^- \quad (14.67)$$

The only stable pentahalide of arsenic is AsF_5 (prepared by reaction 14.68), although $AsCl_5$ can be made at 173 K by treating $AsCl_3$ with Cl_2 under UV radiation. At 298 K,

---

[†] An estimate of $\Delta_f H°([PI_4]^+I^-, s) = +180 \text{ kJ mol}^{-1}$ has been made: see I. Tornieporth-Oetting *et al.* (1990) *Journal of the Chemical Society, Chemical Communications*, p. 132.

AsF_5 is a colourless gas and has a molecular structure similar to **14.23**.

$$AsF_3 + 2SbF_5 + Br_2 \longrightarrow AsF_5 + 2SbBrF_4 \qquad (14.68)$$

Arsenic pentafluoride is a strong F^- acceptor (e.g. reactions 14.52, 14.53 and 14.57) and many complexes containing the octahedral $[AsF_6]^-$ ion are known. One interesting reaction of AsF_5 is with metallic Bi to give $[Bi_5][AsF_6]_3$ which contains the trigonal bipyramidal cluster $[Bi_5]^{3+}$.

Antimony trihalides are low-melting solids, and although these contain trigonal pyramidal molecules, each Sb centre has additional, longer range, intermolecular Sb$\cdots$X interactions. The trifluoride and trichloride are prepared by reacting Sb_2O_3 with concentrated HF or HCl, respectively. SbF_3 is a widely used fluorinating agent, e.g. converting B_2Cl_4 to B_2F_4 (*Section 12.6*), $CHCl_3$ to CHF_2Cl (*equation 13.36*), $COCl_2$ to COClF and COF_2 (*Section 13.8*), $SiCl_4$ to SiF_4 (*Section 13.8*) and $SOCl_2$ to SOF_2 (*Section 15.7*). However, reactions may be complicated by SbF_3 also acting as an oxidizing agent (equation 14.69). Reactions between SbF_3 and MF (M = alkali metal) give salts which include K_2SbF_5 (containing $[SbF_5]^{2-}$, **14.26**), KSb_2F_7 (with discrete SbF_3 and $[SbF_4]^-$, **14.27**), $KSbF_4$ (in which the anion is $[Sb_4F_{16}]^{4-}$, **14.28**) and $CsSb_2F_7$ (containing $[Sb_2F_7]^-$, **14.29**).

$$3C_6H_5PCl_2 + 4SbF_3 \longrightarrow 3C_6H_5PF_4 + 2SbCl_3 + 2Sb \qquad (14.69)$$

(14.26)   (14.27)

(14.28)   (14.29)

Antimony pentafluoride (mp 280 K, bp 422 K) is prepared from SbF_3 and F_2, or by reaction 14.70. In the solid state, SbF_5 is tetrameric (Figure 14.9a) and the presence of Sb–F–Sb bridges accounts for the very high viscosity of the liquid. Antimony pentachloride (mp 276 K, bp 352 K) is prepared from the elements, or by reaction of Cl_2 with $SbCl_3$; trigonal bipyramidal molecules exist in the liquid state but the solid contains dimers.

$$SbCl_5 + 5HF \longrightarrow SbF_5 + 5HCl \qquad (14.70)$$

We have already illustrated the role of SbF_5 as an extremely powerful fluoride acceptor (e.g. reactions 8.44, 8.55, 14.54,

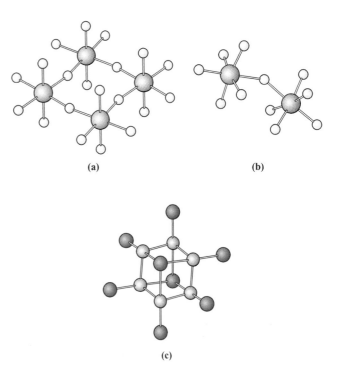

(a)   **(b)**

(c)

Fig. 14.9 The solid state structures of (a) $\{SbF_5\}_4$, (b) $[Sb_2F_{11}]^-$ (X-ray diffraction) in the *tert*-butyl salt [S. Hollenstein *et al.* (1993) *J. Am. Chem. Soc.*, vol. 115, p. 7240] and (c) $[As_6I_8]^{2-}$ (X-ray diffraction) in [{$MeC(CH_2PPh_2)_3$}$NiI]_2[As_6I_8]$ [P. Zanello *et al.* (1990) *J. Chem. Soc., Dalton Trans.*, p. 3761]. The bridge Sb–F bonds in $\{SbF_5\}_4$ and $[Sb_2F_{11}]^-$ are ≈ 15 pm longer than the terminal bonds. Colour code: Sb and As, blue; F, pale grey; I, black.

14.55 and 14.66), and similarly, $SbCl_5$ is one of the strongest chloride acceptors known (e.g. reactions 14.58 and 14.71). Reactions of SbF_5 and $SbCl_5$ with alkali metal fluorides and chlorides yield compounds of the type $M[SbF_6]$ and $M[SbCl_6]$.

$$SbCl_5 + AlCl_3 \longrightarrow [AlCl_2]^+[SbCl_6]^- \qquad (14.71)$$

Whereas the addition of Cl^- to $SbCl_5$ invariably gives $[SbCl_6]^-$, acceptance of F^- by SbF_5 may be accompanied by further association by the formation of Sb–F–Sb bridges. Thus, products may contain $[SbF_6]^-$, $[Sb_2F_{11}]^-$ (Figure 14.9b) or $[Sb_3F_{16}]^-$ in which each Sb centre is octahedrally sited. The strength with which SbF_5 can accept F^- has led to the isolation of salts of some unusual cations, including $[O_2]^+$, $[XeF]^+$, $[Br_2]^+$, $[ClF_2]^+$ and $[NF_4]^+$. Heating $Cs[SbF_6]$ and CsF (molar ratio 1:2) at 573 K for 45 h produces $Cs_2[SbF_7]$; vibrational spectroscopic and theoretical results are consistent with the $[SbF_7]^{2-}$ ion having a pentagonal bipyramidal structure.

When $SbCl_3$ is partially oxidized by Cl_2 in the presence of CsCl, dark blue Cs_2SbCl_6 precipitates; black $[NH_4]_2[SbBr_6]$ can be similarly obtained. Since these compounds are diamagnetic, they cannot contain Sb(IV), and are, in fact, mixed oxidation state species containing $[SbX_6]^{3-}$ and $[SbX_6]^-$; the dark colours of the compounds arise from absorption of light associated with electron-transfer between the two anions. The solid state structures of Cs_2SbCl_6 and

[NH$_4$]$_2$[SbBr$_6$] show similar characteristics, e.g. in [NH$_4$]$_2$[SbBr$_6$], two distinct octahedral anions are present, [SbBr$_6$]$^-$ (Sb−Br = 256 pm) and [SbBr$_6$]$^{3-}$ (Sb−Br = 279 pm); the lone pair in the Sb(III) species appears to be stereochemically inactive.

A number of high nuclearity halo-anions of As and Sb are known which contain doubly and triply bridging X$^-$, e.g. [As$_6$I$_8$]$^{2-}$ (Figure 14.9c), [As$_8$I$_{28}$]$^{4-}$, [Sb$_5$I$_{18}$]$^{3-}$ and [Sb$_6$I$_{22}$]$^{4-}$.

Bismuth halides

The trihalides BiF$_3$, BiCl$_3$, BiBr$_3$ and BiI$_3$ are all well characterized, but BiF$_5$ is the only Bi(V) halide known; all are solids at 298 K. In the vapour phase, the trihalides have molecular (trigonal pyramidal) structures. In the solid state, β-BiF$_3$ contains nine-coordinate Bi(III) centres, BiCl$_3$ and BiBr$_3$ have molecular structures but with an additional five long Bi⋅⋅⋅X contacts, and in BiI$_3$, the Bi atoms occupy octahedral sites in an hcp array of I atoms. The trihalides can be formed by combination of the elements at high temperature; each trihalide is hydrolyzed by water to give BiOX, insoluble compounds with layer structures. The reaction of BiF$_3$ with F$_2$ at 880 K yields BiF$_5$ which is a powerful fluorinating agent. Heating BiF$_5$ with an excess of MF (M = Na, K, Rb or Cs) at 503–583 K for 4 days produces M$_2$[BiF$_7$]; the reactions are carried out under a low pressure of F$_2$ to prevent reduction of Bi(V) to Bi(III). Treatment of BiF$_5$ with an excess of FNO at 195 K yields [NO]$_2$[BiF$_7$], but this is thermally unstable and forms [NO][BiF$_6$] when warmed to room temperature. The [BiF$_7$]$^{2-}$ ion has been assigned a pentagonal bipyramidal structure on the basis of vibrational spectroscopic and theoretical data.

The trihalides are Lewis acids and form donor–acceptor complexes with a number of ethers, e.g. *fac*-[BiCl$_3$(THF)$_3$] and the macrocyclic ligand complexes shown in Figure 14.10. Reactions with halide ions give species such as

[BiCl$_5$]$^{2-}$ (square pyramidal), [BiBr$_6$]$^{3-}$ (octahedral), [Bi$_2$Cl$_8$]$^{2-}$ (**14.30**) and [Bi$_2$I$_9$]$^{3-}$ (**14.31**). Bismuth(III) also forms some higher nuclearity halide complexes, e.g. [Bi$_4$Cl$_{16}$]$^{4-}$, as well as the polymeric species [{BiX$_4$}$_n$]$^{n-}$ and [{BiX$_5$}$_n$]$^{2n-}$; in each case, the Bi atoms are octahedrally sited.

(14.30)

(14.31)

14.8 Oxides of nitrogen

As in group 14, the first element of group 15 stands apart in forming oxides in which $(p–p)\pi$ bonding is important. Table 14.6 lists selected properties of nitrogen oxides, excluding NO$_3$ which is an unstable radical; NO$_2$ exists in equilibrium with N$_2$O$_4$.

Dinitrogen monoxide, N$_2$O

Dinitrogen monoxide (Table 14.6) is usually prepared by decomposition of solid ammonium nitrate (equation 14.72, compare reaction 14.6) but the aqueous solution reaction 14.73 is useful for obtaining a purer product. For further detail on the oxidation of NH$_2$OH to N$_2$O, see *Section 14.5*.

$$NH_4NO_3 \xrightarrow{450-520\,K} N_2O + 2H_2O \qquad (14.72)$$

$$NH_2OH + HNO_2 \longrightarrow N_2O + 2H_2O \qquad (14.73)$$

Dinitrogen monoxide has a faint, sweet odour. It dissolves in water to give a neutral solution, but does not react to any significant extent; the position of equilibrium 14.74 is far to the left.

$$N_2O + H_2O \rightleftharpoons H_2N_2O_2 \qquad (14.74)$$

$$\overset{-}{N}\!\!=\!\!=\!\!\overset{+}{N}\!\!=\!\!=\!\!O$$
$$113\,\text{pm} \quad 119\,\text{pm}$$

(14.32)

$$\overset{+}{N}\!\!\equiv\!\!\overset{-}{N}\!\!-\!\!O$$

(14.33)

Dinitrogen monoxide is a non-toxic gas which is fairly unreactive at 298 K. The N$_2$O molecule is linear, and the bonding can be represented as in structure **14.32**, although

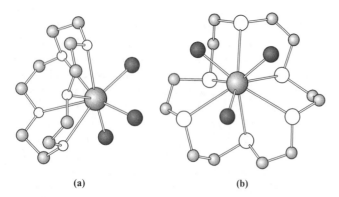

Fig. 14.10 The structures (X-ray diffraction) of (a) [BiCl$_3$(15-crown-5)] [N.W. Alcock *et al.* (1993) *Acta Crystallogr., Sect. B*, vol. 49, p. 507] and (b) [BiCl$_3$L] where L = 1,4,7,10,13,16-hexathiacyclooctadecane [G.R. Willey *et al.* (1992) *J. Chem. Soc., Dalton Trans.*, p. 1339]. Note the high coordination numbers of the Bi(III) centres. Hydrogen atoms have been omitted. Colour code: Bi, blue; O and S, white; Cl, black; C, grey.

Table 14.6 Selected data for the oxides of nitrogen.

	N_2O	NO	N_2O_3	NO_2	N_2O_4	N_2O_5
Name	Dinitrogen monoxide[‡]	Nitrogen monoxide[‡]	Dinitrogen trioxide	Nitrogen dioxide	Dinitrogen tetraoxide	Dinitrogen pentaoxide
Melting point / K	182	109	173	–	262	303
Boiling point / K	185	121	277 dec.	–	294	305 sublimes
Physical appearance	Colourless gas	Colourless gas	Blue solid or liquid	Brown gas	Colourless solid or liquid, but see text	Colourless solid, stable below 273 K
$\Delta_f H^\circ(298\,K) / kJ\,mol^{-1}$	82.1 (g)	90.2 (g)	50.3 (l) 83.7 (g)	33.2 (g)	−19.5 (l) 9.2 (g)	−43.1 (s)
Dipole moment of gas phase molecule / D	0.16	0.16	–	0.315	–	–
Magnetic properties	Diamagnetic	Paramagnetic	Diamagnetic	Paramagnetic	Diamagnetic	Diamagnetic

[‡] N_2O and NO are commonly called nitrous oxide and nitric oxide, respectively.

the bond lengths suggest some contribution from resonance structure **14.33**. One application of N_2O is as a general anaesthetic ('laughing gas'), but its major use is in the preparation of whipped cream. Its reactivity is higher at elevated temperatures; N_2O supports combustion, and reacts with $NaNH_2$ at 460 K (equation 14.40). This reaction is used commercially to prepare NaN_3, a precursor to other azides such as $Pb(N_3)_2$ which is used as a detonator.

Nitrogen monoxide, NO

Nitrogen monoxide (Table 14.6) is made industrially from NH_3 (equation 14.75), and on a laboratory scale by reducing HNO_3 in the presence of H_2SO_4 (reaction 14.76 or 14.77).

$$4NH_3 + 5O_2 \xrightarrow{\text{1300 K, Pt catalyst}} 4NO + 6H_2O \qquad (14.75)$$

$$[NO_3]^- + 3Fe^{2+} + 4H^+ \longrightarrow NO + 3Fe^{3+} + 2H_2O \qquad (14.76)$$

$$2[NO_3]^- + 6Hg + 8H^+ \longrightarrow 2NO + 3[Hg_2]^{2+} + 4H_2O \qquad (14.77)$$

Reaction 14.76 is the basis of the brown ring test for $[NO_3]^-$. After the addition of an equal volume of aqueous $FeSO_4$ to the test solution, cold concentrated H_2SO_4 is added slowly to form a separate, lower layer. If $[NO_3]^-$ is present, NO is liberated, and a brown ring forms between the two layers. The brown colour is due to the formation of $[Fe(NO)(H_2O)_5]^{2+}$, an example of one of many *nitrosyl complexes* in which NO acts as a ligand (see *Section 20.4*).

$$N \overset{\bullet}{=\!=\!=} O$$
$$115\ pm$$
$$(14.34)$$

Structure **14.34** shows that NO is a radical; unlike NO_2, it does not dimerize unless cooled to low temperature, and in the diamagnetic solid, a dimer with a long N−N bond (218 pm) is present. It is probable that a dimer is intermediate in the reactions 14.78, for which reaction rates decrease with increasing temperature.

$$\left.\begin{array}{ll} 2NO + Cl_2 \longrightarrow 2ClNO & \text{Rate} \propto (P_{NO})^2 (P_{Cl_2}) \\ 2NO + O_2 \longrightarrow 2NO_2 & \text{Rate} \propto (P_{NO})^2 (P_{O_2}) \end{array}\right\} \quad (14.78)$$

The reaction with O_2 is important in the manufacture of nitric acid (Section 14.9), but NO can also be oxidized

---

BOX 14.5   **RESOURCES, ENVIRONMENTAL AND BIOLOGICAL**

Nitrogen monoxide in biology

Research into the role played by NO in biological systems is an active area, and in 1992, *Science* named NO 'Molecule of the Year'. The small molecular dimensions of NO mean that it readily diffuses through cell walls. It acts as a messenger-molecule in biological systems, and appears to have an active role in mammalian functions such as the regulation of blood pressure, muscle relaxation and neuro-transmission. A remarkable property exhibited by NO is that it appears to be cytotoxic (i.e. it is able to specifically destroy particular cells) and it affects the ability of the body's immune system to kill tumour cells.

Further reading

A.R. Butler (1995) *Chemistry & Industry*, p. 828 – 'The biological roles of nitric oxide'.

P.L. Feldman, O.W. Griffith and D.J. Stuehr (1993) *Chemical & Engineering News*, December 20 issue, p. 26 – 'The surprising life of nitric oxide'.

E. Palmer (1999) *Chemistry in Britain*, January issue, p. 24 – 'Making the love drug'.

R.J.P. Williams (1995) *Chemical Society Reviews*, vol. 24, p. 77 – 'Nitric oxide in biology: Its role as a ligand'.

See also *Box 28.2*: How the blood-sucking *Rhodnius prolixus* utilizes NO.

NO_x: tropospheric pollutant

'NO_x' (pronounced 'NOX') is a combination of nitrogen oxides arising from both natural (soil emissions and lightning) and man-made sources. The major man-made culprits are vehicle and aircraft exhausts and large industrial power (e.g. electricity-generating) plants; NO_x also contributes to waste effluent in some industrial processes such as the manufacture of adipic acid, and processes are being developed to reduce these pollutant levels. In the closing years of the 20th century, a better awareness of our environment led to the regulation of exhaust emissions; regulated emissions are CO, hydrocarbons and NO_x, as well as particulate matter. The effects of NO_x in the troposphere (0–12 km altitude above the Earth's surface) are to increase $OH^{\bullet}$ and O_3 concentrations. While O_3 in the upper atmosphere acts as a barrier against UV radiation, increased levels at lower altitudes are detrimental to human lung tissue.

For leads into the literature, see:

I. Folkins and G. Brasseur (1992) *Chemistry & Industry*, p. 294 – 'The chemical mechanisms behind ozone depletion'.

M.G. Lawrence and P.J. Crutzen (1999) *Nature*, vol. 402, p. 167 – 'Influence of NO_x emissions from ships on tropospheric photochemistry and climate'.

L. Ross Raber (1997) *Chemical & Engineering News*, April 14 issue, p. 10 – 'Environmental Protection Agency's Air Standards: Pushing too far, too fast?'

See also: *Boxes 8.4* and *10.3*: non-polluting electric vehicles. *Section 26.6* with *Figure 26.13*: catalytic converters.

---

directly to HNO_3 by acidified $[MnO_4]^-$. The reduction of NO depends on the reducing agent, e.g. with SO_2, the product is N_2O, but reduction with tin and acid gives NH_2OH. Although NO is thermodynamically unstable with respect to its elements (Table 14.6), it does not decompose at an appreciable rate below 1270 K, and so does not support combustion well. The positive value of $\Delta_f H^{\circ}$ means that at high temperatures, the formation of NO is favoured, and this is significant during combustion of motor and aircraft fuels where NO is one of several oxides formed; the oxides are collectively described by NO_x (see *Box 14.6*) and contribute to the formation of smogs over large cities.

Reactions 14.57 and 14.58 showed the formation of salts containing the $[NO]^+$ (nitrosyl) cation. Many salts are known, e.g. $[NO][HSO_4]$ is an intermediate in the lead-chamber process for the manufacture of H_2SO_4. X-Ray diffraction data for $[NO]^+$ salts confirm an N−O distance of 106 pm, i.e. less than in NO (115 pm), and a molecular orbital treatment of the bonding (see end of chapter *problem 14.16*) is consistent with this observation. In going from NO to $[NO]^+$ there is an increase in the NO vibration frequency (1876 to $\approx 2300 \text{ cm}^{-1}$), in keeping with an increase in bond strength. All nitrosyl salts are decomposed by water (equation 14.79).

$$[NO]^+ + H_2O \longrightarrow HNO_2 + H^+ \qquad (14.79)$$

Dinitrogen trioxide, N_2O_3

Dinitrogen trioxide (Table 14.6 and Figure 14.11) is obtained as a dark blue liquid in reaction 14.80 at low temperatures, but even at 195 K, extensive dissociation back to NO and N_2O_4 occurs.

$$2NO + N_2O_4 \rightleftharpoons 2N_2O_3 \qquad (14.80)$$

Dinitrogen trioxide is water-soluble and is the *acid anhydride* of HNO_2, nitrous acid (equation 14.81).

$$N_2O_3 + H_2O \longrightarrow 2HNO_2 \qquad (14.81)$$

> An *acid anhydride* is formed when one or more molecules of acid lose one or more molecules of water.

Dinitrogen tetraoxide, N_2O_4, and nitrogen dioxide, NO_2

Dinitrogen tetraoxide and nitrogen dioxide (Table 14.6 and Figure 14.11) exist in equilibrium 14.82, and must be discussed together.

$$N_2O_4 \rightleftharpoons 2NO_2 \qquad (14.82)$$

The solid is colourless and is diamagnetic, consistent with the presence of only N_2O_4. Dissociation of this dimer gives the brown NO_2 radical. Solid N_2O_4 melts to give a yellow liquid, the colour arising from the presence of a little NO_2. At 294 K (bp), the brown vapour contains 15% NO_2; the colour of the vapour darkens as the temperature is raised, and at 413 K dissociation of N_2O_4 is almost complete. Above 413 K, the colour lightens again as NO_2 dissociates to NO and O_2. Laboratory scale preparations of NO_2 or N_2O_4 are usually by thermal decomposition of *dry* lead(II) nitrate (equation 14.83); if the brown gaseous NO_2 is cooled to ≈ 273 K, N_2O_4 condenses as a yellow liquid.

$$2Pb(NO_3)_2(s) \xrightarrow{\Delta} 2PbO(s) + 4NO_2(g) + O_2(g) \qquad (14.83)$$

Dinitrogen tetraoxide is a powerful oxidizing agent (for example, see *Box 8.3*) which attacks many metals, including Hg, at 298 K. The reaction of NO_2 or N_2O_4 with water gives a 1:1 mixture of nitrous and nitric acids (equation 14.84),

Fig. 14.11 The molecular structures of N_2O_3 (with resonance structures), NO_2, N_2O_4 and N_2O_5; molecules of N_2O_3, N_2O_4 and N_2O_5 are planar. The N−N bonds in N_2O_3 and N_2O_4 are particularly long (compare with N_2H_4, Figure 14.3). N atoms are shown in blue.

although nitrous acid disproportionates (see below). Because of the formation of these acids, atmospheric NO_2 is corrosive and contributes to 'acid rain' (see *Box 15.5*). In concentrated H_2SO_4, N_2O_4 yields the nitrosyl and nitryl cations (equation 14.85). The reactions of N_2O_4 with halogens were described in *Section 14.7*, and uses of N_2O_4 as a non-aqueous solvent were outlined in *Section 8.11*.

$$2NO_2 + H_2O \longrightarrow HNO_2 + HNO_3 \qquad (14.84)$$

$$N_2O_4 + 3H_2SO_4$$
$$\longrightarrow [NO]^+ + [NO_2]^+ + [H_3O]^+ + 3[HSO_4]^- \qquad (14.85)$$

The nitryl cation **14.35** is linear, compared with the bent structures of NO_2 (Figure 14.11) and of $[NO_2]^-$ ($\angle O-N-O = 115°$).

$$\underset{115\,pm}{O \!=\!\!=\! \overset{+}{N} \!=\!\!=\! O}$$

(14.35)

Dinitrogen pentaoxide, N_2O_5

Dinitrogen pentaoxide (Table 14.6 and Figure 14.11) is the acid anhydride of HNO_3 and is prepared by reaction 14.86.

$$2HNO_3 \xrightarrow[\text{(dehydrating agent)}]{P_4O_{10}} N_2O_5 + H_2O \qquad (14.86)$$

It forms colourless deliquescent crystals (see *Section 11.5*) but slowly decomposes above 273 K to give N_2O_4 and O_2. In the solid state, N_2O_5 consists of $[NO_2]^+$ and $[NO_3]^-$ ions, but the vapour contains planar molecules (Figure 14.11). A molecular

form of the solid can be formed by sudden cooling of the vapour to 93 K. Dinitrogen pentaoxide reacts violently with water, yielding HNO_3, and is a powerful oxidizing agent (e.g. reaction 14.87).

$$N_2O_5 + I_2 \longrightarrow I_2O_5 + N_2 \qquad (14.87)$$

14.9 Oxoacids of nitrogen

Hyponitrous acid, $H_2N_2O_2$

An aqueous solution of sodium hyponitrite can be made from organic nitrites by reaction 14.88, and addition of Ag^+ leads to the precipitation of $Ag_2N_2O_2$. Treatment of this salt with anhydrous HCl in dry diethyl ether leads to the formation of hyponitrous acid.

$$RONO + NH_2OH + 2EtONa$$
$$\longrightarrow Na_2N_2O_2 + ROH + 2EtOH \qquad (14.88)$$

Free $H_2N_2O_2$ is a weak acid and decomposes spontaneously into N_2O and H_2O. Spectroscopic data indicate that it has structure **14.36**.

$$\underset{HO}{\overset{}{\diagdown}} N \!=\!\!=\! N \underset{}{\overset{OH}{\diagup}}$$

(14.36)

Nitrous acid, HNO_2

Nitrous acid is known only in solution and in the vapour phase and in the latter, it has structure **14.37**. It is a weak acid ($pK_a = 3.37$), but is unstable with respect to disproportionation in solution (equation 14.89). It may be prepared *in situ* by reaction 14.90, the reagents being chosen so as to give an insoluble metal salt; $AgNO_2$ is insoluble but other metal nitrites are soluble in water.

$$\angle O\text{–}N\text{–}O = 111°$$
$$\angle H\text{–}O\text{–}N = 102°$$

(14.37)

$$3HNO_2 \longrightarrow 2NO + HNO_3 + H_2O \qquad (14.89)$$

$$Ba(NO_2)_2 + H_2SO_4 \xrightarrow{\text{aqu}} BaSO_4 + 2HNO_2 \qquad (14.90)$$

Sodium nitrite is an important reagent in the preparation of diazonium compounds, e.g. reaction 14.91 in which HNO_2 is prepared *in situ*. Alkali metal nitrates yield the nitrites when heated alone or, better, with Pb (reaction 14.92).

$$PhNH_2 \xrightarrow{\text{NaNO}_2, \text{ HCl}, <273\text{ K}} [PhN_2]^+Cl^- \qquad (14.91)$$

$$NaNO_3 + Pb \xrightarrow{\Delta} NaNO_2 + PbO \qquad (14.92)$$

Nitrous acid can be oxidized to $[NO_3]^-$ by powerful oxidants such as acidified $[MnO_4]^-$. The products of the reduction of HNO_2 depend on the reducing agent:

- NO is formed with I^- or Fe^{2+};
- N_2O is produced with Sn^{2+};

- NH_2OH results from reduction by SO_2;
- NH_3 is formed with Zn in alkaline solution.

Kinetic rather than thermodynamic control over a reaction is illustrated by the fact that, in dilute solution, HNO_2, but not HNO_3, oxidizes I^- to I_2. Equations 14.93 show that the values of E°_{cell} for these redox reactions are similar; nitrous acid is a faster, rather than a more powerful, oxidizing agent than dilute nitric acid.

$$\left. \begin{array}{ll} I_2 + 2e^- \rightleftharpoons 2I^- & E^\circ = +0.54\,\text{V} \\ [NO_3]^- + 3H^+ + 2e^- \rightleftharpoons HNO_2 + H_2O & E^\circ = +0.93\,\text{V} \\ HNO_2 + H^+ + e^- \rightleftharpoons NO + H_2O & E^\circ = +0.98\,\text{V} \end{array} \right\}$$

(14.93)

Nitric acid, HNO_3, and its derivatives

Nitric acid is an important industrial chemical and is manufactured on a large scale in the Haber–Bosch process closely tied to NH_3 production; the first step is the oxidation of NH_3 to NO (equation 14.20). After cooling, NO is mixed with air and absorbed in a countercurrent of water. The reactions involved are summarized in scheme 14.94; this produces HNO_3 in a concentration of $\approx 60\%$ by weight and it can be concentrated to 68% by distillation.

$$\left. \begin{array}{l} 2NO + O_2 \rightleftharpoons 2NO_2 \\ 2NO_2 \rightleftharpoons N_2O_4 \\ N_2O_4 + H_2O \longrightarrow HNO_3 + HNO_2 \\ 2HNO_2 \longrightarrow NO + NO_2 + H_2O \\ 3NO_2 + H_2O \longrightarrow 2HNO_3 + NO \end{array} \right\}$$

(14.94)

Pure nitric acid can be made in the laboratory by adding H_2SO_4 to KNO_3 and distilling the product *in vacuo*. It is a

---

BOX 14.7   APPLICATIONS

Commercial demand for HNO_3

The industrial production of nitric acid (scheme 14.94) is carried out on a large scale: in the last decade of the 20th century, output climbed in the US from 8.0 Mt in 1988 to 9.4 Mt in 1998, and these figures are surpassed by European manufacturers by a factor of ≈ 2. Nitric acid and ammonia manufacturing are closely linked. About 80% of all HNO_3 produced is destined for conversion into fertilizers, with $[NH_4][NO_3]$ being a key product:

$$NH_3 + HNO_3 \longrightarrow [NH_4][NO_3]$$

Ammonium nitrate has other important applications: about 25% of the manufactured output is used directly in explosives, but its ready accessibility makes it a target for misuse, e.g. in the Oklahoma City bombing in 1995. The potentially explosive nature of $[NH_4][NO_3]$ also makes it a high-risk chemical for transportation.

Nitric acid is usually produced as an aqueous solution containing 50–70% HNO_3 by weight, and this is highly suitable for use in the fertilizer industry. However, for applications of HNO_3 as a nitrating agent in the production of, for example, explosives, acid containing >98% HNO_3 by weight is needed. Ordinary distillation is not appropriate because HNO_3 and H_2O form an azeotrope (see text). Alternative methods are dehydration using concentrated H_2SO_4, or adapting the oxidation of NH_3 (equation 14.20 and first step in scheme 14.94) to include a final step:

$$2N_2O_4 + O_2 + 2H_2O \rightleftharpoons 4HNO_3$$

See also *Box 14.2*: Ammonia: an industrial giant.

BOX 14.8   RESOURCES, ENVIRONMENTAL AND BIOLOGICAL

Nitrates and nitrites in waste water

Levels of $[NO_3]^-$ in waste water are controlled by legislation, limits being recommended by the World Health Organization, the Environmental Protection Agency (in the US) and the European Community. Nitrites, because of their toxicity, must also be removed. Methods of nitrate removal include anion exchange, reverse osmosis (see *Box 15.3*), and denitrification. The last process is a biological one in which certain anaerobic bacteria reduce $[NO_3]^-$ and $[NO_2]^-$ to N_2:

$$2[NO_3]^- + 12H^+ + 10e^- \rightleftharpoons N_2 + 6H_2O$$

$$2[NO_2]^- + 8H^+ + 6e^- \rightleftharpoons N_2 + 4H_2O$$

Other methods of removing $[NO_2]^-$ involve oxidation:

$$[NO_2]^- + [OCl]^- \longrightarrow [NO_3]^- + Cl^-$$

$$[NO_2]^- + H_2O_2 \longrightarrow [NO_3]^- + H_2O$$

with the $[NO_3]^-$ then being removed as described above. Nitrite can also be removed by reduction using urea or sulfamic acid:

$$[NO_2]^- + H_2NSO_3H \longrightarrow N_2 + [HSO_4]^- + H_2O$$

For related information, see *Box 15.3*: Purification of water.

---

colourless liquid, but must be stored below 273 K to prevent slight decomposition (equation 14.95) which gives the acid a yellow colour.

$$4HNO_3 \longrightarrow 4NO_2 + 2H_2O + O_2 \qquad (14.95)$$

Ordinary concentrated HNO_3 is the *azeotrope* containing 68% by weight of HNO_3 and boiling at 393 K; photochemical decomposition occurs by reaction 14.95. Fuming HNO_3 is orange owing to the presence of an excess of NO_2.

> An *azeotrope* is a mixture of two liquids that distils unchanged, the composition of liquid and vapour being the same. Unlike a pure substance, the composition of the azeotropic mixture depends on pressure.

In aqueous solution, HNO_3 acts as a strong acid which attacks most metals, often more rapidly if a trace of HNO_2 is present. Exceptions are Au and the *platinum-group metals* (see *Section 22.9*); Fe and Cr are passivated by concentrated HNO_3. Equations 8.8–8.10 illustrate HNO_3 acting as a base.

Tin, arsenic and a few *d*-block metals are converted to their oxides when treated with HNO_3, but others form nitrates. Only Mg, Mn and Zn liberate H_2 from *very dilute* nitric acid. If the metal is a more powerful reducing agent than H_2, reaction with HNO_3 reduces the acid to N_2, NH_3, NH_2OH or N_2O; other metals liberate NO or NO_2 (e.g. reactions 14.96 and 14.97).

$$3Cu(s) + 8HNO_3(aq)$$
dilute

$$\longrightarrow 3Cu(NO_3)_2(aq) + 4H_2O(l) + 2NO(g) \qquad (14.96)$$

$$Cu(s) + 4HNO_3(aq)$$
conc.

$$\longrightarrow Cu(NO_3)_2(aq) + 2H_2O(l) + 2NO_2(g) \qquad (14.97)$$

Nitrate salts of all metals and cations such as $[NH_4]^+$ are soluble in water. Alkali metal nitrates decompose on heating to the nitrite (reaction 14.98, see also equation 14.92); the decomposition of NH_4NO_3 depends on the temperature (equations 14.6 and 14.72). Most metal nitrates decompose

to the oxide when heated (reaction 14.99), but silver and mercury(II) nitrates give the respective metal (equation 14.100)

$$2KNO_3 \xrightarrow{\Delta} 2KNO_2 + O_2 \qquad (14.98)$$

$$2Cu(NO_3)_2 \xrightarrow{\Delta} 2CuO + 4NO_2 + O_2 \qquad (14.99)$$

$$2AgNO_3 \xrightarrow{\Delta} 2Ag + 2NO_2 + O_2 \qquad (14.100)$$

Many organic and inorganic compounds are oxidized by concentrated HNO_3, although nitrate ion in aqueous solution is usually a very *slow* oxidizing agent (see above). *Aqua regia* contains free Cl_2 and ONCl and attacks Au (reaction 14.101) and Pt with the formation of chloro complexes.

$$Au + HNO_3 + 4HCl \longrightarrow HAuCl_4 + NO + 2H_2O \qquad (14.101)$$

> *Aqua regia* is a mixture of concentrated nitric and hydrochloric acids.

Concentrated HNO_3 oxidizes I_2, P_4 and S_8 to HIO_3, H_3PO_4 and H_2SO_4 respectively.

The molecular structure of HNO_3 is depicted in Figure 14.12a; differences in N–O bond distances are readily understood in terms of the resonance structures shown. The nitrate ion has a trigonal planar structure and the equivalence of the bonds may be rationalized using valence bond or molecular theory (Figure 14.12b). The hydrogen atom in HNO_3 can be replaced by fluorine by treating dilute HNO_3 or KNO_3 with F_2. The product, fluorine nitrate, **14.38**, is an explosive gas which reacts slowly with H_2O but rapidly with aqueous alkali (equation 14.102).

$$\begin{array}{c} {}^-O \qquad\quad F \\ \diagdown \qquad \diagup \\ {}^+N — O \\ \diagup\diagup \\ O \end{array}$$

(14.38)

$$2FONO_2 + 4[OH]^-$$

$$\longrightarrow 2[NO_3]^- + 2F^- + 2H_2O + O_2 \qquad (14.102)$$

Fig. 14.12 (a) The gas phase planar structure of HNO_3, and appropriate resonance structures. (b) The molecular structure of the planar $[NO_3]^-$ anion; the equivalence of the three N—O bonds can be rationalized by valence bond theory (one of three resonance structures is shown) or by MO theory (partial π-bonds are formed by overlap of N and O $2p$ atomic orbitals and the π-bonding is delocalized over the NO_3-framework). N atoms are shown in blue.

The reaction of $NaNO_3$ with Na_2O at 570 K leads to the formation of Na_3NO_4 (sodium orthonitrate); K_3NO_4 may be prepared similarly. X-Ray diffraction data confirm that the $[NO_4]^{3-}$ ion is tetrahedral with N—O bond lengths of 139 pm, consistent with single bond character. Structure **14.39** gives a valence bond picture of the bonding; compare this with $[PO_4]^{3-}$ (**14.40**) where $(p-d)\pi$-bonding *may* be possible (see the discussion of bonding in Section 14.3). The free acid H_3NO_4 is not known.

(14.39)   (14.40)

14.10 Oxides of phosphorus, arsenic, antimony and bismuth

Each of the group 15 elements from P to Bi forms two oxides, E_2O_3 (or E_4O_6) and E_2O_5 (or E_4O_{10}), the latter becoming less stable as the group is descended:

- E_2O_5 (E = P, As, Sb, Bi) are acidic;
- P_4O_6 is acidic;
- As_4O_6 and Sb_4O_6 are amphoteric;
- Bi_2O_3 is basic.

In addition, the discussion below introduces several other oxides of phosphorus.

Oxides of phosphorus

Phosphorus(III) oxide, P_4O_6, is obtained by burning white phosphorus in a restricted supply of O_2. It is a colourless, volatile solid (mp 297 K, bp 447 K) with molecular structure **14.41**; the P—O bond distances (165 pm) are consistent with single bonds, and the angles P—O—P and O—P—O are 128° and 99° respectively. The oxide is soluble in diethyl ether or benzene, but reacts with cold water (equation 14.103).

$$P_4O_6 + 6H_2O \longrightarrow 4H_3PO_3 \qquad (14.103)$$

The most important oxide of phosphorus is P_4O_{10} (phosphorus(V) oxide), commonly called *phosphorus pentoxide*. It can be made directly from P_4 (equation 14.8) or by oxidizing P_4O_6. In the vapour phase, phosphorus(V) oxide contains P_4O_{10} molecules with structure **14.42**; the P—O and P=O bond distances are 160 and 140 pm, consistent with single and double bond character, respectively. When the vapour is condensed rapidly, a volatile and extremely hygroscopic solid is obtained which also contains P_4O_{10} molecules. If this solid is heated in a closed vessel for several hours and the melt maintained at a high temperature before being allowed to cool, the solid obtained is macromolecular. Three polymorphic forms exist at ordinary pressure and temperature, with the basic building block being unit **14.43**; only three of the four O atoms are available for interconnecting the PO_4 units via P—O—P bridges. Phosphorus(V) oxide has a great affinity for water (equation 14.104), and is the anhydride of the wide range of oxoacids described in Section 14.11. It is used as a drying agent (see *Box 11.4*).

(14.41)

(14.42)

(14.43)

$$P_4O_{10} + 6H_2O \longrightarrow 4H_3PO_4 \qquad (14.104)$$

Three other oxides of phosphorus, P_4O_7 (**14.44**), P_4O_8 (**14.45**) and P_4O_9 (**14.46**) have structures that are related to those of P_4O_6 and P_4O_{10}.

(14.44)

(14.45)

(14.46)

These oxides are mixed P(III)P(V) species, each centre bearing a terminal oxo group being oxidized to P(V). For example, P_4O_8 is made by heating P_4O_6 in a sealed tube at 710 K, the other product being red phosphorus.

Oxides of arsenic, antimony and bismuth

The normal combustion products of As and Sb are As(III) and Sb(III) oxides (equation 14.11). The vapour and high temperature solid polymorph of each oxide contains E_4O_6 (E = As or Sb) molecules structurally related to **14.41**. Lower temperature polymorphs have layer structures containing trigonal pyramidal As or Sb atoms. Condensation

of As_4O_6 vapour above 520 K leads to the formation of As_2O_3 glass. Arsenic(III) oxide is an important precursor in arsenic chemistry and is made industrially from the sulfide (Section 14.2). Dissolution of As_2O_3 in water gives a very weakly acidic solution, and it is probable that the species present is $As(OH)_3$ (*arsenous acid*) although this has never been isolated; crystallization of aqueous solutions yields As_2O_3. Arsenic(III) oxide dissolves in aqueous alkali to give salts containing the $[AsO_2]^-$ ion, and in aqueous HCl with the formation of $AsCl_3$. The properties of Sb_2O_3 in water and aqueous alkali or HCl resemble those of As_2O_3.

Bismuth(III) oxide occurs naturally as *bismite*, and is formed when Bi combines with O_2 on heating. In contrast to earlier members of group 15, molecular species are not observed for Bi_2O_3, and the structure is more like that of a typical *metal* oxide.

Arsenic(V) oxide is most readily made by reaction 14.105 than by direct oxidation of the elements. The route makes use of the fact that As_2O_5 is the acid anhydride of arsenic acid, H_3AsO_4. In the solid state, As_2O_5 has a lattice structure consisting of As–O–As linked octahedral AsO_6 and tetrahedral AsO_4-units.

$$As_2O_3 \xrightarrow{\text{conc HNO}_3} 2H_3AsO_4 \xrightarrow{\text{dehydration}} As_2O_5 + 3H_2O$$
$$(14.105)$$

Antimony(V) oxide may be made by reacting Sb_2O_3 with O_2 at high temperatures and pressures. It crystallizes with a lattice structure in which the Sb atoms are octahedrally sited with respect to six O atoms. Bismuth(V) oxide is poorly characterized, and its formation requires the action of strong oxidants (e.g. alkaline hypochlorite) on Bi_2O_3.

14.11 Oxoacids of phosphorus

Table 14.7 lists selected oxoacids of phosphorus. This is an important group of compounds, but the acids are difficult to classify in a straightforward manner. It should be remembered that the basicity of each acid corresponds to the number of OH-groups, *and not simply to the total number of hydrogen atoms*, e.g. H_3PO_3 and H_3PO_2 are dibasic and monobasic respectively (Table 14.7). Diagnostic absorptions in the IR spectra of H_3PO_3 and H_3PO_2 confirm the presence of P–H bonds; the P-attached hydrogens do not ionize in aqueous solution.

Phosphinic acid, H_3PO_2

The reaction of white phosphorus with aqueous alkali (equation 14.9) produces the phosphinate (or hypophosphite) ion, $[H_2PO_2]^-$. By using $Ba(OH)_2$ as alkali, precipitating the Ba^{2+} ions as $BaSO_4$, and evaporating the aqueous solution, white deliquescent crystals of H_3PO_2 can be obtained. In

Table 14.7 Selected oxoacids of phosphorus; older names that are still in common use are given in parentheses.

Formula	Name	Structure	pK_a values
H_3PO_2	Phosphinic acid (hypophosphorous acid)		pK_a = 1.24
H_3PO_3	Phosphonic acid (phosphorous acid)		p$K_a(1)$ = 2.00; p$K_a(2)$ = 6.59
H_3PO_4	Phosphoric acid (orthophosphoric acid)		p$K_a(1)$ = 2.21; p$K_a(2)$ = 7.21; p$K_a(3)$ = 12.67
$H_4P_2O_6$	Hypophosphoric acid		p$K_a(1)$ = 2.2; p$K_a(2)$ = 2.8; p$K_a(3)$ = 7.3; p$K_a(4)$ = 10.0
$H_4P_2O_7$	Diphosphoric acid (pyrophosphoric acid)		p$K_a(1)$ = 0.85; p$K_a(2)$ = 1.49; p$K_a(3)$ = 5.77; p$K_a(4)$ = 8.22
$H_5P_3O_{10}$	Triphosphoric acid		p$K_a(1) \leq 0$ p$K_a(2)$ = 0.89; p$K_a(3)$ = 4.09; p$K_a(4)$ = 6.98; p$K_a(5)$ = 9.93

aqueous solution, H_3PO_2 is a fairly strong monobasic acid (equation 14.106 and Table 14.7).

$$H_3PO_2 + H_2O \rightleftharpoons [H_3O]^+ + [H_2PO_2]^- \qquad (14.106)$$

Phosphinic acid and its salts are reducing agents, and $NaH_2PO_2 \cdot H_2O$ is used industrially in a non-electrochemical reductive process which plates nickel on to, for example, steel. When heated, H_3PO_2 disproportionates according to equation 14.107, the products being determined by reaction temperature.

$$\left. \begin{array}{l} 3H_3PO_2 \xrightarrow{\Delta} PH_3 + 2H_3PO_3 \\ \text{or} \\ 2H_3PO_2 \xrightarrow{\Delta} PH_3 + H_3PO_4 \end{array} \right\} \qquad (14.107)$$

Phosphonic acid, H_3PO_3

Phosphonic acid (commonly called *phosphorous acid*) may be crystallized from the solution obtained by adding ice-cold

water to P_4O_6 (equation 14.103) or PCl_3 (equation 14.61). Pure H_3PO_3 forms colourless, deliquescent crystals (mp 343 K) and in the solid state, molecules of the acid (Table 14.7) are linked by hydrogen bonds to form a three-dimensional network. In aqueous solution, it is dibasic (equations 14.108 and 14.109).

$$H_3PO_3(aq) + H_2O \rightleftharpoons [H_3O]^+ + [H_2PO_3]^- \qquad (14.108)$$

$$[H_2PO_3]^- + H_2O \rightleftharpoons [H_3O]^+ + [HPO_3]^{2-} \qquad (14.109)$$

Salts containing the $[HPO_3]^{2-}$ ion are called *phosphonates*. Although the name 'phosphite' remains in common use, it is a possible source of confusion since esters of type $P(OR)_3$ are also called phosphites, e.g. $P(OEt)_3$ is triethylphosphite.

Phosphonic acid is a reducing agent, but disproportionates when heated (equation 14.110).

$$4H_3PO_3 \xrightarrow{470\,K} PH_3 + 3H_3PO_4 \qquad (14.110)$$

Hypophosphoric acid, $H_4P_2O_6$

The reaction between red phosphorus and $NaOCl$ or $NaClO_2$ yields $Na_2H_2P_2O_6$, which can be converted in aqueous solution into the dihydrate of the free acid which is best formulated as $[H_3O]_2[H_2P_2O_6]$. Dehydration using P_4O_{10} gives $H_4P_2O_6$. The first indication of a P–P bonded dimer (i.e. rather than H_2PO_3) came from the observation that the acid was diamagnetic, and X-ray diffraction data for the salt $[NH_4]_2[H_2P_2O_6]$ have confirmed this structural feature. One resonance structure for $[H_2P_2O_6]^{2-}$ is **14.47**, but the fact that all four terminal P–O bonds are 157 pm in length indicates that the negative charges are delocalized. The acid is thermodynamically unstable with respect to disproportionation and reaction 14.111 occurs slowly in aqueous solution. For this reason, $H_4P_2O_6$ cannot be made by reduction of H_3PO_4 or by oxidation of H_3PO_3 in aqueous media. Hence the need to use a precursor (i.e. elemental phosphorus) in which the P–P bond is already present.

$$\begin{array}{ccc} O & & OH \\ \| & & / \\ {}^-O-P-P-O^- \\ / & & \| \\ HO & & O \end{array}$$

(14.47)

$$H_4P_2O_6 + H_2O \longrightarrow H_3PO_3 + H_3PO_4 \qquad (14.111)$$

Phosphoric acid, H_3PO_4, and its derivatives

Phosphoric acid is made from phosphate rock (equation 14.112) or by hydration of P_4O_{10} (equation 14.104).

$$\underset{\text{conc.}}{Ca_3(PO_4)_2 + 3H_2SO_4} \longrightarrow 2H_3PO_4 + 3CaSO_4 \qquad (14.112)$$

The pure acid forms deliquescent, colourless crystals (mp 315 K). It has a molecular structure (Table 14.7) with P–O and P=O bond distances of 157 and 152 pm; this difference is significantly less than in P_4O_{10} (structure **14.42**) and is the result of extensive hydrogen bonding in the crystalline state which links H_3PO_4 molecules into a layered network. On standing, crystalline H_3PO_4 rapidly forms a viscous liquid. In this and in the commercially available 85% (by weight with water) acid, extensive hydrogen bonding is responsible for the syrupy nature of the acid. In dilute aqueous solutions, acid molecules are hydrogen bonded to water molecules rather than to each other.

Phosphoric acid is very stable and has no oxidizing properties except at very high temperatures. Aqueous H_3PO_4 is a tribasic acid (Table 14.7) and salts containing $[H_2PO_4]^-$, $[HPO_4]^{2-}$ and $[PO_4]^{3-}$ (**14.40**) can be isolated. Thus, three Na^+ salts can be prepared under suitable neutralization conditions; ordinary sodium phosphate is $Na_2HPO_4\cdot12H_2O$, and the common K^+ salt is KH_2PO_4. Sodium phosphates are extensively used for buffering aqueous solutions, and tri-n-butyl phosphate is a valuable solvent for the extraction of metal ions from aqueous solution (see *Box 6.3*).

When H_3PO_4 is heated at 510 K, it is dehydrated to diphosphoric acid (equation 14.113). Comparison of the structures of these acids (Table 14.7) shows that water is eliminated with concomitant P–O–P bridge formation. Further heating yields triphosphoric acid (equation 14.114).

$$2H_3PO_4 \xrightarrow{\Delta} H_4P_2O_7 + H_2O \qquad (14.113)$$

$$H_3PO_4 + H_4P_2O_7 \xrightarrow{\Delta} H_5P_3O_{10} + H_2O \qquad (14.114)$$

Such species containing P–O–P bridges are commonly called *condensed phosphates* and equation 14.115 shows the general condensation process.

(14.115)

Phosphate fertilizers: essential to crops but are they damaging our lakes?

As we pointed out in Box 14.2, worldwide demand for fertilizers is enormous. Phosphorus is an essential plant nutrient and up to 90% (depending on the country) of phosphate rock (see *Section 14.2*) that is mined is consumed in the manufacture of phosphorus-containing fertilizers. Insoluble phosphate rock is treated with concentrated H_2SO_4 to generate soluble *superphosphate* fertilizers containing $Ca(H_2PO_4)_2$ mixed with $CaSO_4$ and other sulfates; reaction between phosphate rock and H_3PO_4 gives *triple superphosphate*, mainly $Ca(H_2PO_4)_2$. Ammonium phosphate fertilizers are valuable sources of *both* N and P. Environmentalists are concerned about the effects that phosphates and polyphosphates from fertilizers and detergents have on the natural balance of lake populations. Phosphates in run-off water which flows into lakes contribute to the excessive growth of algae (*eutrophication*), the presence of which depletes the lakes of O_2, thereby affecting fish and other water-life. However, the issue of phosphates in lakes is not clear cut: recent field studies indicate that adding phosphates to acid lakes (the result of acid rain pollution) stimulates plant growth which in turn leads to a production of $[OH]^-$ which neutralizes excess acid.

Further reading

W. Davison, D.G. George and N.J.A. Edwards (1995) *Nature*, vol. 377, p. 504 – 'Controlled reversal of lake acidification by treatment with phosphate fertilizer'.

The *controlled* hydrolysis of P_4O_{10} is sometimes useful as a means of preparing condensed phosphoric acids. In principle, the condensation of phosphate ions (e.g. reaction 14.116) should be favoured at low pH, but in practice such reactions are usually slow.

$$2[PO_4]^{3-} + 2H^+ \rightleftharpoons [P_2O_7]^{4-} + H_2O \qquad (14.116)$$

Clearly, the number of OH groups in a particular unit determines the extent of the condensation processes. In condensed phosphate anion formation, chain-terminating end groups (**14.48**) are formed from $[HPO_4]^{2-}$, chain members (**14.49**) from $[H_2PO_4]^-$, and cross-linking groups (**14.50**) from H_3PO_4.

(14.48)   (14.49)

(14.50)

In free condensed acids such as $H_5P_3O_{10}$, different phosphorus environments can be distinguished by ^{31}P NMR spectroscopy or chemical methods:

- the pK_a values for successive proton dissociations depend on the position of the OH group; terminal P atoms carry one strongly and one weakly acidic proton, while each P atom in the body of the chain bears one strongly acidic group;
- cross-linking P—O—P bridges are hydrolyzed by water much faster then other such units.

The simplest condensed phosphoric acid, $H_4P_2O_7$, is a solid at 298 K and can be obtained from reaction 14.113 or, in a purer form, by reaction 14.117. It is a stronger acid than H_3PO_4 (Table 14.7).

$$5H_3PO_4 + POCl_3 \longrightarrow 3H_4P_2O_7 + 3HCl \qquad (14.117)$$

The sodium salt $Na_4P_2O_7$ is obtained by heating Na_2HPO_4 at 510 K; note the electronic and structural relationship between $[P_2O_7]^{4-}$ (in which the terminal P—O bond distances are equal) and $[Si_2O_7]^{6-}$, **13.17**. In aqueous solution, $[P_2O_7]^{4-}$ is very slowly hydrolyzed to $[PO_4]^{3-}$, and the two ions can be distinguished by chemical tests, e.g. addition of Ag^+ ions precipitates white $Ag_4P_2O_7$ or pale yellow Ag_3PO_4.

The acid referred to as 'metaphosphoric acid' with an empirical formula of HPO_3 is actually a sticky mixture of polymeric acids, obtained by heating H_3PO_4 and $H_4P_2O_7$ at ≈ 600 K. More is known about the salts of these acids than about the acids themselves. For example, $Na_3P_3O_9$ can be isolated by heating NaH_2PO_4 at 870–910 K and maintaining the melt at 770 K to allow water vapour to escape. It contains the cyclic $[P_3O_9]^{3-}$ ion (*cyclo*-triphosphate ion, Figure 14.13a) which has a chair conformation. In alkaline solution,

(a)   (b)

(c)   (d)

Fig. 14.13 Schematic representations of the structures of (a) $[P_3O_9]^{3-}$, (b) $[P_3O_{10}]^{5-}$ and (c) $[P_4O_{12}]^{4-}$. (d) The structure of $[P_6O_{18}]^{6-}$ (X-ray diffraction) in the compound $[Et_4N]_6[P_6O_{18}]\cdot 4H_2O$ [M.T. Averbuch-Pouchot *et al.* (1991) *Acta Crystallogr., Sect. C*, vol. 47, p. 1579]. Compare these structures with those of the isoelectronic silicates, see Figure 13.16 and associated text. Colour code: P, blue; O, pale grey.

Biological significance of phosphates

Phosphates play an enormously important role in biological systems. The genetic substances deoxyribonucleic acid (DNA) and ribonucleic acid (RNA) are phosphate esters (see *Figure 9.9*). Bones and teeth are constructed from *collagen* (fibrous protein) and single crystals of *hydroxyapatite*, $Ca_5(OH)(PO_4)_3$. Tooth decay involves acid attack on the phosphate, but the addition of fluoride ion to water supplies facilitates the formation of fluoroapatite which is more resistant to decay.

$$Ca_5(OH)(PO_4)_3 + F^- \longrightarrow Ca_5F(PO_4)_3 + [OH]^-$$

All living cells contain *adenosine triphosphate*, ATP:

Hydrolysis results in the loss of a phosphate group and converts ATP to ADP (adenosine diphosphate), releasing energy which is used for functions such as cell growth and muscle movement. The reaction can be written in a simplified form as:

$$[ATP]^{4-} + 2H_2O \longrightarrow [ADP]^{3-} + [HPO_4]^{2-} + [H_3O]^+$$

and, at the standard state usually employed in discussions of biochemical processes (pH 7.4 and $[CO_2] = 10^{-5}$ M), $\Delta G \approx -40$ kJ per mole of reaction. Conversely, energy released by, for example, the oxidation of carbohydrates, can be used to convert ADP to ATP (see *Section 28.4*); thus ATP is continually being reformed, ensuring a continued supply of stored energy in the body.

Further reading

J.J.R. Fraústo da Silva and R.J.P. Williams (1991) *The Biological Chemistry of the Elements*, Clarendon Press, Oxford.
C.K. Mathews and K.E. van Holde (1996) *Biochemistry*, 2nd edn, Benjamin/Cummings, New York, Chapter 3.

$[P_3O_9]^{3-}$ hydrolyzes to $[P_3O_{10}]^{5-}$ (triphosphate ion, Figure 14.13b). The salts $Na_5P_3O_{10}$ and $K_5P_3O_{10}$ (along with several hydrates) are well characterized and $Na_5P_3O_{10}$ (manufactured by reaction 14.118) is used in detergents where it acts as a water softener; uses of polyphosphates as sequestering agents were mentioned in *Sections 11.7* and *11.8*. The parent acid $H_5P_3O_{10}$ has not been prepared in a pure form, but solution titrations allow pK_a values to be determined (Table 14.7).

$$2Na_2HPO_4 + NaH_2PO_4 \xrightarrow{550-650\,K} Na_5P_3O_{10} + 2H_2O$$

$$(14.118)$$

The salt $Na_4P_4O_{12}$ may be prepared by heating $NaHPO_4$ with H_3PO_4 at 670 K and slowly cooling the melt. Alternatively, the volatile form of P_4O_{10} may be treated with ice-cold aqueous NaOH and $NaHCO_3$. Figure 14.13c shows the structure of $[P_4O_{12}]^{4-}$, in which the P_4O_4-ring adopts a chair conformation. Several salts of the $[P_6O_{18}]^{6-}$ ion (Figure 14.13d) are also well characterized; the Na^+ salt is made by heating NaH_2PO_4 at ≈ 1000 K.

The discussion above illustrates how changes in the conditions of heating Na_2HPO_4 or NaH_2PO_4 cause product variation. Carefully controlled conditions are needed to obtain long-chain polyphosphates. Depending on the relative orientations of the PO_4-units, several modifications can be made. Cross-linked polyphosphates (some of which are glasses) can be made by heating NaH_2PO_4 with P_4O_{10}.

14.12 Oxoacids of arsenic, antimony and bismuth

Although 'arsenous acid' ($As(OH)_3$ or H_3AsO_3) has not been isolated, aqueous solutions of As_2O_3 (see *Section 14.10*) probably contain this acid; there is little evidence for the existence of an acid of formula $As(O)OH$. However, several arsenite and meta-arsenite salts containing $[AsO_3]^{3-}$ and $[AsO_2]^-$ respectively have been isolated. Sodium meta-arsenite, $NaAsO_2$ (commercially available), contains Na^+ ions and infinite chains, **14.51**, with trigonal pyramidal As(III) centres.

(14.51)

Arsenic acid, H_3AsO_4, is obtained by dissolving As_2O_5 in water or by oxidation of As_2O_3 using nitric acid (reaction 14.105). Values of $pK_a(1) = 2.25$, $pK_a(2) = 6.77$ and

$pK_a(3) = 11.60$ for H_3AsO_4 show that it is of similar acidic strength to phosphoric acid (Table 14.7). Salts derived from H_3AsO_4 and containing the $[AsO_4]^{3-}$, $[HAsO_4]^{2-}$ and $[H_2AsO_4]^{-}$ ions can be prepared under appropriate conditions. In acidic solution, H_3AsO_4 acts as an oxidizing agent and the pH-dependence of the ease of oxidation or reduction is understood in terms of half-equation 14.119 and the relevant discussion in *Section 7.2*.

$$H_3AsO_4 + 2H^+ + 2e^- \rightleftharpoons H_3AsO_3 + H_2O$$
$$E^\circ = +0.56\,V \qquad (14.119)$$

Condensed polyarsenate ions are kinetically much less stable with respect to hydrolysis (i.e. cleavage of As–O–As bridges) than condensed polyphosphate ions, and only monomeric $[AsO_4]^{3-}$ exists in aqueous solution. Thus, $Na_2H_2As_2O_7$ can be made by dehydrating NaH_2AsO_4 at 360 K; further dehydration (410 K) yields $Na_3H_2As_3O_{10}$ and, at 500 K, polymeric $(NaAsO_3)_n$ is formed. In the solid state, the latter contains infinite chains of tetrahedral AsO_4 units linked by As–O–As bridges. All these condensed arsenates revert to $[AsO_4]^{3-}$ on adding water.

Oxoacids of Sb(III) are not stable, and few antimonite salts are well characterized. Meta-antimonites include $NaSbO_2$ which can be prepared as the trihydrate from Sb_2O_3 and aqueous NaOH; the anhydrous salt has a polymeric structure. No oxoacids of Sb(V) are known, and neither is the tetrahedral anion '$[SbO_4]^{3-}$'. However, well-defined antimonates can be obtained, for example, by dissolving antimony(V) oxide in aqueous alkali and crystallizing the product. Some antimonates contain the octahedral $[Sb(OH)_6]^{-}$ ion, e.g. $Na[Sb(OH)_6]$, originally formulated as $Na_2H_2Sb_2O_7 \cdot 5H_2O$, and $[Mg(H_2O)_6][Sb(OH)_6]_2$, with the old formula of $Mg(SbO_3)_2 \cdot 12H_2O$. The remaining antimonates should be considered as mixed metal oxides; their solid state structures consist of lattices in which Sb(V) centres are octahedrally coordinated by six O atoms and connected by Sb–O–Sb bridges, e.g. $NaSbO_3$, $FeSbO_4$, $ZnSb_2O_6$ and $FeSb_2O_6$ (Figure 14.14).

No oxoacids of Bi are known, although some bismuthate salts are well characterized. Sodium bismuthate is an insoluble, orange solid, obtained by fusing Bi_2O_3 with NaOH in air or with Na_2O_2. It is a very powerful oxidizing agent, e.g. in the presence of acid, it oxidizes Mn(II) to $[MnO_4]^{-}$, and liberates Cl_2 from hydrochloric acid. Like antimonates, some of the bismuthates are better considered as mixed metal oxides. An example is the Bi(III)–Bi(V) compound $K_{0.4}Ba_{0.6}BiO_{3-x}$ ($x \approx 0.02$) which has a perovskite lattice (*Figure 5.23*) and is of interest as a Cu-free superconductor at 30 K (see *Section 27.4*).

14.13 Phosphazenes

Phosphazenes are a group of P(V)/N(III) compounds featuring chain or cyclic structures, and are oligomers of the hypothetical $N\equiv PR_2$. The reaction of PCl_5 with NH_4Cl in a chlorinated solvent (e.g. C_6H_5Cl) gives a mixture of colourless solids of formula $(NPCl_2)_n$ in which the predominant species have $n = 3$ or 4. The compounds $(NPCl_2)_3$ and $(NPCl_2)_4$ are readily separated by distillation under reduced pressure. Although equation 14.120 summarizes the overall reaction, the mechanism is complicated; there is some evidence to support the scheme in Figure 14.15 which illustrates the formation of the trimer.

$$nPCl_5 + nNH_4Cl \longrightarrow (NPCl_2)_n + 4nHCl \qquad (14.120)$$

Reaction 14.120 can be adapted to produce $(NPBr_2)_n$ or $(NPMe_2)_n$ by using PBr_5 or Me_2PCl_3 (in place of PCl_5) respectively. The fluoro derivatives $(NPF_2)_n$ ($n = 3$ or 4) are not made directly, but are prepared by treating $(NPCl_2)_n$ with NaF suspended in MeCN or $C_6H_5NO_2$.

(14.52)

(14.53)

The Cl atoms in $(NPCl_2)_3$, **14.52**, and $(NPCl_2)_4$, **14.53**, readily undergo nucleophilic substitutions, e.g. the following groups can be introduced:

- F using NaF (see above);
- NH_2 using liquid NH_3;
- NMe_2 using Me_2NH;
- N_3 using LiN_3;
- OH using H_2O;
- Ph using LiPh.

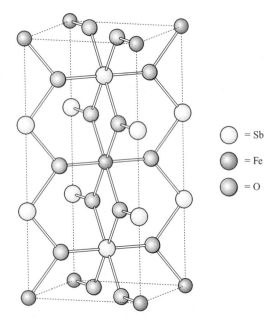

Fig. 14.14 The unit cell of $FeSb_2O_6$ which has a *trirutile* lattice; compare with the rutile unit cell in Figure 5.21.

= Sb

= Fe

= O

$$3PCl_5 + NH_4Cl \xrightarrow{-4HCl} [Cl_3P=N=PCl_3]^+ [PCl_6]^-$$

$$[NH_4]^+ + [PCl_6]^- \longrightarrow Cl_3P=NH + 3HCl$$

$$[Cl_3P=N=PCl_3]^+ + Cl_3P=NH \xrightarrow{-HCl} [Cl_3P=N-PCl_2=N=PCl_3]^+$$

$$\downarrow \begin{array}{c} Cl_3P=NH \\ -HCl \end{array}$$

$$[Cl_3P=N-PCl_2=N-PCl_2=N=PCl_3]^+ \xrightarrow{-[PCl_4]^+}$$

Fig. 14.15 Proposed reaction scheme for the formation of the cyclic phosphazene $(NPCl_2)_3$.

Two substitution pathways are observed. If the group that first enters *decreases* the electron density on the P centre (e.g. F replaces Cl), the second substitution occurs at the *same* P atom. If the electron density *increases* (e.g. NMe$_2$ substitutes for Cl), then the second substitution site is a different P centre.

$$Cl_3P = N \underbrace{\left[P \begin{array}{c} Cl \\ | \\ | \\ Cl \end{array} = N \right]}_{n} PCl_4$$

(14.54)

Small amounts of linear polymers, **14.54**, are also produced in reaction 14.120, and their yield can be increased by using excess PCl$_5$. Polymers with molecular masses of $\leq 20\,000$ are formed when $(NPCl_2)_3$ is heated at 520–620 K in CCl_4 in the presence of O_2; $(NPCl_2)_4$ polymerizes much more slowly. The Cl atoms in the polymers are readily replaced, and this is a route to some commercially important materials. Treatment with sodium alkoxides, NaOR, yields linear polymers $[NP(OR)_2]_n$ which have water-resistant properties, and when $R = CH_2CF_3$, the polymers are inert enough for use in the construction of artificial blood vessels and organs. Many phosphazene polymers are used in fire-resistant materials (see *Box 16.1*).

The structures of $(NPCl_2)_3$, $(NPCl_2)_4$, $(NPF_2)_3$ and $(NPF_2)_4$ are shown in Figure 14.16. In each of $(NPF_2)_3$ and $(NPF_2)_4$, the P_nN_n ring is planar and it is roughly planar in $(NPCl_2)_3$. However, in $(NPCl_2)_4$ the ring is non-planar and two forms exist: the metastable form with a boat conformation, and the stable form with a chair conformation (Figure 14.16b). Although structures **14.52** and **14.53** indicate

	X = Cl	X = F
P—N / pm	158	156
P—X / pm	199	152
∠ P–N–P / deg	121	120
∠ N–P–N / deg	118	121
∠ X–P–X / deg	102	99

Planar

Boat

Chair

(a)   (b)

Fig. 14.16 (a) Structural parameters for the phosphazenes $(NPX_2)_3$ (X = Cl or F); (b) schematic representations of the P_4N_4 ring conformations in $(NPF_2)_4$ (planar) and $(NPCl_2)_4$ (boat and chair).

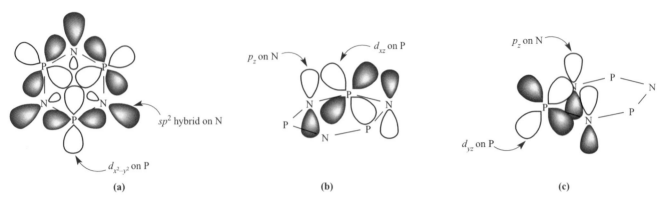

Fig. 14.17 Traditional bonding descriptions for phosphazenes invoke (*p–d*) orbital interactions between nitrogen and phosphorus (although see comments in the text). Three possible descriptions (which may be mutually operative) are shown for the P_3N_3 ring; the *d*-orbital labels are given according to Cartesian axes sets which are *local* at each P centre. (a) Orbital overlap within the plane of the P_3N_3 ring; (b) and (c) orbital overlap perpendicular to the plane of the P_3N_3 ring, and shown for only part of the π-system for clarity.

double and single bonds in the rings, crystallographic data show that the P–N bond lengths in a given ring are equal. Data for $(NPCl_2)_3$ and $(NPF_2)_3$ are given in Figure 14.16a; in $(NPF_2)_4$, $d(P-N) = 151$ pm, and in the boat and chair conformers of $(NPCl_2)_4$, $d(P-N) = 157$ and 156 pm respectively. The P–N bond distances are significantly shorter than expected for a P–N single bond (e.g. 177 pm in the anion in $Na[H_3NPO_3]$), and bonding models for cyclic phosphazenes generally invoke delocalized $(p-d)π$-bonding. However, generalized bonding models should be viewed with some caution, since it is not easy to provide rationalizations for *all* the experimentally observed facts. For example, not all rings are planar, and, while $\angle N-P-N \approx 120°$ in $(NPCl_2)_3$, $(NPCl_2)_4$, $(NPF_2)_3$ and $(NPF_2)_4$, $\angle P-N-P \approx 120°$ in $(NPCl_2)_3$ and $(NPF_2)_3$, 131° in $(NPCl_2)_4$ (both conformers), and 147° in $(NPF_2)_4$. Figure 14.17 illustrates possible N–P $(p-d)$ orbital overlap capable of enhancing the bonding provided by the N–P σ-framework. How important these contributions are to any bonding scheme remains unclear in the light of the uncertain role played by the phosphorus *d*-orbitals (see *Section 14.3*).

A number of phosphazenes containing two or more heterocycles are known, e.g. compound **14.55** is a byproduct in reaction 14.120, and **14.56** is formed when $(NPCl_2)_3$ is treated with sodium 2,6-*tert*-butylphenoxide.

(14.55)

(14.56)

14.14 Sulfides and selenides

Sulfides and selenides of phosphorus

Sulfur–nitrogen compounds are described in *Section 15.10*, and in this section we look at the molecular sulfides formed by phosphorus. Although their structures (Figure 14.18) appear to be closely related to those of the oxides (Section 14.10), there are some notable differences, e.g. P_4O_6 and P_4S_6 (until 1991, a missing member of the sulfide family) are not isostructural. The bond distances *within* the cages of all the sulfides indicate single P–P and P–S bonds; the data for P_4S_3 shown in Figure 14.18 are typical. The terminal P–S bonds are shorter than those in the cage (e.g. 191 versus 208 pm in P_4S_{10}) consistent with double versus single bond character as indicated in Figure 14.18. Only some of the sulfides are prepared by direct combination of the elements. Above 570 K, white phosphorus combines with sulfur to give P_4S_{10} which is the most useful of the phosphorus sulfides. It is a thiating agent (i.e. one that introduces sulfur into a system) in organic reactions, and is a precursor to organothiophosphorus compounds. The reaction of red phosphorus with sulfur above 450 K yields P_4S_3, and P_4S_7 can also be made by direct combination under appropriate conditions. The remaining sulfides in Figure 14.18 are made by one of the general routes:

- abstraction of sulfur using PPh_3 (e.g. reaction 14.121);
- treatment of a phosphorus sulfide with sulfur (e.g. reaction 14.122);
- treatment of a phosphorus sulfide with phosphorus (e.g. reaction 14.123);
- reaction of α- (**14.57**) or β-$P_4S_3I_2$ (**14 58**) with $(Me_3Sn)_2S$ (reaction 14.124).

There is ^{31}P NMR spectroscopic evidence that P_4S_8 has been prepared by treating P_4S_9 with PPh_3.

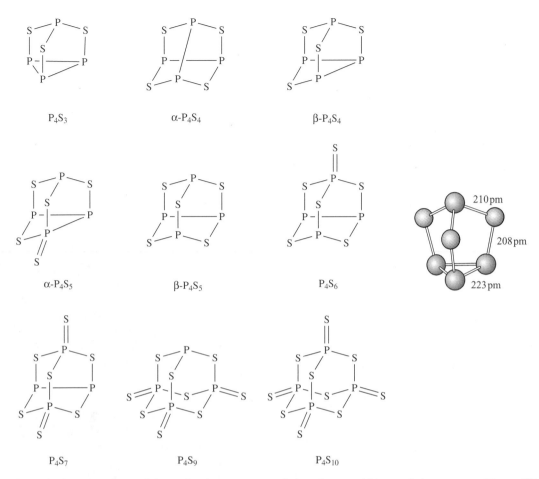

Fig. 14.18 Schematic representations of the molecular structures of phosphorus sulfides, and the structure (X-ray diffraction) of P_4S_3 [L.Y. Goh *et al.* (1995) *Organometallics*, vol. 14, p. 3886]. Colour code: S, blue; P, grey.

(14.57)

(14.58)

$$P_4S_7 + Ph_3P \longrightarrow P_4S_6 + Ph_3P{=}S \qquad (14.121)$$

$$P_4S_3 \xrightarrow{\text{excess sulfur}} P_4S_9 \qquad (14.122)$$

$$P_4S_{10} \xrightarrow{\text{red phosphorus}} \alpha\text{-}P_4S_5 \qquad (14.123)$$

$$\beta\text{-}P_4S_3I_2 + (Me_3Sn)_2S \longrightarrow \beta\text{-}P_4S_4 + 2Me_3SnI \qquad (14.124)$$

Phosphorus sulfides ignite easily, and P_4S_3 is used in 'strike anywhere' matches; it is combined with $KClO_3$, and the compounds inflame when subjected to friction. Whereas P_4S_3 is stable to water, other phosphorus sulfides are slowly hydrolyzed (e.g. reaction 14.125).

$$P_4S_{10} + 16H_2O \longrightarrow 4H_3PO_4 + 10H_2S \qquad (14.125)$$

We have already noted (Section 14.10) that, although sometimes referred to as 'phosphorus pentoxide', phosphorus(V) oxide does not exist as P_2O_5 molecules. In contrast, the vapour of phosphorus(V) sulfide contains some P_2S_5 molecules (although decomposition of the vapour to S, P_4S_7 and P_4S_3 also occurs). In 1991, it was shown that P_2Se_5 and P_4Se_{10} are distinct species. Both can be made by direct combination of P and Se under appropriate conditions; P_2Se_5 is also formed by the decomposition of P_3Se_4I, and P_4Se_{10} from the reaction of P_4Se_3 and selenium at 620 K. Structure **14.59** has been confirmed by X-ray diffraction for P_2Se_5, while P_4Se_{10} is isostructural with P_4S_{10} and P_4O_{10}.

(14.59)

Arsenic, antimony and bismuth sulfides

Arsenic and antimony sulfide ores are major sources of the group 15 elements (see *Section 14.2*). In the laboratory, As_2S_3 and As_2S_5 are usually precipitated from aqueous solutions of arsenite or arsenate; reaction 14.126 proceeds with dilute acid, but at 273 K and high acid concentrations, As_2S_5 is formed.

$$2[AsO_4]^{3-} + 6H^+ + 5H_2S \longrightarrow As_2S_3 + 2S + 8H_2O \quad (14.126)$$

Solid As_2S_3 has the same layer structure as the low temperature polymorph of As_2O_3, but it vaporizes to give As_4S_6 molecules (see below). As_2S_5 exists in crystalline and vitreous forms, but structural details are not known. Both As_2S_3 and As_2S_5 are readily soluble in alkali metal sulfide solutions with the formation of thioarsenites and thioarsenates (e.g. equation 14.127); acids decompose these salts, reprecipitating the sulfides.

$$As_2S_3 + 3S^{2-} \longrightarrow 2[AsS_3]^{3-} \quad (14.127)$$

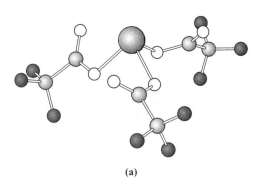

(14.60)

The sulfides As_4S_3 (*dimorphite*), As_4S_4 (*realgar*) and As_2S_3 (*orpiment*) occur naturally; the last two are red and golden-yellow respectively and were used as pigments in early times.[†] The arsenic sulfides As_4S_3, α-As_4S_4, β-As_4S_4 and β-As_4S_5 are structural analogues of the phosphorus sulfides in Figure 14.18, but As_4S_6 is structurally related to P_4O_6 and As_4O_6 rather than to P_4S_6. Structure **14.60** shows another representation of the structure of α-As_4S_4; the bond distances are in accord with As–As and As–S single bonds, and this view of the cage allows a comparison with S_4N_4 (see *Section 15.10*).

The only well-characterized binary sulfide of Sb is the naturally occurring Sb_2S_3 (*stibnite*), which has a double-chain structure in which each Sb(III) is pyramidally sited with respect to three S atoms. The sulfide can be made by direct combination of the elements. A metastable red form can be precipitated from aqueous solution, but reverts to the stable black form on heating. Like As_2S_3, Sb_2S_3 dissolves in alkali metal sulfide solutions (see equation 14.127). Bismuth(III) sulfide, Bi_2S_3, is isostructural with Sb_2S_3; it does not dissolve in alkali metal sulfide solutions.

14.15 Aqueous solution chemistry

Many aspects of the aqueous solution chemistry of the group 15 elements have already been covered:

- acid–base properties of NH_3, PH_3, N_2H_4, HN_3 (Section 14.5);
- redox behaviour of nitrogen compounds (Section 14.5 and Figure 14.4);
- the *brown ring test* for nitrate ion (Section 14.8);
- oxoacids (Sections 14.9, 14.11 and 14.12);
- condensed phosphates (Section 14.11);
- lability of condensed arsenates (Section 14.12);
- sequestering properties of polyphosphates (Section 14.11).

In this section we focus on the formation of aqueous solution species by Sb(III) and Bi(III). Solutions of Sb(III) contain either hydrolysis products or complex ions. The former are commonly written as $[SbO]^+$, but by analogy with Bi(III) (see below), this is surely oversimplified. Complexes are formed with ligands such as oxalate, tartrate or trifluoroacetate ions, and it is usual to observe an arrangement of donor atoms about the Sb atom that reflects the presence of a stereochemically active lone pair of electrons; e.g. in $[Sb(O_2CCF_3)_3]$, the Sb(III) centre is in a trigonal pyramidal environment (Figure 14.19a).

The cation $[Bi_6(OH)_{12}]^{6+}$ is the dominant species in highly acidic aqueous media. The six Bi(III) centres are

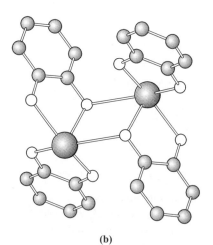

(a)

(b)

Fig. 14.19 The structures (X-ray diffraction) of (a) (*R*)-$[Sb(O_2CCF_3)_3]$ [D.P. Bullivant *et al.* (1980) *J. Chem. Soc., Dalton Trans.*, p. 105] and (b) $[Bi_2(C_6H_4O_2)_4]^{2-}$, crystallized as a hydrated ammonium salt [G. Smith *et al.* (1994) *Aust. J. Chem.*, vol. 47, p. 1413]. Atom colours: Sb and Bi, blue; O, white, F, black, C, grey.

[†] For wider discussions of inorganic pigments, see: R.J.H. Clark (1995) *Chemical Society Reviews*, vol. 24, p. 187 – 'Raman microscopy: Application to the identification of pigments on medieval manuscripts'; R.J.H. Clark and P.J. Gibbs (1997) *Chemical Communications*, p. 1003 – 'Identification of lead(II) sulfide and pararealgar on a 13th century manuscript by Raman microscopy'.

arranged in an octahedron, but at non-bonded separations $(Bi \cdots Bi = 370 \, pm)$, and each of the twelve Bi–Bi edges is supported by a bridging hydroxo ligand. In more alkaline solutions, $[Bi_6O_6(OH)_3]^{3+}$ is formed, and ultimately, $Bi(OH)_3$ is precipitated. As for Sb(III), it is often the case that the coordination geometry is influenced by the presence of a stereochemically active lone pair; e.g. in the catecholate complex $[Bi_2(C_6H_4O_2)_4]^{2-}$ (Figure 14.19b), each Bi atom is in a square-based pyramidal environment. Figure 14.10 showed the structures of two complexes of $BiCl_3$ with macrocyclic ligands.

Summary

This chapter has introduced the extensive descriptive chemistry of the group 15 elements in which kinetic control plays an important part. Metallic character and stabilities of lower oxidation states increase in descending the group. The chemistry of nitrogen differs markedly from that of the later elements, and that of phosphorus (like nitrogen, a non-metal) is also quite distinct.

The following terms were introduced in this chapter. Do you know what they mean?
❑ chemiluminescent reaction
❑ acid anhydride
❑ azeotrope

Suggested further reading

D.E.C. Corbridge (1995) *Phosphorus*, 5th edn, Elsevier, Amsterdam – An up-to-date review of all aspects of phosphorus chemistry.

N.N. Greenwood and A. Earnshaw (1997) *Chemistry of the Elements*, 2nd edn, Butterworth-Heinemann, Oxford – Chapters 11–13 give a detailed account of the chemistries of the group 15 elements.

N.C. Norman, ed. (1998) *Chemistry of Arsenic, Antimony and Bismuth*, Blackie, London – A series of articles covering both inorganic and organometallic aspects of the later group 15 elements.

J. Novosad (1994) 'Phosphorus: Inorganic chemistry' in *Encyclopedia of Inorganic Chemistry*, ed. R.B. King, Wiley, Chichester, vol. 6, p. 3144 – An overview which includes information on ^{31}P NMR spectroscopy.

H.H. Sisler (1994) 'Nitrogen: Inorganic chemistry' in *Encyclopedia of Inorganic Chemistry*, ed. R.B. King, Wiley, Chichester, vol. 5, p. 2516 – A well-referenced, up-to-date account.

A.F. Wells (1984) *Structural Inorganic Chemistry*, 5th edn, Clarendon Press, Oxford – Chapters 18–20 give detailed accounts of the structures of compounds of the group 15 elements.

Specialized topics

J.C. Bottaro (1996) *Chemistry & Industry*, p. 249 – 'Recent advances in explosives and solid propellants'.

K. Dehnicke and J. Strähle (1992) *Angewandte Chemie International Edition, in English*, vol. 31, p. 955 – 'Nitrido complexes of the transition metals'.

P. Ettmayer and W. Lengauer (1994) 'Nitrides: Transition metal solid state chemistry' in *Encyclopedia of Inorganic Chemistry*, ed. R.B. King, Wiley, Chichester, vol. 5, p. 2498.

D.P. Gates and I. Manners (1997) *J. Chem. Soc., Dalton Trans.*, p. 2525 – 'Main-group-based rings and polymers'.

A.C. Jones (1997) *Chemical Society Reviews*, vol. 26, p. 101 – 'Developments in metal-organic precursors for semiconductor growth from the vapour phase'.

S.T. Oyama (1996) *The Chemistry of Transition Metal Carbides and Nitrides*, Kluwer, Dordrecht.

H.G. von Schnering and W. Hönle (1994) 'Phosphides: Solid state chemistry' in *Encyclopedia of Inorganic Chemistry*, ed. R.B. King, Wiley, Chichester, vol. 6, p. 3106.

Problems

14.1 What are the formal oxidation states of N or P in the following species? (a) N_2; (b) $[NO_3]^-$; (c) $[NO_2]^-$; (d) NO_2; (e) NO; (f) NH_3; (g) NH_2OH; (h) P_4; (i) $[PO_4]^{3-}$; (j) P_4O_6; (k) P_4O_{10}.

14.2 Using bond enthalpy terms from Tables 13.2 and 14.3, estimate values of $\Delta_r H^\circ$ for the following reactions:

(a) $2N_2 \longrightarrow N_4$ (tetrahedral structure);
(b) $2P_2 \longrightarrow P_4$ (tetrahedral structure);
(c) $2C_2H_2 \longrightarrow C_4H_4$ (tetrahedrane, with a tetrahedral C_4 core).

14.3 Give a brief account of allotropy among the group 15 elements.

14.4 Write equations for the reactions of: (a) water with Ca_3P_2; (b) aqueous NaOH with NH_4Cl; (c) aqueous NH_3 with $Mg(NO_3)_2$; (d) AsH_3 with an excess of I_2 in aqueous solution; (e) PH_3 with KNH_2 in liquid NH_3.

14.5 Explain why (a) a dilute aqueous solution of NH_3 smells of the gas whereas dilute HCl does not retain the acrid odour of gaseous HCl, and (b) ammonium carbamate is used in smelling salts.

14.6 If (at 298 K) pK_b for NH_3 is 4.75, show that pK_a for $[NH_4]^+$ is 9.25.

14.7 Give the relevant half-equations for the oxidation of NH_2OH to HNO_3 by $[BrO_3]^-$, and write a balanced equation for the overall process.

14.8 (a) Write a balanced equation for the preparation of NaN_3 from $NaNH_2$ with $NaNO_3$. (b) Suggest a route for preparing the precursor $NaNH_2$. (c) How might NaN_3 react with $Pb(NO_3)_2$ in aqueous solution?

14.9 (a) We noted that $[N_3]^-$ is isoelectronic with CO_2. Give three other species that are also isoelectronic with $[N_3]^-$. (b) Describe the bonding in $[N_3]^-$ in terms of an MO picture.

14.10 Refer to Figure 14.7. (a) By considering a number of unit cells of NiAs connected together, confirm that the coordination number of each Ni atom is 6. (b) How does the information contained in the unit cell of NiAs confirm the stoichiometry of the compound?

14.11 Suggest how you might confirm the conformation of N_2H_4 in (a) the gas phase and (b) the liquid phase.

14.12 (a) Discuss structural variation among the phosphorus(III) and phosphorus(V) halides, indicating where stereochemical non-rigidity is possible. (b) On what basis is it appropriate to compare the lattice of $[PCl_4][PCl_6]$ with that of CsCl?

14.13 What might you expect to observe (at 298 K) in the ^{19}F NMR spectra of solutions containing (a) $[PF_6]^-$ and (b) $[SbF_6]^-$. Data needed are in Table 14.2.

14.14 Suggest products for the reactions between: (a) $SbCl_5$ and PCl_5; (b) KF and AsF_5; (c) NOF and SbF_5; (d) HF and SbF_5.

14.15 (a) Draw the structures of $[Sb_2F_{11}]^-$ and $[Sb_2F_7]^-$, and rationalize them in terms of VSEPR theory. (b) Suggest likely structures for the $[\{BiX_4\}_n]^{n-}$ and $[\{BiX_5\}_n]^{2n-}$ oligomers mentioned in Section 14.7.

14.16 By using an MO approach, rationalize why, in going from NO to $[NO]^+$, the bond order increases, bond distance decreases and NO vibrational wavenumber increases.

14.17 $25.0 \, cm^3$ of a 0.0500 M solution of sodium oxalate $(Na_2C_2O_4)$ reacted with $24.8 \, cm^3$ of a solution of $KMnO_4$, **A**, in the presence of excess H_2SO_4. $25.0 \, cm^3$ of a 0.0494 M solution of NH_2OH in H_2SO_4 was boiled with an excess of iron(III) sulfate solution, and when the reaction was complete, the iron(II) produced was found to be equivalent to $24.65 \, cm^3$ of solution **A**. The product **B** formed from the NH_2OH in this reaction can be assumed not to interfere with the determination of iron(II). What can you deduce about the identity of **B**?

14.18 Write a brief account that supports the statement that: 'all the oxygen chemistry of phosphorus(V) is based on the tetrahedral PO_4 unit.'

14.19 Figure 14.14 shows a unit cell of $FeSb_2O_6$. (a) How is this unit cell related to the rutile lattice type? (b) Why can the solid state structure of $FeSb_2O_6$ not be described in terms of a single unit cell of the rutile lattice? (c) What is the coordination environment of each atom type? (d) Confirm the stoichiometry of this compound using only the information provided in the unit cell diagram.

14.20 How may NMR spectroscopy be used:

(a) to distinguish between solutions of $Na_5P_3O_{10}$ and $Na_6P_4O_{13}$;
(b) to determine whether F atoms exchange rapidly between non-equivalent sites in AsF_5;
(c) to determine the positions of the NMe_2 groups in $P_3N_3Cl_3(NMe_2)_3$?

14.21 Deduce what you can about the nature of the following reactions.

(a) One mole of NH_2OH reacts with two moles of Ti(III) in the presence of excess alkali, and the Ti(III) is converted to Ti(IV).
(b) When Ag_2HPO_3 is warmed in water, all the silver is precipitated as metal.
(c) When one mole of H_3PO_2 is treated with excess I_2 in acidic solution, one mole of I_2 is reduced; on making the solution alkaline, a second mole of I_2 is consumed.

14.22 Predict the structures of (a) $[NF_4]^+$; (b) $[N_2F_3]^+$; (c) NH_2OH; (d) $SPCl_3$; (e) PCl_3F_2.

14.23 Suggest syntheses for each of the following from $K^{15}NO_3$: (a) $Na^{15}NH_2$, (b) $^{15}N_2$ and (c) $[^{15}NO][AlCl_4]$.

14.24 Suggest syntheses for each of the following from $Ca_3(^{32}PO_4)_2$: (a) $^{32}PH_3$, (b) $H_3^{32}PO_3$ and (c) $Na_3^{32}PS_4$.

14.25 $25.0 \, cm^3$ of a 0.0500 M solution of sodium oxalate reacted with $24.7 \, cm^3$ of a solution of $KMnO_4$, **C**, in the presence of excess H_2SO_4. $25.0 \, cm^3$ of a 0.0250 M solution of N_2H_4 when treated with an excess of alkaline $[Fe(CN)_6]^{3-}$ solution gave $[Fe(CN)_6]^{4-}$ and a product **D**. The $[Fe(CN)_6]^{4-}$ formed was reoxidized to $[Fe(CN)_6]^{3-}$ by $24.80 \, cm^3$ of solution **C**, and the presence of **D** did not influence this determination. What can you deduce about the identity of **D**?

14.26 Comment on the fact that $AlPO_4$ exists in several forms, each of which has a structure which is also that of a form of silica.

Chapter 15

The group 16 elements

TOPICS

- Occurrence, extraction and uses
- Physical properties and bonding considerations
- The elements
- Hydrides
- Metal sulfides, polysulfides, polyselenides and polytellurides

- Halides, oxohalides and complex halides
- Oxides
- Oxoacids and their salts
- Compounds of sulfur and selenium with nitrogen
- Aqueous solution chemistry of sulfur, selenium and tellurium

1	2		13	14	15	**16**	17	18
H								He
Li	Be		B	C	N	**O**	F	Ne
Na	Mg		Al	Si	P	**S**	Cl	Ar
K	Ca	*d*-block	Ga	Ge	As	**Se**	Br	Kr
Rb	Sr		In	Sn	Sb	**Te**	I	Xe
Cs	Ba		Tl	Pb	Bi	**Po**	At	Rn
Fr	Ra							

15.1 Introduction

The group 16 elements – oxygen, sulfur, selenium, tellurium and polonium – are called the *chalcogens*

Oxygen occupies so central a position in any treatment of inorganic chemistry that discussions of many of its compounds are dealt with under other elements. The decrease in non-metallic character down the group is easily recognized in the elements:

- oxygen exists only as two gaseous allotropes (O_2 and O_3);
- sulfur has many allotropes, all of which are insulators;
- the stable forms of selenium and tellurium are semiconductors;
- polonium is a metallic conductor.

Knowledge of the chemistry of Po and its compounds is limited because of the absence of a stable isotope and the difficulty of

working with ^{210}Po, the most readily available isotope. Polonium-210 is produced from ^{209}Bi by an (n,γ) reaction (see *Section 2.4*) followed by β-decay of the product. It is an intense α-emitter ($t_{\frac{1}{2}} = 138$ days) liberating $520 \, kJ \, g^{-1} \, h^{-1}$, and is a lightweight source of energy in space satellites. However, this large energy loss causes many compounds of Po to decompose; Po decomposes water, making studies of chemical reactions in aqueous solution difficult. Polonium is a metallic conductor and crystallizes in a simple cubic lattice. It forms volatile, readily hydrolyzed halides $PoCl_2$, $PoCl_4$, $PoBr_2$, $PoBr_4$ and PoI_4 and complex ions $[PoX_6]^{2-}$ (X = Cl, Br, I). Polonium(IV) oxide is formed by reaction between Po and O_2 at 520 K; it adopts a fluorite lattice (see *Figure 5.18*) and is sparingly soluble in aqueous alkali. The observed properties are those expected by extrapolation from Te.

15.2 Occurrence, extraction and uses

Occurrence

Dioxygen makes up 21% of the Earth's atmosphere, and 49% of the Earth's crust is composed of O-containing compounds, e.g. water, limestone, silica, silicates, bauxite and haematite. It is a component of innumerable compounds and is essential to life, being converted to CO_2 during respiration. Native sulfur occurs in deposits around volcanoes and hot springs, and sulfur-containing minerals include *iron pyrites* (*fool's gold*, FeS_2), *galena* (PbS), *sphalerite* or *zinc blende* (ZnS), *cinnabar* (HgS), *realgar* (As_4S_4), *orpiment* (As_2S_3), *stibnite* (Sb_2S_3), *molybdenite* (MoS_2) and *chalcocite* (Cu_2S). Selenium and tellurium are relatively rare; Se occurs

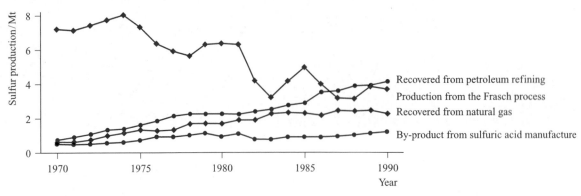

Fig. 15.1 Production of sulfur in the US from 1970 to 1990; note the increasing importance of recovery methods which have now overtaken the Frasch process as a source of sulfur [data: U.S. Geological Survey].

in only a few minerals, while Te is usually combined with other metals, e.g. in *sylvanite* (AgAuTe$_2$).

Extraction

Sulfur is obtained commercially from natural underground deposits by the Frasch process in which super-heated water (440 K under pressure) is used to melt the sulfur, and compressed air is then applied to force it to the surface.

Conventional mining methods are impossible because of quicksands above the sulfur beds. *Roll sulfur* produced by the Frasch process is ≈99% pure. Other sources of sulfur compounds are *anhydrite* (CaSO$_4$), and SO$_2$ (obtained as a by-product in the processing of metal sulfide ores). Sulfur is also obtained from H$_2$S (which occurs in concentrations of up to 30% in some natural gases) by reaction 15.1 and is recovered from petroleum refining processes. Figure 15.1 illustrates the growing importance of sulfur-recovery, and

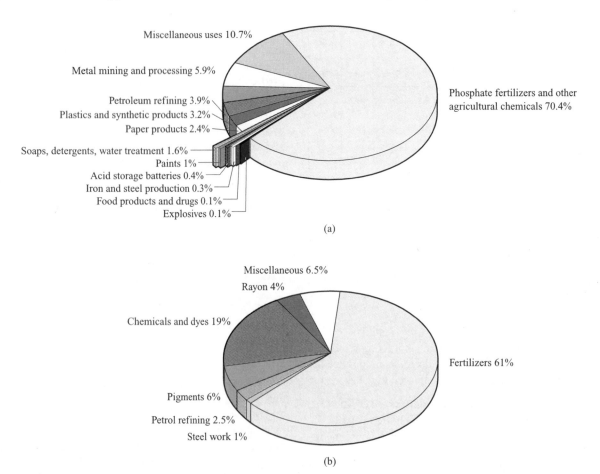

Fig. 15.2 (a) Uses of sulfur in the US in 1990; its role tends to be as a reagent, e.g. as sulfuric acid, in the production of the given product [data: U.S. Geological Survey]. (b) Uses of sulfuric acid [data: *Encyclopedia of Inorganic Chemistry* (1994) ed. R.B. King, Wiley, New York, vol. 7, p. 3968].

Photocopying with selenium

The photoreceptive properties of selenium are responsible for its role in photocopiers: the technique of *xerography* developed rapidly in the latter half of the 20th century. Amorphous selenium or As_2Se_3 (a better photoreceptor than Se) is deposited by a vaporization technique to provide a thin film ($\approx 50\,\mu m$ thick) on an Al drum which is then installed in a photocopier. At the start of a photocopying run, the Se or As_2Se_3 film is charged by a high-voltage corona discharge. Exposure of the Se film to light, with the image to be copied present in the light beam, creates a latent image which is produced in the form of regions of differing electrostatic potential. The image is developed using powdered toner which distributes itself over the 'electrostatic image'. The latter is then transferred to paper (again electrostatically) and fixed by heat treatment. An Se or As_2Se_3-coated photoreceptor drum has a lifetime of $\approx 100\,000$ photocopies. Spent drums are recycled, with some of the main recycling units being in Canada, Japan, the Philippines and several European countries. Once the mainstay of the photocopying industry, Se is gradually being replaced by organic photoreceptors.

in *Box 11.2* we discussed the importance of desulfurization processes.

$$2H_2S + O_2 \xrightarrow{\text{activated carbon or alumina catalyst}} 2S + 2H_2O$$
(15.1)

Commercial sources of Se and Te are flue dusts deposited during the refining of, for example, copper sulfide ores.

Uses

The chief use of O_2 is as a fuel (e.g. for oxyacetylene and hydrogen flames), as a supporter of respiration under special conditions (e.g. in air- and spacecraft), and in steel manufacturing.

Sulfur, mainly in the form of sulfuric acid, is an enormously important industrial chemical; sulfuric acid ranks first among US chemicals and its consumption in a given nation is an indicator of that country's industrial development. Figure 15.2 illustrates applications of sulfur and sulfuric acid (see also *Boxes 8.4* and *10.3*). Sulfur is usually present in the form of an industrial *reagent*, (e.g. in H_2SO_4 in the production of superphosphate fertilizers described in Section 14.2) and it is not necessarily present in the end product; 88% of the uses shown in Figure 15.2a consume H_2SO_4.

An important property of Se is its ability to convert light into electricity, and the element is used in photoelectric cells, photographic exposure meters and photocopiers (see *Box 15.1*). Below its melting point, Se is a semiconductor. Tellurium is used to harden lead, and other applications stem from its semiconducting properties, e.g. cadmium telluride has recently been incorporated into solar cells (see *Box 13.3*). However, uses of Te are limited, partly because Te compounds are readily absorbed by the body and excreted in the breath and perspiration as foul-smelling organic derivatives.

15.3 Physical properties and bonding considerations

Table 15.1 lists selected physical properties of the group 16 elements. The trend in electronegativity values has important consequences as regards the ability of O−H bonds to form hydrogen bonds; this pattern follows that in group 15. There is thermochemical evidence for weak O−H····S bonds, e.g. between phenol and *n*-butyl sulfide in CCl_4 solution.

In comparing Table 15.1 with analogous tables in Chapters 10–14, we should note the importance of *anion*, rather than cation, formation. With the possible exception of PoO_2, there is no evidence that group 16 compounds contain *monatomic cations*; thus Table 15.1 lists values only of the *first* ionization energies to illustrate the expected decrease on descending the group. Electron affinity data for oxygen show that reaction 15.2 for E = O is highly endothermic, and O^{2-} ions exist in ionic lattices only because of the high lattice energies of metal oxides (see *Section 5.16*).

$$\left.\begin{array}{l} E(g) + 2e^- \longrightarrow E^{2-}(g) \qquad (E = O, S) \\ \Delta_r H^\circ(298\,K) = \Delta_{EA}H^\circ_1(298\,K) + \Delta_{EA}H^\circ_2(298\,K) \end{array}\right\}$$
(15.2)

Reaction 15.2 for E = S is also endothermic (Table 15.1), but less so than for O since the repulsion between electrons is less in the larger anion. However, the energy needed to compensate for this endothermic step tends not to be available since lattice energies for sulfides are much lower than those of the corresponding oxides because of the much greater radius of the S^{2-} ion. Consequences of this are that:

- high oxidation state oxides (e.g. MnO_2) often have no sulfide analogues;
- corresponding metal oxides and sulfides are often not isostructural;

Table 15.1 Some physical properties of the group 16 elements and their ions.

Property	O	S	Se	Te	Po
Atomic number, Z	8	16	34	52	84
Ground state electronic configuration	$[He]2s^2 2p^4$	$[Ne]3s^2 3p^4$	$[Ar]3d^{10}4s^2 4p^4$	$[Kr]4d^{10}5s^2 5p^4$	$[Xe]4f^{14}5d^{10}6s^2 6p^4$
Enthalpy of atomization, $\Delta_a H^\circ(298\,K)/kJ\,mol^{-1}$	249‡	277	227	197	≈ 146
Melting point, mp / K	54	388	494	725	527
Boiling point, bp / K	90	718	958	1263*	1235
Standard enthalpy of fusion, $\Delta_{fus}H^\circ(mp)/kJ\,mol^{-1}$	0.44	1.72	6.69	17.49	–
First ionization energy, IE_1 / kJ mol^{-1}	1314	999.6	941.0	869.3	812.1
$\Delta_{EA}H_1^\circ(298\,K)/kJ\,mol^{-1}$ **	−141	−201	−195	−190	−183
$\Delta_{EA}H_2^\circ(298\,K)/kJ\,mol^{-1}$ **	+798	+640			
Covalent radius, r_{cov} / pm	73	103	117	135	–
Ionic radius, r_{ion} for X^{2-} / pm	140	184	198	211	–
Pauling electronegativity, χ^P	3.4	2.6	2.6	2.1	2.0
NMR active nuclei (% abundance, nuclear spin)	^{17}O (0.04, $I = \frac{5}{2}$)	^{33}S (0.76, $I = \frac{3}{2}$)	^{77}Se (7.6, $I = \frac{1}{2}$)	^{123}Te (0.9, $I = \frac{1}{2}$) ^{125}Te (7.0, $I = \frac{1}{2}$)	

‡ For oxygen, $\Delta_a H^\circ = \frac{1}{2} \times$ Dissociation energy of O_2.

* For amorphous Te.

** $\Delta_{EA}H_1^\circ(298\,K)$ is the enthalpy change associated with the process $X(g) + e^- \longrightarrow X^-(g) \approx -\Delta U(0\,K)$; see Section 1.10. $\Delta_{EA}H_2^\circ(298\,K)$ refers to the process $X^-(g) + e^- \longrightarrow X^{2-}(g)$.

- agreement between calculated and experimental values of lattice energies (see *Section 5.15*) for many *d*-block metal sulfides is much poorer than for oxides, indicating significant covalent contributions to the bonding.

Similar considerations apply to selenides and tellurides.

Some bond enthalpy terms for compounds of the group 16 elements are given in Table 15.2. In discussing groups 14 and 15, we emphasized the importance of $(p-p)\pi$ bonding for the first element in each group, and the effect of the restriction of an octet of electrons in the valence shell. These factors are also responsible for some of the differences between O and its heavier congeners. For example:

- there are no stable sulfur analogues of CO and NO (although CS_2 and OCS are well known);
- the highest fluoride of oxygen is OF_2, but the later elements form SF_6, SeF_6 and TeF_6.

Values in Table 15.2 illustrate the particular weakness of the O−O and O−F bonds and this can be rationalized in terms

Table 15.2 Some covalent bond enthalpy terms (kJ mol^{-1}) for bonds involving oxygen, sulfur, selenium and tellurium.

O−O	O=O	O−H	O−C	O−F	O−Cl
146	498	464	359	190‡	205‡
S−S	S=S	S−H	S−C	S−F	S−Cl
266	427	366	272	326‡	255‡
Se−Se		Se−H		Se−F	Se−Cl
192		276		285‡	243‡
		Te−H		Te−F	
		238		335‡	

‡ Values for O−F, S−F, Se−F, Te−F, O−Cl, S−Cl and Se−Cl derived from OF_2, SF_6, SeF_6, TeF_6, OCl_2, S_2Cl_2 and $SeCl_2$ respectively.

of lone pair repulsions (see *Figure 14.1*). Note that O−H and O−C bonds are much stronger than S−H and S−C bonds.

NMR active nuclei and isotopes as tracers

Despite its low abundance (Table 15.1), ^{17}O has been used in studies of hydrated ions in aqueous solution.

The isotope ^{18}O is present to an extent of 0.2% in naturally occurring oxygen and is commonly used as a (non-radioactive) tracer for the element. The usual tracer for sulfur is ^{35}S, which is made by an (n,p) reaction on ^{35}Cl; ^{35}S is a β-emitter with $t_{\frac{1}{2}} = 87$ days.

15.4 The elements

Dioxygen

Dioxygen is obtained industrially by the liquefaction and fractional distillation of air, and is stored and transported as a liquid. Convenient laboratory preparations of O_2 are the electrolysis of aqueous alkali using Ni electrodes, and decomposition of H_2O_2 (equation 15.3). A mixture of $KClO_3$ and MnO_2 used to be sold as 'oxygen mixture' (equation 15.4) and the thermal decompositions of many other oxo salts (e.g. KNO_3, $KMnO_4$ and $K_2S_2O_8$) produce O_2.

$$2H_2O_2 \xrightarrow{MnO_2 \text{ or Pt catalyst}} O_2 + 2H_2O \qquad (15.3)$$

$$2KClO_3 \xrightarrow{\Delta, \, MnO_2 \text{ catalyst}} 3O_2 + 2KCl \qquad (15.4)$$

Caution! Chlorates are potentially explosive.

Dioxygen is a colourless gas, but condenses to a pale blue liquid or solid. Its bonding was described in *Sections 1.12*

BOX 15.2   **CHEMICAL AND THEORETICAL BACKGROUND**

Accurate determination of the O–O bond distance in $[O_2]^-$

Textbook discussions of MO theory of homonuclear diatomic molecules often consider the trends in bond distances in $[O_2]^+$, O_2, $[O_2]^-$ and $[O_2]^{2-}$ (see Chapter 1, problem 1.26) in terms of the occupancy of molecular orbitals. However, the determination of the bond distance in the superoxide ion $[O_2]^-$ has not been straightforward owing to disorder problems in the solid state and, as a result, the range of reported values for $d(O–O)$ is large. A cation-exchange method in liquid NH_3 has recently been used to isolate the salt $[1,3-(NMe_3)_2C_6H_4][O_2]_2 \cdot 3NH_3$ from $[NMe_4][O_2]$. In the solid state, each $[O_2]^-$ ion is fixed in a particular orientation by virtue of a hydrogen-bonded network: $N–H\cdots O$ between solvate NH_3 and $[O_2]^-$, and

$C–H\cdots O$ between cation methyl groups and $[O_2]^-$. Structural parameters for the hydrogen bonds indicate that the interactions are very weak; consequently, the length of the bond in the $[O_2]^-$ anion ought not to be significantly perturbed by their presence. In $[1,3-(NMe_3)_2C_6H_4][O_2]_2 \cdot 3NH_3$, there are two crystallographically independent anions with O–O distances of 133.5 and 134.5 pm.

Further reading

H. Seyeda and M. Jansen (1998) *Journal of the Chemical Society, Dalton Transactions*, p. 875.

and *1.13*. In all phases, it is paramagnetic with a *triplet* ground state (see *Box 20.5*), i.e. the two unpaired electrons have the same spin. In this state, O_2 is a powerful oxidizing agent (see *equation 7.28* and associated discussion) but, fortunately, the kinetic barrier is often high; if it were not, almost all organic chemistry would have to carried out in closed systems. However, a *singlet* state, O_2^*, in which the two unpaired electrons have opposite spins (despite being in different molecular orbitals) lies only $92 \, kJ \, mol^{-1}$ above the ground state. This excited state can be generated photochemically by irradiation of O_2 in the presence of an organic dye as sensitizer, or non-photochemically by reactions such as 15.5.[†]

$$H_2O_2 + NaOCl \longrightarrow O_2^* + NaCl + H_2O \tag{15.5}$$

Singlet O_2 is short-lived, but extremely reactive, combining with many organic compounds, e.g. in reaction 15.6, O_2^* acts as a dienophile in a Diels–Alder reaction.

$$\tag{15.6}$$

At high temperatures, O_2 combines with most elements, exceptions being the halogens and noble gases, and N_2 unless under special conditions. Reactions with the group 1 metals are of particular interest, oxides, peroxides, superoxides and suboxides being possible products. Bond lengths in O_2, $[O_2]^-$ and $[O_2]^{2-}$ are 121, 134 and 149 pm (see *Box 15.2*), consistent with a weakening of the bond caused by increased occupation of the π^* MOs (see *Figure 1.23*).

The first ionization energy of O_2 is $1168 \, kJ \, mol^{-1}$ and it may be oxidized by very powerful oxidizing agents such as PtF_6 (equation 15.7). The bond distance of 112 pm in $[O_2]^+$ is in keeping with the trend for O_2, $[O_2]^-$ and $[O_2]^{2-}$. Other

salts include $[O_2]^+[SbF_6]^-$ (made from irradiation of O_2 and F_2 in the presence of SbF_5, or from O_2F_2 and SbF_5) and $[O_2]^+[BF_4]^-$ (equation 15.8).

$$O_2 + PtF_6 \longrightarrow [O_2]^+[PtF_6]^- \tag{15.7}$$

$$2O_2F_2 + 2BF_3 \longrightarrow 2[O_2]^+[BF_4]^- + F_2 \tag{15.8}$$

The chemistry of O_2 is an enormous topic, and examples of its reactions can be found throughout this book; its biological role is discussed in *Chapter 28*.

Ozone

Ozone, O_3, is usually prepared in up to 10% concentration by the action of a silent electrical discharge between two concentric metallized tubes in an apparatus called an *ozonizer*; electrical discharges in thunderstorms convert O_2 into ozone. The action of UV radiation on O_2, or heating O_2 above 2750 K followed by rapid quenching, also produces O_3. In each of these processes, O atoms are produced and combine with O_2 molecules. Pure ozone can be separated from reaction mixtures by fractional liquefaction; the liquid is blue and boils at 163 K to give a perceptibly blue gas with a characteristic 'electric' smell. Molecules of O_3 are bent (Figure 15.3). Ozone absorbs strongly in the UV region, and its presence in the upper atmosphere of the Earth is essential in protecting the planet's surface from over-exposure to UV-radiation from the Sun (see *Box 13.6*).

Ozone is highly endothermic (equation 15.9); the pure liquid is dangerously explosive, and the gas is a very powerful oxidizing agent (equation 15.10).

$$\tfrac{3}{2}O_2(g) \longrightarrow O_3(g) \qquad \Delta_f H^{\circ}(O_3, g, 298 \, K) = +142.7 \, kJ \, mol^{-1} \tag{15.9}$$

$$O_3(g) + 2H^+(aq) + 2e^- \rightleftharpoons O_2(g) + H_2O(l)$$

$$E^{\circ} = +2.07 \, V \tag{15.10}$$

The value of E° in equation 15.10 refers to pH = 0 (see *Box 7.1*), and at higher pH, E diminishes: +1.65 V at pH = 7, and

---

[†]For an introduction to singlet state O_2, see: C.E. Wayne and R.P. Wayne (1996) *Photochemistry*, Oxford University Press, Oxford.

Fig. 15.3 The structures of O_3 and $[O_3]^-$, and contributing resonance structures in O_3. The O–O bond order in O_3 is taken to be 1.5.

+1.24 V at pH = 14. The presence of high concentrations of alkali stabilizes O_3 both thermodynamically and kinetically. Ozone is much more reactive than O_2 (hence the use of O_3 in water purification); reactions 15.11–15.13 typify this high reactivity, as does its reaction with alkenes to give ozonides.

$$O_3 + S + H_2O \longrightarrow H_2SO_4 \qquad (15.11)$$

$$O_3 + 2I^- + H_2O \longrightarrow O_2 + I_2 + 2[OH]^- \qquad (15.12)$$

$$4O_3 + PbS \longrightarrow 4O_2 + PbSO_4 \qquad (15.13)$$

Potassium ozonide, KO_3 (formed in reaction 15.14), is an unstable, red salt which contains the paramagnetic $[O_3]^-$ ion (Figure 15.3). Ozonide salts are known for all the alkali metals, LiO_3 and NaO_3 being isolated for the first time in 1996. The compounds $[Me_4N][O_3]$ and $[Et_4N][O_3]$ have also recently been prepared, using reactions of the type shown in equation 15.15. Ozonides are explosive, but $[Me_4N][O_3]$ is relatively stable, decomposing above 348 K (see also *Sections 10.6 and 10.8*).

$$2KOH + 5O_3 \longrightarrow 2KO_3 + 5O_2 + H_2O \qquad (15.14)$$

$$CsO_3 + [Me_4N][O_2] \xrightarrow{\text{liquid } NH_3} CsO_2 + [Me_4N][O_3] \qquad (15.15)$$

Sulfur: allotropes

The allotropy of sulfur is complicated, and we describe only the best-established species. The tendency for catenation (see *Section 13.3*) by sulfur is high and leads to formation of both rings of varying sizes and chains. Allotropes of

known structure include cyclic S_6, S_7, S_8, S_9, S_{10}, S_{11}, S_{12}, S_{18} and S_{20} (all with puckered rings, e.g. Figures 15.4a–c) and fibrous sulfur (*catena*-S_∞, Figure 15.4d). In most of these, the S–S bond distances are 206 ± 1 pm, indicative of single bond character; the S–S–S bond angles lie in the range 102–108°. The ring conformations of S_6 (chair) and S_8 (crown) are readily envisaged but other rings have more complicated conformations; the structure of S_7 (Figure 15.4b) is noteworthy because of the wide range of S–S bond lengths (199–218 pm) and angles (101.5–107.5°). The energies of interconversion between the cyclic forms are very small.

The most stable allotrope is orthorhombic sulfur (the α-form and standard state of the element) which occurs naturally as large yellow crystals in volcanic areas. At 367.2 K, the α-form transforms reversibly into monoclinic sulfur (β-form). Both the α- and β-forms contain S_8 rings; the density of the α-form is 2.07 g cm^{-3}, compared with 1.94 g cm^{-3} for the β-form in which the packing of the rings is less efficient. However, if single crystals of the α-form are rapidly heated to 385 K, they melt before the α $\longrightarrow$ β transformation occurs. If crystallization takes place at 373 K, the S_8 rings adopt the structure of the β-form, but the crystals must be cooled rapidly to 298 K; on standing at 298 K, a β $\longrightarrow$ α transition occurs within a few weeks. β-Sulfur melts at 401 K, but this is not a true melting point, since some breakdown of S_8 rings take place causing the melting point to be depressed.

Rhombohedral sulfur (the ρ-form) comprises S_6 rings and is obtained by the ring closure reaction 15.16. It decomposes in light to S_8 and S_{12}.

$$S_2Cl_2 + H_2S_4 \xrightarrow{\text{dry diethyl ether}} S_6 + 2HCl \qquad (15.16)$$

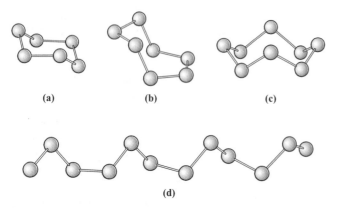

Fig. 15.4 Schematic representations of the structures of some allotropes of sulfur: (a) S_6, (b) S_7, (c) S_8 and (d) *catena*-S_∞ (the chain continues at each end).

Similar ring closures starting from H_2S_x (**15.1**) and S_yCl_2 (**15.2**) lead to larger rings, but a more recent strategy makes use of $[(C_5H_5)_2TiS_5]$ (**15.3**) which is prepared by reaction 15.17 and contains a coordinated $[S_5]^{2-}$ ligand. The Ti(IV) complex reacts with S_yCl_2 to give *cyclo*-S_{y+5}, allowing synthesis of a series of sulfur allotropes. All the *cyclo*-allotropes are soluble in CS_2.

(15.3)

$$2NH_3 + H_2S + \tfrac{1}{2}S_8 \longrightarrow [NH_4]_2[S_5]$$

$$\xrightarrow{[(C_5H_5)_2TiCl_2]} [(C_5H_5)_2TiS_5] \qquad (15.17)$$

By rapidly quenching molten sulfur at 570 K in ice-water, fibrous sulfur (which is insoluble in water) is produced. Fibrous sulfur, *catena*-S_∞, contains infinite, helical chains (Figures 3.15a and 15.4d) and slowly reverts to α-sulfur on standing. α-Sulfur melts to a mobile yellow liquid which darkens in colour as the temperature is raised. At 433 K, the viscosity increases enormously as S_8 rings break by homolytic S—S bond fission, giving diradicals which react together to form polymeric chains containing $\leq 10^6$ atoms. The viscosity reaches a maximum at ≈473 K, and then decreases up to the boiling point (718 K); at this point the liquid contains a mixture of rings and shorter chains. The vapour above liquid sulfur at 473 K consists mainly of S_8 rings, but at higher temperatures, smaller molecules predominate, and above 873 K, paramagnetic S_2 (a diradical like O_2) becomes the main species. Dissociation into atoms occurs above 2470 K.

Sulfur: reactivity

Sulfur is a reactive element. It burns in air with a blue flame to give SO_2, and reacts with F_2, Cl_2 and Br_2 (equation 15.18). For the syntheses of other halides and oxides, see *Sections 15.7* and *15.8*.

$$S_8 \begin{cases} \xrightarrow{F_2} SF_6 \\ \xrightarrow{Cl_2} S_2Cl_2 \\ \xrightarrow{Br_2} S_2Br_2 \end{cases} \qquad (15.18)$$

Sulfur does not react directly with I_2, but in the presence of SbF_5, the salt $[S_7I][SF_6]$ is produced; the cation $[S_7I]^+$ possesses structure **15.4**. When treated with hot aqueous alkali, sulfur forms a mixture of polysulfides, $[S_x]^{2-}$, and polythionates (**15.5**), while oxidizing agents convert it to H_2SO_4.

(15.4)   (15.5)

Saturated hydrocarbons are dehydrogenated when heated with sulfur, and further reaction with alkenes occurs; an application of this reaction is in the vulcanization of rubber in which soft rubber is toughened by cross-linking of the polyisoprene chains, making it suitable for use in, for example, tyres. The reactions of sulfur with CO or $[CN]^-$ yield OCS (**15.6**) or the thiocyanate ion (**15.7**), while treatment with sulfites gives thiosulfates (equation 15.19).

(15.6)   (15.7)

$$Na_2SO_3 + \tfrac{1}{8}S_8 \xrightarrow{H_2O, 373\,K} Na_2S_2O_3 \qquad (15.19)$$

The oxidation of S_8 by AsF_5 or SbF_5 in liquid SO_2 (see *Section 8.5*) yields salts containing the cations $[S_4]^{2+}$, $[S_8]^{2+}$ (Figure 15.5a) and $[S_{19}]^{2+}$. In reaction 15.20, AsF_5 acts as an oxidizing agent *and* a fluoride acceptor (equation 15.21).

$$S_8 + 3AsF_5 \xrightarrow{liquid\ SO_2} [S_8][AsF_6]_2 + AsF_3 \qquad (15.20)$$

$$\left.\begin{array}{l} AsF_5 + 2e^- \longrightarrow AsF_3 + 2F^- \\ AsF_5 + F^- \longrightarrow [AsF_6]^- \end{array}\right\} \qquad (15.21)$$

Two-electron oxidation of S_8 is accompanied by a change in ring conformation (Figure 15.5b); the S—S bond distances around the ring in $[S_8]^{2+}$ are 204 pm (compare 206 pm in S_8) but of significance is the relatively close approach of the two S atoms connected by the hashed line in Figure 15.5b. (Compare the bonding in $[S_8]^{2+}$ with that in S_4N_4, Section 15.10.) Their separation of 283 pm is considerably shorter than in S_8 (470 pm) and indicates some contribution to the bonding by a resonance form with a *transannular* interaction (Figure 15.5c); the second resonance form shown in Figure 15.5c is one of eight structures in which the S=S bond is in different positions around the ring. The $[S_4]^{2+}$ cation is square (S—S = 198 pm) with delocalized bonding. In $[S_{19}]^{2+}$ (Figure 15.5d) two 7-membered, puckered rings are connected by a 5-atom chain; the positive charge may be considered to be localized on two S centres.

A cyclic species has an *annular* form, and a *transannular* interaction is one between atoms across a ring.

Selenium and tellurium

Selenium possesses several allotropes; crystalline, red monoclinic selenium exists in three forms, each containing Se_8 rings with the crown-conformation of S_8 (Figure 15.4c); black selenium consists of larger polymeric rings, and the thermodynamically stable allotrope is grey selenium. Tellurium has only one crystalline form which is a silvery-white metallic-looking solid. In both grey Se and Te, the atoms form infinite, helical chains, the axes of which lie parallel to each other. The red allotropes of Se can be obtained by rapid cooling of molten Se and extraction into CS_2. The photoconductivity of Se (see *Box 15.1*) and Te arises because,

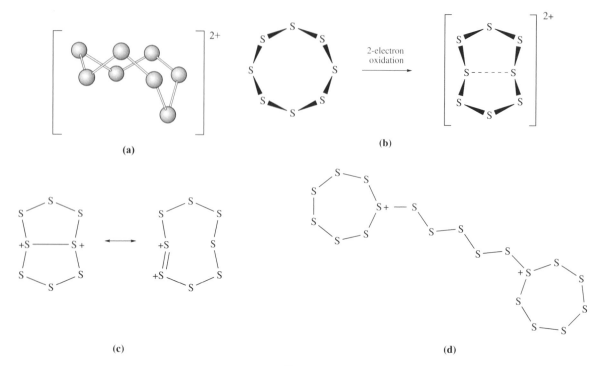

Fig. 15.5 (a) Schematic representation of the structure of $[S_8]^{2+}$. (b) The change in conformation of the ring during oxidation of S_8 to $[S_8]^{2+}$. (c) Resonance structures for $[S_8]^{2+}$ with and without transannular bonds; other similar structures will also contribute. (d) Schematic representation of the structure of $[S_{19}]^{2+}$; the rings are puckered.

in the solid, the band gap of $160 \, kJ \, mol^{-1}$ is small enough for the influence of visible light to cause the promotion of electrons from the filled, bonding MOs to the unoccupied, antibonding MOs (see *Section 5.8*). Although *cyclo*-Te_8 is not known as an allotrope of the element, it has been characterized in the salt $Cs_3[Te_{22}]$ which has the composition $[Cs^+]_3[Te_6^{3-}][Te_8]_2$.

Although less reactive, Se and Te are chemically similar to sulfur. This resemblance extends to the formation of cations such as $[Se_4]^{2+}$, $[Te_4]^{2+}$, $[Se_8]^{2+}$ and $[Te_8]^{2+}$. The salt $[Se_8][AsF_6]_2$ can be made in an analogous manner to $[S_8][AsF_6]_2$ in liquid SO_2 (equation 15.20), whereas reaction 15.22 is carried out in fluorosulfonic acid (see *Section 8.9*). Recent methods use metal halides (e.g. $ReCl_4$ and WCl_6) as oxidizing agents, e.g. the formation of $[Te_8]^{2+}$ (equation 15.23). Reaction 15.24 (in AsF_3 solvent) produces $[Te_6]^{4+}$, **15.8**, which has no S or Se analogue.

$$4Se + S_2O_6F_2 \xrightarrow{HSO_3F} [Se_4][SO_3F]_2 \xrightarrow{Se, HSO_3F} [Se_8][SO_3F]_2$$
$$(15.22)$$

$$2ReCl_4 + 15Te + TeCl_4 \xrightarrow{\Delta. \, sealed \, tube} 2[Te_8][ReCl_6]$$
$$(15.23)$$

$$6Te + 6AsF_5 \xrightarrow{AsF_3} [Te_6][AsF_6]_4 + 2AsF_3 \quad (15.24)$$

The structures of $[Se_4]^{2+}$, $[Te_4]^{2+}$ and $[Se_8]^{2+}$ mimic those of their S analogues, but $[Te_8]^{2+}$ exists in two forms. In $[Te_8][ReCl_6]$, $[Te_8]^{2+}$ is structurally similar to $[S_8]^{2+}$ and

$[Se_8]^{2+}$, but in $[Te_8][WCl_6]_2$, the cation has the bicyclic structure, i.e. resonance structure **15.9** is dominant.

a lies in the range 266–269 pm
b lies in the range 306–315 pm

(15.8)

(15.9)

15.5 Hydrides

Water, H_2O

Aspects of the chemistry of water have already been covered as follows:

- the properties of H_2O (*Section 6.2*);
- acids, bases and ions in aqueous solution (*Chapter 6*);
- 'heavy water', D_2O (*Section 9.3*);
- comparison of the properties of H_2O and D_2O (*Table 9.2*);
- hydrogen bonding (*Section 9.6*).

Water purification is discussed in Box 15.3.

BOX 15.3

Purification of water

The simplest method for the removal of all solid solutes from water is by distillation, but because of the high boiling point and enthalpy of vaporization (Table 6.1), this method is expensive. If the impurities are ionic, ion exchange is an effective (and relatively cheap) means of purification. The treatment involves the passage of water down a column of an organic resin containing acidic groups (e.g. $-SO_3H$) and then down a similar column containing basic groups (e.g. $-NR_3OH$):

$$Resin-SO_3H + M^+ + X^- \longrightarrow Resin-SO_3M + H^+ + X^-$$

$$Resin-NR_3OH + H^+ + X^- \longrightarrow Resin-NR_3X + H_2O$$

After treatment, *deionized water* is produced. The resins are reactivated by treatment with dilute H_2SO_4 and Na_2CO_3 solutions respectively. Reverse osmosis at high pressures is also an important process in water purification, with cellulose acetate as the usual membrane; the latter prevents the passage of dissolved solutes or insoluble impurities. The removal of nitrates is highlighted in Box 14.8.

The purification of drinking water is a complicated industrial process; water may be abundant on the Earth, but impurities such as microorganisms, particulate materials and chemicals usually make it unfit for human consumption. Coagulation and separation methods are used to remove many particles. Aluminium and iron(III) salts are widely used in the coagulation stages, and the treatment relies upon the formation of polymeric species in solution. Pre-polymerized coagulants are now available and include poly-aluminium silicate sulfate (PASS) and polyferric sulfate (PFS). About 67% of all $Al_2(SO_4)_3$ produced ($Al_2(SO_4)_3$ output in the US was 1.25 Mt in 1998) goes into water treatment processes, with the paper manufacturing industry consuming about a half of this amount.

Further reading

J.-Q. Jiang and N.J.D. Graham (1997) *Chemistry & Industry* p. 388 – 'Pre-polymerized inorganic coagulants for treating water and waste water'.

A.A. Delyannis and E.A. Delyannis (1979) Gmelin Handbook of Inorganic Chemistry, *O: Water Desalting*, Supplement Volume 1, System-Number 3, Springer-Verlag, Berlin.

Hydrogen peroxide, H_2O_2

The oldest method for the preparation of H_2O_2 is reaction 15.25. The hydrolysis of peroxodisulfate (produced by electrolytic oxidation of $[HSO_4]^-$ at high current densities using Pt electrodes) has also been an important route to H_2O_2 (equation 15.26).

$$BaO_2 + H_2SO_4 \longrightarrow BaSO_4 + H_2O_2 \qquad (15.25)$$

$$2[NH_4][HSO_4] \xrightarrow[-H_2]{\text{electrolytic oxidation}} [NH_4]_2[S_2O_8]$$

$$\xrightarrow{H_2O} 2[NH_4][HSO_4] + H_2O_2 \qquad (15.26)$$

Nowadays, H_2O_2 is manufactured by the oxidation of 2-ethylanthraquinol (or a related alkyl derivative). The H_2O_2 formed is extracted into water and the organic product is reduced back to starting material; the process is summarized in the catalytic cycle in Figure 15.6.

Some physical properties of H_2O_2 are given in Table 15.3; like water, it is strongly hydrogen-bonded. Pure or strongly concentrated aqueous solutions of H_2O_2 readily decompose (equation 15.27) in the presence of alkali, heavy metal ions, or heterogeneous catalysts (e.g. Pt or MnO_2), and traces of complexing agents (e.g. 8-hydroxyquinoline, **15.10**) or adsorbing materials (e.g. sodium stannate, Na_2SnO_3) are often added as stabilizers.

$$H_2O_2(l) \longrightarrow H_2O(l) + \tfrac{1}{2}O_2(g)$$

$$\Delta_r H^\circ (298 \, K) = -98 \, kJ \text{ per mole of } H_2O_2 \qquad (15.27)$$

Fig. 15.6 The catalytic cycle used in the industrial manufacture of hydrogen peroxide; O_2 is converted to H_2O_2 during the oxidation of the organic alkylanthraquinol. The organic product is reduced by H_2 in a Pd- or Ni-catalyzed reaction. Such cycles are discussed in detail in Chapter 26.

Table 15.3 Selected properties of H_2O_2.

Property	
Physical appearance at 298 K	Colourless (very pale blue) liquid
Melting point / K	272.6
Boiling point / K	425 (decomposes)
$\Delta_f H^o(298\,K)/kJ\,mol^{-1}$	−187.8
$\Delta_f G^o(298\,K)/kJ\,mol^{-1}$	−120.4
Dipole moment / debye	1.57
O−O bond distance (gas phase) / pm	147.5
∠O−O−H (gas phase) / deg	95

(15.10)

Mixtures of H_2O_2 and organic or other readily oxidized materials are dangerously explosive; H_2O_2 mixed with hydrazine has been used as a rocket propellant. Major uses of H_2O_2 include as a bleach, antiseptic, in water pollution control and for the manufacture of sodium peroxoborate (see *Section 12.7*) and peroxocarbonates (see *Section 13.9*).

Figure 15.7 shows the gas phase structure of H_2O_2 and bond parameters are listed in Table 15.3. The internal dihedral angle is sensitive to the surroundings (i.e. the extent of hydrogen bonding) being 111° in the gas phase, 90° in the solid state and 180° in the adduct $Na_2C_2O_4 \cdot H_2O_2$. In this last example, H_2O_2 has a *trans*-planar conformation and the O lone pairs appear to interact with the Na^+ ions. Values of the dihedral angle in organic peroxides, ROOR, show wide variations ($\approx 80° - 145°$).

In aqueous solution, H_2O_2 is partially ionized (equation 15.28), and in alkaline solution, is present as the $[HO_2]^-$ ion.

$$H_2O_2 + H_2O \rightleftharpoons [H_3O]^+ + [HO_2]^-$$
$$K_a = 2.4 \times 10^{-12} \quad (298\,K) \quad (15.28)$$

Hydrogen peroxide is a powerful oxidizing agent as is seen from the standard reduction potential (at pH = 0) in equation 15.29; e.g. it oxidizes I^- to I_2, SO_2 to H_2SO_4 and (in alkaline solution) Cr(III) to Cr(VI). Powerful oxidants such as $[MnO_4]^-$ and Cl_2 will oxidize H_2O_2 (equations 15.30–15.32), and in alkaline solution, H_2O_2 is a good reducing agent (half-equation 15.33).

$$H_2O_2 + 2H^+ + 2e^- \rightleftharpoons 2H_2O \qquad E^o = +1.78\,V \quad (15.29)$$

$$O_2 + 2H^+ + 2e^- \rightleftharpoons H_2O_2 \qquad E^o = +0.70\,V \quad (15.30)$$

$$2[MnO_4]^- + 5H_2O_2 + 6H^+ \longrightarrow 2Mn^{2+} + 8H_2O + 5O_2$$
$$(15.31)$$

$$Cl_2 + H_2O_2 \longrightarrow 2HCl + O_2 \quad (15.32)$$

$$O_2 + 2H_2O + 2e^- \rightleftharpoons H_2O_2 + 2[OH]^- \qquad E^o = -0.15\,V$$
$$(15.33)$$

Tracer studies using ^{18}O show that in these redox reactions $H_2(^{18}O)_2$ is converted to $(^{18}O)_2$, confirming that no oxygen from the solvent (which is not labelled) is incorporated and the O−O bond is not broken.

Deprotonation of H_2O_2 gives $[OOH]^-$ and loss of a second proton yields the peroxide ion, $[O_2]^{2-}$. In addition to peroxide salts such as those of the alkali metals (see *Section 10.6*), many peroxo complexes are known. Figure 15.8 shows two such complexes, one of which also contains the $[OOH]^-$ ion in a bridging mode; typical O−O bond distances for coordinated peroxo groups are $\approx 140 - 148$ pm. Further peroxo complexes are described elsewhere in this book, e.g. *Figure 21.9* and accompanying discussion.

Hydrides H_2E (E = S, Se, Te)

Selected physical data for hydrogen sulfide, selenide and telluride are listed in Table 15.4 and illustrated in *Figures 9.6* and *9.7*. Hydrogen sulfide is more toxic that HCN, but because H_2S has a very characteristic odour of rotten eggs, its presence is easily detected. It is a natural product of decaying S-containing matter, and is present in coal pits, gas wells and sulfur springs. Where it occurs in natural gas deposits, H_2S is removed by reversible absorption in a solution of an organic base and is converted to S by controlled oxidation. In the laboratory, H_2S was historically prepared by reaction 15.34 in a Kipp's apparatus. The hydrolysis of calcium or barium sulfides (e.g. equation 15.35) produces purer H_2S, but the gas is also commercially available in small cylinders.

$$FeS(s) + 2HCl(aq) \longrightarrow H_2S(g) + FeCl_2(aq) \quad (15.34)$$

$$CaS + 2H_2O \longrightarrow H_2S + Ca(OH)_2 \quad (15.35)$$

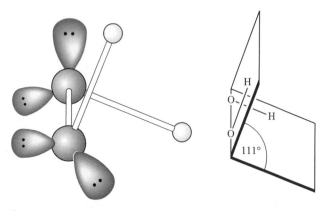

Fig. 15.7 The gas phase structure of H_2O_2. The angle shown as 111° is the *internal dihedral angle*, the angle between the planes containing each OOH-unit; see Table 15.3 for other bond parameters.

(a)

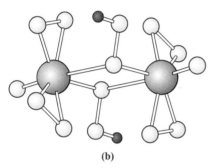

(b)

Fig. 15.8 The structures (X-ray diffraction) of
(a) $[V(O_2)_2(O)(bpy)]^-$ in the hydrated ammonium salt
[H. Szentivanyi *et al.* (1983) *Acta Chem. Scand., Ser. A*, vol.
37, p. 553] and (b) $[Mo_2(O_2)_4(O)_2(\mu\text{-}OOH)_2]^{2-}$ in the
pyridinium salt [J.-M. Le Carpentier *et al.* (1972) *Acta
Crystallogr., Sect. B*, vol. 28, p. 1288]. The H atoms in the
second structure were not located but have been added here
for clarity. Colour code: V and Mo, blue; O, pale grey; N,
pale blue; C, grey; H, black.

Hydrogen selenide may be prepared by reaction 15.36, and a
similar reaction can be used to make H_2Te.

$$Al_2Se_3 + 6H_2O \longrightarrow 3H_2Se + 2Al(OH)_3 \qquad (15.36)$$

The enthalpies of formation of H_2S, H_2Se and H_2Te
(Table 15.4) indicate that the sulfide can be prepared by

direct combination of H_2 and sulfur (boiling), and is more
stable with respect to decomposition into its elements than
H_2Se or H_2Te.

Like H_2O, the hydrides of the later elements in group 16
have bent structures but the angles of $\approx 90°$ (Table 15.4)
are significantly less than that in H_2O (105°). This suggests
that the E−H bonds (E = S, Se or Te) involve p character
from the central atom (i.e. little or no contribution from
the valence s orbital).

In aqueous solution, the hydrides behave as weak acids
(Table 15.4 and Section 6.5). The second ionization constant
of H_2S is $\approx 10^{-19}$ and, thus, metal sulfides are hydrolyzed in
aqueous solution. The only reason that many metal sulfides
can be isolated by the action of H_2S on solutions of their salts
is that the sulfides are extremely insoluble. For example, a
qualitative test for H_2S is its reaction with aqueous lead
acetate (equation 15.37).

$$H_2S + Pb(O_2CCH_3)_2 \longrightarrow \underset{\text{Black ppt.}}{PbS} + 2CH_3CO_2H \qquad (15.37)$$

Sulfides such as CuS, PbS, HgS, CdS, Bi_2S_3, As_2S_3, Sb_2S_3
and SnS have solubility products (see *Sections 6.9* and
6.10) less than $\approx 10^{-30}$ and can be precipitated by H_2S in
the presence of dilute HCl. The acid suppresses ionization
of H_2S, lowering the concentration of S^{2-} in solution.
Sulfides such as ZnS, MnS, NiS and CoS with solubility
products in the range $\approx 10^{-15}$ to 10^{-30} are precipitated only
from neutral or alkaline solutions.

Protonation of H_2S to $[H_3S]^+$ can be achieved using the
superacid HF/SbF_5 (see *Section 8.9*). The salt $[H_3S][SbF_6]$
is a white crystalline solid which reacts with quartz glass;
vibrational spectroscopic data for $[H_3S]^+$ are consistent
with a trigonal pyramidal structure like that of $[H_3O]^+$.

Polysulfanes

Polysulfanes are compounds of the general type H_2S_x where
$x \geq 2$ (see structure **15.1**). Sulfur dissolves in aqueous
solutions of group 1 or 2 metal sulfides (e.g. Na_2S) to yield
polysulfide salts, (e.g. Na_2S_x). Acidification of such solutions

Table 15.4 Selected data for H_2S, H_2Se and H_2Te.

	H_2S	H_2Se	H_2Te
Name[‡]	Hydrogen sulfide	Hydrogen selenide	Hydrogen telluride
Physical appearance and general characteristics	Colourless gas; offensive smell of rotten eggs; toxic	Colourless gas; offensive smell; toxic	Colourless gas; offensive smell; toxic
Melting point / K	187.5	207	224
Boiling point / K	214	232	271
$\Delta_{vap}H°$(bp) / kJ mol^{-1}	18.7	19.7	19.2
$\Delta_f H°$(298 K) / kJ mol^{-1}	−20.6	+29.7	+99.6
$pK_a(1)$	7.04	4.0	3.0
$pK_a(2)$	19	–	–
E−H bond distance / pm	134	146	169
∠H−E−H / deg	92	91	90

‡ The IUPAC names of sulfane, selane and tellane are rarely used.

gives a mixture of polysulfanes as a yellow oil, which can be fractionally distilled to yield H_2S_x ($x = 2$–6). An alternative method of synthesis, particularly useful for polysulfanes with $x > 6$, is by condensation reactions 15.38.

$$2H_2S + S_nCl_2 \longrightarrow H_2S_{n+2} + 2HCl \qquad (15.38)$$

(15.11)

The structure of H_2S_2 (**15.11**) resembles that of H_2O_2 (Figure 15.7) with an internal dihedral angle of $91°$ in the gas phase. All polysulfanes are thermodynamically unstable with respect to decomposition to H_2S and S. Their use in the preparation of *cyclo*-S_n species was described in Section 15.4.

15.6 Metal sulfides, polysulfides, polyselenides and polytellurides

Sulfides

Descriptions of metal sulfides already covered include:

- the zinc blende and wurtzite lattices (*Section 5.11*, *Figures 5.18* and *5.20*);
- precipitation of metal sulfides using H_2S (*Section 15.5*);
- sulfides of the group 14 metals (*Section 13.11*);
- sulfides of the group 15 elements (*Section 14.14*).

The group 1 and 2 metal sulfides possess the antifluorite and NaCl lattices respectively (see *Section 5.11*), and appear to be typical ionic salts. However, the adoption of the NaCl lattice (e.g. by PbS and MnS) cannot be regarded as a criterion for ionic character as we discussed in *Section 13.11*. Most *d*-block metal monosulfides crystallize with the NiAs lattice (e.g. FeS, CoS, NiS) (see *Figure 14.7*) or the zinc blende or wurtzite structure (e.g. ZnS, CdS, HgS) (see *Figures 5.18* and *5.20*). Metal disulfides may adopt the CdI_2 lattice (e.g. TiS_2 and SnS_2 with metal(IV) centres), but others such as FeS_2 (iron pyrites) contain $[S_2]^{2-}$ ions. The latter are formally analogous to peroxides and may be considered to be salts of H_2S_2.

The blue paramagnetic $[S_2]^-$ ion is an analogue of the superoxide ion and has been detected in solutions of alkali metal sulfides in acetone or dimethyl sulfoxide. Simple salts containing $[S_2]^-$ are not known, but the blue colour of the silicate mineral *ultramarine* is due to the presence of the radical anions $[S_2]^-$ and $[S_3]^-$ (see *Box 15.4*).

Polysulfides

Polysulfide ions $[S_x]^{2-}$ are not prepared by deprotonation of the corresponding polysulfanes. Instead, methods of synthesis include reactions 15.17 and 15.39, and that of H_2S with S suspended in NH_4OH solution which yields a mixture of $[NH_4]_2[S_4]$ and $[NH_4]_2[S_5]$.

$$2Cs_2S + S_8 \xrightarrow{\text{aq medium}} 2Cs_2[S_5] \qquad (15.39)$$

(15.12)

Polysulfides of the *s*-block metals are well established. The $[S_3]^{2-}$ ion is bent (**15.12**), but as the chain length increases, it develops a helical twist, rendering it chiral

Ultramarine blues

The soft, metamorphic mineral *lapis lazuri* (or *lazurite*) was prized by Ancient Egyptians for its blue colour and was cut, carved and polished for ornamental uses. Deposits of the mineral occur in, for example, Iran and Afghanistan. Powdering lapis lazuli produces the pigment *ultramarine*, although for commercial purposes, synthetic ultramarine is now manufactured by heating together kaolinite (see *Box 13.8*), Na_2CO_3 and sulfur. Lapis lazuri is related to the aluminosilicate mineral *sodalite*, $Na_8[Al_6Si_6O_{24}]Cl_2$, which contains a zeolite framework (the sodalite or SOD lattice type). The cavities in the zeolite framework contain Na^+ cations and Cl^- anions. Partial or full replacement of Cl^- by the radical anions $[S_2]^-$ and $[S_3]^-$ results in the formation of ultramarines, and the chalcogenide ions give rise to the blue pigmentation. The relative amounts of $[S_2]^-$ and $[S_3]^-$ present determine the colour of the pigment: in the UV–VIS spectrum, $[S_2]^-$ absorbs at $370\,nm$ and $[S_3]^-$ at $595\,nm$. In artificial ultramarines, this ratio can be controlled, so producing a range of colours through from blues to greens.

Further reading

D. Reinen and G.-G. Linder (1999) *Chemical Society Reviews*, vol. 28, p. 75 – 'The nature of the chalcogen colour centres in ultramarine-type solids'.

✳ To view lapis lazuli and the SOD zeolite structure, use the Web links listed on the Website that accompanies this book accessed via http://www.booksites.net/housecroft

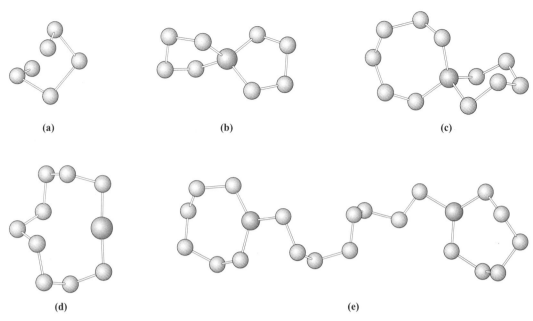

Fig. 15.9 The structures (X-ray diffraction) of (a) $[S_6]^{2-}$ in the salt $[H_3NCH_2CH_2NH_3][S_6]$ [P. Bottcher *et al.* (1984) *Z. Naturforsch., Teil B*, vol. 39, p. 416], (b) $[Zn(S_4)_2]^{2-}$ in the tetraethylammonium salt [D. Coucouvanis *et al.* (1985) *Inorg. Chem.*, vol. 24, p. 24], (c) $[Mn(S_5)(S_6)]^{2-}$ in the $[Ph_4P]^+$ salt [D. Coucouvanis *et al.* (1985) *Inorg. Chem.*, vol. 24, p. 24], (d) $[AuS_9]^-$ in the $[AsPh_4]^+$ salt [G. Marbach *et al.* (1984) *Angew. Chem. Int. Ed., Engl.*, vol. 23, p. 246], and (e) $[(S_6)Cu(\mu\text{-}S_8)Cu(S_6)]^{4-}$ in the $[Ph_4P]^+$ salt [A. Müller *et al.* (1984) *Angew. Chem. Int. Ed., Engl.*, vol. 23, p. 632]. Colour code: metal atoms are shown in grey, S atoms in blue.

(Figure 15.9a). The coordination chemistry of these anions leads to some complexes such as those in Figures 15.9 and 22.21b. For chains containing four or more S atoms, the $[S_x]^{2-}$ ligand often chelates to one metal centre or bridges between two centres; the structure of $[AuS_9]^-$ (Figure 15.9d) illustrates a case where a long chain is required to satisfy the fact that the Au(I) centre favours a linear arrangement of donor atoms.

Polyselenides and polytellurides

Although Se and Te analogues of polysulfanes do not extend beyond the poorly characterized H_2Se_2 and H_2Te_2, the chemistries of polyselenides, polytellurides and their metal complexes are well established. Equations 15.40–15.43 illustrate preparations of salts of $[Se_x]^{2-}$ and $[Te_x]^{2-}$; see *Section 10.8* for details of crown ethers and cryptands.

$$3Se + K_2Se_2 \xrightarrow{\text{DMF}} K_2[Se_5] \qquad (15.40)$$

$$4Se + K_2Se_2 + 2[Ph_4P]Br \longrightarrow [Ph_4P]_2[Se_6] + 2KBr \qquad (15.41)$$

$$3Se + K_2Se_2 \xrightarrow{\text{DMF, 15-crown-5}} [K(15\text{-crown-5})]_2[Se_5] \qquad (15.42)$$

$$2K + 3Te \xrightarrow{\text{1.2-diaminoethane, crypt-222}} [K(\text{crypt-222})]_2[Te_3] \qquad (15.43)$$

(15.13)

(15.14)

Structurally, the smaller polyselenide and polytelluride ions resemble their polysulfide analogues, e.g. $[Te_5]^{2-}$ has structure **15.13** with a helically twisted chain. The structures of higher anions are less simple, e.g. $[Te_8]^{2-}$ (**15.14**) can be considered in terms of $[Te_4]^{2-}$ and $[Te_5]^{2-}$ ligands bound to a Te^{2+} centre. Similarly, $[Se_{11}]^{2-}$ can be described in terms of two $[Se_5]^{2-}$ ligands chelating to a Se^{2+} centre. The coordination chemistry of the $[Se_x]^{2-}$ and $[Te_x]^{2-}$ chain anions has developed significantly since 1990; examples include $[(Te_4)Cu(\mu\text{-}Te_4)Cu(Te_4)]^{4-}$ and $[(Se_4)_2In(\mu\text{-}Se_5)In(Se_4)_2]^{4-}$ (both of which have bridging and chelating ligands), octahedral $[Pt(Se_4)_3]^{2-}$ with chelating $[Se_4]^{2-}$ ligands, and $[Zn(Te_3)(Te_4)]^{2-}$.

15.7 Halides, oxohalides and complex halides

In contrast to the trend found in earlier groups, the stability of the lowest oxidation state ($+2$) of the central atom in the halides of the group 16 elements *decreases* down the group. This is well exemplified in the halides discussed in this section. Our discussion is confined to the fluorides of O, and the fluorides and chlorides of S, Se and Te; the bromides and iodides of the later elements are similar to their chloride analogues. Compounds of O with Cl, Br and I are described in *Section 16.8*.

Oxygen fluorides

Oxygen difluoride, OF_2 (**15.15**), is highly toxic and may be prepared by reaction 15.44. Selected properties are given in Table 15.5. Although OF_2 is formally the anhydride of hypofluorous acid, HOF, only reaction 15.45 occurs with water and this is very slow at 298 K. With concentrated alkali, decomposition is much faster, and with steam, it is explosive.

(15.15)

$$2NaOH + 2F_2 \longrightarrow OF_2 + 2NaF + H_2O \qquad (15.44)$$

$$H_2O + OF_2 \longrightarrow O_2 + 2HF \qquad (15.45)$$

Pure OF_2 can be heated to 470 K without decomposition, but it reacts with many elements (to form fluorides and oxides) at, or slightly above, room temperature. When subjected to UV radiation in an argon matrix at 4 K, the $OF^\bullet$ radical is formed (equation 15.46) and on warming, the radicals combine to give dioxygen difluoride, O_2F_2.

$$OF_2 \xrightarrow{\text{UV radiation}} OF^\bullet + F^\bullet \qquad (15.46)$$

Dioxygen difluoride may also be made by the action of a high voltage discharge on a mixture of O_2 and F_2 at 77–90 K and 1–3 kPa pressure. Selected properties of O_2F_2 are listed in Table 15.5. The low temperature decomposition of O_2F_2 initially yields $O_2F^\bullet$ radicals. Even at low temperatures, O_2F_2 is an extremely powerful fluorinating agent, e.g. it inflames with S at 93 K, and reacts with BF_3 (equation 15.8) and SbF_5 (reaction 15.47).

$$2O_2F_2 + 2SbF_5 \longrightarrow 2[O_2]^+[SbF_6]^- + F_2 \qquad (15.47)$$

The molecular shape of O_2F_2 (**15.16**) resembles that of H_2O_2 (Figure 15.7) although the internal dihedral angle is smaller (87°). The very long O–F bond probably accounts for the ease of dissociation into $O_2F^\bullet$ and $F^\bullet$. Structures **15.17** show appropriate valence bond representations which reflect the long O–F and short O–O bonds; compare the O–O bond distance with those for O_2 and derived ions (Section 15.4) and H_2O_2 (Table 15.3).

(15.16)   (15.17)

Sulfur fluorides and oxofluorides

Table 15.5 lists some properties of the most stable fluorides of sulfur. The fluorides SF_2 and S_2F_4 can be prepared from the reaction of SCl_2 and HgF_2 at elevated temperatures; both are highly unstable. Disulfur difluoride exists as two isomers; S_2F_2 (**15.18**) and $F_2S=S$ (**15.19**), with S_2F_2 (made from AgF and S at 398 K) readily isomerizing to $F_2S=S$. The structure of S_2F_2 is like that of O_2F_2, with an internal dihedral angle of 88°. The S–S bond distances in both isomers are very short (compare $\approx$206 pm for a single S–S bond) and imply multiple bond character; for S_2F_2, contributions from resonance structures analogous to those shown for O_2F_2 are therefore important. Both isomers are

Table 15.5 Selected physical properties of oxygen and sulfur fluorides.

Property	OF_2	O_2F_2	S_2F_2	$F_2S=S$	SF_4	SF_6	S_2F_{10}
Physical appearance and general characteristics	Colourless (very pale yellow) gas; explosive and toxic	Yellow solid below 119 K; decomposes above 223 K	Colourless gas; extremely toxic	Colourless gas	Colourless gas; toxic; reacts violently with water	Colourless gas; highly stable	Colourless liquid; extremely toxic
Melting point / K	49	119	140	108	148	222 (under pressure)	220
Boiling point / K	128	–	288	262	233	subl. 209	303
Δ_fH°(298 K) / kJ mol^{-1}	+24.7	+18.0			–763.2	–1220.5	
Dipole moment / D	0.30	1.44			0.64	0	0
E–F bond distance / pm‡	141	157.5	163.5	160	164.5 (ax) 154.5 (eq)	156	156

$\ddagger$ For other structural data, see text.

unstable with respect to disproportionation into SF_4 and S, and are extremely reactive, attacking glass and being rapidly hydrolyzed by water and alkali (e.g. equation 15.48).

(15.18)   (15.19)

$$2S=SF_2 + 2[OH]^- + H_2O \longrightarrow \tfrac{1}{4}S_8 + [S_2O_3]^{2-} + 4HF$$

$$(15.48)$$

Sulfur tetrafluoride, SF_4, is best prepared by reaction 15.49. It is commercially available and is used as a selective fluorinating agent, e.g. it converts carbonyl groups into CF_2 groups without destroying any unsaturation in the molecule. Representative reactions are shown in Figure 15.10; SF_4 hydrolyzes rapidly and must be handled in moisture-free conditions.

$$3SCl_2 + 4NaF \xrightarrow{\text{MeCN, 350 K}} SF_4 + S_2Cl_2 + 4NaCl$$

$$(15.49)$$

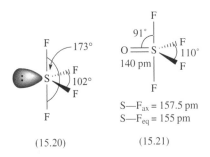

(15.20)   (15.21)

The structure of SF_4, **15.20**, is derived from a trigonal bipyramid. The $S-F_{ax}$ and $S-F_{eq}$ bond distances are quite different (Table 15.5) and, while providing an approximate description of the molecular structure, the VSEPR model (see *Section 1.19*) is an oversimplification. Oxidation by O_2 in the absence of a catalyst to form SOF_4 is slow; the structure of SOF_4, **15.21**, is related to that of SF_4, but with $S-F_{ax}$ and $S-F_{eq}$ bond distances that are close in value.

(15.22)

Among the sulfur fluorides, SF_6, **15.22**, stands out for its high stability and chemical inertness. It can be made by burning S in F_2, and is commercially available, being widely used as an electrical insulator. Its lack of reactivity (e.g. it is unaffected by steam at 770 K or molten alkalis) is kinetic rather than thermodynamic in origin; the value of $\Delta_r G^{\circ}$ for reaction 15.50 certainly indicates thermodynamic spontaneity.

$$SF_6 + 3H_2O \longrightarrow SO_3 + 6HF$$

$$\Delta_r G^{\circ}(298\,\text{K}) = -221\,\text{kJ mol}^{-1} \quad (15.50)$$

The $S-F$ bond in SF_6 is undoubtedly strong (Table 15.2) but the extent to which the S $3d$ orbitals participate in bonding remains a matter for debate as we discussed in *Section 4.7*.

The preparation of SF_6 from S and F_2 produces small amounts of S_2F_{10} and the yield can be optimized by controlling the reaction conditions. An alternative route is reaction 15.51. Selected properties of S_2F_{10} are given in Table 15.5.

$$2SF_5Cl + H_2 \xrightarrow{h\nu} S_2F_{10} + 2HCl \quad (15.51)$$

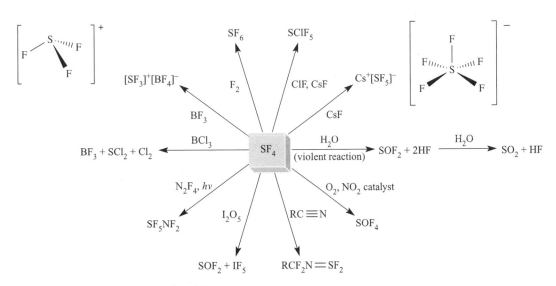

Fig. 15.10 Selected reactions of sulfur tetrafluoride.

(15.23)

Molecules of S_2F_{10} have the staggered structure **15.23**; the S–S bond length of 221 pm is significantly longer than the single bonds in elemental S (206 pm). It disproportionates when heated (equation 15.52) and is a powerful oxidizing agent; an interesting reaction is that with NH_3 to yield $N{\equiv}SF_3$ (see structure **15.52**).

$$S_2F_{10} \xrightarrow{420\,K} SF_4 + SF_6 \qquad (15.52)$$

Many compounds containing SF_5 groups are now known, including $SClF_5$ and SF_5NF_2 (Figure 15.10). In accord with the relative strengths of the S–Cl and S–F bonds (Table 15.2), reactions of $SClF_5$ usually involve cleavage of the S–Cl bond (e.g. reaction 15.53).

$$2SClF_5 + O_2 \xrightarrow{h\nu} F_5SOOSF_5 + Cl_2 \qquad (15.53)$$

∠F–S–F = 92°
∠F–S–O = 106°

(15.24)

Sulfur forms several oxofluorides, and we have already mentioned SOF_4. Thionyl fluoride (or sulfinyl fluoride),[†] SOF_2 (**15.24**), is a colourless gas (bp 229 K), prepared by fluorinating $SOCl_2$ using SbF_3. It reacts with F_2 to give SOF_4, and is slowly hydrolyzed by water (see Figure 15.10). Sulfuryl fluoride (or sulfonyl fluoride), SO_2F_2 (**15.25**), is a colourless gas (bp 218 K) which is made by reactions 15.54 or 15.55.

$$SO_2Cl_2 + 2NaF \longrightarrow SO_2F_2 + 2NaCl \qquad (15.54)$$

$$Ba(SO_3F)_2 \xrightarrow{\Delta} SO_2F_2 + BaSO_4 \qquad (15.55)$$

∠F–S–F = 97°
∠O–S–O = 123°

(15.25)

Although unaffected by water, SO_2F_2 is hydrolyzed by concentrated aqueous alkali. A series of sulfuryl fluorides

---

[†] The names thionyl or sulfinyl signify the presence of an SO group; sulfonyl or sulfuryl show that an SO_2 group is present.

is known, including FSO_2OSO_2F and FSO_2OOSO_2F. The latter compound is prepared by reaction 15.56; fluorosulfonic acid (see *Section 8.9*) is related to the intermediate in this reaction.

$$SO_3 + F_2 \xrightarrow{AgF_2,\,450\,K} FSO_2OF \xrightarrow{SO_3} FSO_2OOSO_2F \qquad (15.56)$$

The dissociation of FSO_2OOSO_2F at 393 K produces the brown paramagnetic radical $FSO_2O^{\bullet}$, selected reactions of which are shown in scheme 15.57.

$\xrightarrow{C_2F_4} FSO_2OCF_2CF_2OSO_2F$
$\xrightarrow{KI} K[I(OSO_2F)_4]$
$\xrightarrow{Cl_2} 2ClOSO_2F$

(15.57)

The reaction of F_2 with sulfate ion yields $[FSO_4]^-$ which can be isolated as the caesium salt and is an extremely powerful oxidizing agent (equation 15.58).

$$[FSO_4]^- + 2H^+ + 2e^- \rightleftharpoons [HSO_4]^- + HF \quad E^{\circ} \approx +2.5\,V \qquad (15.58)$$

Sulfur chlorides and oxochlorides

The range of sulfur chlorides and oxochlorides (which are all hydrolyzed by water) is far more restricted than that of the corresponding fluorides, and there are no stable chloro-analogues of SF_4, SF_6 and S_2F_{10}. One example of a high oxidation state chloride is $SClF_5$, prepared as shown in Figure 15.10.

Disulfur dichloride, S_2Cl_2, is a fuming orange liquid (mp 193 K, bp 409 K) which is toxic and has a repulsive smell. It is manufactured by passing Cl_2 through molten S, and further chlorination yields SCl_2 (a dark-red liquid, mp 195 K, dec. 332 K); both are used industrially for the manufacture of $SOCl_2$ (reactions 15.59) and S_2Cl_2 for the vulcanization of rubber. Pure SCl_2 is unstable with respect to equilibrium 15.60.

$$\left.\begin{array}{l} 2SO_2 + S_2Cl_2 + 3Cl_2 \longrightarrow 4SOCl_2 \\ SO_3 + SCl_2 \longrightarrow SOCl_2 + SO_2 \end{array}\right\} \qquad (15.59)$$

$$2SCl_2 \rightleftharpoons S_2Cl_2 + Cl_2 \qquad (15.60)$$

Internal dihedral angle = 84°

(15.26)

The structure of S_2Cl_2, **15.26**, resembles that of S_2F_2, while SCl_2 is a bent molecule (S–Cl = 201 pm, ∠Cl–S–Cl = 103°). Decomposition of both chlorides by water yields a complex

Table 15.6   Selected properties of the fluorides of selenium and tellurium.

Property	SeF$_4$	SeF$_6$	TeF$_4$	TeF$_6$
Physical appearance and general characteristics	Colourless fuming liquid; toxic; violent hydrolysis	White solid at low temp.; colourless gas; toxic	Colourless solid; highly toxic	White solid at low temp.; colourless gas; foul smelling; highly toxic
Melting point / K	263.5	subl. 226	403	subl. 234
Boiling point / K	375	–	dec. 467	
$\Delta_f H^\circ$(298 K) / kJ mol^{-1}		−1117.0		−1318.0
E−F bond distance for gas phase molecules / pm ‡	Se−F$_{ax}$ = 176.5 Se−F$_{eq}$ = 168	169	Te−F$_{ax}$ = 190 Te−F$_{eq}$ = 179	181.5

‡ For other structural data, see text.

mixture containing S, SO$_2$, H$_2$S$_5$O$_6$ and HCl. Equation 15.16 showed the use of S$_2$Cl$_2$ in the formation of an S$_n$ ring. Condensation of S$_2$Cl$_2$ with polysulfanes (equation 15.61) gives rise to chlorosulfanes that can be used, for example, in the formation of various sulfur rings (see structures **15.1** and **15.2**).

$$ClS\!-\!SCl + H\!-\!\!\overset{(S)_x}{\frown}\!\!-H + ClS\!-\!SCl$$
$$\longrightarrow ClS_{x+4}Cl + 2HCl \quad (15.61)$$

Thionyl chloride, SOCl$_2$ (prepared, for example, according to equations 15.59 or 15.62), and sulfonyl chloride, SO$_2$Cl$_2$ (prepared by reaction 15.63) are colourless, fuming liquids: SOCl$_2$, bp 351 K, SO$_2$Cl$_2$, bp 342 K. Their ease of hydrolysis by water accounts for their fuming nature, e.g. equation 15.64.

$$SO_2 + PCl_5 \longrightarrow SOCl_2 + POCl_3 \quad (15.62)$$

$$SO_2 + Cl_2 \xrightarrow{\text{activated charcoal}} SO_2Cl_2 \quad (15.63)$$

$$SOCl_2 + H_2O \longrightarrow SO_2 + 2HCl \quad (15.64)$$

The structural parameters shown for SOCl$_2$, **15.27**, and SO$_2$Cl$_2$, **15.28**, are for the gas phase molecules.

144 pm $\diagdown$ S $\overset{\text{207 pm}}{\cdots\cdots\!/\!/\!/\!\text{Cl}}$
O $\diagup$ $\qquad$ Cl
∠Cl–S–Cl = 97°
∠Cl–S–O = 108°

(15.27)

140 pm $\diagdown$ S $\overset{\text{O}\;\;\text{201 pm}}{\cdots\cdots\!/\!/\!/\!\text{Cl}}$
O $\diagup$ $\qquad$ Cl
∠Cl–S–Cl = 100°
∠O–S–O = 123.5°

(15.28)

Both thionyl and sulfonyl chlorides are available commercially. Thionyl chloride is used for the preparation of acyl chlorides (equation 15.65) and anhydrous metal chlorides (i.e. removing water of crystallization by reaction 15.64), while SO$_2$Cl$_2$ is a chlorinating agent.

$$RCO_2H + SOCl_2 \xrightarrow{\Delta} RC(O)Cl + SO_2 + HCl \quad (15.65)$$

Halides of selenium and tellurium

In contrast to the various fluorides formed by S, well-characterized fluorides of Se and Te are known only for oxidation states +4 and +6. Table 15.6 lists selected properties of SeF$_4$, SeF$_6$, TeF$_4$ and TeF$_6$. Selenium tetrafluoride is a good fluorinating agent; it is a liquid at 298 K and (compared to SF$_4$) is relatively convenient to handle. It is prepared by reacting SeO$_2$ with SF$_4$; combination of F$_2$ and Se yields SeF$_6$ which is thermally stable and relatively inert. The tellurium fluorides are similarly prepared, TeF$_4$ from TeO$_2$ and SF$_4$ (or SeF$_4$), and TeF$_6$ from the elements. In the liquid and gas phases, SeF$_4$ contains discrete molecules (Figure 15.11a) but in the solid state, significant intermolecular interactions are present. However, these are considerably weaker than in TeF$_4$ in which the formation of Te−F−Te bridges leads to a polymeric structure in the crystal (Figure 15.11b). Fluorine-19 NMR spectroscopic studies of liquid SeF$_4$ have shown that the molecules are stereochemically non-rigid (see *Section 2.11*). The structures of SeF$_6$ and TeF$_6$ are regular octahedra. Tellurium hexafluoride is hydrolyzed by water to telluric acid, H$_6$TeO$_6$, and undergoes a number of exchange reactions such as reaction 15.66. It is also a fluoride acceptor, reacting with alkali metal fluorides and [Me$_4$N]F under anhydrous conditions (equation 15.67).

$$TeF_6 + Me_3SiNMe_2 \longrightarrow Me_2NTeF_5 + Me_3SiF \quad (15.66)$$

$$\left.\begin{array}{l} TeF_6 + [Me_4N]F \xrightarrow{\text{MeCN, 233 K}} [Me_4N][TeF_7] \\[4pt] [Me_4N][TeF_7] + [Me_4N]F \xrightarrow{\text{MeCN, 273 K}} [Me_4N]_2[TeF_8] \end{array}\right\}$$
$$(15.67)$$

The [TeF$_7$]$^-$ ion has a pentagonal bipyramidal structure (**15.29**) although in the solid state, the equatorial F atoms deviate slightly from the mean-equatorial plane. In [TeF$_8$]$^{2-}$, **15.30**, vibrational spectroscopic data are consistent with the Te centre being in a square-antiprismatic environment.

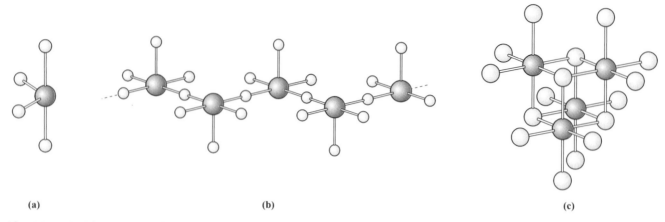

Fig. 15.11 (a) The structure of SeF_4 in the gas and liquid phases; (b) in the solid state, TeF_4 consists of polymeric chains; (c) the structure of the molecular Se_4Cl_{16}-unit present in the crystal lattice of $SeCl_4$. Colour code: S, Se and Te, blue; F and Cl, pale grey.

Te—F_{ax} = 179 pm
Te—F_{eq} = 183–190 pm

(15.29)   (15.30)

In contrast to S, Se and Te form stable tetrachlorides, formed by direct combination of the elements. Both the tetrachlorides are solids ($SeCl_4$, colourless, subl. 469 K; $TeCl_4$ yellow, mp 497 K, bp 653 K) which contain tetrameric units, depicted in Figure 15.11c for $SeCl_4$. The E—Cl (E = Se or Te) bonds within the cubane-core are significantly longer than the terminal E—Cl bonds; e.g. Te—Cl = 293 (core) and 231 (terminal) pm. Thus, the structure may also be described in terms of $[ECl_3]^+$ and Cl^- ions.

> A *cubane* contains a central cubic (or near-cubic) arrangement of atoms.

The $[SeCl_3]^+$ and $[TeCl_3]^+$ cations are also formed in reactions with Cl^- acceptors, e.g. reaction 15.68.

$$SeCl_4 + AlCl_3 \longrightarrow [SeCl_3]^+ + [AlCl_4]^- \qquad (15.68)$$

Both $SeCl_4$ and $TeCl_4$ are readily hydrolyzed by water, but with group 1 metal chlorides in the presence of concentrated HCl, yellow complexes such as $K_2[SeCl_6]$ and $K_2[TeCl_6]$ are formed. Reaction 15.69 is an alternative route to $[TeCl_6]^{2-}$, while $[SeCl_6]^{2-}$ is formed when $SeCl_4$ is dissolved in molten $SbCl_3$ (equation 15.70).

$$TeCl_4 + 2^tBuNH_2 + 2HCl \longrightarrow 2[^tBuNH_3]^+ + [TeCl_6]^{2-}$$

$$\qquad (15.69)$$

$$2SbCl_3 + SeCl_4 \rightleftharpoons 2[SbCl_2]^+ + [SeCl_6]^{2-} \qquad (15.70)$$

The $[ECl_6]^{2-}$ (E = Se or Te) ions are *regular octahedral*, rather than the distorted structure (with a stereochemically active lone pair) expected on the basis of VSEPR theory. Two explanations for this observation can be considered, and we use $[TeCl_6]^{2-}$ as the example.

- The 5s electrons in the valence shell of Te are not used in Te—Cl bonding, and remain stereochemically inactive; using low-lying 6s and 5d orbitals, an sp^3d^2 hybridization scheme accommodates 12 electrons in the valence shell of Te, sufficient to give Te—Cl single bonds.
- A molecular orbital scheme for $[TeCl_6]^{2-}$ can be developed that uses only the valence shell 5p orbitals; combined with six Cl 3p orbitals, this leads to *three* bonding, three non-bonding and three antibonding MOs in $[TeCl_6]^{2-}$, and a Te—Cl bond order of 0.5 (a similar bonding scheme was described for XeF_2 in *Section 4.7*).

In the case of $[SeCl_6]^{2-}$ and $[TeCl_6]^{2-}$, there are no available experimental data to distinguish between the two bonding schemes.[†]

Tellurium forms a series of subhalides, e.g. Te_3Cl_2 and Te_2Cl, the structures of which can be related to the helical chains in elemental Te. When Te is oxidized to Te_3Cl_2, oxidation of one in three Te atoms occurs to give polymer **15.31**.

(15.31)

[†] For discussion of regular versus distorted structures in $[TeCl_6]^{2-}$ and $[TeBr_6]^{2-}$, see: L.-J. Baker, C.E.F. Rickard and M.J. Taylor (1995) *Polyhedron*, vol. 14, p. 401.

15.8 Oxides

Oxides of sulfur

The most important oxides of sulfur are SO_2 and SO_3, but there are also a number of unstable oxides. Among these are S_2O (**15.32**) and S_8O (**15.33**), made by reactions 15.71 and 15.72; the oxides S_nO ($n = 6$–10) can be prepared by reaction 15.73, exemplified for S_8O.

(15.32) (15.33)

$$SOCl_2 + Ag_2S \xrightarrow{430\,K} S_2O + 2AgCl \qquad (15.71)$$

$$HS_7H + SOCl_2 \longrightarrow S_8O + 2HCl \qquad (15.72)$$

$$S_8 \xrightarrow{CF_3C(O)OOH} S_8O \qquad (15.73)$$

Sulfur dioxide is manufactured on a large scale by burning sulfur (the most important process) or H_2S, by roasting sulfide ores (e.g. equation 15.74), or reducing $CaSO_4$ (equation 15.75). Desulfurization processes to limit SO_2 emissions (see *Box 11.2*) and reduce acid rain (see *Box 15.5*) are now in use. In the laboratory, SO_2 may be prepared by, for example, reaction 15.76, and it is commercially available in cylinders. Selected physical properties of SO_2 are listed in Table 15.7.

$$4FeS_2 + 11O_2 \xrightarrow{\Delta} 2Fe_2O_3 + 8SO_2 \qquad (15.74)$$

$$CaSO_4 + C \xrightarrow{>1620\,K} CaO + SO_2 + CO \qquad (15.75)$$

$$Na_2SO_3 + 2HCl \xrightarrow{\text{conc}} SO_2 + 2NaCl + H_2O \qquad (15.76)$$

At 298 K, SO_2 is a liquid and a good solvent (see *Section 8.5*). Sulfur dioxide has a molecular structure (**15.34**). Unlike O, S is not restricted to an octet of electrons in the valence shell and the S–O bonds are commonly taken to be double bonds; this is in contrast to the bond order of 1.5 for the O–O bonds in O_3 (Figure 15.3).

(15.34)

Sulfur dioxide reacts with O_2 (see below), F_2 and Cl_2 (equation 15.77).

$$SO_2 + X_2 \longrightarrow SO_2X_2 \qquad (X = F, Cl) \qquad (15.77)$$

BOX 15.5	RESOURCES, ENVIRONMENTAL AND BIOLOGICAL

The contribution of SO_2 to acid rain

Despite being recognized as far back as the 1870s, the environmental problems associated with 'acid rain' came to the fore in the 1960s with the decline of fish stocks in European and North American lakes. Two of the major contributors towards acid rain are SO_2 and NO_x. (In Section 26.6, we discuss the use of catalytic converters to combat pollution due to nitrogen oxides, NO_x.) Although SO_2 emissions arise from natural sources such as volcanic eruptions, artificial sources contribute ≈90% of the sulfur in the atmosphere. Fossil fuels such as coal contain ≈2–3% sulfur and combustion produces SO_2, and the gas is released when metal sulfide ores are roasted in the production of metals such as Co, Ni, Cu (equation 21.6) and Zn. Once released, SO_2 dissolves in the atmospheric water vapour, forming H_2SO_3 and H_2SO_4. Acid formation may take several days and involves multistage reactions, the outcome of which is:

$$2SO_2 + O_2 + 2H_2O \longrightarrow 2H_2SO_4$$

By the time acid rain falls to the Earth's surface, the pollutants can have travelled long distances from their industrial sources so, for example, prevailing winds in Europe may carry SO_2 from the UK, France and Germany to Scandinavia.

The effects of acid rain can be devastating. The pH of lakes and streams is lowered, although the composition of the bedrock is significant, and in some cases provides a natural buffering effect. A second effect is that acid rain penetrating the bedrock can react with aluminosilicate minerals, or can leach heavy metal ions from the bedrock; as the acid rain makes its ways through the bedrock and into waterways, it carries with it the metal pollutants. Acidified and polluted waters not only kill fish, but also affect the food chain. Acid rain falling on soils may be neutralized if the soil is alkaline, but otherwise the lowering of the pH and the leaching of plant nutrients has devastating effects on vegetation. The effects of acid rain on some building materials are all around us: crumbling gargoyles on ancient churches are a sad reminder of pollution by acid rain.

International legalization to reduce acidic gas emissions has been in operation since the 1980s, and recent environmental studies indicate some improvement in the state of Western European and North American streams and lakes. There is, however, a long way to go.

For related information: see *Box 11.2*: Desulfurization processes to limit SO_2 emissions; *Box 15.6*: Volcanic emissions.

Further reading

J.L. Stoddard *et al.* (1999) *Nature*, vol. 401, p. 575 – 'Regional trends in aquatic recovery from acidification in North America and Europe'.

Table 15.7 Selected physical properties of SO_2 and SO_3.

Property	SO_2	SO_3
Physical appearance and general characteristics	Colourless, dense gas; pungent smell	Volatile white solid, or a liquid
Melting point / K	198	290
Boiling point / K	263	318
$\Delta_{vap}H^\circ$(bp) / kJ mol^{-1}	24.9	40.7
$\Delta_f H^\circ$(298 K) / kJ mol^{-1}	−296.8 (SO_2, g)	−441.0 (SO_3, l)
Dipole moment / D	1.63	0
S−O bond distance / pm ‡	143	142
∠O−S−O / deg ‡	119	120

‡ Gas phase parameters; for SO_3, data refer to the monomer.

In aqueous solution, it is converted to only a small extent to sulfurous acid; aqueous solutions of H_2SO_3 contain significant amounts of dissolved SO_2 (see *equations 6.18–6.20*). Sulfur dioxide is a weak reducing agent in acidic solution, and a slightly stronger one in basic media (equations 15.78 and 15.79).

$$[SO_4]^{2-}(aq) + 4H^+(aq) + 2e^- \rightleftharpoons H_2SO_3(aq) + H_2O(l)$$

$$E^\circ = +0.17\,V \qquad (15.78)$$

$$[SO_4]^{2-}(aq) + H_2O(l) + 2e^- \rightleftharpoons [SO_3]^{2-}(aq) + 2[OH]^-(aq)$$

$$E^\circ_{[OH^-]=1} = -0.93\,V \qquad (15.79)$$

Thus, aqueous solutions of SO_2 are oxidized to sulfate by many oxidizing agents (e.g. I_2, $[MnO_4]^-$, $[Cr_2O_7]^{2-}$ and Fe^{3+} in acidic solutions). However, if the concentration of H^+ is very high, $[SO_4]^{2-}$ can be reduced to SO_2 as in, for example, reaction 15.80; the dependence of E° on $[H^+]$ was detailed in *Section 7.2*.

$$Cu + 2H_2SO_4 \xrightarrow{\quad} SO_2 + CuSO_4 + 2H_2O \qquad (15.80)$$
$$\text{conc}$$

In the presence of concentrated HCl, SO_2 will itself act as an oxidizing agent; in reaction 15.81, the Fe(III) produced is then complexed by Cl$^-$.

$$\left.\begin{array}{l} SO_2 + 4H^+ + 4Fe^{2+} \xrightarrow{\quad} S + 4Fe^{3+} + 2H_2O \\ Fe^{3+} + 4Cl^- \xrightarrow{\quad} [FeCl_4]^- \end{array}\right\} \qquad (15.81)$$

The oxidation of SO_2 by atmospheric O_2 (equation 15.82) is very slow, but is catalyzed by V_2O_5 (see *Section 26.6*). This is the first step in the *Contact process* for the manufacture of sulfuric acid; operating conditions are crucial since equilibrium 15.82 shifts further towards the left-hand side as the temperature is raised, although the yield can be increased somewhat by use of high pressures of air. In practice, the industrial catalytic process operates at $\approx$750 K and achieves conversion factors >98%.

$$2SO_2 + O_2 \rightleftharpoons 2SO_3 \qquad \Delta_r H^\circ = -96\,kJ \text{ per mole of } SO_2 \qquad (15.82)$$

Gaseous SO_3 is removed from the reaction mixture by passage through concentrated H_2SO_4, in which it dissolves to form *oleum* (see *Section 15.9*). Absorption into water to yield H_2SO_4 directly is not a viable option; SO_3 reacts vigorously and very exothermically with H_2O, forming a thick mist. On a small scale, SO_3 can be prepared by heating oleum.

(15.35)

Table 15.7 lists selected physical properties of SO_3. In the gas phase, it is an equilibrium mixture of monomer (planar molecules, **15.35**) and trimer. Solid SO_3 is polymorphic, with all forms containing SO_4-tetrahedra sharing two oxygen atoms. Condensation of the vapour at low temperatures yields γ-SO_3 which contains trimers (Figure 15.12a); crystals of γ-SO_3 have an ice-like appearance. The bond lengths of 140 and 160 pm are consistent with localized double and single bond character respectively. In the presence of traces of water, white crystals of β-SO_3 form; β-SO_3 consists of polymeric chains (Figure 15.12b), as does α-SO_3 in which the chains are arranged into layers in the solid state lattice. Differences in the thermodynamic properties of the different polymorphs are very small, although they do react with water at different rates. Sulfur trioxide is very reactive and representative reactions are given in scheme 15.83.

(15.83)

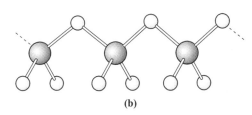

(a)

(b)

Fig. 15.12 The structures of solid state polymorphs of sulfur trioxide contains tetrahedral SO_4 units: (a) γ-SO_3 consists of trimeric units and (b) α- and β-SO_3 contain polymeric chains. Colour code: S, blue; O, pale grey.

Oxides of selenium and tellurium

Selenium and tellurium dioxides are white solids obtained by direct combination of the elements; the polymorph of TeO_2 so formed is α-TeO_2, whereas β-TeO_2 occurs naturally as the mineral *tellurite*. Both forms of TeO_2 contain structural units **15.36** which are connected by shared O atoms into a three-dimensional lattice in α-TeO_2, and into a sheet structure in the β-form. The structure of SeO_2 consists of chains (**15.37**) in which the Se centres are in trigonal pyramidal environments. Whereas SeO_2 sublimes at 588 K, TeO_2 is an involatile solid (mp 1006 K). In the gas phase, SeO_2 is monomeric with structure **15.38**. The trends in structures of the dioxides of S, Se and Te and their associated properties

(e.g. mp, volatility) reflect the increase in metallic character on descending group 16.

(15.36)   (15.37)   (15.38)

Selenium dioxide is very toxic and is readily soluble in water to give selenous acid, H_2SeO_3. It is readily reduced, e.g. by hydrazine, and is used as an oxidizing agent in organic reactions. The α-form of TeO_2 is sparingly soluble in water, giving H_2TeO_3, but is soluble in aqueous alkali. Like SeO_2, TeO_2 is a good oxidizing agent. Selenium trioxide is a white, hygroscopic solid. It is difficult to prepare, being thermodynamically unstable with respect to SeO_2 and O_2 ($\Delta_f H^\circ$(298 K): $SeO_2 = -225$; $SeO_3 = -184\,kJ\,mol^{-1}$). It may be made by reaction of SO_3 with K_2SeO_4 (a salt of selenic acid). Selenium trioxide decomposes at 438 K, is soluble in water, and is a stronger oxidizing agent than SO_3. In the solid state, tetramers (**15.39**) are present.

(15.39)

BOX 15.6

RESOURCES, ENVIRONMENTAL AND BIOLOGICAL

Volcanic emissions

The eruption of a volcano is accompanied by emissions of water vapour (>70% of the volcanic gases), CO_2 and SO_2 plus lower levels of CO, sulfur vapour and Cl_2. Carbon dioxide contributes to the 'greenhouse' effect, and it has been estimated that volcanic eruptions produce $\approx$112 million tonnes of CO_2 per year. Levels of CO_2 in the flume of a volcano can be monitored by IR spectroscopy. Sulfur dioxide emissions are particularly damaging to the environment, since they result in the formation of acid rain. Sulfuric acid aerosols persist as suspensions in the atmosphere for long periods after an eruption. The Mount St. Helens eruption occurred in May 1980. Towards the end of the eruption, the level of SO_2 in the volcanic plume was $\approx$2800 tonnes per day, and an emission rate of $\approx$1600 tonnes per day was

measured in July 1980. Emissions of SO_2 (diminishing with time after the major eruption) continued for over two years, being boosted periodically by further volcanic activity.
Related discussions: see *Box 11.2*; *Box 13.7*; *Box 15.5*.

Further reading

T. Casadevall, W. Rose, T. Gerlach, L.P. Greenland, J. Ewert, R. Wunderman and R. Symonds (1983) *Science*, vol. 221, p. 1383 – 'Gas emissions and eruptions of Mount St. Helens through 1982'.

L.L. Malinconico, Jr. (1979) *Nature*, vol. 278, p. 43 – 'Fluctuations in SO_2 emission during recent eruptions of Etna'.

Tellurium trioxide (the α-form) is formed by dehydrating telluric acid (equation 15.84); it is an orange solid which is insoluble in water but dissolves in aqueous alkali, and is a very powerful oxidizing agent. On heating above 670 K, TeO_3 decomposes to TeO_2 and O_2. The solid state structure of TeO_3 is a three-dimensional lattice in which each Te(VI) centre is octahedrally sited and connected by bridging O atoms.

$$H_6TeO_6 \longrightarrow TeO_3 + 3H_2O \qquad (15.84)$$

15.9 Oxoacids and their salts

By way of introduction, we note some generalities:

- oxoacid chemistry of sulfur resembles the complicated system of phosphorus;

- there are structural analogies between sulfates and phosphates, although fewer condensed sulfates are known;
- redox processes involving sulfur oxoanions are often slow, and thermodynamic data alone do not give a very good picture of their chemistry (compare similar situations for nitrogen- and phosphorus-containing oxoanions);
- selenium and tellurium have a relatively simple oxoacid chemistry.

Structures and pK_a values for important sulfur oxoacids are give in Table 15.8.

Dithionous acid, $H_2S_2O_4$

Although we show the structure of dithionous acid in Table 15.8, only its salts are known and these are powerful reducing agents. Dithionite is prepared by reduction of sulfite in aqueous solution (equation 15.85) by Zn or Na amalgam and possesses eclipsed structure **15.40**.

Table 15.8 Selected oxoacids of sulfur.[‡]

Formula	Name (IUPAC systematic name, acid nomenclature)	Structure[*]	pK_a values (298 K)
$H_2S_2O_4$	Dithionous acid (tetraoxodisulfuric acid)		$pK_a(1) = 0.35$; $pK_a(2) = 2.45$
H_2SO_3	Sulfurous acid[**] (trioxodisulfuric acid)		$pK_a(1) = 1.82$; $pK_a(2) = 6.92$
H_2SO_4	Sulfuric acid (tetraoxosulfuric acid)		$pK_a(2) = 1.92$
$H_2S_2O_7$	Disulfuric acid (μ-oxo-hexaoxodisulfuric acid)		$pK_a(1) = 3.1$
$H_2S_2O_8$	Peroxodisulfuric acid (μ-peroxo-hexaoxodisulfuric acid)		
$H_2S_2O_3$	Thiosulfuric acid (trioxothiosulfuric acid)		$pK_a(1) = 0.6$; $pK_a(2) = 1.74$

[‡] Commonly used names have been included in this table; for systematic name and comments on uses of traditional names, see: *IUPAC: Nomenclature of Inorganic Chemistry* (*Recommendations 1990*), ed. G.J. Leigh, Blackwell Scientific Publications, Oxford.
[*] See text; not all the acids can be isolated.
[**] See text for comment on structure of conjugate base.

BOX 15.7

APPLICATIONS

SO$_2$ and sulfites in wine

During the fermentation process in the manufacture of wine, SO$_2$ or K$_2$S$_2$O$_5$ is added to the initial wine pressings to kill microorganisms, the presence of which results in spoilage of the wine. Molecular SO$_2$ is only used for large scale wine production, while K$_2$S$_2$O$_5$ is the common additive in small scale production. In acidic solution, $[S_2O_5]^{2-}$ undergoes the following reactions:

$$[S_2O_5]^{2-} + H_2O \rightleftharpoons 2[HSO_3]^-$$

$$[HSO_3]^- + H^+ \rightleftharpoons SO_2 + H_2O$$

The overall equilibrium system for aqueous SO$_2$ is:

$$SO_2 + H_2O \rightleftharpoons H^+ + [HSO_3]^- \rightleftharpoons H^+ + [SO_3]^{2-}$$

(These equilibria are discussed more fully with equations 6.18–6.20.) The position of equilibrium is pH-dependent; for the fermentation process, the pH is in the range 2.9–3.6. Only *molecular* SO$_2$ is active against microorganisms.

The first (i.e. yeast) fermentation step is followed by a bacterial fermentation step (malolactic fermentation) in which malic acid is converted to lactic acid. After this stage, SO$_2$ is added to stabilize the wine against oxidation. Adding SO$_2$ too early destroys the bacteria that facilitate

malolactic fermentation. Malolactic fermentation is usually only important in red wine production.

The addition of SO$_2$ to white and red wines is handled differently. Red wines contain anthocyanin pigments, and these react with $[HSO_3]^-$ or $[SO_3]^{2-}$ resulting in a partial loss of the red coloration. Clearly, this must be avoided and means that addition of SO$_2$ to red wine must be carefully controlled. On the other hand, significantly more SO$_2$ can be added to white wine. Red wine, therefore, is less well protected by SO$_2$ against oxidation and spoilage by microorganisms than white wine, and it is essential to ensure that sugar and malic acid (food for the microbes) are removed from red wine before bottling. Red wine does possess a higher phenolic content than white wine, and this acts as a built-in anti-oxidant.

Wines manufactured in the US carry a '*contains sulfites*' statement on the label. Some people are allergic to sulfites, and one possible substitute for SO$_2$ is the enzyme lysozyme. Lysozyme attacks lactic bacteria, and is used in cheese manufacture. However, it is not able to act as an anti-oxidant. A possible solution (not yet adopted by the wine industry) would be to mount a combined offensive: adding lysozyme and a reduced level of SO$_2$.

(15.40)

$$2[SO_3]^{2-} + 2H_2O + 2e^- \rightleftharpoons 4[OH]^- + [S_2O_4]^{2-}$$
$$E^{\circ} = -1.12 \, V \qquad (15.85)$$

The very long S–S bond in $[S_2O_4]^{2-}$ (compare $r_{cov}(S) = 103$ pm) shows it to be particularly weak and this is consistent with the observation that ^{35}S undergoes rapid exchange between $[S_2O_4]^{2-}$ and SO$_2$ in neutral or acidic solution. The presence of the $[SO_2]^-$ radical anion in solutions of Na$_2$S$_2$O$_4$ has been demonstrated by ESR spectroscopy. In aqueous solutions, $[S_2O_4]^{2-}$ is oxidized by air but in the absence of air, it undergoes reaction 15.86.

$$2[S_2O_4]^{2-} + H_2O \longrightarrow [S_2O_3]^{2-} + 2[HSO_3]^- \qquad (15.86)$$

Sulfurous and disulfurous acids, H$_2$SO$_3$ and H$_2$S$_2$O$_5$

Neither 'sulfurous acid' (see also *Section 15.8*) nor 'disulfurous acid' has been isolated as a free acid. Salts containing the sulfite ion, $[SO_3]^{2-}$, are well established (e.g. Na$_2$SO$_3$ and

K$_2$SO$_3$ are commercially available) and are quite good reducing agents (equation 15.79). Applications of sulfites include those as food preservatives, e.g. an additive in wines (see *Box 15.7*). The $[SO_3]^{2-}$ ion has a trigonal pyramidal structure with delocalized bonding (S–O = 151 pm, $\angle$O–S–O = 106°). There is evidence from ^{17}O NMR spectroscopic data that protonation of $[SO_3]^{2-}$ occurs to give a mixture of isomers as shown in equilibrium 15.87.

$$\left[H-OSO_2 \right]^- \rightleftharpoons \left[H-SO_3 \right]^- \qquad (15.87)$$

Although the $[HSO_3]^-$ ion exists in solution, and salts such as NaHSO$_3$ (used as a bleaching agent) may be isolated, evaporation of a solution of NaHSO$_3$ which has been saturated with SO$_2$ results in the formation of Na$_2$S$_2$O$_5$ (equation 15.88).

$$2[HSO_3]^- \rightleftharpoons H_2O + [S_2O_5]^{2-} \qquad (15.88)$$

(15.41)

The $[S_2O_5]^{2-}$ ion is the only known derived anion of disulfurous acid and possesses structure **15.41** with a long, weak S–S bond.

Dithionic acid, $H_2S_2O_6$

Dithionic acid is another sulfur oxoacid that is only known in aqueous solution (in which it behaves as a strong acid) or in the form of salts containing the dithionate, $[S_2O_6]^{2-}$, ion. Such salts can be isolated as crystalline solids and Figure 15.13 shows the presence of a long S–S bond; the anion possesses a staggered conformation in the solid state. The dithionate ion can be prepared by controlled oxidation of $[SO_3]^{2-}$ (equations 15.89 and 15.90), but *not* by the reduction of $[SO_4]^{2-}$ (equation 15.91). The $[S_2O_6]^{2-}$ can be isolated as the soluble salt BaS_2O_6, which is easily converted into salts of other cations.

$$[S_2O_6]^{2-} + 4H^+ + 2e^- \rightleftharpoons 2H_2SO_3 \quad E^\circ = +0.56\,V \quad (15.89)$$

$$MnO_2 + 2[SO_3]^{2-} + 4H^+ \longrightarrow Mn^{2+} + [S_2O_6]^{2-} + 2H_2O \quad (15.90)$$

$$2[SO_4]^{2-} + 4H^+ + 2e^- \rightleftharpoons [S_2O_6]^{2-} + 2H_2O$$
$$E^\circ = -0.22\,V \quad (15.91)$$

The $[S_2O_6]^{2-}$ ion is not easily oxidized or reduced, but in acidic solution it slowly decomposes according to equation 15.92, consistent with there being a weak S–S bond.

$$[S_2O_6]^{2-} \longrightarrow SO_2 + [SO_4]^{2-} \quad (15.92)$$

Sulfuric acid, H_2SO_4

Sulfuric acid is by far the most important of the oxoacids of sulfur and is manufactured on a huge scale by the *Contact process* (see *Figure 26.12*). The first stages of this process (conversion of SO_2 to SO_3 and formation of oleum) were described in Section 15.8; the oleum is finally diluted with water to give H_2SO_4. Pure H_2SO_4 is a colourless, liquid with a high viscosity caused by extensive intermolecular hydrogen bonding. Its self-ionization and use as a non-aqueous solvent were described in *Section 8.8*, and selected properties given in *Table 8.6*. Gas phase H_2SO_4 molecules have C_2 symmetry (Figure 15.13b) with S–O bond distances that reflect double and single bond character (**15.42**). In the sulfate ion, all four S–O bond distances are equal (149 pm) because of charge delocalization, and in the $[HSO_4]^-$, the

S–OH bond distance is 156 pm and the remaining S–O bonds are of equal length (147 pm).

$$\begin{array}{c} O \\ \| \; 143\,pm \\ 154\,pm \; \diagdown\, S \diagup \; O \\ O \diagup \; \big| \; \diagdown\, O \\ 97\,pm \; \big| \qquad \diagup \\ H \qquad O \\ H \end{array}$$

(15.42)

In aqueous solution, H_2SO_4 acts as a strong acid (equation 15.93) but the $[HSO_4]^-$ ion is a fairly weak acid (equation 15.94 and Table 15.8). Two series of salts are formed and can be isolated, e.g. $KHSO_4$ and K_2SO_4.

$$H_2SO_4 + H_2O \longrightarrow [H_3O]^+ + [HSO_4]^- \quad (15.93)$$

$$[HSO_4]^- + H_2O \rightleftharpoons [H_3O]^+ + [SO_4]^{2-} \quad (15.94)$$

Dilute aqueous H_2SO_4 (typically 2 M) neutralizes bases (e.g. equation 15.95), and reacts with electropositive metals, liberating H_2, and metal carbonates (equation 15.96).

$$H_2SO_4(aq) + 2KOH(aq) \longrightarrow K_2SO_4(aq) + 2H_2O(l) \quad (15.95)$$

$$H_2SO_4(aq) + CuCO_3(s) \longrightarrow CuSO_4(aq) + H_2O(l) + CO_2(g) \quad (15.96)$$

Commercial applications of sulfate salts are numerous, e.g. $(NH_4)_2SO_4$ as a fertilizer, $CuSO_4$ in fungicides, $MgSO_4$ as a laxative, and hydrated $CaSO_4$ (see *Boxes 11.2 and 11.7*); uses of H_2SO_4 were shown in Figure 15.2b.

Concentrated H_2SO_4 is a good oxidizing agent (e.g. reaction 15.80) and a powerful dehydrating agent (see *Box 11.4*); its reaction with HNO_3 is important for organic nitrations (equation 15.97).

$$HNO_3 + 2H_2SO_4 \longrightarrow [NO_2]^+ + [H_3O]^+ + 2[HSO_4]^- \quad (15.97)$$

Fluoro- and chlorosulfonic acids, HSO_3F and HSO_3Cl

Fluoro- and chlorosulfonic acids, HSO_3F and HSO_3Cl, are obtained as shown in reaction 15.83, and their structures are related to that of H_2SO_4 with one OH group replaced by F or Cl. Both are colourless liquids at 298 K, and fume in moist air; HSO_3Cl reacts explosively with water. They are commercially available; HSO_3F has wide applications in *superacid* systems (see *Section 8.9*) and as a fluorinating agent, while HSO_3Cl is used as a chlorosulfonating agent.

Polyoxoacids with S–O–S units

Although K^+ salts of the polysulfuric acids $HO_3S(OSO_2)_nOSO_3H$ ($n = 2, 3, 5, 6$) have been obtained by the reaction of SO_3 with K_2SO_4, the free acids cannot be isolated. Disulfuric and trisulfuric acids are present in

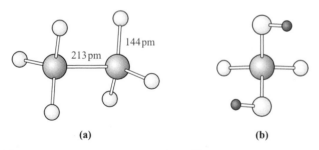

(a)   (b)

Fig. 15.13 (a) The structure of $[S_2O_6]^{2-}$ showing the staggered conformation; from the salt $[Zn\{H_2NNHC(O)Me\}_3][S_2O_6]\cdot2.5H_2O$ [I.A. Krol *et al.* (1981) *Koord. Khim.*, vol. 7, p. 800]; (b) the C_2 structure of gas phase H_2SO_4. Colour code: S, blue; O, pale grey; H, black.

oleum, i.e. when SO_3 is dissolved in concentrated H_2SO_4. The salt $[NO_2]_2[S_3O_{10}]$ has also been prepared and structurally characterized. Structure **15.43** shows $[S_3O_{10}]^{2-}$ as a representative of this group of polyoxoanions.

(15.43)

Peroxosulfuric acids, $H_2S_2O_8$ and H_2SO_5

The reaction between cold, anhydrous H_2O_2 and chlorosulfonic acid yields peroxomonosulfuric acid, H_2SO_5, and peroxodisulfuric acid, $H_2S_2O_8$ (scheme 15.98). Conversion of $H_2S_2O_8$ (Table 15.8) to H_2SO_5 (**15.44**) occurs by controlled hydrolysis.

(15.44)

$$H_2O_2 \xrightarrow[-HCl]{ClSO_3H} H_2SO_5 \xrightarrow[-HCl]{ClSO_3H} H_2S_2O_8$$
$$H_2S_2O_8 + H_2O \xrightarrow{273\,K} H_2SO_5 + H_2SO_4$$

(15.98)

Both acids are crystalline solids at 298 K. Few salts of H_2SO_5 are known, but those of $H_2S_2O_8$ are easily made by anodic oxidation of the corresponding sulfates in acidic solution at low temperatures and high current densities. Peroxodisulfates are strong oxidizing agents (equation 15.99), and oxidations are often catalyzed by Ag^+, with Ag(II) species being formed as intermediates. In acidic solutions, $[S_2O_8]^{2-}$ oxidizes Mn^{2+} to $[MnO_4]^-$, and Cr^{3+} to $[CrO_4]^{2-}$.

$$[S_2O_8]^{2-} + 2e^- \rightleftharpoons 2[SO_4]^{2-} \qquad E^\circ = +2.01\,V \quad (15.99)$$

Peroxodisulfuric acid smells of ozone, and when $K_2S_2O_8$ is heated, a mixture of O_2 and O_3 is produced.

Thiosulfuric acid, $H_2S_2O_3$, and polythionates

Thiosulfuric acid may be prepared under *anhydrous* conditions by reaction 15.100, or by treatment of lead thiosulfate (PbS_2O_3) with H_2S, or sodium thiosulfate with HCl. The free acid is very unstable, decomposing at 243 K or upon contact with water.

$$H_2S + HSO_3Cl \xrightarrow{low\ temp} H_2S_2O_3 + HCl \qquad (15.100)$$

A representation of the structure of thiosulfuric acid is given in Table 15.8, but the conditions of reaction 15.100 may suggest protonation at sulfur, i.e. $(HO)(HS)SO_2$. Thiosulfate salts are far more important than the acid; crystallization of the aqueous solution from reaction 15.101 yields $Na_2S_2O_3 \cdot 5H_2O$.

$$Na_2SO_3 + S \xrightarrow{in\ aqueous\ solution} Na_2S_2O_3 \qquad (15.101)$$

(15.45)

The thiosulfate ion, **15.45**, is a very good complexing agent for Ag^+ and $Na_2S_2O_3$ is widely used in photography for removing unchanged AgBr from exposed photographic film (equation 15.102). In the complex ion $[Ag(S_2O_3)_3]^{5-}$, each thiosulfate ion coordinates to Ag^+ through a sulfur donor atom.

$$AgBr + 3Na_2S_2O_3 \longrightarrow Na_5[Ag(S_2O_3)_3] + NaBr \qquad (15.102)$$

Most oxidizing agents (including Cl_2 and Br_2) slowly oxidize $[S_2O_3]^{2-}$ to $[SO_4]^{2-}$, and $Na_2S_2O_3$ is used to remove excess Cl_2 in bleaching processes. In contrast, I_2 rapidly oxidizes $[S_2O_3]^{2-}$ to tetrathionate; reaction 15.103 is of great importance in titrimetric analysis.

$$2[S_2O_3]^{2-} + I_2 \longrightarrow [S_4O_6]^{2-} + 2I^-$$
$$[S_4O_6]^{2-} + 2e^- \rightleftharpoons 2[S_2O_3]^{2-} \qquad E^\circ = +0.08\,V$$
$$I_2 + 2e^- \rightleftharpoons 2I^- \qquad E^\circ = +0.54\,V$$

(15.103)

Polythionates contain ions of type $[S_nO_6]^{2-}$ and may be prepared by condensation reactions such as those in scheme 15.104, but some ions must be made by specific routes. Polythionate ions are structurally similar and have two $\{SO_3\}^-$ groups connected by a sulfur chain (**15.46** shows $[S_5O_6]^{2-}$); solid state structures for a number of salts show chain conformations are variable. In aqueous solution, polythionates slowly decompose to H_2SO_4, SO_2 and sulfur.

$$SCl_2 + 2[HSO_3]^- \longrightarrow [S_3O_6]^{2-} + 2HCl$$
$$S_2Cl_2 + 2[HSO_3]^- \longrightarrow [S_4O_6]^{2-} + 2HCl$$

(15.104)

Some compounds are known in which S atoms in a polythionate are replaced by Se or Te, e.g. $Ba[Se(SSO_3)_2]$ and $Ba[Te(SSO_3)_2]$. Significantly, Se and Te *cannot* replace the terminal S atoms, presumably because in their highest oxidation states, they are too powerfully oxidizing and attack the remainder of the chain.

(15.46)

Oxoacids of selenium and tellurium

Selenous acid, H_2SeO_3, may be crystallized from aqueous solutions of SeO_2 and gives rise to two series of salts containing the $[HSeO_3]^-$ and $[SeO_3]^{2-}$ ions. In aqueous solution, it behaves as a weak acid: $pK_a(1) \approx 2.46$, $pK_a(2) \approx 7.31$. Heating salts of $[HSeO_3]^-$ generates diselenites containing ion **15.47**. Tellurous acid, H_2TeO_3, is not as stable as H_2SeO_3 and is usually prepared in aqueous solution where it acts as a weak acid: $pK_a(1) \approx 2.48$, $pK_a(2) \approx 7.70$. Most tellurite salts contain the $[TeO_3]^{2-}$ ion.

(15.47)

Oxidation of H_2SeO_3 with 30% aqueous H_2O_2 yields selenic acid, H_2SeO_4, which may be crystallized from the solution. In some ways it resembles H_2SO_4, being fully dissociated in aqueous solution with respect to loss of the first proton; for the second step, $pK_a = 1.92$. It is a more powerful oxidant than H_2SO_4, e.g. it liberates Cl_2 from concentrated HCl. The formula, H_6TeO_6 or $Te(OH)_6$, and properties of telluric acid contrast with those of selenic acid. In the solid, octahedral molecules (**15.48**) are present and in solution, it behaves as a weak acid: $pK_a(1) = 7.68$, $pK_a(2) = 11.29$. Typical salts include those containing $[Te(O)(OH)_5]^-$ and $[Te(O)_2(OH)_4]^{2-}$ and, in 1990, the presence of the $[TeO_4]^{2-}$ ion was confirmed in the solid state structure of $Rb_6[TeO_5][TeO_4]$.

(15.48)

15.10 Compounds of sulfur and selenium with nitrogen

Sulfur–nitrogen compounds

Sulfur–nitrogen chemistry is an area that has seen major developments over the last few decades, in part because of the conductivity of the polymer $(SN)_x$. The following discussion is necessarily selective, and more detailed accounts are listed at the end of the chapter. Probably the best known of the sulfur–nitrogen compounds is tetrasulfur tetranitride, S_4N_4. It is obtained from reaction 15.105 as a diamagnetic orange solid (mp 451 K) which explodes when heated or struck; pure samples are very sensitive. It is hydrolyzed slowly by water (in which it is insoluble) and rapidly by warm alkali (equation 15.106).

$$6S_2Cl_2 + 16NH_3 \xrightarrow{CCl_4, \, 320\,K} S_4N_4 + 12NH_4Cl + S_8$$

(15.105)

$$S_4N_4 + 6[OH]^- + 3H_2O \longrightarrow [S_2O_3]^{2-} + 2[SO_3]^{2-} + 4NH_3$$

(15.106)

$\angle N–S–N = 104°$
$\angle S–N–S = 113°$

(15.49)

The structure of S_4N_4, **15.49**, is a cradle-like ring in which pairs of S atoms are brought within weak bonding distance of one another (compare to $[S_8]^{2+}$, Figure 15.5). The S–N bond distances in S_4N_4 indicate delocalized π-bonding, and resonance structures **15.50** are among those which can be drawn.

(15.50)

Figure 15.14 gives selected reactions of S_4N_4; some lead to products containing S–N rings in which the transannular interactions of S_4N_4 are lost. Reduction (at N) gives tetrasulfur tetraimide, $S_4N_4H_4$, which has a crown-shaped ring with equal S–N bond lengths. Tetrasulfur tetraimide is one of a number of compounds in which S atoms in S_8 are replaced by NH groups with retention of the crown-conformation; S_7NH, $S_6N_2H_2$, $S_5N_3H_3$ (along with S_4N_4 and S_8) are all obtained by treating S_2Cl_2 with NH_3. No members of this family with adjacent NH groups in the ring are known.

(15.51)

$\angle F–S–F = 94°$
$\angle N–S–F = 122°$

(15.52)

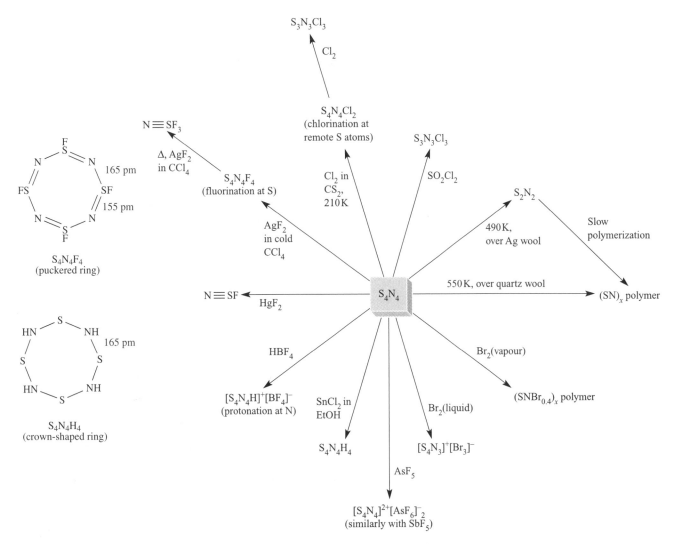

Fig. 15.14 Selected reactions of S_4N_4; the rings in $S_4N_4H_4$ and $S_4N_4F_4$ are non-planar.

Halogenation of S_4N_4 (at S) may degrade the ring depending on X_2 or the conditions (Figure 15.14). The ring in $S_4N_4F_4$ has a puckered conformation quite different from that in $S_4N_4H_4$. Fluorination of S_4N_4 under appropriate conditions (Figure 15.14) yields thiazyl fluoride, NSF, **15.51**, or thiazyl trifluoride NSF_3, **15.52**, which contain S≡N triple bonds. Both are pungent gases at room temperature, and NSF slowly trimerizes to $S_3N_3F_3$; note that $S_4N_4F_4$ is not made from the monomer. The structures of $S_3N_3Cl_3$ (**15.53**) and $S_3N_3F_3$ are similar; the rings exhibit only slight puckering and the S–N bond distances are equal in $S_3N_3Cl_3$ and approximately equal in the fluoro analogue. Structure **15.53** is one of two resonance forms that can be drawn. Oxidation of S_4N_4 with AsF_5 or SbF_5 gives $[S_4N_4][EF_6]_2$ (E = As or Sb) containing $[S_4N_4]^{2+}$. This has the planar structure **15.54** in many of its salts, but $[S_4N_4]^{2+}$ can also adopt a planar structure with alternating bond distances, or a puckered conformation. The $[S_4N_3]^+$ cation (prepared as shown in Figure 15.14) has the planar structure **15.55** with delocalized bonding.

(15.53)

∠S–N–S = 151°
∠N–S–N = 120°

(15.54)

206 pm

S–N in the range
152–160 pm

(15.55)

The S_4N_4 cage can be degraded to S_2N_2 (Figure 15.14) which is isoelectronic with $[S_4]^{2+}$ (see Section 15.4); S_2N_2 is planar with delocalized bonding (S−N = 165 pm), two resonance structures of which are shown in **15.56**. At room temperature, this converts to the lustrous golden-yellow, fibrous polymer $(SN)_x$, which can also be prepared from S_4N_4. The polymer decomposes explosively at 520 K, but can be sublimed *in vacuo* at $\approx$410 K. It is a remarkable material, being covalently bonded but showing metallic properties: a one-dimensional pseudo-metal. It has an electrical conductance about one-quarter of that of mercury in the direction of the polymer chains, and at 0.3 K it becomes a superconductor. However, the explosive nature of S_4N_4 and S_2N_2 limits commercial production of $(SN)_x$, and new routes to $(SN)_x$ or related polymers are goals of current research. In the solid state, X-ray diffraction data indicate that the S−N bond lengths in $(SN)_x$ alternate (159 and 163 pm) but highly precise data are still not available; the closest interchain distances are non-bonding S−S contacts of 350 pm. Structure **15.57** gives a representation of the polymer chain (note the relationship to its monomer precursor **15.56**) and the conductivity can be considered to arise from the unpaired electrons on sulfur occupying a half-filled conduction band (see *Section 5.8*).

(15.56)

∠S–N–S = 119°   ∠ N–S–N = 106°

(15.57)

The reactions of S_7NH with $SbCl_5$ in liquid SO_2, or $S_3N_3Cl_3$ with $SbCl_5$ and sulfur in $SOCl_2$, lead to the formation of the salt $[NS_2][SbCl_6]$ containing the $[NS_2]^+$ ion, (**15.58**) which is isoelectronic (in terms of valence electrons) with $[NO_2]^+$ (see *structure* **14.35**).

146 pm

(15.58)

Tetraselenium tetranitride

Among the compounds formed by Se and N, we mention only the Se analogue of S_4N_4 which can be prepared from $SeCl_4$ or $(Et_2O)_2SeO$ and NH_3. Tetraselenium tetranitride, Se_4N_4, forms orange, hygroscopic crystals and is highly explosive. The structure of Se_4N_4 is like that of S_4N_4 (**15.49**) with Se−N bond lengths of 180 pm and cross-cage Se····Se separa-

tions of 276 pm (compare with $r_{cov}(Se) = 117$ pm). The reactivity of Se_4N_4 has not been as fully explored as that of S_4N_4.

15.11 Aqueous solution chemistry of sulfur, selenium and tellurium

As we saw earlier in the chapter, the redox reactions between compounds of S in different oxidation states are often slow, and values of E^o for half-reactions are invariably obtained from thermochemical information or estimated on the basis of observed chemistry. The data in Figure 15.15 illustrate the relative redox properties of some S, Se and Te-containing species. Points to note are:

- the greater oxidizing powers of selenate and tellurate than of sulfate;
- the similarities between the oxidizing powers of sulfate, selenite and tellurite;
- the instabilities in aqueous solution of H_2Se and H_2Te.

Further, there is little difference in free energy between the various oxidation state species of sulfur, a fact that is doubtless involved in the complicated oxoacid and oxoanion chemistry of sulfur. We have already discussed some aspects of the aqueous solution chemistry of the group 16 elements:

- the ionization of the hydrides (*Sections 6.5 and 15.5*);
- formation of metal sulfides (*Section 15.6*);
- formation of polysulfide ions, e.g. $[S_5]^{2-}$ (*equation 15.39*);
- oxoacids and their salts (*Section 15.9*);
- the oxidizing power of $[S_2O_8]^{2-}$ (*equation 15.99*).

There is no cation chemistry in aqueous solution for the group 16 elements. The coordination to metal ions of oxoanions such as $[SO_4]^{2-}$ and $[S_2O_3]^{2-}$ is well established (e.g. see equation 15.102).

Fig. 15.15 Potential diagrams for sulfur, selenium and tellurium at pH = 0.

Summary

This chapter has introduced the descriptive chemistry of the group 16 elements (the chalcogens). The inaccessibility of the heaviest element, polonium, means that it has been little studied. The chemistry of oxygen differs markedly from that of the later elements, being influenced substantially by its restriction to an octet of electrons, the weakness of the O−O single bond and the small size of the O^{2-} ion, and participation in hydrogen bonding. The complicated chemistry of sulfur is greatly influenced by kinetic factors, especially in its extensive series of compounds with oxygen and with nitrogen. The usual increase in basic character with increase in atomic number appears for selenium and tellurium, but only polonium seems to form a stable cation.

The following terms were introduced in this chapter.
Do you know what they mean?
❏ annular
❏ transannular interaction
❏ cubane

Suggested further reading

N.N. Greenwood and A. Earnshaw (1997) *Chemistry of the Elements*, 2nd edn, Butterworth-Heinemann, Oxford – Chapters 14–16 cover the chalcogens in detail.

D.T. Sawyer (1994) 'Oxygen: Inorganic chemistry' in *Encyclopedia of Inorganic Chemistry*, ed. R.B. King, Wiley, Chichester, vol. 6, p. 2947.

A.F. Wells (1984) *Structural Inorganic Chemistry*, 5th edn, Clarendon Press, Oxford – Chapters 11–17 cover the structures of a large number of compounds of the group 16 elements.

J.D. Woollins (1994) 'Sulfur: Inorganic chemistry' in *Encyclopedia of Inorganic Chemistry*, ed. R.B. King, Wiley, Chichester, vol. 7, p. 3954.

Sulfur-nitrogen compounds

N.N. Greenwood and A. Earnshaw (1997) *Chemistry of the Elements*, 2nd edn, Butterworth-Heinemann, Oxford, pp. 721–746.

S. Parsons and J. Passmore (1994) *Accounts of Chemical Research*, vol. 27, p. 101 – 'Rings, radicals and synthetic metals: The chemistry of $[SNS]^+$'.

J.M. Rawson and J.J. Longridge (1997) *Chemical Society Reviews*, vol. 26, p. 53 – 'Sulfur–nitrogen chains: rational and irrational behaviour'.

Specialized topics

J. Beck (1994) *Angewandte Chemie, International Edition in English*, vol. 33, p. 163 – 'New forms and functions of tellurium: From polycations to metal halide tellurides'.

P. Kelly (1997) *Chemistry in Britain*, vol. 33, number 4, p. 25 – 'Hell's angel: A brief history of sulfur'.

Problems

15.1 (a) Write down, in order, the names and symbols of the elements in group 16; check your answer by reference to the first page of this chapter. (b) Give a *general* notation showing the ground state electronic configuration of each element.

15.2 The formation of ^{210}Po from ^{209}Bi is described in Section 15.1. Write an equation to show this nuclear reaction.

15.3 Write half equations to show the reactions involved during the electrolysis of aqueous alkali.

15.4 By considering the reactions $8E(g) \longrightarrow 4E_2(g)$ and $8E(g) \longrightarrow E_8(g)$ for E = O and E = S, show that the formation of diatomic molecules is favoured for oxygen, whereas ring formation is favoured for sulfur. [Data: see *Table 15.2*.]

15.5 (a) Use the values of E° for reactions 15.29 and 15.30 to show that H_2O_2 is thermodynamically unstable with respect to decomposition into H_2O and O_2. (b) '20 Volume' H_2O_2 is so called because 1 volume of the solution liberates 20 volumes of O_2 when it decomposes. If the volumes are measured at 273 K and 1 bar pressure, what is the concentration of the solution expressed in grams of H_2O_2 per dm^3?

15.6 Suggest products for the following reactions; data needed: see Appendix 11. (a) H_2O_2 and Ce^{4+} in acidic solution; (b) H_2O_2 and I^- in acidic solution.

15.7 Hydrogen peroxide oxidizes $Mn(OH)_2$ to MnO_2. (a) Write an equation for this reaction. (b) What secondary reaction will occur?

15.8 Predict the structures of (a) H_2Se; (b) $[H_3S]^+$; (c) SO_2; (d) SF_4; (e) SF_6; (f) S_2F_2.

15.9 (a) Explain why the reaction of SF_4 with BF_3 yields $[SF_3]^+$, whereas the reaction with CsF gives $Cs[SF_5]$. (b) Suggest how SF_4 might react with a carboxylic acid, RCO_2H.

15.10 Discuss the trends in (a) the O−O bond lengths in O_2 (121 pm), $[O_2]^+$ (112 pm), H_2O_2 (147.5 pm), $[O_2]^{2-}$ (149 pm) and O_2F_2 (122 pm), and (b) the S−S bond distances in S_6 (206 pm), S_2 (189 pm), $[S_4]^{2+}$ (198 pm), H_2S_2 (206 pm), S_2F_2 (189 pm), S_2F_{10} (221 pm) and S_2Cl_2 (193 pm). [Data: $r_{cov}(S) = 103$ pm.]

15.11 Comment on the following values of gas phase dipole moments: SeF_6, 0 D; SeF_4, 1.78 D; SF_4, 0.64 D; SCl_2, 0.36 D; $SOCl_2$, 1.45 D; SO_2Cl_2, 1.81 D.

15.12 The ^{125}Te NMR spectrum of $[Me_4N][TeF_7]$ (298 K in MeCN) consists of a binomial octet ($J = 2876$ Hz), while the ^{19}F NMR spectrum exhibits a singlet with two (superimposed over the singlet), very low intensity doublets ($J = 2876$ and 2385 Hz respectively). Rationalize these observations. [Data: see *Table 15.1*; ^{19}F, 100%, $I = \frac{1}{2}$.]

15.13 In the following series of compounds or ions, identify those that are isoelectronic (with respect to the valence electrons) and those that are also isostructural: (a) $[SiO_4]^{4-}$, $[PO_4]^{3-}$, $[SO_4]^{2-}$; (b) CO_2, SiO_2, SO_2, TeO_2, $[NO_2]^+$; (c) SO_3, $[PO_3]^-$, SeO_3; (d) $[P_4O_{12}]^{4-}$, Se_4O_{12}, $[Si_4O_{12}]^{8-}$.

15.14 (a) Give the structures of SO_3 and $[SO_3]^{2-}$ and rationalize the difference between them. (b) Outline the properties of aqueous solutions of SO_2 and discuss the species that can be derived from them.

15.15 (a) Draw the structures of S_7NH, $S_6N_2H_2$, $S_5N_3H_3$ and $S_4N_4H_4$, illustrating isomerism where appropriate.
(b) Write a brief account of the preparation and reactivity of S_4N_4, giving the structures of the products formed in the reactions described.

15.16 Discuss the interpretation of each of the following observations.

(a) When metallic Cu is heated with concentrated H_2SO_4, in addition to $CuSO_4$ and SO_2, some CuS is formed.
(b) The $[TeF_5]^-$ ion is square pyramidal.
(c) Silver nitrate gives a white precipitate with aqueous sodium thiosulfate; the precipitate dissolves in an excess of $[S_2O_3]^{2-}$. If the precipitate is heated with water, it turns black, and the supernatant liquid then gives a white precipitate with acidified aqueous $Ba(NO_3)_2$.

15.17 Interpret the following experimental results.

(a) Sodium dithionite, $Na_2S_2O_4$ (0.0261 g) was added to excess of ammoniacal $AgNO_3$ solution; the precipitated silver was removed by filtration, and

dissolved in nitric acid. The resulting solution was found to be equivalent to $30.0\,\text{cm}^3$ 0.1 M thiocyanate solution.
(b) A solution containing 0.0725 g of $Na_2S_2O_4$ was treated with $50.0\,\text{cm}^3$ 0.0500 M iodine solution and acetic acid. After completion of the reaction, the residual I_2 was equivalent to $23.75\,\text{cm}^3$ 0.1050 M thiosulfate.

15.18 The action of concentrated H_2SO_4 on urea, $(H_2N)_2CO$, results in the production of a white crystalline solid **X** of formula H_3NO_3S. This is a monobasic acid. On treatment with sodium nitrite and dilute hydrochloric acid at 273 K, one mole of **X** liberates one mole of N_2, and on addition of aqueous $BaCl_2$, the resulting solution yields one mole of $BaSO_4$ per mole of **X** taken initially. Deduce the structure of **X**.

15.19 Write a brief account of the oxoacids of sulfur, paying particular attention to which species are isolable.

15.20 Give the structures of S_2O, $[S_2O_3]^{2-}$, NSF, NSF_3, $[NS_2]^+$ and S_2N_2 and rationalize their shapes.

The group 17 elements

TOPICS

■ Occurrence, extraction and uses

■ Physical properties

■ The elements

■ Hydrogen halides

■ Interhalogen compounds and polyhalogen ions

■ Oxides and oxofluorides of chlorine, bromine and iodine

■ Oxoacids and their salts

■ Aqueous solution chemistry

1	2		13	14	15	16	**17**	18
H								He
Li	Be		B	C	N	O	F	Ne
Na	Mg		Al	Si	P	S	Cl	Ar
K	Ca		Ga	Ge	As	Se	Br	Kr
Rb	Sr	*d*-block	In	Sn	Sb	Te	I	Xe
Cs	Ba		Tl	Pb	Bi	Po	At	Rn
Fr	Ra							

16.1 Introduction

The group 17 elements are called the *halogens*.

Fluorine, chlorine, bromine and iodine

The chemistry of fluorine, chlorine, bromine and iodine is probably better understood than that of any other group of elements except the alkali metals. This is partly because much of the chemistry of the halogens is that of singly bonded atoms or singly charged anions, and partly because of the wealth of structural and physicochemical data available for most of their compounds. The fundamental principles of inorganic chemistry are often illustrated by discussing properties of the halogens and halide compounds, and topics already discussed include:

- electron affinities of the halogens (*Section 1.10*);
- valence bond theory for F_2 (*Section 1.12*);
- molecular orbital theory for F_2 (*Section 1.13*);
- electronegativities of the halogens (*Section 1.15*);
- dipole moments of hydrogen halides (*Section 1.16*);

- bonding in HF by molecular orbital theory (*Section 1.17*);
- VSEPR model (which works well for many halide compounds, *Section 1.19*);
- application of the packing-of-spheres model, solid state structure of F_2 (*Section 5.3*);
- ionic radii (*Section 5.10*);
- ionic lattices: NaCl, CsCl, CaF_2, antifluorite, CdI_2 (*Section 5.11*);
- lattice energies: comparisons of experimental and calculated values for metal halides (*Section 5.15*);
- estimation of fluoride ion affinities (*Section 5.16*);
- estimation of standard enthalpies of formation and disproportionation, illustrated using halide compounds (*Section 5.16*);
- halogen halides as Brønsted acids (*Section 6.4*);
- energetics of hydrogen halide dissociation in aqueous solution (*Section 6.5*);
- solubilities of metal halides (*Section 6.9*);
- common-ion effect, exemplified by AgCl (*Section 6.10*);
- stability of complexes containing hard and soft metal ions and ligands, illustrated with halides of Fe(III) and Hg(II) (*Section 6.13*);
- redox half-cells involving silver halides (*Section 7.3*);
- non-aqueous solvents: liquid HF (*Section 8.7*);
- non-aqueous solvents: BrF_3 (*Section 8.10*);
- reactions of halogens with H_2 (*Section 9.4, equations 9.19–9.21*);
- hydrogen bonding involving halogens (*Section 9.6*).

In *Sections 10.5, 11.5, 12.6, 13.8, 14.7* and *15.7* we have discussed the halides of the group 1, 2, 13, 14, 15 and 16 elements respectively. Fluorides of the noble gases are discussed in *Sections 17.4* and *17.5*, and of the *d*- and *f*-block metals in *Chapters 21, 22* and *24*. In this chapter, we discuss the halogens themselves, their oxides and oxoacids, interhalogen compounds and polyhalide ions.

Astatine

Astatine is the heaviest member of group 17 and is known only in the form of radioactive isotopes, all of which have short half-lives. The longest lived isotope is ^{210}At ($t_{\frac{1}{2}} = 8.1$ h). Several isotopes are present naturally as transient products of the decay of uranium and thorium minerals; ^{218}At is formed from the β-decay of ^{218}Po, but the path competes with decay to ^{214}Pb (the dominant decay, see *Figure 2.3*). Other isotopes are artificially prepared, e.g. ^{211}At (an α-emitter) from the nuclear reaction $^{209}_{83}$Bi(α,2n)$^{211}_{85}$At, and may be separated by vacuum distillation. In general, At is chemically similar to iodine. Tracer studies (which are the only sources of information about the element) show that At$_2$ is less volatile than I$_2$, is soluble in organic solvents, and is reduced by SO$_2$ to At$^-$ which can be coprecipitated with AgI or TlI. Hypochlorite, [ClO]$^-$, or peroxodisulfate,

[S$_2$O$_8$]$^{2-}$, oxidizes astatine to an anion that is carried by [IO$_3$]$^-$ (e.g. coprecipitation with AgIO$_3$) and is therefore probably [AtO$_3$]$^-$. Less powerful oxidizing agents such as Br$_2$ also oxidize astatine, probably to [AtO]$^-$ or [AtO$_2$]$^-$.

16.2 Occurrence, extraction and uses

Occurrence

The major natural sources of fluorine are the minerals *fluor-spar* (*fluorite*, CaF$_2$), *cryolite* (Na$_3$[AlF$_6$]) and *fluorapatite*, (Ca$_5$F(PO$_4$)$_3$) (see *Section 14.2* and *Box 14.10*), although the importance of cryolite lies in its being an *aluminium* ore (see *Section 12.2*). Sources of chlorine are closely linked to

---

BOX 16.1   **APPLICATIONS**

Flame retardants

The incorporation of flame retardants into consumer products is big business: it is forecast that the value of chemicals sold for use in flame-retardants in the US in 2000 will be over 900 million US dollars. In Europe, the predicted split of income in 2003 between the three main categories of flame retardants is shown in the pie chart below. The halogen-based chemicals are dominated by the perbrominated ether (C$_6$Br$_5$)$_2$O (used in television and computer casings), tetra-bromobisphenol A, Me$_2$C{4-(2,6-Br$_2$C$_6$H$_2$OH)}$_2$ (used in printed circuit boards) and an isomer of hexabromocyclodecane (used in polystyrene foams and some textiles). Concerns about the side-effects of bromine-based flame retardants (including hormone-related effects and possible production of bromodioxins) are now resulting in their (at least, partial) withdrawal.

Phosphorus-based flame retardants include tris(1,3-dichloroisopropyl) phosphate, used in polyurethane foams and polyester resins. Once again, there is debate concerning toxic side-effects of such products: although these flame retardants may save lives, they produce noxious fumes during a fire.

Many inorganic compounds are used as flame retardants; for example

- Sb$_2$O$_3$ is used in PVC, and in aircraft and motor vehicles; scares that Sb$_2$O$_3$ in cot mattresses may be the cause of 'cot deaths' appear to have subsided.
- Ph$_3$Sb(OC$_6$Cl$_5$)$_2$ is added to polypropene;
- borates, exemplified by:

are used in polyurethane foams, polyesters and polyester resins;
- ZnSnO$_3$ has applications in PVC, thermoplastics, polyester resins and certain resin-based gloss paints.

Tin-based flame retardants appear to have a great potential future: they are non-toxic, apparently producing none of the hazardous side-effects of the widely used phosphorus-based materials.

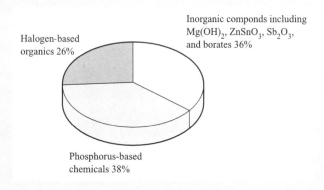

Inorganic compounds including Mg(OH)$_2$, ZnSnO$_3$, Sb$_2$O$_3$, and borates 36%

Halogen-based organics 26%

Phosphorus-based chemicals 38%

[Data: *Chemistry in Britain* (1998) vol. 34, June issue, p. 20.]

Further reading

M.S. Reisch (1997) *Chemical & Engineering News*, 24 Feb., p. 19 – 'Flame retardants sales heat up'.
C. Martin (1998) *Chemistry in Britain*, vol. 34, June issue, p. 20 – 'In the line of fire'.

BOX 16.2

Iodine: from cattle feed supplements to catalytic uses

The annual output of iodine is significantly lower than that of chlorine or bromine, but, nonetheless, it has a wide range of important applications.

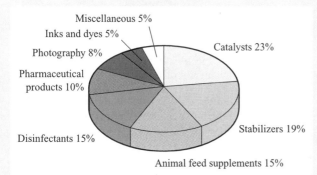

The major catalytic uses involve the complex *cis*-$[Rh(CO)_2I_2]^-$ in the Monsanto acetic acid and Tennessee–Eastman acetic anhydride processes, discussed in detail in Section 26.3. Application of iodine as a stabilizer includes its incorporation into nylon used in carpet and tyre manufacture. Iodized animal feed supplements are responsible for reduced instances of goitre (enlarged thyroid gland) which are otherwise prevalent in regions where the iodine content of soil and drinking water is low; iodized hen feeds increase

egg production. Iodine is usually added to feeds in the form of $[H_3NCH_2CH_2NH_3]I_2$, KI, $Ca(IO_3)_2$ or $Ca(IO_4)_2$. Uses of iodine as a disinfectant range from wound antiseptics to maintaining germ-free swimming pools and water supplies. We have already mentioned the use of ^{131}I as a medical radioisotope (Box 2.3), and photographic applications of AgI are highlighted in Box 22.13. Among dyes that have a high iodine content is erythrosine B (food red-colour additive E127) which is added to carbonated soft drinks, gelatins and cake icings.

Erythrosine B (58% iodine; $\lambda_{max} = 525$ nm)

those of Na and K (see *Section 10.2*): *rock salt* (NaCl), *sylvite* (KCl) and *carnallite* ($KCl \cdot MgCl_2 \cdot 6H_2O$). Seawater is one source of Br_2, but significantly higher concentrations of Br^- are present in salt lakes and natural brine wells (see *Box 16.3*). The natural abundance of iodine is less than that of the lighter halogens; it occurs as iodide ion in seawater and is taken up by seaweed, from which it may be extracted. Impure Chile saltpetre (*caliche*) contains up to 1% sodium iodate and this has become an important source of I_2; brines associated with oil and salt wells are of increasing importance.

Extraction

Most fluorine-containing compounds are made using HF, the latter being prepared from fluorite by reaction 16.1; in 1995, $\approx$74% of CaF_2 consumed in the US was converted into HF. Hydrogen fluoride is also recycled from Al manufacturing processes and from petroleum alkylation processes, and re-enters the supply chain. Difluorine is strongly oxidizing and must be prepared industrially by electrolytic oxidation of F^- ion. The electrolyte is a mixture of anhydrous molten KF and HF, and the electrolysis cell contains a steel or copper cathode, ungraphitized carbon anode, and a Monel metal (Cu/Ni) diaphragm which is perforated below the surface of the electrolyte, but not

above it, thus preventing the H_2 and F_2 products from recombining. As electrolysis proceeds, the HF content of the melt is renewed by adding dry gas from cylinders.

$$CaF_2 + H_2SO_4 \xrightarrow[\text{conc}]{} CaSO_4 + 2HF \qquad (16.1)$$

We have already described the Downs process for extracting Na from NaCl (*Figure 10.1*) and this is also the method of manufacturing Cl_2 (see *Box 10.4*), the tenth most important industrial chemical in the US. The manufacture of Br_2 involves oxidation of Br^- by Cl_2, with air being swept through the system to remove Br_2. Similarly, I^- in brines is oxidized to I_2. The extraction of I_2 from $NaIO_3$ involves controlled reduction by SO_2; complete reduction yields NaI.

Uses

The nuclear fuel industry (see *Section 2.5*) uses large quantities of F_2 in the production of UF_6 for fuel enrichment processes and this is now the major use of F_2. Industrially, the most important F-containing compounds are HF, BF_3, CaF_2 (as a flux in metallurgy), synthetic cryolite (see *reaction 12.35*) and chlorofluorocarbons (CFCs, see *Box 13.6*).

Figure 16.1a summarizes the major uses of chlorine. Chlorinated organic compounds, including 1,2-dichloro-ethene and vinyl chloride for the polymer industry, are

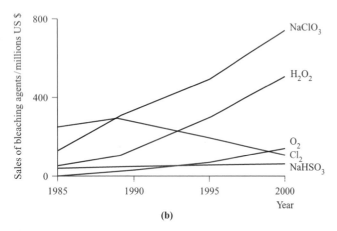

Fig. 16.1 (a) Industrial uses of Cl_2 in Western Europe in 1994 [data: *Chemistry & Industry* (1995) p. 832]. (b) The role of Cl_2 among bleaching agents in the paper and pulp industry is declining in the US, while that of ClO_2, which is prepared from $NaClO_3$, continues to grow (see Section 16.8); data for 1996–2000 are projected trends [data: *Chem. & Eng. News* (1997) 29 Sept. issue, p. 15].

hugely important. Dichlorine is widely used as a bleach and disinfectant, and drinking-water supplies are often chlorinated to make the water safer for human consumption. A disadvantage of Cl_2 as a bleach in the paper and pulp industry is the resultant toxic effluents, and the use of alternative bleaching agents is now encouraged (Figure 16.1b).[†]

The manufacture of bromine- and iodine-containing organic compounds is a primary application of these halogens. Other uses include those of iodide salts (e.g. KI) and silver bromide in the photographic industry (see *Box 22.13*), bromine-based organic compounds as flame retardants (see *Box 16.1*), and solutions of I_2 in aqueous KI as disinfectants for wounds. Iodine is essential for life and a deficiency results in a swollen thyroid gland; 'iodized salt' (NaCl with added I^-) provides us with iodine-supplement; we highlight uses of iodine in *Box 16.2*.

[†] For a discussion of methods of cleaning up contaminated groundwater, including the effects of contamination by chlorinated solvent waste, see: B. Ellis and K. Gorder (1997) *Chemistry & Industry*, p. 95.

16.3 Physical properties and bonding considerations

Table 16.1 lists selected physical properties of the group 17 elements (excluding astatine). Most of the differences between fluorine and the later halogens can be attributed to the:

- restriction of the valence shell of fluorine to an octet;
- relatively small size of the F atom and F^- ion;
- low dissociation energy of F_2 (Figures 14.1 and 16.2);
- higher oxidizing power of F_2;
- high electronegativity of fluorine.

The last factor is *not* a rigidly defined quantity. However, it is useful in rationalizing such observations as the anomalous physical properties of, for example, HF (see *Section 9.6*), the strength of F-substituted carboxylic acids, the deactivating effect of the CF_3 group in electrophilic aromatic substitutions, and the non-basic character of NF_3 and $(CF_3)_3N$.

The fact that F obeys the octet rule means that it forms no high oxidation state compounds (e.g. there are no analogues of $HClO_3$ and Cl_2O_7). When F is attached to another atom, Y, the Y–F bond is usually *stronger* than the corresponding Y–Cl bond (e.g. Tables 13.2, 14.3 and 15.2). If atom Y possesses no lone pairs, or has lone pairs but a large r_{cov}, then the Y–F bond is much stronger than the corresponding Y–Cl bond (e.g. C–F versus C–Cl, Table 13.2). Consequences of the small size of the F atom are that high coordination numbers can be achieved in molecular fluorides YF_n, and good overlap of atomic orbitals between Y and F leads to short, strong bonds, reinforced by ionic contributions when the difference in electronegativities of Y and F is large. The volatility of covalent F-containing compounds (e.g. fluorocarbons, see *Section 13.8*) originates in the weakness of the intermolecular van der Waals or London dispersion forces. This, in turn, can be correlated with the low polarizability and small size of the F atom. The small ionic radius of F^- leads to high coordination numbers in saline fluorides, high lattice energies and highly negative values of $\Delta_f H^o$ for these compounds, as well as a large negative standard enthalpy and entropy of hydration of the ion (Table 16.1).

In Section 15.3, we pointed out the importance of anion, rather than cation, formation in group 15. As expected, this is even more true in group 16. Table 16.1 lists values of the first ionization energies simply to show the expected decrease down the group. Although none of the halogens has yet been shown to form a discrete and stable monocation, complexed or solvated I^+ is established, e.g. in $[I(py)_2]^+$ (Figure 16.3), $[Ph_3PI]^+$ (see Section 16.4) and, apparently, in solutions obtained from reaction 16.2.

$$I_2 + AgClO_4 \xrightarrow{Et_2O} AgI + IClO_4 \qquad (16.2)$$

The corresponding Br- and Cl-containing species are less stable, though they are probably involved in aromatic

Table 16.1 Some physical properties of fluorine, chlorine, bromine and iodine.

Property	F	Cl	Br	I
Atomic number, Z	9	17	35	53
Ground state electronic configuration	$[He]2s^2 2p^5$	$[Ne]3s^2 3p^5$	$[Ar]3d^{10}4s^2 4p^5$	$[Kr]4d^{10}5s^2 5p^5$
Enthalpy of atomization, $\Delta_a H^\circ(298\,K)/kJ\,mol^{-1}$ ‡	79	121	112	107
Melting point, mp/K	53.5	172	266	387
Boiling point, bp/K	85	239	332	457.5
Standard enthalpy of fusion of X_2, $\Delta_{fus}H^\circ(mp)/kJ\,mol^{-1}$	0.51	6.40	10.57	15.52
Standard enthalpy of vaporization of X_2, $\Delta_{vap}H^\circ(bp)/kJ\,mol^{-1}$	6.62	20.41	29.96	41.57
First ionization energy, $EI_1/kJ\,mol^{-1}$	1681	1251	1140	1008
$\Delta_{EA}H_1^\circ(298\,K)/kJ\,mol^{-1}$ *	−328	−349	−325	−295
$\Delta_{hyd}H^\circ(X^-, g)/kJ\,mol^{-1}$	−504	−361	−330	−285
$\Delta_{hyd}S^\circ(X^-, g)/J\,K^{-1}\,mol^{-1}$	−150	−90	−70	−50
$\Delta_{hyd}G^\circ(X^-, g)/kJ\,mol^{-1}$	−459	−334	−309	−270
Standard reduction potential, $E^\circ(X_2/2X^-)/V$	+2.87	+1.36	+1.09	+0.54
Covalent radius, r_{cov}/pm	71	99	114	133
Ionic radius, r_{ion} for X^-/pm **	133	181	196	220
van der Waals radius, r_v/pm	135	180	195	215
Pauling electronegativity, χ^P	4.0	3.2	3.0	2.7

‡ For each element X, $\Delta_a H^\circ = \frac{1}{2} \times$ Dissociation energy of X_2.
* $\Delta_{EA}H_1^\circ(298\,K)$ is the enthalpy change associated with the process $X(g) + e^- \longrightarrow X^-(g) \approx -$(electron affinity); see *Section 1.10*.
** Values of r_{ion} refer to a coordination number of 6 in the solid state.

bromination and chlorination reactions in aqueous media.

The electron affinity of F is out of line with the trend observed for the later halogens (Table 16.1). Addition of an electron to the small F atom is accompanied by greater electron–electron repulsion than is the case for Cl, Br and I, and this probably explains why the process is less exothermic than might be expected on chemical grounds.

As we consider the chemistry of the halogens, it will be clear that there is an increasing trend towards higher oxidation states down the group; this is well exemplified among the interhalogen compounds (Section 16.7).

NMR active nuclei and isotopes as tracers

Although F, Cl, Br and I all possess spin active nuclei, in practice only ^{19}F (100%, $I = \frac{1}{2}$) is used routinely. Fluorine-19 NMR spectroscopy is a valuable tool in the elucidation of structures and reaction mechanisms of F-containing compounds; see case studies 1 and 5 and the discussion of stereochemically non-rigid species in *Section 2.11*.

Artificial isotopes of F include ^{18}F (β^+ emitter, $t_{\frac{1}{2}} = 1.83\,h$) and ^{20}F ($\beta^-$ emitter, $t_{\frac{1}{2}} = 11.0\,s$). The former is the longest lived radioisotope of F and may be used as a radioactive tracer. The ^{20}F isotope has application in F dating of bones and teeth; these usually contain apatite (see *Section 14.2* and *Box 14.10*) which is slowly converted to fluorapatite when the mineral is buried in the soil. By using the technique of *neutron activation analysis*, naturally occurring ^{19}F is converted to ^{20}F by neutron bombardment; the radioactive decay of the latter is then monitored, allowing the amount of ^{19}F originally present in the sample to be determined.

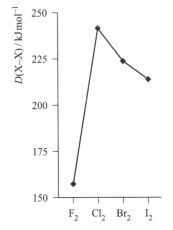

Fig. 16.2 The trend in X−X bond energies for the first four

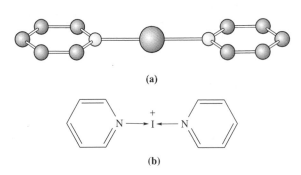

(a)

(b)

Fig. 16.3 (a) The structure of $[I(py)_2]^+$ (determined by X-ray crystallography) from the salt $[I(py)_2][I_3]\cdot2I_2$ [O. Hassel *et al.* (1961) *Acta Chem. Scand.*, vol. 15, p. 407], and (b) a representation of the bonding in the cation.

16.4 The elements

Difluorine

Difluorine is a pale-yellow gas with a characteristic smell similar to that of O_3 or Cl_2. It is extremely corrosive, being easily the most reactive element known. Difluorine is handled in Teflon or special steel vessels, although glass (see below) apparatus can be used if the gas is freed of HF by passage through sodium fluoride (equation 16.3).

$$NaF + HF \longrightarrow Na[HF_2] \qquad (16.3)$$

The synthesis of F_2 cannot be carried out in aqueous media because F_2 decomposes water, liberating ozonized oxygen (i.e. O_2 containing O_3); the oxidizing power of F_2 is apparent from the E^o value listed in Table 16.1. The decomposition of a few high oxidation state metal fluorides generates F_2, but the only efficient alternative to the electrolytic method used industrially (see Section 16.2) is reaction 16.4. However, F_2 is commercially available in cylinders, making laboratory synthesis generally unnecessary.

$$K_2[MnF_6] + 2SbF_5 \xrightarrow{420\,K} 2K[SbF_6] + MnF_2 + F_2 \quad (16.4)$$

Difluorine combines directly with all elements except O_2, N_2 and the lighter noble gases; reactions tend to be very violent. Combustion in compressed F_2 (*fluorine bomb calorimetry*) is a suitable method for determining values of $\Delta_f H^o$ for many binary metal fluorides. However, many metals are passivated by the formation of a layer of non-volatile metal fluoride. Silica is thermodynamically unstable with

respect to reaction 16.5, but, unless the SiO_2 is powdered, the reaction is slow provided that HF is absent; the latter sets up the chain reaction 16.6.

$$SiO_2 + 2F_2 \longrightarrow SiF_4 + O_2 \qquad (16.5)$$

$$\left.\begin{array}{l} SiO_2 + 4HF \longrightarrow SiF_4 + 2H_2O \\ 2H_2O + 2F_2 \longrightarrow 4HF + O_2 \end{array}\right\} \qquad (16.6)$$

The high reactivity of F_2 arises partly from the low bond dissociation energy (Figure 16.2) and partly from the strength of the bonds formed with other elements (see *Section 16.3*).

Dichlorine, dibromine and diiodine

Dichlorine is a pale green-yellow gas with a characteristic odour. Inhalation causes irritation of the respiratory system and liquid Cl_2 burns the skin. Reaction 16.7 can be used for small scale synthesis, but, like F_2, Cl_2 may be purchased in cylinders for laboratory use.

$$MnO_2 + \underset{conc}{4HCl} \longrightarrow MnCl_2 + Cl_2 + 2H_2O \qquad (16.7)$$

Dibromine is a dark-orange, volatile liquid (the only liquid non-metal at 298 K) but is often used as the aqueous solution 'bromine water'. Skin contact with liquid Br_2 results in burns, and Br_2 vapour has an unpleasant smell and causes eye and respiratory irritation. At 298 K, I_2 forms dark purple crystals which sublime readily at 1 bar pressure into a purple vapour.

In the crystalline state, Cl_2, Br_2 or I_2 molecules are arranged in layers as represented in Figure 16.4. The molecules Cl_2 and Br_2 have *intra*molecular distances which are the same as in the vapour (compare these distances with r_{cov}, Table 16.1). *Inter*molecular distances for Cl_2 and Br_2 are also listed in

---

BOX 16.3   **RESOURCES, ENVIRONMENTAL AND BIOLOGICAL**

Bromine: resources and commercial demand

World reserves of bromine in seawater, salt lakes and natural brine wells are plentiful. The major producers of Br_2 draw on brines from Arkansas and Michigan in the US, and from the Dead Sea in Israel, and the chart below indicates the extent to which these countries dominate the world market.

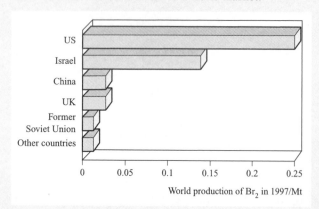

[Data: *Chemistry & Industry* (1999), p. 547.]

Environmental issues, however, are likely to have a dramatic effect on the commercial demand for Br_2. We have already mentioned the call to phase out some (or all) bromine-based flame retardants (Box 16.1). If a change to other types of flame retardants does become a reality, it would mean a massive cut in the demand for Br_2. The commercial market for Br_2 has already been hit by the switch from leaded to unleaded motor vehicle fuels. Leaded fuels contain 1,2-$C_2H_4Br_2$ as an additive to facilitate the release of lead (formed by decomposition of the anti-knock agent Et_4Pb) as a volatile bromide. 1,2-Dibromoethane is also used as a nematocide and fumigant, and CH_3Br is a widely applied fumigant for soil. Bromomethane, however, falls in the category of a potential ozone depleter (see *Box 13.6*) and its use will be phased out in industrialized countries by 2005, and in developing countries by 2015.

Further reading

B. Reuben (1999) *Chemistry & Industry*, p. 547 – 'An industry under threat?'

	Intramolecular distance for molecule in the gaseous state / pm	Intramolecular distance, a / pm	Intermolecular distance *within* a layer, b / pm	Intermolecular distance *between* layers / pm
Cl	199	198	332	374
Br	228	227	331	399
I	267	272	350	427

Fig. 16.4 Part of the solid state structures of Cl_2, Br_2 and I_2 in which molecules are arranged in stacked layers, and pertinent intramolecular and intermolecular distance data.

Figure 16.4; the distances within a layer are shorter than $2r_v$ (Table 16.1), suggesting some degree of interaction between the X_2 molecules. The shortest *inter*molecular $X\cdots X$ distance *between* layers is significantly longer. In solid I_2, the *intra*-molecular I–I bond distance is longer than in a gaseous molecule, and the lowering of the bond order (i.e. decrease in intramolecular bonding) is offset by a degree of *inter*-molecular bonding within each layer (Figure 16.4). It is significant that solid I_2 possesses a metallic lustre and exhibits appreciable electrical conductivity at higher temperatures; under very high pressure I_2 becomes a metallic conductor.

Chemical reactivity decreases steadily from Cl_2 to I_2, notably in reactions of the halogens with H_2, P_4, S_8 and most metals. The values of E^o in Table 16.1 indicate the decrease in oxidizing power along the series $Cl_2 > Br_2 > I_2$, and this trend is the basis of the methods of extraction of Br_2 and I_2 described in Section 16.2. Notable features of the chemistry of iodine which single it out among the halogens are that it is more easily:

- oxidized to high oxidation states;
- converted to stable salts containing I in the +1 oxidation state (e.g. Figure 16.3).

Charge transfer complexes

A *charge transfer complex* is one in which a donor and acceptor interact *weakly* together with some transfer of electronic charge, usually facilitated by the acceptor.

The observed colours of the halogens arise from an electronic transition from the highest occupied π^* MO to the lowest unoccupied σ^* MO (see *Figure 1.23*). The HOMO–LUMO energy gap decreases in the order $F_2 > Cl_2 > Br_2 > I_2$, leading to a progressive shift in the absorption maximum from the near-UV to the red region of the visible spectrum. Dichlorine, dibromine and diiodine dissolve unchanged in many organic solvents (e.g. saturated hydrocarbons, CCl_4). However in, for example, ethers, ketones and pyridine, which contain donor atoms, Br_2 and I_2 (and Cl_2 to smaller extent) form *charge transfer complexes* with the halogen σ^* MO acting as the acceptor orbital. In the extreme, complete transfer of charge could lead to heterolytic bond fission as in the formation of $[I(py)_2]^+$ (Figure 16.3 and equation 16.8).

$$2py + 2I_2 \longrightarrow [I(py)_2]^+ + [I_3]^- \qquad (16.8)$$

Solutions of I_2 in the above donor solvents are brown or yellow. Even benzene acts as a donor, forming charge transfer complexes with I_2 and Br_2; the colours of these solutions are noticeably different from those of I_2 or Br_2 in cyclohexane (a non-donor). Whereas amines, ketones and similar compounds donate electron density through a σ lone pair, benzene uses its π-electrons; this is apparent in the relative orientations of the donor (benzene) and acceptor (Br_2) molecules in Figure 16.5b. That solutions of the charge transfer complexes are coloured means that they absorb in the visible region of the spectrum (≈ 400–750 nm), but the electronic spectrum also contains an intense absorption in the UV region (≈ 230–330 nm) arising from an electronic transition from the solvent–X_2 occupied bonding MO to vacant anti-bonding MO. This is the so-called *charge transfer band*. Many charge transfer complexes can be isolated in the solid state and examples are given in Figure 16.5. In complexes in which the donor is weak, e.g. C_6H_6, the X–X bond distance is unchanged (or nearly so) by complex formation. Elongation as in $1,2,4,5$-$(EtS)_4C_6H_2\cdot(Br_2)_2$ (compare the Br–Br distance in Figure 16.5c with that for free Br_2, in Figure 16.4) is consistent with the involvement of a good donor; it has been estimated from theoretical calculations that -0.25 negative charges are transferred from $1,2,4,5$-$(EtS)_4C_6H_2$ to Br_2. Different degrees of charge transfer are also reflected in the relative magnitudes of $\Delta_r H$ given in equation 16.9. Further evidence for the weakening of the X–X bond comes from vibrational spectroscopic data, e.g. a shift for $\nu(X–X)$ from 215 cm^{-1} in I_2 to 204 cm^{-1} in $C_6H_6\cdot I_2$.

$$\left. \begin{array}{ll} C_6H_6 + I_2 \longrightarrow C_6H_6\cdot I_2 & \Delta_r H = -5\,\text{kJ mol}^{-1} \\ C_2H_5NH_2 + I_2 \longrightarrow C_2H_5NH_2\cdot I_2 & \Delta_r H = -31\,\text{kJ mol}^{-1} \end{array} \right\}$$
$$(16.9)$$

(16.1)

Figure 16.5d shows the solid state structure of $Ph_3P\cdot Br_2$; $Ph_3P\cdot I_2$ has a similar structure (I–I = 316 pm). In CH_2Cl_2

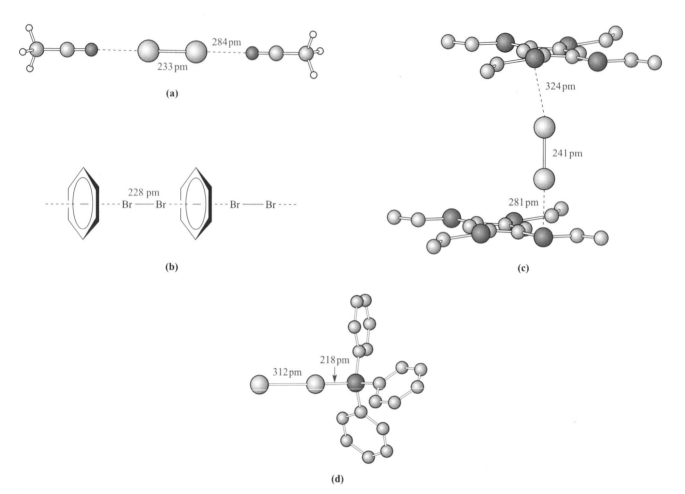

Fig. 16.5 Some examples of charge transfer complexes involving Br_2; the crystal structure of each has been determined by X-ray diffraction: (a) $2MeCN \cdot Br_2$ [K.-M. Marstokk *et al.* (1968) *Acta Crystallogr., Sect. B*, vol. 24, p. 713]; (b) schematic representation of the chains structure of $C_6H_6 \cdot Br_2$; (c) $1,2,4,5-(EtS)_4C_6H_2 \cdot (Br_2)_2$ in which Br_2 molecules are sandwiched between layers of $1,2,4,5-(EtS)_4C_6H_2$ molecules; interactions involving only one Br_2 molecule are shown and H atoms are omitted [H. Bock *et al.* (1996) *J. Chem. Soc., Chem. Commun.*, p. 1529]; (d) $Ph_3P \cdot Br_2$ [N. Bricklebank *et al.* (1992) *J. Chem. Soc., Chem. Commun.*, p. 355]. Colour code: Br, blue; C, grey; N, S or P, black; H, small grey.

solution, $Ph_3P \cdot Br_2$ ionizes to give $[Ph_3PBr]^+Br^-$ and, similarly, Ph_3PI_2 forms $[Ph_3PI]^+I^-$ or, in the presence of excess I_2, $[Ph_3PI]^+[I_3]^-$. The formation of complexes of this type is not easy to predict:

- the reaction of Ph_3Sb with Br_2 or I_2 is an oxidative addition yielding Ph_3SbX_2, **16.1**;
- Ph_3AsBr_2 is an As(V) compound, whereas $Ph_3As \cdot I_2$, $Me_3As \cdot I_2$ and $Me_3As \cdot Br_2$ are charge transfer complexes of the type shown in Figure 16.5d.[†]

Clathrates

Dichlorine, dibromine and diiodine are sparingly soluble in water. By freezing aqueous solutions of Cl_2 and Br_2, solid hydrates of approximate composition $X_2 \cdot 8H_2O$ may be obtained. These crystalline solids (known as *clathrates*) consist of a hydrogen-bonded structures with X_2 molecules occupying cavities in the lattice. An example is $1,3,5-(HO_2C)_3C_6H_3 \cdot 0.16Br_2$; the hydrogen-bonded lattice of pure $1,3,5-(HO_2C)_3C_6H_3$ was described in *Box 9.3*.

> A *clathrate* is a *host–guest compound*, a molecular assembly in which the guest molecules occupy cavities in the lattice of the guest species.

16.5 Hydrogen halides

All the hydrogen halides, HX, are gases at 298 K with sharp, acid smells. Selected properties are given in Table 16.2. Direct combination of H_2 and X_2 to form HX (see *equations 9.19–9.21* and accompanying discussion) can be used synthetically only for the chloride and bromide. Hydrogen fluoride is

---

[†] For insight into the complexity of this problem, see for example: N. Bricklebank, S.M. Godfrey, H.P. Lane, C.A. McAuliffe, R.G. Pritchard and J.-M. Moreno (1995) *Journal of the Chemical Society, Dalton Trans.*, p. 3873.

Table 16.2 Selected properties of the hydrogen halides.

Property	HF	HCl	HBr	HI
Physical appearance at 298 K	Colourless gas	Colourless gas	Colourless gas	Colourless gas
Melting point / K	189	159	186	222
Boiling point / K	293	188	207	237.5
$\Delta_{fus}H^\circ$(mp) / kJ mol^{-1}	4.6	2.0	2.4	2.9
$\Delta_{vap}H^\circ$(bp) / kJ mol^{-1}	34.0	16.2	18.0	19.8
$\Delta_{f}H^\circ$(298 K) / kJ mol^{-1}	−273.3	−92.3	−36.3	+26.5
$\Delta_{f}G^\circ$(298 K) / kJ mol^{-1}	−275.4	−95.3	−53.4	+1.7
Bond dissociation energy / kJ mol^{-1}	570	432	366	298
Bond length / pm	92	127.5	141.5	161
Dipole moment / D	1.83	1.11	0.83	0.45

prepared by treating suitable fluorides with concentrated H_2SO_4 (e.g. reaction 16.10) and analogous reactions are also a convenient means of making HCl. Analogous reactions with bromides and iodides result in partial oxidation of HBr or HI to Br_2 or I_2 (reaction 16.11), and synthesis is thus by reaction 16.12 with PX_3 prepared *in situ*.

$$CaF_2 + 2H_2SO_4 \xrightarrow{conc} 2HF + Ca(HSO_4)_2 \qquad (16.10)$$

$$2HBr + H_2SO_4 \xrightarrow{conc} Br_2 + 2H_2O + SO_2 \qquad (16.11)$$

$$PX_3 + 3H_2O \longrightarrow 3HX + H_3PO_3 \qquad X = Br \text{ or } I \qquad (16.12)$$

Some aspects of the chemistry of the hydrogen halides have already been covered:

- liquid HF (*Section 8.7*);
- solid state structure of HF (*Figure 9.8*);
- hydrogen bonding and trends in boiling points, melting points and $\Delta_{vap}H^\circ$ (*Section 9.6*);
- formation of the $[HF_2]^-$ ion (*Section 8.7; equation 9.25* and accompanying discussion);
- Brønsted acid behaviour in aqueous solution and energetics of acid dissociation (*Sections 6.4 and 6.5*).

Hydrogen fluoride is an important reagent for the introduction of F into organic and other compounds (e.g. *reaction 13.36* in the production of CFCs). It differs from the other hydrogen halides in being a weak acid in aqueous solution ($pK_a = 3.45$). This is in part due to the high H−F bond dissociation enthalpy (*Table 6.2* and *Section 6.5*). At high concentrations, the acid strength increases owing to the stabilization of F^- by formation of $[HF_2]^-$, **16.2** (*scheme 16.13* and *Table 9.4*).

$$\left[F\!-\!\!-\!H\!-\!\!-\!F \right]^-$$
$$(16.2)$$

$$\left. \begin{array}{l} HF(l) + H_2O(l) \rightleftharpoons [H_3O]^+(aq) + F^-(aq) \\[2mm] F^-(aq) + HF(aq) \rightleftharpoons [HF_2]^-(aq) \quad K = \dfrac{[HF_2^-]}{[HF][F^-]} = 0.2 \end{array} \right\}$$
$$(16.13)$$

The formation of $[HF_2]^-$ is also observed when HF reacts with group 1 metal fluorides; $M[HF_2]$ salts are stable at room temperature. Analogous compounds are formed with HCl, HBr and HI only at low temperatures.

16.6 Metal halides: structures and energetics

All the halides of the alkali metals have NaCl or CsCl structures (*Figures 5.15* and *5.16*) and their formation may be considered in terms of the Born−Haber cycle (see *Section 5.14*). In *Section 10.5*, we discussed trends in lattice energies of these halides, and showed that lattice energy is proportional to $1/(r_+ + r_-)$. We can apply this relationship to see why, for example, CsF is the best choice of alkali metal fluoride to effect the halogen exchange reaction 16.14.

$$(16.14)$$

In the absence of solvent, the energy change associated with reaction 16.14 involves:

- the difference between the C−Cl and C−F bond energy terms (*not* dependent on M);
- the difference between the electron affinities of F and Cl (*not* dependent on M);
- the difference in lattice energies between MF and MCl (dependent on M).

The last difference is approximately proportional to the expression:

$$\frac{1}{(r_{M^+} + r_{Cl^-})} - \frac{1}{(r_{M^+} + r_{F^-})}$$

which is always negative because $r_{F^-} < r_{Cl^-}$; the term approaches zero as r_{M^+} increases. Thus, reaction 16.14 is favoured most for $M^+ = Cs^+$.

A few other monohalides possess the NaCl or CsCl structure, e.g. AgF, AgCl, and we have already discussed (*Section 5.15*) that these compounds exhibit significant covalent character. The same is true for CuCl, CuBr, CuI and AgI which possess the wurtzite structure (*Figure 5.20*).

Most metal difluorides crystallize with CaF_2 (*Figure 5.18*) or rutile (*Figure 5.21*) lattices, and for most of these, a simple ionic model is appropriate (e.g. CaF_2, SrF_2, BaF_2, MgF_2, MnF_2 and ZnF_2). With slight modification, this model also holds for other d-block difluorides. Chromium(II) chloride adopts a *distorted* rutile lattice, but other first row d-block metal dichlorides, dibromides and diiodides possess $CdCl_2$ or CdI_2 lattices (see *Figure 5.22* and accompanying discussion). For these dihalides, neither purely electrostatic nor purely covalent models are satisfactory. Dihalides of the heavier d-block metals are considered in *Chapter 22*.

Metal trifluorides are crystallographically more complex than the difluorides, but symmetrical three-dimensional structures are commonly found, and many contain octahedral (sometimes distorted) metal centres, e.g. AlF_3 (*Section 12.6*), VF_3 and MnF_3. For trichlorides, tribromides and triiodides, layer structures predominate. Among the tetrafluorides, a few have lattice structures, e.g. the two polymorphs of ZrF_4 possess, respectively, corner-sharing square antiprismatic and dodecahedral ZrF_8 units. Most metal tetrahalides are either volatile molecular species (e.g. $SnCl_4$, $TiCl_4$) or contain rings or chains with M–F–M bridges (e.g. SnF_4, **13.11**); metal–halogen bridges are longer than terminal bonds. Metal pentahalides may possess chain or rings structures (e.g. NbF_5, RuF_5, SbF_5, *Figure 14.9a*) or molecular structures (e.g. $SbCl_5$), while metal hexahalides are molecular and octahedral (e.g. UF_6, MoF_6, WF_6, WCl_6). In general, an increase in oxidation state results in a structural change along the series three-dimensional ionic $\longrightarrow$ layer or polymer $\longrightarrow$ molecular.

For metals exhibiting variable oxidation states, the relative thermodynamic stabilities of two halides that contain a common halide ion but differ in the oxidation state of the metal (e.g. AgF and AgF_2) can be assessed using Born–Haber cycles. In such a reaction as 16.15, if the increase in ionization energies (e.g. $M \longrightarrow M^+$ *versus* $M \longrightarrow M^{2+}$) is approximately offset by the difference in lattice energies of the compounds, the two metal halides will be of about equal stability. This commonly happens with d-block metal halides.

$$AgX + \tfrac{1}{2}X_2 \longrightarrow AgX_2 \qquad (16.15)$$

16.7 Interhalogen compounds and polyhalogen ions

Interhalogen compounds

Properties of interhalogen compounds are listed in Table 16.3. All are prepared by direct combination of elements,

and where more than one product is possible, the outcome of the reaction is controlled by temperature and relative proportions of the halogens. Reactions of F_2 with the later halogens at ambient temperature and pressure gives ClF, BrF_3 or IF_5, but increased temperatures gives ClF_3, ClF_5, BrF_5 and IF_7. Table 16.3 shows clear trends among the four families of compounds XY, XY_3, XY_5 and XY_7:

- F is always in oxidation state -1;
- highest oxidation states for X reached are Cl < Br < I;
- combination of the later halogens with *fluorine* leads to the highest oxidation state compounds.

The structural families are **16.3**–**16.6** and are consistent with the VSEPR model, although in *Section 1.19*, we used ClF_3 to illustrate the limitations of VSEPR theory. Angle α in **16.4** is 87.5° in ClF_3 and 86° in BrF_3. In each of ClF_5, BrF_5 and IF_5, the X atom lies just below the plane of the four F atoms; in **16.5**, $\approx 90°$ (Cl) $> \alpha > 81°$ (I). Among the interhalogens, 'ICl_3' is unusual in being dimeric and possesses structure **16.7**; the planar I environments are consistent with VSEPR theory.

T-shaped

(16.3)   (16.4)

Square-based pyramid   Pentagonal bipyramid

(16.5)   (16.6)

(16.7)

In a series XY_n in which the oxidation state of X increases, the X–Y bond enthalpy term decreases, e.g. for the Cl–F bonds in ClF, ClF_3 and ClF_5, they are 257, 172 and 153 kJ mol^{-1} respectively.

The most stable of the diatomic molecules are ClF and ICl; at 298 K, IBr dissociates somewhat into its elements, while BrCl is substantially dissociated (Table 16.3). Bromine monofluoride readily disproportionates (equation 16.16), while reaction 16.17 is facile enough to render IF unstable at room temperature.

Table 16.3 Properties of interhalogen compounds.

Compound	Appearance at 298 K	Melting point / K	Boiling point / K	$\Delta_f H^\circ(298\,\text{K})^{***}$ / kJ mol^{-1}	Dipole moment for gas phase molecule / D	Bond distances in gas phase molecules / pm[§]
ClF	Colourless gas	117	173	−50.3	0.89	163
BrF	Pale brown gas	≈240*	≈293*	−58.5	1.42	176
BrCl	‡	–	–	+14.6	0.52	214
ICl	Red solid	300 (α) 287 (β)	≈373**	−23.8	1.24	232
IBr	Black solid	313	389**	−10.5	0.73	248.5
ClF$_3$	Colourless gas	197	285	−163.2	0.6	160 (eq), 170 (ax)
BrF$_3$	Yellow liquid	282	399	−300.8	1.19	172 (eq), 181 (ax)
IF$_3$	Yellow solid	245 (dec)	–	≈ −500	–	–
I$_2$Cl$_6$	Orange solid	337 (sub)	–	−89.3	0	238 (terminal)[§§] 268 (bridge)
ClF$_5$	Colourless gas	170	260	−255	–	172 (basal), 162 (apical)
BrF$_5$	Colourless liquid	212.5	314	−458.6	1.51	178 (basal), 168 (apical)
IF$_5$	Colourless liquid	282.5	373	−864.8	2.18	187 (basal), 185 (apical)
IF$_7$	Colourless gas	278 (sub)	–	−962	0	186 (eq), 179 (ax)

[‡] Exists only in equilibrium with dissociation products: $2\text{BrCl} \rightleftharpoons \text{Br}_2 + \text{Cl}_2$.
[*] Significant disproportionation means values are approximate.
[**] Some dissociation: $2\text{IX} \rightleftharpoons \text{I}_2 + \text{X}_2$ (X = Cl, Br).
[***] Values quoted for the state observed at 298 K.
[§] See structures **16.3–16.7**.
[§§] Solid state (X-ray diffraction) data.

$$3\text{BrF} \longrightarrow \text{Br}_2 + \text{BrF}_3 \qquad (16.16)$$

$$5\text{IF} \longrightarrow 2\text{I}_2 + \text{IF}_5 \qquad (16.17)$$

In general, the diatomic interhalogens exhibit properties intermediate between their parent halogens. However, where the electronegativities of X and Y differ significantly, the X–Y bond is stronger than the mean of the X–X and Y–Y bond strengths (see *equations 1.32 and 1.33*). Consistent with this is the observation that, if $\chi^P(X) \ll \chi^P(Y)$, the X–Y bond lengths (Table 16.3) are shorter than the mean of $d(\text{X–X})$ and $d(\text{Y–Y})$. In the solid state, both α- and β-forms of ICl have chain structures; in each form, two ICl environments are present (e.g. in α-ICl, I–Cl distances are 244 or 237 pm) and there are significant intermolecular interactions with I····Cl separations of 300–308 pm. Solid IBr has a similar structure (**16.8**) although it differs from ICl in that ICl contains I····Cl, I····I and Cl····Cl intermolecular contacts, whereas IBr has only I····Br contacts. Compare these structures with those in Figure 16.2.

(16.8)

Chlorine monofluoride (which is commercially available) acts as a powerful fluorinating and oxidizing agent (e.g. reaction 16.18); oxidative addition to SF$_4$ was shown in Figure 15.10. It may behave as a fluoride donor (equation 16.19) or acceptor (equation 16.20). The structures of $[\text{Cl}_2\text{F}]^+$ (**16.9**) and $[\text{ClF}_2]^-$ (**16.10**) can be rationalized using the VSEPR model. Iodine monochloride and monobromide are less reactive than ClF, but of importance is the fact that, *in polar solvents*, ICl is a source of I$^+$ and iodinates aromatic compounds.

$$\text{W} + 6\text{ClF} \longrightarrow \text{WF}_6 + 3\text{Cl}_2 \qquad (16.18)$$

$$2\text{ClF} + \text{AsF}_5 \longrightarrow [\text{Cl}_2\text{F}]^+[\text{AsF}_6]^- \qquad (16.19)$$

$$\text{ClF} + \text{CsF} \longrightarrow \text{Cs}^+[\text{ClF}_2]^- \qquad (16.20)$$

$$\left[\begin{array}{c} \text{Cl} \\ \diagdown \\ \text{Cl} \qquad \text{F} \end{array}\right]^+ \qquad \left[\text{F} \!—\! \text{Cl} \!—\! \text{F}\right]^-$$

(16.9)   (16.10)

With the exception of I$_2$Cl$_6$, the higher interhalogens contain F and are extremely reactive, exploding or reacting violently with water or organic compounds; ClF$_3$ even ignites asbestos. Despite these hazards, they are valuable fluorinating agents, e.g. the highly reactive ClF$_3$ converts metals, metal chlorides and metal oxides to metal fluorides.

One of its main uses is in nuclear fuel reprocessing (see *Section 2.5*) for the formation of UF_6 (reaction 16.21).

$$U + 3ClF_3 \xrightarrow{\Delta} UF_6 + 3ClF \qquad (16.21)$$

Reactivity decreases in the general order $ClF_n > BrF_n > IF_n$, and within a series having common halogens, the compound with the highest value of n is the most reactive, e.g. $BrF_5 > BrF_3 > BrF$. In line with these trends is the use of IF_5 as a relatively mild fluorinating agent in organic chemistry.

We have already discussed the self-ionization of BrF_3 and its use as a non-aqueous solvent (see *Section 8.10*). There is some evidence for the self-ionization of IF_5 (equation 16.22), but little to support similar processes for other interhalogens.

$$2IF_5 \rightleftharpoons [IF_4]^+ + [IF_6]^- \qquad (16.22)$$

Reactions 16.19 and 16.20 showed the fluoride donor and acceptor abilities of ClF. All the higher interhalogens undergo similar reactions, although ClF_5 does not form stable complexes at 298 K with alkali metal fluorides but does react with CsF or $[Me_4N]F$ at low temperatures to give salts containing $[ClF_6]^-$. Examples are given in equations 8.42 and 16.23–16.27.

$$NOF + ClF_3 \longrightarrow [NO]^+[ClF_4]^- \qquad (16.23)$$

$$CsF + IF_7 \longrightarrow Cs^+[IF_8]^- \qquad (16.24)$$

$$IF_3 \xrightarrow{[Me_4N]F} [Me_4N]^+[IF_4]^- \xrightarrow{[Me_4N]F} [Me_4N]^+_2[IF_5]^{2-}$$
$$(16.25)$$

$$ClF_3 + AsF_5 \longrightarrow [ClF_2]^+[AsF_6]^- \qquad (16.26)$$

$$IF_5 + 2SbF_5 \longrightarrow [IF_4]^+[Sb_2F_{11}]^- \qquad (16.27)$$

The choice of a large cation (e.g. Cs^+, $[NMe_4]^+$) for stabilizing $[XY_n]^-$ anions follows from lattice energy considerations; see also *Boxes 10.5* and *23.2*. Thermal decomposition of salts of $[XY_n]^-$ leads to the halide salt of highest lattice energy, e.g. reaction 16.28.

$$Cs[ICl_2] \xrightarrow{\Delta} CsCl + ICl \qquad (16.28)$$

Whereas $[IF_6]^+$ can be made by treating IF_7 with a fluoride acceptor (e.g. AsF_5), $[ClF_6]^+$ or $[BrF_6]^+$ must be made from ClF_5 or BrF_5 using an extremely powerful oxidizing agent in the presence of an F^- acceptor (e.g. reaction 16.29). Reaction 16.30 further illustrates the use of a noble gas fluoride in interhalogen synthesis; unlike reaction 16.25, this route to $[Me_4N][IF_4]$ avoids the use of the thermally unstable IF_3.

$$[Kr_2F_3]^+[AsF_6]^- + BrF_5 \longrightarrow [BrF_6]^+[AsF_6]^- + 2Kr + F_2$$
$$(16.29)$$

$$2XeF_2 + [Me_4N]I \xrightarrow{242\ K;\ warm\ to\ 298\ K} [Me_4N][IF_4] + 2Xe$$
$$(16.30)$$

Table 16.4 Structures of selected interhalogens and derived anions and cations. Each is consistent with VSEPR theory.

Shape	Examples
Linear	$[ClF_2]^-$, $[IF_2]^-$, $[ICl_2]^-$, $[IBr_2]^-$
Bent	$[ClF_2]^+$, $[BrF_2]^+$, $[ICl_2]^+$
T-shaped	ClF_3, BrF_3, ICl_3
Square planar	$[ClF_4]^-$, $[BrF_4]^-$, $[IF_4]^-$, $[ICl_4]^-$
Disphenoidal, **16.11**	$[ClF_4]^+$, $[BrF_4]^+$, $[IF_4]^+$
Square-based pyramidal	ClF_5, BrF_5, IF_5
Pentagonal planar	$[IF_5]^{2-}$
Octahedral	$[ClF_6]^+$, $[BrF_6]^+$, $[IF_6]^+$
Pentagonal bipyramidal	IF_7
Square antiprismatic	$[IF_8]^-$

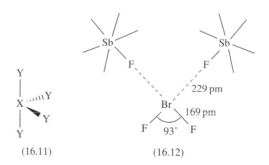

(16.11)   (16.12)

On the whole, the observed structures of interhalogen anions and cations (Table 16.4) are in accord with VSEPR theory, but $[BrF_6]^-$ is regular octahedral, and arguments reminiscent of those used in *Section 15.7* to rationalize the structures of $[SeCl_6]^{2-}$ and $[TeCl_6]^{2-}$ appertain. Raman spectroscopic data suggest that $[ClF_6]^-$ is isostructural with $[BrF_6]^-$. On the other hand, the vibrational spectrum of $[IF_6]^-$ shows it is not regular octahedral; however, on the ^{19}F NMR time-scale, $[IF_6]^-$ is stereochemically non-rigid. The difference between $[BrF_6]^-$ and $[IF_6]^-$ is not easy to rationalize.

Of particular interest in Table 16.4 is $[IF_5]^{2-}$. Only two examples of pentagonal planar XY_n species are known, the other being $[XeF_5]^-$ (see *Section 17.4*). In salts such as $[BrF_2][SbF_6]$, $[ClF_2][SbF_6]$ and $[BrF_4][Sb_2F_{11}]$, there is significant cation–anion interaction; diagram **16.12** focuses on the Br environment on the solid state structure of $[BrF_2][SbF_6]$.

Bonding in $[XY_2]^-$ ions

In *Section 4.7*, we used molecular orbital theory to describe the bonding in XeF_2, and developed a picture which gave a bond order of $\frac{1}{2}$ for each $Xe-F$ bond. In terms of valence electrons, XeF_2 is isoelectronic with $[ICl_2]^-$, $[IBr_2]^-$, $[ClF_2]^-$ and related anions, and all have linear structures. The bonding in these anions can be viewed as being similar to that in XeF_2, and thus suggests weak $X-Y$ bonds; note that this is in contrast to the localized picture that emerges from a structure such as **16.13**. Evidence for weak bonds comes from the $X-Y$ bond lengths (e.g. 255 pm in $[ICl_2]^-$ compared with 232 in ICl) and from $X-Y$ bond stretching wavenumbers (e.g. 267 and 222 cm^{-1} for the symmetric and asymmetric stretches of $[ICl_2]^-$ compared with 384 cm^{-1} in ICl).

(16.13)

Polyhalogen cations

In addition to the interhalogen cations described above, homonuclear cations $[Br_2]^+$, $[I_2]^+$, $[Cl_3]^+$, $[Br_3]^+$, $[I_3]^+$, $[Br_5]^+$, $[I_5]^+$ and $[I_4]^{2+}$ are well established. Other ions such as $[Cl_2]^+$ and $[I_7]^+$ exist but are not well characterized. Cations $[Br_2]^+$ and $[I_2]^+$ can be obtained by oxidation of the corresponding halogen (equations 16.31, 16.32 and 8.15).

$$Br_2 + SbF_5 \xrightarrow{BrF_5} [Br_2]^+[Sb_3F_{16}]^- \qquad (16.31)$$

$$2I_2 + S_2O_6F_2 \xrightarrow{HSO_3F} 2[I_2]^+[SO_3F]^- \qquad (16.32)$$

On going from X_2 to the corresponding $[X_2]^+$, the bond shortens consistent with the loss of an electron from an antibonding orbital (see *Figure 1.20*). In $[Br_2]^+[Sb_3F_{16}]^-$, the $Br-Br$ distance is 215 pm, and in $[I_2]^+[Sb_2F_{11}]^-$ the $I-I$ bond length is 258 pm (compare values of X_2 in Figure 16.4). Correspondingly, the stretching wavenumber increases, e.g. 368 cm^{-1} in $[Br_2]^+$ compared with 320 cm^{-1} in Br_2. The cations are paramagnetic, and $[I_2]^+$ dimerizes at 193 K to give $[I_4]^{2+}$ (**16.14**); the structure has been determined for the salt $[I_4][Sb_3F_{16}][SbF_6]$ and exhibits significant cation–anion interaction.

The cations $[Cl_3]^+$, $[Br_3]^+$ and $[I_3]^+$ are bent (**16.15**) as expected from VSEPR theory, and the $X-X$ bond lengths are similar to those in gaseous X_2, consistent with single bonds. Reactions 16.33 and 16.34 may be used to prepare salts of $[Br_3]^+$ and $[I_3]^+$, and use of a higher concentration of I_2 in the I_2 / AsF_5 reaction leads to the formation of $[I_5]^+$ (see *reaction 8.15*). The $[I_5]^+$ and $[Br_5]^+$ ions are structurally

(16.14)

(16.15) $X = Cl, Br, I$

(16.16) $X = Br, I$

similar (**16.16**) with $d(X-X)_{terminal} < d(X-X)_{non\text{-}terminal}$, e.g. in $[I_5]^+$, the distances are 264 and 289 pm.

$$3Br_2 + 2[O_2]^+[AsF_6]^- \longrightarrow 2[Br_3]^+[AsF_6]^- + 2O_2 \qquad (16.33)$$

$$3I_2 + 3AsF_5 \xrightarrow{\text{in liquid } SO_2} 2[I_3]^+[AsF_6]^- + AsF_3 \qquad (16.34)$$

Polyhalide anions

Of the group 17 elements, iodine forms the largest range of homonuclear polyhalide ions: $[I_3]^-$, $[I_4]^{2-}$, $[I_5]^-$, $[I_7]^-$, $[I_8]^{2-}$, $[I_9]^-$, $[I_{10}]^{4-}$, $[I_{12}]^{2-}$, $[I_{16}]^{2-}$, $[I_{16}]^{4-}$, $[I_{22}]^{4-}$ and $[I_{29}]^{3-}$. Attempts to make $[F_3]^-$ have failed, but $[Cl_3]^-$ and $[Br_3]^-$ are well established, and $[Br_4]^{2-}$ and $[Br_8]^{2-}$ have also been reported. The $[I_3]^-$ ion is formed when I_2 is dissolved in aqueous solutions containing iodide ion. It has a linear structure, and in the solid state, the two $I-I$ bond lengths may be equal (e.g. 290 pm in $[Ph_4As][I_3]$) or dissimilar (e.g. 283 and 303 pm in $Cs[I_3]$). The latter indicates something approaching to an $[I-I\cdots I]^-$ entity (compare $I-I = 266$ pm in I_2), and in the higher polyiodide ions, different $I-I$ bond distances point to the structures being described in terms of association between I_2, I^- and $[I_3]^-$ units as examples in Figure 16.6 show. This reflects their origins, since the higher polyiodides are formed upon crystallization of solutions containing I_2 and I^-. Details of the solid state structures of the anions are cation-dependent; e.g. although usually V-shaped, linear $[I_5]^-$ has also been observed in the solid state.

Fig. 16.6 The structures (X-ray diffraction) of (a) $[I_5]^-$ in $[Fe(S_2CNEt_2)_3][I_5]$ [C.L. Raston *et al.* (1980) *J. Chem. Soc., Dalton Trans.*, p. 1928], (b) $[I_7]^-$ in $[Ph_4P][I_7]$ [R. Poli *et al.* (1992) *Inorg. Chem.*, vol. 31, p. 3165], and (c) $[I_8]^{2-}$ in $[C_{10}H_8S_8]_2[I_3][I_8]_{0.5}$ [M.A. Beno *et al.* (1987) *Inorg. Chem.*, vol. 26, p. 1912].

Fewer studies of polybromide ions have been carried out. Many salts involving $[Br_3]^-$ are known, and the association in the solid state of $[Br_3]^-$ and Br^- has been observed to give rise to the linear species **16.17**. The $[Br_8]^{2-}$ ion is structurally analogous to $[I_8]^{2-}$ (Figure 16.6) with Br–Br bond distances that indicate association between Br_2 and $[Br_3]^-$ units in the crystal.

$$\left[Br\!-\!\!-\!Br\!-\!\!-\!\!-\!\!-\!Br\!-\!\!-\!\!-\!\!-\!\!-\!Br\!-\!\!-\!Br\!-\!\!-\!\!-\!\!-\!\!-\!Br\!-\!\!-\!\!-\!\!-\!\!-\!Br \right]^{3-}$$

(16.17)

16.8 Oxides and oxofluorides of chlorine, bromine and iodine

Oxides

Oxygen fluorides were described in *Section 15.7*. Iodine is the only halogen to form an oxide which is *thermodynamically stable* with respect to decomposition into its elements (equation 16.35). The chlorine and bromine oxides are hazardous materials with a tendency to explode.

$$I_2 + \tfrac{5}{2}O_2 \longrightarrow I_2O_5 \qquad \Delta_f H^\circ(298\,\text{K}) = -158.1\,\text{kJ mol}^{-1}$$

(16.35)

Chlorine oxides, although not difficult to prepare, are all liable to decompose explosively. Far less is known about the oxides of Br (which are very unstable) than those of Cl and iodine, although recently Br_2O_3 (**16.18**) and Br_2O_5 (**16.19**) have been unambiguously prepared (equation 16.36) and structurally characterized. The Br(V) centres are trigonal pyramidal and in Br_2O_5, the BrO_2 groups are eclipsed.

$$Br_2 \xrightarrow[\text{Brown}]{O_3,\,195\,\text{K}} Br_2O_3 \xrightarrow[\text{Orange}]{O_3,\,195\,\text{K}} Br_2O_5$$
$$\text{Colourless}$$

(16.36)

(16.18)   (16.19)

(16.20)

Dichlorine monoxide, Cl_2O (**16.20**), is obtained as a yellow-brown gas by action of Cl_2 on mercury(II) oxide or moist sodium carbonate (equations 16.37 and 16.38); it liquefies at $\approx 277\,\text{K}$, and explodes on warming. It hydrolyzes to hypochlorous acid (equation 16.39), and is formally the anhydride of this acid (see *Section 14.8*).

$$2Cl_2 + 3HgO \longrightarrow Cl_2O + Hg_3O_2Cl_2 \tag{16.37}$$
$$2Cl_2 + 2Na_2CO_3 + H_2O \longrightarrow 2NaHCO_3 + 2NaCl + Cl_2O \tag{16.38}$$
$$Cl_2O + H_2O \longrightarrow 2HOCl \tag{16.39}$$

Cl–O = 147 pm; ∠O–Cl–O = 117°

(16.21)

Chlorine dioxide, ClO_2 (**16.21**) is a yellow gas (bp 283 K), and is produced in the highly dangerous reaction between potassium chlorate, $KClO_3$, and concentrated H_2SO_4. Reaction 16.40 is a safer method of synthesis, and reaction 16.41 is used industrially; ClO_2 is used to bleach flour and wood pulp and for water treatment. Its application as a bleach in the paper industry is increasing in, for example, North America, since the use of Cl_2 is now discouraged on environmental grounds (see *Figure 16.1*).

$$2KClO_3 + 2H_2C_2O_4 \longrightarrow K_2C_2O_4 + 2ClO_2 + 2CO_2 + 2H_2O \tag{16.40}$$
$$2NaClO_3 + SO_2 + H_2SO_4 \longrightarrow 2NaHSO_4 + 2ClO_2 \tag{16.41}$$

Despite being a radical, ClO_2 shows no tendency to dimerize. It dissolves unchanged in water, but is slowly hydrolyzed to HCl and $HClO_3$, a reaction that involves the $ClO^\bullet$ radical. In alkaline solution, hydrolysis is rapid (equation 16.42). Ozone reacts with ClO_2 at 273 K to form Cl_2O_6, a dark red liquid which is also made by reaction 16.43.

$$2ClO_2 + 2[OH]^- \longrightarrow [ClO_3]^- + [ClO_2]^- + H_2O \tag{16.42}$$
$$ClO_2F + HClO_4 \longrightarrow Cl_2O_6 + HF \tag{16.43}$$

(16.22)

Reaction 16.43, and the hydrolysis of Cl_2O_6 to chlorate and perchlorate, suggest that it has structure **16.22** and is the mixed anhydride of $HClO_3$ and $HClO_4$; the IR spectrum of matrix-isolated Cl_2O_6 is consistent with two inequivalent Cl centres. The solid contains $[ClO_2]^+$ and $[ClO_4]^-$ ions. Cl_2O_6 is unstable with respect to decomposition into ClO_2 and $O_2^\bullet$, and, with H_2O, reaction 16.44 occurs. The oxide $ClOClO_3$ is the mixed acid anhydride of HOCl and $HClO_4$, and is made by reaction 16.45.

$$Cl_2O_6 + H_2O \longrightarrow HClO_4 + HClO_3 \tag{16.44}$$
$$Cs[ClO_4] + ClSO_3F \longrightarrow Cs[SO_3F] + ClOClO_3 \tag{16.45}$$

The anhydride of perchloric acid is Cl_2O_7 (**16.23**), an oily, explosive liquid (bp $\approx 353\,\text{K}$), which is made by dehydrating $HClO_4$ using phosphorus(V) oxide at low temperatures.

(16.23)

In contrast to Br_2O_5 which is thermally unstable, I_2O_5 is stable to 573 K. It is a white, hygroscopic solid, prepared by dehydration of iodic acid; the reaction is reversed when I_2O_5 dissolves in water (equation 16.46). I_2O_5 is used in analysis for CO (see *equation 13.50*).

$$I_2O_5 + H_2O \longrightarrow 2HIO_3 \qquad (16.46)$$

In the solid state, I_2O_5 is structurally related to Br_2O_5 (**16.19**), with the difference that it has a staggered conformation, probably as a result of extensive intermolecular interactions ($I \cdots O \leq 223$ pm).

Oxofluorides

Several families of halogen oxides with $X-F$ bonds exist: FXO_2 (X = Cl, Br, I), FXO_3 (X = Cl, Br, I), F_3XO (X = Cl, Br, I), F_3XO_2 (X = Cl, I) and F_5IO; the thermally unstable FClO is also known. Their structures are consistent with VSEPR theory (**16.24–16.29**).

(16.24)   (16.25)   (16.26)

(16.27)   (16.28)   (16.29)

Chloryl fluoride, $FClO_2$, is a colourless gas (bp 267 K) and can be prepared by reacting F_2 with ClO_2. It hydrolyzes to $HClO_3$ and HF, and acts as a fluoride donor towards SbF_5 (equation 16.47) and a fluoride acceptor with CsF (equation 16.48).

$$FClO_2 + SbF_5 \longrightarrow [ClO_2]^+[SbF_6]^- \qquad (16.47)$$

$$CsF + FClO_2 \longrightarrow Cs^+[F_2ClO_2]^- \qquad (16.48)$$

Perchloryl fluoride, $FClO_3$ (bp 226 K, $\Delta_f H^\circ(298\,K) = -23.8$ kJ mol^{-1}) is surprisingly stable and decomposes only above 673 K. It can be prepared by reaction 16.49, or by treating $KClO_3$ with F_2.

$$KClO_4 + 2HF + SbF_5 \longrightarrow FClO_3 + KSbF_6 + H_2O \quad (16.49)$$

Alkali attacks $FClO_3$ only slowly, even at 500 K. Perchloryl fluoride is a mild fluorinating agent and has been used in the preparation of fluorinated steroids; it is also a powerful oxidizing agent at elevated temperatures, e.g. it oxidizes SF_4 to SF_6. Reaction 16.50 illustrates its reaction with an organic nucleophile. In contrast to $FClO_2$, $FClO_3$ does not behave as a fluoride donor or acceptor.

$$C_6H_5Li + FClO_3 \longrightarrow LiF + C_6H_5ClO_3 \qquad (16.50)$$

The reaction between F_2 and Cl_2O at low temperatures yields F_3ClO (mp 230 K, bp 301 K, $\Delta_f H^\circ(g, 298\,K) = -148$ kJ mol^{-1}) which decomposes at 570 K to ClF_3 and O_2. Reactions of F_3ClO with CsF and SbF_5 show its ability to accept or donate F^-, producing $[F_4ClO]^-$ and $[F_2ClO]^+$ respectively.

The only representative of the F_5XO family of oxofluorides is F_5IO, produced when IF_7 reacts with water; it does not readily undergo further reaction with H_2O. One reaction of note is that with $[Me_4N]F$ in which the pentagonal bipyramidal ion $[F_6IO]^-$ is formed; X-ray diffraction data show that the oxygen atom is in an axial site and that the equatorial F atoms are essentially coplanar, in contrast to the puckering observed in IF_7 (see *Section 1.19*).

16.9 Oxoacids and their salts

Hypofluorous acid, HOF

Fluorine is unique among the halogens in forming no species in which it has a formal oxidation state other than -1. The only known oxoacid is hypofluorous acid, HOF, which is unstable and does not ionize in water but reacts according to equation 16.51; no salts are known. It is obtained by passing F_2 over ice at 230 K (equation 16.52) and condensing the gas produced. At 298 K, HOF decomposes rapidly (equation 16.53).

$$HOF + H_2O \longrightarrow H_2O_2 + HF \qquad (16.51)$$

$$F_2 + H_2O \xrightarrow{230\,K} HOF + HF \qquad (16.52)$$

$$2HOF \longrightarrow 2HF + O_2 \qquad (16.53)$$

Oxoacids of chlorine, bromine and iodine

Table 16.5 lists the families of oxoacids known for Cl, Br and I. The hypohalous acids, HOX, are obtained in aqueous solution by reaction 16.54 (compare reactions 16.37 and 16.39).

$$2X_2 + 3HgO + H_2O \longrightarrow Hg_3O_2X_2 + 2HOX \qquad (16.54)$$

All are unknown as isolated compounds, but act as weak acids in aqueous solutions (pK_a values: HOCl, 4.53; HOBr, 8.69; HOI, 10.64). Hypochlorite salts such as NaOCl, KOCl and $Ca(OCl)_2$ (equation 16.55) can be isolated; NaOCl can be crystallized from a solution obtained by

Table 16.5 Oxoacids of chlorine, bromine and iodine.

Oxoacids of chlorine		Oxoacids of bromine		Oxoacids of iodine	
Hypochlorous acid	HOCl	Hypobromous acid	HOBr	Hypoiodous acid	HOI
Chlorous acid	HOClO ($HClO_2$)				
Chloric acid	HOClO$_2$ ($HClO_3$)	Bromic acid	HOBrO$_2$ ($HBrO_3$)	Iodic acid	HOIO$_2$ (HIO$_3$)
Perchloric acid	HOClO$_3$ ($HClO_4$)	Perbromic acid	HOBrO$_3$ ($HBrO_4$)	Periodic acid	HOIO$_3$ (HIO$_4$)
				Orthoperiodic acid	(HO)$_5$IO (H$_5$IO$_6$)

electrolyzing aqueous NaCl in such a way that the Cl_2 liberated at the anode mixes with the NaOH produced at the cathode. Hypochlorites are powerful oxidizing agents and in the presence of alkali convert $[IO_3]^-$ to $[IO_4]^-$, Cr^{3+} to $[CrO_4]^{2-}$, and even Fe^{3+} to $[FeO_4]^{2-}$. Bleaching powder is a non-deliquescent mixture of $CaCl_2$, $Ca(OH)_2$ and $Ca(OCl)_2$ and is manufactured by the action of Cl_2 on $Ca(OH)_2$; NaOCl is a bleaching agent and disinfectant.

$$2CaO + 2Cl_2 \longrightarrow Ca(OCl)_2 + CaCl_2 \qquad (16.55)$$

All hypohalites are unstable with respect to disproportionation (equation 16.56); at 298 K, the reaction is slow for $[OCl]^-$, fast for $[OBr]^-$ and very fast for $[OI]^-$. Sodium hypochlorite disproportionates in hot aqueous solution (equation 16.57), and the passage of Cl_2 through *hot* aqueous alkali yields chlorate and chloride salts rather than hypochlorites. Hypochlorite solutions decompose by reaction 16.58 in the presence of cobalt(II) compounds as catalysts.

$$3[OX]^- \longrightarrow [XO_3]^- + 2X^- \qquad (16.56)$$

$$3NaOCl \longrightarrow NaClO_3 + 2NaCl \qquad (16.57)$$

$$2[OX]^- \longrightarrow 2X^- + O_2 \qquad (16.58)$$

Like HOCl, chlorous acid, $HClO_2$, is not isolable but is known in aqueous solution and is prepared by reaction 16.59; it is weak acid ($pK_a = 2.0$). Sodium chlorite (used as a bleach) is made by reaction 16.60; the chlorite ion has the bent structure **16.30**.

$$Ba(ClO_2)_2 + H_2SO_4(aq) \longrightarrow 2HClO_2(aq) + BaSO_4(s)$$
suspension

$$(16.59)$$

$$Na_2O_2 + 2ClO_2 \longrightarrow 2NaClO_2 + O_2 \qquad (16.60)$$

Cl—O = 157 pm; ∠O–Cl–O = 111°

(16.30)

Alkaline solutions of chlorites persist unchanged over long periods, but in the presence of acid, a complex decomposition occurs which is summarized in equation 16.61.

$$5HClO_2 \longrightarrow 4ClO_2 + H^+ + Cl^- + 2H_2O \qquad (16.61)$$

Chloric and bromic acids, $HClO_3$ and $HBrO_3$, are both strong acids but cannot be isolated as pure compounds.

The aqueous acids can be made by reaction 16.62 (compare with reaction 16.59).

$$Ba(XO_3)_2 + H_2SO_4 \longrightarrow BaSO_4 + 2HXO_3 \qquad (X = Cl, Br)$$
$$(16.62)$$

Iodic acid HIO_3 is a stable, white solid at room temperature, and is produced by reacting I_2O_5 with water (equation 16.46) or by the oxidation of I_2 with nitric acid. Crystalline iodic acid contains trigonal HIO_3 molecules connected by extensive hydrogen bonding. In aqueous solution it is a fairly strong acid ($pK_a = 0.77$).

Chlorates are strong oxidizing agents; commercially, $NaClO_3$ is an important bleaching agent (Figure 16.1b) and is used as a weedkiller, and $KClO_3$ has applications in fireworks and safety matches. Chlorates are produced by electrolysis of brine at 340 K, allowing the products to mix efficiently; chlorate salts are crystallized from the mixture. Bromates are made by, for example, reaction 16.63 under alkaline conditions. Reaction 16.64 is a convenient synthesis of KIO_3.

$$KBr + 3KOCl \longrightarrow KBrO_3 + 3KCl \qquad (16.63)$$

$$2KClO_3 + I_2 \longrightarrow 2KIO_3 + Cl_2 \qquad (16.64)$$

Potassium bromate and iodate are commonly used in volumetric analysis. Very pure KIO_3 is easily obtained, and reaction 16.65 is used as a source of I_2 for the standardization of thiosulfate solutions (reaction 15.103).

$$[IO_3]^- + 5I^- + 6H^+ \longrightarrow 3I_2 + 3H_2O \qquad (16.65)$$

X = Cl, Br, I

(16.31)

Halate ions are trigonal pyramidal (**16.31**) although, in the solid state, some metal halates do not contain discrete ions.[†] The thermal decomposition of alkali metal chlorates follows reaction 16.66, but in the presence of a suitable catalyst, $KClO_3$ decomposes to give O_2 (equation 15.4). Some iodates (e.g. KIO_3) decompose when heated to iodide and O_2, but others (e.g. $Ca(IO_3)_2$) give oxide, I_2 and O_2; bromates

[†] For further discussion, see: A.F. Wells (1984) *Structural Inorganic Chemistry*, 5th edn, Clarendon Press, Oxford, pp. 327–337.

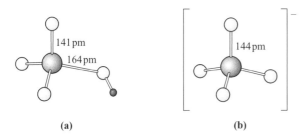

Fig. 16.7 Structures of (a) perchloric acid in which one Cl–O bond is unique, and (b) perchlorate ion in which all Cl–O bonds are equivalent. Colour code: Cl atoms, blue; O, pale grey; H, black.

behave similarly and the interpretation of these observations is a difficult problem in energetics and kinetics.

$$4[ClO_3]^- \longrightarrow 3[ClO_4]^- + Cl^- \quad \textit{Caution: risk of explosion!}$$
$$(16.66)$$

Perchloric acid is the only oxoacid of Cl that can be isolated, and its structure is shown in Figure 16.7a. It is a colourless liquid (bp 363 K with some decomposition), made by heating $KClO_4$ with concentrated H_2SO_4 under reduced pressure. Pure perchloric acid is liable to explode when heated or in the presence of organic material, but in dilute solution, $[ClO_4]^-$ is very difficult to reduce despite the reduction potentials (which provide thermodynamic but not kinetic data) shown in equations 16.67 and 16.68. Zinc, for example, merely liberates H_2, and iodide ion has no action. Reduction to Cl^- can be achieved by Ti(III) in acidic solution or by Fe(II) in the presence of alkali.

$$[ClO_4]^- + 2H^+ + 2e^- \rightleftharpoons [ClO_3]^- + H_2O$$
$$E^\circ = +1.19\,V \quad (\text{at pH}\,0) \quad (16.67)$$

$$[ClO_4]^- + 8H^+ + 8e^- \rightleftharpoons Cl^- + 4H_2O$$
$$E^\circ = +1.39\,V \quad (\text{at pH}\,0) \quad (16.68)$$

Perchloric acid is an extremely strong acid in aqueous solution (see *Table 6.3*). Although $[ClO_4]^-$ (Figure 16.7b) does form complexes with metal cations, the tendency to do so is less than for other common anions. Consequently, $NaClO_4$ solution is a standard medium for the investigation of ionic equilibria in aqueous systems. Alkali metal perchlorates can be obtained by disproportionation of chlorates (equation 16.66) under carefully controlled conditions; traces of impurities can catalyze decomposition to chloride and O_2. *Perchlorate salts are potentially explosive and must be handled with particular care*; mixtures of ammonium perchlorate and aluminium are standard missile propellants, e.g. in the space shuttle. When heated, $KClO_4$ gives KCl and O_2, apparently without intermediate formation of $KClO_3$. Silver perchlorate, like silver salts of some other very strong acids (e.g. $AgBF_4$, $AgSbF_6$ and AgO_2CCF_3), is soluble in many organic solvents including C_6H_6 and Et_2O owing to complex formation between Ag^+ and the organic molecules.

Perbromic acid and perbromate salts have been known since 1968. The best method of preparation of perbromate is by reaction 16.69. Cation exchange (see *Section 10.6*) can be used to give $HBrO_4$, but the anhydrous acid has not been isolated.

$$[BrO_3]^- + F_2 + 2[OH]^- \longrightarrow [BrO_4]^- + 2F^- + H_2O \quad (16.69)$$

Potassium perbromate has been structurally characterized and contains tetrahedral $[BrO_4]^-$ ions (Br–O = 161 pm). Thermochemical data show that $[BrO_4]^-$ (half-reaction 16.70) is a slightly stronger oxidizing agent than $[ClO_4]^-$ or $[IO_4]^-$ under the same conditions. However, oxidations by $[BrO_4]^-$ (as for $[ClO_4]^-$) are slow in dilute neutral solution, but more rapid at higher acidities.

$$[BrO_4]^- + 2H^+ + 2e^- \rightleftharpoons [BrO_3]^- + H_2O$$
$$E^\circ = +1.76\,V \quad (\text{at pH}\,0) \quad (16.70)$$

Several different periodic acids and periodates are known; Table 16.5 lists periodic acid, HIO_4 and orthoperiodic acid, H_5IO_6 (compare with H_6TeO_6, *Section 15.9*). Oxidation of KIO_3 by hot alkaline hypochlorite yields $K_2H_3IO_6$ which is converted to KIO_4 by nitric acid; treatment with concentrated alkali yields $K_4H_2I_2O_{10}$, and dehydration of this at 353 K leads to $K_4I_2O_9$. Apart from $[IO_4]^-$ (**16.32**) and $[IO_5]^{3-}$ and $[HIO_5]^{2-}$ (which are square-based pyramidal), periodic acids and periodate ions feature octahedral I centres, e.g. H_5IO_6 (**16.33**), $[H_2I_2O_{10}]^{4-}$ (**16.34**) and $[I_2O_9]^{4-}$ (**16.35**).

I–O = 178 pm

(16.32)

I=O = 178 pm
I–O = 189 pm

(16.33)

I–O(terminal) = 181 pm
I–O(bridge) = 200 pm
I–OH = 198 pm

(16.34)

I–O(terminal) = 177 pm
I–O(bridge) = 201 pm

(16.35)

The relationships between these ions may be expressed by equilibria 16.71, and aqueous solutions of periodates are therefore not simple systems.

$$\left.\begin{array}{l} [H_3IO_6]^{2-} + H^+ \rightleftharpoons [IO_4]^- + 2H_2O \\ 2[H_3IO_6]^{2-} \rightleftharpoons 2[HIO_5]^{2-} + 2H_2O \\ 2[HIO_5]^{2-} \rightleftharpoons [H_2I_2O_{10}]^{4-} \rightleftharpoons [I_2O_9]^{4-} + H_2O \end{array}\right\} \quad (16.71)$$

Orthoperiodic acid is obtained by electrolytic oxidation of iodic acid, or by adding concentrated nitric acid to $Ba_5(IO_6)_2$, prepared by reaction 16.72.

$$5Ba(IO_3)_2 \xrightarrow{\Delta} Ba_5(IO_6)_2 + 4I_2 + 9O_2 \quad (16.72)$$

Heating H_6IO_6 dehydrates it, first to $H_4I_2O_9$, and then to HIO_4. In aqueous solution, both H_6IO_6 ($pK_a = 3.3$) and HIO_4 ($pK_a = 1.64$) behave as rather weak acids. Periodate oxidizes iodide (equation 16.73) *rapidly* even in neutral solution (compare the actions of chlorate and bromate); it liberates ozonized O_2 from hot acidic solution, and oxidizes Mn(II) to $[MnO_4]^-$.

$$[IO_4]^- + 2I^- + H_2O \longrightarrow [IO_3]^- + I_2 + 2[OH]^- \quad (16.73)$$

16.10 Aqueous solution chemistry

In this section, we are mainly concerned with redox processes in aqueous solution; see *Section 16.1* for a list of relevant topics already covered in the book. Values of E^o for half-reactions 16.74 can be measured directly for X = Cl, Br and I (Table 16.1) and their magnitudes are determined by the X–X bond energies (Figure 16.2), the electron affinities of the halogen atoms (Table 16.1) and the standard free energies of hydration of the halide ions (Table 16.1). This can be seen from scheme 16.75; for X = Br and I, an additional vaporization stage is needed for the element.

$$\tfrac{1}{2}X_2 + e^- \rightleftharpoons X^- \quad (16.74)$$

$$\tfrac{1}{2}X_2 \longrightarrow X(g) \longrightarrow X^-(g) \longrightarrow X^-(aq) \quad (16.75)$$

Dichlorine is a more powerful oxidizing agent in aqueous media than Br_2 or I_2, partly because of a more negative enthalpy of formation of the anion but, more importantly, because the Cl^- ion (which is smaller than Br^- or I^-) interacts more strongly with solvent molecules. (In solid salt formation, the lattice energy factor similarly explains why chloride salts are more exothermic than corresponding bromides or iodides.)

Since F_2 liberates ozonized O_2 from water, the value of E^o for half-reaction 16.74 has no physical reality, but a value of $+2.87$ V can be estimated by comparing the energy changes for each step in scheme 16.75 for X = F and Cl, and hence deriving the difference in E^o for half-equation 16.74 for X = F and Cl. Most of the difference between these E^o values arises from the much more negative value of $\Delta_{hyd}G^o$ of the smaller F^- ion (Table 16.1).

Diiodine is much more soluble in aqueous iodide solutions than in water. At low concentrations of I_2, equation 16.76 describes the system; K can be found be partitioning I_2 between the aqueous layer and a solvent immiscible with water (e.g. CCl_4).

$$I_2 + I^- \rightleftharpoons [I_3]^- \qquad K \approx 10^2 \ (298\,K) \quad (16.76)$$

Potential diagrams (partly calculated from thermochemical data) for Cl, Br and I are given in Figure 16.8. Because several of the oxoacids are weak, the effects of $[H^+]$ on values of some of the reduction potentials are quite complicated. For example, the disproportionation of hypochlorite

Fig. 16.8 Potential diagrams for chlorine, bromine and iodine at pH = 0.

to chlorate and chloride could be written as equilibrium 16.77 without involving protons.

$$3[OCl]^- \rightleftharpoons [ClO_3]^- + 2Cl^- \tag{16.77}$$

However, the fact that HOCl is a weak acid, while $HClO_3$ and HCl are strong ones (see *Table 6.3*) means that, in the presence of hydrogen ions, $[OCl]^-$ is protonated and this affects the position of equilibrium 16.77: HOCl is more stable with respect to disproportionation than $[OCl]^-$. On the other hand, the disproportionation of chlorate into perchlorate and chloride is realistically represented by equilibrium 16.78. From the data in Figure 16.8, this reaction is easily shown to be thermodynamically favourable; nevertheless, the reaction does not occur in aqueous solution owing to some undetermined kinetic factor.

$$4[ClO_3]^- \rightleftharpoons 3[ClO_4]^- + Cl^- \tag{16.78}$$

Another example of the limitations of the data in Figure 16.8 is the inference that O_2 should oxidize I^- and Br^- at pH 0. Further, the fact that Cl_2 rather than O_2 is evolved when hydrochloric acid is electrolyzed is a consequence of the high overpotential for O_2 evolution at most surfaces. Despite some limitations, Figure 16.8 does provide some useful information: for example, the more powerful oxidizing properties of periodate and perbromate than of perchlorate when these species are being reduced to halate ions, and the more weakly oxidizing powers of iodate and iodine than of the other halates or halogens respectively.

The fact that Figure 16.8 refers only to *specific conditions* is well illustrated by considering the stability of I(I). Hypoiodous acid is unstable with respect to disproportionation into $[IO_3]^-$ and I_2, and is therefore not formed when $[IO_3]^-$ acts as an oxidant in aqueous solution. However, in hydrochloric acid, HOI undergoes reaction 16.79.

$$HOI + 2HCl \longrightarrow [ICl_2]^- + H^+ + H_2O \tag{16.79}$$

Under these conditions, the potential diagram becomes:

$$[IO_3]^- \xrightarrow{\;+1.23\;} [ICl_2]^- \xrightarrow{\;+1.06\;} I_2$$

and I(I) is now stable with respect to disproportionation.

Summary

This chapter has introduced the descriptive chemistry of the group 17 elements (the halogens); much relevant discussion has appeared in previous chapters. We have said little about astatine, all isotopes of which are radioactive. The chemistry of F is quite distinct from that of Cl, Br and I because of the restriction of the valence shell of F to an octet, the relatively small size of the F atom and F^- ion, the low dissociation energy of F_2, the higher oxidizing power of F_2, and the high electronegativity of F. In terms of bonding models, the interhalogens are of interest in exemplifying VSEPR theory.

The following terms were introduced in this chapter. Do you know what they mean?
- ❑ ozonized oxygen
- ❑ charge transfer complex
- ❑ charge transfer band
- ❑ clathrate
- ❑ polyhalide ion

Suggested further reading

D.D. DesMarteau, C.W. Bauknight, Jr. and T.E. Mlsna (1994) 'Fluorine: Inorganic chemistry' in *Encyclopedia of Inorganic Chemistry*, ed. R.B. King, Wiley, Chichester, vol. 3, p. 1223 – A review which includes data on ^{19}F NMR spectroscopy.

N.N. Greenwood and A. Earnshaw (1997) *Chemistry of the Elements*, 2nd edn, Butterworth-Heinemann, Oxford – Chapter 17 covers the halogens in detail.

J. Shamir (1994) 'Chlorine, bromine, iodine & astatine: Inorganic chemistry' in *Encyclopedia of Inorganic Chemistry*, ed. R.B. King, Wiley, Chichester, vol. 2, p. 646 – An overview of the heavier halogens.

A.G. Sharpe (1990) *Journal of Chemical Education*, vol. 67, p. 309 – A review of the solvation of halide ions and its chemical significance.

A.F. Wells (1984) *Structural Inorganic Chemistry*, 5th edn, Clarendon Press, Oxford – Chapter 9 gives a detailed account of inorganic halide structures.

A.A. Woolf (1981) *Advances in Inorganic Chemistry and Radiochemistry*, vol. 24, p. 1 – A review of the thermochemistry of fluorine compounds.

Special topics

E.H. Appelman (1973) *Accounts of Chemical Research*, vol. 6, p. 113 – 'Nonexistent compounds: Two case histories'; deals with the histories of the perbromates and hypofluorous acid.

A.J. Blake, F.A. Devillanova, R.O. Gould, W.S. Li, V. Lippolis, S. Parsons, C. Radek and M. Schröder (1998) *Chemical Society Reviews*, vol. 27, p. 195 – 'Template self-assembly of polyiodide networks'.

K. Seppelt (1997) *Accounts of Chemical Research*, vol. 30, p. 111 – 'Bromine oxides'.

Problems

16.1 (a) What is the collective name for the group 17 elements? (b) Write down, in order, the names and symbols of these elements; check your answer by reference to the first two pages of this chapter. (c) Give a *general* notation showing the ground state electronic configuration of each element.

16.2 (a) Write equations to show the reactions involved in the extraction of Br_2 and I_2 from brines. (b) What reactions occur in the Downs process, and why must the products of the process be kept apart? (c) In the electrolysis cell used for the industrial preparation of F_2, a diaphragm is used to separate the products. Give an equation for the reaction that would occur in the absence of the diaphragm and describe the nature of the reaction.

16.3 For a given atom Y, the Y−F bond is usually stronger than the corresponding Y−Cl bond. An exception is when Y is oxygen (Table 15.3). Suggest a reason for this observation.

16.4 Briefly discuss the trends in boiling points and values of $\Delta_{vap}H^\circ$ listed in Table 16.2 for the hydrogen halides.

16.5 Use values of r_{cov} (Table 16.1) to estimate the X−Y bond lengths of ClF, BrF, BrCl, ICl and IBr. Compare the answers with values in Figure 16.2 and Table 16.3, and comment on the validity of the method of calculation.

16.6 Suggest products for the following reactions (which are not balanced):

(a) $AgCl + ClF_3 \longrightarrow$
(b) $ClF + BF_3 \longrightarrow$
(c) $CsF + IF_5 \longrightarrow$
(d) $SbF_5 + ClF_5 \longrightarrow$
(e) $Me_4NF + IF_7 \longrightarrow$
(f) $K[BrF_4] \overset{\Delta}{\longrightarrow}$

16.7 Discuss the role of halide acceptors in the formation of interhalogen cations and anions.

16.8 Predict the structures of (a) $[ICl_4]^-$, (b) $[BrF_2]^+$, (c) $[ClF_4]^+$, (d) IF_7, (e) I_2Cl_6, (f) $[IF_6]^+$, (g) BrF_5.

16.9 (a) Assuming *static* structures, what would you expect to see in the ^{19}F NMR spectra of BrF_5 and $[IF_6]^+$. (b) Do you expect these spectra to be temperature-dependent?

16.10 Discuss the interpretation of each of the following observations:

(a) Al_2Cl_6 and I_2Cl_6 are not isostructural.
(b) Thermal decomposition of $[Bu_4N][ClHI]$ yields $[Me_4N]I$ and HCl.
(c) 0.01 M solutions of I_2 in *n*-hexane, benzene, ethanol and pyridine are violet, purple, brown and yellow respectively. When 0.001 mol of pyridine are added to $100 \, cm^3$ of each of the solutions of I_2 in *n*-hexane, benzene and ethanol, all become yellow.

16.11 Suggest likely structures for (a) $[F_2ClO_2]^-$, (b) $FBrO_3$, (c) $[ClO_2]^+$, (d) $[F_4ClO]^-$.

16.12 (a) Give equations to show the effect of temperature on the reaction between Cl_2 and aqueous NaOH.
(b) In neutral solution 1 mol $[IO_4]^-$ reacts with excess I^- to produce 1 mol I_2. On acidification of the resulting solution, a further 3 mol I_2 are liberated. Derive equations for the reactions which occur under these conditions.
(c) In strongly alkaline solution containing an excess of barium ions, a solution containing 0.01587 g of I^- was treated with 0.1 M $[MnO_4]^-$ until a pink colour persisted in the solution; $10.0 \, cm^3$ was required. Under these conditions, $[MnO_4]^-$ was converted into the sparingly soluble $BaMnO_4$. What is the product of the oxidation of iodide?

16.13 (a) Give descriptions of the bonding in ClO_2 and $[ClO_2]^-$ (**16.21** and **16.30**), and rationalize the differences in Cl−O bond lengths. (b) Rationalize why $KClO_4$ and $BaSO_4$ are isomorphous.

16.14 Suggest products for the following (which are not balanced):

(a) $[ClO_3]^- + Fe^{2+} + H^+ \longrightarrow$
(b) $[IO_3]^- + [SO_3]^{2-} \longrightarrow$
(c) $[IO_3]^- + Br^- + H^+ \longrightarrow$

16.15 Describe in outline how you would attempt:

(a) to determine the equilibrium constant and standard enthalpy change for the aqueous solution reaction:

$$Cl_2 + H_2O \rightleftharpoons HCl + HOCl$$

(b) to show that the oxide I_4O_9 (reported to be formed by reaction between I_2 and O_3) reacts with water according to the reaction:

$$I_4O_9 + 9H_2O \longrightarrow 18HIO_3 + I_2$$

(c) to show that when alkali metal atoms and Cl_2 interact in a solidified noble gas matrix at very low temperatures, the ion $[Cl_2]^-$ is formed.

16.16 Discuss the interpretation of each of the following observations:

(a) Although the hydrogen bonding in HF in stronger than that in H_2O, water has much the high boiling point.
(b) Silver chloride and silver iodide are soluble in saturated aqueous KI, but insoluble in saturated aqueous KCl.

16.17 Explain why:

(a) $[NH_4]F$ has the wurtzite structure, unlike other ammonium halides which possess the CsCl or NaCl lattice depending on temperature.
(b) $[PH_4]I$ is the most stable of the $[PH_4]^+X^-$ halides with respect to decomposition to PH_3 and HX.

Chapter 17 The group 18 elements

TOPICS

■ Occurrence, extraction and uses

■ Physical properties

■ Compounds of xenon

■ Compounds of krypton and radon

1	2		13	14	15	16	17	**18**
H								He
Li	Be		B	C	N	O	F	Ne
Na	Mg		Al	Si	P	S	Cl	Ar
K	Ca	*d*-block	Ga	Ge	As	Se	Br	Kr
Rb	Sr		In	Sn	Sb	Te	I	Xe
Cs	Ba		Tl	Pb	Bi	Po	At	Rn
Fr	Ra							

17.1 Introduction

The group 18 elements (helium, neon, argon, krypton, xenon and radon) are called the *noble gases*.

This section gives a brief, partly historical, introduction to the group 18 elements, the ground state electronic configurations of which tend to suggest chemical inertness. Until 1962, the chemistry of the noble gases was restricted to a few very unstable species such as $[HHe]^+$, $[He_2]^+$, $[ArH]^+$, $[Ar_2]^+$ and $[HeLi]^+$ formed by the combination of an ion and an atom under highly energetic conditions, and detected spectroscopically. Molecular orbital theory provides a simple explanation of why diatomic species such as He_2 and Ne_2 are not known. As we showed for He_2 in *Section 1.13*, bonding and antibonding MOs are fully occupied. However, in a monocation such as $[Ne_2]^+$, the highest energy MO is *singly* occupied, meaning that there is a *net bonding* interaction. Thus, the bond energies in $[He_2]^+$, $[Ne_2]^+$ and $[Ar_2]^+$ are 126, 67

and $104 \, kJ \, mol^{-1}$, respectively, but no stable compounds containing these cations have been isolated. Although $[Xe_2]^+$ has been known for some years and characterized by Raman spectroscopy ($\nu(XeXe) = 123 \, cm^{-1}$), it was only in 1997 that $[Xe_2][Sb_4F_{21}]$ (prepared from $[XeF][Sb_2F_{11}]$ and HF/SbF_5, see *Section 8.9*) was crystallographically characterized. Discrete $[Xe_2]^+$ ions (**17.1**) are present in the solid state of $[Xe_2][Sb_4F_{21}]$, although there are weak Xe····F interactions. The Xe–Xe bond is extremely long, the longest recorded homonuclear bond between main group elements.

$$\left[\underset{309 \, pm}{Xe\rule{2cm}{0.4pt}Xe} \right]^+$$

(17.1)

When H_2O is frozen in the presence of Ar, Kr or Xe at high pressures, clathrates (see *Box 13.5* and *Section 16.4*) of limiting composition $Ar \cdot 6H_2O$, $Kr \cdot 6H_2O$ and $Xe \cdot 6H_2O$ are obtained. The noble gas atoms are guests within hydrogen-bonded host lattices. Other noble gas-containing clathrates include $3.5Xe \cdot 8CCl_4 \cdot 136D_2O$ and $0.866Xe \cdot 3[1,4-(OH)_2C_6H_4]$ (Figure 17.1). Although this type of system is well established, it must be stressed that no *chemical* change has occurred to the noble gas atoms upon formation of the clathrate.

The first indication that Xe was not chemically inert came in 1962 from work of Neil Bartlett when the reaction between Xe and PtF_6 gave a compound formulated as '$XePtF_6$' (see *Section 5.16*). A range of species containing Xe chemically bonded to other elements (usually F or O) is now known. Compounds of Kr are limited to KrF_2 and its derivatives. In principle, there should be many more compounds of Rn. However, the longest lived isotope, ^{222}Rn, has a half-life of 3.8 d and is an intense α-emitter (which leads to decomposition of its compounds), and, in practice, information about the chemistry of Rn is very limited.

Fig. 17.1 Part of the solid state lattice of tris(β-hydroquinone) xenon clathrate showing the arrangement of the xenon atoms in cavities formed between hydrogen-bonded organic molecules [T. Birchall *et al.* (1989) *Acta Cryst., Sect. C*, vol. 45, p. 944]. Colour code: Xe, blue; C, grey; O, white; H, black.

17.2 Occurrence, extraction and uses

Occurrence

After hydrogen, He is the second most abundant element in the universe. It occurs to an extent of ≤7% by volume in natural gas from sources in the US and Canada, and this origin is doubtless from the radioactive decay of heavier elements (see *Section 2.3*). Helium is also found in various minerals containing α-emitting unstable isotopes. Helium was first detected spectroscopically in the Sun's atmosphere where it is formed by nuclear fusion (see *Section 2.8*). Argon is present to the extent of 0.94% by volume in the Earth's atmosphere, while neon and krypton are present to extents of only 1 part per 65 000, and 1 part per million (1 ppm), respectively; xenon is even rarer. Radon is formed by decay of ^{226}Ra in the ^{238}U decay chain (see *Figure 2.3*), and poses a serious health hazard in uranium mines, being linked to cases of lung cancer.[†]

Extraction

In terms of commercial production, He and Ar are the two most important noble gases. Helium is extracted from natural gas by liquefaction of other gases present (He has the lowest boiling point of all the elements), leaving gaseous He which is removed by pumping. Neon is extracted as a by-product when air is liquefied, being left behind as the only gas. Argon has almost the same boiling point as O_2 (Ar, 87 K; O_2, 90 K) and the two gases remain together during the fractionation of liquid air. The O_2/Ar mixture can be partially separated by further fractionation; the crude Ar is mixed with H_2 and sparked to removed O_2 as H_2O, excess H_2 being removed by passage over hot CuO. Krypton and xenon are usually separated from O_2 by selective absorption on charcoal.

Uses

Helium and argon are used widely to provide inert atmospheres, e.g. for arc-welding (see *Box 17.1*) and during the growth of single Si or Ge crystals for the semiconductor industry (see *Box 5.2*). Argon is also used in laboratory inert atmosphere ('dry' or 'glove') boxes for handling air-sensitive compounds. Being very light and non-inflammable, He is used to inflate the tyres of large aircraft, and in balloons including weather-balloons and NASA's unmanned, suborbital research balloons. Liquid He is an important coolant and is used in highfield NMR spectrometers including those used in medical imaging (see *Box 2.5*). The superconductivity of metals cooled to the temperature of liquid He suggests that the latter may become important in power transmission. An O_2/He mixture is used in place of O_2/N_2 for deep-sea divers; He is much less soluble in blood than N_2, and does not cause 'the bends'

[†] Development of lung cancer apparently associated with radon emissions is a more general cause for concern: P. Phillips, T. Denman and S. Barker (1997) *Chemistry in Britain*, vol. 33, number 1, p. 35 – 'Silent, but deadly'.

Protective inert gases for metal arc-welding

The high temperature conditions under which metal arc-welding takes place would, in the absence of protective gases, lead to reaction between molten metal and atmospheric gases including O_2 and N_2. Noble gases such as He and Ar are an obvious choice for the protective blanket, but these may be mixed with an active ingredient such as CO_2 (or H_2) to provide an oxidizing (or reducing) component to the protective layer. Of He and Ar, the latter is of greater industrial importance and is used in welding CrNi alloy steels and a range of metals. Argon is denser than He (1.78 versus $0.18\,g\,cm^{-3}$ at 273 K) and so gives better protection. High-purity Ar (>99.99%) is commercially available and such levels of purity are essential when dealing with metals such as Ti, Ta and Nb which are extremely prone to attack by O_2 or N_2 during arc-welding.

Xenon as a 21st century space fuel

In October 1998, at the start of its New Millennium Program, NASA launched a new space probe called *Deep Space One* (DS1), designed to test new technologies with potential applications in future solar exploration. One of the revolutionary technologies on this flight was a xenon-based ion propulsion system, ten times more efficient than any other used prior to the DS1 mission. The system operates by using a solar power source, and ionizes Xe gas contained in a chamber, at one end of which is a pair of metal grids charged at 1280 V. A xenon-ion beam is produced as ions are ejected through the grids at $\approx 145\,000\,km\,h^{-1}$, and the resultant thrust is used to propel DS1 through space. Since the fuel is Xe gas (and only 81 kg are required for approximately a two year mission), an advantage of the system, in addition to the efficient thrust, is that DS1 is smaller and lighter than previous unmanned spacecraft.

✱ *Further information*: http://nmp.jpl.nasa.gov/ds1

when the pressure is released on surfacing. Helium is also used as a heat-transfer agent in gas-cooled nuclear reactors, for which it has the advantages of being non-corrosive and of not becoming radioactive under irradiation. Neon, krypton and xenon are used in electric discharge signs (e.g. for advertising) and Ar is contained in metal filament bulbs to reduce evaporation from the filament.

17.3 Physical properties

Some physical properties of the group 18 elements are listed in Table 17.1. Of particular significance is the fact that the noble gases have the highest ionization energies of the elements in their respective periods (*Figure 1.15*), but there is

Table 17.1 Some physical properties of the group 18 elements (noble gases).

Property	He	Ne	Ar	Kr	Xe	Rn
Atomic number, Z	2	10	18	36	54	86
Ground state electronic configuration	$1s^2$	$[He]2s^22p^6$	$[Ne]3s^23p^6$	$[Ar]3d^{10}4s^24p^6$	$[Kr]4d^{10}5s^25p^6$	$[Xe]4f^{14}5d^{10}6s^26p^6$
Melting point, mp / K	$-^{\ddagger}$	24.5	84	116	161	202
Boiling point, bp / K	4.2	27	87	120	165	211
Standard enthalpy of fusion, $\Delta_{fus}H^{\circ}(mp)\,/\,kJ\,mol^{-1}$	–	0.34	1.12	1.37	1.81	–
Standard enthalpy of vaporization, $\Delta_{vap}H^{\circ}(bp)\,/\,kJ\,mol^{-1}$	0.08	1.71	6.43	9.08	12.62	18.0
First ionization energy, $IE_1\,/\,kJ\,mol^{-1}$	2372	2081	1521	1351	1170	1037
Van der Waals radius, r_v / pm	99	160	191	197	214	–

‡ Helium cannot be solidified under any conditions of temperature and pressure.

a decrease in values on descending the group (*Figure 5.25*). The extremely low values of $\Delta_{fus}H^\circ$ and $\Delta_{vap}H^\circ$ correspond to the weak van der Waals interactions between the atoms, and the increase in values of $\Delta_{vap}H^\circ$ down the group is due to increased interatomic interactions as atomic size and polarizability increase.

The properties of He deserve special note; it can diffuse through rubber and most glasses. Below 2.18 K, ordinary liquid ^{4}He (but not ^{3}He) is transformed into liquid He(II) which has the remarkable properties of a thermal conductivity 600 times that of copper, and a viscosity approaching zero; it forms films only a few hundred atoms thick which flow up and over the side of the containing vessel.

NMR active nuclei

In the characterization of Xe-containing compounds, use is made of the fact that ^{129}Xe, with a natural abundance of 26.4%, has $I = \frac{1}{2}$. Although direct observation of ^{129}Xe is possible, the observation of satellite peaks in, for example, ^{19}F NMR spectra of xenon fluorides, is a valuable diagnostic tool as we illustrated for $[XeF_5]^{2-}$ in case study 5, *Section 2.11*. For a potential clinical application of ^{129}Xe, see *Box 2.5*.

17.4 Compounds of xenon

Fluorides

The most stable Xe compounds are the colourless fluorides XeF_2, XeF_4 and XeF_6 (Table 17.2). Upon irradiation with UV light, Xe reacts with F_2 at ambient temperature to give XeF_2; the rate of formation is increased by using HF as a catalyst and pure XeF_2 can be prepared by this method. Xenon difluoride may also be made by action of an electrical discharge on a mixture of Xe and F_2, or by passing these gases through a short nickel tube at 673 K. The latter method gives a mixture of XeF_2 and XeF_4, and the yield of XeF_4 is optimized by using a 1:5 $Xe:F_2$ ratio. With an NiF_2 catalyst, the reaction proceeds at a lower temperature, and even at 393 K, XeF_6 can be formed under these same conditions. It is not possible to prepare XeF_4 free of XeF_2 and/or XeF_6; similarly, XeF_6 always forms with contamination by the lower fluorides.

Separation of XeF_4 from a mixture involves preferential complexation of XeF_2 and XeF_6 (equation 17.1) and the XeF_4 is then removed *in vacuo*, while separation of XeF_6 involves reaction 17.2 followed by thermal decomposition of the complex.

$$\left.\begin{array}{r}XeF_2 \\ XeF_4 \\ XeF_6\end{array}\right\} \xrightarrow{\text{excess AsF}_5 \text{ in liq. BrF}_5} \left\{\begin{array}{l}[Xe_2F_3]^+[AsF_6]^- \\ XeF_4 \\ [XeF_5]^+[AsF_6]^-\end{array}\right. \quad (17.1)$$

$$XeF_6 + 2NaF \longrightarrow Na_2[XeF_8] \quad (17.2)$$

All the fluorides sublime *in vacuo*, and all are readily decomposed by water, XeF_2 very slowly, and XeF_4 and XeF_6, rapidly (equations 17.3–17.5 and 17.13).

$$2XeF_2 + 2H_2O \longrightarrow 2Xe + 4HF + O_2 \quad (17.3)$$

$$6XeF_4 + 12H_2O \longrightarrow 2XeO_3 + 4Xe + 24HF + 3O_2 \quad (17.4)$$

$$XeF_6 + 3H_2O \longrightarrow XeO_3 + 6HF \quad (17.5)$$

All three fluorides are powerful oxidizing and fluorinating agents, the relative reactivities being $XeF_6 > XeF_4 > XeF_2$. The difluoride is available commercially and is widely used for fluorinations, e.g. equations 16.30, 17.6 and 17.7. At 298 K, XeF_6 reacts with silica (preventing the handling of XeF_6 in silica glass apparatus, equation 17.8) and with H_2, while XeF_2 and XeF_4 do so only when heated.

$$S + 3XeF_2 \xrightarrow{\text{anhydrous HF}} SF_6 + 3Xe \quad (17.6)$$

$$2Ir + 5XeF_2 \xrightarrow{\text{anhydrous HF}} 2IrF_5 + 5Xe \quad (17.7)$$

$$2XeF_6 + SiO_2 \longrightarrow 2XeOF_4 + SiF_4 \quad (17.8)$$

The structures of the xenon halides are consistent with VSEPR theory. The XeF_2 molecule is linear, but in the solid state, there are significant intermolecular interactions (Figure 17.2a). Square planar XeF_4 molecules also pack in a molecular lattice in the solid state. In the vapour state, the vibrational spectrum of XeF_6 indicates C_{3v} symmetry, i.e. an octahedron distorted by a stereochemically active lone pair in the centre of one face (**17.2**), but the molecule is readily converted into other configurations. Solid XeF_6 is polymorphic, with four crystalline forms, three of which contain tetramers made up of square-pyramidal $[XeF_5]^+$ units (Xe−F = 184 pm) connected by fluoride bridges (Xe−F = 223 and 260 pm) such that the Xe centres form a tetrahedral array (**17.3**). The

Table 17.2 Selected properties of XeF_2, XeF_4 and XeF_6.

Property	XeF_2	XeF_4	XeF_6
Melting point / K	413	390	322
$\Delta_f H^\circ$(298 K) / kJ mol^{-1}	−109	−216	−294
Mean Xe−F bond enthalpy term / kJ mol^{-1}	133$_\ddagger$	131$_\ddagger$	126
Xe−F bond distance / pm	200‡	195‡	189*
Molecular shape	Linear	Square planar	Octahedral

$\ddagger$ Neutron diffraction; * gas phase electron diffraction.

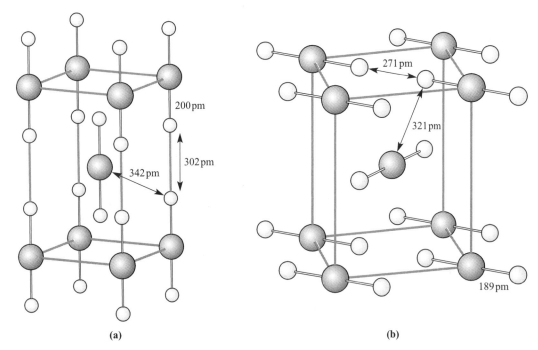

Fig. 17.2 Unit cells of (a) XeF_2 and (b) KrF_2 showing the arrangements and close proximity of molecular units. Colour code: Xe and Kr, blue; F, pale grey.

lowest temperature polymorph contains tetrameric and hexameric units; in the latter, $[XeF_5]^+$ units are connected by fluoride ions, each of which bridges between three Xe centres.

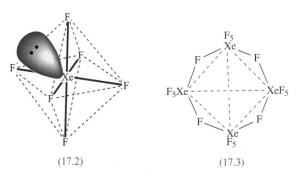

(17.2)   (17.3)

The bonding in XeF_2 and XeF_4 can be described in terms of using only the *s* and *p* valence orbitals (*Section 4.2*) or by invoking delocalized 3c-2e interactions (*Section 4.7*). If the $[XeF]^+$ ion (see below) is taken to contain a single bond, then the fact that its bond distance of 184–190 pm (depending on the salt) is noticeably shorter than those in XeF_2 and XeF_4 (Table 17.2) is consistent with a model of 3c-2e interactions in the xenon fluorides. Further support for low bond orders in XeF_2 and XeF_4 comes from the fact that the strengths of the Xe–F bonds in XeF_2, XeF_4 and XeF_6 are essentially the same (Table 17.2), in contrast to the significant decrease noted (*Section 16.7*) along the series $ClF > ClF_3 > ClF_5$.

Xenon difluoride reacts with F^- acceptors. With pentafluorides such as SbF_5, AsF_5, BrF_5, NbF_5, IrF_5, it forms three types of complex: $[XeF]^+[MF_6]^-$, $[Xe_2F_3]^+[MF_6]^-$ and $[XeF]^+[M_2F_{11}]^-$, although in the solid state, there is

evidence for cation–anion interaction through the formation of Xe–F–M bridges. The $[Xe_2F_3]^+$ cation has structure **17.4**. A number of complexes formed between XeF_2 and metal tetrafluorides have been reported, but structural characterizations are few, e.g. $[XeF]^+[CrF_5]^-$ which has polymeric structure **17.5**.

$$\left[F-Xe \cdots \overset{\overset{\displaystyle F}{\underset{151°}{|}}}{} \cdots \underset{190\,pm}{Xe-F} \right]^+$$

214 pm

(17.4)

(17.5)

Xenon hexafluoride acts as an F^- donor to numerous pentafluorides, giving complexes of types $[XeF_5]^+[MF_5]^-$, $[XeF_5]^+[M_2F_{11}]^-$ (for M = Sb or V) and $[Xe_2F_{11}]^+[MF_5]^-$. The $[XeF_5]^+$ ion (average Xe–F = 184 pm) is isoelectronic

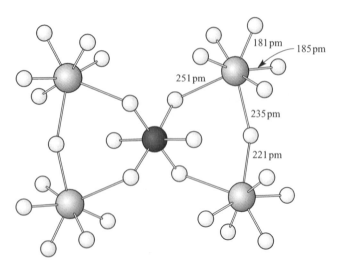

Fig. 17.3 The structure of $[Xe_2F_{11}]_2[NiF_6]$ determined by X-ray diffraction [A. Jesih *et al.* (1989) *Inorg. Chem.*, vol. 28, p. 2911]. The environment about each Xe centre is similar to that in the solid state $[XeF_6]_4$ (**17.3**). Colour code: Xe, blue; Ni, black; F, pale grey.

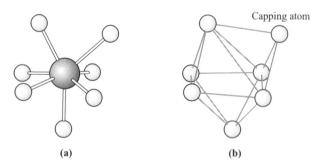

Fig. 17.4 (a) The structure of $[XeF_7]^-$, determined by X-ray diffraction for the caesium salt [A. Ellern *et al.* (1996) *Angew. Chem. Int. Ed. Engl.*, vol. 35, p. 1123]; (b) the capped octahedral arrangement of the F atoms in $[XeF_7]^-$. Colour code: Xe, blue; F, pale grey.

$Cs_2[XeF_8]$ and $Rb_2[XeF_8]$ are the most stable compounds of Xe yet made, and decompose only when heated above 673 K.

$$MF + XeF_6 \longrightarrow M[XeF_7] \xrightarrow{\Delta, -XeF_6} M_2[XeF_8] \qquad (17.9)$$
$$M = Rb, Cs, NO$$

Structural information on $[XeF_7]^-$ has been difficult to obtain because of its ready conversion into $[XeF_8]^{2-}$. However, in 1996, recrystallization of freshly prepared $Cs[XeF_7]$ from liquid BrF_5 yielded crystals suitable for X-ray diffraction studies; the $[XeF_7]^-$ has a capped octahedral structure (Figure 17.4a) with Xe–F = 193 and 197 pm in the octahedron and Xe–F = 210 pm to the capping F atom. The coordination sphere defined by the seven F atoms is shown in Figure 17.4b; the octahedral part is significantly distorted. The reaction between NO_2F and excess XeF_6 gives $[NO_2]^+[Xe_2F_{13}]^-$, the solid state structure of which reveals that the anion can be described as an adduct of $[XeF_7]^-$ and XeF_6 (structure **17.7**). We should mention that there is evidence for the existence of $XeCl_2$ (obtained on condensing the products of a microwave discharge in a mixture of Cl_2 and a large excess of Xe at 20 K) and $XeCl_4$ (detected by ^{129}Xe Mössbauer spectroscopy as a decay product of ^{129}I-labelled $KICl_4 \cdot 2H_2O$), but both are extremely unstable.

and isostructural with IF_5 (**16.5**), but in solid state salts, there is evidence for fluoride bridge formation between cations and anions. The $[Xe_2F_{11}]^+$ cation can be considered as $[F_5Xe\cdots F\cdots XeF_5]^+$ in the same way that $[Xe_2F_3]^+$ can be written as $[FXe\cdots F\cdots XeF]^+$. The compounds $[XeF_5][AgF_4]$ and $[Xe_2F_{11}]_2[NiF_6]$ contain Ag(III) and Ni(IV) respectively, and are prepared from XeF_6, the metal(II) fluoride and KrF_2. In these cases, XeF_6 is not strong enough to oxidize Ag(II) to Ag(III) or Ni(II) to Ni(IV), and KrF_2 is employed as the oxidizing agent. The range of Xe–F bond distances in $[Xe_2F_{11}]_2[NiF_6]$ (Figure 17.3) illustrates the $[F_5Xe\cdots F\cdots XeF_5]^+$ nature of the cation and the longer $F\cdots Xe$ contacts between anion and cations. Xenon tetrafluoride is much less reactive than XeF_2 with F^- acceptors; among the few complexes formed is $[XeF_3]^+[Sb_2F_{11}]^-$. The $[XeF_3]^+$ cation is isostructural with ClF_3 (**16.4**) with bond lengths Xe–F_{eq} = 183 pm and Xe–F_{ax} = 189 pm.

(17.6)

Both XeF_4 and XeF_6 act as F^- acceptors, although the ability of XeF_4 to accept F^- to give $[XeF_5]^-$ has only recently been observed in reactions of CsF and $[Me_4N]F$. The $[XeF_5]^-$ ion (**17.6**) is one of only two pentagonal planar species known, the other being the isoelectronic $[IF_5]^{2-}$ (*Section 16.7*). Equation 17.9 shows the formations of $[XeF_7]^-$ and $[XeF_8]^{2-}$ (which has a square antiprismatic structure); the salts

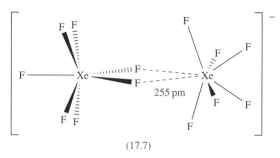

(17.7)

Oxides

Equations 17.4 and 17.5 showed the formation of XeO_3 by hydrolysis of XeF_4 and XeF_6. Solid XeO_3 forms colourless crystals and is dangerously explosive ($\Delta_f H^\circ(298\,K) = +402\,kJ\,mol^{-1}$). The solid contains trigonal pyramidal molecules (**17.8**). Xenon trioxide is only weakly acidic and its aqueous solution is virtually non-conducting. Reactions of XeO_3 and MOH (M = K, Rb, Cs) produce xenates (equation 17.10) which slowly disproportionate in solution (equation 17.11).

Xe—O = 176 pm
$\angle$O–Xe-O = 103°

(17.8)

$$XeO_3 + MOH \longrightarrow M[HXeO_4] \qquad (17.10)$$

$$2[HXeO_4]^- + 2[OH]^- \longrightarrow [XeO_6]^{4-} + Xe + O_2 + 2H_2O$$
$$\text{perxenate}$$

(17.11)

Aqueous $[XeO_6]^{4-}$ is formed when O_3 is passed through a dilute solution of XeO_3 in alkali. Insoluble salts such as $Na_4XeO_6 \cdot 8H_2O$ and Ba_2XeO_6 may be precipitated, but perxenic acid 'H_4XeO_6' (a weak acid in aqueous solution) has not been isolated. The perxenate ion is a powerful oxidant and is rapidly reduced in aqueous acid (equation 17.12); oxidations such as Mn(II) to $[MnO_4]^-$ occur instantly in acidic media at 298 K.

$$[XeO_6]^{4-} + 3H^+ \longrightarrow [HXeO_4]^- + \tfrac{1}{2}O_2 + H_2O \qquad (17.12)$$

Xenon tetraoxide is prepared by the slow addition of concentrated H_2SO_4 to Na_4XeO_6 or Ba_2XeO_6. It is a pale yellow, explosive solid ($\Delta_f H^\circ(298\,K) = +642\,kJ\,mol^{-1}$) which is a very powerful oxidizing agent. Tetrahedral XeO_4 molecules (**17.9**) are present in the gas phase.

O
174 pm

(17.9)

Oxofluorides

Oxofluorides are known for Xe(IV), Xe(VI) and Xe(VIII): $XeOF_2$, $XeOF_4$, XeO_2F_2, XeO_2F_4 and XeO_3F_2. Their structures are consistent with VSEPR theory, see *problem 17.8*. The 1:1 reaction of XeF_4 and H_2O in liquid HF yields $XeOF_2$, isolated as a pale yellow solid which decomposes explosively at 273 K. In contrast to reaction 17.5, *partial* hydrolysis of XeF_6 (equation 17.13) gives $XeOF_4$ (a colourless liquid, mp 227 K), which can be converted to XeO_2F_2 by reaction 17.14. Reaction 17.15 is used to prepare XeO_3F_2 which can be separated *in vacuo*; further reaction between XeO_3F_2 and XeF_6 yields XeO_2F_4.

$$XeF_6 + H_2O \longrightarrow XeOF_4 + 2HF \qquad (17.13)$$

$$XeO_3 + XeOF_4 \longrightarrow 2XeO_2F_2 \qquad (17.14)$$

$$XeO_4 + XeF_6 \longrightarrow XeOF_4 + XeO_3F_2 \qquad (17.15)$$

The stable salts $M[XeO_3F]$ (M = K or Cs) are obtained from MF and XeO_3, and contain infinite chain anions with F^- ions bridging XeO_3 groups. Similar complexes are obtained from CsCl or RbCl with XeO_3 but these contain linked $[XeO_3Cl_2]^{2-}$ anions as shown in **17.10**.

(17.10)

Other compounds of xenon

Members of a series of compounds of the type FXeA where, for example, A^- is $[OClO_3]^-$, $[OSO_2F]^-$, $[OTeF_5]^-$ or $[O_2CCF_3]^-$ have been prepared by the highly exothermic elimination of HF between XeF_2 and HA. Further loss of HF leads to XeA_2 (e.g. equation 17.16). Elimination of HF also drives the reaction of XeF_2 with $HN(SO_3F)_2$ to yield $FXeN(SO_3F)_2$, a relatively rare example of Xe−N bond formation.

$$XeF_2 + HOSO_2F \xrightarrow{-HF} FXeOSO_2F \xrightarrow[-HF]{HOSO_2F} Xe(OSO_2F)_2$$
$$\textbf{(17.11)} \qquad\qquad (17.16)$$

(17.11)

Xenon−carbon bond formation is now quite well exemplified, and many products contain fluorinated aryl substituents, e.g. $(C_6F_5CO_2)Xe(C_6F_5)$, $[(2,6\text{-}F_2C_5H_3N)XeC_6F_5]^+$ (Figure 17.5a), $[(2,6\text{-}F_2C_6H_3)Xe][BF_4]$ (Figure 17.5b), $[(2,6\text{-}F_2C_6H_3)Xe][CF_3SO_3]$ and $[(MeCN)Xe(C_6F_5)]^+$. The degree of interaction between the Xe centre and non-carbon donor (i.e. F, O or N) in these species varies. Some species are best described as containing Xe in a linear environment (e.g. Figure 17.5a) and others tend towards containing an $[RXe]^+$ cation (e.g. Figure 17.5b).

17.5 Compounds of krypton and radon

The only binary compound containing Kr is KrF_2. It is a colourless solid which decomposes >250 K, and is best prepared by UV-irradiation of a mixture of Kr and F_2

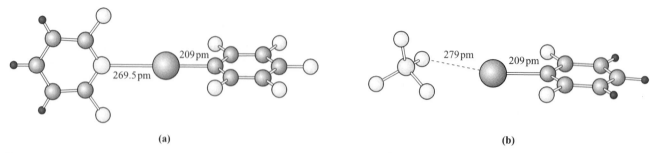

Fig. 17.5 The structures (X-ray diffraction) of (a) $[(2,6\text{-}F_2C_5H_3N)Xe(C_6F_5)]^+$ in the $[AsF_6]^-$ salt [H.J. Frohn *et al.* (1995) *Z. Naturforsch., Teil B*, vol. 50, p. 1799] and (b) $[(2,6\text{-}F_2C_6H_3)Xe][BF_4]$ [T. Gilles *et al.* (1994) *Acta Crystallogr., Sect. C*, vol. 50, p. 411]. Colour code: Xe, blue; N and B, pale blue; C, grey; F, pale grey; H, black.

(4:1 molar ratio) at 77 K. Linear molecules of KrF_2 are arranged in the solid state as shown in Figure 17.2b, and, while still providing close intermolecular contacts, this arrangement contrasts with that adopted by XeF_2 (Figure 17.2a). Krypton difluoride is much less stable than XeF_2. It is rapidly hydrolyzed by water (in an analogous manner to reaction 17.3), and dissociates into Kr and F_2 at 298 K ($\Delta_f H^\circ(298\text{ K}) = +60\text{ kJ mol}^{-1}$). We have already exemplified the use of KrF_2 as a powerful oxidizing agent in the syntheses of $[XeF_5][AgF_4]$ and $[Xe_2F_{11}]_2[NiF_6]$ (*Section 17.4*). Krypton difluoride reacts with SbF_5 to give $[KrF]^+[Sb_2F_{11}]^-$ and $[Kr_2F_3]^+[SbF_6]^-$, and forms salts containing $[KrF]^+$ or $[Kr_2F_3]^+$ with a number of other pentafluorides. Its oxidizing and fluorinating powers are illustrated by its reaction with metallic gold to give $[KrF]^+[AuF_6]^-$.

Few compounds are known that contain Kr bonded to elements other than F. The reactions between KrF_2, $RC\equiv N$ (e.g. R = H, CF_3) and AsF_5 in liquid HF or BrF_5 yield $[(RCN)KrF]^+[AsF_6]^-$ with Kr–N bond formation, and Kr–O bond formation has been observed in the reaction of KrF_2 and $B(OTeF_5)_3$ to give $Kr(OTeF_5)_2$. Crystallographic data for compounds containing Kr (other than clathrates described in Section 17.1) are, as yet, unavailable.

Radon is oxidized by halogen fluorides (e.g. ClF, ClF_3) to the non-volatile RnF_2; the latter is reduced by H_2 at 770 K, and is hydrolyzed by water in a analogous manner to XeF_2 (equation 17.3). As we mentioned in Section 17.1, little chemistry of radon has been explored.

Summary

This chapter has described the properties and chemistry of the noble gases; compound formation is far more extensive for Xe than the other elements in group 18, and, although a more extensive chemistry of radon may be expected, its intense α-radioactivity mitigates against the isolation and characterization of radon-containing compounds.

Suggested further reading

G. Frenking and D. Creme (1990) *Structure and Bonding*, vol. 73, p. 17 – A review: 'The chemistry of the noble gas elements helium, neon and argon'.

N.N. Greenwood and A. Earnshaw (1997) *Chemistry of the Elements*, 2nd edn, Butterworth-Heinemann, Oxford – Chapter 18 covers the noble gases in detail.

J.H. Holloway and E.G. Hope (1999) *Advances in Inorganic Chemistry*, vol. 46, p. 51 – A review of recent developments in noble gas chemistry.

C.K. Jørgensen and G. Frenking (1990) *Structure and Bonding*, vol. 73, p. 1 – A review: 'A historical, spectroscopic and chemical comparison of noble gases'.

B. Žemva (1994) 'Noble gases: Inorganic chemistry' in *Encyclopedia of Inorganic Chemistry*, ed. R.B. King, Wiley, Chichester, vol. 5, p. 2660 – A review of the subject.

Problems

17.1 (a) What is the collective name for the group 18 elements? (b) Write down, in order, the names and symbols of these elements; check your answer by reference to the first page of this chapter. (c) What common feature does the ground state electronic configuration of each element possess?

17.2 Construct MO diagrams for He_2 and $[He_2]^+$ and rationalize why the former is not known but the latter may be detected.

17.3 Confirm that the observed gas phase structures of XeF_2, XeF_4 and XeF_6 are consistent with VSEPR theory.

17.4 Rationalize the structure of $[XeF_8]^{2-}$ (a square antiprism) in terms of VSEPR theory.

17.5 How would you attempt to determine values for (a) $\Delta_f H^\circ(XeF_2, 298\text{ K})$ and (b) the Xe–F bond energy in XeF_2?

17.6 Why is $XeCl_2$ likely to be much less stable than XeF_2?

17.7 How may the standard enthalpy of the unknown salt Xe^+F^- be estimated?

17.8 Predict the structures of $[XeO_6]^{4-}$, $XeOF_2$, $XeOF_4$, XeO_2F_2, XeO_2F_4 and XeO_3F_2.

17.9 Suggest products for the following reactions (which are not necessarily balanced on the left hand sides):

(a) $CsF + XeF_4 \longrightarrow$
(b) $SiO_2 + XeOF_4 \longrightarrow$
(c) $XeF_2 + SbF_5 \longrightarrow$
(d) $XeF_6 + [OH]^- \longrightarrow$
(e) $KrF_2 + H_2O \longrightarrow$

17.10 Write a brief account of the chemistry of the xenon fluorides.

Chapter 18

Organometallic compounds of *s*- and *p*-block elements

1	2		13	14	15	16	17	18
H								He
Li	Be		B	C	N	O	F	Ne
Na	Mg		Al	Si	P	S	Cl	Ar
K	Ca	*d*-block	Ga	Ge	As	Se	Br	Kr
Rb	Sr		In	Sn	Sb	Te	I	Xe
Cs	Ba		Tl	Pb	Bi	Po	At	Rn
Fr	Ra							

18.1 Introduction

This chapter provides an introduction to the large area of the organometallic chemistry of *s*- and *p*-block elements.

An *organometallic* compound contains one or more metal–carbon bonds.

Compounds containing M−C bonds where M is an *s*-block element are readily classified as being organometallic. However, when we come to the *p*-block, the trend from metallic to non-metallic character means that a discussion of strictly

---

BOX 18.1

CHEMICAL AND THEORETICAL BACKGROUND

η-Nomenclature for ligands

In organometallic chemistry in particular, use of the Greek prefix η (eta) is commonly encountered; the letter is accompanied by a superscript number (e.g. η^3). This prefix describes the number of atoms in a ligand which directly interact with the metal centre, the *hapticity* of the ligand. For example, the cyclopentadienyl ligand, $[C_5H_5]^-$ or Cp⁻,

is versatile in its modes of bonding, and examples include the following; note the different ways of representing the η^3- and η^5-modes. This type of nomenclature is also used in coordination chemistry, for example an η^2-peroxo ligand (see *structure* **21.3**).

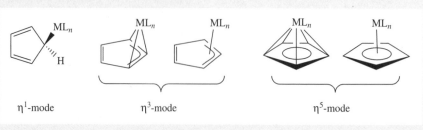

η^1-mode η^3-mode η^5-mode

organometallic compounds would ignore compounds of the semi-metals and synthetically important organoboron compounds. For the purposes of this chapter, we have broadened the definition of an organometallic compound to include species with B−C, Si−C, Ge−C, As−C, Sb−C, Se−C or Te−C bonds. Also relevant to this chapter is the earlier discussion of fullerenes (see *Section 13.4*). Quite often compounds containing, for example, Li−N or Si−N bonds are included in discussions of organometallics, but we have chosen to incorporate these in Chapters 10–16. We do not detail applications of main group organometallic compounds in organic synthesis, but appropriate references are given at the end of the chapter. Abbreviations for the organic substituents mentioned in this chapter are defined in Appendix 2.

18.2 Group 1: alkali metal organometallics

Organic compounds such as terminal alkynes (RC≡CH) which contain relatively acidic hydrogen atoms form salts with the alkali metals, e.g. reactions 18.1, 18.2 and 13.30.

$$2EtC\equiv CH + 2Na \longrightarrow 2Na^+[EtC\equiv C]^- + H_2 \qquad (18.1)$$

$$MeC\equiv CH + K[NH_2] \longrightarrow K^+[MeC\equiv C]^- + NH_3 \qquad (18.2)$$

Similarly, in reaction 18.3, the acidic CH_2 group in cyclopentadiene can be deprotonated to prepare the cyclopentadienyl ligand which is synthetically important in organometallic chemistry (see also *Chapter 23*); Na[Cp] can also be made by direct reaction of Na with C_5H_6.

$$(18.3)$$

$[C_5H_5]^-$ or Cp^-

Colourless alkyl derivatives of Na and K are obtained by *transmetallation* reactions starting from mercury dialkyls (equation 18.4).

$$HgMe_2 + 2Na \longrightarrow 2NaMe + Hg \qquad (18.4)$$

Organolithium compounds are of particular importance among the group 1 organometallics. They may be synthesized by treating an organic halide, RX, with Li (equation 18.5) or by metallation reactions (equation 18.6) using *n*-butyllithium which is commercially available as solutions in hydrocarbon (e.g. hexane) solvents.

$$^nBuCl + 2Li \xrightarrow{\text{hydrocarbon solvent}} {}^nBuLi + LiCl \qquad (18.5)$$

$$^nBuLi + C_6H_6 \longrightarrow {}^nBuH + C_6H_5Li \qquad (18.6)$$

Solvent choices for reactions involving organometallics of the alkali metals are critical. For example, ^nBuLi is decomposed by Et_2O to give ^nBuH, C_2H_4 and LiOEt.

Alkali metal organometallics are extremely reactive and must be handled in air- and moisture-free environments; NaMe, for example, burns explosively in air.[†]

> A *pyrophoric* material is one that burns spontaneously when exposed to air.

Lithium alkyls and aryls are more stable thermally than the corresponding compounds of the heavier group 1 metals (though they ignite spontaneously in air) and mostly differ from them in being soluble in hydrocarbons and other non-polar organic solvents and in being liquids or solids of low melting points. Sodium and potassium alkyls are insoluble in most organic solvents and, when stable enough with respect to thermal decomposition, have fairly high melting points. In the corresponding benzyl and triphenylmethyl compounds, $Na^+[PhCH_2]^-$ and $Na^+[Ph_3C]^-$ (equation 18.7), the negative charge in the organic anions can be delocalized over the aromatic systems; this enhances stability and the salts are red in colour.

$$NaH + Ph_3CH \longrightarrow Na^+[Ph_3C]^- + H_2 \qquad (18.7)$$

Sodium and potassium also form intensely coloured salts with many aromatic compounds (e.g. reaction 18.8). Note that in reactions such as this, the oxidation of the alkali metal involves the transfer of one electron to the aromatic system producing a paramagnetic *radical anion*.

$$(18.8)$$

Naphthalene

> A *radical anion* is an anion that possesses an unpaired electron.

Lithium alkyls are polymeric both in solution and in the solid state. Table 18.1 illustrates the extent to which MeLi, ^nBuLi and ^tBuLi aggregate in solution. In an $(RLi)_4$ tetramer, the Li atoms form a tetrahedral unit, while in an $(RLi)_6$ hexamer, the Li atoms define an octahedron. Figures 18.1a and b show the structure of $(MeLi)_4$; the average Li−Li bond length is 261 pm compared with 267 pm in Li_2 (see *Table 1.6*); the bonding in lithium alkyls is the subject of end of chapter problem 18.2. Figures 18.1c and d show the structure of the Li_6C_6-core of $(LiC_6H_{11})_6$ (C_6H_{11} = cyclohexyl); six Li−Li bond distances lie in the range 295–298 pm while the other six are significantly shorter (238–241 pm). The presence of such aggregates in solution can be determined by using multinuclear NMR spectroscopy. Lithium possesses

---

[†] A useful source of reference is: D.F. Shriver and M.A. Drezdon (1986) *The Manipulation of Air-sensitive Compounds*, Wiley, New York.

Table 18.1 Degree of aggregation of selected lithium alkyls at room temperature (unless otherwise stated).

Compound	Solvent	Species present
MeLi	Hydrocarbons	$(MeLi)_6$
MeLi	Ethers	$(MeLi)_4$
^nBuLi	Hydrocarbons	$(^nBuLi)_6$
^nBuLi	Ethers	$(^nBuLi)_4$
^nBuLi	THF at low temperature	$(^nBuLi)_4 \rightleftharpoons 2(^nBuLi)_2$
^tBuLi	Hydrocarbons	$(^tBuLi)_4$
^tBuLi	Et_2O	Mainly solvated $(^tBuLi)_2$
^tBuLi	THF	Mainly solvated ^tBuLi

two spin-active isotopes (see *Section 2.11* and *Table 10.1*) and the solution structures of lithium alkyls can be studied using ^6Li, ^7Li and ^{13}C NMR spectroscopies as worked example 18.1 illustrates. The alkyls of Na, K, Rb and Cs crystallize with lattice structures (e.g. KMe adopts the NiAs lattice, *Figure 14.7*) or are amorphous solids.

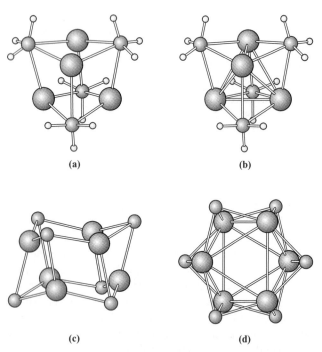

Fig. 18.1 (a) The structure of $(MeLi)_4$ (X-ray diffraction) for the perdeuterated compound [E. Weiss *et al.* (1990) *Chem. Ber.*, vol. 123, p. 79]; the Li atoms define a tetrahedral array while the Li_4C_4-unit can be described as a distorted cube. For clarity, the Li–Li interactions are not been shown in (a) but diagram (b) shows these additional interactions. (c) The Li_6C_6-core of $(LiC_6H_{11})_6$ (X-ray diffraction) [R. Zerger *et al.* (1974) *J. Am. Chem. Soc.*, vol. 96, p. 6048]; the Li_6C_6-core can be considered as a distorted hexagonal prism with Li and C atoms at alternate corners. (d) An alternative view of the structure of the Li_6C_6-core of $(LiC_6H_{11})_6$ which also shows the Li–Li interactions (these were omitted from (c) for clarity); the Li atoms define an octahedral array. Colour code: Li, blue; C, grey; H, pale grey.

Worked example 18.1 NMR spectroscopy of $(^tBuLi)_4$

The structure of $(^tBuLi)_4$ is similar to that of $(MeLi)_4$ shown in Figure 18.1a, but with each H atom replaced by a methyl group. The 75 MHz ^{13}C NMR spectrum of a sample of $(^tBuLi)_4$, prepared from ^6Li metal, consists of two signals, one for the methyl carbons and one for the quaternary carbon atoms. The signal for the quaternary carbons is shown below: (a) at 185 K, and (b) at 299 K. Explain how these signals arise.

[Data: for ^6Li, $I = 1$.]

(a)

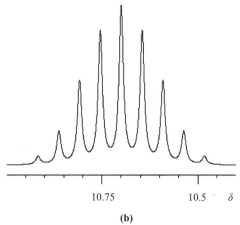

(b)

Firstly, note that the lithium present in the sample is ^6Li, and this is spin active ($I = 1$). The multiplet nature of the signals arises from ^{13}C–^6Li spin–spin coupling.

Multiplicity of signal (number of lines) $= 2nI + 1$

Consider Figure 18.1a with each H atom replaced by an Me group to give $(^tBuLi)_4$. The quaternary C atoms are those bonded to the Li centres, and, *in the static structure*, each ^{13}C nucleus can couple with *three* adjacent and equivalent ^6Li nuclei.

Multiplicity of signal $= (2 \times 3 \times 1) + 1 = 7$

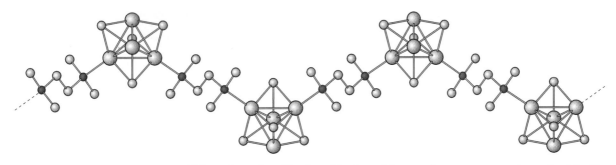

Fig. 18.2 Part of one polymeric chain of $[(^nBuLi)_4 \cdot TMEDA]_\infty$ found in the solid state; the structure was determined by X-ray diffraction. Only the first carbon atom of each ^nBu chain is shown, and all H atoms are omitted for clarity. TMEDA molecules link $(^nBuLi)_4$ units together through the formation of Li–N bonds [N.D.R. Barnett *et al.*, (1993) *J. Am. Chem. Soc.*, vol. 115, p. 1573]. Colour code: Li, blue; C, grey; N, black.

This corresponds to the seven lines (a septet) observed in figure (a) for the low temperature spectrum. Note that the pattern is *non-binomial*. At 299 K, a nonet is observed (non-binomial).

$$\text{Multiplicity of signal} = (2 \times n \times 1) + 1 = 9$$

$$n = 4$$

This means that the molecule is fluxional, and each quaternary ^{13}C nucleus 'sees' four equivalent ^6Li nuclei on the NMR spectroscopic timescale. We can conclude that at 185 K, the molecule possesses a static structure but as the temperature is raised to 299 K, sufficient energy becomes available to allow a fluxional process to occur which exchanges the ^tBu groups.

For a full discussion, see the source of these experimental data: R.D. Thomas, M.T. Clarke, R.M. Jensen and T.C. Young (1986) *Organometallics*, vol. 5, p. 1851.

[For details of NMR spectroscopy: see *Section 2.11*; case study 4 in this section is concerned with a non-binomial multiplet.]

Self-study exercises

1. From the data above, what would you expect to see in the ^{13}C NMR spectrum at 340 K?   [*Ans.* Non-binomial nonet]

2. The ^{13}C NMR spectrum of $(^tBuLi)_4$ at 185 K is called the 'limiting low temperature spectrum'. Explain what this means.

Me₂N   NMe₂

(18.1)

Amorphous alkali metal alkyls such as ^nBuNa are typically insoluble in common solvents, but are solubilized by the chelating ligand TMEDA (**18.1**).[†] Addition of this ligand may break down the aggregates of lithium alkyls to give lower nuclearity complexes, e.g. $[^nBuLi \cdot TMEDA]_2$, **18.2**.

[†] The abbreviation TMEDA stems from the non-IUPAC name N,N,N',N'-tetramethylethylenediamine.

However, detailed studies have revealed that this system is far from simple, and under different conditions, it is possible to isolate crystals of either $[^nBuLi \cdot TMEDA]_2$ or $[(^nBuLi)_4 \cdot TMEDA]_\infty$ (Figure 18.2). In the case of $(MeLi)_4$, the addition of TMEDA does not lead to cluster breakdown, and an X-ray diffraction study of $(MeLi)_4 \cdot 2TMEDA$ confirms the presence of tetramers and amine molecules in the crystal lattice.

Me₂N   NMe₂

Li

Li

Me₂N   NMe₂

(18.2)

Organolithium compounds (in particular MeLi and ^nBuLi) are of great importance as synthetic reagents. In their chemistry they show a general similarly to Grignard reagents (see *Section 18.3*), but are more reactive. Among the many uses of organolithium alkyls and aryls are the conversions of boron trihalides to organoboron compounds (equation 18.9) and similar reactions with other *p*-block halides (e.g. $SnCl_4$).

$$3^nBuLi + BCl_3 \longrightarrow \; ^nBu_3B + 3LiCl \qquad (18.9)$$

Lithium alkyls are important catalysts in the synthetic rubber industry for the stereospecific polymerization of alkenes.

18.3 Group 2 organometallics

Organometallic compounds containing Ca, Sr or Ba are predominantly ionic species which are highly reactive with relative reactivities of compounds being in the order Ca < Sr < Ba. In contrast to the limited studies of organometallic compounds of Ca, Sr and Ba, those of Be

Bonding in cyclopentadienyl complexes

η^1-mode

A bonding description for an $[\eta^1\text{-Cp}]^-$ ligand is straightforward. The M−C single bond is a localized, 2-centre 2-electron interaction:

η^5-mode

If all five carbon atoms of the cyclopentadienyl ring interact

with the metal centre, the bonding is most readily described in terms of an MO scheme. Once the σ-bonding framework of the $[\text{Cp}]^-$ ligand has been formed, there is one $2p$ atomic orbital per C atom remaining, and five combinations are possible. These are shown on the left-hand side of the figure below. Now consider the formation of a compound such as $\text{R}(\eta^5\text{-Cp})\text{Be}$ where R is σ-bonded to the Be centre; for simplicity, let R = H. The valence orbitals of Be are the $2s$ and $2p$ atomic orbitals. For the formation of a $[\text{BeH}]^+$ fragment, we can use an sp hybridization scheme, using one sp hybrid to form the Be−H bond and leaving the other available for bonding to the $[\text{Cp}]^-$ group. The orbitals of the $[\text{BeH}]^+$ fragment are shown on the right-hand side of the figure below.

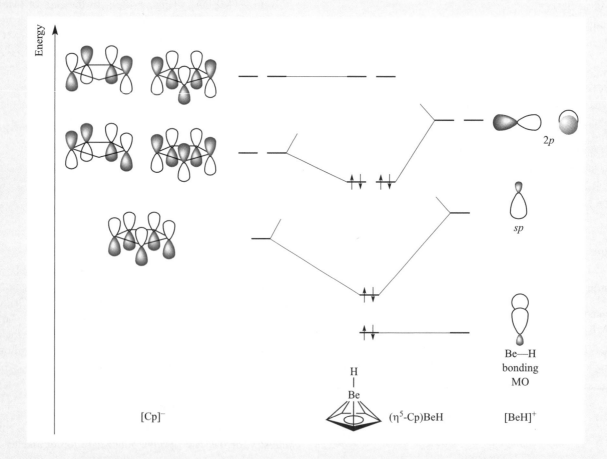

The figure shows the orbital interactions that can occur between the orbitals of $[\text{BeH}]^+$ and three of the MOs of the $[\text{Cp}]^-$ ligand. The remaining two orbitals of the latter remain non-bonding. Finally, consider the number of electrons available. The $[\text{Cp}]^-$ ligand provides six electrons (one per C plus the negative charge). After the formation of the Be−H bond, no more electrons are available from the $[\text{BeH}]^+$ unit. Antibonding MOs have been omitted from the figure for clarity.

In Chapter 22, we describe some cyclopentadienyl complexes of *d*-block metals, e.g. ferrocene, and here, bonding descriptions must take into account the availability of *d*-orbitals on the metal centre.

(Consult Chapter 4 for revision of the MO approach to bonding.)

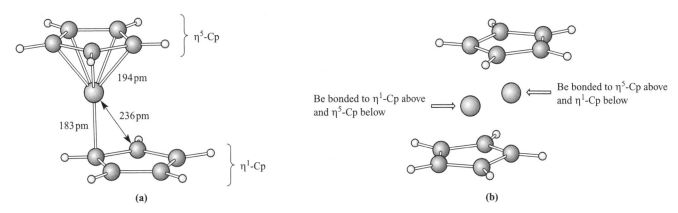

Fig. 18.3 (a) The solid state structure of Cp_2Be determined by X-ray diffraction at 128 K [K.W. Nugent *et al.* (1984) *Aust. J. Chem.*, vol. 37, p. 1601]. (b) The same structure showing the two equivalent sites over which the Be atom is disordered. Colour code: Be, blue; C, grey; H, pale grey.

and Mg are well established and include the synthetically important Grignard reagents.

Beryllium

Beryllium alkyls and aryls are best made by reaction types 18.10 and 18.11 respectively. They are hydrolyzed by water and inflame in air.

$$HgMe_2 + Be \xrightarrow{383\,K} Me_2Be + Hg \qquad (18.10)$$

$$2PhLi + BeCl_2 \xrightarrow{Et_2O} Ph_2Be + 2LiCl \qquad (18.11)$$

In the vapour phase, Me_2Be is monomeric, with a linear C−Be−C unit (Be−C = 170 pm); the bonding was described in *Section 4.2*. The solid state structure is polymeric (**18.3**), and resembles that of $BeCl_2$ (*Figure 11.3b*). However, whereas the bonding in $BeCl_2$ can be described in terms of a localized bonding scheme (*Figure 11.3c*), there are insufficient valence electrons available in $(Me_2Be)_n$ for an analogous bonding picture. Instead, 3c-2e bonds are invoked as described for BeH_2 (see *Figure 9.11* and associated text). Higher alkyls are progressively polymerized to a lesser extent, and the *tert*-butyl derivative is monomeric under all conditions.

(18.3)

$$2Na[Cp] + BeCl_2 \longrightarrow Cp_2Be + 2NaCl \qquad (18.12)$$

Reaction 18.12 leads to the formation of Cp_2Be, and in the solid state, the structure (Figure 18.3a) is in accord with the description $(\eta^1\text{-}Cp)(\eta^5\text{-}Cp)Be$. Electron diffraction and spectroscopic studies of Cp_2Be in the gas phase have provided conflicting views of the structure, but recent data indicate that it resembles that found in the solid state rather than the $(\eta^5\text{-}Cp)_2Be$ originally proposed. In solution, however, the ^1H NMR spectrum shows that *all* proton environments are

equivalent even at 163 K. Furthermore, the solid state structure is not as simple as Figure 18.3a shows; the Be atom is *disordered* over two equivalent sites shown in Figure 18.3b and, thus, the solution NMR spectroscopic data can be interpreted in terms of a fluxional process in which the Be atom moves between these two sites. We consider bonding schemes for complexes containing Cp^- ligands in Box 18.2.

A *crystallographic disorder* involves the position of an atom being described in terms of *partial occupancies* of two or more sites.

Magnesium

Alkyl and aryl magnesium halides (Grignard reagents, represented by the formula RMgX) are extremely well known on account of their uses in synthetic organic chemistry (see further reading list for this chapter). The general preparation of a Grignard reagent (equation 18.13) requires initial activation of the metal, e.g. by addition of I_2.

$$Mg + RX \xrightarrow{Et_2O} RMgX \qquad (18.13)$$

Transmetallation of a suitable organomercury compound is a useful means of preparing pure Grignard reagents (equation 18.14), and transmetallation 18.15 can be used to synthesize compounds of type R_2Mg.

$$Mg + RHgBr \longrightarrow Hg + RMgBr \qquad (18.14)$$

$$Mg + R_2Hg \longrightarrow Hg + R_2Mg \qquad (18.15)$$

Although equations 18.13–18.15 show the magnesium organometallics as simple species, this is an oversimplification. Two-coordination at Mg in R_2Mg is only observed in the solid state when the R groups are especially bulky, e.g. $Mg\{C(SiMe_3)_3\}_2$ (Figure 18.4a). Grignard reagents are generally solvated, and crystal structure data show that the Mg centre is invariably tetrahedrally sited, e.g. in $EtMgBr \cdot 2Et_2O$ (Figure 18.4b) and $PhMgBr \cdot 2Et_2O$. A few examples of 5- and 6-coordination have been observed, e.g. in **18.4** where the macrocyclic ligand imposes the higher

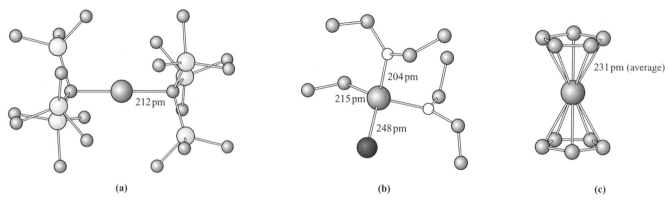

Fig. 18.4 The solid state structures, determined by X-ray diffraction, of (a) $Mg\{C(SiMe_3)_3\}_2$ [S.S. Al-Juaid *et al.* (1994) *J. Organomet. Chem.*, vol. 480, p. 199], (b) $EtMgBr·2Et_2O$ [L.J. Guggenberger *et al.* (1968) *J. Am. Chem. Soc.*, vol. 90, p. 5375], and (c) Cp_2Mg in which each ring is in an η^5-mode and the two rings are mutually staggered [W. Bunder *et al.* (1975) *J. Organomet. Chem.*, vol. 92, p. 1]. Hydrogen atoms have been omitted for clarity; colour code: Mg, dark blue; C, grey; Si, pale blue; Br, black; O, white.

coordination number on the metal centre. Solutions of Grignard reagents may contain several species, e.g. RMgX, R_2Mg, MgX_2, $RMg(\mu\text{-}X)_2MgR$, which are further complicated by solvation. The positions of equilibria between these species are markedly dependent on concentration, temperature and solvent; strongly donating solvents favour monomeric species in which they coordinate to the metal centre.

(18.4)

In comparison to its beryllium analogue, Cp_2Mg has the structure shown in Figure 18.4c, i.e. two η^5-cyclopentadienyl ligands, and is structurally similar to ferrocene (see *Section 23.13*). The reaction between Mg and C_5H_6 yields Cp_2Mg, which is decomposed by water; the compound is therefore often inferred to be an ionic compound and, indeed, significant ionic character is suggested by the long Mg−C bonds in the solid state and also by IR and Raman spectroscopic data.

18.4 Group 13

Boron

We have already discussed the following aspects of organoboron compounds:

- reactions of alkenes with B_2H_6 to give R_3B compounds (see *Figure 12.5*);
- the preparation of $B_4{}^tBu_4$ (*equation 12.31*);
- organoboranes which also contain B−N bonds (*Section 12.8*).

Organoboranes of type R_3B can be prepared by reaction 18.16, or by the hydroboration reaction mentioned above.

$$Et_2O·BF_3 + 3RMgX \longrightarrow R_3B + 3MgXF + Et_2O$$

$$(R = \text{alkyl or aryl}) \quad (18.16)$$

Trialkylboranes are monomeric and inert towards water, but are pyrophoric; the triaryl compounds are less reactive. Both sets of compounds contain planar 3-coordinate B and act as Lewis acids towards amines and carbanions (see also *Sections 12.5* and *12.6*). Reaction 18.17 shows an important example; sodium tetraphenylborate is water-soluble but salts of larger monopositive cations (e.g. K^+) are insoluble, making $Na[BPh_4]$ useful in the precipitation of large metal ions, for example in gravimetric analysis.

$$BPh_3 + NaPh \longrightarrow Na[BPh_4] \quad (18.17)$$

Compounds of the types R_2BCl and $RBCl_2$ can be prepared by transmetallation reactions (e.g. equation 18.18) and are synthetically useful (e.g. reactions 12.53 and 18.19).

$$PhBCl_2 + Ph_3SnCl \longleftarrow BCl_3 + Ph_4Sn$$
$$\longrightarrow Ph_2BCl + Ph_2SnCl_2 \quad (18.18)$$

(18.19)

The bonding in $R_2B(\mu\text{-}H)_2BR_2$ can be described in a similar manner to that in B_2H_6 (see *Section 4.7*). An important

Ziegler–Natta catalysts

Polymerization of alkenes is of great industrial importance, and one key issue is the production of stereoregular polymers; isotactic polypropene has a higher melting point, density and tensile strength than the atactic form. Use of the Ziegler–Natta heterogeneous catalyst controls not only the stereo-specificity of the polymer but also allows the polymerization of $RCH=CH_2$ (R = H or CH_3) to be carried out at 298 K and under atmospheric pressure. This in contrast to early processes which used both higher temperatures and pressures. The Ziegler–Natta catalyst consists of $TiCl_4$ and Et_3Al with Et_2AlCl as co-catalyst, and its development earned K. Ziegler and G. Natta the 1963 Nobel Prize in Chemistry. Ziegler was also responsible for initiating the use of aluminium trialkyls as catalysts for the growth of unbranched alkane chains:

$$Et_3Al + nCH_2=CH_2 \longrightarrow Et_2Al(CH_2CH_2)_nEt$$

Subsequent conversion to long chain alcohols is of commercial importance in the detergent industry.

Ziegler–Natta catalysts are discussed further in Box 23.7 and Section 26.6.

For further discussion of relevant industries, see:

S. Dobson (1995) *Chemistry & Industry*, p. 870 – 'Man-made fibre markets: Recent agitation and change'.
R.G. Harvan (1997) *Chemistry & Industry*, p. 212 – 'Polyethylene: New directions for a commodity thermoplastic'.
D.F. Oxley (1996) *Chemistry & Industry*, p. 535 – 'The world market for polypropylene'.

---

member of this family is **18.5**, commonly known as 9-BBN,[†] which is used for the regioselective reduction of ketones, aldehydes, alkynes and nitriles.

(18.5)

By using bulky organic substituents (e.g. mesityl = 2,4,6-$Me_3C_6H_2$), it is possible to stabilize compounds of type R_2B-BR_2; these should be contrasted with X_2B-BX_2 where X = halogen or NR_2 in which there is $X \longrightarrow B$ π-overlap (see *Sections 12.6* and *12.8*). Two-electron reduction of R_2B-BR_2 gives $[R_2B=BR_2]^{2-}$, an isoelectronic analogue of an alkene. The planar B_2C_4 framework has been confirmed by X-ray diffraction for $Li_2[B_2(2,4,6-Me_3C_6H_2)_3Ph]$, although there is significant interaction between the B=B unit and two Li^+ centres. The shortening of the B−B bond on going from $B_2(2,4,6-Me_3C_6H_2)_3Ph$ (171 pm) to $[B_2(2,4,6-Me_3C_6H_2)_3Ph]^{2-}$ (163 pm) is less than might be expected and this observation is attributed to the large Coulombic repulsion between the two B^- centres.

Aluminium

Aluminium alkyls can be prepared by the transmetallation reaction 18.20, or from Grignard reagents (equation 18.21). On an industrial scale, the direct reaction of Al with a terminal alkene and H_2 (equation 18.22) is employed.

$$2Al + 3R_2Hg \longrightarrow 2R_3Al + 3Hg \qquad (18.20)$$

$$AlCl_3 + 3RMgCl \longrightarrow R_3Al + 3MgCl_2 \qquad (18.21)$$

$$Al + \tfrac{3}{2}H_2 + 3R_2C=CH_2 \longrightarrow (R_2CHCH_2)_3Al \qquad (18.22)$$

Reactions between Al and alkyl halides yield alkyl aluminium halides (equation 18.23); note that **18.6** is in equilibrium with $[R_2Al(\mu-X)_2AlR_2]$ and $[RXAl(\mu-X)_2AlRX]$ via a redistribution reaction, but **18.6** predominates in the mixture.

(18.6)

$$(18.23)$$

$$Al + \tfrac{3}{2}H_2 + 2R_3Al \longrightarrow 3R_2AlH \qquad (18.24)$$

Alkyl aluminium hydrides are obtained by reaction 18.24. These compounds, although unstable to both air and water, are important catalysts for the polymerization of alkenes and other unsaturated organic compounds. Ziegler–Natta catalysts containing trialkyl aluminium compounds are introduced in *Box 18.3*.

(18.7)

Earlier we noted that R_3B compounds are monomeric. In contrast, aluminium trialkyls form dimers. Although this resembles the behaviour of the halides discussed in *Section 12.6*, there are differences in bonding. Trimethylaluminium (mp 313 K) possesses structure **18.7** and so bonding schemes can be developed in like manner as for B_2H_6; the fact that $Al-C_{bridge} > Al-C_{terminal}$ is consistent with 3c-2e bonding in the Al−C−Al bridges, but with 2c-2e terminal bonds.

[†] The systematic name for 9-BBN is 9-borabicyclo[3.3.1]nonane.

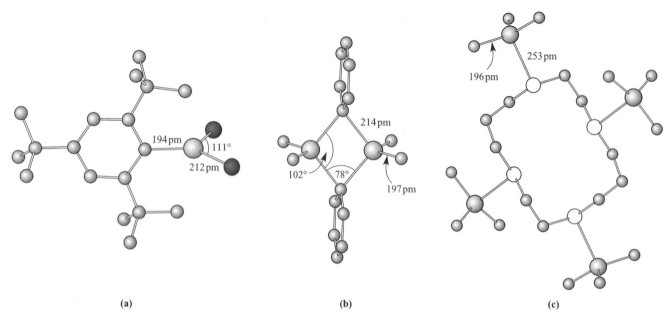

Fig. 18.5 The solid state structures (X-ray diffraction) of (a) $(2,4,6\text{-}^tBu_3C_6H_2)AlCl_2$ [R.J. Wehmschulte *et al.* (1996) *Inorg. Chem.*, vol. 35, p. 3262], (b) $Me_2Al(\mu\text{-}Ph)_2AlMe_2$ [J.F. Malone *et al.* (1972) *J. Chem. Soc., Dalton Trans.*, p. 2649], and (c) the adduct $L\cdot(AlMe_3)_4$ where L is the sulfur-containing macrocyclic ligand 1,4,8,11-tetrathiacyclotetradecane [G.H. Robinson *et al.* (1987) *Organometallics*, vol. 6, p. 887]. Hydrogen atoms are omitted for clarity; colour code: Al, blue; C, grey; Cl, black; S, white.

Equilibria between dimer and monomer exist in solution, with the monomer becoming more favoured as the steric demands of the alkyl group increase. Mixed alkyl halides also dimerize as exemplified in structure **18.6**, but with particularly bulky R groups, the monomer (with trigonal planar Al) is favoured, e.g. $(2,4,6\text{-}^tBu_3C_6H_2)AlCl_2$ (Figure 18.5a). Triphenylaluminium also exists as a dimer, but in the mesityl derivative (mesityl = $2,4,6\text{-}Me_3C_6H_4$), the steric demands of the substituents stabilize the monomer. Figure 18.5b shows the structure of $Me_2Al(\mu\text{-}Ph)_2AlMe_2$, and the orientations of the bridging phenyl groups are the same as in $Ph_2Al(\mu\text{-}Ph)_2AlPh_2$; this orientation is sterically favoured and places each *ipso*-carbon atom in an approximately tetrahedral environment.

> The *ipso*-carbon atom of a phenyl ring is the one to which the substituent is attached; e.g. in PPh_3, the *ipso*-C of each Ph ring is bonded to P.

In dimers containing $RC{\equiv}C$-bridges, a different type of bonding operates. The structure of $Ph_2Al(PhC{\equiv}C)_2AlPh_2$ (**18.8**) shows that the alkynyl bridges lean over towards one of the Al centres. This is interpreted in terms of their behaving as σ,π-ligands: each forms one Al−C σ-bond and interacts with the second Al centre by using the $C{\equiv}C$ π-bond. Thus, each alkynyl group is able to provide three electrons for bridge bonding in contrast to one electron being supplied by an alkyl or aryl group; the bonding is shown schematically in **18.9**.

(18.8)

(18.9)

Trialkylaluminium derivatives behave as Lewis acids, forming a range of adducts, e.g. $R_3N\cdot AlR_3$, $K[AlR_3F]$, $Ph_3P\cdot AlMe_3$ and more exotic complexes such as that shown in Figure 18.5c. Each adduct contains a tetrahedrally sited Al atom. Trialkylaluminium compounds are stronger Lewis acids then either R_3B or R_3Ga, and the sequence for group 13 follows the trend $R_3B < R_3Al > R_3Ga > R_3In > R_3Tl$.

The first R_2Al−AlR_2 derivative was reported in 1988, and was prepared by potassium reduction of the sterically hindered $\{(Me_3Si)_2CH\}_2AlCl$. The Al−Al bond distance in $\{(Me_3Si)_2CH\}_4Al_2$ is 266 pm (compare $r_{cov} = 130$ pm) and the Al_2C_4 framework is *planar*, despite this being a singly bonded compound. A related compound is $(2,4,6\text{-}^iPr_3C_6H_2)_4Al_2$ (Al−Al = 265 pm) and here the Al_2C_4 framework is non-planar (angle between the two AlC_2 planes = 45°). One-electron reduction of Al_2R_4 (R = $(Me_3Si)_2CH$ or $2,4,6\text{-}^iPr_3C_6H_2$) gives the radical anion $[Al_2R_4]^-$ with a formal Al−Al bond order of 1.5. Consistent with the presence of a π-contribution, the

Al–Al bond is shortened upon reduction to 253 pm for R = $(Me_3Si)_2CH$, and 247 pm for R = 2,4,6-^{i}Pr$_3$C$_6$H$_2$; in both anions, the Al$_2$R$_4$ frameworks are essentially planar.

(18.10)

The reaction between cyclopentadiene and Al$_2$Me$_6$ gives CpAlMe$_2$ which is a volatile solid. In the gas phase, it is monomeric with an η^2-Cp bonding mode (**18.10**); this effectively partitions the cyclopentadienyl ring into alkene and allyl parts, since only two of the five π-electrons are donated to the metal centre. In the solid state, the molecules interact to form polymeric chains (Figure 18.6a). The related compound Cp$_2$AlMe is monomeric with an η^2-mode in the solid state (Figure 18.6b). In solution, Cp$_2$AlMe and CpAlMe$_2$ are highly fluxional. A low energy difference between the different modes of bonding of the cyclopentadienyl ligand is also observed in the series of compounds (C$_5$H$_5$)$_3$Al (i.e. Cp$_3$Al), (1,2,4-Me$_3$C$_5$H$_2$)$_3$Al and (Me$_4$C$_5$H)$_3$Al. In solution, even at low temperature, these are stereochemically non-rigid, with negligible energy differences between η^1-, η^2-, η^3- and η^5-modes of bonding. In the solid state, the structural parameters are consistent with the descriptions:

- (η^2-C$_5$H$_5$)($\eta^{1.5}$-C$_5$H$_5$)$_2$Al and (η^2-C$_5$H$_5$)($\eta^{1.5}$-C$_5$H$_5$)(η^1-C$_5$H$_5$)Al for the two independent molecules present in the crystal lattice;
- (η^5-1,2,4-Me$_3$C$_5$H$_2$)(η^1-1,2,4-Me$_3$C$_5$H$_2$)$_2$Al;
- (η^1-Me$_4$C$_5$H)$_3$Al.

These examples serve to indicate the non-predictable nature of these systems, and that subtle balances of steric and electronic effects are in operation.

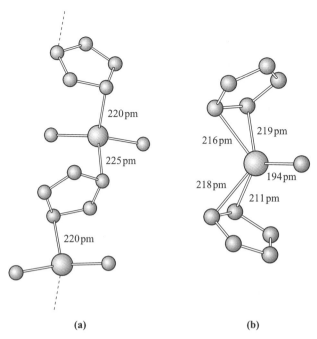

Fig. 18.6 The solid state structures (X-ray diffraction) of (a) polymeric CpAlMe$_2$ [B. Tecle *et al.* (1982) *Inorg. Chem.*, vol. 21, p. 458], and (b) monomeric (η^2-Cp)$_2$AlMe [J.D. Fisher *et al.* (1994) *Organometallics*, vol. 13, p. 3324]. Hydrogen atoms are omitted for clarity; colour code: Al, blue; C, grey.

Gallium, indium and thallium

Since 1980, interest in organometallic compounds of Ga, In and Tl has grown, mainly because of their potential use as precursors to semiconducting materials such as GaAs and InP. Volatile compounds are sought that can be used in the growth of thin films by MOCVD (*metal organic chemical vapour deposition*) or MOVPE (*metal organic vapour phase epitaxy*) techniques (see *Section 27.6*). Precursors include appropriate Lewis base adducts of metal alkyls, e.g. Me$_3$Ga·NMe$_3$ and Me$_3$In·PEt$_3$. Reaction 18.25 is an example of the thermal

---

BOX 18.4

APPLICATIONS

III–V semiconductors

The so-called III–V semiconductors derive their name from the old group numbers for groups 13 and 15, and include AlAs, AlSb, GaP, GaAs, GaSb, InP, InAs and InSb. Of these, GaAs is of the greatest commercial interest. Although Si is probably the most important commercial semiconductor, a major advantage of GaAs over Si is that the charge carrier mobility is much greater. This makes GaAs suitable for high speed electronic devices. Another important difference is that GaAs exhibits a fully allowed electronic transition between valence and conduction bands (i.e. it is a *direct* band gap semiconductor) whereas Si is an *indirect*

band gap semiconductor. The consequence of this difference is that GaAs (and, similarly, the other III–V semiconductors) are more suited than Si for use in optoelectronic devices, since light is emitted more efficiently. The III–Vs have important applications in light-emitting diodes (LEDs). We look in more detail at III–V semiconductors in Section 27.6.

Related information

Box 13.3 – Solar power: thermal and electrical

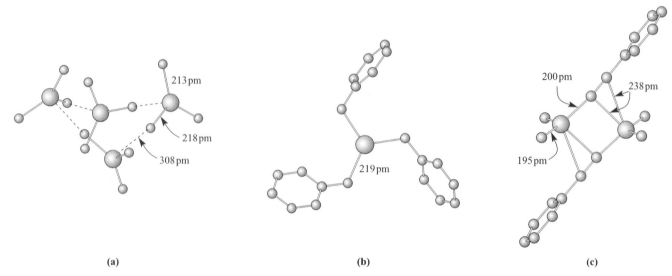

(a)   (b)   (c)

Fig. 18.7 The solid state structures (X-ray diffraction) of (a) Me_3In for which one of the tetrameric units (see text) is shown [A.J. Blake *et al.* (1990) *J. Chem. Soc., Dalton Trans.*, p. 2393], (b) $(PhCH_2)_3In$ [B. Neumuller (1991) *Z. Anorg. Allg. Chem.*, vol. 592, p. 42], and (c) $Me_2Ga(\mu\text{-}C\equiv CPh)_2GaMe_2$ [B. Tecle *et al.* (1981) *Inorg. Chem.*, vol. 20, p. 2335]. Hydrogen atoms are omitted for clarity; colour code: Ga and In, blue; C, grey.

decomposition of gaseous precursors to form a semiconductor which can be deposited in thin films (see *Box 18.4*).

$$Me_3Ga(g) + AsH_3(g) \xrightarrow{1000-1150\,K} GaAs(s) + 3CH_4(g) \tag{18.25}$$

Gallium, indium and thallium trialkyls, R_3M, can be made by use of Grignard reagents (reaction 18.26), RLi (equation 18.27) or R_2Hg (equation 18.28), although a variation in strategy is usually needed to prepare triorganothallium derivatives (e.g. reaction 18.29) since R_2TlX is favoured in reactions 18.26 or 18.27. The Grignard route is valuable for the synthesis of triaryl derivatives. A disadvantage of the Grignard route is that $R_3M\cdot OEt_2$ may be the isolated product.

$$MBr_3 + 3RMgBr \xrightarrow{Et_2O} R_3M + 3MgBr_2 \tag{18.26}$$

$$MCl_3 + 3RLi \xrightarrow{\text{hydrocarbon solvent}} R_3M + 3LiCl \tag{18.27}$$

$$2M + 3R_2Hg \longrightarrow 2R_3M + 3Hg \quad (\textit{not for M = Tl}) \tag{18.28}$$

$$2MeLi + MeI + TlI \longrightarrow Me_3Tl + 2LiI \tag{18.29}$$

Trialkyls and triaryls of Ga, In and Tl are monomeric (trigonal planar metal centres) in solution and the gas phase. In the solid state, monomers are essentially present, but close intermolecular contacts are important in most structures. In trimethylindium, the formation of long $In\cdots C$ interactions (Figure 18.7a) means that the structure can be described in terms of cyclic tetramers; further, each In centre forms an additional weak $In\cdots C$ interaction (356 pm) with the C atom of an adjacent tetramer to give an infinite network. The solid state lattice of Me_3Tl resembles that of Me_3In, with Tl–C distances of 230 and 316 pm within each tetramer, and 330 pm for the weakest $Tl\cdots C$ interactions. The crystal structure of Me_3Ga has not yet been determined. Intermolecular interactions are also observed in, for example, crystalline Ph_3Ga,

Ph_3In and $(PhCH_2)_3In$; Figure 18.7b shows one molecule of $(PhCH_2)_3In$, but each In atom interacts weakly with carbon atoms of phenyl rings of adjacent molecules. Dimer formation is observed in $Me_2Ga(\mu\text{-}C\equiv CPh)_2GaMe_2$ (Figure 18.7c), and the same bonding description that we outlined for $R_2Al(PhC\equiv C)_2AlR_2$ (**18.8** and **18.9**) is appropriate.

Triorganogallium, indium and thallium compounds are air and moisture sensitive. Hydrolysis initially yields the linear $[R_2M]^+$ ion (which can be further hydrolyzed), in contrast to the inertness of R_3B towards water and the formation of $Al(OH)_3$) from R_3Al. The $[R_2Tl]^+$ cation is also present in R_2TlX (X = halide), and the ionic nature of this compound differs from the covalent character of R_2MX for the earlier group 13 elements. Numerous adducts $R_3M\cdot L$ (L = Lewis base) are known in which the metal centre is tetrahedrally sited, e.g. $Me_3Ga\cdot NMe_3$, $Me_3Ga\cdot NCPh$, $Me_3In\cdot OEt_2$, $Me_3In\cdot SMe_2$, $Me_3Tl\cdot PMe_3$, $[Me_4Tl]^-$. In compound **18.11**, donation of the lone pair comes from within the organic moiety; the GaC_3-unit is planar since the ligand is not flexible enough for the usual tetrahedral geometry to be adopted.

(18.11)

As for Al, species of type $[E_2R_4]$ (single E–E bond) and $[E_2R_4]^-$ (E–E bond order 1.5) can be prepared for Ga and In provided that R is especially bulky (e.g. $R = (Me_3Si)_2CH$, $2,4,6\text{-}^iPr_3C_6H_2$), and reduction of $[(2,4,6\text{-}^iPr_3C_6H_2)_4Ga_2]$ to $[(2,4,6\text{-}^iPr_3C_6H_2)_4Ga_2]^-$ is accompanied by a shortening of the Ga–Ga bond from 252 to 234 pm, consistent with an increase in bond order (1 to 1.5). The reaction of Na with

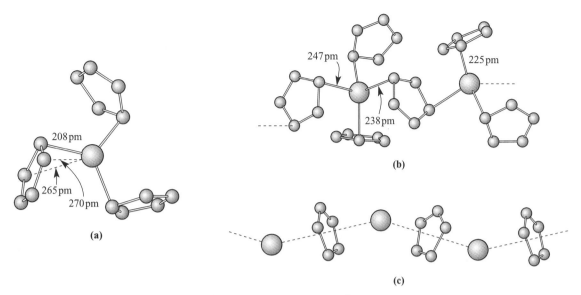

Fig. 18.8 The solid state structures (X-ray diffraction) of (a) monomeric $(\eta^1\text{-Cp})_3\text{Ga}$ [O.T. Beachley *et al.* (1985) *Organometallics*, vol. 4, p. 751], (b) polymeric Cp_3In [F.W.B. Einstein *et al.* (1972) *Inorg. Chem.*, vol. 11, p. 2832], (c) polymeric CpIn [O.T. Beachley *et al.* (1988) *Organometallics*, vol. 7, p. 1051]. Hydrogen atoms are omitted for clarity; colour code: Ga and In, blue; C, grey.

$(2,4,6\text{-}^i\text{Pr}_3\text{C}_6\text{H}_2)\text{GaCl}_2$ in Et_2O gives the Na^+ salt of $[(2,4,6\text{-}^i\text{Pr}_3\text{C}_6\text{H}_2)_2\text{Ga}_2]^{2-}$, **18.12a**, which has been claimed to be the first example of a $\text{Ga}\equiv\text{Ga}$ triple bond. However, there is some controversy as to whether the bond length of 232 pm and bent $\text{C}-\text{Ga}-\text{Ga}-\text{C}$ framework might not better be rationalized in terms of a $\text{Ga}-\text{Ga}$ single bond and presence of lone pairs **18.12b**.[†] The 2,6-dimesitylphenyl substituent is extremely sterically demanding, and reduction of $(2,6\text{-Mes}_2\text{C}_6\text{H}_3)\text{GaCl}_2$ with Na yields $\text{Na}_2[(2,6\text{-Mes}_2\text{C}_6\text{H}_3)_3\text{Ga}_3]$; the $[(2,6\text{-Mes}_2\text{C}_6\text{H}_3)_3\text{Ga}_3]^{2-}$ anion possesses the cyclic structure (**18.13**) and is a 2π-electron aromatic system.

$$\left[\begin{array}{c} R \\ \text{Ga}\equiv\text{Ga} \\ R \end{array} \begin{array}{c} R \\ \\ \end{array} \right]^{2-} \qquad \left[\begin{array}{c} R \\ \ddot{\text{Ga}}-\ddot{\text{Ga}} \\ R \end{array} \begin{array}{c} R \\ \\ \end{array} \right]^{2-}$$

(18.12a)   (18.12b)

$$\left[\begin{array}{c} \text{Mes} \quad\quad \text{Mes} \\ \text{Ga} \\ \text{Mes} \quad\quad\quad\quad \text{Mes} \\ \text{Ga}\text{---}\text{Ga} \\ \text{Mes} \quad \text{Mes} \end{array} \right]^{2-}$$

Mes = $2,4,6\text{-Me}_3\text{C}_6\text{H}_2$

(18.13)

Cyclopentadienyl complexes illustrate the increase in stability of the M(I) oxidation state as group 13 is descended, a consequence of the thermodynamic $6s$ inert pair effect (see *Box 12.3*). Cyclopentadienyl derivatives of Ga(III) which have been prepared (equations 18.30 and 18.31) and structurally characterized include Cp_3Ga and CpGaMe_2.

$$\text{GaCl}_3 + 3\text{Li}[\text{Cp}] \longrightarrow \text{Cp}_3\text{Ga} + 3\text{LiCl} \qquad (18.30)$$

$$\text{Me}_2\text{GaCl} + \text{Na}[\text{Cp}] \longrightarrow \text{CpGaMe}_2 + \text{NaCl} \qquad (18.31)$$

The structure of CpGaMe_2 resembles that of CpAlMe_2 (Figure 18.6a), and Cp_3Ga is monomeric with three $\eta^1\text{-Cp}$ groups bonded to trigonal planar Ga (Figure 18.8a). The In(III) compound Cp_3In is prepared from NaCp and InCl_3, but is structurally different from Cp_3Ga; the solid contains polymeric chains in which each In atom is distorted tetrahedral (Figure 18.6b). The Ga(I) derivative CpGa has been prepared from Cp_2Mg and the metastable GaCl; solution NMR spectroscopic data show that each CH unit is equivalent on the NMR spectroscopic timescale, and on the basis of theoretical data, a monomeric species $(\eta^5\text{-Cp})\text{Ga}$ is proposed. On moving down group 13, the number of M(I) cyclopentadienyl derivatives increases, with a wide range being known for Tl(I). The condensation of In vapour (at 77 K) onto C_5H_6 gives CpIn, and CpTl is readily prepared by reaction 18.32.

$$\text{C}_5\text{H}_6 + \text{TlX} \xrightarrow{\text{KOH/H}_2\text{O}} \text{CpTl} + \text{KX} \qquad \text{e.g. X = halide} \qquad (18.32)$$

Both CpIn and CpTl are monomeric in the gas phase, but in the solid, they possess the polymeric chain structure shown in Figure 18.8c. The cyclopentadienyl derivatives $(\text{C}_5\text{R}_5)\text{M}$ (M = In, Tl) are structurally diverse in the solid state, e.g. for R = PhCH_2 and M = In or Tl, 'quasi-dimers' **18.14** are

[†] For further discussion see: K.W. Klinkhammer (1997) *Angewandte Chemie, International Edition in English*, vol. 36, p. 2320 – 'How can one recognize a triple bond between main group elements?'

present (there may or may not be a meaningful metal–metal interaction), and $(\eta^5\text{-}C_5Me_5)In$ forms hexameric clusters.

PhCH$_2$ substituents omitted

(18.14)

One use of CpTl is as a cyclopentadienyl transfer reagent to *d*-block metal ions, but it can also act as an acceptor of Cp^-, reacting with Cp_2Mg to give $[Cp_2Tl]^-$. This can be isolated as the salt $[CpMgL][Cp_2Tl]$ upon the addition of the chelating ligand $L = Me_2NCH_2CH_2NMeCH_2CH_2NMe_2$. The anion $[Cp_2Tl]^-$ is isoelectronic with Cp_2Sn and possesses a structure in which the η^5-Cp rings are mutually tilted. The structure is as shown in Figure 18.9c for Cp_2Si but with an angle $\alpha = 157°$. Although this ring orientation implies the presence of a stereochemically active lone pair, it has been shown theoretically that there is only a small energy difference ($3.5\,kJ\,mol^{-1}$) between this structure and one in which the η^5-Cp rings are parallel (i.e. as in Figure 18.9a). We return to this scenario in the next section.

18.5 Group 14

Organo-compounds of the group 14 elements include some important commercial products, and we have already discussed *silicones* in *Section 13.10* and *Box 13.10*. Organotin compounds are employed as polyvinylchloride (PVC) stabilizers (against degradation by light and heat), antifouling paints on ships, wood preservatives and agricultural pesticides (see *Box 18.5*). Leaded motor fuels contain the anti-knock agent Et_4Pb, although this use is declining on environmental grounds (see *Figure 13.1*). Several general properties of the organo-derivatives of the group 14 elements, E, are as follows:

- in most compounds, the group 14 element is tetravalent;
- the E−C bonds are generally of low polarity;
- their stability towards all reagents decreases from Si to Pb;
- in contrast to the group 13 organometallics, derivatives of the group 14 elements are less susceptible to nucleophilic attack.

Silicon

Silicon tetraalkyl and tetraaryl derivatives (R_4Si), as well as alkyl or aryl silicon halides (R_nSiCl_{4-n}, $n = 1$–3) can be prepared by reaction types 18.33–18.37. Note that variation

in stoichiometry provides flexibility in synthesis, although the product specificity may be influenced by steric requirements of the organic substituents. Reaction 18.33 is used industrially (the *Rochow process*).

$$n MeCl + Si/Cu \xrightarrow[\text{alloy}]{573\,K} Me_n SiCl_{4-n} \quad (18.33)$$

$$SiCl_4 + 4RLi \longrightarrow R_4Si + 4LiCl \quad (18.34)$$

$$SiCl_4 + RLi \longrightarrow RSiCl_3 + LiCl \quad (18.35)$$

$$SiCl_4 + 2RMgCl \xrightarrow{Et_2O} R_2SiCl_2 + 2MgCl_2 \quad (18.36)$$

$$Me_2SiCl_2 + {}^tBuLi \longrightarrow {}^tBuMe_2SiCl + LiCl \quad (18.37)$$

The structures of these compounds are all similar: monomeric, with tetrahedrally sited Si and resembling their C analogues.

Silicon–carbon bonds are relatively strong (the bond enthalpy term is $318\,kJ\,mol^{-1}$) and R_4Si derivatives possess high thermal stabilities. The stability of the Si−C bond is further illustrated by the fact that chlorination of Et_4Si gives $(ClCH_2CH_2)_4Si$ in contrast to the chlorination of R_4Ge or R_4Sn which yields R_nGeCl_{4-n} or R_nSnCl_{4-n} (see equation 18.44). An important reaction of Me_nSiCl_{4-n} ($n = 1$–3) is hydrolysis to produce *silicones* (e.g. equation 18.38 and see *Section 13.10* and *Box 13.10*).

$$\left.\begin{array}{l} Me_3SiCl + H_2O \longrightarrow Me_3SiOH + HCl \\ 2Me_3SiOH \longrightarrow Me_3SiOSiMe_3 + H_2O \end{array}\right\} \quad (18.38)$$

(18.15)

The reaction of Me_3SiCl with NaCp leads to **18.15** in which the cyclopentadienyl group is η^1. Related η^1-complexes include $(\eta^1\text{-}C_5Me_5)_2SiBr_2$ which reacts with anthracene/potassium to give the diamagnetic *silylene* $(\eta^5\text{-}C_5Me_5)_2Si$; in the solid state, two independent molecules are present (Figure 18.9a,b) which differ in the relative orientations of the cyclopentadienyl rings. In one molecule, the two C_5-rings are parallel and staggered (compare Cp_2Mg) whereas in the other, they are tilted, consistent with the presence of a stereochemically active lone pair (Figure 18.9c).

The reactions between R_2SiCl_2 and alkali metals or alkali metal naphthalides give *cyclo*-$(R_2Si)_n$ by loss of Cl^- and Si−Si bond formation. Bulky R groups favour small rings (e.g. $(2,6\text{-}Me_2C_6H_3)_6Si_3$ and ${}^tBu_6Si_3$) while smaller R substituents encourage the formation of large rings (e.g. $Me_{12}Si_6$, $Me_{14}Si_7$ and $Me_{32}Si_{16}$). The use of bulky substituents for stabilizing $R_2Si=SiR_2$ species was discussed in *Section 13.3*. Reaction 18.39 is designed to provide a specific route to a particular ring-size.

$$Ph_2SiCl_2 + Li(SiPh_2)_5Li \longrightarrow cyclo\text{-}Ph_{12}Si_6 + 2LiCl \quad (18.39)$$

(a)   **(b)**   **(c)**

Fig. 18.9 The solid state structure of $(\eta^5\text{-}C_5Me_5)_2Si$ contains two independent molecules. (a) In the first molecule, the cyclopentadienyl rings are co-parallel, while (b) in the other molecule they are mutually tilted; (c) the tilt angle is measured as angle α, and the tilting indicates the presence of a stereochemically active lone pair [P. Jutzi *et al.* (1986) *Angew. Chem. Int. Ed. Engl.*, vol. 25, p. 164]. Hydrogen atoms are omitted for clarity; colour code: Si, blue; C, grey.

Silylenes, R_2Si (analogues of carbenes), can be formed by a variety of methods, for example, the photolysis of cyclic or linear organopolysilanes. As expected, R_2Si species are highly reactive, undergoing many reactions analogous to those typical of carbenes. Stabilization of R_2Si can be achieved by using sufficiently bulky substituents, and electron diffraction data confirm the bent structure of $\{(Me_3Si)_2HC\}_2Si$ ($\angle C\text{–}Si\text{–}C = 97°$).

Germanium

There are similarities between the methods of preparation of compounds with Ge–C and Si–C bonds, compare reaction 18.40 with 18.33, 18.41 with 18.35, 18.42 with 18.36, and 18.43 with the synthesis of $Me_3Si(\eta^1\text{-}Cp)$.

$$nRCl + Ge/Cu \xrightarrow{\Delta} R_nGeCl_{4-n} \qquad R = \text{alkyl or aryl}$$
$$(18.40)$$

$$GeCl_4 + RLi \longrightarrow RGeCl_3 + LiCl \qquad (18.41)$$

$$GeCl_4 + 4RMgCl \xrightarrow{Et_2O} R_4Ge + 4MgCl_2 \qquad (18.42)$$

$$R_3GeCl + Li[Cp] \longrightarrow R_3Ge(\eta^1\text{-}Cp) + LiCl \qquad (18.43)$$

Tetraalkyl and tetraaryl germanium compounds possess monomeric structures with tetrahedrally sited germanium. They are thermally stable and tend to be chemically inert; halogenation requires a catalyst (equations 18.44 and 18.45). Chlorides can be obtained from the corresponding bromides or iodides by halogen exchange (equation 18.46). The presence of halo-substituents increases reactivity (e.g. equation 18.47) and makes the halo-derivatives synthetically more useful than R_4Ge compounds.

$$2Me_4Ge + SnCl_4 \xrightarrow{AlCl_3} 2Me_3GeCl + Me_2SnCl_2 \qquad (18.44)$$

$$R_4Ge + X_2 \xrightarrow{AlX_3} R_3GeX + RX \quad (X = Br, I) \qquad (18.45)$$

$$R_3GeBr + AgCl \longrightarrow R_3GeCl + AgBr \qquad (18.46)$$

$$R_3GeX \xrightarrow{KOH/EtOH, H_2O} R_3GeOH \qquad (18.47)$$

The availability of Ge(II) halides (see *Section 13.8*) means that the syntheses of $(\eta^5\text{-}C_5R_5)_2Ge$ derivatives does not require a reduction step as was the case for the silicon analogues described above. Reaction 18.48 is a general route to $(\eta^5\text{-}C_5R_5)_2Ge$, which exist as monomers in the solid, solution and vapour states.

$$GeX_2 + 2Na[C_5R_5] \longrightarrow (\eta^5\text{-}C_5R_5)_2Ge + 2NaX$$
$$(X = Cl, Br) \qquad (18.48)$$

X-Ray diffraction studies for Cp_2Ge and $\{\eta^5\text{-}C_5(CH_2Ph)_5\}_2Ge$ confirm the bent structure type illustrated in Figures 18.9b and c for $(\eta^5\text{-}C_5Me_5)_2Si$. Reaction 18.49 generates $[(\eta^5\text{-}C_5Me_5)Ge]^+$.

$$(\eta^5\text{-}C_5Me_5)_2Ge + HBF_4$$
$$\longrightarrow [(\eta^5\text{-}C_5Me_5)Ge][BF_4] + C_5Me_5H \qquad (18.49)$$

Organogermanium(II) compounds are a growing family. Germylenes (R_2Ge) include the highly reactive Me_2Ge which can be prepared by reaction 18.50; photolysis reaction 18.51 shows a general strategy to form R_2Ge.

$$Me_2GeCl_2 + 2Li \xrightarrow{THF} Me_2Ge + 2LiCl \qquad (18.50)$$

$$Me(GeR_2)_{n+1}Me \xrightarrow{h\nu} R_2Ge + Me(GeR_2)_nMe \qquad (18.51)$$

Using very sterically demanding R groups can stabilize the R_2Ge species; thus, compound **18.16** is stable at room temperature. The bent structure of $\{(Me_3Si)_2HC\}_2Ge$ has been confirmed by electron diffraction ($\angle C\text{–}Ge\text{–}C = 107°$).

(18.16)

BOX 18.5   **APPLICATIONS**

Commercial uses of organotin compounds

Organotin(IV) compounds have a wide range of applications, with catalytic and biocidal properties being of prime importance. The compounds below are selected examples; an 'antifouling' agent is typically applied as a component of paint to the underside of a ship's hull to prevent the build-up of, for example, barnacles. Despite the wide use of organotin compounds, there are environmental concerns to be considered in some of the applications.

- $^n Bu_3 Sn(OAc)$ (produced by reacting $^n Bu_3 SnCl$ and NaOAc) is an effective antifouling agent, fungicide and bactericide; it also has applications as a polymerization catalyst.
- $^n Bu_2 Sn(OAc)_2$ (from $^n Bu_2 SnCl_2$ and NaOAc) is used as a polymerization catalyst and a stabilizer for PVC.
- (*cyclo*-$C_6 H_{11}$)$_3$SnOH (formed by alkaline hydrolysis of the corresponding chloride) and (*cyclo*-$C_6 H_{11}$)$_3$Sn(OAc)

(produced by treating (*cyclo*-$C_6 H_{11}$)$_3$SnOH with AcOH) are used widely as insecticides in fruit orchards and vineyards.
- $^n Bu_3 SnOSn^n Bu_3$ (formed by aqueous NaOH hydrolysis of $^n Bu_3 SnCl$) has uses as an algicide, fungicide, antifouling paint and wood-preserving agent.
- $^n Bu_3 SnCl$ (a product of the reaction of $^n Bu_4 Sn$ and $SnCl_4$) is a bactericide and fungicide.
- $Ph_3 SnOH$ (formed by base hydrolysis of $Ph_3 SnCl$) is used as an agricultural fungicide for crops such as potatoes, sugar beet and peanuts.
- $Ph_3 SnOSnPh_3$ (formed by dehydrating $Ph_3 SnOH$) is an antifouling agent.
- The cyclic compound ($^n Bu_2 SnS$)$_3$ (formed by reacting $^n Bu_2 SnCl_2$ with $Na_2 S$) is used as a stabilizer for PVC.

Double bond formation between C and Ge was mentioned in *Section 13.3*, and the formation of Ge=Ge bonds to give digermenes can be achieved (equations 18.52 and 18.53) if particularly bulky substituents (e.g. 2,4,6-$Me_3 C_6 H_2$, 2,6-$Et_2 C_6 H_3$, 2,6-$^i Pr_2 C_6 H_3$) are used to stabilize the system.

$$2RR'GeCl_2 \xrightarrow{\text{LiC}_{10}\text{H}_8, \text{ DME}} RR'Ge{=}GeRR' + 4LiCl$$
$$LiC_{10}H_8 = \text{lithium naphthalide} \qquad (18.52)$$

$$2R_2 Ge\{C(SiMe_3)_3\}_2 \xrightarrow{h\nu} 2\{R_2 Ge{:}\} + (Me_3 Si)_3 CC(SiMe_3)_3$$
$$\downarrow$$
$$R_2 Ge{=}GeR_2 \qquad (18.53)$$

Data for several structurally characterized digermenes confirm a non-planar $Ge_2 C_4$-framework analogous to that observed for distannenes discussed in the next section (see *Figure 18.12*).

Tin

Some features that set organotin chemistry apart from organosilicon or organogermanium chemistries are the:

- greater accessibility of the +2 oxidation state;
- greater range of possible coordination numbers;
- presence of halide bridges (see *Section 13.8*).

Reactions 18.54–18.56 illustrate synthetic approaches to $R_4 Sn$ compounds, and organotin halides can be prepared by routes equivalent to reactions 18.33 and 18.40, redistribution reactions from anhydrous $SnCl_4$ (equation 18.57), or from Sn(II) halides (equation 18.58). Using $R_4 Sn$ in excess in reaction 18.57 gives a route to $R_3 SnCl$. Reaction 18.55

is used industrially for the preparation of tetrabutyltin and tetraoctyltin; commercial applications of organotin compounds are highlighted in *Box 18.5*.

$$4RMgBr + SnCl_4 \longrightarrow R_4 Sn + 4MgBrCl \qquad (18.54)$$

$$3SnCl_4 + 4R_3 Al \xrightarrow{R'_2 O} 3R_4 Sn + 4AlCl_3 \qquad (18.55)$$

$$^n Bu_2 SnCl_2 + 2^n BuCl + 4Na \longrightarrow {}^n Bu_4 Sn + 4NaCl \qquad (18.56)$$

$$R_4 Sn + SnCl_4 \xrightarrow{298\,\text{K}} R_3 SnCl + RSnCl_3 \xrightarrow{500\,\text{K}} 2R_2 SnCl_2 \qquad (18.57)$$

$$SnCl_2 + Ph_2 Hg \longrightarrow Ph_2 SnCl_2 + Hg \qquad (18.58)$$

Tetraorganotin compounds tend to be colourless liquids or solids which are quite stable to attack by water and air, and the ease of cleavage of the Sn–C bonds depends upon the R group, with $Bu_4 Sn$ being relatively stable. In moving to the organotin halides, reactivity increases and the chlorides are useful as precursors to a range of organotin derivatives; Figure 18.10 gives selected reactions of $R_3 SnCl$. The structures of $R_4 Sn$ compounds are all similar with the Sn centre being tetrahedral. However, the presence of halide groups leads to significant variation in solid state structure owing to the possibility of Sn–X–Sn bridge formation. In the solid state, $Me_3 SnF$ molecules are connected into zigzag chains by asymmetric, bent Sn–F–Sn bridges (**18.17**), each Sn being in a trigonal bipyramidal arrangement. The presence of bulky substituents may result in either a straightening of the $\cdots$ Sn–F–Sn–F $\cdots$ backbone (e.g. in $Ph_3 SnF$) or in a monomeric structure (e.g. in $\{(Me_3 Si)_3 C\}Ph_2 SnF$). In $(Me_3 SiCH_2)_3 SnF$ (Figure 18.11a), the $Me_3 SiCH_2$ substituents are very bulky, and the Sn–F distances are much longer than the sum of the covalent radii. Solid state ^{119}Sn NMR spectroscopy and measurements of the ^{119}Sn–^{19}F spin–spin coupling

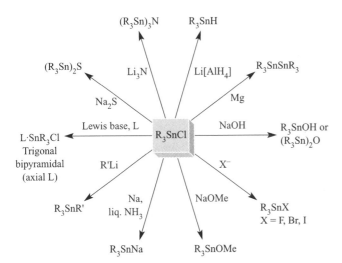

Fig. 18.10 Selected reactions of R_3SnCl; products such as R_3SnH, R_3SnNa and R_3SnSnR_3 are useful starting materials in organotin chemistry.

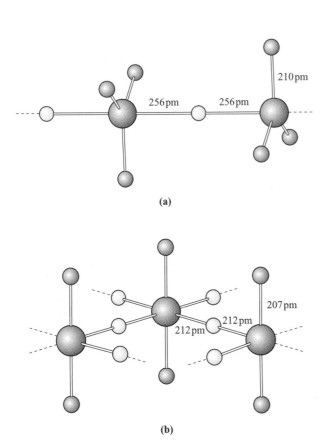

Fig. 18.11 The structures (X-ray diffraction) of (a) $(Me_3SiCH_2)_3SnF$ (only the methylene C atom of each Me_3SiCH_2 group is shown) in which the Sn–F distances are long and indicate the presence of $[(Me_3SiCH_2)_3Sn]^+$ cations interacting with F^- anions to give chains [L.N. Zakharov *et al.* (1983) *Kristallografiya*, vol. 28, p. 271], and (b) Me_2SnF_2 in which Sn–F–Sn bridge formation leads to the generation of sheets [E.O. Schlemper *et al.* (1966) *Inorg. Chem.*, vol. 5, p. 995]. Hydrogen atoms are omitted for clarity; colour code: Sn, blue; C, grey; F, pale grey.

constants provide a useful means of deducing the extent of molecular association in the absence of crystallographic data. Difluoro derivatives R_2SnF_2 tend to contain octahedral Sn in the solid state; in Me_2SnF_2, sheets of interconnected molecules are present (Figure 18.11b). The tendency for association is less for the later halogens (F > Cl > Br > I); thus, $MeSnBr_3$ and $MeSnI_3$ are monomeric, and, in contrast to Me_2SnF_2, Me_2SnCl_2 forms chains of the type shown in **18.18**. Figure 18.10 illustrates the ability of R_3SnCl to act as a Lewis acid; similarly, salts of, for example, $[Me_2SnF_4]^{2-}$ may be prepared and contain discrete octahedral anions.

(18.17)

(18.18)

Tin(II) organometallics of the type R_2Sn which contain Sn–C σ-bonds are stabilized only if R is sterically demanding. Reaction of $SnCl_2$ with $Li[(Me_3Si)_2CH]$ gives $\{(Me_3Si)_2CH\}_2Sn$ which is monomeric in solution and dimeric in the solid state. The dimer (Figure 18.12a) does *not* possess a planar Sn_2C_4 framework (i.e. it is *not* analogous to an alkene) and the Sn–Sn bond distance (276 pm) is too great to be consistent with a normal double bond. A bonding model involving overlap of filled sp^2 hybrids and vacant $5p$ atomic orbitals (Figure 18.12b) has been suggested. Cyclopentadienyl Sn(II) derivatives $(\eta^5\text{-}C_5R_5)_2Sn$ can be prepared by reaction 18.59.

$$2Na[Cp] + SnCl_2 \longrightarrow (\eta^5\text{-}Cp)_2Sn + 2NaCl \qquad (18.59)$$

The structures of $(\eta^5\text{-}C_5R_5)_2Sn$ with various R groups form an interesting series in which the tilt angle α (defined in Figure 18.9 for $(\eta^5\text{-}C_5R_5)_2Si$) increases as the steric demands of R increase: $\alpha = 125°$ for R = H, $144°$ for R = Me, $180°$ for R = Ph. This can be rationalized in terms of a change in hybridization at Sn. In the case of R = H, the Sn centre can be considered to be sp^2 hybridized (with the lone pair in an sp^2 orbital), and for R = Ph, the metal centre can be taken as being sp hybridized (with the lone pair in a $5p$ atomic orbital). Under appropriate conditions, Cp_2Sn reacts with Cp^- to yield $[(\eta^5\text{-}Cp)_3Sn]^-$. Despite the presence of the lone pair in Cp_2Sn, this last reaction shows that it can function as a Lewis *acid*.

(a)

(b)

Fig. 18.12 (a) Schematic representation of the structure of an R_2SnSnR_2 compound which possesses a non-planar Sn_2C_4 framework, and (b) proposed bonding scheme involving sp^2 hybridized tin, and overlap of occupied sp^2 hybrid orbitals with empty $5p$ atomic orbitals to give a weak Sn=Sn double bond.

Lead

Tetraethyllead (made by reaction 18.60 or by electrolysis of $NaAlEt_4$ or EtMgCl using a Pb anode) was formerly widely employed as an anti-knock agent in motor fuels; for environmental reasons, the use of leaded fuels is declining (*Figure 13.1*).

$$4NaPb + 4EtCl \xrightarrow[\text{alloy}]{\approx 373\,K \text{ in an autoclave}} Et_4Pb + 3Pb + 4NaCl$$
(18.60)

Laboratory syntheses of R_4Pb compounds include the use of Grignard reagents (equations 18.61 and 18.62) or organolithium compounds (equations 18.63 and 18.64). High yield routes to $R_3Pb-PbR_3$ involve the reactions of R_3PbLi (see below) with R_3PbCl.

$$2PbCl_2 + 4RMgBr \xrightarrow{Et_2O} 2\{R_2Pb\} + 4MgBrCl$$
$$\downarrow$$
$$R_4Pb + Pb \qquad (18.61)$$

$$3PbCl_2 + 6RMgBr$$
$$\xrightarrow{Et_2O,\ 253\,K} R_3Pb-PbR_3 + Pb + 6MgBrCl \quad (18.62)$$

$$2PbCl_2 + 4RLi \xrightarrow{Et_2O} R_4Pb + 4LiCl + Pb \qquad (18.63)$$

$$R_3PbCl + R'Li \longrightarrow R_3R'Pb + LiCl \qquad (18.64)$$

Alkyllead chlorides can be prepared by reactions 18.65 and 18.66, and these routes are favoured over treatment of R_4Pb with X_2, the outcome of which is hard to control.

$$R_4Pb + HCl \longrightarrow R_3PbCl + RH \qquad (18.65)$$

$$R_3PbCl + HCl \longrightarrow R_2PbCl_2 + RH \qquad (18.66)$$

Compounds of the R_4Pb and R_6Pb_2 families possess monomeric structures with tetrahedral Pb centres as exemplified by the cyclohexyl derivative in Figure 18.13a. The number of Pb derivatives that have been structurally studied is less than for the corresponding Sn-containing compounds. For the organolead halides, the presence of bridging halides is again a common feature giving rise to increased coordination numbers at the metal centre, e.g. in Me_3PbCl (Figure 18.13b). Monomers are favoured if the organic substituents are sterically demanding as in $(2,4,6\text{-}Me_3C_6H_2)_3PbCl$. We mentioned above the use of R_3PbLi reagents; the first structurally characterized member of this group was 'Ph_3PbLi', isolated as the monomeric complex **18.19**.

(18.19)

Tetraalkyl and tetraaryl lead compounds are inert with respect to attack by air and water at room temperature. Thermolysis leads to radical reactions such as those shown in equation 18.67 which will be followed by further radical reaction steps.

$$\left.\begin{array}{l} Et_4Pb \longrightarrow Et_3Pb^{\bullet} + Et^{\bullet} \\ 2Et^{\bullet} \longrightarrow n\text{-}C_4H_{10} \\ Et_3Pb^{\bullet} + Et^{\bullet} \longrightarrow C_2H_4 + Et_3PbH \\ Et_3PbH + Et_4Pb \longrightarrow H_2 + Et_3Pb^{\bullet} + Et_3PbCH_2CH_2^{\bullet} \end{array}\right\}$$
(18.67)

The chloride group in R_3PbCl can be replaced to give a range of R_3PbX species (e.g. $X^- = [N_3]^-$, $[NCS]^-$, $[CN]^-$, $[OR']^-$). Where X^- has the ability to bridge, polymeric structures are observed in the solid state. Both R_3PbN_3 and R_3PbNCS are strong Lewis acids and form adducts such as $[R_3Pb(N_3)_2]^-$. The reaction of Ph_3PbCl with $Na[Cp]$ gives $Ph_3Pb(\eta^1\text{-}Cp)$; structure **18.20** has been confirmed by X-ray diffraction and it is significant that the distance $Pb-C_{Cp} > Pb-C_{Ph}$. This is consistent with a weaker $Pb-C_{Cp}$ bond, and preferential bond cleavage is observed, e.g. in scheme 18.68.

$$Ph_3PbO_2CMe \xleftarrow{MeCO_2H} Ph_3Pb(\eta^1\text{-}Cp) \xrightarrow{PhSH} Ph_3PbSH$$
(18.68)

(18.20)

Fig. 18.13 The solid state structures of (a) $Pb_2(C_6H_{11})_6$ [X-ray diffraction: N. Kleiner *et al.* (1985) *Z. Naturforsch., Teil B*, vol. 40, p. 477], (b) Me_3PbCl [X-ray diffraction: D. Zhang *et al.* (1991) *Z. Naturforsch., Teil A*, vol. 46, p. 337], (c) Cp_2Pb (schematic diagram), and (d) $(\eta^5-Cp)_2Pb(Me_2NCH_2CH_2NMe_2)$ [X-ray diffraction: M.A. Beswick *et al.* (1996) *J. Chem. Soc., Chem. Commun.*, p. 1977]. Hydrogen atoms are omitted for clarity; colour code: Pb, blue; C, grey; F, white; N, black.

Cyclopentadienyl derivatives of Pb(II), $(\eta^5-C_5R_5)_2Pb$, can be prepared by reactions of a Pb(II) salt (e.g. acetate or chloride) with $Na[C_5R_5]$ or $Li[C_5R_5]$. The $(\eta^5-C_5R_5)_2Pb$ compounds are generally sensitive to air, but the presence of bulky R groups increases their stability. The solid state structure of Cp_2Pb consists of polymeric chains (Figure 18.13c), but in the gas phase, discrete $(\eta^5-Cp)_2Pb$ molecules are present which possess the bent structure shown for $(\eta^5-C_5Me_5)_2Si$ in Figure 18.9b. Other $(\eta^5-C_5R_5)_2Pb$ compounds which have been studied in the solid state are monomers. Bent structures (as in Figure 18.9b) are observed for $R = Me$ or $PhCH_2$ for example, but in $\{\eta^5-C_5Me_4(Si^tBuMe_2)\}_2Pb$ where the organic groups are especially bulky, the C_5-rings are coparallel. Discussions of the stereochemically active (or not) role of the lone pair on the Pb(II) centre can be addressed as for similar Sn(II) compounds (see earlier). It has been shown that Cp_2Pb (like Cp_2Sn) can act as a Lewis *acid*; it reacts with the Lewis bases $Me_2NCH_2CH_2NMe_2$ and $4,4'-Me_2bpy$ (**18.21**)

to form the adducts $(\eta^5-Cp)_2Pb·L$ where L is the Lewis base. Figure 18.13d shows the solid state structure of $(\eta^5-Cp)_2Pb·Me_2NCH_2CH_2NMe_2$, and the structure of $(\eta^5-Cp)_2Pb·(4,4'-Me_2bpy)$ is similar. Further evidence for Lewis acid behaviour comes from the reaction of $(\eta^5-Cp)_2Pb$ with $Li[Cp]$ in the presence of a crown ether (see *Section 10.8*), 12-crown-4, which gives $[Li(12-crown-4)]_2[Cp_9Pb_4][Cp_5Pb_2]$. The structures of $[Cp_9Pb_4]^-$ and $[Cp_5Pb_2]^-$ consist of fragments of the polymeric chain of Cp_2Pb (see Figure 18.13c), e.g. in $[Cp_5Pb_2]^-$, one Cp^- ligand bridges between the two Pb(II) centres and the remaining four Cp^- ligands are bonded in an η^5-mode, two to each Pb atom.

(18.21)

18.6 Group 15

Bonding aspects

Our discussion of organometallic compounds of group 15 covers As, Sb and Bi. There is an extensive chemistry of compounds with N–C or P–C bonds, but much of this belongs within the remit of organic chemistry, although amines and phosphines (e.g. R_3E, $R_2E(CH_2)_nER_2$ where $E = N$ or P) are important ligands in inorganic complexes. In both cases, the group 15 element acts as a σ-donor, and in the case of phosphorus, also as a π-acceptor (see *Section 20.4*).

On descending group 15, the E–E and E–C bond enthalpy terms both decrease (e.g. see *Table 14.3*). In Section 14.3, we emphasized differences in bonding between nitrogen and the later elements, and illustrated that $(p-p)\pi$-bonding is important for nitrogen but not for the heavier elements. Thus, nitrogen chemistry provides many compounds of type $R_2N=NR_2$, but for most R groups the analogous $R_2E=ER_2$ compounds ($E = P$, As, Sb or Bi) are unstable with respect to oligomerization to give cyclic compounds such as Ph_6P_6. Only by the use of especially bulky substituents is double bond formation for the later elements made possible, with the steric hindrance preventing oligomerization. Thus, several compounds with P=P, P=As, As=As and P=Sb bonds are now known, and possess *trans*-configurations, e.g. $(Me_3Si)_3CE=E'C(SiMe_3)_3$ ($E=E'$ is P=P, P=As or As=As) and $(2,4,6\text{-}^tBu_3C_6H_2)As=As(2,4,6\text{-}^tBu_3C_6H_2)$ (**18.22**). We return to single bond formation between As, Sb and Bi atoms later.

(18.22)

Arsenic, antimony and bismuth

Organometallic compounds of As(III), Sb(III) and Bi(III) can be prepared from the respective element and organo halides (reaction 18.69) or by use of Grignard reagents (equation 18.70) or organolithium compounds. Treatment of organo halides (e.g. those from reaction 18.69) with R'Li gives RER'_2 or R_2ER' (e.g. equation 18.71).

$$2As + 3RBr \xrightarrow{\text{in presence of Cu, }\Delta} RAsBr_2 + R_2AsBr \quad (18.69)$$

$$EX_3 + 3RMgX \xrightarrow{\text{ether solvent}} R_3E + 3MgX_2 \quad (18.70)$$

$$R_2AsBr + R'Li \longrightarrow R_2AsR' + LiBr \quad (18.71)$$

Metal(V) derivatives, R_5E, cannot be prepared from the corresponding pentahalides, but may be obtained by oxidation of R_3E followed by treatment with RLi (e.g. equation 18.72). In the case of R_5Sb, the reaction of R_3SbCl_2 with RLi yields the salts $Li[SbR_6]$ which hydrolyze to give R_5Sb.

$$R_3As + Cl_2 \longrightarrow R_3AsCl_2 \xrightarrow[-2LiCl]{2RLi} R_5As \quad (18.72)$$

The oxidative addition of R'X (R = alkyl) to R_3E produces $R_3R'EX$, with the tendency of R_3E to undergo this reaction decreasing in the order As > Sb $\gg$ Bi, and I > Br > Cl. Further, conversion of R_3X to $R_3R'EX$ by this route works for R = alkyl or aryl when E = As, but not for R = aryl when E = Sb. Compounds of the type R_3EX_2 are readily prepared as shown in equation 18.72, and R_2EX_3 derivatives can be made by addition of X_2 to R_2EX (E = As, Sb; X = Cl, Br).

(18.23)

Compounds of the type R_3E are sensitive to aerial oxidation but resist attack by water. They are more stable when R = aryl (compared to alkyl), and stability for a given series of triaryl derivatives decreases in the order $R_3As > R_3Sb > R_3Bi$. All R_3E compounds structurally characterized to date are trigonal pyramidal, and the C–E–C angle α in **18.23** decreases for a given R group in the order As > Sb > Bi. Hydrogen peroxide oxidizes Ph_3As to Ph_3AsO for which **18.24** is a bonding representation (see below); Ph_3SbO is similarly prepared or can be obtained by heating $Ph_3Sb(OH)_2$. Triphenylbismuth oxide is made by oxidation of Ph_3Bi or hydrolysis of Ph_3BiCl_2. The ready formation of these oxides should be compared with the relative stability with respect to oxidation of Ph_3P, the ready oxidation of Me_3P, and the use of Me_3NO as an oxidizing agent. Structure **18.25** provides a bonding description in Me_3NO. A comparison of the bonding descriptions shown in **18.24** and **18.25** suggests that a contribution from a resonance structure involving $(p-d)\pi$ bonding is possible for R_3AsO (and likewise for R_3PO, R_3SbO and R_3BiO) but not for R_3NO. We have already discussed the uncertainties surrounding the role of the *d*-orbitals in compounds of the heavier main group elements (see *Sections 13.6* and *14.3*, and the discussion of SF_6 in *Section 4.7*) and, similarly, the relative contributions made by the two resonance structures in **18.24** remains a matter for debate. Triphenylarsenic oxide forms a monohydrate which exists as a hydrogen-bonded dimer (**18.26**) in the solid state; Ph_3SbO crystallizes in several modifications

Fig. 18.14 The solid state structures (X-ray diffraction) of (a) polymeric Me₄SbF in which each Sb(V) centre is distorted octahedral [W. Schwarz *et al.* (1978) *Z. Anorg. Allg. Chem.*, vol. 444, p. 105], (b) Ph₃AsO·BF₃ [N. Burford *et al.* (1990) *Acta Crystallogr.*, *Sect. C*, vol. 46, p. 92], (c) Ph₆As₆ in which the As₆ adopts a chair conformation [A.L. Rheingold *et al.* (1983) *Organometallics*, vol. 2, p. 327], and (d) (η¹-C₅Me₅)₄Sb₄ with methyl groups omitted for clarity [O.M. Kekia *et al.* (1996) *Organometallics*, vol. 15, p. 4104]. Hydrogen atoms are omitted for clarity; colour code: As and Sb, dark blue; C, grey; F, white; B, black; O, pale blue.

which contain either monomers or polymers, and has a range of catalytic uses in organic chemistry, e.g. oxirane polymerization, and reactions between amines and acids to give amides.

(18.24)   (18.25)

(18.26)

The ability of R₃E to act as a Lewis base decreases down group 15. *d*-Block metal complexes involving R₃P ligands are far more numerous than those containing R₃As and R₃Sb, and only a few complexes containing R₃Bi ligands have been structurally characterized, e.g. Cr(CO)₅(BiPh₃) (**18.27**) and [(η⁵-Cp)Fe(CO)₂(BiPh₃)]⁺. Adducts are also formed between R₃E or R₃EO (E = As, Sb) and Lewis acids such as boron trifluoride (Figure 18.14b), and in *Section 16.4*, we described complexes formed between Ph₃E (E = P, As, Sb) and halogens.

(18.27)

Compounds in the R₅E family are usually trigonal bipyramidal (e.g. Ph₅As, Me₅Sb, Me₅Bi) although Ph₅Sb possesses a square-based pyramidal structure. In the solid state, crystal packing forces are probably important in determining the preferred geometry, and in solution, the compounds are stereochemically non-rigid (typical of five-coordinate species; see *Section 2.11*). On heating, R_5E compounds decompose, with the thermal stability decreasing down the group, e.g. Ph_5As is more thermally stable than Ph_5Sb than Ph_5Bi. The decomposition products vary and, for example, Ph_5Sb decomposes to Ph_3Sb and PhPh, while Me_5As gives Me_3As, CH_4 and C_2H_4. Cleavage of an $E-C$ bond in R_5E compounds occurs upon treatment with halogens, Brønsted acids or Ph_3B (equations 18.73–18.75). Both Me_5Sb and Me_5Bi react with MeLi in THF (equation 18.76) to give salts containing the octahedral ions $[Me_6E]^-$ (see also equation 18.72 and accompanying text).

$$Ph_5E + Cl_2 \longrightarrow Ph_4ECl + PhCl \qquad (18.73)$$

$$Ph_5E + HCl \longrightarrow Ph_4ECl + PhH \qquad (18.74)$$

$$Ph_5E + Ph_3B \longrightarrow [Ph_4E]^+[BPh_4]^- \qquad (18.75)$$

$$Me_5Sb + MeLi \xrightarrow{Et_2O,\ THF} [Li(THF)_4]^+[Me_6Sb]^- \qquad (18.76)$$

The monohalides R_4EX tend to be ionic for X = Cl, Br or I, i.e. $[R_4E]^+X^-$, but among the exceptions is Ph_4SbCl which crystallizes as discrete trigonal bipyramidal molecules. The fluorides possess covalent structures; in the solid state Me_4SbF forms polymeric chains (Figure 18.14a) while $MePh_3SbF$ exists as trigonal bipyramidal molecules **18.28**. For the di- and trihalides there is also structural variation, and ionic, discrete molecular and oligomeric structures in the solid state are all exemplified, e.g. Me_3AsBr_2 is ionic and contains the tetrahedral $[Me_3AsBr]^+$ ion, Ph_3BiCl_2 and Ph_3SbX_2 (X = F, Cl, Br or I) are trigonal bipyramidal molecules with axial X atoms, Ph_2SbCl_3 is dimeric (**18.29**), while Me_2SbCl_3 exists in two structural forms, one ionic $[Me_4Sb]^+[SbCl_6]^-$ and the other a covalent dimer.

(18.28)   (18.29)

The family of R_2E-ER_2 compounds has grown significantly since 1980 and those structurally characterized by X-ray diffraction include Ph_4As_2, Ph_4Sb_2 and Ph_4Bi_2. All possess the staggered conformation shown in **18.30** for the C_4E_2 core with values of α and β of 103° and 96° for Ph_4As_2, 94° and 94° for Ph_4Sb_2, and 98° and 91° for Ph_4Bi_2. As expected, the $E-E$ bond length increases: 246 pm in Ph_4As_2, 286 pm in Ph_4Sb_2, and 298 pm in Ph_4Bi_2. Equation 18.77 gives a typical preparative route. Some R_4Sb_2 and R_4Bi_2 (but not R_4As_2) derivatives are *thermochromic*.

$$2R_2BiCl + 2Na \xrightarrow[\text{2. 1,2-dichloroethane}]{\text{1. liquid } NH_3} R_4Bi_2 + 2NaCl \qquad (18.77)$$

(18.30)

> The colour of a *thermochromic* compound is temperature-dependent; the phenomenon is called *thermochromism*.

The reduction of organometal(III) dihalides (e.g. $RAsCl_2$) with sodium or magnesium in THF, or reduction of $RAs(O)(OH)_2$ acids (reaction 18.78) gives *cyclo*-$(RE)_n$, where $n = 3$–6. Figure 18.14c shows the structure of Ph_6As_6 which illustrates the typical trigonal pyramidal environment for the group 15 element. Two crystalline polymorphs of $(\eta^1\text{-}C_5Me_5)_4Sb_4$ are known, differing in details of the molecular geometry and crystal packing; one structure is noteworthy for its acute Sb–Sb–Sb bond angles (Figure 18.14d). Reaction 18.79 is an interesting example of the formation of a *cyclo*-As_3 species, the organic group being tailor-made to encourage the formation of the three-membered ring. A similar reaction occurs with $Me(CH_2SbCl_2)_3$.

$$6PhAs(O)(OH)_2 \xrightarrow{H_3PO_2} Ph_6As_6 \qquad (18.78)$$

$$(18.79)$$

Organometallic chemistry involving cyclopentadienyl ligands is less important in group 15 than for the previous groups we have discussed. We have already mentioned (η^1-$C_5Me_5)_4Sb_4$ (Figure 18.14d). Other compounds for which solid state structures contain η^1-C_5R_5 substituents include (η^1-Cp)$_3$Sb (equation 18.80) and (η^1-$C_5Me_5)AsCl_2$ (Figure 18.15a, prepared by ligand redistribution between (η^1-$C_5Me_5)_3As$ and $AsCl_3$). The derivatives Cp_nSbX_{3-n} (X = Cl, Br, I; $n = 1$, 2) are prepared by treating (η^1-Cp)$_3$Sb with SbX_3, and $CpBiCl_2$ forms in reaction 18.81.

$$Sb(NMe_2)_3 + 3C_5H_6 \xrightarrow{Et_2O,\ 193\,K} (\eta^1\text{-Cp})_3Sb + 3Me_2NH \qquad (18.80)$$

$$BiCl_3 + Na[Cp] \xrightarrow{Et_2O,\ 203\,K} CpBiCl_2 + NaCl \qquad (18.81)$$

In solution the cyclopentadienyl rings in this type of compound are fluxional. In the solid state, crystallographic data (where available) reveal significant variation in bonding modes as examples in Figure 18.15 illustrate. Consideration of the $E-C$ bond distances leads to the designations of η^1 or η^3. Reaction 18.82 gives one of the few η^5-cyclopenta-

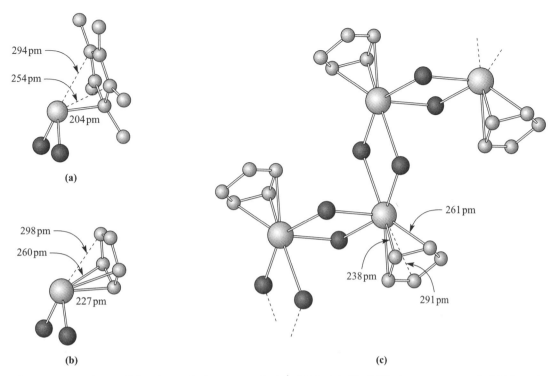

Fig. 18.15 The structures (X-ray diffraction) of (a) monomeric $(\eta^1\text{-}C_5Me_5)AsCl_2$ [E.V. Avtomonov *et al.* (1996) *J. Organomet. Chem.*, vol. 524, p. 253], (b) monomeric $(\eta^3\text{-}C_5H_5)SbCl_2$ [W. Frank (1991) *J. Organomet. Chem.*, vol. 406, p. 331], and (c) polymeric $(\eta^3\text{-}C_5H_5)BiCl_2$ [W. Frank (1990) *J. Organomet. Chem.*, vol. 386, p. 177]. Hydrogen atoms are omitted for clarity; colour code: As and Bi, blue; C, grey; Cl, black.

dienyl derivatives of a heavier group 15 element so far prepared. The $[(\eta^5\text{-}C_5Me_5)_2As]^+$ ion is isoelectronic with $(\eta^5\text{-}C_5Me_5)_2Ge$ and possesses the same tilted bent structure type, illustrated for $(\eta^5\text{-}C_5Me_5)_2Si$ in Figure 18.9.

$$(\eta^1\text{-}C_5Me_5)_2AsF + BF_3 \longrightarrow [(\eta^5\text{-}C_5Me_5)_2As]^+[BF_4]^- \tag{18.82}$$

18.7 Group 16

Our discussion of organo-compounds of group 16 elements is confined to selenium and tellurium (polonium having been little studied, see *Chapter 15*). Of course, there are also vast numbers of organic compounds containing C−O or C−S bonds, and some relevant inorganic topics already covered are:

- oxides and oxoacids of carbon (*Section 13.9*);
- sulfides of carbon (*Section 13.11*).

Selenium and tellurium

The organic chemistry of selenium and tellurium is generally like that of sulfur, and is an expanding area of research. An interesting group of compounds that have no sulfur

analogues is the alkyl and aryl selenium(IV) and tellurium(IV) chlorides, bromides and iodides. Dimethyl selenide and telluride react with Cl_2, Br_2 and I_2 to give Me_2SeX_2 and Me_2TeX_2; aryl compounds can be made similarly. The structure of each compound is based on a trigonal bipyramidal arrangement as shown in structure **18.31**, in accord with VSEPR theory; halogen atoms always occupy the axial sites. What was at one time labelled the β-form of Me_2TeI_2 has now been shown to be $[Me_3Te]^+[MeTeI_4]^-$ with a trigonal pyramidal cation and square-based pyramidal anion; $I-Te \cdots I$ bridges between cations and anions result in each Te centre being in a distorted octahedral environment in the solid state.

$$\begin{array}{c} Cl \\ | \\ :Te\overset{\text{\tiny{IIII}}}{\underset{}{\diagdown}} Me \\ | \\ Cl \end{array}$$

(18.31)

Reaction 18.83 yields the pyrophoric compound Me_4Te which can be oxidized to Me_4TeF_2 using XeF_2; Me_4TeF_2 can be converted to Me_6Te by treatment with Me_2Zn.

$$TeCl_4 + 4MeLi \xrightarrow{Et_2O,\ 195\,K} Me_4Te + 4LiCl \tag{18.83}$$

Summary

Main group organometallic compounds is a large and growing area, and we have only been able to provide an introduction to the topic; a much wider perspective can be obtained by referring to the books and review articles listed below. Much interest in these compounds arises from research into the preparation of suitable precursors for solid state materials and we expand on this topic in Chapter 27.

The following terms were introduced in this chapter. Do you know what they mean?

❑ organometallic compound
❑ pyrophoric
❑ radical anion
❑ disorder (in a crystal structure)
❑ thermochromic

Suggested further reading

General sources

Ch. Elschenbroich and A. Salzer (1992) *Organometallics*, 2nd edn, Wiley-VCH, Weinheim – An excellent text which covers both main group and transition metal organometallic chemistry.

R.B. King, ed. (1994) *Encyclopedia of Inorganic Chemistry*, Wiley, Chichester – The organometallic chemistry of each of the elements discussed in this chapter is surveyed (with many references) in separate articles under the element name.

G. Wilkinson, F.G.A. Stone and E.W. Abel, eds (1982) *Comprehensive Organometallic Chemistry*, Pergamon, Oxford – Volume 1 provides detailed coverage of the organometallic compounds of groups 1, 2 and 13, while volume 2 deals with groups 14 and 15; the reviews include hundreds of literature references up to 1981.

G. Wilkinson, F.G.A. Stone and E.W. Abel, eds (1995) *Comprehensive Organometallic Chemistry II*, Pergamon, Oxford – Volume 1 (ed. C.E. Housecroft) updates the information from the above edition, covering groups 1, 2 and 13 for the period 1982–1994; volume 2 (ed. A.G. Davies) updates the chemistry of groups 14 and 15.

Specialized topics

K.M. Baines and W.G. Stibbs (1996) *Advances in Organometallic Chemistry*, vol. 39, p. 275 – 'Stable doubly bonded compounds of germanium and tin'.

P.J. Brothers and P.P Power (1996) *Advances in Organometallic Chemistry*, vol. 39, p. 1 – 'Multiple bonding involving Al, Ga, In and Tl'.

P. Jutzi (1986) *Advances in Organometallic Chemistry*, vol. 26, p. 217 – 'π-Bonding to main group elements'.

P. Jutzi and N. Burford (1999) *Chemical Reviews*, vol. 99, p. 969 – 'Structurally diverse π-cyclopentadienyl complexes of the main group elements'.

P.R. Markies, O.S. Akkerman, F. Bickelhaupt, W.J.J. Smeets and A.L. Spek (1991) *Advances in Organometallic Chemistry*, vol. 32, p. 147 – 'X-ray structural analysis of organomagnesium compounds'.

N.C. Norman, ed. (1998) *Chemistry of Arsenic, Antimony and Bismuth*, Blackie, London – This book includes chapters dealing with organo-derivatives.

R. Okazaki and R. West (1996) *Advances in Organometallic Chemistry*, vol. 39, p. 231 – 'Chemistry of stable disilenes'.

J.A. Reichl and D.H. Berry (1998) *Advances in Organometallic Chemistry*, vol. 43, p. 197 – 'Recent progress in transition metal-catalyzed reactions of silicon, germanium and tin'.

W.N. Setzer and P. v. R. Schleyer (1985) *Advances in Organometallic Chemistry*, vol. 24, p. 353 – A structural review entitled: 'X-Ray analyses of organolithium compounds'.

D.F. Shriver and M.A. Drezdon (1986) *The manipulation of air-sensitive compounds*, Wiley, New York – An excellent text dealing with inert atmospheres techniques.

L.R. Sita (1995) *Advances in Organometallic Chemistry*, vol. 38, p. 189 – 'Structure/property relationships of polystannanes'.

J.D. Smith (1998) *Advances in Organometallic Chemistry*, vol. 43, p. 267 – 'Organometallic compounds of the heavier alkali metals'.

P.J. Smith, ed. (1998) *Chemistry of Tin*, 2nd edn, Blackie, London – Chapters 4–8 deal in detail with the organometallic chemistry of tin.

Applications

B. Jousseaume and M. Pereyre (1998) in *Chemistry of Tin,* ed. P.J. Smith, 2nd edn, Blackie, London – Chapter 9: 'The uses of organotin compounds in organic synthesis'.

D.S. Matteson (1995) *Stereodirected Synthesis with Organoboranes*, Springer, Berlin.

L.A. Paquette, ed. (1995) *Encyclopedia of Reagents in Organic Synthesis*, Wiley, Chichester – Detailed descriptions of uses of specific main group organometallic compounds are included in this eight-volume encyclopedia.

S.E. Thomas (1991) *Organic Synthesis: The Roles of Boron and Silicon*, Oxford University Press, Oxford.

Problems

18.1 Suggest products of the following reactions:

(a) MeBr + 2Li $\xrightarrow{\text{Et}_2\text{O}}$

(b) Na + $(C_6H_5)_2$ $\xrightarrow{\text{THF}}$

(c) ^{n}BuLi + H_2O $\longrightarrow$

(d) Na + C_5H_6 $\longrightarrow$

18.2 Whether the bonding in lithium alkyls is predominantly ionic or covalent is still a matter for debate. Assuming a covalent model, use a hybrid orbital approach to suggest a bonding scheme for $(MeLi)_4$. Comment on the bonding picture you have described.

18.3 Describe the gas phase and solid state structures of Me_2Be and discuss the bonding in each case. Compare the bonding with that in BeH_2 and $BeCl_2$.

18.4 Suggest products of the following reactions which are *not* necessarily balanced on the left-hand side:

(a) $Mg + C_5H_6 \longrightarrow$

(b) $MgCl_2 + LiR \longrightarrow$

(c) $RBeCl \xrightarrow{\text{LiAlH}_4}$

18.5 The compound $(Me_3Si)_2C(MgBr)_2 \cdot n$THF is monomeric. Suggest a value for n and propose a structure for this Grignard reagent.

18.6 (a) For the equilibrium $Al_2R_6 \rightleftharpoons 2AlR_3$, comment on the fact that values of K are 1.52×10^{-8} for R = Me, and 2.3×10^{-4} for R = Me_2CHCH_2. (b) Describe the bonding in Al_2Me_6, Al_2Cl_6 and $Al_2Me_4(\mu\text{-Cl})_2$.

18.7 Suggest products of the following reactions which are *not* necessarily balanced on the left-hand side:

(a) $Al_2Me_6 + H_2O \longrightarrow$

(b) $AlR_3 + R'NH_2 \longrightarrow$

(c) $Me_3SiCl + Na[C_5H_5] \longrightarrow$

(d) $Me_2SiCl_2 + Li[AlH_4] \longrightarrow$

18.8 (a) Discuss the variation in structure for the group 13 trialkyls and triaryls. (b) Comment on features of interest in the solid state structures of $[Me_2(PhC_2)Ga]_2$ and $[Ph_3Al]_2$.

18.9 The conversion of $(\eta^1\text{-}C_5Me_5)_2SiBr_2$ to $(\eta^5\text{-}C_5Me_5)_2Si$ is achieved using anthracene/potassium. Outline the role of this reagent.

18.10 Suggest the nature of the solid state structures of (a) Ph_2PbCl_2, (b) Ph_3PbCl, (c) $(2,4,6\text{-}Me_3C_6)_3PbCl$, and (d) $[PhPbCl_5]^{2-}$. In each case, state the expected coordination environment of the Pb centre.

18.11 Suggest products when Et_3SnCl reacts with the following reagents: (a) H_2O; (b) $Na[Cp]$; (c) Na_2S; (d) PhLi; (e) Na.

18.12 (a) In what ways do the solid state structures of $(\eta^5\text{-}C_5R_5)_2Sn$ for R = H, Me and Ph differ? Comment on reasons for any differences. (b) In the solid state structure of $(\eta^5\text{-}C_5Me_5)_2Mg$, the two cyclopentadienyl rings are parallel; however, for M = Ca, Sr and Ba, the rings are tilted with respect to one another. Say what you can about this observation.

18.13 The reaction of InBr with an excess of $HCBr_3$ in 1,4-dioxane $(C_4H_8O_2)$ leads to compound **A** which is an adduct of 1,4-dioxane and contains 21.4% In. During the reaction, the indium is oxidized. The ^{1}H NMR spectrum of **A** shows signals at δ 5.36 (singlet) and 3.6 (multiplet) in a ratio 1:8. Treatment of **A** with two molar equivalents of InBr followed by addition of $[Ph_4P]Br$ yields the salt **B** which contains 16.4% In and 34.2% Br. The ^{1}H NMR spectrum of **B** exhibits signals in the range δ 8.01–7.71 and a singlet at δ 0.20 with relative integrals of 60:1. Suggest identities for **A** and **B**.

18.14 Discuss the bonding between the central *p*-block elements in the following compounds and give the expected arrangements of the organic substituents with respect to the central E_2-unit:

(a) $[(2,4,6\text{-}Me_3C_6H_2)_2BB(2,4,6\text{-}Me_3C_6H_2)Ph]^{2-}$;

(b) $[(2,4,6\text{-}^iPr_3C_6H_2)_2GaGa(2,4,6\text{-}^iPr_3C_6H_2)_2]^-$;

(c) $\{(SiMe_3)_2CH\}_2SnSn\{(SiMe_3)_2CH\}_2$;

(d) $^tBu_3GeGe^tBu_3$;

(e) $(Me_3Si)_3CAsAsC(SiMe_3)_3$

18.15 Suggest products when Me_3Sb reacts with the following reagents: (a) B_2H_6; (b) H_2O_2; (c) Br_2; (d) Cl_2 followed by treatment with MeLi; (e) MeI; (f) Br_2 followed by treatment with Na[OEt].

18.16 Write a brief account of how the changes in available oxidation states for elements, E, in groups 13 to 15 affect the families of organoelement compounds of type R_nE that can be formed.

18.17 Give methods of synthesis for the following families of compound, commenting where appropriate on limitations in the choice of R: (a) R_4Ge; (b) R_3B; (c) $(C_5R_5)_3Ga$; (d) *cyclo*-$(R_2Si)_n$; (e) R_5As; (f) R_4Al_2; (g) R_3Sb.

18.18 Give a short account of the structural variation observed for cyclopentadienyl derivatives Cp_nE of the heavier *p*-block elements.

18.19 Write a brief account of the use of sterically demanding substituents in the stabilization of compounds containing E–E and E=E bonds where E is a *p*-block metal or semi-metal.

18.20 Write a short account describing methods of formation of metal-carbon bonds for metals in the *s*- and *p*-block.

Chapter 19

d-Block chemistry: general considerations

1–2	3	4	5	6	7	8	9	10	11	12	13–18
s-block											*p*-block
	Sc	Ti	V	Cr	Mn	Fe	Co	Ni	Cu	Zn	
	Y	Zr	Nb	Mo	Tc	Ru	Rh	Pd	Ag	Cd	
	La	Hf	Ta	W	Re	Os	Ir	Pt	Au	Hg	

- an introduction to coordination complexes (*Section 6.11*);
- redox chemistry in aqueous solution, including potential diagrams (*Chapter 7*);
- geometrical isomerism (*Section 1.20*);
- chiral molecules (*Section 3.8*);
- binary metal hydrides (*Section 9.7*).

19.1 Topic overview

In Chapters 19–23, we discuss the chemistry of the *d*-block metals, covering first some general principles including magnetic and electronic spectroscopic properties. We move then to a systematic coverage of the metals and their compounds, and conclude with a chapter on organometallic chemistry. We have already touched upon some aspects of the *d*-block metals and the following will not be covered again in detail:

- ground state electronic configurations (*Table 1.3*);
- trends in first ionization energies (*Figure 1.15* and *Section 1.10*);
- structures of bulk metals (*Section 5.3*);
- polymorphism (*Section 5.4*);
- metallic radii (*Section 5.5*);
- trends in melting points and $\Delta_a H^\circ$(298 K) (*Section 5.6*);
- alloys and intermetallic compounds (*Section 5.7*);
- metallic bonding including electrical resistivity (*Section 5.8* and *Figure 5.9*);
- aquated cations: formation and acidic properties (*Section 6.7*);
- solubilities of ionic salts and common-ion effect (*Sections 6.9* and *6.10*);
- stability constants for metal complexes (*Section 6.12*);
- selected ligand structures and abbreviations (*Table 6.7*);

19.2 Ground state electronic configurations

d-Block metals versus transition elements

The three rows of *d*-block metals are shown in the schematic periodic table at the beginning of the chapter. The term 'transition elements (metals)' is also widely used, but '*d*-block metal' and 'transition element' are *not* interchangeable. A *transition element* is one for which an atom has an incomplete *d*-subshell, or which gives rise to a cation with an incomplete *d*-subshell,[†] and thus elements in group 12 (which *are* within the *d*-block) are *not* transition elements. The elements in the *f*-block (see *Chapter 24*) are sometimes called *inner transition elements*. Throughout our discussions, we shall use the terms *d*-block and *f*-block metals, so being consistent with the use of the terms *s*-block and *p*-block elements in earlier chapters. Three further points should be noted:

- each group of *d*-block metals consists of three members and is called a *triad*;
- metals of the second and third rows are sometimes called the *heavier d-block metals*;

[†] *IUPAC: Nomenclature of Inorganic Chemistry (Recommendations 1990)*, ed. G.J. Leigh, Blackwell Scientific Publications, Oxford.

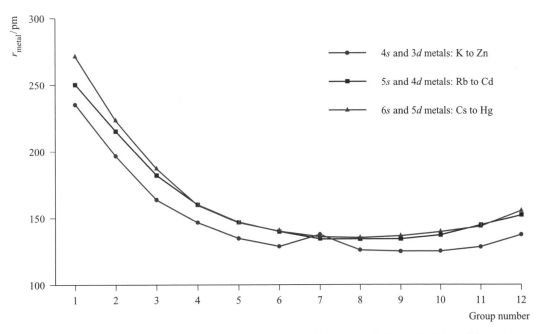

Fig. 19.1 Trends in metallic radii (r_{metal}) across the three rows of *s*- and *d*-block metals K to Zn, Rb to Cd, and Cs to Hg.

• Ru, Os, Rh, Ir, Pd and Pt are collectively known as the *platinum-group metals*.

Electronic configurations

To a first approximation, the observed ground state electronic configurations of the first, second and third row *d*-block metal atoms correspond to the progressive filling of the 3*d*, 4*d* and 5*d* atomic orbitals respectively (*Table 1.3*). However, there are minor deviations from this pattern, e.g. in the first row, the ground state of chromium is $[Ar]4s^1 3d^5$ rather than $[Ar]4s^2 3d^4$. The reasons for these deviations are beyond the scope of this book: we should need to know both the energy difference between the 3*d* and 4*s* atomic orbitals when the nuclear charge is 24 (the atomic number of Cr) and the interelectronic interaction energies for each of the $[Ar]4s^1 3d^5$ and $[Ar]4s^2 3d^4$ configurations. Fortunately, M^{2+} and M^{3+} *ions* of the *first* row *d*-block metals all have electronic configurations of the general form $[Ar]3d^n$, and so the comparative chemistry of these metals is largely concerned with the consequences of the successive filling of the 3*d* orbitals. For metals of the second and third rows, the picture is more complicated, and a systematic treatment of their chemistry cannot be given. The emphasis in this and the next chapter is therefore on the first row metals, but we shall include some material that illustrates ways in which the heavier metals differ from their lighter congeners.

An important point that must not be forgotten is that *d*-block metal atoms are, of course, *many-electron* species, and when we discuss, for example, radial distribution functions of the *nd* atomic orbitals, we refer to hydrogen-like atoms and, therefore, the discussion is extremely approximate.

19.3 Physical properties

In this section, we consider physical properties of the *d*-block metals (see cross references in *Section 19.1* for further details); an extended discussion of properties of the heavier metals is given in *Section 22.1*. Nearly all the *d*-block metals are hard, ductile and malleable, with high electrical and thermal conductivities. With the exceptions of Mn, Zn, Cd and Hg, at room temperature, the metals possess one of the typical metal structures (see *Table 5.2*). The metallic radii (r_{metal}) for 12-coordination (*Table 5.2* and *Figure 19.1*) are much smaller that those of the *s*-block metals of comparable atomic number; *Figure 19.1* also illustrates that values of r_{metal}:

• show little variation across a given row of the *d*-block;
• are greater for second and third row metals than for first row metals:
• are similar for the second and third row metals in a given triad.

This last observation is due to the so-called *lanthanoid contraction* (the steady decrease in size along the 14 lanthanoids between La and Hf, see *Section 24.3*).

Metals of the *d*-block are (with the exception of the group 12 metals) much harder and less volatile than those of the *s*-block. The trends in enthalpies of atomization (*Table 5.2*) are shown in Figure 19.2. Metals in the second and third rows generally possess higher enthalpies of atomization than the corresponding elements in the first row; this is a substantial factor in accounting for the far greater occurrence of metal–metal bonding in compounds of the heavier *d*-block

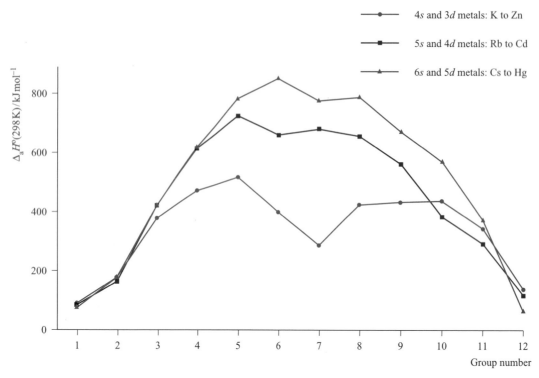

Fig. 19.2 Trends in standard enthalpies of atomization, $\Delta_a H^\circ(298\,\mathrm{K})$, across the three rows of _s_- and _d_-block metals K to Zn, Rb to Cd, and Cs to Hg.

metals compared with their first row congeners. In general, Figure 19.2 shows that metals in the centre of the _d_-block possess higher values of $\Delta_a H^\circ(298\,\mathrm{K})$ than early or late metals. However, one must be careful in comparing metals with different structure types and this is particularly true of manganese (see _Section 5.3_).

The first ionization energies (IE_1) of the _d_-block metals in a given period (_Figure 1.15_ and _Appendix 8_) are higher than those of the preceding _s_-block metals. Figure 1.15 shows that across each of the periods K to Kr, Rb to Xe, and Cs to Rn, the variation in values of IE_1 is small across the _d_-block and far greater among the _s_- and _p_-block elements. Within each period, the overall trend for the _d_-block metals is for the

ionization energies to increase, but many small variations occur. Chemical comparisons between metals from the _s_- and _d_-blocks are complicated by the number of factors involved. Thus, all 3_d_ metals have values of IE_1 (_Figure 1.15_) and IE_2 larger than those of calcium, and all except zinc have higher values of $\Delta_a H^\circ$ (Figure 19.2); these factors make the metals less reactive than calcium. However, since all known M^{2+} ions of the 3_d_ metals are smaller than Ca^{2+}, lattice and solvation energy effects (see _Chapters 5_ and _6_) are more favourable for the 3_d_ metal ions. In practice, it turns out that, in the formation of species containing M^{2+} ions, all the 3_d_ metals are thermodynamically less reactive than calcium, and this is consistent with the standard reduction potentials listed in Table 19.1. However, interpretation of observed chemistry based on these E° data is not always straightforward, since the formation of a coherent surface film of metal oxide often renders a metal less reactive than expected (see _Section 19.4_). A few _d_-block metals are very powerful reducing agents, e.g. E° for the Sc^{3+}/Sc couple ($-2.08\,\mathrm{V}$) is more negative than that for Al^{3+}/Al ($-1.66\,\mathrm{V}$).

19.4 The reactivity of the metals

In Chapters 20 and 21 we shall look at individual elements of the _d_-block in detail. However, a few general points are given here as an overview. In general, the metals are moderately reactive and combine to give binary compounds when

Table 19.1 Standard reduction potentials (298 K) for some metals in the first long period; the concentration of each aqueous solution is $1\,\mathrm{mol\,dm^{-3}}$.

Reduction half-equation	E° / V
$Ca^{2+}(aq) + 2e^- \rightleftharpoons Ca(s)$	-2.87
$Ti^{2+}(aq) + 2e^- \rightleftharpoons Ti(s)$	-1.63
$V^{2+}(aq) + 2e^- \rightleftharpoons V(s)$	-1.18
$Cr^{2+}(aq) + 2e^- \rightleftharpoons Cr(s)$	-0.91
$Mn^{2+}(aq) + 2e^- \rightleftharpoons Mn(s)$	-1.19
$Fe^{2+}(aq) + 2e^- \rightleftharpoons Fe(s)$	-0.44
$Co^{2+}(aq) + 2e^- \rightleftharpoons Co(s)$	-0.28
$Ni^{2+}(aq) + 2e^- \rightleftharpoons Ni(s)$	-0.25
$Cu^{2+}(aq) + 2e^- \rightleftharpoons Cu(s)$	$+0.34$
$Zn^{2+}(aq) + 2e^- \rightleftharpoons Zn(s)$	-0.76

Table 19.2 The visible part of the electromagnetic spectrum.

Colour of light *absorbed*	Approximate wavelength ranges / nm	Colour of light *transmitted*, i.e. complementary colour of the absorbed light	In a 'colour wheel' representation, complementary colours are in opposite sectors
Red	700–620	Green	
Orange	620–580	Blue	
Yellow	580–560	Violet	
Green	560–490	Red	
Blue	490–430	Orange	
Violet	430–380	Yellow	

heated with dioxygen, sulfur or the halogens (e.g. reactions 19.1–19.3), product stoichiometry depending, in part, on the available oxidation states (see below). Combination with H_2, B, C or N_2 may lead to interstitial hydrides (*Section 9.7*), borides (*Section 12.10*), carbides (*Section 13.7*) or nitrides (*Section 14.6*).

$$Os + 2O_2 \xrightarrow{\Delta} OsO_4 \qquad (19.1)$$

$$Fe + S \xrightarrow{\Delta} FeS \qquad (19.2)$$

$$V + \frac{n}{2}X_2 \xrightarrow{\Delta} VX_n \quad (X = F, n = 5; X = Cl, n = 4;$$

$$X = Br, I, n = 3) \quad (19.3)$$

Most *d*-block metals should, on thermodynamic grounds (e.g. Table 19.1), liberate H_2 from acids but, in practice, many do not since they are passivated by a thin surface coating of oxide or by having a high dihydrogen overpotential, or both. The platinum-group metals, silver, gold and mercury (i.e. late, second and third row metals) are, even in the thermodynamic sense, the least reactive metals known; for example, gold is not oxidized by atmospheric O_2 or attacked by acids, except by a 3:1 mixture of concentrated HCl and HNO_3 (*aqua regia*).

19.5 Characteristic properties: a general perspective

In this section, we introduce properties that are characteristic of *d*-block metal compounds. More detailed discussion follows in Chapter 20.

Colour

The colours of *d*-block metal compounds are a characteristic feature of species with ground state electronic configurations other than d^0 and d^{10}. For example,

$[Cr(H_2O)_6]^{2+}$ is sky-blue, $[Mn(H_2O)_6]^{2+}$ very pale pink, $[Co(H_2O)_6]^{2+}$ pink, $[MnO_4]^-$ intense purple and $[CoCl_4]^{2-}$ dark blue. In contrast, salts of Sc(III) (d^0) or Zn(II) (d^{10}) are colourless. The fact that many of the observed colours are of *low intensity* is consistent with the colour originating from electronic '*d–d*' transitions. If we were dealing with an isolated gas phase ion, such transitions would be forbidden by the Laporte selection rule (equation 19.4 where *l* is the orbital quantum number); the pale colours indicate that the probability of a transition occurring is low. Table 19.2 shows relationships between the wavelength of light absorbed and observed colours.

$$\Delta l = \pm 1 \qquad \text{(Laporte selection rule)} \qquad (19.4)$$

The intense colours of species such as $[MnO_4]^-$ have a different origin, namely *charge transfer* absorptions or emissions (see *Section 16.4*). The latter are *not* subject to selection rule 19.4 and are always more intense than electronic transitions between different *d* orbitals. We return to selection rules in *Section 20.6*.

Paramagnetism

The occurrence of *paramagnetic* (see *Sections 20.1* and *20.8*, and the end of *Section 1.12*) compounds of *d*-block metals is common and arises from the presence of unpaired electrons. This phenomenon can be investigated using electron spin resonance (ESR) spectroscopy; it also leads to signal broadening and anomalous chemical shift values in NMR spectra.

Complex formation

d-Block metal ions readily form complexes, with complex formation often being accompanied by a change in colour and sometimes a change in the intensity of colour. Equation 19.5 shows the effect of adding concentrated HCl to aqueous cobalt(II) ions.

$$\underset{\text{pale pink}}{[Co(H_2O)_6]^{2+}} + 4Cl^- \longrightarrow \underset{\text{dark blue}}{[CoCl_4]^{2-}} + 6H_2O \qquad (19.5)$$

Table 19.3 Oxidation states of the *d*-block metals; the most stable states are marked in blue. Tabulation of zero oxidation states refers to its appearance in *compounds* of the metal. In organometallic compounds, oxidation states of less than zero are encountered (see *Chapter 23*). An oxidation state enclosed in [ ] is rare.

Sc	Ti	V	Cr	Mn	Fe	Co	Ni	Cu	Zn
	0	0	0	0	0	0	0	[0]	
		1	1	1	1	1	1	1	
	2	2	2	2	2	2	2	2	2
3	3	3	3	3	3	3	3	3	
	4	4	4	4	4	4	4	[4]	
		5	5	5					
			6	6	6				
				7					

Y	Zr	Nb	Mo	Tc	Ru	Rh	Pd	Ag	Cd
			0	0	0	0	0		
				1		1		1	
	2	2	2	[2]	2	2	2	2	2
3	3	3	3	[3]	3	3		3	
	4	4	4	4	4	4	4		
		5	5	5	5	5			
			6	6	6	6			
				7	7				
					8				

La	Hf	Ta	W	Re	Os	Ir	Pt	Au	Hg
			0	0	0	0	0	[0]	
				1		1		1	1
	2	2	2	2	2	2	2	[2]	2
3	3	3	3	3	3	3		3	
	4	4	4	4	4	4	4		
		5	5	5	5	5	5	5	
			6	6	6	6	6		
				7	7				
					8				

The formation of such complexes is analogous to the formation of those of *s*- and *p*-block metals and discussed in previous chapters, e.g. $[K(18\text{-crown-6})]^+$, $[Be(H_2O)_4]^{2+}$, *trans*-$[SrBr_2(py)_5]$, $[AlF_6]^{3-}$, $[SnCl_6]^{2-}$ and $[Bi_2(C_6H_4O_2)_4]^{2-}$.

Variable oxidation states

The occurrence of variable oxidation states and, often, the interconversion between them, is a characteristic of most *d*-block metals; exceptions are in groups 3 and 12 as Table 19.3 illustrates. A comparison between the available oxidation states for a given metal and the electronic configurations listed in *Table 1.3* is instructive. As expected, metals that display the greatest number of different oxidation states occur in or near the middle of a *d*-block row. Two cautionary notes should be made:

- The apparent oxidation state deduced from a molecular or empirical formula may be misleading, e.g. LaI_2 is a metallic conductor and is best formulated as $La^{3+}(I^-)_2(e^-)$, and $MoCl_2$ contains metal cluster units with metal–metal bonds and is formally $[Mo_6Cl_8]^{4+}(Cl^-)_4$. Indeed, metal–metal bond formation becomes more important for the heavier metals.

- There are many metal compounds in which it is impossible to assign oxidation states unambiguously, e.g. in the complexes $[Ti(bpy)_3]^{n-}$ ($n = 0$, 1, 2), there is evidence that the negative charge is localized on the bpy ligands (see *Table 6.7*) not the metal centres, and in nitrosyl complexes, the NO ligand may donate 1 or 3 electrons (see *Section 20.4*).

19.6 Electroneutrality principle

Pauling's *electroneutrality principle* is an *approximate method* of estimating the charge distribution in molecules and complex ions. It states that the distribution of charge in a molecule or ion is such that the charge on any single atom is within the range +1 to −1 (ideally close to zero).

Let us consider the complex ion $[Co(NH_3)_6]^{3+}$; Figure 19.3a gives a representation of the complex which indicates that the coordinate bonds are formed by lone pair donation from the ligands to Co(III) centre. It implies transfer of

Fig. 19.3 The complex cation $[Co(NH_3)_6]^{3+}$: (a) a conventional diagram showing the donation of lone pairs of electrons from ligand to metal ion; (b) the charge distribution that results from a 100% covalent model of the bonding; (c) the charge distribution that results from a 100% ionic model of the bonding; (d) the approximate charge distribution that results from applying the electroneutrality principle.

charge from ligand to metal, and Figure 19.3b shows the resulting charge distribution. This is clearly unrealistic, since the cobalt(III) centre becomes more negatively charged than would be favourable given its electropositive nature. At the other extreme, we could consider the bonding in terms of a wholly ionic model (Figure 19.3c): the 3+ charge remains localized on the cobalt ion and the six NH_3 ligands remain neutral. However, this model is also flawed; experimental evidence shows that the $[Co(NH_3)_6]^{3+}$ complex ion remains as an entity in aqueous solution, and the electrostatic interactions implied by the ionic model are unlikely to be strong enough to allow this to happen. Thus, neither of the extreme bonding models is appropriate.

If we now apply the electroneutrality principle to $[Co(NH_3)_6]^{3+}$, then, ideally, the net charge on the metal centre should be zero. That is, the Co^{3+} ion may accept a total of *only three electrons* from the six ligands, thus giving the charge distribution shown in Figure 19.3d. The electroneutrality principle results in a bonding description for the $[Co(NH_3)_6]^{3+}$ ion which is 50% ionic (or, 50% covalent).

19.7 Coordination numbers

In this section, we give an overview of the coordination numbers and geometries found within *d*-block metal compounds. It is impossible to give a comprehensive account, and several points should be borne in mind:

- most examples in this section involve mononuclear complexes, and in complexes with more than one metal centre, structural features are often conveniently described in terms of individual metal centres;
- although coordination environments are often described in terms of *regular* geometries such as those in Table 19.4, in practice they are often distorted, for example as a consequence of steric effects;
- detailed discussion of a particular geometry usually involves bond lengths and angles determined in the solid state and these may be affected by crystal packing forces;
- where the energy difference between different possible structures is small (e.g. for 5- and 8-coordinate complexes),

Table 19.4 Coordination geometries; each describes the arrangement of the donor atoms that surround the metal centre. Note that for some coordination numbers, more than one possible arrangement of donor atoms exists.

Coordination number	Arrangement of donor atoms around metal centre	Less common arrangements
2	Linear	
3	Trigonal planar	Trigonal pyramidal
4	Tetrahedral; square planar	
5	Trigonal bipyramidal; square-based pyramidal	
6	Octahedral	Trigonal prismatic
7	Pentagonal bipyramidal	Monocapped trigonal prismatic; monocapped octahedral
8	Dodecahedral; square antiprismatic; hexagonal bipyramidal	Cube; bicapped trigonal prismatic
9	Tricapped trigonal prismatic	

fluxional behaviour in solution may be observed; the small energy difference may also lead to the observation of different structures in the solid state, e.g. in salts of $[Ni(CN)_5]^{3-}$ the shape of the anion depends upon the cation present and in $[Cr(en)_3][Ni(CN)_5]\cdot1.5H_2O$, *both* trigonal bipyramidal and square-based pyramidal structures are present.

We shall not be concerned with ionic lattices in this section. Almost all the examples we discuss involve mononuclear species in which the metal centre is covalently bonded to the atoms in the coordination sphere. The metal–ligand bonding in complexes can generally be considered in terms of σ-donor ligands interacting with a metal centre which acts as a σ-acceptor. This may, in some complexes, be augmented with interactions involving π-donor ligands (with the metal as a π-acceptor) or π-acceptor ligands (with the metal as a π-donor). For a preliminary discussion of stereochemistry, it is not necessary to detail the metal–ligand bonding but we shall find it useful to draw attention to the electronic configuration of the metal centre; the reasons for this will become clear later in the chapter.

The Kepert model

For many years after the classic work of Werner which laid the foundations for the correct formulation of *d*-block metal complexes, it was assumed that a metal in a given oxidation state would have a fixed coordination number and geometry. In the light of the success (albeit not universal success) of VSEPR theory in predicting the shapes of molecular species of the *p*-block elements (see *Section 1.19*), we might reasonably expect the structures of the complex ions $[V(H_2O)_6]^{3+}$ (d^2), $[Mn(H_2O)_6]^{3+}$ (d^4), $[Co(H_2O)_6]^{3+}$ (d^6), $[Ni(H_2O)_6]^{2+}$ (d^8) and $[Zn(H_2O)_6]^{2+}$ (d^{10}) to vary as the electronic configuration of the metal ion changes. However, each of these species has an octahedral arrangement of ligands (**19.1**). Thus, it is clear that VSEPR theory is not applicable to *d*-block metal complexes.

We turn instead to the *Kepert model* in which the metal lies at the centre of a sphere and the ligands are free to move over the surface of the sphere. The ligands are considered to repel one another in a similar manner to the point charges in the VSEPR model; however, unlike the VSEPR model, that of Kepert *ignores non-bonding electrons*. Thus, the coordination geometry of a *d*-block species is considered by Kepert to be *independent* of the ground state electronic configuration of the metal centre, and so ions of type $[ML_n]^{m+}$ and $[ML_n]^{m-}$ have the *same* coordination geometry.

The Kepert model rationalizes the shapes of *d*-block metal complexes $[ML_n]$, $[ML_n]^{m+}$ or $[ML_n]^{m-}$ by considering the repulsions between the groups L. Lone pairs of electrons are ignored. For coordination numbers between two and six, the following arrangements of donor atoms are predicted:

two	linear
three	trigonal planar
four	tetrahedral
five	trigonal bipyramidal *or* square-based pyramidal
six	octahedral

Table 19.4 lists coordination environments associated with coordination numbers between two and nine; not all are predictable using the Kepert model. For example, after considering the repulsions between the cyano ligands in $[Cu(CN)_3]^{2-}$, the coordination sphere would be predicted to be trigonal planar (**19.2**). Indeed, this is what is found experimentally. The other option in Table 19.4 is trigonal pyramidal, but this does not minimize interligand repulsions. One of the most important classes of structure for which the Kepert model does not predict the correct answer is that of the square planar complex, and here electronic effects are usually the controlling factor as we discuss in *Section 20.3*. Another factor that may lead to a breakdown of the Kepert model is the inherent constraint of a ligand. For example:

- the four nitrogen donor atoms of a porphyrin ligand (*Figure 11.8a*) are confined to a square planar array;
- *tripodal ligands* such as **19.3** have limited flexibility which means that the donor atoms are not necessarily free to adopt the positions predicted by Kepert;
- macrocyclic ligands (see *Section 10.8*) are less flexible than open chain ligands.

(19.2)   (19.3)

e.g. X = CMe, N, P

A *tripodal* ligand (e.g. **19.3**) is one containing three arms, each with a donor atom, which radiate from a central atom or group; this central point may itself be a donor atom.

In the remaining part of this section, we give a systematic outline of the occurrence of different coordination numbers and geometries in *solid state d*-block metal complexes. A general word of caution: molecular formulae can be misleading in terms of coordination number, for example in the case of a lattice structure such as CdI_2 (*Figure 5.22*) or in molecular halides or pseudo-halides (e.g. $[CN]^-$) which may contain

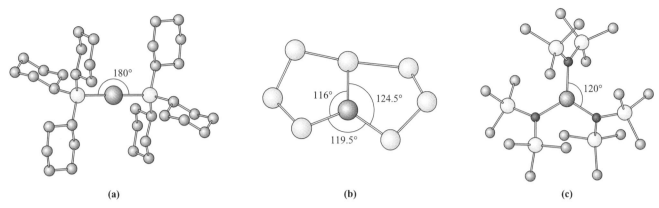

Fig. 19.4 Examples of 2- and 3-coordinate structures (X-ray diffraction data): (a) $[Au\{P(cyclo\text{-}C_6H_{11})_3\}_2]^+$ in the chloride salt [J.A. Muir *et al.* (1985) *Acta Crystallogr.*, *Sect. C*, vol. 41, p. 1174], (b) $[AgTe_7]^{3-}$ in the salt $[Et_4N][Ph_4P]_2[AgTe_7]$ [J.M. McConnachie *et al.* (1993) *Inorg. Chem.*, vol. 32, 3201], (c) $[Fe\{N(SiMe_3)_2\}_3]$ [M.B. Hursthouse *et al.* (1972) *J. Chem. Soc.*, *Dalton Trans.*, p. 2100]. Hydrogen atoms are omitted for clarity; colour code: Fe, Ag and Au, dark blue; C, grey; P and Si, pale grey; Te, pale blue; N, black.

M−X−M bridges and produce oligomers, e.g. α-$PdCl_2$ is polymeric (**19.4**).

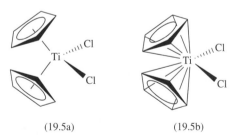

(19.4)

A further ambiguity arises when the bonding mode of a ligand can be described in more than one way. This often happens in organometallic chemistry, for example with cyclopentadienyl ligands as discussed in *Chapter 18*. The nomenclature introduced in *Box 18.1* assists, but there is still the question of whether to consider, for example, an $[\eta^5\text{-}C_5H_5]^-$ ligand as occupying one or five sites in the coordination sphere of a metal atom: thus, the coordination number of the Ti(IV) centre in $[(\eta^5\text{-}C_5H_5)_2TiCl_2]$ may be represented as either **19.5a** or **19.5b**.

(19.5a)   (19.5b)

Coordination number two

Examples of coordination number two are uncommon, being generally restricted to Cu(I), Ag(I), Au(I) and Hg(II), all d^{10} ions. Examples include $[CuCl_2]^-$, $[Ag(NH_3)_2]^+$, $[Au(CN)_2]^-$, $(R_3P)AuCl$, $[Au(PR_3)_2]^+$ (R = alkyl or aryl, Figure 19.4a) and $Hg(CN)_2$, in each of which the metal centre is in a linear environment. However, in the solid state, the Cu(I) centre in $K[Cu(CN)_2]$ is 3-coordinate by virtue of cyano-bridge formation. Bulky amido ligands,

e.g. $[N(SiR_3)_2]^-$, are often associated with *low* coordination numbers for metal centres which usually prefer to be surrounded by a greater number of donor atoms, e.g. $[Fe\{N(SiMePh_2)_2\}_2]$ in which $\angle N\text{-}Fe\text{-}N = 169°$.

Coordination number three

Three-coordinate complexes are not common. Usually, trigonal planar structures are observed, and examples involving d^{10} metal centres include:

- Cu(I) in $[Cu(CN)_3]^{2-}$ (**19.2**), $[Cu(CN)_2]^-$ (see above), $[Cu(SPMe_3)_3]^+$;
- Ag(I) in $[AgTe_7]^{3-}$ (Figure 19.4b), $[Ag(PPh_3)_3]^+$;
- Au(I) in $[Au\{PPh(C_6H_{11})_2\}_3]^+$;
- Hg(II) in $[HgI_3]^-$, $[Hg(SPh_3)_3]^-$;
- Pt(0) in $[Pt(PPh_3)_3]$, $[Pt(P^tBu_2H)_3]$.

Sterically demanding amido ligands have been used to stabilize complexes containing 3-coordinate metal ions, e.g. $[Fe\{N(SiMe_3)_2\}_3]$ (Figure 19.4c). In the solid state, $[Y\{N(SiMe_3)_2\}_3]$ and $[Sc\{N(SiMe_3)_2\}_3]$ possess *trigonal pyramidal* metal centres ($\angle N\text{-}Y\text{-}N = 115°$ and $\angle N\text{-}Sc\text{-}N = 115.5°$), but it is likely that crystal packing effects cause the deviation from planarity. The fact that in the gas phase $[Sc\{N(SiMe_3)_2\}_3]$ contains a trigonal planar Sc(III) centre tends to support this proposal.

p-Block chemistry has a number of examples of T-shaped molecules (e.g. ClF_3) in which stereochemically active lone pairs play a crucial role. *d*-Block metal complexes do not mimic this behaviour, although ligand constraints (e.g. the bite angle of a chelate) may distort a 3-coordinate structure away from the expected trigonal planar structure.

Coordination number four

Four-coordinate complexes are extremely common, with a tetrahedral array of donor atoms being the most frequently observed. The tetrahedron is sometimes 'flattened',

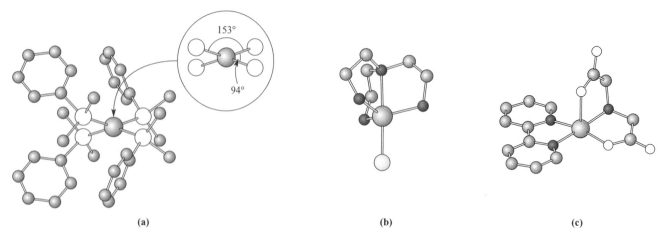

Fig. 19.5 Examples of 4- and 5-coordinate structures (X-ray diffraction data): (a) in $[Rh(PMe_2Ph)_4]^+$, the steric demands of the ligands distort the structure from the square planar structure expected for this d^8 metal centre [J.H. Reibenspies *et al.* (1993) *Acta Crystallogr., Sect. C*, vol. 49, p. 141], (b) $[Zn\{N(CH_2CH_2NH_2)_3\}Cl]^+$ in the $[Ph_4B]^-$ salt [R.J. Sime *et al.* (1971) *Inorg. Chem.*, vol. 10, p. 537], and (c) $[Cu(bpy)\{NH(CH_2CO_2)_2\}]$, crystallized as the hexahydrate [R.E. Marsh *et al.* (1995) *Acta Crystallogr., Sect. B*, vol. 51, p. 300]. Hydrogen atoms are omitted for clarity; colour code: Rh, Zn and Cu, dark blue; C, grey; P and O, white; Cl, pale blue; N, black.

distortions often being attributed to steric or crystal packing effects or, in some cases, electronic effects. Tetrahedral complexes for d^3 ions are not yet known, and for d^4 ions, have only been stabilized with bulky amido ligands, e.g. $[M(NPh_2)_4]$ and $[M\{N(SiMe_3)_2\}_3Cl]$ for M = Hf or Zr. Simple tetrahedral species include:

- d^0: $[VO_4]^{3-}$, $[CrO_4]^{2-}$, $[MoS_4]^{2-}$, $[WS_4]^{2-}$, $[MnO_4]^-$, $[TcO_4]^-$, RuO_4, OsO_4;
- d^1: $[MnO_4]^{2-}$, $[TcO_4]^{2-}$, $[ReO_4]^{2-}$, $[RuO_4]^-$;
- d^2: $[FeO_4]^{2-}$, $[RuO_4]^{2-}$;
- d^5: $[FeCl_4]^-$, $[MnCl_4]^{2-}$;
- d^6: $[FeCl_4]^{2-}$, $[FeI_4]^{2-}$;
- d^7: $[CoCl_4]^{2-}$;
- d^8: $[NiCl_4]^{2-}$, $[NiBr_4]^{2-}$;
- d^9: $[CuCl_4]^{2-}$ (distorted);
- d^{10}: $[ZnCl_4]^{2-}$, $[HgBr_4]^{2-}$, $[CdCl_4]^{2-}$, $[Zn(OH)_4]^{2-}$, $[Cu(CN)_4]^{3-}$, $[Ni(CO)_4]$.

The solid state structures of apparently simple anions may in fact be polymeric (e.g. the presence of fluoride bridges in $[CoF_4]^{2-}$ and $[NiF_4]^{2-}$ leads to a polymer with octahedral metal centres) or may be cation-dependent (e.g. discrete tetrahedral $[MnCl_4]^{2-}$ ions are present in the Cs^+ and $[Me_4N]^+$ salts, but a polymeric structure with Mn−Cl−Mn bridges is adopted by the Na^+ salt).

Square planar complexes are rarer than tetrahedral, and are often associated with d^8 configurations where electronic factors strongly favour a square planar arrangement (see *Section 20.3*), e.g. $[PdCl_4]^{2-}$, $[PtCl_4]^{2-}$, $[AuCl_4]^-$, $[AuBr_4]^-$, $[RhCl(PPh_3)_3]$ and *trans*-$[IrCl(CO)(PPh_3)_2]$. The classification of distorted structures like those in $[Ir(PMePh_2)_4]^+$ and $[Rh(PMe_2Ph)_4]^+$ (Figure 19.5a) may be ambiguous, but in this case, the fact that each metal ion is d^8 suggests that steric crowding causes deviation from a square

planar arrangement (not from a tetrahedral one). The $[Co(CN)_4]^{2-}$ ion is a rare example of a square planar d^7 complex.

Coordination number five

The limiting structures for 5-coordination are the trigonal bipyramid and square-based pyramid. In practice, many structures lie between these two extremes, and we have already emphasized that the energy differences between trigonal bipyramidal and square-based pyramidal structures are often small (see *Section 2.11*). Among simple 5-coordinate complexes are trigonal bipyramidal $[CdCl_5]^{3-}$, $[HgCl_5]^{3-}$ and $[CuCl_5]^{3-}$ (d^{10}) and a series of square-based pyramidal oxo- or nitrido-complexes in which the oxo or nitrido ligand occupies the axial site:

- d^0: $[NbCl_4(O)]^-$;
- d^1: $[V(acac)_2(O)]$, $[WCl_4(O)]^-$ **(19.6)**, $[TcCl_4(N)]^-$ **(19.7)**, $[TcBr_4(N)]^-$;
- d^2: $[TcCl_4(O)]^-$, $[ReCl_4(O)]^-$.

$$\left[\begin{array}{c} O \\ \| \\ Cl\ \substack{\prime\prime\prime\prime\prime} \overset{}{\underset{\displaystyle Cl}{W}} \substack{\cdots\prime\prime\prime} Cl \\ Cl \end{array} \right]^-$$

(19.6)

$$\left[\begin{array}{c} N \\ \||\| \\ Cl\ \substack{\prime\prime\prime\prime} \overset{}{\underset{\displaystyle Cl}{Tc}} \substack{\cdots\prime\prime\prime} Cl \\ Cl \end{array} \right]^-$$

(19.7)

The formulae of some complexes may misleadingly suggest '5-coordinate' metal centres: e.g. Cs_3CoCl_5 is actually $Cs_3[CoCl_4]Cl$.

Five-coordinate structures are found for many compounds with polydentate amine, phosphine or arsine ligands. Of particular interest among these are complexes containing tripodal ligands **(19.3)** in which the central atom is a donor atom;

this makes the ligand ideally suited to occupy one axial and the three equatorial sites of a trigonal bipyramidal complex as in $[CoBr\{N(CH_2CH_2NMe_2)_3\}]^+$, $[Rh(SH)\{P(CH_2CH_2PPh_2)_3\}]$ and $[Zn\{N(CH_2CH_2NH_2)_3\}Cl]^+$ (Figure 19.5b). On the other hand, the conformational constraints of the ligands may result in a preference for a square-based pyramidal complex in the solid state, e.g. $[Cu(bpy)\{NH(CH_2CO_2)_2\}]\cdot 6H_2O$ (Figure 19.5c).

Coordination number six

For many years after Werner's proof from stereochemical studies that many 6-coordinate complexes of chromium and cobalt had octahedral structures (see *Box 21.8*), it was believed that no other form of 6-coordination occurred, and a vast amount of data from X-ray diffraction studies seemed to support this. Eventually, however, examples of trigonal prismatic coordination were confirmed.

The regular or nearly regular octahedral coordination sphere is found for all electronic configurations from d^0 to d^{10}, e.g. $[TiF_6]^{2-}$ (d^0), $[Ti(H_2O)_6]^{3+}$ (d^1), $[V(H_2O)_6]^{3+}$ (d^2), $[Cr(H_2O)_6]^{3+}$ (d^3), $[Mn(H_2O)_6]^{3+}$ (d^4), $[Fe(H_2O)_6]^{3+}$ (d^5), $[Fe(H_2O)_6]^{2+}$ (d^6), $[Co(H_2O)_6]^{2+}$ (d^7), $[Ni(H_2O)_6]^{2+}$ (d^8), $[Cu(NO_2)_6]^{4-}$ (d^9) and $[Zn(H_2O)_6]^{2+}$ (d^{10}). There are distinctions between what we later term *low-spin* and *high-spin* complexes (see *Section 20.1*): where the distinction is meaningful, the examples listed are high-spin complexes, but many octahedral low-spin complexes are also known, e.g. $[Mn(CN)_6]^{3-}$ (d^4), $[Fe(CN)_6]^{3-}$ (d^5), $[Co(CN)_6]^{3-}$ (d^6). Octahedral complexes of d^4 and d^9 metal ions tend to be *tetragonally distorted*, i.e. they are elongated or squashed; this is an electronic effect called *Jahn–Teller distortion* (see *Section 20.3*).

R—C—S⁻
‖
R—C—S⁻

e.g. R = H, Ph

(19.8)

Examples of trigonal prismatic complexes are rare, and include the organometallic d^0 complexes $[WMe_6]$ (Figure 19.6a), $[ReMe_6]$, $[NbMe_6]^-$, $[TaMe_6]^-$ and $[ZrMe_6]^{2-}$. It is difficult to understand this structural preference, but theoretical studies indicate that the origins are electronic. Trigonal prismatic coordination is also favoured in a series of metal d^0, d^1 or d^2 complexes involving ligands **19.8**, e.g. $[Mo(S_2C_2H_2)_3]$ and $[Re(S_2C_2Ph_2)_3]$ (Figure 19.6b). In this case, it has been suggested that the trigonal prism is favoured over the octahedron because it brings the sulfur atoms closer together, allowing a degree of S–S interaction. The relationship between the octahedron and trigonal prism is readily understood in terms of the mutual positions of two triangles as shown in **19.9** (staggered triangles give an octahedron) and **19.10** (eclipsed triangles give a trigonal prism). Thus,

Fig. 19.6 The trigonal prismatic structures of (a) $[WMe_6]$ [V. Pfennig *et al.* (1996) *Science*, vol. 271, p. 626] and (b) $[Re(S_2C_2Ph_2)_3]$, only the *ipso*-C atoms of each Ph ring are shown [R. Eisenberg *et al.* (1966) *Inorg. Chem.*, vol. 5, p. 411]. Hydrogen atoms are omitted from (b); colour code: W and Re, blue; C, grey; S, pale grey; H, black.

there are some complexes that lie part way along this interconversion pathway. An interesting example is $[WL_3]$ where L is ligand **(19.11)**; the structure shows a twist angle of 15° **(19.12)**. At first glance, this appears to be an example of an approximately trigonal prismatic d^6 complex, but theoretical studies suggest that charge localization is such that the ligands are better considered as L^{2-} giving a formally d^0 tungsten centre.[†]

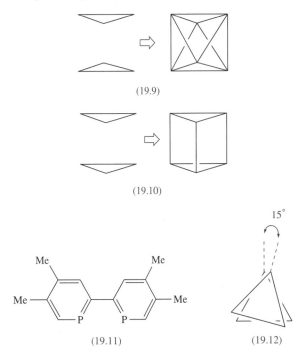

(19.9)

(19.10)

Me — Me

Me — P P — Me

(19.11)

15°

(19.12)

Coordination number seven

High coordination numbers (≥ 7) are observed most frequently for ions of the early, second and third row *d*-block

[†] For a detailed discussion, see: P. Rosa, L. Ricard, P. Le Floch, F. Mathey, G. Sini and O. Eisenstein (1998) *Inorganic Chemistry*, vol. 37, p. 3154.

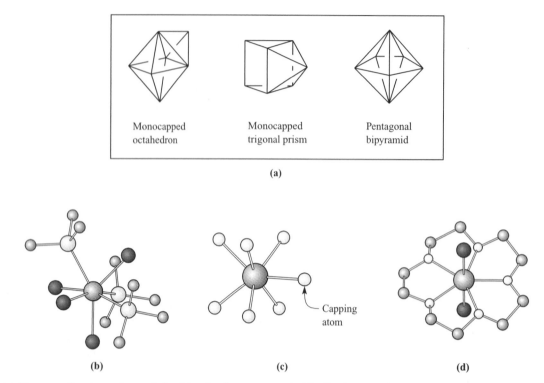

Fig. 19.7 (a) The coordination spheres defined by the donor atoms in idealized 7-coordinate structures. Examples of 7-coordinate complexes (X-ray diffraction data): (b) the capped octahedral structure of [TaCl$_4$(PMe$_3$)$_3$] [F.A. Cotton *et al.* (1984) *Inorg. Chem.*, vol. 23, p. 4046], (c) the capped trigonal prismatic [ZrF$_7$]$^{3-}$ in the guanidinium salt [A.V. Gerasimenko *et al.* (1985) *Koord. Khim.*, vol. 11, p. 566], and (d) the pentagonal bipyramidal cation in [ScCl$_2$(15-crown-5)]$_2$[CuCl$_4$] with the crown ether occupying the equatorial plane [N.R. Strel'tsova *et al.* (1992) *Zh. Neorg. Khim.*, vol. 37, p. 1822]. Hydrogen atoms have been omitted for clarity; colour code: Ta, Zr and Sc, dark blue; C, grey; P, pale blue; Cl, black; F and O, pale grey.

metals and for the lanthanoids and actinoids, i.e. r_{cation} must be relatively large (see *Chapter 24*). Figure 19.7a shows the arrangement of the donor atoms for the three idealized 7-coordinate structures; note that in the capped trigonal prism, the 'cap' is over one of the square faces of the prism. In reality, there is much distortion from these idealized structures, and this is readily apparent for the example of a capped octahedral complex shown in Figure 19.7b. A problem in the chemical literature is that the distortions may lead to ambiguity in the way in which a given structure is described. Among binary metal halides and pseudo-halides, 7-coordinate structures are exemplified by the pentagonal bipyramidal ions [V(CN)$_7$]$^{4-}$ (d^2) and [NbF$_7$]$^{3-}$ (d^1). In the ammonium salt, [ZrF$_7$]$^{3-}$ (d^0) is pentagonal bipyramidal, but in the guanidinium salt, it has a monocapped trigonal prismatic structure (Figure 19.7c). Further examples of monocapped trigonal prismatic complexes are [NbF$_7$]$^{2-}$ and [TaF$_7$]$^{2-}$ (d^0). Seven-coordinate complexes containing oxo ligands may favour pentagonal bipyramidal structures with the oxo group residing in an axial site, e.g. [Nb(O)(ox)$_3$]$^{3-}$, [Nb(O)(H$_2$O)$_2$(ox)$_2$]$^-$ and [Mo(O)(O$_2$)$_2$(ox)]$^{2-}$ (all d^0). In this last example, two peroxo ligands are present, each in an η^2-mode (**19.13**). Macrocyclic ligands containing five donor atoms

(e.g. 15-crown-5) may dictate that the coordination geometry is pentagonal bipyramidal as shown in Figure 19.7d.

$$\left[\begin{array}{c} O \\ \| \\ O \\ \| \\ O\text{—Mo}\cdots\text{O} \\ O\text{=C} \quad\quad O \quad\quad O \\ \quad\quad\quad\quad O \quad\quad O \\ \quad\quad\quad\quad C\text{—O} \\ \quad\quad\quad\quad \| \\ \quad\quad\quad\quad O \end{array} \right]^{2-}$$

(19.13)

Coordination number eight

As the number of vertices in a polyhedron increases, so does the number of possible structures (Figure 19.8a). Probably, the best known eight-vertex polyhedron is the cube, (**19.14**), but this is hardly ever observed as an arrangement of donor atoms in complexes. The few examples include the anions in the actinoid complexes Na$_3$[PaF$_8$], Na$_3$[UF$_8$] and [Et$_4$N]$_4$[U(NCS-*N*)$_8$]. Steric hindrance between ligands

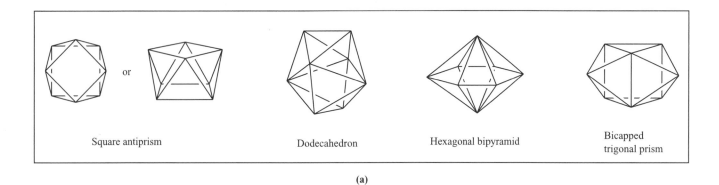

Square antiprism   Dodecahedron   Hexagonal bipyramid   Bicapped trigonal prism

(a)

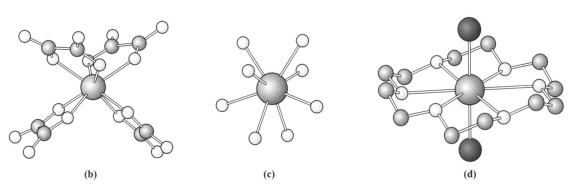

(b)   **(c)**   **(d)**

Fig. 19.8 (a) The coordination spheres defined by the donor atoms in idealized 8-coordinate structures; the left-hand drawing of the square antiprism emphasizes that the two square faces are mutually staggered. Examples of 8-coordinate complexes (X-ray diffraction): (b) the square antiprismatic structure of $[Nb(ox)_4]^{4-}$ in the salt $K_2[H_3NCH_2CH_2NH_3][Nb(ox)_4]\cdot4H_2O$ [F.A. Cotton *et al.* (1987) *Inorg. Chem.*, vol. 26, p. 2889]; (c) the dodecahedral ion $[Y(H_2O)_8]^{3+}$ in the salt $[Y(H_2O)_8]Cl_3\cdot(15\text{-crown-5})$ [R.D. Rogers *et al.* (1986) *Inorg. Chim. Acta*, vol. 116, p. 171]; and (d) $[CdBr_2(18\text{-crown-6})]$ with the macrocyclic ligand occupying the equatorial plane of a hexagonal bipyramid [A. Hazell (1988) *Acta Crystallogr.*, *Sect. C*, vol. 44, p. 88]. Hydrogen atoms have been omitted for clarity; colour code: Nb, Y and Cd, blue; C, grey; Br, black; O, pale grey.

can be reduced by converting a cubic into a square antiprismatic arrangement, i.e. on going from **19.14** to **19.15**.

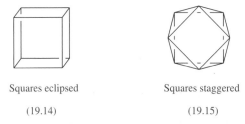

Squares eclipsed   Squares staggered

(19.14)   (19.15)

Square antiprismatic coordination environments occur in $[Zr(acac)_4]$ (d^0) and in the anions in the salts $Na_3[TaF_8]$ (d^0), $K_2[ReF_8]$ (d^1) and $K_2[H_3NCH_2CH_2NH_3][Nb(ox)_4]$ (d^1) (Figure 19.8b). Specifying the counter-ion is important since the energy difference between 8-coordinate structures tends to be small with the result that the preference between two structures may be altered by crystal packing forces in two different salts. Examples are seen in a range of salts of $[Mo(CN)_8]^{3-}$, $[W(CN)_8]^{3-}$, $[Mo(CN)_8]^{4-}$ or $[W(CN)_8]^{4-}$ which possess square antiprismatic or dodecahedral structures depending on the cation. Further examples of dodecahedral complexes include $[Y(H_2O)_8]^{3+}$ (Figure 19.8c) and a number of complexes with didentate ligands: $[Mo(O_2)_4]^{2-}$

(d^0), $[Ti(NO_3)_4]$ (d^0), $[Cr(O_2)_4]^{3-}$ (d^1), $[Mn(NO_3)_4]^{2-}$ (d^5) and $[Fe(NO_3)_4]^-$ (d^5).

The hexagonal bipyramid is a rare coordination environment, but may be favoured in complexes containing a hexadentate macrocyclic ligand, for example $[CdBr_2(18\text{-crown-}6)]$, Figure 19.8d. A bicapped trigonal prism is another option for 8-coordination, but is only rarely observed, e.g. in $[ZrF_8]^{4-}$ (d^0) and $[La(acac)_3(H_2O)_2]\cdot H_2O$ (d^0).

Coordination number nine

The anions $[ReH_9]^{2-}$ and $[TcH_9]^{2-}$ (both d^0) provide examples of 9-coordinate species in which the metal centre is in a tricapped trigonal prismatic environment (see *Figure 9.10c*). A coordination number of nine is most often associated with yttrium, lanthanum and the *f*-block elements; the tricapped trigonal prism is the only regular arrangement of donor atoms yet observed, e.g. in $[Sc(H_2O)_9]^{3+}$, $[Y(H_2O)_9]^{3+}$ and $[La(H_2O)_9]^{3+}$.

Coordination numbers of ten and above

It is always dangerous to draw conclusions on the basis of the non-existence of structure types, but, from data available at

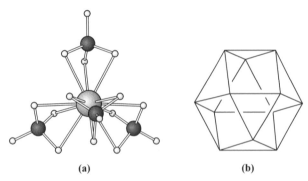

(a)   **(b)**

Fig. 19.9 (a) The structure of $[Hf(BH_4)_4]$ determined by neutron diffraction at low temperature [R.W. Broach *et al.* (1983) *Inorg. Chem.*, vol. 22, p. 1081]. Colour code: Hf, blue; B, black; H, pale grey. (b) The 12-vertex cubeoctahedral coordination sphere of the Hf(IV) centre in $[Hf(BH_4)_4]$.

Fig. 19.10 Classification of types of isomerism in metal complexes.

the present time, it seems that a coordination ≥ 10 is generally confined to the *f*-block metal ions (see *Chapter 24*). Complexes containing $[BH_4]^-$ and related ligands are an exception, e.g. in $[Hf(BH_4)_4]$ and $[Zr(MeBH_3)_4]$ the ligands are tridentate (see *structure* **12.5**) and the metal centres are 12-coordinate. Figure 19.9 shows the structure of $[Hf(BH_4)_4]$ and the cubeoctahedral arrangement of the 12 hydrogen atoms around the metal centre. The same coordination environment is found in $[Zr(MeBH_3)_4]$.

19.8 Isomerism in *d*-block metal complexes

In this book so far, we have not mentioned isomerism very often, and most references have been to *trans* and *cis* isomers, e.g. *trans*-$[CaI_2(THF)_4]$ (*Section 11.5*) and the *trans*- and *cis*-isomers of N_2F_2 (*Section 14.7*). These are *geometrical isomers*, and our previous discussion of this topic (see *Section 1.20*) will not be elaborated further here.

In this section, we shall be concerned with other types of isomerism exhibited by *d*-block metal complexes, and we use a classification that goes back to the work of Werner (Figure 19.10).

Structural isomerism: ionization isomers

Ionization isomers result from the interchange of an anionic ligand within the first coordination sphere with an anion outside the coordination sphere.

Examples are ionization isomers are violet $[Co(NH_3)_5Br][SO_4]$ (prepared by reaction scheme 19.6) and red $[Co(NH_3)_5(SO_4)]Br$ (prepared by reaction sequence 19.7). These isomers can be readily distinguished by appropriate qualitative tests for *ionic* sulfate or bromide, respectively. The isomers are also easily distinguished by

IR spectroscopy; free and coordinated sulfate ions give rise to one or three IR active S−O stretching vibrations respectively.

$$CoBr_2 \xrightarrow{[NH_4]Br,\ NH_3,\ O_2} [Co(NH_3)_5(H_2O)]Br_3$$

$$\downarrow \Delta$$

$$[Co(NH_3)_5Br]Br_2$$

$$\downarrow Ag_2SO_4$$

$$[Co(NH_3)_5Br][SO_4] \qquad (19.6)$$

$$[Co(NH_3)_5Br]Br_2 \xrightarrow{conc\ H_2SO_4} [Co(NH_3)_5(SO_4)][HSO_4]$$

$$\downarrow BaBr_2$$

$$[Co(NH_3)_5(SO_4)]Br \quad (19.7)$$

Structural isomerism: hydration isomers

Hydration isomers result from the interchange of H_2O and another ligand between the first coordination sphere and the ligands outside it.

The classic example of hydrate isomerism is that of the compound of formula $CrCl_3 \cdot 6H_2O$. Green crystals of chromium(III) chloride formed from a hot solution obtained by reducing chromium(VI) oxide with concentrated hydrochloric acid are $[Cr(H_2O)_4Cl_2]Cl \cdot 2H_2O$. When this is dissolved in water, the chloride ions in the complex are slowly replaced by water to give blue-green $[Cr(H_2O)_5Cl]Cl_2 \cdot H_2O$ and finally, violet $[Cr(H_2O)_6]Cl_3$. The complexes can be distinguished by precipitation of the *free* chloride ion using aqueous silver nitrate.

Structural isomerism: coordination isomerism

Coordination isomers are possible only for salts in which both cation and anion are complex ions; the isomers arise from interchange of ligands between the two metal centres.

Examples of coordination isomers are:

- $[Co(NH_3)_6][Cr(CN)_6]$ and $[Cr(NH_3)_6][Co(CN)_6]$;
- $[Co(NH_3)_6][Co(NO_2)_6]$ and $[Co(NH_3)_4(NO_2)_2][Co(NH_3)_2(NO_2)_4]$;
- $[Pt^{II}(NH_3)_4][Pt^{IV}Cl_6]$ and $[Pt^{IV}(NH_3)_4Cl_2][Pt^{II}Cl_4]$.

Structural isomerism: linkage isomerism

Linkage isomers may arise when one or more of the ligands can coordinate to the metal ion in more than one way, e.g. in $[SCN]^-$ (**19.16**), both the N and S atoms are potential donor sites.

$$\left[S = C = N \right]^-$$

(19.16)

Thus, the complex $[Co(NH_3)_5(NCS)]^{2+}$ has two isomers which are distinguished by using the following nomenclature:

- in $[Co(NH_3)_5(NCS-N)]^{2+}$, the thiocyanate ligand coordinates through the nitrogen donor atom;
- in $[Co(NH_3)_5(NCS-S)]^{2+}$, the thiocyanate ion is bonded to the metal centre through the sulfur atom.

Scheme 19.8 shows how linkage isomers of $[Co(NH_3)_5(NO_2)]^{2+}$ can be prepared.

$$[Co(NH_3)_5Cl]Cl_2 \xrightarrow{\text{dil } NH_3(aq)} [Co(NH_3)_5(H_2O)]Cl_3$$

$$\Bigg\downarrow NaNO_2 \qquad\qquad \Bigg\downarrow NaNO_2, \text{ conc HCl}$$

$$\underset{\text{red}}{[Co(NH_3)_5(NO_2-O)]Cl_2} \underset{UV}{\overset{\text{warm HCl or spontaneous}}{\rightleftharpoons}} \underset{\text{yellow}}{[Co(NH_3)_5(NO_2-N)]Cl_2}$$

(19.8)

In this example, the complexes $[Co(NH_3)_5(NO_2-O)]^{2+}$ and $[Co(NH_3)_5(NO_2-N)]^{2+}$ can be distinguished by using IR spectroscopy. For the *O*-bonded ligand, characteristic absorption bands at 1065 and $1470\,cm^{-1}$ are observed, while for the *N*-bonded ligand, the corresponding vibrational wavenumbers are 1310 and $1430\,cm^{-1}$.

Structural isomerism: polymerization isomerism

'Polymerization' isomerism is a rather unfortunate term since we are not dealing with polymeric structures *per se*.

Polymerization isomers denote complexes which have the same empirical formulae but different molecular masses.

Examples of polymerization isomers are:

- $[PtCl_2(NH_3)_2]$ and $[Pt(NH_3)_4][PtCl_4]$;
- $[Co(NH_3)_3(NO_2)_3]$ and $[Co(NH_3)_6][Co(NO_2)_6]$.

Stereoisomerism: geometrical isomers

Distinguishing between *cis*- and *trans*-isomers of a square planar complex or between *mer*- and *fac*-isomers of an octahedral complex is most unambiguously confirmed by structural determinations using single crystal X-ray diffraction. Vibrational spectroscopy (applications of which were introduced in *Section 3.7*) may also be of assistance. For example, Figure 19.11 illustrates that the asymmetric stretch for the $PtCl_2$ unit in $[Pt(NH_3)_2Cl_2]$ is IR active for both the *trans*- and *cis*-isomers, but the symmetric stretch is IR active only for the *cis*-isomer. In square planar complexes containing phosphine ligands, the ^{31}P NMR spectrum may be particularly diagnostic, as is illustrated in *Box 19.1*.

The existence of ions or molecules in different structures (e.g. trigonal bipyramidal and square-based pyramidal $[Ni(CN)_5]^{3-}$) is just a special case of geometrical isomerism. In the cases of, for example, tetrahedral and square planar $[NiBr_2(PBzPh_2)_2]$ (Bz = benzyl), the two forms can be distinguished by the fact that they exhibit different magnetic properties as we shall discuss in *Section 20.8*. To complicate matters, square planar $[NiBr_2(PBzPh_2)_2]$ may exist as either *trans*- or *cis*-isomers.

Stereoisomerism: optical isomers

Optical isomerism is concerned with *chirality*, and some important terms relating to chiral complexes are defined in Box 19.2. The simplest case of optical isomerism among *d*-block complexes involves a metal ion surrounded by three didentate ligands, for example $[Cr(acac)_3]$ or $[Co(en)_3]^{3+}$ (Figures 3.15b and 19.12). These are examples of *tris-chelate complexes*. Pairs of enantiomers such as Δ- and Λ-$[Cr(acac)_3]$ or Δ- and Λ-$[Co(en)_3]Cl_3$ differ only in their action on polarized light. However, for ionic complexes such as $[Co(en)_3]^{3+}$, there is the opportunity to form salts with a chiral counter-ion A^-. These salts now contain two different types of chirality: the Δ- or Λ-chirality at the metal centre and the (+) or (−) chirality of the anion. Four combinations are possible of which the pair {Δ-(+)} and {Λ-(−)} is enantiomeric as is the pair {Δ-(−)} and {Λ-(+)}. However, with a given anion chirality, the pair of salts {Δ-(−)} and {Λ-(−)} are diastereomers and may differ in the packing of the ions in the solid state, and separation by fractional crystallization is often possible.

Bis-chelate octahedral complexes such as $[Co(en)_2Cl_2]^+$ exist in both *cis* and *trans*-isomers depending on the arrangement of the chloro ligands. In addition, the *cis* isomer (but

Fig. 19.11 The *trans*- and *cis*-isomers of the square planar complex $[PtCl_2(NH_3)_2]$ can be distinguished by IR spectroscopy. The selection rule for an IR active vibration is that it must lead to a *change in molecular dipole moment* (see *Section 3.7*).

BOX 19.1   **CHEMICAL AND THEORETICAL BACKGROUND**

Trans- and *cis*-isomers of square planar complexes: an NMR spectroscopic probe

In Section 2.11 we described how *satellite peaks* may arise in some NMR spectra. In square planar platinum(II) complexes containing two phosphine (PR_3) ligands, the ^{31}P NMR spectrum of the complex provides valuable information about the *cis* or *trans* arrangement of the ligands. The isotope ^{195}Pt ($I = \frac{1}{2}$) constitutes 33.8% of naturally occurring platinum; in a ^{31}P NMR spectrum of a complex such as $[PtCl_2(PPh_3)_2]$, there is spin–spin coupling between the ^{31}P and ^{195}Pt nuclei which gives rise to satellite peaks.

If the PR_3 ligands are mutually *trans*, the value of $J_{PPt} \approx 2000\text{–}2500\,Hz$, but if the ligands are *cis*, the coupling constant is much larger, $\approx 3000\text{–}3500\,Hz$. While the values vary somewhat, comparison of the ^{31}P NMR spectra of *cis* and *trans* isomers of a given complex enables the isomers to be assigned. For example, for *cis*- and *trans*-$[PtCl_2(P^nBu_3)_2]$, values of J_{PPt} are 3508 and 2380 Hz, respectively; the figure shows a 162 MHz ^{31}P NMR spectrum of *cis*-$[PtCl_2(P^nBu_3)_2]$, simulated using experimental data; (the chemical shift reference is 85% aq H_3PO_4).

Similar diagnostic information can be obtained from NMR spectroscopy for square planar complexes containing metal centres with spin-active isotopes. For example, rhodium is *monotopic* (i.e. 100% of one isotope) with ^{103}Rh

having $I = \frac{1}{2}$. In square planar rhodium(I) complexes containing two phosphine ligands, values of J_{PRh} are $\approx 160\text{–}190\,Hz$ for a *cis*-arrangement and $\approx 70\text{–}90\,Hz$ for a *trans*-arrangement. Thus, the ^{31}P NMR spectrum of a complex of the type $[RhCl(PR_3)_2L]$ (L = neutral ligand) exhibits a *doublet* with a coupling constant characteristic of a particular isomer.

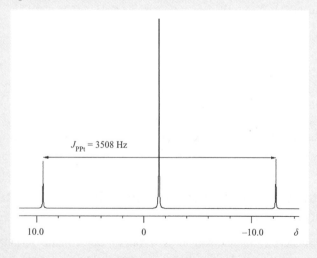

BOX 19.2   **CHEMICAL AND THEORETICAL BACKGROUND**

Definitions and notation for chiral complexes

Chirality was introduced in Section 3.8 and Box 3.2. Here, we add to this introduction and collect together some terms that are frequently encountered in discussing optically active complexes.

Enantiomers are a pair of stereoisomers that are non-superimposable mirror images.

Diastereomers are stereoisomers that are not an enantiomeric pair.

(+) and (−) prefixes: the specific rotation (see *Section 3.8*) of enantiomers is equal and opposite, and a useful means of distinguishing between enantiomers is to denote the *sign* of $[\alpha]_D$. Thus, if two enantiomers of a compound A have $[\alpha]_D$ values of $+12°$ and $-12°$, they are labelled (+)-A and (−)-A.

***d* and *l* prefixes**: sometimes (+) and (−) are denoted by *dextro-* and *laevo-* (derived from the Latin for right and left) and these refer to right- and left-handed rotation of the plane of polarized light respectively; *dextro* and *laevo* are generally abbreviated to *d* and *l*.

The $+/-$ or d/l notation has nothing to do with the *absolute configuration* of an enantiomer (the arrangement of the substituents or ligands) for which the following prefixes are used.

***R* and *S* prefixes**: the convention for labelling chiral carbon atoms (tetrahedral with four different groups attached) uses the Cahn–Ingold–Prelog notation. The four groups attached to the chiral carbon atom are prioritized according to the atomic number of the attached atoms, highest priority being assigned to highest atomic number, and the molecule then viewed down the C–X vector, where X has the lowest priority. The *R*- and *S*-labels for the enantiomers refer to a clockwise (rectus) and anticlockwise (sinister) sequence of the prioritized atoms, working from high to low. Example: CHClBrI, view down the C–H bond:

This notation is used for chiral organic ligands, and also for tetrahedral complexes.

Δ and Λ prefixes: enantiomers of octahedral complexes containing three equivalent didentate ligands (tris-chelate complexes) are among those which are distinguished using Δ (delta) and Λ (lambda) prefixes. The octahedron is viewed down a 3-fold axis, and the chelates then define either a right- or left-handed helix. The enantiomer with

right-handedness is labelled Δ, and that with left-handedness is Λ.

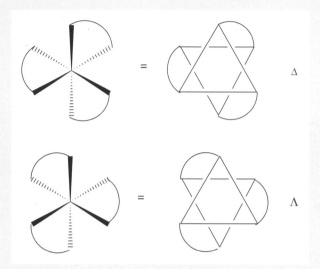

δ and λ-prefixes: actually, the situation with chelating ligands is often more complicated. Consider the formation of a 1,2-diaminoethane ring. The 5-membered ring is not planar but adopts an envelope conformation. This is most easily seen by taking a Newman projection along the C–C bond of the ligand; clearly, two enantiomers are possible and are distinguished by the prefixes δ and λ.

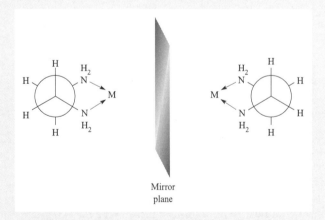

Mirror plane

Depending upon the metal ion and the number and nature of chelate rings, these enantiomers may sometimes be separable giving additional complications.

For detailed information on the nomenclature of chiral complexes, see:

IUPAC: *Nomenclature of Inorganic Chemistry (Recommendations 1990)*, ed. G.J. Leigh, Blackwell Scientific Publications, Oxford, p. 182.

A. von Zelewsky (1996) *Stereochemistry of Coordination Compounds*, Wiley, Chichester.

See also *Box 3.2*.

Fig. 19.12 The octahedral complex $[Co(en)_3]^{3+}$ contains three didentate ligands and therefore possesses optical isomers, i.e. the isomers are non-superimposable mirror images of each other.

not the *trans*) possesses optical isomers. The first purely inorganic complex to be *resolved* into its optical isomers was $[CoL_3]^{6+}$ **(19.17)** in which each L^+ ligand is the complex *cis*-$[Co(NH_3)_4(OH)_2]^+$ which chelates through the two *O*-donor atoms.

(19.17)

Summary

In this chapter, we have introduced some of general features of *d*-block metal chemistry, with particular emphasis on coordination numbers and isomerism. The connection between electronic configuration and stereochemistry is clearly more complicated than in the case of the *s*- and *p*-block elements. For the *d*-block metals, additional factors operate and we go on to discuss these in the next chapter.

The following terms were introduced in this chapter. Do you know what they mean?

- ❑ *d*-block metal
- ❑ transition element
- ❑ platinum-group metal
- ❑ electroneutrality principle
- ❑ Kepert model
- ❑ tripodal ligand
- ❑ structural isomerism
- ❑ stereoisomerism
- ❑ ionization isomerism
- ❑ hydration isomerism
- ❑ linkage isomerism
- ❑ polymerization isomerism
- ❑ geometrical isomerism
- ❑ optical isomerism

Suggested further reading

M. Gerloch and E.C. Constable (1994) *Transition Metal Chemistry: The Valence Shell in d-Block Chemistry*, VCH, Weinheim – An introductory and very readable text.

C.E. Housecroft (1999) *The Heavier d-Block Metals: Aspects of Inorganic and Coordination Chemistry*, Oxford University Press, Oxford – A short textbook which highlights differences between the first-row and the heavier *d*-block metals.

J.E. Huheey, E.A. Keiter and R.L. Keiter (1993) *Inorganic Chemistry: Principles of Structure and Reactivity*, 4th edn, Harper Collins, New York – Chapter 12 gives a good account of coordination numbers and geometries.

D.L. Kepert (1987) 'Coordination numbers and geometries' in *Comprehensive Coordination Chemistry*, eds G. Wilkinson, R.D. Gillard and J.A. McCleverty, Pergamon, Oxford, vol. 1, chapter 2, p. 95 – A review which surveys coordination numbers and geometries.

J.A. McCleverty (1999) *Chemistry of the First-row Transition Metals*, Oxford University Press, Oxford – A valuable introduction to metals and solid compounds, solution species, high and low oxidation state species and bio-transition metal chemistry.

D. Venkataraman, Y. Du, S.R. Wilson, K.A. Hirsch, P. Zhang and J.S. Moore, (1997) *Journal of Chemical Education*, vol. 74, p. 915 – An article entitled: 'A coordination geometry table of the *d*-block elements and their ions'.

M.J. Winter (1994) *d-Block Chemistry*, Oxford University Press, Oxford – An introductory text concerned with the principles of the *d*-block metals.

Problems

Ligand abbreviations: see *Table 6.7*

19.1 Comment on (a) the observation of variable oxidation states among elements of the *s*- and *p*-blocks, and (b) the statement that 'variable oxidation states are a characteristic feature of any *d*-block metal'.

19.2 (a) Write down, in order, the metals that comprise the first row of the *d*-block and give the ground state valence electronic configuration of each element.
(b) Which triads of metals make up groups 4, 8 and 11?
(c) Which metals are collectively known as the platinum-group metals?

19.3 Comment on the reduction potential data in Table 19.1.

19.4 By referring to relevant sections earlier in the book, write a brief account of the formation of hydrides, borides, carbides and nitrides of the *d*-block metals.

19.5 Give a brief overview of properties that characterize a *d*-block metal.

19.6 Suggest why (a) high coordination numbers are not usual for first row *d*-block metals, (b) in early *d*-block metal complexes the combination of a high oxidation state and high coordination number is common, and (c) in first row *d*-block metal complexes, high oxidation states are stabilized by fluoro or oxo ligands.

19.7 For each of the following complexes, give the oxidation state of the metal and its d^n configuration:

(a) $[Mn(CN)_6]^{4-}$; (b) $[FeCl_4]^{2-}$; (c) $[CoCl_3(py)_3]$; (d) $[ReO_4]^-$; (e) $[Ni(en)_3]^{2+}$; (f) $[Ti(H_2O)_6]^{3+}$; (g) $[VCl_6]^{3-}$; (h) $[Cr(acac)_3]$.

19.8 Within the Kepert model, what geometries do you associate with the following coordination numbers: (a) two; (b) three; (c) four; (d) five; (e) six?

19.9 Show that the trigonal bipyramid, square-based pyramid, square antiprism and dodecahedron belong to the point groups D_{3h}, C_{4v}, D_{4d} and D_{2d} respectively.

19.10. (a) In the solid state, $Fe(CO)_5$ possesses a trigonal bipyramidal structure. How many carbon environments are there? (b) Explain why only one signal is observed in the ^{13}C NMR spectrum of solutions of $Fe(CO)_5$, even at low temperature.

19.11 Structures **19.18–19.20** show bond angle data (determined by X-ray diffraction) for some complexes with low coordination numbers. Comment on these data, suggesting reasons for deviations from regular geometries.

(19.18)   (19.19)

(19.20)

19.12 Suggest a structure for the complex $[CuCl(\mathbf{19.21})]^+$ assuming that all donor atoms are coordinated to the Cu(II) centre.

(19.21)

19.13 What chemical tests would you use to distinguish between (a) $[Co(NH_3)_5Br][SO_4]$ and $[Co(NH_3)_5(SO_4)]Br$, and (b) $[CrCl_2(H_2O)_4]Cl \cdot 2H_2O$ and $[CrCl(H_2O)_5]Cl_2 \cdot H_2O$? (c) What is the relationship between these pairs of compounds? (d) What isomers are possible for $[CrCl_2(H_2O)_4]^+$?

19.14 (a) Give formulae for compounds that are coordination isomers of the salt $[Co(bpy)_3]^{3+}[Fe(CN)_6]^{3-}$. (b) What other types of isomerism could be exhibited by any of the complex ions noted down in your answer to part (a)?

19.15 What isomers would you expect to exist for the platinum(II) compounds:

(a) $[Pt(H_2NCH_2CHMeNH_2)_2]Cl_2$, and

(b) $[Pt(H_2NCH_2CMe_2NH_2)(H_2NCH_2CPh_2NH_2)]Cl_2$?

19.16 How many different forms of $[Co(en)_3]^{3+}$ are possible in principle? Indicate how they are related as enantiomers or diastereomers.

19.17 State the types of isomerism that may be exhibited by the following complexes, and draw structures of the isomers: (a) $[Co(en)_2(ox)]^+$, (b) $[Cr(ox)_2(H_2O)_2]^-$, (c) $[PtCl_2(PPh_3)_2]$, (d) $[PtCl_2(Ph_2PCH_2CH_2PPh_2)]$ and (e) $[Co(en)(NH_3)_2Cl_2]^{2+}$.

19.18 Using spectroscopic methods, how would you distinguish between the pairs of isomers (a) *cis*- and *trans*-$[PdCl_2(PPh_3)_2]$, (b) *cis*- and *trans*-$[PtCl_2(PPh_3)_2]$ and (c) *fac*- and *mer*-$[RhCl_3(PMe_3)_3]$.

19.19 Comment on the possibility of isomer formation for each of the following complexes (the ligand tpy is 2,2':6',2''-terpyridine, **19.22**): (a) $[Ru(py)_3Cl_3]$; (b) $[Ru(bpy)_2Cl_2]^+$; (c) $[Ru(tpy)Cl_3]$.

(19.22)

d-Block chemistry: coordination complexes

TOPICS

- Bonding in *d*-block metal complexes: valence bond theory
- Bonding in *d*-block metal complexes: crystal field theory
- Spectrochemical series
- Crystal field stabilization energy

- Bonding in *d*-block metal complexes: molecular orbital theory
- Electronic spectra
- Nephelauxetic effect
- Magnetic properties
- Thermodynamic aspects

20.1 Introduction

In this chapter, we discuss complexes of the *d*-block metals and we consider bonding theories that rationalize experimental facts such as electronic spectra and magnetic properties. Most of our discussion centres on first row *d*-block metals for which theories of bonding are most successful. The bonding in *d*-block metal complexes is not fundamentally different from that in other compounds, and we shall show applications of valence bond theory, the electrostatic model and molecular orbital theory. A key point is that three of the five *d* orbitals for a given principal quantum number have their lobes directed *in between* the *x*, *y* and *z* axes, while the other two are directed *along* these axes (Figure 20.1).[†] As a consequence of this difference, the *d* orbitals in the presence of ligands are split into groups of different energies, the type of splitting and the magnitude of the energy differences depending on the arrangement and nature of the ligands. Magnetic properties and electronic spectra, both of which are observable properties, reflect the splitting of *d* orbitals.

High- and low-spin states

In Section 19.5, we stated that paramagnetism is a characteristic of some *d*-block metal compounds. In Section 20.8 we consider magnetic properties in detail, but for now, let us simply state that magnetic data allow us to determine the number of unpaired electrons. In an isolated first row *d*-block metal

ion, the 3*d* orbitals are degenerate and the electrons occupy them according to Hund's rules: e.g. diagram **20.1** shows the arrangement of six electrons.

(20.1)

However, magnetic data for a range of octahedral d^6 *complexes* show that they fall into two categories: paramagnetic or diamagnetic. The former are called *high-spin complexes* and correspond to those in which, despite the *d* orbitals being split, there are still four unpaired electrons. The diamagnetic d^6 complexes are termed *low-spin* and correspond to those in which electrons are doubly occupying three orbitals, leaving two unoccupied.

20.2 Bonding in *d*-block metal complexes: valence bond theory

Hybridization schemes

Although VB theory (see *Sections 1.11, 1.12* and *4.2*) in the form developed by Pauling in the 1930s is not much used now in discussing *d*-block metal complexes, the terminology and many of the ideas have been retained and some knowledge of the theory remains essential. In *Section 4.2*, we described the use of sp^3d, sp^3d^2 and sp^2d hybridization schemes in trigonal pyramidal, square-based pyramidal, octahedral and square planar molecules. These same hybridization schemes can be used to describe the bonding in *d*-block metal complexes (Table 20.1); an *empty* hybrid orbital on the metal centre can accept a pair of electrons

---

[†] Although we refer to the *d* orbitals in these 'pictorial' terms, it is important not to lose sight of the fact that these orbitals are *not real* but merely mathematical solutions of the Schrödinger wave equation (see *Section 1.5*).

(a)

(b)

Fig. 20.1 (a) The six M–L vectors of an octahedral complex $[ML_6]^{n+}$ can be defined to lie along the x, y and z axes. (b) The five *d*-orbitals; the d_{z^2} and $d_{x^2-y^2}$ atomic orbitals point directly along the axes, but the d_{xy}, d_{yz} and d_{xz} atomic orbitals point between them.

from a ligand to form a σ-bond. The choice of particular p or d atomic orbitals may depend on the definition of the axes with respect to the molecular framework, e.g. in linear ML_2, the M–L vectors are defined to lie along the z axis. We have included the cube in Table 20.1 only to point out the required use of an f orbital.

Applying VB theory

We illustrate the applications of VB theory by first considering octahedral complexes of Cr(III) (d^3) and Fe(III) (d^5). The atomic orbitals required for hybridization in an octahedral complex are the $3d_{z^2}$, $3d_{x^2-y^2}$, $4s$, $4p_x$, $4p_y$ and

$4p_z$ (Table 20.1); these orbitals must be *unoccupied* so as to be available to accept six pairs of electrons from the ligands. The Cr^{3+} ion has three unpaired electrons and these are accommodated in the $3d_{xy}$, $3d_{xz}$ and $3d_{yz}$ orbitals:

Vacant orbitals available to accept ligand electrons

Table 20.1 Hybridization schemes for the σ-bonding frameworks of different geometrical configurations of ligand donor atoms.

Coordination number	Arrangement of donor atoms	Orbitals hybridized	Hybrid orbital description	Example
2	Linear	s, p_z	sp	$[Ag(NH_3)_2]^+$
3	Trigonal planar	s, p_x, p_y	sp^2	$[HgI_3]^-$
4	Tetrahedral	s, p_x, p_y, p_z	sp^3	$[FeBr_4]^{2-}$
4	Square planar	$s, p_x, p_y, d_{x^2-y^2}$	sp^2d	$[Ni(CN)_4]^{2-}$
5	Trigonal bipyramidal	$s, p_x, p_y, p_z, d_{z^2}$	sp^3d	$[CuCl_5]^{3-}$
5	Square-based pyramidal	$s, p_x, p_y, p_z, d_{x^2-y^2}$	sp^3d	$[Ni(CN)_5]^{3-}$
6	Octahedral	$s, p_x, p_y, p_z, d_{z^2}, d_{x^2-y^2}$	sp^3d^2	$[Co(NH_3)_6]^{3+}$
6	Trigonal prismatic	$s, d_{xy}, d_{yz}, d_{xz}, d_{z^2}, d_{x^2-y^2}$	sd^5	$[Mo(S_2C_2Ph_2)_3]$
		or	or	
		$s, p_x, p_y, p_z, d_{xz}, d_{yz}$	sp^3d^2	
7	Pentagonal bipyramidal	$s, p_x, p_y, p_z, d_{xy}, d_{x^2-y^2}, d_{z^2}$	sp^3d^3	$[V(CN)_7]^{4-}$
7	Monocapped trigonal prismatic	$s, p_x, p_y, p_z, d_{xy}, d_{xz}, d_{z^2}$	sp^3d^3	$[NbF_7]^{2-}$
8	Cubic	$s, p_x, p_y, p_z, d_{xy}, d_{xz}, d_{yz}, f_{xyz}$	sp^3d^3f	$[PaF_8]^{3-}$
8	Dodecahedral	$s, p_x, p_y, p_z, d_{z^2}, d_{xy}, d_{xz}, d_{yz}$	sp^3d^4	$[Mo(CN)_8]^{4-}$
8	Square antiprismatic	$s, p_x, p_y, p_z, d_{xy}, d_{xz}, d_{yz}, d_{x^2-y^2}$	sp^3d^4	$[TaF_8]^{3-}$
9	Tricapped trigonal prismatic	$s, p_x, p_y, p_z, d_{xy}, d_{xz}, d_{yz}, d_{z^2}, d_{x^2-y^2}$	sp^3d^5	$[ReH_9]^{2-}$

BOX 20.1 CHEMICAL AND THEORETICAL BACKGROUND

Valence bond theory: a historical note

In the early use of VB theory, complexes in which the electronic configuration of the metal ion was the same as that of the free gaseous atom were called *ionic complexes*, while those in which the electrons had been paired up as far as possible were called *covalent complexes*. Later, first row metal complexes in which ligand electrons entered 3*d* orbitals (as in $[Fe(CN)_6]^{3-}$) were termed *inner orbital complexes*, and those in which 4*d* orbitals were occupied (as in $[FeF_6]^{3-}$) were *outer orbital complexes*. The various terms, all of which may be encountered, relate to one another as follows:

- high-spin complex = ionic complex = outer orbital complex;
- low-spin complex = covalent complex = inner orbital complex.

With the electrons from the ligands included and a hybridization scheme applied for an octahedral complex, the diagram becomes:

This diagram is appropriate for all octahedral Cr(III) complexes because the three 3*d* electrons always occupy different orbitals.

For octahedral Fe(III) complexes, we must account for the existence of both high- and low-spin complexes. The electronic configuration of the free Fe^{3+} ion is:

For a low-spin octahedral complex such as $[Fe(CN)_6]^{3-}$, we can represent the electronic configuration by means of the following diagram where the electrons shown in blue are donated by the ligands:

For a high-spin octahedral complex such as $[FeF_6]^{3-}$, the five 3*d* electrons occupy the five 3*d* atomic orbitals (as in the free ion shown above) and the two *d* orbitals required for the sp^3d^2 hybridization scheme must come from the 4*d* set. With the ligand electrons included, valence bond theory describes the bonding as follows leaving three empty 4*d* atomic orbitals (not shown):

This scheme, however, is unrealistic because the 4*d* orbitals are at a significantly higher energy than the 3*d* atomic orbitals.

Nickel(II) (d^8) forms paramagnetic tetrahedral and octahedral complexes, and diamagnetic square planar complexes. Bonding in a tetrahedral complex can be represented as follows (ligand electrons are shown in blue):

and an octahedral complex can be described by the diagram:

in which, the three empty 4*d* atomic orbitals are not shown. For diamagnetic square planar complexes, valence bond theory gives the following picture:

Valence bond theory may rationalize stereochemical and magnetic properties, but only at a simplistic level. It can say *nothing* about electronic spectroscopic properties nor about the kinetic inertness (see *Section 25.2*) that is a characteristic of the low-spin d^6 configuration. Furthermore, the model implies a distinction between high- and low-spin complexes that is actually misleading. Finally, it cannot tell us *why* certain ligands are associated with the formation of high- (or low-) spin complexes. We therefore move on to alternative approaches to the bonding.

20.3 Crystal field theory

A second approach to the bonding in complexes of the *d*-block metals is *crystal field theory*. This is an *electrostatic model* and simply uses the ligand electrons to create an electric field around the metal centre. Ligands are considered as point charges and there are *no* metal–ligand covalent interactions.

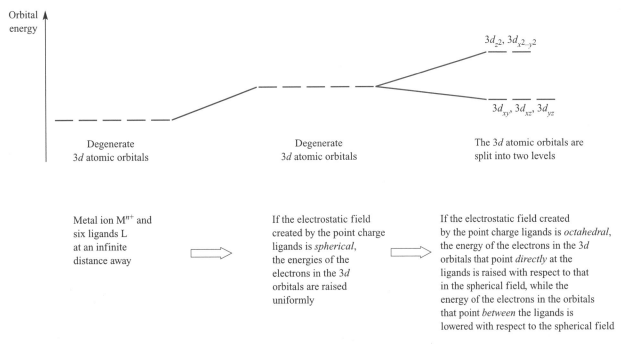

Fig. 20.2 The changes in the energies of the electrons occupying the *d* orbitals of an M^{n+} ion when the latter is in an octahedral crystal field. The energy changes are shown in terms of the orbital energies.

The octahedral crystal field

Consider a first row metal cation surrounded by six ligands placed on the Cartesian axes at the vertices of an octahedron (Figure 20.1a). Each ligand is treated as a negative point charge and there is an electrostatic attraction between the metal ion and ligands. However, there is also a repulsive interaction between electrons in the *d* orbitals and the ligand point charges. *If* the electrostatic field (the *crystal field*) were spherical, then the energies of the five 3*d* orbitals would be raised (destabilized) by the same amount. However, since the d_{z^2} and $d_{x^2-y^2}$ atomic orbitals point *directly at* the ligands while the d_{xy}, d_{yz} and d_{xz} atomic orbitals point *between* them, the d_{z^2} and $d_{x^2-y^2}$ atomic orbitals are destabilized to a greater extent than the d_{xy}, d_{yz} and d_{xz} atomic orbitals (Figure 20.2). Thus, with respect to their energy in a spherical field (the *barycentre*, a kind of 'centre of gravity'), the d_{z^2} and $d_{x^2-y^2}$ atomic orbitals are destabilized while the d_{xy}, d_{yz} and d_{xz} atomic orbitals are stabilized.

> *Crystal field theory* is an electrostatic model which predicts that the *d* orbitals in a metal complex are not degenerate. The pattern of splitting of the *d* orbitals depends on the crystal field, this being determined by the arrangement and type of ligands.

The two sets of *d* orbitals are labelled e_g and t_{2g} (see *Box 20.2*), and the energy separation between them is Δ_{oct} ('delta oct') or $10Dq$ (Figure 20.3). The overall stabilization

---

BOX 20.2 **CHEMICAL AND THEORETICAL BACKGROUND**

Some symmetry labels

The two sets of *d* orbitals in an octahedral field are labelled e_g and t_{2g} (Figure 20.3). In a tetrahedral field (Figure 20.8), the labels become *e* and t_2. The symbols *t* and *e* refer to the degeneracy of the level:

- a triply degenerate level is labelled *t*;
- a doubly degenerate level is labelled *e*.

The subscript *g* means *gerade* (German for 'even') and the subscript *u* means *ungerade* (German for 'odd'). Gerade and ungerade designate the behaviour of the wavefunction

under the operation of *inversion*, and denote the *parity* (even or odd) of an orbital. The *u* and *g* labels are *only* applicable if the system possesses a centre of symmetry (centre of inversion) and thus are used for the octahedral field, but not for the tetrahedral one. We shall simply accept the subscript 2 as an additional label which provides further symmetry information. A further symmetry label is needed later in the chapter:

- *a* and *b* designate a singly degenerate level.

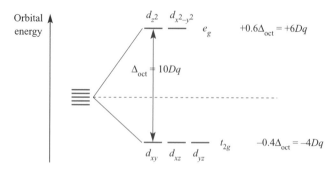

Fig. 20.3 Splitting of the *d* orbitals in an octahedral crystal field, with the energy changes measured with respect to the barycentre.

of the t_{2g} orbitals equals the overall destabilization of the e_g set. Thus, orbitals in the e_g set are raised by $0.6\Delta_{oct}$ with respect to the barycentre while those in the t_{2g} set are lowered by $0.4\Delta_{oct}$. Figure 20.3 also shows these energy differences in terms of $10Dq$. Both Δ_{oct} and $10Dq$ notations are in common use, but we use Δ_{oct} in this book. The stabilization and destabilization of the t_{2g} and e_g sets, respectively, are given *in terms of* Δ_{oct}. The magnitude of Δ_{oct} is determined by the *strength of the crystal field*, the two extremes being called *weak field* and *strong field* (equation 20.1).

$$\Delta_{oct}(\text{weak field}) < \Delta_{oct}(\text{strong field}) \qquad (20.1)$$

It is a merit of crystal field theory that, in principle at least, values of Δ_{oct} can be evaluated from electronic spectroscopic data (see *Section 20.6*). Consider the d^1 complex $[\text{Ti}(H_2O)_6]^{3+}$ for which the ground state is represented by diagram **20.2** or the notation $t_{2g}{}^1 e_g{}^0$.

$$\begin{array}{c} \underline{\quad}\ \ \underline{\quad}\ \ e_g \\[4pt] \underline{\uparrow}\ \ \underline{\quad}\ \ \underline{\quad}\ \ t_{2g} \end{array}$$

(20.2)

The absorption spectrum of the ion exhibits one broad band for which $\lambda_{max} = 20\,300\,\text{cm}^{-1}$ corresponding to an energy change of $243\,\text{kJ mol}^{-1}$ (Figure 20.4). The absorption results from a change in electronic configuration from $t_{2g}{}^1 e_g{}^0$ to $t_{2g}{}^0 e_g{}^1$, and the value of λ_{max} (see *Figure 20.15*) gives a measure of Δ_{oct}. For systems with more than one *d* electron, the evaluation of Δ_{oct} is more complicated; it is important to remember that Δ_{oct} is an *experimental* quantity.

Factors governing the magnitude of Δ_{oct} (Table 20.2) are the identity and oxidation state of the metal ion and the nature of the ligands. We shall see later that Δ parameters are also defined for other ligand arrangements (e.g. Δ_{tet}). For octahedral complexes, Δ_{oct} increases along the following *spectrochemical series* of ligands; the $[NCS]^-$ ion may coordinate through the *N*- or *S*-donor (distinguished in blue below) and accordingly, it has two positions in the series:

$$I^- < Br^- < [NCS]^- < Cl^- < F^- < [OH]^- < [ox]^{2-}$$
$$\approx H_2O < [NCS]^- < NH_3 < en < bpy < phen < [CN]^- \approx CO$$

weak field ligands ⎯⎯⎯⎯⎯⎯⎯⎯⎯→ strong field ligands

increasing Δ_{oct}

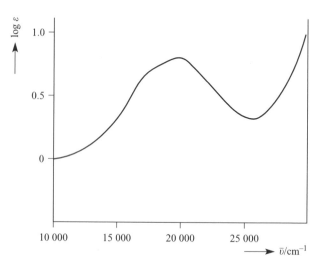

Fig. 20.4 The electronic spectrum of $[\text{Ti}(H_2O)_6]^{3+}$ in aqueous solution.

The spectrochemical series is reasonably general. Ligands with the same donor atoms are close together in the series. If we consider octahedral complexes of *d*-block metal ions, a number of points arise which can be illustrated by the following examples:

- the complexes of Cr(III) listed in Table 20.2 illustrate the effects of different ligand field strengths for a given M^{n+} ion;
- the complexes of Fe(II) and Fe(III) in Table 20.2 illustrate that for a given ligand and a given metal, Δ_{oct} increases with increasing oxidation state;
- where analogous complexes exist for a series of M^{n+} metals ions (constant *n*) in a triad, Δ_{oct} increases significantly down the triad (e.g. Figure 20.5);
- for a given ligand and a given oxidation state, Δ_{oct} varies *irregularly* across the first row of the *d*-block, e.g. over the range 8000 to $14\,000\,\text{cm}^{-1}$ for the $[M(H_2O)_6]^{2+}$ ions.

Table 20.2 Values of Δ_{oct} for selected *d*-block metal complexes.

Complex	Δ / cm^{-1}	Complex	Δ / cm^{-1}
$[TiF_6]^{3-}$	17 000	$[Fe(ox)_3]^{3-}$	14 100
$[Ti(H_2O)_6]^{3+}$	20 300	$[Fe(CN)_6]^{3-}$	35 000
$[V(H_2O)_6]^{3+}$	17 850	$[Fe(CN)_6]^{4-}$	33 800
$[V(H_2O)_6]^{2+}$	12 400	$[CoF_6]^{3-}$	13 100
$[CrF_6]^{3-}$	15 000	$[Co(NH_3)_6]^{3+}$	22 900
$[Cr(H_2O)_6]^{3+}$	17 400	$[Co(NH_3)_6]^{2+}$	10 200
$[Cr(H_2O)_6]^{2+}$	14 100	$[Co(en)_3]^{3+}$	24 000
$[Cr(NH_3)_6]^{3+}$	21 600	$[Co(H_2O)_6]^{3+}$	18 200
$[Cr(CN)_6]^{3-}$	26 600	$[Co(H_2O)_6]^{2+}$	9 300
$[MnF_6]^{2-}$	21 800	$[Ni(H_2O)_6]^{2+}$	8 500
$[Fe(H_2O)_6]^{3+}$	13 700	$[Ni(NH_3)_6]^{2+}$	10 800
$[Fe(H_2O)_6]^{2+}$	9 400	$[Ni(en)_3]^{2+}$	11 500

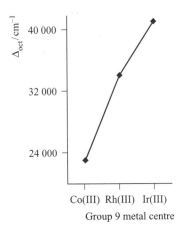

Fig. 20.5 The trend in values of Δ_{oct} for the complexes $[M(NH_3)_6]^{3+}$ where M = Co, Rh, Ir.

Trends in values of Δ_{oct} lead to the conclusion that metal ions can be placed in a spectrochemical series which is independent of the ligands:

$$Mn(II) < Ni(II) < Co(II) < Fe(III) < Cr(III) < Co(III) < Ru(III)$$
$$< Mo(III) < Rh(III) < Pd(II) < Ir(III) < Pt(IV)$$

increasing field strength

Spectrochemical series are empirical generalizations and simple crystal field theory *cannot* account for the magnitudes of Δ_{oct} values.

Crystal field stabilization energy: high- and low-spin octahedral complexes

We now consider the effects of different numbers of electrons occupying the d orbitals in an octahedral crystal field. For a d^1 system, the ground state corresponds to the configuration t_{2g}^{1} (**20.1**). With respect to the barycentre, there is a stabilization energy of $-0.4\Delta_{oct}$ (Figure 20.3); this is the so-called *crystal field stabilization energy*, *CFSE*.[†] For a d^2 ion, the ground state configuration is t_{2g}^{2} and the CFSE $= -0.8\Delta_{oct}$ (equation 20.2); a d^3 ion (t_{2g}^{3}) has a CFSE $= -1.2\Delta_{oct}$.

$$CFSE = -(2 \times 0.4)\Delta_{oct} = -0.8\Delta_{oct} \qquad (20.2)$$

e_g e_g

t_{2g} t_{2g}

(20.3)   (20.4)

For a d^4 ion, two arrangements are available: the four electrons may occupy the t_{2g} set with the configuration t_{2g}^{4} (**20.3**), or may singly occupy four d orbitals, $t_{2g}^{3}e_g^{1}$ (**20.4**). Configuration **20.3** corresponds to a low-spin arrangement, and **20.4** to a high-spin case. The preferred configuration is

that with the lower energy and depends on whether it is energetically preferable to pair the fourth electron or promote it to the e_g level. Two terms contribute to the electron-pairing energy, P, which is the energy required to transform two electrons with parallel spin in different degenerate orbitals into spin-paired electrons in the same orbital:

- the loss in the *exchange energy* (see *Box 1.7*) which occurs upon pairing the electrons;
- the coulombic repulsion between the spin-paired electrons.

Inequalities 20.3 and 20.4 show the requirements for high- or low-spin configurations. Inequality 20.3 holds when the crystal field is weak, whereas expression 20.4 is true for a strong crystal field. Figure 20.6 summarizes the preferences for low- and high-spin d^5 octahedral complexes.

For high-spin: $\Delta_{oct} < P$ (20.3)

For low-spin: $\Delta_{oct} > P$ (20.4)

Table 20.3 lists ground state electronic configurations and values of CFSE for d^n configurations; in the third column of the table, the pairing energy term is added where applicable. In each of d^4 to d^7, the difference in energy between the high- and low-spin configurations can be given in terms of the inequalities 20.3 and 20.4.

We can now relate types of ligand with a preference for high- or low-spin complexes. Strong field ligands such as $[CN]^-$ favour the formation of low-spin complexes while weak field ligands such as halides tend to favour high-spin complexes. However, we cannot predict whether high- or low-spin complexes will be formed unless we have accurate values of Δ_{oct} and P. On the other hand, with some experimental knowledge in hand, we can make some comparative predictions: if we know from magnetic data that $[Co(H_2O)_6]^{3+}$ is low-spin, then from the spectrochemical series we can say that $[Co(ox)_3]^{3-}$ and $[Co(CN)_6]^{3-}$ will be low-spin. The only common high-spin cobalt(III) complex is $[CoF_6]^{3-}$.

Jahn–Teller distortions

Octahedral complexes of d^9 and high-spin d^4 ions are often distorted, e.g. CuF_2 (the solid state structure of which contains octahedrally sited Cu^{2+} centres, see *Section 21.12*) and $[Cr(H_2O)_6]^{2+}$, so that two metal–ligand bonds (axial) are different lengths from the remaining four (equatorial). This is shown in structures **20.5** (elongated octahedron) and **20.6** (compressed octahedron).[‡] For a high-spin d^4 ion, one of the e_g orbitals contains one electron while the other is vacant. If the singly occupied orbital is in the d_{z^2}, most of the electron density in this orbital will be

---

[†] The sign convention used here for CFSE follows the thermodynamic convention.

[‡] Other distortions may arise and these are exemplified for Cu(II) complexes in Section 21.12.

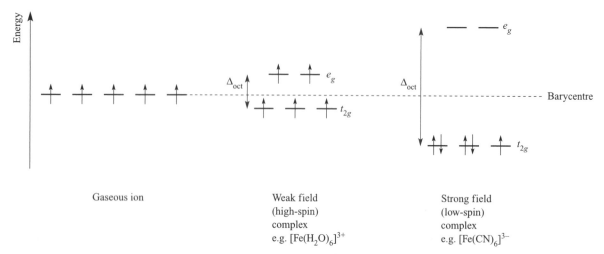

Fig. 20.6 The occupation of the 3*d* orbitals in weak and strong field Fe^{3+} (d^5) complexes.

concentrated between the cation and the two ligands on the *z* axis. Thus, there will be greater electrostatic repulsion associated with these ligands than with the other four and the complex suffers elongation (**20.5**). Conversely, occupation of the $d_{x^2-y^2}$ orbital would lead to elongation along the *x* and *y* axes as in structure **20.6**. A similar argument can be put forward for the d^9 configuration in which the two orbitals in the e_g set are occupied by one and two electrons respectively. Electron-density measurements confirm that the electronic configuration of the Cr^{2+} ion in $[Cr(H_2O)_6]^{2+}$ is *approximately* $d_{xy}^{\ 1}d_{yz}^{\ 1}d_{xz}^{\ 1}d_{z^2}^{\ 1}$. The corresponding effect when the t_{2g} set is unequally occupied is expected to be very much smaller since the orbitals are not pointing directly at the ligands; this expectation is usually, but not invariably, confirmed experimentally. Distortions of this kind are called *Jahn–Teller distortions*.

Bond length $a > e$
(20.5)

Bond length $a < e$
(20.6)

The *Jahn–Teller theorem* states that any non-linear molecular system in a degenerate electronic state will be unstable and will undergo distortion to form a system of lower symmetry and lower energy, thereby removing the degeneracy.

Table 20.3 Octahedral crystal field stablization energies (CFSE) for d^n configurations, and energy terms including pairing energies where applicable. High- and low-spin octahedral complexes are shown only where the distinction is appropriate.

d^n	High-spin = weak field			Low-spin = strong field		
	Electronic configuration	CFSE	Stabilization energy including pairing energy	Electronic configuration	CFSE	Stabilization energy including pairing energy
d^1	$t_{2g}^{\ 1}e_g^{\ 0}$	$-0.4\Delta_{oct}$				
d^2	$t_{2g}^{\ 2}e_g^{\ 0}$	$-0.8\Delta_{oct}$				
d^3	$t_{2g}^{\ 3}e_g^{\ 0}$	$-1.2\Delta_{oct}$				
d^4	$t_{2g}^{\ 3}e_g^{\ 1}$	$-0.6\Delta_{oct}$		$t_{2g}^{\ 4}e_g^{\ 0}$	$-1.6\Delta_{oct}$	$-1.6\Delta_{oct} + P$
d^5	$t_{2g}^{\ 3}e_g^{\ 2}$	0		$t_{2g}^{\ 5}e_g^{\ 0}$	$-2.0\Delta_{oct}$	$-2.0\Delta_{oct} + 2P$
d^6	$t_{2g}^{\ 4}e_g^{\ 2}$	$-0.4\Delta_{oct}$	$-0.4\Delta_{oct} + P$	$t_{2g}^{\ 6}e_g^{\ 0}$	$-2.4\Delta_{oct}$	$-2.4\Delta_{oct} + 3P$
d^7	$t_{2g}^{\ 5}e_g^{\ 2}$	$-0.8\Delta_{oct}$	$-0.8\Delta_{oct} + 2P$	$t_{2g}^{\ 6}e_g^{\ 1}$	$-1.8\Delta_{oct}$	$-1.8\Delta_{oct} + 3P$
d^8	$t_{2g}^{\ 6}e_g^{\ 2}$	$-1.2\Delta_{oct}$	$-1.2\Delta_{oct} + 3P$			
d^9	$t_{2g}^{\ 6}e_g^{\ 3}$	$-0.6\Delta_{oct}$	$-0.6\Delta_{oct} + 4P$			
d^{10}	$t_{2g}^{\ 6}e_g^{\ 4}$	0	$5P$			

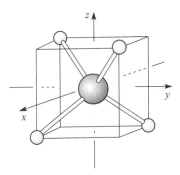

Fig. 20.7 The relationship between a tetrahedral ML_4 complex and a cube; the cube is readily related to a Cartesian axis set. The ligands lie *between* the x, y and z axes; compare this to an octahedral complex where the ligands lie on the axes.

The tetrahedral crystal field

So far we have restricted the discussion to octahedral complexes; we now turn to the tetrahedral crystal field. Figure 20.7 (see also *Figure 4.6*) shows a convenient way of relating a tetrahedron to a Cartesian axis set. With the complex in this orientation, none of the metal d orbitals points exactly at the ligands, but the d_{xy}, d_{yz} and d_{xz} orbitals come nearer to doing so than the d_{z^2} and $d_{x^2-y^2}$ orbitals. For a regular tetrahedron, the splitting of the d orbitals is inverted compared with that for a regular octahedral structure, and the energy difference (Δ_{tet}) is smaller. If all other things are equal (and of course, they never are), the relative splittings Δ_{oct} and Δ_{tet} are related by equation 20.5.

$$\Delta_{tet} = \tfrac{4}{9}\Delta_{oct} \approx \tfrac{1}{2}\Delta_{oct} \qquad (20.5)$$

Figure 20.8 compares crystal field splitting for octahedral and tetrahedral fields; remember, the subscript g in the symmetry labels (see *Box 20.2*) is not needed in the tetrahedral case.

Since Δ_{tet} is significantly smaller than Δ_{oct}, tetrahedral complexes are high-spin. Also, since smaller amounts of energy are needed for an $t_2 \longleftarrow e$ transition (tetrahedral) than for an $e_g \longleftarrow t_{2g}$ transition (octahedral), corresponding octahedral and tetrahedral complexes often have different colours. (The notation for electronic transitions is given in *Box 20.3*.)

The square planar crystal field

A square planar arrangement of ligands can be formally derived from an octahedral array by removal of two *trans*-ligands (Figure 20.9). If we remove the ligands lying along the z axis, then the d_{z^2} orbital is greatly stabilized; the energies of the d_{yz} and d_{xz} orbitals are also lowered, although to a smaller extent. The resultant ordering of the metal d orbitals is shown at the left-hand side of Figure 20.10. The fact that square planar d^8 complexes such as $[Ni(CN)_4]^{2-}$ are diamagnetic is a consequence of the relatively large energy difference between the d_{xy} and $d_{x^2-y^2}$ orbitals. Worked example 20.1 shows an experimental means (other than single crystal X-ray diffraction) by which square planar and tetrahedral d^8 complexes can be distinguished.

Worked example 20.1 Square planar and tetrahedral d^8 complexes

The d^8 complexes $[Ni(CN)_4]^{2-}$ and $[NiCl_4]^{2-}$ are square planar and tetrahedral respectively. Will these complexes be paramagnetic or diamagnetic?

Consider the splitting diagrams shown in Figures 20.8 and 20.10. For $[Ni(CN)_4]^{2-}$ and $[NiCl_4]^{2-}$, the eight electrons occupy the d-orbitals as follows:

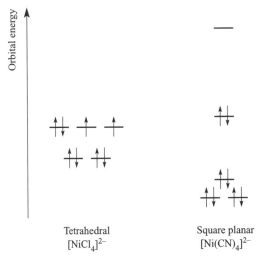

Thus, $[NiCl_4]^{2-}$ is paramagnetic while $[Ni(CN)_4]^{2-}$ is diamagnetic.

BOX 20.3   **CHEMICAL AND THEORETICAL BACKGROUND**

Notation for electronic transitions

For electronic transitions caused by the absorption and emission of energy, the following notation is used:

Emission: (high energy level) $\longrightarrow$ (low energy level)

Absorption: (high energy level) $\longleftarrow$ (low energy level)

For example, to denote an electronic transition from the e to t_2 level in a tetrahedral complex, the notation should be $t_2 \longleftarrow e$.

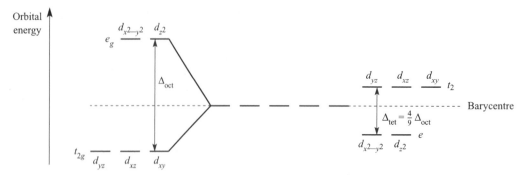

Fig. 20.8 Crystal field splitting diagrams for octahedral (left-hand side) and tetrahedral (right-hand side) fields. The splittings are referred to a common barycentre.

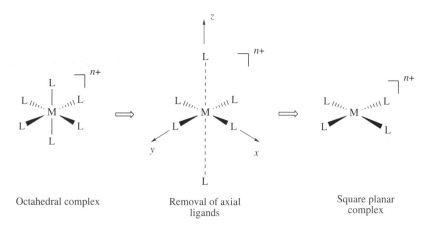

Fig. 20.9 A square planar complex can be derived from an octahedral complex by the removal of two ligands, e.g. those on the *z* axis; the intermediate stage is a Jahn–Teller distorted (elongated) octahedral complex.

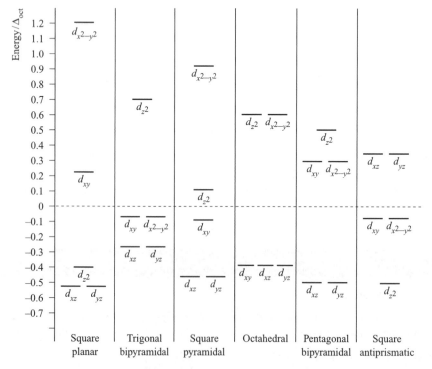

Fig. 20.10 Crystal field splitting diagrams for some common fields referred to a common barycentre; splittings are given with respect to Δ_{oct}. For tetrahedral splitting, see *Figure 20.8*.

Self-study exercises

No specific answers are given here, but the answer to each question is closely linked to the theory in worked example 20.1.

1. The complexes $NiCl_2(PPh_3)_2$ and $PdCl_2(PPh_3)_2$ are paramagnetic and diamagnetic respectively. What does this tell you about their structures?

2. The anion $[Ni(SPh)_4]^{2-}$ is tetrahedral. Explain why it is paramagnetic.

3. Diamagnetic *trans*-$[NiBr_2(PEtPh_2)_2]$ converts to a form which is paramagnetic. Suggest a reason for this observation.

Although $[NiCl_4]^{2-}$ is tetrahedral and paramagnetic, $[PdCl_4]^{2-}$ and $[PtCl_4]^{2-}$ are square planar and diamagnetic. This difference is a consequence of the larger crystal field splitting observed for second and third row metal ions compared to their first row congener; Pd(II) and Pt(II) complexes are invariably square planar (but see *Box 20.6*).

Other crystal fields

Figure 20.10 shows crystal field splittings for some common geometries with the relative splittings of the *d* orbitals referred to Δ_{oct}. By using these splitting diagrams, it is possible to rationalize the magnetic properties of a given complex (see *Section 20.8*). However, a word of caution: Figure 20.10 refers to ML_x complexes containing *like* ligands, and so *only* applies to simple complexes.

Crystal field theory: uses and limitations

Crystal field theory can bring together structures, magnetic properties and electronic properties, and we shall expand upon the last two topics later in the chapter. Trends in CFSEs provide some understanding of thermodynamic and kinetic aspects of *d*-block metal complexes (see *Sections 20.9–20.11* and *25.4*). Crystal field theory is surprisingly useful when one considers its simplicity. However, it has limitations. For example, although we can interpret the contrasting magnetic properties of high- and low-spin octahedral complexes on the basis of the positions of weak- and strong-field ligands in the spectrochemical series, crystal field theory provides no explanation as to *why* particular ligands are placed where they are in the series.

20.4 Molecular orbital theory: octahedral complexes

In this section, we consider a third approach to the bonding in metal complexes: the use of molecular orbital theory. In contrast to crystal field theory, the molecular orbital model considers covalent interactions between the metal centre and ligands.

Complexes with *no* metal–ligand π-bonding

We illustrate the application of MO theory to *d*-block metal complexes first by considering an octahedral complex such as $[Co(NH_3)_6]^{3+}$ in which metal–ligand σ-bonding is dominant. In the construction of an MO energy level diagram for such a complex, many approximations are made and the result is only *qualitatively* accurate. Even so, the results are useful to an understanding of metal–ligand bonding.

By following the procedures that we detailed in Chapter 4, an MO diagram can be constructed to describe the bonding in an $[ML_6]^{n+}$ complex. For a first row metal, the valence shell atomic orbitals are 3*d*, 4*s* and 4*p*. Each ligand provides one orbital and combinations of these orbitals are shown in Figure 20.11. Six ligand group orbitals (see *Section 4.4*) can be constructed and their symmetries are matched to those of the metal *s* and *p* atomic orbitals and to the d_{z^2} and $d_{x^2-y^2}$ orbitals (Figure 20.11). Combinations of the metal and ligand orbitals generate six bonding and six antibonding molecular orbitals. The metal d_{xy}, d_{yz} and d_{xz} atomic orbitals are non-bonding (Figure 20.12). The overlap between the ligand and metal *s* and *p* orbitals is greater than that involving the metal *d* orbitals, and so the a_{1g} and t_{1u} MOs are stabilized to a greater extent than the e_g MOs. The non-bonding d_{xy}, d_{yz} and d_{xz} atomic orbitals are the t_{2g} set of orbitals in the complex. In an octahedral complex *with no π-bonding*, the energy difference between the t_{2g} and e_g^* levels corresponds to Δ_{oct} in crystal field theory (Figure 20.12).

Having constructed the MO diagram in Figure 20.12, we are able to describe the bonding in a range of octahedral, σ-bonded complexes. For example:

- in low-spin $[Co(NH_3)_6]^{3+}$, 18 electrons (6 from Co^{3+} and 2 from each ligand) occupy the a_{1g}, t_{1u}, e_g and t_{2g} MOs;
- in high-spin $[CoF_6]^{3-}$, 18 electrons are available, 12 occupy the a_{1g}, t_{1u} and e_g MOs, 4 the t_{2g} level, and 2 the e_g^* level.

Whether a complex is high- or low-spin depends upon the energy separation of the t_{2g} and e_g^* levels. *Notionally*, in a σ-bonded octahedral complex, the 12 electrons supplied by the ligands are considered to occupy the a_{1g}, t_{1u} and e_g orbitals. Occupancy of the t_{2g} and e_g^* levels corresponds to the number of valence electrons of the metal ion, just as in crystal field theory. The molecular orbital model of bonding in octahedral complexes gives much the same results as crystal field theory. It is when we move to complexes with M−L π-bonding that distinctions between the models emerge.

Complexes with metal–ligand π-bonding

The metal d_{xy}, d_{yz} and d_{xz} atomic orbitals (the t_{2g} set) are non-bonding in an $[ML_6]^{n+}$, σ-bonded complex (Figure 20.12) and these orbitals may overlap with ligand orbitals of the correct symmetry to give π-interactions (Figure 20.13). Of the three types of interaction shown in Figure 20.13, (a) and (b) are commonly encountered (see below). Although π-bonding between metal and ligand *d* orbitals is

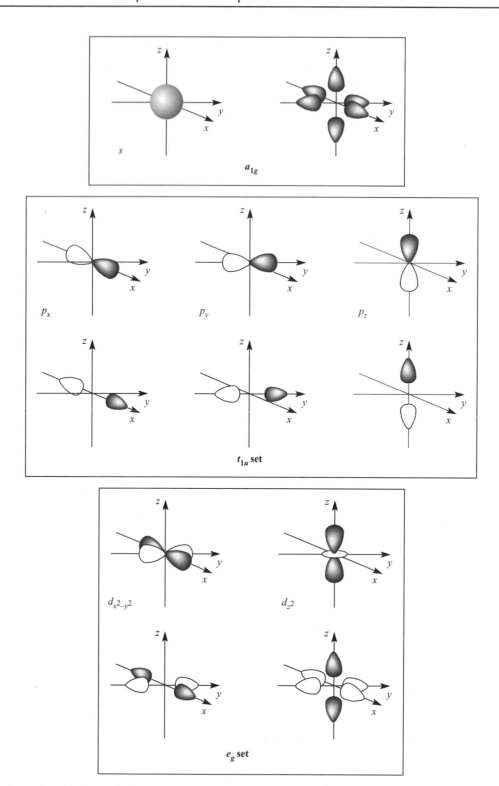

Fig. 20.11 Metal atomic orbitals matched with appropriate ligand group orbitals for an octahedral complex with only σ-bonding.

often considered for interactions between metals and phosphine ligands (e.g. PR_3 or PF_3), the role of the phosphorus *d* orbitals is a matter of some debate and involvement of ligand σ^* acceptor orbitals has also been suggested. This parallels the controversy surrounding the role of main group element *d* orbitals that we have already discussed in

Sections 4.7, 13.6 and *14.3.*[†] Two types of ligand must be differentiated: π-donor and π-acceptor ligands.

[†] For further discussion, see: A.G. Orpen and N.G. Connelly (1985) *Journal of the Chemical Society, Chemical Communications*, p. 1310. See also the discussion of *negative hyperconjugation* at the end of Section 13.6.

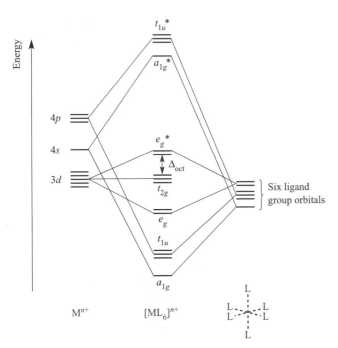

Fig. 20.12 An approximate MO diagram for the formation of $[ML_6]^{n+}$ (where M is a first row metal) using the ligand group orbital approach; the orbitals are shown pictorially in Figure 20.11. The bonding only involves M−L σ-interactions.

A *π-donor ligand* donates electrons to the metal centre in an interaction that involves a filled ligand orbital and an empty metal orbital; a *π-acceptor ligand* accepts electrons from the metal centre in an interaction that involves a filled metal orbital and an empty ligand orbital.

π-Donor ligands include Cl^-, Br^- and I^- and the metal–ligand π-interaction involves transfer of electrons from filled ligand p orbitals to the metal centre (Figure 20.13a). Examples of π-acceptor ligands are CO, N_2, NO and alkenes, and the metal–ligand π-bonds arise from the *back donation* of electrons from the metal centre to vacant antibonding

orbitals on the ligand (for example, Figure 20.13b). π-Acceptor ligands can stabilize low oxidation state metal complexes (see *Chapter 23*). Figure 20.14 shows partial MO diagrams which describe metal–ligand π-interactions in octahedral complexes; the metal s and p orbitals (see *Figure 20.12*) have been omitted. Figure 20.14a shows the interaction between a metal ion and six π-donor ligands; electrons are omitted from the diagram, and we return to them later. The ligand group π-orbitals (see *Box 20.4*) are filled and lie above, but relatively close to, the ligand σ-orbitals, and interaction with the metal d_{xy}, d_{yz} and d_{xz} atomic orbitals leads to bonding (t_{2g}) and antibonding (t_{2g}^*) MOs. The energy separation between the t_{2g}^* and e_g^* levels corresponds to Δ_{oct}. Figure 20.14b shows the interaction between a metal ion and six π-acceptor ligands. The vacant ligand π^*-orbitals lie significantly higher in energy than the ligand σ-orbitals. Orbital interaction leads to bonding (t_{2g}) and antibonding (t_{2g}^*) MOs as before, but now the t_{2g}^* MOs are at high energy and Δ_{oct} is identified as the energy separation between the t_{2g} and e_g^* levels (Figure 20.14b).

Although Figures 20.12 and 20.14 are qualitative, they reveal important differences between octahedral $[ML_6]^{n+}$ complexes containing σ-donor, π-donor and π-acceptor ligands:

- Δ_{oct} *decreases* in going from a σ-complex to one containing π-donor ligands;
- for a complex with π-donor ligands, increased π-donation stabilizes the t_{2g} level and destabilizes the t_{2g}^*, thus decreasing Δ_{oct};
- Δ_{oct} values are relatively large for complexes containing π-acceptor ligands, and such complexes are likely to be low-spin;
- for a complex with π-acceptor ligands, increased π-acceptance stabilizes the t_{2g} level, increasing Δ_{oct}.

The above points are consistent with the positions of the ligands in the spectrochemical series; π-donors such as I^- and Br^- are weak field, while π-acceptor ligands such as CO and $[CN]^-$ are strong field ligands.

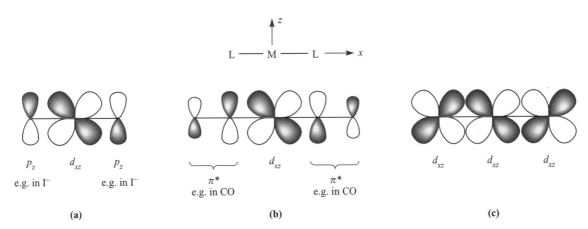

Fig. 20.13 π-Bond formation in a linear L−M−L unit in which the metal and ligand donor atoms lie on the x axis: (a) between metal d_{xz} and ligand p_z orbitals as for L = I^-, an example of a π-acceptor ligand; (b) between metal d_{xz} and ligand π^* orbitals as for L = CO, an example of a π-donor ligand; (c) between metal d_{xz} and ligand d_{xz} orbitals as may occur for PR_3 ligands (*but see text*).

BOX 20.4

The t_{2g} set of ligand π-orbitals for an octahedral complex

Figure 20.14 shows *three* ligand group π-orbitals and the reader may wonder how these arise from the combination of six ligands, especially since we show a simplistic view of the π-interactions in Figure 20.13. In an octahedral $[ML_6]^{n+}$ complex with six π-donor or acceptor ligands lying on the x, y and z axes, each ligand provides *two* π-orbitals, e.g. for ligands on the x axis, both p_y and p_z orbitals are available for π-bonding. Now consider just one plane containing four ligands of the octahedral complex, e.g. the xz plane. Diagram (a) below shows a ligand group orbital (LGO) comprising the p_z orbitals of two ligands and the p_x orbitals of the other two. Diagram (b) shows how the LGO in (a) combines with the metal d_{xz} orbital to give a bonding MO, while (c) shows the antibonding combination.

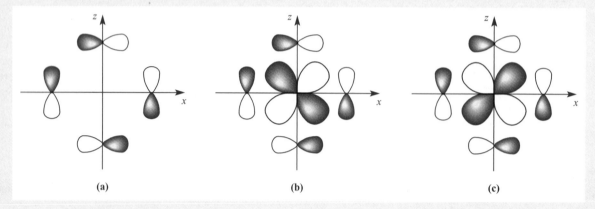

Three LGOs of the type shown in (a) can be constructed, one in each plane, and these can, respectively, overlap with the metal d_{xy}, d_{yz} and d_{xz} atomic orbitals to give the t_{2g} and t_{2g}^{*} MOs shown in Figure 20.14.

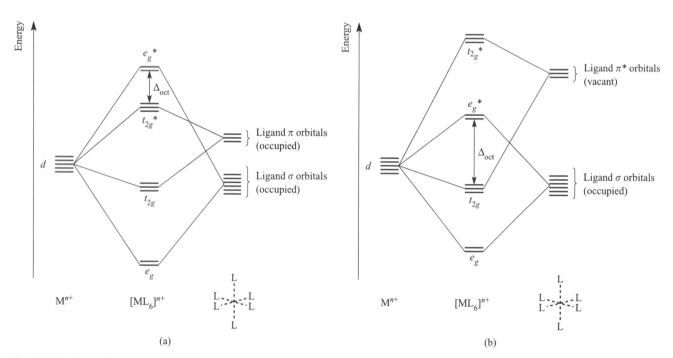

Fig. 20.14 Approximate, partial MO diagrams for metal–ligand π-bonding in an octahedral complex: (a) with π-donor ligands and (b) with π-acceptor ligands. In addition to the MOs shown, σ-bonding in the complex involves the a_{1g} and t_{1u} MOs (see *Figure 20.12*). Electrons are omitted from the diagram, because we are dealing with a general M^{n+} ion. Compared to Figure 20.12, the energy scale is expanded.

Let us complete this section by considering the occupancies of the MOs in Figures 20.14a and 20.14b. Six π-donor ligands provide 18 electrons (12 σ and 6 π-electrons) and these can *notionally* be considered to occupy the a_{1g}, t_{1u}, e_g and t_{2g} orbitals of the complex. The occupancy of the t_{2g}^* and e_g^* levels corresponds to the number of valence electrons of the metal ion. Six π-acceptor ligands provide 12 electrons (the π-ligand orbitals are empty) and, *formally*, we can place these in the a_{1g}, t_{1u} and e_g orbitals of the complex. The number of electrons supplied by the metal centre then corresponds to the occupancy of the t_{2g} and e_g^* levels. Since occupying *antibonding* MOs is detrimental to metal–ligand bond formation, it follows that, for example, octahedral complexes with π-accepting ligands will not be favoured for metal centres with d^7, d^8, d^9 or d^{10} configurations. This last point brings us to back to some fundamental observations in experimental inorganic chemistry: d-block metal organometallic and related complexes tend to obey the *effective atomic number rule* or *18-electron rule*. We return to the 18-electron rule in *Chapter 23*.

A low oxidation state organometallic complex contains π-acceptor ligands and the metal centre tends to acquire 18 electrons in its valence shell (the *18-electron rule*), thus filling the valence orbitals, e.g. Cr in $Cr(CO)_6$, Fe in $Fe(CO)_5$, and Ni in $Ni(CO)_4$.

Worked example 20.2   18-Electron rule

Show that $Cr(CO)_6$ obeys the 18-electron rule.

The Cr(0) centre has six valence electrons.
Each CO ligand is a two-electron donor.
The total electron count at the metal centre in $Cr(CO)_6 = 6 + (6 \times 2) = 18$.

Self-study exercises

1. Show that each of the metal centres in each of the following obey the 18-electron rule: (a) $Fe(CO)_5$; (b) $Ni(CO)_4$; (c) $[Mn(CO)_5]^-$; (d) $Mo(CO)_6$.

2. (a) How many electrons does a PPh_3 ligand donate? (b) Use your answer to (a) to confirm that the Fe centre in $Fe(CO)_4(PPh_3)$ obeys the 18-electron rule.

3. What is the oxidation state of each metal centre in the complexes in question (1)?   [*Ans.* (a) 0; (b) 0; (c) -1; (d) 0]

In applying the 18-electron rule, one clearly needs to know the number of electrons donated by a ligand, e.g. CO is a 2-electron donor. An ambiguity arises over NO groups in complexes. Nitrosyl complexes fall into two classes:

● NO as a 3-electron donor: crystallographic data show linear M−N−O (observed range $\angle$M−N−O = 165–180°) and

short M−N and N−O bonds indicating multiple character; IR spectroscopic data give ν(NO) in the range 1650–1900 cm^{-1}; the bonding mode is represented as **20.7** with the N atom taken to be *sp* hybridized.

● NO as a 1-electron donor: crystallographic data reveal a bent M−N−O group (observed range $\angle$M−N−O $\approx$ 120–140°), and N−O bond length typical of a double bond; IR spectroscopic data show ν(NO) in the range 1525–1690 cm^{-1}; the bonding mode is represented as **20.8** with the N atom considered as sp^2 hybridized.

(20.7)   (20.8)

Although the 18-electron rule is quite widely obeyed for low oxidation state organometallic compounds containing π-acceptor ligands, it is useless for higher oxidation state metals. This is clear from examples of octahedral complexes cited in *Section 19.7*, and can be rationalized in terms of the smaller energy separations between bonding and antibonding orbitals illustrated in Figures 20.12 and 20.14a compared with that in Figure 20.14b. We could extend our arguments to complexes such as $[CrO_4]^{2-}$ and $[MnO_4]^-$ showing how π-donor ligands help to stabilize high oxidation state complexes. However, for a valid discussion of these examples, we need to construct new MO diagrams appropriate to tetrahedral species. To do so would not provide much more insight than we have gained from considering the octahedral case, and interested readers are directed to more specialized texts.[†]

20.5 Ligand field theory

Although we shall not be concerned with the mathematics of ligand field theory, it is important to comment upon it briefly since we shall be using ligand field stabilization energies (LFSEs) later in this chapter.

Ligand field theory is an extension of crystal field theory which is freely parameterized rather than taking a localized field arising from point charge ligands.

Ligand field, like crystal field, theory is *confined* to the role of d orbitals, but unlike crystal field model, the ligand field approach is *not* a purely electrostatic model. It is a freely parameterized model, and uses Δ_{oct} and *Racah parameters*

[†] For application of MO theory to geometries other than octahedral, see Chapter 9 in: J.K. Burdett (1980) *Molecular Shapes: Theoretical Models of Inorganic Stereochemistry*, Wiley, New York.

(to which we return later) which are obtained from electronic spectroscopic (i.e. *experimental*) data. Most importantly, although (as we showed in the last section) it is possible to approach the bonding in *d*-block metal complexes by using molecular orbital theory, it is *incorrect* to state that ligand field theory is simply the application of MO theory.[†]

20.6 Electronic spectra

Spectral features

A characteristic feature of many *d*-block metal complexes is their colours which arise because they absorb light in the visible region (see *Figure 20.4*). Studies of electronic spectra of metal complexes provide information about structure and bonding, although interpretation of the spectra is not always straightforward. Absorptions arise from transitions between electronic energy levels:

● transitions between metal-centred orbitals possessing *d*-character ('*d*–*d*' transitions);
● transitions between metal- and ligand-centred MOs which transfer charge from metal-to-ligand or ligand-to-metal.

Charge transfer (CT) gives rise to intense absorptions, whereas '*d*–*d*' bands are much weaker. In some spectra, CT absorptions mask bands due to '*d*–*d*' transitions, although CT absorptions (as well as ligand-centred $n-\pi^*$ and $\pi-\pi^*$ bands) often occur at higher energies than '*d*–*d*' absorptions.

> MLCT = metal-to-ligand charge transfer
> LMCT = ligand-to-metal charge transfer

$$\bar{\nu} = \frac{1}{\lambda} = \frac{\nu}{c}$$

400 nm corresponds to 25 000 cm^{-1}; 200 nm corresponds to 50 000 cm^{-1}

Absorption bands in electronic spectra are usually broad; the absorption of a photon of light occurs in $\approx 10^{-18}$ s whereas molecular vibrations and rotations occur more slowly. Therefore, an electronic transition is a 'snapshot' of a molecule in a particular vibrational and rotational state, and it follows that the electronic spectrum will record a range of energies corresponding to different vibrational and rotational states. Absorption bands are described in terms of λ_{max} corresponding to the absorption maximum

---

[†] For a more detailed introduction to ligand field theory, see: M. Gerloch and E.C. Constable (1994) *Transition Metal Chemistry: the Valence Shell in d-Block Chemistry*, VCH, Weinheim, pp. 117–120; also see the further reading list at the end of the chapter.

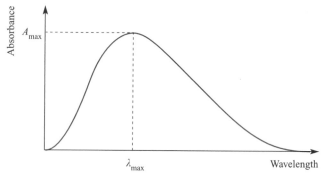

Fig. 20.15 Absorptions in the electronic spectrum of a molecule or molecular ion are often broad, and cover a range of wavelengths. The absorption is characterized by values of λ_{max} and ε_{max} (see *equation 20.6*).

A_{max} (Figure 20.15); the wavelength, λ_{max} is usually given in nm, but the position of the absorption may also be described in terms of wavenumbers, $\bar{\nu}$ (cm^{-1}). The molar extinction coefficient ε_{max} of an absorption must also be quoted; ε_{max} indicates how intense an absorption is and is related to A_{max} by equation 20.6 where c is the concentration of the solution and ℓ is the pathlength (in cm) of the spectrometer cell.

$$\varepsilon_{max} = \frac{A_{max}}{c \times \ell} \qquad (\varepsilon_{max} \text{ in dm}^3 \text{ mol}^{-1} \text{ cm}^{-1}) \qquad (20.6)$$

Values of ε_{max} range from close to zero (a very weak absorption) to >10 000 dm^3 mol^{-1} cm^{-1} (an intense absorption).

Some important points (for which explanations will be given later in the section) are that the electronic spectra of:

● d^1, d^4, d^6 and d^9 complexes consist of one absorption;
● d^2, d^3, d^7 and d^8 complexes consist of three absorptions;
● d^5 complexes consist of a series of very weak, relatively sharp absorptions.

Selection rules

Electronic transitions obey the following selection rules.

> *Spin selection rule*: $\Delta S = 0$
>
> Transitions may occur from singlet to singlet, or triplet to triplet states and so on, but a change in spin multiplicity is *forbidden*.
>
> *Laporte selection rule*: There must be a change in parity:
>
> allowed transitions: $g \leftrightarrow u$
> forbidden transitions: $g \leftrightarrow g$ $u \leftrightarrow u$
>
> This leads to the selection rule:
>
> $\Delta l = \pm 1$
>
> and, thus, *allowed* transitions are $s \longrightarrow p, p \longrightarrow d, d \longrightarrow f$; *forbidden* transitions are $s \longrightarrow s, p \longrightarrow p, d \longrightarrow d, f \longrightarrow f$, $s \longrightarrow d, p \longrightarrow f$ etc.

Table 20.4 Typical ε_{max} values for electronic absorptions; a large ε_{max} corresponds to an intense absorption and, if the absorption is in the visible region, a highly coloured complex.

Type of transition	Typical ε_{max} / $dm^3\,mol^{-1}\,cm^{-1}$	Example
Spin-forbidden 'd–d'	<1	$[Mn(H_2O)_6]^{2+}$ (high-spin d^5)
Laporte-forbidden, spin-allowed 'd–d'	1–10	Centrosymmetric complexes, e.g. $[Ti(H_2O)_6]^{3+}$ (d^1)
	10–1000	Non-centrosymmetric complexes, e.g. $[NiCl_4]^{2-}$
Charge transfer (fully allowed)	1000–50 000	$[MnO_4]^-$

Since these selection rules *must* be *strictly obeyed*, why do many d-block metal complexes exhibit 'd–d' bands in their electronic spectra?

A spin-forbidden transition becomes 'allowed' if, for example, a singlet state mixes to some extent with a triplet state. This is possible by *spin–orbit coupling* (see *Box 20.5*) but for first row metals, the degree of mixing is small and so bands associated with 'spin-forbidden' transitions are very weak (Table 20.4). Spin-allowed 'd–d' transitions remain Laporte-forbidden and their observation is explained by a mechanism called '*vibronic coupling*'. An octahedral complex possesses a centre-of-symmetry, but molecular vibrations result in its temporary loss. At an instant when the molecule does *not* possess a centre-of-symmetry, mixing of d and p orbitals can occur. Since the lifetime of the vibration ($\approx 10^{-13}$ s) is longer than that of an electronic transition ($\approx 10^{-18}$ s), a 'd–d' transition involving an orbital of mixed pd character can occur although the absorption is still relatively weak (Table 20.4). In a molecule which is non-centrosymmetric (e.g. tetrahedral), p–d mixing can occur to a greater extent and so the probability of 'd–d' transitions is greater than in a centrosymmetric complex. This leads to tetrahedral complexes being more intensely coloured than octahedral complexes.

Charge transfer transitions are allowed by the selection rules, and as a result, the intensities of CT absorptions are much greater than those of 'd–d' bands (Table 20.4).

Worked example 20.3 Spin-allowed and spin-forbidden transitions

Explain why an electronic transition for high-spin $[Mn(H_2O)_6]^{2+}$ is spin-forbidden, but for $[Co(H_2O)_6]^{2+}$ is spin-allowed.

$[Mn(H_2O)_6]^{2+}$ is high-spin d^5 Mn(II):

A transition from a t_{2g} to e_g orbital is impossible without breaking the spin selection rule: $\Delta S = 0$, which means that S must remain the same.

$[Co(H_2O)_6]^{2+}$ is a high-spin d^7 Co(II) complex:

A transition from a t_{2g} to e_g orbital can occur without violating the spin selection rule.

NB: Transitions in both complexes are Laporte-forbidden.

Self-study exercises

1. Write down the spin selection rule. [*Ans.* see text]

2. What is the d^n configuration and the spin multiplicity of the ground state of a (a) Ti^{3+} and (b) V^{3+} ion?
[*Ans.* (a) d^1; doublet; (b) d^2; triplet]

3. Why is a transition from a t_{2g} to e_g orbital spin allowed in $[V(H_2O)_6]^{3+}$? [*Ans.* triplet to triplet; see question 1]

Electronic spectra of octahedral and tetrahedral complexes

Electronic spectroscopy is a complicated topic and we shall restrict our discussion to high-spin complexes. The spectrum of $[Ti(H_2O)_6]^{3+}$ (Figure 20.4) exhibits one broad band; close inspection shows the presence of a shoulder indicating that the absorption is actually two closely spaced bands arising from a Jahn–Teller effect in the excited state. The term symbol for the ground state of Ti^{3+} (d^1, one electron for which $m_l = 2$, $L = 2$, $S = \frac{1}{2}$) is ^2D. In an octahedral field, this is split into $^2T_{2g}$ and a ^2E_g terms separated by energy Δ_{oct}, the magnitude of which increases with increasing field strength (Figure 20.16). The electronic spectrum of Ti^{3+} arises from a transition from the T_{2g} to E_g term; the energy of the transition depends on the field strength of the ligands in the octahedral Ti(III) complex.

For the d^9 configuration (e.g. Cu^{2+}) in an octahedral field (actually, a rare occurrence because of Jahn–Teller effects which lower the symmetry), the ground state of the free ion (^2D) is again split into $^2T_{2g}$ and a ^2E_g terms, but, in contrast to the d^1 ion (Figure 20.16), the ^2E_g term is lower than the $^2T_{2g}$ term. The d^9 and d^1 configurations are related by a *positive hole* concept: d^9 is derived from a

Term symbols for free atoms and ions

This Box provides an introduction to term symbols for free atoms and ions; for the most part, use of these symbols is confined to this chapter. In Section 1.7, we showed how to assign a set of quantum numbers to a given electron. For many purposes, this level of discussion is adequate. However, for an understanding of term symbols, a more detailed study is required; before studying this background information, readers may wish to review Box 1.4.

Quantum numbers for multi-electron species

Consider worked example 1.7; in the answer, we ignored a complication. In assigning quantum numbers to the four $2p$ electrons, how do we distinguish between the possibilities of placing the last electron in an orbital with $m_l = +1$, 0 or -1? This, and related questions, can be answered only by considering the interaction of electrons, primarily by means of the *coupling* of magnetic fields generated by their spin or orbital motion: hence the importance of spin and orbital angular momentum (see *Section 1.6*).

As we have seen, for any system containing more than one electron, the energy of an electron in a shell of principal quantum number n depends on the value of l and this also determines the orbital angular momentum of the electrons:

$$\text{orbital angular momentum} = \left[\sqrt{l(l+1)}\right]\frac{h}{2\pi}$$

We assume that the energy of a multi-electron species and its orbital angular momentum is determined by a *resultant orbital quantum number*, L, which is obtained directly from the values of l for the individual electrons; since the orbital angular momentum has magnitude *and* $(2l+1)$ spatial orientations (the number of values of m_l), *vectorial* summation of individual l values is necessary. Since the value of m_l for any electron denotes the component of its orbital angular momentum along the z axis, $m_l(h/2\pi)$, algebraic summation of m_l values for individual electrons gives the resultant orbital magnetic quantum number M_L and the component of the resultant orbital angular momentum along the z axis $M_L(h/2\pi)$. Just as m_l may have the $(2l+1)$ values l, $(l-1)\ldots0\ldots-(l-1), -l$, so M_L may have the $(2L+1)$ values L, $(L-1)\ldots0\ldots-(L-1), -L$, and if, for the multi-electron system, we can find all possible values of M_L, this tells us the value of L. Energy states for which $L = 0, 1, 2, 3, 4 \ldots$ are known as $S, P, D, F, G \ldots$ terms respectively, the letters corresponding to $s, p, d, f, g \ldots$ used to denote orbitals for which $l = 0, 1, 2, 3, 4 \ldots$ in the one-electron case. The resultant orbital angular momentum is given by:

$$\text{orbital angular momentum for a multi-electron species} = \left[\sqrt{L(L+1)}\right]\frac{h}{2\pi}$$

The resultant spin quantum number, S, from which the resultant spin angular momentum is determined by:

$$\text{spin angular momentum for a multi-electron species} = \left[\sqrt{S(S+1)}\right]\frac{h}{2\pi}$$

M_S is obtained by algebraic summation of the m_s values for individual electrons. One electron with $s = \frac{1}{2}$ obviously has $S = \frac{1}{2}$ with $M_S = +\frac{1}{2}$ or $-\frac{1}{2}$; M_S for the multi-electron system is analogous to m_s for the one-electron species. Two electrons lead to $S = 0$ ($m_s = +\frac{1}{2}$ and $-\frac{1}{2}$ giving $M_S = 0$) or $S = 1$ ($m_s = +\frac{1}{2}$ and $+\frac{1}{2}$, or $+\frac{1}{2}$ and $-\frac{1}{2}$, or $-\frac{1}{2}$ and $-\frac{1}{2}$ giving $M_S = 1$, 0 or -1). In general, for any value of S, there are $(2S+1)$ values of M_S: $S, (S-1)\ldots0\ldots-(S-1), -S$.

> The quantity $(2S+1)$ is the *multiplicity* of the term where S is the total spin quantum number. Terms for which $(2S+1) = 1, 2, 3, 4\ldots$ (corresponding to $S = 0, \frac{1}{2}, 1, \frac{3}{2}\ldots$) are called *singlets, doublets, triplets, quartets*...

The use of S for the resultant spin quantum number, a term for which $L = 0$, *and* screening constants is unfortunate, but is firmly established and in practice rarely causes confusion.

Finally, we have the resultant inner quantum number, J, also called the total angular momentum quantum number since the total angular momentum is given by:

$$\text{total angular momentum for a multi-electron species} = \left[\sqrt{J(J+1)}\right]\frac{h}{2\pi}$$

where J is compounded vectorially from L and S, i.e. algebraically from M_L and M_S. Quantum number J can take values $(L+S), (L+S-1)\ldots|L-S|$, the last symbol denoting the modulus of the quantity (i.e. only the magnitude, and not the sign is involved). Like j (see *Box 1.4*) for a single electron, J for the multi-electron system must be positive or zero. There are thus $(2S+1)$ possible values of J for $S < L$, and $(2L+1)$ possible values for $L < S$.

This method of obtaining J from L and S is based on LS (or *Russell–Saunders*) *coupling*. Although it is the only form of coupling of orbital and spin angular momentum that we shall consider in this book, it is not valid for all elements (especially those with high atomic numbers). In an alternative method of coupling, l and s for all the individual electrons are first combined to give j, and the individual j values are combined in a j–j *coupling* scheme.

Term symbols

If we know $(2S+1)$, L and J for an energy state, we can write the full *term symbol*. This is done by writing the symbol of the value of L (i.e. $S, P, D\ldots$) with the value of $(2S+1)$ as a left-superscript and the value of J as a right-subscript. Thus, the electronic ground state of carbon is $^3P_0$ ('triplet P zero') denoting $L = 1$, $(2S+1) = 3$ (i.e. $S = 1$) and $J = 0$. Different values of J denote different *levels* within the term, i.e. $^{(2S+1)}L_{J_1}$, $^{(2S+1)}L_{J_2}$ ..., the levels having different energies. Inorganic chemists often omit the value of J and refer to a $^{(2S+1)}L$ term; we shall usually follow this practice in this book.

Now we look in detail at the electronic ground states of atoms with $Z = 1$ to 10.

Hydrogen (Z = 1)

A hydrogen atom has an electronic configuration of $1s^1$; for the electron, $l = 0$ so L must be 0 and, therefore, we have an S term. The total spin quantum number $S = \frac{1}{2}$ so $(2S + 1) = 2$ (a doublet term). The only possible value of J is $\frac{1}{2}$, and so the term symbol for the hydrogen atom is $^2S_{1/2}$.

Helium (Z = 2)

For helium ($1s^2$), both electrons have $l = 0$, so $L = 0$. Two electrons both with $n = 1$ and $l = 0$ must have $m_s = +\frac{1}{2}$ and $-\frac{1}{2}$, so $S = 0$ and $(2S + 1) = 1$ (a singlet term). The only value of J is 0, and so the term symbol is $^1S_0$. Thus, the ns^2 configuration, having $L = 0$, $S = 0$, and $J = 0$ will contribute nothing to the term symbol in lithium and later atoms. The same conclusion can be drawn for any np^6 configuration and the reader is left to confirm this statement.

Lithium (Z = 3) and beryllium (Z = 4)

Atomic lithium has the electronic configuration $1s^2 2s^1$, and its term symbol is the same as that for hydrogen, $^2S_{1/2}$. Similarly, the term symbol for beryllium ($1s^2 2s^2$) is the same as that for helium, $^1S_0$.

Boron (Z = 5)

For boron ($1s^2 2s^2 2p^1$) we need only consider the p electron for reasons outlined above. For this, $l = 1$ so $L = 1$ (a P term); $S = \frac{1}{2}$ and so $(2S + 1) = 2$ (a doublet term). J can take values $(L + S), (L + S - 1) \ldots |L - S|$, and so $J = \frac{3}{2}$ or $\frac{1}{2}$. The term symbol for boron may be $^2P_{3/2}$ or $^2P_{1/2}$.

Carbon (Z = 6)

For carbon ($1s^2 2s^2 2p^2$), only the p electrons need be considered, and each has $l = 1$. Values of m_l may be +1, 0 or −1, and the algebraic sum of m_l for the individual electrons gives values of $L = 2$, 1 or 0 (D, P or S terms respectively). The two electrons may be spin-paired or have parallel spins and so $S = 0$ or 1, giving $(2S + 1) = 1$ (singlet term) or 3 (triplet term). It might seem that J could be 3, 2, 1 or 0, but this is not so. If, for example, the two electrons each have $n = 2$, $l = 1$ and $m_l = 1$ (giving $L = 2$), they cannot both have $m_s = +\frac{1}{2}$ as this would violate Pauli's principle. The only allowed combinations of m_l and m_s (and corresponding values of M_L and M_S) for two p electrons with the same value of n are shown in the table; such combinations are called *microstates*.

Table: Microstates for two electrons in an np level: values of m_l and m_s (represented as paired or unpaired electrons) and resultant values of M_L, M_S and M_J.

$m_l = +1$	$m_l = 0$	$m_l = -1$	M_L	M_S	M_J
↑↓			2	0	2
	↑↓		0	0	0
		↑↓	−2	0	−2
↑	↑		1	1	2
↑		↑	0	1	1
	↑	↑	−1	1	0
↓	↓		1	−1	0
	↓	↓	0	−1	−1
	↓	↓	−1	−1	−2
↑	↓		1	0	1
↓	↑		1	0	1
↑		↓	0	0	0
↓		↑	0	0	0
	↑	↓	−1	0	−1
	↓	↑	−1	0	−1

Inspection of the table reveals the following:

- the 15 microstates can be grouped into three sets, with the proviso that no set can contain a repetition;
- there is a set of 5 microstates with $M_L = 2$, 1, 0, −1, −2 and $M_S = 0$ (and thus $M_J = 2$, 1, 0, −1, −2) corresponding to $L = 1$ and a D term; moreover, since $S = 0$ (singlet) and $J = 2$, the term symbol is $^1D_2$;
- there is a set of 9 microstates with $M_L = 1$, 0, −1 and $M_S = 1$, 0, −1 which can be assigned the term ^3P (because $L = 1$ and $S = 1$);
- further examination of this last set of microstates reveals that it can be subdivided into a set of 5 with $J = 2$ (term symbol $^3P_2$), a set of 3 with $J = 1$ ($^3P_1$), and a single entry with $J = 0$ ($^3P_0$);
- one entry in the table remains unaccounted for and has $M_L = 0$, $M_S = 0$ and $M_J = 0$, corresponding to the term $^1S_0$.

We have, of course, no means of telling which entry with $M_L = 0$ and $M_S = 0$ should be assigned to which term (or similarly, how entries with $M_L = 1$ and $M_S = 0$, or $M_L = -1$ and $M_S = 0$ should be assigned). Indeed, it is *not meaningful* to do so.

Of the five terms that we have denoted for carbon ($^1D_2$, $^3P_2$, $^3P_1$, $^3P_0$ and $^1S_0$), the one with the lowest energy is $^3P_0$ and this is the electronic ground state. The others are excited states; notice that *Hund's rules do not always apply to excited states*.

Nitrogen to neon (Z = 7−10)

A similar treatment for the nitrogen atom shows that the $2p^3$ configuration gives rise to ^4S, ^2P and ^2D terms. For the $2p^4$ configuration (oxygen), we introduce a useful simplification by considering it as $2p^6$ plus two positrons which annihilate two of the electrons. Since positrons differ from electrons only in charge, the terms arising from the np^4 and np^2 configurations are the same. Similarly, np^5 is equivalent to np^1. This positron or *positive hole* concept is very useful and can be extended to *nd* configurations.

Relative energies of terms and levels

In regard to relative energies of terms, we state all of Hund's rules in a formal way. It is found from analysis of spectroscopic data that, provided that Russell–Saunders coupling holds:

- the term having the highest spin multiplicity (highest value of S) is the most stable (lowest energy);
- if two or more terms have the same value of S, the term having the higher value of L is the more stable;
- for all terms having the same values of S and L, the level with the lowest value of J is the most stable if the sub-shell is less than half-filled, and the level with the highest value of J is most stable if the sub-shell is more than half-filled (if the sub-shell is half-filled and S has the highest possible

value, $L = 0$ and $J = S$).

Thus for the terms corresponding to the electronic configuration np^2 ($^1D_2$, $^3P_2$, $^3P_1$, $^3P_0$ and $^1S_0$, see above, that of lowest energy is ^3P and the level of lowest energy is $^3P_0$.

Further reading

For a detailed, but readable, account of state symbols which includes *j–j coupling*, see: M. Gerloch (1986) *Orbitals, Terms and States*, Wiley, Chichester.

A good introduction to term symbols is included in Chapter 13 of: P.W. Atkins (1998) *Physical Chemistry*, 6th edn, Oxford University Press, Oxford.

A related topic not covered in this book is term symbols for diatomic molecules; an extremely good summary is: M.L. Campbell (1996) *Journal of Chemical Education*, vol. 73, p. 749.

d^{10} configuration by replacing one electron by a positive hole; thus, whereas the d^1 configuration contains one electron, d^9 contains one 'hole' (see *Box 20.5*). For a d^9 ion in an octahedral field, the splitting diagram is an inversion of that for the octahedral d^1 ion. This relationship is shown in Figure 20.17 (an *Orgel diagram*) where the right-hand side describes the octahedral d^1 case and the left-hand side, the octahedral d^9 ion.

Just as there is a relationship between the d^1 and d^9 configurations, there is a similar relationship between the d^4 and d^6 configurations. Further, we can relate the four configurations in an octahedral field as follows. In the weak-field limit, a d^5 ion is high-spin and spherically symmetric, and in this latter regard, d^0, d^5 and d^{10} configurations are analogous. Addition

of one electron to the high-spin d^5 ion to give a d^6 configuration mimics going from a d^0 to d^1 configuration; likewise, going from d^5 to d^4 by adding a positive hole mimics going from d^{10} to d^9. The result is that the Orgel diagram for octahedral d^1 and d^6 ions are the same, as are the diagrams for octahedral d^4 and d^9 (Figure 20.17).

Figure 20.17 also shows that the diagram for a d^1 or d^9 ion is inverted by going from an octahedral to tetrahedral field. Because the Orgel diagram uses a single representation for octahedral *and* tetrahedral fields, it is not possible to indicate that $\Delta_{tet} = \frac{4}{9}\Delta_{oct}$. Tetrahedral d^4 and d^6 ions can also be represented on the same Orgel diagram.

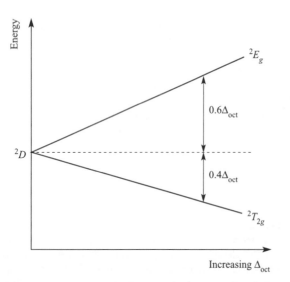

Fig. 20.16 Energy level diagram for a d^1 ion in an octahedral field.

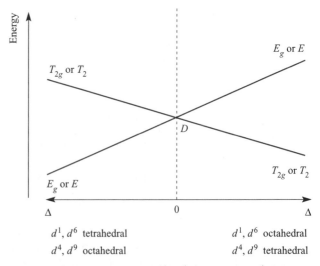

Fig. 20.17 Orgel diagram for d^1, d^4 (high-spin), d^6 (high-spin) and d^9 ions in octahedral (for which T_{2g} and E_g labels are relevant) and tetrahedral (E and T_2 labels) fields. In contrast to Figure 20.16, multiplicities are not stated because they depend on the d^n configuration.

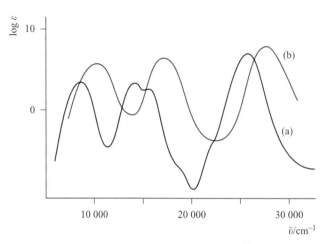

Fig. 20.19 Electronic spectra of (a) $[Ni(H_2O)_6]^{2+}$ and (b) $[Ni(NH_3)_6]^{2+}$ in aqueous solution showing three absorption bands.

Fig. 20.18 Orgel diagram for d^2, d^3, d^7 and d^8 ions (high-spin) in octahedral (for which T_{1g}, T_{2g} and A_{2g} labels are relevant) and tetrahedral (T_1, T_2 and A_2 labels) fields. Multiplicities are not stated because they depend on the d^n configuration, e.g. for the octahedral d^2 ion, $^3T_{1g}$, $^3T_{2g}$ and $^3A_{2g}$ labels are appropriate.

Finally, Figure 20.17 shows that for each of the octahedral and tetrahedral d^1, d^4, d^6 and d^9 ions, only one electronic transition from a ground to excited state is possible:

- for octahedral d^1 and d^6, the transition is $E_g \longleftarrow T_{2g}$
- for octahedral d^4 and d^9, the transition is $T_{2g} \longleftarrow E_g$
- for tetrahedral d^1 and d^6, the transition is $T_2 \longleftarrow E$
- for tetrahedral d^4 and d^9, the transition is $E \longleftarrow T_2$

Each transition is *spin-allowed* (no change in total spin, S) and the electronic spectrum of each ion exhibits one absorption. For sake of completeness, the notation for the transitions given above should include spin multiplicities, $2S + 1$ (see *Box 20.5*), e.g. for octahedral d^1, the notation is $^2E_g \longleftarrow {}^2T_{2g}$, and for high-spin, octahedral d^4, $^5T_{2g} \longleftarrow {}^5E_g$.

In an analogous manner to grouping d^1, d^4, d^6 and d^9 ions, we can consider together d^2, d^3, d^7 and d^8 ions in octahedral and tetrahedral fields. Two terms arise for the d^2 ion: ^3F (ground state) and ^3P (excited state); deriving terms from a table of microstates is dealt with in the next subsection. In an octahedral field, the ^3P term does not split, and is labelled $^3T_{1g}$. The ^3F term splits into $^3T_{1g}$, $^3T_{2g}$ and $^3A_{2g}$ terms. The $^3T_{1g}(F)$ term corresponds to a $t_{2g}^2 e_g^0$ arrangement and is triply degenerate because there are three ways of placing two electrons (with parallel spins) in any two of the d_{xy}, d_{yz} and d_{xz} orbitals. The $^3A_{2g}$ term corresponds to $t_{2g}^0 e_g^2$ arrangement (singly degenerate). The $^3T_{2g}$ and $^3T_{1g}(P)$ terms equate with a $t_{2g}^1 e_g^1$

configuration; the lower energy $^3T_{2g}$ term arises from placing two electrons in orbitals lying in mutually perpendicular planes, e.g. $(d_{xy})^1(d_{z^2})^1$, while the higher energy $^3T_{1g}(P)$ term arises from placing two electrons in orbitals lying in the same plane e.g. $(d_{xy})^1(d_{x^2-y^2})^1$. The energies of the $^3T_{1g}(F)$, $^3T_{2g}$, $^3A_{2g}$ and $^3T_{1g}(P)$ terms are shown on the right-hand side of Figure 20.18; note the effect of increasing field strength. Starting from this diagram and using the same arguments as for the d^1, d^4, d^6 and d^9 ions, we can derive the complete Orgel diagram shown in Figure 20.18. At increased field strengths, the lines describing the $T_{1g}(F)$ and $T_{1g}(P)$ terms (or T_1, depending on whether we are dealing with octahedral or tetrahedral cases) curve away from one another; there is interaction between terms of the same symmetry and they are not allowed to cross (the *non-crossing rule*). From Figure 20.18, we can see why three absorptions are observed in the electronic spectra of d^2, d^3, d^7 and d^8 octahedral and tetrahedral complexes; the transitions are from the ground to excited states, and are all spin-allowed. Transitions are possible from one excited state to another, but their probability is so low that they can be ignored. Figure 20.19 illustrates spectra for octahedral nickel(II) (d^8) complexes.

For the high-spin d^5 configuration, all transitions are *spin-forbidden* and 'd–d' transitions that are observed are between the ^6S ground state and quartet states (three unpaired electrons). Associated absorptions are extremely weak.

For a proper interpretation of electronic spectral features, interelectronic interaction must be taken into account and parameters additional to Δ_{oct} are needed to quantify the description of the spectrum. There are *Racah parameters* which we shall meet again when we describe Tanabe–Sugano diagrams. The evaluation of Δ_{oct} from electronic spectra is, therefore, more difficult for d^2 (and related d^n) than for d^1 (and d^6 etc.) ions and some uncertainty is often associated with reported values.

Table 20.5 Table of microstates for two electrons in an *nd* level ($l = 2$): values of m_l (read down each column) and resultant values of M_L; each electron has $m_s = \frac{1}{2}$. The microstates are grouped so as to show the values of M_L corresponding to ^3F and ^3P terms.

m_l											
-2			↑		↑	↑				↑
-1		↑		↑		↑		↑		
0		↑		↑	↑			↑		
$+1$	↑							↑	↑	↑
$+2$	↑	↑	↑	↑						
M_L	$+3$	$+2$	$+1$	0	-1	-2	-3	$+1$	0	-1

$$^3F \qquad\qquad\qquad\qquad\qquad ^3P$$

Worked example 20.4   Electronic spectra

The electronic spectrum of an aqueous solution of $[\text{Ni(en)}_3]^{2+}$ exhibits broad absorptions with $\lambda_{\text{max}} \approx 325$, 550 and 900 nm. (a) Suggest assignments for the electronic transitions. (b) Which bands are in the visible region?

(a) $[\text{Ni(en)}_3]^{2+}$ is a Ni(II), d^8 complex. From the Orgel diagram in Figure 20.18 the three transitions can be assigned; *lowest wavelength* corresponds to *highest energy* transition:

900 nm assigned to $^3T_{2g} \longleftarrow {}^3A_{2g}$
550 nm assigned to $^3T_{1g}(F) \longleftarrow {}^3A_{2g}$
325 nm assigned to $^3T_{1g}(P) \longleftarrow {}^3A_{2g}$

(b) Visible region spans $\approx$400–750 nm, so only the 550 nm absorption falls in this range (see *Table 19.2*).

Self-study exercises

1. Of the three absorptions in $[\text{Ni(en)}_3]^{2+}$, which is closest to the UV end of the spectrum?   [*Ans.* Look at Appendix 4]

2. Does the notation $^3T_{2g} \longleftarrow {}^3A_{2g}$ indicate an absorption or emission band?   [*Ans.* Look at Box 20.3]

3. Why are the three transitions for $[\text{Ni(en)}_3]^{2+}$ (a) spin-allowed, and (b) Laporte-forbidden?

Microstates

In discussing the spectra of octahedral d^2 complexes, we have emphasized the physical principles underlying the number of bands in the electronic spectrum. We now present a more formal treatment involving the construction of a table of microstates (see *Box 20.5*). Table 20.5 shows all possible combinations of two unpaired *d* electrons; for parallel spins, the terms must be triplets ($S = 1$) and in addition to the ^3F ground state, the only excited state is the ^3P. It can be shown from group theory that in an octahedral field, D and F, but not S and P, terms split.

Term	Components in an octahedral field
S	A_{1g}
P	T_{1g}
D	$E_g + T_{2g}$
F	$A_{2g} + T_{2g} + T_{1g}$
G	$A_{1g} + E_g + T_{2g} + T_{1g}$

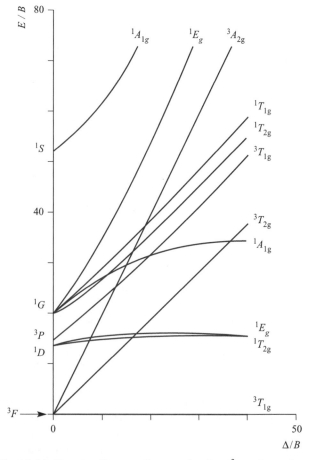

Fig. 20.20 Tanabe–Sugano diagram for the d^2 configuration in an octahedral field.

Similar splittings occur in a tetrahedral field, but the g labels given above are no longer applicable. For a d^2 ion in an octahedral field, the possible triplet terms are therefore $^3A_{2g}$, $^3T_{2g}$ and $^3T_{1g}$ derived from the triplet ^3F ground term, and the $^3T_1$ derived from the triplet ^3P excited term.

Tanabe–Sugano diagrams

A more advanced treatment of the energies of electronic states is found in *Tanabe–Sugano diagrams*. The energy of the ground state is taken to be zero for all field strengths, and the energies of all other terms and their components are plotted with respect to the ground term; if there is a change in ground term as the field strength increases, a discontinuity appears in the diagram. Figure 20.20 shows the Tanabe–Sugano diagram for the d^2 configuration in an octahedral field; notice that the energy and field strength are both expressed in terms of the Racah parameter B. Racah parameters take into account interelectronic repulsion: there are three parameters A, B and C, although we shall only mention B. Application of Tanabe–Sugano diagrams is illustrated in worked example 20.5.

Worked example 20.5   Application of Tanabe–Sugano diagrams

Aqueous solutions of $[V(H_2O)_6]^{3+}$ show absorptions at 17 200 and 25 600 cm^{-1} assigned to the $^3T_{2g} \longleftarrow {}^3T_{1g}(F)$ and $^3T_{1g}(P) \longleftarrow {}^3T_{1g}(F)$ transitions. Estimate values of B and Δ_{oct} for $[V(H_2O)_6]^{3+}$.

$[V(H_2O)_6]^{3+}$ is a d^2 ion and the Tanabe–Sugano diagram in Figure 20.20 is therefore appropriate. An important point to recognize is that with the diagram provided, only approximate values of B and Δ_{oct} can be obtained.

Let the transition energies be $E_2 = 25\,600$ cm^{-1} and $E_1 = 17\,200$ cm^{-1}.

Values of transition energies cannot be read directly from the Tanabe–Sugano diagram, but ratios of energies can be obtained since:

$$\frac{\left(\dfrac{E_2}{B}\right)}{\left(\dfrac{E_1}{B}\right)} = \frac{E_2}{E_1}$$

From the observed absorption data:

$$\frac{E_2}{E_1} = \frac{25\,600}{17\,200} = 1.49$$

We now proceed by trial-and-error, looking for the value of $\dfrac{\Delta_{oct}}{B}$ which corresponds to a ratio

$$\frac{\left(\dfrac{E_2}{B}\right)}{\left(\dfrac{E_1}{B}\right)} = 1.49.$$

Trial points:

$$\text{when } \frac{\Delta_{oct}}{B} = 20, \qquad \frac{\left(\dfrac{E_2}{B}\right)}{\left(\dfrac{E_1}{B}\right)} \approx \frac{32}{18} = 1.78$$

$$\text{when } \frac{\Delta_{oct}}{B} = 30, \qquad \frac{\left(\dfrac{E_2}{B}\right)}{\left(\dfrac{E_1}{B}\right)} \approx \frac{41}{28} = 1.46$$

$$\text{when } \frac{\Delta_{oct}}{B} = 29, \qquad \frac{\left(\dfrac{E_2}{B}\right)}{\left(\dfrac{E_1}{B}\right)} \approx \frac{40.0}{26.9} = 1.49$$

This is an *approximate* answer but we are now able to estimate B and Δ_{oct} as follows:

- when $\dfrac{\Delta_{oct}}{B} = 29$, we have $\dfrac{E_2}{B} \approx 40.0$, and since $E_2 = 25\,600$ cm^{-1}, $B \approx 637$ cm^{-1};
- when $\dfrac{\Delta_{oct}}{B} = 29$, $\dfrac{E_1}{B} \approx 26.9$, and since $E_1 = 17\,200$ cm^{-1}, $B \approx 640$ cm^{-1}.

Substitution of the value of B into $\dfrac{\Delta_{oct}}{B} = 29$ gives an estimate of $\Delta_{oct} \approx 18\,600$ cm^{-1}.

Accurate methods involving mathematical expressions can be used and these can be found in the advanced texts listed at the end of the chapter.

Self-study exercises

1. Why are the two values of B obtained above self-consistent?

2. For $[Ti(H_2O)_6]^{3+}$, a value of Δ_{oct} can be determined directly from λ_{max} in the electronic spectrum. Why is this not possible for $[V(H_2O)_6]^{3+}$, and for most other octahedral ions?

20.7 Evidence for metal–ligand covalent bonding

The nephelauxetic effect

In metal complexes, there is evidence for sharing of electrons between metal and ligand. Pairing energies are lower in complexes than in gaseous M^{n+} ions indicating that interelectronic repulsion is less in complexes and that the *effective* size of the metal orbitals has increased; this is the *nephelauxetic effect*.

Nephelauxetic means (electron) 'cloud expanding'.

For complexes with a common metal ion, it is found that the nephelauxetic effect of ligands varies according to a series

Table 20.6 Selected values of *h* and *k* which are used to parameterize the nephelauxetic series; worked example 20.6 shows their application.

Metal ion	*k*	Ligands	*h*
Co(III)	0.35	6 Br^-	2.3
Rh(III)	0.28	6 Cl^-	2.0
Co(II)	0.24	6 $[CN]^-$	2.0
Fe(III)	0.24	3 en	1.5
Cr(III)	0.21	6 NH_3	1.4
Ni(II)	0.12	6 H_2O	1.0
Mn(II)	0.07	6 F^-	0.8

independent of metal ion:

$$F^- < H_2O < NH_3 < en < [ox]^{2-} < [NCS]^- < Cl^- < [CN]^- < Br^- < I^-$$

increasing nephelauxetic effect →

A nephelauxetic series for metal ions (independent of ligands) is as follows:

$$Mn(II) < Ni(II) \approx Co(II) < Mo(II) < Re(IV) < Fe(III)$$
$$< Ir(III) < Co(III) < Mn(IV)$$

increasing nephelauxetic effect →

The nephelauxetic effect can be parameterized and the values shown in Table 20.6 used to estimate the reduction in electron–electron repulsion upon complex formation. In equation 20.7, the interelectronic repulsion in the complex is the Racah parameter *B* (see *Section 20.6*); B_0 is the interelectronic repulsion in the gaseous M^{n+} ion.

$$\frac{B_0 - B}{B_0} \approx h_{\text{ligands}} \times k_{\text{metal ion}} \quad (20.7)$$

Worked example 20.6   The nephelauxetic series

Using data in Table 20.6, estimate the reduction in the interelectronic repulsion in going from the gaseous Fe^{3+} ion to $[FeF_6]^{3-}$.

The reduction in interelectronic repulsion is given by:

$$\frac{B_0 - B}{B_0} \approx h_{\text{ligands}} \times k_{\text{metal ion}}$$

In Table 20.6, values of *h* refer to a set of six ligands. For $[FeF_6]^{3-}$:

$$\frac{B_0 - B}{B_0} \approx 0.8 \times 0.24 = 0.192$$

Therefore, the reduction in interelectronic repulsion in going from the gaseous Fe^{3+} ion to $[FeF_6]^{3-}$ is $\approx 19\%$.

Self-study exercises

Refer to Table 20.6.

1. Show that the reduction in interelectronic repulsion in going

from the gaseous Ni^{2+} ion to $[NiF_6]^{4-}$ is $\approx 10\%$.

2. Estimate the reduction in interelectronic repulsion on going from gaseous Rh^{3+} to $[Rh(NH_3)_6]^{3+}$.   [*Ans.* $\approx 39\%$]

ESR spectroscopy

Further proof of electron sharing comes from electron spin resonance (ESR) spectroscopy.[†] Unpaired electrons behave as magnets, and may align themselves parallel or antiparallel to a magnetic field. The former configuration has a slightly lower energy than the latter, and transitions from one configuration to the other can be observed by applying radio-frequency radiation. If the metal ion carrying the unpaired electron is linked to a ligand containing nuclei with the nuclear spin quantum number, $I \neq 0$, hyperfine splitting of the ESR signal is observed showing that the orbital occupied by the electron has both metal and ligand character, i.e. there is metal–ligand covalent bonding. An example is the ESR spectrum of $Na_2[IrCl_6]$ (paramagnetic low-spin d^5) recorded for a solid solution in $Na_2[PtCl_6]$ (diamagnetic low-spin d^6); this was a classic ESR experiment, reported in 1953.[‡]

20.8 Magnetic properties

Magnetic susceptibility and the spin-only formula

We begin the discussion of magnetochemistry with the so-called *spin-only formula*, an *approximation* that has limited, but useful, applications.

Paramagnetism arises from unpaired electrons. Each electron has a magnetic moment with one component associated with the spin angular momentum of the electron and (except when the quantum number $l = 0$) a second component associated with the orbital angular momentum. For many complexes of first row *d*-block metal ions we can ignore the second component and the magnetic moment, μ, can be regarded as being determined by the number of unpaired electrons, *n* (equations 20.8 and 20.9). The two equations are related because the total spin quantum number $S = \frac{n}{2}$.

$$\mu(\text{spin-only}) = 2\sqrt{S(S + 1)} \quad (20.8)$$

$$\mu(\text{spin-only}) = \sqrt{n(n + 2)} \quad (20.9)$$

The *effective magnetic moment*, μ_{eff}, can be obtained from the experimentally measured *molar magnetic susceptibility*, χ_{m},

---

[†] For an introduction to ESR (also referred to as EPR) spectroscopy, see: R.V. Parish (1990) *NMR, NQR, EPR and Mössbauer Spectroscopy in Inorganic Chemistry*, Ellis Horwood, Chichester.
[‡] See: J. Owen and K.W.H. Stevens (1953) *Nature*, vol. 171, p. 836.

BOX 20.6

CHEMICAL AND THEORETICAL BACKGROUND

A paramagnetic, tetrahedral Pd(II) complex

In Section 20.3, we discussed why Pd(II) and Pt(II) complexes are square planar, basing our arguments on electronic factors. Nickel(II), on the other hand, forms both square planar and tetrahedral complexes. A square planar geometry for the coordination sphere of a d^8 ion can readily be distinguished from a tetrahedral one by measuring the magnetic moment: square planar complexes are diamagnetic, while tetrahedral ones have two unpaired electrons. There is a temptation to ignore the possibility of geometries other than square planar for 4-coordinate Pd(II) and Pt(II) complexes, but in rare cases, tetrahedral coordination is observed

and gives rise to a paramagnetic complex. An example is $[PdCl_2L]$ where ligand L is the ferrocenyl (see Section 23.13) derivative shown below. The ligand coordinates through the two O-donors and in the complex (shown below with the Pd atom in blue), the O−Pd−O bond angle is 104° and the Cl−Pd−Cl angle is 120°. It is thought that the steric effects of ligand L force the metal coordination environment to be tetrahedral. The effective magnetic moment of $[PdCl_2L]$ is $2.48\,\mu_B$; although this is lower than is typical for a tetrahedral Ni(II) complex, it is still consistent with two unpaired electrons (Table 20.7).

Ligand L   The structure of $[PdCl_2L]$

and is expressed in Bohr magnetons (μ_B or BM) where $1\mu_B = eh/4\pi m_e = 9.27 \times 10^{-24}\,J\,T^{-1}$. Equation 20.10 gives the relationship between μ_{eff} and χ_m; using SI units for the constants, this expression reduces to equation 20.11 in which χ_m is in $cm^3\,mol^{-1}$. In the laboratory, the continued use of Gaussian units in magnetochemistry means that *irrational susceptibility* is the measured quantity and equation 20.12 is therefore usually applied.[†]

$$\mu_{eff} = \sqrt{\frac{3k\chi_m T}{L\mu_0\mu_B^{\,2}}} \qquad (20.10)$$

where k = Boltzmann constant; L = Avogadro number; μ_0 = vacuum permeability; T = temperature in kelvin.

$$\mu_{eff} = 0.7977\sqrt{\chi_m T} \qquad (20.11)$$

$$\mu_{eff} = 2.828\sqrt{\chi_m T} \quad \text{(for use with Gaussian units)} \quad (20.12)$$

Several methods can be used to measure χ_m: e.g. the *Gouy balance* (Figure 20.21), the *Faraday balance* (which operates in a similar manner to the Gouy balance) and a more recent technique using a *SQUID* (see *Section 27.4*). The Gouy method makes use of the interaction between unpaired electrons and a magnetic field; a diamagnetic material is repelled

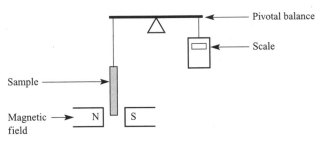

Fig. 20.21 Schematic representation of a Gouy balance.

[†] Units in magnetochemistry are non-trivial; for detailed information, see: I. Mills, T. Cvitaš, K. Homann, N. Kallay and K. Kuchitsu (1993) *IUPAC: Quantities, Units and Symbols in Physical Chemistry*, 2nd edn, Blackwell Science, Oxford.

Magnetic susceptibility

It is important to distinguish between the magnetic susceptibilities χ, χ_g and χ_m.

- Volume susceptibility is χ and is dimensionless.

- Gram susceptibility is $\chi_g = \dfrac{\chi}{\rho}$ where ρ is the density of the sample; the units of χ_g are $m^3\,kg^{-1}$.

- Molar susceptibility is $\chi_m = \chi_g M$ (where M is the molecular mass of the compound) and has SI units of $m^3\,mol^{-1}$.

by a magnetic field whereas a paramagnetic material is attracted into it. The compound for study is placed in a glass tube, suspended from a balance on which the weight of the sample is recorded. The tube is placed so that one end of the sample lies at the point of maximum magnetic flux in an electromagnetic field while the other end is at a point of low flux. Initially the magnet is switched off, but upon applying a magnetic field, paramagnetic compounds are drawn into it by an amount that depends on the number of unpaired electrons. The change in weight caused by the movement of the sample into the field is recorded, and from the associated force it is possible to calculate the magnetic susceptibility of the compound. The effective magnetic moment is then derived using equation 20.12.

For metal complexes in which the spin quantum number S is the same as for the isolated gaseous metal ion, the spin-only formula (equation 20.8 or 20.9) can be applied to find the number of unpaired electrons. Table 20.7 lists examples in which measured values of μ_{eff} correlate fairly well with those derived from the spin-only formula; note that all the metal ions are from the *first row* of the *d*-block. Use of the spin-only formula allows the number of unpaired electrons to be determined and gives information about, for example, oxidation state of the metal and whether the complex is low- or high-spin.

Worked example 20.7   Magnetic moments: spin-only formula

At room temperature, the observed value of μ_{eff} for $[Cr(en)_3]Br_2$ is 4.75 μ_B. Is the complex high- or low-spin? (Ligand abbreviations: see Table 6.7.)

$[Cr(en)_3]Br_2$ contains the octahedral $[Cr(en)_3]^{2+}$ complex, and a Cr^{2+} (d^4) ion. Low-spin will have two unpaired electrons ($n = 2$), and high-spin, four ($n = 4$).

Assume that the spin-only formula is valid (first row metal, octahedral complex):

$$\mu(\text{spin-only}) = \sqrt{n(n+2)}$$

For low-spin: $\mu(\text{spin-only}) = \sqrt{8} = 2.83$

For high-spin: $\mu(\text{spin-only}) = \sqrt{24} = 4.90$

The latter is close to the observed value, and is consistent with a high-spin complex.

Self-study exercises

1. Given that the observed value of μ_{eff} for $[Cr(en)_3]^{2+}$ is 4.75 μ_B, deduce the number of unpaired electrons.   [*Ans.* 4]

Table 20.7 Spin-only values of μ_{eff} compared with approximate ranges of observed magnetic moments for high-spin complexes of first row *d*-block ions.

Metal ion	d^n configuration	S	μ_{eff}(spin-only) / μ_B	Observed values of μ_{eff} / μ_B
Sc^{3+}, Ti^{4+}	d^0	0	0	0
Ti^{3+}	d^1	$1/2$	1.73	1.7–1.8
V^{3+}	d^2	1	2.83	2.8–3.1
V^{2+}, Cr^{3+}	d^3	$3/2$	3.87	3.7–3.9
Cr^{2+}, Mn^{3+}	d^4	2	4.90	4.8–4.9
Mn^{2+}, Fe^{3+}	d^5	$5/2$	5.92	5.7–6.0
Fe^{2+}, Co^{3+}	d^6	2	4.90	5.0–5.6
Co^{2+}	d^7	$3/2$	3.87	4.3–5.2
Ni^{2+}	d^8	1	2.83	2.9–3.9
Cu^{2+}	d^9	$1/2$	1.73	1.9–2.1
Zn^{2+}	d^{10}	0	0	0

2. At 298 K, the observed value of μ_{eff} for $[Cr(NH_3)_6]Cl_2$ is 4.85 μ_B. Confirm that the complex is high-spin.

3. At 300 K, the observed value of μ_{eff} for $[V(NH_3)_6]Cl_2$ is 3.9 μ_B. Confirm that this corresponds to what is expected for an octahedral, d^3 complex.

Spin and orbital contributions to the magnetic moment

By no means all paramagnetic complexes obey the spin-only formula and caution must be exercised in its use. It is often the case that moments arising from *both* the spin and orbital angular momenta contribute to the observed magnetic moment. Details of the Russell–Saunders coupling scheme to obtain the total angular momentum quantum number, J, from quantum numbers L and S are given in *Box 20.5*, along with notation for state symbols $^{(2S+1)}L_J$. The energy difference between adjacent states with J values of J' and $(J' + 1)$ is given by the expression $(J' + 1)\lambda$ where λ is called the *spin–orbit coupling constant*. For the d^2 configuration, for example, the ^3F state in an octahedral field is split into states $^3F_2$, $^3F_3$ and $^3F_4$, the energy differences between successive pairs being 3λ and 4λ respectively. In a magnetic field, each state with a different J value splits again to give $(2J + 1)$ different levels separated by $g_J\mu_B B_0$ where g_J is a constant called the Landé splitting factor and B_0 is the magnetic field; it is the very small energy differences between these levels with which ESR (also called electron paramagnetic resonance, EPR) spectroscopy is concerned.[†] The overall splitting pattern for a d^2 ion is shown in Figure 20.22.

The value of λ varies from a fraction of a cm^{-1} for the very lightest atoms to a few thousand cm^{-1} for the heaviest ones. The extent to which states of different J values are populated at ambient temperature depends on how large their separation is compared with the thermal energy available, kT; at 300 K, $kT \approx 200$ cm^{-1} or 2.6 kJ mol^{-1}. It can be shown theoretically that if the separation of energy levels is large, the magnetic moment is given by equation 20.13. Strictly, this applies only to free-ion energy levels, but it gives values for the magnetic moments of lanthanoid ions (for which λ is typically 1000 cm^{-1}) that are in good agreement with observed values (see *Section 24.4*).

$$\left.\begin{array}{l} \mu_{\text{eff}} = g_J\sqrt{J(J + 1)} \\[2ex] \text{where} \quad g_J = 1 + \dfrac{S(S + 1) - L(L + 1) + J(J + 1)}{2J(J + 1)} \end{array}\right\}$$

(20.13)

For d-block metal ions, equation 20.13 gives results that correlate poorly with experimental data (Tables 20.7 and 20.8). For many (but not all) first row metal ions, λ is very

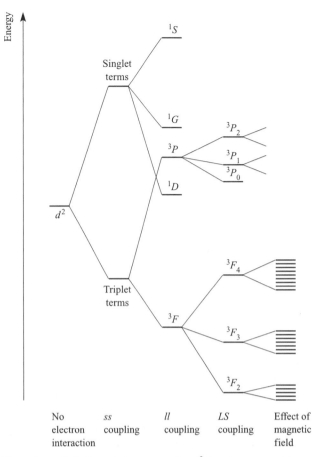

Fig. 20.22 Splitting of the terms of a d^2 ion (not to scale).

small and the spin and orbital angular momenta of the electrons operate independently. For this case, the van Vleck formula (equation 20.14) has been derived; strictly, equation 20.14 applies to free ions but, in a complex ion, the crystal field partly or fully *quenches* the orbital angular momentum. Data in Tables 20.7 and 20.8 reveal a poor fit between observed values of μ_{eff} and those calculated from equation 20.14.

$$\mu_{\text{eff}} = \sqrt{4S(S + 1) + L(L + 1)}$$

(20.14)

If there is *no contribution* from orbital motion, then equation 20.14 reduces to equation 20.15 which is the spin-only formula we met earlier. Any ion for which $L = 0$ (e.g. high-spin d^5 Mn^{2+} or Fe^{3+} in which each orbital with $m_l = +2$, $+1$, 0, -1, -2 is singly occupied giving $L = 0$) should, therefore, obey equation 20.15.

$$\mu_{\text{eff}} = \sqrt{4S(S + 1)} = 2\sqrt{S(S + 1)}$$

(20.15)

However, some other complex ions also obey the spin-only formula (Tables 20.7 and 20.8). In order for an electron to have orbital angular momentum, it must be possible to transform the orbital it occupies into an entirely equivalent and degenerate orbital by rotation; the electron is then effectively rotating about the axis used for the rotation of the orbital. In an octahedral complex, for example, the three t_{2g} orbitals can be interconverted by rotations through 90°; thus, an electron in a t_{2g} orbital has orbital angular momentum. The e_g

---

[†] For an introduction to ESR spectroscopy, see Chapter 5 in: R.V. Parish (1990) *NMR, NQR, EPR and Mössbauer Spectroscopy in Inorganic Chemistry*, Ellis Horwood, Chichester.

Table 20.8 Calculated magnetic moments for first row d-block metal ions in high-spin complexes at ambient temperatures. Compare these values with those observed (Table 20.7).

Metal ion	Ground term	μ_{eff} / μ_B calculated from equation 20.13	μ_{eff} / μ_B calculated from equation 20.14	μ_{eff} / μ_B calculated from equation 20.15
Ti^{3+}	$^2D_{3/2}$	1.55	3.01	1.73
V^{3+}	$^3F_2$	1.63	4.49	2.83
V^{2+}, Cr^{3+}	$^4F_{3/2}$	0.70	5.21	3.87
Cr^{2+}, Mn^{3+}	$^5D_0$	0	5.50	4.90
Mn^{2+}, Fe^{3+}	$^6S_{5/2}$	5.92	5.92	5.92
Fe^{2+}, Co^{3+}	$^5D_4$	6.71	5.50	4.90
Co^{2+}	$^4F_{9/2}$	6.63	5.21	3.87
Ni^{2+}	$^3F_4$	5.59	4.49	2.83
Cu^{2+}	$^2D_{5/2}$	3.55	3.01	1.73

orbitals, having different shapes, cannot be interconverted and so electrons in e_g orbitals never have angular momentum. There is, however, another factor that needs to be taken into account: if all the t_{2g} orbitals are singly occupied, an electron in, say, the d_{xz} orbital cannot be transferred into the d_{xy} or d_{yz} orbital because these already contain an electron having the same spin quantum number as the incoming electron. If all the t_{2g} orbitals are doubly occupied, electron transfer is also impossible. It follows that in high-spin octahedral complexes, orbital contributions to the magnetic moment are important only for the configurations t_{2g}^{1}, t_{2g}^{2}, $t_{2g}^{4}e_g^{2}$ and $t_{2g}^{5}e_g^{2}$. For tetrahedral complexes, it is similarly shown that the configurations that give rise to an orbital contribution are $e^2t_2^{1}$, $e^2t_2^{2}$, $e^4t_2^{4}$ and $e^4t_2^{5}$. These results lead us to the conclusion that an octahedral high-spin d^7 complex should have a magnetic moment greater than the spin-only value of 3.87 μ_B but a tetrahedral d^7 complex should not. However, the observed values of μ_{eff} for $[Co(H_2O)_6]^{2+}$ and $[CoCl_4]^{2-}$ are 5.0 and 4.4 μ_B respectively, i.e. *both* complexes have magnetic moments greater than μ(spin-only). The third factor involved is *spin-orbit coupling*.

Spin-orbit coupling is a complicated subject and we can give only a brief mention here; for further details of spin-orbit coupling, see the advanced texts listed in the end of chapter reading list. As a result of a mixing of states (which we have so far ignored), theory brings us to equation 20.16; this modifies the μ(spin-only) formula to take into account spin-orbit coupling and, although dependent on Δ_{oct}, applies also to tetrahedral complexes.

$$\mu_{\text{eff}} = \mu(\text{spin-only})\left(1 - \frac{\alpha\lambda}{\Delta_{\text{oct}}}\right) \quad (20.16)$$

where λ = spin-orbit coupling constant and α = constant that depends on the ground term: 4 for an A ground state, and 2 for an E ground state. The simple approach of equation 20.16 is not applicable to ions with a T ground state.

Some values of λ are given in Table 20.9. Note that λ is positive for less than half-filled shells and negative for shells that are more than half-filled. Thus, spin-orbit coupling leads to:

- $\mu_{\text{eff}} > \mu(\text{spin-only})$ for d^6, d^7, d^8 and d^9 ions;
- $\mu_{\text{eff}} < \mu(\text{spin-only})$ for d^1, d^2, d^3 and d^4 ions.

Worked example 20.8 Magnetic moments: spin-orbit coupling

Calculate a value for μ_{eff} for $[Ni(en)_3]^{2+}$ taking into account spin-orbit coupling. Compare your answer with μ(spin-only) and the value of 3.16 μ_B observed experimentally for $[Ni(en)_3][SO_4]$. [Data: see *Tables 20.2 and 20.9*.]

Octahedral Ni(II) (d^8) has a $^3A_{2g}$ ground state. Equation needed:

$$\mu_{\text{eff}} = \mu(\text{spin-only})\left(1 - \frac{4\lambda}{\Delta_{\text{oct}}}\right)$$

$$\mu(\text{spin-only}) = \sqrt{n(n+2)} = \sqrt{8} = 2.83$$

From Table 20.2: $\Delta_{\text{oct}} = 11\,500\,\text{cm}^{-1}$
From Table 20.9: $\lambda = -315\,\text{cm}^{-1}$

$$\mu_{\text{eff}} = 2.83\left(1 + \frac{4 \times 315}{11\,500}\right) = 3.14\,\mu_B$$

Table 20.9 Spin-orbit coupling coefficients for selected first row d-block metal ions.

Metal ion	Ti^{3+}	V^{3+}	Cr^{3+}	Mn^{3+}	Fe^{2+}	Co^{2+}	Ni^{2+}	Cu^{2+}
d^n configuration	d^1	d^2	d^3	d^4	d^6	d^7	d^8	d^9
λ/cm^{-1}	155	105	90	88	-102	-177	-315	-830

The calculated value is significantly larger than μ(spin-only) as expected for a d^n configuration with a more than half-full shell; it agrees well with the experimental value.

Self-study exercises

Use data in Tables 20.2 and 20.9.

1. Calculate a value for μ_{eff} for $[\mathrm{Ni(NH_3)_6}]^{2+}$ taking into account spin–orbit coupling. [*Ans.* 3.16 μ_B]

2. Estimate a value for μ_{eff} for $[\mathrm{Ni(H_2O)_6}]^{2+}$ taking into account spin–orbit coupling. [*Ans.* 3.25 μ_B]

An important point is that spin–orbit coupling is generally large for second and third row d-block metal ions and this leads to large discrepancies between μ(spin-only) and observed values of μ_{eff}. The d^1 complexes *cis*-$[\mathrm{NbBr_4(NCMe)_2}]$ and *cis*-$[\mathrm{TaCl_4(NCMe)_2}]$ illustrate this clearly: room temperature values of μ_{eff} are 1.27 and 0.45 μ_B respectively compared to μ(spin-only) = 1.73 μ_B.

The effects of temperature on μ_{eff}

So far, we have ignored the effects of temperature on μ_{eff}. If a complex obeys the Curie Law (equation 20.17), then μ_{eff} is independent of temperature; this follows from a combination of equations 20.11 and 20.17.

$$\chi = \frac{C}{T} \qquad (20.17)$$

where C = Curie constant; T = temperature in K. However, the Curie Law is rarely obeyed and so it is essential to state the temperature at which a value of μ_{eff} has been measured. For second and third row d-block metal ions in particular, quoting *only* a room temperature value of μ_{eff} is usually meaningless; when spin–orbit coupling is large, μ_{eff} is highly dependent on T. For a given electronic configuration, the influence of temperature on μ_{eff} can be seen from a *Kotani plot* of μ_{eff} against $-kT/\lambda$ where k is the Boltzmann constant, T is the temperature in K, and λ is the spin–orbit coupling constant. Remember that λ is small for first row metal ions, is large for a second row metal ion, and is even larger for a third row ion. Figure 20.23 shows a Kotani plot for a t_{2g}^4 configuration; four points are indicated on the curve and correspond to typical values of μ_{eff}(298 K) for complexes of Cr(II) and Mn(III) from the first row, and Ru(IV) and Os(IV) from the second and third rows respectively. Points to note from these data are:

- since the points corresponding to μ_{eff}(298 K) for the first row metal ions lie on the near-horizontal part of the curve, changing the temperature has little effect on μ_{eff};
- since the points relating to μ_{eff}(298 K) for the heavier metal ions lie on parts of the curve with steep gradients, μ_{eff} is sensitive to changes in temperature; this is especially true for Os(IV).

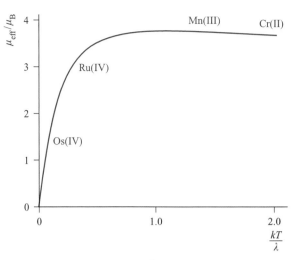

Fig. 20.23 Kotani plot for a t_{2g}^4 configuration; λ is the spin–orbit coupling constant. Typical values of μ_{eff}(298 K) for Cr(II), Mn(III), Ru(IV) and Os(IV) are indicated on the curve.

Spin crossover

The choice between a low- and high-spin configuration for d^4, d^5, d^6 and d^7 complexes is not always unique and a *spin crossover* sometimes occurs; this may be initiated by a change in pressure (e.g. a low to high-spin crossover for $[\mathrm{Fe(CN)_5(NH_3)}]^{3-}$ at high pressure) or temperature (e.g. octahedral $[\mathrm{Fe(phen)_2(NCS}$-$N)_2]$, octahedral $[\mathrm{Fe(20.9)_2}]$ and the square-based pyramidal complex **20.10** undergo low-to-high spin crossovers at 175, 391 and 180 K respectively). The change in the value of μ_{eff} which accompanies the spin crossover may be gradual or abrupt (Figure 20.24).[†]

* = coordination site

(20.9)

(20.10)

[†] An application of spin crossover is described in 'Molecules with short memories': O. Kahn (1999) *Chemistry in Britain*, vol. 35, number 2, p. 24.

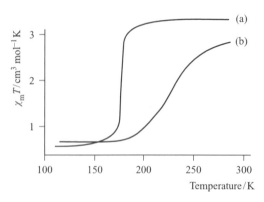

phen = 1,10-phenanthroline

btz = 2,2′-bi-4,5-dihydrothiazine

Fig. 20.24 The dependence of the observed values of μ_{eff} on temperature for (a) $[\text{Fe(phen)}_2(\text{NCS-}N)_2]$ where low-to-high spin crossover occurs abruptly at 175 K, and (b) $[\text{Fe(btz)}_2(\text{NCS-}N)_2]$ where low-to-high spin crossover occurs more gradually. Ligand abbreviations are defined in the figure [data: J.-A. Real *et al.* (1992) *Inorg. Chem.*, vol. 31, p. 4972].

Ferromagnetism, antiferromagnetism and ferrimagnetism

Whenever we have mentioned magnetic properties so far, we have assumed that metal centres have no interaction with each other (Figure 20.25a). This is true for substances where the paramagnetic centres are well separated from each other by diamagnetic species; such systems are said to be *magnetically dilute*. When the paramagnetic species are very close together (as in the bulk metal) or are separated by a species that can transmit magnetic interactions (as in many oxides, fluorides or chlorides), the metal centres may interact (*couple*) with one another. The interaction may give rise to *ferromagnetism* or *antiferromagnetism* (Figures 20.25b and 20.25c).

> In a *ferromagnetic* material, large domains of magnetic dipoles are aligned in the same direction; in an *antiferromagnetic* material, neighbouring magnetic dipoles are aligned in opposite directions.

Ferromagnetism leads to greatly enhanced paramagnetism as in iron metal at temperatures of up to 1041 K (the

Curie temperature, T_{C}) above which thermal energy is sufficient to overcome the alignment and normal paramagnetic behaviour prevails. Antiferromagnetism occurs below the *Néel temperature*, T_{N}; as the temperature decreases, less thermal energy is available and the paramagnetic susceptibility falls rapidly. The classic example of antiferromagnetism is MnO which has a NaCl-type lattice and a Néel temperature of 118 K. Neutron diffraction is capable of distinguishing between sets of atoms having opposed magnetic moments and reveals that the unit cell of MnO at 80 K is double the one at 293 K; this indicates that in the conventional unit cell (*Figure 5.15*), metal atoms at adjacent corners have opposed moments at 80 K and that the cells must be stacked to produce the 'true' unit cell. More complex behaviour may occur if some moments are systematically aligned so as to oppose others, but relative numbers or relative values of the moments are such as to lead to a finite resultant magnetic moment: this is *ferrimagnetism* and is represented schematically in Figure 20.25d.

When a bridging ligand facilitates the coupling of electron spins on adjacent metal centres, the mechanism is one of *superexchange*. This is shown schematically in diagram

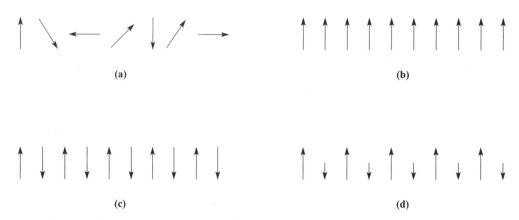

Fig. 20.25 Representations of (a) paramagnetism, (b) ferromagnetism, (c) antiferromagnetism and (d) ferrimagnetism.

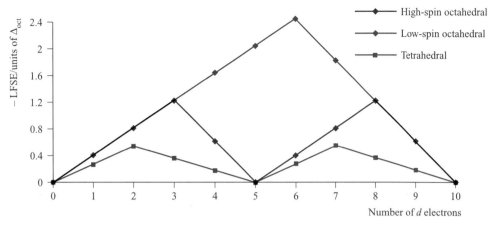

Fig. 20.26 Ligand field stabilization energies as a function of Δ_{oct} for high- and low-spin octahedral systems and for tetrahedral systems; Jahn–Teller effects for d^4 and d^9 configurations have been ignored.

20.11 in which the unpaired metal electrons are represented in blue.

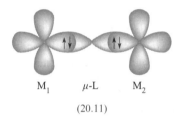

(20.11)

In a *superexchange* pathway, the unpaired electron on the first metal centre, M_1, interacts with a spin-paired pair of electrons on the bridging ligand with the result that the unpaired electron on M_2 is aligned in an antiparallel manner with respect to that on M_1.

20.9 Thermodynamic aspects: ligand field stabilization energies (LFSE)

Trends in LFSE

So far, we have considered Δ_{oct} (or Δ_{tet}) only as a quantity derived from electronic spectroscopy and representing the energy required to transfer an electron from a t_{2g} to an e_g level (or from an e to t_2 level). However, chemical significance can be attached to these values. Table 20.3 showed the variation in crystal field stabilization energies (CFSE) for high- and low-spin octahedral systems; these trends are restated in Figure 20.26 where they are compared with that for a tetrahedral field, Δ_{tet} being expressed as a fraction of Δ_{oct} (see *equation 20.5*). Note the change from CFSE to LFSE in moving from Table 20.3 to Figure 20.26; this reflects the fact that we are now dealing with *ligand field theory* and *ligand field stabilization energies*. In Table 20.3 we included pairing energies; if we are considering only the *extra* stability that results from the splitting of the d orbitals

in high-spin complexes, we can to a first approximation ignore the pairing energy term, since even in the free gaseous d^6–d^{10} ions, pairing is inevitable and so only the *difference* in pairing energy between that in the free ion and in the complex is involved. For low-spin octahedral complexes of d^4–d^7 ions, Figure 20.26 reveals a large increase in LFSE, but this is *partly* offset by pairing energy considerations. In practice it is often very difficult to interpret the electronic spectra of low-spin d^4–d^7 complexes unambiguously, and so we shall have little to say about these cases. However, the stabilization of the low-spin d^6 configuration does have important consequences in the reaction kinetics of these systems and we return to this in Chapter 25.

In the discussion that follows, we consider relationships between observed trends in LFSE values and selected thermodynamic properties of high-spin compounds of the d-block metals.

Lattice energies and hydration energies of M^{n+} ions

Figure 20.27 shows a plot of experimental lattice energy data for metal(II) chlorides of first row d-block elements; in each salt, the metal ion is high-spin and lies in an octahedral environment in the solid state.[†] The 'double-hump' in Figure 20.27 is reminiscent of that in Figure 20.26, albeit with respect to a reference line which shows a general increase in lattice energy as the period is crossed. Similar plots can be obtained for species such as MF_2, MF_3 and $[MF_6]^{3-}$, but for each series, only limited data are available and complete trends cannot be studied.

Water is a weak-field ligand and $[M(H_2O)_6]^{2+}$ ions of the first row metals are high-spin. The relationship between absolute enthalpies of hydration of M^{2+} ions (see *Section*

[†] Strictly, a purely electrostatic model does not hold for chlorides, but we include them because more data are available than for fluorides for which the electrostatic model is more appropriate.

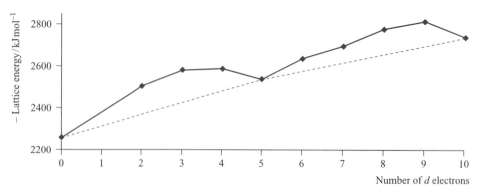

Fig. 20.27 Lattice energies (derived from Born–Haber cycle data) for MCl_2 where M is a first row *d*-block metal; the point for d^0 corresponds to $CaCl_2$. Data are not available for scandium where the stable oxidation state is +3.

6.9) and d^n configuration is shown in Figure 20.28, and again we see the 'double-humped' appearance of Figures 20.26 and 20.27.

For each plot in Figures 20.27 and 20.28, deviations from the reference line joining the d^0, d^5 and d^{10} points may be taken as measures of 'thermochemical LFSE' values. In general, the agreement between these values and those calculated from the values of Δ_{oct} derived from electronic spectroscopic data are fairly close. For example, for $[Ni(H_2O)_6]^{2+}$, the values of LFSE(thermochemical) and LFSE(spectroscopic) are 120 and $126 \, kJ \, mol^{-1}$ respectively; the latter comes from an evaluation of $1.2\Delta_{oct}$ where Δ_{oct} is determined from the electronic spectrum of $[Ni(H_2O)_6]^{2+}$ to be $8500 \, cm^{-1}$. We have to emphasize that this level of agreement is fortuitous; if we look more closely at the problem, we note that only *part* of the measured hydration enthalpy can be attributed to the first coordination sphere of six H_2O molecules, and, moreover, the definitions of LFSE(thermochemical) and LFSE(spectroscopic) are not strictly equivalent. In conclusion, we must make the very important point that, interesting and useful though discussions of 'double-humped' graphs are in dealing with trends in the thermodynamics of high-spin complexes, they are never more than *approximations*. It is crucial to remember

that LFSE terms are only *small parts* of the total interaction energies (generally <10%).

Octahedral versus tetrahedral coordination: spinels

Figure 20.26 indicates that, if all other factors are equal, d^0, high-spin d^5 and d^{10} ions should have no preference between tetrahedral and octahedral coordination, and that the strongest preference for octahedral coordination should be found for d^3 and d^8 ions. In practice, other factors do operate. For example, the smaller size of tetrahedral complexes results in higher lattice and solvation energies; thus, although Ni^{2+} (d^8) does not form tetrahedral complexes in aqueous solution, it does so in melts and non-aqueous media.

The distribution of metal ions between tetrahedral and octahedral sites in a spinel (see *Box 12.6*) can be rationalized in terms of LFSEs. In a normal spinel $A^{II}B_2^{III}O_4$ the tetrahedral sites are occupied by the A^{2+} ions and the octahedral sites by B^{3+} ions: $(A^{II})^{tet}(B^{III})_2^{oct}O_4$. In an inverse spinel, the distribution is $(B^{III})^{tet}(A^{II}B^{III})^{oct}O_4$. For spinel itself, $A = Mg$, $B = Al$. If at least one of the cations is from the *d*-block, the inverse structure is frequently (though by no means always) observed: $Zn^{II}Fe_2^{III}O_4$,

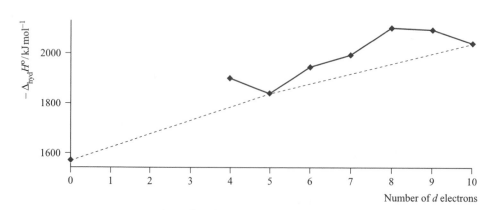

Fig. 20.28 Absolute enthalpies of hydration of the M^{2+} ions of the first row metals; the point for d^0 corresponds to Ca^{2+}. Data are not available for Sc^{2+}, Ti^{2+} and V^{2+}.

$Fe^{II}Cr_2^{III}O_4$ and $Mn^{II}Mn_2^{III}O_4$ are normal spinels while $Ni^{II}Ga_2^{III}O_4$, $Co^{II}Fe_2^{III}O_4$ and $Fe^{II}Fe_2^{III}O_4$ are inverse spinels. To account for these observations we first note the following:

- the Madelung constants for the spinel and inverse spinel lattices are usually nearly equal;
- the charges on the metal ions are independent of environment (an assumption);
- Δ_{oct} values for complexes of M^{3+} ions are significantly greater than for corresponding complexes of M^{2+} ions.

Consider compounds with normal spinel structures: in $Zn^{II}Fe_2^{III}O_4$ (d^{10} and d^5), LFSE = 0 for each ion; in $Fe^{II}Cr_2^{III}O_4$ (d^6 and d^3), Cr^{3+} has a much greater LFSE in an octahedral site than does high-spin Fe^{2+}; in $Mn^{II}Mn_2^{III}O_4$ (d^5 and d^4), only Mn^{3+} has any LFSE and this is greater in an octahedral than a tetrahedral site. Now consider some inverse spinels: in $Ni^{II}Ga_2^{III}O_4$, only Ni^{2+} (d^8) has any LFSE and this is greater in an octahedral site; in each of $Co^{II}Fe_2^{III}O_4$ (d^7 and d^5) and $Fe^{II}Fe_2^{III}O_4$ (d^6 and d^5), LFSE = 0 for Fe^{3+} and so the preference is for Co^{2+} and Fe^{2+} respectively to occupy octahedral sites. While this argument is impressive, we must note that observed structures do not always agree with LFSE expectations, e.g. $Fe^{II}Al_2^{III}O_4$ is a normal spinel.

20.10 Thermodynamic aspects: the Irving–Williams series

In aqueous solution, water is replaced by other ligands (equation 20.18, and see *Table 6.7*) and the position of equilibrium will be related to the difference between two LFSEs, since Δ_{oct} is ligand-dependent.

$$[Ni(H_2O)_6]^{2+} + [EDTA]^{4-} \rightleftharpoons [Ni(EDTA)]^{2-} + 6H_2O$$
$$(20.18)$$

Table 20.10 lists overall stability constants (see *Section 6.12*) for $[M(en)_3]^{2+}$ and $[M(EDTA)]^{2-}$ high-spin complexes for d^5 to d^{10} first row M^{2+} ions. For a given ligand and cation charge, ΔS^o should be nearly constant along the series and the variation in $\log \beta_6$ should approximately parallel the trend in values of $-\Delta H^o$. Table 20.10 shows that the trend from d^5 to d^{10} follows a 'single-hump', with the ordering of $\log \beta_6$ for the high-spin ions being:

$$Mn^{2+} < Fe^{2+} < Co^{2+} < Ni^{2+} < Cu^{2+} > Zn^{2+}$$

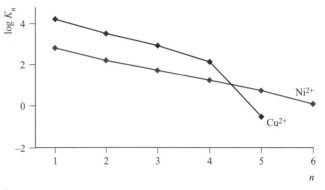

Fig. 20.29 Stepwise stability constants ($\log K_n$) for the displacement of H_2O by NH_3 from $[Ni(H_2O)_6]^{2+}$ (d^8) and $[Cu(H_2O)_6]^{2+}$ (d^9).

This is called the *Irving–Williams series* and is observed for a wide range of ligands. The trend is a 'hump' that peaks at Cu^{2+} (d^9) and not at Ni^{2+} (d^8) as might be expected from a consideration of LFSEs (Figure 20.26); while the variation in LFSE values is a contributing factor, it is not the sole arbiter. Trends in stability constants should bear a relationship to trends in ionic radii (see *Appendix 6*); the pattern in values of r_{ion} for 6-coordinate high-spin ions is:

$$Mn^{2+} > Fe^{2+} > Co^{2+} > Ni^{2+} < Cu^{2+} < Zn^{2+}$$

We might expect r_{ion} to decrease from Mn^{2+} to Zn^{2+} as Z_{eff} increases, but once again we see a dependence on the d^n configuration with Ni^{2+} being smallest. In turn, this predicts the highest value of $\log \beta_6$ for Ni^{2+}. Why, then, are copper(II) complexes so much more stable than might be expected? The answer lies in the Jahn–Teller distortion that a d^9 complex suffers; the six metal–ligand bonds are not of equal length and thus the concept of a 'fixed' ionic radius for Cu^{2+} is not valid. In an elongated complex (structure **20.5**) such as $[Cu(H_2O)_6]^{2+}$, there are four short and two long Cu–O bonds. Plots of stepwise stability constants for the displacement of H_2O by NH_3 ligands in $[Cu(H_2O)_6]^{2+}$ and $[Ni(H_2O)_6]^{2+}$ are shown in Figure 20.29; for the first four substitution steps, complex stability is greater for Cu^{2+} than Ni^{2+}, reflecting the formation of four short (strong) Cu–N bonds. The value of $\log K_5$ for Cu^{2+} is consistent with the formation of a weak (axial) Cu–N bond; $\log K_6$ cannot be measured in aqueous solution. The magnitude of the overall stability constant for complexation of Cu^{2+} is dominated by values of K_n for the first four steps and the thermodynamic favourability of these displacement steps is

Table 20.10 Overall stability constants for selected d-block metal complexes.

Metal ion	Mn^{2+}	Fe^{2+}	Co^{2+}	Ni^{2+}	Cu^{2+}	Zn^{2+}
$\log \beta_6$ for $[M(en)_3]^{2+}$	5.7	9.5	13.8	18.6	18.7	12.1
$\log \beta_6$ for $[M(EDTA)]^{2-}$	13.8	14.3	16.3	18.6	18.7	16.1

responsible for the position of Cu^{2+} in the Irving–Williams series.

20.11 Thermodynamic aspects: oxidation states in aqueous solution

In the preceding sections, we have, with some degree of success, attempted to rationalize irregular trends in some thermodynamic properties of the first row *d*-block metals. Now we consider the variation in E° values for equilibrium 20.19 (Table 19.1 and Figure 20.30); the more negative the value of E°, the less easily M^{2+} is reduced.

$$M^{2+}(aq) + 2e^{-} \rightleftharpoons M(s) \qquad (20.19)$$

This turns out to be a difficult problem. Water is relatively easily oxidized or reduced, and the range of oxidation states on which measurements can be made under aqueous conditions is therefore restricted, e.g. Sc(II) and Ti(II) would liberate H_2. Values of $E^{\circ}(M^{2+}/M)$ are related to energy changes accompanying the processes:

$$M(s) \longrightarrow M(g) \qquad \textit{atomization } (\Delta_{a}H^{\circ})$$

$$M(g) \longrightarrow M^{2+}(g) \qquad \textit{ionization } (IE_1 + IE_2)$$

$$M^{2+}(g) \longrightarrow M^{2+}(aq) \qquad \textit{hydration } (\Delta_{hyd}H^{\circ})$$

In crossing the first row of the *d*-block, the general trend is for $\Delta_{hyd}H^{\circ}$ to become more negative (Figure 20.28). There is also a successive increase in the sum of the first two ionization energies albeit with discontinuities at Cr and Cu (Figure 20.31). Values of $\Delta_{a}H^{\circ}$ vary erratically and over a wide range with a particularly low value for zinc (*Table 5.2*). The net effect of all these factors is an irregular variation in values of $E^{\circ}(M^{2+}/M)$ across the row, and it is clearly not worth while discussing the relatively small variations in LFSEs.

Consider now the variations in $E^{\circ}(M^{3+}/M^{2+})$ across the row. The enthalpy of atomization is no longer relevant and we are concerned only with trends in the third ionization energy (Table 20.11) and the hydration energies of M^{2+} and M^{3+}. Experimental values for $E^{\circ}(M^{3+}/M^{2+})$ (Table 20.11) are restricted to middle of the series; Sc(II) and Ti(II) would reduce water while Ni(III), Cu(III) and Zn(III) would oxidize it. In general, larger values of IE_3

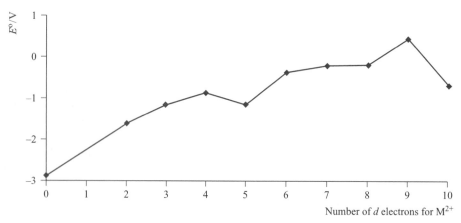

Fig. 20.30 The variation in values of $E^{\circ}(M^{2+}/M)$ as a function of d^{n} configuration for the first row metals; the point for d^{0} corresponds to M = Ca.

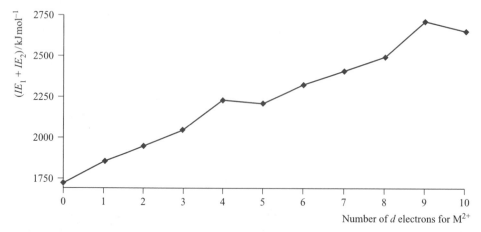

Fig. 20.31 The variation in the sum of the first and second ionization energies as a function of d^{n} configuration for the first row metals; the point for d^{0} corresponds to M = Ca.

Table 20.11 Standard reduction potentials for the equilibrium $M^{3+}(aq) + e^- \rightleftharpoons M^{2+}(aq)$ and values of the third ionization energies.

M	V	Cr	Mn	Fe	Co
E°/V	-0.26	-0.41	$+1.54$	$+0.77$	$+1.92$
$IE_3/\text{kJ mol}^{-1}$	2827	2992	3252	2962	3232

correspond to more positive E° values; this suggests that a steady increase in the difference between the hydration energies of M^{3+} and M^{2+} (which would become larger as the ions become smaller) is outweighed by the variation in IE_3. The only pair of metals for which the change in E° appears out of step is vanadium and chromium. The value of IE_3 for Cr is 165 kJ mol^{-1} greater than for V and so it is harder to oxidize *gaseous* Cr^{2+} than V^{2+}. In aqueous solution however, Cr^{2+} is a more powerful reducing agent than V^{2+}. These oxidations correspond to changes in electronic configuration of $d^3 \longrightarrow d^2$ for V and $d^4 \longrightarrow d^3$ for Cr. The V^{2+}, V^{3+}, Cr^{2+} and Cr^{3+} hexaaqua ions are high-spin; oxidation of V^{2+} is accompanied by a *loss* of LFSE (Table 20.3) while there is a *gain* in LFSE (i.e. more negative) upon oxidation of Cr^{2+} (minor consequences of the Jahn–Teller effect are ignored). Using values of Δ_{oct} from Table 20.2, these changes in LFSE are expressed as follows:

Change in LFSE on oxidation of V^{2+} is

$$-(1.2 \times 12\,400) \text{ to } -(0.8 \times 17\,850)$$

$$= -14\,880 \text{ to } -14\,280 \text{ cm}^{-1}$$

$$= +600 \text{ cm}^{-1}$$

Change in LFSE on oxidation of Cr^{2+} is

$$-(0.6 \times 14\,100) \text{ to } -(1.2 \times 17\,400)$$

$$= -8460 \text{ to } -20\,880 \text{ cm}^{-1}$$

$$= -12\,420 \text{ cm}^{-1}$$

The gain in LFSE upon formation of Cr^{3+} corresponds to $\approx 149 \text{ kJ mol}^{-1}$ and largely cancels out the effect of the third ionization energy. Thus, the apparent anomaly of $E^{\circ}(Cr^{3+}/Cr^{2+})$ can be mostly accounted for in terms of LFSE effects – a considerable achievement in view of the simplicity of the theory.

Summary

This chapter has provided an introduction to *d*-block metal chemistry with an emphasis on bonding and physical properties. Crystal and ligand field theories are central to the interpretation of electronic spectra of *d*-block metal complexes, and trends in a number of thermodynamic quantities can be rationalized in terms of variations in ligand field stabilization energies. Magnetic properties of metal complexes can be understood in some cases by use of the spin-only formula, but this approximation has limited applications and for the most part, the effects of spin–orbit coupling must be considered.

The following terms were introduced in this chapter. Do you know what they mean?

- ☐ high-spin
- ☐ low-spin
- ☐ outer orbital complex
- ☐ inner orbital complex
- ☐ crystal field theory
- ☐ Δ_{oct}, $\Delta_{\text{tet}} \cdots$
- ☐ weak field ligand
- ☐ strong field ligand
- ☐ spectrochemical series
- ☐ crystal field stabilization energy (CFSE)
- ☐ pairing energy
- ☐ Jahn–Teller distortion
- ☐ π-donor ligand
- ☐ π-acceptor ligand
- ☐ 18-electron rule
- ☐ 'd–d' transition
- ☐ charge transfer absorption
- ☐ λ_{max} and ε_{max} for an absorption band
- ☐ selection rule: $\Delta S = 0$
- ☐ selection rule: $\Delta l = \pm 1$
- ☐ vibronic coupling
- ☐ Orgel diagram
- ☐ table of microstates
- ☐ Tanabe–Sugano diagram
- ☐ nephelauxetic effect
- ☐ Racah parameter, B
- ☐ magnetic susceptibility
- ☐ effective magnetic moment
- ☐ spin-only formula
- ☐ Gouy balance
- ☐ Russell–Saunders coupling
- ☐ spin–orbit coupling constant
- ☐ Curie Law
- ☐ Kotani plot
- ☐ spin crossover
- ☐ ferromagnetism
- ☐ antiferromagnetism
- ☐ ferrimagnetism
- ☐ superexchange
- ☐ ligand field stabilization energy (LFSE)

Suggested further reading

Texts that complement the present treatment

I.B. Bersuker (1996) *Electronic Structure and Properties of Transition Metal Compounds*, Wiley, New York.

M. Gerloch and E.C. Constable (1994) *Transition Metal Chemistry: the Valence Shell in d-Block Chemistry*, VCH, Weinheim.

J.E. Huheey, E.A. Keiter and R.L. Keiter (1993) *Inorganic Chemistry*, 4th edn, Harper Collins, New York, Chapter 11.

W.L. Jolly (1991) *Modern Inorganic Chemistry*, 2nd edn, McGraw-Hill, New York, Chapters 15, 17 and 18.

S.F.A. Kettle (1996) *Physical Inorganic Chemistry*, Spektrum, Oxford.

Crystal and ligand field theories, electronic spectra and magnetism: advanced texts

B.N. Figgis (1966) *Introduction to Ligand Fields*, Interscience, New York.

M. Gerloch (1983) *Magnetism and Ligand Field Analysis*, Cambridge University Press, Cambridge.

M. Gerloch and R.C. Slade (1973) *Ligand Field Parameters*, Cambridge University Press, Cambridge.

A.B.P. Lever (1984) *Inorganic Electronic Spectroscopy*, 2nd edn, Elsevier, Amsterdam.

F.A. Mabbs and D.J. Machin (1973) *Magnetism and Transition Metal Complexes*, Chapman and Hall, London.

Problems

20.1 Outline how you would apply crystal field theory to explain why the five *d*-orbitals in an octahedral complex are not degenerate. Include in your answer an explanation of the 'barycentre'.

20.2 The absorption spectrum of $[Ti(H_2O)_6]^{3+}$ exhibits a band with $\lambda_{max} = 510\,nm$. What colour of light is absorbed and what colour will aqueous solutions of $[Ti(H_2O)_6]^{3+}$ appear?

20.3 Draw the structures of the following ligands, highlight the donor atoms and give the likely modes of bonding (e.g. monodentate): (a) en; (b) bpy; (c) $[CN]^-$; (d) $[N_3]^-$; (e) CO; (f) phen; (g) $[ox]^{2-}$; (h) $[NCS]^-$; (i) PMe_3.

20.4 Arrange the following ligands in order of increasing field strength: Br^-, F^-, $[CN]^-$, NH_3, $[OH]^-$, H_2O.

20.5 For which member of the following pairs of complexes would Δ_{oct} be the larger and why: (a) $[Cr(H_2O)_6]^{2+}$ and $[Cr(H_2O)_6]^{3+}$; (b) $[CrF_6]^{3-}$ and $[Cr(NH_3)_6]^{3+}$; (c) $[Fe(CN)_6]^{4-}$ and $[Fe(CN)_6]^{3-}$; (d) $[Ni(H_2O)_6]^{2+}$ and $[Ni(en)_3]^{2+}$; (e) $[MnF_6]^{2-}$ and $[ReF_6]^{2-}$; (f) $[Co(en)_3]^{3+}$ and $[Rh(en)_3]^{3+}$?

20.6 (a) Explain why there is no distinction between low and high spin arrangements for an octahedral d^8 metal ion. (b) Discuss the factors that contribute to the preference for forming either a high- or low-spin d^4 complex. (c) How would you distinguish experimentally between the two configurations in (b)?

20.7 Verify the CFSE values in Table 20.3.

20.8 In each of the following complexes, rationalize the number of observed unpaired electrons (stated after the formula): (a) $[Mn(CN)_6]^{4-}$ (1); (b) $[Mn(CN)_6]^{2-}$ (3); (c) $[Cr(en)_3]^{2+}$ (4); (d) $[Fe(ox)_3]^{3-}$ (5); (e) $[Pd(CN)_4]^{2-}$ (0); (f) $[CoCl_4]^{2-}$ (3); (g) $[NiBr_4]^{2-}$ (2).

20.9 (a) Explain the forms of the *d* orbital splitting diagrams for trigonal bipyramidal and square pyramidal complexes of formula ML_5 shown in Figure 20.10. (b) What would you expect concerning the magnetic properties of such complexes of Ni(II)?

20.10 (a) What do you understand by the *nephelauxetic effect*? (b) Place the following ligands in order of increasing nephelauxetic effect: H_2O, I^-, F^-, en, $[CN]^-$, NH_3.

20.11 Discuss each of the following observations:

(a) The $[CoCl_4]^{2-}$ ion is a regular tetrahedron but $[CuCl_4]^{2-}$ has a flattened tetrahedral structure.

(b) The electronic spectrum of $[CoF_6]^{3-}$ contains two bands with maxima at $11\,500$ and $14\,500\,cm^{-1}$.

20.12 (a) Set up a table of microstates to show that the ground term for the d^1 ion is the singlet ^2D. What are the components of this term in a tetrahedral field? (b) Repeat the process for a d^2 ion and show that the ground and excited terms are the ^3F and ^3P. What are the components of these terms in tetrahedral and octahedral fields?

20.13 (a) On Figure 20.19, convert the wavenumber scale to nm. (b) Which part of the scale corresponds to the visible range? (c) What would you predict are the colours of $[Ni(H_2O)_6]^{2+}$ and $[Ni(NH_3)_6]^{2+}$. (d) Are the spectra in Figure 20.19 consistent with the relative positions of H_2O and NH_3 in the spectrochemical series?

20.14 (a) How many '*d*–*d*' bands would you expect to find in the electronic spectrum of an octahedral Cr(III) complex? (b) Account for the observation that the colour of *trans*-$[Co(en)_2F_2]^+$ is less intense than those of *cis*-$[Co(en)_2F_2]^+$ and *trans*-$[Co(en)_2Cl_2]^+$.

20.15 Find x in the formulae of the following complexes by determining the oxidation state of the metal from the experimental values of μ_{eff}: (a) $[VCl_x(bpy)]$, $1.77\,\mu_B$; (b) $K_x[V(ox)_3]$, $2.80\,\mu_B$; (c) $[Mn(CN)_6]^{x-}$, $3.94\,\mu_B$. What assumption have you made and how valid is it?

20.16 Explain why in high-spin octahedral complexes, orbital contributions to the magnetic moment are only important for d^1, d^2, d^6 and d^7 configurations.

20.17 The observed magnetic moment for $K_3[TiF_6]$ is $1.70\,\mu_B$. (a) Calculate μ(spin-only) for this complex. (b) How can you improve on this estimate?

20.18 Comment on the observations that octahedral Ni(II) complexes have magnetic moments in the range 2.9–3.4 μ_B, tetrahedral Ni(II) complexes have moments up to $\approx 4.1\ \mu_B$, and square planar Ni(II) complexes are diamagnetic.

20.19 (a) Using data from Appendix 6, plot a graph to show how the ionic radii of high-spin, 6-coordinate M^{2+} ions of the first row of the d-block vary with the d^n configuration. Comment on factors that contribute to observed trend. (b) Briefly discuss other properties of these metal ions that show related trends.

20.20 Values of Δ_{oct} for $[Ni(H_2O)_6]^{2+}$ and high-spin $[Mn(H_2O)_6]^{3+}$ have been evaluated spectroscopically as 8500 and 21 000 cm^{-1} respectively. Assuming that these values also hold for the corresponding oxide lattices, predict whether $Ni^{III}Mn_2^{III}O_4$ should have the normal or inverse spinel structure. What factors might make your prediction unreliable?

20.21 Discuss each of the following observations:

(a) Although Co^{2+}(aq) forms the tetrahedral complex $[CoCl_4]^{2-}$ on treatment with concentrated HCl, Ni^{2+}(aq) does not form a similar complex.

(b) E^o for the half-reaction: $[Fe(CN)_6]^{3-} + e^- \rightleftharpoons [Fe(CN)_6]^{4-}$ depends on the pH of the solution, being most positive in strongly acidic medium.

(c) E^o for the Mn^{3+}/Mn^{2+} couple is much more positive than that for Cr^{3+}/Cr^{2+} or Fe^{3+}/Fe^{2+}.

d-Block metal chemistry: the first row metals

TOPICS

■ Occurrence, extraction and uses

■ Physical properties

■ Inorganic chemistry

1–2	3	4	5	6	7	8	9	10	11	12	13–18
s-block											*p*-block
	Sc	Ti	V	Cr	Mn	Fe	Co	Ni	Cu	Zn	
	Y	Zr	Nb	Mo	Tc	Ru	Rh	Pd	Ag	Cd	
	La	Hf	Ta	W	Re	Os	Ir	Pt	Au	Hg	

21.1 Introduction

The chemistry of the first row *d*-block metals is best considered separately from that of the second and third row metals for several reasons, including the following:

• the chemistry of the first member of a triad is distinct from that of the two heavier metals, e.g. Zr and Hf have similar chemistries but that of Ti differs;

• electronic spectra and magnetic properties of many complexes of the first row metals can often be rationalized using crystal or ligand field theory, but effects of spin–orbit coupling are more important for the heavier metals (see *Sections 20.7* and *20.8*);

• complexes of the heavier metal ions show a wider range of coordination numbers than those of their first row congeners;

• trends in oxidation states (*Table 19.3*) are not consistent for all members of a triad, e.g. although the *maximum* oxidation state of Cr, Mo and W is +6, its stability is greater for Mo and W than for Cr;

• metal–metal bonding is more important for the heavier metals than for those in the first row.

The emphasis of this chapter is on inorganic and coordination chemistry; organometallic complexes are discussed in *Chapter 23*.

21.2 Occurrence, extraction and uses

Scandium occurs as a rare component in a range of minerals. Its main source is *thortveitite* $(Sc,Y)_2Si_2O_7$ (a rare mineral found in Scandinavia), and it can also be extracted from residues in uranium processing. Uses of scandium are limited; it is a component in high-intensity lights.

The main ore of *titanium* is ilmenite $(FeTiO_3)$, and it also occurs as three forms of TiO_2 (*anatase*, *rutile* and *brookite*) and *perovskite* $(CaTiO_3$, *Figure 5.23*). The structures of anatase, rutile and brookite differ as follows: whereas the lattice of rutile (*Figure 5.21*) is based on an hcp array of O^{2-} ions with half the octahedral holes occupied by Ti(IV) centres, those of anatase and brookite contains ccp arrays of O^{2-} ions. Titanium is present in meteorites, and rock samples from the Apollo 17 lunar mission contain $\approx 12\%$ of the metal. Production of Ti involves conversion of rutile or ilmenite to $TiCl_4$ (by heating in a stream of Cl_2 at 1200 K in the presence of coke) followed by reduction using Mg. Titanium(IV) oxide is also purified via $TiCl_4$ in the 'chloride process' (see *Box 21.3*). Titanium metal is resistant to corrosion at ambient temperatures, and is lightweight and strong, making it valuable as a component in alloys, e.g. in aircraft construction. Superconducting magnets (used, for example, in MRI equipment, see *Box 2.5*) contain NbTi multicore conductors.

Vanadium occurs in several minerals including *vanadinite* $(Pb_5(VO_4)_3Cl)$, *carnotite* $(K_2(UO_2)_2(VO_4)_2 \cdot 3H_2O)$, *roscoelite* (a vanadium-containing mica) and *patronite* (VS_4). It also occurs in phosphate rock (see *Section 14.2*) and in some crude oils. It is not mined directly and extraction of vanadium is associated with that of other metals. Roasting vanadium ores with Na_2CO_3 gives water-soluble $NaVO_3$ and from solutions of this salt, the sparingly soluble $[NH_4][VO_3]$ can be precipitated. This is heated to give V_2O_5, reduction of which with Ca yields V. The steel

Chromium: resources and recycling

About 95% of the world's reserves of chromium ore lie in South Africa and Zimbabwe, and the bar chart below illustrates the dominance of South Africa in world chromite output.

Industrial nations in Europe and North America must rely on a supply of chromium ore from abroad, the US consuming ≈13% of world output. Because chromium is such a vital metal to the economy, government stockpiles in the US are considered an important strategy to ensure supplies during periods of military activity. Chromium ore is converted to chromium ferroalloys (for stainless steel and other alloys), chromite-containing refractory materials and chromium-based chemicals. The most important commercial applications of the latter are for pigments, leather tanning and wood preservation.

Recycling of stainless steel scrap as a source of Cr is an important secondary source, with the amount being recycled in the US rising from 19% in 1993 to ≈25% in 1997.

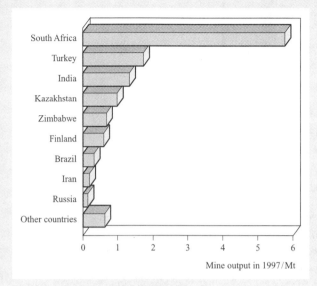

[Data: U.S. Geological Survey.]

industry consumes about 85% of world-supplies of V and *ferrovanadium* (used for toughening steels) is made by reducing a mixture of V_2O_5 and Fe_2O_3 with Al; steel–vanadium alloys are used for spring and high-speed cutting-tool steels. Vanadium(V) oxide is used as a catalyst in the oxidations of SO_2 to SO_3 and of naphthalene to phthalic acid.

The major ore of **chromium** is *chromite* ($FeCr_2O_4$) which has a normal spinel structure (see *Box 12.6* and *Section 20.9*). Chromite is reduced with carbon to produce *ferrochromium* for the steel industry; stainless steels contain ≈13% Cr to increase their corrosion resistance. For the production of Cr metal, chromite is fused with Na_2CO_3 in the presence of air (equation 21.1) to give water-soluble Na_2CrO_4 and insoluble Fe_2O_3. Extraction with water followed by acidification with H_2SO_4 gives a solution from which $Na_2Cr_2O_7$ can be crystallized. Equations 21.2 and 21.3 show the final two stages of production.

$$4FeCr_2O_4 + 8Na_2CO_3 + 7O_2$$
$$\longrightarrow 8Na_2CrO_4 + 2Fe_2O_3 + 8CO_2 \quad (21.1)$$

$$Na_2Cr_2O_7 + 2C \xrightarrow{\Delta} Cr_2O_3 + Na_2CO_3 + CO \quad (21.2)$$

$$Cr_2O_3 + 2Al \xrightarrow{\Delta} Al_2O_3 + 2Cr \quad (21.3)$$

The corrosion resistance of Cr leads to its widespread use as a protective coating (*chromium plating*); the metal is deposited by electrolyzing aqueous $Cr_2(SO_4)_3$, produced by dissolving Cr_2O_3 in H_2SO_4. After the steel industry, the next major consumer of Cr (≈25%) is the chemical industry; applications include pigments (e.g. chrome yellow), tanning agents, mordants, catalysts and oxidizing agents. Chromite

is used as a refractory material (see *Section 11.6*) e.g. in refractory bricks and furnace linings. Chromium compounds are toxic; chromates are corrosive to skin.

Several oxides of **manganese** occur naturally, the most important being *pyrolusite* (β-MnO_2); South Africa holds 80% of the world's ore reserves but mine production in China, South Africa and Ukraine is currently at similar levels. Little recycling of Mn currently takes place. Manganese nodules containing up to 24% of the metal have been discovered on the ocean bed. The main use of the element is in the steel industry; pyrolusite is mixed with Fe_2O_3 and reduced with coke to give *ferromanganese* (≈80% Mn). Almost all steels contain some Mn; those with a high Mn content (up to 12%) possess very high resistance to shock and wear and are suitable for crushing, grinding and excavating machinery. Manganese metal is produced by the electrolysis of $MnSO_4$ solutions. Manganese(IV) oxide is used in dry cell batteries. Figure 21.1 shows the Leclanché cell (the 'acid' cell); in the long-life 'alkaline' version, NaOH or KOH replaces NH_4Cl. The strong oxidizing power of $KMnO_4$ makes this an important chemical (see *Box 21.4*); Mn is an essential trace element for plants and small amounts of $MnSO_4$ are added to fertilizers.

Iron is the most important of all metals and is the fourth most abundant element in the Earth's crust. The Earth's core is believed to consist mainly of iron and it is the main constituent of metallic meteorites. The chief ores are *haematite* (α-Fe_2O_3), *magnetite* (Fe_3O_4), *siderite* ($FeCO_3$), *goethite* (α-Fe(O)OH) and *lepidocrocite* (γ-Fe(O)OH). While *iron pyrites* (FeS_2) and *chalcopyrite* ($CuFeS_2$) are common, their high sulfur contents render them unsuitable for Fe

Fig. 21.1 Schematic representation of the dry battery cell ('acid' version).

production. Pure Fe (made by reduction of the oxides with H_2) is reactive and rapidly corrodes; finely divided iron is pyrophoric. Although *pure* iron is not of commercial importance, steel production is carried out on a huge scale (see *Section 5.7*, *Box 5.1* and *Box 7.3*). α-Iron(III) oxide is used as a polishing and grinding agent and in the formation of ferrites (see *Section 21.9*); iron oxides are important commercial pigments: α-Fe_2O_3 (red), γ-Fe_2O_3 (red-brown), Fe_3O_4 (black) and $Fe(O)OH$ (yellow). Iron is of immense biological importance (see *Chapter 28*), and is present in, for example, haemoglobin and myoglobin (O_2-carriers), ferredoxins and cytochromes (redox processes), ferritin (iron storage), acid phosphatase (hydrolysis of phosphates), superoxide dismutases (O_2 dismutation) and nitrogenase (nitrogen fixation). A deficiency of iron in the body causes anaemia (see *Box 21.7*), while an excess causes haemochromatosis.

Cobalt occurs as a number of sulfide and arsenide ores including *cobaltite* (CoAsS) and *skutterudite* ((Co,Ni)As₃) which contains planar As_4-units). Production of the metal generally relies on the fact that it often occurs in ores of other metals (e.g. Ni, Cu and Ag) and the final processes involve reduction of Co_3O_4 with Al or C followed by electrolytic refining. Pure Co is brittle but it is commercially

important in special steels, alloyed with Al, Fe and Ni (*Alnico* is a group of carbon-free alloys) in permanent magnets, and in the form of hard, strong, corrosion-resistant non-ferrous alloys (e.g. with Cr and W) which are important in the manufacture of jet-engines and aerospace components. Cobalt compounds are widely used as pigments (blue hues in porcelain, enamels and glass, see *Box 21.9*), catalysts and as additives to animal feeds. Vitamin B_{12} is a cobalt complex, and a range of enzymes require B_{12} coenzymes. The artificial isotope ^{60}Co is used as a tracer (see *Box 2.3*). Like cobalt, *nickel* occurs as sulfide and arsenide minerals, e.g. *pentlandite*, $(Ni,Fe)_9S_8$. Roasting such ores in air gives nickel oxide which is then reduced to the metal using carbon. The metal is refined electrolytically or by conversion to $Ni(CO)_4$ followed by thermal decomposition (equation 21.4). This is the *Mond process* which is based on the fact that Ni forms a carbonyl derivative more readily than any other metal.

$$Ni + 4CO \underset{423-573\,K}{\overset{323\,K}{\rightleftharpoons}} Ni(CO)_4 \qquad (21.4)$$

Nickel is used extensively in alloys, notably in stainless steel, other corrosion-resistant alloys such as *Monel metal*, and coinage metals. Electroplated Ni provides a protective coat for other metals. Nickel has widespread use in batteries; recently, this has included the production of 'environmentally friendly' nickel–metal hydride batteries (see *Section 9.7*) which out-perform NiCd cells (equation 21.5) as rechargeable sources of power in portable appliances.

$$\left.\begin{array}{ll} \textit{Anode}: & Cd + 2[OH]^- \longrightarrow Cd(OH)_2 + 2e^- \\ \textit{Cathode}: & NiO(OH) + H_2O + e^- \longrightarrow Ni(OH)_2 + [OH]^- \end{array}\right\} \qquad (21.5)$$

Nickel is an important catalyst, e.g. for the hydrogenation of unsaturated organic compounds and in the water–gas shift reaction (see *Section 9.4*). *Raney nickel* is prepared by treating a NiAl alloy with NaOH and is a spongy material (pyrophoric when dry) which is a highly active catalyst.

BOX 21.2   **RESOURCES, ENVIRONMENTAL AND BIOLOGICAL**

Copper: resources and recycling

The estimated resources of copper on the Earth's surface are 2.3 billion tonnes, which includes bedrock minerals and deep-sea nodules. Among metals, consumption of Cu is exceeded only by steel and Al. The recovery of Cu from scrap metal is an essential part of copper-based industries, e.g. in 1998 in the US, recycled metal constituted ≈34% of the Cu supply. Worldwide mine production in 1998 was 11.4 Mt, with 30% originating from Chile and 17% from the US. Recycling of the metal is important for environmental reasons: dumping of waste leads to pollution, e.g. of water supplies. In the electronics industry solutions of NH_3–NH_4Cl in the presence of O_2 are used to etch Cu in printed circuit boards. The resulting Cu(II) waste is subjected to a solvent extraction process, first by treating with an

organic solvent XH which is a compound of the type $RR'C(OH)C(NOH)R''$, the conjugate base of which can function as a ligand:

$[Cu(NH_3)_4]^{2+}(aq) + 2XH(org)$

$\qquad\qquad \longrightarrow CuX_2(org) + 2NH_3(aq) + 2NH_4{}^+(aq)$

where aq and org represent the aqueous and organic phases respectively. Treatment with H_2SO_4 follows:

$CuX_2 + H_2SO_4 \longrightarrow CuSO_4 + 2XH$

and then Cu is reclaimed by electrolytic methods:

At the cathode: $Cu^{2+}(aq) + 2e^- \longrightarrow Cu(s)$

Copper is, by a considerable margin, the most noble of the first row metals and occurs native in small deposits in several countries. The chief ore is *chalcopyrite* ($CuFeS_2$) ($\approx 80\%$ of world copper production); others include *chalcanthite* ($CuSO_4 \cdot 5H_2O$), *atacamite* ($Cu_2Cl(OH)_3$), *cuprite* (Cu_2O) and *malachite* ($Cu_2(OH)_2CO_3$). Polished malachite is widely used for decorative purposes. The first step in Cu production is to roast chalcopyrite in a limited air supply to give Cu_2S and FeO; the latter is removed by combination with silica to form a slag, and Cu_2S is converted to Cu by reaction 21.6.

$$Cu_2S + O_2 \longrightarrow 2Cu + SO_2 \qquad (21.6)$$

Electrolytic purification of Cu is carried out by constructing a cell with impure Cu as the anode, clean Cu as the cathode and $CuSO_4$ as electrolyte; during electrolysis, Cu is transferred from anode to cathode yielding high purity metal (e.g. suitable for electrical wiring, a major use) and a deposit under the anode from which metallic Ag and Au can be extracted. Recycling of copper is now important (see *Box 21.2*). Being corrosion-resistant, Cu is in demand for water and steam piping and is used on the exterior of buildings, e.g. roofing and flashing, where long-term exposure results in a green patina of basic copper sulfate or carbonate. Alloys of Cu such as brass (Cu/Zn) (see *Section 5.7*), bronze (Cu/Sn), nickel silver (Cu/Zn/Ni) and coinage metal (Cu/Ni) are commercially important. Copper(II) sulfate is used extensively as a fungicide. Copper has a vital biochemical role, e.g. in cytochrome oxidase (involved in reduction of O_2 to H_2O) and haemocyanin (an O_2-carrying copper protein in arthropods). Copper compounds have numerous catalytic uses, and analytical applications include the biuret test and use of Fehling's solution (see *Section 21.12*).

The principal ores of *zinc* are *sphalerite* (*zinc blende*, ZnS, see *Figure 5.18*), *calamine* (*hemimorphite*, $Zn_4Si_2O_7(OH)_2 \cdot H_2O$) and *smithsonite* ($ZnCO_3$). Extraction from ZnS involves roasting in air to give ZnO followed by reduction with carbon. Zinc is more volatile (bp 1180 K) than most metals and can be separated by rapid chilling (to prevent reversing the reaction) and purified by distillation or electrolysis. Recycling of Zn has grown in importance, providing a secondary source of the metal. Figure 21.2 summarizes major uses of Zn. It is used to galvanize steel (see *Section 5.7* and *Box 7.3*) and Zn alloys are commercially important, e.g. brass (Cu/Zn) and nickel silver (Cu/Zn/Ni). Dry cell batteries use zinc as the anode (see *Figure 21.1*). Zinc oxide is used as a polymer stabilizer and an emollient in zinc ointment, and in the production of Zn_2SiO_4 for television screens; its major use is in the rubber industry where it lowers the vulcanization temperature and facilitates faster vulcanization (see *Section 15.4*). Both ZnO and ZnS are used as white pigments, although for most purposes TiO_2 is superior (see *Box 21.3* and *Section 27.5*).

21.3 Physical properties: an overview

A lot of the physical data for the first row metals has been discussed earlier, but Table 21.1 summarizes selected physical properties. Additional data are tabulated as follows:

- metal lattice types (*Table 5.2*);
- values of ionic radii, r_{ion}, which depend on charge, geometry and whether the ion is high- or low-spin (*Appendix 6*);
- standard reduction potentials, $E^{\circ}(M^{2+}/M)$ and $E^{\circ}(M^{3+}/M^{2+})$ (see *Tables 19.1* and *20.11*).

For electronic spectroscopic data (e.g. Δ_{oct} and spin–orbit coupling constants) and magnetic moments, relevant sections in *Chapter 20* should be consulted.

21.4 Group 3: scandium

The metal

In its chemistry, Sc shows a greater similarity to Al than to the heavier group 3 metals; E° values are given for comparison in equation 21.7.

$$M^{3+}(aq) + 3e^- \rightleftharpoons M(s) \quad \begin{cases} M = Al, \ E^{\circ} = -1.66\,V \\ M = Sc, \ E^{\circ} = -2.03\,V \end{cases} \quad (21.7)$$

Scandium metal dissolves in both acids and alkalis, and combines with halogens; it reacts with N_2 at high temperatures to give ScN which is hydrolyzed by water. Scandium normally shows one stable oxidation state in its compounds, Sc(III), but lower halides, e.g. ScCl (a layer structure ClScScCl . . . ClScScCl), Sc_7Cl_{10} (containing double chains of edge-sharing octahedral Sc_6 clusters) and Sc_7Cl_{12} (with Sc_6 clusters with Sc–Sc bonds and formulated as $Sc^{3+}[Sc_6Cl_{12}]^{3-}$) are obtained from reactions between $ScCl_3$ and Sc at high temperatures.

Scandium(III)

Direct combination of the elements gives anhydrous ScF_3 (water-insoluble white solid), $ScCl_3$ and $ScBr_3$ (soluble white solids) and ScI_3 (moisture-sensitive yellow solid). The

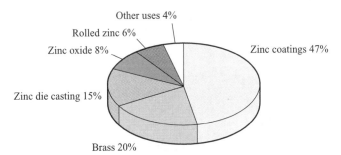

Fig. 21.2 Major uses of zinc [data: J. Emsley (1992) *Zinc*, Hobsons, Cambridge].

Other uses 4%
Rolled zinc 6%
Zinc oxide 8%
Zinc die casting 15%
Zinc coatings 47%
Brass 20%

Table 21.1 Selected physical properties of the metals of the first row of the *d*-block.

Property	Sc	Ti	V	Cr	Mn	Fe	Co	Ni	Cu	Zn
Atomic number, Z	21	22	23	24	25	26	27	28	29	30
Physical appearance of pure metal	Soft; silver-white; tarnishes in air	Hard; lustrous silver coloured	Soft; ductile; bright white	Hard; blue-white	Hard; lustrous silver-blue	Quite soft; malleable; lustrous, white	Hard; brittle; lustrous blue-white	Hard; malleable and ductile; grey-white	Malleable and ductile; reddish	Brittle at 298 K; malleable 373–423 K; lustrous blue-white
Melting point / K	1814	1941	2183	2180	1519	1811	1768	1728	1358	693
Boiling point / K	3104	3560	3650	2945	2235	3023	3143	3005	2840	1180
Ground state valence electronic configuration (core = [Ar]):										
Atom	$4s^2 3d^1$	$4s^2 3d^2$	$4s^2 3d^3$	$4s^1 3d^5$	$4s^2 3d^5$	$4s^2 3d^6$	$4s^2 3d^7$	$4s^2 3d^8$	$4s^1 3d^{10}$	$4s^2 3d^{10}$
M^+	$4s^1 3d^1$	$4s^2 3d^1$	$3d^4$	$3d^5$	$4s^1 3d^5$	$4s^1 3d^6$	$3d^8$	$3d^9$	$3d^{10}$	$4s^1 3d^{10}$
M^{2+}	$3d^1$	$3d^2$	$3d^3$	$3d^4$	$3d^5$	$3d^6$	$3d^7$	$3d^8$	$3d^9$	$3d^{10}$
M^{3+}	[Ar]	$3d^1$	$3d^2$	$3d^3$	$3d^4$	$3d^5$	$3d^6$	$3d^7$	$3d^8$	$3d^9$
Enthalpy of atomization, $\Delta_a H^\circ (298 \text{ K}) / \text{kJ mol}^{-1}$	378	470	514	397	283	418	428	430	338	130
First ionization energy, $IE_1 / \text{kJ mol}^{-1}$	633.1	658.8	650.9	652.9	717.3	762.5	760.4	737.1	745.5	906.4
Second ionization energy, $IE_2 / \text{kJ mol}^{-1}$	1235	1310	1414	1591	1509	1562	1648	1753	1958	1733
Third ionization energy, $IE_3 / \text{kJ mol}^{-1}$	2389	2653	2828	2987	3248	2957	3232	3395	3555	3833
Metallic radius, $r_{metal} / \text{pm}^\ddagger$	164	147	135	129	137	126	125	125	128	137
Electrical resistivity $(\rho) \times 10^8 / \Omega\,\text{m}$ (at 273 K)*	56**	39	18.1	11.8	143	8.6	5.6	6.2	1.5	5.5
	Sc	**Ti**	**V**	**Cr**	**Mn**	**Fe**	**Co**	**Ni**	**Cu**	**Zn**

‡ Metallic radius for 12-coordinate atom.
* See equation 5.3 for relationship between electrical resistivity and resistance.
** At 290–300 K.

fluoride crystallizes with the ReO_3 lattice (Figure 21.3) in which each Sc centre is octahedrally sited; compare this lattice with that of perovskite (Figure 5.23). In each of $ScCl_3$, $ScBr_3$ and ScI_3, the Sc atoms occupy octahedral sites in an hcp array of halogen atoms (i.e. a BiI_3-type lattice). On reaction with MF (M = Na, K, Rb, NH_4), ScF_3 forms water-soluble complexes $M_3[ScF_6]$ containing octahedral $[ScF_6]^{3-}$.

Addition of aqueous alkali to solutions of Sc(III) salts precipitates ScO(OH) which is isostructural with AlO(OH);

in the presence of excess $[OH]^-$, ScO(OH) redissolves as $[Sc(OH)_6]^{3-}$. Dehydration of ScO(OH) yields Sc_2O_3.

The coordination chemistry of Sc(III) is far more limited than that of the other first row *d*-block metal ions and is generally restricted to hard donors such as N and O. Coordination numbers of six are favoured, e.g. $[ScF_6]^{3-}$, $[Sc(bpy)_3]^{3+}$, *mer*-$[ScCl_3(H_2O)_3]$, *mer*-$[ScCl_3(THF)_3]$ and $[Sc(acac)_3]$. Among complexes with higher coordination numbers are $[ScF_7]^{4-}$ (pentagonal bipyramid), $[ScCl_2(15\text{-}crown\text{-}5)]^+$ (Figure 19.7d), $[Sc(NO_3)_5]^{2-}$ (see end of *Section 8.11*) and $[Sc(H_2O)_9]^{3+}$ (tricapped trigonal prism); bulky amido ligands stabilize low coordination numbers, e.g. $[Sc\{N(SiMe_3)_2\}_3]$.

21.5 Group 4: titanium

The metal

Titanium does not react with alkalis (cold or hot) and does not dissolve in mineral acids at room temperature. It is attacked by hot HCl forming Ti(III) and H_2, and hot

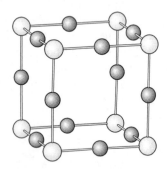

Fig. 21.3 Unit cell of ReO_3; Re atoms are shown in pale grey and O atoms in blue. This structure type is adopted by ScF_3 and FeF_3.

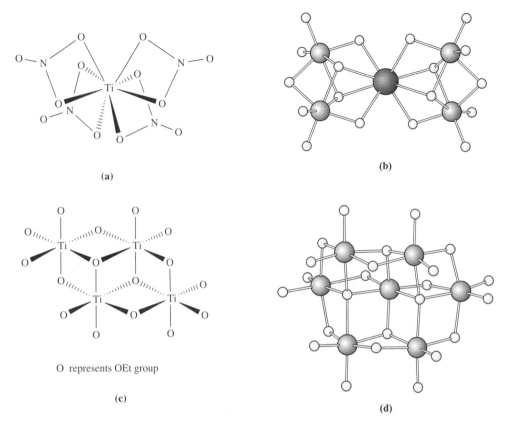

(a)

(b)

(c)

O represents OEt group

(d)

Fig. 21.4 (a) The structure of Ti(NO$_3$)$_4$ showing the dodecahedral environment of the titanium atom; compare with Figure 19.8; (b) the structure of [Ca{Ti$_2$(OEt)$_9$}$_2$] (X-ray diffraction); Et groups are omitted, colour code: Ca, black; Ti, blue; O, pale grey [E.P. Turevskaya *et al.* (1994) *J. Chem. Soc., Chem. Comm.*, p. 2303]; (c) the tetrameric structure of [Ti(OEt)$_4$] i.e. [Ti$_4$(OEt)$_{16}$] with ethyl groups omitted for clarity; (d) the structure of [Ti$_7$(μ_4-O)$_2$(μ_3-O)$_2$(OEt)$_{20}$] determined by X-ray diffraction; Et groups are omitted, colour code: Ti, blue; O, pale grey [R. Schmid *et al.* (1991) *J. Chem. Soc., Dalton Trans.*, p. 1999].

HNO$_3$ oxidizes the metal to hydrous TiO$_2$. Titanium reacts with most non-metals at elevated temperatures; with C, O$_2$, N$_2$ and halogens X$_2$, it forms TiC, TiO$_2$ (see *Figure 5.21*), TiN (see *Section 14.6*) and TiX$_4$ respectively. With H$_2$, it forms 'TiH$_2$' but this has a wide non-stoichiometric range, e.g. TiH$_{1.7}$ (see *Section 9.7*). The binary hydrides, carbide (see *Section 13.7*), nitride and borides (see *Section 12.10*) are all inert, high-melting, refractory materials.

In its compounds, Ti exhibits oxidation states of +4 (by far the most stable), +3, +2 and, rarely, 0.

Titanium(IV)

Titanium(IV) halides can be formed from the elements; industrially, TiCl$_4$ is prepared by reacting TiO$_2$ with Cl$_2$ in the presence of carbon and this reaction is also used in the purification of TiO$_2$ in the 'chloride process' (see *Box 21.3*). Titanium(IV) fluoride is a hygroscopic, white solid which forms HF on hydrolysis. The vapour contains tetrahedral TiF$_4$ molecules, but the solid state structure has not been confirmed. Both TiCl$_4$ and TiBr$_4$ hydrolyze more readily than TiF$_4$; at 298 K, TiCl$_4$ is a colourless liquid (mp 249 K, bp 409 K) and TiBr$_4$ a yellow solid. The tetraiodide is a

red-brown hygroscopic solid which sublimes *in vacuo* at 473 K to a red vapour. Tetrahedral molecules are present in the solid and vapour phases of TiCl$_4$, TiBr$_4$ and TiI$_4$. Each tetrahalide acts as a Lewis acid; TiCl$_4$ is the most important, being used with AlCl$_3$ in Ziegler–Natta catalysts for alkene polymerization (see *Box 18.3* and *Section 26.6*) and as a catalyst in a variety of other organic reactions. The Lewis acidity of TiCl$_4$ is seen in complex formation; it combines with tertiary amines and phosphines to give octahedral complexes such as [TiCl$_4$(NMe$_3$)$_2$] and [TiCl$_4$(PEt$_3$)$_2$]. Salts containing [TiCl$_6$]$^{2-}$ are best made in thionyl chloride solution since they are hydrolyzed by water; salts of [TiF$_6$]$^{2-}$ can be prepared in aqueous media. With the diarsine **21.1**, the dodecahedral complex [TiCl$_4$(**21.1**)$_2$] is formed. The reaction of N$_2$O$_5$ with TiCl$_4$ yields anhydrous [Ti(NO$_3$)$_4$] in which the Ti(IV) centre is in a dodecahedral environment (Figure 21.4a).

(21.1)

Commercial demand for TiO$_2$

Titanium dioxide has wide industrial applications as a brilliant white pigment, for example in paints and plastics. This commercial application arises from the fact that fine particles scatter incident light extremely strongly; even crystals of TiO$_2$ possess a very high refractive index ($\mu = 2.6$ for rutile, 2.55 for anatase). Historically, Pb(II) compounds were used as pigments in paints but the associated health hazards make lead undesirable; TiO$_2$ has negligible risks. Two manufacturing methods are used:

- the *sulfate process* produces TiO$_2$ in the form of rutile and anatase;
- the *chloride process* produces rutile.

The raw material for the sulfate process is ilmenite, FeTiO$_3$; treatment with H$_2$SO$_4$ at 420–470 K yields Fe$_2$(SO$_4$)$_3$ and TiOSO$_4$. The Fe$_2$(SO$_4$)$_3$ is reduced and separated as FeSO$_4$·7H$_2$O by a crystallization process. Hydrolysis of TiOSO$_4$ yields hydrated TiO$_2$ which is subsequently dehydrated to give TiO$_2$. This is in the form of *anatase* unless seed crystals of *rutile* are introduced in the final stages of production. Rutile ore occurs naturally in, for example, apatite veins in Norway, and is the raw material for the chloride process. Initially, TiO$_2$ ore is converted to TiCl$_4$ by treatment with Cl$_2$ and C at 1200 K. Oxidation by O$_2$ at $\approx$1500 K yields pure rutile.

Originally, the sulfate process was the more industrially important process, but since the early 1990s, the chloride process has been favoured on both financial and environmental grounds. The chart below shows TiO$_2$ production in megatonnes per year with projected output to 2000.

Titanium dioxide is a semiconductor and is an excellent photocatalyst for the photomineralization of water, i.e. the

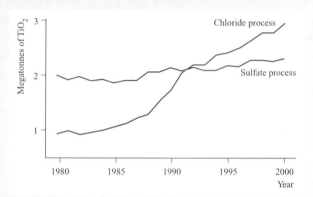

[Data: *Chemistry & Industry* (1997) p. 51; converted to SI units.]

degradation of pollutants in water is catalyzed by TiO$_2$ in the presence of UV radiation. Pollutants which can be successfully destroyed include a wide range of hydrocarbons and halogenated organic compounds as well as some herbicides, pesticides and dyes. The semiconducting properties of TiO$_2$ have also led to its being used as a gas sensor for detection of Me$_3$N emitted from decaying fish. Other uses of TiO$_2$ include applications in cosmetics and ceramics, and in anodes for various electrochemical processes.

Further reading

A. Mills, R.H. Davies and D. Worsley (1993) *Chemical Society Reviews*, vol. 22, p. 417.
See also *Section 27.5*.

We describe the commercial importance of TiO$_2$ in Box 21.3, and its rutile lattice was shown in *Figure 5.21*. Although it may be formulated as Ti^{4+}(O^{2-})$_2$, the very high value of the sum of the first four ionization energies of the metal (8797 kJ mol^{-1}) makes the validity of the ionic model doubtful. Dry TiO$_2$ is difficult to dissolve in acids, but the hydrous form (precipitated by adding base to solutions of Ti(IV) salts) dissolves in HF, HCl and H$_2$SO$_4$ giving fluoro, chloro and sulfato complexes respectively. There is no simple aqua ion of Ti^{4+}. The reaction of TiO$_2$ with CaO at 1620 K gives the *titanate* CaTiO$_3$; other members of this group include BaTiO$_3$ and FeTiO$_3$ (ilmenite). The structure type depends on the size of M^{2+}; if it is large (e.g. M = Ca), a perovskite lattice is favoured (*Figure 5.23*) but if M^{2+} is similar in size to Ti(IV), a corundum structure (see *Section 12.7*), in which M(II) and Ti(IV) replace two Al(III) centres, is preferred, e.g. ilmenite. The M^{II}TiO$_3$ titanates are *mixed oxides* and do *not* contain [TiO$_3$]$^{2-}$ ions. Above 393 K, BaTiO$_3$ has the perovskite lattice, but at lower temperatures

it transforms successively into three phases, each of which is a *ferroelectric*, i.e. the phase has an electric dipole moment even in the absence of an external magnetic field. This arises because the small Ti(IV) centre tends to lie off-centre in the octahedral O$_6$-hole (*Figure 5.23*). Application of an electric field causes all such ions to be drawn to the same side of the holes and leads to a great increase in specific permittivity; thus, barium titanates are used in capacitors. Application of pressure to one side of a BaTiO$_3$ crystal causes the Ti^{4+} ions to migrate generating an electric current (the piezoelectric effect, see *Section 13.9*), and this property makes BaTiO$_3$ suitable for use in electronic devices such as microphones. Interest in perovskite-phases such as BaTiO$_3$ and CaTiO$_3$ has led to investigations of solid state materials such as [M{Ti$_2$(OEt)$_9$}$_2$] (M = Ba or Ca) (Figure 21.4b) derived from reactions of alkoxides of Ti(IV) and Ba or Ca. Titanium alkoxides are widely used in waterproofing fabrics and in heat-resistant paints. Thin films of TiO$_2$ are used in capacitors and can be deposited using Ti(IV) alkoxides such as [Ti(OEt)$_4$]. The ethoxide is prepared from TiCl$_4$ and Na[OEt] (or from TiCl$_4$, dry NH$_3$ and EtOH) and has a tetrameric structure (Figure 21.4c) in which each Ti is octahedrally sited. Larger structures which retain TiO$_6$ 'building-blocks' can be assembled; reaction of [Ti(OEt)$_4$] with anhydrous EtOH at 373 K gives [Ti$_{16}$O$_{16}$(OEt)$_{32}$], while [Ti$_7$O$_4$(OEt)$_{20}$] (Figure 21.4d) is the product if basic CuCO$_3$ is present. Similar structures are observed for vanadates (*Section 21.6*), molybdates and tungstates (*Section 22.7*).

(21.2)

(21.3)

The reaction of TiO$_2$ and TiCl$_4$ at 1320 K in a fluidized bed produces [Cl$_3$Ti(μ-O)TiCl$_3$] which reacts with [Et$_4$N]Cl to give [Et$_4$N]$_2$[TiOCl$_4$]. The [TiOCl$_4$]$^{2-}$ ion (**21.2**) has a square-based pyramidal structure with the oxo ligand in the apical position. A number of peroxo complexes of Ti(IV) are known and include products of reactions between TiO$_2$ in 40% HF and 30% H$_2$O$_2$; at pH 9 the product is [TiF$_2$(η^2-O$_2$)$_2$]$^{2-}$ while at pH 6, [TiF$_5$(η^2-O$_2$)]$^{3-}$ is formed. The dinuclear species [Ti$_2$F$_6$(μ-F)$_2$(η^2-O$_2$)$_2$]$^{4-}$ (**21.3**) is made by treating [TiF$_6$]$^{2-}$ with 6% H$_2$O$_2$ at pH 5.

Titanium(III)

Titanium(III) fluoride is prepared by passing H$_2$ and HF over Ti or its hydride at 970 K; it is a blue solid (mp 1473 K) with a structure related to ReO$_3$ (Figure 21.3). The trichloride exists in four forms (α, β, γ and δ); the α-form (a violet solid) is prepared by reducing TiCl$_4$ with H$_2$ above 770 K and has a layer structure with Ti atoms in octahedral sites. The brown β-form is prepared by heating TiCl$_4$ with trialkyl aluminium compounds; it is fibrous and contains face-sharing TiCl$_6$ octahedra. The trichloride is commercially available; it is used as a catalyst in alkene polymerization and is a powerful reducing agent. In air, TiCl$_3$ is readily oxidized, and disproportionates above 750 K (equation 21.8).

$$2TiCl_3 \longrightarrow TiCl_4 + TiCl_2 \qquad (21.8)$$

Titanium tribromide is made by heating TiBr$_4$ with Al, or by reaction of BBr$_3$ with TiCl$_3$; it is a grey solid with a layer structure analogous to α-TiCl$_3$. Reduction of TiI$_4$ with Al gives violet TiI$_3$; both TiBr$_3$ and TiI$_3$ disproportionate when heated >600 K. The magnetic moment of TiF$_3$ (1.75 μ_B at 300 K) is consistent with one unpaired electron per metal centre. However, magnetic data for TiCl$_3$, TiBr$_3$ and TiI$_3$ indicate significant Ti−Ti interactions in the solid state; for TiCl$_3$, the magnetic moment at 300 K is 1.31 μ_B and TiBr$_3$ is only weakly paramagnetic.

When aqueous solutions of Ti(IV) are reduced by Zn, the purple aqua ion [Ti(H$_2$O)$_6$]$^{3+}$ is obtained (see *equation 6.35* and *Figure 20.4*). This is a powerful reductant (equation 21.9) and is used in titrimetric analyses of Fe(III) and nitro groups (reduced to NH$_2$ groups); aqueous solutions of Ti(III) must be protected from aerial oxidation.

$$[TiO]^{2+}(aq) + 2H^+(aq) + e^- \rightleftharpoons Ti^{3+}(aq) + H_2O(l)$$
$$E^\circ = +0.1 \text{ V} \qquad (21.9)$$

In alkaline solution (partly because of the involvement of H$^+$ in redox equilibrium 21.9, and partly because of the low solubility of the product) Ti(III) compounds liberate H$_2$ from H$_2$O and are oxidized to TiO$_2$. In the absence of air, alkali precipitates hydrous Ti$_2$O$_3$ from solutions of TiCl$_3$. Dissolution of this oxide in acids gives salts containing [Ti(H$_2$O)$_6$]$^{3+}$, e.g. [Ti(H$_2$O)$_6$]Cl$_3$ and CsTi(SO$_4$)$_2\cdot$12H$_2$O, the latter being isomorphous with other alums (see *Section 12.9*).

Titanium(III) oxide is made by reducing TiO$_2$ with Ti at high temperatures; it is a purple-black, insoluble solid with the corundum structure (see *Section 12.7*) and exhibits a transition from semiconductor to metallic character on heating above 470 K or doping with, for example, V(III). Uses of Ti$_2$O$_3$ include those in thin film capacitors.

Complexes of Ti(III) usually have octahedral structures, e.g. [TiF$_6$]$^{3-}$, [TiCl$_6$]$^{3-}$, [Ti(CN)$_6$]$^{3-}$, *trans*-[TiCl$_4$(THF)$_2$]$^-$, *trans*-[TiCl$_4$(py)$_2$]$^-$, *mer*-[TiCl$_3$(THF)$_3$], *mer*-[TiCl$_3$(py)$_3$] and [Ti{(H$_2$N)$_2$CO-*O*}$_6$]$^{3+}$, and magnetic moments close to the spin-only values. Examples of 7-coordinate complexes include [Ti(EDTA)(H$_2$O)]$^-$ and [Ti(H$_2$O)$_3$(ox)$_2$]$^-$.

Low oxidation states

Titanium(II) chloride, bromide and iodide can be prepared by thermal disproportionation of TiX$_3$ (equation 21.8) or by reaction 21.10. They are red or black solids which adopt the CdI$_2$ lattice (*Figure 5.22*).

$$TiX_4 + Ti \xrightarrow{\Delta} 2TiX_2 \qquad (21.10)$$

With water, TiCl$_2$, TiBr$_2$ and TiI$_2$ react violently liberating H$_2$; thus, there is no aqueous chemistry of Ti(II).

Titanium(II) oxide is manufactured by heating TiO$_2$ and Ti *in vacuo*; it is a black solid and a metallic conductor which adopts an NaCl lattice with one-sixth of both anion and cation sites unoccupied. A commercial use of TiO is in electrochromic systems (see *Box 22.4*). Conducting properties of the first row metal(II) oxides are compared in *Section 27.3*.

Reduction of TiCl$_3$ with Na/Hg, or of TiCl$_4$ with Li in THF and 2,2′-bipyridine leads to violet [Ti(bpy)$_3$]. Formally this contains Ti(0), but results of MO calculations and spectroscopic studies indicate that electron-delocalization occurs such that the complex should be considered as [Ti^{3+}(bpy$^-$)$_3$]; see also the end of *Section 19.5* and discussion of [W(**19.11**)$_3$] in *Section 19.7*.

21.6 Group 5: vanadium

The metal

In many ways, V metal is similar to Ti. Vanadium is a powerful reductant (equation 21.11) but is passivated by an oxide film.

$$V^{2+} + 2e^- \rightleftharpoons V \qquad E^o = -1.18\,V \qquad (21.11)$$

The metal is insoluble in non-oxidizing acids (except HF) and alkalis, but is attacked by HNO_3, *aqua regia* and peroxodisulfate solutions. On heating, V reacts with halogens (equation 21.12) and combines with O_2 to give V_2O_5, and with B, C and N_2 to yield solid state materials (see *Sections 12.10, 13.7* and *14.6*).

$$V \begin{cases} \xrightarrow{F_2} VF_5 \\ \xrightarrow{Cl_2} VCl_4 \\ \xrightarrow{X_2} VX_3 \quad (X = Br\ or\ I) \end{cases} \qquad (21.12)$$

The normal oxidation states of vanadium are +5, +4, +3 and +2; 0 occurs in a few compounds with π-acceptor ligands, e.g. $V(CO)_6$ (see *Chapter 23*).

Vanadium(V)

The only binary halide of vanadium(V) is VF_5 (equation 21.12); it is a volatile white solid which is readily hydrolyzed and is a powerful fluorinating agent. In the gas phase, VF_5 exists as trigonal bipyramidal molecules but the solid has a polymeric structure (**21.4**). The salts $K[VF_6]$. and $[Xe_2F_{11}][VF_6]$ are made by reacting VF_5 with KF or XeF_6 (at 250 K) respectively.

(21.4)

The oxohalides VOX_3 (X = F or Cl) are made by halogenation of V_2O_5; reaction of VOF_3 with $(Me_3Si)_2O$ yields VO_2F, and treatment of $VOCl_3$ with Cl_2O gives VO_2Cl. The oxohalides are hygroscopic and hydrolyze readily. Both VO_2F and VO_2Cl decompose on heating (equation 21.13).

$$3VO_2X \xrightarrow{\Delta} VOX_3 + V_2O_5 \qquad (X = F\ or\ Cl) \qquad (21.13)$$

Pure V_2O_5 is an orange or red powder depending on its state of division, and is manufactured by heating $[NH_4][VO_3]$ (equation 21.14).

$$2[NH_4][VO_3] \xrightarrow{\Delta} V_2O_5 + H_2O + 2NH_3 \qquad (21.14)$$

Vanadium(V) oxide is amphoteric, being sparingly soluble in water but dissolving in alkalis to give a wide range of vanadates, and in strong acids to form complexes of $[VO_2]^+$. The species present in vanadium(V)-containing solutions depend on the pH:

pH 14	$[VO_4]^{3-}$
	$[VO_3(OH)]^{2-}$ in equilibrium with $[V_2O_7]^{4-}$
	$[V_4O_{12}]^{4-}$
pH 6	$[H_nV_{10}O_{28}]^{(6-n)-}$
	V_2O_5
pH 0	$[VO_2]^+$

This dependence can be expressed in terms of a series of equilibria such as equations 21.15–21.21.

$$[VO_4]^{3-} + H^+ \rightleftharpoons [VO_3(OH)]^{2-} \qquad (21.15)$$

$$2[VO_3(OH)]^{2-} \rightleftharpoons [V_2O_7]^{4-} + H_2O \qquad (21.16)$$

$$[VO_3(OH)]^{2-} + H^+ \rightleftharpoons [VO_2(OH)_2]^- \qquad (21.17)$$

$$4[VO_2(OH)_2]^- \rightleftharpoons [V_4O_{12}]^{4-} + 4H_2O \qquad (21.18)$$

$$10[V_3O_9]^{3-} + 15H^+ \rightleftharpoons 3[HV_{10}O_{28}]^{5-} + 6H_2O \qquad (21.19)$$

$$[HV_{10}O_{28}]^{5-} + H^+ \rightleftharpoons [H_2V_{10}O_{28}]^{4-} \qquad (21.20)$$

$$[H_2V_{10}O_{28}]^{4-} + 14H^+ \rightleftharpoons 10[VO_2]^+ + 8H_2O \qquad (21.21)$$

Isopolyanions (*homopolyanions*) are complex metal oxoanions (polyoxometallates) of type $[M_xO_y]^{n-}$, e.g. $[V_{10}O_{28}]^{6-}$ and $[Mo_6O_{19}]^{2-}$. A *heteropolyanion* contains a hetero atom, e.g. $[PW_{12}O_{40}]^{3-}$.

The formation of polyoxometallates is a characteristic of V, Mo, W (see *Section 22.7*) and, to a lesser extent, Nb, Ta and Cr. Characterization of solution species is aided by ^{17}O and ^{51}V NMR spectroscopies, and solid state structures for a range of salts are known. The structural chemistry of V_2O_5 and vanadates is complicated and only a brief survey is given here. The structure of V_2O_5 consists of chains of edge-sharing square-based pyramids; each V centre is bonded to one O at 159 pm (apical site and not shared), one O at 178 pm (shared with one other V) and two O at 188 pm and one at 202 pm (shared with two other V atoms). Salts of $[VO_4]^{3-}$ (*orthovanadates*) contain discrete tetrahedral ions, and those of $[V_2O_7]^{4-}$ (*pyrovanadates*) also contain discrete anions (Figure 21.5a); $[V_2O_7]^{4-}$ is isoelectronic and isostructural with $[Cr_2O_7]^{2-}$. The ion $[V_4O_{12}]^{4-}$ has a cyclic structure (Figure 21.5b). Anhydrous salts of $[VO_3]^-$ (*metavanadates*) contain infinite chains of vertex sharing VO_4 units (Figures 21.5c and d). However, this structure type is not common to all metavanadates, e.g. in $KVO_3 \cdot H_2O$ and $Sr(VO_3)_2 \cdot 4H_2O$ each V is bonded to five O atoms in a double-chain structure. The

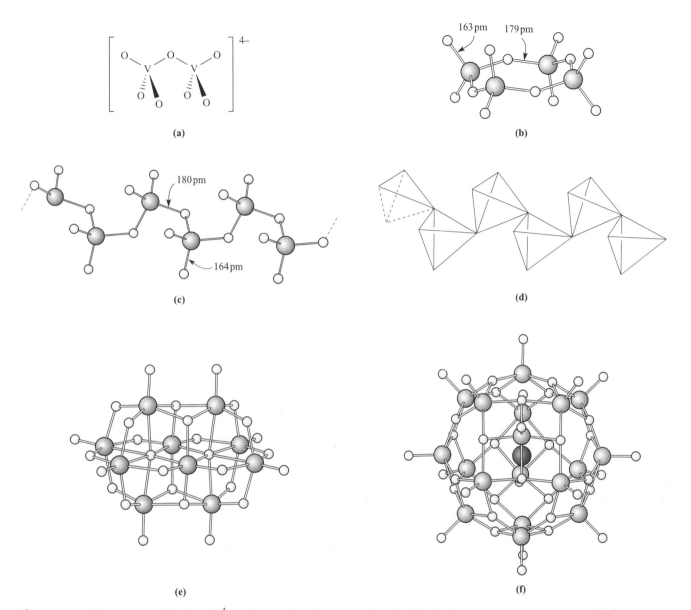

Fig. 21.5 (a) The structure of the $[V_2O_7]^{4-}$ anion consists of two tetrahedral units sharing a common oxygen atom; (b) the structure of $[V_4O_{12}]^{4-}$ in the salt $[Ni(bpy)_3]_2[V_4O_{12}]\cdot11H_2O$ (X-ray diffraction) [G.-Y. Yang *et al.* (1998) *Acta Crystallogr., Sect. C*, vol. 54, p. 616]; (c) infinite chains of corner-sharing tetrahedral VO_4 units are present in anhydrous metavanadates; this shows part of one chain in $[n\text{-}C_6H_{13}NH_3][VO_3]$ (an X-ray diffraction determination); colour code: V, blue; O, pale grey [P. Roman *et al.*, (1991) *Mater. Res. Bull.*, vol. 26, p. 19]; (d) the structure of the metavanadate shown in (c) can be represented as a chain of corner-sharing tetrahedra, each tetrahedron representing a VO_4 unit; (e) the structure of $[V_{10}O_{28}]^{6-}$ in the salt $[^iPrNH_3]_6[V_{10}O_{28}]\cdot4H_2O$ (X-ray diffraction); colour code: V, blue; O, pale grey [M.-T. Averbuch-Pouchot *et al.* (1994) *Eur. J. Solid State Inorg. Chem.*, vol. 31, p. 351]; (f) in $[Et_4N]_5[V_{18}O_{42}I]$ (X-ray diffraction), the $[V_{18}O_{42}]^{4-}$ ion contains square-based pyramidal VO_5 units and the cage encapsulates I^-; colour code: V, blue; O, pale grey; I, black [A. Müller *et al.* (1997) *Inorg. Chem.*, vol. 36, p. 5239].

$[V_{10}O_{28}]^{6-}$ anion exists in solution (at appropriate pH) and has been characterized in the solid state in, for example, $[H_3NCH_2CH_2NH_3]_3[V_{10}O_{28}]\cdot6H_2O$ and $[^iPrNH_3]_6[V_{10}O_{28}]\cdot4H_2O$ (Figure 21.5e). It consists of 10 VO_6 octahedral units with 2 μ_6-O, 4 μ_3-O, 14 μ-O and 8 terminal O atoms. Crystalline salts of $[HV_{10}O_{28}]^{5-}$, $[H_2V_{10}O_{28}]^{4-}$ and $[H_3V_{10}O_{28}]^{3-}$ have also been isolated and the anions retain the framework shown in Figure 21.5e. Examples of isopolyanions of vanadium with open ('bowl-shaped') structures are known, e.g. $[V_{12}O_{32}]^{4-}$, and

these may act as 'hosts' to small molecules, e.g. in $[Ph_4P]_4[V_{12}O_{32}]\cdot4MeCN\cdot4H_2O$, one MeCN molecule resides partially within the cavity of the anion.

Reduction of yellow $[VO_2]^+$ in acidic solution yields successively blue $[VO]^{2+}$, green V^{3+} and violet V^{2+}. The potential diagram in Figure 21.6 shows that all oxidation

$$[VO_2]^+ \xrightarrow{+0.99} [VO]^{2+} \xrightarrow{+0.34} V^{3+} \xrightarrow{-0.26} V^{2+} \xrightarrow{-1.18} V$$

Fig. 21.6 Potential diagram for vanadium at pH 0.

states of vanadium in aqueous solution are stable with respect to disproportionation.

Vanadium(IV)

The highest chloride of vanadium is VCl_4 (equation 21.12); it is a toxic, red-brown liquid (bp 247 K, bp 421 K) and the liquid and vapour phases contain tetrahedral molecules (**21.5**). It readily hydrolyzes to $VOCl_2$ (see below), and at 298 K, slowly decomposes (equation 21.22). The reaction of VCl_4 with anhydrous HF gives lime-green VF_4 (solid at 298 K) which is also formed with VF_5 when V reacts with F_2. On heating, VF_4 disproportionates (equation 21.23) in contrast to the behaviour of VCl_4 (equation 21.22).

(21.5)

$$2VCl_4 \longrightarrow 2VCl_3 + Cl_2 \qquad (21.22)$$

$$2VF_4 \xrightarrow{\geq 298\,K} VF_5 + VF_3 \qquad (21.23)$$

The structure of VF_4 probably consists of fluorine bridged VF_6-units. Reaction between VF_4 and KF in anhydrous HF gives $K_2[VF_6]$ containing octahedral $[VF_6]^{2-}$. Vanadium(IV) bromide is known but decomposes at 250 K to VBr_3 and Br_2.

The green oxochloride $VOCl_2$ (prepared from V_2O_5 and VCl_3) is polymeric and has a temperature dependent magnetic moment (1.40 μ_B at 296 K, 0.95 μ_B at 113 K); it decomposes on heating (equation 21.24).

$$2VOCl_2 \xrightarrow{650\,K} VOCl_3 + VOCl \qquad (21.24)$$

(21.6)

Figure 21.6 shows that vanadium(V) is quite a powerful oxidant, and only mild reducing agents (e.g. SO_2) are needed to convert V(V) to V(IV). In aqueous solution, V(IV) is present as the hydrated vanadyl ion $[VO]^{2+}$ (**21.6**) of which many salts are known. Anhydrous $V(O)SO_4$ is manufactured by reducing a solution of V_2O_5 in H_2SO_4 with $H_2C_2O_4$; the blue solid has a polymeric structure with vertex-sharing VO_6 octahedra linked by sulfate groups. The hydrate $V(O)SO_4 \cdot 5H_2O$ contains octahedrally sited V(IV) involving one oxo ligand (V−O = 159 pm) and five other O atoms (from sulfate and four H_2O) at 198–222 pm. The reaction of V_2O_5 and Hacac (see *Table 6.7*) gives blue $[VO(acac)_2]$ which has a square-based pyramidal

structure (**21.7**); this readily forms complexes with *N*-donor ligands which occupy the site *trans* to the oxo ligand. The salt $[NH_4]_2[VOCl_4]$ can be obtained by crystallization of a solution of $VOCl_3$ and $[NH_4]Cl$ in hydrochloric acid; the $[VOCl_4]^{2-}$ ion has a square-based pyramidal structure with the oxo ligand in the apical site. This preference is seen throughout related derivatives containing the $[VO]^{2+}$ unit; its presence is detected by a characteristic IR spectroscopic absorption around 980 cm^{-1} (the corresponding value for a V−O single bond is $\approx 480\,cm^{-1}$).

(21.7)

Vanadium(IV) oxide, VO_2, is prepared by heating V_2O_5 with $H_2C_2O_4$. It crystallizes with a rutile lattice (*Figure 5.21*) which is distorted at 298 K such that pairs of V(IV) centres are alternately 262 and 317 pm apart; the shorter distance is consistent with metal–metal bonding. This polymorph is an insulator, but above 343 K, the electrical conductivity and magnetic susceptibility of VO_2 increase as the regular rutile lattice is adopted. Vanadium(IV) oxide is blue but shows thermochromic behaviour (see *Section 18.6*). It is amphoteric, dissolving in non-oxidizing acids to give $[VO]^{2+}$ and in alkalis to form homopolyanions such as $[V_{18}O_{42}]^{12-}$, the Na^+ and K^+ salts of which can be isolated by heating $V(O)SO_4$ and MOH (M = Na or K) in water at pH 14 in an inert atmosphere. The structure of $[V_{18}O_{42}]^{12-}$ consists of square-based pyramidal VO_5-units, the apical O atoms of which are terminal (i.e. V=O units) while basal O atoms are involved in V−O−V bridges to build an almost spherical cage. Related anions such as $[V_{18}O_{42}]^{4-}$, $[V_{18}O_{42}]^{5-}$ and $[V_{18}O_{42}]^{6-}$ formally contain V(IV) and V(V) centres. The cavity in $[V_{18}O_{42}]^{n-}$ is able to accommodate an *anionic* guest as in $[V_{18}O_{42}I]^{5-}$ (Figure 21.5f) or $[H_4V_{18}O_{42}X]^{9-}$ (X = Cl, Br, I).

Vanadium(III)

The trihalides VF_3, VCl_3, VBr_3 and VI_3 are all known. The yellow-green, insoluble trifluoride is made from V and HF at 500 K. Vanadium(III) chloride is a violet, hygroscopic solid which dissolves in water without decomposition to give $[V(H_2O)_6]Cl_3$. Anhydrous VCl_3 is made by decomposition of VCl_4 at 420 K (equation 21.22), but above 670 K, it disproportionates to VCl_4 and VCl_2. Reaction of VCl_3 with BBr_3, or V with Br_2, yields VBr_3, a green-black, water-soluble solid which disproportionates to VBr_2 and VBr_4. The brown, hygroscopic VI_3 is made from V with I_2, and decomposes above 570 K to VI_2 and I_2. All the trihalides adopt a lattice in which the V(III) centres occupy octahedral sites in an hcp array of halogen atoms (i.e. a BiI_3 lattice).

(21.8)

Vanadium(III) forms a variety of octahedral complexes, e.g. *mer*-[VCl$_3$(THF)$_3$)] and *mer*-[VCl$_3$(^{t}BuNC-*C*)$_3$], which have magnetic moments close to the spin-only value for a d^2 ion. The $[VF_6]^{3-}$ ion is present in simple salts such as K_3VF_6, but various lattice structures are observed in other salts. The reaction of CsCl with VCl_3 at 1000 K produces $Cs_3[V_2Cl_9]$; $[V_2Cl_9]^{3-}$ (**21.8**) is isomorphous with $[Cr_2Cl_9]^{3-}$ and consists of two face-sharing octahedra with *no* metal–metal interaction. Examples of complexes with higher coordination numbers are known, e.g. $[V(CN)_7]^{4-}$ (pentagonal bipyramidal) made from VCl_3 and KCN in aqueous solution and isolated as the K^+ salt.

The oxide V_2O_3 (which, like Ti_2O_3, adopts the corundum lattice, see *Section 12.7*) is made by partial reduction of V_2O_5 using H_2, or by heating (1300 K) V_2O_5 with vanadium. It is a black solid which, on cooling, exhibits a metal–insulator transition at 155 K. The oxide is exclusively basic, dissolving in acids to give $[V(H_2O)_6]^{3+}$. The hydrated oxide may be precipitated by adding alkali to green solutions of vanadium(III) salts. The $[V(H_2O)_6]^{3+}$ ion is present in alums such as $[NH_4]V(SO_4)_2 \cdot 12H_2O$ formed by electrolytic reduction of $[NH_4][VO_3]$ in sulfuric acid.

Vanadium(II)

Green VCl_2 is made from VCl_3 and H_2 at 770 K and is converted to blue VF_2 by reaction with HF and H_2; VCl_2 can also be obtained from VCl_3 as described above, and similarly brown-red VBr_2 and violet VI_2 can be produced from VBr_3 and VI_3, respectively. Vanadium(II) fluoride is antiferromagnetic and crystallizes with a rutile lattice (*Figure 5.21*); VCl_2, VBr_2 and VI_2 (all paramagnetic) possess CdI_2 layer structures (*Figure 5.22*). The dihalides are water soluble.

Vanadium(II) is present in aqueous solution as the violet, octahedral $[V(H_2O)_6]^{2+}$ ion; it can be prepared by reduction of vanadium in higher oxidation states electrolytically or using zinc amalgam. It is strongly reducing, being rapidly oxidized on exposure to air. Compounds such as Tutton salts contain $[V(H_2O)_6]^{2+}$; e.g. $K_2V(SO_4)_2 \cdot 6H_2O$ is made by adding K_2SO_4 to an aqueous solution of VSO_4 and forms violet crystals.

A *Tutton salt* has the general formula $[M^I]_2M^{II}(SO_4)_2 \cdot 6H_2O$ (compare with an *alum*, Section 12.9).

Vanadium(II) oxide is a grey, metallic solid and is obtained by reduction of higher oxides at high temperatures. It is non-stoichiometric, varying in composition from $VO_{0.8}$ to $VO_{1.3}$, and possesses an NaCl (*Figure 5.15*) or defect NaCl lattice (see *Section 27.2*). Conducting properties of the first row metal(II) oxides are compared in *Section 27.3*.

Vanadium(II) complexes are relatively rare. They include $[V(CN)_6]^{4-}$, the K^+ salt of which is made by reducing $K_4[V(CN)_7]$ with K metal in liquid NH_3; the magnetic moment of 3.5 μ_B is close to the spin-only value of 3.87 μ_B. Octahedral $[V(NCMe)_6]^{2+}$ has been isolated in the $[ZnCl_4]^{2-}$ salt from the reaction of VCl_3 with Et_2Zn in MeCN. Treatment of $VCl_2 \cdot 4H_2O$ with phen gives $[V(phen)_3]Cl_2$, for which $\mu_{eff} = 3.82\,\mu_B$ (300 K), consistent with octahedral d^3.

21.7 Group 6: chromium

The metal

At ordinary temperatures, Cr metal is resistant to chemical attack (although it dissolves in dilute HCl and H_2SO_4). This inertness is kinetic rather than thermodynamic in origin as the Cr^{2+}/Cr and Cr^{3+}/Cr couples in Figure 21.7 show. Nitric acid renders Cr passive; alkalis have no action. At higher temperatures the metal is reactive: it decomposes steam and combines with O_2, halogens, and most other non-metals. Borides, carbides and nitrides (see *Sections 12.10*, *13.7* and *14.6*) exist in various phases (e.g. CrN, Cr_2N, Cr_3N, Cr_3N_2) and are inert materials (e.g. CrN is used in wear-resistant coatings). The black sulfide Cr_2S_3 is formed by direct combination of the elements on heating; a range of other sulfides is known, but methods of synthesis vary.

The main oxidation states of chromium are +6, +3 and +2. A few compounds of Cr(V) and Cr(IV) are known, but are unstable with respect to disproportionation. Chromium(0) is stabilized by π-acceptor ligands (see *Chapter 23*).

Chromium(VI)

The only halide of chromium(VI) to have been reported is yellow CrF_6, produced by fluorination of the metal at 670 K and 200 bar followed by rapid chilling. The material so-prepared reacts violently in moist air and decomposes at 173 K into CrF_5 and F_2; whether or not CrF_6 actually exists and, if it *does*, the question of whether it is octahedral

Fig. 21.7 Potential diagram for chromium at pH 0.

or trigonal prismatic, are controversial topics.[†] The oxohalides CrO_2F_2 and CrO_2Cl_2 are much more stable; fluorination of CrO_3 with SeF_4, SF_4 or HF yields CrO_2F_2 (violet crystals, mp 305 K), while CrO_2Cl_2 (red liquid, mp 176 K, bp 390 K) is prepared by heating a mixture of $K_2Cr_2O_7$, KCl and concentrated H_2SO_4. Chromyl chloride is an oxidant and chlorinating agent. It has a molecular structure (**21.9**) and is light-sensitive and readily hydrolyzed (equation 21.25); if it is added to a concentrated KCl solution, $K[CrO_3Cl]$ precipitates. Structure **21.10** shows the $[CrO_3Cl]^-$ ion.

Fig. 21.8 Structures (X-ray diffraction) of (a) $[Cr_2O_7]^{2-}$ in the 2-amino-5-nitropyridinium salt [J. Pecaut *et al.* (1993) *Acta Crystallogr.*, *Sect. B*, vol. 49, p. 277], and (b) $[Cr_3O_{10}]^{2-}$ in the guanidinium salt [A. Stepien *et al.* (1977) *Acta Crystallogr.*, *Sect. B*, vol. 33, p. 2924]. Colour code: Cr, blue; O, pale grey.

(21.9)

∠O–Cr–O = 108.5°
∠Cl–Cr–Cl = 113°

(21.10)

∠O–Cr–O = 111°
∠Cl–Cr–O = 106°

$$2CrO_2Cl_2 + 3H_2O \longrightarrow [Cr_2O_7]^{2-} + 4Cl^- + 6H^+ \quad (21.25)$$

Chromium(VI) oxide ('chromic acid'), CrO_3, separates as a purple-red solid when concentrated H_2SO_4 is added to a solution of a dichromate(VI) salt; it is a powerful oxidant with uses in organic synthesis. It melts at 471 K and at slightly higher temperatures, decomposes to Cr_2O_3 and O_2 with CrO_2 formed as an intermediate. The solid state structure of CrO_3 consists of chains of corner-sharing tetrahedral CrO_4 units (as in Figure 21.5d).

Chromium(VI) oxide dissolves in base to give yellow solutions of $[CrO_4]^{2-}$; this is a weak base and forms $[HCrO_4]^-$ and then H_2CrO_4 as the pH is lowered (H_2CrO_4: p$K_a(1) = 0.74$; p$K_a(2) = 6.49$). In solution, these equilibria are complicated by the formation of orange dichromate(VI), $[Cr_2O_7]^{2-}$ (equation 21.26).

$$2[HCrO_4]^- \rightleftharpoons [Cr_2O_7]^{2-} + H_2O \quad (21.26)$$

Further condensation occurs at high $[H^+]$ to give $[Cr_3O_{10}]^{2-}$ and $[Cr_4O_{13}]^{2-}$. The structures (determined for solid state salts) of $[Cr_2O_7]^{2-}$ and $[Cr_3O_{10}]^{2-}$ are shown in Figure 21.8; like $[CrO_4]^{2-}$, they contain tetrahedral CrO_4 units and the chains in the di- and trinuclear species contain corner-sharing tetrahedra (i.e. as in CrO_3). The $[Cr_4O_{13}]^{2-}$ ion has a related structure. Higher species are not observed and thus chromates do not mimic vanadates in their structural complexity.

Complex formation by Cr(VI) requires strong π-donor ligands such as O^{2-} or $[O_2]^{2-}$. When H_2O_2 is added to an acidified solution of a chromate(VI) salt, the product is a deep violet-blue complex which contains both oxo and peroxo ligands (equation 21.27).

$$[CrO_4]^{2-} + 2H^+ + 2H_2O_2 \longrightarrow [Cr(O)(O_2)_2] + 3H_2O$$
$$(21.27)$$

In aqueous solution, $[Cr(O)(O_2)_2]$ rapidly decomposes to Cr(III) and O_2. An ethereal solution is more stable and, from it, the pyridine adduct $[Cr(O)(O_2)_2(py)]$ may be isolated. In the solid state, $[Cr(O)(O_2)_2(py)]$ contains an approximate pentagonal pyramidal arrangement of oxygen atoms with the oxo ligand in the apical site (Figures 21.9a and b). If each peroxo ligand is considered to occupy one rather than two coordination sites, then the coordination environment is tetrahedral (Figure 21.9c). This and related compounds (which are explosive when dry) have uses as oxidants in organic syntheses. Like other Cr(VI) compounds, $[Cr(O)(O_2)_2(py)]$ has a very small paramagnetic susceptibility (arising from coupling of the diamagnetic ground state with excited states). The action of H_2O_2 on neutral or slightly acidic solutions of $[Cr_2O_7]^{2-}$ (or reaction

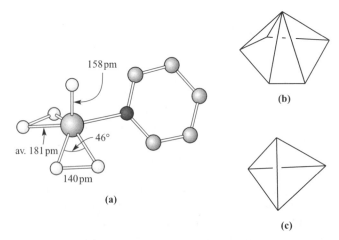

Fig. 21.9 (a) The structure of $[Cr(O)(O_2)_2(py)]$ determined by X-ray diffraction [R. Stomberg (1964) *Ark. Kemi*, vol. 22, p. 29]; colour code: Cr, blue; O, pale grey; N, black; C, grey. The coordination environment can be described as (b) pentagonal pyramidal or (c) tetrahedral (see text).

[†] For detailed discussion see: E. Jacob *et al.* (1990), *Chem. Ber.*, vol. 123, p. 1319; C. Marsden *et al.* (1991), *Inorg. Chem.*, vol. 30, p. 1681; E.G. Hope *et al.* (1991) *Inorg. Chem.*, vol. 30, p. 4875; K. Pierloot *et al.* (1992), *Inorg. Chem.*, vol. 31, p. 5353; A. Neuhaus *et al.* (1992) *Inorg. Chem.*, vol. 31, p. 5355; J. Jacobs *et al.* (1992) *Inorg. Chem.*, vol. 31, p. 5357.

between $[Cr(O)(O_2)_2]$ and alkalis) yields diamagnetic, dangerously explosive, red-violet salts of $[Cr(O)(O_2)_2(OH)]^-$. Imido ligands $[RN]^{2-}$ may formally replace oxo groups in Cr(VI) species, e.g. $[Cr(N^tBu)_2Cl_2]$ is structurally related to CrO_2Cl_2.

Chromium(VI) in acidic solution is a powerful oxidizing agent (equation 21.28), albeit often slow. Both $Na_2Cr_2O_7$ and $K_2Cr_2O_7$ are manufactured on a large scale; $K_2Cr_2O_7$ is less soluble than $Na_2Cr_2O_7$. Both are widely used as oxidants in organic syntheses; commercial applications include those in tanning, corrosion inhibitors, wood preservatives and insecticides. Potassium dichromate(VI) is used in titrimetric analysis (e.g. reaction 21.29) and the colour change accompanying reduction of $[Cr_2O_7]^{2-}$ to Cr^{3+} is the basis for some types of breathalyzer units in which ethanol is oxidized to acetaldehyde. Sodium chromate(VI), also an important oxidant, is manufactured by reaction 21.30.

$$[Cr_2O_7]^{2-} + 14H^+ + 6e^- \rightleftharpoons 2Cr^{3+} + 7H_2O$$
$$\underset{\text{orange}}{\phantom{[Cr_2O_7]^{2-}}} \qquad \underset{\text{green}^\dagger}{\phantom{2Cr^{3+}}}$$

$$E^o = +1.33\,V \qquad (21.28)$$

$$[Cr_2O_7]^{2-} + 14H^+ + 6Fe^{2+} \longrightarrow 2Cr^{3+} + 7H_2O + 6Fe^{3+}$$
$$(21.29)$$

$$Na_2Cr_2O_7 + 2NaOH \longrightarrow 2Na_2CrO_4 + H_2O \qquad (21.30)$$

Chromium(VI) compounds are highly toxic (suspected carcinogens) and must be stored away from combustible materials; violent reactions occur with some organic compounds.

Chromium(V) and chromium(IV)

Unlike CrF_6, CrF_5 is well established. It is a red, volatile solid (mp 303 K), formed by direct combination of the elements at $\approx 570\,K$. The vapour is yellow and contains distorted trigonal bipyramidal CrF_5 molecules; the solid state structure consists of infinite chains supported by $Cr-F-Cr$ bridges. It is a strong oxidizing and fluorinating agent. For Cr(V), the fluoride is the only halide known. Chromium(IV) fluoride is made by reaction of Cr and F_2, but reports of its colour vary (colourless, green, blue and brown) with some descriptions being affected by the presence of impurities. The vapour contains tetrahedral molecules. Chromium(IV) chloride and bromide have been prepared but are unstable.

Chromium(IV) oxide, CrO_2, is usually made by controlled decomposition of CrO_3. It is a brown-black solid which has the rutile structure and is a metallic conductor (compare with VO_2). It is ferromagnetic and is widely used in magnetic recording tapes.

When an acidic solution in which $[Cr_2O_7]^{2-}$ is oxidizing propan-2-ol is added to aqueous $MnSO_4$, MnO_2 is precipitated, although acidified $[Cr_2O_7]^{2-}$ alone does not effect

this oxidation. This observation is evidence for the participation of Cr(V) or Cr(IV) in dichromate(VI) oxidations. Under suitable conditions, it is possible to isolate salts of $[CrO_4]^{3-}$ and $[CrO_4]^{4-}$. For example, dark blue Sr_2CrO_4 is produced by heating $SrCrO_4$, Cr_2O_3 and $Sr(OH)_2$ at 1270 K, and dark green Na_3CrO_4 results from reaction of Na_2O, Cr_2O_3 and Na_2CrO_4 at 770 K.

Complexes of chromium(V) may be stabilized by π-donor ligands, e.g. $[CrF_6]^-$, $[CrOF_4]^-$, $[CrOF_5]^{2-}$ and $[Cr(N^tBu)Cl_3]$. Peroxo complexes containing $[Cr(O_2)_4]^{3-}$ are obtained by reaction of chromate(V) with H_2O_2 in alkaline solution; $[Cr(O_2)_4]^{3-}$ has a dodecahedral structure. These salts are explosive but are less dangerous than the Cr(VI) peroxo complexes. The explosive Cr(IV) peroxo complex $[Cr(O_2)_2(NH_3)_3]$ (**21.11**) is formed when $[Cr_2O_7]^{2-}$ reacts with aqueous NH_3 and H_2O_2; a related complex is $[Cr(O_2)_2(CN)_3]^{3-}$.

(21.11)

Chromium(III)

The +3 oxidation state is the most stable for chromium in its compounds and octahedral coordination dominates for Cr(III) centres. Table 20.3 shows the large LFSE associated with the octahedral d^3 configuration, and Cr(III) complexes are generally kinetically inert (see *Section 25.2*).

Anhydrous $CrCl_3$ (red-violet solid, mp 1425 K) is made from the metal and Cl_2, and is converted to green CrF_3 by heating with HF at 750 K. Solid CrF_3 is isostructural with VF_3, and $CrCl_3$ adopts a BiI_3 lattice. The dark-green tribromide and triiodide can be prepared from Cr and the respective halogen and are isostructural with $CrCl_3$. Chromium(III) trifluoride is sparingly soluble and may be precipitated as the hexahydrate. The formation of and hydrate isomerism in $CrCl_3 \cdot 6H_2O$ were described in *Section 19.8*. Although pure $CrCl_3$ is insoluble in water, addition of a trace of Cr(II) (e.g. $CrCl_2$) results in dissolution; the fast redox reaction between Cr(III) in the $CrCl_3$ lattice and Cr(II) in solution is followed by rapid substitution of Cl^- by H_2O at the solid surface since Cr(II) is labile (see *Chapter 25*).

Chromium(III) oxide is made by combination of the elements at high temperature, by reduction of CrO_3, or by reaction 21.31. It has the corundum structure (*Section 12.7*) and is semiconducting and antiferromagnetic ($T_N = 310\,K$). Commercially Cr_2O_3 is used in abrasives and is an important green pigment; the dihydrate (*Guignet's green*) is used in paints. Traces of Cr(III) in Al_2O_3 give rise to the red colour of rubies (see *Section 12.2*).

---

† The green colour is due to a sulfato complex H^+ being supplied as sulfuric acid; $[Cr(H_2O)_6]^{3+}$ is violet, see later.

$$[NH_4]_2[Cr_2O_7] \xrightarrow{\Delta} Cr_2O_3 + N_2 + 4H_2O \qquad (21.31)$$

Large numbers of mononuclear, octahedral Cr(III) complexes are known with magnetic moments close to the spin-only value of 3.87 μ_B (*Table 20.7*); the electronic spectra of octahedral d^3 complexes contain three absorptions due to '*d–d*' transitions (see *Figure 20.18*). Selected examples of octahedral chromium(III) complexes are [Cr(acac)$_3$], [Cr(ox)$_3$]$^{3-}$, [Cr(en)$_3$]$^{3+}$, [Cr(bpy)$_3$]$^{3+}$, *cis*- and *trans*-[Cr(en)$_2$F$_2$]$^+$, *trans*-[CrCl$_2$(MeOH)$_4$]$^+$, [Cr(CN)$_6$]$^{3-}$ and [Cr(NH$_3$)$_2$(S$_5$)$_2$]$^-$ ([S$_5$]$^{2-}$ is didentate, see *Figure 15.9* for related structures). Complex halides include [CrF$_6$]$^{3-}$, [CrCl$_6$]$^{3-}$ and [Cr$_2$Cl$_9$]$^{3-}$. Violet Cs$_3$[Cr$_2$Cl$_9$] is made by reaction 21.32; [Cr$_2$Cl$_9$]$^{3-}$ is isostructural with [V$_2$Cl$_9$]$^{3-}$ (**21.8**) and magnetic data are consistent with the presence of three unpaired electrons per Cr(III) centre, i.e. *no* Cr–Cr interaction.

$$3CsCl + 2CrCl_3 \xrightarrow{\text{in a melt}} Cs_3[Cr_2Cl_9] \qquad (21.32)$$

Pale violet [Cr(H$_2$O)$_6$]$^{3+}$ is obtained in aqueous solution when [Cr$_2$O$_7$]$^{2-}$ is reduced by SO$_2$ or by EtOH and H$_2$SO$_4$ below 200 K. The commonest salt containing [Cr(H$_2$O)$_6$]$^{3+}$ is chrome alum, KCr(SO$_4$)$_2$·12H$_2$O; [Cr(H$_2$O)$_6$]$^{3+}$ has been structurally characterized in the solid state in a number of salts, e.g. [Me$_2$NH$_2$][Cr(H$_2$O)$_6$][SO$_4$]$_2$ (av. Cr–O = 196 pm). From aqueous solutions of Cr(III) salts, alkali precipitates Cr$_2$O$_3$ which dissolves to give [Cr(OH)$_6$]$^{3-}$. The hexaaqua ion is quite acidic (p$K_a \approx 4$) and hydroxo-bridged species are present in solution (see *equation 6.38* and accompanying discussion); Figure 21.10 shows the structure of [Cr$_2$(H$_2$O)$_8$(μ-OH)$_2$]$^{4+}$. Addition of NH$_3$ to aqueous solutions of [Cr(H$_2$O)$_6$]$^{3+}$ results in the slow formation of ammine complexes; it is preferable to use Cr(II) precursors since substitution is faster (see *Chapter 25*). The dinuclear complex **21.12** is reversibly converted to the oxo-bridged **21.13** in the presence of alkali (equation 21.33).

$$\left[\begin{array}{c} L \quad L \qquad L \quad L \\ L-Cr \overset{H}{\underset{O}{\diagdown}} Cr-L \\ L \quad L \qquad L \quad L \end{array} \right]^{5+} + [OH]^-$$

∠Cr–O–Cr = 166°

L = NH$_3$

(**21.12**)

$$\rightleftharpoons \left[\begin{array}{c} L \quad L \qquad L \quad L \\ L-Cr-O-Cr-L \\ L \quad L \qquad L \quad L \end{array} \right]^{4+} + H_2O \qquad (21.33)$$

∠Cr–O–Cr = 180°

L = NH$_3$

(**21.13**)

The two Cr(III) (d^3) centres in complex **21.13** are antiferromagnetically coupled and this is rationalized in terms

Fig. 21.10 The structure of [Cr$_2$(H$_2$O)$_8$(μ-OH)$_2$]$^{4+}$ determined by X-ray diffraction for the mesitylene-2-sulfonate salt; the *non-bonded* Cr····Cr separation is 301 pm [L. Spiccia *et al.* (1987) *Inorg. Chem.*, vol. 26, p. 474]. Colour code: Cr, blue; O, pale blue; H, black.

of $(d-p)\pi$ bonding involving Cr d and O p orbitals. Weak antiferromagnetic coupling also occurs between the Cr(III) centres in trinuclear complexes of type [Cr$_3$L$_3$(μ-O$_2$CR)$_6$(μ_3-O)]$^+$ (Figure 21.11).

Chromium(II)

Anhydrous CrF$_2$, CrCl$_2$ and CrBr$_2$ are made by reacting Cr with HX (X = F, Cl, Br) at >850 K; CrI$_2$ is formed by heating the elements. The fluoride and chloride adopt distorted rutile structures (*Figure 5.21*), while CrBr$_2$ and CrI$_2$ crystallize with distorted CdI$_2$ lattices (*Figure 5.22*); the distortions arise from the Jahn–Teller effect (high-spin d^4, see *Section 20.3*). Crystals of CrCl$_2$ are colourless but dissolve in water to give blue solutions of the strongly reducing hexaaqua ion. Solutions of [Cr(H$_2$O)$_6$]$^{2+}$ are usually obtained by dissolving Cr in acids or by reduction (Zn amalgam or electrolytically) of Cr(III)-containing solutions. Hydrated salts such as Cr(ClO$_4$)$_2$·6H$_2$O, CrCl$_2$·4H$_2$O and CrSO$_4$·7H$_2$O may be isolated from solution, but cannot be dehydrated without decomposition.

For the Cr^{3+}/Cr^{2+} couple, $E^\circ = -0.41$ V, and Cr(II) compounds slowly liberate H$_2$ from water, as well as undergo oxidation by O$_2$ (see *worked example 7.3*). The potential diagram in Figure 21.7 shows that Cr(II) compounds are just stable with respect to disproportionation. The study of the oxidation of Cr^{2+} species has played an important role in establishing the mechanisms of redox reactions (see *Chapter 25*).

Complexes of Cr(II) include halide anions such as [CrX$_3$]$^-$, [CrX$_4$]$^{2-}$, [CrX$_5$]$^{3-}$ and [CrX$_6$]$^{4-}$. Despite the range of formulae, the Cr(II) centres in the solids are usually octahedrally sited, e.g. [CrCl$_3$]$^-$ consists of chains of distorted face-sharing octahedra, the distortion being a Jahn–Teller effect. Some of these salts show interesting magnetic properties. For example, salts of [CrCl$_4$]$^{2-}$ show *ferromagnetic coupling* (as opposed to antiferromagnetic coupling which is a more common phenomenon, see *Section 20.8*) at low temperatures, with T_C values in the range 40–60 K; communication between the metal centres is by virtue of Cr–Cl–Cr bridging interactions.

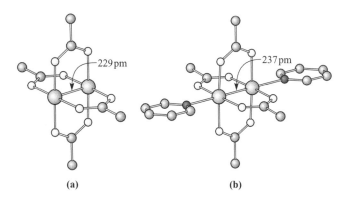

Fig. 21.12 The structures (X-ray diffraction) of (a) [Cr$_2$(μ-O$_2$CMe)$_4$] [F.A. Cotton *et al.* (1977) *J. Am. Chem. Soc.*, vol. 99, p. 4704] and (b) [Cr$_2$(py)$_2$(μ-O$_2$CMe)$_4$] [F.A. Cotton *et al.* (1980) *Inorg. Chem.*, vol. 19, p. 328]. Hydrogen atoms are omitted for clarity; colour code: Cr, blue; O, pale grey; C, grey; N, black.

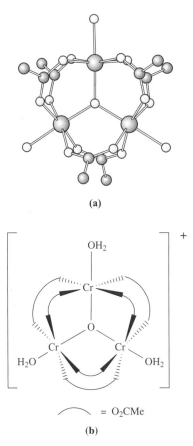

Fig. 21.11 A representative member of the [Cr$_3$L$_3$(μ-O$_2$CR)$_6$(μ$_3$-O)]$^+$ family of complexes: (a) the structure of [Cr$_3$(H$_2$O)$_3$(μ-O$_2$CMe)$_6$(μ$_3$-O)]$^+$ (X-ray diffraction) in the hydrated chloride salt [C.E. Anson *et al.* (1997) *Inorg. Chem.*, vol. 36, p. 1265], and (b) a schematic representation of the same complex. In (a), the H atoms are omitted for clarity; colour code: Cr, blue; O, pale grey; C, grey.

Chromium–chromium multiple bonds

Chromium(II) carboxylates are dimers of general formula [Cr$_2$(μ-O$_2$CR)$_4$] or [Cr$_2$L$_2$(μ-O$_2$CR)$_4$] and are examples of *d*-block metal complexes that involve metal–metal multiple bonding. For example, red [Cr$_2$(H$_2$O)$_2$(μ-O$_2$CMe)$_4$] is precipitated when aqueous CrCl$_2$ is added to saturated aqueous Na[MeCO$_2$]. Figure 21.12 shows the structures of [Cr$_2$(μ-O$_2$CMe)$_4$] and [Cr$_2$(py)$_2$(μ-O$_2$CMe)$_4$]. Such compounds are *diamagnetic*, possess *short* Cr–Cr bonds (cf. 258 pm in Cr metal), and have eclipsed ligand conformations. These properties are consistent with the Cr(II) *d* electrons being involved in *quadruple* bond formation. For the bridging ligands in [Cr$_2$(μ-O$_2$CMe)$_4$] to be eclipsed is less surprising than in complexes with monodentate ligands, e.g. [Re$_2$Cl$_8$]$^{2-}$ (see *Section 22.8*), but the observation is a key feature in the description of the metal–metal quadruple bond. The bonding in [Cr$_2$(μ-O$_2$CMe)$_4$] can be described as shown in Figure 21.13. The Cr atoms are defined to lie on the *z* axis, and each Cr atom uses four (*s*, *p$_x$*, *p$_y$* and *d$_{x^2-y^2}$*) of its nine atomic orbitals to form Cr–O bonds. Now allow

mixing of the *p$_z$* and *d$_{z^2}$* orbitals to give two hybrid orbitals directed along the *z* axis; each Cr atom has four orbitals available for metal–metal bonding: *d$_{xz}$*, *d$_{yz}$*, *d$_{xy}$* and one *p$_z$d$_{z^2}$* hybrid, with the second *p$_z$d$_{z^2}$* hybrid being non-bonding and pointing outwards from the Cr–Cr-unit (see below). Figure 21.13a shows that overlap of the metal *p$_z$d$_{z^2}$* hybrid orbitals leads to σ-bond formation, while *d$_{xz}$*–*d$_{xz}$* and *d$_{yz}$*–*d$_{yz}$* overlap gives a degenerate pair of π-orbitals. Finally, overlap of the *d$_{xy}$* orbitals gives rise to a δ-bond. The degree of overlap follows the order σ > π > δ and Figure 21.13b shows an approximate energy level diagram for the σ, π, δ, σ*, π* and δ* MOs. Each Cr(II) centre provides four electrons for Cr–Cr bond formation and these occupy the MOs in Figure 21.13b to give a σ^{2}π^{4}δ^2 configuration, i.e. a quadruple bond. A consequence of this bonding picture is that the δ-component forces the two CrO$_4$-units to be eclipsed. The red colour of [Cr$_2$(μ-O$_2$CMe)$_4$] (λ_{max} = 520 nm, see *Table 19.2*) can be understood in terms of the δ–δ* energy gap and a σ^{2}π^{4}δ^{1}δ^{*1} ⟵ σ^{2}π^{4}δ^2 transition.

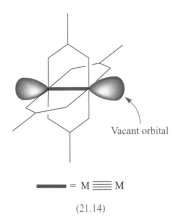

Vacant orbital

━━━ = M ≡≡≡ M

(21.14)

This bonding description for [Cr$_2$(μ-O$_2$CMe)$_4$] leaves a non-bonding, outward pointing *p$_z$d$_{z^2}$* hybrid orbital per Cr

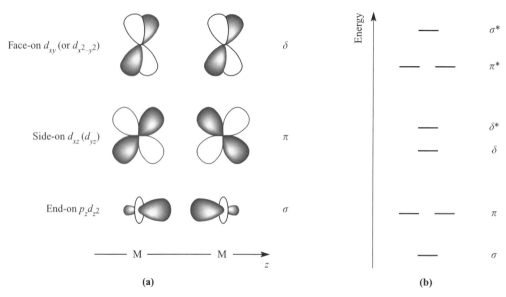

Fig. 21.13 (a) The formation of σ, π and δ components of a metal–metal quadruple bond by overlap of appropriate metal orbitals. *Both* the d_{xz} and d_{yz} atomic orbitals are used to form π-bonds, but *either* the d_{xy} or $d_{x^2-y^2}$ atomic orbital is used for δ-bond formation, the choice being arbitrary depending on which has been 'chosen' for Cr–O bond formation. (b) Approximate energy levels of the metal–metal bonding and antibonding MOs.

atom (**21.14**); complex formation with donors such as H_2O and pyridine (Figure 21.12b) occurs by donation of a lone pair of electrons into each vacant orbital.

21.8 Group 7: manganese

The metal

Metallic Mn is slowly attacked by water and dissolves readily in acids (e.g. equation 21.34). The finely divided metal is pyrophoric in air, but the bulk metal is not attacked unless heated (equation 21.35). At elevated temperatures, it combines with most non-metals, e.g. N_2 (equation 21.36), halogens (equation 21.37), C, Si and B (see *Sections 12.10, 13.7* and *14.6*).

$$Mn + 2HCl \longrightarrow MnCl_2 + H_2 \qquad (21.34)$$

$$3Mn + 2O_2 \xrightarrow{\Delta} Mn_3O_4 \qquad (21.35)$$

$$3Mn + N_2 \xrightarrow{\Delta} Mn_3N_2 \qquad (21.36)$$

$$Mn + Cl_2 \xrightarrow{\Delta} MnCl_2 \qquad (21.37)$$

Manganese exhibits the widest range of oxidation states of any of the first row *d*-block metals; the lowest states are stabilized by π-acceptor ligands, usually in organometallic complexes (see *Chapter 23*). However, dissolution of Mn powder in air-free aqueous NaCN gives the Mn(I) complex $Na_5[Mn(CN)_6]$.

Section 21.8 describes Mn(II)–Mn(VII) species and a potential diagram for manganese was given in *Figure 7.2*.

On going from Cr to Mn, there is an abrupt change in the stability with respect to oxidation of M^{2+} (equation 21.38); the difference in E° values arises from the much higher third ionization energy of Mn (see *Table 21.1*). All oxidation states above Mn(II) are powerful oxidizing agents.

$$M^{3+}(aq) + e^- \rightleftharpoons M^{2+}(aq) \quad \begin{cases} M = Mn, \ E^{\circ} = +1.54\,V \\ M = Cr, \ E^{\circ} = -0.41\,V \end{cases}$$

$$(21.38)$$

Manganese(VII)

Binary halides of Mn(VII) have not been isolated. The oxohalides MnO_3F and MnO_3Cl may be made by reacting $KMnO_4$ with HSO_3X (X = F or Cl) at low temperature; both are powerful oxidants and decompose explosively at room temperature. Both MnO_3F and MnO_3Cl have molecular (C_{3v}) structures. The oxo and imido, $[RN]^{2-}$, groups are isoelectronic, and compounds of the type $Mn(NR)_3Cl$ have been prepared by reacting $MnCl_3$ with $RNH(SiMe_3)$; the chloride group in $Mn(NR)_3Cl$ can be substituted by a range of anions (Figure 21.14).

Manganese(VII) chemistry is dominated by the manganate(VII) ion (permanganate). The potassium salt, $KMnO_4$, is a strong oxidizing agent and is corrosive to human tissue; it is manufactured on a large scale (see *Box 21.4*) by conversion of MnO_2 to K_2MnO_4 followed by electrolytic oxidation. In analytical chemistry, Mn determination involves oxidation of Mn(II) to $[MnO_4]^-$ by bismuthate, periodate or peroxodisulfate. Solid $KMnO_4$ forms dark purple-black crystals and is isostructural with $KClO_4$; tetrahedral $[MnO_4]^-$ ions have equivalent bonds (Mn–O = 163 pm). Aqueous solutions of $KMnO_4$ deposit MnO_2 on standing. Although $KMnO_4$ is

KMnO$_4$: a powerful oxidant at work

About 0.05 Mt per year of KMnO$_4$ are manufactured world-wide. Although this amount does not compete with those of inorganic chemicals such as CaO, NH$_3$, TiO$_2$ and the major mineral acids, the role of KMnO$_4$ as an oxidizing agent is nonetheless extremely important. In addition to oxidations of organic compounds in industrial manufacturing processes, KMnO$_4$ is used in water purification where it is preferable to Cl$_2$ for two reasons: it does not affect the taste of the water,

and MnO$_2$ (produced on reduction) is a coagulant for particulate impurities. The oxidizing power of KMnO$_4$ is also applied to the removal of impurities, for example in the purification of MeOH, EtOH, MeCO$_2$H and NC(CH$_2$)$_4$CN (a precursor in nylon manufacturing). Some commercial bleaching processes use KMnO$_4$, e.g. bleaching some cotton fabrics, jute fibres and beeswax.

insoluble in benzene, the addition of the cyclic ether 18-crown-6 results in the formation of the soluble [K(18-crown-6)][MnO$_4$] (see *Section 10.8*). Potassium permanganate is intensely coloured owing to charge-transfer (O $\longrightarrow$ Mn). It also shows weak temperature-independent paramagnetism arising from the coupling of the diamagnetic ground state of [MnO$_4$]$^-$ with paramagnetic excited states under the influence of a magnetic field.

The free acid HMnO$_4$ can be obtained by low temperature evaporation of its aqueous solution (made by ion exchange).

e.g. X = Cl, O$_2$CMe, OC$_6$F$_5$

(a)

(b)

Fig. 21.14 (a) Examples of manganese(VII) imido complexes. (b) The structure of [Mn(N^{t}Bu)$_3$(O$_2$CMe)] determined by X-ray diffraction [A.A. Danopoulos *et al.* (1994) *J. Chem. Soc.*, *Dalton Trans.*, p. 1037]. Hydrogen atoms are omitted for clarity; colour code: Mn, blue; N, black; O, pale grey; C,

It is a violent oxidizing agent and explodes above 273 K. The anhydride of HMnO$_4$ is Mn$_2$O$_7$, made by the action of concentrated H$_2$SO$_4$ on pure KMnO$_4$. It is a green, hygroscopic, highly explosive liquid, unstable above 263 K (equation 21.39) and has molecular structure **21.15**.

$$2Mn_2O_7 \xrightarrow{>263\,K} 4MnO_2 + 3O_2 \qquad (21.39)$$

$$\angle Mn\text{–}O\text{–}Mn = 121°$$

(21.15)

Equations 21.40–21.42 show reductions of [MnO$_4$]$^-$ to Mn(VI), Mn(IV) and Mn(II) respectively.

$$[MnO_4]^-(aq) + e^- \rightleftharpoons [MnO_4]^{2-}(aq)$$
$$E^\circ = +0.56\,V \qquad (21.40)$$

$$[MnO_4]^-(aq) + 4H^+ + 3e^- \rightleftharpoons MnO_2(s) + 2H_2O$$
$$E^\circ = +1.69\,V \qquad (21.41)$$

$$[MnO_4]^-(aq) + 8H^+ + 5e^- \rightleftharpoons Mn^{2+}(aq) + 4H_2O$$
$$E^\circ = +1.51\,V \qquad (21.42)$$

The H$^+$ concentration plays an important part in influencing which reduction takes place (see *Section 7.2*). Although many reactions of KMnO$_4$ can be understood by considering redox potentials, kinetic factors are also important. Permanganate at pH 0 should oxidize water, but in practice the reaction is extremely slow. It should also oxidize [C$_2$O$_4$]$^{2-}$ at room temperature, but reaction 21.43 is extremely slow unless Mn^{2+} is added or the temperature is raised.

$$2[MnO_4]^- + 16H^+ + 5[C_2O_4]^{2-}$$
$$\longrightarrow 2Mn^{2+} + 8H_2O + 10CO_2 \qquad (21.43)$$

Many studies have been made on the mechanism of such reactions and, as in oxidations by [Cr$_2$O$_7$]$^{2-}$, it has been shown that intermediate oxidation states are involved.

Manganese(VI)

No binary halides of Mn(VI) have been isolated, and the only oxohalide is MnO_2Cl_2 (**21.16**). It is prepared by reducing $KMnO_4$ with SO_2 at low temperature in HSO_3Cl, and is a brown liquid which readily hydrolyzes and decomposes at 240 K.

(21.16)

Salts of dark green $[MnO_4]^{2-}$ are made by fusing MnO_2 with group 1 metal hydroxides in the presence of air, or by reaction 21.44. This oxidation may be reversed by reaction 21.45.

$$4[MnO_4]^- + 4[OH]^- \longrightarrow 4[MnO_4]^{2-} + 2H_2O + O_2 \quad (21.44)$$

$$2[MnO_4]^{2-} + Cl_2 \longrightarrow 2[MnO_4]^- + 2Cl^- \quad (21.45)$$

Manganate(VI) is unstable with respect to disproportionation (equation 21.46) in the presence of even weak acids such as H_2CO_3 and is therefore not formed in the reduction of acidified $[MnO_4]^-$.

$$3[MnO_4]^{2-} + 4H^+ \longrightarrow 2[MnO_4]^- + MnO_2 + 2H_2O \quad (21.46)$$

The $[MnO_4]^{2-}$ ion is tetrahedral (Mn$-$O $= 166$ pm), and K_2MnO_4 is isomorphous with K_2CrO_4 and K_2SO_4. At 298 K, the magnetic moment of K_2MnO_4 is $1.75\,\mu_B$ (d^1). The tetrahedral anion $[Mn(N^tBu)_4]^{2-}$ (an imido analogue of $[MnO_4]^{2-}$) is made by treating $Mn(N^tBu)_3Cl$ with $Li[NH^tBu]$.

Manganese(V)

Binary halides of Mn(V) have not been isolated. The only oxohalide is $MnOCl_3$ (**21.17**) which is made by reacting $KMnO_4$ with $CHCl_3$ in HSO_3Cl. Above 273 K, $MnOCl_3$ decomposes, and in moist air, it hydrolyzes to $[MnO_4]^{3-}$. Salts of $[MnO_4]^{3-}$ are blue and moisture-sensitive; the most accessible are $K_3[MnO_4]$ and $Na_3[MnO_4]$, made by reduction of $[MnO_4]^-$ in concentrated aqueous KOH or NaOH at 273 K. Solutions of $[MnO_4]^{3-}$ must be strongly alkaline to prevent disproportionation which occurs readily in weakly alkaline (equation 21.47) or acidic (equation 21.48) media.

(21.17)

$$2[MnO_4]^{3-} + 2H_2O \longrightarrow [MnO_4]^{2-} + MnO_2 + 4[OH]^- \quad (21.47)$$

$$3[MnO_4]^{3-} + 8H^+ \longrightarrow [MnO_4]^- + 2MnO_2 + 4H_2O \quad (21.48)$$

The tetrahedral structure of $[MnO_4]^{3-}$ has been confirmed in the solid state in $Na_{10}Li_2(MnO_4)_4$; the Mn$-$O bonds are longer (170 pm) than in manganate(VI) or manganate(VII). Magnetic moments of $[MnO_4]^{3-}$ salts are typically $\approx 2.8\,\mu_B$.

Manganese(IV)

The only binary halide of Mn(IV) is MnF_4, prepared from the elements. It is an unstable blue solid which decomposes at ambient temperatures (equation 21.49).

$$2MnF_4 \longrightarrow 2MnF_3 + F_2 \quad (21.49)$$

Manganese(IV) oxide is polymorphic and often markedly non-stoichiometric. Only the high temperature β-form has the stoichiometry MnO_2, and adopts the rutile structure (*Figure 5.21*). It acts as an oxidizing agent when heated with concentrated acids (e.g. reaction 21.50).

$$MnO_2 + 4HCl \xrightarrow[\text{conc}]{\Delta} MnCl_2 + Cl_2 + 2H_2O \quad (21.50)$$

Hydrated forms of MnO_2 are extremely insoluble and are often obtained as dark black-brown precipitates in redox reactions involving $[MnO_4]^-$ (equation 21.41) when the $[H^+]$ is insufficient to allow reduction to Mn^{2+}.

The reaction of Mn_2O_3 with $CaCO_3$ at 1400 K yields Ca_2MnO_4 which formally contains $[MnO_4]^{4-}$. However, Ca_2MnO_4 crystallizes with a layer structure in which each Mn(IV) centre is in an octahedral MnO_6 environment, and isolated $[MnO_4]^{4-}$ ions are not present.

The coordination chemistry of Mn(IV) is limited. Mononuclear complexes include $[MnF_6]^{2-}$, obtained by fluorinating mixtures of chlorides or by reducing $[MnO_4]^-$ with H_2O_2 in aqueous HF. Reaction 21.51 shows the first practicable non-electrolytic method of producing F_2.

$$K_2[MnF_6] + 2SbF_5 \xrightarrow{\Delta} MnF_2 + 2K[SbF_6] + F_2 \quad (21.51)$$

The structure of K_2MnF_6 is a lattice prototype for some AB_2X_6 systems (e.g. Cs_2FeF_6 and K_2PdF_6). It is best considered as a close-packed array of K^+ and F^- ions in an alternating cubic-hexagonal sequence; the Mn^{4+} centres occupy some of the octahedral holes such that they are surrounded by six F^- ions giving $[MnF_6]^{2-}$ ions present in the lattice. Closely related lattice types are K_2GeF_6 and K_2PtCl_6 in which the K^+ and X^- ions in each compound form hcp or ccp arrays respectively.[†]

Of significance is a series of multinuclear manganese complexes (Figure 21.15) which have been studied as models for the enzyme *Photosystem II* which converts H_2O to O_2 represented by equation 21.52.

$$2H_2O \longrightarrow O_2 + 4H^+ + 4e^- \quad (21.52)$$

[†] For detailed descriptions of these lattice types, see A.F. Wells (1984) *Structural Inorganic Chemistry* 5th edn, Oxford University Press, Oxford, p. 458.

(a)

(b)

Fig. 21.15 An example of a manganese(III),(IV) cluster as a model for *Photosystem II*. (a) The $Mn_4(\mu_3\text{-}Cl)(\mu_3\text{-}O)_3$ core of $[Mn_4Cl_3(\mu_3\text{-}Cl)(\mu_3\text{-}O)_3(\mu\text{-}O_2CMe)_3(py)_3]$, the structure of which has been determined by X-ray diffraction [D.N. Hendrickson *et al.* (1992) *J. Am. Chem. Soc.*, vol. 114, p. 2455] and (b) a schematic representation of the same complex. Colour code in (a): Mn, blue; O, pale grey; Cl, black.

Although the structure of *Photosystem II* has not been established by crystallography, it is known that four Mn centres, connected by μ-O or μ-OH ligands, are involved; donor atoms from amino acid residues probably occupy other Mn coordination sites. Electron transfer is proposed to involve the four metal centres undergoing a sequence of redox steps, the fully reduced and oxidized states being $\{Mn^{II}Mn^{III}_3\}$ and $\{Mn^{III}Mn^{IV}_3\}$ respectively.

Manganese(III)

The only binary halide of Mn(III) is the red-purple MnF_3 which is made by the action of F_2 on Mn(II) halides at 520 K. It is thermally stable but is immediately hydrolyzed by water. The solid state structure of MnF_3 is related to those of TiF_3, VF_3, CrF_3, FeF_3 and CoF_3 but is Jahn–Teller distorted (high-spin d^4); there are, however, *three* pairs of Mn–F distances (179, 191 and 209 pm) rather than the distortions shown in structures **20.5** and **20.6**. At room temperature, the magnetic moment of MnF_3 is 4.94 μ_B, but on cooling, it becomes antiferromagnetic ($T_N = 43$ K) (see *Section 20.8*).

The black oxide Mn_2O_3 (the α-form) is obtained when MnO_2 is heated at 1070 K or (in the hydrous form) by oxidation of Mn(II) in alkaline media. At higher temperatures, it

forms Mn_3O_4, a normal spinel ($Mn^{II}Mn^{III}_2O_4$, see *Box 12.6*) but with the Mn(III) centres being Jahn–Teller distorted. The Mn atoms in α-Mn_2O_3 are in distorted octahedral MnO_6 sites (elongated, diagram **20.5**); the structure differs from the corundum lattice adopted by Ti_2O_3, V_2O_3 and Cr_2O_3. Whereas Mn_2O_3 is *antiferromagnetic* below 80 K, Mn_3O_4 is *ferrimagnetic* below 43 K.

Most complexes of Mn(III) are octahedral, high-spin systems being Jahn–Teller distorted. The red aqua ion $[Mn(H_2O)_6]^{3+}$ can be obtained by electrolytic oxidation of aqueous Mn^{2+} and is present in the alum $CsMn(SO_4)_2\cdot12H_2O$; this, surprisingly, shows no Jahn–Teller distortion, at least down to 78 K. In aqueous solution, $[Mn(H_2O)_6]^{3+}$ is appreciably hydrolyzed (see *Section 6.7*) and polymeric cations are present. It is also unstable with respect to disproportionation (equation 21.53) as expected from the potentials in *Figure 7.2*; it is less unstable in the presence of high concentrations of Mn^{2+} or H^+ ions.

$$2Mn^{3+} + 2H_2O \longrightarrow Mn^{2+} + MnO_2 + 4H^+ \qquad (21.53)$$

The Mn^{3+} ion is stabilized by hard ligands including F^-, $[PO_4]^{3-}$, $[SO_4]^{2-}$ or $[C_2O_4]^{2-}$; the pink colour sometimes seen before the end of the permanganate–oxalate titration (equation 21.43) is due to an oxalato complex of Mn(III). The salt $Na_3[MnF_6]$ is made by heating NaF with MnF_3, and reaction of MnO_2 with KHF_2 in aqueous HF gives $K_3[MnF_6]$; both salts are violet and have magnetic moments of 4.9 μ_B (298 K) consistent with the spin-only value for high-spin d^4. Reaction of NaF with MnF_3 in aqueous HF yields pink $Na_2[MnF_5]$ which contains chains of distorted octahedral Mn(III) centres (**21.18**) in the solid state. Salts of $[MnF_4]^-$ also crystallize with the Mn centres in distorted octahedral sites, e.g. $CsMnF_4$ has a layer structure. However, in salts of $[MnCl_5]^{2-}$ for which solid state data are available, discrete square-based pyramidal anions are present. Contrasting structures are also observed in the related complexes $[Mn(N_3)(acac)_2]$ and $[Mn(NCS-N)(acac)_2]$; whereas the azido ligand presents two nitrogen donors to adjacent Mn(III) centres to produce a chain polymer, the thiocyanate ligand binds only through the hard *N*-donor leaving the soft *S*-donor uncoordinated (Figure 21.16). The complex $[Mn(acac)_3]$ (obtained from $MnCl_2$ and $[acac]^-$ followed by oxidation with $KMnO_4$) is also of structural interest; it is dimorphic, crystallizing in one form with an elongated octahedral coordination sphere (**20.5**) while in the other it is compressed (**20.6**).

$$\left[\begin{array}{c} F \quad F \\ | \\ \text{- - - -} F \text{—} Mn \text{- - - - -} \\ | \\ F \quad F \end{array} \right]^{2n-}_n$$

(21.18)

The only well-known low-spin complex of Mn(III) is the dark red $K_3[Mn(CN)_6]$, made from KCN and $K_3[MnF_6]$

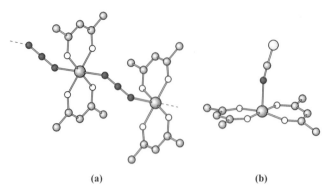

Fig. 21.16 The structures (X-ray diffraction) of the Mn(III) complexes (a) [Mn(N₃)(acac)₂] which forms polymeric chains [B.R. Stults _et al._ (1975) _Inorg. Chem._, vol. 14, p. 722] and (b) [Mn(NCS-_N_)(acac)₂] [B.R. Stults _et al._ (1979) _Inorg. Chem._, vol. 18, p. 1847]. Hydrogen atoms are omitted for clarity; colour code: Mn, dark blue; C, grey; O, white; N, black; S, pale blue.

or by oxidation of $K_4[Mn(CN)_6]$ using 3% H_2O_2. As expected for low-spin d^4, $[Mn(CN)_6]^{3-}$ has a regular octahedral structure ($Mn-C = 198$ pm).

Manganese(II)

Manganese(II) salts are obtained from MnO_2 by a variety of methods; the soluble $MnCl_2$ and $MnSO_4$ result from heating MnO_2 with the appropriate concentrated acid (equations 21.50 and 21.54). The sulfate is commercially made by this route (MnO_2 being supplied as the mineral pyrolusite) and is commonly encountered as the hydrate $MnSO_4 \cdot 5H_2O$.

$$2MnO_2 + 2H_2SO_4 \xrightarrow[\text{conc}]{\Delta} 2MnSO_4 + O_2 + 2H_2O \qquad (21.54)$$

Insoluble $MnCO_3$ is obtained by precipitation from solutions containing Mn^{2+}; however, the carbonate so obtained contains hydroxide. Pure $MnCO_3$ can be made by reaction of manganese(II) acetate or hydroxide with supercritical CO_2 (see _Box 8.1_).

Manganese(II) salts are characteristically very pale pink or colourless. For the d^5 Mn^{2+} ion in an octahedral high-spin complex, '_d–d_' transitions are both spin- and Laporte-forbidden (see _Section 20.6_). Although the electronic spectrum of $[Mn(H_2O)_6]^{2+}$ does contain several absorptions, they are all weaker by a factor of $\approx 10^2$ than those arising from spin-allowed transitions of other first row metal ions. The weak absorptions observed for Mn^{2+} arise from promotion of an electron to give various excited states containing only three unpaired electrons.

All four halides of Mn(II) are known. Hydrates of MnF_2 and $MnBr_2$ are prepared from $MnCO_3$ and aqueous HF or HBr and the anhydrous salts are then obtained by dehydration. The chloride is prepared by reaction 21.50, and MnI_2 results from direct combination of the elements. The fluoride adopts a rutile lattice (_Figure 5.21_) in the solid state, while $MnCl_2$, $MnBr_2$ and MnI_2 possess the CdI_2 layer structure (_Figure 5.22_).

Fig. 21.17 Part of one of the infinite chains of face-sharing octahedra present in the lattice of $[Me_2NH_2][MnCl_3]$; the structure was determined by X-ray diffraction [R.E. Caputo _et al._ (1976) _Phys. Rev. B_, vol. 13, p. 3956]. Colour code: Mn, blue; Cl, black.

The reduction of a higher oxide of manganese (e.g. MnO_2 or Mn_2O_3) with H_2 at elevated temperature gives MnO, which is also obtained by thermal decomposition of manganese(II) oxalate. Green MnO adopts an NaCl lattice and its antiferromagnetic behaviour was discussed in _Section 20.8_. (The conductivity of metal(II) oxides is described in _Section 27.3_.) It is a basic oxide, insoluble in water but dissolving in acids to give pale pink solutions containing $[Mn(H_2O)_6]^{2+}$. The oxidation of Mn(II) compounds in acidic solution requires a powerful oxidant such as periodate, but in alkaline media, oxidation is easier because hydrous Mn_2O_3 is far less soluble than $Mn(OH)_2$. Thus, when alkali is added to a solution of a Mn(II) salt in the presence of air, the white precipitate of $Mn(OH)_2$ that initially forms rapidly darkens owing to atmospheric oxidation.

Large numbers of Mn(II) complexes exist. This oxidation state is stable with respect to both oxidation and reduction, and in high-spin complexes, the lack of any LFSE means that Mn^{2+} does not favour a particular arrangement of ligand donor atoms. Manganese(II) halides form a range of complexes. Reaction of MnF_2 with MF (e.g. M = Na, K, Rb) gives $M[MnF_3]$ salts which adopt the perovskite structure (_Figure 5.23_); discrete $[MnF_3]^-$ ions are not present. Heating a 1:2 ratio of MnF_2:KF at 950 K gives $K_2[MnF_4]$ which has an extended lattice structure containing MnF_6 octahedra connected by $Mn-F-Mn$ bridges. Discrete anions are, again, _not_ present in salts of $[MnCl_3]^-$, e.g. $[Me_2NH_2][MnCl_3]$ crystallizes with infinite chains of face-sharing $MnCl_6$ octahedra (Figure 21.17). Structural determinations for several compounds which appear to be salts of $[MnCl_5]^{3-}$ reveal significant cation-dependence; the green-yellow Cs_3MnCl_5 contains discrete tetrahedral $[MnCl_4]^{2-}$ and Cl^- ions, whereas pink $[(H_3NCH_2CH_2)_2NH_2][MnCl_5]$ has an extended lattice containing corner-sharing $MnCl_6$ octahedra. The salt $K_4[MnCl_6]$ contains discrete octahedral anions, and in green-yellow $[Et_4N]_2[MnCl_4]$ and $[PhMe_2(PhCH_2)N]_2[MnCl_4]$, isolated tetrahedral anions are present. Tetrahedral complexes are more intensely coloured than octahedral ones (see _Section 20.6_) and in these last examples, the presence of the tetrahedral $[MnCl_4]^{2-}$ ion leads to rather more intensely coloured complexes.

The reactions of $MnCl_2$, $MnBr_2$ and MnI_2 with, for example, _N_-, _O_-, _P_- or _S_-donor ligands has led to the isolation of a wide variety of complexes. A range of coordination geometries are observed as the following examples show ($H_2pc = $ **21.19**; Hpz = **21.20**; tpy = **19.22**):

- tetrahedral: $[MnCl_2(OPPh_3)_2]$, $[Mn(N_3)_4]^{2-}$, $[Mn(Se_4)_2]^{2-}$;
- square planar: $[Mn(pc)]$;
- trigonal bipyramidal: $[MnBr_2\{OC(NHMe)_2\}_3]$, $[MnBr\{N(CH_2CH_2NMe_2)_3\}]^+$, $[MnI_2(THF)_3]$;
- octahedral: *trans*-$[MnBr_2(Hpz)_4]$, *cis*-$[Mn(bpy)_2(NCS-N)_2]$, *cis*-$[MnCl_2(HOCH_2CH_2OH)_2]$, $[MnI(THF)_5]^+$, *mer*-$[MnCl_3(H_2O)_3]^-$, $[Mn(tpy)_2]^{2+}$, $[Mn(EDTA)]^{2-}$;
- 7-coordinate: $[Mn(EDTA)(H_2O)]^{2-}$;
- square-antiprism: $[Mn(\textbf{21.21})_2]^{2+}$.

H_2pc = phthalocyanine

(21.19)

Hpz = 1*H*-pyrazole

(21.20)   (21.21)

The only common low-spin complex of Mn(II) is the blue, efflorescent $K_4[Mn(CN)_6]\cdot 3H_2O$ ($\mu_{eff} = 2.18\,\mu_B$) which is prepared in aqueous solution from $MnCO_3$ and KCN. Yellow $Na_4[Mn(CN)_6]\cdot 10H_2O$ can be made in a similar manner, but is unstable in air. Conversion of $K_4[Mn(CN)_6]$ to $K_3[Mn(CN)_6]$ occurs readily, the presence of the cyano ligands significantly destabilizing Mn(II) with respect to Mn(III) (see *Section 7.3*).

Efflorescence is the loss of water from a hydrated salt.

21.9 Group 8: iron

The metal

Finely divided Fe is pyrophoric in air, but the bulk metal oxidizes in dry air only when heated. In moist air, Fe rusts forming a hydrated oxide $Fe_2O_3\cdot xH_2O$. Rusting is an electrochemical process (equation 21.55) and occurs only in

the presence of O_2, H_2O and an electrolyte; the latter may be water, but is more effective if it contains dissolved SO_2 (e.g. from industrial pollution) or NaCl (e.g. from sea-spray or salt-treated roads). Diffusion of the ions formed in reaction 21.55 deposits $Fe(OH)_2$ at places between the points of attack and this is further oxidized to hydrated iron(III) oxide (see *Box 7.3*).

$$\left.\begin{array}{l} 2Fe \longrightarrow 2Fe^{2+} + 4e^- \\ O_2 + 2H_2O + 4e^- \longrightarrow 4[OH]^- \end{array}\right\} \quad (21.55)$$

Iron reacts with halogens at 470–570 K to give FeF_3, $FeCl_3$, $FeBr_3$ and FeI_2, respectively. The metal dissolves in dilute mineral acids to yield Fe(II) salts, but concentrated HNO_3 and other powerful oxidizing agents make it passive; it is unaffected by alkalis. When powdered iron and sulfur are heated together, FeS is produced. The formation of iron carbides and alloys is crucial to the steel industry (see *Box 5.1* and *Section 5.7*).

Most of the chemistry of Fe involves Fe(II) or Fe(III), with Fe(IV) and Fe(VI) known in a small number of compounds; Fe(V) is rare. Lower formal oxidation states occur with π-acceptor ligands (see *Chapter 23*).

Iron(VI), iron(V) and iron(IV)

The highest oxidation states of iron are found in compounds of $[FeO_4]^{2-}$, $[FeO_4]^{3-}$, $[FeO_4]^{4-}$ and $[FeO_3]^{2-}$ although these free ions are not necessarily present. Salts of $[FeO_4]^{2-}$ can be made by hypochlorite oxidation of Fe(III) salts in the presence of alkali (equation 21.56); they contain discrete tetrahedral ions and are paramagnetic with magnetic moments corresponding to two unpaired electrons. The Na^+ and K^+ salts are deep red-purple and are soluble in water; aqueous solutions decompose (equation 21.57) but alkaline solutions are stable. Ferrate(VI) is a powerful oxidant (equation 21.58).

$$Fe_2O_3 + 3[OCl]^- + 4[OH]^- \longrightarrow 2[FeO_4]^{2-} + 3Cl^- + 2H_2O \quad (21.56)$$

$$4[FeO_4]^{2-} + 6H_2O \longrightarrow 4FeO(OH) + 8[OH]^- + 3O_2 \quad (21.57)$$

$$[FeO_4]^{2-} + 8H^+ + 3e^- \rightleftharpoons Fe^{3+} + 4H_2O \qquad E^\circ = +2.20\,V \quad (21.58)$$

The reaction of K_2FeO_4 with KOH in O_2 at 1000 K gives K_3FeO_4, a rare example of an Fe(V) salt.

Iron(IV) ferrates include Na_4FeO_4 (made from Na_2O_2 and $FeSO_4$), Sr_2FeO_4 (prepared by heating Fe_2O_3 and SrO in the presence of O_2) and Ba_2FeO_4 (made from BaO_2 and $FeSO_4$). Typically, these are mixed metal oxides with extended lattice structures, but Na_4FeO_4 contains discrete $[FeO_4]^{4-}$ ions. In aqueous solution Na_4FeO_4 disproportionates (equation 21.59).

$$3Na_4FeO_4 + 5H_2O \longrightarrow Na_2FeO_4 + Fe_2O_3 + 10NaOH \quad (21.59)$$

The super-ion battery

The MnO_2–Zn dry battery is a major contributor to the commercial supply of batteries; in the long-life 'alkaline' version, the lifetime of the battery is mainly dependent on the lifetime of the MnO_2 cathode. Prolonging the lifetimes of batteries which are used, for example, in implanted pacemakers has obvious advantages, and the use of the Fe(VI) compounds K_2FeO_4 and $BaFeO_4$ as cathodic materials has recently been investigated with promising results. The so-called 'super-ion battery' contains K_2FeO_4 and $BaFeO_4$ as a replacement for MnO_2 in the alkaline dry battery. The reduction of Fe(VI) to Fe(III) provides a high capacity source of cathodic charge and the $[FeO_4]^{2-}$-for-MnO_2 cathode replacement leads to an increase in the energy capacity of

the battery of more than 50%. The cell reaction of the super-ion battery is:

$$2K_2FeO_4 + 3Zn \longrightarrow Fe_2O_3 + ZnO + 2K_2ZnO_2$$

and a further advantage of the system is that it is rechargeable.

Further reading

S. Licht, B. Wang and S. Ghosh (1999) *Science*, vol. 285, p. 1039 – 'Energetic iron(VI) chemistry: The super-ion battery'.

Compounds formally containing $[FeO_3]^{2-}$ are actually mixed metal oxides; $CaFeO_3$, $SrFeO_3$ and $BaFeO_3$ crystallize with the perovskite structure (*Figure 5.23*).

Attempts to stabilize Fe in high oxidation states using fluoro ligands have met with limited success. The reaction of Cs_2FeO_4 with F_2 (40 bar, 420 K) gives Cs_2FeF_6 along with $CsFeF_4$ and Cs_3FeF_6. In the solid state, Cs_2FeF_6 adopts the K_2MnF_6 lattice (see *Section 21.8*, Mn(IV)). There is current interest in the coordination chemistry of Fe(IV) since Fe(IV) intermediates are present in bioinorganic processes involving haem peroxidases (see *Chapter 28*); however, the number of Fe(IV) complexes so far isolated and structurally characterized is small. The coordination environment is octahedral or square-based pyramidal, and ligands that stabilize Fe(IV) include dithiocarbamates (Figure 21.18a), dithiolates as in $[Fe(PMe_3)_2(1,2-S_2C_6H_4)_2]$, porphyrins and phthalocyanines. Ligand **21.22** stabilizes

both Fe(IV) and Fe(V) with the four donor atoms occupying the basal sites in the square-based pyramidal complexes shown in Figures 21.18b and c; $[Fe(\mathbf{21.22})(P^nPr_3)]$ is oxidized by I_2 to $[Fe(\mathbf{21.22})I]$. Magnetic moments at 298 K for the two complexes (d^2 and d^1) are 2.76 and 2.19 μ_B respectively.

(21.22)

Iron(III)

The old name for iron(III) is *ferric*. Iron(III) fluoride, chloride and bromide are made by heating Fe with the halogen. The

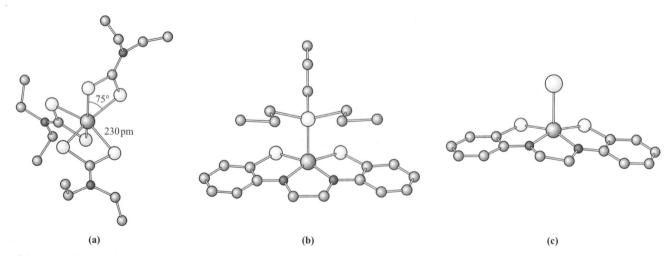

(a)   (b)   (c)

Fig. 21.18 The structures (X-ray diffraction) of some high oxidation state iron complexes: (a) the iron(IV) complex $[Fe(S_2CNEt_2)_3]^+$ in the $[I_5]^-$ salt [C.L. Raston *et al.* (1980) *J. Chem. Soc., Dalton Trans.*, p. 1928], (b) the iron(IV) species $[Fe(\mathbf{21.22})(P^nPr_3)]$ [D. Sellmann *et al.* (1997) *Angew. Chem., Int. Ed. Engl.*, vol. 36, p. 1201] and (c) the iron(V) complex $[Fe(\mathbf{21.22})I]$ [D. Sellmann *et al.* (1997) *Angew. Chem., Int. Ed. Engl.*, vol. 36, p. 1734]. Hydrogen atoms are omitted; colour code: Fe, dark blue; S and P, pale grey; C, grey; N, black; I, pale blue.

fluoride is a white, involatile solid isostructural with ScF_3 (Figure 21.3). In the solid state, $FeCl_3$ adopts the BiI_3 lattice but the gas phase (bp 588 K) contains discrete molecules, dimers below 970 K and monomers above 1020 K. Anhydrous $FeCl_3$ forms hygroscopic dark green or black crystals; it dissolves in water to give strongly acidic solutions (see below) from which the orange-brown hydrate $FeCl_3 \cdot 6H_2O$ (properly formulated as trans-$[FeCl_2(H_2O)_4]Cl \cdot 2H_2O$) can be crystallized. The trichloride is a useful precursor in Fe(III) chemistry, and both anhydrous $FeCl_3$ and $FeBr_3$ are used as Lewis acid catalysts in organic synthesis. Anhydrous $FeBr_3$ forms deliquescent, red-brown, water-soluble crystals; the solid adopts a BiI_3 structure, but in the gas phase, molecular dimers are present. Until 1990, the ready decomposition of FeI_3 (reaction 21.60) had precluded its isolation, but it has now been made by reaction 21.61 under inert conditions.

$$2FeI_3 \longrightarrow 2FeI_2 + I_2 \tag{21.60}$$

$$2Fe(CO)_4I_2 + I_2 \xrightarrow{h\nu} 2FeI_3 + 8CO \tag{21.61}$$

Iron(III) oxide exists in a number of forms. The paramagnetic α-form (a red-brown solid or grey-black crystals) occurs as the mineral *haematite* and adopts a corundum lattice (see *Section 12.7*) with octahedrally sited Fe(III) centres. The β-form is produced by hydrolyzing $FeCl_3 \cdot 6H_2O$, or by chemical vapour deposition (CVD, see *Section 27.6*) at 570 K from iron(III) trifluoroacetylacetonate; on annealing at 770 K, a β $\longrightarrow$ α phase change occurs. The γ-form is obtained by careful oxidation of Fe_3O_4 and crystallizes with a lattice in which the O^{2-} ions adopt a ccp array and the Fe^{3+} ions randomly occupy octahedral and tetrahedral holes; it is ferromagnetic and is widely used in magnetic recording tapes. Iron(III) oxide is insoluble in water but can be dissolved with difficulty in acids. Several hydrates of Fe_2O_3 exist, and when Fe(III) salts are dissolved in alkali, the red-brown gelatinous precipitate that forms is *not* $Fe(OH)_3$ but $Fe_2O_3 \cdot H_2O$ (also written as $Fe(O)OH$). The precipitate is soluble in acids giving $[Fe(H_2O)_6]^{3+}$, and in concentrated aqueous alkalis giving $[Fe(OH)_6]^{3-}$. Several forms of $Fe(O)OH$ exist and consist of chain structures with edge-sharing FeO_6 octahedra; the minerals *goethite* and *lepidocrocite* are α- and γ-$Fe(O)OH$ respectively.

Mixed metal oxides derived from Fe_2O_3 and of general formula $M^{II}Fe^{III}_2O_4$ or $M^IFe^{III}O_2$ are commonly known as *ferrites* despite the absence of discrete oxoanions. They include compounds of commercial importance by virtue of their magnetic properties, e.g. electromagnetic devices for information storage; for discussion of the magnetic properties of mixed metal oxides, see *Chapter 27*. Spinel and inverse spinel lattices adopted by $M^{II}Fe^{III}_2O_4$ oxides were described in *Box 12.6* and *Section 20.9*, e.g. $MgFe_2O_4$ and $NiFe_2O_4$ are inverse spinels while $MnFe_2O_4$ and $ZnFe_2O_4$ are normal spinels. Some oxides of the $M^IFe^{III}O_2$ type adopt structures that are related to NaCl (e.g. $LiFeO_2$ in which the Li^+ and Fe^{3+} ions occupy Na^+ sites and O^{2-} ions occupy Cl^- sites, *Figure 5.15*). Among the $M^IFe^{III}O_2$ group of compounds, $CuFeO_2$ and $AgFeO_2$ are noteworthy in being semiconductors. Other ferrites exist with more complex structures: permanent magnets are made using $BaFe_{12}O_{19}$, and the *iron garnet* family includes $Y_3Fe_5O_{12}$ (yttrium iron garnet, YIG) which is used as a microwave filter in radar equipment.

When Fe_2O_3 is heated at 1670 K, it converts to black Fe_3O_4 ($Fe^{II}Fe^{III}_2O_4$) which also occurs as the mineral *magnetite*, and possesses an inverse spinel structure (see *Box 12.6*). Its ferrimagnetic behaviour (see *Figure 20.25*) makes Fe_3O_4 commercially important, e.g. it is used in magnetic toner in photocopiers. Mixtures of Fe_3O_4 and γ-Fe_2O_3 are used in magnetic recording tape, and this market competes with that of CrO_2 (see *Section 21.7*).

The chemistry of Fe(III) is well researched and among many commercially available starting materials are the chloride (see above), perchlorate, sulfate and nitrate. *Hazard*: *Perchlorates are potentially explosive.* Anhydrous $Fe(ClO_4)_3$ is a yellow solid, but it is commercially available as a hydrate $Fe(ClO_4)_3 \cdot xH_2O$ with variable water content; the hydrate is prepared from aqueous $HClO_4$ and $Fe_2O_3 \cdot H_2O$ and, depending on contamination with chloride, may be pale violet (<0.005% chloride content) or yellow. Iron(III) sulfate (made by oxidation of $FeSO_4$ with concentrated H_2SO_4) is

---

| BOX 21.6 | RESOURCES, ENVIRONMENTAL AND BIOLOGICAL |

Rusticles are destroying RMS *Titanic*

Green, iron-rich structures called *rusticles* growing from the sunken hull of RMS *Titanic* are gradually destroying what remains of the ship. Rusticles contain colonies of metallophilic bacteria; the structures are composed externally of *lepidocrocite*, γ-$Fe(O)OH$, and internally of *goethite*, α-$Fe(O)OH$. The rate at which the bacteria are converting the ship's hull into rusticles is alarming and the phenomenon is a topic of current research.

Further reading

W. Wells and H. Mann (1997) *Resources of Environmental Biotechnology*, vol. 1, p. 271.

purchased as the hydrate $Fe_2(SO_4)_3 \cdot 5H_2O$. The nitrate is available as $Fe(NO_3)_3 \cdot 9H_2O$ (correctly formulated as $[Fe(H_2O)_6][NO_3]_3 \cdot 3H_2O$) which forms colourless or pale violet deliquescent crystals; it is made by reaction of iron oxides with concentrated HNO_3. The violet hexahydrate, $Fe(NO_3)_3 \cdot 6H_2O$ (correctly written as $[Fe(H_2O)_6][NO_3]_3$), can be obtained by reaction of $Fe_2O_3 \cdot H_2O$ with HNO_3. The octahedral $[Fe(H_2O)_6]^{3+}$ ion is also present in crystals of the violet alum $[NH_4]Fe(SO_4)_2 \cdot 12H_2O$ (see *Section 12.9*). These Fe(III) salts are all water-soluble, dissolving to give brown-yellow solutions due to hydrolysis of $[Fe(H_2O)_6]^{3+}$ (*equations 6.36* and *6.37*); solution species include $[(H_2O)_5FeOFe(H_2O)_5]^{4+}$ (**21.23**) which has a *linear* Fe−O−Fe bridge indicative of $(d-p)\pi$ bonding involving Fe *d* and O *p* orbitals. The magnetic moment of $5.82\,\mu_B$ for $[Fe(H_2O)_6]^{3+}$ is close to the spin-only value for high-spin d^5.

(21.23)

The $[Fe(CN)_6]^{3-}$ ion (Figure 21.19a) contains low-spin Fe(III) ($\mu_{eff} = 2.25\,\mu_B$) and is made by oxidation of $[Fe(CN)_6]^{4-}$, e.g. by reaction 21.62 or electrolytically. The cyano ligands in $[Fe(CN)_6]^{3-}$ are more labile than in $[Fe(CN)_6]^{4-}$ and cause the former to be more toxic than the latter.

$$2K_4[Fe(CN)_6] + Cl_2 \longrightarrow 2K_3[Fe(CN)_6] + 2KCl \qquad (21.62)$$

Fig. 21.19 Examples of iron(III) cyano complexes: (a) the structure of $[Fe(CN)_6]^{3-}$ in the salt $Cs[NH_4]_2[Fe(CN)_6]$ (X-ray diffraction) [D. Babel (1982) *Z. Naturforsch., Teil B*, vol. 37, p. 1534], (b) one-eighth of the unit cell of $KFe[Fe(CN)_6]$ (the K^+ ions occupy the cavities and are omitted from the figure), and (c) the structure (determined by X-ray diffraction) of $[Fe(CN)_2(TPP)]$ where $H_2TPP = 5,10,15,20$-tetraphenyl-$21H,23H$-porphyrin (see *Figure 11.8* for parent porphyrin) [W.R. Scheidt *et al.* (1980) *J. Am. Chem. Soc.*, vol. 102, p. 3017]. Hydrogen atoms are omitted from (c); colour code in (a) and (c): Fe, blue; N, black, C, grey.

The ruby-red salt $K_3[Fe(CN)_6]$ (potassium hexacyano-ferrate(III) or ferricyanide) is commercially available. It is an oxidizing agent although $[Fe(CN)_6]^{3-}$ is less powerful an oxidant than $[Fe(H_2O)_6]^{3+}$ (see *Section 7.3*). Addition of $[Fe(CN)_6]^{3-}$ to aqueous Fe^{2+} gives the deep blue complex *Turnbull's blue* and this reaction is used as a qualitative test for Fe^{2+}. Conversely, if $[Fe(CN)_6]^{4-}$ is added to aqueous Fe^{3+}, the deep blue complex *Prussian blue* is produced. Both Prussian blue and Turnbull's blue are hydrated salts of formula $Fe_4^{III}[Fe^{II}(CN)_6]_3 \cdot xH_2O$ ($x \approx 14$), and related to them is $KFe[Fe(CN)_6]$, *soluble Prussian blue*. In the solid state, these complexes possess extended lattices containing cubic arrangements of Fe^{n+} centres linked by $[CN]^-$ bridges. Figure 21.19b shows part of the unit cell of $KFe[Fe(CN)_6]$; each Fe^{n+} is in an octahedral environment, either FeC_6 or FeN_6. Turnbull's blue, Prussian blue and Berlin green ($Fe^{III}[Fe^{III}(CN)_6]$) have been widely used in inks and dyes.

Figure 21.19b shows the ability of $[CN]^-$ to act as a bridging ligand and a number of polymeric materials containing either Fe(III) or Fe(II) as well as other metal centres have been made utilizing this property. An example is $[Ni(en)_2]_3[Fe(CN)_6]_2 \cdot 2H_2O$, the solid state structure of which (Figure 21.20) consists of interconnected helical chains in which octahedral Ni^{2+} and Fe^{3+} centres are connected by bridging $[CN]^-$ ligands. The latter facilitate electronic communication between the metal centres resulting in a ferromagnetic material.

Large numbers of Fe(III) complexes are known, and octahedral coordination is common. Examples of simple complexes (see *Table 6.7* for ligand abbreviations) include:

- high-spin octahedral: $[Fe(H_2O)_6]^{3+}$, $[FeF_6]^{3-}$, $[Fe(ox)_3]^{3-}$, $[Fe(acac)_3]$;
- low-spin octahedral: $[Fe(CN)_6]^{3-}$, $[Fe(bpy)_3]^{3+}$, $[Fe(phen)_3]^{3+}$, $[Fe(en)_3]^{3+}$;
- 7-coordinate: $[Fe(EDTA)(H_2O)]^-$.

The octahedral complex $[Fe(NH_3)_6]^{3+}$ can be prepared in liquid NH_3, but it has low stability in aqueous solutions, decomposing with loss of NH_3. Both bpy and phen stabilize Fe(II) more than they do Fe(III); this is ascribed to the existence of relatively low lying π^* MOs on the ligands allowing them to function as π-acceptors. In aqueous solution, both $[Fe(bpy)_3]^{3+}$ and $[Fe(phen)_3]^{3+}$ are more readily reduced than the hexaaqua ion (equations 21.63 and 21.64).

$$[Fe(bpy)_3]^{3+} + e^- \rightleftharpoons [Fe(bpy)_3]^{2+} \quad E^\circ = +1.03\,V$$
blue   red

$$(21.63)$$

$$[Fe(phen)_3]^{3+} + e^- \rightleftharpoons [Fe(phen)_3]^{2+} \quad E^\circ = +1.12\,V$$
blue   red

$$(21.64)$$

The addition of thiocyanate to aqueous solutions of Fe^{3+} produces a blood-red coloration due to the formation of $[Fe(H_2O)_5(SCN\text{-}N)]^{2+}$. Complete exchange of ligands to give $[Fe(SCN\text{-}N)_6]^{3-}$ is best carried out in non-aqueous media.

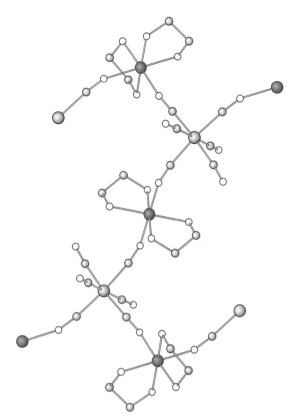

Fig. 21.20 Part of the polymeric structure (X-ray diffraction) of $[Ni(en)_2]_3[Fe(CN)_6]_2 \cdot 2H_2O$ in which Fe^{3+} ions are in $Fe(CN\text{-}C)_6$ environments and Ni^{2+} ions are in $Ni(CN\text{-}N)_2(en)_2$ sites [M. Ohba *et al.* (1994) *J. Am. Chem. Soc.*, vol. 116, p. 11566]. Hydrogen atoms are omitted; colour code: Fe, blue; Ni, black; N, white; C, grey.

Iron(III) favours *O*-donor ligands and stable complexes such as the green $[Fe(ox)_3]^{3-}$ and red $[Fe(acac)_3]$ are commonly encountered. Iron(III) porphyrinato complexes are of relevance for modelling haem-proteins (see *Section 28.3*) and there is interest in reactions of these complexes with, for example, CO, O_2, NO and $[CN]^-$. The N_4-donor set of a porphyrinato ligand is confined to a plane and this restriction forces the Fe(III) centre to be in a square planar environment with respect to the macrocycle. Other ligands may then enter in axial sites above and below the FeN_4-plane to give either square-based pyramidal (Figure 21.19c) or octahedral complexes.

Low coordination numbers can be stabilized by interaction with amido ligands, e.g. $[Fe\{N(SiMe_3)_2\}_3]$ (*Figure 19.4c*).

Iron(II)

The old name for iron(II) is *ferrous*. Anhydrous FeF_2, $FeCl_2$ and $FeBr_2$ can be prepared by reaction 21.65, while FeI_2 is made by direct combination of the elements.

$$Fe + 2HX \xrightarrow{\Delta} FeX_2 + H_2 \quad (X = F, Cl, Br) \qquad (21.65)$$

Iron(II) fluoride is a sparingly soluble, white solid with a distorted rutile structure (*Figure 5.21*); the environment around

BOX 21.7

APPLICATIONS

Iron complexes fight anaemia

In Chapter 28, the crucial role that iron plays in biological systems is discussed in detail. Anaemia, in which the body suffers from a deficiency of iron, leads to a general state of lethargy and weakness. Iron is usually administered orally to a patient as iron supplement tablets containing an Fe(II) or Fe(III) salt. Iron(II) salts are more typical because of they exhibit better solubilities than Fe(III) salts at physiological pH, but Fe(III) has the advantage that, unlike Fe(II), it is not susceptible to oxidation in aqueous solution. Among compounds which are in common use are iron(III) chloride, iron(II) sulfate, iron(II) fumarate, iron(II) succinate and iron(II) gluconate; the structures of fumaric acid, succinic acid and gluconic acid are shown here.

Fumaric acid

Succinic acid

Gluconic acid

the high-spin Fe(II) centre (d^6) is surprisingly irregular with 4F at 212 pm and 2F at 198 pm. In the gas phase, FeF_2 is monomeric. Iron(II) chloride forms white, hygroscopic, water-soluble crystals and adopts a $CdCl_2$ lattice (see *Section 5.11*). In the gas phase of $FeCl_2$, monomers and dimers are present. The pale green hydrate $FeCl_2 \cdot 4H_2O$, properly formulated as octahedral $[FeCl_2(H_2O)_4]$, is a convenient precursor in Fe(II) chemistry. The hexahydrate (which loses water readily) can be obtained by recrystallizing $FeCl_2$ from water below 285 K. Iron(II) bromide is a deliquescent, yellow or brown solid and adopts a CdI_2 structure. It is very soluble in water and forms hydrates $FeBr_2 \cdot xH_2O$ where $x = 4$, 6 or 9 depending on crystallization conditions. Dark-violet FeI_2 has a CdI_2 layer structure, and is hygroscopic and light-sensitive; it forms a green tetrahydrate. All the halides or their hydrates are commercially available, as are salts such as the perchlorate, sulfate and $[NH_4]_2Fe[SO_4]_2 \cdot 6H_2O$. Iron(II) sulfate is a common source of Fe(II) and is available as the blue-green $FeSO_4 \cdot 7H_2O$, an old name for which is *green vitriol*. Like most hydrated Fe(II) salts, it dissolves in water to give $[Fe(H_2O)_6]^{2+}$, the electronic spectrum and magnetic moment of which are consistent with high-spin d^6. The salt $[NH_4]_2Fe[SO_4]_2 \cdot 6H_2O$ is an important source of Fe^{2+} since, in the solid state, it is kinetically more stable towards oxidation than most Fe(II) salts.

Iron(II) oxide is a black, insoluble solid with an NaCl lattice above its Curie temperature (200 K); the lattice suffers defects because it is always deficient in Fe (see *Section 27.2*). Below 200 K, FeO undergoes a phase change and becomes antiferromagnetic. It can be made *in vacuo* by thermal decomposition of iron(II) oxalate but the product must be cooled rapidly to prevent disproportionation (equation 21.66).

$$4FeO \longrightarrow Fe_3O_4 + Fe \qquad (21.66)$$

White $Fe(OH)_2$ is precipitated by adding alkali to solutions of Fe(II) salts but it rapidly absorbs O_2, turning dark green, then brown. The products are a mixed Fe(II)Fe(III) hydroxide and $Fe_2O_3 \cdot H_2O$. Iron(II) hydroxide dissolves in acids, and from concentrated NaOH solutions, the blue-green $Na_4[Fe(OH)_6]$ can be crystallized.

An interesting distinction between iron(II) oxides and sulfides is that whereas FeO has an analogue in FeS, there is no peroxide analogue of FeS_2 (*iron pyrites*). The sulfide FeS is made by heating together the elements; it is found in lunar rock samples and adopts an NiAs lattice (*Figure 14.7*). Reaction of FeS with hydrochloric acid used to be a familiar laboratory synthesis of H_2S (*equation 15.34*). Iron pyrites is $Fe^{2+}(S_2)^{2-}$ and contains low-spin Fe(II) in a distorted NaCl structure.

The coordination chemistry of Fe(II) is well developed and only a brief introduction to simple species is given here. Iron(II) halides combine with gaseous NH_3 to give salts of $[Fe(NH_3)_6]^{2+}$ but this decomposes in aqueous media, precipitating $Fe(OH)_2$. In aqueous solutions, $[Fe(H_2O)_6]^{2+}$ is unstable with respect to oxidation, although as we saw above, double salts such as $[NH_4]_2Fe[SO_4]_2 \cdot 6H_2O$ are more stable. Displacement of the ligands in $[Fe(H_2O)_6]^{2+}$ leads to a range of complexes. We have already discussed the stabilization of Fe(II) by bpy and phen (equations 21.63 and 21.64); oxidation of red $[Fe(phen)_3]^{2+}$ to blue $[Fe(phen)_3]^{3+}$ is more difficult than that of $[Fe(H_2O)_6]^{2+}$ to $[Fe(H_2O)_6]^{3+}$, and hence arises the use of $[Fe(phen)_3][SO_4]$ as a redox indicator. Both $[Fe(phen)_3]^{2+}$ and $[Fe(bpy)_3]^{2+}$ are low-spin d^6 and diamagnetic; $[Fe(CN)_6]^{4-}$ is also low-spin. The latter, like $[Fe(CN)_6]^{3-}$ (Figure 21.19a), is octahedral but the Fe–C bonds in the Fe(II) species are shorter (192 pm) than those in the Fe(III) complex. Taken together with the fact that the C–N bond lengths and stretching frequencies differ little between $[Fe(CN)_6]^{4-}$ and $[Fe(CN)_6]^{3-}$, this provides convincing evidence for stronger π-bonding in the lower oxidation state complex. There are many known *mono*substitution products of $[Fe(CN)_6]^{4-}$. Sodium nitropentacyanoferrate(II) (*sodium nitroprusside*),

$Na_2[Fe(CN)_5(NO)]\cdot2H_2O$, is made by reaction 21.67 or 21.68; among its uses are those as an anti-hypertensive drug (it acts as a vasodilator through release of NO) and as a standard reference for ^{57}Fe Mössbauer spectroscopy.

$$[Fe(CN)_6]^{4-} + 4H^+ + [NO_3]^-$$
$$\longrightarrow [Fe(CN)_5(NO)]^{2-} + CO_2 + [NH_4]^+ \quad (21.67)$$

$$[Fe(CN)_6]^{4-} + H_2O + [NO_2]^-$$
$$\longrightarrow [Fe(CN)_5(NO)]^{2-} + [CN]^- + 2[OH]^- \quad (21.68)$$

Nitrogen monoxide is a radical (structure **14.34**), but $Na_2[Fe(CN)_5(NO)]$ is diamagnetic; the N–O distance of 113 pm is shorter, and the stretching wavenumber of $1947\,cm^{-1}$ higher, than in free NO. Thus, the complex is formulated as containing an $[NO]^+$ ligand. The addition of S^{2-} to $[Fe(CN)_5(NO)]^{2-}$ produces the red $[Fe(CN)_5(NOS)]^{4-}$ and this is the basis of a sensitive test for S^{2-}. Similarly, reaction with $[OH]^-$ gives $[Fe(CN)_5(NO_2)]^{4-}$ (equation 21.69). For details of $[Fe(NO)(H_2O)_5]^{2+}$, see *Section 14.8*.

$$[Fe(CN)_5(NO)]^{2-} + 2[OH]^- \longrightarrow [Fe(CN)_5(NO_2)]^{4-} + H_2O$$
$$(21.69)$$

In addition to the hexaaqua ion, high-spin Fe(II) complexes include $[Fe(en)_3]^{2+}$. Its magnetic moment of $5.45\,\mu_B$ is larger than the spin-only value of $4.90\,\mu_B$ and reflects orbital contributions for the configuration $t_{2g}^4 e_g^2$ (see *Section 20.8*). Although iron(II) favours an octahedral arrangement of donor atoms, there are some tetrahedral complexes, for example $[FeCl_4]^{2-}$ (equation 21.70), $[FeBr_4]^{2-}$, $[FeI_4]^{2-}$ and $[Fe(SCN)_4]^{2-}$.

$$2MCl + FeCl_2 \xrightarrow{\text{in EtOH}} M_2[FeCl_4] \quad \text{(M = group 1 metal)}$$
$$(21.70)$$

The amido complex $[Fe\{N(SiMePh_2)_2\}_2]$ is an unusual example of 2-coordinate Fe(II) (see *Section 19.7*).

21.10 Group 9: cobalt

The metal

Cobalt is less reactive than Fe (e.g. see equation 21.71); Co does not react with O_2 unless heated, although when very finely divided, it is pyrophoric. It dissolves slowly in dilute mineral acids (e.g. reaction 21.72), but concentrated HNO_3 makes it passive; alkalis have no effect on the metal.

$$M^{2+}(aq) + 2e^- \longrightarrow M(s) \quad \begin{cases} M = Fe, \; E^\circ = -0.44\,V \\ M = Co, \; E^\circ = -0.28\,V \end{cases}$$
$$(21.71)$$

$$Co + H_2SO_4 \longrightarrow CoSO_4 + H_2 \quad (21.72)$$

Cobalt reacts at 520 K with F_2 to give CoF_3, but with Cl_2, Br_2 and I_2, CoX_2 is formed. Even when heated, cobalt does not react with H_2 or N_2, but it does combine with B, C (see *Section 13.7*), P, As and S.

The trend in decreasing stability of high oxidation states on going from Mn to Fe continues along the row (*Table 19.3*); Co(IV) is the highest oxidation state but it is of far less importance than Co(III) and Co(II). Cobalt(I) and lower oxidation states are stabilized in organometallic species by π-acceptor ligands (see *Chapter 23*); among Co(I) complexes containing only phosphine ligands is tetrahedral $[Co(PMe_3)_4]^+$.

Cobalt(IV)

Few Co(IV) species have been established. Yellow $Cs_2[CoF_6]$ is obtained by fluorination of a mixture of CsCl and $CoCl_2$ at 570 K; the fact that $[CoF_6]^{2-}$ (d^5) is *low-spin* contrasts with the *high-spin* nature of $[CoF_6]^{3-}$ (d^6) and the difference reflects the increase in Δ_{oct} with increasing oxidation state. Cobalt(IV) oxide (made by oxidizing Co(II) using alkaline hypochlorite) is poorly defined. Several mixed oxides are known: Ba_2CoO_4 and M_2CoO_3 (M = K, Rb, Cs).

Cobalt(III)

There are few *binary* compounds of Co(III) and only a limited number of Co(III) compounds are commercially available. The only binary halide is brown CoF_3 which is isostructural with FeF_3. It is used as a fluorinating agent, e.g. for preparing perfluorinated organics, and is corrosive and an oxidant. The reaction of N_2O_5 with CoF_3 at 200 K gives the dark-green, anhydrous $Co(NO_3)_3$ which has a molecular structure with three didentate $[NO_3]^-$ groups bound to octahedral Co(III).

Although reports of Co_2O_3 are found in the literature, the anhydrous compound probably does not exist. The mixed oxidation state Co_3O_4 ($Co^{II}Co^{III}_2O_4$) is formed when Co is heated in O_2. The insoluble, grey-black Co_3O_4 crystallizes with a normal spinel structure containing high-spin Co^{2+} in tetrahedral holes and low-spin Co^{3+} in octahedral holes; it is, therefore, a worse electrical conductor than Fe_3O_4. A hydrated oxide is precipitated when excess alkali reacts with most Co(III) compounds, or on aerial oxidation of aqueous suspensions of $Co(OH)_2$. Mixed metal oxides $MCoO_2$ where M is an alkali metal can be made by heating mixtures of the oxides and consist of layer lattices built of edge-sharing CoO_6 octahedra with M^+ ions in interlayer sites. Of particular significance is $LiCoO_2$ which is used in lithium-ion batteries (see *Box 10.3*).

The blue, low-spin $[Co(H_2O)_6]^{3+}$ ion can be prepared *in situ* by electrolytic oxidation of aqueous $CoSO_4$ in acidic solution at 273 K. It is a powerful oxidant (equation 21.73) and is unstable in aqueous media, decomposing to Co(II) with the liberation of ozonized O_2. The $[Co(H_2O)_6]^{3+}$ ion is best isolated as the sparingly soluble, blue alum $CsCo(SO_4)_2\cdot12H_2O$, although this decomposes within hours on standing. Complex formation with, for example, bpy, NH_3, RNH_2 or $[CN]^-$ greatly stabilizes Co(III) as equations 21.74–21.77 illustrate.

$$Co^{3+}(aq) + e^- \rightleftharpoons Co^{2+}(aq) \qquad E^\circ = +1.92\,V \qquad (21.73)$$

$$[Co(bpy)_3]^{3+} + e^- \rightleftharpoons [Co(bpy)_3]^{2+}$$
$$E^\circ = +0.31\,V \qquad (21.74)$$

$$[Co(NH_3)_6]^{3+} + e^- \rightleftharpoons [Co(NH_3)_6]^{2+}$$
$$E^\circ = +0.11\,V \qquad (21.75)$$

$$[Co(en)_3]^{3+} + e^- \rightleftharpoons [Co(en)_3]^{2+} \quad E^\circ = -0.26\,V \qquad (21.76)$$

$$[Co(CN)_6]^{3-} + H_2O + e^- \rightleftharpoons [Co(CN)_5(H_2O)]^{3-} + [CN]^-$$
$$E^\circ = -0.83\,V \qquad (21.77)$$

Replacing aqua by ammine ligands, for example, results in a dramatic change in E° (equations 21.73 and 21.75) and shows that the overall stability constant of $[Co(NH_3)_6]^{3+}$ is $\approx 10^{30}$ greater than that of $[Co(NH_3)_6]^{2+}$. Much of this difference arises from LFSEs:

- Δ_{oct} for the ammine complex is far greater than for the aqua complex in both oxidation states (*Table 20.2*);
- both Co(II) complexes are high-spin whereas both Co(III) complexes are low-spin (*Table 20.3*).

Cobalt(III) complexes (d^6) are usually low-spin octahedral and *kinetically inert* (see *Section 25.2*). The latter means that ligands are not labile and so preparative methods of Co(III) complexes usually involve oxidation of the

---

BOX 21.8 — **CHEMICAL AND THEORETICAL BACKGROUND**

Alfred Werner

Alfred Werner (working at the University of Zurich) was awarded the Nobel Prize for Chemistry in 1913 for his pioneering work that began to unravel the previous mysteries of the compounds formed between *d*-block metal ions and species such as H_2O, NH_3 and halide ions. A famous problem that led to Werner's theory of coordination concerns the fact that $CoCl_3$ forms a series of complexes with NH_3:

- violet $CoCl_3 \cdot 4NH_3$
- green $CoCl_3 \cdot 4NH_3$
- purple $CoCl_3 \cdot 5NH_3$
- yellow $CoCl_3 \cdot 6NH_3$

and that addition of $AgNO_3$ precipitates different amounts of AgCl per equivalent of Co(III). Thus, one equivalent of $CoCl_3 \cdot 6NH_3$ reacts with an excess of $AgNO_3$ to precipitate *three* equivalents of AgCl, one equivalent of $CoCl_3 \cdot 5NH_3$ precipitates *two* equivalents of AgCl, while one equivalent of either green or violet $CoCl_3 \cdot 4NH_3$ precipitates only *one* equivalent of AgCl. Werner realized that any Cl^- precipitated was free chloride ion and that any other chloride was held in the compound in some other way. The crucial conclusion that Werner drew was that in all these cobalt(III) compounds, the metal was intimately associated with six ligands (NH_3 molecules or Cl^- ions), and that only the remaining Cl^- ions behaved as 'normal' ions, free to react with Ag^+:

$$Ag^+(aq) + Cl^-(aq) \longrightarrow AgCl(s)$$

Werner referred to the oxidation state of the metal ion as its 'primary valence' and to what we now call the coordination number as its 'secondary valence'. The compounds $CoCl_3 \cdot 6NH_3$, $CoCl_3 \cdot 5NH_3$ and $CoCl_3 \cdot 4NH_3$ were thus reformulated as $[Co(NH_3)_6]Cl_3$, $[Co(NH_3)_5Cl]Cl_2$ and $[Co(NH_3)_4Cl_2]Cl$. This picture contrasted greatly with earlier ideas such as the 'chain theory' of Danish chemist, Sophus Mads Jørgensen.

Werner's studies went on to show that the numbers of ions in solution (determined from conductivity measurements) were consistent with the formulations $[Co(NH_3)_6]^{3+}[Cl^-]_3$,

$[Co(NH_3)_5Cl]^{2+}[Cl^-]_2$ and $[Co(NH_3)_4Cl_2]^+Cl^-$. The fact that $[Co(NH_3)_4Cl_2]Cl$ existed as two *isomers* (the green and violet forms) was a key to the puzzle of the shape of the $[Co(NH_3)_4Cl_2]^+$ complex. The possible *regular* arrangements for six ligands are planar hexagonal, octahedral and trigonal prismatic. There are three ways of arranging the ligands in $[Co(NH_3)_4Cl_2]^+$ in a hexagon:

two ways for an octahedral arrangement (what we now call *cis*- and *trans*-isomers):

and three for a trigonal prismatic arrangement:

From the fact that two isomers of $[Co(NH_3)_4Cl_2]Cl$ had been isolated, Werner concluded that the $[Co(NH_3)_4Cl_2]^+$ had an octahedral structure, and, by analogy, so did other complexes containing six ligands. Werner's work extended well beyond this one system and his contributions to the groundwork of the understanding of coordination chemistry were immense.

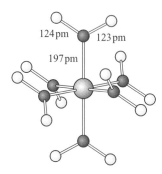

Fig. 21.21 The structure (X-ray diffraction) of $[Co(NO_2-N)_6]^{3-}$ in the salt $Li[Me_4N]_2[Co(NO_2-N)_6]$ [R. Bianchi *et al.* (1996) *Acta Crystallogr.*, *Sect. B*, vol. 52, p. 471]. Colour code: Co, blue; N, black; O, pale grey.

corresponding or related Co(II) species, often *in situ*. For example:

- oxidation by PbO_2 of aqueous Co^{2+} in the presence of excess oxalate gives $[Co(ox)_3]^{3-}$;
- action of excess $[NO_2]^-$ and acid on aqueous Co^{2+} gives $[Co(NO_2-N)_6]^{3-}$ (Figure 21.21); some $[NO_2]^-$ acts as oxidant and NO is liberated;
- reaction between $Co(CN)_2$ and excess KCN in aqueous solution with *in situ* oxidation gives yellow $K_3[Co(CN)_6]$ (the intermediate Co(II) species is $[Co(CN)_5]^{3-}$ or $[Co(CN)_5(H_2O)]^{3-}$, see later);
- reaction of aqueous $CoCl_2$ with bpy and Br_2 gives $[Co(bpy)_3]^{3+}$;
- aerial oxidation of aqueous $CoCl_2$ in the presence of NH_3 and $[NH_4]Cl$ gives purple $[Co(NH_3)_5Cl]Cl_2$ containing cation **21.24**.

$$\left[\begin{array}{c} Cl \\ H_3N\;\cdots\;\overset{|}{\underset{|}{Co}}\;\cdots\;NH_3 \\ H_3N \quad\quad NH_3 \\ NH_3 \end{array} \right]^{2+}$$

$$Co\text{—}N = 197\,pm;\; Co\text{–}Cl = 229\,pm$$

(21.24)

The identity of the product may depend on reaction conditions and in the last example, if charcoal is added as a catalyst, the isolated complex is $[Co(NH_3)_6]Cl_3$ containing the $[Co(NH_3)_6]^{3+}$ ion. Similarly, the preparation of orange-red $[Co(en)_3]Cl_3$ requires careful control of reaction conditions (equation 21.78).

$$[Co(en)_3]Cl_3 \xleftarrow[\text{aqueous soln}]{\text{en, en·HCl, } O_2} CoCl_2$$

$$\xrightarrow[\text{acidic soln}]{\text{en·HCl, } O_2} trans\text{-}[Co(en)_2Cl_2]Cl \quad (21.78)$$

The $[Co(en)_3]^{3+}$ ion is frequently used to precipitate large anions, and the kinetic inertness of the d^6 ion allows its enantiomers to be separated. The green $trans\text{-}[Co(en)_2Cl_2]Cl$

is isolated from reaction 21.78 as the salt $trans\text{-}[Co(en)_2Cl_2]Cl\cdot2H_2O\cdot HCl$ but this loses HCl on heating. It can be converted to the racemic red $cis\text{-}[Co(en)_2Cl_2]Cl$ by heating an aqueous solution and removing the solvent; enantiomers of $cis\text{-}[Co(en)_2Cl_2]^+$ can be separated using a chiral anion such as (1*S*)- or (1*R*)-3-bromocamphor-8-sulfonate. In aqueous solution, one Cl^- ligand in $[Co(en)_2Cl_2]^+$ is replaced by H_2O to give $[Co(en)_2Cl(H_2O)]^{2+}$. Because ligand substitutions in Co(III) complexes are so slow, these species have been the subject of many kinetic studies (see *Chapter 25*).

The $[Co(CN)_6]^{3-}$ ion is so stable that if a solution of $K_3[Co(CN)_5]$ containing excess KCN is heated, H_2 is evolved and $K_3[Co(CN)_6]$ is formed. In this reaction, the hydrido complex $[Co(CN)_5H]^{3-}$ is an intermediate. It can be obtained almost quantitatively (reversible reaction 21.79) and can be precipitated as $Cs_2Na[Co(CN)_5H]$.

$$2[Co^{II}(CN)_5]^{3-} + H_2 \rightleftharpoons 2[Co^{III}(CN)_5H]^{3-} \quad (21.79)$$

The $[Co(CN)_5H]^{3-}$ ion is an effective homogeneous hydrogenation catalyst for alkenes; the process is summarized in equation 21.80, with reaction 21.79 completing the cycle.

$$\left. \begin{array}{l} [Co(CN)_5H]^{3-} + CH_2{=}CHX \longrightarrow [Co(CN)_5CH_2CH_2X]^{3-} \\ [Co(CN)_5CH_2CH_2X]^{3-} + [Co(CN)_5H]^{3-} \\ \quad\quad \longrightarrow CH_3CH_2X + 2[Co(CN)_5]^{3-} \end{array} \right\}$$

(21.80)

$$\left[\begin{array}{ccc} & CN & \\ NC\text{—} & \overset{|}{\underset{|}{Co}} & \text{—}CN \\ NC & | & \\ & CN & \\ & O\text{—}O & \\ & & CN \\ & & | \\ NC\text{—}\overset{|}{\underset{|}{Co}}\text{—}CN \\ & CN & \end{array} \right]^{6-} \quad 145\,pm$$

(21.25)

By aerial oxidation of $[Co^{II}(CN)_5]^{3-}$ in aqueous cyanide solution, it is possible to isolate the diamagnetic peroxo complex $[(CN)_5Co^{III}OOCo^{III}(CN)_5]^{6-}$ (**21.25**) which can be precipitated as the brown potassium salt. Oxidation of $K_6[(CN)_5CoOOCo(CN)_5]$ using Br_2 leads to the paramagnetic, red $K_5[(CN)_5CoOOCo(CN)_5]$. The structure of $[(CN)_5CoOOCo(CN)_5]^{5-}$ resembles that of **21.25**, except that the O–O distance is 126 pm, indicating that oxidation takes place at the peroxo bridge and not at a metal centre. Thus, $[(CN)_5CoOOCo(CN)_5]^{5-}$ is a superoxo complex retaining two Co(III) centres. The ammine complexes $[(H_3N)_5CoOOCo(NH_3)_5]^{4+}$ and $[(H_3N)_5CoOOCo(NH_3)_5]^{5+}$ (which have been isolated as the brown nitrate and green chloride salts respectively) are similar, containing peroxo and superoxo ligands respectively; the peroxo complex is stable in solution only in the presence of >2 M NH_3.

One of the few examples of a high-spin Co(III) complex is $[CoF_6]^{3-}$; the blue K^+ salt (obtained by heating $CoCl_2$, KF and F_2) has a magnetic moment of 5.63 μ_B.

Cobalt(II)

In contrast to Co(III), Co(II) forms a variety of simple compounds and all four Co(II) halides are known. Reaction of anhydrous $CoCl_2$ with HF at 570 K gives sparingly soluble, pink CoF_2 which crystallizes with the rutile structure (see *Figure 5.21*). Blue $CoCl_2$ is made by combination of the elements and has a $CdCl_2$ lattice (see *Section 5.11*). It turns pink on exposure to moisture and readily forms hydrates. The dark pink hexahydrate is commercially available and is a common starting material in Co(II) chemistry; the di- and tetrahydrates can also be crystallized from aqueous solutions of $CoCl_2$, although the latter only with difficulty. Crystalline $CoCl_2 \cdot 6H_2O$ contains *trans*-$[CoCl_2(H_2O)_4]$, connected to the extra water molecules through a hydrogen-bonded network. In contrast, the structure of $CoCl_2 \cdot 4H_2O$ consists of hydrogen-bonded *cis*-$[CoCl_2(H_2O)_4]$ molecules, while $CoCl_2 \cdot 2H_2O$ contains chains of edge-sharing octahedra (structure **21.26**). In aqueous solutions of all forms of $CoCl_2$, the major species are $[Co(H_2O)_6]^{2+}$, $[CoCl(H_2O)_5]^+$ and $[CoCl_4]^{2-}$, with minor amounts of $[CoCl_2(H_2O)_4]$ and $[CoCl_3(H_2O)]^-$. Green $CoBr_2$ (made by heating Co and Br_2) is dimorphic, adopting either the $CdCl_2$ or CdI_2 lattice. It is water-soluble and can be crystallized as the purple-blue dihydrate or red hexahydrate. Heating Co metal with HI produces blue-black CoI_2 which adopts a CdI_2 layer structure; the red hexahydrate can be crystallized from aqueous solutions. Both $CoBr_2 \cdot 6H_2O$ and $CoI_2 \cdot 6H_2O$ contain the octahedral $[Co(H_2O)_6]^{2+}$ ion in the solid state, as do a number of hydrates, e.g. $CoSO_4 \cdot 6H_2O$, $Co(NO_3)_2 \cdot 6H_2O$ and $Co(ClO_4)_2 \cdot 6H_2O$. Aqueous solutions of most simple Co(II) salts contain $[Co(H_2O)_6]^{2+}$ (see below).

(21.26)

Cobalt(II) oxide is an olive-green, insoluble solid but its colour may vary depending on its dispersion. It is best obtained by thermal decomposition of the carbonate or nitrate in the absence of air, and has the NaCl structure; CoO is used as a pigment in glasses and ceramics (see *Box 21.9*). When heated in air at 770 K, CoO converts to Co_3O_4.

The sparingly soluble $Co(OH)_2$ may be pink or blue, with the pink form being the more stable; freshly precipitated blue $Co(OH)_2$ turns pink on standing. The change in colour is presumably associated with a change in coordination about the Co(II) centre. Cobalt(II) hydroxide is amphoteric and dissolves in hot, concentrated alkalis to give salts of $[Co(OH)_4]^{2-}$ (**21.27**).

(21.27)

Whereas the coordination chemistry of Co^{3+} is essentially that of octahedral complexes, that of Co^{2+} is structurally varied since LFSEs for the d^7 configuration do not tend to favour a particular ligand arrangement. The variation in coordination geometries is shown in the following examples:

- linear: $[Co\{N(SiMe_3)_2\}_2]$;
- trigonal planar: $[Co\{N(SiMe_3)_2\}_2(PPh_3)]$; $[Co\{N(SiMe_3)_2\}_3]^-$;
- tetrahedral: $[Co(OH)_4]^{2-}$, $[CoCl_4]^{2-}$, $[CoBr_4]^{2-}$, $[CoI_4]^{2-}$, $[Co(NCS-N)_4]^{2-}$, $[Co(N_3)_4]^{2-}$, $[CoCl_3(NCMe)]^-$;
- square planar: $[Co(CN)_4]^{2-}$, $[Co(pc)]$ ($H_2pc = $ **21.19**);
- trigonal bipyramidal: $[Co\{N(CH_2CH_2PPh_2)_3\}(SMe)]^+$;
- square-based pyramidal: $[Co(CN)_5]^{3-}$;
- octahedral: $[Co(H_2O)_6]^{2+}$, $[Co(NH_3)_6]^{2+}$, $[Co(en)_3]^{2+}$;
- pentagonal bipyramidal: $[Co(15\text{-crown-}5)L_2]^{2+}$ (L = H_2O or MeCN; see **21.32**);
- dodecahedral: $[Co(NO_3)_4]^{2-}$.

BOX 21.9 **APPLICATIONS**

Cobalt blues

Blue glass and ceramic glazes and enamels are in high demand for decorative wear, and the source of colour is very often a cobalt-based pigment. Cobalt(II) oxide is the form that is incorporated into the molten glass, but initial sources vary. Black Co_3O_4 is transformed in $\approx$93% yield to CoO at $\approx$1070 K. Purple $CoCO_3$ can also be used as raw material but has lower conversion yields. Only very small amounts of the oxide are required to obtain a discernible blue pigment. Variations in colour are achieved by combining with other oxides, e.g. purple shades result if manganese oxide is added. Cobalt oxide is also used to counter the

yellow colouring in glazes that arises from iron impurities. Blue pigmentation can also be obtained using $(Zr,V)SiO_4$ (see *Section 27.5*).

While the importance of cobalt-based pigments in ceramics is well established, it has also been shown that thin films of Co_3O_4 provide an effective coating for solar collectors that operate at high temperatures. The properties of black Co_3O_4 that make it suitable for this application are its high solar absorbance and low IR emittance.

Related material: see *Box 13.3* – Solar power: thermal and electrical.

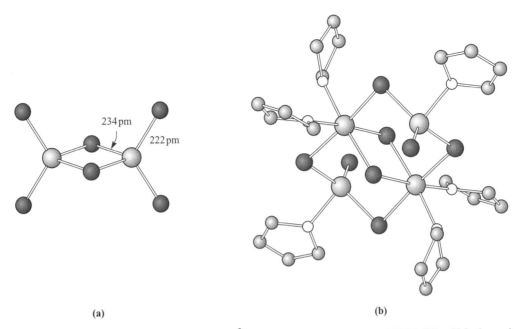

(a)   **(b)**

Fig. 21.22 The structures (X-ray diffraction) of (a) $[Co_2Cl_6]^{2-}$ in the salt $[Co(15-crown)(NCMe)_2][Co_2Cl_6]$; the cation is shown in structure **21.32** [O.K. Kireeva *et al.* (1992) *Polyhedron*, vol. 11, p. 1801], and (b) $[Co_4Cl_2(\mu-Cl)_6(THF)_6]$ [P. Sobota *et al.* (1993) *Polyhedron*, vol. 12, p. 613]. Hydrogen atoms are omitted; colour code: Co, blue; Cl, black; C, grey; O, white.

Aqueous solutions of simple salts usually contain $[Co(H_2O)_6]^{2+}$ but there is evidence for the existence of equilibrium 21.81, although $[Co(H_2O)_6]^{2+}$ is by far the dominant species; speciation in aqueous $CoCl_2$ was discussed earlier.

$$[Co(H_2O)_6]^{2+} \rightleftharpoons [Co(H_2O)_4]^{2+} + 2H_2O \quad (21.81)$$
$$\text{octahedral} \qquad \text{tetrahedral}$$

Whereas $[Co(H_2O)_6]^{2+}$ is a stable complex, $[Co(NH_3)_6]^{2+}$ is easily oxidized (equations 21.73 and 21.75). The same is true of amine complexes; $[Co(en)_3]^{2+}$ can be prepared from $[Co(H_2O)_6]^{2+}$ and en in an inert atmosphere and is usually made *in situ* as required. The $[Co(bpy)_3]^{2+}$ ion is stable enough to be isolated in a number of salts, e.g. orange $[Co(bpy)_3]Cl_2 \cdot 2H_2O \cdot EtOH$ which has been crystallographically characterized (Co–N = 213 pm). Among the stable complexes of Co(II) are tetrahedral $[CoX_4]^{2-}$ (X = Cl, Br, I). Addition of concentrated HCl to solutions of pink $[Co(H_2O)_6]^{2+}$ produces the intensely blue $[CoCl_4]^{2-}$. Many salts of $[CoCl_4]^{2-}$ are known; of note is Cs_3CoCl_5 which is actually $Cs_3[CoCl_4]Cl$ and does *not* contain $[CoCl_5]^{3-}$. Both $[Co(H_2O)_6]^{2+}$ and $[CoCl_4]^{2-}$, like most Co(II) complexes, are high-spin with magnetic moments higher than the spin-only value; typically, for high-spin Co^{2+}, μ_{eff} lies in the range 4.3–5.2 μ_B for octahedral complexes and 4.2–4.8 μ_B for tetrahedral species. Among other tetrahedral complexes is $[Co(NCS-N)_4]^{2-}$, isolated in blue $[Me_4N]_2[Co(NCS-N)_4]$ ($\mu_{eff} = 4.40 \mu_B$) and $K_2[Co(NCS-N)_4] \cdot 4H_2O$ ($\mu_{eff} = 4.38 \mu_B$). The insoluble mercury(II) salt of $[Co(NCS-N)_4]^{2-}$ is the standard calibrant for magnetic susceptibility measurements. By using cation

21.28, it has been possible to isolate a red salt of the octahedral $[Co(NCS-N)_6]^{4-}$.

(21.28)

The ability of chloro ligands to bridge between two metal centres allows the formation of dinuclear species such as $[Co_2Cl_6]^{2-}$ (Figure 21.22a), as well as higher nuclearity complexes such as polymer **21.26**. The complex $[CoCl_2(py)_2]$ exists in two modifications: one is monomer **21.29** containing a tetrahedral Co(II) centre, while the other contains edge-sharing octahedra in polymer **21.30**. Equation 21.82 summarizes the formation of $[CoCl_2(py)_2]$ and $[CoCl_2(py)_2]_n$. Similar tetrahedral–octahedral interconversions are seen for some Ni(II) complexes of type L_2NiX_2 where X^- has the propensity for bridge formation (see *Section 21.11*).

$$CoCl_2 + 2py \longrightarrow [CoCl_2(py)_2] \underset{\Delta,\, 390\,K}{\overset{\text{polymerizes on standing in air}}{\rightleftharpoons}} [CoCl_2(py)_2]_n$$
$$\text{blue} \qquad\qquad\qquad \text{violet}$$
$$(21.82)$$

(21.29)   (21.30)

Heating a solution of $CoCl_2$ in THF at reflux produces the dark blue $[Co_4Cl_2(\mu\text{-}Cl)_6(THF)_6]$ in which bridging chloro ligands support the tetranuclear framework (Figure 21.22b). Two Co(II) centres are octahedrally coordinated and two are in 4-coordinate environments. At 300 K, the magnetic moment is 4.91 μ_B, typical of isolated high-spin Co(II) centres. On lowering the temperature to 4.2 K, the value of μ_{eff} increases to 7.1 μ_B. Such behaviour indicates ferromagnetic coupling between the metal centres which are able to communicate through the bridging ligands (see *Section 20.8*).

Chloride is just one example of a ligand which may coordinate to a metal centre in a terminal or bridging mode; other ligands are equally versatile. For example, $[Co(acac)_2]$ is prepared from $CoCl_2$, Hacac and $Na[O_2CMe]$ in aqueous methanol. In the solid state, the blue anhydrous salt is tetrameric with a structure related to that of the trimer $[\{Ni(acac)_2\}_3]$ (see Figure 21.24b).

Low-spin cyano complexes of Co(II) provide examples of square-based pyramidal and square planar species. The addition of an excess of $[CN]^-$ to aqueous Co^{2+} yields $[Co(CN)_5]^{3-}$. That this is formed in preference to $[Co(CN)_6]^{4-}$ (which has not been isolated) can be understood by considering Figure 20.14b. For the strong-field cyano ligands, Δ_{oct} is large and for a hypothetical octahedral d^7 complex, partial occupancy of the $e_g{}^*$ MOs would be unfavourable since it would impart significant *antibonding* character on the complex. The brown $K_3[Co(CN)_5]$ is paramagnetic, but a violet, diamagnetic salt $K_6[Co_2(CN)_{10}]$ has also been isolated. The $[Co_2(CN)_{10}]^{6-}$ ion, **21.31**, possesses a Co–Co single bond and a staggered conformation; it is isoelectronic and isostructural with $[Mn_2(CO)_{10}]$ (see *Figure 23.10*). By using the large cation $[(Ph_3P)_2N]^+$, it has been possible to isolate a salt of the square planar complex

$[Co(CN)_4]^{2-}$ (Figure 21.23a). This is an unusual example of a square planar Co(II) species where the geometry is *not* imposed by the ligand. In complexes such as $[Co(pc)]$, the phthalocyanine ligand (**21.19**) has a rigid framework and forces the coordination environment to be square planar.

(21.31)

The highest coordination numbers for Co(II) are 7 and 8. The effects of a coordinatively restricted macrocyclic ligand give rise to pentagonal pyramidal structures for $[Co(15\text{-crown-5})(NCMe)_2]^{2+}$ (**21.32**) and $[Co(15\text{-crown-5})(H_2O)_2]^{2+}$. Larger macrocycles are more flexible, and in the complex $[Co(\mathbf{21.33})]^{2+}$, the S_6-donor set is octahedrally arranged. Figure 21.23b shows the solid state structure of $[Co(12\text{-crown-4})(NO_3)_2]$ in which the Co(II) centre is 7-coordinate. In $[Co(NO_3)_4]^{2-}$, a dodecahedral arrangement of donor atoms is observed, although as Figure 21.23c shows, each $[NO_3]^-$ ligand is bound asymmetrically with one oxygen donor interacting more strongly than the other. These nitrato complexes illustrate that caution is sometimes needed in interpreting coordination geometries and a further example concerns $[LCoX]^+$ complexes where L is the tripodal ligand $N(CH_2CH_2PPh_2)_3$. For $X^- = [MeS]^-$ or $[EtO(O)_2S]^-$, the Co(II) centre in $[LCoX]^+$ is 5-coordinate (**21.34**) with a Co–N distance of 213 or 217 pm, respectively. However, for $X^- = Cl^-$, Br^- or I^-, there is only a weak interaction between the nitrogen and metal centre (**21.35**) with Co····N in the range

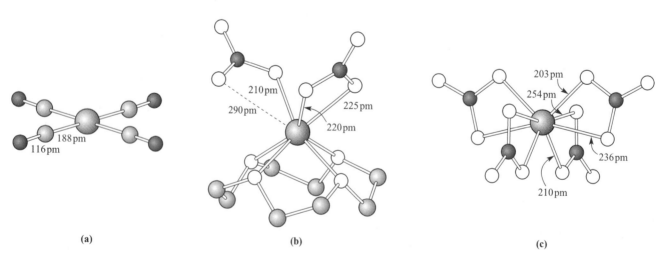

(a)   (b)   (c)

Fig. 21.23 The structures (X-ray diffraction) of (a) $[Co(CN)_4]^{2-}$ in the salt $[(Ph_3P)_2N]_2[Co(CN)_4]\cdot4DMF$; there is also a *weak* interaction with a solvate molecule in an axial site [S.J. Carter *et al.* (1984) *J. Am. Chem. Soc.*, vol. 106, p. 4265]; (b) $[Co(12\text{-crown-4})(NO_3)_2]$ [E.M. Holt *et al.* (1981) *Acta Crystallogr., Sect. B*, vol. 37, p. 1080], and (c) $[Co(NO_3)_4]^{2-}$ in the $[Ph_4As]^+$ salt [J.G. Bergman *et al.* (1966) *Inorg. Chem.*, vol. 5, p. 1208]. Hydrogen atoms are omitted; colour code: Co, blue; N, black; C, grey; O, white.

268–273 pm. We stress that these data refer to the *solid state* and say nothing about solution species.

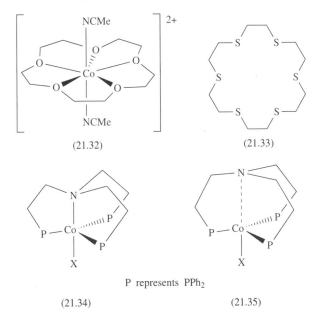

(21.32)

(21.33)

P represents PPh$_2$

(21.34)

(21.35)

21.11 Group 10: nickel

The metal

The reactivity of Ni metal resembles that of Co (e.g. equation 21.83). It is attacked by dilute mineral acids, made passive by concentrated HNO$_3$, and is resistant to aqueous alkalis.

$$M^{2+}(aq) + 2e^- \longrightarrow M(s) \quad \begin{cases} M = Ni, \; E^\circ = -0.25\,V \\ M = Co, \; E^\circ = -0.28\,V \end{cases}$$

(21.83)

The bulk metal is oxidized by air or steam only at high temperatures, but *Raney nickel* (see *Section 21.2*) is pyrophoric. Nickel reacts with F$_2$ to give a coherent coating of NiF$_2$ which prevents further attack; hence the use of nickel and its alloy *Monel metal* in apparatus for handling F$_2$ or xenon fluorides. With Cl$_2$, Br$_2$ and I$_2$, Ni(II) halides are formed. At elevated temperatures, Ni reacts with P, S and B and a range of different phosphide (see *Section 14.6*), sulfide and boride (see *Section 12.10*) phases are known.

Nickel(II) is far the most important oxidation state for the metal (*Table 19.3*). Low oxidation states are most common in organometallic species (*Chapter 23*), but other Ni(0) species include [Ni(PF$_3$)$_4$] and [Ni(CN)$_4$]$^{4-}$. Yellow K$_4$[Ni(CN)$_4$] is made by reduction of K$_2$[Ni(CN)$_4$] in liquid NH$_3$ using excess K, but oxidizes immediately on exposure to air.

Nickel(IV) and nickel(III)

Nickel(IV) is present in only a few species, and its formation requires extremely strong oxidants. The salt [Xe$_2$F$_{11}$]$_2$[NiF$_6$] (*Figure 17.3*) is made from XeF$_2$, KrF$_2$ and NiF$_2$. Octahedral

[NiF$_6$]$^{2-}$ is diamagnetic (low-spin d^6) and the red K$^+$ salt crystallizes with the K$_2$[PtF$_6$] lattice (see Mn(IV), *Section 21.8*). Above 620 K, K$_2$[NiF$_6$] decomposes to K$_3$[NiF$_6$]. Salts of [NiF$_6$]$^{2-}$ are powerful oxidants, and [NF$_4$]$_2$[NiF$_6$] is used as an oxidizing agent in some solid propellants. It decomposes on heating according to equation 21.84. Nickel(IV) fluoride can be prepared from K$_2$[NiF$_6$] and BF$_3$ or AsF$_5$, but is unstable above 208 K (equation 21.85).

$$[NF_4]_2[NiF_6] \longrightarrow 2NF_3 + NiF_2 + 3F_2$$

(21.84)

$$2NiF_4 \longrightarrow 2NiF_3 + F_2$$

(21.85)

Nickel(IV) is present in KNiIO$_6$, formally a salt of [IO$_6$]$^{5-}$ (see *Section 16.9*); it is formed by oxidation of [Ni(H$_2$O)$_6$]$^{2+}$ by [S$_2$O$_8$]$^{2-}$ in the presence of [IO$_4$]$^-$. The structure of KNiIO$_6$ can be considered as an hcp array of O atoms with K, Ni and I occupying octahedral sites.

Impure NiF$_3$ is made by reaction 21.85. It is a black solid, and is a strong fluorinating agent, but decomposes when heated (equation 21.86).

$$2NiF_3 \xrightarrow{\Delta} NiF_2 + F_2$$

(21.86)

Reaction of NiCl$_2$, KCl and F$_2$ produces violet K$_3$[NiF$_6$]. Octahedral [NiF$_6$]$^{3-}$ is low-spin d^7 and shows the expected Jahn–Teller distortion.

The black hydrous oxide Ni(O)OH is obtained by alkaline hypochlorite oxidation of aqueous Ni(II) salts and has widespread use in NiCd rechargeable batteries (equation 21.5). It is a strong oxidizing agent, liberating Cl$_2$ from hydrochloric acid. Mixed metal oxides of Ni(IV) include BaNiO$_3$ and SrNiO$_3$ which are isostructural and contain chains of face-sharing NiO$_6$ octahedra.

Nickel(III) is a very good oxidizing agent, but is stabilized by σ-donor ligands. Complexes include [Ni(1,2-S$_2$C$_6$H$_4$)$_2$]$^-$ (**21.36**) and [NiBr$_3$(PEt$_3$)$_2$] (**21.37**). The latter has a magnetic moment of 1.72 μ_B, indicative of low-spin Ni(III); the solid compound is stable for only a few hours. Other ligands used to stabilize Ni(III) include porphyrins and aza-macrocycles; in [Ni(**21.38**)], each set of three *N*-donors and three *O*-donors is in a *fac*-arrangement about an octahedral Ni(III) centre.

(21.36)

PEt$_3$

Br — Ni$^{\cdots}$Br

Br

PEt$_3$

(21.37)

CO$_2^-$

CO$_2^-$

CO$_2^-$

(21.38)

Nickel(II)

Nickel(II) fluoride is made by fluorination of $NiCl_2$; it is a yellow solid with the rutile lattice (*Figure 5.21*). Both NiF_2 and its green tetrahydrate are commercially available. Anhydrous $NiCl_2$, $NiBr_2$ and NiI_2 are made by direct combination of the elements; $NiCl_2$ and NiI_2 adopt a $CdCl_2$ lattice while $NiBr_2$ has a CdI_2 structure (see *Section 5.11*). The chloride is a useful precursor in Ni(II) chemistry and can be purchased as the yellow anhydrous salt or green hydrate. The hexahydrate contains the $[Ni(H_2O)_6]^{2+}$ ion in the solid state, but the dihydrate (obtained by partial dehydration of $NiCl_2 \cdot 6H_2O$) has a polymeric structure analogous to **21.26**. Anhydrous $NiBr_2$ is yellow and can be crystallized as a number of hydrates; black NiI_2 forms a green hexahydrate.

The water-insoluble, green NiO is obtained by thermal decomposition of $NiCO_3$ or $Ni(NO_3)_2$ and crystallizes with the NaCl lattice; thin amorphous films of NiO exhibiting electrochromic behaviour (see *Box 22.4*) may be deposited by CVD (*chemical vapour deposition*, see *Section 27.6*) starting from $[Ni(acac)_2]$. Nickel(II) oxide is antiferromagnetic ($T_N = 520 \, K$); its conducting properties are discussed in *Section 27.3*. Nickel(II) oxide is basic, reacting with acids, e.g. reaction 21.87.

$$NiO + H_2SO_4 \longrightarrow NiSO_4 + H_2O \qquad (21.87)$$

Oxidation of NiO by hypochlorite yields Ni(O)OH (see earlier). Aerial oxidation converts NiS to Ni(S)OH, a fact that explains why, although NiS is not precipitated in acidic solution, after exposure to air it is insoluble in dilute acid. Addition of $[OH]^-$ to aqueous solutions of Ni^{2+} precipitates green $Ni(OH)_2$ which has a CdI_2 structure; it is used in NiCd batteries (equation 21.5). Nickel(II) hydroxide is insoluble in aqueous NaOH except at very high hydroxide concentrations when it forms soluble $Na_2[Ni(OH)_4]$; $Ni(OH)_2$ is soluble in aqueous NH_3 with formation of $[Ni(NH_3)_6]^{2+}$. The pale green basic carbonate, $2NiCO_3 \cdot 3Ni(OH)_2 \cdot 4H_2O$, forms when Na_2CO_3 is added to aqueous Ni^{2+} and it is this carbonate that is usually bought commercially.

A range of coordination geometries is observed for nickel(II) complexes with coordination numbers from 4 to 6 being common, and 7 rare; octahedral and square planar geometries are most usual. Examples include:

- tetrahedral: $[NiCl_4]^{2-}$, $[NiBr_4]^{2-}$, $[Ni(NCS\text{-}N)_4]^{2-}$;
- square planar: $[Ni(CN)_4]^{2-}$, $[Ni(Hdmg)_2]$ (H_2dmg = dimethylglyoxime);
- trigonal bipyramidal: $[Ni(CN)_5]^{2-}$ (cation-dependent), $[NiCl\{N(CH_2CH_2NMe_2)_3\}]^+$;
- square-based pyramidal: $[Ni(CN)_5]^{2-}$ (cation-dependent);
- octahedral: $[Ni(H_2O)_6]^{2+}$, $[Ni(NH_3)_6]^{2+}$, $[Ni(bpy)_3]^{2+}$, $[Ni(en)_3]^{2+}$, $[Ni(NCS\text{-}N)_6]^{4-}$, $[NiF_6]^{4-}$.

Some structures are complicated by interconversions between square planar and tetrahedral, or square planar and octahedral coordination as we discuss later. In addition,

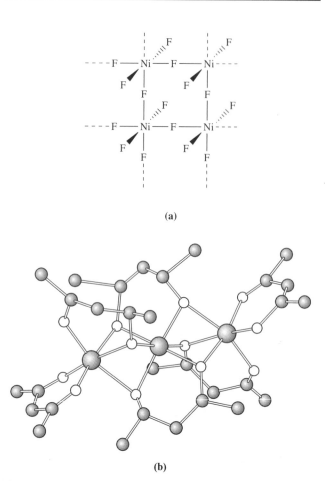

Fig. 21.24 (a) Representation of part of a layer of corner-sharing NiF_6 octahedra in K_2NiF_4. (b) The structure of $[\{Ni(acac)_2\}_3]$ (X-ray diffraction) with H atoms omitted [G.J. Bullen *et al.* (1965) *Inorg. Chem.*, vol. 4, p. 456]. Colour code: Ni, blue; C, grey; O, white.

the potential of some ligands to bridge between metal centres may cause ambiguity. For example, alkali metal salts of $[NiF_3]^-$, $[NiF_4]^{2-}$ and $[NiCl_3]^-$ crystallize with infinite lattices, whereas salts of $[NiCl_4]^{2-}$ and $[NiBr_4]^{2-}$ contain discrete tetrahedral anions. The compounds $KNiF_3$ and $CsNiF_3$ are obtained by cooling melts containing NiF_2 and MHF_2; $KNiF_3$ has a perovskite structure (*Figure 5.23*) and is antiferromagnetic, while $CsNiF_3$ possesses chains of face-sharing NiF_6 octahedra and is ferrimagnetic. A similar chain-structure is adopted by $CsNiCl_3$. The antiferromagnetic K_2NiF_4 contains layers of corner-sharing octahedral NiF_6 units (Figure 21.24a) separated by K^+ ions.

In Section 21.10, we noted that $[Co(acac)_2]$ is tetrameric. Similarly, $[Ni(acac)_2]$ oligomerizes, forming trimers (Figure 21.24b) in which $[acac]^-$ ligands are in chelating and bridging modes. Reaction of $[\{Ni(acac)_2\}_3]$ with aqueous $AgNO_3$ yields $Ag[Ni(acac)_3]$ containing the octahedral $[Ni(acac)_3]^-$ ion.

Solid, hydrated nickel(II) salts and their aqueous solutions usually contain green $[Ni(H_2O)_6]^{2+}$, the electronic spectrum of which was shown in *Figure 20.19* with that of $[Ni(NH_3)_6]^{2+}$. Salts of the latter are typically blue, giving

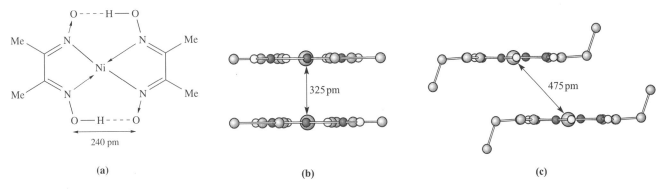

Fig. 21.25 (a) Representation of the square planar structure of bis(dimethylglyoximato)nickel(II), [Ni(Hdmg)$_2$]; (b) in the solid state, molecules of [Ni(Hdmg)$_2$] pack in vertical columns with relatively short Ni····Ni distances [X-ray diffraction data: D.E. Williams *et al.* (1959) *J. Am. Chem. Soc.*, vol. 81, p. 755], but, (c) in bis(ethylmethylglyoximato)nickel(II), the packing is not so efficient [X-ray diffraction data: E. Frasson *et al.* (1960) *Acta Crystallogr.*, vol. 13, p. 893]. Hydrogen atoms are omitted; colour code: Ni, blue; N, black; O, white; C, grey.

violet solutions; in solution, [Ni(NH$_3$)$_6$]$^{2+}$ is stable only in the presence of excess NH$_3$ without which species such as [Ni(NH$_3$)$_4$(H$_2$O)$_2$]$^{2+}$ form. The violet chloride, bromide or perchlorate salts of [Ni(en)$_3$]$^{2+}$ are obtained as racemates, the cation being kinetically labile (see *Section 25.2*). The octahedral complexes *trans*-[Ni(ClO$_4$-*O*,*O'*)$_2$(NCMe)$_2$] (**21.39**) and *trans*-[Ni(ClO$_4$-*O*)$_2$(py)$_4$] illustrate the ability of perchlorate ions to act as didentate or monodentate ligands respectively. The latter complex is discussed again later.

$$(21.39)$$

Magnetic moments of *octahedral* Ni(II) complexes are usually close to the spin-only value of 2.83 μ_B. In contrast, *tetrahedral* complexes possess magnetic moments $\approx 4\,\mu_B$ due to orbital contributions (see *Section 20.8*), and *square planar* complexes such as [Ni(CN)$_4$]$^{2-}$ (equation 21.88) are *diamagnetic*. These differences in magnetic moments are invaluable in providing information about the coordination geometry in a Ni(II) complex.

$$[Ni(H_2O)_6]^{2+} + 4[CN]^- \longrightarrow [Ni(CN)_4]^{2-} + 6H_2O \quad (21.88)$$
$$\text{yellow}$$

The red square planar complex bis(dimethylglyoximato)-nickel(II), [Ni(Hdmg)$_2$]† (Figure 21.25a), is used for gravimetric analysis of nickel; Ni(II) is precipitated along with Pd(II) when the ligand H$_2$dmg in weakly ammoniacal solution is used as a reagent. The specificity for Ni^{2+} arises from the low solubility of [Ni(Hdmg)$_2$], *not* its high stability

constant; complexes of type [M(Hdmg)$_2$], where M^{2+} is a first row *d*-block metal ion, have stability constants of the same order. The low solubility of [Ni(Hdmg)$_2$] can be rationalized in terms of its solid state structure. Strong hydrogen bonding links the two ligands (Figure 21.25a) and plays a role in determining a square planar structure; as a consequence of the molecular framework being planar, molecules in the crystal lattice are able to assemble into one-dimensional stacks such that intermolecular Ni····Ni separations are 325 pm (Figure 21.25b; but contrast [Cu(Hdmg)$_2$] in Section 21.12). Bis(ethylmethylglyoximato)nickel(II) has a related structure, but the bulkier ligand forces the molecules to pack less efficiently (Figure 21.25c); the fact that the latter complex is more soluble than [Ni(Hdmg)$_2$] supports a structure–solubility relationship.

For some Ni(II) complexes, there is only a small energy difference between structure types. In *Section 19.7*, we stated that *both* trigonal bipyramidal and square-based pyramidal [Ni(CN)$_5$]$^{3-}$ ions (equation 21.89) are present in crystals of [Cr(en)$_3$][Ni(CN)$_5$]·1.5H$_2$O. In the anhydrous salt, however, the anions are square-based pyramidal. It is impossible to give a simple interpretation of these observations which may be attributed to a 'subtle balance of steric and electronic effects'.

$$[Ni(CN)_4]^{2-} + \text{excess } [CN]^- \longrightarrow [Ni(CN)_5]^{3-} \quad (21.89)$$

The preference between different 4- and 6-coordination geometries for a number of Ni(II) systems is often marginal and examples are as follows.

Octahedral-planar

- [Ni(ClO$_4$)$_2$(py)$_4$] exists in a blue, paramagnetic *trans*-octahedral form and as a yellow diamagnetic salt containing square planar [Ni(py)$_4$]$^{2+}$ ions;
- Salicylaldoxime (2-HOC$_6$H$_4$CH=NOH) reacts with Ni(II) to give colourless crystals of the square planar complex **21.40**, but on dissolving in pyridine, a green solution of the paramagnetic octahedral [Ni(2-OC$_6$H$_4$CH=NOH)$_2$(py)$_2$] forms.

---

† For an introduction to the use of [Ni(Hdmg)$_2$] and related complexes in template syntheses of macrocycles, see: E.C. Constable (1999) *Coordination Chemistry of Macrocyclic Compounds*, Oxford University Press, Oxford (Chapter 4).

(21.40)

Tetrahedral-planar

- Halides of type NiL_2X_2 are generally planar when $L =$ trialkylphosphine, but tetrahedral when L is triarylphosphine; when $X = Br$ and $L = PEtPh_2$ or $P(CH_2Ph)Ph_2$, both forms are known (scheme 21.90).

$$NiBr_2 + PEtPh_2$$

$$\downarrow \text{ethanol}$$

$$[NiBr_2(PEtPh_2)_2]$$
green, tetrahedral
$\mu_{eff} = 3.18 \, \mu_B$

slow isomerization at 298 K $\updownarrow$ dissolve in CS_2

$$[NiBr_2(PEtPh_2)_2]$$
brown, square planar
diamagnetic

(21.90)

Nickel(I)

Nickel(I) is uncommon, but dark red $K_4[Ni_2(CN)_6]$ can be prepared by Na amalgam reduction of $K_2[Ni(CN)_4]$. It is diamagnetic and the anion has structure **21.41** in which the $Ni(CN)_3$ units are mutually perpendicular. The reaction of $K_4[Ni_2(CN)_6]$ with water liberates H_2 and forms $K_2[Ni(CN)_4]$.

(21.41)

21.12 Group 11: copper

The metal

Copper is the least reactive of the first row metals. It is not attacked by non-oxidizing acids in the absence of air (equation 21.91), but it reacts with hot concentrated sulfuric acid (equation 21.92) and with HNO_3 of all concentrations (*equations 14.96* and *14.97*).

$$Cu^{2+} + 2e^- \rightleftharpoons Cu \qquad\qquad E^\circ = +0.34 \, V \qquad (21.91)$$

$$\underset{\text{conc}}{Cu + 2H_2SO_4} \longrightarrow SO_2 + CuSO_4 + 2H_2O \qquad (21.92)$$

BOX 21.10   **CHEMICAL AND THEORETICAL BACKGROUND**

Copper: from antiquity to present day

5000–4000BC	Copper metal used in tools and utensils; heat is used to make the metal malleable.
4000–2000BC	In Egypt, copper is cast into specific shapes; bronze (an alloy with tin) is made; first copper mining in Asia Minor, China and North America.
2000BC–0	Bronze weapons are introduced; bronze is increasingly used in decorative pieces.
0–AD200	Brass (an alloy of copper and zinc) is developed.
AD200–1800	A period of little progress.
1800–1900	Deposits of copper ores in Michigan, US, are mined, increasing dramatically the US output and availability of the metal. The presence of copper in plants and animals is first discovered.
1900–1960	The electrical conducting properties of copper are discovered and as a result, many new applications.
1960 onwards	North American production continues to increase but the world market also reaps the benefits of mining in many other countries, in particular Chile; copper recycling becomes important. From mid-1980s, materials such as $YBa_2Cu_3O_{7-x}$ are discovered to be high-temperature superconductors (see *Section 27.4*).

Table adapted from: R.R. Conry and K.D. Karlin (1994) 'Copper: Inorganic & coordination chemistry' in *Encyclopedia of Inorganic Chemistry*, ed. R.B. King, Wiley, Chichester, vol. 2, p. 829.

In the presence of air, Cu reacts with many dilute acids (the green patina on roofs in cities is basic copper sulfate) and also dissolves in aqueous NH_3 to give $[Cu(NH_3)_4]^{2+}$. When heated strongly, Cu combines with O_2 (equation 21.93).

$$2Cu + O_2 \xrightarrow{\Delta} 2CuO \xrightarrow{>1300\,K} Cu_2O + \tfrac{1}{2}O_2 \qquad (21.93)$$

Heating Cu with F_2, Cl_2 or Br_2 produces the corresponding dihalide.

Copper is the only first row d-block metal to exhibit a *stable* +1 oxidation state. In aqueous solution, Cu(I) is unstable by a relatively small margin with respect to Cu(II) and the metal (equations 21.91, 21.94 and 21.95).

$$Cu^+ + e^- \rightleftharpoons Cu \qquad E^\circ = +0.52\,V \qquad (21.94)$$

$$Cu^{2+} + e^- \rightleftharpoons Cu^+ \qquad E^\circ = +0.15\,V \qquad (21.95)$$

This disproportionation is usually fast, but when aqueous Cu(I) is prepared by reduction of Cu(II) with V(II) or Cr(II), decomposition in the absence of air takes several hours. Copper(I) can be stabilized by the formation of an insoluble compound (e.g. CuCl) or a complex (e.g. $[Cu(CN)_4]^{3-}$) (see *Section 7.4*). The stable oxidation state may depend on reaction conditions: e.g. when Cu powder reacts with aqueous $AgNO_3$, reaction 21.96 takes place, but in MeCN reaction 21.97 occurs.

$$Cu + 2Ag^+ \xrightarrow{aq.\ solution} Cu^{2+} + 2Ag \qquad (21.96)$$

$$Cu + [Ag(NCMe)_4]^+ \longrightarrow [Cu(NCMe)_4]^+ + Ag \qquad (21.97)$$

Copper(0) is rarely stabilized; the unstable $Cu_2(CO)_6$ has been isolated in a matrix at low temperature. The highest oxidation state attained for copper is +4.

Copper(IV) and (III)

Copper(IV) is rare. It exists in the red Cs_2CuF_6 which is made by fluorinating $CsCuCl_3$ at 520 K; the $[CuF_6]^{2-}$ ion is low-spin d^7 and has a Jahn–Teller distorted octahedral structure. Copper(IV) oxide has been prepared in a matrix by vaporizing the metal and co-depositing it with O_2; spectroscopic data are consistent with a linear structure, O=Cu=O.

High pressure fluorination of a mixture of CsCl and $CuCl_2$ gives $Cs_3[CuF_6]$. Green $K_3[CuF_6]$ is similarly prepared and has a magnetic moment of 3.01 μ_B indicative of octahedral Cu(III). The diamagnetic compounds $K[CuO_2]$ and $K_7[Cu(IO_6)_2]$ contain square planar Cu(III) (structures **21.42** and **21.43**).

(21.42)

(21.43)

(21.44)

Ligands that stabilize Cu(III) include 1,2-dithiooxalate. Reaction of $[C_2O_2S_2]^{2-}$ with $CuCl_2$ produces $[Cu^{II}(C_2O_2S_2)_2]^{2-}$, oxidation of which by $FeCl_3$ gives $[Cu^{III}(C_2O_2S_2)_2]^-$ (**21.44**). This readily undergoes a photo-induced two-electron intramolecular transfer, cleaving one of the C–C bonds and releasing two equivalents of SCO.

Probably the most important use of Cu(III) species is in high-temperature superconductors such as $YBa_2Cu_3O_{7-x}$ ($x \approx 0.1$) which are discussed in *Chapter 27*.

Copper(II)

Cupric is the old name for copper(II). Throughout copper(II) chemistry, Jahn–Teller distortions are observed as predicted for an octahedral d^9 ion, although the degree of distortion varies considerably.

White CuF_2 (made, like $CuCl_2$ and $CuBr_2$, from the elements) has a distorted rutile structure (Figure 5.21) with elongated CuF_6-units (193 and 227 pm). In moist air, CuF_2 turns blue as it forms the dihydrate. Copper(II) chloride forms yellow or brown deliquescent crystals and forms the green-blue $CuCl_2\cdot2H_2O$ on standing in moist air. The structure of anhydrous $CuCl_2$ (Figure 21.26a) consists of chains so stacked that each Cu(II) centre is in a distorted octahedral site. In solid $CuCl_2\cdot2H_2O$ (**21.45**), *trans*-square planar molecules are arranged so that there are weak intermolecular Cu····Cl interactions. Above 570 K, $CuCl_2$ decomposes to CuCl and Cl_2. Black $CuBr_2$ has a distorted CdI_2 structure (*Figure 5.22*). Copper(II) iodide is not known.

a = 193; b = 228; c = 291 pm

(21.45)

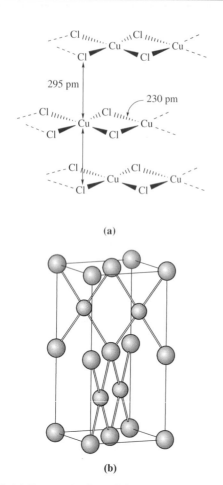

Fig. 21.26 (a) Representation of the solid state structure of CuCl$_2$ in which chains stack to place each Cu(II) centre in a distorted octahedral environment; (b) the *cooperite* (PtS) lattice adopted by CuO with Cu^{2+} (square planar) and O^{2-} (distorted tetrahedral) centres shown in blue and grey respectively.

Black CuO is made by heating the elements (equation 21.93) or by thermal decomposition of solid Cu(NO$_3$)$_2$ or CuCO$_3$ (equation 21.98). Its structure consists of square planar CuO$_4$ units linked by bridging O atoms into chains; these lie in a criss-cross arrangement so that each O atom is in a distorted tetrahedral site. Figure 21.26b shows a unit cell of this lattice which is an example of the *cooperite* (PtS) structure type. Below 225 K, CuO is antiferromagnetic. One use of CuO is as a black pigment in ceramics.

$$CuCO_3 \xrightarrow{\Delta} CuO + CO_2 \qquad (21.98)$$

Blue Cu(OH)$_2$ precipitates when [OH]$^-$ is added to aqueous solutions of Cu^{2+}; Cu(OH)$_2$ dissolves in acids and also in concentrated aqueous alkalis in which an ill-defined hydroxo-species is formed. Copper(II) hydroxide is readily dehydrated to CuO.

Aqueous solutions of Cu^{2+} contain the [Cu(H$_2$O)$_6$]$^{2+}$ ion and this has been isolated in several salts including Cu(ClO$_4$)$_2$·6H$_2$O and the Tutton salt [NH$_4$]$_2$Cu[SO$_4$]$_2$·6H$_2$O (see *Section 21.6*). The solid state structures of both salts reveal distortions of [Cu(H$_2$O)$_6$]$^{2+}$ such that there are *three*

pairs of Cu$-$O distances, e.g. in Cu(ClO$_4$)$_2$·6H$_2$O the Cu$-$O bond lengths are 209, 216 and 228 pm. Crystals of the blue hydrated sulfate CuSO$_4$·5H$_2$O (*blue vitriol*) contain square planar [Cu(H$_2$O)$_4$]$^{2+}$ units with two sulfate O atoms completing the remaining sites in an elongated octahedral coordination sphere. The lattice consists of a hydrogen-bonded assembly which incorporates the non-coordinated H$_2$O molecules. The pentahydrate loses water in stages on heating (equation 21.99) and finally forms the white, hygroscopic anhydrous CuSO$_4$.

$$CuSO_4 \cdot 5H_2O \xrightarrow[-2H_2O]{300\,K} CuSO_4 \cdot 3H_2O$$

$$\xrightarrow[-2H_2O]{380\,K} CuSO_4 \cdot H_2O \xrightarrow[-H_2O]{520\,K} CuSO_4 \qquad (21.99)$$

Copper(II) sulfate and nitrate are commercially available and, in addition to uses as precursors in Cu(II) chemistry, they are used as fungicides, e.g. *Bordeaux mixture* contains CuSO$_4$ and Ca(OH)$_2$ and when added to water forms a basic copper(II) sulfate which acts as the antifungal agent. Copper(II) nitrate is widely used in the dyeing and printing industries. It forms hydrates Cu(NO$_3$)$_2$·xH$_2$O where $x = 2.5$, 3 or 6; the blue hexahydrate readily loses water at 300 K to give green Cu(NO$_3$)$_2$·3H$_2$O. Anhydrous Cu(NO$_3$)$_2$ is made from Cu and N$_2$O$_4$ (*equation 8.75*) followed by decomposition of [NO][Cu(NO$_3$)$_3$] so formed. The solid state structure of α-Cu(NO$_3$)$_2$ consists of Cu(II) centres linked into an infinite lattice by bridging [NO$_3$]$^-$ ligands (**21.46**). At 423 K, the solid volatilizes *in vacuo* giving molecular Cu(NO$_3$)$_2$ (**21.47**).

(21.46)

(21.47)

The salt Cu(O$_2$CMe)$_2$·H$_2$O is dimeric and is structurally similar to [Cr$_2$(H$_2$O)$_2$(μ-O$_2$CMe)$_4$] (see *Figure 21.12* for structure type) but lacks the strong metal–metal bonding. The distance between the two Cu centres of 264 pm is greater than in the bulk metal (256 pm). The magnetic moment of 1.4 μ_B per Cu(II) centre (i.e. less than μ(spin-only) of 1.73 μ_B) suggests that in [Cu$_2$(H$_2$O)$_2$(μ-O$_2$CMe)$_4$] there is only weak antiferromagnetic coupling between the unpaired electrons. On cooling, the magnetic moment decreases. These observations can be explained in terms of the two unpaired electrons giving a singlet ground state ($S = 0$) and a low-lying triplet excited state ($S = 1$) which is thermally populated at 298 K but which becomes less populated as the temperature is lowered (see *Box 20.5* for singlet and triplet states).

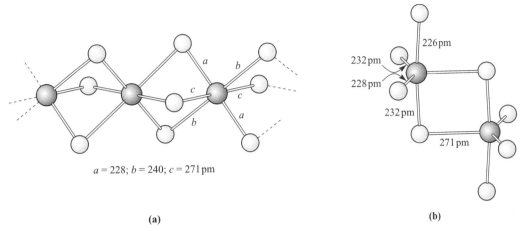

$a = 228; b = 240; c = 271$ pm

(a)

(b)

Fig. 21.27 The structures (X-ray diffraction) of (a) polymeric $[CuCl_3]_n^{n-}$ in the salt $[Me_3NH]_3[CuCl_3][CuCl_4]$; the $[CuCl_4]^{2-}$ ion in this salt is tetrahedral [R.M. Clay *et al.* (1973) *J. Chem. Soc., Dalton Trans.*, p. 595], and (b) the $[Cu_2Cl_8]^{4-}$ ion in the salt $[Rh(en)_3]_2[Cu_2Cl_8]Cl_2 \cdot 2H_2O$ [S.K. Hoffmann *et al.* (1985) *Inorg. Chem.*, vol. 24, p. 1194]. Colour code: Cu, blue; Cl, pale grey.

Vast numbers of copper(II) complexes are known and this discussion covers only simple species; Jahn–Teller distortions are generally observed (d^9 configuration). Halo complexes include $[CuCl_3]^-$, $[CuCl_4]^{2-}$, $[CuCl_5]^{3-}$ but the solid state structures of species possessing these stoichiometries are highly dependent on the counter-ions. For example, $[Ph_4P][CuCl_3]$ contains dimers (**21.48**), whereas $K[CuCl_3]$ and $[Me_3NH]_3[CuCl_3][CuCl_4]$ contain chains of distorted, face-sharing octahedra (Figure 21.27a). The latter salt also contains discrete tetrahedral $[CuCl_4]^{2-}$ ions; $[PhCH_2CH_2NH_2Me]_2[CuCl_4]$ crystallizes in two forms, one with distorted tetrahedral and the other with square planar $[CuCl_4]^{2-}$ ions. The salt $[NH_4]_2[CuCl_4]$ has a polymeric structure containing distorted octahedral Cu(II) centres. Dimeric $[Cu_2Cl_8]^{4-}$ (with edge-sharing trigonal bipyramidal Cu(II) centres) may be stabilized by very bulky cations, e.g. $[M(en)_3]_2[Cu_2Cl_8]Cl_2$ (M = Co, Rh or Ir, Figure 21.27b). The $[CuCl_5]^{3-}$ ion is trigonal bipyramidal in the Cs^+ and $[Me_3NH]^+$ salts, but in $[\mathbf{21.49}][CuCl_5]$, it is square-based pyramidal.

Complexes containing *N*- and *O*-donor ligands are very common, and coordination numbers of 4, 5 and 6 predominate. We have already mentioned the aqua species $[Cu(H_2O)_6]^{2+}$ and $[Cu(H_2O)_4]^{2+}$. When NH_3 is added to aqueous Cu^{2+}, only four aqua ligands in $[Cu(H_2O)_6]^{2+}$ are replaced (see *Section 20.10*), but salts of $[Cu(NH_3)_6]^{2+}$ can be made in liquid NH_3; $[Cu(en)_3]^{2+}$ is formed in very concentrated aqueous solutions of 1,2-diaminoethane. Deep blue aqueous $[Cu(NH_3)_4](OH)_2$ (formed when $Cu(OH)_2$ is dissolved in aqueous NH_3) has the remarkable property of

dissolving cellulose, and if the resulting solution is squirted into acid, the man-made fibre *rayon* is produced as cellulose is precipitated; the precipitation is also used to waterproof canvas. Further examples of complexes with *N*- and *O*-donor ligands are:

- tetrahedral (flattened): $[Cu(NCS-N)_4]^{2-}$; $[CuCl_2(Meim)_2]$ (Figure 21.28a);
- square planar: $[Cu(ox)_2]^{2-}$; *cis*- and *trans*-$[Cu(H_2NCH_2CO_2)_2]$; $[Cu(en)(NO_3-O)_2]$;
- trigonal bipyramidal: $[Cu(NO_3-O)_2(py)_3]$ (equatorial nitrates); $[Cu(CN)\{N(CH_2CH_2NH_2)_3\}]^+$ (axial cyanide);
- square-based pyramidal: $[Cu(NCS-N)(\mathbf{21.50})]^+$ (ligand **21.50** is tetradentate in the basal sites); $[Cu(H_2O)(phen)(\mathbf{21.51})]$ (apical H_2O), $[CuCl_2(H_2O)_2(MeOH)]$ (apical MeOH, *trans* Cl in the basal sites);
- octahedral: $[Cu(HOCH_2CH_2OH)_3]^{2+}$; $[Cu(bpy)_3]^{2+}$; $[Cu(phen)_3]^{2+}$; *trans*-$[CuCl(H_2O)(en)_2]^+$; *trans*-$[Cu(BF_4)_2(en)_2]$.

Jahn–Teller distortions are apparent in many complexes; in $[Cu(bpy)_3]^{2+}$, the distortion is particularly severe with equatorial Cu–N bonds of 203 pm, and axial distances of 223 and 245 pm. The complex *trans*-$[Cu(BF_4)_2(en)_2]$ illustrates the ability of $[BF_4]^-$ to act as a monodentate ligand; the long Cu–F bonds (256 pm) indicate rather weak Cu–F interactions. In Section 21.11, we described the structure of $[Ni(Hdmg)_2]$; $[Cu(Hdmg)_2]$ also exhibits hydrogen bonding between the ligands but, in the solid state, molecules are associated in *pairs* with the coordination sphere being square-based pyramidal (Figure 21.28b).

(21.48)

(21.49)

(21.50)

(21.51)

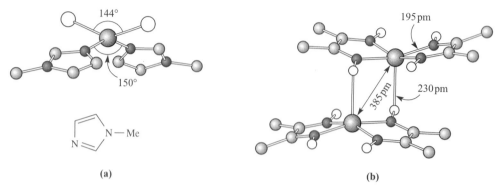

Fig. 21.28 (a) The flattened tetrahedral structure of [CuCl₂(Meim)₂] (determined by X-ray diffraction) and a schematic representation of the *N*-methylimidazole (Meim) ligand [J.A.C. van Ooijen *et al.* (1979) *J. Chem. Soc., Dalton Trans.*, p. 1183]; (b) [Cu(Hdmg)₂] forms dimers in the solid state, in contrast to [Ni(Hdmg)₂], (Figure 21.25); structure determined by X-ray diffraction [A. Vaciago *et al.* (1970) *J. Chem. Soc. A*, p. 218]. Nitrogen atoms are shown in black, Cu atoms in blue, Cl and O atoms in white.

A practical application of the coordination of *N,O*-donors to Cu(II) is the *biuret test* for peptides and proteins. Compounds containing peptide linkages form a violet complex when treated in NaOH solution with a few drops of aqueous CuSO₄. The general form of the complex can be represented by that of **21.52** in which the ligand is the doubly deprotonated form of biuret, H₂NC(O)NHC(O)NH₂.

(21.52)

When a Cu(II) salt is treated with excess KCN at room temperature, cyanogen is evolved and the copper reduced (equation 21.100). However, in aqueous methanol at low temperatures, violet square planar [Cu(CN)₄]²⁻ forms.

$$2Cu^{2+} + 4[CN]^- \longrightarrow 2CuCN(s) + C_2N_2 \qquad (21.100)$$

Some copper-containing complexes are studied as models for bioinorganic systems (see *Chapter 28*).

Copper(I)

Cuprous is the old name for copper(I). The Cu⁺ ion has a d^{10} configuration and salts are diamagnetic and colourless except when the counter-ion is coloured or when charge-transfer absorptions occur in the visible region, e.g. in red Cu₂O.

Copper(I) fluoride is not known; CuCl, CuBr and CuI are white solids (CuCl turns green in moist air) and are made by reduction of a Cu(II) salt in the presence of halide ions, e.g. CuBr forms when SO₂ is bubbled through an aqueous solution of CuSO₄ and KBr. Copper(I) chloride has a zinc blende structure (see *Figure 5.18*). The γ-forms of CuBr and CuI adopt the zinc blende lattice but convert to

the β-forms (wurtzite structure, *Figure 5.20*) at 660 and 690 K respectively. Values of K_{sp}(298 K) for CuCl, CuBr and CuI are 1.72×10^{-7}, 6.27×10^{-9} and 1.27×10^{-12}. Copper(I) iodide precipitates when any Cu(II) salt in added to KI solution (equation 21.101).

$$2Cu^{2+} + 4I^- \longrightarrow 2CuI + I_2 \qquad (21.101)$$

Anions and ligands available in solution strongly influence the relative stabilities of Cu(I) and Cu(II) species. The very low solubility of CuI is crucial to reaction 21.101 which occurs despite the fact that the $E°$ values of the Cu²⁺/Cu⁺ and I₂/I⁻ couples are +0.15 and +0.54 V respectively. However, in the presence of 1,2-diaminoethane or tartrate, which form stable complexes with Cu²⁺, I₂ oxidizes CuI.

Copper(I) hydride is obtained by reduction of Cu(II) salts with H₃PO₂ and crystallizes with the wurtzite lattice. It decomposes when treated with acids, liberating H₂.

Red copper(I) oxide may be made by oxidation of Cu (reaction 21.93), but is more readily obtained by reduction of Cu(II) compounds in alkaline media. When Fehling's solution (Cu²⁺ in aqueous alkaline sodium tartrate) is added to a reducing sugar such as glucose, Cu₂O precipitates; this is a qualitative test for reducing sugars. The solid state structure of Cu₂O is related to that of β-cristobalite (SiO₂, *Figure 5.19c*) but with Cu(I) in linear sites and O²⁻ in tetrahedral sites. Because the Cu₂O framework is particularly open, the crystal consists of *two* interpenetrating frameworks, and the Cu₂O, *cuprite*, lattice is a structural prototype. Copper(I) oxide is used as a red pigment in ceramics, porcelain glazes and glasses; it has fungicidal properties and is added to certain paints as an antifouling agent. It is insoluble in water, but dissolves in aqueous NH₃ to give colourless [Cu(NH₃)₂]⁺ (**21.53**); the solution readily absorbs O₂ and turns blue as [Cu(NH₃)₄]²⁺ forms. In acidic solutions, Cu₂O disproportionates (equation 21.102).

$$Cu_2O + H_2SO_4 \longrightarrow CuSO_4 + Cu + H_2O \qquad (21.102)$$

$$\left[H_3N \longrightarrow Cu \longleftarrow NH_3 \right]^+$$

(21.53)

Complex **21.53** illustrates a linear environment for Cu(I); the most common geometry is tetrahedral, and 3-coordinate species also occur. Halide complexes exhibit great structural diversity and the identity of the cation is often crucial in determining the structure of the anion. For example, $[CuCl_2]^-$ (formed when CuCl dissolves in concentrated HCl) may occur as discrete, linear anions (**21.54**) or as a polymer with tetrahedral Cu(I) centres (**21.55**). Trigonal planar $[CuCl_3]^{2-}$ has been isolated, e.g. in $[Me_4P]_2[CuCl_3]$, but association into discrete, halo-bridged anions is also possible, e.g. $[Cu_2I_4]^{2-}$ (**21.56**), $[Cu_2Br_5]^{3-}$ (Figure 21.29a) and $[Cu_4Br_6]^{2-}$ (Figure 21.29b). Complexation between Cu(I) and $[CN]^-$ can lead to $[Cu(CN)_2]^-$ (polymeric **21.57** as in the K^+ salt), $[Cu(CN)_3]^{2-}$ (trigonal planar, **19.2**) or $[Cu(CN)_4]^{3-}$ (tetrahedral).

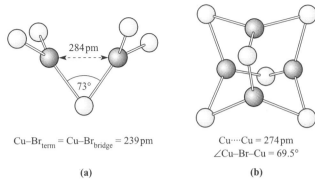

Cu–Br_{term} = Cu–Br_{bridge} = 239 pm

(a)

Cu····Cu = 274 pm
∠Cu–Br–Cu = 69.5°

(b)

Fig. 21.29 The structures (X-ray diffraction) of (a) $[Cu_2Br_5]^{3-}$ in the $[Me_4N]^+$ salt [M. Asplund *et al.* (1985) *Acta Chem. Scand., Ser. A*, vol. 39, p. 47] and (b) $[Cu_4Br_6]^{2-}$ in the $[^nPr_4N]^+$ salt [M. Asplund *et al.* (1984) *Acta Chem. Scand. Ser. A*, vol. 38, p. 725]. In both, the Cu(I) centres are in trigonal planar environments and in $[Cu_4Br_6]^{2-}$, the copper atoms are in a tetrahedral arrangement; the Cu····Cu distances are longer than in the bulk metal. Colour code: Cu, blue; Br, grey.

$$\left[Cl\!\!-\!\!-\!\!-\!\!Cu\!\!-\!\!-\!\!-\!\!Cl \right]^-$$

Cu—Cl = 209 pm

(21.54)

Cu—Cl = 235 pm

(21.55)

$$(21.56)$$

(21.57)

Fig. 21.30 The structure of $[Cu(\mathbf{21.33})]^+$ (ligand **21.33** is an S_6-macrocycle) determined by X-ray diffraction for the $[BF_4]^-$ salt; the Cu^+ is in a distorted tetrahedral environment [J.R. Hartman *et al.* (1986) *J. Am. Chem. Soc.*, vol. 108, p. 1202]. Hydrogen atoms are omitted; colour code: Cu, blue; S, pale grey; C, grey.

Copper(I) is a soft metal centre (*Table 6.9*) and tends to interact with soft donor atoms such as S and P, although complex formation with *O*- and *N*-donor ligands is well documented. Many complexes with *S*-donor ligands are known, and the propensity of sulfur to form bridges leads to many multinuclear complexes, e.g. $[(S_6)Cu(\mu\text{-}S_8)Cu(S_6)]^{4-}$ (*Figure 15.9*), $[Cu_4(SPh)_6]^{2-}$ (which is structurally related to $[Cu_4Br_6]^{2-}$ with $[SPh]^-$ replacing Br^- bridges), and $[\{Cu(S_2O_3)_2\}_n]$ (structurally related to **21.55** with *S*-bonded thiosulfates replacing Cl^- bridges). We have seen several times in this chapter how macrocyclic ligands may impose unusual coordination numbers on metal ions, or, if the ring is large enough, may wrap around a metal ion, e.g. in

$[Co(\mathbf{21.33})]^{2+}$. In $[Cu(\mathbf{21.33})]^+$ (Figure 21.30), the preference for the Cu^+ ion to be tetrahedrally coordinated means that it interacts with only four of the six donor atoms of the macrocycle.

21.13 Group 12: zinc

The metal

Zinc is not attacked by air or water at room temperature, but the hot metal burns in air and decomposes in steam, forming ZnO. Zinc is much more reactive than Cu (compare equations 21.103 and 21.91), liberating H_2 from dilute mineral acids and from alkalis (equation 21.104). With hot concentrated sulfuric acid, reaction 21.105 occurs; the products of reactions with HNO_3 depend on temperature and acid concentration. On heating, Zn reacts with all the halogens to give ZnX_2, and combines with elemental S and P.

$$Zn^{2+} + 2e^- \rightleftharpoons Zn \qquad E^o = -0.76\,V \qquad (21.103)$$

$$Zn + 2NaOH + 2H_2O \longrightarrow Na_2[Zn(OH)_4] + H_2 \qquad (21.104)$$

$$Zn + 2H_2SO_4 \longrightarrow ZnSO_4 + SO_2 + 2H_2O \qquad (21.105)$$
$$\text{hot, conc}$$

The first (Sc) and last (Zn) members of the first row of the *d*-block exhibit a more restricted range of oxidation states than the other metals, and the chemistry of Zn is confined to that of Zn(II). The $[Zn_2]^{2+}$ ion (analogues of which are well established for the heavier group 10 metals) has only been established in a yellow diamagnetic glass obtained by cooling a solution of metallic Zn in molten $ZnCl_2$; it rapidly disproportionates (equation 21.106).

$$[Zn_2]^{2+} \longrightarrow Zn^{2+} + Zn \qquad (21.106)$$

Since the electronic configuration of Zn^{2+} is d^{10}, compounds are colourless and diamagnetic. There is no LFSE associated with the d^{10} ion and, as the discussion below shows, no particular geometry is preferred for Zn^{2+}. There are some similarities with Mg, and many compounds of Zn are isomorphous with their Mg analogues.

Zinc(II)

Binary halides are best made by action of HF, HCl, Br_2 or I_2 on hot Zn; ZnF_2 is also prepared by thermal decomposition of $Zn(BF_4)_2$. The vapours of the halides contain linear molecules. Solid ZnF_2 adopts a rutile structure (*Figure 5.21*) and has a high lattice energy and melting point. Evidence for significant covalent character is apparent in the structures and properties of $ZnCl_2$, $ZnBr_2$ and ZnI_2 which possess layer lattices, have lower melting points than ZnF_2 (Figure 21.31) and are soluble in a range of organic solvents. The water solubility of ZnF_2 is low, but $ZnCl_2$, $ZnBr_2$ and ZnI_2 are highly soluble. Uses of $ZnCl_2$ are varied, e.g. in some fireproofings, wood preservation, combined with NH_4Cl as a soldering flux, an astringent and in deodorants.

Zinc hydride is made by reaction 21.107 (or from LiH and $ZnBr_2$) and is a fairly stable solid at 298 K.

$$ZnI_2 + 2NaH \xrightarrow{\text{THF}} ZnH_2 + 2NaI \qquad (21.107)$$

Zinc is of great commercial significance and ZnO (made from Zn and O_2) is its most important compound (see *Section 21.2*). It is a white solid with the wurtzite structure (*Figure 5.20*) at 298 K. It turns yellow on heating and in this form is a semiconductor owing to loss of oxygen and production of some interstitial Zn atoms. Zinc oxide is amphoteric, dissolving in acids to give solutions containing $[Zn(H_2O)_6]^{2+}$ or derivatives thereof (some anions coordinate to Zn^{2+}), but hydrolysis of $[Zn(H_2O)_6]^{2+}$ occurs to give various solution species resulting from H^+ loss. In alkalis, ZnO forms zincates such as $[Zn(OH)_4]^{2-}$ (**21.58**). This ion also forms when $Zn(OH)_2$ dissolves in aqueous alkalis. Zinc

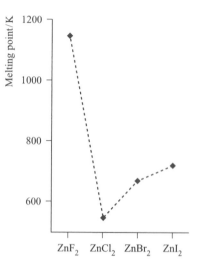

Fig. 21.31 Trend in melting points of the zinc halides.

hydroxide is water-insoluble; there are five polymorphs of which ε-$Zn(OH)_2$ (distorted β-cristobalite lattice, *Figure 5.19c*) is thermodynamically the most stable.

$$\left[\begin{array}{c} OH \\ | \\ HO \diagdown \underset{Zn}{} \overset{\cdots}{\cdots} OH \\ | \\ OH \end{array} \right]^{2-}$$

(21.58)

Zinc sulfide occurs naturally as the minerals *zinc blende* and, more rarely, *wurtzite*; these are structural prototypes (see *Section 5.11*). It is a light-sensitive white solid and, on exposure to cathode- or X-rays, it luminesces or fluoresces and is used in fluorescent paints and radar screens. Adding Cu to ZnO results in a green phosphorescence after exposure to light, and other colour variations are achieved by using different additives. The conversion of ZnS to ZnO by roasting in air is the commercial method of producing the oxide.

Other Zn(II) compounds that are commercially available include the carbonate, sulfate and nitrate. The sulfate is very soluble in water; crystals of $ZnSO_4 \cdot 7H_2O$ form on evaporating solutions from reactions of Zn, ZnO, $Zn(OH)_2$ or $ZnCO_3$ with aqueous H_2SO_4. Dehydration initially occurs on heating, followed by decomposition (equation 21.108).

$$ZnSO_4 \cdot 7H_2O \xrightarrow[-7H_2O]{520\,K} ZnSO_4 \xrightarrow{1020\,K} ZnO + SO_3 \qquad (21.108)$$

Insoluble $ZnCO_3$ occurs naturally as *smithsonite*, but the mineral tends to be coloured owing to the presence of, for example, Fe(II). The carbonate is usually purchased as the basic salt $ZnCO_3 \cdot 2Zn(OH)_2 \cdot xH_2O$ and is used in calamine lotion.

Zinc nitrate can be obtained as one of several hydrates, of which $Zn(NO_3)_2 \cdot 6H_2O$ is the most common; anhydrous $Zn(NO_3)_2$ is made from Zn and N_2O_4 since heating the hydrates yields hydroxy salts. The hexahydrates of $Zn(NO_3)_2$ and $Zn(ClO_4)_2$ contain octahedral $[Zn(H_2O)_6]^{2+}$ in the solid state. Similarly, it is possible to isolate salts containing $[Zn(NH_3)_6]^{2+}$ from reactions done in liquid NH_3, e.g. $ZnCl_2 \cdot 6NH_3$. However, in aqueous solution, $[Zn(NH_3)_6]^{2+}$ exists in equilibrium with tetrahedral $[Zn(NH_3)_4]^{2+}$. Equation 8.25 showed the formation of $[Zn(NH_2)_4]^{2-}$. Basic zinc acetate $[Zn_4(\mu_4\text{-}O)(\mu\text{-}O_2CMe)_6]$ is isostructural with its Be(II) analogue (*Figure 11.6*), but is more readily hydrolyzed in water. Another salt of interest is $Zn(acac)_2 \cdot H_2O$ (**21.59**) in which the coordination of Zn^{2+} is square-based pyramidal.

(21.59)

Our discussion of Zn(II) compounds has introduced complexes including $[Zn(H_2O)_6]^{2+}$, $[Zn(NH_3)_6]^{2+}$, $[Zn(NH_3)_4]^{2+}$, $[Zn(OH)_4]^{2-}$ and $[Zn(acac)_2(H_2O)]$, exemplifying octahedral, tetrahedral and square-based pyramidal coordination. Large numbers of Zn(II) complexes are known (some interest arises from developing models for Zn-containing bioinorganic systems, see *Chapter 28*) and coordination numbers of 4 to 6 are the most common. Zinc(II) is a borderline hard/soft ion and readily complexes with ligands containing a range of donor atoms, e.g. hard *N*- and *O*- and soft *S*-donors.

(21.60)   (21.61)

Tetrahedral $[ZnCl_4]^{2-}$ and $[ZnBr_4]^{2-}$ can be formed from $ZnCl_2$ and $ZnBr_2$ and many salts are known; salts of $[ZnI_4]^{2-}$ are stabilized using large cations. Crystallographic data for '$[ZnCl_3]^-$' salts usually reveal the presence of $[Zn_2Cl_6]^{2-}$, (**21.60**), and in coordinating solvents, tetrahedral $[ZnCl_3(solv)]^-$ is present; salts such as $K[ZnCl_3] \cdot H_2O$ contain $[ZnCl_3(H_2O)]^-$ (**21.61**) in the solid state. A similar picture is true for '$[ZnBr_3]^-$' and '$[ZnI_3]^-$' salts; both $[Zn_2Br_6]^{2-}$ and $[Zn_2I_6]^{2-}$ have been confirmed in the solid state.

Although the structure of $Zn(CN)_2$ is an *anticuprite* lattice (related to *cuprite*, *Section 21.12*, in the same way as *fluorite* and *antifluorite*, *Section 5.11*, are related) with $[CN]^-$ groups bridging between tetrahedral Zn(II) centres, $[Zn(CN)_4]^{2-}$ exists as discrete tetrahedral ions, as do $[Zn(N_3)_4]^{2-}$ and $[Zn(NCS\text{-}N)_4]^{2-}$. Just as it is possible to isolate both $[Zn(NH_3)_4]^{2+}$ and $[Zn(NH_3)_6]^{2+}$, pairs of tetrahedral

$[ZnL_2]^{2+}$ and octahedral $[ZnL_3]^{2+}$ complexes (L = en, bpy, phen) are also known.

(21.62)

Examples of high coordination numbers for Zn^{2+} are rare, but include pentagonal bipyramidal $[Zn(15\text{-crown-}5)(H_2O)_2]^{2+}$ (**21.62**), and dodecahedral $[Zn(NO_3)_4]^{2-}$ (structurally similar to $[Co(NO_3)_4]^{2-}$, Figure 21.23c).

Summary

This chapter has given an introduction to the descriptive chemistry of the first row *d*-block metals. Some of the general principles discussed in Chapters 19 and 20 have now been seen in practice, having been used to account for many structural, spectroscopic and magnetic properties. The range of oxidation states exhibited by the metals reaches a maximum with Mn. In aqueous solution, pH and ligand concentrations affect the relative stabilities of oxidation states to marked extents. The wealth of modern structural data has invalidated earlier simpler views on the correlation of electronic structure, coordination number and coordination geometry. A bewildering number of 'special cases' now exists, and subtle factors often determine the observed coordination geometry.

The following terms were introduced in this chapter.
Do you know what they mean?

❑ isopolyanion
❑ polyoxometallate
❑ homopolyanion
❑ heteropolyanion

Suggested further reading

See also further reading suggested for Chapters 19 and 20.
F.A. Cotton, G. Wilkinson, M. Bochmann and C. Murillo (1999) *Advanced Inorganic Chemistry*, 6th edn, Wiley Interscience, New York – One of the best detailed accounts of the chemistry of the *d*-block metals.
J. Emsley (1998) *The Elements*, 3rd edn, Oxford University Press, Oxford – An invaluable source of data for the elements.
N.N. Greenwood and A. Earnshaw (1997) *Chemistry of the Elements*, 2nd edn, Butterworth-Heinemann, Oxford – A very good account including historical, technological and structural aspects; the metals in each triad are treated together.
R.B. King (ed.) (1994) *Encyclopedia of Inorganic Chemistry*, Wiley, Chichester – Contains an article on the inorganic

and coordination chemistry of each metal with numerous literature citations.

J. McCleverty (1999) *Chemistry of the First-row Transition Metals*, Oxford University Press, Oxford. – An introductory text dealing with the metals Ti to Cu.

M.T. Pope (1994) 'Polyoxoanions' in *Encyclopedia of Inorganic Chemistry*, ed. R.B. King, Wiley, Chichester, vol. 6, p. 3361.

J. Silver (ed.) (1993) *Chemistry of Iron*, Blackie, London – A series of articles covering different facets of the chemistry of iron.

A.F. Wells (1984) *Structural Inorganic Chemistry*, 5th edn, Clarendon Press, Oxford – An excellent source for detailed structural information of, in particular, binary compounds.

Problems

21.1 Write out, in sequence, the first row *d*-block elements and give the valence electronic configuration of each metal and of its M^{2+} ion.

21.2 Comment on the variation in oxidation states of the first row metals. Why are Sc and Zn not classed as *transition metals*?

21.3 In the complex $[Ti(BH_4)_3(MeOCH_2CH_2OMe)]$, the Ti(III) centre is 8-coordinate. Suggest modes of coordination for the ligands.

21.4 Comment on each of the following observations. (a) Li_2TiO_3 forms a continuous range of solid solutions with MgO. (b) When $TiCl_3$ is heated with concentrated aqueous NaOH, H_2 is evolved.

21.5 An acidified solution of $0.1000\,mol\,dm^{-3}$ ammonium vanadate ($25.00\,cm^3$) was reduced by SO_2 and, after boiling off excess reductant, the blue solution remaining was found to require addition of $25.00\,cm^3$ $0.0200\,mol\,dm^{-3}$ $KMnO_4$ to give a pink colour to the solution. Another $25.00\,cm^3$ portion of the vanadate solution was shaken with Zn amalgam and then immediately poured into excess of the ammonium vanadate solution; on titration of the resulting solution with the $KMnO_4$ solution, $74.5\,cm^3$ of the latter were required. Deduce what happened in these experiments.

21.6 Give equations to describe what happens to VBr_3 on heating.

21.7 The magnetic moment of $[NH_4]V(SO_4)_2 \cdot 12H_2O$ is $2.8\,\mu_B$ and the electronic spectrum of an aqueous solution of contains absorptions at 17 800, 25 700 and $34\,500\,cm^{-1}$. Explain these observations.

21.8 Suggest the formula and structure of the mononuclear complex formed between Cr^{3+} and ligand **21.63**. Comment on possible isomerism.

(21.63)

21.9 Use data from Appendix 11 to predict qualitatively the outcome of the following experiment at 298 K: Cr is dissolved in excess of molar $HClO_4$ and the solution is shaken in air.

21.10 Figure 21.32 shows the change in concentration of $[MnO_4]^-$ with time during a reaction with acidified oxalate ions. (a) Suggest a method of monitoring the reaction. (b) Explain the shape of the curve.

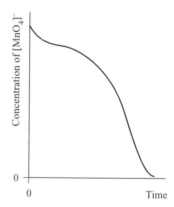

Fig. 21.32 Figure for problem 21.10.

21.11 Comment on the modes of bonding of the ligands in the Mn(II) complexes listed at the end of Section 21.8, drawing attention to any conformational restrictions.

21.12 How would you (a) distinguish between the formulations $Cu^{II}Fe^{II}S_2$ and $Cu^IFe^{III}S_2$ for the mineral *chalcopyrites*, (b) show that Fe^{3+} is a hard cation, and (c) show that the blue compound precipitated when a solution of $[MnO_4]^-$ in concentrated aqueous KOH is reduced by $[CN]^-$ contains Mn(V)?

21.13 Give equations for the following reactions: (a) heating Fe with Cl_2; (b) heating Fe with I_2; (c) solid $FeSO_4$ with conc H_2SO_4; (d) aqueous Fe^{3+} with $[SCN]^-$; (e) aqueous Fe^{3+} with $K_2C_2O_4$; (f) FeO with dil. H_2SO_4; (g) aqueous $FeSO_4$ and NaOH.

21.14 How would you attempt to (a) estimate the crystal field stabilization energy of FeF_2, and (b) determine the overall stability constant of $[Co(NH_3)_6]^{3+}$ in aqueous solution given that the overall formation constant for

$[Co(NH_3)_6]^{2+}$ is 10^5, and:

$$Co^{3+}(aq) + e^- \rightleftharpoons Co^{2+}(aq) \qquad E^\circ = +1.92 \text{ V}$$

$$[Co(NH_3)_6]^{3+}(aq) + e^- \rightleftharpoons [Co(NH_3)_6]^{2+}(aq)$$
$$E^\circ = +0.11 \text{ V}$$

21.15 Suggest why Co_3O_4 adopts a normal rather than inverse spinel structure.

21.16 Give explanations for the following observations. (a) The complex $[Co(en)_2Cl_2]_2[CoCl_4]$ has a room temperature magnetic moment of $3.71\ \mu_{\text{eff}}$. (b) The room temperature magnetic moment of $[CoI_4]^{2-}$ (e.g. $5.01\ \mu_B$ for the $[Et_4N]^+$ salt) is larger than that of salts of $[CoCl_4]^{2-}$.

21.17 (a) When $[CN]^-$ is added to aqueous Ni^{2+} ions, a green precipitate forms; if excess KCN is added, the precipitate dissolves to give a yellow solution and at high concentrations of $[CN]^-$, the solution becomes red. Suggest an explanation for these observations. (b) If the yellow compound from part (a) is isolated and reacted with Na in liquid NH_3, a red, air-sensitive, diamagnetic product can be isolated. Suggest its identity.

21.18 Treatment of an aqueous solution of $NiCl_2$ with $H_2NCHPhCHPhNH_2$ gives a blue complex ($\mu_{\text{eff}} = 3.30\ \mu_B$) which loses H_2O on heating to form a yellow, diamagnetic compound. Suggest explanations for these observations and comment on possible isomerism in the yellow species.

21.19 Give equations for the following reactions: (a) aqueous NaOH with $CuSO_4$; (b) CuO with Cu in concentrated HCl at reflux; (c) Cu with concentrated HNO_3; (d) addition of aqueous NH_3 to a precipitate of $Cu(OH)_2$; (e) $ZnSO_4$ with aqueous NaOH followed by addition of excess NaOH; (f) ZnS with dil. HCl.

21.20 (a) Compare the solid state structures of $[M(Hdmg)_2]$ for M = Ni and Cu and comment on the fact that $[Cu(Hdmg)_2]$ is more soluble in water than is $[Ni(Hdmg)_2]$. (b) Suggest the likely structural features of $[Pd(Hdmg)_2]$.

21.21 Copper(II) chloride is not completely reduced by SO_2 in concentrated HCl solution. Suggest an explanation for this observation and state how you would try to establish if the explanation is correct.

21.22 When the ligands do not sterically control the coordination geometry, do 4-coordinate complexes of (a) Pd(II), (b) Cu(I) and (c) Zn(II) prefer to be square planar or tetrahedral? Explain your answer. In the absence of crystallographic data, how could you distinguish between a square planar or tetrahedral structure for a Ni(II) complex?

21.23 Write down formulae for the following ions: (a) manganate(VII); (b) manganate(VI); (c) dichromate(VI); (d) vanadyl; (e) vanadate (*ortho* and *meta*); (f) hexacyanoferrate(III). Give an alternative name for manganate(VII).

21.24 Give a brief account of the variation in properties of binary oxides of the first row *d*-block metals on going from Sc to Zn.

21.25 Give an overview of the formation of halo complexes of type $[MX_n]^{m-}$ by the first row *d*-block metal ions, noting in particular whether discrete ions are present in the solid state.

21.26 When iron(II) oxalate (oxalate $= ox^{2-}$) is treated with H_2O_2, H_2ox and K_2ox, a green compound **X** is obtained. **X** reacts with aqueous NaOH to give hydrated Fe_2O_3, and is decomposed by light with production of iron(II) oxalate, K_2ox and CO_2. Analysis of **X** shows it contains 11.4% Fe and 53.7% ox^{2-}. Deduce the formula of **X** and write equations for its reaction with alkali and its photochemical decomposition. State, with reasons, whether you would expect **X** to exhibit optical isomerism.

21.27 Dimethyl sulfoxide (DMSO) reacts with cobalt(II) perchlorate in EtOH to give a pink compound **A** which is a 1:2 electrolyte and has a magnetic moment of $4.9\ \mu_B$. Cobalt(II) chloride also reacts with DMSO, but in this case the dark-blue product, **B**, is a 1:1 electrolyte, and the magnetic moment of **B** is $4.6\ \mu_B$ per Co centre. Suggest a formula and structure for **A** and **B**.

21.28 When H_2S is passed into a solution of copper(II) sulfate acidified with H_2SO_4, copper(II) sulfide precipitates. When concentrated H_2SO_4 is heated with metallic Cu, the principal sulfur-containing product is SO_2 but a residue of copper(II) sulfide is also formed. Account for these reactions.

Chapter 22

d-Block metal chemistry: the second and third row metals

1–2	3	4	5	6	7	8	9	10	11	12	13–18
s-block											*p*-block
	Sc	Ti	V	Cr	Mn	Fe	Co	Ni	Cu	Zn	
	Y	Zr	Nb	Mo	Tc	Ru	Rh	Pd	Ag	Cd	
	La	Hf	Ta	W	Re	Os	Ir	Pt	Au	Hg	

22.1 Introduction

Chapter 21 dealt with descriptive chemistry of the first row *d*-block metals and in this chapter, the theme continues with the focus being the second and third row metals (the *heavier metals*). Reasons for discussing the lighter and heavier metals separately were given in Section 21.1.

Lanthanum, La, is commonly classified with the lanthanoids (see *Figure 1.13*) even though 'lanthanoid' means 'like lanthanum' and La is strictly a group 3 metal. Because of the chemical similarity of La to the elements Ce−Lu, we consider them together in Chapter 24; our only mention of La in this chapter is its occurrence.

22.2 Occurrence, extraction and uses

Yttrium and *lanthanum* are similar to the lanthanoids and occur with them in nature. The major yttrium and lanthanum ores are *monazite* (a mixed metal phosphate, $(Ce,La,Nd,Pr,Th,Y...)PO_4$) and *bastnäsite* $((Ce,La,Y...)CO_3F)$; their composition varies, e.g. an 'yttrium-rich' mineral might contain ≤1% Y, a 'lanthanum-rich' one up to 35% La. The extraction of yttrium

involves conversion to YF_3 or YCl_3 followed by reduction with Ca or K respectively; the separation of lanthanoid metals is described in *Section 24.5*. Yttrium is used in the manufacture of phosphors for television tubes (as Y_2O_3 and YVO_4) and corrosion-resistant alloys, and in the formation of yttrium garnets for microwave filters and synthetic gemstones (yttrium aluminium garnets, YAG).

Zirconium is the next most abundant *d*-block metal in the Earth's crust after Fe, Ti and Mn, and is present to quite a large extent in lunar rock samples collected in the Apollo missions. Zirconium and *hafnium* occur naturally together and are hard to separate; Hf is rarer than Zr, 5.3 and 190 ppm, respectively, of the Earth's crust. The main ores are *baddeleyite* (ZrO_2), *zircon* (($Zr,Hf)SiO_4$, <2% Hf) and *alvite* (($Zr,Hf)SiO_4 \cdot xH_2O$, <2% Hf). Extraction of Zr involves reduction of ZrO_2 by Ca, or conversion of ZrO_2 to K_2ZrF_6 (by treatment with K_2SiF_6) followed by reduction; both Zr and Hf can be produced from zircon by reaction sequence 22.1. The mixture of metals so-obtained is used for strengthening steel.

$$MO_2 \xrightarrow{CCl_4,\ 770\,K} MCl_4 \xrightarrow{Mg\ under\ Ar,\ 1420\,K} M$$

$$(M = Zr\ or\ Hf) \qquad (22.1)$$

Zirconium has a high corrosion-resistance and low cross-section for neutron capture (see *Section 2.4*) and is used for cladding fuel rods in water-cooled nuclear reactors. For this application, Zr must be free of Hf which is a very good neutron absorber; the main use of pure Hf is in nuclear reactor control rods. Zirconium and hafnium compounds possess similar lattice energies and solubilities, and their complexes have similar stabilities; this means that separation techniques (e.g. ion exchange, solvent extraction) encounter the same problems as those of the lanthanoids. Very pure metals can be obtained by zone refining (see *Box 5.2*) or by thermal decomposition of the iodides on a hot metal filament. Zirconium compounds have a range of catalytic

BOX 22.1

Environmental catalysts

The platinum-group metals Rh, Pd and Pt play a vital role in keeping the environment devoid of pollutants originating from vehicle exhausts. They are present in catalytic converters (which we discuss in detail in Section 26.6) where they catalyze the conversion of hydrocarbon wastes, CO and NO_x (see Box 14.6) to CO_2, H_2O and N_2. The growth rate of environmental catalyst manufacture by companies such as Johnson Matthey in the UK is propelled by legislative measures for the control of exhaust emissions. Regulations brought in by the US and Europe have already had a major impact on the levels of emissions and have improved the quality of urban air. Action is spreading worldwide: India and China have recently adopted legislative measures.

---

applications. Uses of ZrO_2 are described in Section 22.5; in *Box 14.4*, we highlight applications of Hf and Zr nitrides.

Niobium (formerly known as *columbium*) and *tantalum* occur together in the mineral *columbite* $(Fe,Mn)(Nb,Ta)_2O_6$; when Nb-rich, it is called *niobite* and when Ta-rich, *tantalite*. Fusion of the ore with alkali gives poly-niobates and -tantalates, and further treatment with dilute acid yields Nb_2O_5 and Ta_2O_5. One method of separation utilizes the more basic character of Ta: at a controlled concentration of HF and KF in aqueous solution, the oxides are converted to $K_2[NbOF_5]$ and $K_2[TaF_7]$; the former is more water-soluble than the latter. The modern separation technique is fractional extraction from aqueous HF solution into methyl isobutyl ketone. Niobium is used in the manufacture of tough steels and in frameworks designed for the Gemini space program. Superconducting magnets (e.g. in MRI equipment, see *Box 2.5*) contain NbTi metallic multicore conductors. Tantalum is very high melting (mp 3290 K) and extremely resistant to corrosion by air and water; it is used in corrosion-resistant alloys, e.g. for construction materials in the chemical industry. The inertness of the metal makes it suitable for use in surgical appliances including prostheses.

The German for tungsten is *wolfram*, hence the symbol W. Although *molybdenum* and *tungsten* compounds are usually isomorphous, the elements occur separately. The major Mo-containing ore is *molybdenite* (MoS_2) and the metal is extracted by reaction sequence 22.2. Tungsten occurs in *wolframite* ($(Fe,Mn)WO_4$) and *scheelite* ($CaWO_4$) and scheme 22.3 shows typical extraction processes.

$$MoS_2 \xrightarrow{\Delta\,(870\,K)\,\text{in air}} MoO_3 \xrightarrow{H_2,\,870\,K} Mo \qquad (22.2)$$

$$\left.\begin{array}{l} (Fe,Mn)WO_4 \xrightarrow{Na_2CO_3,\,\text{fusion}} \underset{\text{insoluble}}{(Fe,Mn)_2O_3} + \underset{\text{soluble}}{Na_2WO_4} \\[2em] Na_2WO_4 \xrightarrow{HCl} WO_3 \xrightarrow{H_2,\,870\,K} W \end{array}\right\} \qquad (22.3)$$

Molybdenum is very hard and high melting (mp 2896 K), and tungsten has the highest melting point (3695 K) of all metals (*Table 5.2*). Both metals are used in the manufacture of toughened steels (for which wolframite can be reduced directly by Al); tungsten carbides have extensive use in cutting tools and abrasives. A major use of W metal is in electric light bulb filaments. Molybdenum has an essential role in biological systems (see *Chapter 28*).

Technetium is an artificial element, available as ^{99}Tc (a β-particle emitter, $t_{\frac{1}{2}} = 2.13 \times 10^5$ yr) which is isolated from fission product wastes by oxidation to $[TcO_4]^-$. Separation employs solvent extraction and ion-exchange methods. The $[TcO_4]^-$ ion is the common precursor in technetium chemistry; the metal can be obtained by H_2 reduction of $[NH_4][TcO_4]$ at high temperature. The principal use of Tc compounds is in nuclear medicine where they are important imaging agents (see *Boxes 2.3* and *22.7*). *Rhenium* is rare and occurs in small amounts in Mo ores; during roasting (first step in equation 22.2), volatile Re_2O_7 forms and is deposited in flue dusts. It is dissolved in water and precipitated as $KReO_4$. Rhenium is used in alloys (e.g. ReW alloys in thermocouples) and in filaments for photographic flash equipment and mass spectrometers.

The *platinum-group metals* (Ru, Os, Rh, Ir, Pd and Pt) are rare (Figure 22.1) and expensive, and occur together either

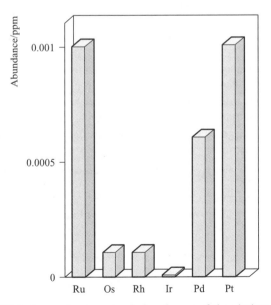

Fig. 22.1 Approximate natural abundances of the platinum-group metals in the Earth's crust.

Treatment of cyanide waste

The toxicity of $[CN]^-$ was brought to public attention early in 2000 when a huge spillage of cyanide (originating from gold extraction processes) entered the River Danube and surrounding rivers in Eastern Europe, devastating fish stocks and other river life.

The high toxicity of $[CN]^-$ makes it essential for cyanide-containing waste produced by industry to be treated. Several methods are used. For dilute solutions of cyanide, destruction using hypochlorite solution is common:

$$[CN]^- + [OCl]^- + H_2O \longrightarrow ClCN + 2[OH]^-$$

$$ClCN + 2[OH]^- \longrightarrow Cl^- + [OCN]^- + H_2O \quad (\text{at pH} > 11)$$

$$[OCN]^- + 2H_2O \longrightarrow [NH_4]^+ + [CO_3]^{2-} \quad (\text{at pH} < 7)$$

The operation must be further modified to take into account the large amounts of Cl^- produced. An alternative method is oxidation by H_2O_2:

$$[CN]^- + H_2O_2 \longrightarrow [OCN]^- + H_2O$$

Older methods such as formation of $[SCN]^-$ or complexation to give $[Fe(CN)_6]^{4-}$ are no longer favoured.

---

native or in sulfide ores of Cu and Ni. Three sites of mineral deposits in the former Soviet Union, Canada and South Africa hold the world's reserves. The main source of **ruthenium** is from wastes from Ni refining, e.g. from *pentlandite*, $(Fe,Ni)S$. **Osmium** and **iridium** occur in *osmiridium*, a native alloy with variable composition: 15–40% osmium and 80–50% iridium. **Rhodium** occurs in *native platinum* and in *pyrrhotite* ores $(Fe_{1-n}S, n = 0–0.2,$ often with $\leq 5\%$ Ni); native platinum is of variable composition but may contain as much as 86% Pt, other constituents being Fe, Ir, Os, Au, Rh, Pd and Cu. The ore is an important source of **palladium** which is also a side-product of Cu and Zn refining. Besides being obtained native, **platinum** is extracted from *sperrylite* $(PtAs_2)$. Extraction and separation methods for the six metals are interlinked, solvent extraction and ion-exchange methods being used.[†] The metals are important heterogeneous catalysts, e.g. Pd for hydrogenation and dehydrogenation, Pt for NH_3 oxidation and hydrocarbon reforming, and Rh and Pt for catalytic converters (see *Section 26.6*). Uses of Ru and Rh include alloying with Pt and Pd to increase their hardness for use in, for example, the manufacture of electrical components (e.g. electrodes and thermocouples) and laboratory crucibles. Osmium and iridium have few commercial uses; they are employed to a limited extent as alloying agents; an IrOs alloy is used in pen-nibs. Palladium is widely used in the electronics industry (in printed circuits and multilayer ceramic capacitors); the ability of Pd to absorb large amounts of H_2 (see *Section 9.7*) leads to its being used in the industrial purification of H_2. Platinum is particularly inert; Pt electrodes[‡] have laboratory applications (e.g. in the standard hydrogen and pH electrodes), and the metal is widely used in electrical wires, thermocouples and jewellery. Platinum-containing compounds

such as *cisplatin* (**22.1a**) and *carboplatin* (**22.1b**) are antitumour drugs, and we discuss these further in *Box 22.10*.

(22.1a)   (22.1b)

Silver and **gold** occur native, and in sulfide, arsenide and telluride ores, e.g. argentite (Ag_2S) and *sylvanite* $((Ag,Au)Te_2)$. Silver is usually worked from the residues of Cu, Ni or Pb refining and, like Au, can be extracted from all its ores by reaction 22.4, the cyano complex being reduced to the metal by Zn.[§]

$$4M + 8[CN]^- + 2H_2O + O_2 \longrightarrow 4[M(CN)_2]^- + 4[OH]^-$$
$$(M = Ag, Au) \quad (22.4)$$

Native gold typically contains 85–95% Au with Ag as the second constituent. Silver is used in soldering-alloys, high capacity batteries, electrical equipment and printed circuits; silver salts are extensively employed in the photographic industry (see *Box 22.13*). Silver iodide (in the form of flares or ground-sited acetone–AgI generators) is used in cloud seeding to control rain patterns in certain regions. Gold has been worked since ancient civilization, not only in the usual yellow form, but as red, purple or blue *colloidal gold*; modern uses of colloidal gold are in electron microscope imaging, staining of microscope slides and as colouring agents, e.g. reduction of Au(III) with $SnCl_2$ yields *purple of Cassius*, used in the manufacture of ruby glass. Uses of gold include coinage, the electronics industry, and jewellery;

[†] For further discussion, see: N.N. Greenwood and A. Earnshaw (1997) *Chemistry of the Elements*, 2nd edn, Butterworth-Heinemann, Oxford, p. 1147.

[‡] Microelectrodes are a relatively new innovation, see: G. Denuault (1996) *Chemistry & Industry*, p. 678.

[§] Extraction of gold, see: J. Barrett and M. Hughes (1997) *Chemistry in Britain*, vol. 33, issue 6, p. 23 – 'A golden opportunity'.

Mercury: a highly toxic, liquid metal

The low melting point (234 K) of Hg results in its being a unique metal. Its high thermal expansion coefficient makes it a suitable liquid for use in thermometers, and it has widespread application in barometers, diffusion pumps and in Hg-switches in electrical apparatus. An older use was in mirrors. Some other metals dissolve in mercury to give *amalgams*; their uses are varied, for example:

- Cd/Hg amalgam is a component in the Weston cell (see equation 22.5);
- Na/Hg amalgam is a convenient source of Na as a reducing agent;
- silver amalgam ($\approx 50\%$ Hg, 35% Ag, 13% Sn, 2% Cu by weight) is used for silver-fillings in dentistry.

Despite these uses, Hg poses a serious health risk, as do its compounds (e.g. Me_2Hg). Mercury has a low enthalpy of vaporization ($59 \, kJ \, mol^{-1}$), and even below its boiling point (630 K) its volatility is high; at 293 K, a drop of liquid Hg vaporizes at a rate of $5.8 \, \mu g \, h^{-1} \, cm^{-2}$, and at its saturation point the surrounding air contains $13 \, mg \, m^{-3}$, a level far in excess of safe limits. Similarly, amalgams are a source of Hg vapour, and those in tooth fillings release toxic vapour directly into the human body; research has shown that brushing teeth and chewing increases the vaporization process. The toxicity is now well established, and steps have been taken in some countries to phase out the use of Hg in dental fillings.

After entering the body as Hg vapour, the metal accumulates in the kidneys, brain and testicles. It is converted to Hg(II) and is readily coordinated by thiol (RSH) donors present in proteins; the end result of mercury poisoning is severe damage to the central nervous system. One of the reasons why the toxicity is so high is that its retention time in body tissue is especially long, ≈ 65 days in the kidneys. The effects of Hg poisoning were referred to by Lewis Carroll in *Alice in Wonderland* – the occupation of the Mad Hatter brought him into regular contact with $Hg(NO_3)_2$ which was used in the manufacture of felt hats.

Further reading

M.J. Vimy (1995) *Chemistry & Industry*, p. 14 – 'Toxic teeth: the chronic mercury poisoning of modern man'.

L. Magos (1988) 'Mercury' in *Handbook on Toxicity of Inorganic Compounds*, ed. H.G. Seiler, H. Sigel and A. Sigel, Dekker, New York, p. 419.

M.B. Blayney, J.S. Winn and D.W. Nierenberg (1997) *Chemical & Engineering News*, vol. 75, May 12 issue, p. 7 – 'Handling dimethyl mercury'.

carat indicates the gold content (24 carat = pure gold). Some gold compounds are used as anti-rheumatic drugs.

Cadmium occurs as the rare mineral *greenockite* (CdS), but the metal is isolated almost entirely from zinc ores, CdS occurring ($<0.5\%$) in ZnS; being more volatile than Zn, Cd can be collected in the first stage of the distillation of the metal. Cadmium (which is toxic) has a relatively low melting point (594 K) and is used as an alloying agent in low-melting alloys. Other important uses are in NiCd batteries (see *equation 21.5*), and in control rods in nuclear reactors (Cd has a large cross-section for capture of neutrons). Cadmium selenide and telluride are semiconductors and are widely employed in the electronics industry; CdTe has potential application in solar cells, although the market currently makes greatest use of Si-based cells.[†] The Weston standard cell (equation 22.5) uses a Cd/Hg amalgam cathode, but use of this cell is declining.

$$Cd(Hg) \, | \, CdSO_4, H_2O \, \vdots \, Hg_2SO_4 \, | \, Hg \qquad (22.5)$$

The symbol Hg is derived from *hydrargyrum* (Latin) meaning 'liquid silver'. The major source of **mercury** is cinnabar (HgS) from which the metal is extracted by roasting in air (equation 22.6).

$$HgS + O_2 \longrightarrow Hg + SO_2 \qquad (22.6)$$

Mercury has many uses but is a cumulative poison (see Box 22.3).

22.3 Physical properties

Some physical properties of the heavier *d*-block metals have already been discussed or tabulated:

- trends in first ionization energies (*Figure 1.15*);
- ionization energies (*Appendix 8*);
- metallic radii (*Table 5.2* and *Figure 19.1*);
- values of $\Delta_a H^\circ$ (*Table 5.2*);
- lattice types (*Table 5.2*);
- an introduction to electronic spectra and magnetism (*Chapter 20*).

For convenience, selected physical properties are listed in Table 22.1.

The electronic configurations of the ground state M(g) atoms change rather irregularly with increasing atomic number, more so than for the first row metals (compare

---

[†] Semiconductors for solar cells, see: 'Getting power from the sun', M. Hammonds (1998) *Chemistry & Industry*, p. 219.

Table 22.1 Selected physical properties of the second and third row *d*-block metals.

Second row	Y	Zr	Nb	Mo	Tc***	Ru	Rh	Pd	Ag	Cd
Atomic number, Z	39	40	41	42	43	44	45	46	47	48
Physical appearance of pure metal	Soft; silver-white	Hard; lustrous; silver-coloured	Soft; shiny; silver-white	Hard; lustrous; silver-coloured; often encountered as grey powder	Silver; often encountered as grey powder	Hard; lustrous; silver-white	Hard; lustrous; silver-white	Grey-white; malleable and ductile; strength increased by cold-working	Lustrous; silver-white	Soft; blue-white; ductile
Ground state valence electronic configuration (core = [Kr])	$5s^2 4d^1$	$5s^2 4d^2$	$5s^1 4d^4$	$5s^1 4d^5$	$5s^2 4d^5$	$5s^1 4d^7$	$5s^1 4d^8$	$5s^0 4d^{10}$	$5s^1 4d^{10}$	$5s^2 4d^{10}$
Melting point / K	1799	2128	2750	2896	2430	2607	2237	1828	1235	594
Boiling point / K	3611	4650	5015	4885	5150	4173	4000	3413	2485	1038
$\Delta_a H^\circ$ (298 K) / kJ mol^{-1}	423	609	721	658	677	651	556	377	285	112
Metallic radius, r_{metal} / pm‡	182	160	147	140	135	134	134	137	144	152
Electrical resistivity $(\rho) \times 10^8$ / Ωm (at 273 K)*	59.6**	38.8	15.2	4.9	–	7.1	4.3	9.8	1.5	6.8

Third row	La	Hf	Ta	W	Re	Os	Ir	Pt	Au	Hg
Atomic number, Z	57	72	73	74	75	76	77	78	79	80
Physical appearance of pure metal	Soft; silver-white; tarnishes in air	Lustrous; silver-coloured; ductile	Hard; shiny; silver-coloured; ductile	Lustrous; silver-white; often encountered as grey powder	Silver-grey; often encountered as grey powder	Very hard; lustrous; blue-white; dense§	Very hard; brittle; lustrous; silver coloured; dense§	Lustrous; silver-coloured; malleable; ductile	Soft; yellow; malleable; ductile	Liquid at 298 K; silver-coloured
Ground state valence electronic configuration (core = [Xe]$4f^{14}$)	$6s^2 5d^1$	$6s^2 5d^2$	$6s^2 5d^3$	$6s^2 5d^4$	$6s^2 5d^5$	$6s^2 5d^6$	$6s^2 5d^7$	$6s^1 5d^9$	$6s^1 5d^{10}$	$6s^2 5d^{10}$
Melting point / K	1193	2506	3290	3695	3459	3306	2719	2041	1337	234
Boiling point / K	3730	5470	5698	5930	5900	5300	4403	4100	3080	630
$\Delta_a H^\circ$ (298 K) / kJ mol^{-1}	423	619	782	850	774	787	669	566	368	61
Metallic radius, r_{metal} / pm‡	188	159	147	141	137	135	136	139	144	155
Electrical resistivity $(\rho) \times 10^8$ / Ωm (at 273 K)*	61.5**	30.4	12.2	4.8	17.2	8.1	4.7	9.6	2.1	94.1

‡ Metallic radius for 12-coordinate atom.
* See equation 5.3 for relationship between electrical resistivity and resistance.
** At 290–300 K.
*** Technetium is radioactive (see text).
§ Osmium and iridium are the densest elements known (22.59 and 22.56 g cm^{-3} respectively).

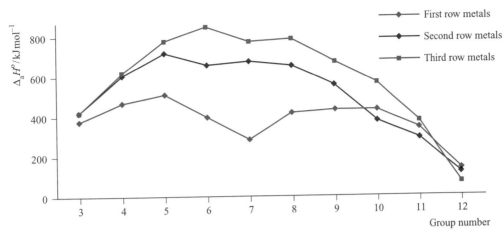

Fig. 22.2 Trends in the values of standard enthalpies of atomization (298 K) of the d-block metals; values are given in Tables 21.1 and 22.2.

Tables 22.1 and 21.1); the nd and $(n + 1)s$ atomic orbitals are closer in energy for $n = 4$ or 5 than for $n = 3$. For ions of the first row metals, the electronic configuration is generally d^n and this gives a certain amount of order to discussions of the properties of M^{2+} and M^{3+} ions. Simple M^{n+} cations of the heavier metals are rare, and it is not possible to discuss their chemistry in terms of simple redox couples (e.g. M^{3+}/M^{2+}) as we did for the most of the first row metals.

The atomic numbers of pairs of second and third row metals (except Sc and La) differ by 32 units and there is an appreciable difference in electronic energy levels and, therefore, electronic spectra and ionization energies. Within a triad, the first ionization energy is generally higher for the third than for the first and second metals, but the reverse is often true for removal of subsequent electrons. Even where a pair of second and third row metal compounds are isostructural, there are often quite significant differences in stability with respect to oxidation and reduction.

Figure 22.2 shows that, with the exception of Hg (group 12), the heavier metals have higher values of $\Delta_a H^\circ$ than their first row congeners. This is a consequence of the greater spatial extent of d-orbitals with an increase in principal quantum number, and greater orbital overlap: $5d$–$5d > 4d$–$4d > 3d$–$3d$. The trend corresponds to the fact that, compared with the first row metals, the heavier metals exhibit many more compounds containing M–M bonds. Figure 22.2 also shows that the highest values of $\Delta_a H^\circ$ occur for metals in the middle of a row. Among the heavier metals, there are numerous multimetal species with metal–metal bonding and these are discussed later in the chapter; many low oxidation-state *metal carbonyl clusters* also exist (see *Chapter 23*).

It is difficult to discuss satisfactorily the relative stabilities of oxidation states (*Table 19.3*). The situation is complicated by the fact that low oxidation states for the heavier metals are stabilized in organometallic complexes, while in non-organometallic species, the stability of *higher* oxidation states tends to *increase* down a group. Consider group 6.

Tungsten forms the stable WF_6 and WCl_6 while the existence of CrF_6 has not been firmly established (see *Section 21.7*); although CrO_3 and chromate(VI) ions are powerful oxidizing agents, WO_3, tungstate(VI) species and molybdenum analogues are not readily reduced. In general, the stability of high oxidation states increases for a given triad in the sequence first row ≪ second row < third row metals. Two factors appear important in the stabilization of third row high oxidation state compounds (e.g. AuF_5 or ReF_7) which have no second or first row counterparts:

- easier promotion of electrons for the $5d$ metals compared with $4d$ or $3d$ metals;
- better orbital overlap for $5d$ orbitals (or those with $5d$ character) than for $4d$ or $3d$ orbitals.

In comparing pairs of compounds such as MoF_6 and WF_6, or RuF_6 and OsF_6, the M–F bonds are stronger for the third than second row metal, and the symmetric stretching wavenumber and force constant are higher. Relativistic effects (see *Box 12.2*) are also important for the third row metals.

Effects of the lanthanoid contraction

Table 22.1 shows that pairs of metals in a triad (Zr and Hf, Nb and Ta etc.) are of similar radii. This is due to the *lanthanoid contraction*: the steady decrease in size along the series of lanthanoid metals Ce–Lu which lie between La and Hf in the third row of the d-block. The similarity extends to values of r_{ion} (where meaningful) and r_{cov} (e.g. the M–O distances in the high temperature forms of ZrO_2 and HfO_2 differ by less than 1 pm) and to many pairs of second and third row compounds being isomorphous. Properties that depend mainly on atom or ion size (e.g. lattice energies, solvation energies, complex stability constants) are nearly the same for corresponding pairs of $4d$ and $5d$ metal compounds. Pairs of metals often occur naturally together (e.g. Zr and Hf, Nb and Ta) and are difficult to separate (*Section 22.2*).

Coordination numbers

Consistent with the increase in size in going from a first row to later metals in a triad, the heavier metals tend to show higher coordination numbers. The common range is 4 to 9 with the highest numbers being especially prevalent for metals in groups 3–5.

NMR active nuclei

Several of the metals have spin-active nuclei and this sometimes allows *direct* observation using NMR spectroscopy, e.g. ^{89}Y has a shift range >1000 ppm and ^{89}Y NMR spectroscopy is valuable for characterizing yttrium-containing compounds. In general, it is more convenient to make use of the coupling of metal nuclei to more easily observed nuclei such as ^{1}H, ^{13}C or ^{31}P. Some examples of nuclei with $I = \frac{1}{2}$ are ^{89}Y (100% abundant), ^{103}Rh (100%), ^{183}W (14.3%), ^{107}Ag (51.8%), ^{109}Ag (48.2%), ^{195}Pt (33.8%). Coupling to low abundant isotopes gives rise to satellite peaks (see *Section 2.11* and *Figure 2.12*).

22.4 Group 3: yttrium

The metal

Bulk yttrium metal is passivated by an oxide layer and is quite stable in air; metal turnings ignite if heated >670 K (equation 22.7). Yttrium reacts with halogens (equation 22.8) and combines with most other non-metals. The reaction between Y and H_2 under pressure was described in *Box 9.4*. Yttrium reacts slowly with cold water and dissolves in dilute acids (half-equation 22.9), liberating H_2.

$$4Y + 3O_2 \xrightarrow{\Delta} 2Y_2O_3 \qquad (22.7)$$

$$2Y + 3X_2 \xrightarrow{\Delta} 2YX_3 \qquad (X = F, Cl, Br, I) \qquad (22.8)$$

$$Y^{3+} + 3e^- \rightleftharpoons Y \qquad E^\circ = -2.37\,V \qquad (22.9)$$

The chemistry of yttrium is that of the +3 oxidation state, the formation of lower hydrides being an exception.

Yttrium(III)

The halides YF_3, YCl_3, YBr_3 and YI_3 are white solids; the fluoride is water-insoluble, but YCl_3, YBr_3 and YI_3 are soluble. In solid YF_3, each Y atom is 9-coordinate (distorted tricapped trigonal prismatic), while both YCl_3 and YI_3 have layer structures (e.g. YI_3 adopts the BiI_3 lattice) with 6-coordinate Y centres. Yttrium(III) chloride forms a hexahydrate which is correctly formulated as $[YCl_2(H_2O)_6]^+Cl^-$. Reaction of YCl_3 with KCl gives $K_3[YCl_6]$ containing the octahedral $[YCl_6]^{3-}$ ion.

The white oxide Y_2O_3 is insoluble in water but dissolves in acids; it is used in ceramics, optical glasses and refractory materials (see also *Section 22.2*). Yttrium(III) hydroxide is a colourless solid, in which each Y^{3+} ion is in a tricapped trigonal prismatic YO_9 environment. The hydroxide is water-insoluble and exclusively basic.

Hpic = picric acid

(22.2)

The coordination chemistry of Y^{3+} is a developing field and coordination numbers of 6 to 9 are usual. Crystalline salts containing the aqua ions $[Y(H_2O)_8]^{3+}$ (dodecahedral, *Figure 19.8c*) and $[Y(H_2O)_9]^{3+}$ (tricapped trigonal prismatic) have been structurally characterized. The Y^{3+} ion is 'hard' and in its complexes favours hard *N*- and *O*- donors, e.g. *trans*-$[YCl_4(THF)_2]^-$ (octahedral), *trans*-$[YCl_2(THF)_5]^+$ (pentagonal bipyramidal), $[Y(H_2O)_7(pic)]^{2+}$ (8-coordinate, Hpic = **22.2**), $[Y(NO_3\text{-}O,O')_3(H_2O)_3]$ (irregular 9-coordinate) and $[Y(NO_3)_5]^{2-}$ (see end of *Section 8.11*). Reaction 22.10 yields a rare example of 3-coordinate Y(III); in the solid state, $[Y\{N(SiMe_3)_2\}_3]$ has a trigonal pyramidal rather than planar structure but this is probably caused by crystal packing effects (see *Section 19.7*).

$$YCl_3 + 3Na[N(SiMe_3)_2] \longrightarrow [Y\{N(SiMe_3)_2\}_3] + 3NaCl \qquad (22.10)$$

22.5 Group 4: zirconium and hafnium

The metals

In a finely divided form, Hf and Zr metals are pyrophoric, but the bulk metals are passivated; the high corrosion-resistance of Zr is due to the formation of a dense layer of inert ZrO_2. The metals are not attacked by dilute acids (except HF) unless heated, and aqueous alkalis have no effect even when hot. At elevated temperatures, Hf and Zr combine with most non-metals (e.g. equation 22.11).

$$MCl_4 \xleftarrow{Cl_2,\Delta} M \xrightarrow{O_2,\Delta} MO_2 \qquad (M = Hf\ or\ Zr) \qquad (22.11)$$

More is known about the chemistry of Zr than Hf, the former being more readily available (see *Section 22.2*).

Much of the chemistry concerns Zr(IV) and Hf(IV), the lower oxidation states being less stable with respect to oxidation than the first group member, Ti(III). In aqueous solutions, only M(IV) is stable although not as M^{4+}, even though tables of data may quote half-equation 22.12;

solution species (see below) depend upon conditions, e.g. in basic solution, $MO(OH)_2$ is present.

$$M^{4+} + 4e^- \rightleftharpoons M \quad \begin{cases} M = Zr, & E^\circ = -1.70\,V \\ M = Hf, & E^\circ = -1.53\,V \end{cases} \quad (22.12)$$

Stabilization of low oxidation states of Zr and Hf by π-acceptor ligands is discussed in *Chapter 23*.

Zirconium(IV) and hafnium(IV)

The halides MX_4 (M = Zr, Hf; X = F, Cl, Br, I), formed by direct combination of the elements, are white solids with the exception of orange-yellow ZrI_4 and HfI_4. The solids possess infinite lattices ($ZrCl_4$, $ZrBr_4$, ZrI_4 and HfI_4 contain chains of edge-sharing octahedra) but the vapours contain tetrahedral molecules. Zirconium(IV) fluoride is dimorphic; the α-form consists of a network of F-bridged ZrF_8 square antiprisms and converts ($> 720\,K$) to β-ZrF_4 in which each Zr centre is dodecahedrally sited. Ultra-pure ZrF_4 for use in optical fibres and IR spectrometer parts is made by treatment of $[Zr(BH_4)_4]$ (see *Section 12.5*) with HF and F_2. The chlorides, bromides and iodides are water-soluble, but hydrolyze to MOX_2. Water reacts with ZrF_4 to give $[F_3(H_2O)_3Zr(\mu\text{-}F)_2Zr(H_2O)_3F_3]$. Both ZrF_4 and $ZrCl_4$ form highly electrically conducting materials with graphite, e.g. the reaction of ZrF_4, F_2 and graphite gives

$C_nF(ZrF_4)_m$ ($n = 1$–100, $m = 0.0001$–0.15). The Lewis acidity of the halides is seen in the formation of complexes such as $HfCl_4\cdot2L$ (L = NMe_3, THF) and in the use of $ZrCl_4$ as a Lewis acid catalyst.

Oxides of Zr(IV) and Hf(IV) are produced by direct combination of the elements or by heating MCl_4 with H_2O followed by dehydration. The white oxides are isostructural and adopt lattice structures in which Zr and Hf centres are 7-coordinate. Zirconium(IV) oxide is inert, and is used as an opacifier in ceramics and enamels and as an additive to synthetic apatites (see *Section 14.2* and *Box 14.10*) used in dentistry; pure ZrO_2 undergoes a phase change at $1370\,K$ which results in cracking of the material, and for use in, for example, refractory materials, the higher temperature cubic phase is stabilized by adding MgO or CaO. Crystals of *cubic zirconia* (see *Section 27.2*) are commercially important as artificial diamonds. The addition of $[OH]^-$ to any water-soluble Zr(IV) compound produces the white amorphous $ZrO_2\cdot xH_2O$; there is no true hydroxide.

In aqueous acidic solution, Zr(IV) compounds are present as partly hydrolyzed species, e.g. $[Zr_3(OH)_4]^{8+}$ and $[Zr_4(OH)_8]^{8+}$. From solutions of $ZrCl_4$ in dilute HCl, '$ZrOCl_2\cdot8H_2O$' can be crystallized; this is tetrameric, $[Zr_4(OH)_8(H_2O)_{16}]Cl_8\cdot12H_2O$, and contains $[Zr_4(OH)_8(H_2O)_{16}]^{8+}$ (Figure 22.3a) in which each Zr is dodecahedrally sited.

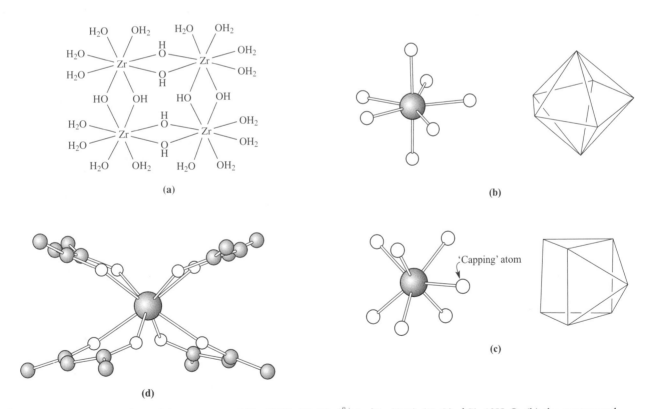

Fig. 22.3 (a) Representation of the structure of $[Zr_4(OH)_8(H_2O)_{16}]^{8+}$ in $[Zr_4(OH)_8(H_2O)_{16}]Cl_8\cdot12H_2O$, (b) the pentagonal bipyramidal structure (X-ray diffraction) of $[ZrF_7]^{3-}$ in $[H_3N(CH_2)_2NH_2(CH_2)_2NH_3][ZrF_7]$ [V.V. Tkachev *et al.* (1993) *Koord. Khim.*, vol. 19, p. 288], (c) the monocapped trigonal prismatic structure (X-ray diffraction) of $[ZrF_7]^{3-}$ in the guanidinium salt [A.V. Gerasimenko *et al.* (1985) *Koord. Khim.*, vol. 11, p. 566], and (d) the square antiprismatic structure (X-ray diffraction) of $[Zr(acac)_4]$ [W. Clegg (1987) *Acta Crystallogr.*, *Sect. C*, vol. 43, p. 789]. Hydrogen atoms in (d) have been omitted; colour code: Zr, blue; C, grey; O, white.

The high coordination numbers exhibited in some apparently simple compounds of Zr(IV) and Hf(IV) extend to their complexes (e.g. see *Figure 19.9*), hard fluoride and oxygen-donor ligands being favoured, e.g.:

- pentagonal bipyramidal: $[ZrF_7]^{3-}$ (Figure 22.3b, e.g. Na^+, K^+ salts, structure is cation-dependent), $[HfF_7]^{3-}$, (e.g. K^+ salt, equation 22.13), $[F_4(H_2O)Zr(\mu\text{-}F)_2Zr(H_2O)F_4]^{2-}$;
- capped trigonal prismatic: $[ZrF_7]^{3-}$ (Figure 22.3c, e.g. $[NH_4]^+$ salt, structure is cation-dependent);
- dodecahedral: $[Zr(NO_3\text{-}O,O')_4]$ (equation 22.14), $[Zr(ox)_4]^{4-}$;
- square antiprismatic: $[Zr(acac)_4]$ (Figure 22.3d).

$$HfF_4 + 3KF \xrightarrow{\Delta \text{ in sealed Pt tube}} K_3[HfF_7] \qquad (22.13)$$

$$ZrCl_4 + 4N_2O_5$$
$$\xrightarrow{\text{anhydrous conditions}} [Zr(NO_3\text{-}O,O')_4] + 4NO_2Cl \quad (22.14)$$

$$ZrCl_4 + 2CsCl \xrightarrow{1070 \text{ K, in sealed } SiO_2 \text{ tube}} Cs_2[ZrCl_6] \quad (22.15)$$

The $[ZrCl_6]^{2-}$ ion (equation 22.15) is octahedral; colourless $Cs_2[ZrCl_6]$ adopts a $K_2[PtCl_6]$ lattice (see Mn(IV) in *Section 21.8*) and is used as an image intensifier in X-ray imaging. A number of oxo-complexes with square-based pyramidal structures are known, e.g. $[M(O)(ox)_2]^{2-}$ (M = Hf, Zr, **22.3**) and $[Zr(O)(bpy)_2]^{2+}$. Lower coordination numbers are stabilized by amido ligands, e.g. tetrahedral $[M(NPh_2)_4]$ and $[M\{N(SiMe_3)_2\}_3Cl]$ (M = Hf, Zr).

$$
\left[
\begin{array}{c}
O \\
\| \\
O{=}\!\!=\!\!=\!\!M\!\!=\!\!=\!\!=O \\
\end{array}
\right]^{2-}
$$

M = Hf, Zr

(22.3)

Lower oxidation states of zirconium and hafnium

The blue or black halides ZrX_3, ZrX_2 and ZrX (X = Cl, Br, I) are obtained by reduction of ZrX_4, e.g. heating Zr and $ZrCl_4$ in a sealed Ta tube gives ZrCl or $ZrCl_3$ depending on temperature. The corresponding hafnium chlorides are prepared similarly, e.g. equations 22.16 and 22.17.

$$Hf + HfCl_4 \xrightarrow{1070 \text{ K, in sealed Ta tube}} HfCl \qquad (22.16)$$

$$Hf + HfCl_4 \xrightarrow{720 \text{ K, in sealed Ta tube}} HfCl_3 \qquad (22.17)$$

The monohalides have layer structures consisting of sheets of metal and halogen atoms sequenced XMMX...XMMX... and are metallic conductors in a direction *parallel* to the layers; compare this with the conductivity of graphite (see *Section 13.4*). The di- and trihalides disproportionate (equations 22.18 and 22.19).

$$2MCl_2 \longrightarrow M + MCl_4 \qquad (M = Hf, Zr) \qquad (22.18)$$

$$2MCl_3 \longrightarrow MCl_2 + MCl_4 \qquad (M = Hf, Zr) \qquad (22.19)$$

There is no aqueous chemistry of M(I), M(II) and M(III).

Zirconium clusters

In this section, we introduce the first group of cluster compounds of the heavier *d*-block metals in which the external ligands are halides. Octahedral M_6 frameworks are present in most of these clusters, but, in contrast to similar group 5 and 6 species (*Sections 22.6* and *22.7*), most zirconium clusters are stabilized by an *interstitial atom*, e.g. Be, B, C or N.

Heating Zr powder, $ZrCl_4$ and carbon in a sealed Ta tube above 1000 K produces $Zr_6Cl_{14}C$; under similar reaction conditions and with added alkali metal halides, clusters such as $Cs_3[Zr_6Br_{15}C]$, $K[Zr_6Br_{13}Be]$ and $K_2[Zr_6Br_{15}B]$ have been made. In the solid state, these octahedral Zr_6 clusters are connected by bridging halide ligands to

(a)   (b)

Fig. 22.4 The structures (determined by X-ray diffraction) of (a) $[Zr_6Cl_{14}(P^nPr_3)_4]$ [F.A. Cotton *et al.* (1992) *Angew. Chem., Int. Ed. Engl.*, vol. 31, p. 1050] and (b) $[Zr_5Cl_{12}(\mu\text{-H})_2(\mu_3\text{-H})_2(PMe_3)_5]$ [F.A. Cotton *et al.* (1994) *J. Am. Chem. Soc.*, vol. 116, p. 4364]. Colour code: Zr, blue; Cl, pale grey; P, black; Me and ^nPr groups are omitted. The inset in (b) shows the μ-H and μ_3-H positions (black dots) with respect to the Zr_5 framework.

generate extended lattices. The formulae may be written to show the connectivity, e.g. writing $[Zr_6Br_{15}B]^{2-}$ as $[\{Zr_6(\mu\text{-}Br)_{12}B\}Br_{6/2}]^{2-}$ indicates that Zr_6 clusters are connected into a 3D-network by six Br atoms, three 'belonging' to each cluster.[†]

In contrast to the high temperature syntheses of these Zr_6X clusters (X = B, C, N), reduction of $ZrCl_4$ by Bu_3SnH followed by addition of PR_3 gives *discrete* clusters such as $[Zr_6Cl_{14}(P^nPr_3)_4]$ (Figure 22.4a, Zr–Zr = 331–337 pm) and $[Zr_5Cl_{12}(\mu\text{-}H)_2(\mu_3\text{-}H)_2(PMe_3)_5]$ (Figure 22.4b, Zr–Zr = 320–354 pm). Varying the reaction conditions leads to clusters such as $[Zr_6Cl_{14}(PMe_3)H_4]$, $[Zr_6Cl_{18}H_4]^{3-}$ and $[Zr_6Cl_{18}H_5]^{4-}$.[‡]

22.6 Group 5: niobium and tantalum

The metals

The properties of Nb and Ta (and of pairs of corresponding compounds) are similar. At high temperatures, both are attacked by O_2 (equation 22.20) and the halogens (equation 22.21) and combine with most non-metals.

$$4M + 5O_2 \xrightarrow{\Delta} 2M_2O_5 \quad (M = Nb, Ta) \quad (22.20)$$

$$2M + 5X_2 \xrightarrow{\Delta} 2MX_5$$
$$(M = Nb, Ta; X = F, Cl, Br, I) \quad (22.21)$$

The metals are passivated by the formation of oxide coatings, giving them high corrosion resistance; they are inert towards non-oxidizing acids, HF and HF/HNO_3 being two of the few reagents to attack them under ambient conditions. Fused alkalis react with Nb and Ta at high temperatures.

The chemistry of Nb and Ta is predominantly that of the +5 oxidation state. The heavier group 5 metals differ from V (see *Section 21.6*) in the relative instability of their lower oxidation states, their failure to form simple ionic compounds, and the inertness of the M(V) oxides. In contrast to V, it is not meaningful to assign ionic radii to Nb and Ta in their lower oxidation states since they tend to form hexanuclear clusters with metal–metal bonding (see later). For M(V), radii of 64 pm are usually tabulated for 'Nb^{5+}' and 'Ta^{5+}', but these are unreal quantities since Nb(V) and Ta(V) compounds are essentially covalent.

Niobium(V) and tantalum(V)

Niobium and tantalum halides (white MF_5, yellow MCl_5, yellow-red MBr_5 and yellow-brown MI_5) are volatile,

air- and moisture-sensitive solids made by reaction 22.21. The chlorides and bromides are also made by halogenation of M_2O_5; NbI_5 is produced commercially by reaction of $NbCl_5$, I_2 and HI, and TaI_5 by treating $TaCl_5$ with BI_3. Each halide is monomeric (trigonal bipyramidal) in the gas phase, but the solid fluorides are tetrameric (**22.4**), while solid MCl_5, MBr_5 and MI_5 consist of dimers (**22.5**). The M–F–M bridges in tetramer **22.4** are linear and the M–F$_{bridge}$ bonds are longer (and weaker) than M–F$_{terminal}$ (206 *vs* 177 pm from M = Nb). Similarly, in dimer **22.5**, M–X$_{bridge}$ > M–X$_{terminal}$.

M = Nb, Ta

(22.4)

M = Nb, Ta; X = Cl, Br, I

(22.5)

The halides NbF_5, TaF_5, $NbCl_5$ and $TaCl_5$ are useful starting materials in the chemistry of these metals. They are Friedel–Crafts catalysts and the Lewis acidity of NbF_5 and TaF_5 is apparent in reaction 22.22 (which takes place in non-aqueous media, see *Section 8.10*), in the formation of related salts and other complexes (see later), and in the ability of a TaF_5/HF mixture to act as a superacid (see *Sections 8.7* and *8.9*).

$$MF_5 + BrF_3 \longrightarrow [BrF_2]^+[MF_6]^- \quad (M = Nb, Ta) \quad (22.22)$$

The oxohalides MOX_3 and MO_2X (M = Nb, Ta; X = F, Cl, Br, I) are prepared by halogenation of M_2O_5, or reaction of MX_5 with O_2 under controlled conditions. The oxohalides are monomeric in the vapour and polymeric in the solid; $NbOCl_3$ is representative with monomer (**22.6**) and polymer (**22.7**) which contains oxygen-bridged Nb_2Cl_6-units. Oxoanions include octahedral $[NbOF_5]^{2-}$, $[MOCl_5]^{2-}$ (M = Nb, Ta), $[MOCl_4]^-$ (equation 22.23), and $[Ta_2OX_{10}]^{2-}$ (X = F, Cl; Figure 22.5a). The linearity of the bridge in $[Ta_2OX_{10}]^{2-}$ indicates multiple bond character (refer to Figure 22.15).

$$MOCl_3 + ONCl \longrightarrow [NO]^+[MOCl_4]^- \quad (M = Nb, Ta) \quad (22.23)$$

[†] The nomenclature is actually more complicated, but more informative, e.g. R.P. Ziebarth and J.D. Corbett (1985) *Journal of the American Chemical Society*, vol. 107, p. 4571.

[‡] For a discussion of the location of H atoms in these and related Zr_6 cages, see: L. Chen, F.A. Cotton and W.A. Wojtczak (1997) *Inorganic Chemistry*, vol. 36, p. 4047.

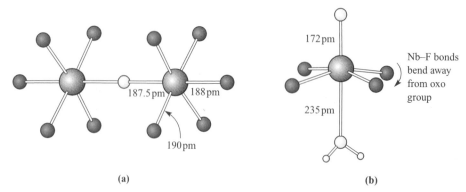

Fig. 22.5 The structures (determined by X-ray diffraction for the $[Et_4N]^+$ salts) of (a) $[Ta_2OF_{10}]^{2-}$ [J.C. Dewan *et al.* (1977) *J. Chem. Soc., Dalton Trans.*, p. 978] and (b) $[Nb(H_2O)(O)F_4]^-$ [N.G. Furmanova *et al.* (1992) *Kristallografiya*, vol. 37, p. 136]. Colour code: Ta and Nb, blue; F, black; O and H, pale grey.

The structure of $[Nb(H_2O)(O)F_4]^-$ (Figure 22.5b) shows how oxo and aqua ligand O atoms can be distinguished from the Nb–O bond lengths; it is not always possible to locate H atoms in X-ray diffraction studies.

$$(22.6) \qquad (22.7)$$

Hydrolysis of $TaCl_5$ with H_2O produces the hydrated oxide $Ta_2O_5 \cdot xH_2O$; $Nb_2O_5 \cdot xH_2O$ is best formed by boiling $NbCl_5$ in aqueous HCl. Heating the hydrates yields anhydrous Nb_2O_5 and Ta_2O_5 which are dense, inert white solids. Various polymorphs of Nb_2O_5 exist, with NbO_6 octahedra being the usual structural unit; the lattices of both metal(V) oxides are complicated networks. Uses of Nb_2O_5 include those as a catalyst, in ceramics and in humidity sensors. Both Nb_2O_5 and Ta_2O_5 are insoluble in acids except concentrated HF, but dissolve in molten alkalis. If the resultant melts are dissolved in water, salts of niobates (precipitated below $\approx$pH 7) and tantalates (precipitated below $\approx$pH 10) can be isolated, e.g. $K_8[Nb_6O_{19}] \cdot 16H_2O$ and $[Et_4N]_6[Nb_{10}O_{28}] \cdot 6H_2O$. The $[Nb_6O_{19}]^{8-}$ ion consists of six MO_6 octahedral units with shared O atoms; it is isoelectronic and isostructural with $[Mo_6O_{19}]^{2-}$ and $[W_6O_{19}]^{2-}$ (see *Figure 22.8c*). The $[Nb_{10}O_{28}]^{6-}$ ion is isostructural with $[V_{10}O_{28}]^{6-}$ and contains octahedral building-blocks as in $[Nb_6O_{19}]^{8-}$.

Heating Nb_2O_5 or Ta_2O_5 with group 1 or group 2 metal carbonates at high temperatures (e.g. Nb_2O_5 with Na_2CO_3 at 1650 K in a Pt crucible) yields mixed metal oxides such as $LiNbO_3$, $NaNbO_3$, $LiTaO_3$, $NaTaO_3$ and $CaNb_2O_6$. The $M'MO_3$ compounds crystallize with perovskite lattices (*Figure 5.23*), and exhibit ferroelectric and piezoelectric properties (see *Section 13.9*) which lead to uses in electro-optical and acoustic devices.

The coordination chemistry of Nb(V) and Ta(V) is well developed and there is a close similarity in the complexes formed by the two metals; complexes with hard donors are favoured. Although 6-, 7- and 8-coordinate complexes are the most common, lower coordination numbers are observed, e.g. in $[Ta(NEt_2)_5]$ (trigonal bipyramid), $[Nb(NMe_2)_5]$ and $[NbOCl_4]^-$ (square-based pyramidal). The Lewis acidity of the pentahalides, especially NbF_5 and TaF_5, leads to the formation of salts such as $Cs[NbF_6]$ and $K[TaF_6]$ (octahedral anions), $K_2[NbF_7]$ and $K_2[TaF_7]$ (capped trigonal prismatic anions), $Na_3[TaF_8]$ and $Na_3[NbF_8]$ (square antiprismatic anions) and $[^nBu_4N][M_2F_{11}]$ (equation 22.24 and structure **22.8**).

$$MF_5 \xrightarrow{[^nBu_4N][BF_4]} [^nBu_4N][MF_6] \xrightarrow{MF_5} [^nBu_4N][M_2F_{11}]$$
$$(M = Nb, Ta) \qquad (22.24)$$

$$M = Nb, Ta$$
$$(22.8)$$

Other complexes include:

- octahedral: $[Nb(H_2O)(O)F_4]^-$ (Figure 22.5b), $[Nb(NCS-N)_6]^-$, $[NbF_5(OEt_2)]$, *mer*-$[NbCl_3(O)(NCMe)_2]$;
- intermediate between octahedral and trigonal prismatic: $[Nb(SCH_2CH_2S)_3]^-$;
- pentagonal bipyramidal: $[Nb(H_2O)_2(O)(ox)_2]^-$ (**22.9**); $[Nb(O)(ox)_3]^{3-}$ (oxo ligand in an axial site);
- dodecahedral: $[Nb(\eta^2\text{-}O_2)_4]^{3-}$, $[Nb(\eta^2\text{-}O_2)_2(ox)_2]^{3-}$;
- square antiprismatic: $[Ta(\eta^2\text{-}O_2)_2F_4]^{3-}$.

(For explanation of the η-nomenclature, see *Box 18.1*.)

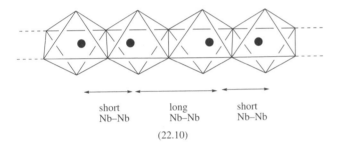

(22.9)

Niobium(IV) and tantalum(IV)

With the exception of TaF_4, all halides of Nb(IV) and Ta(IV) are known. They are dark solids, prepared by reducing the respective MX_5 by heating with metal M or Al. Niobium(IV) fluoride is paramagnetic (d^1) and isostructural with SnF_4 (**13.11**). In contrast, MCl_4, MBr_4 and MI_4 are diamagnetic (or weakly paramagnetic) consistent with the pairing of metal atoms in the solid state. The structures of $NbCl_4$ and NbI_4 are known in detail, and consist of edge-sharing distorted NbX_6 octahedra (**22.10**) with alternating Nb–Nb distances (303 and 379 pm in $NbCl_4$; 331 and 436 pm in NbI_4).

short long short
Nb–Nb Nb–Nb Nb–Nb

(22.10)

The tetrahalides are readily oxidized in air (e.g. NbF_4 to NbO_2F) and disproportionate on heating (reaction 22.25).

$$2TaCl_4 \xrightarrow{\Delta} TaCl_5 + TaCl_3 \qquad (22.25)$$

Blue-black NbO_2 is formed by reduction of Nb_2O_5 at 1070 K using H_2 or NH_3; it has a rutile structure, distorted by pairing of Nb atoms (Nb–Nb = 280 pm).

A range of Nb(IV) and Ta(IV) complexes are formed by reactions of MX_4 (X = Cl, Br, I) with Lewis bases containing *N*-, *P*-, *As*-, *O*- or *S*-donors, or by reduction of MX_5 in the presence of a ligand. Coordination numbers are typically 6, 7 or 8, for example some structures confirmed for the *solid state* are:

- octahedral: *trans*-$[TaCl_4(PEt_2)_2]$, *cis*-$[TaCl_4(PMe_2Ph)_2]$;
- capped octahedral: $[TaCl_4(PMe_3)_3]$ (Figure 19.7b);
- capped trigonal prismatic: $[NbF_7]^{3-}$ (equation 22.26);
- dodecahedral: $[Nb(CN)_8]^{4-}$;
- square antiprismatic: $[Nb(ox)_4]^{4-}$.

$$4NbF_5 + Nb + 15KF \longrightarrow 5K_3[NbF_7] \qquad (22.26)$$

Lower oxidation state halides

Of the lower oxidation state compounds of Nb and Ta, we focus on halides. The dark brown or black trihalides $NbCl_3$, $NbBr_3$, NbI_3, $TaCl_3$ and $TaBr_3$ are quite inert solids with polymeric structures; fluoride analogues have not been fully established.

● = Cl

(22.11)

A range of halides with M_3 or M_6 frameworks exists, but all have lattice structures with the metal cluster units connected by bridging halides. The structure of Nb_3Cl_8 is represented in **22.11**, but of the Cl atoms shown, six are shared between two adjacent units, and three between three. Alternatively, the structure can be considered in terms of an hcp array of Cl atoms with three-quarters of the octahedral holes occupied by Nb atoms such that they form Nb_3 triangles. Reduction of Nb_3I_8 (structurally analogous to Nb_3Cl_8) with Nb in a sealed tube at 1200 K yields Nb_6I_{11}; the formula can be written as $[Nb_6I_8]I_{6/2}$ indicating that $[Nb_6I_8]^{3+}$ units are connected by iodides shared between two clusters. (The ionic formulation is a formalism.) The $[Nb_6I_8]^{3+}$ cluster consists of an octahedral Nb_6-core, each face of which is iodo-capped (Figure 22.6a). The clusters are connected into a network by bridges (Figure 22.6c). Two other families of halides are M_6X_{14} (e.g. Nb_6Cl_{14}, Ta_6Cl_{14}, Ta_6I_{14}) and M_6X_{15} (e.g. Nb_6F_{15}, Ta_6Cl_{15}, Ta_6Br_{15}). Their formulae can be written as $[M_6X_{12}]X_{4/2}$ or $[M_6X_{12}]X_{6/2}$ showing that they contain cluster units $[M_6X_{12}]^{2+}$ and $[M_6X_{12}]^{3+}$ respectively (Figure 22.6b). The clusters are connected either into a 3-dimensional network (M_6X_{15}, Figure 22.6c) or 2-dimensional sheet (M_6X_{14}, Figure 22.6d).

Magnetic data show that the subhalides exhibit metal–metal bonding. The magnetic moment of Nb_3Cl_8 is 1.86 μ_B per Nb_3-unit (298 K) indicating one unpaired electron. This can be rationalized as follows:

- 3 Nb atoms provide 15 electrons (Nb s^2d^3);
- 8 Cl atoms provide 8 electrons (this is irrespective of the Cl bonding mode because bridge formation invokes coordinate bonds using Cl lone pairs);
- the total number of valence electrons is 23;
- 22 electrons are used in 8 Nb–Cl and 3 Nb–Nb single bonds;
- one electron is left over.

Compounds of the type M_6X_{14} are diamagnetic, while M_6X_{15} compounds have magnetic moments corresponding to one unpaired electron per M_6-cluster. If we consider M_6X_{14} to contain an $[M_6X_{12}]^{2+}$ unit, there are 8 pairs of valence electrons remaining after allocation of 12 M–X bonds, giving a bond order of two-thirds per M–M edge

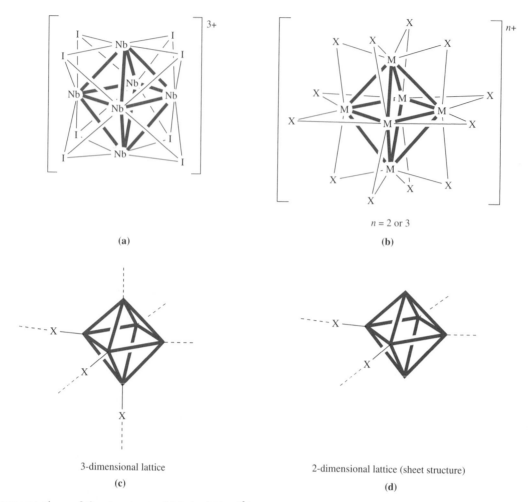

Fig. 22.6 Representations of the structures of (a) the $[Nb_6I_8]^{3+}$ unit found in Nb_6I_{11} and (b) the $[M_6X_{12}]^{n+}$ ($n = 2$ or 3) unit found in compounds of type M_6X_{14} and M_6X_{15} for M = Nb or Ta, X = halide. The cluster units are connected into (c) a 3D-network or (d) a 2D-sheet by bridging halides (see text).

(12 edges). In M_6X_{15}, after allocating electrons to 12 M–X single bonds, the $[M_6X_{12}]^{3+}$ unit has 15 valence electrons for M–M bonding; the observed paramagnetism indicates that one unpaired electron remains unused. The magnetic

moment (per hexametal unit) of Ta_6Br_{15}, for example, is temperature-dependent: $\mu_{eff} = 2.17\,\mu_B$ at 623 K, 1.73 μ_B at 222 K and 1.34 μ_B at 77 K.

There is also a family of discrete clusters $[M_6X_{18}]^{n-}$ (M = Nb, Ta; X = Cl, Br, I); for example, the reaction of Nb_6Cl_{14} with KCl at 920 K produces $K_4[Nb_6Cl_{18}]$. The $[Nb_6Cl_{18}]^{4-}$ ion is oxidized by I_2 to $[Nb_6Cl_{18}]^{3-}$ or by Cl_2 to $[Nb_6Cl_{18}]^{2-}$. The $[M_6X_{18}]^{n-}$ ions are structurally similar (Figure 22.7) and relationships between the structure of this discrete ion, that of the $[M_6Cl_{12}]^{n+}$ ion (Figure 22.6b) and of the Zr_6 clusters (e.g. Figure 22.4a) are clear.

22.7 Group 6: molybdenum and tungsten

The metals

The properties of Mo and W are similar. Both have very high melting points and enthalpies of atomization (*Table 5.2* and Figure 22.2). The metals are not attacked in air at 298 K, but react with O_2 at high temperatures to give MO_3, and are

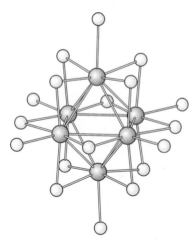

Fig. 22.7 The structure (X-ray diffraction) of $[Nb_6Cl_{18}]^{3-}$ in the $[Me_4N]^+$ salt [F.W. Koknat *et al.* (1974) *Inorg. Chem.*, vol. 13, p. 295]. Colour code: Nb, blue; Cl, pale grey.

readily oxidized by the halogens (see later); even at 298 K, oxidation to M(VI) occurs with F_2 (equation 22.27). Sulfur reacts with Mo or W (e.g. equation 22.28); other sulfide phases of Mo are produced under different conditions.

$$M + 3F_2 \longrightarrow MF_6 \qquad (M = Mo, W) \qquad (22.27)$$

$$M + 2S \xrightarrow{\Delta} MS_2 \qquad (M = Mo, W) \qquad (22.28)$$

The metals are inert towards most acids but are rapidly attacked by fused alkalis in the presence of oxidizing agents.

Molybdenum and tungsten exhibit a range of oxidation states (*Table 19.3*) although simple mononuclear species are not known for all states. The extensive chemistry of Cr(II) and Cr(III) (see *Section 21.7*) has no counterpart in the chemistries of the heavier group 6 metals, and, in contrast to Cr(VI), Mo(VI) and W(VI) are poor oxidizing agents. Since $W^{3+}(aq)$ is not known, no reduction potential for the W(VI)/W(III) couple can be given, but equations 22.29 and 22.30 compare the Cr and Mo systems at pH 0.

$$[Cr_2O_7]^{2-} + 14H^+ + 6e^- \rightleftharpoons 2Cr^{3+} + 7H_2O \quad E^\circ = +1.33\,V$$
$$(22.29)$$

$$H_2MoO_4 + 6H^+ + 3e^- \rightleftharpoons Mo^{3+} + 4H_2O \qquad E^\circ = +0.1\,V$$
$$(22.30)$$

Molybdenum and tungsten compounds are usually isomorphous and essentially isodimensional.

Molybdenum(VI) and tungsten(VI)

The hexafluorides are formed by reaction 22.27, or by reactions of MoO_3 with SF_4 (sealed vessel, 620 K) and WCl_6 with HF or SbF_3. Both MoF_6 (colourless liquid, bp 307 K)

and WF_6 (pale yellow, volatile liquid, bp 290 K) have molecular structures (**22.12**) and are readily hydrolyzed. The only other hexahalides that are well established are the dark blue WCl_6 and WBr_6. The former is made by heating W or WO_3 with Cl_2 and has an octahedral, molecular structure; WBr_6 (also molecular) is best made by reaction 22.31. Both WCl_6 and WBr_6 readily hydrolyze. Reactions of WF_6 with Me_3SiCl, or WCl_6 with F_2, yield mixed halo-derivatives, e.g. *cis*- and *trans*-WCl_2F_4 and *mer*- and *fac*-WCl_3F_3.

$$M = Mo, W$$
$$(22.12)$$

$$W(CO)_6 + 3Br_2 \longrightarrow WBr_6 + 6CO \qquad (22.31)$$

Oxohalides MOX_4 (M = Mo, X = F, Cl; M = W, X = F, Cl, Br) and MO_2X_2 (M = Mo, W; X = F, Cl, Br) can be made by a variety of routes, e.g. equation 22.32. Reactions of MO_3 with CCl_4 yield MO_2Cl_2; WO_2Cl_2 decomposes on heating (equation 22.33). The oxohalides readily hydrolyze.

$$\left.\begin{array}{l} M + O_2 + F_2 \\ MOCl_4 + HF \\ MO_3 + F_2 \end{array}\right\} \longrightarrow MOF_4 \quad (M = Mo, W) \qquad (22.32)$$

$$2WO_2Cl_2 \xrightarrow{450-550\,K} WO_3 + WOCl_4 \qquad (22.33)$$

The solids are polymeric, e.g. $MoOF_4$ contains chains of $MoOF_5$ octahedra linked by Mo−F−Mo bridges; in

---

BOX 22.4   **APPLICATIONS**

Electrochromic 'smart' windows

In an electrochromic cell, the passage of charge causes an electrode to change colour; reversing the flow of charge reverses the colour change. By laying the electrode on the surface of glass, electrochromic windows or mirrors can be created, their use being to reduce light transmission when

light intensity exceeds comfortable limits. The design of electrochromic windows is a current topic of research by major glass manufacturers, with WO_3 playing a vital role. An example of a window design is as follows:

$Li_xV_2O_5$ is a non-stoichiometric compound ($x \approx 1$) and acts as a Li atom store. When a small potential is applied across the cell, Li^+ ions migrate from the lithium polymer electrolyte into the WO_3 layer forming a tungsten bronze (see equation 22.42 and discussion). Its formation results in a colour change from colourless to blue.

Further reading

M. Green (1996) *Chemistry & Industry*, p. 643 – 'The promise of electrochromic systems'.
R.J. Mortimer (1997) *Chemical Society Reviews*, vol. 26, p. 147 – 'Electrochromic materials'.

MoS$_2$: a solid lubricant

After purification and conversion into appropriate grade powders, the mineral *molybdenite*, MoS$_2$, has widespread commercial applications as a solid lubricant. It is applied to reduce wear and friction, and is able to withstand high temperature working conditions. The lubricating properties

are a consequence of the solid state layer structure (compare with graphite), in which there are only weak van der Waals forces operating between S–Mo–S slabs. Applications of MoS$_2$ lubricants range from engine oils and greases (used in engineering equipment) to coatings on sliding fitments.

WOCl$_4$, W–O–W bridges are present. The layer structure of WO$_2$Cl$_2$ is related to that of SnF$_4$ (**13.11**); each layer comprises bridged WO$_4$Cl$_2$ units (**22.13**) and the lattice is able to act as an intercalation host.

• = Cl

(22.13)

The most important compounds of Mo(VI) and W(VI) are the oxides and the molybdate and tungstate anions. White MoO$_3$ (mp 1073 K) is usually made by reaction 22.34, and yellow WO$_3$ (mp 1473 K) by dehydration of tungstic acid (see below). Both oxides are commercially available.

$$MoS_2 \xrightarrow{\text{roast in air}} MoO_3 \qquad (22.34)$$

The structure of MoO$_3$ consists of layers of linked MoO$_6$ octahedra; the arrangement of the MoO$_6$ units is complex and results in a unique 3D-lattice. Several polymorphs of WO$_3$ exist, all based on the ReO$_3$ lattice (*Figure 21.3*). Thin films of WO$_3$ are used in electrochromic windows. Neither oxide reacts with acids, but in aqueous alkali, [MO$_4$]$^{2-}$ or polyoxometallate ions are produced. The chemistry of molybdates and tungstates is complicated and an active area of research; uses of the homo- and heteropoly-anions are extremely varied.[†] The simplest molybdate(VI) and tungstate(VI) are [MoO$_4$]$^{2-}$ and [WO$_4$]$^{2-}$, many salts of which are known. Alkali metal salts such as Na$_2$MoO$_4$ and Na$_2$WO$_4$ (commercially available as the dihydrates and useful starting materials in this area of chemistry) are made by dissolving MO$_3$ (M = Mo, W) in aqueous alkali metal hydroxide. From *strongly acidic* solutions of these molybdates and tungstates, it is possible to isolate yellow 'molybdic acid' and 'tungstic acid'. Crystalline molybdic and tungstic acids are formulated as MoO$_3$·2H$_2$O and

WO$_3$·2H$_2$O, and possess layer structures consisting of corner-sharing MO$_5$(H$_2$O) octahedra with additional H$_2$O molecules residing between the layers. In crystalline salts, the [MO$_4$]$^{2-}$ ions are discrete and tetrahedral. In acidic media and dependent on the pH, condensation occurs to give polyanions, e.g. reaction 22.35.

$$7[MoO_4]^{2-} + 8H^+ \longrightarrow [Mo_7O_{24}]^{6-} + 4H_2O \qquad pH \approx 5$$

(22.35)

The structure of [Mo$_7$O$_{24}$]$^{6-}$ is shown in Figure 22.8a, and structural features, common to other polynuclear molybdates and tungstates, are that:

- the cage is supported by oxygen bridges and there is no metal–metal bonding;
- the cage is constructed from octahedral MO$_6$-units connected by shared oxygen atoms.

As a consequence of this last point, the structures may be represented in terms of linked octahedra, in much the same way that silicate structures are depicted by linked tetrahedra (see *structure* **13.16** and *Figure 13.15*). Figure 22.8b shows such a representation for [Mo$_7$O$_{24}$]$^{6-}$; each vertex corresponds to an O atom in Figure 22.8a. By controlling the pH or working in non-aqueous media, salts of other molybdates and tungstates can be isolated. One of the simplest is [M$_6$O$_{19}$]$^{2-}$ (M = Mo, W) which is isostructural with [M$_6$O$_{19}$]$^{8-}$ (M = Nb, Ta) and possesses the *Lindqvist structure* (Figure 22.8c). For tungsten, the solution system is more complicated than for molybdenum, and involves equilibria with W$_7$, W$_{10}$, W$_{11}$ and W$_{12}$ species; the lowest nuclearity anion, [W$_7$O$_{24}$]$^{6-}$, is isostructural with [Mo$_7$O$_{24}$]$^{6-}$. Salts can be isolated by careful control of pH, and under non-aqueous conditions salts of polytungstates unknown in aqueous solution can be crystallized.

Heteropolyanions have been well studied and have many applications, e.g. as catalysts. Two families are important:

- the α-Keggin anions,[‡] [XM$_{12}$O$_{40}$]$^{n-}$ (M = Mo, W; e.g. X = P or As, n = 3; X = Si, n = 4; X = B, n = 5);

[†] For overviews of applications, see: J.T. Rhule, C.L. Hill and D.A. Judd (1998) *Chemical Reviews*, vol. 98, p. 327; D.E. Katsoulis (1998) *Chemical Reviews*, vol. 98, p. 359.

[‡] The prefix α distinguishes the structural type discussed here from other isomers; the first example, [PMo$_{12}$O$_{40}$]$^{3-}$, was reported in 1826 by Berzelius, and was structurally elucidated using X-ray diffraction in 1933 by J.F. Keggin.

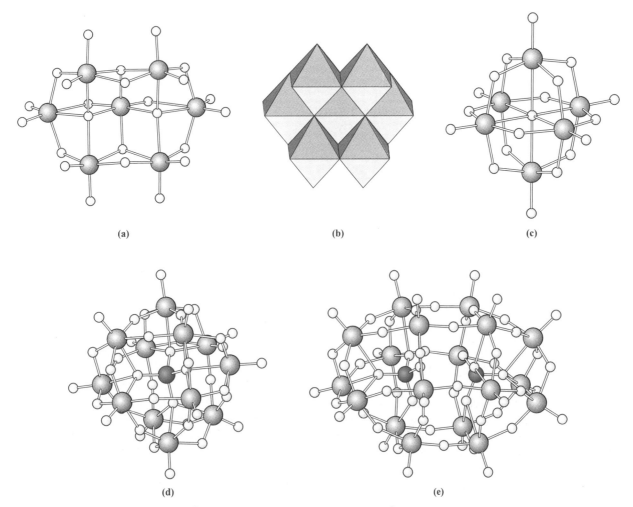

(a) (b) (c)

(d) (e)

Fig. 22.8 (a) The structure of $[Mo_7O_{24}]^{6-}$ in the $[H_3N(CH_2)_2NH_2(CH_2)_2NH_3]^{3+}$ salt [P. Roman *et al.* (1992) *Polyhedron*, vol. 11, p. 2027]; (b) the $[Mo_7O_{24}]^{6-}$ ion represented in terms of seven octahedral building blocks (these can be generated in diagram (a) by connecting the O atoms); (c) the structure of $[W_6O_{19}]^{2-}$ determined for the $[W(CN^tBu)_7]^{2+}$ salt [W.A. LaRue *et al.* (1980) *Inorg. Chem.*, vol. 19, p. 315]; (d) the structure of the α-Keggin ion $[SiMo_{12}O_{40}]^{4-}$ in the guanidinium salt (the Si atom is shown in black) [H. Ichida *et al.* (1980) *Acta Crystallogr.*, *Sect. B*, vol. 36, p. 1382]; (e) the structure of $[H_3S_2Mo_{18}O_{62}]^{5-}$ (in the $[^nBu_4N]^+$ salt) formed by reducing the α-Dawson anion $[S_2Mo_{18}O_{62}]^{4-}$ (the S atoms are shown in black; H atoms are omitted) [R. Neier *et al.* (1995) *J. Chem. Soc.*, *Dalton Trans.*, p. 2521]. Colour code: Mo and W, blue; O, pale grey; heteroatoms, black.

BOX 22.6   **APPLICATIONS**

Catalytic applications of MoO₃ and molybdates

Molybdenum-based catalysts are used to facilitate a range of organic transformations including benzene to cyclohexane, ethylbenzene to styrene, and propene to acetone.

Acrylonitrile (used in the manufacture of acrylic fibres, resins and rubbers) is produced commercially on a large scale by the reaction:

$$CH_2=CHCH_3 + NH_3 + \tfrac{3}{2}O_2$$

$$\xrightarrow{\text{bismuth-molybdate catalyst}} CH_2=CHCN + 3H_2O$$

The bismuth-molybdate catalyst functions by providing intimately associated Bi–O and Mo=O sites. The Bi–O sites are involved in abstracting α-hydrogen (see structure **23.36**) while the Mo=O sites interact with the incoming

alkene, and are involved in activation of NH₃ and in C–N bond formation.

In Box 11.2, we described methods of desulfurizing emission gases. A combination of MoO₃ and CoO supported on activated alumina acts as an effective catalyst for the desulfurization of petroleum and coal-based products. This catalyst system has wide application in a process that contributes significantly to reducing SO₂ emissions.

Further reading

R.K. Grasselli (1986) *Journal of Chemical Education*, vol. 63, p. 216 – 'Selective oxidation and ammoxidation of olefins by heterogeneous catalysis'.

- the α-Dawson anions, $[X_2M_{18}O_{62}]^{n-}$ (M = Mo, W; e.g. X = P or As, $n = 6$).

Equations 22.36 and 22.37 show typical syntheses of α-Keggin ions; all ions are structurally similar (Figure 22.8d) with the hetero-atom tetrahedrally sited in the centre of the polyoxometallate cage. The construction of the cage from oxygen-linked, octahedral MO_6-units is apparent by studying Figure 22.8d.

$$[HPO_4]^{2-} + 12[WO_4]^{2-} + 23H^+ \longrightarrow [PW_{12}O_{40}]^{3-} + 12H_2O \quad (22.36)$$

$$[SiO_3]^{2-} + 12[WO_4]^{2-} + 22H^+ \longrightarrow [SiW_{12}O_{40}]^{4-} + 11H_2O \quad (22.37)$$

α-Dawson anions of Mo are formed spontaneously in solutions containing $[MoO_4]^{2-}$ and phosphates or arsenates at appropriate pH, but formation of corresponding tungstate species is slower and requires an excess of phosphate or arsenate. The α-Dawson cage structure can be viewed as the condensation of two α-Keggin ions with loss of six WO_3-units (compare Figures 22.8d and 22.8e). The structure shown in Figure 22.8e is that of $[H_3S_2Mo_{18}O_{62}]^{5-}$, a protonated product of the four-electron reduction of the α-Dawson ion $[S_2Mo_{18}O_{62}]^{4-}$; apart from bond length changes, the cage remains unaltered by the addition of electrons. Similarly, reduction of α-Keggin ions occurs without gross structural changes. Reduction converts some of the M(VI) to M(V) centres and is accompanied by a change in colour to intense blue; hence, reduced Keggin and Dawson anions are called *heteropoly blues*.

Heteropolyanions with incomplete cages, *lacunary* anions, may be made under controlled pH conditions, e.g. at pH ≈ 1, $[PW_{12}O_{40}]^{3-}$ can be prepared (equation 22.36) while at pH ≈ 2, $[PW_{11}O_{39}]^{7-}$ can be formed. Lacunary ions act as ligands by coordination through terminal *O*-atoms; and complexes include $[PMo_{11}VO_{40}]^{4-}$, $[(PW_{11}O_{39})Ti(\eta^5-C_5H_5)]^{4-}$ and $[(PW_{11}O_{39})Rh_2(O_2CMe)_2(DMSO)_2]^{5-}$.

The formation of mononuclear complexes by Mo(VI) and W(VI) is limited. Simple complexes include octahedral $[WOF_5]^-$ and *cis*-$[MoF_4O_2]^{2-}$; salts of $[MoF_7]^-$ (equation 22.38) and $[MoF_8]^{2-}$ (equation 22.39) have recently been isolated.

$$MoF_6 + [Me_4N]F \xrightarrow{MeCN} [Me_4N][MoF_7] \quad (22.38)$$

$$MoF_6 + 2KF \xrightarrow{in\ IF_5} K_2[MoF_8] \quad (22.39)$$

The peroxo ligand, $[O_2]^{2-}$, forms a range of complexes with Mo(VI) and W(VI), e.g. $[M(O_2)_2(O)(ox)]^{2-}$ (M = Mo, W) is pentagonal bipyramidal (22.14) and $[Mo(O_2)_4]^{2-}$ is dodecahedral. Some peroxo complexes of Mo(VI) catalyze the epoxidation of alkenes.

(22.14)

Molybdenum(V) and tungsten(V)

The known pentahalides are yellow MoF_5, yellow WF_5, black $MoCl_5$, dark green WCl_5 and black WBr_5, all solids at 298 K. The pentafluorides are made by heating MoF_6 with Mo (or WF_6 with W, equation 22.40), but both disproportionate on heating (equation 22.41).

$$5WF_6 + W \xrightarrow{\Delta} 6WF_5 \quad (22.40)$$

$$2MF_5 \xrightarrow{\Delta,\ T\ K} MF_6 + MF_4 \quad (22.41)$$

M = Mo, $T > 440$ K; M = W, $T > 320$ K

Direct combination of the elements under controlled conditions gives $MoCl_5$ and WCl_5. The pentafluorides MoF_5 and WF_5 are tetrameric in the solid, isostructural with NbF_5 and TaF_5 (22.4); $MoCl_5$ and WCl_5 are dimeric and structurally similar to $NbCl_5$ and $TaCl_5$ (22.5). Each pentahalide is paramagnetic, indicating little or no metal–metal interaction.

Tungsten bronzes contain M(V) and M(VI) and are formed by vapour phase reduction of WO_3 by alkali metals, reduction of Na_2WO_4 by H_2 at 800–1000 K, or by reaction 22.42.

$$\frac{x}{2}Na_2WO_4 + \frac{3-2x}{3}WO_3 + \frac{x}{6}W \xrightarrow{1120\ K} Na_xWO_3 \quad (22.42)$$

Tungsten bronzes are inert materials M_xWO_3 ($0 < x < 1$) with defect perovskite structures (*Figure 5.23*). Their colour depends on x: golden for $x \approx 0.9$, red for $x \approx 0.6$, violet for $x \approx 0.3$. Bronzes with $x > 0.25$ exhibit metallic conductivity owing to a band-like structure associated with W(V) and W(VI) centres in the lattice; those with $x < 0.25$ are semiconductors (see *Section 5.8*). Similar compounds are formed by Mo, Ti and V.

Our discussion of complexes of Mo(V) and W(V) is restricted to selected, mononuclear species; octahedral coordination is common. Halo anions include $[MoF_6]^-$ (equation 22.43), $[WF_6]^-$, $[MoCl_6]^-$ and $[WF_8]^{3-}$ (equation 22.44).

$$MoF_6\ (in\ excess) \xrightarrow{KI\ in\ liquid\ SO_2} K[MoF_6] \quad (22.43)$$

$$W(CO)_6 \xrightarrow{KI\ in\ IF_5} K_3[WF_8] \quad (22.44)$$

Treatment of WCl_5 with concentrated HCl leads to $[WOCl_5]^{2-}$; $[WOBr_5]^{2-}$ forms when $[W(O)_2(ox)_2]^{3-}$ reacts with aqueous HBr. Dissolution of $[MoOCl_5]^{2-}$ in aqueous acid produces yellow $[Mo_2O_4(H_2O)_6]^{2+}$ (22.15) which is diamagnetic, consistent with a Mo−Mo single bond. A number of complexes $[MOCl_3L_2]$ are known, e.g. $WOCl_3(THF)_2$ (a useful starting material since the THF ligands are labile), $[WOCl_3(PEt_3)_2]$ (**22.16**) and $[MoOCl_3(bpy)]$. High coordination numbers are observed in $[Mo(CN)_8]^{3-}$ and $[W(CN)_8]^{3-}$, formed by oxidation of $[M(CN)_8]^{4-}$ using Ce^{4+} or $[MnO_4]^{-}$; the coordination geometries are cation-dependent, illustrating the small energy difference between dodecahedral or square antiprismatic structures.

$$
\left[
\begin{array}{c}
\text{(22.15)}
\end{array}
\right]^{2+}
\qquad
\text{(22.16)}
$$

Molybdenum(IV) and tungsten(IV)

Binary halides MX_4 are established for $M = Mo$, W and $X = F$, Cl and Br; WI_4 exists but is not well characterized. Equations 22.45 and 22.46 show representative syntheses.

$$MoO_3 \xrightarrow{H_2,\ 720\,K} MoO_2 \xrightarrow{CCl_4,\ 520\,K} MoCl_4 \qquad (22.45)$$

$$WCl_6 \xrightarrow{W(CO)_6,\ reflux\ in\ chlorobenzene} WCl_4 \qquad (22.46)$$

Tungsten(IV) fluoride is polymeric, and a polymeric structure for MoF_4 is consistent with Raman spectroscopic data. Three polymorphs of $MoCl_4$ exist: α-$MoCl_4$ has the $NbCl_4$ structure (**22.10**) and, at 520 K, transforms to the β-form containing cyclic Mo_6Cl_{24} units; the structure of the third polymorph is unknown. Tungsten(IV) chloride (structurally like α-$MoCl_4$) is a useful starting material in W(IV) and lower oxidation state chemistry. All the tetrahalides are air- and moisture-sensitive.

Reduction of MO_3 by H_2 yields MoO_2 and WO_2 which adopt rutile lattices (*Figure 5.21*), distorted (as in NbO_2) by pairing of metal centres; in MoO_2, Mo−Mo distances are 251 and 311 pm. The oxides do not dissolve in non-oxidizing acids. Molybdenum(IV) sulfide (equation 22.28) has a layer structure and is used as a lubricant (see *Box 22.5*).

Molybdenum(IV) is stabilized in acidic solution as red $[Mo_3(\mu_3\text{-O})(\mu\text{-O})_3(H_2O)_9]^{4+}$ (Figure 22.9) which is formed by reduction of $Na_2[MoO_4]$ or oxidation of $[Mo_2(H_2O)_8]^{4+}$.

The halo-complexes $[MX_6]^{2-}$ (M = Mo, W; X = F, Cl, Br) are known although $[WF_6]^{2-}$ has been little studied. By adjusting the conditions of reaction 22.43 (i.e. taking a 1:2 molar ratio MoF_6:I^-, and removing I_2 as it is formed), $K_2[MoF_6]$ can be isolated. Salts of $[MoCl_6]^{2-}$ can be made starting from $MoCl_5$, e.g. $[NH_4]_2[MoCl_6]$ by heating $MoCl_5$ with NH_4Cl. Many salts of $[WCl_6]^{2-}$ are known (e.g. reaction 22.47) but the ion decomposes on contact with water.

$$2M[WCl_6] \xrightarrow{550\,K} M_2[WCl_6] + WCl_6 \quad \text{(M = group 1 metal)} \qquad (22.47)$$

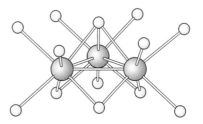

Fig. 22.9 The structure of $[Mo_3(\mu_3\text{-O})(\mu\text{-O})_3(H_2O)_9]^{4+}$ determined by X-ray diffraction for the hydrated $[4\text{-MeC}_6H_4SO_3]^{-}$ salt; H atoms are omitted from the terminally bound H_2O ligands. Mo−Mo distances are in the range 247–249 pm. [D.T. Richens *et al.* (1989) *Inorg. Chem.*, vol. 28, p. 1394]. Colour code: Mo, blue; O, pale grey.

Reduction of H_2WO_4 using Sn in HCl in the presence of K_2CO_3 leads to $K_4[W_2(\mu\text{-O})Cl_{10}]$; the anion is structurally like $[Ta_2(\mu\text{-O})F_{10}]^{2-}$ (Figure 22.5a).

Octahedral geometries are common for complexes of Mo(IV) and W(IV), syntheses of which often involve ligand-mediated reduction of the metal centre, e.g. reactions 22.48 and 22.49.

$$WOCl_4 + 3Ph_3P \longrightarrow \textit{trans}\text{-}[WCl_4(PPh_3)_2] + Ph_3PO \qquad (22.48)$$

$$MoCl_5 \xrightarrow{\text{excess py or bpy}} \begin{cases} MoCl_4(py)_2 \\ MoCl_4(bpy) \end{cases} \qquad (22.49)$$

The salt $K_4[Mo(CN)_8]\cdot 2H_2O$ was the first example (in 1939) of an 8-coordinate (dodecahedral) complex. However, studies on a range of salts of $[Mo(CN)_8]^{4-}$ and $[W(CN)_8]^{4-}$ reveal cation dependence, both dodecahedral and square antiprismatic anions being found. The $K_4[M(CN)_8]$ salts are formed by reactions of K_2MoO_4, KCN and KBH_4 in the presence of acetic acid; the $[M(CN)_8]^{4-}$ ions are kinetically inert with respect to ligand substitution (see *Section 25.2*), but can be oxidized to $[M(CN)_8]^{3-}$ as described earlier.

Molybdenum(III) and tungsten(III)

All the binary halides of Mo(III) and W(III) are known except for WF_3. The Mo(III) halides are made by reducing a halide of a higher oxidation state. Reduction of $MoCl_5$ with H_2 at 670 K gives $MoCl_3$ which has a layer lattice similar to $CrCl_3$ but distorted and rendered diamagnetic by pairing of metal atoms (Mo−Mo = 276 pm). The 'W(III) halides', prepared by controlled halogenation of a lower halide, contain cluster units (see equations 22.55 and 22.56).

In contrast to Cr(III) (see *Section 21.7*), mononuclear complexes of Mo(III) and W(III) (especially the latter) are rare, there being an increased tendency for M−M bonding for the M(III) state. Electrolytic reduction of MoO_3 in concentrated HCl yields $[MoCl_5(H_2O)]^{2-}$ and $[MoCl_6]^{3-}$, the red K^+ salts of which are stable in dry air but are readily hydrolyzed to $[Mo(H_2O)_6]^{3+}$, one of the few simple aqua ions of the heavier metals. By changing the reaction conditions, $[Mo_2Cl_9]^{3-}$ is formed in place of $[MoCl_6]^{3-}$, but reduction of WO_3 in concentrated HCl always gives

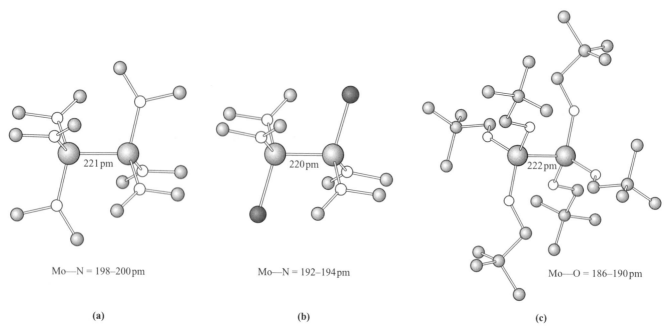

Mo—N = 198–200 pm   Mo—N = 192–194 pm   Mo—O = 186–190 pm

(a)   (b)   (c)

Fig. 22.10 The staggered structures (X-ray diffraction) of (a) $Mo_2(NMe_2)_6$ [M.H. Chisholm *et al.* (1976) *J. Am. Chem. Soc.*, vol. 98, p. 4469], (b) $Mo_2Cl_2(NMe_2)_4$ [M. Akiyama *et al.* (1977) *Inorg. Chem.*, vol. 16, p. 2407] and (c) $Mo_2(OCH_2{}^tBu)_6$ [M.H. Chisholm *et al.* (1977) *Inorg. Chem.*, vol. 16, p. 1801]. Hydrogen atoms are omitted for clarity; colour code: Mo, blue; N and O, white; C, grey; Cl, black.

$[W_2Cl_9]^{3-}$, and $[WX_6]^{3-}$ has not been isolated. Both $[MoF_6]^{3-}$ and $[MoCl_6]^{3-}$ are paramagnetic with magnetic moments close to $3.8\,\mu_B$ ($\approx \mu$(spin-only) for d^3). The $[M_2X_9]^{3-}$ ions adopt structure **22.17**; magnetic data and M–M distances (from crystalline salts) are consistent with metal–metal bonding. The $[W_2Cl_9]^{3-}$ ion is diamagnetic, indicating a $W\equiv W$ triple bond consistent with the short bond length of 242 pm; oxidation (equation 22.50) to $[W_2Cl_9]^{2-}$ causes the W–W bond to lengthen to 254 pm consistent with a lower bond order of 2.5.

$$2[W_2Cl_9]^{3-} + Cl_2 \longrightarrow 2[W_2Cl_9]^{2-} + 2Cl^- \qquad (22.50)$$

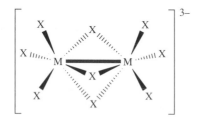

M = Mo, W; X = Cl, Br (see text)

(22.17)

In $Cs_3[Mo_2X_9]$, the Mo–Mo bond lengths are 266 pm (X = Cl) and 282 pm (X = Br); these data and magnetic moments at 298 K of $0.6\,\mu_B$ (X = Cl) and $0.8\,\mu_B$ (X = Br) per Mo, indicate significant Mo–Mo interaction but with a bond order <3. Contrast this with $[Cr_2X_9]^{3-}$ in which there is no Cr–Cr bonding (*Section 21.7*).

In Mo(III) and W(III) chemistry, $Mo\equiv Mo$ and $W\equiv W$ triple bonds ($\sigma^2\pi^4$, see *Figure 21.13*) are common, and derivatives with amido and alkoxy ligands have received much attention, e.g. as precursors for solid state materials. Reaction of $MoCl_3$ (or $MoCl_5$) or WCl_4 with $LiNMe_2$ gives $Mo_2(NMe_2)_6$ or $W_2(NMe_2)_6$ respectively. Both possess staggered structures (Figure 22.10a) with M–M bond lengths of 221 (Mo) and 229 pm (W) typical of triple bonds. The orientations of the NMe_2 groups in the solid state suggest that the M–N bonds contain metal *d*–nitrogen *p* π-contributions. A staggered conformation, short Mo–Mo bond and shortened Mo–N bonds are also observed in $Mo_2Cl_2(NMe_2)_4$ (Figure 22.10b); this and the W analogue are made by reacting $M_2(NMe_2)_6$ with Me_3SiCl. The air- and moisture-sensitive $M_2(NMe_2)_6$ and $M_2Cl_2(NMe_2)_4$ (M = Mo, W) are precursors for many derivatives including alkoxy compounds (equation 22.51 and Figure 22.10c); $[W_2(OR)_6]$ compounds are less stable than their Mo analogues. An extensive chemistry of alkoxy derivatives has been developed.[†]

$$Mo_2(NMe_2)_6 \xrightarrow{\text{ROH}} Mo_2(OR)_6 \qquad (22.51)$$

(R must be bulky, e.g. ${}^tBu$, ${}^iPr$, $CH_2{}^tBu$)

Molybdenum(II) and tungsten(II)

With the exception of organometallic and cyano complexes, few mononuclear species are known for Mo(II) and W(II). The pentagonal bipyramidal $[Mo(CN)_7]^{5-}$ ion is formed by reducing $[MoO_4]^{2-}$ using H_2S in the presence of $[CN]^-$; in

---

[†] For example, see M.H. Chisholm, D.M. Hoffman and J.C. Huffman (1985) *Chemical Society Reviews*, vol. 14, p. 69; M.H. Chisholm (1995) *Chemical Society Reviews*, vol. 24, 79; M.H. Chisholm (1996) *J. Chem. Soc., Dalton Trans.*, p. 1781.

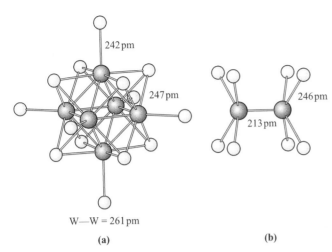

Fig. 22.11 The structures (X-ray diffraction) of (a) $[Mo_6Cl_{14}]^{2-}$ in the $[Ph_4P]^+$ salt [M.A. White *et al.* (1994) *Acta Crystallogr.*, *Sect. C*, vol. 50, p. 1087] and (b) $[Mo_2Cl_8]^{4-}$ in the compound $[H_3NCH_2CH_2NH_3]_2[Mo_2Cl_8]\cdot 2H_2O$ [J.V. Brenic *et al.* (1969) *Inorg. Chem.*, vol. 8, p. 2698]. Colour code: Mo, blue; Cl, pale grey.

the capped trigonal prismatic $[MoBr(CN^tBu)_6]^+$ (equation 22.52), the Br^- ligand occupies the capping site.

$$[MoBr_2(CO)_4] + 6\,^tBuNC \longrightarrow [MoBr(CN^tBu)_6]Br + 4CO \tag{22.52}$$

The binary M(II) (M = Mo, W) halides are made by thermally decomposing higher halides. The structures of the dihalides consist of $[M_6X_8]^{4+}$ clusters (structurally like $[Nb_6I_8]^{3+}$, Figure 22.6a) with each M atom bonded to an additional X atom; the clusters are connected into a 2D-layer lattice by M–X–M bridges, i.e. $[M_6X_8]X_2X_{4/2}$. Reactions such as 22.53 and 22.54 produce salts containing discrete $[M_6X_{14}]^{2-}$ ions (Figure 22.11a). The diamagnetism of $[M_6X_{14}]^{2-}$ is consistent with M–M bonding, and M–M single bonds can be allocated by following a similar electron-counting procedure as for $[Nb_6X_{12}]^{2+}$ and $[Ta_6X_{12}]^{2+}$ (see *Section 22.6*).

$$MoCl_2 \xrightarrow{\text{Et}_4\text{NCl in dil. HCl, EtOH}} [Et_4N]_2[Mo_6Cl_{14}] \tag{22.53}$$

$$MoBr_2 \xrightarrow{\text{CsBr, IBr, 325 K}} Cs_2[Mo_6Br_{14}] \tag{22.54}$$

While the Mo(II) halides are not readily oxidized, WCl_2 and WBr_2 (equations 22.55 and 22.56) are oxidized to give products containing $[W_6(\mu\text{-}Cl)_{12}]^{6+}$ or $[W_6(\mu_3\text{-}Br)_8]^{6+}$ clusters, terminal halides and bridging $[Br_4]^{2-}$ units. The formulae of the products indicate whether the clusters are discrete or linked.

$$[W_6Cl_8]Cl_2Cl_{4/2} \xrightarrow{\text{Cl}_2,\ 373\ K} [W_6Cl_{12}]Cl_6 \tag{22.55}$$

$$[W_6Br_8]Br_2Br_{4/2} \xrightarrow{\text{Br}_2,\ T<420\ K} \begin{cases} [W_6Br_8]Br_6 \\ [W_6Br_8]Br_4(Br_4)_{2/2} \\ [W_6Br_8]Br_2(Br_4)_{4/2} \end{cases} \tag{22.56}$$

Compounds containing an $\{Mo\equiv Mo\}^{4+}$ unit are well exemplified in Mo(II) chemistry, although species containing $W\equiv W$ bonds are difficult to make. A description of a $Mo\equiv Mo$ quadruple bond in terms of σ, π and δ components is analogous to that of a $Cr\equiv Cr$ bond (see *Section 21.7* and *Figure 21.13*), and the effect of the δ-component in forcing the ligands to be eclipsed is illustrated by the structure of $[Mo_2Cl_8]^{4-}$ (Figure 22.11b). This is made in the reaction sequence 22.57, the intermediate acetate $Mo_2(\mu\text{-}O_2CMe)_4$ (**22.18**) being a useful synthon in this area of chemistry, e.g. reaction 22.58. Replacement of the Cl^- ligands in $[Mo_2Cl_8]^{4-}$ yields a range of derivatives; equation 22.59 gives examples, one of which involves concomitant oxidation of the $\{Mo\equiv Mo\}^{4+}$ core. Derivatives containing $[MeSO_3]^-$ or $[CF_3SO_3]^-$ bridges are useful precursors and can be used to prepare the highly reactive $[Mo_2(NCMe)_8]^{4+}$ (equation 22.60).

= quadruple bond

(22.18)

$$Mo(CO)_6 \xrightarrow{\text{MeCO}_2\text{H}} Mo_2(\mu\text{-}O_2CMe)_4 \xrightarrow{\text{KCl in HCl}} [Mo_2Cl_8]^{4-} \tag{22.57}$$

$$Mo_2(\mu\text{-}O_2CMe)_4 \xrightarrow{\text{LiMe in Et}_2\text{O}} Li_4[Mo_2Me_8]\cdot 4Et_2O \tag{22.58}$$

$$[Mo_2Cl_8]^{4-} \xrightarrow[\text{in presence of O}_2]{[\text{HPO}_4]^{2-}} [Mo_2(\mu\text{-}HPO_4)_4]^{2-}$$

$$\downarrow {[\text{SO}_4]^{2-}} \tag{22.59}$$

$$[Mo_2(\mu\text{-}SO_4)_4]^{4-}$$

$$[Mo_2(\mu\text{-}O_3SCF_3)_2(H_2O)_4]^{2+} \xrightarrow{\text{MeCN}} [Mo_2(NCMe)_8]^{4+} \tag{22.60}$$

Each Mo centre in these Mo_2-derivatives possesses a vacant orbital (as in structure **21.14**), but forming Lewis base adducts is not facile; '$[Mo_2(\mu\text{-}O_2CMe)_4(H_2O)_2]$' has not been isolated, although the oxidized species $[Mo_2(\mu\text{-}SO_4)_4(H_2O)_2]^{3-}$ and $[Mo_2(\mu\text{-}HPO_4)_4(H_2O)_2]^{2-}$ are known. An unstable adduct $[Mo_2(\mu\text{-}O_2CMe)_4(py)_2]$ results from addition of pyridine to $[Mo_2(\mu\text{-}O_2CMe)_4]$, and a more stable one by using $[Mo_2(\mu\text{-}O_2CCF_3)_4]$.

Not all the derivatives mentioned above contain $Mo\equiv Mo$ bonds, e.g. oxidation occurs in the reaction with $[HPO_4]^{2-}$ (equation 22.59). Table 22.2 lists the Mo–Mo bond lengths

Table 22.2   Mo−Mo bond lengths and orders in selected dimolybdenum species.

Compound or ion[‡]	Mo−Mo bond distance / pm	Mo−Mo bond order	Notes
$Mo_2(\mu\text{-}O_2CMe)_4$	209	4.0	Structure **22.18**
$Mo_2(\mu\text{-}O_2CCF_3)_4$	209	4.0	Analogous to **22.18**
$[Mo_2Cl_8]^{4-}$	214	4.0	Figure 22.11b
$[Mo_2(\mu\text{-}SO_4)_4]^{4-}$	211	4.0	Analogous to **22.18**
$[Mo_2(\mu\text{-}SO_4)_4(H_2O)_2]^{3-}$	217	3.5	Contains axial H_2O ligands
$[Mo_2(\mu\text{-}HPO_4)_4(H_2O)_2]^{2-}$	223	3.0	Contains axial H_2O ligands

[‡] Data for anions refer to K^+ salts; contrast Figure 22.11b where the $[Mo_2Cl_8]^{4-}$ parameters refer to the $[H_3NCH_2CH_2NH_3]^{2+}$ salt.

in selected compounds, and the bond orders follow from the energy level diagram in Figure 21.13b; e.g. $[Mo_2Cl_8]^{4-}$ has a $\sigma^2\pi^4\delta^2$ configuration (Mo≡Mo), but $[Mo_2(HPO_4)_4]^{2-}$ and $[Mo_2(HPO_4)_4(H_2O)_2]^{2-}$ are $\sigma^2\pi^4$ (Mo≡Mo).

22.8 Group 7: technetium and rhenium

The metals

The heavier group 7 metals, Tc and Re, are less reactive than Mn. Technetium does not occur naturally (see *Section 22.2*). The bulk metals tarnish slowly in air, but more finely divided Tc and Re burn in O_2 (equation 22.61) and react with the halogens (see below). Reactions with sulfur give TcS_2 and ReS_2.

$$4M + 7O_2 \xrightarrow{650\,K} 2M_2O_7 \quad (M = Tc, Re) \quad (22.61)$$

The metals dissolve in oxidizing acids (e.g. conc HNO_3) to give $HTcO_4$ (pertechnetic acid) and $HReO_4$ (perrhenic acid), but are insoluble in HF or HCl.

Technetium and rhenium exhibit oxidation states from 0 to +7 (*Table 19.3*), although M(II) and lower states are stabilized by π-acceptor ligands such as CO and will not be considered further in this section. The chemistry of Re is better developed than that of Tc, but interest in the latter has expanded with the current use of its compounds in nuclear medicine. There are significant differences between the chemistries of Mn and the heavier group 7 metals:

- a comparison of the potential diagrams in Figure 22.12 with that for Mn (*Figure 7.2*) shows that $[TcO_4]^-$ and $[ReO_4]^-$ are significantly more stable with respect to reduction than $[MnO_4]^-$;

- the heavier metals have less cationic chemistry than manganese;

- a tendency for M−M bond formation leads to higher nuclearity species being important for the heavier metals.

High oxidation states of technetium and rhenium: M(VII), M(VI) and M(V)

Rhenium reacts with F_2 to give yellow ReF_6 and ReF_7 depending on conditions, and ReF_5 is made by reaction 22.62. Direct combination of Tc and F_2 leads to TcF_6 and TcF_5; TcF_7 is not known.

$$ReF_6 \xrightarrow{\text{over W wire, 870 K}} ReF_5 \quad (22.62)$$

For the later halogens, combination of the elements at appropriate temperatures affords $TcCl_6$, $ReCl_6$, $ReCl_5$ and $ReBr_5$. The high oxidation state halides are volatile solids which are hydrolyzed by water to $[MO_4]^-$ and MO_2 (e.g. equation 22.63).

$$3ReF_6 + 10H_2O \longrightarrow 2HReO_4 + ReO_2 + 18HF \quad (22.63)$$

(22.19)

(22.20)

The fluorides ReF_7, ReF_6 and TcF_6 are molecular with pentagonal bipyramidal, **22.19**, and octahedral structures; $ReCl_6$ is probably a molecular monomer, but $ReCl_5$ (a useful precursor in Re chemistry) is a dimer (**22.20**).

$$[TcO_4]^- \xrightarrow{+0.78} TcO_2 \xrightarrow{(+0.90)} Tc^{3+} \xrightarrow{+0.30} Tc^{2+} \xrightarrow{+0.40} Tc$$
$$+0.62$$

$$[ReO_4]^- \xrightarrow{+0.51} ReO_2 \xrightarrow{(+0.16)} Re^{3+} \xrightarrow{+0.30} Re$$
$$+0.37$$

Fig. 22.12 Potential diagrams for technetium and rhenium in aqueous solution at pH 0; compare with the diagram for manganese in Figure 7.2.

Oxohalides are well represented:

- M(VII): $ReOF_5$, TcO_2F_3, ReO_2F_3, ReO_3F, TcO_3Cl, TcO_3Br, TcO_3I, ReO_3Cl, ReO_3Br;
- M(VI): $TcOF_4$, $ReOF_4$, ReO_2F_2, $TcOCl_4$, $ReOCl_4$, $ReOBr_4$;
- M(V): $ReOF_3$, $TcOCl_3$.

They are prepared by reacting oxides with halogens, or halides with O_2, or by reactions such as 22.64 and 22.65.

$$ReCl_5 + 3Cl_2O \longrightarrow ReO_3Cl + 5Cl_2 \qquad (22.64)$$

$$\left.\begin{array}{l} Tc_2O_7 + 4HF \longrightarrow 2TcO_3F + [H_3O]^+ + [HF_2]^- \\ TcO_3F + XeF_6 \longrightarrow TcO_2F_3 + XeOF_4 \end{array}\right\} \qquad (22.65)$$

Few oxohalides have been structurally elucidated; $ReOCl_4$ is molecular (**22.21**), while TcO_2F_3 is polymeric with oxo groups *trans* to bridging F atoms (**22.22**). A number of imido analogues ($[RN]^{2-}$ is isoelectronic with O^{2-}) have been structurally characterized and include tetrahedral $Re(N^tBu)_3Cl$ and trigonal bipyramidal $Re(N^tBu)_2Cl_3$ (**22.23**). Reduction of $Tc(NAr)_3I$ using Na leads to the dimer $[(ArN)_2Tc(\mu-NAr)_2Tc(NAr)_2]$ when $Ar = 2,6-Me_2C_6H_3$, but when Ar is the bulkier $2,6-^iPr_2C_6H_3$, the product is $Tc_2(NAr)_6$ with an ethane-like configuration; each dimer contains a Tc–Tc single bond.

(22.21)

(22.22)

(22.23)

The yellow, volatile oxides M_2O_7 (M = Tc, Re) form when the metals burn in O_2; the volatility of Re_2O_7 is utilized in manufacturing Re (see *Section 22.2*). In the solid and vapour states, Tc_2O_7 is molecular with a linear Tc–O–Tc

(22.24)

bridge (**22.24**). In the vapour, Re_2O_7 has a similar structure but the solid adopts a complex layer lattice. The oxides are the anhydrides of $HTcO_4$ and $HReO_4$ and dissolve in water (equation 22.66) to give solutions containing $[TcO_4]^-$ (pertechnetate) and $[ReO_4]^-$ (perrhenate); pertechnetate and perrhenate salts are the commonest starting materials in Tc and Re chemistries.

$$M_2O_7 + H_2O \longrightarrow 2HMO_4 \qquad (M = Tc, Re) \qquad (22.66)$$

Pertechnetic and perrhenic acids are strong acids; crystalline $HReO_4$ (yellow), $HReO_4 \cdot H_2O$ and $HTcO_4$ (dark red) have been isolated; crystalline $HReO_4 \cdot H_2O$ consists of a hydrogen-bonded network of $[H_3O]^+$ and $[ReO_4]^-$ ions. The acids react with H_2S to precipitate M_2S_7 (equation 22.67) in striking contrast to the reduction of $[MnO_4]^-$ to Mn^{2+} by H_2S.

$$2HMO_4 + 7H_2S \longrightarrow M_2S_7 + 8H_2O \qquad (M = Tc, Re)$$
$$(22.67)$$

The $[TcO_4]^-$ and $[ReO_4]^-$ ions are tetrahedral and isostructural with $[MnO_4]^-$. Whereas $[MnO_4]^-$ is intense purple due to a charge transfer absorption in the visible region, $[ReO_4]^-$ is colourless because the corresponding CT band is in the UV region; salts of $[TcO_4]^-$ are also usually colourless.

Rhenium(VI) oxide, ReO_3, is made by reducing Re_2O_7 with CO; TcO_3 has not been isolated. Red ReO_3 crystallizes with a cubic lattice (*Figure 21.3*) and is a metallic-like electrical conductor owing to delocalization of the d^1 electrons. No reaction between ReO_3 and H_2O, dilute acids or alkalis occurs, but reaction 22.68 occurs with concentrated alkalis.

$$3ReO_3 + 2[OH]^- \xrightarrow{\Delta} 2[ReO_4]^- + ReO_2 + H_2O \qquad (22.68)$$

For the +5 oxidation state, only the blue Re_2O_5 is known but it is unstable with respect to disproportionation.

Technetium(VII) and rhenium(VII) form a series of hydride complexes (neutron diffraction data are essential for accurate H location) including the tricapped trigonal prismatic $[TcH_9]^{2-}$, $[ReH_9]^{2-}$ (see *Section 9.7*) and $[ReH_7(Ph_2PCH_2CH_2PPh_2-P,P')]$. Some hydrido complexes contain coordinated η^2-H_2 (**22.25**) with a 'stretched' H–H bond, e.g. in $[ReH_5(\eta^2-H_2)\{P(4-C_6H_4Me)_3\}_2]$ two H atoms are separated by 136 pm, the next shortest H····H separation being 175 pm.

(22.25)

A few halo complexes are known for M(VI) and M(V): square antiprismatic $[ReF_8]^{2-}$ (formed from KF and ReF_6), $[ReF_6]^-$ (from reduction of ReF_6 with KI in liquid SO_2), $[TcF_6]^-$ (from TcF_6 and CsCl in IF_5) and $[Re_2F_{11}]^-$ (**22.26**, as the $[Re(CO)_6]^+$ salt formed when an excess of ReF_6 reacts with $Re_2(CO)_{10}$ in anhydrous HF).

Eclipsed; linear bridge

(22.26)

Complexes of M(VII), M(VI) and M(V) (M = Tc, Re) are dominated by oxo and nitrido species, with octahedral and square-based pyramidal (oxo or nitrido ligand in the apical site) structures being common. Complexes of M(V) outnumber those of the higher oxidation states, with square-based pyramidal structures usually favoured. Examples include:

- octahedral M(VII): *fac*-[ReO$_3$L]$^+$ (L = **22.27**, tridentate), *fac*-[ReO$_3$Cl(phen)], [TcNCl(η^2-O$_2$)$_2$]$^-$;
- octahedral M(VI): [ReOCl$_5$]$^-$, *trans*-[TcN(H$_2$O)Br$_4$]$^-$, *mer*-[TcNCl$_3$(bpy)];
- square-based pyramidal M(VI): [TcNCl$_4$]$^-$, [TcNBr$_4$]$^-$;
- octahedral M(V): [ReOCl$_5$]$^{2-}$, [ReOCl$_4$(py)]$^-$, *trans*-[TcO$_2$(en)$_2$]$^+$, *trans*-[TcO$_2$(py)$_4$]$^+$, *cis*-[TcNBr(bpy)$_2$]$^+$;
- square-based pyramidal M(V): [ReOCl$_4$]$^-$, [TcOCl$_4$]$^-$, [TcO(ox)$_2$]$^-$.

X = NH or S

(22.27)

The development of technetium agents for imaging the brain, heart and kidneys has prompted the study of a range of Tc(V) oxo complexes, many of which are square-based pyramidal and contain a tetradentate ligand, often a mixed *S*- and *N*-donor; the oxo ligand occupies the apical site. Complexes **22.28** and **22.29** (in their ^{99m}Tc forms, see *Boxes 2.3* and *22.7*) are examples of radiopharmaceuticals used as kidney and brain imaging agents respectively.[†]

(22.28)   (22.29)

Technetium(IV) and rhenium(IV)

The reaction of Tc$_2$O$_7$ with CCl$_4$ at 670 K (or heating Tc and Cl$_2$) gives TcCl$_4$ as a moisture sensitive, red solid. The halides ReX$_4$ (X = F, Cl, Br, I) are all known; blue ReF$_4$ forms when ReF$_5$ is reduced by H$_2$ over a Pt gauze, and black ReCl$_4$ is made by heating ReCl$_5$ and Re$_3$Cl$_9$. Solid TcCl$_4$ and ReCl$_4$ are polymeric but not isostructural; TcCl$_4$ adopts chain-structure **22.30** and has a magnetic moment of 3.14 μ_B (298 K) per Tc(IV) centre. In ReCl$_4$, dimers are linked into zigzag chains by chloro-bridges (**22.31**) and the short Re—Re distance is consistent with metal–metal bonding (compare **22.31** with **22.17**).

(22.30)

Re–Re = 273 pm

(22.31)

The oxides TcO$_2$ and ReO$_2$ are made by thermal decomposition of [NH$_4$][MO$_4$] or reduction of M$_2$O$_7$ by M or H$_2$. Both adopt rutile lattices (*Figure 5.21*), distorted by pairing of metal centres as in MoO$_2$. With O$_2$, TcO$_2$ is oxidized to Tc$_2$O$_7$, and with H$_2$ at 770 K, reduction of TcO$_2$ to the metal occurs.

Reduction of KReO$_4$ using I$^-$ in concentrated HCl produces K$_4$[Re$_2$(μ-O)Cl$_{10}$]; the anion has a linear Re—O—Re bridge with Re—O π-character (Re—O = 186 pm) and is structurally related to [W$_2$(μ-O)Cl$_{10}$]$^{4-}$ and [Ru$_2$(μ-O)Cl$_{10}$]$^{4-}$ (see *Figure 22.15*). The octahedral complexes [MX$_6$]$^{2-}$ (M = Tc, Re; X = F, Cl, Br, I) are all known and are probably the most important M(IV) complexes. The ions [MX$_6$]$^{2-}$ (X = Cl, Br, I) are formed by reducing [MO$_4$]$^-$ (e.g. by I$^-$) in concentrated HX. Reactions of [MBr$_6$]$^{2-}$ with HF yield [MF$_6$]$^{2-}$. The chloro complexes (e.g. as K$^+$ or [Bu$_4$N]$^+$ salts) are useful starting materials in Tc and Re chemistries, but both are readily hydrolyzed in water. In aqueous solution, [TcCl$_6$]$^{2-}$ is in equilibrium with [TcCl$_5$(H$_2$O)]$^-$, and complete hydrolysis gives

---

[†] For review articles, see: R.C. Elder and K. Tepperman (1994) 'Metal-based drugs & imaging agents' in *Encyclopedia of Inorganic Chemistry*, ed. R.B. King, Wiley, Chichester, vol. 4, p. 2165; K. Schwochau (1994) *Angewandte Chemie International Edition in English*, vol. 33, p. 2258 – 'Technetium radiopharmaceuticals: Fundamentals, synthesis, structure and development'; J.R. Dilworth and S.J. Parrott (1998) *Chemical Society Reviews*, vol. 27, p. 43 – 'The biomedical chemistry of technetium and rhenium'.

TcO_2. Halide exchange between $[ReI_6]^{2-}$ and HCl leads to *fac*-$[ReCl_3I_3]^{2-}$, *cis*- and *trans*-$[ReCl_4I_2]^{2-}$ and $[ReCl_5I]^{2-}$. In most complexes, octahedral coordination for Re(IV) and Tc(IV) is usual, e.g. *cis*-$[TcCl_2(acac)_2]$, *trans*-$[TcCl_4(PMe_3)_2]$, $[Tc(NCS-N)_6]^{2-}$, $[Tc(ox)_3]^{2-}$, *trans*-$[ReCl_4(PPh_3)_2]$, $[ReCl_5(H_2O)]^-$, $[ReCl_5(PEt_3)]^-$, $[ReCl_4(bpy)]$ and *cis*-$[ReCl_4(THF)_2]$.

Technetium(III) and rhenium(III)

For the +3 oxidation state, metal–metal bonding becomes important. Rhenium(III) halides (X = Cl, Br, I) are trimeric, M_3X_9; no Tc(III) halides nor ReF_3 are known. Rhenium(III) chloride is an important precursor in Re(III) chemistry and is made by heating $ReCl_5$. Its structure (Figure 22.13a) consists of an Re_3 triangle (Re−Re = 248 pm), each edge being chloro-bridged; the terminal Cl atoms lie above and below the metal framework. In the solid, two-thirds of the terminal Cl atoms are involved in weak bridging interactions to Re atoms of adjacent molecules. Rhenium(III) chloride is diamagnetic, and Re=Re double bonds are allocated to the metal framework, i.e. the (formally) $\{Re_3\}^{9+}$ core contains 12 valence electrons (Re, s^2d^5) which are used for metal–metal bonding. Lewis bases react with Re_3Cl_9 (or $Re_3Cl_9(H_2O)_3$) to give complexes of type $Re_3Cl_9L_3$ (Figure 22.13b); $Re_3Cl_9(H_2O)_3$ can be isolated from aqueous solutions of the chloride at 273 K. Equation 22.69 shows further examples of Lewis base additions.

$$Re_3Cl_9(py)_3 \xleftarrow{\text{py}} Re_3Cl_9 \xrightarrow{PR_3} Re_3Cl_9(PR_3)_3 \qquad (22.69)$$

The reaction of MCl with Re_3Cl_9 gives $M[Re_3Cl_{10}]$, $M_2[Re_3Cl_{11}]$ or $M_3[Re_3Cl_{12}]$ depending upon the conditions, for example reactions 22.70 and 22.71. Figure 22.13c shows the structure of the $[Re_3Cl_{12}]^{3-}$ ion.

$$Re_3Cl_9 \xrightarrow{\text{excess CsCl, conc HCl}} Cs_3[Re_3Cl_{12}] \qquad (22.70)$$

$$Re_3Cl_9 \xrightarrow{[Ph_4As]Cl, \text{dil HCl}} [Ph_4As]_2[Re_3Cl_{11}] \qquad (22.71)$$

The diamagnetic $[Re_2Cl_8]^{2-}$ was the first example of a species containing a metal–metal quadruple bond. It is made by reducing $[ReO_4]^-$ using H_2 or $[HPO_2]^{2-}$ and is isostructural with $[Mo_2Cl_8]^{4-}$ (Figure 22.11b) with a Re−Re distance of 224 pm. Salts of $[Re_2Cl_8]^{2-}$ are blue (λ_{max} = 700 nm) arising from a $\sigma^2\pi^4\delta^1\delta^{*1} \leftarrow \sigma^2\pi^4\delta^2$ transition (*Figure 21.13*). Reactions of $[Re_2Cl_8]^{2-}$ include ligand displacements and redox processes. With Cl_2, $[Re_2Cl_9]^-$ is formed (i.e. oxidation and Cl^- addition); this ion is isoelectronic and isostructural with $[W_2Cl_9]^{3-}$ (**22.17**), with Re−Re = 270 pm. Reaction 22.72 shows the reaction of carboxylates with $[Re_2Cl_8]^{2-}$; the reaction can be reversed by treatment with HCl.

$$[Re_2Cl_8]^{2-} + 4[RCO_2]^- \longrightarrow [Re_2(\mu\text{-}O_2CR)_4Cl_2] + 6Cl^- \qquad (22.72)$$
$$(22.32)$$

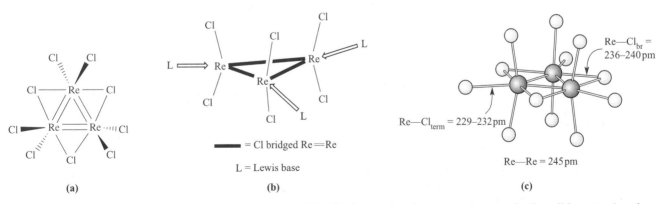

= quadruple bond

(22.32)

When $[Re_2Cl_8]^{2-}$ reacts with phosphines (equation 22.73), the $\{Re_2\}^{6+}$ core with a $\sigma^2\pi^4\delta^2$ configuration (Re≡Re) is reduced to a $\{Re_2\}^{4+}$ unit ($\sigma^2\pi^4\delta^2\delta^{*2}$, Re≡Re). The change might be expected to lead to an increase in the Re−Re bond length, but in fact it stays the same (224 pm); the introduction of the bridging ligands must counter the decrease in bond order by 'clamping' the Re atoms together.

$$[Re_2Cl_8]^{2-} + 2Ph_2PCH_2CH_2PPh_2$$
$$\xrightarrow{-Cl^-} [Re_2Cl_4(\mu\text{-}Ph_2PCH_2CH_2PPh_2)_2] \qquad (22.73)$$

reduction

Fig. 22.13 Schematic representations of (a) the structure of Re_3Cl_9 (interactions between units occur in the solid, see text) and (b) the sites of addition of Lewis bases to Re_3Cl_9. (c) The structure (X-ray diffraction) of $[Re_3Cl_{12}]^{3-}$ in the $[Me_3NH]^+$ salt [M. Irmler *et al.* (1991) *Z. Anorg. Allg. Chem.*, vol. 604, p. 17]; colour code: Re, blue; Cl, pale grey.

Technetium-99m labelling using [Tc(H₂O)₃(CO)₃]⁺

Diagnostic imaging agents incorporating ^{99m}Tc labels were mentioned in Box 2.3. In developing new techniques of tumour imaging with radioisotopes, one goal is to label single-chain antibody fragments which may efficiently target tumours. Recent research has shown that the complex $[^{99m}\text{Tc}(\text{H}_2\text{O})_3(\text{CO})_3]^+$ (which can be made easily by the reaction of $[^{99m}\text{TcO}_4]^-$ with CO at 1 bar pressure, at pH 11 in aqueous NaCl solution) can be used to label single-chain antibody fragments which carry *C*-terminal histidine tags. High activities are achieved ($90\,\text{mCi}\,\text{mg}^{-1}$), and *in vivo*, the

technetium-labelled fragments are very stable. The new technique appears to have a high potential for application in clinical medicine.

Further reading

R. Waibel *et al.* (1999) *Nature Biotechnology*, vol. 17, p. 897 – 'Stable one-step technetium-99m labelling of His-tagged recombinant proteins with a novel Tc(I)-carbonyl complex'.

The $[\text{Tc}_2\text{Cl}_8]^{2-}$ ion is also known (Tc–Tc = 215 pm) but is less stable than $[\text{Re}_2\text{Cl}_8]^{2-}$; interestingly, the paramagnetic $[\text{Tc}_2\text{Cl}_8]^{3-}$ ($\sigma^2\pi^4\delta^2\delta^{*1}$, Tc–Tc = 211 pm, eclipsed ligands) is easier to isolate than $[\text{Tc}_2\text{Cl}_8]^{2-}$ ($\sigma^2\pi^4\delta^2$, Tc–Tc = 215 pm, eclipsed ligands). The *increase* in Tc–Tc distance of 4 pm in going from $[\text{Tc}_2\text{Cl}_8]^{3-}$ to $[\text{Tc}_2\text{Cl}_8]^{2-}$ is not readily rationalized. Reduction of the $\{\text{Tc}_2\}^{6+}$ core occurs when $[\text{Tc}_2\text{Cl}_8]^{2-}$ undergoes reaction 22.74; the product (also made from $\text{Tc}^{II}_2\text{Cl}_4(\text{PR}_3)_4$ and $\text{HBF}_4\cdot\text{OEt}_2$) is expected to have a staggered arrangement of ligands consistent with the change from $\sigma^2\pi^4\delta^2$ to $\sigma^2\pi^4\delta^2\delta^{*2}$, and this has been confirmed for the related $[\text{Tc}_2(\text{NCMe})_8(\text{OSO}_2\text{CF}_3)_2]^{2+}$.

$$[\text{Tc}_2\text{Cl}_8]^{2-} \xrightarrow{\text{HBF}_4\cdot\text{OEt}_2,\,\text{in MeCN}} [\text{Tc}_2(\text{NCMe})_{10}]^{4+} \qquad (22.74)$$

Mononuclear complexes of Re(III) and Tc(III) are quite well exemplified (often with π-acceptor ligands stabilizing the +3 oxidation state) and octahedral coordination is usual, e.g. $[\text{Tc}(\text{acac})_2(\text{NCMe})_2]^+$, $[\text{Tc}(\text{acac})_3]$, $[\text{Tc}(\text{NCS-}N)_6]^{3-}$, *mer*-$[\text{Tc}(\text{Ph}_2\text{PCH}_2\text{CH}_2\text{CO}_2)_3]$, *mer,trans*-$[\text{ReCl}_3(\text{NCMe})(\text{PPh}_3)_2]$. Seven-coordination has been observed in $[\text{ReBr}_3(\text{CO})_2(\text{bpy})]$ and $[\text{ReBr}_3(\text{CO})_2(\text{PMe}_2\text{Ph})_2]$. Simple aqua ions such as $[\text{Tc}(\text{H}_2\text{O})_6]^{3+}$ are not known, although, stabilized by CO, it has been possible to prepare the Tc(I) species $[\text{Tc}(\text{H}_2\text{O})_3(\text{CO})_3]^+$ (see *Box 22.7*).

22.9 Group 8: ruthenium and osmium

The metals

Like all platinum-group metals, Ru and Os are relatively noble. Osmium powder reacts slowly with O_2 at 298 K to give the volatile OsO_4 (the bulk metal requires heating to 670 K); Ru is passivated by a coating of non-volatile RuO_2 and reacts further with O_2 only >870 K. Both metals react with F_2 and Cl_2 when heated (see below), and are attacked

by mixtures of HCl and oxidizing agents, and by molten alkalis.

Table 19.3 shows the range of oxidation states exhibited by the group 8 metals. In this section we consider oxidation states from +2 to +8; the lower states are stabilized by π-acceptor ligands and are covered in Chapter 23. Consistent with trends seen for earlier metals and in contrast to the first group member, Ru and Os form some compounds with metal–metal multiple bonds.

High oxidation states of ruthenium and osmium: M(VIII), M(VII) and M(VI)

The only binary halides formed for the high oxidation states are RuF_6 (equation 22.75) and OsF_6 (equation 22.76); the formation of OsF_7 has been claimed but not proven.

$$2\text{RuF}_5 + \text{F}_2 \xrightarrow{500\,\text{K},\,50\,\text{bar}} 2\text{RuF}_6 \qquad (22.75)$$

$$\text{Os} + 3\text{F}_2 \xrightarrow{500\,\text{K},\,1\,\text{bar}} \text{OsF}_6 \qquad (22.76)$$

Ruthenium(VI) fluoride is an unstable brown solid; OsF_6 is a volatile yellow solid with a molecular (octahedral) structure. Several oxofluorides of Os(VIII), Os(VII) and Os(VI) are known, but RuOF_4 is the only example for Ru; all are very moisture-sensitive. Red *cis*-OsO_2F_4 forms when OsO_4 reacts with HF and KrF_2 at 77 K. Yellow OsO_3F_2 (made from F_2 and OsO_4) is also molecular in the gas phase (**22.33**) but is polymeric in the solid with Os–F–Os bridges connecting *fac*-octahedral units. Heating OsO_3F_2 with F_2 gives OsOF_5 and OsOF_4.

(22.33)

M = Ru, Os; *d* = 171 pm (gas phase)

(22.34)

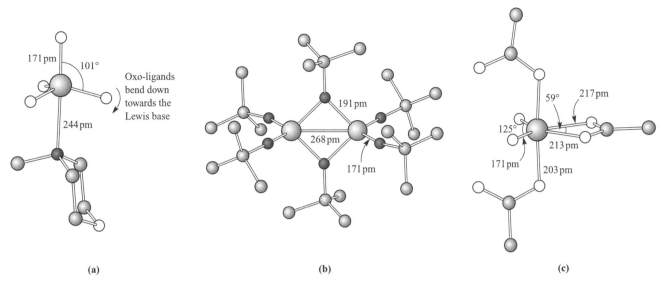

Fig. 22.14 The structures (X-ray diffraction) of (a) the adduct formed between *N*-methylmorpholine and OsO$_4$ [A.J. Bailey *et al.* (1997) *J. Chem. Soc., Dalton Trans.*, p. 3245], (b) [Os$_2$(N^{t}Bu)$_4$(μ-N^{t}Bu)$_2$]$^{2+}$ in the [BF$_4$]$^-$ salt [A.A. Danopoulos *et al.* (1991) *J. Chem. Soc., Dalton Trans.*, p. 269] and (c) [OsO$_2$(O$_2$CMe)$_3$]$^-$ in the solvated K$^+$ salt [T. Behling *et al.* (1982) *Polyhedron*, vol. 1, p. 840]. Colour code: Os, blue; O, white; C, grey; N, black.

Both Ru and Os form toxic, volatile, yellow oxides MO$_4$ (RuO$_4$ mp 298 K, bp 403 K; OsO$_4$ mp 313 K, bp 403 K)[†] but RuO$_4$ is more readily reduced than OsO$_4$. Osmium(VIII) oxide ('osmic acid') is made from Os and O$_2$ (see above), but the formation of RuO$_4$ requires acidified [IO$_4$]$^-$ or [MnO$_4$]$^-$ oxidation of RuO$_2$ or RuCl$_3$. Both tetraoxides have penetrating ozone-like odours; they are sparingly soluble in water but soluble in CCl$_4$. The oxides are isostructural with molecular structures (**22.34**). Ruthenium(VIII) oxide is thermodynamically unstable with respect to RuO$_2$ and O$_2$ (equation 22.77) and is liable to explode; it is a very powerful oxidant, reacting violently with organic compounds. Osmium(VIII) oxide is used as an oxidizing agent in organic synthesis (e.g. converting alkenes to 1,2-diols) and as a biological stain, but its ease of reduction and its volatility make it dangerous to the eyes. Reaction 22.77 occurs on heating.

$$MO_4 \xrightarrow{\Delta} MO_2 + O_2 \qquad (M = Ru, Os) \qquad (22.77)$$

Osmium(VIII) oxide forms adducts with Lewis bases such as Cl$^-$, 4-phenylpyridine and *N*-morpholine; the adducts are distorted trigonal bipyramidal with the oxo ligands in the equatorial and one axial site (Figure 22.14a).

When RuO$_4$ dissolves in aqueous alkali, O$_2$ is evolved and [RuO$_4$]$^-$ forms; in concentrated alkali, reduction proceeds to [RuO$_4$]$^{2-}$ (equation 22.78); K$_2$RuO$_4$ can also be made by fusing Ru with KNO$_3$ and KOH.

$$4[RuO_4]^- + 4[OH]^- \longrightarrow 4[RuO_4]^{2-} + 2H_2O + O_2 \qquad (22.78)$$

$$\left[\begin{array}{c} OH \\ O = Ru = O \\ O \\ OH \end{array} \right]^{2-}$$

(22.35)

$$\left[\begin{array}{c} O \\ O = Os \begin{array}{c} OH \\ OH \end{array} \\ O \end{array} \right]^{2-}$$

(22.36)

Both [RuO$_4$]$^-$ and [RuO$_4$]$^{2-}$ are powerful oxidants but can be stabilized in solution by pH control under non-reducing conditions. In solid state salts, [RuO$_4$]$^-$ has a flattened tetrahedral structure (Ru−O = 173 pm), but crystals of 'K$_2$[RuO$_4$]·H$_2$O' are actually K$_2$[RuO$_3$(OH)$_2$] containing anion **22.35**. In contrast to its action on RuO$_4$, alkali reacts with OsO$_4$ to give *cis*-[OsO$_4$(OH)$_2$]$^{2-}$ (**22.36**) which is reduced to *trans*-[OsO$_2$(OH)$_4$]$^{2-}$ by EtOH. Anion **22.36** has been isolated in crystalline Na$_2$[OsO$_4$(OH)$_2$]·2H$_2$O. Reaction 22.79 gives K[Os(N)O$_3$] which contains tetrahedral [Os(N)O$_3$]$^-$ (**22.37**), isoelectronic and isostructural with OsO$_4$. The IR spectrum of [Os(N)O$_3$]$^-$ contains bands at 871 and 897 cm^{-1} ($\nu_{Os=O}$) and 1021 cm^{-1} ($\nu_{Os\equiv N}$); this compares with absorptions at 954 and 965 cm^{-1} for OsO$_4$.

$$OsO_4 + NH_3 + KOH \longrightarrow K[Os(N)O_3] + 2H_2O \qquad (22.79)$$

$$OsO_4 \xrightarrow{(Me_3Si)NH^tBu} Os(N^tBu)_4 \qquad (22.80)$$

Reaction 22.80 gives an imido analogue of OsO$_4$; the tetrahedral shape is retained and the Os−N bond lengths

$$\left[\begin{array}{c} N \\ \| \\ O = Os = O \\ O \end{array} \right]^-$$

(22.37)

[†] The literature contains differing values for OsO$_4$: see Y. Koda (1986) *Journal of the Chemical Society, Chemical Communications*, p. 1347.

of 175 pm are consistent with double bonds. Sodium amalgam reduces $Os(N^tBu)_4$ to the Os(VI) dimer $Os_2(N^tBu)_4(\mu-N^tBu)_2$ (Os–Os = 310 pm) and subsequent oxidation gives $[Os_2(N^tBu)_4(\mu-N^tBu)_2]^{2+}$ (Figure 22.14b), a rare example of an Os(VII) complex. Trigonal planar $Os(NAr)_3$ is stabilized against dimerization if Ar is very bulky, e.g. $2,6\text{-}^iPr_2C_6H_3$.

Complexes of M(VIII) and M(VII) are few, e.g. $[Os(N)O_3]^-$ (see above), but are well exemplified for M(VI), particularly for M = Os, with oxo, nitrido or imido ligands commonly present, e.g.

- tetrahedral: $[OsO_2(S_2O_3\text{-}S)_2]^{2-}$;
- square-based pyramidal: $[RuNBr_4]^-$, $[OsNBr_4]^-$;
- octahedral: $[OsO_2(O_2CMe)_3]^-$ (distorted, Figure 22.14c), *trans*-$[OsO_2Cl_4]^{2-}$, *trans*-$[RuO_2Cl_4]^{2-}$ (see *Section 8.12*), $[OsO_2Cl_2(py)_2]$ (**22.38**), *trans*-$[OsO_2(en)_2]^{2+}$;

(22.38)

The reaction of FNO with OsF_6 gives $[NO]^+[OsF_7]^-$, but other simple halo complexes for oxidation states $\geq +6$ are not known.

Ruthenium(V), (IV) and osmium(V), (IV)

Green RuF_5 and OsF_5 (readily hydrolyzed solids) are made by reactions 22.81 and 22.82 and are tetrameric like NbF_5 (**22.4**) but with non-linear bridges. Black $OsCl_5$ is the only other halide of the M(V) state and is made by reducing and chlorinating OsF_6 with BCl_3. It is dimeric, analogous to $NbCl_5$ (**22.5**).

$$2Ru + 5F_2 \xrightarrow{570K} 2RuF_5 \tag{22.81}$$

$$OsF_6 \xrightarrow{I_2, IF_5, 328K} OsF_5 \tag{22.82}$$

For the M(IV) state, RuF_4, OsF_4, $OsCl_4$ (two polymorphs) and $OsBr_4$ are known and are polymeric; the fluorides are made by reducing higher fluorides, and $OsCl_4$ and $OsBr_4$ by combining the elements at high temperature and, for $OsBr_4$, high pressure.

In contrast to iron, the lowest oxides formed by the heavier group 8 metals are for the M(IV) state. Both RuO_2 and OsO_2 adopt the rutile lattice (*Figure 5.21*); these oxides are far less important than RuO_4 and OsO_4.

The electrochemical oxidation of $[Ru(H_2O)_6]^{2+}$ in aqueous solution produces a Ru(IV) species. Its formulation as $[Ru_4O_6(H_2O)_{12}]^{4+}$ (or a protonated form depending on pH) is consistent with ^{17}O NMR spectroscopic data and of the two proposed structures **22.39a** and **22.39b**, the latter is supported by EXAFS studies (see *Box 26.2*).

(22.39a)

(22.39b)

Octahedral halo complexes of Ru(V) and Os(V) are represented by $[MF_6]^-$ (M = Ru, Os) and $[OsCl_6]^-$. For the +4 oxidation state, all the $[MX_6]^{2-}$ ions are known except $[RuI_6]^{2-}$. Various synthetic routes are used, e.g. $[RuCl_6]^{2-}$ can be made by heating Ru, Cl_2 and an alkali metal chloride, or by oxidizing $[RuCl_6]^{3-}$ with Cl_2. The salt $K_2[RuCl_6]$ has a magnetic moment (298 K) of 2.8 μ_B, close to μ(spin-only) for a low-spin d^4 ion but the value is temperature dependent; for $K_2[OsCl_6]$, the value of 1.5 μ_B arises from the greater spin–orbit coupling constant for the $5d$ metal ion (see *Figure 20.23* and discussion). Mixed halo complexes are produced by halogen exchange; in reaction 22.83, the products are formed by stepwise substitution, the position of F^- entry being determined by the stronger *trans*-effect (see *Section 25.3*) of the chloro ligand.

$$[OsCl_6]^{2-} \xrightarrow{BrF_3} [OsCl_5F]^{2-} + cis\text{-}[OsCl_4F_2]^{2-}$$
$$+ fac\text{-}[OsCl_3F_3]^{2-} + cis\text{-}[OsCl_2F_4]^{2-}$$
$$+ [OsClF_5]^{2-} + [OsF_6]^{2-} \tag{22.83}$$

Reaction of RuO_4 in aqueous HCl in the presence of KCl gives K^+ salts of $[Ru^{IV}_2OCl_{10}]^{4-}$, $[Ru^{III}Cl_5(H_2O)]^{2-}$ and $[Ru^{III}Cl_6]^{3-}$. Each Ru(IV) centre in $[Ru_2OCl_{10}]^{4-}$ is octahedrally sited and the Ru–O–Ru bridge is linear (Figure 22.15a); salts of $[Ru_2OCl_{10}]^{4-}$ are diamagnetic. This is rationalized by considering the formation of two 3-centre π-interactions (Figure 22.15b) involving the d_{xz} and d_{yz} atomic orbitals of the two low-spin Ru(IV) centres (each of configuration $d_{xy}^2 d_{xz}^1 d_{yz}^1$) and the filled p_x and p_y atomic orbitals of the O atom. In addition to the π and π^* MOs, four non-bonding MOs result from combinations of the d_{xy}, d_{xz} and d_{yz} orbitals (Figure 22.15b); these are fully occupied in $[Ru_2OCl_{10}]^{4-}$. The same MO diagram can be

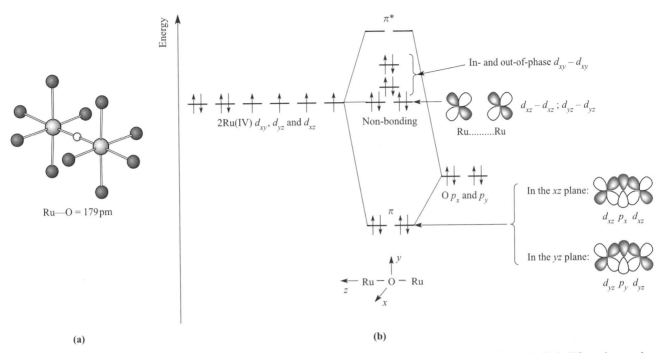

Fig. 22.15 (a) The structure of $[Ru_2(\mu\text{-}O)Cl_{10}]^{4-}$ determined by X-ray diffraction for the histaminium salt [I.A. Efimenko *et al.* (1994) *Koord. Khim.*, vol. 20, p. 294]; colour code: Ru, blue; Cl, black; O, pale grey. (b) A partial MO diagram for the interaction between the d_{xy}, d_{xz} and d_{yz} atomic orbitals of the Ru(IV) centres and the p_x and p_y atomic orbitals of the O atom to give two bonding, two antibonding and four non-bonding MOs; the non-bonding MOs are derived from combinations of d orbitals with no oxygen contribution. Relative orbital energies are approximate, and the non-bonding MOs lie close together.

used to describe the bonding in the related anions $[Os_2OCl_{10}]^{4-}$ (two d^4 metal centres), $[W_2OX_{10}]^{4-}$ (X = Cl, Br; d^2), $[Re_2OCl_{10}]^{4-}$ (d^3) and $[Ta_2OX_{10}]^{2-}$ (X = F, Cl; d^0); changes in d^n configuration only affect the occupancy of the non-bonding MOs, leaving the metal–oxygen π-bonding MOs occupied. The diamagnetic $[Ru_2(\mu\text{-}N)Cl_8(H_2O)_2]^{3-}$ (**22.40**) is a nitrido-bridged analogue of $[Ru_2OCl_{10}]^{4-}$, and Ru–N distances of 172 pm indicate strong π-bonding; it is made by reducing $[Ru(NO)Cl_5]^{2-}$ with $SnCl_2$ in HCl.

$$\left[\begin{array}{c} \text{Cl} \quad\quad\quad \text{Cl} \\ | \quad\text{Cl} \quad\quad | \quad\text{Cl} \\ H_2O - Ru ---- N === Ru - OH_2 \\ | \quad\quad\quad\quad | \\ \text{Cl} \quad\quad\quad \text{Cl} \\ \text{Cl} \quad\quad\quad \text{Cl} \end{array} \right]^{3-}$$

(22.40)

$$\left[\begin{array}{c} O \\ \| \\ Et_2NC - S \quad\quad S \\ S - Ru \quad S \quad CNEt_2 \\ S \quad\quad S \\ C \\ N \\ Et_2 \end{array} \right]^{-}$$

(22.41)

Although the coordination chemistry of Ru(IV) and Os(IV) is quite varied, halo complexes are dominant;

complexes of Os(IV) outnumber those of Ru(IV). Hexahalo complexes are common precursors (Figure 22.16). Apart from those already described, examples with mixed ligands include octahedral *trans*-$[OsBr_4(AsPh_3)_2]$, $[OsX_4(acac)]^-$ (X = Cl, Br, I), $[OsX_4(ox)]^{2-}$ (X = Cl, Br, I), *cis*-$[OsCl_4(NCS\text{-}N)_2]^{2-}$ and *trans*-$[OsCl_4(NCS\text{-}S)_2]^{2-}$. Two complexes of special note for their stereochemistries are the pentagonal bipyramidal $[RuO(S_2CNEt_2)_3]^-$ (**22.41**) and the square planar *trans*-$[Ru(PMe_3)_2(NAr)_2]$ in which Ar is the bulky $2,6\text{-}^iPr_2C_6H_3$.

Ruthenium(III) and osmium(III)

All the binary halides RuX_3 are known but for Os, only $OsCl_3$ and OsI_3 have been established; OsF_4 is the lowest fluoride of Os. Reduction of RuF_5 with I_2 gives RuF_3, a brown solid isostructural with FeF_3. Reactions 22.84–22.86 show preparations of $RuCl_3$, $RuBr_3$ and RuI_3; the chloride is commercially available as a hydrate of variable composition '$RuCl_3 \cdot xH_2O$' and is an important starting material in Ru(III) and Ru(II) chemistry.

$$Ru_3(CO)_{12} \xrightarrow{Cl_2, \, 630\,K \text{ under } N_2} \beta\text{-RuCl}_3 \xrightarrow{Cl_2, \, >720\,K} \alpha\text{-RuCl}_3 \tag{22.84}$$

$$Ru \xrightarrow{Br_2, \, 720\,K, \, 20\,bar} RuBr_3 \tag{22.85}$$

$$RuO_4 \xrightarrow{aq\,HI} RuI_3 \tag{22.86}$$

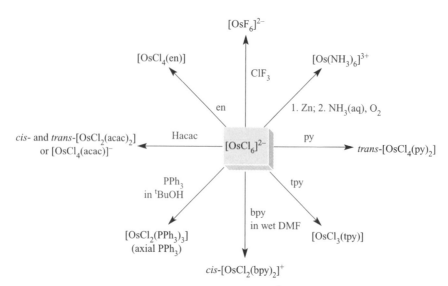

Fig. 22.16 Representative complex-forming reactions starting from $[OsCl_6]^{2-}$. Note that reduction to Os(III) occurs in three reactions, and to Os(II) in one. See Table 6.7 and structure **19.22** for ligand abbreviations.

The α-forms of $RuCl_3$ and $OsCl_3$ are isostructural with α-$TiCl_3$ (see *Section 21.5*), while β-$RuCl_3$ has the same structure as $CrCl_3$ (see *Section 21.7*); infinite lattices with octahedral Ru(III) are adopted by $RuBr_3$ and RuI_3.

There are no binary oxides or oxoanions for Ru(III), Os(III) or lower oxidation states. No simple aqua ion of Os(III) has been established, but octahedral $[Ru(H_2O)_6]^{3+}$ can be obtained by aerial oxidation of $[Ru(H_2O)_6]^{2+}$ and has been isolated in the alum (see *Section 12.9*) $CsRu(SO_4)_2\cdot12H_2O$ and the salt $[Ru(H_2O)_6][4\text{-}MeC_6H_4SO_3]_3\cdot3H_2O$. The Ru—O bond length of 203 pm is shorter than in $[Ru(H_2O)_6]^{2+}$ (212 pm). In aqueous solution, $[Ru(H_2O)_6]^{3+}$ is acidic (compare equation 22.87 with equation 6.36 for Fe^{3+}) and it is less readily reduced than $[Fe(H_2O)_6]^{3+}$ (equation 22.88).

$$[Ru(H_2O)_6]^{3+} + H_2O \rightleftharpoons [Ru(H_2O)_5(OH)]^{2+} + [H_3O]^+$$
$$pK_a \approx 2.4 \qquad (22.87)$$

$$[M(H_2O)_6]^{3+} + e^- \rightleftharpoons [M(H_2O)_6]^{2+}$$
$$\begin{cases} M = Ru, \quad E^\circ = +0.23\,V \\ M = Fe, \quad E^\circ = +0.77\,V \end{cases} \qquad (22.88)$$

Substitution in Ru(III) complexes (low-spin d^5) is slow (see *Chapter 25*) and all members of the series $[RuCl_n(H_2O)_{6-n}]^{(n-3)-}$, including isomers, have been characterized. Aerial oxidation of $[Ru(NH_3)_6]^{2+}$ (see below) gives $[Ru(NH_3)_6]^{3+}$ (equation 22.89).

$$[Ru(NH_3)_6]^{3+} + e^- \rightleftharpoons [Ru(NH_3)_6]^{2+} \qquad E^\circ = +0.24\,V$$
$$(22.89)$$

Halo complexes $[MX_6]^{3-}$ are known for M = Ru, X = F, Cl, Br, I and M = Os, X = Cl, Br, I. The anions $[RuCl_5(H_2O)]^{2-}$ and $[RuCl_6]^{3-}$ are made in the same reaction as $[Ru_2OCl_{10}]^{4-}$ (see above); in aqueous solution $[RuCl_6]^{3-}$ is rapidly aquated to $[RuCl_5(H_2O)]^{2-}$. The anion $[Ru_2Br_9]^{3-}$ can be made by treating $[RuCl_6]^{3-}$ with

HBr; this and $[Ru_2Cl_9]^{3-}$ (reaction 22.90) adopt structure **22.42** but the Ru—Ru distances of 273 (Cl) and 287 pm (Br) along with magnetic moments of 0.86 (Cl) and 1.18 μ_B (Br) suggest a degree of Ru—Ru bonding.

$$\begin{bmatrix} \\ X\text{---}M\text{---}X \cdots M\text{---}X \\ \end{bmatrix}^{3-}$$

M = Ru, X = Cl, Br
M = Os, X = Br

(22.42)

$$K_2[RuCl_5(H_2O)] \xrightarrow{520\,K} K_3[Ru_2Cl_9] \qquad (22.90)$$

$$[OsCl_6]^{2-} \xrightarrow{MeCO_2H,\,(MeCO)_2O} [Os_2(\mu\text{-}O_2CMe)_4Cl_2]$$
$$\xrightarrow[X=Cl,\,Br,\,I]{HX(g),\,EtOH} [Os_2X_8]^{2-} \qquad (22.91)$$

The anions $[Os_2X_8]^{2-}$ (X = Cl, Br, I) are made in reaction sequence 22.91. In diamagnetic $[Os_2X_8]^{2-}$ and $[Os_2(\mu\text{-}O_2CMe)_4Cl_2]$, the electronic configuration of the Os_2-unit is (from *Figure 21.13*) $\sigma^2\pi^4\delta^2\delta^{*2}$ corresponding to an Os≡Os triple bond. Since the δ^* MO is occupied, the influence of the δ-bond is lost and so no electronic factor restricts the orientation of the ligands (cf. the M≡M bond). Crystal structures for several salts of $[Os_2Cl_8]^{2-}$ show different ligand arrangements (Figure 22.17) and this is true also for $[Os_2Br_8]^{2-}$; for $[Os_2I_8]^{2-}$, steric factors appear to favour a staggered arrangement. Ruthenium(III) forms a number of acetate complexes which differ from the Os(III) species in equation 22.91. The reaction of $RuCl_3\cdot xH_2O$ with $MeCO_2H$ and $MeCO_2Na$ yields paramagnetic $[Ru_3(H_2O)_3(\mu\text{-}O_2CMe)_6(\mu_3\text{-}O)]^+$ (structurally analogous to the Cr(III) species in *Figure 21.11*)

Fig. 22.17 Differences in energy between the arrangement of the chloro ligands (staggered, eclipsed or somewhere in between) in $[Os_2Cl_8]^{2-}$ are small. The solid state structure of $[Os_2Cl_8]^{2-}$ (viewed along the Os−Os bond) in (a) the $[^nBu_4N]^+$ salt [P.A. Agaskar *et al.* (1986) *J. Am. Chem. Soc.*, vol. 108, p. 4850] ($[Os_2Cl_8]^{2-}$ is also staggered in the $[Ph_3PCH_2CH_2PPh_3]^{2+}$ salt), (b) the $[MePh_3P]^+$ salt [F.A. Cotton *et al.* (1990) *Inorg. Chem.*, vol. 29, p. 3197] and (c) the $[(Ph_3P)_2N]^+$ salt (structure determined at 83 K) [P.E. Fanwick *et al.* (1986) *Inorg. Chem.*, vol. 25, p. 4546]. In the $[(Ph_3P)_2N]^+$ salt, an eclipsed conformer is also present.

which is reduced by PPh_3 to give the mixed-valence complex $[Ru_3(PPh_3)_3(\mu\text{-}O_2CMe)_6(\mu_3\text{-}O)]$.

Both Ru(III) and Os(III) form a range of octahedral complexes with ligands other than those already mentioned; Ru(III) complexes outnumber those of Os(III), the reverse of the situation for the M(IV) state reflecting the relative stabilities Os(IV) > Ru(IV) but Ru(III) > Os(III). Examples of mononuclear complexes include $[Ru(acac)_3]$, $[Ru(ox)_3]^{3-}$, $[Ru(en)_3]^{3+}$, *cis*-$[RuCl(H_2O)(en)_2]^{2+}$, *cis*-$[RuCl_2(bpy)_2]^+$, $[RuCl_4(bpy)]^-$, *trans*-$[RuCl(OH)(py)_4]^+$, *mer*-$[RuCl_3(DMSO-S)_2(DMSO-O)]$, $[Ru(NH_3)_5(py)]^{3+}$, *mer*-$[OsCl_3(py)_3]$, $[Os(acac)_3]$, $[Os(en)_3]^{3+}$, *trans*-$[OsCl_2(PMe_3)_4]^+$ and *trans*-$[OsCl_4(PEt_3)_2]^-$.

Ruthenium(II) and osmium(II)

Binary halides of Ru(II) and Os(II) are not well characterized and there are no oxides. Heating the metal with S gives MS_2 (M = Ru, Os) which contain $[S_2]^{2-}$ and adopt a pyrites structure (see *Section 21.9*). Most of the chemistry of Ru(II) and Os(II) concerns complexes, all of which are diamagnetic, low-spin d^6 and, with a few exceptions, octahedral. We saw in *Section 20.3* that values of Δ_{oct} (for a set of related complexes) are greater for second and third row metals than for the first member of the triad, and low-spin complexes are favoured. A vast number of Ru(II) complexes are known and we can give only a brief introduction.

The hydrido anions $[RuH_6]^{4-}$ and $[OsH_6]^{4-}$ (analogous to $[FeH_6]^{4-}$, *Figure 9.10b*) are formed by heating the metal with MgH_2 or BaH_2 under a pressure of H_2. There are no simple halo complexes; H_2 or electrochemical reduction of $RuCl_3\cdot xH_2O$ in MeOH produces blue solutions (*ruthenium blues*) which, despite their synthetic utility for preparing Ru(II) complexes, have not been fully characterized. The blue species present have been variously formulated, but cluster anions seem likely.

Substitution reactions involving Ru(II) or Os(II) are affected by the kinetic inertness of the low-spin d^6 ion (see *Section 25.2*),

and methods of preparation of M(II) complexes often start from higher oxidation states, e.g. $RuCl_3\cdot xH_2O$ or $[OsCl_6]^{2-}$. Reducing aqueous solutions of $RuCl_3\cdot xH_2O$ from which Cl^- has been precipitated by Ag^+ produces $[Ru(H_2O)_6]^{2+}$; there is no Os(II) analogue. In air, $[Ru(H_2O)_6]^{2+}$ readily oxidizes (equation 22.88) but is present in Tutton salts (see *Section 21.6*) $M_2Ru(SO_4)_2\cdot 6H_2O$ (M = Rb, NH_4); its structure has been determined in the salt $[Ru(H_2O)_6][4\text{-}MeC_6H_4SO_3]_2$ (see discussion of $[Ru(H_2O)_6]^{3+}$). Under 200 bar pressure of N_2, $[Ru(H_2O)_6]^{2+}$ reacts to give $[Ru(H_2O)_5(N_2)]^{2+}$. The related $[Ru(NH_3)_5(N_2)]^{2+}$ (which can be isolated as the chloride salt and is structurally similar to **22.44**) is formed either by reaction scheme 22.92 or by N_2H_4 reduction of aqueous solutions of $RuCl_3\cdot xH_2O$.[†]

$$[Ru(NH_3)_5(H_2O)]^{3+} \xrightarrow{Zn/Hg} [Ru(NH_3)_5(H_2O)]^{2+}$$

$$\xrightarrow[-H_2O]{N_2,\,100\,bar} [Ru(NH_3)_5(N_2)]^{2+} \quad (22.92)$$

The cation $[(H_3N)_5Ru(\mu\text{-}N_2)Ru(NH_3)_5]^{4+}$ (**22.43**) forms when $[Ru(NH_3)_5(H_2O)]^{2+}$ reacts with $[Ru(NH_3)_5(N_2)]^{2+}$, or when aqueous $[Ru(NH_3)_5Cl]^{2+}$ is reduced by Zn amalgam under N_2. Reduction of $[OsCl_6]^{2-}$ with N_2H_4 gives $[Os(NH_3)_5(N_2)]^{2+}$ (**22.44**) which can be oxidized or converted to the bis(N_2) complex (equation 22.93); note the presence of the π-acceptor ligand to stabilize Os(II).

$$[Os(NH_3)_5(N_2)]^{2+} \begin{cases} \xrightarrow[-2H_2O]{HNO_2} cis\text{-}[Os(NH_3)_4(N_2)_2]^{2+} \\ \\ \xrightarrow{Ce^{4+}} [Os(NH_3)_5(N_2)]^{3+} \end{cases} \quad (22.93)$$

N–N = 112 pm   Ru–N$_{bridge}$ = 193 pm
Ru–N(NH$_3$) = 212–214 pm

(22.43)

N–N = 112 pm   Os–N(N$_2$) = 184 pm
Os–N(NH$_3$) = 214–215 pm

(22.44)

† Much of the interest in complexes containing N_2 ligands arises from the possibility of reducing the ligand to NH_3: see Y. Nishibayashi, S. Iwai and M. Hidai (1998) *Science*, vol. 279, p. 540.

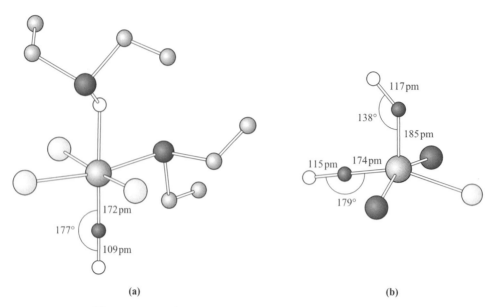

Fig. 22.18 The structures (X-ray diffraction) of (a) [RuBr₃(Et₂S)(Et₂SO)(NO)] [R.K. Coll *et al.* (1987) *Inorg. Chem.*, vol. 26, p. 106] and (b) [RuCl(NO)₂(PPh₃)₂] (only the P atoms of the PPh₃ groups are shown) [C.G. Pierpont *et al.* (1972) *Inorg. Chem.*, vol. 11, p. 1088]. Hydrogen atoms are omitted in (a); colour code: Ru, dark blue; Br and Cl, pale blue; O, white; S and P, black; C, grey.

Most dinitrogen complexes decompose when gently heated, but those of Ru, Os and Ir can be heated to 370–470 K. Although the bonding in a terminal, linear M−N≡N unit can be described in a similar manner to a terminal M−C≡O unit, the bridging modes of N₂ and CO are different; coordination of CO to metals is described in *Section 23.2*.

The complex $[Ru(NH_3)_6]^{2+}$ (which oxidizes in air, equation 22.89) is made by reacting $RuCl_3 \cdot xH_2O$ with Zn dust in concentrated NH_3 solution. The analogous Os(II) complex may be formed in liquid NH_3 but is unstable. The reaction of HNO_2 with $[Ru(NH_3)_6]^{2+}$ gives the nitrosyl complex $[Ru(NH_3)_5(NO)]^{3+}$ in which the Ru−N−O angle is close to 180°. Numerous mononuclear nitrosyl complexes

of ruthenium are known; in each of $[Ru(NH_3)_5(NO)]^{3+}$, $[RuCl_5(NO)]^{2-}$, $[RuCl(bpy)_2(NO)]^{2+}$, *mer,trans*-$[RuCl_3(PPh_3)_2(NO)]$ and $[RuBr_3(Et_2S)(Et_2SO)(NO)]$ (Figure 22.18a), the Ru−N−O unit is linear and an Ru(II) state is formally assigned. Without prior knowledge of structural and spectroscopic properties of nitrosyl complexes (see *Section 20.4*), the oxidation state of the metal centre remains ambiguous, for example in $[RuCl(NO)_2(PPh_3)_2]$ (Figure 22.18b). Stable ruthenium nitrosyl complexes are formed during the extraction processes for the recovery of uranium and plutonium from nuclear wastes, and are difficult to remove; ^{106}Ru is a fission product from uranium and plutonium and the use of HNO_3 and TBP (see *Box 6.3*) in the extraction process facilitates the formation of Ru(NO)-containing complexes.

The tris-chelates $[Ru(en)_3]^{2+}$, $[Ru(bpy)_3]^{2+}$ (*Figure 9.2*) and $[Ru(phen)_3]^{2+}$ are made in a similar manner to $[Ru(NH_3)_6]^{2+}$. The $[Ru(bpy)_3]^{2+}$ complex is widely studied as a photosensitizer; it absorbs light at 452 nm to give an excited singlet state $^1\{[Ru(bpy)_3]^{2+}\}^*$ (Figure 22.19) which results from transfer of an electron for the Ru(II) centre to a bpy π^* orbital, i.e. the excited state may be considered to contain Ru(III), two bpy and one [bpy]⁻. The singlet excited state rapidly decays to a triplet excited state,[†] the lifetime of

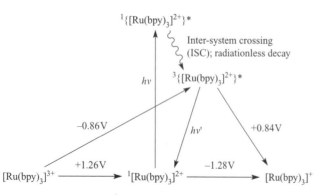

Fig. 22.19 $[Ru(bpy)_3]^{2+}$ (low-spin d^6 is a singlet state) absorbs light to give an excited state which rapidly decays to a longer-lived excited state, $^3\{[Ru(bpy)_3]^{2+}\}^*$. This state can decay by emission or can undergo electron transfer. Standard reduction potentials are given for one-electron processes involving $[Ru(bpy)_3]^{2+}$ and $^3\{[Ru(bpy)_3]^{2+}\}^*$.

[†] For a detailed review, see: A. Juris, V. Balzani, F. Barigelletti, S. Campagna, P. Belser and A. von Zelewsky (1988) *Coordination Chemistry Reviews*, vol. 84, p. 85 – 'Ru(II) polypyridine complexes: Photophysics, photochemistry, electrochemistry and chemiluminescence'. For an introduction to photochemical principles, see C.E. Wayne and R.P. Wayne (1996) *Photochemistry*, Oxford University Press Primer Series, Oxford.

which in aqueous solution at 298 K is 600 ns, long enough to allow redox activity to occur. The standard reduction potentials in Figure 22.19 show that the excited $^3\{[Ru(bpy)_3]^{2+}\}^*$ state is both a better oxidant *and* reductant than the ground $[Ru(bpy)_3]^{2+}$ state; in neutral solution, for example, H_2O can be oxidized or reduced by the excited complex. In practice, the system only works in the presence of a *quenching agent* such as methyl viologen (paraquat), $[MV]^{2+}$ (**22.45**) and a sacrificial donor, D, ($[EDTA]^{4-}$ is often used) which reduces $[Ru(bpy)_3]^{3+}$ to $[Ru(bpy)_3]^{2+}$. Scheme 22.94 summarizes the use of $[Ru(bpy)_3]^{2+}$ as a photosensitizer in the photolytic production of H_2 from H_2O (see *Section 9.4*). However, this and similar reaction schemes are not yet suited to commercial application.

(22.45)

$$[Ru] = [Ru(bpy)_3]$$

(22.94)

Many low oxidation state complexes of Ru and Os including those of Ru(II) and Os(II) are stabilized by PR_3 (π-acceptor) ligands. Treatment of $RuCl_3 \cdot xH_2O$ with PPh_3 in EtOH/HCl at reflux gives *mer*-$[RuCl_3(PPh_3)_3]$ or, with excess PPh_3 in MeOH at reflux, $[RuCl_2(PPh_3)_3]$. Reaction with H_2 converts $[RuCl_2(PPh_3)_3]$ to $[HRuCl(PPh_3)_3]$ which is a hydrogenation catalyst for alk-1-enes (see *Section 26.3*). Both $[RuCl_2(PPh_3)_3]$ and $[HRuCl(PPh_3)_3]$ have square-based pyramidal structures (**22.46** and **22.47**).

(22.46)

(22.47)

Mixed-valence ruthenium complexes

Equation 22.91 showed the formation of the Os(III) complex $[Os_2(\mu\text{-}O_2CMe)_4Cl_2]$. For ruthenium, the scenario is different, and in reaction 22.95, the product is an Ru(II)/Ru(III) polymer (**22.48**).

$$RuCl_3 \cdot xH_2O \xrightarrow{MeCO_2H, (MeCO)_2O} [Ru_2(\mu\text{-}O_2CMe)_4Cl]_n$$

(22.95)

= multiple bond (see text)

(22.48)

Complex **22.48** formally possesses a $\{Ru_2\}^{5+}$ core and from Figure 21.13 we would predict a configuration of $\sigma^2\pi^4\delta^2\delta^{*2}\pi^{*1}$. However, the observed paramagnetism corresponding to three unpaired electrons is consistent with the π^* level lying at lower energy than the δ^*, i.e. $\sigma^2\pi^4\delta^2\pi^{*2}\delta^{*1}$. This reordering is reminiscent of the $\sigma - \pi$ crossover amongst first row diatomics (*Figure 1.23*) and illustrates the importance of utilizing *experimental facts* when constructing and interpreting qualitative MO diagrams.

The *Creutz–Taube* cation $[(H_3N)_5Ru(\mu\text{-pz}')Ru(NH_3)_5]^{5+}$ (pz' = pyrazine) is a member of the series of cations **22.49** (equation 22.96).

(22.96)

$n = 4, 5, 6$

(22.49)

When the charge is 4+ or 6+, the complexes are Ru(II)/Ru(II) or Ru(III)/Ru(III) species respectively. For $n = 5$, a mixed-valence Ru(II)/Ru(III) species might be formulated but spectroscopic and structural data show the Ru centres are equivalent with charge delocalization across the pyrazine bridge. Such electron transfer (see *Section 25.5*) is not observed in all related species. For example, $[(bpy)_2ClRu(\mu\text{-pz}')RuCl(bpy)_2]^{3+}$ exhibits an intervalence charge transfer absorption in its electronic spectrum indicating an Ru(II)/Ru(III) formulation; $[(H_3N)_5Ru^{III}(\mu\text{-pz}')Ru^{II}Cl(bpy)_2]^{4+}$ is similar.

22.10 Group 9: rhodium and iridium

The metals

Rhodium and iridium are unreactive metals; they react with O_2 or the halogens only at high temperatures (see below) and neither is attacked by *aqua regia*. The metals dissolve in fused alkalis. For Rh and Ir, the range of oxidation states (*Table 19.3*) and the stabilities of the highest ones are less than for Ru and Os. The most important states are Rh(III) and Ir(III), i.e. d^6 which is invariably low-spin giving diamagnetic and kinetically inert complexes (see *Section 25.2*).

High oxidation states of rhodium and iridium: M(VI) and M(V)

Rhodium(VI) and iridium(VI) occur only in black RhF_6 and yellow IrF_6, formed by heating the metals with F_2 under pressure and quenching the volatile products; RhF_6 is the least stable platinum-group metal hexafluoride. Both RhF_6 and IrF_6 are octahedral monomers. The pentafluorides are made by direct combination of the elements (equation 22.97) or by reduction of MF_6, and are moisture-sensitive (reaction 22.98) and very reactive. They are tetramers, structurally analogous to NbF_5 (**22.4**).

$$2RhF_5 \xleftarrow{\text{M = Rh, 520 K, 6 bar}} 2M + 5F_2 \xrightarrow{\text{M = Ir, 650 K}} 2IrF_5 \tag{22.97}$$

$$IrF_5 \xrightarrow{H_2O} IrO_2 \cdot xH_2O + HF \tag{22.98}$$

For M(V) and M(VI), no binary compounds with the heavier halogens and no oxides are known.

Salts of octahedral $[MF_6]^-$ (M = Rh, Ir) can be made in HF or interhalogen solvents (reaction 22.99). On treatment with water, they liberate O_2 forming Rh(IV) and Ir(IV) compounds.

$$RhF_5 + KF \xrightarrow{\text{HF or IF}_5} K[RhF_6] \tag{22.99}$$

A number of Ir(V) hydrido complexes are known, e.g. $[IrH_5(PMe_3)_2]$.

Rhodium(IV) and iridium (IV)

The unstable fluorides are the only established halides of Rh(IV) and Ir(IV), and no oxohalides are known. The reaction of $RhBr_3$ or $RhCl_3$ with BrF_3 yields RhF_4; IrF_4 is made by reduction of IrF_6 or IrF_5 with Ir, but above 670 K, IrF_4 disproportionates (equation 22.100). Before 1965, reports of 'IrF_4' were erroneous and actually described IrF_5.

$$8IrF_5 + 2Ir \xrightarrow{670\,K} 10IrF_4 \xrightarrow{>670\,K} 5IrF_3 + 5IrF_5 \tag{22.100}$$

Iridium(IV) oxide forms when Ir is heated with O_2 and is the only well-established oxide of Ir. It is also made by controlled hydrolysis of $[IrCl_6]^{2-}$ in alkaline solution. Heating Rh and O_2 gives Rh_2O_3 (see below) unless the reaction is carried out under high pressure, in which case RhO_2 is obtained. Rutile lattices (*Figure 5.21*) are adopted by RhO_2 and IrO_2.

The series of paramagnetic (low-spin d^7) halo anions $[MX_6]^{2-}$ with M = Rh, X = F, Cl and M = Ir, X = F, Cl, Br, can be made, but the Ir(IV) species are the more stable; $[RhF_6]^{2-}$ and $[RhCl_6]^{2-}$ (equations 22.101 and 22.102) are hydrolyzed to RhO_2 by an excess of H_2O. White alkali metal salts of $[IrF_6]^{2-}$ are made by reaction 22.103; $[IrF_6]^{2-}$ is stable in neutral or acidic solution but decomposes in alkali.

$$2KCl + RhCl_3 \xrightarrow{BrF_3} K_2[RhF_6] \tag{22.101}$$

$$CsCl + [RhCl_6]^{3-} \xrightarrow{Cl_2, \text{aqu.}} Cs_2[RhCl_6] \tag{22.102}$$

$$M[IrF_6] \xrightarrow{H_2O} M_2[IrF_6] + IrO_2 + O_2$$
$$(M = Na, K, Rb, Cs) \tag{22.103}$$

Salts of $[IrCl_6]^{2-}$ are common starting materials in Ir chemistry. Alkali metal salts are made by chlorinating a mixture of MCl and Ir; $Na_2[IrCl_6] \cdot 3H_2O$, $K_2[IrCl_6]$ and $H_2[IrCl_6] \cdot xH_2O$ (*chloroiridic acid*) are commercially available. The $[IrCl_6]^{2-}$ ion is quantitatively reduced (equation 22.104) by KI or $[C_2O_4]^{2-}$ and is used as an oxidizing agent in some organic reactions. In alkaline solution, $[IrCl_6]^{2-}$ decomposes, liberating O_2 but the reaction is reversed in strongly acidic solution (see *Section 7.2*). In its reactions, $[IrCl_6]^{2-}$ is often reduced to Ir(III) (scheme 22.105), but reaction with Br^- yields $[IrBr_6]^{2-}$.

$$[IrCl_6]^{2-} + e^- \rightleftharpoons [IrCl_6]^{3-} \qquad E^\circ = +0.87\,V \tag{22.104}$$

$$[IrCl_6]^{2-} \begin{cases} \xrightarrow{[CN]^-} [Ir(CN)_6]^{3-} \\ \xrightarrow{NH_3} [Ir(NH_3)_5Cl]^{2+} \xrightarrow{NH_3} [Ir(NH_3)_6]^{3+} \\ \xrightarrow{Et_2S} [IrCl_3(SEt_2)_3] \end{cases} \tag{22.105}$$

Octahedral coordination is usual for Ir(IV). Complexes with *O*-donors are relatively few, and include $[Ir(OH)_6]^{2-}$ (the red K^+ salt is made by heating $Na_2[IrCl_6]$ with KOH), $[Ir(NO_3)_6]^{2-}$ (formed by treating $[IrBr_6]^{2-}$ with N_2O_5) and $[Ir(ox)_3]^{2-}$ (made by oxidizing $[Ir(ox)_3]^{3-}$). Complexes with group 15 donors include $[IrCl_4(phen)]$, $[IrCl_2H_2(P^iPr_3)_2]$ (**22.50**) and *trans*-$[IrBr_4(PEt_3)_2]$.

(22.50)

Rhodium(III) and iridium(III)

Binary halides MX_3 for M = Rh, Ir and X = Cl, Br and I can be made by heating the appropriate elements. Reactions 22.106 and 22.107 show routes to MF_3; direct reaction of M and F_2 leads to higher fluorides (e.g. equation 22.97).

$$RhCl_3 \xrightarrow{F_2, 750\,K} RhF_3 \tag{22.106}$$

$$Ir + IrF_6 \xrightarrow{750\,K} 2IrF_3 \qquad (22.107)$$

Anhydrous $RhCl_3$ and α-$IrCl_3$ adopt layer structures and are isomorphous with $AlCl_3$; brown α-$IrCl_3$ converts to the red β-form at 870–1020 K. Water-soluble $RhCl_3 \cdot 3H_2O$ (dark red) and $IrCl_3 \cdot 3H_2O$ (dark green) are commercially available, being common starting materials in Rh and Ir chemistry. Figure 22.20 shows selected complex formations starting from $IrCl_3 \cdot 3H_2O$. In particular, note the formation of $[Ir(bpy)_2(bpy\text{-}C,N)]^{2+}$: this contains a 2,2'-bipyridine ligand which has undergone *orthometallation*. As the insert in Figure 22.20 illustrates, deprotonation of 2,2'-bipyridine in the 6'-position occurs to give the $[bpy\text{-}C,N]^-$ ligand. This leaves an uncoordinated N atom which can be protonated as is observed in $[Ir(bpy)_2(Hbpy\text{-}C,N)]^{3+}$.

The oxide Ir_2O_3 is known only as an impure solid. Rhodium(III) oxide is well characterized, and is made by heating the elements at ordinary pressure or by thermal decomposition of $Rh(NO_3)_3$ (equation 22.108). Several polymorphs of Rh_2O_3 are known; α-Rh_2O_3 has a corundum structure (see *Section 12.7*).

$$4Rh(NO_3)_3 \cdot 6H_2O$$
$$\xrightarrow{1000\,K} 2Rh_2O_3 + 24H_2O + 12NO_2 + 3O_2 \quad (22.108)$$

In the presence of aqueous $HClO_4$, the octahedral $[Rh(H_2O)_6]^{3+}$ can be formed although it hydrolyzes (equation 22.109). Crystalline $Rh(ClO_4)_3 \cdot 6H_2O$ contains $[Rh(H_2O)_6]^{3+}$, i.e. it should be formulated as $[Rh(H_2O)_6][ClO_4]_3$. The $[Ir(H_2O)_6]^{3+}$ ion exists in aqueous solutions in the presence of concentrated $HClO_4$. The hexaaqua ions are present in the crystalline alums $CsM(SO_4)_2 \cdot 12H_2O$ (M = Rh, Ir).

$$[Rh(H_2O)_6]^{3+} + H_2O \rightleftharpoons [Rh(H_2O)_5(OH)]^{2+} + [H_3O]^+$$
$$pK_a = 3.33 \qquad (22.109)$$

When $Rh_2O_3 \cdot H_2O$ is dissolved in a limited amount of aqueous HCl, $RhCl_3 \cdot 3H_2O$ (better written as $[RhCl_3(H_2O)_3]$) forms. All members of the series $[RhCl_n(H_2O)_{6-n}]^{(3-n)+}$ ($n = 0$–6) are known and can be made in solution by reaction of $[Rh(H_2O)_6]^{3+}$ with Cl^- or by substitution starting from $[RhCl_6]^{3-}$ (see *Problem 25.8*). Interconversions involving $[Rh(H_2O)_6]^{3+}$ and $[RhCl_6]^{3-}$ are given in scheme 22.110.

$$[RhCl_6]^{3-}$$

$$Rh_2O_3 \cdot H_2O \underset{[OH]^-}{\overset{H^+}{\rightleftharpoons}} [Rh(H_2O)_6]^{3+} \qquad (22.110)$$

Reduction of $[IrCl_6]^{2-}$ by SO_2 yields $[IrCl_6]^{3-}$ (Figure 22.20) which hydrolyzes in H_2O to $[IrCl_5(H_2O)]^{2-}$ (isolated as the green $[NH_4]^+$ salt), $[IrCl_4(H_2O)_2]^-$ and $[IrCl_3(H_2O)_3]$.

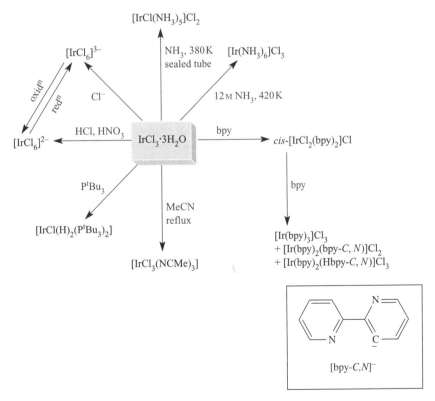

Fig. 22.20 Selected reactions of $IrCl_3 \cdot xH_2O$. In the complexes $[Ir(bpy)_2(bpy\text{-}C,N)]^{2+}$ and $[Ir(bpy)_2(Hbpy\text{-}C,N)]^{3+}$, the ligands coordinating in a *C,N*-mode have undergone *orthometallation* in which a C–H bond has been broken and a C^- coordination site formally created.

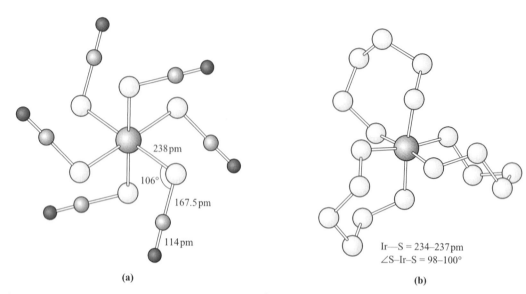

Fig. 22.21 The structures (X-ray diffraction) of (a) $[Ir(NCS\text{-}S)_6]^{3-}$ in the $[Me_4N]^+$ salt [J.-U. Rohde *et al.* (1998) *Z. Anorg. Allg. Chem.*, vol. 624, p. 1319] and (b) $[Ir(S_6)_3]^{3-}$ in the $[NH_4]^+$ salt [T.E. Albrecht-Schmitt *et al.* (1996) *Inorg. Chem.*, vol. 35, p. 7273]. Colour code: Ir, blue; S, pale grey; C, grey; N, black.

Reaction of $[Ir(H_2O)_6]^{3+}$ with **[22.51]**Cl_3 in aqueous Cs_2SO_4 produces **[22.51]**$[IrCl_2(H_2O)_4][SO_4]_2$ containing cation **22.52** for which $pK_a(1) = 6.31$.

(22.51) (22.52)

Routes to ammine complexes of Ir(III) were shown in Figure 22.20. For Rh(III), it is more difficult to form $[Rh(NH_3)_6]^{3+}$ than $[Rh(NH_3)_5Cl]^{2+}$ (equation 22.111). Reaction of $RhCl_3 \cdot 3H_2O$ with Zn dust and aqueous NH_3 gives $[Rh(NH_3)_5H]^{2+}$.

$$RhCl_3 \cdot 3H_2O \xrightarrow{\text{aq } NH_3, EtOH} [RhCl(NH_3)_5]Cl_2$$

$$\xrightarrow[\text{sealed tube}]{NH_3, 373\,K} [Rh(NH_3)_6]Cl_3 \qquad (22.111)$$

Large numbers of octahedral Rh(III) and Ir(III) complexes exist, and common precursors include $[IrCl_6]^{3-}$ (e.g. Na^+, K^+ or $[NH_4]^+$ salts), $[RhCl(NH_3)_5]Cl_2$, $[Rh(H_2O)(NH_3)_5][ClO_4]_3$ (made by treating $[RhCl(NH_3)_5]Cl_2$ with $AgClO_4$) and *trans*-$[RhCl_2(py)_4]^+$ (made from $RhCl_3 \cdot 3H_2O$ and pyridine). Schemes 22.112 and 22.113 give selected examples.

$$[IrCl_6]^{3-} \begin{cases} \xrightarrow{\text{en, }\Delta} [Ir(en)_3]^{3+} \\ \xrightarrow{\text{py in EtOH}} \textit{trans-}[IrCl_2(py)_4]^+ \\ \xrightarrow{ox^{2-},\,\Delta} [Ir(ox)_3]^{3-} \\ \xrightarrow{\text{conc } HNO_3,\,\Delta} [Ir(NO_3)_6]^{3-} \end{cases} \quad (22.112)$$

$$[RhCl_2(py)_4]^+ \begin{cases} \xrightarrow{\text{en, }H_2O} \textit{cis-}[RhCl(en)_2(py)]^{2+} \\ \xrightarrow{\text{EtOH, }\Delta} \textit{mer-}[RhCl_3(py)_3] \\ \xrightarrow{\text{in } CHCl_3,\,\Delta} [RhCl_3(py)_2]_n \\ \xrightarrow{NH_3} [RhCl(NH_3)_5]^{2+} \\ \xrightarrow{ox^{2-},\,\Delta} [Rh(NH_3)_4(ox)]^+ \end{cases}$$

$$(22.113)$$

Rhodium(III) and iridium(III) form complexes with both hard and soft donors and examples (in addition to those above and in Figure 22.20) include:

- *N*-donors: $[Ir(NO_2)_6]^{3-}$, *cis*-$[RhCl_2(bpy)_2]^+$, $[Rh(bpy)_2(phen)]^{3+}$, $[Rh(bpy)_3]^{3+}$, $[Rh(en)_3]^{3+}$;
- *O*-donors: $[Rh(acac)_3]$, $[Ir(acac)_3]$, $[Rh(ox)_3]^{3-}$;
- *P*-donors: *fac*- and *mer*-$[IrH_3(PPh_3)_3]$, $[RhCl_4(PPh_3)_2]^-$, $[RhCl_2(H)(PPh_3)_2]$;
- *S*-donors: $[Ir(NCS\text{-}S)_6]^{3-}$ (Figure 22.21a), *mer*-$[IrCl_3(SEt_3)_3]$, $[Ir(S_6)_3]^{3-}$ (Figure 22.21b).

Both metal ions form $[M(CN)_6]^{3-}$. Linkage isomerization is exhibited by $[Ir(NH_3)_5(NCS)]^{2+}$, i.e. both $[Ir(NH_3)_5(NCS\text{-}N)]^{2+}$ and $[Ir(NH_3)_5(NCS\text{-}S)]^{2+}$ can be isolated. The nitrite ligand in $[Ir(NH_3)_5(NO_2)]^{2+}$ undergoes a change from *O*- to *N*-coordination in alkaline solution.

Rhodium(II) and iridium(II)

Mononuclear Rh(II) and Ir(II) complexes are relatively rare. The chemistry of Rh(II) is quite distinct from that of Ir(II) since dimers of type $[Rh_2(\mu\text{-}L)_4]$ and $[Rh_2(\mu\text{-}L)_4L'_2]$ are well known but have no Ir analogues. The best-known dimers contain carboxylate bridges (Figures 22.22a and 22.22b); other bridging ligands include $[RC(O)NH]^-$ and

(a)

(b)

231 pm

204 pm

Rh—Rh = 238.5 pm

(c)

Fig. 22.22 Schematic representations of two families of Rh(II) carboxylate dimers: (a) $[Rh_2(\mu\text{-}O_2CR)_4]$ and (b) $[Rh_2(\mu\text{-}O_2CR)_4L_2]$. (c) The structure of $[Rh_2(\mu\text{-}O_2CMe)_4(H_2O)_2]$ (H atoms omitted) determined by X-ray diffraction [F.A. Cotton *et al.* (1971) *Acta Crystallogr.*, *Sect. B*, vol. 27, p. 1664]; colour code Rh, blue; C, grey; O, white.

$[RC(O)S]^-$. The dimers $[Rh_2(\mu\text{-}O_2CMe)_4L_2]$ (L = MeOH or H_2O) are made by reactions 22.114 and 22.115; the axial ligands can be removed by heating *in vacuo*, or replaced (e.g. reaction 22.116).

$RhCl_3 \cdot 3H_2O$

$$\xrightarrow[\text{MeOH reflux}]{\text{MeCO}_2\text{H / Na[MeCO}_2]} [Rh_2(\mu\text{-}O_2CMe)_4(MeOH)_2] \tag{22.114}$$

$$[NH_4]_3[RhCl_6] \xrightarrow[\text{EtOH, H}_2\text{O}]{\text{MeCO}_2\text{H}} [Rh_2(\mu\text{-}O_2CMe)_4(H_2O)_2] \tag{22.115}$$

$$[Rh_2(\mu\text{-}O_2CMe)_4(H_2O)_2] \begin{cases} \xrightarrow{\text{py}} [Rh_2(\mu\text{-}O_2CMe)_4(py)_2] \\ \xrightarrow{\text{SEt}_2} [Rh_2(\mu\text{-}O_2CMe)_4(SEt_2)_2] \end{cases} \tag{22.116}$$

(22.53)   (22.54)

Figure 22.22c shows the structure of $[Rh_2(\mu\text{-}O_2CMe)_4(H_2O)_2]$, and related complexes are similar. If the axial ligand has a second donor atom appropriately oriented, polymeric chains in which L' bridges $[Rh_2(\mu\text{-}L)_4]$ units can result, e.g. when $L' = $ **22.53** or **22.54**. Each dimer formally contains a $\{Rh_2\}^{4+}$ core which (from *Figure 21.13*) has a $\sigma^2\pi^4\delta^2\delta^{*2}\pi^{*4}$ configuration, and a Rh–Rh *single* bond. Compare this to the multiply bonded Mo(II), Re(III) and Os(III) dimers discussed earlier.

Rhodium(I) and iridium(I)

The +1 oxidation state is stabilized by π-acceptor ligands such as phosphines, with square planar, and to a lesser extent, trigonal bipyramidal coordination being favoured. Being low oxidation state species, it may be appropriate to consider the bonding in terms of the 18-electron rule (*Section 20.4*). In fact, most Rh(I) complexes are square planar, 16-electron species and some such as $[RhCl(PPh_3)_3]$ (*Wilkinson's catalyst*, **22.55**) have important applications in homogeneous catalysis (see *Chapter 26*). Preparation of $[RhCl(PPh_3)_3]$ involves reduction of Rh(III) by PPh_3 (equation 22.117). Other $[RhCl(PR_3)_3]$ complexes are made by routes such as 22.118; alkene complexes like that in this reaction are described in *Chapter 23*. Starting from $[RhCl(PPh_3)_3]$, it is possible to make a variety of square planar complexes in which phosphine ligands remain to stabilize the Rh(I) centre, e.g. scheme 22.119.

$$\begin{array}{c} PPh_3 \\ | \\ Cl — Rh — PPh_3 \\ | \\ PPh_3 \end{array}$$

(22.55)

$$RhCl_3 \cdot 3H_2O + 6PPh_3 \xrightarrow{\text{EtOH, reflux}} [RhCl(PPh_3)_3] + \dots \tag{22.117}$$

$$+6PR_3 \longrightarrow 2[RhCl(PR_3)_3] + 4C_2H_4$$

$$R \neq Ph \tag{22.118}$$

$$[RhCl(PPh_3)_3] \begin{cases} \xrightarrow{NaBH_4} [RhH(PPh_3)_3] \\[2mm] \xrightarrow[MeCN]{[Me_3O][BF_4]} [Rh(MeCN)(PPh_3)_3][BF_4] \\[2mm] \xrightarrow{RCO_2H} [Rh(O_2CR-O)(PPh_3)_3] \end{cases}$$

$$\tag{22.119}$$

Treatment of $[RhCl(PR_3)_3]$ with $TlClO_4$ yields the perchlorate salt of the trigonal planar cation $[Rh(PPh_3)_3]^+$. The square planar Ir(I) complex *trans*-$[IrCl(CO)(PPh_3)_2]$ (*Vaska's compound*, **22.56**) is strictly organometallic since it contains an Ir–C bond, but it is an important precursor in Ir(I) chemistry. Both *trans*-$[IrCl(CO)(PPh_3)_2]$ and $[RhCl(PPh_3)_3]$ undergo many oxidative addition reactions (see *Chapter 23*) in which the M(I) centre is oxidized to M(III).

$$(22.56)$$

22.11 Group 10: palladium and platinum

The metals

At 298 K, bulk Pd and Pt are resistant to corrosion. Palladium is more reactive than Pt, and at high temperatures is attacked by O_2, F_2 and Cl_2 (equation 22.120).

$$PdO \xleftarrow{O_2, \Delta} Pd \xrightarrow{Cl_2, \Delta} PdCl_2 \tag{22.120}$$

Palladium dissolves in hot oxidizing acids (e.g. HNO_3), but both metals dissolve in *aqua regia* and are attacked by molten alkali metal oxides. The absorption of H_2 by Pd was described at the end of *Section 9.7*.

The dominant oxidation states are M(II) and M(IV), but the M(IV) state is more stable for Pt than Pd. Within a given oxidation state, resemblances are close with the exception of their behaviour towards oxidizing and reducing agents. In comparing the chemistries of Ni(II), Pd(II) and Pt(II), structural similarities between low-spin square planar complexes are observed, but octahedral and tetrahedral high-spin Ni(II) complexes have only a few

parallels in Pd(II) chemistry and effectively none among Pt(II) species (see *Box 20.6*).

The highest oxidation states: M(VI) and M(V)

The M(VI) and M(V) states are confined to platinum fluorides (reactions 22.121 and 22.122); PtF_5 readily disproportionates to PtF_4 and PtF_6.

$$PtCl_2 \xrightarrow{F_2, 620 K} PtF_5 \tag{22.121}$$

$$Pt \xrightarrow[\text{2. rapid quenching}]{\text{1. } F_2, 870 K} PtF_6 \tag{22.122}$$

Platinum(V) fluoride is a tetramer (structurally like **22.4**), while PtF_6 is monomeric and Jahn–Teller distorted octahedral (d^4). The hexafluoride is a very powerful oxidant (equation 22.123, see *Section 5.16*) and attacks glass.

$$O_2 + PtF_6 \longrightarrow [O_2]^+[PtF_6]^- \tag{22.123}$$

The fluorides PdF_5 and PdF_6 have not been confirmed, but $[PdF_6]^-$ can be made by reaction 22.124.

$$PdF_4 + KrF_2 + O_2 \longrightarrow [O_2]^+[PdF_6]^- + Kr \tag{22.124}$$

Palladium(IV) and platinum(IV)

The only tetrahalide of Pd(IV) is PdF_4, a diamagnetic, red solid made from the elements at 570 K. The paramagnetic compound 'PdF_3' (also formed from Pd and F_2) is actually $Pd^{II}[Pd^{IV}F_6]$; both Pd centres are octahedrally sited in the solid, and while $[PdF_6]^{2-}$ (Pd–F = 190 pm) is diamagnetic, the Pd(II) centre (Pd–F = 217 pm) has two unpaired electrons. All the Pt(II) halides are known, and $PtCl_4$ and $PtBr_4$ are formed by reactions of the halogens with Pt. Treatment of $PtCl_2$ with F_2 ($T < 475\,K$) gives PtF_4 (compare reaction 22.121). In $PtCl_4$, $PtBr_4$ and PtI_4, the metal is octahedrally sited as shown in **22.57**; in PdF_4 and PtF_4, the connectivity is similar but results in a 3D-structure.

$$X = Cl, Br, I$$
$$(22.57)$$

Hydrated PtO_2 is made by hydrolyzing $[PtCl_6]^{2-}$ in boiling aqueous Na_2CO_3; heating converts it to the black anhydrous oxide. Above 920 K, PtO_2 decomposes to the elements. The hydrated oxide dissolves in NaOH as

$Na_2[Pt(OH)_6]$ and in aqueous HCl as $H_2[PtCl_6]$ (*chloroplatinic acid*); the latter is an important starting material and has catalytic applications. Water hydrolyzes $H_2[PtCl_6]$ to $H[PtCl_5(H_2O)]$ and $[PtCl_4(H_2O)_2]$; the reaction is reversed by adding HCl.

In their complexes, Pd(IV) and Pt(IV) are low-spin, octahedral and diamagnetic (d^6). The full range of halo complexes $[MX_6]^{2-}$ is known (e.g. equations 22.125–22.127), in contrast to PdF_4 being the only neutral Pd(IV) halide. The $[MX_6]^{2-}$ ions are stabilized by large cations.

$$M \xrightarrow{\text{aqua regia; KCl}} K_2[MCl_6] \qquad M = Pd, Pt \qquad (22.125)$$

$$K_2[PtCl_6] \xrightarrow{BrF_3} K_2[PtF_6] \qquad (22.126)$$

$$PtCl_4 \xrightarrow{HCl} H_2[PtCl_6] \begin{cases} \xrightarrow{KCl} K_2[PtCl_6] \\ \xrightarrow{KI} K_2[PtI_6] \end{cases} \qquad (22.127)$$

The greater kinetic inertness (see *Section 25.2*) of the Pt(IV) complexes is illustrated by the fact that $K_2[PdF_6]$ is decomposed in the air by moisture, but $K_2[PtF_6]$ can be crystallized from boiling water even though $[PtF_6]^{2-}$ is *thermodynamically* unstable with respect to hydrolysis. For details of $K_2[PtH_6]$, see *Section 9.7*.

(22.58)

The variety of Pd(IV) complexes is far less than that of Pt(IV), and their syntheses usually involve oxidation of a related Pd(II) species, e.g. reaction 22.128. When a chelating ligand such as bpy or $Me_2PCH_2CH_2PMe_2$ is present, the complex is constrained to being *cis*, e.g. **22.58**. The Pd(IV) complexes are of limited stability.

$$\text{\textit{trans}-}[PdCl_2(NH_3)_2] + Cl_2 \longrightarrow \text{\textit{trans}-}[PdCl_4(NH_3)_2] \qquad (22.128)$$

Platinum(IV) forms a wide range of thermodynamically and kinetically inert, octahedral complexes, and ammine complexes, for example, have been known since the days of Werner (see *Box 21.8*). In liquid NH_3 at 230 K, $[NH_4]_2[PtCl_6]$ is converted to $[Pt(NH_3)_6]Cl_4$; *trans*-$[PtCl_2(NH_3)_4]^{2+}$ is made by oxidative addition of Cl_2 to $[Pt(NH_3)_4]^{2+}$, and, as for Pd, oxidative addition is a general strategy for Pt(II) $\longrightarrow$ Pt(IV) conversions. Amine complexes include the optically active $[Pt(en)_3]^{4+}$ and *cis*-$[PtCl_2(en)_2]^{2+}$, both of which can be resolved. Although the range of ligands coordinating to Pt(IV) covers soft and hard donors (see *Table 6.9*), some ligands such as phosphines tend to reduce Pt(IV) to Pt(II).

(22.59)

Of note is $[Pt(NH_3)_4]_2[\mathbf{22.59}]\cdot 9H_2O$, formed when aqueous $[Pt(NH_3)_4][NO_3]_2$ reacts with $K_3[Fe(CN)_6]$. Localized Fe(II) and Pt(IV) centres in $[\mathbf{22.59}]^{4-}$ have been assigned on the basis of magnetic, electrochemical and ESR spectroscopic data, and the complex exhibits an intense Fe(II) $\longrightarrow$ Pt(IV) charge-transfer absorption at 470 nm.

Palladium(III), platinum(III) and mixed valence complexes

We saw above that 'PdF$_3$' is the mixed valence $Pd[PdF_6]$, and this cautionary note extends to some other apparently Pd(III) and Pt(III) species. For example, both PtCl$_3$ and PtCl$_3$ are mixed valence compounds. The compounds of empirical formulae $Pt(NH_3)_2Br_3$ (**22.60**) and $Pt(NEtH_2)_4Cl_3\cdot H_2O$ (*Wolffram's red salt*, **22.61**) contain halide-bridged chains; extra Cl^- in the lattice of the latter balance the 4+ charge. Such mixed-valence compounds possess intense colours due to intervalence charge-transfer absorptions. Partially oxidized $[Pt(CN)_4]^{2-}$ salts are described under platinum(II).

octahedral Pt(IV)   square planar Pt(II)

L = NH_3

(22.60)

(b)

$a = 283\,\mathrm{pm}$
$b = 291\,\mathrm{pm}$

(a)

N⌣O = conjugate base of glutarimide

(c)

Fig. 22.23 (a) The structure (X-ray diffraction) of the cation in the platinum blue $[Pt_4(en)_4(\mu\text{-}L)_4][NO_3]_5\cdot H_2O$ where $HL = \alpha$-pyridone. Hydrogen atoms have been omitted; colour code: Pt, blue; N, black; O, white; C, grey [T.V. O'Halloran *et al.* (1984) *J. Am. Chem. Soc.*, vol. 106, p. 6427]. (b) Examples of *N,O*-donor ligands present in platinum blues. (c) Schematic representation of the platinum blue $[Pt_4(NH_3)_8(\mu\text{-}L)_4]^{5+}$ where $HL = $ glutarimide.

octahedral Pt(IV)   square planar Pt(II)

$L = NEtH_2$

(22.61)

Palladium(III) dimers that are structurally related to the Rh(II) dimers discussed above (Figure 22.22) include $[Pd_2(\mu\text{-}SO_4\text{-}O,O')_4(H_2O)_2]^{2-}$ and $[Pd_2(\mu\text{-}O_2CMe)_4(H_2O)_2]^{2+}$ each of which formally contains a $\{Pd_2\}^{6+}$ core (isoelectronic with $\{Rh_2\}^{4+}$) and a Pd–Pd single bond.

The *platinum blues*[†] are mixed valence complexes containing discrete Pt_n chains (Figure 22.23). They are formed by

hydrolysis of *cis*-$[PtCl_2(NH_3)_2]$ or *cis*-$[PtCl_2(en)]$ in aqueous $AgNO_3$ (i.e. replacing Cl^- by H_2O and precipitating AgCl) followed by treatment with *N,O*-donors such as pyrimidines, uracils or the compounds shown in Figure 22.23b. Figures 22.23a and 22.23c show two examples; each formally contains a $\{Pt_4\}^{9+}$ core which can be considered as $(Pt^{III})(Pt^{II})_3$. ESR spectroscopic data show that the unpaired electron is delocalized over the Pt_4-chain. Interest in platinum blues lies in the fact that some exhibit anti-tumour activity.

Palladium(II) and platinum(II)

In *Section 20.3*, we discussed the increase in crystal field splitting on progressing down group 10, and explained why Pd(II) and Pt(II) complexes favour a square planar arrangement of donor atoms (but see *Box 20.6*). In this section, the discussion of Pd(II) and Pt(II) compounds reiterates these points.

All the halides of Pd(II) and Pt(II) except PtF_2 are known. Reaction of Pd and F_2 gives 'PdF_3' (see above) which is reduced to violet PdF_2 by SeF_4. Unusually for Pd(II), PdF_2 is paramagnetic and each Pd(II) centre is octahedrally sited in a rutile lattice (*Figure 5.21*). The other dihalides are

[†] For related complexes, see: C. Tejel, M.A. Ciriano and L.A. Oro (1999) *Chemistry – A European Journal*, vol. 5, p. 1131 – 'From platinum blues to rhodium and iridium blues'.

BOX 22.8   **APPLICATIONS**

Detecting CO with PdCl$_2$

In Box 13.9, we looked at the use of semiconductors such as SnO$_2$ as sensors for gases including CO and hydrocarbons. For workers who may be exposed to CO, a Pd-based detector that can be worn as a badge has been developed; the presence of CO is conveniently signalled by a colour change. The detector contains hydrated PdCl$_2$ and CuCl$_2$. Carbon monoxide reduces Pd(II) to Pd(0):

$$PdCl_2 \cdot 2H_2O + CO \longrightarrow Pd + CO_2 + 2HCl + H_2O$$

and the production of Pd metal causes the chemical patch to darken. When the detector is removed from a CO-containing zone and enters an oxidizing environment, the Pd metal is oxidized to Pd(II) by CuCl$_2$, with the CuCl$_2$ being

regenerated by atmospheric oxidation of Cu(I):

$$Pd + 2CuCl_2 \cdot 2H_2O \longrightarrow PdCl_2 \cdot 2H_2O + 2CuCl$$
$$4CuCl + 4HCl + O_2 + 6H_2O \longrightarrow 4CuCl_2 \cdot 2H_2O$$

The reaction sequence requires the presence of H$_2$O and HCl in the detector: the support for the chemical patch is silica gel which absorbs moisture, CaCl$_2$ acts as a Cl$^-$ source, and a strong acid is added in the form of, for example, H$_8$[SiMo$_{12}$O$_{42}$]$\cdot$28H$_2$O.

Related information

See the Wacker process: Figure 26.2 and associated text.

---

diamagnetic (low-spin d^8) and contain square planar M(II) centres in polymeric structures. Heating Pd and Cl$_2$ gives PdCl$_2$; the α-form is a polymer (**22.62**) and above 820 K, α-PdCl$_2$ converts to the β-form which contains hexameric units **22.63**. Palladium(II) bromide is made from the elements, and PdI$_2$ by heating PdCl$_2$ with HI. Direct combination of Pt and a halogen affords PtCl$_2$, PtBr$_2$ and PtI$_2$; PtCl$_2$ is dimorphic like PdCl$_2$.

(22.62)   (22.63)

Black PdO, formed by heating Pd and O$_2$, is the only well-established oxide of Pd. Thus, the heavier group 10 metals contrast in their oxide formation, PtO$_2$ being the only well-characterized oxide of Pt. Dissolution of PdO in perchloric acid gives [Pd(H$_2$O)$_4$][ClO$_4$]$_2$ containing a diamagnetic, square planar tetraaqua ion. The [Pt(H$_2$O)$_4$]$^{2+}$ ion is made by treating [PtCl$_4$]$^{2-}$ with AgClO$_4$. Both aqua ions are considerably better oxidizing agents than aqueous Ni^{2+} (equation 22.129), but neither [Pd(H$_2$O)$_4$]$^{2+}$ nor [Pt(H$_2$O)$_4$]$^{2+}$ is very stable.

$$M^{2+} + 2e^- \rightleftharpoons M \begin{cases} M = Ni, \; E^\circ = -0.25\,V \\ M = Pd, \; E^\circ = +0.95\,V \\ M = Pt, \; E^\circ = +1.18\,V \end{cases} \quad (22.129)$$

Palladium(II) and platinum(II) form a wealth of square planar complexes; a tendency for Pt–Pt interactions (i.e. for the heaviest group 10 metal) is quite often observed.

The mechanisms of substitution reactions in Pt(II) complexes and the *trans-effect* have been much-studied and we return to this in *Section 25.3*. However, for the discussion that follows, it is important to note that mutually *trans* ligands exert an effect on one another, and this dictates the order in which ligands are displaced and, therefore, the products of substitution reactions. *A word of caution:* do not confuse *trans*-effect with *trans*-influence (see *Box 22.9*).

Important families of complexes with monodentate ligands include [MX$_4$]$^{2-}$ (e.g. X = Cl, Br, I, CN, SCN-*S*), [MX$_2$L$_2$] (e.g. X = Cl, Br; L = NH$_3$, NR$_3$, RCN, py, PR$_3$, SR$_2$; or X = CN; L = PR$_3$) and [ML$_4$]$^{2+}$ (e.g. L = PR$_3$, NH$_3$, NR$_3$, MeCN). For [MX$_2$L$_2$], *trans*- and *cis*-isomers may exist and, in the absence of X-ray diffraction data, IR spectroscopy can be used to distinguish between *cis*- and *trans*-[MX$_2$Y$_2$] species (*Figure 19.11*). Isomers of, for example, [PtCl$_2$(PR$_3$)$_2$] can also be distinguished using ^{31}P NMR spectroscopy (*Box 19.1*). Equations 22.130–22.132 show the formation of some ammine complexes, the choice of route for *cis*- and *trans*-[PtCl$_2$(NH$_3$)$_2$] being a manifestation of the *trans*-effect (see above).

Isomerization of *cis*- to *trans*-[PtCl$_2$(NH$_3$)$_2$] occurs in solution.

$$[PtCl_6]^{2-} \xrightarrow{SO_2;\; HCl} H_2PtCl_4 \xrightarrow{conc\, NH_3,\, \Delta} [Pt(NH_3)_4]Cl_2 \quad (22.130)$$

$$[PtCl_4]^{2-} + 2NH_3 \xrightarrow{aq\, soln} cis\text{-}[PtCl_2(NH_3)_2] + 2Cl^- \quad (22.131)$$

$$[Pt(NH_3)_4]^{2+} + 2HCl$$
$$\xrightarrow{aq\, soln} trans\text{-}[PtCl_2(NH_3)_2] + 2[NH_4]^+ \quad (22.132)$$

BOX 22.9   **CHEMICAL AND THEORETICAL BACKGROUND**

The *trans*-influence

Consider a square planar complex which contains a *trans* L–M–L' arrangement:

Ligands L and L' compete with each other for electron density because formation of M–L and M–L' bonds uses the same metal orbitals, e.g. d_{z^2} and p_z if L and L' lie on the z axis. The existence of a *ground state trans*-influence (i.e. the influence that L has on the M–L' bond) is established by comparing the solid state structural, and vibrational and NMR spectroscopic data for series of related complexes. Structural data are exemplified by the following series of square planar Pt(II) complexes; H^- exerts a strong *trans*-influence and as a consequence, the Pt–Cl bond in $[PtClH(PEtPh_2)_2]$ is relatively long and weak:

| Pt–Cl / pm | 231.6 | $a = 232.7$ $b = 230.5$ | 237 | 242 |

IR and ^{1}H NMR spectroscopic data for a series of square planar complexes *trans*-$[PtXH(PEt_3)_2]$ are as follows:

X^-	$[CN]^-$	I^-	Br^-	Cl^-
$\bar{\nu}(Pt-H)$ / cm^{-1}	2041	2156	2178	2183
δ (^{1}H for Pt–H)	−7.8	−12.7	−15.6	−16.8

Values of $\bar{\nu}(Pt-H)$ show that the Pt–H bond is weakest for $X^- = [CN]^-$ and the *trans*-influence of the X- ligands follows the order $[CN]^- > I^- > Br^- > Cl^-$. The signal for the hydride in the ^{1}H NMR spectrum moves to lower frequency (higher field) with a decrease in the *trans*-influence of X^-. The *trans*-influence is not unique to square planar complexes, and may be observed wherever ligands are mutually *trans*, e.g. in octahedral species.

The *trans*-influence is *not* the same as the *trans*-effect. The first is a ground state phenomenon, while the second is a kinetic effect (see Section 25.3).

Pt–Pt = 324 pm

(22.64)

Magnus's green salt, $[Pt(NH_3)_4][PtCl_4]$, is a polymerization isomer (see *Section 19.8*) of $[PtCl_2(NH_3)_2]$ and is prepared by precipitation from colourless $[Pt(NH_3)_4]Cl_2$ and pink $[PtCl_4]^{2-}$. It contains chains of alternating cations and anions (**22.64**) with significant Pt–Pt interactions, and this structural feature leads to the change in colour in going from the constituent ions to the solid salt. However the Pt–Pt distance is not as short as in the partially oxidized $[Pt(CN)_4]^{2-}$ (see below), and $[Pt(NH_3)_4][PtCl_4]$ is not a metallic conductor.

The paucity of complexes with *O*-donors arises because Pd(II) and Pt(II) are soft metal centres (see *Table 6.9*); we noted above the instability of the tetraaqua ions. Reaction of $[PtCl_4]^{2-}$ with KOH and excess Hacac gives monomeric $[Pt(acac)_2]$. Palladium(II) and platinum(II) acetates are trimeric and tetrameric respectively. The Pd atoms in $[Pd(O_2CMe)_2]_3$ are arranged in a triangle with each Pd····Pd (non-bonded, 310–317 pm) bridged by two $[MeCO_2]^-$ ligands giving square planar coordination. In $[Pt(O_2CMe)_2]_4$, the Pt atoms form a square (Pt–Pt bond lengths = 249 pm) with two $[MeCO_2]^-$ bridging each edge. Palladium(II) acetate is an important industrial catalyst for the conversion of ethene to vinyl acetate.

BOX 22.10

APPLICATIONS

Platinum-containing drugs for cancer treatment

Cisplatin is the square planar complex *cis*-[PtCl$_2$(NH$_3$)$_2$] (**22.1a**) and its capacity to act as an anti-tumour drug has been known since the 1960s. It is used to treat bladder and cervical tumours, as well as testicular and ovarian cancers, but patients may suffer from side-effects of nausea and kidney damage. Carboplatin (**22.1b**) shows similar anti-tumour properties and has the advantage of producing fewer side-effects than cisplatin. The drugs operate by inter-acting with guanine (G) bases in strands of DNA (Figure 9.9), with the *N*-donors of the nucleotide base coordinating to the Pt(II) centre; GG intrastrand cross-links are formed by cisplatin. A recent Pt(II) complex to enter into clinical trials is the triplatinum species shown on the right. Preliminary results show that the complex is significantly more active than cisplatin, and is capable of forming interstrand crosslinks involving three base pairs in DNA.

X = (CH$_2$)$_6$

Further reading

J. Reedijk (1996), *Chemical Communications*, p. 801 – 'Improved understanding in platinum antitumour chemistry'.

The cyano complexes [Pd(CN)$_4$]$^{2-}$ and [Pt(CN)$_4$]$^{2-}$ are very stable; K$_2$[Pt(CN)$_4$]·3H$_2$O can be isolated from the reaction of K$_2$[PtCl$_4$] with KCN in aqueous solution. Aqueous solutions of K$_2$[Pt(CN)$_4$] are colourless, but the hydrate forms yellow crystals; similarly, other salts are colourless in solution but form coloured crystals. The colour change arises from stacking (non-eclipsed) of square planar anions in the solid, although the Pt····Pt separations are significantly larger (e.g. 332 pm in yellow-green Ba[Pt(CN)$_4$]·4H$_2$O and 309 pm in violet Sr[Pt(CN)$_4$]·3H$_2$O) than in Pt metal (278 pm). When K$_2$[Pt(CN)$_4$] is partially oxidized with Cl$_2$ or Br$_2$, bronze complexes of formula K$_2$[Pt(CN)$_4$]X$_{0.3}$· 2.5H$_2$O (X = Cl, Br) are obtained. These contain isolated X$^-$ ions and stacks of staggered [Pt(CN)$_4$]$^{2-}$ ions (**22.65**)

with short Pt–Pt separations, and are good one-dimensional metallic conductors. The conductivity arises from electron delocalization along the Pt$_n$ chain, the centres no longer being localized Pt(II) after partial oxidation. Some salts do contain non-stacked [Pt(CN)$_4$]$^{2-}$ ions, e.g. [PhNH$_3$]$_2$[Pt(CN)$_4$].

Zeise's salt, K[PtCl$_3$(η^2-C$_2$H$_4$)], is a well-known organo-metallic Pt(II) complex and is discussed in *Section 23.10*.

22.12 Group 11: silver and gold

The metals

Silver and gold are generally inert, and are not attacked by O$_2$ or non-oxidizing acids. Silver dissolves in HNO$_3$ and liberates H$_2$ from concentrated HI owing to the formation of stable iodo complexes. Where sulfide (e.g. as H$_2$S) is present, Ag tarnishes as a surface coating of Ag$_2$S forms. Gold dissolves in concentrated HCl in the presence of oxidizing agents due to the formation of chloro complexes (equation 22.133).

$$[AuCl_4]^- + 3e^- \rightleftharpoons Au + 4Cl^- \qquad E^o_{[Cl^-]=1} = +1.00\,V$$

(22.133)

Both metals react with halogens (see below), and gold dissolves in liquid BrF$_3$, forming [BrF$_2$]$^+$[AuF$_4$]$^-$. The dissolution of Ag and Au in cyanide solutions in the presence of air is used in their extraction from crude ores (equation 22.4).

Stable oxidation states for the group 11 metals differ: in contrast to the importance of Cu(II) and Cu(I), silver has only one stable oxidation state, Ag(I), and for gold, Au(III) and Au(I) are dominant, with Au(III) being the

L = CN$^-$
a = 288 pm

(22.65)

BOX 22.11   **RESOURCES, ENVIRONMENTAL AND BIOLOGICAL**

Recycling materials: silver and gold

With precious metals such as silver and gold, recycling is an important way of conserving resources. In 1997, the US industrial turnover of silver was ≈5000 t, with 1360 t of this recovered from photographic scrap (mainly fixer solutions),

spent catalysts, jewellery manufacturing waste and miscellaneous silver-containing materials. About 18% of the gold supply in the US is recovered metal.

more stable. Relativistic effects (discussed in *Box 12.2*) are considered to be important in stabilizing Au(III). As we have already noted, discussing oxidation states of the heavy *d*-block metals in terms of independently obtained physicochemical data is usually impossible owing to the absence of *IE* values and the scarcity of simple ionic compounds or aqua ions. Data are more plentiful for Ag than for many of the heavier metals, and some comparisons with Cu are possible. Although the enthalpy of atomization of Ag < Cu (Table 22.3), the greater ionic radius for the silver ion along with relevant ionization energies (Table 22.3) make Ag more noble than Cu (equations 22.134–22.136). Gold is more noble still (equation 22.137).

$$Ag^+ + e^- \rightleftharpoons Ag \qquad E^\circ = +0.80\,V \qquad (22.134)$$

$$Cu^+ + e^- \rightleftharpoons Cu \qquad E^\circ = +0.52\,V \qquad (22.135)$$

$$Cu^{2+} + e^- \rightleftharpoons Cu^+ \qquad E^\circ = +0.34\,V \qquad (22.136)$$

$$Au^+ + e^- \rightleftharpoons Au \qquad E^\circ = +1.69\,V \qquad (22.137)$$

Gold(V) and silver (V)

Gold(V) is found only in AuF_5 and $[AuF_6]^-$ (scheme 22.138); AuF_5 is highly reactive and (from Raman and Mössbauer spectroscopic data) possesses chains of *cis*-F-bridged octahedral units.

$$Au + O_2 + 3F_2 \xrightarrow[\text{8 bar}]{650K,} [O_2][AuF_6] \xrightarrow[\text{reactor}]{\text{hot/cold}} AuF_5 + O_2 + \tfrac{1}{2}F_2$$

$$\downarrow KrF_2, \text{anhyd HF}$$

$$AuF_5 + KrF_2 \xleftarrow{300\,K} [KrF][AuF_6] \,(22.138)$$

Gold(III) and silver (III)

For gold, AuF_3, $AuCl_3$ and $AuBr_3$ are known, but AgF_3 is the only high oxidation state halide of silver. It is made in

anhydrous HF by treating $K[AgF_4]$ with BF_3, $K[AgF_4]$ being prepared from fluorination of a KCl and AgCl mixture. Red AgF_3 is diamagnetic (d^8) and isostructural with AuF_3; diamagnetic $K[AuF_4]$ presumably contains a square planar anion. Gold(III) fluoride is made from Au with F_2 (1300 K, 15 bar) or by reaction 22.139. It is a polymer consisting of helical chains. Part of one chain is shown in **22.66**; the coordination of each Au(III) is square planar but F atoms above and below the plane interact weakly.

(22.66)

$$Au \xrightarrow{BrF_3} [BrF_2][AuF_4] \xrightarrow{330\,K} AuF_3 \qquad (22.139)$$

Red $AuCl_3$ and brown $AuBr_3$ (made by direct combination of the elements) are diamagnetic, planar dimers (**22.67**). In hydrochloric acid, $AuCl_3$ forms $[AuCl_4]^-$ which reacts with Br^- to give $[AuBr_4]^-$, but with I^- to yield AuI and I_2. '*Chloroauric acid*' ($HAuCl_4 \cdot xH_2O$), its bromo analogue, $K[AuCl_4]$ and $AuCl_3$ are commercially available and are valuable starting materials in Au(III) and Au(I) chemistry.

a = 234 pm, *b* = 224 pm

(22.67)

The hydrated oxide $Au_2O_3 \cdot H_2O$ is precipitated by alkali from solutions of $Na[AuCl_4]$, and reacts with an excess of $[OH]^-$ to give $[Au(OH)_4]^-$. While $Au_2O_3 \cdot H_2O$ is the only established oxide of gold, Ag_2O_3 is less important than Ag_2O (see below).

Limited numbers of Ag(III) complexes are known; examples are paramagnetic CsK_2AgF_6 and diamagnetic $K[AgF_4]$. Numerous gold(III) complexes have been made,

Table 22.3 Selected physical data for Cu and Ag.

Quantity	Cu	Ag
$IE_1 / kJ\,mol^{-1}$	745.5	731.0
$IE_2 / kJ\,mol^{-1}$	1958	2073
$IE_3 / kJ\,mol^{-1}$	3555	3361
$\Delta_a H^\circ(298\,K) / kJ\,mol^{-1}$	338	285

and square planar coordination (d^8 metal centre) predominates. Haloanions $[AuX_4]^-$ (X = F, Cl, Br, see above) can be made by oxidation of Au metal (e.g. equation 22.139), and the unstable $[AuI_4]^-$ by treating $[AuCl_4]^-$ with anhydrous, liquid HI. Other simple complexes include $[Au(CN)_4]^-$ (from $[AuCl_4]^-$ with $[CN]^-$), $[Au(NCS-S)_4]^-$, $[Au(N_3)_4]^-$ and $[Au(NO_3-O)_4]^-$ (equation 22.140). Phosphino complexes of type R_3PAuCl_3 can be made by oxidative addition of Cl_2 to Ph_3PAuCl.

$$Au \xrightarrow{N_2O_5} [NO_2][Au(NO_3)_4] \xrightarrow{KNO_3} K[Au(NO_3)_4]$$

(22.140)

Most compounds which may appear to contain gold(II) are mixed valence compounds, e.g. 'AuCl$_2$' is actually the tetramer $(Au^I)_2(Au^{III})_2Cl_8$ (**22.68**), and $CsAuCl_3$ is $Cs_2[AuCl_2][AuCl_4]$. Both compounds contain square planar Au(III) and linear Au(I), and their dark colours arise from charge transfer between Au(I) and Au(III).

(22.68)

Silver(II)

Brown AgF_2 is obtained by reacting F_2 and Ag at 520 K, but is instantly decomposed by H_2O; it is paramagnetic (Ag^{2+}, d^9) but the magnetic moment of $1.07\,\mu_B$ reflects antiferromagnetic coupling. In solid AgF_2, the environments of Ag^{2+} centres are Jahn–Teller distorted (elongated) octahedral, Ag–F = 207 and 259 pm.

The black solid of composition AgO which is precipitated when $AgNO_3$ is warmed with persulfate solution is diamagnetic and contains Ag(I) (with two O nearest neighbours) and Ag(III) (4-coordinate). However, when AgO dissolves in aqueous $HClO_4$, the paramagnetic $[Ag(H_2O)_4]^{2+}$ ion is formed. This (equation 22.141), AgO and Ag(II) complexes are very powerful oxidizing agents, e.g. AgO converts Mn(II) to $[MnO_4]^-$ in acidic solution.

$$Ag^{2+} + e^- \rightleftharpoons Ag^+ \qquad E^\circ = +1.98\,V \qquad (22.141)$$

Silver(II) complexes can be made in aqueous solution from Ag(I) salts by using a powerful oxidizing agent in the presence of an appropriate ligand. They are paramagnetic and usually square planar. Examples include $[Ag(py)_4]^{2+}$, $[Ag(bpy)_2]^{2+}$ and $[Ag(bpy)(NO_3-O)_2]$.

Gold(I) and silver(I)

Many Ag(I) salts are familiar laboratory reagents; they are nearly always anhydrous and (except for AgF, $AgNO_3$ and $AgClO_4$) are usually sparingly soluble in water. Topics already covered are:

- solubilities of Ag(I) halides (*Section 6.9*);
- common-ion effect, exemplified with AgCl (*Section 6.10*);
- half-cells involving Ag(I) halides (*Section 7.3*);
- Frenkel defects illustrated by the structure of AgBr (*Section 5.17*).

Yellow AgF can be made from the elements or by dissolving AgO in HF; it adopts an NaCl lattice (*Figure 5.15*) as do AgCl and AgBr. Precipitation reactions 22.142 are used to prepare AgCl (white), AgBr (pale yellow) and AgI (yellow); for K_{sp} values, see *Table 6.4*.

$$AgNO_3(aq) + X^-(aq) \longrightarrow AgX(s) + [NO_3]^-(aq)$$

$$(X = Cl, Br, I) \qquad (22.142)$$

Silver(I) iodide is polymorphic. The stable form at 298 K and 1 bar pressure, γ-AgI, has a zinc blende structure (*Figure 5.18*). At high pressures, this converts to δ-AgI with an NaCl structure, the Ag–I distance increasing from 281 to 304 pm. Between 409 and 419 K, the β-form exists with a wurtzite lattice (*Figure 5.20*). Above 419 K, α-AgI becomes a fast ion electrical conductor (see *Section 27.3*), the conductivity at the transition temperature increasing by a factor of ≈4000. In this form, the I^- ions occupy positions in a CsCl structure (*Figure 5.16*) but the much smaller Ag^+ ions move freely between sites of 2-, 3- or 4-coordination

---

BOX 22.12
APPLICATIONS

Bactericidal effects of silver sols

Solutions of silver sols (i.e. colloidal dispersions of Ag in aqueous solution) have some application as bactericidal agents. The active agent is Ag^+, with the metabolism of bacteria being disrupted by its presence. A silver sol exhibits a large metal surface area, and oxidation by atmospheric O_2 occurs to some extent to give Ag_2O. While this is only sparingly soluble in water, the concentration of Ag^+ in solution is sufficient to provide the necessary bactericidal effects. Over-exposure to Ag, however, results in argyria: this is a darkening of the skin, caused by absorption of metallic Ag, which cannot be medically reversed.

Silver(I) halides in photochromic glasses and photography

Silver(I) halides darken in light owing to photochemical decomposition. If the halogen produced is kept in close proximity to the finely divided metal which is also formed, the process may be reversed when the source light is cut off; hence the use of AgCl in photochromic glasses.

The light-sensitivity of silver halides is fundamental to all types of photography. In a black-and-white photographic film, the emulsion coating consists of a layer of gelatine containing AgBr (or to a lesser extent, AgCl or AgI) in suspension. Exposure of the film produces some submicroscopic particles of Ag; on addition of an organic reducing agent, more AgBr is reduced, the rate of reduction depending on the intensity of illumination during the exposure period. Thus, the parts of the film which were the most strongly

illuminated become the darkest. It is this intensification of the latent image first formed that allows the use of short exposure times and causes silver halides to occupy their unique position in photography. Unchanged AgX is washed out using aqueous $[S_2O_3]^{2-}$:

$$AgX + 2[S_2O_3]^{2-} \longrightarrow [Ag(S_2O_3)_2]^{3-} + X^-$$

and the photographic negative is converted into a print by allowing light to pass through it on to AgBr-containing photographic paper. This is again treated with thiosulfate.

The recent innovation of digital cameras may well have the effect of considerably decreasing the use of silver halides in photography.

---

among the easily deformed I^- ions. The high temperature form of Ag_2HgI_4 shows similar behaviour.

Gold(I) fluoride is not known, but yellow AuCl, AuBr and AuI can be made by reactions 22.143 and 22.144; overheating AuCl and AuBr results in decomposition to the elements. Crystalline AuCl, AuBr and AuI possess zig-zag chain structures (**22.69**). The halides disproportionate when treated with H_2O; disproportionation of Au(I) (equation 22.145) does not parallel that of Cu(I) to Cu and Cu(II).

Au–I = 262 pm; ∠Au-I-Au = 72°

(22.69)

$$AuX_3 \xrightarrow{\Delta} AuX + X_2 \qquad (X = Cl, Br) \qquad (22.143)$$

$$2Au + I_2 \xrightarrow{\Delta} 2AuI \qquad (22.144)$$

$$3Au^+ \longrightarrow Au^{3+} + 2Au \qquad (22.145)$$

Silver(I) oxide is precipitated by adding alkali to solutions of Ag(I) salts. It is a brown solid which decomposes above 423 K. Aqueous suspensions of Ag_2O are alkaline and absorb atmospheric CO_2. In alkalis, Ag_2O dissolves forming $[Ag(OH)_2]^-$. No gold(I) oxide has been confirmed.

For gold(I), linear coordination is usual, although Au····Au interactions in the solid state are a common feature (Figure 22.24a); trigonal planar and tetrahedral complexes are also found. For Ag(I), linear and tetrahedral complexes are common, but the metal ion can tolerate a range of environments and coordination numbers from 2 to 6 (the latter is rare) are well established. Both Ag(I) and Au(I) favour soft donor atoms, and there is a wide variety of complexes with M–P and M–S bonds, including some thiolate complexes with intriguing structures (Figure 15.9d and Figure 22.24b).

Dissolving Ag_2O in aqueous NH_3 gives the linear $[Ag(NH_3)_2]^+$, but in liquid NH_3 tetrahedral $[Ag(NH_3)_4]^+$ forms. Trigonal planar $[Ag(NH_3)_3]^+$ can be isolated as the nitrate. Equation 22.4 showed the formation of $[M(CN)_2]^-$ (M = Ag, Au) during metal extraction; the complexes are also made by dissolving MCN in aqueous KCN. Both AgCN and AuCN are linear polymers (**22.70**), and $[M(CN)_2]^-$ are discrete, linear ions.

(22.70)

Dissolution of AgX in aqueous halide solutions produces $[AgX_2]^-$ and $[AgX_3]^{2-}$. In aqueous solutions, $[AuX_2]^-$ (X = Cl, Br, I) are unstable with respect to disproportionation but can be stabilized by adding excess X^- (equation 22.146).

$$3[AuX_2]^- \rightleftharpoons [AuX_4]^- + 2Au + 2X^- \qquad (22.146)$$

Routes to Au(I) complexes often involve reduction of Au(III) as illustrated by the formation of R_3PAuCl and R_2SAuCl species (equations 22.147 and 22.148).

$$[AuCl_4]^- + 2R_3P \longrightarrow [R_3PAuCl] + R_3PCl_2 + Cl^- \qquad (22.147)$$

$$[AuCl_4]^- + 2R_2S + H_2O$$
$$\longrightarrow [R_2SAuCl] + R_2SO + 2H^+ + 3Cl^- \qquad (22.148)$$

Molecules of R_3PAuCl and R_2SAuCl (for which many examples with different R groups are known) contain linear Au(I), but aggregation in the solid state by virtue of Au····Au interactions (similar to those in Figure 22.24a) is often observed. Other than the expectation of a linear Au(I) environment, structures may be hard to predict. For example in $[Au_2(CS_3)_2]^{2-}$ (made from $[Au(SH)_2]^-$ and CS_2), there is a short Au–Au contact, but in $[Au_2(TeS_3)_2]^{2-}$ (made from AuCN and $[TeS_3]^{2-}$), the Au(I) centres are out of bonding range (Figures 22.24c–d).

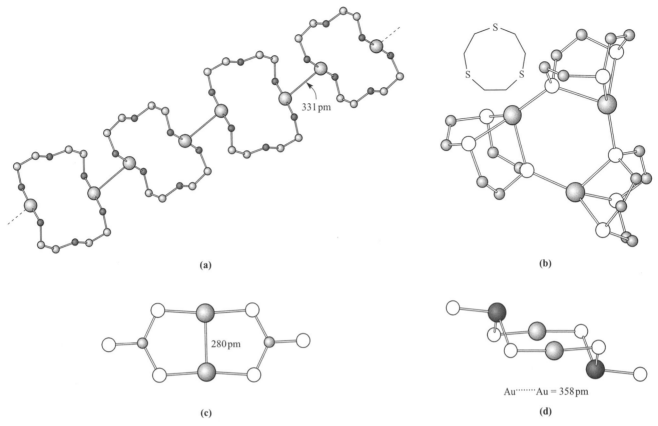

Fig. 22.24 The structures (X-ray diffraction) of (a) the cation in $[Au_2(H_2NCH_2CH_2NHCH_2CH_2NH_2)_2][BF_4]_2$ (part of one chain in which dimers are connected by weak Au—Au interactions is shown) [J. Yau *et al.* (1995) *J. Chem. Soc., Dalton Trans.*, p. 2575], (b) the trimer $[Ag_3L_3]^{3+}$ where L is the sulfur-containing macrocycle shown in the inset [H.-J. Kuppers *et al.* (1987) *Angew. Chem., Int. Ed. Engl.*, vol. 26, p. 575], (c) the planar $[Au_2(CS_3)_2]^{2-}$ (from the $[(Ph_3P)_2N]^+$ salt) [J. Vicente *et al.* (1995) *J. Chem. Soc., Chem. Commun.*, p. 745], and (d) $[Au_2(TeS_3)_2]^{2-}$ (from the $[Me_4N]^+$ salt) [D.-Y. Chung *et al.* (1995) *Inorg. Chem.*, vol. 34, p. 4292]. Hydrogen atoms are omitted for clarity; colour code: Au, blue; C, grey; N and Te, black; S, white.

22.13 Group 12: cadmium and mercury

The metals

Cadmium is chemically very like Zn, and any differences are attributable to the larger sizes of the Cd atom and Cd^{2+} ion. Among the group 12 metals, Hg is distinct. It does bear some resemblance to Cd, but in many respects is very like Au and Tl. It has been suggested that the relative inertness of Hg towards oxidation is a manifestation of the thermodynamic 6*s* inert pair effect (see *Box 12.3*).

Cadmium is a reactive metal and dissolves in non-oxidizing and oxidizing acids, but unlike Zn, it does not dissolve in aqueous alkali. In moist air, Cd slowly oxidizes, and when heated in air, it forms CdO. When heated, Cd reacts with the halogens and sulfur.

Mercury is less reactive than Zn and Cd. It is attacked by oxidizing (but not non-oxidizing) acids, with products dependent on conditions, e.g. with dilute HNO_3, Hg forms $Hg_2(NO_3)_2$ (containing $[Hg_2]^{2+}$, see below) but with concentrated HNO_3, the product is $Hg(NO_3)_2$. Reaction of the metal with hot concentrated H_2SO_4 gives $HgSO_4$ and SO_2. Mercury reacts with the halogens (equations 22.149 and 22.150); it combines with O_2 at 570 K to give HgO, but at higher temperatures HgO decomposes back to the elements, and, if sulfur is present, HgS is produced rather than the oxide.

$$Hg + X_2 \xrightarrow{\Delta} HgX_2 \qquad (X = F, Cl, Br) \qquad (22.149)$$

$$3Hg + 2I_2 \xrightarrow{\Delta} HgI_2 + Hg_2I_2 \qquad (22.150)$$

Mercury dissolves many metals to give *amalgams* (see *Box 22.3*); in the Na—Hg system, for example, Na_3Hg_2, NaHg and $NaHg_2$ have been characterized.

For cadmium, the +2 oxidation state is of most importance, but compounds of Hg(I) and Hg(II) are both well known. Mercury is unique among the group 12 metals in forming a stable $[M_2]^{2+}$ ion. Although there is evidence for $[Zn_2]^{2+}$ and $[Cd_2]^{2+}$ in metal–metal halide melts, and $Cd_2[AlCl_4]$ has been isolated from a molten mixture of Cd, $CdCl_2$ and $AlCl_3$, it is not possible to obtain $[Zn_2]^{2+}$ and $[Cd_2]^{2+}$ in aqueous solution. Force constants (60, 110 and 250 N m^{-1} for M = Zn, Cd and Hg, calculated from Raman spectra of $[M_2]^{2+}$) show that the bond in $[Hg_2]^{2+}$ is stronger than those in $[Zn_2]^{2+}$ and $[Cd_2]^{2+}$. However, given

Table 22.4 Selected physical data for the group 12 metals.

Quantity	Zn	Cd	Hg
$IE_1 / kJ\,mol^{-1}$	906.4	867.8	1007
$IE_2 / kJ\,mol^{-1}$	1733	1631	1810
$\Delta_a H^\circ(298\,K) / kJ\,mol^{-1}$	130	112	61
$E^\circ\,(M^{2+}/M)\,/\,V$	−0.76	−0.40	+0.85
r_{ion} for M^{2+} / pm‡	74	95	101

‡ For Hg, the value of based on the structure of HgF_2, one of the few mercury compounds with a typical ionic lattice.

that Hg has the lowest value of $\Delta_a H^\circ$ of all the *d*-block metals (*Table 5.2*), the stability of $[Hg_2]^{2+}$ (**22.71**) is difficult to rationalize. Other polycations of mercury are known; $[Hg_3]^{2+}$ (**22.72**) is formed as the $[AlCl_4]^-$ salt in molten Hg, $HgCl_2$ and $AlCl_3$, and $[Hg_4]^{2+}$ (**22.73**) is produced as the $[AsF_6]^-$ salt from reaction of Hg with AsF_5 in liquid SO_2.

$$\left[\,Hg\!-\!\!-\!Hg\,\right]^{2+}$$
253 pm
(22.71)

$$\left[\,Hg\!-\!\!-\!Hg\!-\!\!-\!Hg\,\right]^{2+}$$
255 pm
(22.72)

$$\left[\,Hg\!-\!\!-\!Hg\!\overset{262\ pm}{-\!\!-\!Hg}\!-\!\!-\!Hg\,\right]^{2+}$$
259 pm
(22.73)

Ionization energies decrease from Zn to Cd but increase from Cd to Hg (Table 22.4). Whatever the origin of the high ionization energies for Hg, it is clear that they far outweigh the small change in $\Delta_a H^\circ$ and make Hg a noble metal. The reduction potentials in Table 22.4 reveal the relative electropositivities of the group 12 metals.

Since much of the chemistries of Cd and Hg are distinct, we shall deal with the two metals separately. In making this decision, we are effectively saying that the consequences of the lanthanoid contraction are of minor significance for the heavier metals of the last group of the *d*-block.

Cadmium(II)

All four Cd(II) halides are known. The action of HF on $CdCO_3$ gives CdF_2, and of gaseous HCl on Cd (720 K) yields $CdCl_2$; $CdBr_2$ and CdI_2 are formed by direct combination of the elements. White CdF_2 adopts a CaF_2 lattice (*Figure 5.18*), while $CdCl_2$ (white), $CdBr_2$ (pale yellow) and CdI_2 (white) have layer structures (see *Section 5.11*). The fluoride is sparingly soluble in water, while the other halides are readily soluble, giving solutions containing aquated Cd^{2+} and a range of halo complexes, e.g. CdI_2 dissolves to give an equilibrium mixture of $[Cd(H_2O)_6]^{2+}$, $[Cd(H_2O)_5I]^+$, $[CdI_3]^-$ and $[CdI_4]^{2-}$, while 0.5 M aqueous $CdBr_2$ contains $[Cd(H_2O)_6]^{2+}$, $[Cd(H_2O)_5Br]^+$, $[Cd(H_2O)_5Br_2]$, $[CdBr_3]^-$ and $[CdBr_4]^{2-}$. In contrast to Zn^{2+}, the stability of halo complexes of Cd^{2+} increases from F^- to I^-, i.e. Cd^{2+} is a softer metal centre than Zn^{2+} (*Table 6.9*).

Cadmium(II) oxide (formed by heating Cd in O_2, and varying in colour from green to black) adopts an NaCl lattice. It is insoluble in H_2O and alkalis, but dissolves in acids, i.e. CdO is more basic than ZnO. Addition of dilute alkali to aqueous solutions of Cd^{2+} precipitates white $Cd(OH)_2$, and this dissolves only in concentrated alkali to give $[Cd(OH)_4]^{2-}$ (contrast $[Zn(OH)_4]^{2-}$, **21.58**). Equation 21.5 showed the role of $Cd(OH)_2$ in NiCd cells. Yellow CdS (the stable α-form has a wurtzite lattice, *Figure 5.20*) is commercially important as a pigment and phosphor; CdSe and CdTe are semiconductors (see *Section 22.2*).

In aqueous solutions, $[Cd(H_2O)_6]^{2+}$ is present but is quite acidic (equation 22.151); in concentrated solutions, aquated $[Cd_2(OH)]^{3+}$ is present.

$$[Cd(H_2O)_6]^{2+} + H_2O \rightleftharpoons [Cd(H_2O)_5(OH)]^+ + H^+$$
(22.151)

In aqueous NH_3, tetrahedral $[Cd(NH_3)_4]^{2+}$ is present, but at high concentrations, $[Cd(NH_3)_6]^{2+}$ forms. The lack of LFSE for Cd^{2+} (d^{10}) means that a range of coordination geometries is observed; coordination numbers of 4, 5 and 6 are most common, but higher coordination numbers can be forced upon the metal centre by using macrocyclic ligands. Examples of complexes include:

- tetrahedral: $[CdCl_4]^{2-}$, $[Cd(NH_3)_4]^{2+}$, $[Cd(en)_2]^{2+}$;
- trigonal bipyramidal: $[CdCl_5]^{3-}$;
- octahedral: $[Cd(DMSO\text{-}O)_6]^{2+}$, $[Cd(en)_3]^{2+}$, $[Cd(acac)_3]^-$, $[CdCl_6]^{4-}$;
- hexagonal bipyramidal: $[CdBr_2(18\text{-crown-}6)$ (see *Section 10.8*).

As we have seen previously, formulae can be deceptive in terms of structure, e.g. $[Cd(NH_3)_2Cl_2]$ is polymeric with octahedral Cd^{2+} and bridging halo ligands.

Mercury(II)

All four Hg(II) halides can be prepared from the elements. A fluorite structure (*Figure 5.18*) is adopted by HgF_2 (Hg−F = 225 pm); it is completely hydrolyzed by H_2O (equation 22.152).

$$HgF_2 + H_2O \longrightarrow HgO + 2HF$$
(22.152)

The chloride and bromide are volatile solids, soluble in H_2O (in which they are un-ionized), EtOH and Et_2O. The solids possess molecular lattices containing HgX_2 units packed to give distorted octahedral Hg(II) centres (two long Hg−X contacts to adjacent molecules). Below 400 K, HgI_2 is red with a layer structure, and above 400 K, is yellow with HgI_2 molecules assembled into a lattice with distorted octahedral metal centres. The vapours contain linear HgX_2 molecules with bond distances of 225, 244 and 261 pm for X = Cl, Br and I respectively. Figure 22.25 shows the trend in solubilities of the halides; for HgI_2, $K_{sp} = 2.82 \times 10^{-29}$.

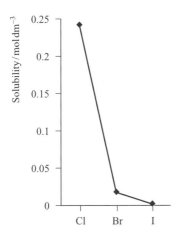

Fig. 22.25 The trend in solubilities of Hg(II) halides in water; HgF_2 decomposes.

(22.74)

Mercury(II) oxide exists in yellow (formed by heating Hg in O_2 or by thermal decomposition of $Hg(NO_3)_2$) and red (prepared by precipitation from alkaline solutions of Hg^{2+}) forms; both have infinite chain structures (**22.74**) with linear Hg(II). The thermal decomposition of HgO (equation 22.153) led to the discovery of O_2 by Priestley in 1774.

$$2HgO \xrightarrow{>670\,K} 2Hg + O_2 \qquad (22.153)$$

Although the oxide dissolves in acids, it is only weakly basic; in aqueous solution, Hg(II) salts that are ionized (e.g. $Hg(NO_3)_2$ and $HgSO_4$) are hydrolyzed to a considerable extent and many basic salts are formed, e.g. $HgO \cdot HgCl_2$ and $[O(HgCl)_3]Cl$ (a substituted oxonium salt).

In its complexes, Hg(II) (d^{10}) exhibits coordination numbers of 2 to 6. Like Cd^{2+}, Hg^{2+} is a soft metal centre, and coordination to S-donors is especially favoured. Complex chlorides, bromides and iodides are formed in aqueous solution, and the tetrahedral $[HgI_4]^{2-}$ is particularly stable. A solution of $K_2[HgI_4]$ (*Nessler's reagent*) gives a characteristic brown compound, $[Hg_2N]^+I^-$, on treatment with NH_3 and is used in determination of NH_3. In solid $[Hg_2N]I$, the $[Hg_2N]^+$ cations assemble into an infinite network related to that of β-cristobalite (*Figure 5.19c*) and containing linear Hg(II). Reaction 22.154 shows the formation of its hydroxide.

$$2HgO + NH_3 \xrightarrow{aq} Hg_2N(OH) + H_2O \qquad (22.154)$$

The salt $[Hg(NH_3)_2]Cl_2$ (equation 22.155) contains linear $[Hg(NH_3)_2]^{2+}$ ions and dissolves in aqueous NH_3 to give $[Hg(NH_2)]Cl$ which contains polymeric chains (**22.75**).

$$HgCl_2 + 2NH_3(g) \longrightarrow [Hg(NH_3)_2]Cl_2 \qquad (22.155)$$

(22.75)

Examples of Hg(II) complexes illustrating different coordination environments (see *Table 6.7* for ligand abbreviations) include:

- linear: $[Hg(NH_3)_2]^{2+}$, $[Hg(CN)_2]$, $[Hg(py)_2]^{2+}$, $[Hg(SEt)_2]$;
- trigonal planar: $[HgI_3]^-$;
- tetrahedral: $[Hg(en)_2]^{2+}$, $[Hg(NCS-S)_4]^{2-}$, $[HgI_4]^{2-}$, $[Hg(S_4\text{-}S,S')_2]^{2-}$, $[Hg(Se_4\text{-}Se,Se')_2]^{2-}$, $[Hg(phen)_2]^{2+}$;
- trigonal bipyramidal: $[HgCl_2(tpy)]$, $[HgCl_2(dien)]$, $[HgCl_5]^{3-}$;
- square-based pyramidal: $[Hg(H_2O)L]^{2+}$ (L = **22.76**);
- octahedral: $[Hg(en)_3]^{2+}$, *fac*-$[HgL_2]^{2+}$ (L = **22.77**);
- square antiprism: $[Hg(NO_2\text{-}O,O')_4]^{2-}$.

(22.76)   (22.77)

Mercury(I)

The chemistry of Hg(I) is that of the $[Hg_2]^{2+}$ unit which contains an Hg–Hg single bond (**22.71**). The general method of preparation of Hg(I) compounds is by the action of Hg metal on Hg(II) compounds, e.g. reaction 22.156 in which Hg_2Cl_2 (*calomel*) is freed from $HgCl_2$ by washing with hot water. The *standard calomel electrode* is a reference electrode (equation 22.157) consisting of a Pt wire dipping into Hg in contact with Hg_2Cl_2 and immersed in 1 M KCl solution. This electrode is more convenient than the standard hydrogen electrode which requires a source of purified gas.

$$HgCl_2 + Hg \xrightarrow{\Delta} Hg_2Cl_2 \qquad (22.156)$$

$$Hg_2Cl_2 + 2e^- \rightleftharpoons 2Hg + 2Cl^-$$
$$E^o = +0.268\,V \text{ (in 1 M aq KCl)} \qquad (22.157)$$

Potential diagrams for Hg are shown in scheme 22.158, and the data in acidic solution illustrate that the disproportionation of Hg(I) (equation 22.159) has a small and positive ΔG^o value at 298 K.

$$\left.
\begin{array}{l}
Hg^{2+} \xrightarrow{+0.91} [Hg_2]^{2+} \xrightarrow{+0.77} Hg \qquad pH = 0 \\
\qquad\qquad\qquad +0.84 \\[4pt]
HgO \xrightarrow{+0.10} Hg \qquad\qquad\qquad pH = 14
\end{array}
\right\} \quad (22.158)$$

$[Hg_2]^{2+} \rightleftharpoons Hg^{2+} + Hg \quad K = 6 \times 10^{-3}\,(298\,K) \qquad (22.159)$

Reagents that form insoluble Hg(II) salts or stable Hg(II) complexes upset equilibrium 22.159 and decompose Hg(I) salts, e.g. addition of $[OH]^-$, S^{2-} or $[CN]^-$ results in formation of Hg and HgO, HgS or $[Hg(CN)_4]^{2-}$, and the Hg(I) compounds Hg_2O, Hg_2S and $Hg_2(CN)_2$ are not known. Mercury(II) forms more stable complexes than the larger $[Hg_2]^{2+}$ and relatively few Hg(I) compounds are known. The most important are the halides (**22.78**). Whereas Hg_2F_2 decomposes to Hg, HgO and HF on contact with water, the later halides are sparingly soluble.

$$X \longrightarrow Hg \longrightarrow Hg \longrightarrow X$$

X	Hg–Hg / pm
F	251
Cl	253
Br	258
I	269

(22.78)

Other Hg(I) salts include $Hg_2(NO_3)_2$, Hg_2SO_4 and $Hg_2(ClO_4)_2$; the nitrate is commercially available as the dihydrate, the solid state structure of which contains $[(H_2O)HgHg(OH_2)]^{2+}$ cations. Scheme 22.160 summarizes some reactions of $Hg_2(NO_3)_2$.

$$Hg_2(NO_3)_2 \begin{cases} \xrightarrow{\text{KSCN}} Hg_2(SCN)_2 \xrightarrow{\text{decomp.}} Hg(SCN)_2 + Hg \\ \xrightarrow{\text{HN}_3} Hg_2(N_3)_2 \text{ (explosive)} \\ \xrightarrow{\text{H}_2\text{SO}_4} Hg_2SO_4 \end{cases}$$

(22.160)

Summary

The descriptive chemistry of the second and third row *d*-block metals has been the topic of this chapter. The chemistry of each pair of metals in a group differs significantly from that of the first member of the group, but the effects of the lanthanoid contraction mean that there are similarities between pairs such as Hf and Zr, Nb and Ta, Mo and W etc. This trend is broken in group 12 (Cd and Hg). Metal–

metal bonding, high coordination numbers, metal halo clusters and polyoxometallates are four characteristics that feature in the chemistry of the heavier *d*-block metals.

Suggested further reading

See also further reading suggested for Chapters 19 and 20.

F.A. Cotton, G. Wilkinson, M. Bochmann and C. Murillo (1999) *Advanced Inorganic Chemistry*, 6th edn, Wiley Interscience, New York – One of the best detailed accounts of the chemistry of the *d*-block metals.

F.A. Cotton and R.A. Walton (1993) *Multiple Bonds between Metal Atoms*, 2nd edn, Oxford University Press, Oxford.

S.A. Cotton (1997) *Chemistry of Precious Metals*, Blackie, London – Covers descriptive inorganic chemistry (including σ-bonded organometallic complexes) of the heavier group 8, 9, 10 and 11 metals.

J. Emsley (1998) *The Elements*, 3rd edn, Oxford University Press, Oxford – An invaluable source of data for the elements.

N.N. Greenwood and A. Earnshaw (1997) *Chemistry of the Elements*, 2nd edn, Butterworth-Heinemann, Oxford – A very good account including historical, technological and structural aspects; the metals in each triad are treated together.

C.E. Housecroft (1999) *The Heavier d-Block Metals: Aspects of Inorganic and Coordination Chemistry*, Oxford University Press, Oxford – An introductory text including chapters on aqueous solution species, structure, M–M bonded dimers and clusters, and polyoxometallates.

R.B. King, ed. (1994) *Encyclopedia of Inorganic Chemistry*, Wiley, Chichester – Contains an article on the inorganic and coordination chemistry of each metal with numerous literature citations.

M.T. Pope (1994) 'Polyoxoanions' in *Encyclopedia of Inorganic Chemistry*, ed. R.B. King, Wiley, Chichester, vol. 6, p. 3361.

M.T. Pope and A. Müller, eds. (1994) *Polyoxometalates: From Platonic Solids to Anti-Retroviral Activity*, Kluwer Academic Publishers, Dordrecht.

E.A. Seddon and K.R. Seddon (1984) *The Chemistry of Ruthenium*, Elsevier, Amsterdam – An excellent, well-referenced account of the chemistry of Ru.

H. Schmidbauer (1999) *Gold: Chemistry, Biochemistry and Technology*, Wiley, New York – An overview of the chemistry of gold including applications.

A.F. Wells (1984) *Structural Inorganic Chemistry*, 5th edn, Clarendon Press, Oxford – An excellent source for detailed structural information of, in particular, binary compounds.

Problems

22.1 (a) Write out the first row *d*-block metals in sequence and then complete each triad of metals. (b) Between which two metals is the series of lanthanoid metals?

22.2 Briefly discuss trends in (a) metallic radii and (b) values of $\Delta_a H^\circ (298\,K)$ for the *d*-block metals.

22.3 (a) Estimate the value of $\Delta_f H^\circ (WCl_2)$ assuming it to be an ionic compound. Comment on any assumptions made. Data needed in addition to those in Tables 21.1,

22.1 or the Appendices: $\Delta_f H^\circ (CrCl_2) = -397\,kJ\,mol^{-1}$. (b) What does your answer to (a) tell you about the likelihood of WCl_2 being ionic?

22.4 Comment on the following observations:

(a) The density of HfO_2 (9.68 g cm^{-3}) is much greater than that of ZrO_2 (5.73 g cm^{-3}).

(b) NbF_4 is paramagnetic but $NbCl_4$ and $NbBr_4$ are essentially diamagnetic.

22.5 Suggest products in the following reactions: (a) CsBr heated with $NbBr_5$ at 383 K; (b) KF and TaF_5 melted together; (c) NbF_5 with bpy at 298 K. (d) Comment on the structures of the group 5 metal halides in the starting materials and give possible structures for the products.

22.6 Comment on the observation that $K_3[Cr_2Cl_9]$ is strongly paramagnetic but $K_3[W_2Cl_9]$ is diamagnetic.

22.7 (a) Interpret the formula $[Mo_6Cl_8]Cl_2Cl_{4/2}$ in structural terms, and show that the formula is consistent with the stoichiometry $MoCl_2$. (b) Show that the $[W_6Br_8]^{4+}$ cluster can be considered to contain W–W single bonds.

22.8 Give a short account of Tc(V) and Re(V) oxo species.

22.9 Briefly summarize similarities and differences between Mn and Tc chemistries.

22.10 Draw the structure of $[Re_2Cl_8]^{2-}$; discuss the metal–metal bonding in the anion and its consequences on ligand orientation.

22.11 Suggest reasons for the variation in Re–Re bond lengths in the following species: $ReCl_4$ (273 pm), Re_3Cl_9 (249 pm), $[Re_2Cl_8]^{2-}$ (224 pm), $[Re_2Cl_9]^-$ (270 pm) and $[Re_2Cl_4(\mu\text{-}Ph_2PCH_2CH_2PPh_2)_2]$ (224 pm).

22.12 When $K_2[OsCl_4]$ is heated with NH_3 under pressure, compound **A** of composition $Os_2Cl_5H_{24}N_9$ is isolated. Treatment of a solution of **A** with HI precipitates a compound in which three of the five chlorines have been replaced by iodine. Treating 1 mmol of **A** with KOH releases 9 mmol NH_3. Compound **A** is diamagnetic and none of the stronger absorption bands in the IR spectrum is Raman active. Suggest a structure for **A** and account for the diamagnetism.

22.13 Give an account of the halides of Ru and Os.

22.14 (a) Give an account of the methods of synthesis of Rh(IV) and Ir(IV) halides and halo anions.

(b) Reaction of $[IrCl_6]^{2-}$ with PPh_3 and $Na[BH_4]$ in EtOH gives $[IrH_3(PPh_3)_3]$. Give the structures of the isomers of this complex and suggest how you would distinguish between them using NMR spectroscopy.

22.15 When $RhBr_3$ in the presence of $MePh_2As$ is treated with H_3PO_2, a monomeric compound **X** is formed. **X** contains 2 Br and 3 $MePh_2As$ per Rh, and is a non-electrolyte. Its IR spectrum has a band at 2073 cm^{-1}, and the corresponding band if the complex is made using D_3PO_2 in a deuterated solvent is 1483 cm^{-1}. Spectrophotometric titration of **X** with Br_2 shows that one molecule of **X**

reacts with one molecule of Br_2; treating the product with excess mineral acid regenerates $RhBr_3$. What can you conclude about the products?

22.16 (a) Compare the structures of β-$PdCl_2$ and $[Nb_6Cl_{12}]^{2+}$.

(b) Discuss, with examples, the existence (or not) of Pt(III) species.

(c) Discuss the variation in stereochemistries of Ni(II), Pd(II) and Pt(II) complexes.

22.17 (a) Describe the methods by which *cis*- and *trans*-$[PtCl_2(NH_3)_2]$ can be distinguished from each other and from $[Pt(NH_3)_4][PtCl_4]$. (b) Another possible isomer would be $[(H_3N)_2Pt(\mu\text{-}Cl)_2Pt(NH_3)_2]Cl_2$. What diagnostic data would enable you to rule out its formation?

22.18 Suggest products in the reactions of $K_2[PtCl_4]$ with (a) excess KI; (b) aqueous NH_3; (c) phen; (d) tpy; (e) excess KCN. What are the expected structures of these products? [Ligand abbreviations: see Table 6.7 and **19.22**.]

22.19 Complexes of the type $[PtCl_2(R_2P(CH_2)_nPR_2)]$ may be monomeric or dimeric. Suggest factors that might influence this preference and suggest structures for the complexes.

22.20 Comment on each of the following observations:

(a) Unlike $[Pt(NH_3)_4][PtCl_4]$, $[Pt(EtNH_2)_4][PtCl_4]$ has an electronic absorption spectrum that is the sum of those of the constituent ions.

(b) AgI is readily soluble in saturated aqueous $AgNO_3$, but AgCl is not.

(c) When $Hg(ClO_4)_2$ is shaken with liquid Hg, the ratio [Hg(I)]/[Hg(II)] in the resulting solution is independent of the value of [Hg(II)].

22.21 Discuss the variation in stable oxidation states for the group 11 metals, using examples of metal halides, oxides and complexes to illustrate your answer.

22.22 'The group 12 metals differ significantly from the *d*-block metals from groups 4-11.' Discuss this statement.

22.23 Studies of the heavier *d*-block metals are often used to introduce students to (a) metal–metal bonding, (b) high coordination numbers, (c) metal halo clusters and (d) polyoxometallates. Write an account of each topic, and include examples that illustrate why the first row metals are not generally as relevant as their heavier congeners for such discussing these topics.

Chapter 23 Organometallic compounds of *d*-block elements

1–2	3	4	5	6	7	8	9	10	11	12	13–18
s-block											*p*-block
	Sc	Ti	V	Cr	Mn	Fe	Co	Ni	Cu	Zn	
	Y	Zr	Nb	Mo	Tc	Ru	Rh	Pd	Ag	Cd	
	La	Hf	Ta	W	Re	Os	Ir	Pt	Au	Hg	

23.1 Introduction

Organometallic chemistry of the *s*- and *p*-block elements was described in *Chapter 18*, and we now extend the discussion to organometallic compounds containing *d*-block metals. This topic covers a huge area of chemistry, and we can only provide an introduction to it, emphasizing the fundamental families of complexes and reactions.

> An *organometallic* compound contains one or more metal–carbon bonds.

Many compounds described in *Chapter 18* contained σ-bonds or π-interactions between a metal centre and a cyclopentadienyl ligand. We also introduced examples of 3-electron donor bridging ligands, e.g. halides (**18.6**) and alkynyls (**18.9**), and 2-electron alkene donors, e.g. **18.10**.

Hapticity of a ligand

The *hapticity of a ligand* is the number of atoms that are directly bonded to the metal centre (see *Boxes 18.1* and

18.2). Structures **23.1a** and **23.1b** show two representations of an $[\eta^5\text{-}C_5H_5]^-$ (cyclopentadienyl, Cp^-) ligand. For clarity in the diagrams in this chapter, we adopt **23.1b** and similar representations for π-ligands such as $\eta^3\text{-}C_3H_5$ and $\eta^6\text{-}C_6H_6$.

(23.1a)   (23.1b)

23.2 Common types of ligand: bonding and spectroscopy

In this section, we introduce some of the most common ligands found in organometallic complexes. Many other ligands are related to those discussed below, and bonding descriptions can be developed by comparison with the ligands chosen for detailed coverage.

σ-Bonded alkyl, aryl and related ligands

In complexes such as WMe_6, $TiMe_4$ and $MeMn(CO)_5$, the $M-C_{Me}$ bond can be described as a localized 2c-2e interaction, i.e. it parallels that for the $[\eta^1\text{-}Cp]^-$ ligand (see *Box 18.2*). The same bonding description is applicable to the $Fe-C_{Ph}$ bond in **23.2** and the $Fe-C_{CHO}$ bond in **23.3**.

$$M \longrightarrow C \equiv O \longrightarrow z$$

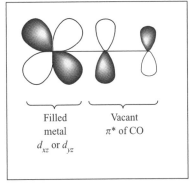

Vacant metal orbital, e.g. $sp_z d_{z^2}$ hybrid | HOMO of CO; σ-donor MO

Filled metal d_{xz} or d_{yz} | Vacant π^* of CO

CO-to-M donation
(a)

M-to-CO back-donation
(b)

Fig. 23.1 Components of metal–carbonyl bonding: (a) the M−CO σ-bond, and (b) the M−CO π-interaction which leads to back-donation of charge from metal to carbonyl. The orbital labels are examples, and assume that the M, C and O atoms lie on the z axis.

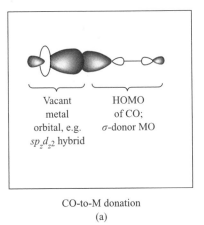

(23.2)

(23.3)

(23.5)   (23.6)   (23.7)   (23.8)

μ-CO μ_3-CO

Carbonyl ligands

The bonding in octahedral $M(CO)_6$ complexes was described in *Section 20.4* using a molecular orbital approach, but it is also convenient to give a simple picture to describe the bonding in one M−CO interaction. Figure 23.1a shows the σ-interaction between the highest occupied molecular orbital of CO (which has predominantly C character) and a vacant orbital on the metal centre (e.g. an $sp_z d_{z^2}$ hybrid). As a result of this interaction, electronic charge is donated from the CO ligand to the metal. Figure 23.1b shows the π-interaction that leads to *back-donation* of charge from metal to ligand; compare Figure 23.1b with Figure 20.13b. This 'donation/back-donation' bonding picture is the *Dewar–Chatt–Duncanson model*. Carbon monoxide is a *weak σ-donor* and a *strong π-acceptor* (or *π-acid*) and population of the CO π^* MO weakens and lengthens the C−O bond while also enhancing M−C bonding. Resonance structures **23.4** for the MCO unit also indicate a lowering of the C−O bond order as compared with free CO.

$$:M \!-\! C \!\equiv\! O: \quad \longleftrightarrow \quad M \!=\! C \!=\! O:$$

(23.4)

In multinuclear metal species, CO ligands may adopt terminal (**23.5**) or bridging (**23.6** and **23.7**) modes. Other modes are known, e.g. semi-bridging (part way between **23.5** and **23.6**) and mode **23.8**.

Evidence for a lowering of the C−O bond order on coordination comes from structural and spectroscopic data. In the IR spectrum of free CO, an absorption at $2143\,\text{cm}^{-1}$ is assigned to ν_{CO} (ν designates a stretching mode, see *Section 3.7*), and typical changes in vibrational wavenumber on going to metal carbonyl complexes are illustrated in Figure 23.2. Absorptions due to ν_{CO} are strong and easily observed. The lower the value of $\bar{\nu}_{CO}$, the weaker the C−O bond and this indicates greater back-donation of charge from metal to CO. Table 23.1 lists data for two sets of isoelectronic metal carbonyl complexes. On going from $Ni(CO)_4$ to $[Co(CO)_4]^-$ to $[Fe(CO)_4]^{2-}$, the additional negative charge is delocalized onto the ligands, causing a decrease in $\bar{\nu}_{CO}$. A similar effect is seen along the series $[Mn(CO)_6]^+$, $Cr(CO)_6$ and $[V(CO)_6]^-$. The increased back-donation is also reflected in values of $\bar{\nu}_{MC}$, e.g. $416\,\text{cm}^{-1}$ for $[Mn(CO)_6]^+$, and $441\,\text{cm}^{-1}$ for $Cr(CO)_6$.[†]

Carbonyl ligand environments can also be investigated using ^{13}C NMR spectroscopy, although systems are often fluxional (e.g. $Fe(CO)_5$, see *structure 2.2* and discussion)

The interplay of donation and back-donation of electronic charge between a metal and π-acceptor ligand is an example of a *synergic effect*.

[†] For a detailed discussions of IR spectroscopy in metal carbonyls, see: F.A. Cotton and G. Wilkinson (1988) *Advanced Inorganic Chemistry*, 5th edn, Wiley Interscience, New York, p. 1034; P.S. Braterman (1975) *Metal Carbonyl Spectra*, Academic Press, New York.

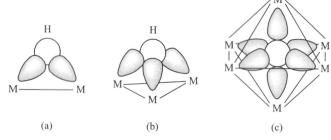

Fig. 23.2 Approximate regions in the IR spectrum in which $\bar{\nu}_{CO}$ is observed for different carbonyl modes; there is often overlap between the regions, e.g. see *Table 23.1*.

and information about specific CO environments may therefore be masked. Some useful points are that:

- typical ^{13}C NMR shifts for metal carbonyl ^{13}C nuclei are δ +170 to +240;

- within a series of analogous compounds containing metals from a given triad, the ^{13}C NMR signals for the CO ligands shift to lower frequency, e.g. in the ^{13}C NMR spectra of $Cr(CO)_6$, $Mo(CO)_6$ and $W(CO)_6$, signals are at δ +211, +201 and +191 respectively;

- for a given metal, signals for μ-CO ligands occur at higher frequency (more positive δ value) than those for terminal carbonyls.

In keeping with the weakening of the C–O bond on going from free CO to coordinated CO, X-ray diffraction data show a lengthening of the C–O bond. In CO, the C–O bond length is 112.8 pm, whereas typical values in metal carbonyls for terminal and μ-CO are 117 and 120 pm respectively.

Hydride ligands

The term *hydride ligand* suggests $H^{\delta-}$ and is consistent with the charge distribution expected for an H atom attached to an electropositive metal centre. However, the properties of H ligands depend on environment and in many organometallic complexes, hydrido ligands behave as protons, being removed by base (equation 23.1) or introduced by treatment with acid (reaction 23.2).

$$HCo(CO)_4 + H_2O \longrightarrow [Co(CO)_4]^- + [H_3O]^+ \qquad (23.1)$$

$$[HFe(CO)_4]^- + H^+ \longrightarrow H_2Fe(CO)_4 \qquad (23.2)$$

Hydride ligands can adopt terminal, bridging or (in metal clusters) interstitial modes of bonding (**23.9–23.12**). A localized 2c-2e M–H bond is an appropriate description for a terminal hydride, delocalized 3c-2e or 4c-2e interactions describe μ-H and μ_3-H interactions respectively (Figures 23.3a and 23.3b), and a 7c-2e interaction is appropriate for an interstitial hydride in an octahedral cage (Figure 23.3c).

Fig. 23.3 Overlap of the H 1*s* atomic orbital with (a) two or (b) three appropriate metal hybrid orbitals to form μ-H and μ_3-H bridges. (c) For an interstitial H atom within an octahedral M_6-cage, a delocalized description involves the overlap of the H 1*s* atomic orbital with six appropriate metal orbitals to give a 7c-2e interaction.

H $\vert$ M	H $\triangle$ M——M	μ_3-H	Interstitial, μ_6-H
μ-H	μ-H	μ_3-H	Interstitial, μ_6-H
(23.9)	(23.10)	(23.11)	(23.12)

Locating hydride ligands by X-ray diffraction is difficult. X-Rays are diffracted by *electrons* and the electron density in the region of the M–H bond is dominated by the heavy atoms. Neutron diffraction can be used, but this is an expensive and less readily available technique. In IR spectra, absorptions due to ν_{MH} modes are generally weak. Proton NMR spectroscopy is the routine way of observing metal hydrides. In ^{1}H NMR spectra, signals due to metal hydrides usually occur in the approximate range δ −8 to −30, although it is not easy to distinguish between terminal and bridging modes. The chemical shifts of interstitial hydrides are less diagnostic, and may occur at high frequency, e.g. δ +16.4 in $[(\mu_6\text{-H})Ru_6(CO)_{18}]^-$. Spin–spin coupling to spin-active metal nuclei such as ^{103}Rh (100% abundant, $I = \frac{1}{2}$), ^{183}W (14.3%, $I = \frac{1}{2}$) or ^{195}Pt (33.8%, $I = \frac{1}{2}$) gives valuable structural information, as does that to nuclei such as ^{31}P. Typical values of J_{PH} for a *cis*-arrangement (**23.13**) are 10–15 Hz, compared with ≈30 Hz for *trans*-coupling (**23.14**).

cis	*trans*
(23.13)	(23.14)

Table 23.1 Infrared spectroscopic data: values of $\bar{\nu}_{CO}$ for isoelectronic sets of tetrahedral $M(CO)_4$ and octahedral $M(CO)_6$ complexes.

Complex	$Ni(CO)_4$	$[Co(CO)_4]^-$	$[Fe(CO)_4]^{2-}$	$[Mn(CO)_6]^+$	$Cr(CO)_6$	$[V(CO)_6]^-$
$\bar{\nu}_{CO}$ / cm^{-1}	2060	1890	1790	2101	1981	1859

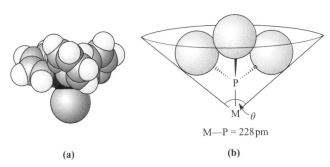

(a)

(b)

M—P = 228 pm

Fig. 23.4 (a) A space-filling diagram of an FePPh$_3$ unit; the phenyl groups adopt a 'paddle-wheel' arrangement; colour code: Fe, blue; P, black; C, grey; H, pale grey. (b) Schematic representation of the measurement of the Tolman cone angle, θ, for a PR$_3$ ligand; each circle represents the spatial extent of an R group.

Examples of stereochemically non-rigid hydride complexes are common (e.g. in the tetrahedral cluster [H$_3$Ru$_4$(CO)$_{12}$]$^-$) and variable temperature NMR spectroscopic studies are routinely carried out.

Phosphine and related ligands

Monodentate organophosphines may be tertiary (PR$_3$), secondary (PR$_2$H) or primary (PRH$_2$) and are *terminally* bound; PF$_3$ behaves similarly. Bridging modes can be adopted by [PR$_2$]$^-$ (**23.15**) or [PR]$^{2-}$ (**23.16**) ligands. Phosphines are σ-donor and π-acceptor ligands (see *Section 20.4*) and related to them are arsines (AsR$_3$), stibines (SbR$_3$) and phosphites (P(OR)$_3$). The extent of σ-donation and π-acceptance depends on the substituents, e.g. PR$_3$ (R = alkyl) is a poor π-acceptor, whereas PF$_3$ is a poor σ-donor and as strong a π-acceptor as CO. The π-accepting properties of some PR$_3$ ligands follow the ordering:

$$PF_3 > P(OPh)_3 > P(OMe)_3 > PMe_3 > P^tBu_3$$

(23.15)   (23.16)

Infrared spectroscopic data can be used to determine this sequence: a ligand *trans* to a CO affects the M ⟶ CO back-donation and, therefore, $\bar{\nu}_{CO}$, e.g. in octahedral Mo(CO)$_3$(PF$_3$)$_3$, $\bar{\nu}_{CO}$ = 2090 and 2055 cm^{-1}, compared with 1937 and 1841 cm^{-1} in Mo(CO)$_3$(PPh$_3$)$_3$.

The steric requirements of a PR$_3$ ligand depend on the R groups. Ligands such as PPh$_3$ (Figure 23.4a) or P^{t}Bu$_3$ are sterically demanding while others such as PMe$_3$ are less so. The steric requirements are assessed using the *Tolman cone angle*,[†] found by estimating the angle of a cone that has the metal atom at its apex and encompasses the PR$_3$ ligand

---

[†] For a full discussion, see C.A. Tolman (1977) *Chemical Reviews*, vol. 77, p. 313.

Table 23.2 Tolman cone angles for selected phosphine and phosphite ligands.

Ligand	Tolman cone angle / deg	Ligand	Tolman cone angle / deg
P(OMe)$_3$	107	PPh$_3$	145
PMe$_3$	118	P(4-MeC$_6$H$_4$)$_3$	145
PMe$_2$Ph	122	P^{i}Pr$_3$	160
PHPh$_2$	126	P(3-MeC$_6$H$_4$)$_3$	165
P(OPh)$_3$	128	P(*cyclo*-C$_6$H$_{11}$)$_3$	170
PEt$_3$	132	P^{t}Bu$_3$	182
P^{n}Bu$_3$	132	P(2-MeC$_6$H$_4$)$_3$	194
PMePh$_2$	136	P(2,4,6-Me$_3$C$_6$H$_2$)$_3$	212

taking the van der Waals surfaces of the H atoms as its boundary (Figure 23.4b). Table 23.2 lists Tolman cone angles for selected ligands.

The variation in electronic and steric effects in PR$_3$ and related ligands can significantly alter the reactivities of complexes in a series in which the only variant is the phosphine ligand. Many polydentate phosphines are known, two of the more common being bis(diphenylphosphino)methane (dppm, **23.17**) and bis(diphenylphosphino)ethane (dppe, **23.18**). The modes of bonding of polydentate phosphines depend on the flexibility of the backbone of the ligand. For example, dppm is ideally suited to bridge between two adjacent M centres, whereas dppe is found in chelating and bridging modes, or may act as a monodentate ligand with one P atom uncoordinated. Assigning the bonding mode is often aided by ^{31}P NMR spectroscopy; coordination of a P atom shifts its ^{31}P NMR resonance to higher frequency, e.g. the signal in the ^{31}P NMR spectrum of free PPh$_3$ is at δ −6, compared with δ +20.6 for W(CO)$_5$(PPh$_3$).

Ph$_2$P⁀PPh$_2$ Ph$_2$P⁀⁀PPh$_2$

(23.17)   (23.18)

π-Bonded organic ligands

Alkenes, R$_2$C=CR$_2$, tend to bond to metal centres in a 'side-on' (i.e. η^2) manner and behave as 2-electron donors. The metal–ligand bonding can be described in terms of the Dewar–Chatt–Duncanson model (Figure 23.5). The C=C π-bonding MO acts as an electron donor while the π^* MO is an electron acceptor. Populating the π^* MO leads to:

- C−C bond lengthening, e.g. 133.9 pm in C$_2$H$_4$ *vs* 144.5 pm in (η^5-Cp)Rh(η^2-C$_2$H$_4$)(PMe$_3$);
- a lowering of the absorption due to $\nu_{C=C}$ in the vibrational spectrum, e.g. 1623 cm^{-1} in free C$_2$H$_4$ *vs* 1551 cm^{-1} in Fe(CO)$_4$(η^2-C$_2$H$_4$).

The extent of back-donation to R$_2$C=CR$_2$ is influenced by the nature of R, and is enhanced by electron-withdrawing groups such as CN. In the extreme, the π-contribution to the C−C bond is completely removed and the complex

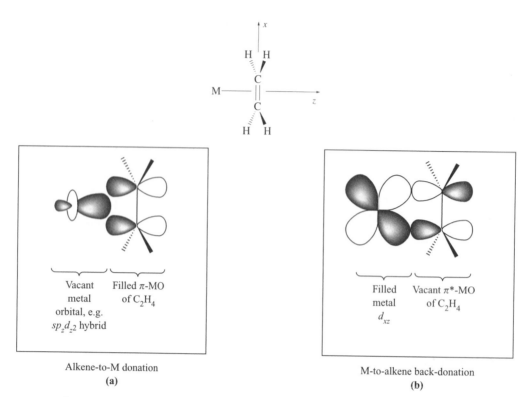

Fig. 23.5 Components of metal–alkene bonding: (a) donation of electrons from the alkene π-MO to a suitable metal d orbital or hybrid, and (b) back-donation of electrons from metal to alkene π^* MO.

becomes a *metallacyclopropane* ring. Structures **23.19a** and **23.19b** show limiting bonding schemes. In **23.19a**, alkene $\longrightarrow$ M donation of charge is dominant, while in **23.19b**, π-back-donation has fully populated the alkene π^*-MO, reducing the C–C bond order to one. On going from

23.19a to **23.19b**, the alkene C atoms rehybridize from sp^2 to sp^3, M–C σ-bonds are formed, and the alkene substituents bend away from the metal (Figure 23.6a). Comparisons of X-ray diffraction data for series of complexes provide evidence for these structural changes.

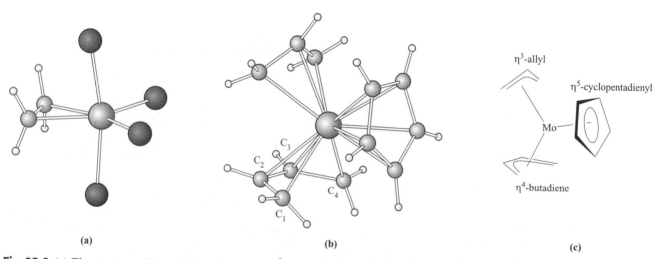

Fig. 23.6 (a) The structure (X-ray diffraction) of $Ru(\eta^2\text{-}C_2H_4)(PMe_3)_4$ illustrating the non-planarity of the coordinated ethene ligand (C–C = 144 pm); only the P atoms of the PMe_3 ligands are shown [W.-K. Wong *et al.* (1984) *Polyhedron*, vol. 3, p. 1255], (b) the structure (X-ray diffraction) of $Mo(\eta^3\text{-}C_3H_5)(\eta^4\text{-}CH_2CHCHCH_2)(\eta^5\text{-}C_5H_5)$ [L.-S. Wang *et al.* (1997) *J. Am. Chem. Soc.*, vol. 119, p. 4453], and (c) a schematic representation of $Mo(\eta^3\text{-}C_3H_5)(\eta^4\text{-}CH_2CHCHCH_2)(\eta^5\text{-}C_5H_5)$. Colour code: Ru and Mo, blue; C, grey; P, black; H, pale grey.

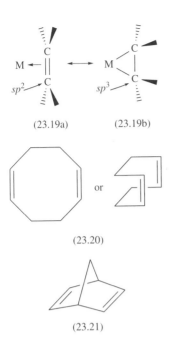

(23.19a)   (23.19b)

or

(23.20)

(23.21)

The bonding description for a coordinated alkene can be extended to other unsaturated organic ligands. Polyalkenes may be non-conjugated or conjugated. In complexes of non-conjugated systems (e.g. cycloocta-1,5-diene (cod), **23.20**, or 2,5-norbornadiene (nbd), **23.21**), the metal–ligand bonding

is analogous to isolated alkene groups. For complexes of conjugated polyenes such as buta-1,3-diene, a delocalized bonding picture is appropriate. Figure 23.7a shows the four π-molecular orbitals of buta-1,3-diene, and in Figure 23.7b, their symmetries are matched to available metal orbitals. Two combinations lead to ligand $\longrightarrow$ M donations, and two to M $\longrightarrow$ ligand back-donation. The interactions involving ψ_2 and ψ_3 weaken bonds C_1–C_2 and C_3–C_4, while strengthening C_2–C_3. The extent of ligand donation or metal back-donation depends on the metal, substituents on the diene, and other ligands present. Structure **23.22** shows the C–C bond lengths in free buta-1,3-diene, and examples of complexes include $Fe(CO)_3(\eta^4\text{-}C_4H_6)$ in which all three C–C bonds in the coordinated diene are 145 pm, and $Mo(\eta^3\text{-}C_3H_5)(\eta^4\text{-}C_4H_6)(\eta^5\text{-}C_5H_5)$ (Figure 23.6b and c) in which the butadiene ligand has C–C bond lengths of 142 (C_1–C_2), 138 (C_2–C_3) and 141 pm (C_3–C_4). Just as for alkene coordination, we can draw two limiting resonance structures (**23.23**) for a buta-1,3-diene (or other 1,3-diene) complex.

145 pm

136 pm

(23.22)   (23.23)

(a)

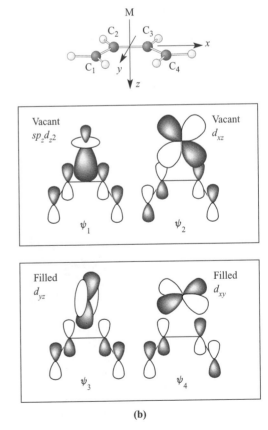

(b)

Fig. 23.7 (a) The four π-MOs of buta-1,3-diene (the energy scale is arbitrary); (b) axis definition for a metal–buta-1,3-diene complex and the combinations of metal and ligand orbitals that lead to transfer of charge from a 1,3-diene to metal (top diagram) and from metal to 1,3-diene (lower diagram).

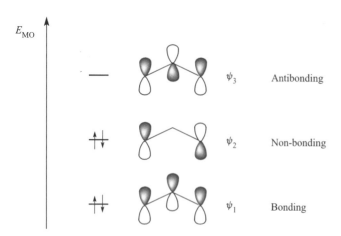

Fig. 23.8 The three π-MOs of the allyl anion, $[C_3H_5]^-$ (the energy scale is arbitrary).

$$H_2C \overset{\overset{\displaystyle H}{\underset{|}{C}}}{\underset{-}{\diagdown}} CH_2$$

(23.24)

The allyl ligand, $[C_3H_5]^-$ (**23.24**), coordinates in an η^3-mode, using the two occupied π-MOs (bonding and non-bonding) as donors and the π^* MO as an acceptor (Figure 23.8). Allyl can also be considered as $[C_3H_5]^{\cdot}$ (see later). Similar schemes can be developed for cyclobutadiene (η^4-C_4H_4), cyclopentadienyl (η^5-C_5H_5, see *Box 18.2*), benzene (η^6-C_6H_6) and related ligands as we discuss later in the chapter.

In solution, complexes with π-bonded organic ligands are often fluxional, with rotation of the ligand being a common dynamic process (see *structure* **23.43**, *Figure 23.17* and *scheme 23.77*). Variable temperature NMR spectroscopy is used to study such phenomena.

Dinitrogen

The molecules N_2 and CO are isoelectronic, and the bonding description in Figure 23.1 can be qualitatively applied to N_2 complexes (see *Section 22.9*), although it must be remembered that the MOs of N_2 have equal atomic orbital contributions from each atom. Complexes of N_2 are not as stable as those of CO, and far fewer examples are known. Terminal $M-N\equiv N$ units are linear (like a terminal $M-C\equiv O$), but bridging N_2 ligands do not mimic bridging CO groups (compare structure **22.43** with **23.6**).

Dihydrogen

We have already mentioned dihydrogen complexes of Re (e.g. **22.25**) and noted the presence of a 'stretched' H−H bond. Other examples include $W(CO)_3(\eta^2\text{-}H_2)(P^iPr_3)_2$ and $[OsH(\eta^2\text{-}H_2)\{P(OEt)_3\}_4]^+$. The H_2 molecule only has available a σ-MO (electron donor orbital) and σ^*-MO (acceptor). Both metal–ligand interactions shown in Figure 23.9 weaken the H−H bond, and coordination readily leads to H−H cleavage (see *Section 23.7*).

23.3 The 18-electron rule

In *Section 20.4*, we applied molecular orbital theory to octahedral complexes containing π-acceptor ligands and gave a

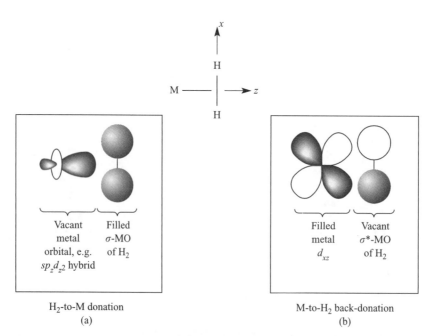

Fig. 23.9 Components of metal–dihydrogen bonding: (a) H_2-to-M donation using the H_2 σ-bonding MO, and (b) M-to-H_2 back-donation into the H_2 σ^*-MO. The axis set is defined in the top diagram.

rationale for the fact that *low oxidation state organometallic complexes tend to obey the 18-electron rule*. This rule often breaks down for early and late *d*-block metals as examples later in the chapter show: 16-electron complexes are common for e.g. Rh(I), Ir(I), Pd(0) and Pt(0). The majority of organometallic compounds with metals from the middle of the *d*-block obey the 18-electron rule and its application is useful, for example, in checking proposed structures. For electron-counting purposes, it is convenient to treat all ligands as *neutral* entities as this avoids the need to assign an oxidation state to the metal centre. However, one must not lose sight of the fact that this is a *formalism*. For example, in the synthesis of cyclopentadienyl derivatives, a common precursor is the salt $Na^+[Cp]^-$. Ferrocene, Cp_2Fe, may be formulated as an Fe(II) compound containing $[Cp]^-$ ligands, but for electron-counting, it is convenient to consider the combination of an Fe(0) centre (group 8, 8 valence electrons) and two neutral $Cp^•$ ligands (5-electron donor) giving an 18-electron complex (**23.25**).

Electron count:

Fe(0) = 8 valence electrons
2 $Cp^•$ = 2×5 valence electrons
Total = 18 electrons

(23.25)

The number of valence electrons for a zero-valent metal centre is equal to the group number (e.g. Cr, 6; Fe, 8; Rh, 9); some commonly encountered ligands[†] donate the following numbers of valence electrons:

- 1-electron donor: $H^•$ (in any bonding mode), and terminal $Cl^•$, $Br^•$, $I^•$, $R^•$ (e.g. R = alkyl or Ph) or $RO^•$;
- 2-electron donor: CO, PR_3, $P(OR)_3$, $R_2C=CR_2$ (η^2-alkene), $R_2C:$ (carbene);
- 3-electron donor: η^3-$C_3H_5{}^•$ (allyl radical), RC (carbyne), μ-$Cl^•$, μ-$Br^•$, μ-$I^•$, μ-$R_2P^•$;
- 4-electron donor: η^4-diene (e.g. **23.22**), η^4-C_4R_4 (cyclobutadienes);
- 5-electron donor: η^5-$C_5H_5{}^•$ (as in **23.25**), μ_3-$Cl^•$, μ_3-$Br^•$, μ_3-$I^•$, μ_3-$RP^•$;
- 6-electron donor: η^6-C_6H_6 (and other η^6-arenes, e.g. η^6-C_6H_5Me);
- 1- or 3-electron donor: NO (see discussion at the end of *Section 20.4*).

(23.26)

Counting electrons provided by bridging ligands, metal–metal bonds and net charges requires care. When bridging between two metal centres, an $X^•$ (X = Cl, Br, I) or $R_2P^•$ ligand uses the unpaired electron and one lone pair to give an interaction formally represented by structure **23.26**, i.e. one electron is donated to M, and two to M′. In a doubly bridged species such as $(CO)_2Rh(\mu$-$Cl)_2Rh(CO)_2$, the μ-Cl atoms are equivalent as are the Rh atoms, and the two Cl bridges together contribute three electrons to each Rh. A bridging $H^•$ provides only one electron *in total*, shared between the metal atoms it bridges, e.g. in $[HFe_2(CO)_8]^-$, (**23.27**). Example **23.27** also illustrates that the formation of a M−M single bond provides each M atom with one extra electron; an M=M double bond contributes two electrons to each metal.

Electron count:

(23.27)

Fe(0) = 8 electrons
3 terminal CO = 3×2 electrons
2 μ-CO = 2×1 electron per Fe
Fe–Fe bond = 1 electron per Fe
H = $^1/_2$ electron per Fe
1− charge = $^1/_2$ electron per Fe
Total = 18 electrons per Fe

Worked example 23.1   18-Electron rule

Confirm that the Cr centre in $[(\eta^6$-$C_6H_6)Cr(CO)_3]$ obeys the 18-electron rule, but Rh in $[(CO)_2Rh(\mu$-$Cl)_2Rh(CO)_2]$ does not.

Cr(0) (group 6) contributes 6 electrons
η^6-C_6H_6 contributes 6 electrons
3 CO contribute $3 \times 2 = 6$ electrons

Total = 18 electrons

Rh(0) (group 9) contributes 9 electrons
μ-Cl contributes 3 electrons (1 to one Rh and 2 to the other Rh)
2 CO contribute $2 \times 2 = 4$ electrons

Total per Rh = 16 electrons

Self-study exercises

1. Confirm that the Fe centres in $H_2Fe(CO)_4$ and $[(\eta^5$-$C_5H_5)Fe(CO)_2]^-$ obey the 18-electron rule.

2. Show that $Fe(CO)_4(\eta^2$-$C_2H_4)$, $HMn(CO)_3(PPh_3)_2$ and $[(\eta^6$-$C_6H_5Br)Mn(CO)_3]^+$ contain 18-electron metal centres.

3. Show that $[Rh(PMe_3)_4]^+$ contains a 16-electron metal centre. Comment on whether violation of the 18-electron rule is expected.

[†] Notation for bridging ligands: see *Section 6.7*.

Metal–metal bonding in multinuclear species is not always clear-cut. *Solely on the basis of the 18-electron rule*, suggest whether $(\eta^5\text{-Cp})\text{Ni}(\mu\text{-PPh}_2)_2\text{Ni}(\eta^5\text{-Cp})$ might be expected to contain a metal–metal bond.

The formula is instructive in terms of drawing a structure, except in respect of any M–M bond. Thus, we can draw as an initial structure:

Now count the valence electrons around each metal centre: Ni(0) (group 10) contributes 10 electrons; $\eta^5\text{-Cp}^{\bullet}$ gives 5 electrons. Two $\mu\text{-PPh}_2$ contribute 3 electrons, 3 per Ni. Total per Ni = 18 electrons.
Conclusion: each Ni atom obeys the 18-electron rule and no Ni–Ni bond is required.

Note that in all such examples, a prediction about the presence or not of the M–M bond *assumes* that the 18-electron rule is obeyed. Bridging ligands often play a very important role in supporting a dimetal framework.

Self-study exercises

1. Show that an M–M single bond is expected in $M_2(CO)_{10}$ (M = Mn, Tc, Re) on the basis of the 18-electron rule.
   [*Hint*: see Figure 23.10b]

2. The presence of an Fe–Fe bond in $[(\eta^5\text{-Cp})(CO)\text{Fe}(\mu\text{-CO})_2\text{Fe}(CO)(\eta^5\text{-Cp})]$ has been a controversial topic. *Solely on the basis of the 18-electron rule*, show that an Fe–Fe bond is expected. What does your conclusion depend on? Are your assumptions infallible?

23.4 Metal carbonyls: synthesis, physical properties and structure

Table 23.3 lists many of the stable, neutral, *d*-block metal carbonyl compounds containing ≤ 6 metal atoms. A range of unstable carbonyls has been obtained by *matrix isolation*: the action of CO on metal atoms in a noble gas matrix at very low temperatures or the photolysis of stable metal carbonyls

Table 23.3 Neutral, low nuclearity ($\leq M_6$) metal carbonyls of the *d*-block metals (dec. = decomposes).

Group number	5	6	7	8	9	10
First row metals	$V(CO)_6$ Dark-blue solid; paramagnetic; dec. 343 K	$Cr(CO)_6$ White solid; sublimes *in vacuo*; dec. 403 K	$Mn_2(CO)_{10}$ Yellow solid; mp 427 K	$Fe(CO)_5$ Yellow liquid; mp 253 K; bp 376 K	$Co_2(CO)_8$ Air-sensitive, orange-red solid; mp 324 K	$Ni(CO)_4$ colourless, volatile liquid; highly toxic vapour; bp 316 K
				$Fe_2(CO)_9$ Golden crystals; mp 373 K (dec.)	$Co_4(CO)_{12}$ Air-sensitive, black solid	
				$Fe_3(CO)_{12}$ Dark green solid; dec. 413 K	$Co_6(CO)_{16}$ Black solid; slowly dec. in air	
Second row metals		$Mo(CO)_6$ White solid; sublimes *in vacuo*	$Tc_2(CO)_{10}$ White solid; slowly dec. in air; mp 433 K	$Ru(CO)_5$ Colourless liquid; mp 251 K; dec. in air at 298 K to $Ru_3(CO)_{12}$ + CO	$Rh_4(CO)_{12}$ Red solid; >403 K dec. to $Rh_6(CO)_{16}$	
				$Ru_3(CO)_{12}$ Orange solid; mp 427 K; sublimes *in vacuo*	$Rh_6(CO)_{16}$ Black solid; dec. >573 K	
Third row metals		$W(CO)_6$ White solid; sublimes *in vacuo*	$Re_2(CO)_{10}$ White solid; mp 450 K	$Os(CO)_5$ Yellow liquid; mp 275 K	$Ir_4(CO)_{12}$ Slightly air sensitive yellow solid; mp 443 K	
				$Os_3(CO)_{12}$ Yellow solid; mp 497 K	$Ir_6(CO)_{16}$ Red solid	

BOX 23.1

Manufacturing iron powder from Fe(CO)$_5$

For applications in magnetic cores for electronic components, iron powder is manufactured by the thermal decomposition of Fe(CO)$_5$:

$$Fe(CO)_5(g) \longrightarrow Fe(s) + 5CO(g)$$

Commercial-scale decomposition takes place in special externally heated vessels in which Fe particles act as nucleation centres. As Fe(CO)$_5$ decomposes, Fe deposits on these particles causing them to grow (up to 8 μm in diameter). Manufacturing processes must control particle size (e.g. by controlling the concentration of Fe(CO)$_5$ in the reactor) and the C, N and O content of the particles (C and O arise from CO decomposition, while N comes from NH$_3$, added to reduce the C and O).

---

under similar conditions. Among species made this way are Ti(CO)$_6$, Pd(CO)$_4$, Pt(CO)$_4$, Cu$_2$(CO)$_6$, Ag$_2$(CO)$_6$, Cr(CO)$_4$, Mn(CO)$_5$, Fe(CO)$_4$, Fe(CO)$_3$ and Ni(CO)$_3$ (the last five being fragments formed by decomposition of stable carbonyls). In the rest of this section, we discuss compounds isolable at ordinary temperatures.

Synthesis and physical properties

The carbonyls Ni(CO)$_4$ and Fe(CO)$_5$ are the only ones normally obtained by action of CO on the finely divided metal. Formation of Ni(CO)$_4$ (equation 21.4) occurs at 298 K and 1 bar pressure, but Fe(CO)$_5$ is made under 200 bar CO at 420–520 K. Most other simple metal carbonyls are prepared by *reductive carbonylation*, i.e. action of CO and a reducing agent (which may be excess CO) on a metal oxide, halide or other compound (e.g. reactions 23.3–23.10). Yields are often poor and we have not attempted to write stoichiometric reactions; for the preparation of [Tc(H$_2$O)$_3$(CO)$_3$]$^+$, see *Box 22.7*.

$$VCl_3 + Na + CO \xrightarrow[\text{in diglyme}]{420\,K,\,150\,bar} [Na(diglyme)_2][V(CO)_6]$$

$$\Big\downarrow \text{HCl, Et}_2\text{O} \qquad (23.3)$$

$$V(CO)_6$$

$$CrCl_3 + CO + Li[AlH_4] \xrightarrow[\text{in Et}_2\text{O}]{390\,K,\,70\,bar} Cr(CO)_6 + LiCl + AlCl_3 \qquad (23.4)$$

$$MoCl_5 + CO + AlEt_3 \xrightarrow{373\,K,\,200\,bar} Mo(CO)_6 + AlCl_3 \qquad (23.5)$$

$$WCl_6 + Fe(CO)_5 \xrightarrow{373\,K} W(CO)_6 + FeCl_2 \qquad (23.6)$$

$$OsO_4 + CO \xrightarrow{520\,K,\,350\,bar} Os(CO)_5 + CO_2 \qquad (23.7)$$

$$Co(O_2CMe)_2 \cdot 4H_2O \xrightarrow[\text{in acetic anhydride}]{CO/H_2\,(4:1),\,200\,bar,\,430\,K} Co_2(CO)_8 \qquad (23.8)$$

$$RuCl_3 \cdot xH_2O + CO \xrightarrow[\text{in MeOH}]{400\,K,\,50\,bar} Ru_3(CO)_{12} \qquad (23.9)$$

$$OsO_4 + CO \xrightarrow[\text{in MeOH}]{400\,K,\,\leq 200\,bar} Os_3(CO)_{12} \qquad (23.10)$$

Diiron nonacarbonyl, Fe$_2$(CO)$_9$, is usually made by photolysis of Fe(CO)$_5$ (equation 23.11) while Fe$_3$(CO)$_{12}$ is obtained by several methods, e.g. oxidation of [HFe(CO)$_4$]$^-$ using MnO$_2$.

$$2Fe(CO)_5 \xrightarrow{h\nu} Fe_2(CO)_9 + CO \qquad (23.11)$$

Some metal carbonyls including M(CO)$_6$ (M = Cr, Mo, W), Fe(CO)$_5$, Fe$_2$(CO)$_9$, Fe$_3$(CO)$_{12}$, Ru$_3$(CO)$_{12}$, Os$_3$(CO)$_{12}$ and Co$_2$(CO)$_8$ are commercially available. All carbonyls are thermodynamically unstable with respect to oxidation in air, but the rates of oxidation vary: Co$_2$(CO)$_8$ reacts under ambient conditions, Fe(CO)$_5$ and Ni(CO)$_4$ are also easily oxidized (their vapours forming explosive mixtures with air), but M(CO)$_6$ (M = Cr, Mo, W) do not oxidize unless heated. Table 23.3 lists some physical properties of some of the more common metal carbonyls. Note the increased importance of M−M bonding as one descends groups 8 and 9: e.g. whereas Co$_2$(CO)$_8$ is stable, Rh$_2$(CO)$_8$ is unstable with respect to Rh$_4$(CO)$_{12}$. The latter can also be formed by reaction 23.12, and above 400 K, it decomposes to Rh$_6$(CO)$_{16}$. Reactions 23.13 and 23.14 are routes to Ir$_4$(CO)$_{12}$ and Ir$_6$(CO)$_{16}$.

$$(CO)_2Rh(\mu\text{-Cl})_2Rh(CO)_2 \xrightarrow[\text{in hexane, NaHCO}_3]{CO,\,1\,bar,\,298\,K} Rh_4(CO)_{12}$$
$$(23.12)$$

$$Na_3[IrCl_6] \xrightarrow[\text{2. base}]{1.\,CO,\,1\,bar,\,in\,MeOH\,under\,reflux} Ir_4(CO)_{12}$$
$$(23.13)$$

$$[Et_4N]_2[Ir_6(CO)_{15}] \xrightarrow{CF_3SO_3H\,under\,CO} Ir_6(CO)_{16} \qquad (23.14)$$

Metal carbonyl *clusters* containing four or more metal atoms are made by a variety of methods; osmium in particular forms a range of binary compounds and pyrolysis of Os$_3$(CO)$_{12}$ yields a mix of products (equation 23.15) which can be separated by chromatography.

$$Os_3(CO)_{12} \xrightarrow{483\,K} Os_5(CO)_{16}$$
$$+ \underset{\text{major product}}{Os_6(CO)_{18}} + Os_7(CO)_{21} + Os_8(CO)_{23}$$
$$(23.15)$$

Large cations for large anions: 2

Metal cluster anions are usually stabilized in salts that contain large cations, common choices are $[Ph_4P]^+$, $[Ph_4As]^+$, $[^nBu_4N]^+$ and $[(Ph_3P)_2N]^+$. Compatibility between cation and anion sizes is important. The figure below shows part of the packing diagram for the salt $[Ph_4P]_2[Ir_8(CO)_{22}]$; the

ions are shown in space-filling form with the H atoms of the Ph rings omitted for clarity. The diagram illustrates how well the large cations pack with the cluster anions, and this is essential for the stabilization and crystallization of such salts.

Colour coding: Ir atoms in blue; C atoms in grey; O atoms in pale grey; P atoms in black. [Data from: F. Demartin *et al.* (1981) *J. Chem. Soc., Chem. Commun.*, p. 528].

See also: *Box 10.5* – Large cations for large anions: 1.

Metal carbonyl anions can be derived by reduction, e.g. reactions 23.16–23.21; dimers such as $Mn_2(CO)_{10}$ and $Co_2(CO)_8$ undergo simple cleavage of the M−M bond, but in other cases, reduction is accompanied by an increase in metal nuclearity. In reactions 23.16 and 23.21 $Na[C_{10}H_8]$ (sodium naphthalide) is made from Na and naphthalene; both $Na[C_{10}H_8]$ and $K[C_{10}H_8]$ are powerful reducing agents.

$$Fe(CO)_5 \xrightarrow{Na[C_{10}H_8]} Na_2[Fe(CO)_4] \qquad (23.16)$$

$$Mn_2(CO)_{10} + 2Na \longrightarrow 2Na[Mn(CO)_5] \qquad (23.17)$$

$$Co_2(CO)_8 + 2Na \longrightarrow 2Na[Co(CO)_4] \qquad (23.18)$$

$$Ru_3(CO)_{12} \xrightarrow{Na,\ THF,\ \Delta} [Ru_6(CO)_{18}]^{2-} \qquad (23.19)$$

$$Os_3(CO)_{12} \xrightarrow{Na,\ diglyme,\ \Delta} [Os_6(CO)_{18}]^{2-} \qquad (23.20)$$

$$Ni(CO)_4 \xrightarrow{Na[C_{10}H_8]} [Ni_5(CO)_{12}]^{2-} + [Ni_6(CO)_{12}]^{2-}$$
$$(23.21)$$

The salt $Na_2[Fe(CO)_4]$ (equation 23.16) is *Collman's reagent* which has numerous synthetic applications; it is very air-sensitive and is best prepared *in situ*. In reactions 23.19–23.21, Na^+ salts are the initial products, but the large cluster anions are isolated as salts of large cations such as $[(Ph_3P)_2N]^+$, $[Ph_4P]^+$ or $[Ph_4As]^+$ (see *Box 23.2*).

Structures

Mononuclear metal carbonyls possess the following structures:

- tetrahedral: $Ni(CO)_4$ (Ni−C = 184 pm), $[Co(CO)_4]^-$ (Co−C = 175 pm), $[Fe(CO)_4]^{2-}$ (Figure 23.10a);
- trigonal bipyramidal: $Fe(CO)_5$ (Fe−C_{axial} = 181 pm, Fe−C_{equ} = 183 pm);
- octahedral: $V(CO)_6$ (V−C = 200 pm), $Cr(CO)_6$ (Cr−C = 192 pm), $Mo(CO)_6$ (Mo−C = 206 pm), $W(CO)_6$ (W−C = 207 pm).

With the exception of $V(CO)_6$, each obeys the 18-electron rule. The 17-electron count in $V(CO)_6$ suggests the possibility of dimerization to '$V_2(CO)_{12}$' containing a V−V bond, but this is sterically unfavourable. A mononuclear carbonyl of Mn would, like $V(CO)_6$, be a radical, but now, dimerization occurs and $Mn_2(CO)_{10}$ is the lowest nuclearity neutral binary carbonyl of Mn. A similar situation arises for cobalt: '$Co(CO)_4$' is a 17-electron species and the lowest nuclearity binary carbonyl is $Co_2(CO)_8$. The group 7 dimers $Mn_2(CO)_{10}$, $Tc_2(CO)_{10}$ and $Re_2(CO)_{10}$ are isostructural and have staggered arrangements of carbonyls (Figure 23.10b); the M−M bond is unbridged and longer (Mn−Mn = 290 pm, Tc−Tc = 303 pm, Re−Re = 304 pm) than twice the metallic radius (see *Tables 21.1* and *22.1*). In $Fe_2(CO)_9$ (Figure 23.10c), three CO ligands bridge between

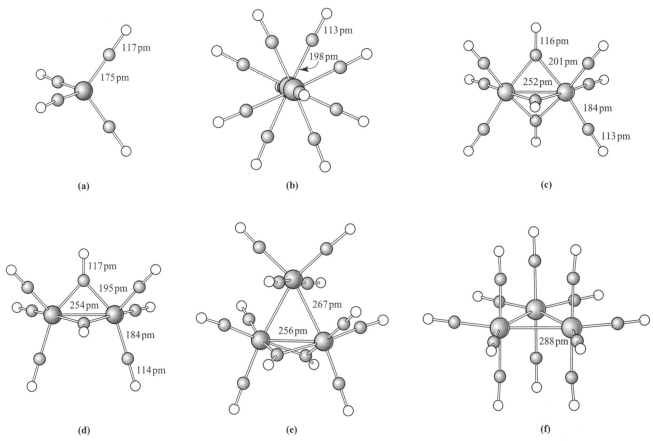

Fig. 23.10 The structures (X-ray diffraction) of (a) [Fe(CO)$_4$]$^{2-}$ in the K$^+$ salt [R.G. Teller *et al.* (1977) *J. Am. Chem. Soc.*, vol. 99, p. 1104], (b) Re$_2$(CO)$_{10}$ showing the staggered configuration also adopted by Mn$_2$(CO)$_{10}$ and Tc$_2$(CO)$_{10}$ [M.R. Churchill *et al.* (1981) *Inorg. Chem.*, vol. 20, p. 1609], (c) Fe$_2$(CO)$_9$ [F.A. Cotton *et al.* (1974) *J. Chem. Soc., Dalton Trans.*, p. 800], (d) Co$_2$(CO)$_8$ [P.C. Leung *et al.* (1983) *Acta Crystallogr., Sect. B*, vol. 39, p. 535], (e) Fe$_3$(CO)$_{12}$ [D. Braga *et al.* (1994) *J. Chem. Soc., Dalton Trans.*, p. 2911], and (f) Os$_3$(CO)$_{12}$ which is isostructural with Ru$_3$(CO)$_{12}$ [M.R. Churchill *et al.* (1977) *Inorg. Chem.*, vol. 16, p. 878]. Colour code: metals, blue; C, grey; O, white.

the Fe centres; each Fe atom obeys the 18-electron rule if an Fe−Fe bond is present and this is consistent with the observed diamagnetism of the complex. Even so, many theoretical studies have been carried out to investigate the presence (or not) of Fe−Fe bonding in Fe$_2$(CO)$_9$. Figure 23.10d shows the *solid state* structure of Co$_2$(CO)$_8$. When solid Co$_2$(CO)$_8$ is dissolved in hexane, the IR spectrum changes; the spectrum of the solid contains bands assigned to terminal *and* bridging CO ligands, but in hexane, only absorptions due to terminal carbonyls are seen. This is explained by the equilibrium in scheme **23.28** and solid state ^{13}C NMR spectroscopic data show that terminal-bridge CO exchange occurs even in *solid* Co$_2$(CO)$_8$.

$$OC \overset{\overset{\displaystyle O}{\underset{\displaystyle \parallel}{C}}}{\underset{\underset{\displaystyle O}{\underset{\displaystyle \parallel}{C}}}{\overset{OC\,\text{'''''}}{\underset{OC}{Co}}\!-\!\underset{CO}{\overset{CO}{Co}}}} \rightleftharpoons \begin{matrix}OC & CO & CO \\ | & | & | \\ OC\!-\!Co\!-\!Co\!-\!CO \\ | & | & | \\ OC & OC & CO\end{matrix}$$

(23.28)

Each group 8 metal forms a trinuclear binary carbonyl M$_3$(CO)$_{12}$ containing a triangular framework of metal atoms. However, the arrangement of CO ligands in Fe$_3$(CO)$_{12}$ (Figure 23.10e) differs from that in Ru$_3$(CO)$_{12}$ and Os$_3$(CO)$_{12}$ (Figure 23.10f). The latter contain equilateral M$_3$ triangles and four terminal CO per metal, whereas in the solid state, Fe$_3$(CO)$_{12}$ contains an isosceles Fe$_3$ triangle with one Fe−Fe edge (the shortest) bridged by two CO ligands. Each M atom in Fe$_3$(CO)$_{12}$, Ru$_3$(CO)$_{12}$ and Os$_3$(CO)$_{12}$ obeys the 18-electron rule. The solution ^{13}C NMR spectrum of Fe$_3$(CO)$_{12}$ exhibits one resonance even as low as 123 K showing that the molecule is fluxional. The process can be described in terms of exchange of terminal and bridging CO ligands, or by considering the tilting of the Fe$_3$-unit within a shell of CO ligands. X-Ray data collected at several temperatures show that Fe$_3$(CO)$_{12}$ also undergoes a dynamic process in the solid state. This illustrates that the CO$_{\text{term}}$−CO$_{\text{bridge}}$ (CO$_{\text{term}}$ = terminal CO ligand) exchange is a low energy process and is one of many such examples.

The group 9 carbonyls Co$_4$(CO)$_{12}$ and Rh$_4$(CO)$_{12}$ (Figure 23.11a) are isostructural; three μ-CO ligands are

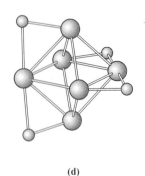

(a)   **(b)**   **(c)**   **(d)**

Fig. 23.11 The structures (X-ray diffraction) of (a) $Rh_4(CO)_{12}$ which is isostructural with $Co_4(CO)_{12}$ [C.H. Wei (1969) *Inorg. Chem.*, vol. 8, p. 2384], (b) $Ir_4(CO)_{12}$ [M.R. Churchill *et al.* (1978) *Inorg. Chem.*, vol. 17, p. 3528], (c) the red isomer of $Ir_6(CO)_{16}$ and (d) the black isomer of $Ir_6(CO)_{16}$ [L. Garlaschelli *et al.* (1984) *J. Am. Chem. Soc.*, vol. 106, p. 6664]. In (a) and (b), O atoms have been omitted for clarity. In (c) and (d), the terminal CO and the O atoms of the bridging CO ligands have been omitted; each Ir has two CO_{term}. Colour code: metals, blue; C, grey.

arranged around the edges of one face of the M_4 tetrahedron. In $Ir_4(CO)_{12}$, all ligands are terminal (Figure 23.11b). Each group 9 metal forms a hexanuclear carbonyl, $M_6(CO)_{16}$, in which the metal atoms form an octahedral cluster; in $Co_6(CO)_{16}$, $Rh_6(CO)_{16}$ and the red isomer of $Ir_6(CO)_{16}$, each M atom has two CO_{term} and there are four μ_3-CO as shown in Figure 23.11c. A black isomer of $Ir_6(CO)_{16}$ has been isolated and in the solid state has 12 CO_{term} and 4 μ-CO (Figure 23.11d). Other octahedral carbonyl clusters include $[Ru_6(CO)_{18}]^{2-}$ and $[Os_6(CO)_{18}]^{2-}$,

but in contrast $Os_6(CO)_{18}$ has a bicapped tetrahedral structure (**23.29**). This is an example of a *condensed polyhedral* cluster.

(23.29)

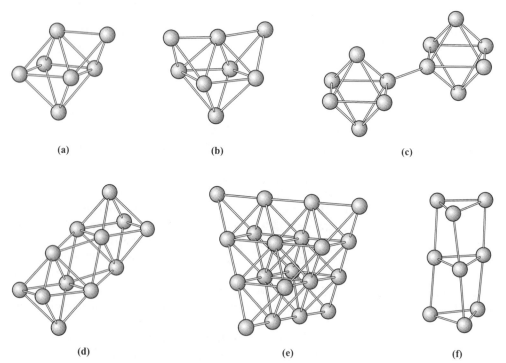

(a)   **(b)**   **(c)**

(d)   **(e)**   **(f)**

Fig. 23.12 The structures (X-ray diffraction) of the metal-cores in (a) $Os_7(CO)_{21}$ [C.R. Eady *et al.* (1977) *J. Chem. Soc., Chem. Commun.*, p. 385], (b) $[Os_8(CO)_{22}]^{2-}$ in the $[(Ph_3P)_2N]^+$ salt [P.F. Jackson *et al.* (1980) *J. Chem. Soc., Chem. Commun.*, p. 60], (c) $[Rh_{12}(CO)_{30}]^{2-}$ in the $[Me_4N]^+$ salt [V.G. Albano *et al.* (1969) *J. Organomet. Chem.*, vol. 19, p. 405], (d) $[Ir_{12}(CO)_{26}]^{2-}$ in the $[Ph_4P]^+$ salt [R.D. Pergola *et al.* (1987) *Inorg. Chem.*, vol. 26, p. 3487], (e) $[Os_{20}(CO)_{40}]^{2-}$ in the $[^nBu_4P]^+$ salt [L.H. Gade *et al.* (1994) *J. Chem. Soc., Dalton Trans.*, p. 521], and (f) $[Pt_9(CO)_{18}]^{2-}$ in the $[Ph_4P]^+$ salt [J.C. Calabrese *et al.* (1974) *J. Am. Chem. Soc.*, vol. 96, p. 2614].

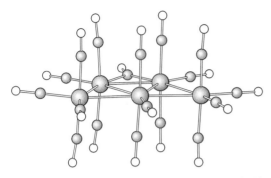

Fig. 23.13 The structure (X-ray diffraction) of $Os_5(CO)_{18}$ in which the Os atoms form a planar 'raft' [W. Wang *et al.* (1992) *J. Chem. Soc., Chem. Commun.*, p. 1737]. Colour code: Os, blue; C, grey; O, white.

> In a *condensed polyhedral cluster*, two or more polyhedral cages are fused together through atom, edge or face-sharing.

The syntheses of high-nuclearity metal carbonyl clusters are not readily generalized,[†] and we focus only on the structures of selected species. For seven or more metal atoms, metal carbonyl clusters tend to be composed of condensed (or less often, linked) tetrahedral or octahedral units. The group 10 metals form a series of clusters containing stacked triangles, e.g. $[Pt_9(CO)_{18}]^{2-}$ and $[Pt_{15}(CO)_{30}]^{2-}$. Figure 23.12 shows the metal cores of representative clusters; in $[Os_{20}(CO)_{40}]^{2-}$ the Os atoms form a ccp arrangement. Some metal carbonyls possess 'raft' structures, i.e. the metal atoms form planar arrangements of edge-sharing triangles, e.g. $Os_5(CO)_{18}$ (Figure 23.13).

23.5 The isolobal principle and application of Wade's rules

In *Section 12.11*, we introduced *Wade's rules* to rationalize the structures of borane clusters. This method of counting electrons can be extended to simple organometallic clusters by making use of the *isolobal relationship* between cluster fragments.

> Two cluster fragments are *isolobal* if they possess the same frontier orbital characteristics: same symmetry, same number of electrons available for cluster bonding, and *approximately* the same energy.

Figure 23.14 shows the frontier MOs (i.e. those close to and including the HOMO and LUMO) of BH and C_{3v}

[†] For further details, see for example: C.E. Housecroft (1996) *Metal–Metal Bonded Carbonyl Dimers and Clusters*, Oxford University Press, Oxford.

$M(CO)_3$ (M = Fe, Ru, Os) fragments. In *Box 12.9*, we considered how the frontier orbitals of six BH combined to give the cluster bonding MOs in $[B_6H_6]^{2-}$ (a process that can be extended to other clusters), and now we look at why it is that BH and some organometallic fragments can be regarded as being similar in terms of cluster bonding. The points to note in Figure 23.14 are that the BH and $M(CO)_3$ fragments have three frontier MOs with matching symmetries and containing the same number of electrons; the ordering of the MOs is not important. The BH and C_{3v} $M(CO)_3$ (M = Fe, Ru, Os) fragments are *isolobal* and their relationship allows BH units in borane clusters to be replaced (in theory and sometimes in practice, although syntheses are not as simple as this formal replacement suggests) by $Fe(CO)_3$, $Ru(CO)_3$ or $Os(CO)_3$ fragments. Thus, for example, we can go from $[B_6H_6]^{2-}$ to $[Ru_6(CO)_{18}]^{2-}$. Wade's rules categorize $[B_6H_6]^{2-}$ as a 7 electron pair *closo*-cluster, and similarly, $[Ru_6(CO)_{18}]^{2-}$ is a *closo*-species; both are predicted (and have in practice) octahedral cages.

Moving to the left or right of group 8 removes or adds electrons to the frontier MOs shown in Figure 23.14. Removing or adding a CO ligand removes or adds two electrons. (The frontier MOs also change, but this is unimportant if we are simply counting electrons.) Changing the ligands similarly alters the number of electrons available. Equation 23.22 shows how the number of electrons provided by a given fragment can be determined and Table 23.4 applies this to selected fragments. These numbers are used *within the Wade approach*, also known as *polyhedral skeletal electron pair theory* (PSEPT).

$$x = v + n - 12 \tag{23.22}$$

where x = number of cluster-bonding electrons provided by a fragment, v = number of valence electrons from the metal atom and n = number of valence electrons provided by the ligands.

Worked example 23.3   Application of Wade's rules (PSEPT)

(a) Rationalize why $Rh_4(CO)_{12}$ has a tetrahedral core.
(b) What class of cluster is $Ir_4(CO)_{12}$?

[If you are unfamiliar with Wade's rules, first review *Section 12.11*.]
(a) Break the formula of $Rh_4(CO)_{12}$ down into convenient units and determine the number of cluster-bonding electrons.

- Each $\{Rh(CO)_3\}$-unit provides 3 cluster-bonding electrons.
- Total number of electrons available in $Rh_4(CO)_{12} = (4 \times 3) = 12$ electrons = 6 pairs.
- Thus, $Rh_4(CO)_{12}$ has 6 pairs of electrons with which to bond 4 cluster units.
- There are $(n + 2)$ pairs of electrons for n vertices, and so $Rh_4(CO)_{12}$ is a *nido*-cage; the parent deltahedron is a

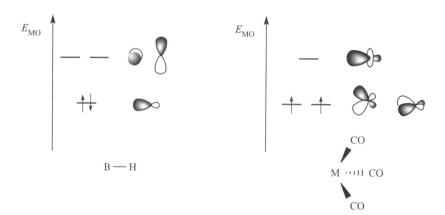

Fig. 23.14 The frontier MOs of a BH unit and a C_{3v} (i.e. 'conical') $M(CO)_3$ (M = Fe, Ru, Os) group. For the BH unit, the occupied MO is an *sp* hybrid; for $M(CO)_3$, the orbitals are represented by *pd* or *spd* hybrids. These orbitals combine with those of other cluster fragments to give cluster-bonding, non-bonding and antibonding MOs (see *Box 12.9*).

trigonal bipyramid, and thus $Rh_4(CO)_{12}$ is expected to be tetrahedral.

(b) Rh and Ir are both in group 9 and so $Ir_4(CO)_{12}$ is also a *nido*-cluster.

This example illustrates an important point about the use of such electron counting schemes: *no information about the positions of the ligands can be obtained*. Although Wade's rules rationalize why $Rh_4(CO)_{12}$ and $Ir_4(CO)_{12}$ both have tetrahedral cores, they say nothing about the fact that the ligand arrangements are different (Figures 23.11a and b).

Self-study exercises

1. Using Wade's approach, rationalize why $Co_4(CO)_{12}$ has a tetrahedral core.

2. Using PSEPT, rationalize why $[Fe_4(CO)_{13}]^{2-}$ has a tetrahedral Fe_4 core.

3. The cluster $Co_2(CO)_6C_2H_2$ has a tetrahedral Co_2C_2 core. How many electrons does each CH unit contribute to cluster bonding? [*Ans.* 3]

The diversity of cage structures among metal clusters is greater than that of boranes; Wade's rules were developed

for boranes and extension of the rules to rationalize the structures of high nuclearity metal clusters is limited. Boranes tend to adopt rather open structures, and there are few examples of BH units in capping positions. However, application of the *capping principle* does allow satisfactory rationalization of some condensed cages such as $Os_6(CO)_{18}$ (**23.29**).

> Within the remit of Wade's rules (PSEPT), the addition of one or more *capping units* to a deltahedral cage requires no additional bonding electrons; a capping unit is a cluster fragment placed over the *triangular* face of a central cage.

Worked example 23.4 Application of Wade's rules (PSEPT): the capping principle

Rationalize why $Os_6(CO)_{18}$ adopts structure **23.29** rather than an octahedral cage.

- $Os_6(CO)_{18}$ can be broken down into 6 $Os(CO)_3$ fragments.
- Each $\{Os(CO)_3\}$-unit provides 2 cluster-bonding electrons.
- Total number of electrons available in $Os_6(CO)_{18} = (6 \times 2) = 12$ electrons = 6 pairs.

Table 23.4 The number of electrons (*x* in equation 23.22) provided for cluster bonding by selected fragments; $\eta^5\text{-}C_5H_5 = \eta^5\text{-}Cp$.

Cluster fragment	Group 6: Cr, Mo, W	Group 7: Mn, Tc, Re	Group 8: Fe, Ru, Os	Group 9: Co, Rh, Ir
$M(CO)_2$	−2	−1	0	1
$M(CO)_3$	0	1	2	3
$M(CO)_4$	2	3	4	5
$M(\eta^5\text{-}C_5H_5)$	−1	0	1	2
$M(\eta^6\text{-}C_6H_6)$	0	1	2	3
$M(CO)_2(PR_3)$	0	1	2	3

- Thus, $Os_6(CO)_{18}$ has 6 pairs of electrons with which to bond 6 cluster units.
- This corresponds to a monocapped structure, the parent deltahedron being one with 5 vertices, i.e. a trigonal bipyramid:

- The monocapped trigonal bipyramid is the same as a bicapped tetrahedron (**23.29**).
- A *closo*-octahedral cage requires 7 pairs of electrons, and $Os_6(CO)_{18}$ has insufficient electrons for this structure.

Self-study exercises

1. The Os_7 core of $Os_7(CO)_{21}$ is a capped octahedron. Show that this is consistent with the PSEPT capping principle.

2. Use the capping principle to account for the fact that $[Os_8(CO)_{22}]^{2-}$ has a bicapped octahedral structure.

Using the isolobal principle, one can relate clusters that contain fragments having analogous orbital properties. Some isolobal pairs of metal carbonyl and hydrocarbon fragments are:

- $Co(CO)_3$ (C_{3v}) and CH (provides 3 orbitals and 3 electrons);
- $Fe(CO)_4$ (C_{2v}) and CH_2 (provides 2 orbitals and 2 electrons);
- $Mn(CO)_5$ (D_{4h}) and CH_3 (provides 1 orbital and 1 electron).

C_{3v} $\qquad$ C_{2v} $\qquad$ D_{4h}

Thus, for example, $Co_4(CO)_{12}$, $Co_3(CO)_9CH$, $Co_2(CO)_6C_2H_2$ and C_4H_4 form an isolobal series. Isolobal relationships have a *theoretical* premise and tell us nothing about methods of cluster synthesis.

23.6 Total valence electron counts in *d*-block organometallic clusters

The structures of many polynuclear organometallic species are not conveniently described in terms of Wade's rules, and an alternative approach is to consider the *total valence electron count*, also called the *Mingos cluster valence electron count*.

Single cage structures

Each low oxidation state *metal cluster* cage possesses a characteristic number of valence electrons (ve) as Table 23.5 shows. We shall not describe the MO basis for these numbers, but merely apply them to rationalize observed structures. Look back to Section 23.2 for the numbers of electrons donated by ligands. Any organometallic complex with a triangular M_3-framework requires 48 valence electrons, for example:

- $Ru_3(CO)_{12}$ has $(3 \times 8) + (12 \times 2) = 48$ ve;
- $H_2Ru_3(CO)_8(\mu\text{-PPh}_2)_2$ has $(2 \times 1) + (3 \times 8) + (8 \times 2) + (2 \times 3) = 48$ ve;
- $H_3Fe_3(CO)_9(\mu_3\text{-CMe})$ has $(3 \times 1) + (3 \times 8) + (9 \times 2) + (1 \times 3) = 48$ ve.

Similarly, clusters with tetrahedral or octahedral cages require 60 or 86 valence electrons respectively, for example:

- $Ir_4(CO)_{12}$ has $(4 \times 9) + (12 \times 2) = 60$ ve;
- $(\eta^5\text{-Cp})_4Fe_4(\mu_3\text{-CO})_4$ has $(4 \times 5) + (4 \times 8) + (4 \times 2) = 60$ ve;
- $Rh_6(CO)_{16}$ has $(6 \times 9) + (16 \times 2) = 86$ ve;
- $Ru_6(CO)_{17}C$ has $(6 \times 8) + (17 \times 2) + 4 = 86$ ve.

The last example is of a cage containing an *interstitial atom* (see structure **23.12**) and *contributes all of its valence electrons* to cluster bonding. An interstitial C atom contributes four electrons, a B atom, three, an N or P atom, five, and so on.

> **Worked example 23.5 An application of total valence electrons counts**
>
> Suggest what change in cluster structure might accompany the reaction:
>
> $$[Co_6(CO)_{15}N]^- \longrightarrow [Co_6(CO)_{13}N]^- + 2CO$$
>
> Both clusters contain an interstitial N atom which contributes 5 electrons to cluster bonding. The negative charge contributes one electron.
> Total valence electron count for
>
> $$[Co_6(CO)_{15}N]^- = (6 \times 9) + (15 \times 2) + 5 + 1 = 90$$
>
> Total valence electron count for
>
> $$[Co_6(CO)_{13}N]^- = (6 \times 9) + (13 \times 2) + 5 + 1 = 86$$
>
> i.e. the loss of 2 CO ligands corresponds to a loss of 4 electrons, and a change from a trigonal prism to octahedral Co_6-cage.

Self-study exercises

1. $[Ru_6(CO)_{17}B]^-$ and $[Os_6(CO)_{18}P]^-$ contain interstital B and P atoms respectively. Account for the fact that while

Table 23.5 Characteristic total valence electron counts for selected low oxidation state metal clusters.

Cluster framework	Diagrammatic representation of the cage	Valence electron count
Triangle		48
Tetrahedron		60
Butterfly, or planar raft of four atoms		62
Square		64
Trigonal bipyramid		72
Square-based pyramid		74
Octahedron		86
Trigonal prism		90

$[Ru_6(CO)_{17}B]^-$ has an octahedral M_6 core, $[Os_6(CO)_{18}P]^-$ adopts a trigonal prismatic core. [*Ans.* 86 ve; 90 ve]

2. Rationalize why $Os_3(CO)_{12}$ has a triangular Os_3 core but in $Os_3(CO)_{12}Br_2$, the Os atoms are in a linear arrangement. [*Ans.* 48 ve; 50 ve]

Condensed cages

Structure **23.29** showed one type of *condensed cluster*. The sub-cluster units are connected either through shared M atoms, M−M edges or M_3 faces. The total valence electron count for a condensed structure is equal to the total number of electrons required by the sub-cluster units *minus* the electrons associated with the shared unit. The numbers to *subtract* are:

- 18 electrons for a shared M atom;
- 34 electrons for a shared M−M edge;
- 48 electrons for a shared M_3 face.

Examples of these families of condensed polyhedral clusters are $Os_5(CO)_{19}$ (atom sharing, **23.30**), $H_2Os_5(CO)_{16}$ (edge-sharing, **23.31**) and $H_2Os_6(CO)_{18}$ (face-sharing, **23.32**).

(23.30)   (23.31)   (23.32)

Worked example 23.6   Electron counts in condensed cluster structures

$Os_6(CO)_{18}$ has structure **23.29**, i.e. three face-sharing tetrahedra. Show that this structure is consistent with the number of valence electrons available.

can be represented as three face-sharing tetrahedra:

Valence electron count for three tetrahedra $= 3 \times 60 = 180$
For each shared face, subtract 48 electrons.
Valence electron count for the bicapped tetrahedron $= 180 − (2 \times 48) = 84$

The number of valence electrons available in $Os_6(CO)_{18} = (6 \times 8) + (18 \times 2) = 84$

Thus, the observed structure is consistent with the number of valence electrons available.

Self-study exercises

1. While $[Os_6(CO)_{18}]^{2−}$ has an octahedral Os_6 core, that in $Os_6(CO)_{18}$ is a capped trigonal bipyramid. Use total valence electron counts to rationalize this difference.

2. The core in $Os_5(CO)_{19}$ is shown in structure **23.30**. Show that this shape is consistent with the total valence electron count of the cluster.

Limitations of total valence counting schemes

For some clusters such as Rh_x species, the number of electrons available may not match the number apparently required by the structure adopted. Two examples are $[Rh_5(CO)_{15}]^−$ and $[Rh_9(CO)_{19}]^{3−}$. The former possesses 76 valence electrons and yet has a trigonal bipyramidal Rh_5-core for which 72 electrons are usual. However, a look at the Rh−Rh bond lengths reveals that six edges are in the range 292–303 pm, while three are 273–274 pm, indicating that the extra electrons have caused bond lengthening. In $[Rh_9(CO)_{19}]^{3−}$, 122 electrons are available but the Rh_9-core consists of two face-sharing octahedra for which 124 electrons are required by the scheme outlined above.[†]

These are but two examples of the limitations of electron counting schemes. As more clusters are structurally characterized, further exceptions arise providing yet more challenges for the theorist.

23.7 Types of organometallic reactions

In this section, we introduce the main types of ligand transformations that take place at metal centres in organometallic compounds:

- ligand substitution;
- oxidative addition (including orthometallation);
- reductive elimination;
- alkyl and hydrogen migration;
- β-hydrogen elimination;
- α-hydrogen abstraction.

Substitution of CO ligands

The substitution of a CO ligand by another two-electron donor (e.g. PR_3) may occur by photochemical or thermal activation, either by direct reaction of the metal carbonyl and incoming ligand, or by first replacing a CO by a more labile ligand such as THF or MeCN. An example of the latter is the formation of $Mo(CO)_5(PPh_3)$ (equation 23.23) which is most effectively carried out by first making the THF (**23.33**) adduct *in situ*.

[†] For detailed discussion, see: D.M.P. Mingos and D.J. Wales (1990) *Introduction to Cluster Chemistry*, Prentice Hall, Englewood Cliffs, NJ.

(23.33)

$$Mo(CO)_6 \xrightarrow[-CO]{\text{in THF}, h\nu} Mo(CO)_5(THF)$$

$$\xrightarrow[-THF]{PPh_3} Mo(CO)_5(PPh_3) \quad (23.23)$$

The substitution steps are *dissociative* (see *Chapter 25*). The outgoing ligand leaves, creating a 16-electron metal centre which is *coordinatively unsaturated*. The entry of a new 2-electron ligand restores the 18-electron count. Competition between ligands for coordination to the 16-electron centre may be countered by having the incoming ligand (L in equation 23.24) present in excess.

(23.24)

In reaction 23.25, the incoming ligand provides four electrons and displaces two CO ligands. Multiple substitution by two-electron donors is exemplified by reaction 23.26.

$$Fe(CO)_5 + \text{\raisebox{0pt}{/\!\!\!\!\diagup}} \xrightarrow[\text{or } \Delta, 20\,bar]{h\nu}$$

$$Fe(CO)_3(\eta^4\text{-}CH_2CHCHCH_2) + 2CO \quad (23.25)$$

$$Fe(CO)_5 + \text{excess } PMe_3 \xrightarrow{h\nu} Fe(CO)_3(PMe_3)_2 + 2CO$$

(23.34) (23.26)

(23.34)

Oxidative addition

Oxidative addition reactions are very important in organometallic synthesis. Oxidative addition involves:

- the addition of a molecule XY with cleavage of the X−Y single bond (equation 23.27), addition of a multiply bonded species with reduction in the bond order and formation of a metallacycle (equation 23.28), addition of a C−H bond in an *orthometallation* step (equation 23.29) or a similar addition;
- oxidation of the metal centre by two units;
- increase in metal coordination number by two.

(23.27)

(23.28)

(23.29)

Addition of O_2 to give an η^2-peroxo complex is related to reaction type 23.28. Each addition in equations 23.27–23.29 occurs *at a 16-electron metal centre*, taking it to an 18-electron centre in the product. Most commonly, the precursor has a d^8 or d^{10} configuration, e.g. Rh(I), Ir(I), Pd(0), Pd(II), Pt(0), Pt(II), and the metal must have an accessible higher oxidation state, e.g. Rh(III). If the starting compound contains an 18-electron metal centre, oxidative addition cannot occur without loss of a two-electron ligand as in reaction 23.30.

$$Os(CO)_5 + H_2 \xrightarrow{\Delta,\,80\,bar} H_2Os(CO)_4 + CO \quad (23.30)$$

Many examples of the addition of small molecules (e.g. H_2, HX, RX) are known. The reverse of oxidative addition is *reductive elimination*, e.g. reaction 23.31 in which an acyl substituent is converted to an aldehyde.

$$H_2Co\{C(O)R\}(CO)_3 \longrightarrow HCo(CO)_3 + RCHO \quad (23.31)$$

Oxidative addition initially gives a *cis*-addition product, but ligand rearrangements can occur and the isolated product may contain the added groups mutually *cis* or *trans*.

Alkyl and hydrogen migrations

Reaction 23.32 is an example of *alkyl migration*.

(23.32)

The reaction is also called *CO insertion* since the incoming CO molecule seems to have been inserted into the Mn−C$_{Me}$ bond: this name is misleading. If reaction 23.32

(23.35)

is carried out using ^{13}CO, *none* of the incoming ^{13}CO ends up in the acyl group nor in the position *trans* to the acyl group; the isolated product is **23.35**. Reaction 23.32 involves the *intramolecular* transfer of an alkyl group to the C atom of a CO group which is *cis* to the original alkyl site; the incoming CO occupies the coordination site vacated by the alkyl group as shown in scheme 23.33.

(23.33)

The migration is reversible and the *decarbonylation* reaction has been studied with the ^{13}C-labelled compound; the results are shown in Figure 23.15. The distribution of the products is consistent with the migration of the Me group, and not with a mechanism that involves movement of the 'inserted' CO. The reaction products can be monitored using ^{13}C NMR spectroscopy.

The 'insertion of CO' into $M-C_{alkyl}$ bonds is well exemplified in organometallic chemistry, and one industrial example

(equation 23.34) is a step in the Monsanto process for the production of acetic acid (see *Section 26.3*).

$$[Rh(Me)(CO)_2I_3]^- + CO \longrightarrow [Rh(CMeO)(CO)_2I_3]^-$$

(23.34)

Alkyl migrations are not confined to the formation of acyl groups, and, for example, 'alkene insertion' involves the conversion of a coordinated alkene to a σ-bonded alkyl group. Equation 23.35 shows the migration of an *H atom*; related alkyl migrations occur and result in *carbon chain growth*.

(23.35)

β-Hydrogen elimination

(23.36)

The reverse of reaction 23.35 is a *β-elimination* step. It involves the transfer of a β-H atom (structure **23.36**) from the alkyl group to the metal and the conversion of the σ-alkyl group to a π-bonded alkene, i.e. a C−H bond is activated. The first step is thought to involve a cyclic intermediate **23.37** with an *agostic* M−H−C interaction.

Fig. 23.15 The distribution of products from the decarbonylation of $Mn(CO)_4(^{13}CO)\{C(O)Me\}$ provides evidence for the migration of the Me group rather than movement of a CO molecule.

Homogeneous catalysts

Many of the reaction types discussed in Section 23.7 are represented in the catalytic processes described in Chapter 26. Unsaturated (16-electron) metal centres play an important role in catalytic cycles; selected catalysts or catalyst precursors are summarized below.

Homogeneous catalyst	Catalytic application
$RhCl(PPh_3)_3$	Alkene hydrogenation
cis-$[Rh(CO)_2I_2]^-$	Monsanto acetic acid synthesis; Tennessee–Eastman acetic anhydride process
$HCo(CO)_4$	Hydroformylation; alkene isomerization
$HRh(CO)_4$	Hydroformylation (only for certain branched alkenes)[‡]
$HRh(CO)(PPh_3)_3$	Hydroformylation
$[Ru(CO)_2I_3]^-$	Homologation of carboxylic acids
$[HFe(CO)_4]^-$	Water–gas shift reaction
$(\eta^5\text{-}C_5H_5)_2TiMe_2$	Alkene polymerization
$(\eta^5\text{-}C_5H_5)_2ZrH_2$	Hydrogenation of alkenes and alkynes
$Pd(PPh_3)_4$	Many laboratory applications including the Heck reaction

[‡] $HRh(CO)_4$ is more active than $HCo(CO)_4$ in hydroformylation, but shows a lower regioselectivity (see equation 26.5 and discussion).

(23.37)

$$L_nMCH_2CH_2R \longrightarrow L_nMH(\eta^2\text{-}RCH=CH_2)$$

$$\longrightarrow L_nMH + RCH=CH_2 \qquad (23.36)$$

β-Elimination is responsible for the decomposition of some metal alkyl complexes (equation 23.36), but the reaction may be hindered or prevented by:

- steric hindrance;
- having a coordinatively *saturated* metal centre as in $(\eta^5\text{-}C_5H_5)Fe(CO)_2Et$;
- preparing an alkyl derivative which does not possess a β-hydrogen atom.

Examples of σ-bonded alkyl groups that cannot undergo β-elimination because they lack a β-H atom are Me, CH_2CMe_3, CH_2SiMe_3 and CH_2Ph. Thus, methyl derivatives cannot decompose by a β-elimination route and are usually more stable than their ethyl analogues. This does not mean that methyl derivatives are necessarily stable; the co-ordinatively unsaturated $TiMe_4$ decomposes at 233 K, but the stability can be increased by the formation of six-coordinate adducts such as $Ti(bpy)Me_4$ and $Ti(Me_2PCH_2CH_2PMe_2)Me_4$.

α-Hydrogen abstraction

Early *d*-block metal complexes containing one or two α-hydrogen atoms (see **23.36**) may undergo α-*hydrogen*

abstraction to yield *carbene* (*alkylidene*, **23.38**) or *carbyne* (*alkylidyne*, **23.39**) complexes. The solid state structure of the product of reaction 23.37 confirms differences in the Ti–C bond lengths: 225 pm for $Ti\text{-}C_{alkyl}$ and 205 pm for $Ti\text{-}C_{carbene}$.

$$L_nM=CR_2 \qquad\qquad L_nM\equiv CR$$

(23.38) (23.39)

$$TaCl_5 \xrightarrow{Zn(CH_2{}^tBu)_2} Ta(CH_2{}^tBu)_3Cl_2 \xrightarrow[-LiCl,\ -CMe_4]{LiCH_2{}^tBu}$$

$$({}^tBuCH_2)_3Ta = C \overset{\displaystyle H}{\underset{\displaystyle {}^tBu}{\big\langle}} \qquad (23.37)$$

Abstraction of a second α-H atom gives a carbyne complex (e.g. reaction 23.38). Other routes to carbenes and carbynes are described in *Section 23.12*.

$$WCl_6 \xrightarrow[-CMe_4,\ -LiCl]{LiCH_2CMe_3} W(\equiv CCMe_3)(CH_2CMe_3)_3 \qquad (23.38)$$

Summary

A basic knowledge of the reaction types described in this section allows us to proceed to a discussion of the chemistry of selected organometallic complexes and (in *Chapter 26*) catalysis. Oxidative additions and alkyl migrations in particular are very important in the catalytic processes used in the manufacture of many organic chemicals; selected important organometallic compounds used as catalysts are summarized in *Box 23.3*.

23.8 Metal carbonyls: selected reactions

Reactions 23.16–23.21 illustrated conversions of neutral carbonyl compounds to carbonylate anions. Reduction by Na is typically carried out using Na/Hg amalgam; with Na in liquid NH_3, highly reactive anions can be formed (equations 23.39 and 23.40).

$$Cr(CO)_4(Me_2NCH_2CH_2NMe_2\text{-}N,N')$$

$$\xrightarrow{\text{Na, liquid } NH_3} Na_4[Cr(CO)_4] \qquad (23.39)$$

$$M_3(CO)_{12} \xrightarrow[\text{low temp.}]{\text{Na, liquid } NH_3} Na_2[M(CO)_4]$$

$$M = Ru, Os \qquad (23.40)$$

The IR spectra (see *Section 23.2*) of highly charged anions exhibit absorptions for the terminal CO ligands in regions usually characteristic of bridging carbonyls, e.g. 1680 and $1471\,cm^{-1}$ for $[Mo(CO)_4]^{4-}$.

The action of alkali on $Fe(CO)_5$ (equation 23.41) gives $[HFe(CO)_4]^-$ (**23.40**); nucleophilic attack by $[OH]^-$ on a CO ligand is followed by Fe–H bond formation and elimination of CO_2. The $[HFe(CO)_4]^-$ ion has a variety of synthetic uses.

$$Fe(CO)_5 + 3NaOH \xrightarrow{H_2O} Na[HFe(CO)_4] + Na_2CO_3 + H_2O$$
$$(23.41)$$

(23.40)

Hydrido ligands can be introduced by various routes including protonation (equations 23.2 and 23.42), reaction with H_2 (reactions 23.43 and 23.44) and action of $[BH_4]^-$ (reactions 23.45 and 23.46).

$$Na[Mn(CO)_5] \xrightarrow{H_3PO_4, \text{ in THF}} HMn(CO)_5 \qquad (23.42)$$

$$Mn_2(CO)_{10} + H_2 \xrightarrow{200\,\text{bar},\,470\,K} 2HMn(CO)_5 \qquad (23.43)$$

$$Ru_3(CO)_{12} + H_2 \xrightarrow{\text{in boiling octane}} (\mu\text{-}H)_4Ru_4(CO)_{12}$$
$$(23.44)$$

$$Cr(CO)_6 \xrightarrow{Na[BH_4]} [(OC)_5Cr(\mu\text{-}H)Cr(CO)_5]^- \qquad (23.45)$$

$$Ru_3(CO)_{12} \xrightarrow{Na[BH_4] \text{ in THF}} [HRu_3(CO)_{11}]^- \qquad (23.46)$$
$$(\mathbf{23.41})$$

(23.41)

Reactions 23.47–23.49 illustrate preparations of selected metal carbonyl halides (see *Section 23.9*) from binary carbonyls.

$$Fe(CO)_5 + I_2 \longrightarrow Fe(CO)_4I_2 + CO \qquad (23.47)$$

$$Mn_2(CO)_{10} + X_2 \longrightarrow 2Mn(CO)_5X \qquad X = Cl, Br, I \quad (23.48)$$

$$M(CO)_6 + [Et_4N]X \longrightarrow [Et_4N][M(CO)_5X] + CO$$

$$M = Cr, Mo, W; \ X = Cl, Br, I \qquad (23.49)$$

Large numbers of derivatives are formed by displacement of CO by other ligands (see equations 23.23–23.26 and discussion). Whereas substitution by *tertiary* phosphine ligands gives terminal ligands, the introduction of a *secondary* or *primary* phosphine into a multinuclear carbonyl complex creates the possibility of oxidative addition of a P–H bond to a second metal centre and the formation of a bridging phosphido ligand (reaction 23.50).

(23.50)

We saw earlier that CO displacement can be carried out photolytically or thermally, and that activation of the starting compound (as in reaction 23.23) may be necessary. In multinuclear compounds, activation of one site can control the degree of substitution, e.g. $Os_3(CO)_{11}(NCMe)$ is used as an *in situ* intermediate during the formation of mono-substituted derivatives (equation 23.51).

$$Os_3(CO)_{12} \xrightarrow{MeCN, Me_3NO} Os_3(CO)_{11}(NCMe)$$

$$\xrightarrow{L \ (\text{e.g. } PR_3)} Os_3(CO)_{11}L \qquad (23.51)$$

In the first step of the reaction, Me_3NO oxidizes CO to CO_2, liberation of which leaves a vacant coordination site that is occupied temporarily by the labile MeCN ligand. This method can be applied to higher nuclearity clusters to achieve control over otherwise complex reactions.

Displacement of CO by a nitrosyl ligand (see end of *Section 20.4*) alters the electron count and, for an 18-electron centre to be retained, one-for-one ligand substitution cannot occur. Reaction 23.52 shows the conversion of octahedral $Cr(CO)_6$ to tetrahedral $Cr(NO)_4$ in which NO is a 3-electron donor.

$$Cr(CO)_6 + \text{excess NO} \longrightarrow Cr(NO)_4 + 6CO \qquad (23.52)$$

Table 23.6 Selected properties of $HMn(CO)_5$, $H_2Fe(CO)_4$ and $HCo(CO)_4$.

Property	$HMn(CO)_5$	$H_2Fe(CO)_4$	$HCo(CO)_4$
Physical appearance at 298 K	Colourless liquid	Yellow liquid	Yellow liquid
Stability	Stable up to 320 K	Dec. ≥ 253 K	Dec. >247 K (mp)
pK_a values	15.1	p$K_a(1) = 4.4$;	<0.4
		p$K_a(2) = 14.0$	
$\delta\ ^1H$	-10.7	-11.2	-7.9

Reactions of metal carbonyls with unsaturated organic ligands are discussed in later sections.

23.9 Metal carbonyl hydrides and halides

Methods of preparing selected hydrido complexes were given in equations 23.2 and 23.42–23.46. Selected properties of the mononuclear complexes $HMn(CO)_5$, $H_2Fe(CO)_4$ and $HCo(CO)_4$ are give in Table 23.6. Metal hydrides play an important role in organometallic chemistry, and scheme 23.53 illustrates some ligand transformations involving M—H bonds; $HCo(CO)_4$ is an industrial catalyst (see *Section 26.3*).

$$\text{(23.53)}$$

Methods of forming carbonyl halides include starting from binary metal carbonyls (equations 23.47–23.49) or metal halides (equations 23.54–23.56).

$$RhCl_3 \cdot xH_2O \xrightarrow{\text{CO, 373 K}} (CO)_2Rh(\mu\text{-Cl})_2Rh(CO)_2 \quad \text{(23.54)}$$

Vaska's compound

$$\text{(23.55)}$$

$$\text{(23.56)}$$

The 16-electron halide complexes *cis*-$[Rh(CO)_2I_2]^-$ and *trans*-$[Ir(CO)Cl(PPh_3)_2]$ (Vaska's compound) undergo many oxidative addition reactions and have important catalytic applications (see *Chapter 26*). The product of reaction 23.56 is a catalyst precursor for alkene hydrogenation. Vaska's compound readily takes up O_2 to give the peroxo complex **23.42**.

$$\text{(23.42)}$$

23.10 Alkyl, aryl, alkene and alkyne complexes

σ-Bonded alkyl and aryl ligands

Simple σ-bonded organic derivatives of low oxidation state *d*-block metals are generally more reactive than analogous main group metal species. The origin is kinetic rather than thermodynamic: e.g. the $Ti-C_{Me}$ bond is stronger than $Pb-C_{Me}$ or $Sn-C_{Me}$, but the availability of vacant $3d$ atomic orbitals in titanium alkyl complexes means that they readily undergo β-elimination to give alkene complexes (see *Section 23.7*).

Alkyl and aryl derivatives can be made by reactions such as 23.57–23.62, the latter being an example of an oxidative addition to a 16-electron complex.

$$TiCl_4 + 4LiMe \xrightarrow{\text{in Et}_2O,\ 193\ K} TiMe_4 + 4LiCl \quad \text{(23.57)}$$

$$WCl_6 + 3Al_2Me_6 \longrightarrow WMe_6 + 3Al_2Me_4Cl_2 \quad \text{(23.58)}$$

$$Na_2[Fe(CO)_4] + EtBr \longrightarrow Na[Fe(CO)_4Et] + NaBr \quad \text{(23.59)}$$

$$Na[Mn(CO)_5] + PhCH_2Cl \longrightarrow Mn(CO)_5(CH_2Ph) + NaCl \quad \text{(23.60)}$$

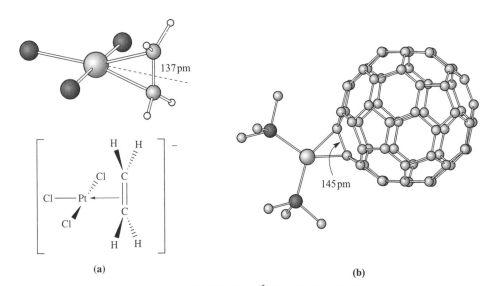

Fig. 23.16 (a) The structure of the anion in Zeise's salt, $K[PtCl_3(\eta^2\text{-}C_2H_4)]$. The Pt(II) centre can be regarded as being square planar as indicated in the schematic representation [neutron diffraction: R.A. Love *et al.* (1975) *Inorg. Chem.*, vol. 14, p. 2653]. (b) The structure of $Pd(\eta^2\text{-}C_{60})(PPh_3)_2$; for clarity, only the *ipso*-C atoms of the Ph rings are shown [X-ray diffraction: V.V. Bashilov *et al.* (1993) *Organometallics*, vol. 12, p. 991]. Colour code: Pt and Pd, blue; C, grey; Cl and P, black.

$$Li[Mn(CO)_5] + PhC(O)Cl \xrightarrow{-LiCl} Mn(CO)_5\{C(O)Ph\}$$

$$\xrightarrow{\Delta} Mn(CO)_5Ph + CO \quad (23.61)$$

$$cis\text{-}[Rh(CO)_2I_2]^- + MeI \longrightarrow mer\text{-}[Rh(CO)_2I_3Me]^- \quad (23.62)$$

Hexamethyltungsten was the first example of a discrete trigonal prismatic complex (*Figure 19.6a*); it is highly reactive in air and is potentially explosive.

Alkene ligands

Alkene complexes are often made by displacement of CO or halide ion by an alkene. The formation of *Zeise's salt*, $K[PtCl_3(\eta^2\text{-}C_2H_4)]$ (reaction 23.63) is catalyzed by $SnCl_2$ with $[PtCl_3(SnCl_3)]^{2-}$ being the intermediate. The $[PtCl_3(\eta^2\text{-}C_2H_4)]^-$ ion (Figure 23.16a) contains a square planar (or pseudo-square planar) Pt(II) centre and in the solid state, the ethene ligand lies perpendicular to the coordination 'square plane', thereby minimizing steric interactions.

$$K_2[PtCl_4] + H_2C=CH_2$$

$$\xrightarrow{SnCl_2} K[PtCl_3(\eta^2\text{-}C_2H_4)] + KCl \quad (23.63)$$

Addition to 16-electron metal complexes is exemplified by reaction 23.64; ethene readily dissociates from $Ir(CO)Cl(\eta^2\text{-}C_2H_4)(PPh_3)_2$, but the related complex $Ir(CO)Cl(\eta^2\text{-}C_2(CN)_4)(PPh_3)_2$ is very stable.

$$trans\text{-}Ir(CO)Cl(PPh_3)_2 + R_2C=CR_2$$

$$\longrightarrow Ir(CO)Cl(\eta^2\text{-}C_2R_4)(PPh_3)_2 \quad (23.64)$$

Recent additions to the family of alkene complexes are fullerene derivatives such as $Rh(CO)(\eta^2\text{-}C_{60})(H)(PPh_3)_2$ and $Pd(\eta^2\text{-}C_{60})(PPh_3)_2$ (Figure 23.16b). The C_{60} cage (see *Section 13.4*) functions as a polyene with localized C=C bonds, and in $C_{60}\{Pt(PEt_3)_2\}_6$, six C=C bonds (remote

from one another) in the C_{60} cage have undergone addition. Reaction 23.65 illustrates C_{60}-for-ethene substitution (the 16-electron centre is retained), and reaction 23.66 shows addition to Vaska's compound (a 16 to 18-electron conversion).

$$Pt(\eta^2\text{-}C_2H_4)(PPh_3)_2 + C_{60} \xrightarrow{-C_2H_4} Pt(\eta^2\text{-}C_{60})(PPh_3)_2 \quad (23.65)$$

$$trans\text{-}Ir(CO)Cl(PPh_3)_2 + C_{60}$$

$$\longrightarrow Ir(CO)Cl(\eta^2\text{-}C_{60})(PPh_3)_2 \quad (23.66)$$

Reactions of alkenes with metal carbonyl clusters may give simple substitution products such as $Os_3(CO)_{11}(\eta^2\text{-}C_2H_4)$ or may involve oxidative addition of one or more C–H bonds. Reaction 23.67 illustrates the reaction of $Ru_3(CO)_{12}$ with $RHC=CH_2$ (R = alkyl) to give isomers of $H_2Ru_3(CO)_9(RCCH)$ in which the organic ligand acts as a 4-electron donor (one π- and two σ-interactions).

(23.67)

In solution, alkene complexes are often fluxional, with rotation occurring as shown in Figure 23.17. The model compound in the figure contains a mirror plane passing through M, L and L′. The limiting *low* temperature ^{1}H spectrum shows one resonance for H_1 and H_4, and another due to H_2 and H_3, i.e. a static picture of the molecule. On raising the temperature, the molecule gains sufficient energy for the alkene to rotate and the limiting *high* temperature spectrum contains one resonance since H_1, H_2, H_3 and H_4 become equivalent on the NMR timescale. In $(\eta^5\text{-Cp})\text{Rh}(\eta^2\text{-C}_2\text{H}_4)_2$, two alkene proton signals are observed as 233 K (the different H environments are blue and black respectively in **23.43**); at 373 K, the proton environments become equivalent on the NMR spectroscopic timescale as each alkene ligand rotates about the metal–ligand coordinate bond.

Fig. 23.18 The structure (X-ray diffraction) of $\text{Co}_2(\text{CO})_6(\text{C}_2\text{Ph}_2)$ [D. Gregson *et al.* (1983) *Acta Crystallogr., Sect. C*, vol. 39, p. 1024]. Hydrogen atoms are omitted for clarity; colour code: Co, blue; C, grey; O, pale grey.

(23.43)

Coordinated alkenes may be displaced by other ligands (equation 23.65). Unlike free alkenes, which undergo electrophilic additions, *coordinated* alkenes are more susceptible to nucleophilic attack and many reactions of catalytic importance involve this pathway (see *Chapter 26*). Reaction 23.68 shows that addition of a nucleophile, R⁻, leads to a σ-bonded complex; the mechanism may involve direct attack at one alkene C atom, or attack at the $M^{\delta+}$ centre followed by alkyl migration (see *Section 23.7*).

Alkyne ligands

Many mono- and polynuclear organometallic complexes involving alkyne ligands are known. An alkyne RC≡CR has *two* fully occupied π-MOs and is potentially a 2- or 4-electron donor. The bonding in a monometallic alkyne complex can be described in a similar manner to that in an alkene complex (see *Section 23.2*), but allowing for the participation of the two orthogonal π-MOs. A typical C≡C bond length in a free alkyne is 120 pm and, in complexes, this lengthens to ≈124–137 pm depending on the mode of bonding. In **23.44**, the C–C bond length (124 pm) is consistent with a weakened triple bond; the alkyne lies perpendicular to the PtCl₃-plane and occupies one site in the square planar coordination sphere of the Pt(II) centre. A similar example is $[\text{PtCl}_3(\eta^2\text{-C}_2\text{Ph}_2)]^-$ (equation 23.69). In **23.45**, the alkyne acts as a 4-electron donor, forming a metallacycle; the C–C bond length (132 pm) is consistent with a double bond. A decrease in the alkyne C–C–C$_R$ bond angle accompanies the change in bonding mode on going from **23.44** to **23.45**. The addition of an alkyne to $\text{Co}_2(\text{CO})_8$ (equation 23.70) results in the formation of an Co_2C_2 cluster (Figure 23.18) in which the alkyne C–C bond is lengthened to 136 pm.

(23.68)

$$[\text{Pt}_2\text{Cl}_6]^{2-} + 2\text{PhC≡CPh} \longrightarrow 2[\text{PtCl}_3(\eta^2\text{-C}_2\text{Ph}_2)]^- \qquad (23.69)$$

Fig. 23.17 Schematic representation of the rotation of an $\eta^2\text{-C}_2\text{H}_4$ ligand in a complex MLL′$(\eta^2\text{-C}_2\text{H}_4)$. The complex is viewed along a line connecting the centre of the C–C bond and the M atom; the M atom is shown in blue. Because L and L′ are *different*, rotation of the alkene interchanges the environments of H_1 and H_4 with H_2 and H_3.

$$Co_2(CO)_8 + PhC\equiv CPh \longrightarrow Co_2(CO)_6(C_2Ph_2) + 2CO \tag{23.70}$$

L = 4-MeC$_6$H$_4$NH$_2$

(23.44) (23.45)

The reactions between alkynes and multinuclear metal carbonyls give various product-types, with alkyne coupling and alkyne–CO coupling often being observed, e.g. reaction 23.71 in which the organic ligand in the product is a 6-electron donor (2 σ and 2 π-interactions). Predicting the outcome of such reactions is difficult.

$$Fe_2(CO)_9 + 2MeC\equiv CMe \xrightarrow{-2CO} (OC)_3Fe \longrightarrow Fe(CO)_3 \tag{23.71}$$

In solution, π-bonded alkynes of the type in structure **23.44** undergo rotations analogous to those of alkenes.

23.11 Allyl and buta-1,3-diene complexes

Allyl and related ligands

π-Allyl and related complexes can be prepared by reactions such as 23.72–23.76; the last two reactions illustrate formation of allyl ligands by deprotonation of coordinated propene, and protonation of coordinated buta-1,3-diene respectively. Reactions 23.73 and 23.74 are examples of pathways that go via σ-bonded intermediates (e.g. **23.46**) which eliminate CO.

(23.46)

$$NiCl_2 + 2C_3H_5MgBr$$
$$\xrightarrow{\text{in Et}_2O, \ 263\,K} Ni(\eta^3\text{-}C_3H_5)_2 + 2MgBrCl \tag{23.72}$$

$$Na[Mn(CO)_5] + H_2C=CHCH_2Cl$$
$$\longrightarrow Mn(\eta^3\text{-}C_3H_5)(CO)_4 + CO + NaCl \tag{23.73}$$

$$Na[(\eta^5\text{-}Cp)Mo(CO)_3] + H_2C=CHCH_2Cl$$
$$\xrightarrow{h\nu,\ \text{in THF}} (\eta^5\text{-}Cp)Mo(\eta^3\text{-}C_3H_5)(CO)_2 + NaCl + CO \tag{23.74}$$

$$[PdCl_4]^{2-} \xrightarrow{MeCH=CH_2}$$

(23.75)

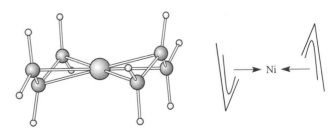

(23.76)

In Figure 23.8 we showed the three π-MOs that the π-allyl ligand uses in bonding to a metal centre. In Mo(η^3-C$_3$H$_5$)(η^4-C$_4$H$_6$)(η^5-C$_5$H$_5$) (Figure 23.6b) the central and two outer Mo–C bond lengths in the Mo(η^3-C$_3$) unit are different (221 and 230 pm respectively). This is a typical observation, e.g. 198 and 203 pm respectively for the central and outer Ni–C bonds in Ni(η^3-C$_3$H$_5$)$_2$ (Figure 23.19). In the latter, the two allyl ligands are staggered. In Figures 23.6b and 23.19, note the orientations of the H atoms with respect to the metal centres; the two H atoms in each terminal CH$_2$ group of an η^3-C$_3$H$_5$ ligand are non-equivalent. In solution, however, they are often

Fig. 23.19 The structure (neutron diffraction at 100 K) of Ni(η^3-C$_3$H$_5$)$_2$ and a schematic representation of the complex [R. Goddard *et al.* (1985) *Organometallics*, vol. 4, p. 285]. Colour code: Ni, blue; C, grey; H, pale grey.

equivalent on the NMR spectroscopic timescale, and this can be rationalized in terms of the $\eta^3-\eta^1-\eta^3$ (i.e. $\pi-\sigma-\pi$) rearrangement shown in scheme 23.77. An $\eta^3-\eta^1$ rearrangement also features in some reactions of allyl ligands.

Bond rotation
about single bonds

(23.77)

Buta-1,3-diene and related ligands

Photolysis of $Fe(CO)_5$ with buta-1,3-diene gives complex **23.47**, an orange liquid which loses CO at room temperature

to give **23.48**, an air-stable yellow solid. The coordinated diene is difficult to hydrogenate, and does not undergo Diels–Alder reactions characteristic of conjugated dienes. Structural data for **23.48** confirm that the Fe atom is equidistant from each C atom of the ligand; bonding schemes for the metal–ligand interaction were discussed in *Section 23.2*.

Iron tricarbonyl complexes of 1,3-dienes (e.g. cyclohexa-1,3-diene) play an important role in organic synthesis; the complexes are stable under a variety of reaction conditions, and iron carbonyls are inexpensive. The $Fe(CO)_3$ group acts as a protecting group for the diene functionality (e.g. against additions to the C=C bonds), allowing reactions to

---

BOX 23.4   **APPLICATIONS**

Molecular wires

As we move into the new millennium, the ability of chemists to design molecules for electronic applications is becoming a reality: 'molecular wires' are a topical area of research. A molecular wire is a molecule capable of transporting charge carriers from one end of the wire to the other. Molecules with extended conjugated systems are prime candidates for molecular wires since the conjugation provides the necessary electronic communication between atomic centres. The molecule must also possess a small band gap.

(For details on charge carriers and band gaps, see *Section 5.9*.)

Although commercial applications are still future goals, much progress has been made in the design of potential molecular wires. Molecules so far studied have included organic molecules with conjugated alkyne functionalities, porphyrins connected by alkyne units, chains of connected thiophenes, and a number of organometallic complexes. Examples of the latter are shown below:

n = 2, 3, 4, 6, 8, 10

For a detailed review which also assesses the means of testing the properties of molecular wires, see: M.D. Ward (1996) *Chemistry & Industry*, p. 568 – 'Current developments in molecular wires'.

be carried out on other parts of the organic molecule as illustrated by reaction 23.78.

$$(23.78)$$

The presence of the $Fe(CO)_3$ group also permits reactions with *nucleophiles* to be carried out at the diene functionality with control of the stereochemistry, the nucleophile being able to attack only on the side of the coordinated diene *remote* from the metal centre. The organic ligand can be removed in the final step of the reaction.[†]

23.12 Carbene and carbyne complexes

In *Section 23.7*, we introduced carbene and carbyne complexes when we discussed α-hydrogen abstraction. Equations 23.37 and 23.38 exemplified methods of preparation. Carbenes can also be made by nucleophilic attack on a carbonyl C atom followed by alkylation (equation 23.79).

$$(23.79)$$

Compounds of the type formed in reactions such as 23.79 are called *Fischer-type carbenes*; they possess a low oxidation state metal, a heteroatom (O in this example) and an *electrophilic* carbene centre (i.e. subject to attack by nucleophiles, e.g. reaction 23.80). Resonance pair **23.49** gives a bonding description for a Fischer-type carbene complex.

$$(23.80)$$

$$(23.49)$$

$$(23.50)$$

In contrast, *Schrock-type carbenes* are made by reactions such as 23.37, contain an early *d*-block metal in a high oxidation state, and show nucleophilic character (i.e. susceptible to attack by electrophiles, e.g. reaction 23.81). Resonance pair **23.50** describes a Schrock-type carbene complex.

$$(\eta^5\text{-Cp})_2\text{MeTa}{=}\text{CH}_2 + \text{AlMe}_3$$
$$\longrightarrow (\eta^5\text{-Cp})_2\text{Me}\overset{+}{\text{Ta}}{-}\text{CH}_2\overset{-}{\text{Al}}\text{Me}_3 \qquad (23.81)$$

The $M-C_{carbene}$ bonds in both Fischer- and Schrock-type complexes are *longer* than typical $M-C_{CO(term)}$ bonds, but *shorter* than typical $M-C$ single bonds, e.g. in $(OC)_5Cr{=}C(OMe)Ph$, $Cr-C_{carbene} = 204\,pm$ and $Cr-C_{CO} = 188\,pm$. This implies some degree of $(d-p)\pi$-character as indicated by resonance structures **23.49** and **23.50**. The π-system can be extended to the heteroatom in the Fischer-type system as shown in diagram **23.51**.

$$(23.51)$$

One route to a carbyne (alkylidyne) complex was reaction 23.38; equation 23.82 illustrates the initial method of Fischer. The abstraction of an α-H atom from a Schrock-type carbene yields the corresponding carbyne complex (equation 23.83).

[†] For detailed discussion of the use of metal carbonyl 1,3-diene complexes in organic synthesis, see: *Comprehensive Organometallic Chemistry II*, eds E.W. Abel, F.G.A. Stone and G. Wilkinson (1995), Pergamon, Oxford, vol. 12 – Chapter 6.2, W.A. Donaldson, p. 623, and Chapter 6.3, A.J. Pearson, p. 637; L.R. Cox and S.V. Ley (1998) *Chemical Society Reviews*, vol. 27, p. 301 – 'Tricarbonyl complexes: An approach to acyclic stereocontrol'.

$$(23.82)$$

$$(23.83)$$

An M—C$_{carbyne}$ bond is typically shorter than an M—C$_{CO(term)}$ bond, e.g. structure **23.52**. The multiple bonding can be considered in terms of an *sp* hybridized C$_{carbyne}$ atom, with one M—C σ-interaction (using an *sp$_z$* hybrid) and two π-interactions (using the C$_{carbyne}$ 2p_x and 2p_y atomic orbitals overlapping with the metal d_{xz} and d_{yz} atomic orbitals).

$$(23.52)$$

Alkylidyne (carbyne) complexes containing μ$_3$-CR groups interacting with a triangle of metal atoms include Co$_3$(CO)$_9$(μ$_3$-CMe) and H$_3$Ru$_3$(CO)$_9$(μ$_3$-CMe), and we considered the bonding in such compounds in terms of the

isolobal principle in *Section 23.5*. Reactions 23.84 and 23.85 illustrate methods of introducing μ$_3$-CR groups into clusters; the precursor in reaction 23.85 is unsaturated and contains an Os=Os bond which undergoes additions. The intermediate in this reaction contains a bridging carbene group which undergoes oxidative addition of a C—H bond on heating.

$$Co_2(CO)_8 \xrightarrow{MeCCl_3} Co_3(CO)_9CMe \qquad (23.84)$$

$$H_2Os_3(CO)_{10} \xrightarrow[-N_2]{CH_2N_2} H_2Os_3(CO)_{10}(\mu\text{-}CH_2)$$

$$\xrightarrow[-CO]{\Delta} H_3Os_3(CO)_9(\mu_3\text{-}CH) \qquad (23.85)$$

In mononuclear carbyne complexes, the M≡C bond undergoes addition reactions, e.g. addition of HCl and alkynes.

23.13 Complexes containing η^5-cyclopentadienyl ligands

The cyclopentadienyl ligand was discussed in *Chapter 18* and in *Sections 23.1* and *23.2*. Now we look at examples of some of its more important *d*-block metal complexes.

In a *sandwich complex*, the metal centre lies between two π-bonded hydrocarbon (or derivative) ligands. Complexes of the type (η^5-Cp)$_2$M are called *metallocenes*.

Ferrocene and other metallocenes

The best known cyclopentadienyl complex is the *sandwich compound* ferrocene, (η^5-Cp)$_2$Fe; it is a diamagnetic,

(23.53)

orange solid (mp 393 K) which obeys the 18-electron rule (scheme **23.25**). In the gas phase, the two cyclopentadienyl rings are *eclipsed* (**23.53**) but the solid exists in several phases in which the rings are co-parallel but in different orientations. Solving the structure has been hampered by disorder problems. The barrier to rotation of the two rings is low and at 298 K, there is motion even in the solid state. In derivatives of ferrocene with substituents on the Cp rings, the barrier to rotation is higher, and in $(\eta^5\text{-}C_5Me_5)_2Fe$, the two C_5 rings are staggered in both the gas and solid states. The bonding in $(\eta^5\text{-}Cp)_2Fe$ can be described in terms of the interactions between the π-MOs of the ligands (see *Box 18.2*) and the metal 3d atomic orbitals (see *problem 23.19*). Ferrocene is oxidized (e.g. by I_2 or $FeCl_3$) to the paramagnetic, blue ferrocenium ion, $[(\eta^5\text{-}Cp)_2Fe]^+$. Equation 23.86 gives E° relative to the standard hydrogen electrode, but the Fc^+/Fc couple is commonly used as a convenient internal, secondary reference electrode (i.e. E° is defined as 0 V for reference purposes).

$$[(\eta^5\text{-}Cp)_2Fe]^+ + e^- \rightleftharpoons (\eta^5\text{-}Cp)_2Fe$$
$$\underset{Fc^+}{} \qquad \qquad \underset{Fc}{}$$
$$E^\circ = +0.40 \text{ V} \qquad (23.86)$$

Metallocenes of the first row metals are known for V(II), Cr(II), Mn(II), Fe(II), Co(II) and Ni(II); reaction 23.87 is a general synthesis for all except $(\eta^5\text{-}Cp)_2V$ where the starting chloride is VCl_3. Reaction 23.88 gives an alternative synthesis for ferrocene and nickelocene.

$$MCl_2 + 2Na[Cp] \longrightarrow (\eta^5\text{-}Cp)_2M + 2NaCl \qquad (23.87)$$

$$MCl_2 + 2C_5H_6 + 2Et_2NH \longrightarrow (\eta^5\text{-}Cp)_2M + 2[Et_2NH_2]Cl$$
$$(M = Fe, Ni) \qquad (23.88)$$

The complexes $(\eta^5\text{-}Cp)_2V$ (air-sensitive, violet solid), $(\eta^5\text{-}Cp)_2Cr$ (air-sensitive, red solid), $(\eta^5\text{-}Cp)_2Mn$ (brown solid, pyrophoric when finely divided), $(\eta^5\text{-}Cp)_2Co$ (very air-sensitive, black solid) and $(\eta^5\text{-}Cp)_2Ni$ (green solid) are paramagnetic; $(\eta^5\text{-}Cp)_2Cr$ and $(\eta^5\text{-}Cp)_2Ni$ have two unpaired electrons. The 19-electron complex $(\eta^5\text{-}Cp)_2Co$ is readily oxidized to the 18-electron $[(\eta^5\text{-}Cp)_2Co]^+$, yellow salts of which are air-stable. Nickelocene is a 20-electron complex and in its reactions (Figure 23.20), often relieves this situation, forming 18-electron complexes. Manganocene, unlike the other metallocenes, is dimorphic; the room

Fig. 23.20 Selected reactions of nickelocene, $(\eta^5\text{-}Cp)_2Ni$.

temperature form is polymeric and structurally similar to $(\eta^5\text{-}Cp)_2Pb$ (Figure 18.13c), while the high temperature form is structurally related to ferrocene.

The chemistry of ferrocene dominates that of the other metallocenes; it is commercially available and large numbers of derivatives are known. The rings in $(\eta^5\text{-}Cp)_2Fe$ possess aromatic character, and selected reactions are shown in Figure 23.21. That protonation occurs at the Fe(II) centre is indicated by the appearance of a signal at δ −2.1 in the ^1H NMR spectrum of $[(\eta^5\text{-}Cp)_2FeH]^+$. The exchange of an η^5-Cp ring for an η^6-arene ligand (equation 23.89) is accompanied by a change in overall charge so that the Fe atom remains an 18-electron centre.

(23.89)

$(\eta^5\text{-}Cp)_2Fe_2(CO)_4$ and derivatives

Reactions between metal carbonyls and cyclopentadiene usually yield mixed ligand complexes, e.g. $Fe(CO)_5$ reacts with C_5H_6 to give $(\eta^5\text{-}Cp)_2Fe_2(CO)_4$. Two isomers of $(\eta^5\text{-}Cp)_2Fe_2(CO)_4$ exist, *cis* (**23.54**) and *trans* (**23.55**), and both have been confirmed in the solid state. The Fe–Fe bond length (253 pm) is consistent with a single bond giving each Fe centre 18 electrons.

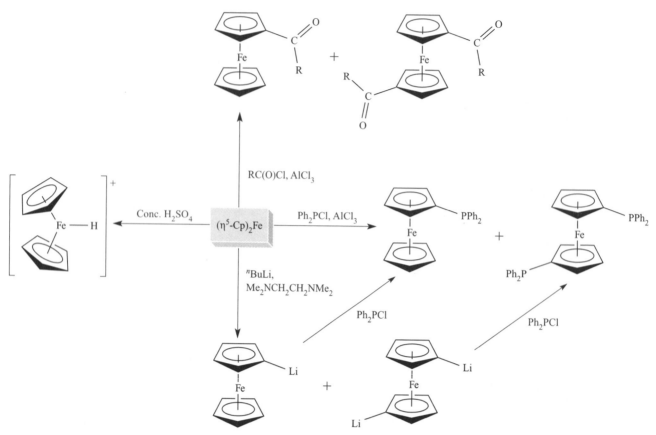

Fig. 23.21 Selected reactions of ferrocene, $(\eta^5\text{-Cp})_2\text{Fe}$.

(23.54)   (23.55)

In solution at 298 K, both the *cis*- and *trans*-forms are present and the terminal and bridging ligands exchange by an intramolecular process. Above 308 K, *cis* → *trans* isomerism occurs, probably through an unbridged intermediate; the *cis*-isomer can be obtained by crystallization at low temperatures.

The dimer $(\eta^5\text{-Cp})_2\text{Fe}_2(\text{CO})_4$ is commercially available and is a valuable starting material in organometallic chemistry. Reactions with Na or halogens cleave the Fe–Fe bond (equations 23.90 and 23.91) giving useful organometallic reagents, reactions of which are exemplified in equations 23.92–23.95.

$$(\eta^5\text{-Cp})_2\text{Fe}_2(\text{CO})_4 + 2\text{Na} \longrightarrow 2\text{Na}[(\eta^5\text{-Cp})\text{Fe}(\text{CO})_2]$$
$$(23.90)$$

$$(\eta^5\text{-Cp})_2\text{Fe}_2(\text{CO})_4 + X_2 \longrightarrow 2(\eta^5\text{-Cp})\text{Fe}(\text{CO})_2X$$
$$X = \text{Cl, Br, I} \quad (23.91)$$

$$\text{Na}[(\eta^5\text{-Cp})\text{Fe}(\text{CO})_2] + \text{RCl} \longrightarrow (\eta^5\text{-Cp})\text{Fe}(\text{CO})_2\text{R} + \text{NaCl}$$
$$\text{e.g. R = alkyl} \quad (23.92)$$

$$\text{Na}[(\eta^5\text{-Cp})\text{Fe}(\text{CO})_2]$$

$$\text{CH}_2=\text{CHCH}_2\text{Br}$$
$$-\text{NaBr}$$

$$h\nu \mid -\text{CO}$$

$$(23.93)$$

$$(\eta^5\text{-Cp})\text{Fe}(\text{CO})_2\text{Br}$$

$$\text{Na[Co(CO)4]}$$
$$-\text{NaBr}$$

$$(23.94)$$

BOX 23.6　　　　　　　　　　　　　　　　　　　　　　　　　　　　　　　　　　　　　　**APPLICATIONS**

Enantioselectivity in the preparation of the herbicide *S*-Metolachlor

Each year, Novartis manufactures ≈10 000 tonnes of the herbicide *S*-Metolachlor, the synthesis of which is shown in the scheme below. The key to enantioselectivity is the first step of *asymmetric hydrogenation*.

The hydrogenation step is catalyzed by an iridium(I) complex containing the chiral ferrocenyl bisphosphine ligand shown below. The ligand coordinates to the Ir(I) centre through the two *P*-donor atoms, and the complete catalyst

system comprises Ir(I), the ferrocenyl ligand, I^- and H_2SO_4. The 80% ee is not as high as would be required for, say, chiral drug synthesis, but is adequate for the production of a herbicide. Chiral catalysts play a vital role in directing asymmetric syntheses, and the % ee is highly sensitive to the choice of chiral ligand; 'ee' is explained in *Section 26.3*. The manufacture of *S*-Metolachlor provides an example of an industrial application of one *particular* chiral ferrocenyl bisphosphine ligand.

Further reading

More about asymmetric syntheses in Section 26.3.

A. Togni (1996) *Angewandte Chemie, International Edition in English*, vol. 35, p. 1475 – 'Planar–chiral ferrocenes: Synthetic methods and applications'.

D.L. Lewis *et al.* (1999) *Nature*, vol. 401, p. 898 – 'Influence of environmental changes on degradation of chiral pollutants in soils'.

23.14 Complexes containing η^6- and η^7-ligands

η^6-Arene ligands

Arenes such as benzene and toluene can act as 6π-electron donors as illustrated in equations 23.89 and 23.95. A wide range of arene complexes exists, and sandwich complexes

Zirconocene derivatives as catalysts

The development of Ziegler–Natta-type catalysts (see *Box 18.3* and *Section 26.6*) has, since the 1980s, included the use of zirconocene derivatives. In the presence of methylaluminoxane $[MeAl(\mu\text{-}O)]_n$ as a co-catalyst, compounds **A**, **B** and **C** (shown below) are active catalysts for propene polymerization. Note that **A** and **B** are chiral because of the relative orientations of the two halves of the organic ligand. A racemic mixture of **A** facilitates the formation of *isotactic* polypropene, while use of catalyst **C** results in *syndiotactic* polypropene (see footnote in *Section 26.6* for

definitions of syndiotactic, isotactic and atactic). If $(\eta^5\text{-}Cp)_2ZrCl_2$ is used, *atactic* polypropene is produced. Such catalysts are commercially available. The active species in the catalyst system is a cation of the general type $[Cp_2ZrR]^+$, and such cations are now being used directly as polymerization catalysts. Formed by protonolysis, oxidative Zr–R cleavage, or abstraction of R^- from Cp_2ZrR_2 (e.g. R = Me), $[Cp_2ZrR]^+$ reagents are active catalysts *without* the need for addition of the methylaluminoxane co-catalyst.

A   **B**   **C**

Zirconocene derivatives are used to catalyze a range of organic hydrogenation and C–C bond-forming reactions. In the presence of methylaluminoxane, chiral complex **A** catalyzes asymmetric hydrogenations (see *Section 26.3*), with the active species being a cationic zirconium hydrido complex.

Further reading

Encyclopedia of Reagents in Organic Synthesis (1995), ed. L.A. Paquette, Wiley, Chichester, vol. 4, p. 2445.
Comprehensive Organometallic Chemistry II (1995), eds E.W. Abel, F.G.A. Stone and G. Wilkinson, Pergamon, Oxford, vol. 4, p. 618.

---

can be made by co-condensation of metal and arene vapours (equation 23.96) or by reaction 23.97.

$$Cr(g) + 2C_6H_6(g) \xrightarrow[\text{warm to 298 K}]{\text{co-condense on to surface at 77 K;}}$$

$$(\eta^6\text{-}C_6H_6)_2Cr \quad (23.96)$$

$$\left.\begin{array}{l} 3CrCl_3 + 2Al + AlCl_3 + 6C_6H_6 \\ \quad \longrightarrow 3[(\eta^6\text{-}C_6H_6)_2Cr]^+[AlCl_4]^- \\ 2[(\eta^6\text{-}C_6H_6)_2Cr]^+ + 4[OH]^- + [S_2O_4]^{2-} \\ \quad \longrightarrow 2(\eta^6\text{-}C_6H_6)_2Cr + 2H_2O + 2[SO_3]^{2-} \end{array}\right\} \quad (23.97)$$

The group 6 metals form air-sensitive 18-electron complexes $(\eta^6\text{-}C_6H_6)_2M$ (M = Cr, Mo, W). In the solid state, the two benzene rings in $(\eta^6\text{-}C_6H_6)_2Cr$ are eclipsed (**23.56**); the C–C bonds are equal in length (142 pm) and slightly longer than in free benzene (140 pm). The bonding can be described in terms of the interaction between the π-MOs of the ligands (Figure 23.22) and the metal 3*d* atomic orbitals,

with the occupied ligand π-MOs acting as donors and the vacant MOs functioning as acceptors.

(23.56)

Surprisingly, the brown Cr complex is easily oxidized by I_2 to the 17-electron, air-stable yellow $[(\eta^6\text{-}C_6H_6)_2Cr]^+$. The ease of oxidation precludes $(\eta^6\text{-}C_6H_6)_2Cr$ from undergoing electrophilic substitution reactions; electrophiles oxidize $(\eta^6\text{-}C_6H_6)_2Cr$ to $[(\eta^6\text{-}C_6H_6)_2Cr]^+$ which does not react further. The lithiated derivative $(\eta^6\text{-}C_6H_5Li)_2Cr$ can be made by reaction of $(\eta^6\text{-}C_6H_6)_2Cr$ with ^nBuLi (compare with the lithiation of ferrocene, Figure 23.21) and is a precursor to other derivatives.

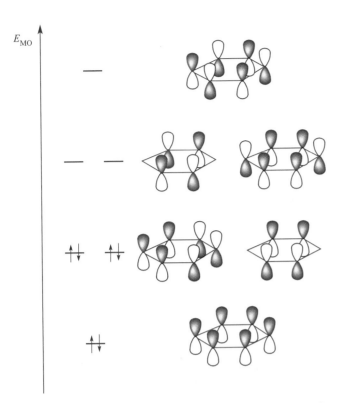

Fig. 23.22 The π-molecular orbitals of C_6H_6; the energy scale is arbitrary.

(23.57)

The reaction of $Cr(CO)_6$ or $Cr(CO)_3(NCMe)_3$ with benzene gives the *half-sandwich complex* $(\eta^6\text{-}C_6H_6)Cr(CO)_3$ (**23.57**), and related complexes can be made similarly. The $Cr(CO)_3$ unit in $(\eta^6\text{-arene})Cr(CO)_3$ complexes withdraws electrons from the arene ligand making it *less* susceptible to electrophilic attack than the free arene, but *more* susceptible to attack by nucleophiles (reaction 23.98).

$$(\eta^6\text{-}C_6H_5Cl)Cr(CO)_3 + NaOMe$$

$$\longrightarrow (\eta^6\text{-}C_6H_5OMe)Cr(CO)_3 + NaCl \qquad (23.98)$$

As in $(\eta^6\text{-}C_6H_6)_2Cr$, the benzene ligand in $(\eta^6\text{-}C_6H_6)Cr(CO)_3$ can be lithiated (equation 23.99) and then derivatized (scheme 23.100). The reactivity of half-sandwich complexes is not confined to sites within the π-bonded ligand: equation 23.101 illustrates substitution of a CO ligand for PPh_3.

$$(\eta^6\text{-}C_6H_6)Cr(CO)_3$$

$$\xrightarrow[\text{with } Me_2NCH_2CH_2NMe_2]{^nBuLi} (\eta^6\text{-}C_6H_5Li)Cr(CO)_3 \qquad (23.99)$$

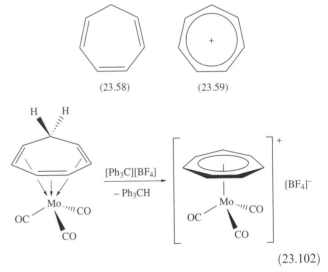

(23.100)

$$(\eta^6\text{-}C_6H_6)Cr(CO)_3 + PPh_3$$

$$\xrightarrow{h\nu} (\eta^6\text{-}C_6H_6)Cr(CO)_2(PPh_3) + CO \qquad (23.101)$$

Cycloheptatriene and derived ligands

Cycloheptatriene (**23.58**) can act as a 6π-electron donor, and in its reaction with $Mo(CO)_6$, it forms $(\eta^6\text{-}C_7H_8)Mo(CO)_3$. The solid state structure of this complex (Figure 23.23a) confirms that the ligand coordinates as a triene, the ring being folded with the CH_2 group bent away from the metal centre. Reaction 23.102 shows the abstraction of H^- from coordinated $\eta^6\text{-}C_7H_8$ to give the planar $[\eta^7\text{-}C_7H_7]^+$ ion, **23.59** (the cycloheptatrienylium cation),[†] which has an aromatic π-system and retains the ability of cycloheptatriene to act as a 6π-electron donor.

(23.58)   (23.59)

(23.102)

The planarity of the $[C_7H_7]^+$ ligand has been confirmed in the structure of the ion $[(\eta^7\text{-}C_7H_7)Mo(CO)_3]^+$ (Figure 23.23b); all the C–C bond lengths are close to 140 ppm in contrast to the variation observed in $(\eta^6\text{-}C_7H_8)Mo(CO)_3$ (Figure 23.23a).

[†] The non-systematic name for the cycloheptatrienylium cation is the *tropylium cation*.

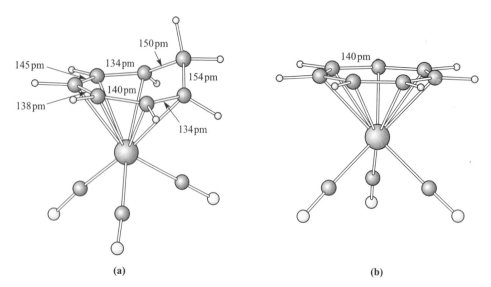

Fig. 23.23 The structures (X-ray diffraction) of (a) $[(\eta^6\text{-}C_7H_8)Mo(CO)_3]$ [J.D. Dunitz *et al.* (1960) *Helv. Chim. Acta*, vol. 43, p. 2188] and (b) $[(\eta^7\text{-}C_7H_7)Mo(CO)_3]^+$ in the $[BF_4]^-$ salt [G.R. Clark *et al.* (1973) *J. Organomet. Chem.*, vol. 50, p. 185]. Colour code: Mo, blue; C, grey; O and H, pale grey.

In the complex $(\eta^4\text{-}C_7H_8)Fe(CO)_3$, cycloheptatriene acts as a diene, giving the Fe(0) centre its required 18 electrons. Equation 23.103 shows that deprotonation generates a coordinated $[C_7H_7]^-$ ligand which bonds in an η^3 manner, allowing the metal to retain 18 electrons. At room temperature, the $[C_7H_7]^-$ ligand is fluxional, and on the NMR spectroscopic timescale, the $Fe(CO)_3$ unit effectively 'visits' every carbon atom.

23.15 Complexes containing the η^4-cyclobutadiene ligand

Cyclobutadiene, C_4H_4, is anti-aromatic (i.e. it does not have $4n + 2$ π-electrons) and readily polymerizes. However, it can be stabilized by coordination to a low oxidation state metal centre. Yellow crystalline $(\eta^4\text{-}C_4H_4)Fe(CO)_3$ is made by reaction 23.104 and its solid state structure (Figure 23.24a) shows that (in contrast to the free ligand in which the double bonds are localized) the C−C bonds in coordinated C_4H_4 are of equal length.

$$(23.103)$$

$$(23.104)$$

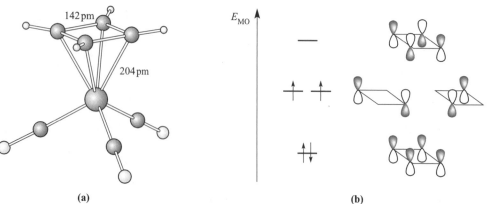

Fig. 23.24 (a) The structure (X-ray diffraction) of $(\eta^4\text{-}C_4H_4)Fe(CO)_3$ [P.D. Harvey *et al.* (1988) *Inorg. Chem.*, vol. 27, p. 57], and (b) the π-molecular orbitals of C_4H_4 in which the ligand geometry is as in its complexes, i.e. a *square* C_4-framework. Colour code: Fe, blue; C, grey; O and H, pale grey.

A C_4H_4 ligand with the geometry found in its complexes, i.e. a *square* C_4-framework, has the π-MOs shown in Figure 23.24b and is *paramagnetic*. However, $(\eta^4\text{-}C_4H_4)Fe(CO)_3$ is *diamagnetic* and this provides evidence for pairing of electrons between ligand and metal: a C_{3v} $Fe(CO)_3$ fragment also has two unpaired electrons (Figure 23.14).

Cyclobutadiene complexes can also be formed by the cycloaddition of alkynes as in reaction 23.105.

$$2PdCl_2(NCPh)_2 + 4PhC\equiv CPh$$

$$\longrightarrow (\eta^4\text{-}C_4Ph_4)ClPd(\mu\text{-}Cl)_2PdCl(\eta^4\text{-}C_4Ph_4) \quad (23.105)$$

In its reactions, *coordinated* cyclobutadiene exhibits aromatic character, undergoing electrophilic substitution, e.g. Friedel–Crafts acylation. A synthetic application of $(\eta^4\text{-}C_4H_4)Fe(CO)_3$ in organic chemistry is as a stable source of cyclobutadiene; oxidation releases the ligand making it available for reaction with, for example, alkynes as in scheme 23.106.

Dewar-benzene
derivative

(23.106)

Summary

This chapter has introduced the fundamentals of organometallic chemistry of low oxidation state *d*-block metals: bonding principles, structures, reactivity and uses of spectroscopic methods to probe ligand environment and fluxionality. The reaction types detailed in *Section 23.7* and the ability of some metal centres to exist as either 16- or 18- electron centres are crucial to the use of some organometallic compounds as homogeneous catalysts; this subject is developed in *Chapter 26*.

The following terms were introduced in this chapter.
Do you know what they mean?

❑ organometallic compound
❑ hapticity of a ligand
❑ Dewar–Chatt–Duncanson model
❑ synergic effect
❑ Tolman cone angle
❑ 18-electron rule
❑ condensed polyhedral cluster

❑ isolobal principle
❑ polyhedral skeletal electron pair theory (PSEPT)
❑ capping principle (within Wade's rules)
❑ total valence electron counts (for metal frameworks)
❑ reductive carbonylation
❑ ligand substitution
❑ oxidative addition
❑ orthometallation
❑ reductive elimination
❑ alkyl and hydrogen migration
❑ CO insertion
❑ β-hydrogen elimination
❑ agostic $M-H-C$ interaction
❑ α-hydrogen abstraction
❑ carbene (alkylidene)
❑ carbyne (alkylidyne)
❑ sandwich complex
❑ metallocene
❑ half-sandwich complex

Suggested further reading

M. Bochmann (1994) *Organometallics 1: Complexes with Transition Metal–Carbon σ-Bonds*, Oxford University Press, Oxford – This and the companion book (see below) give a concise introduction to organometallic chemistry.

M. Bochmann (1994) *Organometallics 2: Complexes with Transition Metal–Carbon π-Bonds*, Oxford University Press, Oxford – see above.

Ch. Elschenbroich and A. Salzer (1992) *Organometallics*, 2nd edn, Wiley-VCH, Weinheim – An excellent text which covers both main group and transition metal organometallic chemistry.

R.B. King, ed. (1994) *Encyclopedia of Inorganic Chemistry*, Wiley, Chichester – The organometallic chemistry of each *d*-block metal discussed is covered in separate articles under the element name.

S. Komiya, ed. (1997) *Synthesis of Organometallic Compounds: A Practical Guide*, Wiley-VCH, Weinheim – A book emphasizing methods of synthesis and handling of air-sensitive compounds.

P.L. Pauson (1993) 'Organo-iron compounds' in *Chemistry of Iron*, ed. J. Silver, Blackie Academic, Glasgow, p. 73 – A good summary of ferrocene chemistry and of other organo-iron complexes.

A. Togni and R.L. Halterman, eds (1998) *Metallocenes*, Wiley-VCH, Weinheim – A two-volume book covering synthesis, reactivity and applications of metallocenes.

A. Togni and T. Hayashi, eds (1995) *Ferrocenes. Homogeneous Catalysis, Organic Synthesis, Materials Science*, Wiley-VCH, Weinheim – Excellent survey of ferrocene applications.

G. Wilkinson, F.G.A. Stone and E.W. Abel, eds (1982) *Comprehensive Organometallic Chemistry*, Pergamon, Oxford – A series of volumes reviewing the literature up to ≈1981.

G. Wilkinson, F.G.A. Stone and E.W. Abel, eds (1995) *Comprehensive Organometallic Chemistry II*, Pergamon, Oxford – An update of the previous set of volumes which provides an excellent entry into the literature.

Organometallic clusters of the *d*-block metals

C.E. Housecroft (1996) *Metal–Metal Bonded Carbonyl Dimers and Clusters*, Oxford University Press, Oxford.

D.M.P. Mingos and D.J. Wales (1990) *Introduction to Cluster Chemistry*, Prentice Hall, Englewood Cliffs, NJ.

D.F. Shriver, H.D. Kaesz and R.D. Adams, eds. (1990) *The Chemistry of Metal Cluster Complexes*, VCH, New York.

M.D. Vargas and J.N. Nicholls (1986) 'High nuclearity clusters: Their syntheses and reactivity' in *Advances in Inorganic Chemistry and Radiochemistry*, vol. 30, p. 123.

Fluxionality in organometallic complexes and uses of NMR spectroscopy

J.W. Faller (1994) 'Stereochemical nonrigidity of organometallic complexes' in *Encyclopedia of Inorganic Chemistry*, ed. R.B. King, Wiley, Chichester, vol. 7, p. 3914.

B.E. Mann (1982) 'Non-rigidity in organometallic compounds' in *Comprehensive Organometallic Chemistry*, eds G. Wilkinson, F.G.A. Stone and E.W. Abel, Pergamon, Oxford, vol 3, p. 89.

W. von Phillipsborn (1999) *Chemical Society Reviews*, vol. 28, p. 95 – 'Probing organometallic structure and reactivity by transition metal NMR spectroscopy'.

Problems

23.1 (a) Explain the meaning of the following notations: μ-CO; μ_4-PR; η^5-C_5Me_5; η^4-C_6H_6; μ_3-H. (b) Why can the cyclopentadienyl and CO ligands be regarded as being versatile in their bonding modes? (c) Is PPh_3 a 'versatile ligand'?

23.2 What is a synergic effect, and how does it relate to metal–carbonyl bonding?

23.3 Comment on the following:

(a) Infrared spectra of $[V(CO)_6]^-$ and $Cr(CO)_6$ show absorptions at 1859 and 1981 cm^{-1} respectively assigned to ν_{CO}, and 460 and 441 cm^{-1} assigned to ν_{MC}.

(b) The Tolman cone angles of PPh_3 and $P(4\text{-}MeC_6H_4)_3$ are both 145°, but that of $P(2\text{-}MeC_6H_4)_3$ is 194°.

(c) Before reaction with PPh_3, $Ru_3(CO)_{12}$ may be treated with Me_3NO in MeCN.

(d) In the complex $[Os(en)_2(\eta^2\text{-}C_2H_4)(\eta^2\text{-}C_2H_2)]^{2+}$ the $Os-C_{ethyne}-H_{ethyne}$ bond angle is 127°.

23.4 (a) Draw a structure corresponding to the formula $[(CO)_2Ru(\mu\text{-H})(\mu\text{-CO})(\mu\text{-Me}_2PCH_2PMe_2)_2Ru(CO)_2]^+$. (b) The ^1H NMR spectrum of this complex contains a quintet centred at δ −10.2. Assign the signal and explain the origin of the observed multiplicity.

23.5 Rationalize the following observations.

(a) On forming $[IrBr(CO)(\eta^2\text{-}C_2(CN)_4)(PPh_3)_2]$, the C−C bond in $C_2(CN)_4$ lengthens from 135 to 151 pm.

(b) During the photolysis of $Mo(CO)_5(THF)$ with PPh_3, a signal in the ^{31}P NMR spectrum at δ −6 disappears and is replaced by one at δ +37.

(c) On going from $Fe(CO)_5$ to $Fe(CO)_3(PPh_3)_2$, absorptions in the IR spectrum at 2025 and 2000 cm^{-1} are replaced by bands at 1944, 1886 and 1881 cm^{-1}.

23.6 Draw out a bonding scheme (similar to that in Figure 23.7b) for the interaction of an η^3-allyl ligand with a low oxidation state metal centre.

23.7 Show that the metal centres in the following complexes obey the 18-electron rule:

(a) $(\eta^5\text{-Cp})Rh(\eta^2\text{-}C_2H_4)(PMe_3)$;

(b) $(\eta^3\text{-}C_3H_5)_2Rh(\mu\text{-Cl})_2Rh(\eta^3\text{-}C_3H_5)_2$;

(c) $Cr(CO)_4(PPh_3)_2$;

(d) $Fe(CO)_3(\eta^4\text{-}CH_2CHCHCH_2)$;

(e) $Fe_2(CO)_9$;

(f) $[HFe(CO)_4]^-$;

(g) $[(\eta^5\text{-Cp})CoMe(PMe_3)_2]^+$;

(h) $RhCl(H)_2(\eta^2\text{-}C_2H_4)(PPh_3)_2$.

23.8 Reaction of $Fe(CO)_5$ with $Na_2[Fe(CO)_4]$ in THF gives a salt $Na_2[A]$ and CO. The Raman spectrum of $[Et_4N]_2[A]$ shows an absorption at 160 cm^{-1} assigned to an unbridged Fe−Fe bond. Suggest an identity and structure for $[A]^{2-}$.

23.9 Suggest possible structures for the cation in $[Fe_2(NO)_6][PF_6]_2$ and state how you would attempt to distinguish between them experimentally.

23.10 Comment on the following observations:

(a) In the IR spectrum of free $MeCH=CH_2$, $\bar{\nu}_{C=C}$ comes at 1652 cm^{-1}, but in the complex $K[PtCl_3(\eta^2\text{-MeCH=CH}_2)]$, the corresponding absorption is at 1504 cm^{-1}.

(b) At 303 K, the ^1H NMR spectrum of $(\eta^5\text{-Cp})(\eta^1\text{-Cp})Fe(CO)_2$ shows two singlets.

23.11 Use Wade's rules (PSEPT) to suggest structures for $Os_7(CO)_{21}$ and $[Os_8(CO)_{22}]^{2-}$.

23.12 For each of the following clusters, confirm that the total valence electron count is consistent with the metal cage framework adopted: (a) $[Ru_6(CO)_{18}]^{2-}$, octahedron; (b) $H_4Ru_4(CO)_{12}$, tetrahedron; (c) $Os_5(CO)_{16}$, trigonal bipyramid; (d) $Os_4(CO)_{16}$, square; (e) $Co_3(CO)_9(\mu_3\text{-CCl})$, triangle; (f) $H_2Os_3(CO)_9(\mu_3\text{-PPh})$, triangle; (g) $HRu_6(CO)_{17}B$, octahedron; (h) $Co_3(\eta^5\text{-Cp})_3(CO)_3$, triangle; (i) $Co_3(CO)_9Ni(\eta^5\text{-Cp})$, tetrahedron.

23.13 (a) $Os_5(CO)_{18}$ has a metal framework consisting of three edge-sharing triangles (a *raft* structure). Show that the valence electron count for this raft is consistent with the number available. (b) Figure 23.25 shows the metal core of $[Ir_8(CO)_{22}]^{2-}$. What would be an appropriate electron-counting scheme for this cluster?

23.14 Suggest products in the following reactions, and give likely structures for the products: (a) $Fe(CO)_5$ irradiated with C_2H_4; (b) $Re_2(CO)_{10}$ with Na/Hg; (c) $Na[Mn(CO)_5]$ with ONCl; (d) $Na[Mn(CO)_5]$ with H_3PO_4; (e) $Ni(CO)_4$ with PPh_3.

Fig. 23.25 Figure for problem 23.13b.

23.15 In Section 23.7, we stated that the distribution of the products in Figure 23.15 is consistent with the migration of the Me group, and not with a mechanism that involves movement of the 'inserted' CO. Confirm that this is true by determining the distribution of products for the latter mechanism and comparing it with that for the former mechanism.

23.16 Illustrate, with examples, what is meant by (a) oxidative addition, (b) reductive elimination, (c) α-hydrogen abstraction, (d) β-hydrogen elimination, (e) alkyl migration and (f) orthometallation.

23.17 Discuss the following statements:

(a) Complexes $Fe(CO)_3L$ where L is a 1,3-diene have applications in organic synthesis.
(b) The fullerenes C_{60} and C_{70} form a range of organometallic complexes.
(c) $Mn_2(CO)_{10}$ and C_2H_6 are related by the isolobal principle.

23.18 Explain why scheme 23.77 is invoked to explain the equivalence of the H atoms in each terminal CH_2 group

of al η^3-allyl ligand, rather than a process involving rotation about the metal–ligand coordination axis.

23.19 With reference to Box 18.2, develop a bonding scheme for $(\eta^5$-Cp$)_2$Fe.

23.20 Suggest products in the following reactions: (a) excess $FeCl_3$ with $(\eta^5$-Cp$)_2$Fe; (b) $(\eta^5$-Cp$)_2$Fe with PhC(O)Cl in the presence of $AlCl_3$; (c) $(\eta^5$-Cp$)_2$Fe with toluene in the presence of Al and $AlCl_3$; (d) $(\eta^5$-Cp$)Fe(CO)_2Cl$ with $Na[Co(CO)_4]$.

23.21 In the reaction of ferrocene with MeC(O)Cl and $AlCl_3$, how could one distinguish between the products $Fe(\eta^5$-$C_5H_4C(O)Me)_2$ and $(\eta^5$-Cp$)Fe(\eta^5$-$C_5H_4C(O)Me)$ by methods other than elemental analysis and X-ray crystallography?

23.22 The reaction of $[(C_6Me_6)RuCl_2]_2$ (**A**) with C_6Me_6 in the presence of $AgBF_4$ gives $[(C_6Me_6)_2Ru][BF_4]_2$ containing cation **B**. Treatment of this compound with Na in liquid NH_3 yields a neutral Ru(0) complex, **C**. Suggest structures for **A**, **B** and **C**.

23.23 (a) Suggest structures for the complexes $LFe(CO)_3$ where L = 2,5-norbornadiene (**23.21**) or cycloheptatriene.
(b) How is the bonding mode of the cycloheptatriene ligand affected on going from $LFe(CO)_3$ to $LMo(CO)_3$?
(c) For L = cycloheptatriene, what product would you expect from the reaction of $LMo(CO)_3$ and $[Ph_3C][BF_4]$?

23.24 Describe the bonding in $(\eta^4$-$C_4H_4)Fe(CO)_3$, accounting for the diamagnetism of the complex.

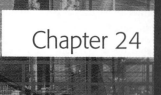

Chapter 24

The *f*-block metals: lanthanoids and actinoids

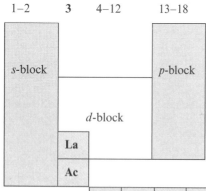

Ce	Pr	Nd	Pm	Sm	Eu	Gd	Tb	Dy	Ho	Er	Tm	Yb	Lu
Th	Pa	U	Np	Pu	Am	Cm	Bk	Cf	Es	Fm	Md	No	Lr

24.1 Introduction

In this chapter we look at *f*-block metals and their compounds. There are two series of metals: the *lanthanoids* (the 14 elements that follow lanthanum in the periodic table) and the *actinoids* (the 14 elements following actinium).[†] The lanthanoids and actinoids (Table 24.1) are collectively known as the *inner transition metals*, while scandium, yttrium, lanthanum and the lanthanoids are together called the *rare earth metals*. Although La and Ac are strictly group 3 metals, the chemical similarity of La to the elements Ce−Lu, and of Ac to Th−Lr, means that La is commonly classified with the lanthanoids, and Ac with the actinoids.

> The symbol Ln is often used to refer generically to the elements La−Lu.

The lanthanoids resemble each other much more closely than do the members of a row of *d*-block metals. The chemistry of the actinoids is more complicated, and in addition, only Th and U have naturally occurring isotopes. Studies of the *transuranium elements* (those with $Z > 92$) require specialized techniques. The occurrence of artificial isotopes among the *f*-block elements can be seen from *Appendix 5*: all the actinoids are unstable with respect to radioactive decay (see *Section 24.9*), although the half-lives of the most abundant isotopes of thorium and uranium (^{232}Th and

---

[†] The IUPAC recommends the names lanthanoid and actinoid in preference to lanthanide and actinide; the ending '-ide' usually implies a negatively charged ion.

Table 24.1 Lanthanum, actinium and the f-block elements. Ln is used as a general symbol for the metals La–Lu.

Element name	Symbol	Z	Ground state electronic configuration				Radius / pm	
			Ln	Ln^{2+}	Ln^{3+}	Ln^{4+}	Ln	$Ln^{3+\ddagger}$
Lanthanum	La	57	$[Xe]6s^2 5d^1$	$[Xe]5d^1$	$[Xe]4f^0$		188	116
Cerium	Ce	58	$[Xe]4f^1 6s^2 5d^1$	$[Xe]4f^2$	$[Xe]4f^1$	$[Xe]4f^0$	183	114
Praseodymium	Pr	59	$[Xe]4f^3 6s^2$	$[Xe]4f^3$	$[Xe]4f^2$	$[Xe]4f^1$	182	113
Neodymium	Nd	60	$[Xe]4f^4 6s^2$	$[Xe]4f^4$	$[Xe]4f^3$		181	111
Promethium	Pm	61	$[Xe]4f^5 6s^2$	$[Xe]4f^5$	$[Xe]4f^4$		181	109
Samarium	Sm	62	$[Xe]4f^6 6s^2$	$[Xe]4f^6$	$[Xe]4f^5$		180	108
Europium	Eu	63	$[Xe]4f^7 6s^2$	$[Xe]4f^7$	$[Xe]4f^6$		199	107
Gadolinium	Gd	64	$[Xe]4f^7 6s^2 5d^1$	$[Xe]4f^7 5d^1$	$[Xe]4f^7$		180	105
Terbium	Tb	65	$[Xe]4f^9 6s^2$	$[Xe]4f^9$	$[Xe]4f^8$	$[Xe]4f^7$	178	104
Dysprosium	Dy	66	$[Xe]4f^{10} 6s^2$	$[Xe]4f^{10}$	$[Xe]4f^9$	$[Xe]4f^8$	177	103
Holmium	Ho	67	$[Xe]4f^{11} 6s^2$	$[Xe]4f^{11}$	$[Xe]4f^{10}$		176	102
Erbium	Er	68	$[Xe]4f^{12} 6s^2$	$[Xe]4f^{12}$	$[Xe]4f^{11}$		175	100
Thulium	Tm	69	$[Xe]4f^{13} 6s^2$	$[Xe]4f^{13}$	$[Xe]4f^{12}$		174	99
Ytterbium	Yb	70	$[Xe]4f^{14} 6s^2$	$[Xe]4f^{14}$	$[Xe]4f^{13}$		194	99
Lutetium	Lu	71	$[Xe]4f^{14} 6s^2 5d^1$	$[Xe]4f^{14} 5d^1$	$[Xe]4f^{14}$		173	98

Element name	Symbol	Z	Ground state electronic configuration			Radius / pm	
			M	M^{3+}	M^{4+}	$M^{3+\,*}$	$M^{4+\,*}$
Actinium	Ac	89	$[Rn]6d^1 7s^2$	$[Rn]5f^0$		111	99
Thorium	Th	90	$[Rn]6d^2 7s^2$	$[Rn]5f^1$	$[Rn]5f^0$		94
Protactinium	Pa	91	$[Rn]5f^2 7s^2 6d^1$	$[Rn]5f^2$	$[Rn]5f^1$	104	90
Uranium	U	92	$[Rn]5f^3 7s^2 6d^1$	$[Rn]5f^3$	$[Rn]5f^2$	103	89
Neptunium	Np	93	$[Rn]5f^4 7s^2 6d^1$	$[Rn]5f^4$	$[Rn]5f^3$	101	87
Plutonium	Pu	94	$[Rn]5f^6 7s^2$	$[Rn]5f^5$	$[Rn]5f^4$	100	86
Americium	Am	95	$[Rn]5f^7 7s^2$	$[Rn]5f^6$	$[Rn]5f^5$	98	85
Curium	Cm	96	$[Rn]5f^7 7s^2 6d^1$	$[Rn]5f^7$	$[Rn]5f^6$	97	85
Berkelium	Bk	97	$[Rn]5f^9 7s^2$	$[Rn]5f^8$	$[Rn]5f^7$	96	83
Californium	Cf	98	$[Rn]5f^{10} 7s^2$	$[Rn]5f^9$	$[Rn]5f^8$	95	82
Einsteinium	Es	99	$[Rn]5f^{11} 7s^2$	$[Rn]5f^{10}$	$[Rn]5f^9$		
Fermium	Fm	100	$[Rn]5f^{12} 7s^2$	$[Rn]5f^{11}$	$[Rn]5f^{10}$		
Mendelevium	Md	101	$[Rn]5f^{13} 7s^2$	$[Rn]5f^{12}$	$[Rn]5f^{11}$		
Nobelium	No	102	$[Rn]5f^{14} 7s^2$	$[Rn]5f^{13}$	$[Rn]5f^{12}$		
Lawrencium	Lr	103	$[Rn]5f^{14} 7s^2 6d^1$	$[Rn]5f^{14}$	$[Rn]5f^{13}$		

$\ddagger$ Ionic radius is for a 8-coordinate ion.

$*$ Ionic radius is for a 6-coordinate ion.

^{238}U, $t_{\frac{1}{2}} = 1.4 \times 10^{10}$ and 4.5×10^9 yr respectively) are so long that for many purposes their radioactivity can be neglected.

The *transuranium elements* are those with atomic number higher than that of uranium ($Z > 92$).

24.2 f-Orbitals and oxidation states

For an f-orbital, the quantum numbers are $n = 4$ or 5, $l = 3$ and $m_l = +3, +2, +1, 0, -1, -2, -3$; a set of f-orbitals is seven-fold degenerate. f-Orbitals are *ungerade*.

A set of f-orbitals is seven-fold degenerate and there is more than one way to represent them. The *cubic set* is commonly used and is readily related to tetrahedral, octahedral and cubic ligand fields. The cubic set comprises the f_{x^3}, f_{y^3}, f_{z^3}, f_{xyz}, $f_{z(x^2 - y^2)}$, $f_{y(z^2 - x^2)}$ and $f_{x(z^2 - y^2)}$ atomic orbitals, and Figure 24.1 shows representations of the f_{z^3} and f_{xyz} orbitals and indicates how the remaining five atomic orbitals are related to them.[†] Figure 24.1c emphasizes how the directionalities of the lobes of the f_{xyz} atomic orbital relate to the corners of a cube. Each f orbital contains three nodal planes.

[†] Three-dimensional representations of the f orbitals can be viewed using the following World Wide Websites: http://wulff.mit.edu/orbs and http://yip5.chem.wfu.edu/yip/organic/pchem-atoms.html

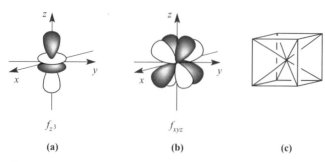

f$_{z^3}$ *f*$_{xyz}$

(a)   (b)   (c)

Fig. 24.1 The 'cubic set' of *f*-orbitals: (a) f_{z^3}, (b) f_{xzy} and (c) a diagram to highlight the fact that the lobes of the f_{xyz} orbital point towards the corners of a cube. The f_{x^3} and f_{y^3} orbitals are like f_{z^3} but point along the *x* and *y* axes respectively. The $f_{z(x^2-y^2)}$, $f_{y(z^2-x^2)}$ and $f_{x(z^2-y^2)}$ look like the f_{xzy} atomic orbital but, with respect to the latter, are rotated by 45° about the *z*, *y* and *x* axes respectively.

The valence shell of a lanthanoid element contains 4*f* orbitals and that of an actinoid, 5*f* atomic orbitals. The ground state electronic configurations of the *f*-block elements are listed in Table 24.1. A 4*f* atomic orbital has no radial node, whereas a 5*f* atomic orbital has one radial node (see *Section 1.6*). A crucial difference between the 4*f* and 5*f* orbitals is the fact that the 4*f* atomic orbitals are deeply buried and 4*f* electrons are not available for covalent bonding. Neither, usually, is ionization beyond the M^{3+} ion energetically favourable, and this leads to a characteristic +3 oxidation state across the whole row from La to Lu.

> The elements La to Lu are characterized by the +3 oxidation state, and the chemistry is mostly that of the Ln^{3+} ion.

The known oxidation states of the actinoids are shown in Table 24.2. The existence of at least two oxidation states for nearly all these metals implies that the successive ionization energies (see *Appendix 8*) probably differ by less than they do for the lanthanoids. For the higher oxidation states, covalent bonding must certainly be involved. This may occur either because the 5*f* atomic orbitals extend further from the nucleus than do the 4*f* atomic orbitals and are available for bonding, or because the energy separations between the 5*f*, 6*d*, 7*s* and 7*p* atomic orbitals are sufficiently small that appropriate valence states for covalent bonding are readily

attained. Evidence that 5*f* atomic orbitals have a greater spatial extent than 4*f* atomic orbitals comes from the fine structure of the ESR spectrum of UF$_3$ (in a CaF$_2$ lattice) which arises from interaction of the electron spin of the U^{3+} ion and the F$^-$ ions; NdF$_3$ (the corresponding lanthanoid species) shows no such effect.

Table 24.2 shows that a wide range of oxidation states is exhibited by the earlier actinoids, but from Cm to Lr, the elements resemble the lanthanoids. This follows from the lowering in energy of the 5*f* atomic orbitals on crossing the period and the stabilization of the 5*f* electrons.

24.3 Atom and ion sizes

The lanthanoid contraction

> The *lanthanoid contraction* is the steady decrease in size along the series of elements La–Lu.

The overall decrease in atomic and ionic radii (Table 24.1) from La to Lu has major consequences on the chemistry of the third row of *d*-block metals (see *Section 22.3*). The contraction is similar to that observed in a period of *d*-block metals and is attributed to the same effect: the imperfect shielding of one electron by another in the same sub-shell. However, the shielding of one 4*f* electron by another is less than for one *d* electron by another, and as the nuclear charge increases from La to Lu, there is a fairly regular decrease in the size of the 4*f^n* sub-shell.

The ionic radii for the lanthanoids in Table 24.1 refer to 8-coordinate ions, and those for the actinoids to 6-coordination. The values should only be used as a guide; they increase with increasing coordination number and are by no means absolute values.

Coordination numbers

In *Section 19.7*, we looked at coordination numbers up to 9. The large size of the lanthanoid and actinoid metals means that in their complexes, high coordination numbers (>6) are common. The splitting of the degenerate set of *f* orbitals in crystal fields in small ($\Delta_{oct} \approx 1$ kJ mol^{-1}) and crystal field

Table 24.2 Oxidation states of actinium and the actinoids. The most stable states are shown in blue.

Ac	Th	Pa	U	Np	Pu	Am	Cm	Bk	Cf	Es	Fm	Md	No	Lr
						2			2	2	2	2	2	
3			3	3	3	3	3	3	3	3	3	3	3	3
	4	4	4	4	4	4	4	4	4					
		5	5	5	5	5								
			6	6	6	6								
				7	7									

stabilization considerations are of minor importance in lanthanoid and actinoid chemistry. Preferences between different coordination numbers and geometries tend to be controlled by steric effects.

24.4 Spectroscopic and magnetic properties

Electronic spectra and magnetic moments: lanthanoids

The reader should refer to *Box 20.5* for term symbols for free atoms and ions. The interpretation of the electronic spectra of $4f^n$ ions is based on the principles outlined for *d*-block metal ions (*Section 20.6*) but there are important differences. For the lanthanoids, spin–orbit coupling is more important than crystal field splitting, and terms differing only in J values are sufficiently different in energy to be separated in the electronic spectrum. Further, since $l = 3$ for an *f* electron, m_l may be 3, 2, 1, 0, −1, −2 or −3, giving rise to high values of L for some f^n ions: e.g. for the configuration f^2, application of Hund's rules gives the ground state (with $L = 5$, $S = 1$) as $^3H_4$. Since *S*, *P*, *D*, *F* and *G* terms are also possible,

many of them with different positive values of J, the number of possible transitions is large, even after taking into account the limitations imposed by selection rules. As a result, spectra of Ln^{3+} ions often contain large numbers of absorptions. Since the 4*f* electrons are well shielded and not affected by the environment of the ion, bands arising from *f*–*f* transitions are sharp (rather than broad like *d*–*d* absorptions) and their positions in the spectrum are little affected by complex formation. Intensities of the absorptions are low, indicating that the probabilities of the *f*–*f* transitions are low, i.e. little *d*–*f* mixing. Absorptions due to 4*f*–5*d* transitions are broad and *are* affected by ligand environment. Small amounts of some lanthanoid salts are used in phosphors for television tubes (see *luminescence* below) because of the sharpness of their electronic transitions.

> In the electronic spectra of lanthanoid metal ions, absorptions due to *f*–*f* transitions are *sharp*, but bands due to 4*f*–5*d* transitions are *broad*.

Typical colours of Ln^{3+} ions in aqueous solution are listed in Table 24.3. Usually (but not invariably) f^n and f^{14-n} species have similar colours.

The bulk magnetic moments (see *Section 20.8*) of Ln^{3+} ions are given in Table 24.3. In general, experimental

---

BOX 24.1   APPLICATIONS

Neodymium lasers

The word *laser* stands for 'light amplification by stimulated emission of radiation'. A laser produces beams of monochromated, very intense radiation in which the radiation waves are coherent. The principle of a laser is that of *stimulated emission*: an excited state can decay spontaneously to the ground state by emitting a photon, but in a laser, the emission is stimulated by an incoming photon of the same energy as the emission. The advantages of this over spontaneous emission are that (i) the energy of emission is exactly defined, (ii) the radiation emitted is in phase with the radiation used to stimulate it, and (iii) the emitted radiation is coherent with the stimulating radiation. Further, because their properties are identical, the *emitted as well as the stimulating* radiation can stimulate further decay, and so on, i.e. the stimulating radiation has been *amplified*.

A neodymium laser consists of a YAG rod (see *Section 22.2*) containing a low concentration of Nd^{3+}; at each end of the rod is a mirror, one of which can also transmit radiation. An initial irradiation from an external source *pumps* the system, exciting the Nd^{3+} ions which then spontaneously relax to the longer-lived $^4F_{3/2}$ excited state (see diagram). That the lifetime of the $^4F_{3/2}$ is relatively long is essential, allowing there to be a *population inversion* of ground and excited states. Decay to the $^4I_{11/2}$ state is the *laser transition*, and is stimulated by a photon of the correct energy. As the

diagram shows, the neodymium laser is a *four level laser*. The mirror-system in the laser allows the radiation to be reflected between the ends of the rod until a high-intensity beam is eventually emitted. The wavelength of the emission from the neodymium laser is usually 1064 nm (i.e. in the infrared), but *frequency doubling* can give lasers emitting at 532 nm.

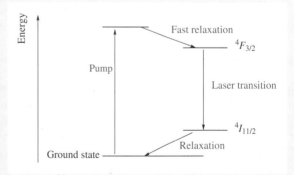

Further reading

P.W. Atkins (1999) *Physical Chemistry*, 6th edn, Oxford University Press, Oxford, p. 507.

Table 24.3   Colours of aqua complexes of La^{3+} and Ln^{3+}, and observed and calculated magnetic moments for the M^{3+} ions.

Metal ion	Colour	Ground state electronic configuration	Magnetic moment, μ (298 K) / μ_B	
			Calculated	Observed
La^{3+}	Colourless	$[Xe]4f^0$	0	0
Ce^{3+}	Colourless	$[Xe]4f^1$	2.54	2.3–2.5
Pr^{3+}	Green	$[Xe]4f^2$	3.58	3.4–3.6
Nd^{3+}	Lilac	$[Xe]4f^3$	3.62	3.5–3.6
Pm^{3+}	Yellow	$[Xe]4f^4$	2.68	2.7
Sm^{3+}	Yellow	$[Xe]4f^5$	0.84	1.5–1.6
Eu^{3+}	Pale pink	$[Xe]4f^6$	0	3.4–3.6
Gd^{3+}	Colourless	$[Xe]4f^7$	7.94	7.8–8.0
Tb^{3+}	Pale pink	$[Xe]4f^8$	9.72	9.4–9.6
Dy^{3+}	Yellow	$[Xe]4f^9$	10.63	10.4–10.5
Ho^{3+}	Yellow	$[Xe]4f^{10}$	10.60	10.3–10.5
Er^{3+}	Rose-red	$[Xe]4f^{11}$	9.57	9.4–9.6
Tm^{3+}	Pale green	$[Xe]4f^{12}$	7.63	7.1–7.4
Yb^{3+}	Colourless	$[Xe]4f^{13}$	4.50	4.4–4.9
Lu^{3+}	Colourless	$[Xe]4f^{14}$	0	0

values agree well with those calculated from formulae based on the assumption of Russell–Saunders coupling and large spin–orbit coupling constants, as a consequence of which only the states of lowest J value are populated. This is *not* true for Eu^{3+}, and not quite true for Sm^{3+}. For Eu^{3+}, the spin–orbit coupling constant λ is $\approx 300\,cm^{-1}$, only slightly greater than kT ($\approx 200\,cm^{-1}$); the ground state of the f^6 ion is $^7F_0$ (which is diamagnetic, since $J = 0$), but the states $^7F_1$ and $^7F_2$ are also populated to some extent and give rise to the observed magnetic moment. As expected, at low temperatures, the moment of Eu^{3+} approaches zero. The variation of μ with n (number of unpaired electrons) in Table 24.3 arises from the operation of Hund's third rule (see *Box 20.5*): $J = L - S$ for a shell less than half full but $J = L + S$ for a shell more than half full; accordingly, J and g_J (see *Equation 20.13*) for ground states are both larger in the second half than the first half of the lanthanoid series.

Luminescence of lanthanoid complexes

Irradiation with UV light of many Ln^{3+} complexes causes them to fluoresce; in some species, low temperatures are required to observe the phenomenon. Their fluorescence leads to the use of lanthanoids in phosphors in televisions and fluorescent lighting. The origin of the fluorescence is $4f$–$4f$ transitions, no transitions being possible for f^0, f^7 (spin-forbidden) and f^{14}. Irradiation produces Ln^{3+} in an excited state which decays to the ground state either with emission of energy (observed as fluorescence) or by a non-radiative pathway. The ions that are commercially important for their emitting properties are Eu^{3+} (red emission) and Tb^{3+} (green emission).

Electronic spectra and magnetic moments: actinoids

The spectroscopic and magnetic properties of actinoids are complicated and we mention them only briefly. Absorptions due to $5f$–$5f$ transitions are weak, but they are somewhat broader and more intense (and considerably more dependent on the ligands present) than those due to $4f$–$4f$ transitions. The interpretation of electronic spectra is made difficult by the large spin–orbit coupling constants (about twice those of the lanthanoids) as a result of which the Russell–Saunders coupling scheme partially breaks down.

Magnetic properties show an overall similarity to those of the lanthanoids in the variation of magnetic moment with the number of unpaired electrons, but values for isoelectronic lanthanoid and actinoid species, e.g. Np(VI) and Ce(III), Np(V) and Pr(III), Np(IV) and Nd(III), are lower for the actinoids, indicating partial quenching of the orbital contribution by crystal field effects.

24.5 Sources of the lanthanoids and actinoids

Occurrence and separation of the lanthanoids

All the lanthanoids except Pm occur naturally. The most stable isotope of promethium, ^{147}Pm (β-emitter, $t_{\frac{1}{2}} = 2.6\,yr$) is formed as a product of the fission of heavy nuclei and is obtained in mg amounts from products of nuclear reactors.

Bastnäsite and *monazite* are the main ores for La and the lanthanoids. All the metals (excluding Pm) can be

BOX 24.2

APPLICATIONS

Commercial demand for the rare earth metals

Bastnäsite deposits in China and the US provide the major commercial source of the rare earth metals, and these countries dominate world output:

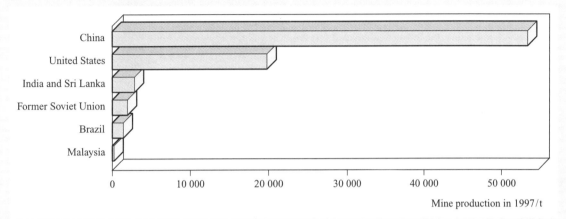

Mine production in 1997/t

[Data: U.S. Geological Survey.]

The demand for the rare earth metals increased over the last two decades of the 20th century, and the demand for cerium oxides in motor vehicle catalytic converters (see equations 26.36 and 26.37) has been a major contributing factor. In 1997, the US consumed $\approx 19\,400\,t$ of rare earth metals with 48% of this total being in catalytic converters. Petroleum refining catalysts accounted for a further 17%, with other applications being in glass polishing and ceramics (14%), permanent magnets (12%), alloys (7%), phosphors (1%) and miscellaneous uses (1%). Only small amounts of the rare earth metals are recycled, most originating from scrapped permanent magnets.

---

obtained from *monazite*, a mixed phosphate $(Ce,La,Nd,Pr,Th,Y\ldots)PO_4$. *Bastnäsite*, $(Ce,La\ldots)CO_3F$, is a source of the lighter lanthanoids. The first step in extraction of the metals from monazite is removal of phosphate and thorium. The ore is heated with caustic soda, and, after cooling, Na_3PO_4 is dissolved in water. The residual hydrated Th(IV) and Ln(III) oxides are treated with hot, aqueous HCl; ThO_2 is not dissolved, but the Ln(III) oxides give a solution of MCl_3 (M = La, Ce...) which is then purified. Starting from bastnäsite, the ore is treated with dilute HCl to remove $CaCO_3$, and then converted to an aqueous solution of MCl_3 (M = La, Ce...). The similarity in ion size and properties of the lanthanoids makes separation difficult. Modern methods of separating the lanthanoids involve solvent extraction using $(^nBuO)_3PO$ (see *Box 6.3*) or ion exchange (see *Section 10.6*).

A typical cation-exchange resin is sulfonated polystyrene or its Na^+ salt. When a solution containing Ln^{3+} ions is poured on to a resin-column, the cations exchange with the H^+ or Na^+ ions (equation 24.1).

$$Ln^{3+}(aq) + 3H^+(resin) \rightleftharpoons Ln^{3+}(resin) + 3H^+(aq) \quad (24.1)$$

The equilibrium distribution coefficient between the resin and the aqueous solution ($[Ln^{3+}(resin)]/[Ln^{3+}(aq)]$) is large for all the ions, but is nearly constant. The resin-bound Ln^{3+} ions are now removed using a complexing agent such as $EDTA^{4-}$ (see *equation 6.75*). The formation constants of the $EDTA^{4-}$ complexes of the Ln^{3+} ions increase regularly from $10^{15.3}$ for La^{3+} to $10^{19.2}$ for Lu^{3+}. If a column on which all the Ln^{3+} ions have been absorbed is eluted with dilute aqueous H_4EDTA, and the pH adjusted to 8 using NH_3, Lu^{3+} is preferentially complexed, then Yb^{3+}, and so on. By using a long ion-exchange column, 99.9% pure components can be separated (Figure 24.2).

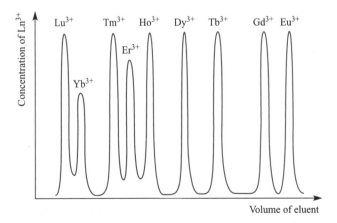

Fig. 24.2 A representation of the order in which $EDTA^{4-}$ complexes of the heavier lanthanoids are eluted from a cation-exchange column.

Detecting smoke with ^{241}Am

Commercial smoke detectors may function using a photo-electric detector or an ionization chamber. An ionization detector consists of two plates across which a voltage (supplied by a battery) is applied (see diagram). One plate has a hole in it, and under the hole lies a small quantity (typically 2×10^{-4} g) of ^{241}Am, an α-particle emitter with $t_{\frac{1}{2}} = 432$ yr. As the α-particles enter the chamber, they ionize atmospheric gas molecules resulting in X^+ ions which are attracted to the negatively charged plate, and electrons which migrate to the positively charged plate. A current flows which is calibrated to correspond to a smoke-free zone.

When smoke enters the chamber, the current changes as ions interact with the smoke particles. The sensor is equipped with an alarm which is triggered when a change in current is detected.

^{241}Am source

The actinoids

With the exception of Th and U, the actinoids are man-made, produced by nuclear reactions (see *Chapter 2*). Radiation hazards of all but Th and U lead to technical difficulties in studying actinoid compounds, and conventional experimental techniques are not generally applicable.

Uranium and thorium are isolated from natural sources. Thorium is extracted from monazite as ThO_2 (see above), and the most important source of uranium is *pitchblende* (U_3O_8). The uranium ore is heated with H_2SO_4 in the presence of an oxidizing agent to give the sulfate salt of the uranyl cation, $[UO_2]^{2+}$, which is separated on an anion-exchange resin, eluting with HNO_3 to give $[UO_2][NO_3]_2$. After further work-up, the uranium is precipitated as the oxo-peroxo complex $UO_2(O_2) \cdot 2H_2O$ or as 'yellow cake' (approximate composition $[NH_4]_2[U_2O_7]$). Thermal decomposition gives yellow UO_3 which is converted to UF_4 (reactions 2.19 and 2.20); reduction with Mg yields U metal.

The isotopes ^{227}Ac and ^{231}Pa can be isolated from the decay products of ^{235}U in pitchblende, but are better synthesized by nuclear reactions 24.2 and 24.3.

$$^{226}Ra \xrightarrow{(n,\gamma)} {}^{227}Ra \xrightarrow{-\beta^-} {}^{227}Ac \qquad (24.2)$$

$$^{230}Th \xrightarrow{(n,\gamma)} {}^{231}Th \xrightarrow{-\beta^-} {}^{231}Pa \qquad (24.3)$$

Equation 2.17 showed the syntheses of ^{239}Np and ^{239}Pu; lengthy irradiation of ^{239}Pu in a nuclear pile leads to the successive formation of small quantities of ^{240}Pu, ^{241}Pu, ^{242}Pu and ^{243}Pu. The last is a β^--emitter ($t_{\frac{1}{2}} = 5$ h) and decays to ^{243}Am ($t_{\frac{1}{2}} = 7400$ yr) which gives ^{244}Cm by sequence 24.4.

$$^{243}Am \xrightarrow{(n,\gamma)} {}^{244}Am \xrightarrow{-\beta^-} {}^{244}Cm \qquad (24.4)$$

Both ^{243}Am and ^{244}Cm are available on a 100 g scale, and multiple neutron capture followed by β^--decay yields milligram amounts of ^{249}Bk, ^{252}Cf, ^{253}Es and ^{254}Es, plus microgram amounts of ^{257}Fm. Synthesis of the heaviest

actinoids was detailed in *Section 2.6*. In *Box 24.3*, we highlight an everyday use of ^{241}Am.

24.6 Lanthanoid metals

Lanthanum and the lanthanoids, except Eu, crystallize in one or both of the close-packed structures; Eu has a bcc lattice and the value of r_{metal} given in Table 24.1 can be adjusted to 205 pm for 12-coordination (see *Section 5.5*). It is important to notice in Table 24.1 that Eu and Yb have much larger metallic radii than the other lanthanoids, implying that Eu and Yb (which have well defined lower oxidation states) contribute fewer electrons to M–M bonding. This is consistent with the lower values of $\Delta_a H^\circ$: Eu and Yb, 177 and 152 kJ mol^{-1} respectively, compared with the other lanthanoids (206–430 kJ mol^{-1}). The lowest in this range (206 kJ mol^{-1}) belongs to Sm; like Eu and Yb, Sm also has a well-defined lower oxidation state, but unlike Eu and Yb, Sm shows no anomaly in its metallic radius. Further, Eu and Yb, but not Sm, form blue solutions in liquid NH_3 due to reaction 24.5 (see *Section 8.6*).

$$Ln \xrightarrow{\text{liquid } NH_3} Ln^{2+} + 2e^- \text{(solv)} \qquad Ln = Eu, Yb \qquad (24.5)$$

All the lanthanoids are soft white metals. The later metals are passivated by an oxide coating and are kinetically more inert than the earlier metals. Values of E° for half-reaction 24.6 lie in the range -2.0 to -2.4 V, and the small variation indicates that variations in $\Delta_a H^\circ$, *IE* values and $\Delta_{\text{hyd}} H^\circ$ (which are considerable) effectively cancel out.

$$Ln^{3+} + 3e^- \rightleftharpoons Ln \qquad (24.6)$$

All the metals liberate H_2 from dilute acids or steam. They burn in air to give Ln_2O_3 with the exception of Ce which forms CeO_2. When heated, lanthanoids react with H_2 to give a range of compounds between metallic (i.e. conducting) hydrides LnH_2 (best formulated as $Ln^{3+}(H^-)_2(e^-)$) and saline hydrides LnH_3; non-stoichiometric hydrides are typified by 'GdH$_3$' which actually has compositions in the range $GdH_{2.85-3}$. Europium forms only EuH_2. The alloy $LaNi_5$ is a potential 'hydrogen storage vessel' (see *Section 9.7*) since it reversibly absorbs H_2 (equation 24.7).

$$LaNi_5 \underset{413 \text{ K}, -H_2}{\overset{H_2}{\rightleftharpoons}} LaNi_5H_x \qquad x \approx 6 \qquad (24.7)$$

The carbides Ln_2C_3 and LnC_2 are formed when the metals are heated with carbon. The LnC_2 carbides adopt the same structure as CaC_2 (see *Section 13.7*), but the C–C bonds (128 pm) are significantly lengthened (119 pm in CaC_2). They are metallic conductors and are best formulated as $Ln^{3+}[C_2]^{2-}(e^-)$. Lanthanoid borides were discussed in *Section 12.10*; halides are described below.

24.7 Inorganic compounds and coordination complexes of the lanthanoids

The discussion in this section is necessarily selective. Most of the chemistry concerns the +3 oxidation state, with Ce(IV) being the only stable +4 state (equation 24.8).

$$Ce^{4+} + e^- \rightleftharpoons Ce^{3+} \qquad E^{\circ} = +1.72 \text{ V} \qquad (24.8)$$

The +2 oxidation state is well defined for Eu, Sm and Lu. The estimated E° values for the Sm^{3+}/Sm^{2+} and Yb^{3+}/Yb^{2+} couples are -1.5 and -1.1 V respectively, indicating that Sm(II) and Yb(II) are highly unstable with respect to oxidation even by water. For the Eu^{3+}/Eu^{2+} couple, the E° value (-0.35 V) is similar to that for Cr^{3+}/Cr^{2+} (-0.41 V) and colourless Eu(II) solutions can be used for chemical studies if air is excluded.

Halides

Reactions of F_2 with Ln give LnF_3 for all the metals and, for Ce, Pr and Tb, also LnF_4; CeF_4 can also be made by reaction 24.9.

$$CeO_2 \xrightarrow{F_2, \text{ XeF}_2, \text{ ClF}_3} CeF_4 \qquad (24.9)$$

With Cl_2, Br_2 and I_2, LnX_3 are formed. Reaction 24.10 gives metallic diiodides with high electrical conductivities; as with the dihydrides above, these diiodides are actually $Ln^{3+}(I^-)_2(e^-)$. Saline LnX_2 (Ln = Sm, Eu, Yb; X = F, Cl, Br, I) can be formed by reducing LnX_3 (e.g. with H_2).

$$Ln + 2LnI_3 \longrightarrow 3LnI_2 \qquad Ln = La, Ce, Pr, Gd \qquad (24.10)$$

The solid state structures of LnX_3 contain Ln(III) centres with high coordination numbers, and as $r_{M^{3+}}$ decreases across the row, the coordination number decreases. The chlorides $LnCl_3$ for Ln = La to Gd, possess the UCl_3 lattice; this is a structural prototype containing tricapped

trigonal prismatic (*Figure 9.10c*) metal centres. For Ln = Tb to Lu, $LnCl_3$ adopts an $AlCl_3$ layer lattice with octahedral Ln(III).

Compounds such as $KCeF_4$, $NaNdF_4$ and Na_2EuCl_5 are made by fusion of group 1 metal fluorides and LnF_3. These are *double salts* and do not contain complex anions. Several discrete hexahalo anions of Ln(II) are known, e.g. $[YbI_6]^{4-}$.

Hydroxides and oxides

Lanthanum hydroxide, though sparingly soluble, is a strong base and absorbs CO_2, giving the carbonate. Base strength and solubility decrease on crossing the lanthanoid series, and $Yb(OH)_3$ and $Lu(OH)_3$ dissolve in hot concentrated NaOH (equation 24.11).

$$Ln(OH)_3 + 3[OH]^- \longrightarrow 2[Ln(OH)_6]^{3-} \qquad Ln = Yb, Lu \qquad (24.11)$$

Cerium(III) hydroxide is a white solid, and in air slowly forms yellow $Ce(OH)_4$. Most oxides Ln_2O_3 are formed by thermal decomposition of oxoacid salts, e.g. reaction 24.12, but Ce, Pr and Tb give higher oxides by this method and H_2 is used to reduce the latter to Ln_2O_3.

$$4Ln(NO_3)_3 \xrightarrow{\Delta} 2Ln_2O_3 + 12NO_2 + 3O_2 \qquad (24.12)$$

Complexes of Ln(III)

The Ln^{3+} ions are hard and show a preference for F^- and *O*-donor ligands, e.g. in complexes with $[EDTA]^{4-}$ (Section 24.5), $[Yb(OH)_6]^{3-}$ (equation 24.11) and in β-diketonate complexes (*Box 24.4*). In their aqua complexes, the Ln^{3+} ions are typically 9-coordinate, and a tricapped trigonal prismatic structure has been confirmed in crystalline salts such as $[Pr(H_2O)_9][OSO_3Et]_3$ and $[Ho(H_2O)_9][OSO_3Et]_3$. High coordination numbers are the norm in complexes of Ln^{3+}, with the highest exhibited by the early lanthanoids; examples include:[†]

- 12-coordinate: $[La(NO_3\text{-}O,O')_6]^{3-}$, $[La(H_2O)_2(NO_3\text{-}O,O')_5]^{2-}$;
- 11-coordinate: $[La(H_2O)_5(NO_3\text{-}O,O')_3]$, $[Ce(H_2O)_5(NO_3\text{-}O,O')_3]$, $[Ce(15\text{-crown-}5)(NO_3\text{-}O,O')_3]$, $[La(15\text{-crown-}5)(NO_3\text{-}O,O')_3]$ (Figure 24.3a);
- 10-coordinate: $[Ce(CO_3\text{-}O,O')_5]^{6-}$ (Ce^{4+}, Figure 24.3b), $[Nd(NO_3\text{-}O,O')_5]^{2-}$, $[Eu(18\text{-crown-}6)(NO_3\text{-}O,O')_2]^+$;
- 9-coordinate: $[Ln(EDTA)(H_2O)_3]^-$ (Ln = La, Ce, Nd, Sm, Eu, Gd, Tb, Dy, Ho), $[CeCl_2(18\text{-crown-}6)(H_2O)]^+$, $[PrCl(18\text{-crown-}6)(H_2O)_2]^{2+}$, $[LaCl_3(18\text{-crown-}6)]$ (Figure 24.3c), $[Eu(tpy)_3]^{3+}$;
- 8-coordinate: $[Pr(NCS\text{-}N)_8]^{5-}$ (between cubic and square antiprismatic);
- 7-coordinate: (**24.1**);

[†] Ligand abbreviations: see *Table 6.7* and structure **19.22**.

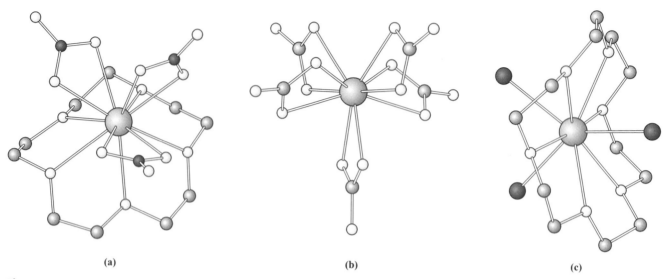

Fig. 24.3 The structures (X-ray diffraction) of (a) [La(15-crown-5)(NO$_3$-*O,O'*)$_3$] [R.D. Rogers *et al.* (1990) *J. Crystallogr. Spectrosc. Res.*, vol. 20, p. 389], (b) [Ce(CO$_3$-*O,O'*)$_5$]$^{6-}$ in the guanidinium salt [R.E. Marsh *et al.* (1988) *Acta Crystallogr., Sect. B*, vol. 44, p. 77], and (c) [LaCl$_3$(18-crown-6)] [R.D. Rogers *et al.* (1993) *Inorg. Chem.*, vol. 32, p. 3451]. Hydrogen atoms have been omitted for clarity; colour code: metal atom, blue; C, grey; N and Cl, black; O, pale grey.

- 6-coordinate: *cis*-[GdCl$_4$(THF)$_2$]$^-$, [YbI$_6$]$^{4-}$, [Ln(β-ketonate)$_3$] (see *Box 24.4*).

The variation found in coordination geometries for a given high coordination number is consistent with the argument that spatial requirements of a ligand and coordination restrictions of multidentate ligands are controlling factors, the 4*f* atomic orbitals being deeply buried and playing little role in metal–ligand bonding. Thus, the 4*f^n* configuration is not a controlling influence on the coordination number. Recent development of MRI contrast agents (see *Box 2.5*) has made studies of Gd^{3+} complexes containing polydentate ligands with *O*- and *N*-donors important.

$$\left[\begin{array}{c} \text{(structure 24.1)} \end{array} \right]^+$$

(24.1)

BOX 24.4

APPLICATIONS

Lanthanoid shift reagents in NMR spectroscopy

The magnetic field experienced by a proton is very different from that of the applied field when a paramagnetic metal centre is present, and results in the δ range over which the ^{1}H NMR spectroscopic signals appear being larger than in a spectrum of a related diamagnetic complex. Signals for protons *close* to the paramagnetic metal centre are significantly *shifted* and this has the effect of 'spreading out' the spectrum. Values of coupling constants are generally not much changed.

^{1}H NMR spectra of large organic compounds or of mixtures of diastereomers, for example, are often difficult to interpret and assign due to overlapping of signals. This is particularly true when the spectrum is recorded on a lowfield (e.g. 100 or 250 MHz) instrument. Paramagnetic lanthanoid complexes have application as *NMR shift reagents*. The addition of a small amount of a shift reagent to a solution of an organic compound can lead to an equilibrium being established between the free and coordinated organic species. The result is that signals due to the organic species which originally overlapped, spread out, and the spectrum becomes

easier to interpret. The europium(III) complex shown below is a commercially available shift reagent (Resolve-Al^{TM}), used, for example, to resolve mixtures of diastereomers.

See also: Box 2.5 for application of Gd(III) complexes as MRI contrast agents.

Lower coordination numbers can be stabilized by using aryloxy or amido ligands, for example:

- 5-coordinate: (**24.2**), (**24.3**);
- 3-coordinate: $[Nd\{N(SiMe_3)_2\}_3]$.

In the solid state, $[Nd\{N(SiMe_3)_2\}_3]$ is trigonal pyramidal but this may be a consequence of crystal packing forces (see *Section 19.7*).

(24.2)

(24.3)

24.8 Organometallic complexes of the lanthanoids

Organolanthanoid chemistry is a rapidly expanding research area. In contrast to the extensive carbonyl chemistry of the *d*-block metals (see *Sections 23.4* and *23.9*), lanthanoid metals do not form complexes with CO under normal conditions. Unstable carbonyls such as $Nd(CO)_6$ have been prepared by matrix isolation. Organolanthanoids are air- and moisture-sensitive and some are pyrophoric; handling the compounds under inert atmospheres is essential.[†]

σ-Bonded complexes

Reaction 24.13 shows a general method of forming Ln–C σ-bonds.

$$LnCl_3 + 3LiR \longrightarrow LnR_3 + 3LiCl \qquad (24.13)$$

In the presence of excess LiR and with R groups that are not too sterically demanding, reaction 24.13 may proceed

---

[†] For details of inert atmosphere techniques, see: D.F. Shriver and M.A. Drezdon (1986) *The Manipulation of Air-sensitive Compounds*, Wiley, New York.

---

further to give $[LnR_4]^-$ or $[LnR_6]^{3-}$ (equations 24.14 and 24.15).

$$YbCl_3 + 4^tBuLi \xrightarrow{\text{THF, 218 K}} [Li(THF)_3]^+[Yb^tBu_4]^- + 3LiCl$$
$$(24.14)$$

$$LuCl_3 + 6MeLi \xrightarrow{\text{DME, 195 K}} [Li(DME)]_3^+[LuMe_6]^{3-} + 3LiCl$$
$$\text{DME} = 1,2\text{-dimethoxyethane} \qquad (24.15)$$

In the solid state, $[LuMe_6]^{3-}$ is octahedral (Lu–C = 253 pm) and analogues for all the lanthanoids except Eu are known. In these reactions, a coordinating solvent such as DME or $Me_2NCH_2CH_2NMe_2$ (TMED) is needed to stabilize the product with a *solvated* Li^+ ion. In some cases, the solvent coordinates to the lanthanoid, e.g. $TmPh_3(THF)_3$ (reaction 24.16 and structure **24.4**).

$$Tm + HgPh_2 \xrightarrow[\text{in presence of TmCl}_3]{\text{in THF,}} TmPh_3(THF)_3 \qquad (24.16)$$

Tm–C = 242 pm

(24.4)

Complexes containing σ-bonded $-C\equiv CR$ groups have been prepared by a number of routes, e.g. reaction 24.17.

$$[Lu^tBu_4(THF)_4]^- \xrightarrow[-^tBuH]{HC\equiv C^tBu \text{ in THF}} [Lu(C\equiv C^tBu)_4]^-$$
$$(24.17)$$

Cyclopentadienyl complexes

Many organolanthanoids contain cyclopentadienyl ligands and reaction 24.18 is a general route to Cp_3Ln.

$$LnCl_3 + 3NaCp \longrightarrow Cp_3Ln + 3NaCl \qquad (24.18)$$

The solid state structures of Cp_3Ln compounds vary with Ln, e.g. Cp_3Tm and Cp_3Yb are monomeric, while Cp_3La, Cp_3Pr and Cp_3Lu are polymeric. Adducts with donors such as THF, pyridine and MeCN are readily formed, e.g. tetrahedral $(\eta^5\text{-Cp})_3Tb(NCMe)$ and $(\eta^5\text{-Cp})_3Dy(THF)$, and trigonal bipyramidal $(\eta^5\text{-Cp})_3Pr(NCMe)_2$ (axial MeCN groups). The complexes $(\eta^5\text{-}C_5Me_5)_3Sm$ and $(\eta^5\text{-}C_5Me_5)_3Nd$ are one-electron reductants: $(\eta^5\text{-}C_5Me_5)_3Ln$ reduces $Ph_3P=Se$ to PPh_3 and forms $(\eta^5\text{-}C_5Me_5)_2Ln(\mu\text{-Se})_nLn(\eta^5\text{-}C_5Me_5)_2$ (Ln = Sm, $n = 1$; Ln = Nd, $n = 2$) and $(C_5Me_5)_2$. The reducing ability is attributed to the severe steric congestion in $(\eta^5\text{-}C_5Me_5)_3Ln$, the reducing agent being the $[C_5Me_5]^-$ ligand.

By altering the $LnCl_3$:NaCp ratio in reaction 24.18, $(\eta^5\text{-}C_5H_5)_2LnCl$ and $(\eta^5\text{-}C_5H_5)LnCl_2$ can be isolated.

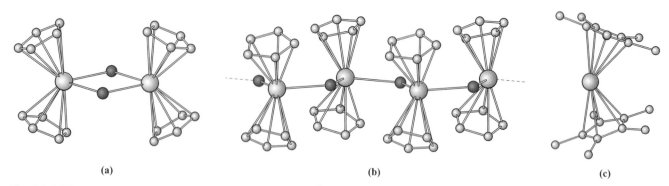

Fig. 24.4 The structures (X-ray diffraction) of (a) dimeric $[(\eta^5\text{-}C_5H_5)_2ErCl]_2$ [W. Lamberts *et al.* (1987) *Inorg. Chim. Acta*, vol. 134, p. 155], (b) polymeric $(\eta^5\text{-}C_5H_5)_2DyCl$ [W. Lamberts *et al.* (1987) *Inorg. Chim. Acta*, vol. 132, p. 119] and (c) the bent metallocene $(\eta^5\text{-}C_5Me_5)Sm$ [W.J. Evans *et al.* (1986) *Organometallics*, vol. 5, p. 1285]. Hydrogen atoms are omitted for clarity; colour code: metal atom, blue; C, grey; Cl, black.

Crystallographic data reveal more complex structures than these formulae suggest: e.g. $(\eta^5\text{-}C_5H_5)ErCl_2$ and $(\eta^5\text{-}C_5H_5)YbCl_2$ crystallize from THF as adducts **24.5**, $(\eta^5\text{-}C_5H_5)_2YbCl$ and $(\eta^5\text{-}C_5H_5)_2ErCl$ are dimeric (Figure 24.4a), and $(\eta^5\text{-}C_5H_5)_2DyCl$ consists of polymeric chains (Figure 24.4b).

Ln = Er, Yb

(24.5)

Schemes 24.19 and 24.20 show some reactions of $[(\eta^5\text{-}C_5H_5)_2LuCl]_2$ and $[(\eta^5\text{-}C_5H_5)_2YbCl]_2$; coordinating solvents are often incorporated into the products and can cause bridge cleavage as in reaction 24.21.

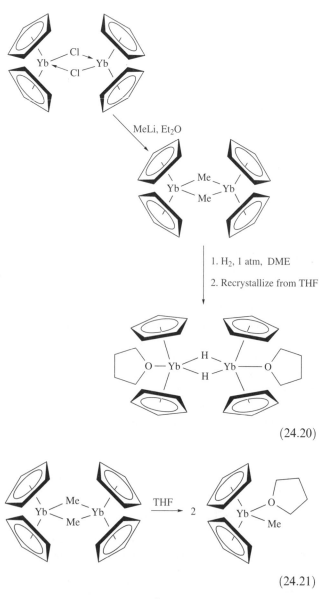

MeLi, Et$_2$O

1. H$_2$, 1 atm, DME

2. Recrystallize from THF

(24.20)

NaH, THF

(24.19)

THF

2

(24.21)

Compounds of the type $(\eta^5\text{-}Cp)_2LnR$ (isolated as THF adducts) can be made directly from $LnCl_3$, e.g. for Lu in reaction 24.22.

$$LuCl_3 + 2NaCp + LiR \xrightarrow{\text{THF, low temperature}} (\eta^5\text{-Cp})_2LuR$$

$$R = CH_2Ph, CH_2{}^tBu, 4\text{-MeC}_6H_4 \qquad (24.22)$$

Lanthanoid(II) metallocenes exist for Sm, Eu and Yb and are stabilized by using the bulky $[C_5Me_5]^-$ ligand (equations 24.23–24.25). The products are obtained as solvates; the desolvated metallocenes have *bent* structures in the solid state (Figure 24.4c) rather than a ferrocene-like structure.

$$2Na[C_5Me_5] + YbCl_2 \xrightarrow{\text{THF}} (\eta^5\text{-C}_5Me_5)_2Yb + 2NaCl$$
$$(24.23)$$

$$2K[C_5Me_5] + SmI_2 \xrightarrow{\text{THF}} (\eta^5\text{-C}_5Me_5)_2Sm + 2KI \qquad (24.24)$$

$$2C_5Me_5H + Eu \xrightarrow{\text{liquid NH}_3} (\eta^5\text{-C}_5Me_5)_2Eu + H_2 \qquad (24.25)$$

Bis(arene) derivatives

The co-condensation at 77 K of $1,3,5\text{-}^tBu_3C_6H_3$ with Ln metal vapour yields the bis(arene) derivatives $(\eta^6\text{-}1,3,5\text{-}^tBu_3C_6H_3)_2Ln$. The complexes are thermally stable for Ln = Nd, Tb, Dy, Ho, Er and Lu, but unstable for Ce, Eu, Tm and Yb.

Complexes containing the η^8-cyclooctatetraenyl ligand

In *Chapter 23*, we described organometallic sandwich and half-sandwich complexes containing π-bonded ligands with hapticities ≤ 7, e.g. $[(\eta^7\text{-C}_7H_7)Mo(CO)_3]^+$. The larger sizes of the lanthanoids permit the formation of sandwich complexes with the planar, octagonal $[C_8H_8]^{2-}$ ligand (see equation 24.46). Lanthanoid(III) chlorides react with $K_2C_8H_8$ to give $[(\eta^8\text{-C}_8H_8)_2Ln]^-$ (Ln = La, Ce, Pr, Sm, Tb, Yb). Cerium also forms $(\eta^8\text{-C}_8H_8)_2Ce$ (**24.6**), an analogue of uranocene (see *Section 24.11*),

and for the lanthanoids with a stable +2 oxidation state, K^+ salts of $[(\eta^8\text{-C}_8H_8)_2Ln]^{2-}$ (Ln = Sm, Eu, Yb) are isolable.

(24.6)

24.9 The actinoid metals

The artificial nature (see *Section 24.5*) of all but two of the actinoid metals affects the extent of knowledge of their properties, and this is reflected in the varying amounts of information that we give for each metal. The instability of the actinoids with respect to radioactive decay has already been mentioned, and Table 24.4 lists data for the longest-lived isotope of each element. All the actinoids are highly toxic, the ingestion of long-lived α-emitters such as ^{231}Pa being extremely hazardous; lethal doses are extremely small.

Actinium is a soft metal which glows in the dark. It is readily oxidized to Ac_2O_3 in moist air, and liberates H_2 from H_2O. *Thorium* is relatively stable in air, but is attacked slowly by H_2O and rapidly by steam or dilute HCl. On heating, Th reacts with H_2 to give ThH_2, halogens to give ThX_4, and N_2 and C to give nitrides and carbides; it forms alloys with a range of metals (e.g. Th_2Zn, $CuTh_2$). *Protactinium* is ductile and malleable, is not corroded by air, but reacts with O_2, H_2 and halogens when heated (scheme 24.26), and with concentrated HF, HCl and H_2SO_4.

$$Pa \begin{cases} \xrightarrow{O_2, \Delta} Pa_2O_5 \\ \xrightarrow{H_2, \Delta} PaH_3 \\ \xrightarrow{I_2, \Delta} PaI_5 \end{cases} \qquad (24.26)$$

Table 24.4 Half-lives and decay modes of the longest-lived isotopes of actinium and the actinoids.

Longest-lived isotope	Half-life	Decay mode	Longest-lived isotope	Half-life	Decay mode
^{227}Ac	21.8 yr	β^-	^{247}Bk	1.4×10^3 yr	α, γ
^{232}Th	1.4×10^{10} yr	α, γ	^{251}Cf	9.0×10^2 yr	α, γ
^{231}Pa	3.3×10^4 yr	α, γ	^{252}Es	1.3 yr	α
^{238}U	4.5×10^9 yr	α, γ	^{257}Fm	100 d	α, γ
^{237}Np	2.1×10^6 yr	α, γ	^{258}Md	52 d	α
^{244}Pu	8.2×10^7 yr	α, γ	^{259}No	58 min	α
^{243}Am	7.4×10^3 yr	α, γ	^{262}Lr	3 min	α
^{247}Cm	1.6×10^7 yr	α, γ			

Uranium corrodes in air; it is attacked by water and dilute acids but not alkali. Scheme 24.27 gives selected reactions. With O_2, UO_2 is produced, but on heating, U_3O_8 forms.

$$U \begin{cases} \xrightarrow{H_2, \Delta} UH_3 \\ \xrightarrow{F_2, \Delta} UF_6 \\ \xrightarrow{Cl_2, \Delta} UCl_4 + UCl_5 + UCl_6 \\ \xrightarrow{H_2O, 373\,K} UO_2 \end{cases}$$
(24.27)

Neptunium is a reactive metal which quickly tarnishes in air. It reacts with dilute acids liberating H_2, but is not attacked by alkali. Despite the fact that the critical mass (see *Section 2.5*) of *plutonium* is $<0.5\,kg$ and it is extremely toxic, its uses as a nuclear fuel and explosive make it a much-studied element. It reacts with O_2, steam and acids, but is inert towards alkali; on heating, Pu combines with many non-metals to give, for example, PuH_2, PuH_3, $PuCl_3$, PuO_2 and Pu_3C_2. *Americium* is a very intense α- and γ-emitter. It tarnishes slowly in dry air, reacts with steam and acids, and on heating forms binary compounds with a range of non-metals. *Curium* corrodes rapidly in air; only minute quantities can be handled ($<20\,mg$ in controlled conditions). *Berkelium* and *californium* behave similarly to Cm, being attacked by air and acids, but not by alkali. Curium and the later elements are handled only in specialized research laboratories.

In the remaining sections, we focus on the chemistries of thorium and uranium, the actinoids for which the most extensive chemistries have been developed. Potential diagrams for Np, Pu and Am are included in Figure 24.6.

24.10 Inorganic compounds and coordination complexes of thorium and uranium

Thorium

The chemistry of thorium largely concerns Th(IV) and, in aqueous solution, there is no evidence for any other oxidation state. The E^o value for the Th^{4+}/Th couple is $-1.9\,V$.

Thorium(IV) halides are made by direct combination of the elements. White ThF_4, $ThCl_4$ and $ThBr_4$, and yellow ThI_4 crystallize with lattices in which Th(IV) is 8-coordinate. Reaction of ThI_4 with Th yields ThI_2 and ThI_3 (both polymorphic) which are metallic conductors and are formulated as $Th^{4+}(I^-)_2(e^-)_2$ and $Th^{4+}(I^-)_3(e^-)$ respectively. Thallium(IV) fluoride is insoluble in water and aqueous alkali metal fluoride solutions, but a large number of double or complex fluorides can be made by direct combination of their constituents. Their structures are complicated, e.g. $[NH_4]_3[ThF_7]$ and $[NH_4]_4[ThF_8]$ contain infinite

$[ThF_7]_n^{3n-}$ chains consisting of edge-sharing tricapped trigonal prismatic Th(IV). Thorium(IV) chloride is soluble in water, and a range of salts containing the discrete, octahedral $[ThCl_6]^{2-}$ is known (reaction 24.28).

$$ThCl_4 + 2MCl \longrightarrow M_2ThCl_6 \qquad e.g.\,M = K, Rb, Cs$$
(24.28)

White ThO_2 is made by thermal decomposition of $Th(ox)_2$ or $Th(NO_3)_4$ and adopts a CaF_2 lattice (*Figure 5.18*). It is precipitated in neutral or even weakly acidic solution. Nowadays, ThO_2 has application as a Fischer–Tropsch catalyst, but its property of emitting a blue glow when heated led to an earlier use in incandescent gas mantles. As expected from the high formal charge on the metal centre, aqueous solutions of Th(IV) salts contain hydrolysis products such as $[ThOH]^{3+}$, $[Th(OH)_2]^{2+}$. The addition of alkali to these solutions gives a gelatinous, white precipitate of $Th(OH)_4$ which is converted to ThO_2 at $>700\,K$.

Coordination complexes of Th(IV) characteristically exhibit high coordination numbers, and hard donors such as oxygen are preferred, for example:

- 12-coordinate: $[Th(NO_3\text{-}O,O')_6]^{2-}$ (Figure 24.5), $[Th(NO_3\text{-}O,O')_5(OPMe_3)_2]^-$;
- 10-coordinate: $[Th(CO_3\text{-}O,O')_5]^{6-}$;
- 9-coordinate (tricapped trigonal prismatic): $[ThCl_2(H_2O)_7]^{2+}$;
- 8-coordinate (dodecahedral): $[ThCl_4(OSPh_2)_4]$, α-$[Th(acac)_4]$, $[ThCl_4(THF)_4]$;
- 8-coordinate (square antiprismatic): β-$[Th(acac)_4]$;
- 8-coordinate (cubic): $[Th(NCS\text{-}N)_8]^{4-}$;
- 7-coordinate: $[ThCl_4(NMe_3)_3]$.

Lower coordination numbers can be stabilized by using amido or aryloxy ligands. In reaction 24.29, the bis(silyl)amido ligands are too bulky to allow the last chloro group to be replaced. Reactions 24.30 and 24.31 illustrate that steric control dictates whether Th(OR)$_4$ is

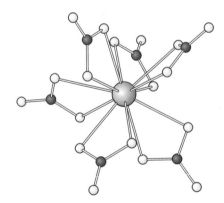

Fig. 24.5 The structure (X-ray diffraction) of the 12-coordinate $[Th(NO_3\text{-}O,O')_6]^{2-}$ in the 2,2'-bipyridinium salt [M.A. Khan *et al.* (1984) *Can. J. Chem.*, vol. 62, p. 850]. Colour code: Th, blue; N, black; O, pale grey.

stabilized with or without other ligands in the coordination sphere.

$$ThCl_4 + 3LiN(SiMe_3)_2 \longrightarrow \underset{\text{tetrahedral}}{ThCl\{N(SiMe_3)_2\}_3} + 3LiCl$$

(24.29)

$$ThI_4 + 4KO^tBu \xrightarrow{\text{pyridine/THF}} \textit{cis-}\underset{\text{octahedral}}{Th(py)_2(O^tBu)_4} + 4KCl$$

(24.30)

$$ThI_4 + 4KOC_6H_3\text{-}2,6\text{-}^tBu_2 \longrightarrow \underset{\text{tetrahedral}}{Th(OC_6H_3\text{-}2,6\text{-}^tBu_2)_4} + 4KI$$

(24.31)

Uranium

Uranium exhibits oxidation states from +3 to +6, although U(IV) and U(VI) are the more common. The key starting point for the preparation of many uranium compounds is UO_2 and scheme 24.32 shows the syntheses of fluorides and chlorides. The fluoride UF_5 is made by controlled reduction of UF_6 but readily disproportionates to UF_4 and UF_6.

(24.32)

Uranium hexafluoride is a colourless, volatile solid with a vapour pressure of 1 bar at 329 K; it is of great importance in the separation of uranium isotopes (see *Section 2.5*). The solid and vapour consist of octahedral UF_6 molecules (U–F = 199 pm). The hexafluoride is immediately hydrolyzed by H_2O (equation 24.33) and is a vigorous fluorinating agent. Treatment of UF_6 with BCl_3 gives the unstable, molecular UCl_6.

$$UF_6 + H_2O \longrightarrow UOF_4 + 2HF$$

(24.33)

The sparingly soluble, green UF_4 is an inert solid (mp 1309 K) with a lattice structure containing 8-coordinate U(IV). Solid UCl_4 also contains 8-coordinate U, but UCl_5 is a dimer (**24.7**); the latter disproportionates on heating. The halides accept X^- to give complexes such as $NaUF_7$, Cs_2UCl_6 and $[NH_4]_4UF_8$; the alkali metal salts adopt lattice structures with U–F–U interactions giving U in high coordination environments.

(24.7)

(24.8)   (24.9)

(24.10)

The oxide UO_3 is polymorphic and all forms decompose to the mixed oxidation state U_3O_8 on heating. Most acids dissolve UO_3 to give yellow solutions containing the uranyl ion (**24.8**), present as a complex, e.g. in aqueous solution, **24.8** exists as an aqua ion and the perchlorate salt of the pentagonal bipyramidal $[UO_2(H_2O)_5]^{2+}$ has been isolated. The $[UO_2]^{2+}$ ion is also present in many solid compounds including the alkaline earth uranates (e.g. $BaUO_4$) which are best described as mixed metal oxides. Uranyl salts of oxoacids include $[UO_2][NO_3]_2 \cdot 6H_2O$ (see *Box 6.3*), $[UO_2][MeCO_2]_2 \cdot 2H_2O$ and $[UO_2][CF_3SO_3]_2 \cdot 3H_2O$, and coordination of the oxoanions and water commonly places the U(IV) centre in a 7- or 8-coordinate environment as in **24.9** and **24.10**. In aqueous solution, the $[UO_2]^{2+}$ ion is partially hydrolyzed to species such as $[U_2O_5]^{2+}$ and $[U_3O_8]^{2+}$. The $[UO_2]^{2+}$ ion is hard and forms a more stable complex with F^- than with the later halides. Figure 24.6 gives a potential diagram for uranium at pH = 0. Reduction of $[UO_2]^{2+}$ first gives $[UO_2]^+$, but this is somewhat unstable with respect to the disproportionation reaction 24.34. Since protons are involved in this reaction, the position of the equilibrium is pH-dependent. Uranium(V) can be stabilized with respect to disproportionation by complexing with F^- as $[UF_6]^-$.

$$2[UO_2]^+ + 4H^+ \rightleftharpoons [UO_2]^{2+} + U^{4+} + 2H_2O$$

(24.34)

Uranium metal liberates H_2 from acids to give the claret-coloured U^{3+} which is a powerful reducing agent (Figure 24.6). The U^{4+} ion is rapidly oxidized to $[UO_2]^{2+}$ by Cr(VI), Ce(IV) or Mn(VII) but oxidation by air is slow. The U^{4+}/U^{3+} and $[UO_2]^{2+}/[UO_2]^+$ redox couples are reversible, but the $[UO_2]^+/U^{4+}$ couple is not: the first two involve only electron transfer, but the last couple involves a structural reorganization around the metal centre.

Whereas the coordination chemistry of thorium is concerned with only the +4 oxidation state, that of uranium covers oxidation states +3 to +6. For U(VI), the linear $[UO_2]^{2+}$ unit is generally present and *trans*-octahedral, pentagonal bipyramidal and hexagonal bipyramidal complexes are usual. For other oxidation states, the coordination polyhedron is essentially determined by the spatial requirements of the ligands rather than electronic factors, and the large size of the U centre allows high coordination numbers to

$$[UO_2]^{2+} \xrightarrow{+0.06} [UO_2]^{+} \xrightarrow{+0.62} U^{4+} \xrightarrow{-0.61} U^{3+} \xrightarrow{-1.80} U$$

$$\xrightarrow{+0.33}$$

$$[NpO_6]^{5-} \xrightarrow{+2.0} [NpO_2]^{2+} \xrightarrow{+1.24} [NpO_2]^{+} \xrightarrow{+0.64} Np^{4+} \xrightarrow{+0.15} Np^{3+} \xrightarrow{-1.86} Np$$

$$\xrightarrow{+0.94}$$

$$[PuO_2]^{2+} \xrightarrow{+1.02} [PuO_2]^{+} \xrightarrow{+1.04} Pu^{4+} \xrightarrow{+1.01} Pu^{3+} \xrightarrow{-2.03} Pu$$

$$\xrightarrow{+1.03}$$

$$\xrightarrow{+1.03}$$

$$[AmO_2]^{2+} \xrightarrow{+1.60} [AmO_2]^{+} \xrightarrow{+0.82} Am^{4+} \xrightarrow{+2.62} Am^{3+} \xrightarrow{-2.05} Am$$

$$\xrightarrow{+1.72}$$

$$\xrightarrow{+1.68}$$

Fig. 24.6 Potential diagram for uranium at pH = 0, and comparative diagrams for Np, Pu and Am.

be attained. Complexes involving different oxidation states and coordination numbers include:

- 14-coordinate: $[U(\eta^3\text{-}BH_4)_4(THF)_2]$;
- 12-coordinate: $[U(NO_3\text{-}O,O')_6]^{2-}$, $[U(\eta^3\text{-}BH_3Me)_4]$;[†]
- 11-coordinate: $[U(\eta^3\text{-}BH_4)_2(THF)_5]^{+}$;
- 9-coordinate: $[UCl_3(18\text{-crown-}6)]^{+}$, $[UBr_2(H_2O)_5(MeCN)_2]^{+}$, $[U(H_2O)(ox)_4]^{4-}$;
- 8-coordinate: $[UCl_3(DMF)_5]^{+}$, $[UCl(DMF)_7]^{3+}$, $[UCl_2(acac)_2(THF)_2]$, $[UO_2(18\text{-crown-}6)]^{2+}$, $[UO_2(NO_3\text{-}O,O')_2(NO_3\text{-}O)_2]^{2-}$, $[UO_2(\eta^2\text{-}O_2)_3]^{4-}$;
- 7-coordinate: $[UO_2Cl(THF)_4]^{+}$, $[UO_2(OSMe_2)_5]^{2+}$;
- 6-coordinate: *trans*-$[UO_2Cl_4]^{2-}$.

Lower coordination numbers are observed in alkoxy derivatives having sterically demanding substituents, e.g. $U(OC_6H_3\text{-}2,6\text{-}^tBu_2)_4$ (preparation analogous to reaction 24.31). The U(III) complex $U(OC_6H_3\text{-}2,6\text{-}^tBu_2)_3$ is probably monomeric; it is oxidized to $UX(OC_6H_3\text{-}2,6\text{-}^tBu_2)_3$ (X = Cl, Br, I; oxidant = PCl$_5$, CBr$_4$, I$_2$ respectively) in which tetrahedral U(IV) has been confirmed for X = I. Coordinating solvents are often non-innocent and tend to lead to increased coordination numbers as in reaction 24.35.

$$UCl_4 + 2LiOC^tBu_3 \xrightarrow{THF} \quad (24.35)$$

24.11 Organometallic complexes of thorium and uranium

Although organometallic complexes are known for all the early actinoids, compounds of Th and U far exceed those of the other metals. In addition to radioactive properties, organoactinoids are air-sensitive and inert atmospheres techniques are required for their handling.

σ-Bonded complexes

Some difficulty was originally encountered in preparing homoleptic σ-bonded alkyl or aryl complexes of the actinoids, but (as for the lanthanoids, Section 24.8) use of the chelate TMED was the key to stabilizing the Li$^+$ salt of $[ThMe_7]^{3-}$ (equation 24.36 and Figure 24.7a). Similarly, hexaalkyls of type $Li_2UR_6 \cdot 7TMED$ have been isolated.

$$ThCl_4 + \text{excess MeLi} \xrightarrow{EtO, TMED} [Li(TMED)]_3[ThMe_7] \quad (24.36)$$

$(Me_3Si)_2HC—U—CH(SiMe_3)_2$
$CH(SiMe_3)_2$

(24.11)

Bulky alkyl groups are also a stabilizing influence as illustrated by the isolation of $U\{CH(SiMe_3)_2\}_3$ (reaction 24.37). The solid contains *trigonal pyramidal* molecules (**24.11**) although the deviation from planarity may be due to crystal packing effects.

$$U(OC_6H_3\text{-}2,6\text{-}^tBu_2)_3 + 3LiCH(SiMe_3)_2$$
$$\longrightarrow U\{CH(SiMe_3)_2\}_3 + 3LiOC_6H_3\text{-}2,6\text{-}^tBu_2 \quad (24.37)$$

[†] η^3-$[BH_3Me]^-$ is like η^3-$[BH_4]^-$: see structure **12.5**.

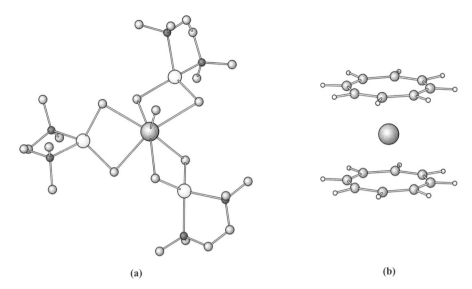

(a)   **(b)**

Fig. 24.7 The structures (X-ray diffraction) of (a) $[Li(TMED)]_3[ThMe_7]$ showing the role of the TMED in stabilizing the structure (H atoms are omitted) [H. Lauke *et al.* (1984) *J. Am. Chem. Soc.*, vol. 106, p. 6841] and (b) $(\eta^8\text{-}C_8H_8)_2Th$ [A. Avdeef *et al.* (1972) *Inorg. Chem.*, vol. 11, p. 1083]. Colour code: Th, blue; Li, pale blue; C, grey; N, black; H, small pale grey.

Alkyl derivatives are more stable if the actinoid is also bound to cyclopentadienyl ligands and reactions 24.38–24.40 show general methods of synthesis where M = Th or U.

$$(\eta^5\text{-}Cp)_3MCl + RLi \xrightarrow{Et_2O} (\eta^5\text{-}Cp)_3MR + LiCl \qquad (24.38)$$

$$(\eta^5\text{-}Cp)_3MCl + RMgX \xrightarrow{THF} (\eta^5\text{-}Cp)_3MR + MgClX \qquad (24.39)$$

$$(\eta^5\text{-}C_5Me_5)_2MCl_2 + 2RLi \xrightarrow{Et_2O} (\eta^5\text{-}C_5Me_5)_2MR_2 + 2LiCl \qquad (24.40)$$

Cyclopentadienyl derivatives

Cyclopentadienyl derivatives are plentiful among organometallic complexes of Th(IV), Th(III), U(IV) and U(III), and reactions 24.41–24.44 give methods of synthesis for the main families of compounds (M = Th, U).

$$MCl_4 + 4KCp \xrightarrow{C_6H_6} (\eta^5\text{-}Cp)_4M + 4KCl \qquad (24.41)$$

$$MX_4 + 3NaCp \xrightarrow{THF} (\eta^5\text{-}Cp)_3MX + 3NaX$$
$$X = Cl, Br, I \qquad (24.42)$$

$$MX_4 + TlCp \xrightarrow{THF} (\eta^5\text{-}Cp)MX_3(THF)_2 + TlX$$
$$X = Cl, Br \qquad (24.43)$$

$$(\eta^5\text{-}Cp)_3MCl + NaC_{10}H_8$$
$$\xrightarrow{THF} (\eta^5\text{-}Cp)_3M(THF) + NaCl + C_{10}H_8 \qquad (24.44)$$
$$(NaC_{10}H_8 = \text{sodium naphthalide})$$

Compounds of type $(\eta^5\text{-}Cp)_2MX_2$ are usually subject to a redistribution reaction such as 24.45 unless sterically hindered as in $(\eta^5\text{-}C_5Me_5)_2ThCl_2$ and $(\eta^5\text{-}C_5Me_5)_2UCl_2$.

$$2(\eta^5\text{-}Cp)_2UCl_2 \xrightarrow{THF} (\eta^5\text{-}Cp)_3UCl + (\eta^5\text{-}Cp)UCl_3(THF)_2 \qquad (24.45)$$

M = Th, U

(24.12)

Colourless $(\eta^5\text{-}Cp)_4Th$ and red $(\eta^5\text{-}Cp)_4U$ are monomeric in the solid state with pseudo-tetrahedral structures **24.12** (Th–C = 287 pm, U–C = 281 pm). Tetrahedral structures are also observed for $(\eta^5\text{-}Cp)_3MX$ and $(\eta^5\text{-}Cp)_3M(THF)$ derivatives, while $(\eta^5\text{-}Cp)MX_3(THF)_2$ is octahedral. How to describe the metal–ligand bonding in these and other Cp derivatives of the actinoids is the subject of much theoretical debate. The current picture suggests involvement of the metal $6d$ atomic orbitals with the $5f$ orbitals being fairly unperturbed; relativistic effects (see *Box 12.2*) also work in favour of a bonding role for the $6d$ rather than $5f$ atomic orbitals. Covalent contributions to the bonding appear to be present in Th(IV) and U(IV) cyclopentadienyl complexes, but for Th(III) and U(III), it is suggested that the bonding is mainly ionic.

A range of organometallic species can be made starting from $(\eta^5\text{-}Cp)_3ThCl$ and $(\eta^5\text{-}Cp)_3UCl$, and Figure 24.8 shows selected reactions of $(\eta^5\text{-}Cp)_3UCl$. The heterometallic complex $(\eta^5\text{-}Cp)_3UFe(CO)_2(\eta^5\text{-}Cp)$ contains an unbridged U–Fe bond.

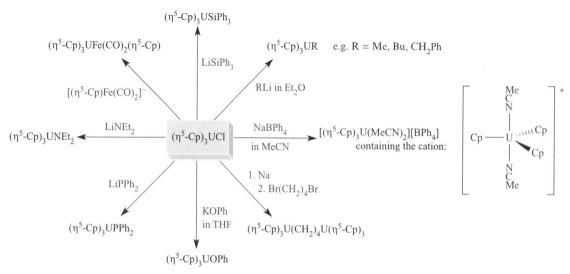

Fig. 24.8 Selected reactions of $(\eta^5\text{-Cp})_3\text{UCl}$.

Complexes containing the η^8-cyclooctatetraenyl ligand

As we have seen, the large U(IV) and Th(IV) centres accommodate up to four η^5-Cp$^-$ ligands, and ferrocene-like complexes are not observed. However, with the large $[C_8H_8]^{2-}$ ligand (reaction 24.46), sandwich complexes are formed by reaction 24.47.

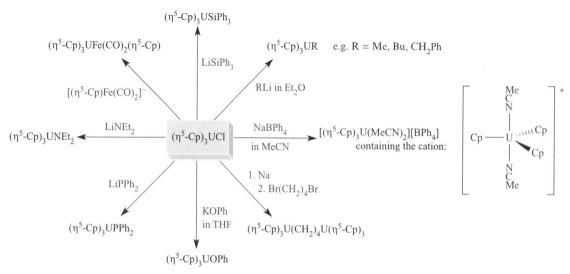

$$(24.46)$$

$$MCl_4 + 2K_2C_8H_8 \longrightarrow (\eta^8\text{-C}_8\text{H}_8)_2\text{M} + 4\text{KCl}$$

$$M = \text{Th, U} \qquad (24.47)$$

The green $(\eta^8\text{-C}_8\text{H}_8)_2\text{U}$ (*uranocene*) and yellow $(\eta^8\text{-C}_8\text{H}_8)_2\text{Th}$ (*thorocene*) are isostructural (Figure 24.7b, mean Th–C = 270 pm and U–C = 265 pm). Bonding in these metallocenes is much studied by theorists, with arguments mirroring those discussed above for cyclopentadienyl derivatives. Uranocene is flammable in air, but does not react with H_2O at 298 K; $(\eta^8\text{-C}_8\text{H}_8)_2\text{Th}$ is air-sensitive, is attacked by protic reagents and explodes when red hot.

Reaction of $ThCl_4$ with $(\eta^8\text{-C}_8\text{H}_8)_2\text{Th}$ in THF yields the half-sandwich $(\eta^8\text{-C}_8\text{H}_8)\text{ThCl}_2(\text{THF})_2$ **24.13**, but the

analogous U(IV) species is made by reaction 24.48, and the iodo derivative by reaction 24.49.

$$UCl_4 + C_8H_8 + 2NaH$$

$$\xrightarrow{\text{THF}} (\eta^8\text{-C}_8\text{H}_8)\text{UCl}_2(\text{THF})_2 + 2\text{NaCl} + H_2 \quad (24.48)$$

$$(\eta^8\text{-C}_8\text{H}_8)_2\text{U} + I_2 \xrightarrow{\text{THF}} (\eta^8\text{-C}_8\text{H}_8)\text{UI}_2(\text{THF})_2 + C_8H_8 \quad (24.49)$$

The halides are useful synthons in this area of chemistry, e.g. reactions 24.50 and 24.51.

$$(\eta^8\text{-C}_8\text{H}_8)\text{UCl}_2(\text{THF})_2 + 2\text{NaN(SiMe}_3)_2$$

$$\xrightarrow{\text{THF}} (\eta^8\text{-C}_8\text{H}_8)\text{U}\{\text{N(SiMe}_3)_2\}_2 + 2\text{NaCl} \quad (24.50)$$

$$(\eta^8\text{-C}_8\text{H}_8)\text{UI}_2(\text{THF})_2 + 3\text{LiCH}_2\text{SiMe}_3$$

$$\xrightarrow{\text{THF}} [\text{Li(THF)}_3]^+[(\eta^8\text{-C}_8\text{H}_8)\text{U}\{\text{CH}_2\text{SiMe}_3\}_3]^- \quad (24.51)$$

Summary

In this chapter, we have given an introduction to the chemistry of the *f*-block elements: the lanthanoids (4*f* metals) and actinoids (5*f*). The +3 oxidation state dominates for the lanthanoids, but for the actinoids, wider variation in oxidation state (+2 to +7) is observed. The large sizes of the 4*f* and 5*f* metals and their ions lead to high coordination numbers. Of the 5*f* metals, only Th and U occur naturally, and the heaviest actinoids can only be made and handled in very small amounts. The majority of studies of the chemistry of the 5*f* metals concern Th and U.

The following terms have been introduced in this chapter. Do you know what they mean?

❏ lanthanoid

❑ actinoid
❑ transuranium element
❑ *f*-orbital
❑ lanthanoid contraction

Suggested further reading

S.A. Cotton (1991) *Lanthanides and Actinides*, Macmillan, London – A good general introduction to the *f*-block elements.

D.C. Hoffmann and D.M. Lee (1999) *Journal of Chemical Education*, vol. 76, p. 331 – 'Chemistry of the heaviest elements – One atom at a time' is an excellent article covering the development and future prospects of 'atom-at-a-time' chemistry of the transuranium elements.

S.F.A. Kettle (1996) *Physical Inorganic Chemistry*, Spektrum, Oxford – Chapter 11 gives an excellent introduction to orbital, spectroscopic and magnetic properties of the *f*-block elements.

G.T. Seaborg (1995) *Accounts of Chemical Research*, vol. 28, p. 257 – A review by one of the pioneers and Nobel Prize winner in the field: 'Transuranium elements: Past, present and future'.

G.T. Seaborg, J.J. Katz and L.R. Moss (1986) *The Chemistry of the Actinide Elements*, 2nd edn, Kluwer, Dordrecht – An excellent account of the actinoid metals.

G.T. Seaborg and W.D. Loveland (1990) *The Elements Beyond Uranium*, Wiley, New York – A text that covers syntheses of the elements, properties, experimental techniques and applications.

G.L. Soloveichik (1994) 'Actinides: Inorganic chemistry' in *Encyclopedia of Inorganic Chemistry*, ed. R.B. King, Wiley, Chichester, vol. 1, p. 2 – A well-referenced review.

Structure and Bonding (1985) vol. 59/60 – A volume containing a series of articles covering different aspects of the actinoids.

Organometallic complexes

D.L. Clark and A.P. Sattelberger (1994) 'Actinides: Organometallic chemistry' in *Encyclopedia of Inorganic Chemistry*, ed. R.B. King, Wiley, Chichester, vol. 1, p. 19 – A well-referenced review.

F.G.N. Cloke (1995) 'Zero oxidation state complexes of scandium, yttrium and the lanthanide elements' in *Comprehensive Organometallic Chemistry II*, eds G. Wilkinson, F.G.A. Stone and E.W. Abel, Pergamon, Oxford, vol. 4, p. 1.

S.A. Cotton (1997) *Coordination Chemistry Reviews*, vol. 160, p. 93 – 'Aspects of the lanthanide-carbon σ-bond'.

F.T. Edelmann (1995) 'Scandium, yttrium and the lanthanide and actinide elements' in *Comprehensive Organometallic Chemistry II*, eds G. Wilkinson, F.G.A. Stone and E.W. Abel, Pergamon, Oxford, vol. 4, p. 11.

W.J. Evans (1985) *Advances in Organometallic Chemistry*, vol. 24, p. 131 – 'Organometallic lanthanide chemistry'.

T.J. Marks and R.D. Ernst (1982) 'Scandium, yttrium and the lanthanides and actinides' in *Comprehensive Organometallic Chemistry*, eds G. Wilkinson, F.G.A. Stone and E.W. Abel, Pergamon, Oxford, vol. 3, p. 173.

C.J. Schaverien (1994) *Advances in Organometallic Chemistry*, vol. 36, p. 283 – 'Organometallic chemistry of the lanthanides'.

Problems

24.1 (a) What is the *lanthanoid contraction*? (b) Explain how the lanthanoids can be separated from their ores.

24.2 Use Hund's rules to derive the ground state of the Ce^{3+} ion, and calculate its magnetic moment. (The spin–orbit coupling constant for Ce^{3+} is $1000 \, cm^{-1}$ and so the population of states other than the ground state can be neglected at 298 K.)

24.3 Show that the stability of a lanthanoid dihalide LnX_2 with respect to disproportionation into LnX_3 and Ln is greatest for X = I.

24.4 How would you attempt to show that a given lanthanoid diiodide, LnI_2, has saline rather than metallic character?

24.5 Comment on each of the following observations:

(a) ΔH° for the formation of $[Ln(EDTA)(H_2O)_x]^-$ (x = 2 or 3) in aqueous solution is nearly constant for all Ln and is almost zero.

(b) The value of E° for the Ce(IV)/Ce(III) couple (measured at pH 0) decreases along the series of acids $HClO_4$, HNO_3, H_2SO_4, HCl.

(c) $BaCeO_3$ has a perovskite structure.

24.6 Comment on the observations that the electronic spectra of lanthanoid complexes contain many absorptions some of which are weak and sharp and similar to those of the gas phase metal ions, and some of which are broad and are affected by the ligands present.

24.7 Discuss the variation in coordination numbers among complexes of the 4*f* metals.

24.8 The reactions of $Ln(NCS)_3$ with $[NCS]^-$ under varying conditions lead to discrete anions such as $[Ln(NCS)_6]^{3-}$, $[Ln(NCS)_7(H_2O)]^{4-}$ and $[Ln(NCS)_7]^{4-}$. What can you say about possible structures for these species?

24.9 (a) Give a brief account of the formation of Ln–C σ-bonds and of complexes containing cyclopentadienyl ligands, and comment on the roles of coordinating solvents. (b) Suggest products for the reactions of $SmCl_3$ and SmI_2 with $K_2C_8H_8$.

24.10 (a) By considering Figure 24.6, suggest a method for the separation of Am from U, Np and Pu. (b) What would you expect to happen when a solution of $NpO_2(ClO_4)_2$ in 1 M $HClO_4$ is shaken with Zn amalgam and the resulting liquid decanted from the amalgam and aerated?

24.11 $25.00 \, cm^3$ of a solution **X** containing $21.4 \, g \, U(VI) \, dm^{-3}$ was reduced with Zn amalgam, decanted from the amalgam, and after being aerated for 5 min, was titrated with $0.1200 \, mol \, dm^{-3}$ Ce(IV) solution; $37.5 \, cm^3$ of the latter was required for reoxidation of the uranium to

U(VI). Solution **X** (100 cm^3) was then reduced and aerated as before, and treated with excess of dilute aqueous KF. The resulting precipitate (after drying at 570 K) weighed 2.826 g. Dry O$_2$ was passed over the precipitate at 1070 K, after which the solid product weighed 1.386 g. This product as dissolved in water, the fluoride in the solution precipitated as PbClF, 2.355 g being obtained. Deduce what you can concerning the chemical changes in these experiments.

24.12 Suggest likely products in the following reactions: (a) UF$_4$ with F$_2$ at 570 K; (b) Pa$_2$O$_5$ with SOCl$_2$ followed by heating with H$_2$; (c) UO$_3$ with H$_2$ at 650 K; (d) heating UCl$_5$; (e) UCl$_3$ with NaOC$_6$H$_2$-2,4,6-Me$_3$.

24.13 What structural features would you expect in the solid state of (a) Cs$_2$[NpO$_2$(acac)$_3$], (b) Np(BH$_4$)$_4$, (c) the guanidinium salt of [ThF$_3$(CO$_3$)$_3$]$^{5-}$, (d) Li$_3$[LuMe$_6$]·3DME, (e) Sm{CH(SiMe$_3$)$_2$}$_3$, and (f) a complex that analyzes as having the composition [UO$_2$][CF$_3$SO$_3$]$_2$·2(18-crown-6)·5H$_2$O.

24.14 Identify isotopes **A–F** in the following sequence of nuclear reactions:

(a) ^{238}U $\xrightarrow{(n, \gamma)}$ **A** $\xrightarrow{-\beta^-}$ **B** $\xrightarrow{-\beta^-}$ **C**

(b) **D** $\xrightarrow{-\beta^-}$ **E** $\xrightarrow{(n, \gamma)}$ ^{242}Am $\xrightarrow{-\beta^-}$ **F**

24.15 Identify the starting isotopes **A–F** in each of the following syntheses of transactinoid elements:

(a) **A** + $^4_2$He ⟶ $^{256}_{101}$Md + n

(b) **B** + $^{16}_8$O ⟶ $^{255}_{102}$No + 5n

(c) **C** + $^{11}_5$B ⟶ $^{256}_{103}$Lr + 4n

(d) **D** + $^{18}_8$O ⟶ $^{261}_{104}$Rf + 5n

(e) **E** + $^{18}_8$O ⟶ $^{263}_{106}$Sg + 4n

24.16 Discuss the following statements:

(a) Thorium forms iodides of formulae ThI$_2$, ThI$_3$ and ThI$_4$.
(b) In the solid state, salts of [UO$_2$]$^{2+}$ contain a linear cation.
(c) Reactions of NaOR with UCl$_4$ lead to monomeric U(OR)$_4$ complexes.

24.17 (a) What Th-containing products would you expect from the reactions of (η^5-Cp)$_3$ThCl with (i) Na[(η^5-Cp)Ru(CO)$_2$], (ii) LiCHMeEt, (iii) LiCH$_2$Ph?
(b) What advantage does (η^5-C$_5$Me$_5$)$_2$ThCl$_2$ have over (η^5-Cp)$_2$ThCl$_2$ as a starting material?
(c) How might (η^5-C$_5$Me$_5$)UI$_2$(THF)$_3$ react with K$_2$C$_8$H$_8$?

24.18 (a) Suggest a method of preparing U(η^3-C$_3$H$_5$)$_4$.
(b) How might U(η^3-C$_3$H$_5$)$_4$ react with HCl?
(c) (η^5-C$_5$Me$_5$)(η^8-C$_8$H$_8$)ThCl is dimeric, but its THF adduct is a monomer. Draw the structures of these compounds, and comment on the role of coordinating solvents in stabilizing other monomeric organothorium and organouranium complexes.

24.19 Discuss the following:

(a) Many actinoid oxides are non-stoichiometric, but few lanthanoid oxides are.
(b) The ion [NpO$_6$]$^{5-}$ can only be made in aqueous solution if the solution is strongly alkaline.
(c) A solution containing Pu(IV) undergoes negligible disproportionation in the presence of an excess of molar H$_2$SO$_4$.

24.20 Give a short account of aspects of the organometallic compounds formed by the lanthanoids and actinoids and highlight major differences between families of organometallic complexes of the *d*- and *f*-block metals.

d-Block metal complexes: reaction mechanisms

25.1 Introduction

We have already touched on some aspects of inorganic reaction mechanisms: *kinetically inert* metal centres such as Co(III) (*Section 21.10*) and organometallic reaction types (*Section 23.7*). Now, we discuss in more detail the mechanisms of ligand substitution and electron-transfer reactions in coordination complexes; for the substitution reactions, we confine our attention to square planar and octahedral complexes for which kinetic data are plentiful.

A proposed mechanism *must* be consistent with all experimental facts. A mechanism cannot be proven since another mechanism may also be consistent with the experimental data.

25.2 Ligand substitutions: some general points

In a ligand substitution reaction:

$$ML_xX + Y \longrightarrow ML_xY + X$$

X is the *leaving group* and Y is the *entering group*.

Kinetically inert and labile complexes

Metal complexes that undergo reactions with $t_{\frac{1}{2}} \leq 1$ min are described as being *kinetically labile*. If the reaction takes significantly longer than this, the complex is *kinetically inert*.

There is no connection between the *thermodynamic* stability of a complex and its lability towards substitution. For example, values of $\Delta_{hyd}G^o$ for Cr^{3+} and Fe^{3+} are almost equal, yet $[Cr(H_2O)_6]^{3+}$ (d^3) undergoes substitution slowly and $[Fe(H_2O)_6]^{3+}$ (high-spin d^5) rapidly. Similarly, although the overall formation constant of $[Hg(CN)_4]^{2-}$ is greater than that of $[Fe(CN)_6]^{4-}$, the Hg(II) complex rapidly exchanges $[CN]^-$ with isotopically labelled cyanide while exchange is extremely slow for $[Fe(CN)_6]^{4-}$. The kinetic inertness of d^3 and low-spin d^6 octahedral complexes is in part associated with crystal field effects (see *Section 25.4*).

Stoichiometric equations say nothing about mechanism

The processes that occur in a reaction are not necessarily obvious from the stoichiometric equation. For example, reaction 25.1 might suggest a mechanism involving the direct substitution of coordinated $[CO_3]^{2-}$ by H_2O.

$$[(H_3N)_5Co(CO_3)]^+ + 2[H_3O]^+$$
$$\longrightarrow [(H_3N)_5Co(H_2O)]^{3+} + CO_2 + 2H_2O \quad (25.1)$$

However, use of $H_2^{18}O$ as solvent shows that all the oxygen in the aqua complex is derived from carbonate, and scheme 25.2 shows the proposed pathway of the reaction.

$$[(H_3N)_5Co(OCO_2)]^+ + [H_3O]^+ \longrightarrow \left[(H_3N)_5Co - O \begin{array}{c} H \\ \\ CO_2 \end{array} \right]^{2+} + H_2O$$

$$\downarrow$$

$$[(H_3N)_5Co(OH)]^{2+} + CO_2$$

$$\downarrow [H_3O]^+$$

$$[(H_3N)_5Co(H_2O)]^{3+} + H_2O$$

$$(25.2)$$

Types of substitution mechanism

In inorganic substitutions, the limiting mechanisms are *dissociative* (*D*) in which the intermediate has a lower coordination number than the starting complex (equation 25.3) and *associative* (*A*) in which the intermediate has a higher coordination number (equation 25.4).[†]

> *Dissociative* and *associative* reaction mechanisms involve two-step pathways and an *intermediate*.

$$
\left.\begin{array}{l}
ML_xX \longrightarrow \underset{\text{intermediate}}{ML_x} + \underset{\substack{\text{leaving}\\\text{group}}}{X} \\[2.5em]
\underset{\substack{\text{entering}\\\text{group}}}{ML_x + Y} \longrightarrow ML_xY
\end{array}\right\} \quad dissociative\ (D) \quad (25.3)
$$

$$
\left.\begin{array}{l}
ML_xX + \underset{\substack{\text{entering}\\\text{group}}}{Y} \longrightarrow \underset{\text{intermediate}}{ML_xXY} \\[2.5em]
ML_xXY \longrightarrow ML_xY + \underset{\substack{\text{leaving}\\\text{group}}}{X}
\end{array}\right\} \quad associative\ (A) \quad (25.4)
$$

> An *intermediate* occurs at a local energy minimum; it can be detected and, sometimes, isolated. A *transition state* occurs at an energy maximum, and cannot be isolated.

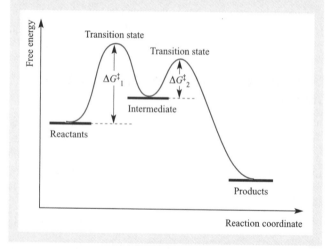

In most metal complex substitution pathways, bond formation between the metal and entering group is thought to be *concurrent* with bond cleavage between the metal and leaving group (equation 25.5). This is the *interchange* (*I*) mechanism.

$$
ML_xX + \underset{\substack{\text{entering}\\\text{group}}}{Y} \longrightarrow \underset{\substack{\text{transition}\\\text{state}}}{Y\cdots ML_x\cdots X} \longrightarrow ML_xY + \underset{\substack{\text{leaving}\\\text{group}}}{X}
$$

$$(25.5)$$

In an *I* mechanism, there is no intermediate but various transition states are possible. Two types of interchange mechanisms can be identified:

- *dissociative interchange* (*I*$_d$) in which bond breaking dominates over bond formation;
- *associative interchange* (*I*$_a$) in which bond formation dominates over bond breaking.

In an *I*$_a$ mechanism, the reaction rate shows a dependence on the entering group; in an *I*$_d$ mechanism, the rate shows only a very small dependence on the entering group. It is usually difficult to distinguish between *A* and *I*$_a$, *D* and *I*$_d$, and *I*$_a$ and *I*$_d$ processes.

> An *interchange* (*I*) mechanism is a concerted process in which there is *no intermediate* species with a coordination number different from that of the starting complex.

Activation parameters

The diagram which distinguishes between a transition state and an intermediate also shows the free energy of activation, $\Delta G^{\ddagger}$, for each step in the two-step reaction path. Enthalpies and entropies of activation, $\Delta H^{\ddagger}$ and $\Delta S^{\ddagger}$, obtained from temperature dependence of rate constants, can shed light on mechanisms. Equation 25.6 gives the relationship between the rate constant, temperature and activation parameters.

$$
\ln\left(\frac{k}{T}\right) = \frac{-\Delta H^{\ddagger}}{RT} + \ln\left(\frac{k'}{h}\right) + \frac{\Delta S^{\ddagger}}{R} \quad (25.6)
$$

where k = rate constant, T = temperature (K), $\Delta H^{\ddagger}$ = enthalpy of activation (J mol^{-1}), $\Delta S^{\ddagger}$ = enthalpy of activation (J K^{-1} mol^{-1}), R = molar gas constant, k' = Boltzmann constant, h = Planck constant.[‡]

From equation 25.6, a plot of $\ln(k/T)$ against $1/T$ (an *Eyring plot*) is linear; the activation parameters $\Delta H^{\ddagger}$ and $\Delta S^{\ddagger}$ can be determined as shown in Figure 25.1.

Values of $\Delta S^{\ddagger}$ are particularly useful in distinguishing between associative and dissociative mechanisms. A *large negative value* of $\Delta S^{\ddagger}$ is indicative of an *associative* mechanism, i.e. there is a decrease in entropy as the entering group associates with the starting complex. However, caution is needed; solvent reorganization can result in negative values of $\Delta S^{\ddagger}$ even for a dissociative mechanism, and hence the qualifier that $\Delta S^{\ddagger}$ should be *large* and negative to indicate an associative pathway.

The pressure dependence of rate constants leads to a measure of the *volume of activation*, $\Delta V^{\ddagger}$ (equation 25.7).

[†] The terminology for inorganic substitution mechanisms is not the same as for organic nucleophilic substitutions. Since readers will already be familiar with the S$_N$1 (unimolecular) and S$_N$2 (bimolecular) notation, it may be helpful to note that the *D* mechanism corresponds to S$_N$1, and *I*$_a$ to S$_N$2.

[‡] Physical constants: see back inside cover of this book.

Table 25.1 Activation parameters for substitution in selected square planar complexes (see *Table 6.7* for ligand abbreviations).

Reactants	$\Delta H^{\ddagger} \, / \, \mathrm{kJ \, mol^{-1}}$	$\Delta S^{\ddagger} \, / \, \mathrm{J \, K^{-1} \, mol^{-1}}$	$\Delta V^{\ddagger} \, / \, \mathrm{cm^3 \, mol^{-1}}$
$[Pt(dien)Cl]^+ + H_2O$	+84	−63	−10
$[Pt(dien)Cl]^+ + [N_3]^-$	+65	−71	−8.5
trans-$[PtCl_2(PEt_3)_2]$ + py	+14	−25	−14
trans-$[PtCl(NO_2)(py)_2]$ + py	+12	−24	−9

or, in integrated form:

$$\frac{d(\ln k)}{dP} = \frac{-\Delta V^{\ddagger}}{RT} \left.\begin{array}{c} \\ \\ \\ \ln\left(\frac{k_{(P_1)}}{k_{(P_2)}}\right) = \frac{-\Delta V^{\ddagger}}{RT}(P_1 - P_2) \end{array}\right\} \quad (25.7)$$

where k = rate constant; P = pressure; $\Delta V^{\ddagger}$ = volume of activation $(\mathrm{cm^3 \, mol^{-1}})$; R = molar gas constant; T = temperature (K).

A reaction in which the transition state has a greater volume than the initial state shows a positive $\Delta V^{\ddagger}$, whereas a negative $\Delta V^{\ddagger}$ corresponds to the transition state being compressed relative to the reactants. After allowance for any change in volume of the solvent (which is important if solvated ions are involved), the sign of $\Delta V^{\ddagger}$ should, in principle, distinguish between an associative and dissociative mechanism.

A negative value of $\Delta V^{\ddagger}$ indicates an associative mechanism; a positive value suggests that the mechanism is dissociative.

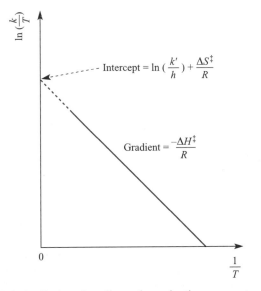

Fig. 25.1 An Eyring plot allows the activation parameters $\Delta H^{\ddagger}$ and $\Delta S^{\ddagger}$ to be determined from the temperature dependence of the rate constant; the dotted part of the line represents an extrapolation. See *equation 25.6* for definitions of quantities.

25.3 Substitution in square planar complexes

Complexes with a d^8 configuration often form square planar complexes (see *Section 20.3*), especially when there is a large crystal field: Rh(I), Ir(I), Pt(II), Pd(II), Au(III). However, 4-coordinate complexes of Ni(II) may be tetrahedral or square planar. The majority of kinetic work on square planar systems has been carried out on Pt(II) complexes because the rate of ligand substitution is conveniently slow. Although data for Pd(II) and Au(III) complexes indicate similarity between their substitution mechanisms and those of Pt(II) complexes, one *cannot assume* a similarity in kinetics among a series of structurally related complexes undergoing similar substitutions.

Rate equations, mechanism and the *trans*-effect

The consensus of opinion, based on a large body of experimental work, is that nucleophilic substitution mechanisms in square planar Pt(II) complexes normally proceed by *associative* mechanisms (A or I_a). Negative values of $\Delta S^{\ddagger}$ and $\Delta V^{\ddagger}$ support this proposal (Table 25.1). The observation that the rate constants for the displacement of Cl^- by H_2O in $[PtCl_4]^{2-}$, $[PtCl_3(NH_3)]^-$, $[PtCl_2(NH_3)_2]$ and $[PtCl(NH_3)_3]^+$ are similar suggests an associative mechanism, since a dissociative pathway would be expected to show a significant dependence on the charge on the complex.

Reaction 25.8 shows the substitution of X by Y in a square planar Pt(II) complex.

$$PtL_3X + Y \longrightarrow PtL_3Y + X \quad (25.8)$$

The usual form of the experimental rate law is given by equation 25.9 indicating that the reaction proceeds simultaneously by two routes.[†]

$$\text{Rate} = -\frac{d[PtL_3X]}{dt} = k_1[PtL_3X] + k_2[PtL_3X][Y] \quad (25.9)$$

Reaction 25.8 would usually be studied under pseudo-first order conditions, with Y (as well as the solvent, S) in vast

---

[†] In rate equations, [ ] stands for 'concentration of' and should not be confused with use of square brackets around formulae of complexes in other contexts. For this reason, we omit [ ] in formulae in most reaction equations in this chapter.

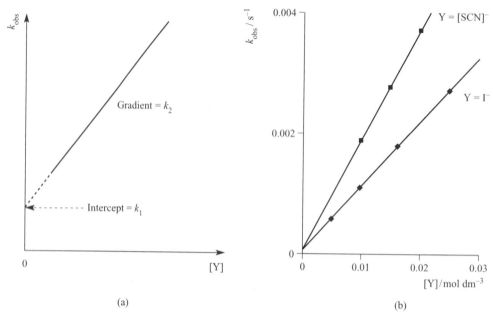

Fig. 25.2 (a) Determination of the k_1 and k_2 rate constants (equation 25.11) from the observed rate data for ligand substitution in a square planar complex; Y is the entering ligand. The dotted part of the line represents an extrapolation. (b) Plots of k_{obs} against concentration of the entering group for the reactions of *trans*-$[PtCl_2(py)_2]$ with $[SCN]^-$ or with I^-; both reactions were carried out in MeOH and so there is a common intercept. [Data from: U. Belluco *et al.* (1965) *J. Am. Chem. Soc.*, vol. 87, p. 241.]

excess. This means that, since $[Y]_t \approx [Y]_0$, and $[S]_t \approx [S]_0$ (where the subscripts represent time *t* and time zero), we can rewrite equation 25.9 in the form of equation 25.10 where k_{obs} is the observed rate constant and is related to k_1 and k_2 by equation 25.11.

$$\text{Rate} = -\frac{d[PtL_3X]}{dt} = k_{obs}[PtL_3X] \quad (25.10)$$

$$k_{obs} = k_1 + k_2[Y] \quad (25.11)$$

Carrying out a series of reactions with various concentrations of Y (always under pseudo-first order conditions) allows k_1 and k_2 to be evaluated (Figure 25.2a). Data plotted in this form, but with different entering groups and the *same solvent*, illustrate the solvent dependence of k_1 since there is a common intercept (Figure 25.2b); if the kinetic runs are repeated using a different solvent, a different common intercept is observed.

The contributions of the two terms in equation 25.9 to the overall rate reflect the relative dominance of one pathway over the other. The k_2 term arises from an associative mechanism involving attack by Y on PtL_3X in the rate-determining step, and when Y is a good nucleophile, the k_2 term is dominant. The k_1 term might appear to indicate a concurrent dissociative pathway. However, experiment shows that the k_1 term becomes dominant if the reaction is carried out in polar solvents, and its contribution diminishes in apolar solvents. This indicates solvent participation, and equation 25.9 is more fully written in the form of equation 25.12 in which S is the solvent. Since S is in vast excess, its concentration is effectively constant during the reaction (i.e. pseudo-first order conditions) and so, comparing equations 25.9 and 25.12, $k_1 = k_3[S]$.

$$\text{Rate} = -\frac{d[PtL_3X]}{dt} = k_3[PtL_3X][S] + k_2[PtL_3X][Y] \quad (25.12)$$

When the solvent is a potential ligand (e.g. H_2O), it competes with the entering group Y in the rate-determining step of the reaction, and X can be displaced by Y or S. Substitution of S by Y then occurs in a *fast* step, i.e. *non-rate determining*. The two competing pathways by which reaction 25.8 occurs are shown in scheme 25.13.

$$
\begin{array}{l}
PtL_3X \;+\; Y \;\xrightarrow{\;k_2\;}\; PtL_3Y \;+\; X \\[4pt]
\text{competes with:} \\[4pt]
PtL_3X \;+\; S \;\xrightarrow{\;k_1\;}\; PtL_3S \;+\; X \\[4pt]
PtL_3S \;+\; Y \;\xrightarrow{\text{fast}}\; PtL_3Y \;+\; S
\end{array}
\quad (25.13)
$$

A further point in favour of both the k_1 and k_2 terms being associative is that *both* rate constants decrease when the steric demands of Y or L increase.

In the majority of reactions, substitution at square planar Pt(II) is *stereoretentive*: the entering group takes the coordination site previously occupied by the leaving group. An A or I_a mechanism involves a 5-coordinate intermediate or transition state and, since the energy difference between different 5-coordinate geometries is small, one would expect rearrangement of the 5-coordinate species to be facile unless, for example, it is sterically hindered (A or I_a) or its lifetime is too short (I_a). The stereochemical retention can be envisaged as shown in Figure 25.3 (in which we ignore any part played by the solvent). Why does Figure 25.3 specifically show a *trigonal bipyramidal* species as the intermediate

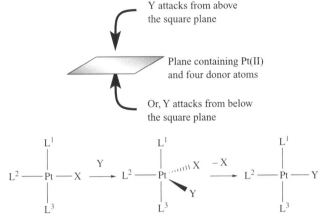

Fig. 25.3 Initial attack by the entering group at a square planar Pt(II) centre is from above or below the plane. Nucleophile Y then coordinates to give a trigonal bipyramidal species which loses X with retention of stereochemistry.

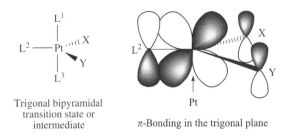

Fig. 25.4 In the trigonal plane of the 5-coordinate transition state or intermediate (see *Figure 25.3*), a π-bonding interaction can occur between a metal *d* orbital (e.g. d_{xy}) and suitable orbitals (e.g. *p* atomic orbitals, or molecular orbitals of π-symmetry) of ligand L^2 (the ligand *trans* to the leaving group), X (the leaving group) and Y (the entering group). Note that ligands may not necessarily contribute to the π-bonding scheme, e.g. NH_3.

or transition state? To answer this, we must consider additional experimental data:

> The choice of leaving group in a square planar complex is determined by the nature of the ligand *trans* to it; this is the *trans-effect* and is *kinetic* in origin.

Reactions 25.14 and 25.15 illustrate the *trans*-effect in operation: *cis* and *trans*-[PtCl₂(NH₃)₂] are prepared *specifically* by different substitution routes.[†]

$$\left[\begin{array}{c} Cl \\ | \\ Cl {\longrightarrow} Pt {\longrightarrow} Cl \\ | \\ Cl \end{array}\right]^{2-} \xrightarrow[- Cl^-]{NH_3} \left[\begin{array}{c} NH_3 \\ | \\ Cl {\longrightarrow} Pt {\longrightarrow} Cl \\ | \\ Cl \end{array}\right]^{-}$$

$$\xrightarrow[- Cl^-]{NH_3} \begin{array}{c} NH_3 \\ | \\ Cl {\longrightarrow} Pt {\longrightarrow} NH_3 \\ | \\ Cl \end{array} \quad (25.14)$$

$$\left[\begin{array}{c} NH_3 \\ | \\ H_3N {\longrightarrow} Pt {\longrightarrow} NH_3 \\ | \\ NH_3 \end{array}\right]^{2+} \xrightarrow[- NH_3]{Cl^-} \left[\begin{array}{c} Cl \\ | \\ H_3N {\longrightarrow} Pt {\longrightarrow} NH_3 \\ | \\ NH_3 \end{array}\right]^{+}$$

$$\xrightarrow[- NH_3]{Cl^-} \begin{array}{c} Cl \\ | \\ H_3N {\longrightarrow} Pt {\longrightarrow} NH_3 \\ | \\ Cl \end{array} \quad (25.15)$$

[†] The use of the terms *trans-effect* and *trans-influence* in different textbooks is not consistent, and may cause confusion; attention should be paid to specific definitions.

One contributing factor to the *trans*-effect is the *trans*-influence (see *Box 22.9*). The second factor, which addresses the *kinetic* origin of the *trans*-effect, is that of shared π-electron density in the 5-coordinate transition state or intermediate as shown in Figure 25.4: ligand L^2 is *trans* to the leaving group, X, in the initial square planar complex and is also *trans* to the entering group, Y, in the final square planar complex (Figure 25.3). These three ligands and the metal centre can communicate electronically through π-bonding *only* if they all lie in the *same plane* in the transition state or intermediate. This implies that the 5-coordinate species must be trigonal bipyramidal rather than square-based pyramidal. If L^2 is a strong π-acceptor (e.g. CO), it will stabilize the transition state by accepting electron density that the incoming nucleophile donates to the metal centre, and will thereby facilitate substitution at the site *trans* to it. The general order of the *trans*-effect (i.e. the ability of ligands to direct *trans*-substitution) spans a factor of about 10^6 in rates and is:

$$H_2O \approx [OH]^- \approx NH_3 \approx py < Cl^- < Br^- < I^- \approx [NO_2]^-$$
$$< Ph^- < Me^- < PR_3 \approx H^- \ll CO \approx [CN]^- \approx C_2H_4$$

Experimental rates of substitution are affected by both the ground state *trans*-influence and the kinetic *trans*-effect, and rationalizing the sequence above in terms of individual factors is difficult. There is no close connection between the relative magnitudes of the *trans*-influence and *trans*-effect. However, the π-bonding scheme in Figure 25.4 does help to explain the very strong *trans*-directing abilities of CO, [CN]⁻ and ethene.

The *trans*-effect is useful in devising syntheses of Pt(II) complexes, e.g. selective preparations of *cis*- and *trans*-isomers of [PtCl₂(NH₃)₂] (schemes 25.14 and 25.15) and of [PtCl₂(NH₃)(NO₂)]⁻ (schemes 25.16 and 25.17).

Table 25.2 Values of n_{Pt} for entering ligands, Y, in reaction 25.18; values are relative to n_{Pt} for MeOH $= 0$ and are measured at 298 K.[‡]

Ligand n_{Pt}	Cl^- 3.04	NH_3 3.07	py 3.19	Br^- 4.18	I^- 5.46	$[CN]^-$ 7.14	PPh_3 8.93

[‡] For further data, see: R.G. Pearson, H. Sobel and J. Songstad (1968) *J. Am. Chem. Soc.*, vol. 90, p. 319.

$$(25.16)$$

$$(25.17)$$

Finally, we should note that a small *cis*-effect does exist, but is usually of far less importance than the *trans*-effect.

Ligand nucleophilicity

If one studies how the rate of substitution by Y in a given complex depends on the entering group, then for most reactions at Pt(II), the rate constant k_2 (equation 25.9) increases in the order:

$$H_2O < NH_3 \approx Cl^- < py < Br^- < I^- < [CN]^- < PR_3$$

This is called the *nucleophilicity sequence* for substitution at square planar Pt(II) and the ordering is consistent with Pt(II) being a soft metal centre (see *Table 6.9*). A *nucleophilicity parameter*, n_{Pt}, is defined by equation 25.19 where k_2' is the rate constant for reaction 25.18 with Y = MeOH (i.e. for Y = MeOH, $n_{Pt} = 0$).

$$trans\text{-}[PtCl_2(py)_2] + Y \longrightarrow trans\text{-}[PtCl(py)_2Y]^+ + Cl^-$$

$$(25.18)$$

(The equation is written assuming Y is a neutral ligand.)

$$n_{Pt} = \log\frac{k_2}{k_2'} \quad \text{or} \quad n_{Pt} = \log k_2 - \log k_2' \quad (25.19)$$

Values of n_{Pt} vary considerably (Table 25.2) and illustrate the dependence of the rate of substitution on the nucleophilicity of the entering group. There is no correlation between n_{Pt} and the strength of the nucleophile as a Brønsted base.

> The *nucleophilicity parameter*, n_{Pt}, describes the dependence of the rate of substitution in a square planar Pt(II) complex on the nucleophilicity of the entering group.

If we now consider substitution reactions of nucleophiles with other Pt(II) complexes, linear relationships are found between values of $\log k_2$ and n_{Pt} as illustrated in Figure 25.5. For the general reaction 25.8 (in which the ligands L do *not* have to be identical), equation 25.20 is defined where s is the *nucleophilicity discrimination factor* and k_2' is the rate constant when the nucleophile is MeOH.

$$\log k_2 = s(n_{Pt}) + \log k_2' \qquad (25.20)$$

For a given substrate, s can be found from the gradient of a line in Figure 25.5; each complex has a characteristic value of s, and selected values are listed in Table 25.3. The relatively small value of s for $[Pt(dien)(H_2O)]^{2+}$ indicates that this complex does not discriminate as much between entering ligands as, for example, does *trans*-$[PtCl_2(PEt_3)_2]$; i.e. $[Pt(dien)(H_2O)]^{2+}$ is generally more reactive towards substitution than other complexes in the table, consistent with the fact that H_2O is a good leaving group.

> The *nucleophilicity discrimination factor*, s, is a characteristic of a given square planar Pt(II) complex and describes how sensitive the complex is to variation in the nucleophilicity of the entering ligand.

25.4 Substitution and racemization in octahedral complexes

Most studies of the mechanism of substitution in octahedral metal complexes have been concerned with Werner-type complexes; organometallic complexes have entered the research field more recently. Among the former, the popular candidates for study have been Cr(III) (d^3) and low-spin Co(III) (d^6) species. These complexes are kinetically inert and their rates of reaction are relatively slow and readily followed by conventional techniques. Both Rh(III) and Ir(III) (both low-spin d^6) undergo very slow substitution

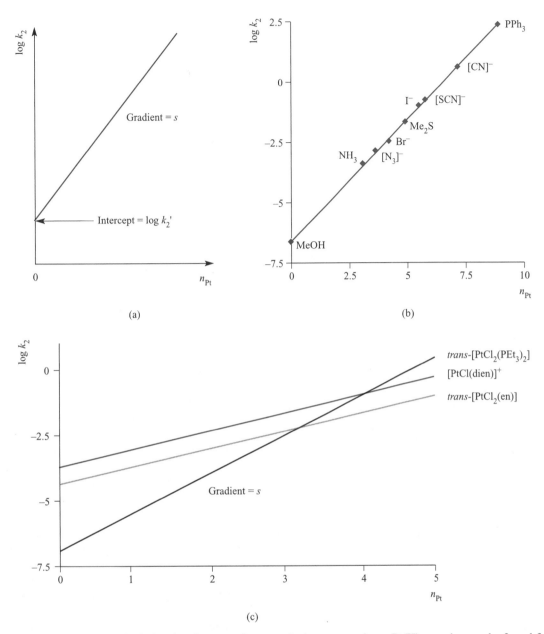

Fig. 25.5 (a) The nucleophilicity discrimination factor, s, for a particular square planar Pt(II) complex can be found from a plot of $\log k_2$ (the second order rate constant, see *equation 25.9*) against n_{Pt} (the nucleophilicity parameter, see *equation 25.19*). Experimental results are plotted in this way in graph (b) which shows data for the reaction of *trans*-$[PtCl_2(py)_2]$ with different nucleophiles in MeOH at 298 or 303 K. [Data from: R.G. Pearson *et al.* (1968) *J. Am. Chem. Soc.*, vol. 90, p. 319.] (c) Plots of $\log k_2$ against n_{Pt} for three square planar Pt(II) complexes; each plot is of the same type as in graph (b). The gradient of each line gives s, the nucleophilicity discrimination factor, for that particular complex. [Data from: U. Belluco *et al.* (1965) *J. Am. Chem. Soc.*, vol. 87, p. 241.]

reactions. There is no universal mechanism by which octahedral complexes undergo substitution, and so care is needed when tackling the interpretation of kinetic data.

Water exchange

The exchange of coordinated H_2O by isotopically labelled water has been investigated for a wide range of octahedral $[M(H_2O)_6]^{n+}$ species (Co^{3+} is not among these because it is unstable in aqueous solution, see *Section 21.10*). Reaction 25.21, where M is an s-, p- or d-block metal, can be studied

by using ^{17}O NMR spectroscopy, and rate constants can thus be determined.

$$[M(H_2O)_6]^{n+} + H_2(^{17}O) \longrightarrow [M(H_2O)_5\{H_2(^{17}O)\}]^{n+} + H_2O \tag{25.21}$$

First order rate constants for the exchange of coordinated water show the following trends:

- within a group of s- or p-block metals, the rate constant increases with increasing cationic radius;

Table 25.3 Nucleophilic discrimination factors, *s*, for selected square planar Pt(II) complexes. (See *Table 6.7* for ligand abbreviations.)

Complex	*s*
trans-[PtCl$_2$(PEt$_3$)$_2$]	1.43
trans-[PtCl$_2$(AsEt$_3$)$_2$]	1.25
trans-[PtCl$_2$(py)$_2$]	1.0
[PtCl$_2$(en)]	0.64
[PtBr(dien)]$^+$	0.75
[PtCl(dien)]$^+$	0.65
[Pt(dien)(H$_2$O)]$^{2+}$	0.44

Table 25.4 Volumes of activation for water exchange reactions (equation 25.21).

Metal ion	High-spin d^n configuration	$\Delta V^{\ddagger}\,/\,cm^3\,mol^{-1}$
V^{2+}	d^3	−4.1
Mn^{2+}	d^5	−5.4
Fe^{2+}	d^6	+3.7
Co^{2+}	d^7	+6.1
Ni^{2+}	d^8	+7.2
Ti^{3+}	d^1	−12.1
V^{3+}	d^2	−8.9
Cr^{3+}	d^3	−9.6
Fe^{3+}	d^5	−5.4

- for cations of similar radii (e.g. Li$^+$, Mg^{2+}, Ga^{3+}), an increase in ionic charge slows down the substitution;
- among M^{2+} ions of the *d*-block, there is *no* correlation between rate constants and ionic size, but trends do indicate a correlation with electronic configuration;
- limited data for M^{3+} ions of the *d*-block also support a correlation between rate constants and electronic configuration.

Table 25.4 lists activation volumes for reaction 25.21 with selected first row *d*-block metal ions. The change from negative to positive values of $\Delta V^{\ddagger}$ indicates a change from associative to dissociative mechanism, and suggests that bond-making becomes less (and bond-breaking more) important on going from a d^3 to d^8 configuration. For the M^{3+} ions in Table 25.4, values of $\Delta V^{\ddagger}$ suggest an associative mechanism. Where data are available, an associative process appears to operate for second and third row metal ions, consistent with the idea that larger metal centres may facilitate association with the entering ligand.

First order rate constants, *k*, for reaction 25.21 vary greatly among the first row *d*-block metals (all high-spin M^{n+} in the hexaaqua ions):

- Cr^{2+} (d^4) and Cu^{2+} (d^9) are kinetically very labile ($k \geq 10^8\,s^{-1}$);
- Cr^{3+} (d^3) is kinetically inert ($k \approx 10^{-3}\,s^{-1}$);
- Mn^{2+} (d^5), Fe^{2+} (d^6), Co^{2+} (d^7) and Ni^{2+} (d^8) are kinetically labile ($k \approx 10^4$ to $10^7\,s^{-1}$);
- V^{2+} (d^2) has $k \approx 10^2\,s^{-1}$, i.e. considerably less labile than the later M^{2+} ions.

Although one can relate some of these trends to CFSE effects as we discuss below, charge effects are also important, e.g. compare [Mn(H$_2$O)$_6$]$^{2+}$ ($k = 2.1 \times 10^7\,s^{-1}$) and [Fe(H$_2$O)$_6$]$^{3+}$ ($k = 1.6 \times 10^2\,s^{-1}$), both of which are high-spin d^5.

The rates of water exchange in high-spin hexaaqua ions follow the sequences:

$$V^{2+} < Ni^{2+} < Co^{2+} < Fe^{2+} < Mn^{2+} < Zn^{2+} < Cr^{2+} < Cu^{2+}$$

and

$$Cr^{3+} \ll V^{3+} < Fe^{3+} < Ti^{3+}$$

For a series of ions of the same charge and about the same size undergoing the same reaction by the same mechanism, we may reasonably suppose that collision frequencies and values of $\Delta S^{\ddagger}$ are approximately constant, and that variations in rate will arise from variation in $\Delta H^{\ddagger}$. Let us assume that the latter arise from loss or gain of CFSE (see *Table 20.3*) on going from the starting complex to the transition state: *a loss of CFSE means an increase in the activation energy for the reaction and hence a decrease in its rate*. The splitting of the *d* orbitals depends on the coordination geometry (Figures 20.8 and 20.10), and we can calculate the change in CFSE on the formation of a transition state. Such calculations make assumptions that are actually unlikely to be valid (e.g. constant M−L bond lengths), but for *comparative* purposes, the results should have some meaning. Table 25.5 lists results of such calculations for high-spin octahedral M^{2+} complexes going to either 5- or 7-coordinate transition states; this provides a model for both dissociative and associative processes. For either model, and despite the simplicity of crystal field theory, there is moderately good qualitative agreement between the calculated order of lability and that observed; Jahn−Teller effects contribute towards the high rates of Cr^{2+} and Cu^{2+}.

The Eigen−Wilkins mechanism

Water exchange is always more rapid than substitutions with other entering ligands. Let us now consider reaction 25.22.

$$ML_6 + Y \longrightarrow products \qquad (25.22)$$

The mechanism may be associative (*A* or *I*$_a$) or dissociative (*D* or *I*$_d$), and it is not at all easy to distinguish between these, even though the rate laws are different. An associative mechanism involves a 7-coordinate intermediate or transition state and, sterically, an associative pathway seems less likely than a dissociative one. Nevertheless, activation volumes do sometimes indicate an associative mechanism (see *Table 25.4*). However, *for most ligand substitutions in octahedral complexes, experimental evidence supports dissociative pathways*. Two limiting cases are often observed for general reaction 25.22:

Table 25.5 Changes in CFSE (ΔCFSE) on converting a high-spin octahedral complex into a square-based pyramidal (for a dissociative process) or pentagonal bipyramidal (for an associative process) transition state, other factors remaining constant (see text).

Metal ion (high-spin)	d^n	ΔCFSE / Δ_{oct}		Metal ion (high-spin)	d^n	ΔCFSE / Δ_{oct}	
		Square-based pyramidal	Pentagonal bipyramidal			Square-based pyramidal	Pentagonal bipyramidal
Sc^{2+}	d^1	+0.06	+0.13	Fe^{2+}	d^6	+0.06	+0.13
Ti^{2+}	d^2	+0.11	+0.26	Co^{2+}	d^7	+0.11	+0.26
V^{2+}	d^3	−0.20	−0.43	Ni^{2+}	d^8	−0.20	−0.43
Cr^{2+}	d^4	+0.31	−0.11	Cu^{2+}	d^9	+0.31	−0.11
Mn^{2+}	d^5	0	0	Zn^{2+}	d^{10}	0	0

Table 25.6 Rate constants, k, for reaction 25.31; see *equation 25.25* for the rate law.

Entering ligand, Y $k \times 10^{-4} / s^{-1}$	NH_3	py	$[MeCO_2]^-$	F^-	$[SCN]^-$
	3	3	3	0.8	0.6

- at high concentrations of Y, the rate of substitution is independent of Y, pointing to a dissociative mechanism;
- at low concentrations of Y, the rate of reaction depends on Y and ML_6, suggesting an associative mechanism.

These apparent contradictions are explained by the *Eigen–Wilkins mechanism*.

> The *Eigen–Wilkins mechanism* applies to ligand substitution in an octahedral complex. An *encounter complex* is first formed between substrate and entering ligand in a pre-equilibrium step, and this is followed by loss of the leaving ligand in the rate-determining step.

Consider reaction 25.22. The first step in the Eigen–Wilkins mechanism is the diffusing together of ML_6 and Y to form a *weakly bound encounter complex* (equilibrium 25.23).

$$ML_6 + Y \xrightleftharpoons{K_E} \underset{\text{encounter complex}}{\{ML_6, Y\}} \tag{25.23}$$

Usually, the rate of formation of $\{ML_6, Y\}$ and the back-reaction to ML_6 and Y are much faster than the subsequent conversion of $\{ML_6, Y\}$ to products. Thus, the formation of $\{ML_6, Y\}$ is a *pre-equilibrium*. The equilibrium constant, K_E, can rarely be determined experimentally, but it can be estimated using theoretical models. The rate determining step in the Eigen–Wilkins mechanism is step 25.24 with a rate constant k; the overall rate law is equation 25.25.

$$\{ML_6, Y\} \xrightarrow{k} \text{Products} \tag{25.24}$$

$$\text{Rate} = k[\{ML_6, Y\}] \tag{25.25}$$

The concentration of $\{ML_6, Y\}$ cannot be measured directly, and we must make use of an estimated value of K_E[†] which is related to $[\{ML_6, Y\}]$ by equation 25.26.

$$K_E = \frac{[\{ML_6, Y\}]}{[ML_6][Y]} \tag{25.26}$$

The *total* concentration of ML_6 and $\{ML_6, Y\}$ in equation 25.23 is measurable because it is the initial concentration of the complex; let this be $[M]_{total}$ (equation 25.27). Thus, we have expression 25.28 for $[ML_6]$.

$$\left.\begin{array}{l} [M]_{total} = [ML_6] + [\{ML_6, Y\}] \\ [M]_{total} = [ML_6] + K_E[ML_6][Y] \\ \qquad = [ML_6](1 + K_E[Y]) \end{array}\right\} \tag{25.27}$$

$$[ML_6] = \frac{[M]_{total}}{1 + K_E[Y]} \tag{25.28}$$

We can now rewrite rate equation 25.25 in the form of equation 25.29 by substituting for $[\{ML_6, Y\}]$ (from equation 25.26) and then for $[ML_6]$ (from equation 25.28).

$$\text{Rate} = \frac{kK_E[M]_{total}[Y]}{1 + K_E[Y]} \tag{25.29}$$

This equation looks complicated, but at *low concentrations of Y* where $K_E[Y] \ll 1$, equation 25.29 approximates to equation 25.30, a second order rate equation in which k_{obs} is the observed rate constant.

$$\text{Rate} = kK_E[M]_{total}[Y] = k_{obs}[M]_{total}[Y] \tag{25.30}$$

Since k_{obs} can be measured experimentally, and K_E can be estimated theoretically, we can estimate k from the expression $k = k_{obs}/K_E$ which follows from equation 25.30. Table 25.6 lists values of k for reaction 25.31 for various entering ligands. The fact that k varies so little is consistent with an

---

[†] K_E can be estimated using an electrostatic approach: for details of the theory, see R.G. Wilkins (1991) *Kinetics and Mechanism of Reactions of Transition Metal Complexes*, 2nd edn, Wiley-VCH, Weinheim, p. 206.

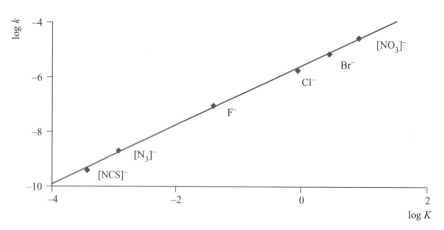

Fig. 25.6 Plot of $\log k$ against $\log K$ for selected leaving groups in reaction 25.33. [Data from: A. Haim (1970), *Inorg. Chem.*, vol. 9, p. 426.]

I_d mechanism. If the pathway were associative, the rate would depend more significantly on the nature of Y.

$$[Ni(H_2O)_6]^{2+} + Y \longrightarrow [Ni(H_2O)_5Y]^{2+} + H_2O \qquad (25.31)$$

> The substitution of an uncharged ligand (e.g. H_2O) by an anionic ligand (e.g. Cl^-) is called *anation*.

At a *high concentration of Y* (e.g. when Y is the solvent), $K_E[Y] \gg 1$, and equation 25.29 approximates to equation 25.32, a first order rate equation with *no dependence* on the entering ligand. The value of k can be measured directly ($k_{obs} = k$).

$$\text{Rate} = k[M]_{total} \qquad (25.32)$$

The water exchange reaction 25.21 exemplifies a case where the entering ligand is the solvent.

Let us now look further at *experimental* trends that are consistent with dissociative (*D* or I_d) mechanisms for substitution in octahedral complexes; I_d is supported in very many instances.

The rate of ligand substitution usually depends on the *nature of the leaving ligand*.

$$[Co(NH_3)_5X]^{2+} + H_2O \rightleftharpoons [Co(NH_3)_5(H_2O)]^{3+} + X^- \qquad (25.33)$$

For reaction 25.33, the rate of substitution increases with X^- in the following order:

$$[OH]^- < NH_3 \approx [NCS]^- < [MeCO_2]^- < Cl^-$$
$$< Br^- < I^- < [NO_3]^-$$

This trend correlates with the M–X bond strength (the stronger the bond, the slower the rate) and is consistent with the rate-determining step involving bond breaking in a dissociative step. We can go one step further: a plot of $\log k$ (where k is the rate constant for the forward reaction 25.33) against $\log K$ (where K is the equilibrium constant for reaction 25.33) is linear *with a gradient of 1.0* (Figure 25.6). Equations 25.34 and 25.35 relate $\log k$ and $\log K$ to

$\Delta G^{\ddagger}$ (free energy of activation) and ΔG (free energy of reaction), respectively. It follows that the linear relationship between $\log k$ and $\log K$ represents a linear relationship between $\Delta G^{\ddagger}$ and ΔG, a so-called *linear free energy relationship* (LFER).[†]

$$\Delta G^{\ddagger} \propto -\log k \qquad (25.34)$$
$$\Delta G \propto -\log K \qquad (25.35)$$

The interpretation of the LFER in Figure 25.6 in mechanistic terms is that the transition state is closely related to the product $[Co(NH_3)_5(H_2O)]^{3+}$, and, therefore, the transition state involves, at most, only a weak $Co\cdots X$ interaction. This is consistent with a dissociative (*D* or I_d) process.

Stereochemistry of substitution

Although most substitutions in octahedral complexes involve *D* or I_d pathways, we consider the stereochemical implications only of the *D* mechanism since this involves a 5-coordinate species which we can readily visualize (equation 25.36).

$$(25.36)$$

Table 25.7 gives the isomer distributions in the products of the reactions of H_2O with *cis*- and *trans*-$[CoCl(en)_2Z]^+$. The specificity of the substitutions in *cis*-$[CoCl(en)_2Z]^+$ extends to other nucleophiles (and to other *cis*-complexes) and is explained in terms of a *square-based pyramidal* species which has a single site for attack by H_2O (Figure 25.7a). Starting from *trans*-$[CoCl(en)_2Z]^+$, the stereochemistry of substitution is not always specific, and similar observations

---

[†] LFERs can also use $\ln k$ and $\ln K$, but it is common practice to use log–log relationships.

Table 25.7 The isomer distributions in the reactions of *cis*- and *trans*-[CoCl(en)₂Z]⁺ with H₂O.

cis-[CoCl(en)₂Z]⁺ + H₂O → [Co(H₂O)(en)₂Z]²⁺ + Cl⁻		*trans*-[CoCl(en)₂Z]⁺ + H₂O → [Co(H₂O)(en)₂Z]²⁺ + Cl⁻	
Y	% of *cis* product	Y	% of *cis* product[‡]
[OH]⁻	100	[OH]⁻	75
Cl⁻	100	Cl⁻	35
[NO₂]⁻	100	[NO₂]⁻	0
[NCS]⁻	100	[NCS]⁻	50–70

[‡] Remaining % is *trans*-product.

are observed for other *trans*-complexes. Figure 25.7b shows that if, after loss of Cl⁻, a *trigonal bipyramidal* species is formed, then attack within the equatorial plane can occur at one of three sites to give *cis*- and *trans*-isomers. Stabilization of the trigonal bipyramidal species by π-bonding involving the *trans*-ligand Z (i.e. as in Figure 25.4) can be

envisaged. The argument certainly rationalizes the observations, but one has to accept an assumption that no rearrangement occurs during the lifetime of the 5-coordinate species. (We have previously discussed rearrangements in 5-coordinate complexes; see for example the end of *Section 2.11*.)

(a)

(b)

Fig. 25.7 Octahedral substitution through (a) a square-based pyramidal species leads to retention of the initial *cis*-isomer but through (b), a trigonal bipyramidal species, may lead to isomerization.

Base-catalyzed hydrolysis

Substitution reactions of Co(III) ammine complexes are catalyzed by $[OH]^-$, and for reaction 25.37, the rate law is equation 25.38.

$$[Co(NH_3)_5X]^{2+} + [OH]^- \longrightarrow [Co(NH_3)_5OH]^{2+} + X^-$$
$$(25.37)$$

$$\text{Rate} = k_{obs}[Co(NH_3)_5X^{2+}][OH^-] \qquad (25.38)$$

That $[OH]^-$ appears in the rate equation shows it has a rate-determining role. However, this is *not* because $[OH]^-$ attacks the metal centre but rather because it deprotonates a co-ordinated NH_3 ligand. Steps 25.39–25.41 show the *conjugate–base mechanism* (*Dcb* or S_N1cb mechanism). A pre-equilibrium is first established, followed by loss of X^- to give the reactive amido species **25.1**, and, finally, formation of the product in a fast step.

$$\begin{bmatrix} & NH_3 & \\ & | & \\ H_2N = Co & \cdots NH_3 \\ & | & \searrow NH_3 \\ & NH_3 & \end{bmatrix}^{2+}$$

(25.1)

$$[Co(NH_3)_5X]^{2+} + [OH]^- \underset{}{\overset{K}{\rightleftharpoons}} [Co(NH_3)_4(NH_2)X]^+ + H_2O$$
$$(25.39)$$

$$[Co(NH_3)_4(NH_2)X]^+ \xrightarrow{k_2} [Co(NH_3)_4(NH_2)]^{2+} + X^-$$
$$(25.40)$$

$$[Co(NH_3)_4(NH_2)]^{2+} + H_2O \xrightarrow{fast} [Co(NH_3)_5(OH)]^{2+}$$
$$(25.41)$$

If the equilibrium constant for equilibrium 25.39 is K, then the rate law consistent with this mechanism is given by equation 25.42. If $K[OH^-] \ll 1$, then equation 25.42 simplifies to equation 25.38 where $k_{obs} = Kk_2$.

$$\text{Rate} = \frac{Kk_2[Co(NH_3)_5X^{2+}][OH^-]}{1 + K[OH^-]} \qquad (25.42)$$

Two observations that are consistent with (but cannot rigidly establish) the conjugate–base mechanism are that:

- if NH_3 is replaced by pyridine or another tertiary amine, base hydrolysis is very much slower;
- the exchange of H (in the NH_3) for D in alkaline D_2O is much faster than the rate of base hydrolysis.

The Green–Taube experiment provides an elegant demonstration that a conjugate–base mechanism operates: when base hydrolysis (with a fixed concentration of $[OH]^-$) of $[Co(NH_3)_5X]^{2+}$ (X = Cl, Br, NO$_3$) is carried out in a mixture of $H_2(^{16}O)$ and $H_2(^{18}O)$, it is found that the ratio of $[Co(NH_3)_5(^{16}OH)]^{2+}$ to $[Co(NH_3)_5(^{18}OH)]^{2+}$ is constant

and independent of X^-. This provides strong evidence that the entering group is H_2O, and not $[OH]^-$, at least in the cases of the leaving groups being Cl^-, Br^- and $[NO_3]^-$.

Isomerization and racemization of octahedral complexes

Although the octahedron is stereochemically rigid, loss of a ligand gives a 5-coordinate species which can undergo Berry pseudo-rotation (see *Figure 2.13*). Although, earlier in this chapter, we discussed cases where the assumption is that such rearrangement does *not* occur, if the lifetime of the intermediate is long enough, it provides a mechanism for isomerization (e.g. equation 25.43). Such isomerization is related to mechanisms already described.

$$trans\text{-}[MX_4Y_2] \xrightarrow{-Y} \{MX_4Y\}$$
$$\xrightarrow{Y} trans\text{-}[MX_4Y_2] + cis\text{-}[MX_4Y_2] \qquad (25.43)$$

Our main concern in this section is the racemization of chiral complexes $M(L-L)_3$ and $cis\text{-}M(L-L)_2XY$ containing symmetrical or asymmetrical chelating ligands, $L-L$, and monodentate ligands, X and Y.

For $[Ni(bpy)_3]^{2+}$ and $[Ni(phen)_3]^{2+}$, the rates of exchange with ^{14}C-labelled ligands are the same as the rates of racemization. This is consistent with a dissociative process (equation 25.44) in which the intermediate is racemic, or racemizes faster than recombination with $L-L$.

$$M(L-L)_3 \underset{-S}{\overset{\text{Solvent, S}}{\rightleftharpoons}} M(L-L)_2S_2 + L-L \qquad (25.44)$$

Such a dissociative mechanism is rare, and kinetic data are usually consistent with an intramolecular process, e.g. for $[Cr(ox)_3]^{3-}$, $[Co(ox)_3]^{3-}$ (low-spin) and $[Fe(bpy)_3]^{2+}$ (low-spin), the rate of racemization exceeds that of ligand exchange.[†] Two intramolecular mechanisms are possible: a twist mechanism, or the cleavage and reformation of the M–L bond of *one end* of the didentate ligand. Alternative twist mechanisms (the *Bailar and Ray–Dutt twists*) for the interconversion of enantiomers of $M(L-L)_3$ are shown in Figure 25.8. Each transition state is a trigonal prism and the mechanisms differ only in which pair of opposing triangular faces twist with respect to each other. The ligands remain coordinated throughout. It is proposed that the racemization of $[Ni(en)_3]^{2+}$ occurs by a twist mechanism.

The second intramolecular mechanism for racemization involves the dissociation of *one* donor atom of a didentate ligand to give a 5-coordinate species which may undergo rearrangement within the time that the donor atom remains uncoordinated. Scheme 25.45 summarizes the available pathways for the interconversion of Δ and Λ enantiomers of $M(L-L)_3$.

[†] Ligand abbreviations: see *Table 6.7*.

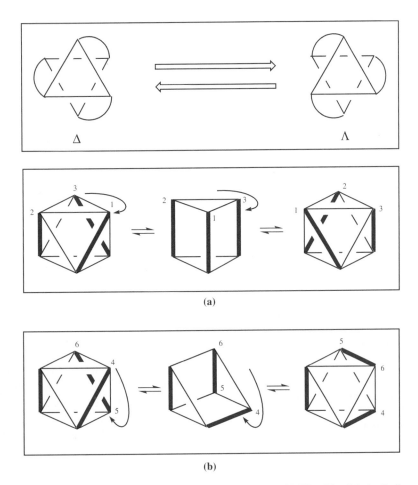

Fig. 25.8 Twist mechanisms for the interconversion of Δ and Λ enantiomers of $M(L-L)_3$: (a) the Bailar twist and (b) the Ray–Dutt twist. The chelating $L-L$ ligands are represented by the blue lines (see also *Box 19.2*).

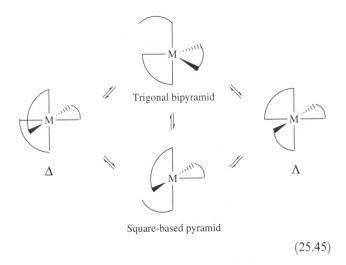

$$(25.45)$$

In aqueous solution, racemization of tris-oxalato complexes is faster than exchange of ox^{2-} by two H_2O ligands, suggesting that the two processes are mechanistically different. For $[Rh(ox)_3]^{3-}$ (**25.2**) the non-coordinated O atoms exchange with ^{18}O (from labelled H_2O) faster than do the coordinated

O atoms, the rate for the latter being comparable to the rate of racemization. This is consistent with a mechanism involving dissociation of one end of the ox^{2-} ligand, both for isotope exchange of the coordinated O, and for racemization.

$$(25.2)$$

If the chelating ligand is asymmetrical (i.e. has two different donor groups), geometrical isomerization is possible as well as racemization making the kinetics of the system more difficult to interpret. Similarly, racemization of complexes of the type *cis*-$M(L-L)_2XY$ is complicated by competing isomerization. The kinetics of these systems are dealt with in more advanced texts.

25.5 Electron-transfer processes

The simplest redox reactions involve *only* the transfer of electrons, and can be monitored by using isotopic tracers, e.g. reaction 25.46.

$$[^{56}Fe(CN)_6]^{3-} + [^{59}Fe(CN)_6]^{4-}$$

$$\longrightarrow [^{56}Fe(CN)_6]^{4-} + [^{59}Fe(CN)_6]^{3-} \qquad (25.46)$$

If $[^{54}MnO_4]^-$ is mixed with unlabelled $[MnO_4]^{2-}$, it is found that however rapidly $[MnO_4]^{2-}$ is precipitated as $BaMnO_4$, incorporation of the label has occurred. In the case of electron transfer between $[Os(bpy)_3]^{2+}$ and $[Os(bpy)_3]^{3+}$, the rate of electron transfer can be measured by studying the loss of optical activity (reaction 25.47).

$$(+)[Os(bpy)_3]^{2+} + (-)[Os(bpy)_3]^{3+}$$

$$\rightleftharpoons (-)[Os(bpy)_3]^{2+} + (+)[Os(bpy)_3]^{3+} \qquad (25.47)$$

Electron-transfer processes fall into two classes, defined by Taube: *outer-sphere* and *inner-sphere mechanisms*.

In an *outer-sphere mechanism*, electron transfer occurs *without a covalent linkage* being formed between the reactants. In an *inner-sphere mechanism*, electron transfer occurs via a *covalently bound bridging ligand*.

In some cases, kinetic data readily distinguish between outer- and inner-sphere mechanisms, but in many reactions, rationalizing the data in terms of a mechanism is not straightforward.

Inner-sphere mechanism

In 1953, Taube (who received the Nobel Prize for Chemistry in 1983) made the classic demonstration of an inner-sphere reaction on a skilfully chosen system (reaction 25.48) in which the reduced forms were substitutionally labile and the oxidized forms were substitutionally inert.

$$[Co(NH_3)_5Cl]^{2+} + [Cr(H_2O)_6]^{2+} + 5[H_3O]^+ \longrightarrow$$

low-spin Co(III)   high-spin Cr(II)
non-labile   labile

$$[Co(H_2O)_6]^{2+} + [Cr(H_2O)_5Cl]^{2+} + 5[NH_4]^+ \qquad (25.48)$$

high-spin Co(II)   Cr(III)
labile   non-labile

All the Cr(III) produced was in the form of $[Cr(H_2O)_5Cl]^{2+}$, and tracer experiments in the presence of excess, unlabelled Cl^- showed that all the chloro ligand in $[Cr(H_2O)_5Cl]^{2+}$ originated from $[Co(NH_3)_5Cl]^{2+}$. Since the Co centre could not have lost Cl^- before reduction, and Cr could not have gained Cl^- after oxidation, the transferred Cl^- must have been bonded to both metal centres during the reaction. Intermediate **25.3** is consistent with these observations.

BOX 25.1   **CHEMICAL AND THEORETICAL BACKGROUND**

Timescales of experimental techniques for studying electron-transfer reactions

In Section 2.11, we discussed fluxional processes in relation to the timescales of NMR and IR spectroscopies. A range of techniques are now available to probe electron-transfer reactions, and the recent development of femtosecond (fs) and picosecond (ps) flash photolysis methods now allow

investigations of extremely rapid reactions. For his studies of transition states of chemical reactions using femtosecond spectroscopy, Ahmed H. Zewail was awarded the 1999 Nobel Prize in Chemistry.

For details of experimental methods, see end-of-chapter reading list.

For information on femtochemistry, see:
J.C. Williamson, J. Cao, H. Ihee, H. Frey and A.H. Zewail (1997) *Nature*, vol. 386, p. 159 – 'Clocking

transient chemical changes by ultrafast electron diffraction'.
A.H. Zewail (1994) *Femtochemistry: Ultrafast Dynamics of the Chemical Bond*, 2 volumes, World Scientific, Singapore.

(25.3)

(25.4)

In the above example, Cl^- is transferred between metal centres; such transfer is often (but not necessarily) observed. In the reaction between $[Fe(CN)_6]^{3-}$ and $[Co(CN)_5]^{3-}$, the intermediate **25.4** (which is stable enough to be precipitated as the Ba^{2+} salt) is slowly hydrolyzed to $[Fe(CN)_6]^{4-}$ and $[Co(CN)_5(H_2O)]^{2-}$ without transfer of the bridging ligand. Common bridging ligands in inner-sphere mechanisms include halides, $[OH]^-$, $[CN]^-$, $[NCS]^-$, pyrazine (**25.5**) and 4,4′-bipyridine (**25.6**); pyrazine acts as an electron-transfer bridge in the Creutz–Taube cation and related species (see structure **22.49** and discussion).

(25.5)　　　　　　(25.6)

> The steps of an *inner-sphere mechanism* are bridge formation, electron transfer and bridge cleavage.

Equations 25.49–25.51 illustrate the inner-sphere mechanism for reaction 25.48; the product $[Co(NH_3)_5]^{2+}$ adds H_2O and then hydrolyzes in a fast step to give $[Co(H_2O)_6]^{2+}$.

$$[Co^{III}(NH_3)_5Cl]^{2+} + [Cr^{II}(H_2O)_6]^{2+} \underset{k_{-1}}{\overset{k_1}{\rightleftharpoons}}$$

$$[(NH_3)_5Co^{III}(\mu\text{-}Cl)Cr^{II}(H_2O)_5]^{4+} + H_2O \qquad (25.49)$$

$$[(NH_3)_5Co^{III}(\mu\text{-}Cl)Cr^{II}(H_2O)_5]^{4+} \underset{k_{-2}}{\overset{k_2}{\rightleftharpoons}}$$

$$[(NH_3)_5Co^{II}(\mu\text{-}Cl)Cr^{III}(H_2O)_5]^{4+} \qquad (25.50)$$

$$[(NH_3)_5Co^{II}(\mu\text{-}Cl)Cr^{III}(H_2O)_5]^{4+} \underset{k_{-3}}{\overset{k_3}{\rightleftharpoons}}$$

$$[Co^{II}(NH_3)_5]^{2+} + [Cr^{III}(H_2O)_5Cl]^{2+} \qquad (25.51)$$

Most inner-sphere processes exhibit second order kinetics overall, and interpreting the data is seldom simple. Any one of bridge formation, electron transfer or bridge cleavage can be rate determining. In the reaction between

Table 25.8 Second order rate constants for reaction 25.52 with different bridging X ligands.

Bridging ligand, X	$k \, / \, dm^3 \, mol^{-1} \, s^{-1}$
F^-	2.5×10^5
Cl^-	6.0×10^5
Br^-	1.4×10^6
I^-	3.0×10^6
$[N_3]^-$	3.0×10^5
$[OH]^-$	1.5×10^6
H_2O	0.1

$[Fe(CN)_6]^{3-}$ and $[Co(CN)_5]^{3-}$, the rate-determining step is the breaking of the bridge, but it is common for the electron transfer to be the rate-determining step. For bridge formation to be rate determining, the substitution required to form the bridge must be slower than electron transfer. This is not so in reaction 25.49: substitution in $[Cr(H_2O)_6]^{2+}$ (high-spin d^4) is very rapid, and the rate-determining step is electron transfer. However, if $[Cr(H_2O)_6]^{2+}$ is replaced by $[V(H_2O)_6]^{2+}$ (d^3), then the rate constant for reduction is similar to that for water exchange. This is also true for the reactions between $[V(H_2O)_6]^{2+}$ and $[Co(NH_3)_5Br]^{2+}$ or $[Co(CN)_5(N_3)]^{3-}$, indicating that the bridging group has little effect on the rate and that the rate-determining step is the ligand substitution required for bridge formation (the rate depending on the *leaving group*, H_2O) (see *Section 25.4*.)

For reaction 25.52 with a range of ligands X, the rate-determining step is electron transfer, and the rates of reaction depend on X (Table 25.8). The increase in k along the series F^-, Cl^-, Br^-, I^- correlates with increased ability of the halide to act as a bridge; k for $[OH]^-$ is similar to that for Br^-, but for H_2O, k is very small and is also pH-dependent. This observation is consistent with H_2O not being the bridging species at all, but rather $[OH]^-$, its availability in solution varying with pH.

$$[Co(NH_3)_5X]^{2+} + [Cr(H_2O)_6]^{2+} + 5[H_3O]^+ \longrightarrow$$

$$[Co(H_2O)_6]^{2+} + [Cr(H_2O)_5X]^{2+} + 5[NH_4]^+ \qquad (25.52)$$

(25.7)

Thiocyanate can coordinate through either the *N*- or *S*-donor, and the reaction of $[Co(NH_3)_5(NCS\text{-}S)]^{2+}$ (**25.7**) with $[Cr(H_2O)_6]^{2+}$ leads to the linkage isomers $[Cr(H_2O)_5(NCS\text{-}N)]^{2+}$ (70%) and $[Cr(H_2O)_5(NCS\text{-}S)]^{2+}$ (30%). The results are explained in terms of different bridge structures. If the free *N*-donor in **25.7** bonds to the Cr(II) centre to give bridge **25.8**, then the reaction proceeds to form $[Cr(H_2O)_5(NCS\text{-}N)]^{2+}$. Alternatively, bridge structure **25.9** gives the green $[Cr(H_2O)_5(NCS-S)]^{2+}$; this is unstable and isomerizes to the purple $[Cr(H_2O)_5(NCS-N)]^{2+}$.

$$\left[(H_3N)_5Co - SCN - Cr(OH_2)_5 \right]^{4+}$$

(25.8)

$$\left[(H_3N)_5Co - \overset{\overset{\displaystyle N}{\overset{\displaystyle \|}{C}}}{S} - Cr(OH_2)_5 \right]^{4+}$$

(25.9)

Conjugated organic anions (e.g. ox^{2-}) lead to faster inner-sphere reactions than non-conjugated anions (e.g. succinate, $^-O_2CCH_2CH_2CO_2^-$). In the reaction of $[Fe(CN)_5(H_2O)]^{3-}$ with $[Co(NH_3)_5(\textbf{25.10})]^{3+}$ in which the spacer X in **25.10** is varied, the reaction is fast when X provides a conjugated bridge allowing efficient electron transfer, and is slower for short, saturated bridges such as CH_2. However, rapid electron transfer is also observed when the spacer is very flexible, even when it is a saturated (insulating) chain. This observation is consistent with the metal centres being brought in closer contact and a change to an outer-sphere mechanism.

X = CH₂, CH₂CH₂, CH₂CH₂CH₂,
CH=CH, C(O)

(25.10)

Outer-sphere mechanism

When *both* reactants in a redox reaction are *kinetically inert*, electron transfer must take place by a *tunnelling* or *outer-sphere* mechanism. For a reaction such as 25.46, $\Delta G^\circ \approx 0$, but activation energy is needed to overcome electrostatic repulsion between ions of like charge, to stretch or shorten bonds so that they are equivalent in the transition state (see below), and to alter the solvent sphere around each complex.

> In a *self-exchange reaction*, the left- and right-hand sides of the equation are identical; only electron transfer, and no net chemical reaction, takes place.

The rates of outer-sphere self-exchange reactions vary considerably as illustrated in Table 25.9. The fastest reactions listed are between two low-spin complexes which differ only in the presence of an extra non-bonding electron in the t_{2g} orbitals of the complex containing the lower oxidation state metal. Such pairs of low-spin complexes have similar bond lengths in their ground states, e.g. Fe−C in $[Fe(CN)_6]^{4-}$ and $[Fe(CN)_6]^{3-}$ are 192 and 195 pm respectively, and Fe−N in $[Fe(bpy)_3]^{2+}$ and $[Fe(bpy)_3]^{3+}$ are 197 and 196 pm respectively. Where one complex is high-spin and one low-spin, there is a significant difference between the metal–ligand bond lengths, e.g. Co−N in $[Co(NH_3)_6]^{2+}$ $(t_{2g}^5 e_g^2)$ and $[Co(NH_3)_6]^{3+}$ (t_{2g}^6) are 211 and 196 pm respectively. The implications of the differing bond lengths and spin states on the mechanism of self-exchange are as follows.

Table 25.9 Second order rate constants, k, for some outer-sphere redox reactions at 298 K in aqueous solution.

	Reaction	$k\,/\,dm^3\,mol^{-1}\,s^{-1}$
No net chemical reaction (self-exchange)	$[Fe(bpy)_3]^{2+} + [Fe(bpy)_3]^{3+} \longrightarrow [Fe(bpy)_3]^{3+} + [Fe(bpy)_3]^{2+}$	$>10^6$
	$[Os(bpy)_3]^{2+} + [Os(bpy)_3]^{3+} \longrightarrow [Os(bpy)_3]^{3+} + [Os(bpy)_3]^{2+}$	$>10^6$
	$[IrCl_6]^{3-} + [IrCl_6]^{2-} \longrightarrow [IrCl_6]^{2-} + [IrCl_6]^{3-}$	$>10^6$
	$[Fe(CN)_6]^{4-} + [Fe(CN)_6]^{3-} \longrightarrow [Fe(CN)_6]^{3-} + [Fe(CN)_6]^{4-}$	10^5
	$[Co(phen)_3]^{2+} + [Co(phen)_3]^{3+} \longrightarrow [Co(phen)_3]^{3+} + [Co(phen)_3]^{2+}$	40
	$[Fe(H_2O)_6]^{2+} + [Fe(H_2O)_6]^{3+} \longrightarrow [Fe(H_2O)_6]^{3+} + [Fe(H_2O)_6]^{2+}$	3
	$[Co(en)_3]^{2+} + [Co(en)_3]^{3+} \longrightarrow [Co(en)_3]^{3+} + [Co(en)_3]^{2+}$	10^{-4}
	$[Co(NH_3)_6]^{2+} + [Co(NH_3)_6]^{3+} \longrightarrow [Co(NH_3)_6]^{3+} + [Co(NH_3)_6]^{2+}$	10^{-6}
Net chemical reaction	$[Os(bpy)_3]^{2+} + [Mo(CN)_8]^{3-} \longrightarrow [Os(bpy)_3]^{3+} + [Mo(CN)_8]^{4-}$	2×10^9
	$[Fe(CN)_6]^{4-} + [Fe(phen)_3]^{3+} \longrightarrow [Fe(CN)_6]^{3-} + [Fe(phen)_3]^{2+}$	10^8
	$[Fe(CN)_6]^{4-} + [IrCl_6]^{2-} \longrightarrow [Fe(CN)_6]^{3-} + [IrCl_6]^{3-}$	4×10^5

Fig. 25.9 The outer-sphere mechanism: when the reactants have differing bond lengths, vibrationally excited states with equal bond lengths must be formed in order to allow electron transfer to occur.

Clearly, the reactants must approach closely for the electron to migrate from reductant to oxidant. However, there is an important restriction imposed by the *Franck–Condon principle*: during electron transfer, the nuclei are essentially stationary and so electron transfer between $[Co(NH_3)_6]^{2+}$ and $[Co(NH_3)_6]^{3+}$ (or between other redox partners with differing bond lengths) can only occur between vibrationally excited states with identical structures (Figure 25.9). This reductant–oxidant pair is called the *encounter* or *precursor complex*.

By the *Franck–Condon principle*, a molecular electronic transition is much faster than a molecular vibration.

The greater the changes in bond lengths required to reach the encounter complex, the slower the rate of electron transfer. Table 25.9 shows that the rate of self-exchange between $[Co(NH_3)_6]^{2+}$ and $[Co(NH_3)_6]^{3+}$ is relatively slow, while that between $[Fe(bpy)_3]^{2+}$ and $[Fe(bpy)_3]^{3+}$ is very fast. Table 25.9 illustrates another point: self-exchange between $[Co(phen)_3]^{2+}$ and $[Co(phen)_3]^{3+}$ is much faster than between $[Co(NH_3)_6]^{2+}$ and $[Co(NH_3)_6]^{3+}$ or $[Co(en)_3]^{2+}$ and $[Co(en)_3]^{3+}$ (all three exchange processes are between high-spin Co(II) and low-spin Co(III)). This is consistent with the ability of phen ligands to use their π-orbitals to facilitate the intermolecular migration of an electron from one ligand to another, and phen complexes tend to exhibit fast rates of self-exchange.

The accepted method of testing for an outer-sphere mechanism is to apply *Marcus–Hush* theory[†] which relates kinetic and thermodynamic data for two self-exchange reactions with data for the *cross-reaction* between the self-exchange partners, e.g. reactions 25.53–25.55.

$$[ML_6]^{2+} + [ML_6]^{3+} \longrightarrow [ML_6]^{3+} + [ML_6]^{2+}$$

Self-exchange 1   (25.53)

$$[M'L_6]^{2+} + [M'L_6]^{3+} \longrightarrow [M'L_6]^{3+} + [M'L_6]^{2+}$$

Self-exchange 2   (25.54)

$$[ML_6]^{2+} + [M'L_6]^{3+} \longrightarrow [ML_6]^{3+} + [M'L_6]^{2+}$$

Cross-reaction   (25.55)

For each self-exchange reaction, $\Delta G^\circ = 0$. The free energy of activation, $\Delta G^\ddagger$, for a self-exchange reaction can be written in terms of four contributing factors (equation 25.56).

$$\Delta G^\ddagger = \Delta_w G^\ddagger + \Delta_o G^\ddagger + \Delta_s G^\ddagger + RT \ln \frac{k'T}{hZ} \quad (25.56)$$

where T = temperature in K; R = molar gas constant; k' = Boltzmann constant; h = Planck constant; Z = effective collision frequency in solution $\approx 10^{11} \, dm^3 \, mol^{-1} \, s^{-1}$.

The contributions in this equation arise as follows:

- $\Delta_w G^\ddagger$ is the free energy associated with bringing the reductant and oxidant together and includes the work done to counter electrostatic repulsions;
- $\Delta_o G^\ddagger$ is the free energy associated with changes in bond distances;
- $\Delta_s G^\ddagger$ arises from rearrangements within the solvent spheres;
- the final term accounts for the loss of translational and rotational free energy on formation of the encounter complex.

Although we shall not delve into the theory, it is possible to calculate the terms on the right-hand side of equation 25.56, and thus to estimate values of $\Delta G^\ddagger$ for self-exchange reactions. The rate constant, k, for the self-exchange can then be calculated using equation 25.57; the results of such calculations have been checked against much experimental data, and the validity of the theory is upheld.

[†] For fuller treatments of Marcus–Hush theory, see end-of-chapter reading list; Rudolph A. Marcus received the Nobel Prize for Chemistry in 1992.

$$k = \kappa Z \, e^{(-\Delta G^{\ddagger}/RT)} \tag{25.57}$$

where κ (the transmission coefficient) ≈ 1; $Z \approx 10^{11} \, dm^3 \, mol^{-1} \, s^{-1}$ (see equation 25.56).

Now consider reactions 25.53–25.55, and let the rate and thermodynamic parameters be designated as follows:

- k_{11} and $\Delta G^{\ddagger}_{11}$ for self-exchange 1;
- k_{22} and $\Delta G^{\ddagger}_{22}$ for self-exchange 2;
- k_{12} and $\Delta G^{\ddagger}_{12}$ for the cross-reaction; the equilibrium constant is K_{12}, and the standard free energy of reaction is ΔG°_{12}.

The Marcus–Hush equation (which we shall not derive) is given by expression 25.58 and applies to outer-sphere mechanisms.

$$k_{12} = (k_{11} k_{22} K_{12} f_{12})^{1/2} \tag{25.58}$$

where f_{12} is defined by the relationship

$$\log f_{12} = \frac{(\log K_{12})^2}{4 \log \left(\dfrac{k_{11} k_{22}}{Z^2} \right)}$$

and Z is the collision frequency (see equation 25.56).

Equation 25.59 gives a logarithmic form of equation 25.58; often, $f \approx 1$ and so $\log f \approx 0$, allowing this term to be neglected in some cases. Thus, equation 25.60 is an approximate form of the Marcus–Hush equation.

$$\log k_{12} = 0.5 \log k_{11} + 0.5 \log k_{22} + 0.5 \log K_{12} + 0.5 \log f_{12} \tag{25.59}$$

$$\log k_{12} \approx 0.5 \log k_{11} + 0.5 \log k_{22} + 0.5 \log K_{12} \tag{25.60}$$

Values of k_{11}, k_{22}, K_{12} and k_{12} can be measured experimentally, or k_{11} and k_{22} theoretically (see above); K_{12} is determined from E_{cell} (see *Section 7.2*). If the value of k_{12} calculated from equation 25.60 agrees with the experimental value, this provides strong evidence that the cross-reaction proceeds by an outer-sphere mechanism. Deviation from equation 25.60 indicates that another mechanism is operative. For example, for the reaction between $[Fe(CN)_6]^{4-}$ and $[IrCl_6]^{2-}$, a calculated value of k_{12} of $7 \times 10^5 \, dm^3 \, mol^{-1} \, s^{-1}$ compares favourably with the observed value (Table 25.9), thereby supporting an outer-sphere mechanism.

Worked example 25.1   Marcus–Hush theory: a test for an outer-sphere mechanism

For the reaction:

$$[Ru(NH_3)_6]^{2+} + [Co(phen)_3]^{3+}$$
$$\longrightarrow [Ru(NH_3)_6]^{3+} + [Co(phen)_3]^{2+}$$

the observed rate constant is $1.5 \times 10^4 \, dm^3 \, mol^{-1} \, s^{-1}$ and the equilibrium constant is 2.6×10^5. The rate constants for the self-exchange reactions $[Ru(NH_3)_6]^{2+}/[Ru(NH_3)_6]^{3+}$ and $[Co(phen)_3]^{2+}/[Co(phen)_3]^{3+}$ are 8.2×10^2 and

$40 \, dm^3 \, mol^{-1} \, s^{-1}$ respectively. Are these data consistent with an outer-sphere mechanism for the cross-reaction?

The approximate form of the Marcus–Hush equation is:

$$k_{12} \approx (k_{11} k_{22} K_{12})^{1/2} \quad \text{(or its log form)}$$

Calculate k_{12} using this equation:

$$k_{12} \approx [(8.2 \times 10^2)(40)(2.6 \times 10^5)]^{1/2}$$
$$\approx 9.2 \times 10^4 \, dm^3 \, mol^{-1} \, s^{-1}$$

This is in quite good agreement with the observed value of $1.5 \times 10^4 \, dm^3 \, mol^{-1} \, s^{-1}$, and suggests that the mechanism is outer-sphere electron transfer.

Self-study exercise

For the reaction given above, use the values of $k_{12} = 1.5 \times 10^4 \, dm^3 \, mol^{-1} \, s^{-1}$, $K_{12} = 2.6 \times 10^5$, and k for the self-exchange reaction $[Ru(NH_3)_6]^{2+}/[Ru(NH_3)_6]^{3+}$ to estimate a value of k for the self-exchange $[Co(phen)_3]^{2+}/[Co(phen)_3]^{3+}$. Comment on the agreement between your value and the observed value of $40 \, dm^3 \, mol^{-1} \, s^{-1}$.

$$[Ans. \approx 1.1 \, dm^3 \, mol^{-1} \, s^{-1}]$$

By using the relationships in equations 25.34 and 25.35, we can write equation 25.60 in terms of free energies (equation 25.61).

$$\Delta G^{\ddagger}_{12} \approx 0.5 \Delta G^{\ddagger}_{11} + 0.5 \Delta G^{\ddagger}_{22} + 0.5 \Delta G^{\circ}_{12} \tag{25.61}$$

In a series of related redox reactions in which one reactant is the same, a plot of $\Delta G^{\ddagger}_{12}$ against ΔG°_{12} is linear with a gradient of 0.5, if an outer-sphere mechanism is operative.

An important application of Marcus–Hush theory is in bioinorganic electron-transfer systems.[†] For example, cytochrome *c* is an electron-transfer metalloprotein (see *Section 28.4*) and contains haem-iron as either Fe(II) or Fe(III). Electron transfer from one Fe centre to another is *long range*, the electron *tunnelling* through the protein. Model systems have been devised to investigate electron transfer between cytochrome *c* and molecular complexes such as $[Ru(NH_3)_6]^{2+}$, and kinetic data are consistent with Marcus theory, indicating outer-sphere processes. For electron transfer in both metalloproteins and the model systems, the distance between the metal centres is significantly greater than for transfer between two simple metal complexes, e.g. up to 2500 pm. The rate of electron transfer decreases exponentially with increasing distance, *r*, between the two metal centres (equation 25.62, where β is a parameter which depends on the molecular environment).

Rate of electron transfer $\propto e^{-\beta r}$ $\qquad$ (25.62)

[†] For further discussion, see: R.G. Wilkins (1991) *Kinetics and Mechanism of Reactions of Transition Metal Complexes*, 2nd edn, Wiley-VCH, Weinheim, p. 285; J.J.R. Fraústo da Silva and R.J.P. Williams (1991) *The Biological Chemistry of the Elements*, Clarendon Press, Oxford, p. 105.

Summary

This chapter has introduced the basic concepts of inorganic reaction kinetics and mechanism, with an emphasis on ligand substitution in square planar and octahedral complexes, racemization of chiral octahedral complexes, and electron-transfer processes. Correlations between electronic structures and CFSE contribute towards our understanding of these processes.

The following terms have been introduced in this chapter. Do you know what they mean?

- ❏ leaving group
- ❏ entering group
- ❏ kinetically inert
- ❏ kinetically labile
- ❏ associative mechanism, A
- ❏ dissociative mechanism, D
- ❏ interchange mechanism, I_a or I_d
- ❏ intermediate
- ❏ transition state
- ❏ rate-determining step
- ❏ fast step
- ❏ activation parameters
- ❏ volume of activation, $\Delta V^{\ddagger}$
- ❏ *trans*-effect
- ❏ nucleophilicity sequence
- ❏ nucleophilicity parameter
- ❏ nucleophilicity discrimination factor
- ❏ Eigen–Wilkins mechanism
- ❏ encounter complex
- ❏ pre-equilibrium
- ❏ anation
- ❏ linear free energy relationship, LFER
- ❏ conjugate–base mechanism, Dcb
- ❏ Bailar twist mechanism
- ❏ Ray–Dutt twist mechanism
- ❏ outer-sphere mechanism
- ❏ inner-sphere mechanism
- ❏ Franck–Condon principle
- ❏ self-exchange mechanism
- ❏ cross-reaction
- ❏ Marcus–Hush theory (fundamental principles)

Suggested further reading

For an introduction to rate laws

P.W. Atkins (1998) *Physical Chemistry*, Oxford University Press, Oxford – Chapters 25–27 give a detailed account.

C.E. Housecroft and E.C. Constable (1997) *Chemistry: An Integrated Approach*, Addison Wesley Longman, Harlow – Chapter 10 gives a basic introduction.

Kinetics and mechanisms of inorganic and organometallic reactions

J.D. Atwood (1997) *Inorganic and Organometallic Reaction Mechanisms*, 2nd edn, Wiley-VCH, Weinheim – One of the most readable texts dealing with coordination and organo-metallic reaction mechanisms.

F. Basolo and R.G. Pearson (1967) *Mechanisms of Inorganic Reactions*, Wiley, New York – A classic book in the field of inorganic mechanisms.

R.D. Cannon (1994) 'Electron-transfer reactions: Theory' in *Encyclopedia of Inorganic Chemistry*, ed. R.B. King, Wiley, Chichester, vol. 3, p. 1098.

J.F. Endicott (1994) 'Electron transfer in coordination compounds' in *Encyclopedia of Inorganic Chemistry*, ed. R.B. King, Wiley, Chichester, vol. 3, p. 1081.

R.B. Jordan (1998) *Reaction Mechanisms of Inorganic and Organometallic Systems*, 2nd edn, Oxford University Press, New York – A detailed text which includes experimental methods, photochemistry and bioinorganic systems.

S.F.A. Kettle (1996) *Physical Inorganic Chemistry*, Spektrum, Oxford – Chapter 14 gives an excellent introduction and includes photokinetics.

A.G. Lappin (1994) *Redox Mechanisms in Inorganic Chemistry*, Ellis Horwood, Chichester – A comprehensive review of redox reactions in inorganic chemistry, including multiple electron transfer and some aspects of bioinorganic chemistry.

M.L. Tobe and J. Burgess (1999) *Inorganic Reaction Mechanisms*, Addison Wesley Longman, Harlow – A comprehensive account of inorganic mechanisms.

R.G. Wilkins (1991) *Kinetics and Mechanism of Reactions of Transition Metal Complexes*, 2nd edn, Wiley-VCH, Weinheim – An excellent and detailed text which includes experimental methods.

More-specialized reviews

M.J. Blandamer, J. Burgess and J.B.F.N. Engberts (1985) *Chemical Society Reviews*, vol.14, p. 237 – 'Activation parameters for chemical reactions in solution'.

R.J. Cross (1985) *Chemical Society Reviews*, vol. 14, p. 197 – 'Ligand substitution reactions in square planar molecules'.

R. van Eldik (1999) *Coordination Chemistry Reviews*, vol. 182, p. 373 – 'Mechanistic studies in coordination chemistry'.

A. Haim (1975) *Accounts of Chemical Research*, vol. 8, p. 264 – 'Role of the bridging ligand in inner-sphere electron-transfer reactions'.

R.A. Marcus (1986) *Journal of Physical Chemistry*, vol. 90, p. 3460 – 'Theory, experiment and reaction rates: a personal view'.

G. McLendon (1988) *Accounts of Chemical Research*, vol. 21, p. 160 – 'Long-distance electron transfer in proteins and model systems'.

W. Preetz, G. Peters and D. Bublitz (1996) *Chemical Reviews*, vol. 96, p. 977 – 'Preparation and investigation of mixed octahedral complexes'.

G. Stochel and R. van Eldik (1999) *Coordination Chemistry Reviews*, vol. 187, p. 329 – 'Elucidation of inorganic reaction mechanisms through volume profile analysis'.

H. Taube (1984) *Science*, vol. 226, p. 1028 – 'Electron transfer between metal complexes: Retrospective' (Nobel Prize in Chemistry lecture).

Problems

25.1 Review what is meant by the following terms:
(a) elementary step, (b) rate-determining step,
(c) activation energy, (d) intermediate, (e) transition
state, (f) rate equation, (g) zero, first and second order
rate laws, (h) nucleophile.

25.2 Sketch reaction profiles for the reaction pathways
described in equations 25.3 and 25.4. Comment on any
significant features including activation energies.

25.3 Discuss evidence to support the proposal that
substitution in square planar complexes is an associative
process.

25.4 Under pseudo-first order conditions, the variation of k_{obs}
with [py] for reaction of square planar
$[Rh(cod)(PPh_3)_2]^+$ (2×10^{-4} mol dm^{-3}, cod $= \textbf{23.20}$)
with pyridine is as follows:

[py] / mol dm^{-3}	0.006 25	0.0125	0.025	0.05
k_{obs} / s^{-1}	27.85	30.06	34.10	42.04

Show that the data are consistent with the reaction
proceeding by two competitive routes, indicate what
these pathways are, and determine values of the rate
constants for each pathway. [Data from: H. Krüger *et al.*
(1987) *J. Chem. Ed.*, vol. 64, p. 262.]

25.5 (a) The *cis*- and *trans*-isomers of $[PtCl_2(NH_3)(NO_2)]^-$
are prepared by reaction sequences 25.16 and 25.17
respectively. Rationalize the observed differences in
products in these routes. (b) Suggest the products of the
reaction of $[PtCl_4]^{2-}$ with PEt$_3$.

25.6 (a) Suggest a mechanism for the reaction:

trans-$[PtL_2Cl_2] + Y \longrightarrow$ *trans*-$[PtL_2ClY]^+ + Cl^-$

(b) If the intermediate in your mechanism is sufficiently
long lived, what complication might arise?

25.7 For the reaction:

$[Co(NH_3)_5(H_2O)]^{3+} + X^- \longrightarrow [Co(NH_3)_5X]^{2+} + H_2O$

it is found that:

$$\frac{d[Co(NH_3)_5X^{2+}]}{dt} = k_{obs}[Co(NH_3)_5(H_2O)^{3+}][X^-]$$

and for $X^- = Cl^-$, $\Delta V^{\ddagger}$ is positive. Rationalize these
data.

25.8 (a) Rationalize the formation of the products in the
following sequence of reactions:

$[Rh(H_2O)_6]^{3+} \xrightarrow[-H_2O]{Cl^-} [RhCl(H_2O)_5]^{2+}$

$\xrightarrow[-H_2O]{Cl^-}$ *trans*-$[RhCl_2(H_2O)_4]^+$

$\xrightarrow[-H_2O]{Cl^-}$ *mer*-$[RhCl_3(H_2O)_3]$

$\xrightarrow[-H_2O]{Cl^-}$ *trans*-$[RhCl_4(H_2O)_2]^-$

(b) Suggest methods of preparing $[RhCl_5(H_2O)]^{2-}$, *cis*-
$[RhCl_4(H_2O)_2]^-$ and *fac*-$[RhCl_3(H_2O)_3]$.

25.9 What reason can you suggest for the sequence
$Co > Rh > Ir$ in the rates of anation of $[M(H_2O)_6]^{3+}$
ions?

25.10 Derive rate law 25.42 for the mechanism shown in steps
25.39–25.41.

25.11 Suggest a mechanism for the possible racemization of
tertiary amines $NR_1R_2R_3$. Is it likely that such molecules
can be resolved?

25.12 The rate of racemization of $[CoL_3]$ where $HL = \textbf{25.11a}$ is
approximately the same as its rate of isomerization into
$[CoL'_3]$ where $HL' = \textbf{25.11b}$. What can you deduce about
the mechanisms of these reactions?

(25.11a)

(25.11b)

25.13 Substitution of H_2O in $[Fe(H_2O)_6]^{3+}$ by thiocyanate is
complicated by proton loss. By considering the reaction
scheme in Figure 25.10, derive an expression for
$-\dfrac{d[SCN^-]}{dt}$ in terms of the equilibrium and rate
constants, $[Fe(H_2O)_6{}^{3+}]$, $[SCN^-]$, $[Fe(H_2O)_5(SCN)^{2+}]$
and $[H^+]$.

25.14 Rationalize the observation that when the reaction:

$[Co(NH_3)_4(CO_3)]^+$

$\xrightarrow{[H_3O]^+, H_2O} [Co(NH_3)_4(H_2O)_2]^{3+} + CO_2$

is carried out in $H_2(^{18}O)$, the water in the complex
contains equal proportions of $H_2(^{18}O)$ and $H_2(^{16}O)$.

25.15 Two twist-mechanisms for the rearrangement of
Δ-$M(L-L)_3$ to Λ-$M(L-L)_3$ are shown in Figure 25.8.
The initial diagrams in (a) and (b) are identical; confirm
that the enantiomers formed in (a) and (b) are also
identical.

$[Fe(H_2O)_6]^{3+} + [SCN]^- \underset{k_{-1}}{\overset{k_1}{\rightleftharpoons}} [Fe(H_2O)_5(SCN)]^{2+} + H_2O$

$\Big\Updownarrow K_1 \qquad\qquad\qquad\qquad\qquad \Big\Updownarrow K_2$

$[Fe(H_2O)_5(OH)]^{2+} + H^+$ $\underset{k_{-2}}{\overset{k_2}{\rightleftharpoons}}$ $[Fe(H_2O)_4(OH)(SCN)]^+$
$+ [SCN]^-$ $+ H^+ + H_2O$

Fig. 25.10 Scheme for problem 25.13.

25.16 The rate constants for racemization (k_r) and dissociation (k_d) of $[FeL_3]^{4-}$ ($H_2L = $ **25.12**) at several temperatures, T, are given in the table. (a) Determine $\Delta H^{\ddagger}$ and $\Delta S^{\ddagger}$ for each reaction. (b) What can you deduce about the mechanism of racemization?

T/K	288	294	298	303	308
$k_r \times 10^5 / s^{-1}$	0.5	1.0	2.7	7.6	13.4
$k_d \times 10^5 / s^{-1}$	0.5	1.0	2.8	7.7	14.0

[Data from: A. Yamagishi (1986) *Inorg. Chem.*, vol. 25, p. 55.]

(25.12)

25.17 The reaction:

$$[Cr(NH_3)_5Cl]^{2+} + NH_3 \longrightarrow [Cr(NH_3)_6]^{3+} + Cl^-$$

in liquid NH_3 is catalyzed by KNH_2. Suggest an explanation for this observation.

25.18 Give an example of a reaction that proceeds by an inner-sphere mechanism. Sketch reaction profiles for inner-sphere electron transfer reactions in which the rate determining step is (a) bridge formation, (b) electron transfer and (c) bridge cleavage. Which profile is most commonly observed?

25.19 Discuss, with examples, the differences between inner- and outer-sphere mechanisms, and state what is meant by a self-exchange reaction.

25.20 Account for the relative values of the rate constants for the following electron transfer reactions in aqueous solution:

Reaction number	Reactants	$k / dm^3\,mol^{-1}\,s^{-1}$
I	$[Ru(NH_3)_6]^{3+} + [Ru(NH_3)_6]^{2+}$	10^4
II	$[Co(NH_3)_6]^{3+} + [Ru(NH_3)_6]^{2+}$	10^{-2}
III	$[Co(NH_3)_6]^{3+} + [Co(NH_3)_6]^{2+}$	10^{-8}

For which reactions is $\Delta G^o = 0$?

25.21 (a) Write equations for electron transfer between $[Fe(CN)_6]^{3-}$ and $[Fe(CN)_6]^{4-}$ (reaction **A**), $[Mo(CN)_8]^{3-}$ and $[Mo(CN)_8]^{4-}$ (reaction **B**), and $[Mo(CN)_8]^{3-}$ and $[Fe(CN)_6]^{4-}$ (reaction **C**) and give the metal oxidation states in each reactant and product. (b) For reactions **A** and **B**, values of k are 7.4×10^2 and $3.0 \times 10^4\,dm^3\,mol^{-1}\,s^{-1}$ respectively; for reaction **C**, $K = 1.0 \times 10^2$. Use these data to estimate k for reaction **C**, and compare it with the experimental value of $3.0 \times 10^4\,dm^3\,mol^{-1}\,s^{-1}$. What conclusions can you draw? State any assumptions made in your calculation.

25.22 The proposed mechanism for the oxidation of ascorbic acid, H_2A, by $[Fe(CN)_6]^{3-}$ is as follows, where the rate-determining step is the one-electron transfer shown in the second step:

$$H_2A \rightleftharpoons H^+ + HA^-$$

$$[Fe(CN)_6]^{3-} + HA^- \xrightarrow{k} [Fe(CN)_6]^{4-} + HA^{\bullet}$$

$$HA^{\bullet} \rightleftharpoons A^{\bullet -} + H^+$$

$$[Fe(CN)_6]^{3-} + A^{\bullet -} \xrightarrow{fast} [Fe(CN)_6]^{4-} + A$$

A value of $k = 1130\,dm^3\,mol^{-1}\,s^{-1}$ has been determined experimentally at 298 K. Under the experimental conditions for studying the kinetics, the values of E for the $[Fe(CN)_6]^{3-}/[Fe(CN)_6]^{4-}$ and $HA^{\bullet}/HA^-$ couples are $+0.45$ and $+0.71$ V respectively. Values of the self-exchange rate constants for reactions between $[Fe(CN)_6]^{3-}$ and $[Fe(CN)_6]^{4-}$, and between $HA^{\bullet}$ and HA^- are 5.8×10^4 and $3 \times 10^5\,dm^3\,mol^{-1}\,s^{-1}$ respectively. (a) Write an equation for the overall reaction between H_2A and $[Fe(CN)_6]^{3-}$. (b) Calculate a value of k using Marcus theory and suggest a mechanism for the reaction. [Data from: L.J.A. Martins *et al.* (1988) *J. Chem. Ed.*, vol. 65, p. 176.]

Homogeneous and heterogeneous catalysis

26.1 Introduction and definitions

Numerous applications of catalysts in small-scale synthesis and the industrial production of chemicals have been described in this book. Now we discuss catalysis in detail, focusing on commercial applications. Catalysts containing *d*-block metals are of immense importance to the chemical industry: they provide cost-effective syntheses, and control the specificity of reactions that might otherwise give mixed products. In 1990 in the US, the value of chemicals (including fuels) produced with at least one manufacturing catalytic step was 890 billion dollars.[†] The search for new catalysts is one of the major driving forces behind organometallic research, and the chemistry in much of this chapter can be understood in terms of the reaction types introduced in *Chapter 23*.

A *catalyst* is a substance that alters the rate of a reaction without appearing in any of the products of that reaction; it may speed up or slow down a reaction. For a reversible reaction, a catalyst alters the rate at which equilibrium is attained; it does *not* alter the position of equilibrium.

The term catalyst is often used to encompass both the *catalyst precursor* and the *catalytically active species*. A catalyst precursor is the substance added to the reaction, but it may undergo loss of a ligand such as CO or PPh_3 before it is available as the catalytically active species.

Although one tends to associate a catalyst with *increasing* the rate of a reaction, a *negative catalyst* slows down a reaction.

Some reactions are internally catalyzed (*autocatalysis*) once the reaction is under way, e.g. in the reaction of $[C_2O_4]^{2-}$ with $[MnO_4]^-$, the Mn^{2+} ions formed catalyze the forward reaction.

In an *autocatalytic reaction*, one of the products is able to catalyze the reaction.

Catalysts fall into two categories, homogeneous and heterogeneous, depending on their relationship to the phase of the reaction in which they are involved.

A *homogeneous catalyst* is in the same phase as the components of the reaction that it is catalyzing; a *heterogeneous catalyst* is in a different phase from the components of the reaction for which it is acting.

26.2 Catalysis: introductory concepts

Free energy profiles for a reaction: catalyzed versus non-catalyzed

A catalyst operates by allowing a reaction to follow a different pathway from that of the non-catalyzed reaction. If the activation barrier is lowered, then the reaction proceeds more rapidly. Figure 26.1 illustrates this for a reaction that follows a single step when it is non-catalyzed, but a two-step path when a catalyst is added. Each step in the catalyzed route has a characteristic free energy of activation, $\Delta G^{\ddagger}$, but

---

[†] For an overview of the growth of catalysis in industry during the 20th century, see: G.W. Parshall and R.E. Putscher (1986) *Journal of Chemical Education*, vol. 63, p. 189.

Fig. 26.1 A schematic representation of the reaction profile of a reaction without and with a catalyst. The pathway for the catalyzed reaction has two steps, and the first step is rate determining.

the step that matters with respect to the rate of reaction is that with the higher barrier; for the catalyzed pathway, the first step is the rate-determining step. (See *Box 26.1* for the relevant equations for and relationship between E_a and $\Delta G^{\ddagger}$.) Values of $\Delta G^{\ddagger}$ for the controlling steps in the catalyzed and non-catalyzed routes are marked in Figure 26.1. A crucial aspect of the catalyzed pathway is that it must not pass through an energy minimum *lower* than the energy of the products – such a minimum would be an 'energy sink', and would lead to the pathway yielding different products from those desired.

Catalytic cycles

A catalyzed reaction pathway is usually represented by a *catalytic cycle*.

A *catalytic cycle* consists of a series of stoichiometric reactions (often reversible) that form a closed loop; the catalyst must be regenerated so that it can participate in the cycle of reactions more than once.

For a catalytic cycle to be efficient, the intermediates must be short-lived. The downside of this for understanding the mechanism is that short lifetimes make studying a cycle difficult. Experimental probes are used to investigate the kinetics of a catalytic process, isolate or trap the intermediates, attempt to monitor intermediates in solution, or devise systems that model individual steps so that the product of the model-step represents an intermediate in the cycle. In the latter, the 'product' can be characterized by conventional techniques (e.g. NMR and IR spectroscopies, X-ray diffraction, mass spectrometry). For many cycles, however, the mechanisms are not firmly established.

We now study one cycle in detail to illustrate the notations. Figure 26.2 shows a simplified catalytic cycle for the Wacker process which converts ethene to acetaldehyde (equation 26.1); the process was developed in the 1950s and although it is not of great industrial significance nowadays, it provides a well-studied example for close examination.

$$CH_2{=}CH_2 + \tfrac{1}{2}O_2 \xrightarrow{\ [PdCl_4]^{2-}\ \text{catalyst}\ } CH_3CHO \qquad (26.1)$$

The *feedstocks* for the industrial process are highlighted along with the final product in Figure 26.2. The catalyst in the Wacker process contains palladium: through most of the cycle, the metal is present as Pd(II) but is reduced to Pd(0) as CH_3CHO is produced. We now work through the cycle, considering each step in terms of the organometallic reaction types discussed in *Section 23.7*.

The first step involves substitution by $CH_2{=}CH_2$ in $[PdCl_4]^{2-}$ (equation 26.2); at the top of Figure 26.2, the arrow notation shows $CH_2{=}CH_2$ entering the cycle and Cl^- leaving. One Cl^- is then replaced by H_2O, but we ignore this in Figure 26.2.

$$[PdCl_4]^{2-} + CH_2{=}CH_2 \longrightarrow [PdCl_3(\eta^2\text{-}C_2H_4)]^- + Cl^- \qquad (26.2)$$

The next step involves nucleophilic attack by H_2O with loss of H^+; recall that coordinated alkenes are susceptible to nucleophilic attack (see *equation 23.68*). In the third step, β-elimination occurs and formation of the Pd−H bond

BOX 26.1   **CHEMICAL AND THEORETICAL BACKGROUND**

Energy and free energy of activation: E_a and $\Delta G^{\ddagger}$

The Arrhenius equation:

$$\ln k = \ln A - \frac{E_a}{RT} \quad \text{or} \quad k = A\,e^{(-E_a/RT)}$$

is often used to relate the rate constant, k, of a reaction to the activation energy, E_a, and to the temperature, T (in K). In this equation, A, is the pre-exponential factor, and R = molar gas constant. The activation energy is often approximated to $\Delta H^{\ddagger}$, but the exact relationship is:

$$E_a = \Delta H^{\ddagger} + RT$$

The free energy of activation, $\Delta G^{\ddagger}$, which is related to the rate constant by the equation:

$$k = \frac{k'T}{h}\,e^{(-\Delta G^{\ddagger}/RT)}$$

where k' = Boltzmann's constant, h = Planck's constant.

In Section 25.2 we discussed activation parameters, including $\Delta H^{\ddagger}$ and $\Delta S^{\ddagger}$, and showed how these can be determined from an Eyring plot (Figure 25.1) which derives from the equation above relating k to $\Delta G^{\ddagger}$.

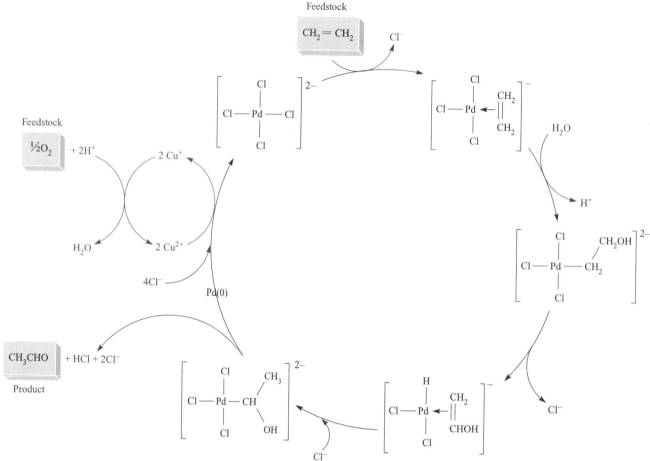

Fig. 26.2 Catalytic cycle for the Wacker process; for simplicity, we have ignored the role of coordinated H_2O, which replaces Cl^- *trans* to the alkene.

results in loss of Cl^-. This is followed by attack by Cl^- with H atom migration to give a σ-bonded $CH(OH)CH_3$ group. Elimination of CH_3CHO, H^+ and Cl^- with reduction of Pd(II) to Pd(0) occurs in the last step. To keep the cycle going, Pd(0) is now oxidized by Cu^{2+} (equation 26.3). The secondary cycle in Figure 26.2 shows the reduction of Cu^{2+} to Cu^+ and reoxidation of the latter by O_2 in the presence of H^+ (equation 26.4).

$$Pd + 2Cu^{2+} + 8Cl^- \longrightarrow [PdCl_4]^{2-} + 2[CuCl_2]^- \quad (26.3)$$

$$2[CuCl_2]^- + \tfrac{1}{2}O_2 + 2HCl \longrightarrow 2CuCl_2 + 2Cl^- + H_2O \quad (26.4)$$

If the whole cycle in Figure 26.2 is considered with species 'in' balanced against species 'out', the *net reaction* is reaction 26.1.

Choosing a catalyst

A reaction is not usually catalyzed by a unique species and a number of criteria must be considered when choosing the most effective catalyst, especially for a commercial process. Moreover, altering a catalyst in an industrial plant already in operation may be costly (e.g. a new plant design may be required) and the change must be guaranteed to be financially viable. Apart from the changes in reaction conditions

that the use of a catalyst may bring about (e.g. pressure and temperature), other factors that must be considered are:

- the concentration of catalyst required;
- the catalytic turnover;
- the selectivity of the catalyst to the desired product;
- how often the catalyst needs renewing.

The *catalytic turnover number* (TON) is the number of moles of product per mole of catalyst; this number indicates the number of catalytic cycles for a given process, e.g. after 2 h, the TON was 2400. The *catalytic turnover frequency* (TOF) is the catalytic turnover per unit time: the number of moles of product per mole of catalyst per unit time, e.g. the TOF was $20 \, \text{min}^{-1}$.

Defining the catalytic turnover number and frequency is not without problems. For example, if there is more than one product, one should distinguish between values of the total TON and TOF for all the catalytic products, and specific values for individual products. The term catalytic turnover number is usually used for batch processes, whereas catalytic turnover frequency is usually applied to continuous processes (flow reactors).

Now we turn to the question of selectivity, and the conversion of propene to an aldehyde provides a good example. Equation 26.5 shows the four possible products that may result from the reaction of propene with CO and H_2 (*hydroformylation*; see also *Section 26.3*).

$$CH_3CH=CH_2 \xrightarrow{CO/H_2} \begin{array}{l} CH_3CH_2CH_2CHO \xrightarrow{H_2} CH_3CH_2CH_2CH_2OH \\ \textit{n-isomer} \\ \\ (CH_3)_2CHCHO \xrightarrow{H_2} (CH_3)_2CHCH_2OH \\ \textit{i-isomer} \end{array}$$

$$(26.5)$$

The following ratios are important:

- the *n:i* ratio of the aldehydes (regioselectivity of the reaction);
- the aldehyde:alcohol ratio for a given chain (chemoselectivity of the reaction).

The choice of catalyst can have a significant effect on these ratios. For reaction 26.5, a cobalt carbonyl catalyst (e.g. $HCo(CO)_4$) gives $\approx 80\%$ C_4-aldehyde, 10% C_4-alcohol and $\approx 10\%$ other products, and an *n:i* ratio $\approx 3:1$. For the same reaction, various rhodium catalysts with phosphine co-catalysts can give an *n:i* ratio of between 8:1 and 16:1, whereas ruthenium cluster catalysts show a high chemoselectivity to aldehydes with the regioselectivity depending on the choice of cluster, e.g. for $Ru_3(CO)_{12}$, $n:i \approx 2:1$, and for $[HRu_3(CO)_{11}]^-$, $n:i \approx 74:1$.

Although a diagram such as Figure 26.2 shows a catalyst being regenerated and passing once more around the cycle, in practice, catalysts eventually become exhausted or are *poisoned*, e.g. by impurities in the feedstock.

26.3 Homogeneous catalysis: industrial applications

We now describe selected homogeneous catalytic processes that are of industrial importance; many more processes are applied in industry and detailed accounts can be found in the suggested reading at the end of the chapter. Two advantages of homogeneous over heterogeneous catalysis are the relatively mild conditions under which many processes operate, and the selectivity that can be achieved. A disadvantage is the need to separate the catalyst at the end of a reaction in order to recycle it, e.g. in the hydroformylation process, volatile $HCo(CO)_4$ can be removed by flash evaporation. The use of polymer supports or biphasic systems (*Section 26.4*) makes catalyst separation easier, and the development of such species is an active area of current research.

Throughout this section, the role of *coordinatively unsaturated 16-electron species* (see *Section 23.7*) and the ability of the metal centre to change coordination number (essential requirements of an active catalyst) should be noted.

Alkene hydrogenation

The most widely used procedures for the hydrogenation of alkenes nearly all employ heterogeneous catalysts, but for certain specialized purposes, homogeneous catalysts are used. Although addition of H_2 to a double bond is thermodynamically favoured (equation 26.6), the kinetic barrier is high and a catalyst is required to permit the reaction to be carried out at a viable rate without the need for high temperatures and pressures.

$$CH_2=CH_2 + H_2 \longrightarrow C_2H_6 \quad \Delta G^o = -101 \text{ kJ mol}^{-1} \quad (26.6)$$

$$\begin{array}{c} Cl \\ | \\ Ph_3P \longrightarrow Rh \longrightarrow PPh_3 \\ | \\ PPh_3 \end{array}$$

$$(26.1)$$

Wilkinson's catalyst (**26.1**) has been widely studied, and in its presence alkene hydrogenation can be carried out at 298 K and 1 bar H_2 pressure. The red, 16-electron Rh(I) complex **26.1** can be prepared from $RhCl_3$ and PPh_3, and is commonly used in benzene/ethanol solution in which it dissociates to some extent (equilibrium 26.7); a solvent molecule (solv) fills the fourth site in $RhCl(PPh_3)_2$ to give $RhCl(PPh_3)_2(\text{solv})$.

$$RhCl(PPh_3)_3 \rightleftharpoons RhCl(PPh_3)_2 + PPh_3 \quad K = 1.4 \times 10^{-4}$$
(26.1)
$$(26.7)$$

The *cis*-oxidative addition of H_2 to $RhCl(PPh_3)_3$ yields an octahedral complex which dissociates giving a coordinatively unsaturated 16-electron species (equation 26.8). The solvated complex $RhCl(PPh_3)_2(\text{solv})$ (formed from $RhCl(PPh_3)_2$ in reaction 26.7) is also involved in the catalytic cycle (but at low concentrations) and probably acts in a similar manner to $RhCl(PPh_3)_3$.

$$RhCl(PPh_3)_3 + H_2 \rightleftharpoons \textit{cis-}RhCl(H)_2(PPh_3)_3$$
$$\text{16-electron} \qquad\qquad\qquad \text{18-electron}$$
$$\rightleftharpoons RhCl(H)_2(PPh_3)_2 + PPh_3 \quad (26.8)$$
$$\text{16-electron}$$

The addition of an alkene to $RhCl(H)_2(PPh_3)_2$ brings alkene and hydrido ligands together on the Rh(I) centre, allowing hydrogen migration, followed by reductive elimination of an alkane. The process is summarized in Figure 26.3, the role of the solvent being ignored. The scheme shown should not be taken as being unique; for example, for some alkenes, experimental data suggest that $RhCl(PPh_3)_2(\eta^2\text{-alkene})$ is an intermediate. Other catalysts that are effective for alkene hydrogenation include $HRuCl(PPh_3)_3$ and $HRh(CO)(PPh_3)_3$ (this precursor loses PPh_3 to become the active catalyst).

Substrates for hydrogenation catalyzed by Wilkinson's catalyst include alkenes, dienes, allenes, terpenes, butadiene rubbers, antibiotics, steroids and prostaglandins. Significantly, ethene actually poisons its own conversion to ethane

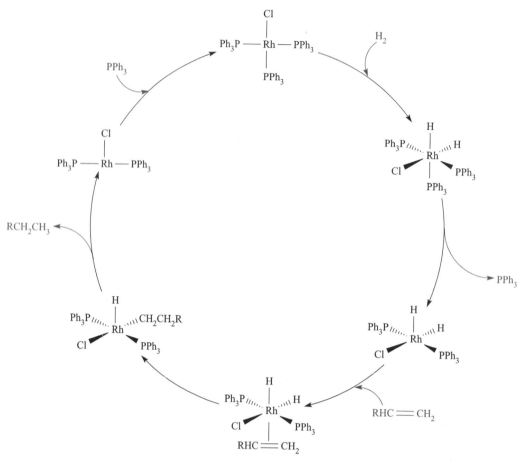

Fig. 26.3 Catalytic cycle for the hydrogenation of RCH=CH₂ using Wilkinson's catalyst, RhCl(PPh₃)₃.

and catalytic hydrogenation using RhCl(PPh₃)₃ cannot be applied in this case. For effective catalysis, the size of the alkene is important. The rate of hydrogenation is hindered by sterically demanding alkenes (Table 26.1); many useful *selective* hydrogenations can be achieved, e.g. reaction 26.9.

$$\text{(26.9)}$$

Biologically active compounds usually have at least one *asymmetric centre* and dramatic differences in the activities of different enantiomers of chiral drugs are commonly observed (see *Box 23.6*). Whereas one enantiomer may be an effective therapeutic drug, the other may be inactive or highly toxic as was the case with thalidomide.[†] *Asymmetric synthesis* is therefore an active field of research.

Asymmetric synthesis is an enantioselective synthesis and its efficiency can be judged from the *enantiomeric excess* (ee):

$$\% \text{ ee} = \left(\frac{|R - S|}{|R + S|} \right) \times 100$$

where R and S = relative quantities of *R* and *S* enantiomers.

An enantiomerically pure compound has 100% enantiomeric excess (100% ee). In *asymmetric catalysis*, the catalyst is chiral.

Table 26.1 Rate constants for the hydrogenation of alkenes (at 298 K in C₆H₆) in the presence of Wilkinson's catalyst.[‡]

Alkene	$k/\times 10^{-2} \text{ dm}^3 \text{ mol}^{-1} \text{ s}^{-1}$
Phenylethene (styrene)	93.0
Dodec-1-ene	34.3
Cyclohexene	31.6
Hex-1-ene	29.1
2-Methylpent-1-ene	26.6
1-Methylcyclohexene	0.6

[†] See for example: 'When drug molecules look in the mirror': E. Thall (1996) *Journal of Chemical Education*, vol. 73, p. 481; 'Counting on chiral drugs': S.C. Stinson (1998) *Chemical & Engineering News*, 21 Sept. issue, p. 83.

[‡] For further data, see: F.H. Jardine, J.A. Osborn and G. Wilkinson (1967) *Journal of the Chemical Society A*, p. 1574.

Table 26.2 Observed % ee of the product of the hydrogenation of $CH_2=C(CO_2H)(NHCOMe)$ using Rh(I) catalysts containing different chiral bisphosphines.

Bisphosphine		
(R,R)-DIOP	(S,S)-BPPM	(R,R)-DIPAMP
% ee (selective to enantiomer R or S) 73 (R)	99 (R)	90 (S)

If hydrogenation of an alkene can, in principle, lead to enantiomeric products, then the alkene is *prochiral* (see *problem 26.4b*). If the catalyst is *achiral* (as $RhCl(PPh_3)_3$ is), then the product of hydrogenation of the prochiral alkene is a racemic mixture: i.e. starting from a prochiral alkene, there is an equal chance that the σ-alkyl complex formed during the catalytic cycle (Figure 26.3) will be an R- or an S-enantiomer. If the catalyst is *chiral*, it should favour the formation of one or other of the R- or S-enantiomers, thereby making the hydrogenation enantioselective. *Asymmetric hydrogenations* can be carried out by modifying Wilkinson's catalyst, introducing a chiral phosphine or chiral didentate bisphosphine, e.g. (R,R)-DIOP (see *Table 26.2*). By varying the chiral catalyst, hydrogenation of a given prochiral alkene proceeds with differing enantiomeric selectivities as exemplified in Table 26.2. An early triumph of the application of asymmetric alkene hydrogenation to drug manufacture was the production of the alanine derivative L-DOPA (**26.2**) which is used in the treatment of Parkinson's disease.[†] The anti-inflammatory drug Naproxen (active in the (S)-form) is prepared by chiral resolution or by asymmetric hydrogenation of a prochiral alkene (reaction 26.10); enantiopurity is essential, since the (R)-enantiomer is a liver toxin.

(26.10)

Monsanto acetic acid synthesis

The conversion of MeOH to $MeCO_2H$ (equation 26.11) is carried out on a huge industrial scale: currently $\approx 3.5\,Mt^{\ddagger}$ of $MeCO_2H$ are produced a year worldwide, and 60% of the world's acetyls are manufactured using the Monsanto process.

$$MeOH + CO \longrightarrow MeCO_2H \qquad (26.11)$$

Before 1970, the BASF process (employing cobalt catalysts) was used commercially, but its replacement by the Monsanto process has brought the advantages of milder conditions and greater selectivity (Table 26.3). The Monsanto process involves two interrelated cycles. In the left-hand cycle in Figure 26.4, MeOH is converted to MeI, which then enters the Rh-cycle by oxidatively adding to the 16-electron complex *cis*-$[Rh(CO)_2I_2]^-$. This addition is the rate-determining step in the process, and so the formation of MeI is critical to the viability of the Monsanto process. The right-hand cycle in Figure 26.4 shows methyl migration to give a species which is shown as 5-coordinate, but an 18-electron species, either dimer **26.4** or $Rh(CO)(COMe)I_3$(solv) where solv = solvent, is more likely. Recent EXAFS (see *Box 26.2*) studies in THF solution indicate a dimer at 253 K, but solvated monomer above 273 K. Addition of CO follows to give an 18-electron, octahedral complex which eliminates

---

[†] For further details, see: W.A. Knowles (1986) *Journal of Chemical Education*, vol. 63, p. 222 – 'Application of organometallic catalysis to the commercial production of L-DOPA'.

[‡] Mt = megatonne; 1 metric tonne $\approx$ 1.1 US ton.

Table 26.3 Major advantages of the Monsanto process over the BASF process for the manufacture of acetic acid (equation 26.11) can be seen from the summary in this table.

Conditions	BASF (Co-based catalyst)	Monsanto (Rh-based catalyst)
Temperature / K	500	453
Pressure / bar	500–700	35
Catalyst concentration / mol dm^{-3}	0.1	0.001
Selectivity / %	90	>99

MeC(O)I. The latter enters the left-hand cycle in Figure 26.4 and is converted to acetic acid.

(26.4)

Optimizing manufacturing processes is essential for financial reasons, and each catalytic process has potential problems that have to be overcome. One difficulty in the Monsanto process is the possible oxidation of cis-[Rh(CO)$_2$I$_2$]$^-$ by HI (reaction 26.12), the product of which easily loses CO, resulting in the loss of the catalyst from the system (equation 26.13). Operating under a pressure of CO prevents this last detrimental step and also has the effect of reversing the effects of reaction 26.12 (equation

26.14). Adding small amounts of H$_2$ prevents oxidation of Rh(I) to Rh(III).

$$[Rh(CO)_2I_2]^- + 2HI \longrightarrow [Rh(CO)_2I_4]^- + H_2 \quad (26.12)$$

$$[Rh(CO)_2I_4]^- \longrightarrow RhI_3(s) + 2CO + I^- \quad (26.13)$$

$$[Rh(CO)_2I_4]^- + CO + H_2O \longrightarrow [Rh(CO)_2I_2]^- + 2HI + CO_2 \quad (26.14)$$

Iridium-based complexes also catalyze reaction 26.11, and the combination of [Ir(CO)$_2$I$_2$]$^-$ with Ru$_2$(CO)$_6$I$_2$(μ-I)$_2$ as a catalyst promoter provides a commercially viable system.

Tennessee–Eastman acetic anhydride process

The Tennessee–Eastman acetic anhydride process converts methyl acetate to acetic anhydride (equation 26.15) and has been in commercial use since 1983.

$$MeCO_2Me + CO \longrightarrow (MeCO)_2O \quad (26.15)$$

Fig. 26.4 The Monsanto acetic acid process involves two interrelated catalytic cycles.

Fig. 26.5 Catalytic cycle for the Tennessee–Eastman acetic anhydride process.

It closely resembles the Monsanto process but uses $MeCO_2Me$ in place of MeOH; *cis*-$[Rh(CO)_2I_2]^-$ remains the catalyst and the oxidative addition of MeI to *cis*-$[Rh(CO)_2I_2]^-$ is still the rate-determining step. One pathway can be described by adapting Figure 26.4, replacing:

- MeOH by $MeCO_2Me$;
- H_2O by $MeCO_2H$;
- $MeCO_2H$ by $(MeCO)_2O$.

However, a second pathway (Figure 26.5) in which LiI replaces HI is found to be extremely important for efficiency of the process; the final product is formed by the reaction of acetyl iodide and lithium acetate. Other alkali metal iodides do not function as well as LiI, e.g. replacing LiI by NaI slows the reaction by a factor of ≈ 2.5.

Hydroformylation (Oxo-process)

Hydroformylation (or the Oxo-process) is the conversion of alkenes to aldehydes (reaction 26.16). It is catalyzed by cobalt and rhodium carbonyl complexes and has been exploited as a manufacturing process since World War II.

$$RCH{=}CH_2 + CO + H_2$$
$$\longrightarrow RCH_2CH_2CHO + RCHMeCHO \qquad (26.16)$$
$$\text{linear (\emph{n}-isomer)} \qquad \text{branched (\emph{i}-isomer)}$$

Cobalt-based catalysts were the first to be employed. Under the conditions of the reaction (370–470 K, 100–400 bar), $Co_2(CO)_8$ reacts with H_2 to give $HCo(CO)_4$ and the latter

is usually represented in catalytic cycles as the precursor to the coordinatively unsaturated (i.e. active) species $HCo(CO)_3$. As equation 26.16 shows, hydroformylation can generate a mixture of linear and branched aldehydes, and the catalytic cycle in Figure 26.6 accounts for both products. All steps (except for the final release of the aldehyde) are reversible. To interpret the catalytic cycle, start with $HCo(CO)_3$ at the top of Figure 26.6. Addition of the alkene is the first step and this is followed by CO addition and accompanying H migration and formation of a σ-bonded alkyl group. At this point, the cycle splits into two routes depending on which C atom is involved in Co–C bond formation. The two pathways are shown as the inner and outer cycles in Figure 26.6. In each, the next step is alkyl migration, followed by oxidative addition of H_2 and the transfer of one H atom to the alkyl group to give elimination of the aldehyde. The inner cycle eliminates a linear aldehyde, while the outer cycle produces a branched isomer. Two major complications in the process are the hydrogenation of aldehydes to alcohols, and alkene isomerization (which is also catalyzed by $HCo(CO)_3$). The first of these problems (see *equation 26.5*) can be controlled by using H_2:CO ratios greater than 1:1 (e.g. 1.5:1). The isomerization problem (regioselectivity) can be addressed by using other catalysts (see below) or can be turned to advantage by purposely preparing mixtures of isomers for separation at a later stage. Scheme 26.17 illustrates the distribution of products formed when oct-1-ene undergoes hydroformylation at 423 K, 200 bar, and with a 1:1 H_2:CO ratio.

Fig. 26.6 Competitive catalytic cycles in the hydroformylation of alkenes to give linear (inner-cycle) and branched (outer-cycle) aldehydes.

(26.17)

Just as we saw that the rate of hydrogenation was hindered by sterically demanding alkenes (Table 26.1), so too is the rate of hydroformylation affected by steric constraints, as is illustrated by the data in Table 26.4.

Other hydroformylation catalysts that are used industrially are $HCo(CO)_3(PBu_3)$ (which, like $HCo(CO)_4$ must lose CO to become coordinatively unsaturated) and $HRh(CO)(PPh_3)_3$ (which loses PPh_3 to give the catalytically active $HRh(CO)(PPh_3)_2$). Data in Table 26.5 compare the operating conditions for, and selectivities of, these catalysts with those of $HCo(CO)_4$. The Rh(I) catalyst is particularly selective towards aldehyde formation, and under certain conditions the $n{:}i$ ratio is as high as 20:1. An excess of PPh_3 prevents reactions 26.18 which occur in the presence of CO; the products are also hydroformylation catalysts but lack the selectivity of $HRh(CO)(PPh_3)_2$. The parent

Table 26.4 Rate constants for the hydroformylation of selected alkenes at 383 K in the presence of the active catalytic species $HCo(CO)_3$.

Alkene	$k\,/\,\times 10^{-5}\,s^{-1}$
Hex-1-ene	110
Hex-2-ene	30
Cyclohexene	10
Oct-1-ene	109
Oct-2-ene	31
2-Methylpent-2-ene	8

Table 26.5 A comparison of the operating conditions for and selectivities of three commercial hydroformylation catalysts.

	$HCo(CO)_4$	$HCo(CO)_3(PBu_3)$	$HRh(CO)(PPh_3)_3$
Temperature / K	410–450	450	360–390
Pressure / bar	250–300	50–100	30
Regioselectivity $n:i$ ratio (see *equation 26.5*)	≈3:1	≈9:1	>10:1
Chemoselectivity (aldehyde predominating over alcohol)	High	Low	High

phosphine complex, $HRh(PPh_3)_3$ is inactive towards hydroformylation, and while $RhCl(PPh_3)_3$ is active, Cl^- acts as an inhibitor.

$$\left.\begin{array}{l} HRh(CO)(PPh_3)_2 + CO \rightleftharpoons HRh(CO)_2(PPh_3) + PPh_3 \\ HRh(CO)_2(PPh_3) + CO \rightleftharpoons HRh(CO)_3 + PPh_3 \end{array}\right\}$$

$$(26.18)$$

Alkene oligomerization

The Shell Higher Olefins Process (SHOP) uses a nickel-based catalyst to oligomerize ethene. The process is designed to be flexible, so that product distributions meet consumer demand. The process is complex, but Figure 26.7 gives a simplified catalytic cycle and indicates the form in which the nickel catalyst probably operates.

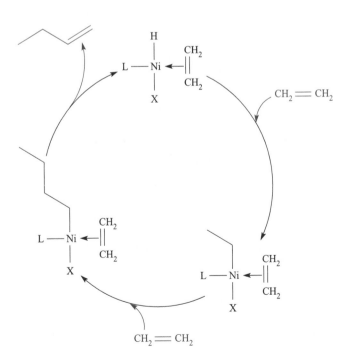

Fig. 26.7 Simplified catalytic cycle illustrating the oligomerization of ethene using a nickel-based catalyst; L = phosphine, X = electronegative group.

26.4 Homogeneous catalyst development

The development of new catalysts is an important research topic, and in this section we briefly introduce some areas of current interest.

Polymer-supported catalysts

Attaching homogeneous metal catalysts to polymer supports retains the advantages of mild operating conditions and selectivity usually found for conventional homogeneous catalysts, while aiming to overcome the difficulties of catalyst separation. Types of support include polymers with a high degree of cross-linking and with large surface areas, and microporous polymers (low degree of cross-linking) which swell when they are placed in solvents. A common method of attaching the catalyst to the polymer is to functionalize the polymer with a ligand that can then be used to coordinate to, and hence bind, the catalytic metal centre. Equation 26.19 gives a schematic representation of the use of a chlorinated polymer to produce phosphine groups supported on the polymer surface.

$$(26.19)$$

Alternatively, some polymers can bind the catalyst directly, e.g. poly-2-vinylpyridine (made from monomer **26.5**) is suitable for application in the preparation of hydroformylation catalysts (equation 26.20).

$$(26.20)$$

Hydroformylation catalysts can also be made by attaching the cobalt or rhodium carbonyl residues to a phosphine-functionalized surface through phosphine-for-carbonyl substitution. The chemo- and regioselectivities observed for the supported homogeneous catalysts are typically quite different from those of their conventional analogues.

While much progress has been made in this area, leaching of the metal into solution (which partly defeats the advantages gained with regard to catalyst separation) is a common problem.

Biphasic catalysis

$$(26.6)$$

Biphasic catalysis addresses the problem of catalyst separation: a water-soluble catalyst is retained in an aqueous layer that is immiscible with the organic medium in which the reaction takes place. Intimate contact between the two solutions is achieved during the catalytic reaction, after which the two liquids are allowed to settle and the catalyst-containing layer separated by decantation. Many homogeneous catalysts are hydrophobic and so it is necessary to introduce ligands that will bind to the metal but that carry hydrophilic substituents. Among ligands that have met with success is **26.6**: e.g. the reaction of an excess of **26.6** with $[Rh_2(nbd)_2(\mu-Cl)_2]$ (nbd = **23.21**) gives a species, probably $[RhCl(\mathbf{26.6})_3]^{3+}$, which catalyzes the hydroformylation of hex-1-ene to aldehydes (at 40 bar, 360 K) in 90% yield with an $n{:}i$ ratio of 4:1. An excess of the ligand in the aqueous phase stabilizes the catalyst and increases the $n{:}i$ ratio to ≈10:1. Much work has been carried out with the P-donor ligand **26.7** which can be introduced into a variety of organometallic complexes by carbonyl or alkene displacement. For example, the water-soluble complex $HRh(CO)(\mathbf{26.7})_3$ is a hydroformylation catalyst precursor; conversion of hex-1-ene to heptanal proceeds with 93% selectivity for the n-isomer, a higher selectivity than is shown by $HRh(CO)(PPh_3)_3$ under conventional homogeneous catalytic conditions. A range of alkene hydrogenations are catalyzed by $RhCl(\mathbf{26.7})_3$ and it is particularly efficient and selective for the hydrogenation of hex-1-ene.

$$(26.7)$$

$$(26.8)$$

Biphasic asymmetric hydrogenation has also been developed using water-soluble chiral bisphosphines such as **26.8** coordinated to Rh(I). With $PhCH{=}C(CO_2H)(NHC(O)Me)$ as substrate, hydrogenation takes place with 87% ee, and similar success has been achieved for related systems.

Although the biphasic catalysts described above appear analogous to those discussed in Section 26.3, it does *not* follow that the mechanisms by which the two catalysts operate for a given reaction are similar.

d-Block organometallic clusters as homogeneous catalysts

Over the past 25 years, much effort has been put into investigating the use of *d*-block organometallic clusters as homogeneous catalysts, and equations 26.21–26.23 give examples of small-scale catalytic reactions. Note that in reaction 26.21, insertion of CO is into the O−H bond; in the Monsanto process using $[Rh(CO)_2I_2]^-$ catalyst, CO insertion is into the C−OH bond (equation 26.11).

$$MeOH + CO \xrightarrow[\text{400 bar, 470 K}]{Ru_3(CO)_{12}} \underset{\text{90\% selectivity}}{HC(O)OMe} \qquad (26.21)$$

$$(26.22)$$

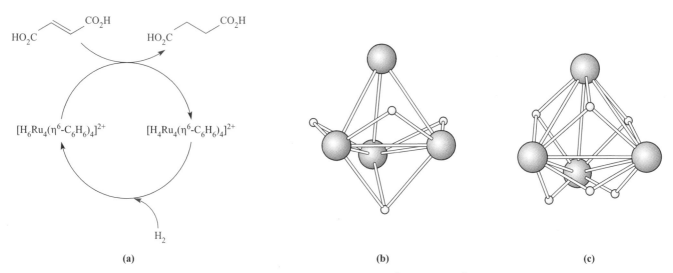

Fig. 26.8 (a) Catalytic cycle for the hydrogenation of fumaric acid by $[H_4(\eta^6\text{-}C_6H_6)_4Ru_4]^{2+}$; (b) H_4Ru_4 core of $[H_4(\eta^6\text{-}C_6H_6)_4Ru_4]^{2+}$ and (c) H_6Ru_4 core of $[H_6(\eta^6\text{-}C_6H_6)_4Ru_4]^{2+}$, both determined by X-ray diffraction [G. Meister *et al.* (1994) *J. Chem. Soc., Dalton Trans.*, p. 3215]. Colour code in (b) and (c): Ru, blue; H, pale grey.

(26.23)

A promising development in the area is the use of *cationic* clusters; $[H_4(\eta^6\text{-}C_6H_6)_4Ru_4]^{2+}$ catalyzes the reduction of fumaric acid, the reaction being selective to the C=C bond and leaving the carboxylic acid units intact (Figure 26.8).

Despite the large of amount of work that has been carried out in the area and the wide range of examples now known,[†] it would appear that no industrial applications of clusters catalysts have yet been found to be viable.

26.5 Heterogeneous catalysis: surfaces and interactions with adsorbates

The majority of industrial catalytic processes involve *heterogeneous catalysis* and Table 26.6 gives selected examples. Conditions are generally harsh, with high temperatures and pressures. Before describing specific industrial applications, we introduce some terminology and discuss the properties of metal surfaces and zeolites that render them useful as heterogeneous catalysts.

We shall mainly be concerned with reactions of gases over heterogeneous catalysts. Molecules of reactants are *adsorbed* on to the catalyst surface, undergo reaction and the products

---

[†] For a well-referenced review of this area, see: G. Süss-Fink and G. Meister (1993) *Advances in Organometallic Chemistry*, vol. 35, p. 41 – 'Transition metal clusters in homogeneous catalysis'.

are *desorbed*. Interaction between the adsorbed species and surface atoms may be of two types:

Physisorption involves weak van der Waals interactions between the surface and adsorbate. *Chemisorption* involves the formation of chemical bonds between surface atoms and the adsorbed species.

The process of adsorption activates molecules, either by cleaving bonds or by weakening them. The dissociation of a diatomic molecule such as H_2 on a metal surface is represented schematically in equation 26.24; bond formation does not have to be with a single metal atom as we illustrate later. Bonds in molecules, e.g. C−H, N−H, are similarly activated.

(26.24)

The balance between the contributing bond energies is a factor in determining whether or not a particular metal will facilitate bond fission in the adsorbate. However, if metal–adsorbate bonds are especially strong, it becomes energetically less favourable for the adsorbed species to leave the surface, and this blocks adsorption sites, reducing catalytic activity.

The adsorption of CO on metal surfaces has been thoroughly investigated. Analogies can be drawn between the interactions of CO with metal atoms on a surface and those in organometallic complexes (see *Section 23.2*), i.e. both terminal and bridging modes of attachment are possible, and IR spectroscopy can be used to study adsorbed CO. Upon interaction with a surface metal atom, the C−O bond is weakened in much the same way that we described in Figure 23.1. The extent of weakening depends not only

---

BOX 26.2   **CHEMICAL AND THEORETICAL BACKGROUND**

Some experimental techniques used in surface science

In much of this book, we have been concerned with studying species that are soluble and subjected to solution techniques such as NMR and electronic spectroscopy, or with structural data obtained from X-ray or neutron diffraction studies of *single crystals* or electron diffraction studies of gases. The investigation of solid surfaces requires specialist techniques, many of which have been developed relatively recently. Selected examples are listed below.

Acronym	Technique	Application and description of technique
AES	Auger electron spectroscopy	Study of surface composition
EXAFS	Extended X-ray absorption fine structure	Estimation of internuclear distances around a central atom
FTIR	Fourier transform infrared spectroscopy	Study of adsorbed species
HREELS	High resolution electron energy loss spectroscopy	Study of adsorbed species
LEED	Low energy electron diffraction	Study of structural features of the surface and of adsorbed species
SIMS	Secondary ion mass spectrometry	Study of surface composition
STM	Scanning tunnelling microscopy	Obtaining images of a surface and adsorbed species at an atomic level
XANES	X-ray absorption near edge spectroscopy	Study of oxidation states of surface atoms
XRD	X-ray diffraction	Investigation of phases and particle sizes
XPS (ESCA)	X-ray photoelectron spectroscopy (electron spectroscopy for chemical analysis)	Study of surface composition and oxidation states of surface atoms

For further details of solid state techniques, see:

J. Evans (1997) *Chemical Society Reviews*, vol. 26, p. 11 – 'Shining light on metal catalysts'.

S.S. Perry and G.A. Somorjai (1994) 'Surfaces' in *Encyclopedia of Inorganic Chemistry*, ed. R.B. King, Wiley, Chichester, vol. 7, p. 4064.

G.A. Somorjai (1994) *Surface Chemistry and Catalysis*, Wiley, New York.

A.R. West (1999) *Basic Solid State Chemistry*, 2nd edn, Wiley, Chichester.

✳ To view selected STM images, use the Web link listed on the Website that accompanies this book.

Table 26.6   Examples of industrial processes that use heterogeneous catalysts.

Industrial manufacturing process	Catalyst system
NH_3 synthesis (Haber process)[‡]	Fe on SiO_2 and Al_2O_3 support
Water–gas shift reaction[*]	Ni, iron oxides
Catalytic cracking of heavy petroleum distillates	Zeolites (see *Section 26.6*)
Catalytic reforming of hydrocarbons to improve octane number[**]	Pt, Pt–Ir and other Pt-group metals on acidic alumina support
Methanation ($CO \longrightarrow CO_2 \longrightarrow CH_4$)	Ni on support
Ethene epoxidation	Ag on support
HNO_3 manufacture (Haber–Bosch process)[***]	Pt–Rh gauzes

[‡] See *Section 14.5*.

[*] See *equation 9.11*.

[**] The octane number is increased by increasing the ratio of branched or aromatic hydrocarbons: straight chain hydrocarbons. The 0–100 octane number scale assigns 0 to *n*-heptane and 100 to 2,2,4-trimethylpentane.

[***] See *Section 14.9*.

---

on the mode of interaction with the surface, but on the surface coverage. In studies of the adsorption of CO on a Pd(111)[†] surface, it is found that the enthalpy of adsorption of CO decreases as more of the surface is covered with adsorbed molecules. An abrupt decrease in the amount of heat evolved per mole of adsorbate is observed when the surface is half-occupied by a *monolayer*; at this point, significant reorganization of the adsorbed molecules is needed to accommodate still more. Changes in the mode of attachment of CO molecules to the surface alter the strength of the C−O bond and the extent to which the molecule is activated.

[†] The notations (111), (110), (101)... are Miller indices and define the crystal planes in the metal lattice.

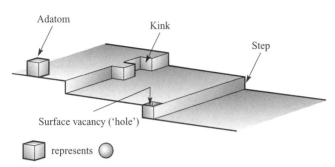

Fig. 26.9 A schematic representation of typical features of a metal surface. [Based on a figure from *Encyclopedia of Inorganic Chemistry* (1994), R. B. King (ed.), vol. 3, p. 1359, John Wiley & Sons: Chichester.]

Diagrams of hcp, fcc or bcc metal lattices such as we showed in *Figure 5.2* imply 'flat' metal surfaces. In practice, a surface contains imperfections such as those illustrated in Figure 26.9. The *kinks* on a metal surface are extremely important for catalytic activity, and their presence increases the rate of catalysis. In a close-packed lattice, sections of 'flat' surface contain M_3 triangles (**26.9**) while a step possesses a line of M_4 'butterflies' (see *Table 23.5*), one of which is highlighted in structure **26.10**. Both can accommodate adsorbed species in sites which can be mimicked by discrete metal clusters; this has led to the *cluster-surface analogy* (see *Section 26.7*).

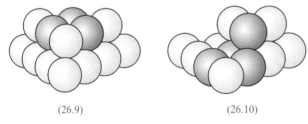

(26.9)   (26.10)

The design of metal catalysts has to take into account not only the available surface, but also the fact that the catalytically active platinum-group metals (see *Section 22.2*) are rare and expensive. There may also be the problem that extended exposure to the metal surface may result in side reactions. In many commercial catalysts including motor vehicle catalytic converters, small metal particles (e.g. 1600 pm in diameter) are dispersed on a support such as γ-alumina (*activated alumina*, see *Section 12.7*) which has a large surface area. Using a support of this type means that a high percentage of the metal atoms are available for catalysis. In some cases, the support itself may beneficially modify the properties of the catalyst; e.g. in hydrocarbon reforming (Table 26.6), the metal and support operate together:

- the platinum-group metal catalyzes the conversion of an alkane to alkene;
- isomerization of the alkene is facilitated by the acidic alumina surface;
- the platinum-group metal catalyzes the conversion of the isomerized alkene to an alkane which is more highly branched than the starting hydrocarbon.

As well as having roles as supports for metals, silica and alumina are used directly as heterogeneous catalysts. A major application is in the catalytic cracking of heavy petroleum distillates; very fine powders of silica and γ-alumina possess a huge surface area of $\approx 900 \, \text{m}^2 \, \text{g}^{-1}$. Large surface areas are a key property of zeolite catalysts (see *Section 13.9*), the selectivity of which can be tuned by varying the sizes, shapes and Brønsted acidity of their cavities and channels; we discuss these properties more fully in Section 26.6.

26.6 Heterogeneous catalysis: commercial applications

In this section, we describe selected commercial applications of heterogeneous catalysts. The examples have been chosen to illustrate a range of catalyst types, as well as the development of motor vehicle catalytic converters.

Alkene polymerization: Ziegler–Natta catalysis

The polymerization of alkenes to yield stereoregular polymers by heterogeneous Ziegler–Natta catalysis (see also *Boxes 18.3* and *23.7*) is of vast importance to the polymer industry. First generation catalysts were made by reacting $TiCl_4$ with Et_3Al to precipitate β-$TiCl_3 \cdot xAlCl_3$ which was converted to γ-$TiCl_3$. While the latter catalyzed the production of isotactic polypropene, its selectivity and efficiency required significant improvement.[†] A change in the method of catalyst preparation generated the δ-form of $TiCl_3$ which is stereoselective below 373 K. The co-catalyst, Et_2AlCl, in these systems is essential, its role being to alkylate Ti atoms on the catalyst surface. In third generation catalysts (used since the 1980s), $TiCl_4$ is supported on $MgCl_2$, and Et_3Al may be used for alkylation. Alkene polymerization is catalyzed at defect sites in the crystal lattice, and the *Cossee–Arlman mechanism* shown in Figure 26.10 is the accepted pathway of the catalytic process. In Figure 26.10, the $TiCl_5$ unit shown at the starting point represents a *surface site* which has a surface Cl atom and a vacant coordination position which renders the Ti centre coordinatively unsaturated. In the first step, the surface Cl atom is replaced by an ethyl group, and it is crucial that the alkyl group is *cis* to the vacant lattice site. Coordination of the alkene then takes place, followed by alkyl migration (see *equations 23.32–23.35*), and repetition of these last two steps results in polymer growth. In propene polymerization, the stereoselective formation of isotactic polypropene is thought to

---

[†] In *isotactic* polypropene, the methyl groups are all on the same side of the carbon chain; the polymer chains pack efficiently to give a crystalline material. Isotactic polypropene is of greater commercial value than the soft and elastic *atactic* polymer in which the Me groups are randomly arranged. Also of commercial importance is *syndiotactic* polypropene in which the Me groups are regularly arranged on either side of the carbon backbone.

Fig. 26.10 A schematic representation of alkene polymerization on the surface of a Ziegler–Natta catalyst; the vacant coordination site must be *cis* to the coordinated alkyl group.

be controlled by the catalyst's surface structure which imposes restrictions on the possible orientations of the coordinated alkene relative to the metal-attached alkyl group.

Fischer–Tropsch carbon chain growth

Scheme 26.25 summarizes the Fischer–Tropsch (FT) reaction, i.e. the conversion of synthesis gas (see *Section 9.4*) into hydrocarbons. A range of catalysts can be used (e.g. Ru, Ni, Fe, Co) but Fe and Co are presently favoured.

$$(26.25)$$

If petroleum is cheap and readily available, the FT process is not commercially viable and in the 1960s, many industrial plants were closed. In South Africa, the *Sasol process* continues to use H_2 and CO as feedstocks. Changes in the availability of oil reserves affect the views of industry as regards its feedstocks, and research interest in the FT reaction continues to be high.

The product distribution, including carbon chain length, of an FT reaction can be controlled by choice of catalyst, reactor design and reaction conditions; the addition of promoters such as group 1 or 2 metal salts (e.g. K_2CO_3) affects the selectivity of a catalyst. The exact mechanism by which the FT reaction occurs is not known, and many model studies have been carried out using discrete metal clusters (see *Section 26.7*). The original mechanism proposed by Fischer and Tropsch involved the adsorption of CO, C–O bond cleavage to give a surface carbide, and hydrogenation to produce CH_2

groups which then polymerized. Various mechanisms have been put forward, and the involvement of a surface-bound CH_3 group has been debated. Any mechanism (or series of pathways) must account for the formation of surface carbide, graphite and CH_4, and the distribution of organic products shown in scheme 26.25. Current opinion favours CO dissociation on the catalyst surface to give surface C and O and, in the presence of adsorbed H atoms (equation 26.24), the formation of surface CH and CH_2 units and release of H_2O. If CO dissociation and subsequent formation of CH_x groups is efficient (as it is on Fe), the build up of CH_x units leads to reaction between them and to the growth of carbon chains. The types of processes that might be envisaged on the metal surface are represented in scheme 26.26. Reaction of the surface-attached alkyl chain would release an alkane; if it undergoes β-elimination, an alkene is released.

$$(26.26)$$

It has also been suggested that vinylic species are involved in FT chain growth and that combination of a surface-bound

CH and CH_2 units to give $CH=CH_2$ may be followed by successive incorporation of CH_2 units alternating with alkene isomerization as shown in scheme 26.27. Release of a terminal alkene results if reaction of the adsorbate is with H instead of CH_2.

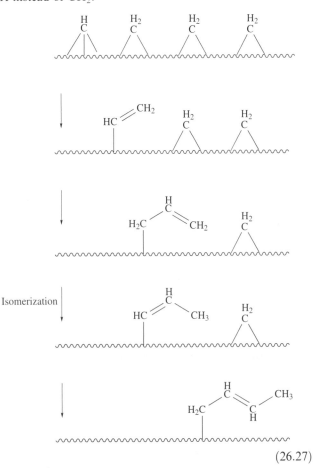

(26.27)

Haber process

Figure 26.11 illustrates the vast scale on which the industrial production of NH_3 is carried out and its growth over the past decade. In 1998, only production of phosphate rock (54.1 Mt), sulfuric acid (52.4 Mt) and lime (24.8 Mt) topped that of ammonia among inorganic chemicals and minerals in the US. In *equation 14.18* and the accompanying discussion, we described the manufacture of NH_3 using a heterogeneous catalyst. Now we focus on the mechanism of the reaction and on catalyst performance.

Without a catalyst, the reaction between N_2 and H_2 occurs only slowly, since the activation barrier for the dissociation of N_2 and H_2 in the gas phase is very high. In the presence of a suitable catalyst such as Fe, dissociation of N_2 and H_2 to give adsorbed atoms is facile, with the energy released by the formation of M−N and M−H bonds more than offsetting the energy required for $N\equiv N$ and H−H fission. The adsorbates then readily combine to form NH_3 which desorbs from the surface. The rate-determining step is the dissociative adsorption of N_2 (equation 26.28); the notation '(ad)' refers to an adsorbed atom.

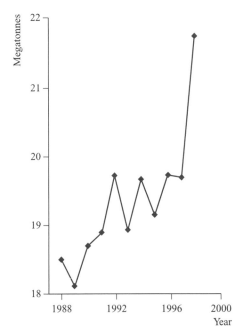

Fig. 26.11 The trend in the industrial output of NH_3 in the US from 1988 to 1998. [Data: *Chemical & Engineering News* (1999), 28 June issue, p. 34 (converted to SI units).]

$$N_2(g) \longrightarrow N(ad) \quad N(ad) \qquad (26.28)$$

Dihydrogen is similarly adsorbed (equation 26.24), and the surface reaction continues as shown in scheme 26.29 with gaseous NH_3 finally being released; activation barriers for each step are relatively low.

$$N(ad) \quad H(ad) \longrightarrow NH(ad)$$

$$NH(ad) \quad H(ad) \longrightarrow NH_2(ad)$$

$$NH_2(ad) \quad H(ad) \longrightarrow NH_3(ad)$$

$$\searrow \; NH_3(g)$$

(26.29)

Metals other than Fe catalyze the reaction between N_2 and H_2, but the rate of formation of NH_3 is metal-dependent. High rates are observed for Fe, Ru and Os. Since the rate-determining step is the chemisorption of N_2, a high activation energy for this step, as is observed for late *d*-block metals (e.g. Co, Rh, Ir, Ni and Pt), slows down the overall formation of NH_3. Early *d*-block metals such as Mo and Re chemisorb N_2 efficiently, but the M−N interaction is strong enough to favour retention of the adsorbed atoms;

this blocks surface sites and inhibits further reaction. The catalyst used industrially is active α-Fe which is produced by reducing Fe_3O_4 mixed with K_2O (an *electronic promoter* which improves catalytic activity), SiO_2 and Al_2O_3 (*structural promoters* which stabilize the catalyst's structure). High purity (often synthetic) magnetite and the catalyst promoters are melted electrically and then cooled; this stage distributes the promoters homogeneously within the catalyst. The catalyst is then ground to an optimum grain size. High purity materials are essential since some impurities poison the catalyst. Dihydrogen for the Haber process is produced as synthesis gas (*Section 9.4*), and contaminants such as H_2O, CO, CO_2 and O_2 are *temporary catalyst poisons*. Reduction of the Haber process catalyst restores its activity, but over-exposure of the catalyst to oxygen-containing compounds decreases the efficiency of the catalyst irreversibly; a 5 ppm CO content in the H_2 supply (see *equations 9.10* and *9.11*) decreases catalyst activity by ≈5% per year. The performance of the catalyst depends critically on the operating temperature of the NH_3 converter, and a 770–790 K range is optimal.

Production of SO_3 in the Contact process

In 1998, production of sulfuric acid in the US (Figure 26.12) was second only to that of phosphate rock among inorganic chemicals and minerals. The oxidation of SO_2 to SO_3 (equation 26.30) is the first step in the Contact process, and in *Section 15.8* we discussed how the yield of SO_3 depends on temperature and pressure. At ordinary temperatures, the reaction is too slow to be commercially viable, while at very high temperatures, equilibrium 26.30 shifts to the left,

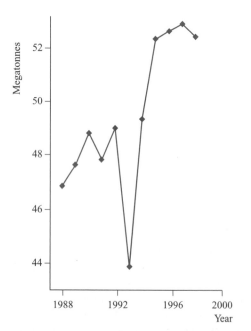

Fig. 26.12 The trend in the industrial output of H_2SO_4 in the US between 1988 and 1998. [Data: *Chemical & Engineering News* (1999), 28 June issue, p. 34 (converted to SI units).]

Fig. 26.13 Sources of NO_x emissions in the US in 1995; the total emission in 1995 was ≈24 million tonnes. [Data: *Chemical & Engineering News* (1997), 14 April issue, p. 10.]

decreasing the yield of SO_3.

$$2SO_2 + O_2 \rightleftharpoons 2SO_3 \qquad \Delta_r H^\circ = -96 \text{ kJ per mole of } SO_2$$
$$(26.30)$$

Use of a catalyst increases the rate of the forward reaction 26.30, and active catalysts are Pt, V(V) compounds and iron oxides. Modern manufacturing plants for SO_3 use a V_2O_5 catalyst on an SiO_2 carrier (which provides a large surface area) with a K_2SO_4 promoter; the catalyst system contains 4–9% by weight of V_2O_5. Passage of the reactants through a series of catalyst beds is required to obtain an efficient conversion of SO_2 to SO_3, and an operating temperature of 690–720 K is optimal. Since oxidation of SO_2 is exothermic and since temperatures >890 K degrade the catalyst, the $SO_2/SO_3/O_2$ mixture must be cooled between leaving one catalyst bed and entering to the next. Although the $V_2O_5/SiO_2/K_2SO_4$ system is introduced as a solid catalyst, the operating temperatures are such that the catalytic oxidation of SO_2 occurs in a liquid melt on the surface of the silica carrier. The reaction mechanism and intermediates have not been established, but the role of the vanadium(V) catalyst can be represented by scheme 26.31.

$$\left. \begin{array}{l} SO_2 + V_2O_5 \rightleftharpoons 2VO_2 + SO_3 \\ \frac{1}{2}O_2 + 2VO_2 \longrightarrow V_2O_5 \end{array} \right\} \qquad (26.31)$$

Catalytic converters

Environmental concerns have grown during the past few decades (see, for example, *Boxes 8.4* and *10.3*), and to the general public, the use of motor vehicle catalytic converters is well known. Regulated exhaust emissions comprise CO, hydrocarbons and NO_x (see *Section 14.8*). The radical NO is one of several species that act as catalysts for the conversion of O_3 to O_2 and is considered to contribute to depletion of the ozone layer. Although industrial processes also contribute to NO_x emissions,[†] the combustion of transport fuels is the major source (Figure 26.13). A typical catalytic converter is ≥90% efficient in reducing emissions, accommodating

---

[†] Shell and Bayer are among companies that have introduced processes to eliminate industrial NO_x emissions: *Chemistry & Industry* (1994) p. 415 – 'Environmental technology in the chemical industry'.

current European regulations which call for a 90% reduction in CO and an 85% decrease in hydrocarbon and NO_x emissions, bringing combined hydrocarbon and NO_x output to $<0.2 \, g \, km^{-1}$.

A catalytic converter consists of a honeycomb ceramic structure coated in finely divided Al_2O_3 (the *washcoat*); fine particles of catalytically active Pt, Pd and Rh are dispersed within the cavities of the washcoat and the whole unit is contained in a stainless steel vessel placed in sequence in the vehicle's exhaust pipe. As the exhaust gases pass through the converter at high temperatures, redox reactions 26.32–26.35 occur. Under legislation, the only acceptable emission products are CO_2, N_2 and H_2O.

$$2CO + O_2 \longrightarrow 2CO_2 \qquad (26.32)$$

$$C_nH_m + \left(n + \frac{m}{4}\right)O_2 \longrightarrow nCO_2 + \frac{m}{2}H_2O$$

$$(C_nH_m = \text{hydrocarbon}) \qquad (26.33)$$

$$2NO + 2CO \longrightarrow 2CO_2 + N_2 \qquad (26.34)$$

$$2NO + 2H_2 \longrightarrow N_2 + 2H_2O \qquad (26.35)$$

Whereas CO and hydrocarbons are oxidized, the destruction of NO_x involves its reduction. Modern catalytic converters have a '3-way' system which promotes both oxidation and reduction; Pd and Pt catalyze reactions 26.32 and 26.33, while Rh catalyzes reactions 26.34 and 26.35.

The efficiency of the catalyst depends, in part, on metal particle size, typically 1000–2000 pm diameter. Over a period of time, the high temperatures needed for the operation of a catalytic converter cause ageing of the metal particles with a loss of their optimal size and a decrease in the efficiency of the catalyst. Constant high temperature running also transforms the Al_2O_3 support into a phase with a lower surface area, again reducing catalytic activity. To counter degradation of the support, group 2 metal oxide stabilizers are added to the alumina; new support materials, such as a high-temperature-resistant fibrous silica–alumina washcoat developed by Toyota in 1998, may eventually replace Al_2O_3. Catalytic converters operate only with unleaded fuels; lead additives bind to the alumina washcoat, deactivating the catalyst.

In order to achieve the regulatory emission standards, it is crucial to control the air:fuel ratio as it enters the catalytic converter: the optimum ratio is 14.7:1. If the air:fuel ratio exceeds 14.7:1, extra O_2 competes with NO for H_2 and the efficiency of reaction 26.35 is lowered. If the ratio is less than 14.7:1, oxidizing agents are in short supply and CO, H_2 and hydrocarbons compete with each other for NO and O_2. The air:fuel ratio is monitored by a sensor fitted in the exhaust pipe; the sensor measures O_2 levels and sends an electronic signal to the fuel injection system or carburettor to adjust the air:fuel ratio as necessary. A recent development in catalytic converter design has been the introduction of a CeO_2/Ce_2O_3 system to store oxygen. During 'lean' periods of vehicle running, O_2 can be 'stored' by reaction 26.36;

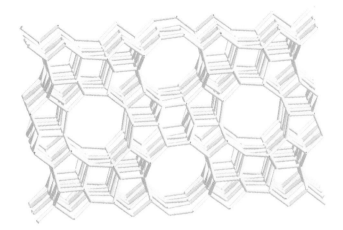

Fig. 26.14 Part of the aluminosilicate framework of synthetic zeolite ZSM-5 (structure-type MFI).

during 'rich' periods when extra oxygen is needed for hydrocarbon and CO oxidation, CeO_2 is reduced (equation 26.37).

$$2Ce_2O_3 + O_2 \longrightarrow 4CeO_2 \qquad (26.36)$$

$$2CeO_2 + CO \longrightarrow Ce_2O_3 + CO_2 \qquad (26.37)$$

A catalytic converter cannot function immediately after the 'cold-start' of an engine; at its 'light-off' temperature (typically 620 K), the catalyst operates at 50% efficiency but during the 90–120 s lead time, exhaust emissions are not controlled. Several methods have been developed to counter this problem, e.g. electrical heating of the catalyst using power from the vehicle's battery.

The development of catalytic converters has recently encompassed the use of zeolites, e.g. Cu-ZSM-5 (a copper-modified ZSM-5 system), but at the present time, and despite some advantages such as low light-off temperatures, zeolite-based catalysts have not shown themselves to be sufficiently durable for their use in catalytic converters to be commercially viable.

Zeolites as catalysts for organic transformations: uses of ZSM-5

For an introduction to zeolites, see *Figure 13.17* and the accompanying discussion. Many natural and synthetic zeolites are known, and it is the presence of well-defined cavities and/or channels, the dimensions of which are comparable with those of small molecules, that makes them invaluable as catalysts and molecular sieves. Zeolites are environmentally 'friendly' and the development of industrial processes in which they can replace less 'acceptable' acid catalysts is advantageous. In this section, we focus on catalytic applications of synthetic zeolites such as ZSM-5 (structure-type code MFI, Figure 26.14); the latter is silicon-rich with composition $Na_n[Al_nSi_{96-n}O_{192}] \cdot \approx 16H_2O$ ($n < 27$).[†] Within the

---

[†] Structures of zeolites can be viewed and manipulated using the Website: http://www.iza-structure.org/

$$(26.12)$$

aluminosilicate framework of ZSM-5 lies a system of inter-linked channels; one set can be seen in Figure 26.14, but the channels are often represented in the form of structure **26.11**. Each channel has an elliptical cross-section (53×56 pm and 51×55 pm) and the *effective pore size* is comparable to the *kinetic molecular diameter* of a molecule such as 2-methylpropane or benzene, leading to the *shape-selective* properties of zeolite catalysts. The effective pore size differs from that determined crystallographically because it takes into account the flexibility of the zeolite framework as a function of temperature; similarly, the kinetic molecular diameter allows for the molecular motions of species entering the zeolite channels or cavities.

$$(26.11)$$

The high catalytic activity of zeolites arises from the Brønsted acidity of Al sites, represented in resonance pair **26.12**; the Si:Al ratio affects the number of such sites and acid strength of the zeolite.

Zeolite catalysts are important in the catalytic cracking of heavy petroleum distillates; their high selectivities and high rates of reactions, coupled with reduced coking effects, are major advantages over the activities of the alumina/silica catalysts that zeolites have replaced. Ultrastable Y (USY) zeolites are usually chosen for catalytic cracking because their use leads to an increase in the gasoline (motor fuels) octane number. It is essential that the catalyst be robust enough to withstand the conditions of the cracking process; both USY and ZSM-5 (used as a co-catalyst because of its shape-selective properties) meet this requirement. The shape-selectivity of ZSM-5 is also crucial to its activity as a catalyst in the conversion of methanol to hydrocarbon fuels; the growth of carbon chains is restricted by the size of the zeolite channels and thereby gives a selective distribution of hydrocarbon products. The MTG (methanol-to-gasoline) process has operated on an industrial scale in New Zealand from 1985, making use of natural gas reserves which can be converted to MeOH and, subsequently, to motor fuels. However, the commercial viability of the process depends on current oil prices. Recent advances have shown zeolites are effective in catalyzing the direct

conversion of synthesis gas to motor fuels. The MTO (methanol-to-olefins) process converts MeOH to C_2–C_4 alkenes and is also catalyzed by ZSM-5. The development of a gallium-modified ZSM-5 catalyst (Ga-ZSM-5) has provided an efficient catalyst for the production of aromatic compounds from mixtures of C_3 and C_4 alkanes (commonly labelled LPG).

Zeolites are replacing acid catalysts in a number of manufacturing processes. One of the most important is the alkylation of aromatics; the Mobil–Badger process for producing C_6H_5Et from C_6H_6 and C_2H_4 provides the precursor for styrene (and hence polystyrene) manufacture. The isomerization of 1,3- to 1,4-dimethylbenzene (xylenes) is also catalyzed on the acidic surface of ZSM-5, presumably with channel shape and size playing an important role in the observed selectivity.

26.7 Heterogeneous catalysis: organometallic cluster models

One of the driving forces behind organometallic cluster research is to model metal-surface catalyzed processes such as the Fischer–Tropsch reaction. The *cluster-surface analogy* assumes that discrete organometallic clusters containing *d*-block metal atoms are realistic models for the bulk metal. In many small clusters, the arrangements of the metal atoms mimic units from close-packed arrays, e.g. the M_3-triangle and M_4-butterfly in structures **26.9** and **26.10**. The success of modelling studies has been limited, but a well established and much-cited result is that shown in Figure 26.15.[†]

$$(26.13)$$

[†] For further details, see M.A. Drezdon, K.H. Whitmire, A.A. Bhatta-charyya, W.-L. Hsu, C.C. Nagel, S.G. Shore and D.F. Shriver (1982) *Journal of the American Chemical Society*, vol. 104, p. 5630 – 'Proton induced reduction of CO to CH_4 in homonuclear and heteronuclear metal carbonyls'.

Fig. 26.15 The proton-induced conversion of a cluster bound CO ligand to CH$_4$: a cluster model for catalyzed hydrogenation of CO on an Fe surface.

Model studies involve transformations of organic fragments which are proposed as surface intermediates, but do not necessarily address a complete sequence as is the case in Figure 26.15. For example, metal-supported ethylidyne units (**26.13**) are proposed as intermediates in the Rh- or Pt-catalyzed hydrogenation of ethene, and there has been much interest in the chemistry of M$_3$-clusters such as H$_3$Fe$_3$(CO)$_9$CR, H$_3$Ru$_3$(CO)$_9$CR and Co$_3$(CO)$_9$CR which contain ethylidyne or other alkylidyne units. In the presence of base, H$_3$Fe$_3$(CO)$_9$CMe undergoes reversible deprotonation and loss of H$_2$ (equation 26.38), perhaps providing a model for an organic fragment transformation on a metal surface.

(26.38)

Summary

In this chapter, we have introduced the concepts of homogeneous and heterogeneous catalysis, and have described examples of industrial processes that rely on catalysis. The discussion has been necessarily selective, but has exemplified some key points such as the use of catalytic cycles, the development of catalysts, the need for industrial processes to be commercially viable, and the move towards using environmentally benign catalytic processes.

The following terms have been introduced in this chapter. Do you know what they mean?
- ❑ catalyst
- ❑ catalyst precursor
- ❑ autocatalytic
- ❑ homogeneous catalyst
- ❑ heterogeneous catalyst
- ❑ catalytic cycle
- ❑ catalytic turnover number
- ❑ catalytic turnover frequency
- ❑ coordinatively unsaturated
- ❑ asymmetric hydrogenation
- ❑ prochiral
- ❑ enantiomeric excess
- ❑ hydroformylation
- ❑ chemoselectivity and regioselectivity (with respect to hydroformylation)
- ❑ biphasic catalysis
- ❑ physisorption
- ❑ chemisorption

- ❑ adsorbate
- ❑ Ziegler–Natta catalysis
- ❑ Fischer–Tropsch reaction
- ❑ catalytic converter
- ❑ zeolite

Suggested further reading

General texts

B. Cornils and W.A. Hermann (eds) (1996) *Applied Homogeneous Catalysis with Organometallic Compounds*, Wiley-VCH, Weinheim (2 volumes) – The first volume covers catalytic processes used in industry; the second volume deals with recent developments and specialized processes.

F.A. Cotton, G. Wilkinson, M. Bochmann and C. Murillo (1999) *Advanced Inorganic Chemistry*, 6th edn, Wiley Interscience, New York – Chapter 22 gives a full account of the homogeneous catalysis of organic reactions by *d*-block metal compounds.

R.J. Farrauto and C.H. Bartholomew (1997) *Fundamentals of Industrial Catalytic Processes*, Kluwer, Dordrecht – Provides a detailed account of catalysts and their industrial applications.

G.W. Parshall and S.D. Ittel (1992) *Homogeneous Catalysis*, 2nd edn, Wiley, New York – Contains an excellent coverage of the applications of homogeneous catalysis in industry.

Homogeneous catalysis

D. Forster and T.W. Dekleva (1986) *Journal of Chemical Education*, vol. 63, p. 204 – 'Catalysis of the carbonylation of alcohols to carboxylic acids': a detailed look at the Monsanto process.

F.H. Jardine (1994) 'Hydrogenation and isomerization of alkenes' in *Encyclopedia of Inorganic Chemistry*, ed. R.B. King, Wiley, Chichester, vol. 3, p. 1471.

P.W. Jolly (1982) 'Nickel catalyzed oligomerization of alkenes and related reactions' in *Comprehensive Organometallic Chemistry*, eds G. Wilkinson, F.G.A. Stone and E.W. Abel, Pergamon, Oxford, vol. 8, p. 615.

H.B. Kagan (1982) 'Asymmetric synthesis using organometallic catalysts' in *Comprehensive Organometallic Chemistry*, eds G. Wilkinson, F.G.A. Stone and E.W. Abel, Pergamon, Oxford, vol. 8, p. 463.

S.W. Polichnowski (1986) *Journal of Chemical Education*, vol. 63, p. 204 – 'Catalysis of the carbonylation of alcohols to carboxylic acids': an account of the elucidation of the mechanism of the Tennessee–Eastman process.

R.L. Pruett (1986) *Journal of Chemical Education*, vol. 63, p. 196

– 'Hydroformylation': A detailed and partly historical account.

G.G. Stanley (1994) 'Carbonylation processes by homogeneous catalysis' in *Encyclopedia of Inorganic Chemistry*, ed. R.B. King, Wiley, Chichester, vol. 2, p. 575 – A well-referenced overview which includes hydroformylation, Monsanto and Tennessee–Eastman processes.

I. Tkatchenko (1982) in *Comprehensive Organometallic Chemistry*, eds G. Wilkinson, F.G.A. Stone and E.W. Abel, Pergamon, Oxford, vol. 8, p. 101 – A detailed account of hydroformylation (with industrial plant flow diagrams) and related processes.

Heterogeneous catalysis

A. Dyer (1994) 'Zeolites', in *Encyclopedia of Inorganic Chemistry*, ed. R.B. King, Wiley, Chichester, vol. 8, p. 4363 – A review of structures, properties and applications of zeolites.

B.L. Goodall (1986) *Journal of Chemical Education*, vol. 63, p. 191 – 'History and current state of the art of propylene polymerization catalysts': an account of the development of Ziegler–Natta catalysts for propene polymerization.

F.H. Ribeiro and G.A. Somorjai (1994) 'Heterogeneous catalysis by metals' in *Encyclopedia of Inorganic Chemistry*, ed. R.B. King, Wiley, Chichester, vol. 3, p. 1359 – A general introduction to concepts and catalyst design.

Industrial processes

R.I. Wijngaarden and K.R. Westerterp (1998) *Industrial Catalysts*, Wiley-VCH, Weinheim – A book that focuses on practical aspects of applying catalysts in industry.

J. Hagen (1999) *Industrial Catalysis*, Wiley-VCH, Weinheim – Covers both homogeneous and heterogeneous catalysis, including catalyst production, testing and development.

Ullman's Encyclopedia of Industrial Inorganic Chemicals and Products (1998) Wiley-VCH, Weinheim – Six volumes with detailed accounts of industrial processes involving inorganic chemicals.

Biphasic catalysis

B. Cornils and W.A. Hermann (eds) (1998) *Aqueous-phase Organometallic Catalysis*: *Concepts and Applications*, Wiley-VCH, Weinheim – An up-to-date account.

P. Kalck and F. Monteil (1992) *Advances in Organometallic Chemistry*, vol. 34, p. 219 – 'Use of water-soluble ligands in homogeneous catalysis'.

D.M. Roundhill (1995) *Advances in Organometallic Chemistry*, vol. 38, p. 155 – 'Organotransition-metal chemistry and homogeneous catalysis in aqueous solution'.

Problems

26.1 (a) Analyze the catalytic cycle shown in Figure 26.16, identifying the types of reactions occurring. (b) Why does this process work best for R′ = vinyl, benzyl or aryl groups?

26.2 The isomerization of alkenes is catalyzed by HCo(CO)$_3$ and Figure 26.17 shows the relevant catalytic cycle. (a) HCo(CO)$_4$ is a catalyst precursor; explain what this

means. (b) Give a fuller description of what is happening in each of the steps shown in Figure 26.17.

26.3 Outline the catalytic processes involved in the manufacture of acetic acid (Monsanto process) and acetic anhydride (Tennessee–Eastman process).

26.4 (a) Of the following alkenes, which are prochiral: PhHC=CHPh, PhMeC=CHPh, H$_2$C=CHPh,

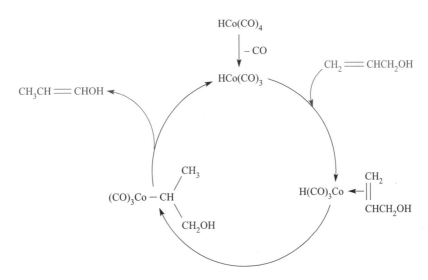

Fig. 26.16 Catalytic cycle for use in problem 26.1.

$H_2C=C(CO_2H)(NHC(O)Me)$? (b) If an asymmetric hydrogenation proceeds with 85% ee favouring the *R*-enantiomer, what is the percentage of each enantiomer formed?

26.5 (a) Assuming some similarity between the mechanism of hydroformylation using $HCo(CO)_4$ and $HRh(CO)(PPh_3)_3$ as catalysts, propose a mechanism for the conversion of $RCH=CH_2$ to RCH_2CH_2CHO and explain what is happening in each step. (b) 'The regioselectivity of the hydroformylation of $RCH=CH_2$ catalyzed by $HRh(CO)(PPh_3)_3$ drops when the temperature is increased.' Explain what is meant by this statement.

26.6 The hydroformylation of pent-2-ene using $Co_2(CO)_8$ as the catalyst was found to give rise to three aldehydes in a ratio 35:12:5. Show how the three products arose, and suggest which was formed in the most and which in the least amount.

26.7 In the catalyzed reaction of $RCH=CH_2$ with H_2, the catalyst precursor is $HRh(CO)(PPh_3)_3$. It is proposed that the first step in the mechanism is the addition of the alkene to the active catalyst, *trans*-$HRh(CO)(PPh_3)_2$. Suggest how the reaction might then proceed and construct an appropriate catalytic cycle.

26.8 (a) Ligand **26.6** is used in biphasic catalysis. The IR spectrum of $Fe(CO)_4(PPh_3)$ shows strong absorptions at 2049, 1975 and 1935 cm^{-1}, while that of $[Fe(CO)_4(\textbf{26.6})]^+$ exhibits bands at 2054, 1983 and 1945 cm^{-1}. What can you deduce from these data?
(b) Which of the complexes $[X][Ru(\textbf{26.14})_3]$ in which $X^+ = Na^+$, $[^nBu_4N]^+$ or $[Ph_4P]^+$ might be suitable candidates for testing in biphasic catalysis using aqueous medium for the catalyst?

(26.14)

26.9 Give a brief discussion of the use of homogeneous catalysis in selected industrial manufacturing processes.

26.10 For the catalyzed hydrocyanation of buta-1,3-diene

$$CH_2=CHCH=CH_2 \xrightarrow{HCN} NC(CH_2)_4CN$$

(a step in the manufacture of nylon-6,6), the catalyst precursor is NiL_4 where $L = P(OR)_3$. Consider the addition of only the first equivalent of HCN. (a) Some values of K for:

$$NiL_4 \rightleftharpoons NiL_3 + L$$

are 6×10^{-10} for $R = 4\text{-MeC}_6H_4$, 3×10^{-5} for $R = {}^iPr$ and 4×10^{-2} for $R = 2\text{-MeC}_6H_4$. Comment on the trend in values and on the relevance of these data to the catalytic process. (b) The first three steps in the proposed catalytic cycle are the addition of HCN to the active

Fig. 26.17 Catalytic cycle for use in problem 26.2.

catalyst, loss of L, and the addition of buta-1,3-diene with concomitant H migration. Draw out this part of the catalytic cycle. (c) Suggest the next step in the cycle, and discuss any complications.

26.11 $H_2Os_3(CO)_{10}$ (**26.15**) catalyzes the isomerization of alkenes:

$$RCH_2CH=CH_2 \longrightarrow E\text{-}RCH=CHMe + Z\text{-}RCH=CHMe$$

(a) By determining the cluster valence electron count for $H_2Os_3(CO)_{10}$ deduce what makes this cluster an effective catalyst. (b) Propose a catalytic cycle that accounts for the formation of the products shown.

(26.15)

26.12 Describe briefly why a clean nickel surface (fcc structure) should not be regarded as comprising a perfect close-packed array of atoms. Indicate the arrangements of atoms that an adsorbate might encounter on the surface, and suggest possible modes of attachment for CO.

26.13 (a) What advantages are there to using Rh supported on $\gamma\text{-}Al_2O_3$ as a catalyst rather than the bulk metal? (b) In a catalytic converter, why is a combination of platinum-group metals used?

26.14 The forward reaction in equation 26.30 is exothermic. What are the effects of (a) increased pressure, and (b) increased temperature on the yield of SO_3? (c) In trying to optimize both the yield and rate of formation of SO_3, what problem does the Contact process encounter and how is it overcome?

26.15 (a) Outline how the gaseous reaction between N_2 and H_2 proceeds in the presence of a heterogeneous catalyst, and state why a catalyst is needed for the commercial production of NH_3. (b) Suggest why V and Pt are poor catalysts for the reaction between N_2 and H_2, and give a possible reason why Os (although it is a good catalyst) is not used commercially.

26.16 (a) Summarize the structural features of importance in a Ziegler–Natta catalyst comprising $TiCl_4$ supported on $MgCl_2$. (b) What is the role of the ethyl aluminium compounds which are added the catalyst? (c) Explain how a Ziegler–Natta catalyst facilitates the conversion of ethene to a representative oligomer.

26.17 Give a brief discussion of the use of heterogeneous catalysis in selected industrial manufacturing processes.

26.18 Comment on each of the following:

(a) Zeolite 5A (effective pore size 430 pm) is used to separate a range of *n*- and *iso*-alkanes.
(b) Zeolite ZSM-5 catalyzes the isomerization of 1,3- to 1,4-$Me_2C_6H_4$ (i.e. *m*- to *p*-xylene), and the conversion of C_6H_6 to EtC_6H_5.

26.19 Summarize the operation of a 3-way catalytic converter, including comments on (a) the addition of cerium oxides, (b) the light-off temperature, (c) optimum air–fuel ratios and (d) catalyst ageing.

Some aspects of solid state chemistry

27.1 Introduction

As we enter the 21st century, there is intense interest in developing new inorganic materials, and solid state and polymer chemistries are 'hot' areas of research. We have already encountered many structural aspects of the solid state and have exemplified applications of solid state materials, e.g. magnetic properties of metal oxides (*Chapters 21* and *22*), semiconductors (*Chapter 5*) and heterogeneous catalysts (*Chapter 26*). The following topics appeared in *Chapter 5*:

- structures of metals;
- polymorphism;
- alloys;
- band theory;
- semiconductors;
- prototype ionic lattices;
- lattice energies and their applications in inorganic chemistry;
- Frenkel and Schottky defects.

With the exception of lattice defects and semiconducting materials, these will not be discussed further here. In *Section 20.8*, we introduced some concepts of magnetism including *ferromagnetism*, *antiferromagnetism* and *ferrimagnetism*; although these properties are important in materials chemistry, it is beyond the scope of this book to take this topic further.

The topics chosen for inclusion in this chapter reflect areas of active interest and elaborate upon some topics that have been given only brief mention in earlier chapters. In describing the chemistries of the *d*- and *f*-block metals in *Chapters 21, 22* and *24*, we included many examples of solid state compounds, and we now look further at electrical conduct-ing and superconducting properties. At various points in the book, we have mentioned colour pigments in ceramic materials when describing applications of inorganic compounds (mainly oxides); Section 27.5 looks at the colouring of ceramics in more detail. We describe *chemical vapour deposition* (CVD) for the formation of thin films of materials and its application in the semiconductor industry, and in the final section, discuss selected inorganic fibres. Throughout the chapter, we emphasize commercial applications in order to exemplify the role that inorganic chemistry plays in technological developments; in contrast to other chapters, applications are not highlighted specifically in boxed material.

27.2 Defects in solid state lattices

Structures of solids are usually described in terms of proto-type lattices (e.g. NaCl, TiO_2) but in reality, almost all crystals are not perfect arrays of ordered atoms or ions but contain *defects*.

Intrinsic defects occur in lattices of pure compounds; *extrinsic defects* result from the addition of dopants.

Types of defect: stoichiometric and non-stoichiometric compounds

The lattice defects that we introduced in Section 5.17 are due to *point defects* in the crystal lattice. A *Schottky defect* arises from vacant lattice sites, but the stoichiometry and electrical neutrality of the compound are retained. Defects that fall in this category include:

- a vacant *atom* site in a metal lattice;

- a vacant *cation and* a vacant *anion* site in an MX lattice (*Figure 5.26*);
- two vacant *cations and* one vacant *anion* site in an M_2X lattice.

In ionic lattices in which there is a significant difference in size between the cation and anion (e.g. AgBr), the smaller ion may occupy a site that is vacant in the ideal lattice. This is a *Frenkel defect* (*Figure 5.27*) and does not affect the stoichiometry or electrical neutrality of the compound.

Many defects result in a compound being *non-stoichiometric*; such defects often occur in crystalline solids of *d*-block metal compounds when the metal exhibits variable oxidation states. Metal oxides and sulfides are particularly prone to non-stoichiometric defects and these often lead to dramatic changes in physical properties even at low levels of crystal imperfection. For example, treatment of TiO_2 with H_2 at 770–970 K results in an O deficiency such that the stoichiometry is $TiO_{1.993}$; the change from TiO_2 to $TiO_{1.993}$ causes a decrease in electrical resistivity by more than five orders of magnitude. Some crystalline metal oxides are extremely difficult to prepare in a stoichiometric form. For example, TiO (see end of *Section 21.5*) exists as a non-stoichiometric compound in the range $TiO_{0.82}$–$TiO_{1.23}$; FeO is always Fe-deficient and is formulated as $Fe_{1-x}O$ ($0.04 < x < 0.11$) occurring naturally in this form in meteorites and oceanic basalt. Thus, in $Fe_{1-x}O$, some Fe^{3+} ions are present to counter what would otherwise be a charge imbalance caused by Fe^{2+} vacancies. The Fe^{3+} centres tend to occupy interstitial sites between vacancies giving well-defined clusters such as **27.1** and **27.2**; the clusters can be described as zones of an Fe_3O_4-type lattice.

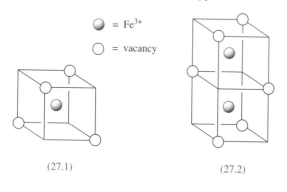

= Fe^{3+}

= vacancy

(27.1)   (27.2)

Metal oxides having a CaF_2 lattice are subject to various non-stoichiometric defects. Uranium(IV) oxide has the stoichiometry UO_{2+x}, i.e. it has an *anion-excess structure* in which excess O^{2-} ions are accommodated in interstitial sites. The addition of dopants to crystalline solids produces defects which can be of commercial significance; dopant cations must be of a similar size to those in the parent lattice. Adding CaO to ZrO_2 stabilizes the cubic form of zirconia (see *Section 22.5*) and prevents a phase change from cubic to monoclinic that would otherwise occur on cooling below 1143 K. The introduction of Ca^{2+} into the ZrO_2 lattice results in replacement of Zr^{4+} by Ca^{2+} and the creation of an O^{2-} vacancy to counter the charge imbalance; the doped zirconia, $Ca_xZr_{1-x}O_{2-x}$, is *anion-deficient*.

The introduction of a dopant may result in a change in the oxidation state of metal sites in the parent lattice; a well-cited example is the doping of NiO with Li_2O in the presence of air/O_2. When an Ni^{2+} ion is replaced by Li^+, electrical neutrality is retained by the oxidation of another Ni^{2+} to Ni^{3+} (Figure 27.1).

Colour centres (F-centres)

Defects that result from the presence of trapped electrons in a crystal lattice cause colour changes. If NaCl is heated in Na vapour, Na atoms enter the NaCl lattice; Na is oxidized to Na^+ (equation 27.1).

$$Na \longrightarrow Na^+ + e^- \qquad (27.1)$$

The electron produced in oxidation step 27.1 remains trapped in the crystal lattice and occupies a lattice site, leaving a Cl^- site vacant. Excitation and subsequent relaxation of the electron results in the emission of radiation in the visible region. The electron centre is known as an *F-centre* (from the German *Farbe* for colour). The origins of F-centres are varied, but their presence has some dramatic consequences. For example, some variants of transparent minerals are coloured owing to the presence of F-centres, e.g. Blue John is a rare blue-purple form of fluorspar and is much prized in jewellery and decorative ornaments.

Thermodynamic effects of crystal defects

At the beginning of this section we stated that almost all

Dope with Li_2O in presence of O_2

Fig. 27.1 NiO possesses an NaCl lattice; doping with Li_2O in the presence of air/O_2 results in the replacement of an Ni^{2+} centre (grey) with an Li^+ ion (black), and the oxidation of one Ni^{2+} to an Ni^{3+} centre (large blue). Oxide ions are shown in blue.

crystals are imperfect. The creation of a defect from an ideal crystal is an endothermic process, but is entropically favoured since a degree of *disorder* is introduced into the otherwise perfectly ordered lattice. The balance between the ΔH and $T\Delta S$ terms (equation 27.2) is therefore important.

$$\Delta G = \Delta H - T\Delta S \qquad (27.2)$$

At temperatures above 0 K, the thermodynamic balance favours the presence of defects, and the minimum value of ΔG is attained at a given equilibrium concentration of defects, the concentration being temperature-dependent.

27.3 Electrical conductivity in ionic solids

Ionic *solids* usually have a high electrical resistance (low conductivity, see *Section 5.8*) and the conductivity is significant only when the compound is molten. The presence of defects in an ionic solid decreases the resistance, e.g. cubic zirconia stabilized with CaO (see above) is a *fast-ion conductor*, the conductivity arising from the migration of O^{2-} ions. An increased concentration of defects can be introduced by heating a solid to a high temperature and then cooling it rapidly. Since more defects are present at high temperatures, the effect of quenching the solid is to 'freeze' the defect concentration present at elevated temperature.

> The presence of defects in a crystal lattice facilitates ion migration and enhances electrical conductivity (i.e. lowers the resistance).

Mechanisms of ion migration can be categorized as follows:

- migration of a cation into a cation vacancy creating a new vacancy into which another cation can migrate, and so on;
- migration of a cation into an interstitial site (as in *Figure 5.27*) creating a vacancy which can be filled by another migrating ion, and so on.

Anion migration could also occur by the first mechanism, but for the second, it is usually the cation that is small enough to occupy an interstitial site, for example, the tetrahedral holes in an NaCl-type lattice.

Sodium and lithium ion conductors

Current developments in battery technology, electrochromic devices (see *Box 22.4*) and research into electrically powered vehicles make use of solid electrolytes (see *Box 10.3*). The sodium/sulfur battery contains a solid β-alumina electrolyte. The name *β-alumina* is misleading since it is prepared by the reaction of Na_2CO_3, $NaNO_3$, $NaOH$ and Al_2O_3 at 1770 K and is a non-stoichiometric compound of approximate

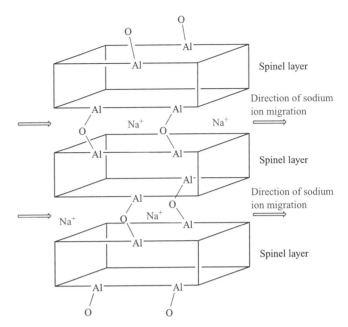

Fig. 27.2 A schematic representation of part of the structure of β-alumina ($Na_2O \cdot 11Al_2O_3$) in which Na^+ ions are mobile between bridged layers of Al_2O_3 spinel lattice.

composition $Na_2Al_{22}O_{34}$ (or $Na_2O \cdot 11Al_2O_3$), always containing an excess of Na^+; we therefore refer to this material as *Na β-alumina*. Equation 27.3 shows the half-reactions that occur in the sodium/sulfur battery; Na^+ ions produced at the anode migrate through the Na β-alumina electrolyte and combine with the polysulfide anions formed at the cathode (equation 27.4). The reactions are reversed when the cell is recharged.

$$\left.\begin{array}{ll} \textit{At the anode:} & Na \longrightarrow Na^+ + e^- \\ \textit{At the cathode:} & nS + 2e^- \longrightarrow [S_n]^{2-} \end{array}\right\} \qquad (27.3)$$

$$2Na^+ + [S_n]^{2-} \longrightarrow Na_2S_n \qquad (27.4)$$

The Na β-alumina acts as a *sodium ion conductor*. The key to this property lies in its structure which consists of spinel-type layers 1123 pm thick, with Na^+ ions occupying the interlayer spaces (Figure 27.2). The conductivity of Na β-alumina ($3\,\Omega^{-1}\,m^{-1}$) arises from the ability of the Na^+ ions to migrate through the gaps between the spinel layers; it therefore conducts in one plane though the crystal in the same way that graphite conducts only in the plane parallel to the carbon-containing planes (*Figure 13.2a*). Although the conductivity of Na β-alumina is small compared with a metal (*Figure 5.9*), it is large compared with typical ionic solids (e.g. $10^{-13}\,\Omega^{-1}\,m^{-1}$ for solid NaCl). The Na^+ ions in Na β-alumina can be replaced by cations such as Li^+, K^+, Cs^+, Rb^+, Ag^+, Tl^+ and H^+. However, the conductivities of these materials are lower than that of Na β-alumina: the match between the size of the Na^+ ions and the interlayer channels in the host lattice leads to the most efficient cation mobility. The conductivities of Na β-alumina and selected cation and anion conductors exhibiting relatively

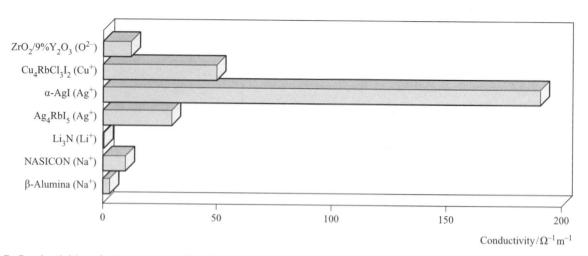

Fig. 27.3 Conductivities of selected ionic solids; the ion given in parentheses after the solid electrolyte is the ion conductor.

high (i.e. in the context of ionic solids) conductivities are compared in Figure 27.3.

Materials of composition $Na_{1+x}Zr_2P_{3-x}Si_xO_{12}$ $(0 \leq x \leq 3)$ are known as *Na super-ionic conductors* (NASICON) and also have potential application in sodium/sulfur batteries. They comprise solid solutions of $NaZr_2(PO_4)_3$ (the host lattice) and $Na_4Zr_2(SiO_4)_3$; the former adopts a lattice composed of corner-sharing ZrO_6 octahedra and PO_4 tetrahedra which generate a network through which channels run. Once $Na_4Zr_2(SiO_4)_3$ is incorporated to give a solid solution, the Na^+ ion conductivity increases and is optimized to $10 \, \Omega^{-1} \, m^{-1}$ when $x \approx 2$.

Solid electrolytes with applications in lithium batteries include Li_7NbO_6, $Li_{12}Ti_{17}O_{40}$, Li_8ZrO_6 and Li_3N which are *Li^+ ion conductors* (see also *Box 10.3*). In Li_7NbO_6, one eighth of the Li^+ sites are vacant, rendering the solid a good ionic conductor. Lithium nitride has the layer structure

shown in *Figure 10.3*. A 1–2% deficiency of Li^+ ions in the hexagonal layers leads to Li^+ conduction *within* these layers, the interlayer Li^+ ion sites remaining fully occupied. Lithium nitride is used as the solid electrolyte in cells containing an Li electrode coupled with a TiS_2, TaS_2 or other metal sulfide electrode. During battery discharge, Li is oxidized to Li^+ ions which flow through the solid Li_3N barrier and enter the MS_2 solid; this acts as a host lattice (equation 27.5), intercalating the Li^+ ions which occupy vacant sites.

$$x Li^+ + TiS_2 + xe^- \underset{\text{charge}}{\overset{\text{discharge}}{\rightleftharpoons}} Li_x TiS_2 \qquad (27.5)$$

d-Block metal(II) oxides

In *Chapter 21*, we described the chemistry of the first row *d*-block metal(II) oxides, but said little about their electrical conductivity properties. The oxides TiO, VO, MnO, FeO,

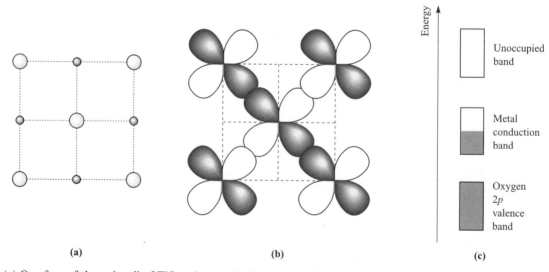

Fig. 27.4 (a) One face of the unit cell of TiO; colour code: Ti, grey; O, blue. (b) Overlap of the Ti t_{2g} orbitals occurs and leads to (c) the formation of a partly filled metal conduction band.

CoO and NiO adopt NaCl lattices but are non-stoichiometric, being metal-deficient as exemplified for TiO and FeO in Section 27.2. In TiO and VO, there is overlap of the metal t_{2g} orbitals giving rise to a partially occupied band (Figure 27.4) and, as a result, TiO and VO are electrically conducting. In contrast, MnO is an insulator at 298 K but is semiconducting (see *Section 5.9*) at higher temperatures; FeO, CoO and NiO behave similarly, their conductivities (which are low at 298 K) increasing with temperature. The conductivity is explained in terms of an *electron-hopping* mechanism, in which an electron moves from an M^{2+} to M^{3+} centre (recall that the metal-deficiency in the non-stoichiometric oxide leads to the presence of M^{3+} sites), effectively creating a positive hole. Heating the oxide in the presence of O_2 results in further M^{2+} ions being oxidized. In turn, this facilitates electron migration through the solid.

27.4 Superconductivity

Superconductors: early examples and basic theory

> A *superconductor* is a material, the electrical resistance of which drops to zero when it is cooled below its *critical temperature*, T_c.

Superconductivity was discovered in 1911 by H. Kamerlingh Onnes (awarded the Nobel Prize for Physics in 1913): on cooling to its critical temperature, T_c, a superconductor loses all electrical resistance and at the same time becomes a perfect diamagnetic material. The latter (the *Meissner effect*) is detected by an unusual phenomenon: if a permanent magnet is placed on top of a superconductor as it is cooled, at T_c the former rises to become suspended in midair above the superconducting material. A range of metals, alloys and metallic compounds are superconductors (Table 27.1). However, to put the practical limitations of working with the materials listed in Table 27.1 (and many others including

Table 27.1 Selected superconducting metals, alloys and metallic compounds.

Element or compound	T_c / K	Element or compound	T_c / K
Al	1.17	$AuPb_2$	3.15
α-Hg	4.15	InPb	6.65
In	3.41	Ir_2Th	6.50
Nb	9.25	Nb_2SnV	9.8
Ru	0.49	CuS	1.62
Sn	3.72	Nb_3Sn	18
Ti	0.40	TiO	0.58
Zn	0.85	SnO	3.81
Al_2Y	0.35	$(SN)_x$	0.26

Table 27.2 Selected high-temperature superconductors with $T_c > 77$ K.

Compound	T_c / K	Element or compound	T_c / K
$YBa_2Cu_3O_7$	93	$Tl_2CaBa_2Cu_2O_8$	119
$YBa_2Cu_4O_8$	80	$Tl_2Ca_2Ba_2Cu_3O_{10}$	128
$Y_2Ba_4Cu_7O_{15}$	93	$TlCaBa_2Cu_2O_7$	103
$Bi_2CaSr_2Cu_2O_8$	92	$TlCa_2Ba_2Cu_3O_8$	110
$Bi_2Ca_2Sr_2Cu_3O_{10}$	110	$Tl_{0.5}Pb_{0.5}Ca_2Sr_2Cu_3O_9$	120

the superconducting fullerides described in *Section 13.4*) into perspective, compare the values of T_c with the boiling points of available coolants, e.g. liquid He (4.2 K), H_2 (20.1 K) and N_2 (77 K).

Superconductivity is usually described in terms of *Cooper pairs* which we describe only at a simple level.[†] A Cooper pair consists of two electrons of opposite spin and momentum that are brought together by a cooperative effect involving the positively charged nuclei in the vibrating crystal lattice. Cooper pairs of electrons (which are present in the Fermi level, see *Section 5.8*) remain as bound pairs below some critical temperature (T_c) and their presence gives rise to resistance-free conductivity. The theory holds for the earliest known superconductors but for the cuprates that we discuss below, while Cooper pairs may still be significant, new theories are required. To date, no complete explanation for the conducting properties of the *high-temperature superconductors* has been forthcoming.

High-temperature superconductors

Since 1987, cuprate superconductors with $T_c > 77$ K have been the centre of intense interest; one of the first to be discovered was $YBa_2Cu_3O_7$ made by reaction 27.6. The oxygen content of the final material depends on reaction conditions (e.g. temperature and pressure).

$$4BaCO_3 + Y_2(CO_3)_3 + 6CuCO_3$$
$$\xrightarrow{1220\,K} 2YBa_2Cu_3O_{7-x} + 13CO_2 \qquad (27.6)$$

Selected *high-temperature superconductors* are listed in Table 27.2; the oxidation state of the Cu centres in $YBa_2Cu_3O_7$ can be inferred by assuming fixed oxidation states of +3, +2 and −2 for Y, Ba and O respectively; the result indicates a mixed Cu(II)/Cu(III) compound. A similar result is obtained for some other materials listed in Table 27.2. High-temperature superconductors have two structural features in common:

- Their structures are related to that of perovskite. *Figure 5.23* showed a 'ball-and-stick' representation of this

---

[†] For a greater depth discussion, see: J. Bardeen, L.N. Cooper and J.R. Schrieffer (1957) *Physics Reviews*, vol. 108, p. 1175; A.R. West (1999) *Basic Solid State Chemistry*, 2nd edn, Wiley-VCH, Weinheim.

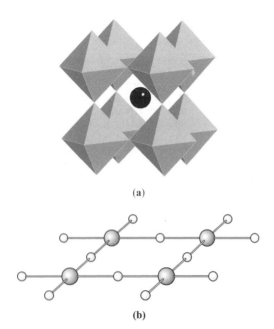

(a)

(b)

Fig. 27.5 (a) A unit cell of perovskite, $CaTiO_3$, using a polyhedral representation for the coordination environments of the Ti centres; an O atom lies at each vertex of the octahedra, and the Ca^{2+} ion is shown in black. (b) Part of a layer of stoichiometry CuO_2 which forms a building block in all cuprate high-temperature superconductors.

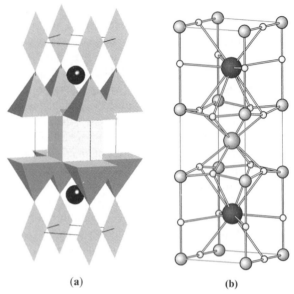

(a) **(b)**

Fig. 27.6 A unit cell of $YBa_2Cu_3O_7$. (a) A representation showing coordination polyhedra for the Cu centres (square planar and square-based pyramidal) and Y^{3+} centre (cubic); the Ba^{2+} ions are shown in black. (b) The unit cell drawn using a 'ball-and-stick' representation; Cu centres are in blue, Ba in black, Y in grey (central ion) and O in pale grey.

lattice; the same structure is depicted in Figure 27.5a, but with the octahedral coordination spheres of the Ti centres shown in polyhedral representation (see *diagram* **13.16** and *Figure 22.8*).

● They always contain layers of stoichiometry CuO_2; these may be planar (Figure 27.5b) or puckered.

The incorporation of the two structural building blocks is illustrated in Figure 27.6 which shows a unit cell of $YBa_2Cu_3O_7$. The unit cell of $YBa_2Cu_3O_7$ can be considered in terms of three stacked perovskite unit cells; taking the prototype perovskite to be $CaTiO_3$, then on going from $CaTiO_3$ to $YBa_2Cu_3O_7$, Ba^{2+} and Y^{3+} ions substitute for Ca^{2+}, while Cu centres substitute for Ti(IV). Compared to the lattice derived by stacking three perovskite unit cells, the structure of $YBa_2Cu_3O_7$ is oxygen-deficient, leading to

the Cu coordination environments being square planar or square-based pyramidal (Figure 27.6a), the Ba^{2+} ions being 10-coordinate (Figure 27.6b), and each Y^{3+} ion being in a cubic environment. The structure is readily described in terms of sheets, and the unit cell in Figure 27.6 can be represented schematically as layer structure **27.3**. Other high-temperature superconductors can be described in similar fashion, e.g. $Tl_2Ca_2Ba_2Cu_3O_{10}$ (containing Tl^{3+}, Ca^{2+} and Ba^{2+} centres) is composed of layer sequence **27.4**. The non-CuO_2 oxide layers in the cuprate superconductors are isostructural with layers from an NaCl-lattice, and so the structures are sometimes described in terms of perovskite and rock salt layers.

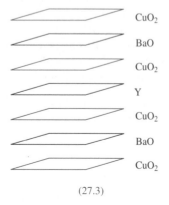

	CuO_2
	BaO
	CuO_2
	Y
	CuO_2
	BaO
	CuO_2

(27.3)

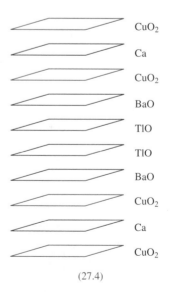

	CuO_2
	Ca
	CuO_2
	BaO
	TlO
	TlO
	BaO
	CuO_2
	Ca
	CuO_2

(27.4)

A full discussion of the bonding and origins of superconductivity in these cuprate materials is beyond the scope of this book, but we can comment on several important points. It is the CuO_2 layers that are responsible for the superconducting properties, while the other layers in the lattice act as sources of electrons; the arrangement of the layers is an important factor in controlling the superconductivity. Taking the square planar Cu centres to be Cu(II) gives a d^9 configuration with the unpaired electron in a $d_{x^2-y^2}$ orbital. The energies of the $3d$ and $2p$ atomic orbitals are sufficiently close to allow significant orbital mixing, and a band structure is appropriate. The half-filled band is then tuned electronically by the effects of the 'electron sinks' which make up the neighbouring layers in the lattice.

Applications of superconductors

Commercial applications of high-temperature superconductors are now established and a host of potential uses should become reality during the 21st century. The majority of magnetic resonance imaging scanners (see *Box 2.5*) rely on superconducting magnets with flux densities of 0.5–2.0 T; presently, NbTi ($T_c = 9.5$ K) multicore conductors are used, but replacement by *high-temperature* superconductors would be financially beneficial.

The combination of two superconductors separated by a thin oxide barrier which is a weak insulator makes up a *Josephson junction*, a device that is very sensitive to magnetic fields. Among applications of Josephson junctions is their role in SQUID (*superconducting quantum interference device*) systems for measuring magnetic susceptibilities. The extreme sensitivity of a SQUID allows it to be used to measure very weak biomagnetic signals such as those originating from the brain, and naval vessels equipped with SQUIDs have increased sensitivity to detect undersea mines.

For the development of applications for superconductors, two obstacles in particular have to be surmounted. The first is that the material must be cooled to low temperatures to attain T_c. As higher temperature superconductors are developed, this has become less of a major drawback, but still militates against the use of superconductors in conventional settings. The second problem is one of fabrication. When prepared as a bulk material, the cuprate superconductors have unacceptably low critical current densities, i.e. the superconductivity is lost after the material has carried only a limited amount of current. The origin of the problem is the presence of grain boundaries in the solid and can be overcome by preparing thin films using, for example, CVD (see *Section 27.6*) or *texturing* the material (i.e. alignment of crystallites) through specialized crystallization techniques or mechanical working. Even with the advances that have been made so far, the application of superconductors for bulk power transmission remains a long way in the future.

27.5 Ceramic materials: colour pigments

A *ceramic* material is a hard, high melting solid which is usually chemically inert.

Ceramic materials are commonplace in everyday life, e.g. floor and wall tiles, crockery, wash-basins, baths and decorative pottery and tiles, and also include the cuprate high-temperature superconductors discussed above. Many ceramic materials consist of metal oxides or silicates, and the addition of white and coloured pigments is a huge industrial concern. In earlier chapters, we mentioned the use of several metal oxides (e.g. CoO and TiO_2, *Boxes 21.3* and *21.9*) as colour pigments. One of the factors that has to be taken into account when choosing a pigment is the need for it to withstand the high firing temperatures involved in the manufacture of ceramics. This is in contrast to the introduction of pigments into, for example, fabrics.

White pigments (opacifiers)

An *opacifier* is a glaze additive that makes an otherwise transparent glaze opaque.

The most important commercial opacifiers in ceramic materials are TiO_2 (in the form of anatase) and $ZrSiO_4$ (zircon). While SnO_2 is also highly suitable, its use is not as cost effective as that of TiO_2 and $ZrSiO_4$, and it is retained only for specialist purposes. Zirconium(IV) oxide is also an excellent opacifier but is more expensive than $ZrSiO_4$. Fine particles of these pigments scatter incident light extremely strongly: the refractive indices of anatase, $ZrSiO_4$, ZrO_2 and SnO_2 are 2.5, 2.0, 2.2 and 2.1 respectively. The firing temperature of the ceramic material determines whether or not TiO_2 is a suitable pigment for a particular application. Above 1120 K, anatase converts to rutile, and although rutile also has a high refractive index ($\mu = 2.6$), the presence of relatively large particles of rutile prevents it from functioning as an effective opacifier. Anatase is therefore useful only if working temperatures do not exceed the phase transition temperature. Zircon is amenable to use at higher firing temperatures; it can be added to the molten glaze and precipitates as fine particles dispersed in the glaze as it is cooled.

Adding colour

Cation substitution in a host lattice such as ZrO_2, TiO_2, SnO_2 or $ZrSiO_4$ is a means of altering the colour of a pigment; the substituting cation must have one or more unpaired electrons so as to give rise to an absorption in the visible region (see *Section 20.6*). Yellow pigments used to colour ceramics include $(Zr,V)O_2$ (which retains the lattice structure of *baddeleyite*, the monoclinic form of ZrO_2 in which the metal is 7-coordinate), $(Sn,V)O_2$ (with a V-doped *cassiterite* lattice) and $(Zr,Pr)SiO_4$ (with a *zircon* lattice doped with ≈5% Pr). Blue pigmentation can be obtained using (Zr,V)SiO$_4$ and this is routinely used when high temperature firing is required; cobalt oxide-based pigments produce a more intense blue coloration than vanadium-doped zirconia, but are unsuitable for use at elevated temperatures. The content of cobalt oxide needed in a blue ceramic is ≈0.4–0.5% Co.

Spinels (AB$_2$O$_4$) (see *Box 12.6*) are an important class of oxide for the manufacture of brown and black pigments for ceramics. The three spinels FeCr$_2$O$_4$, ZnCr$_2$O$_4$ and ZnFe$_2$O$_4$ are structurally related, forming a family in which Fe^{2+} or Zn^{2+} ions occupy tetrahedral sites, while Cr^{3+} or Fe^{3+} ions are octahedrally sited. In nature, cation substitution occurs to produce, for example, black crystals of the mineral *franklinite* (Zn,Mn,Fe)(Fe,Mn)$_2$O$_4$ which has a variable composition. In the ceramics industry, spinels for use as pigments are prepared by heating together suitable metal oxides in appropriate stoichiometric ratios so as to control the cation substitution in a parent spinel lattice. In (Zn,Fe)(Fe,Cr)$_2$O$_4$, a range of brown shades can be obtained by varying the cation site compositions. For the commercial market, reproducibility of shade of colour is, of course, essential.

27.6 Chemical vapour deposition (CVD)

We devote a significant part of this chapter to the method of *chemical vapour deposition*, the development of which has been closely tied to the need to deposit thin films of a range of metals and inorganic materials for use in semiconducting devices, ceramic coatings and electrochromic materials. Table 27.3 lists some applications of selected thin film materials. Part of the challenge of the successful production of thin films is to find suitable molecular precursors, and there is much research interest in this area. We illustrate CVD by focusing on the deposition of specific materials including semiconductors. In any industrial CVD process, reactor design is crucial to the efficiency of the deposition, and it should be recognized that *the diagrams given of CVD reactors are highly schematic*.

> *Chemical vapour deposition* (CVD) is the delivery (by uniform mass transport) of a volatile precursor or precursors to a heated surface on which reaction takes place to deposit a thin film of the solid product; the surface must be hot enough to permit reaction but cool enough to allow solid deposition. Multilayer deposition is also possible. *Metal–organic chemical vapour deposition* (MOCVD) refers specifically to use of metal–organic precursors.

High-purity silicon for semiconductors

Although Ge was the first semiconductor to be used commercially, it is Si that now leads the world market.

Table 27.3 Some applications of selected thin film materials; see also Table 27.5.

Thin film	Applications
Al$_2$O$_3$	Oxidation resistance
AlN	High powered integrated circuits; acoustic devices
CdTe	Solar cells
CeO$_2$	Optical coatings; insulating films
GaAs	Semiconducting devices; electrooptics; (includes solar cells)
GaN	Light-emitting diodes (LED)
GaAs$_{1-x}$P$_x$	Light-emitting diodes (LED)
LiNbO$_3$	Electrooptic ceramic
NiO	Electrochromic devices
Si	Semiconductors, many applications of which include solar cells
Si$_3$N$_4$	Diffusion barriers and inert coatings in semiconducting devices
SiO$_2$	Optical wave guides
SnO$_2$	Sensors for reducing gases, e.g. H$_2$, CO, CH$_4$, NO$_x$
TiC	Wear resistance
TiN	Friction reduction
W	Metal coatings on semiconducting integrated circuits
WO$_3$	Electrochromic windows
ZnS	Infrared windows

Germanium has been replaced, not only by Si, but by a range of recently developed semiconducting materials. All silicon semiconductors are manufactured by CVD. In *Box 5.2*, we described the Czochralski process for obtaining single crystals of pure silicon; the silicon used for the crystal growth must itself be of high purity and a purification stage is needed after the manufacture of Si from SiO$_2$ (reaction 27.7). Crude silicon is first converted to the volatile SiHCl$_3$ which is then converted back to a higher purity grade of Si (equation 27.8) by using CVD.

$$\text{SiO}_2 + 2\text{C} \xrightarrow{\Delta} \text{Si} + 2\text{CO} \qquad (27.7)$$

$$3\text{HCl} + \text{Si} \underset{1400\,\text{K}}{\overset{620\,\text{K}}{\rightleftharpoons}} \text{SiHCl}_3 + \text{H}_2 \qquad (27.8)$$

Figure 27.7 illustrates the industrial CVD procedure: SiHCl$_3$ and H$_2$ pass into the reaction vessel where they come into contact with a high purity silicon surface, electrically heated to 1400 K. Back-reaction 27.8 is highly endothermic and occurs on the Si surface to deposit additional Si (mp = 1687 K); no deposition occurs on the vessel walls because these are kept cold, devoid of the heat energy needed to facilitate the reaction between SiHCl$_3$ and H$_2$. A secondary product of the deposition reaction is SiCl$_4$ (equation 27.9), some of which reacts with H$_2$ to give more SiHCl$_3$. The remainder leaves with the exhaust gases and finds use in the manufacture of silica.

$$4\text{SiHCl}_3 + 2\text{H}_2 \longrightarrow 3\text{Si} + 8\text{HCl} + \text{SiCl}_4 \qquad (27.9)$$

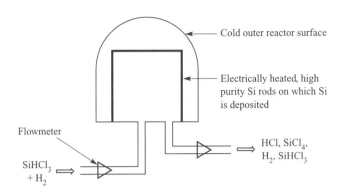

Fig. 27.7 Schematic representation of the CVD setup used to deposit high purity silicon by thermal decomposition of $SiHCl_3$.

A more recently developed CVD process starts with SiH_4 (equation 27.10), which is first prepared from $SiHCl_3$ by scheme 27.11.

$$SiH_4 \xrightarrow{\Delta} Si + 2H_2 \qquad (27.10)$$

$$\left.\begin{array}{l} 2SiHCl_3 \longrightarrow SiH_2Cl_2 + SiCl_4 \\ 2SiH_2Cl_2 \longrightarrow SiH_3Cl + SiHCl_3 \\ 2SiH_3Cl \longrightarrow SiH_4 + SiH_2Cl_2 \end{array}\right\} \qquad (27.11)$$

The high grade silicon produced by CVD is virtually free of B or P impurities, and this is essential despite the fact that doping with B or P is routine. Careful tuning of the properties of n- or p-type semiconductors (see *Section 5.9*) depends on the *controlled* addition of B, Al, P or As during their manufacture.

α-Boron nitride

Thin films of α-BN (which possesses the layer structure shown in *Figure 12.13*) can be deposited by CVD using reactions of NH_3 with volatile boron compounds such as BCl_3 (equation 27.12) or BF_3 at temperatures of $\approx 1000\,K$.

$$BCl_3 + NH_3 \xrightarrow{\Delta} BN + 3HCl \qquad (27.12)$$

An important application of such films is in doping silicon to generate a p-type semiconductor (Figure 27.8). The semiconductor-grade silicon is first oxidized to provide a layer of SiO_2 which is then etched; deposition of a thin film of α-BN provides contact between Si and α-BN within the etched zones. By heating under N_2, B atoms from the film diffuse into the silicon to give the desired p-type semiconductor which is finally plated with a thin film of nickel (see 'metal deposition' below).

α-Boron nitride films have a range of other applications which make use of the material's hardness, resistance to oxidation and insulating properties.

Silicon nitride and carbide

The preparation and structure of Si_3N_4 were discussed at the end of *Section 13.12*. Its uses as a refractory material are widespread, as are its applications in the microelectronics industry and solar cell construction. Thin films of Si_3N_4 can be prepared by reacting SiH_4 or $SiCl_4$ with NH_3 (*equation 13.78*), or $SiCl_4$ with N_2H_4. Films deposited using $(\eta^5\text{-}C_5Me_5)SiH_3$ (**27.5**) as a precursor with a *plasma-enhanced CVD* technique have the advantage of low carbon contamination; the precursor is made by reduction of $(\eta^5\text{-}C_5Me_5)SiCl_3$ using $Li[AlH_4]$ and is an air- and heat-stable volatile compound, ideal for CVD.

(27.5)

Silicon carbide (*carborundum*) has several polymorphs; the β-form adopts the wurtzite lattice (*Figure 5.20*). It is extremely hard, resists wear, withstands very high temperatures, has a high thermal conductivity and a low coefficient of thermal expansion, and has long been used as a refractory material and abrasive powder. Recent development of suitable CVD methods has made possible the deposition of β-SiC of >99.9 % purity, with suitable precursors being alkylsilanes, alkylchlorosilanes, or alkanes with chlorosilanes. Silicon carbide is a IV–IV semiconductor (band gap $= 288\,kJ\,mol^{-1}$) which has particular application for high frequency devices and for systems operating at high

Fig. 27.8 Boron doping of silicon using α-BN.

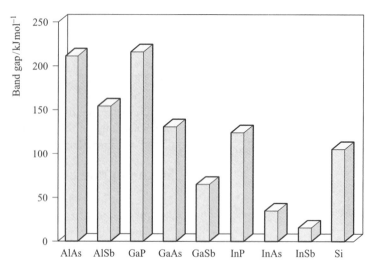

Fig. 27.9 Band gaps (at 298 K) of the III–V semiconductors and of Si.

temperatures. Thin films exhibit excellent reflective properties and are used for manufacturing mirrors for laser radar systems, high energy lasers, synchrotron X-ray equipment and astronomical telescopes. Silicon carbide is also used for blue light-emitting diodes (LEDs); silicon carbide fibres are described in *Section 27.7*.

III–V Semiconductors

The III–V semiconductors comprise AlAs, AlSb, GaP, GaAs, GaSb, InP, InAs and InSb, and of these GaAs is the most important commercially. The band gaps of these materials are compared with that of Si in Figure 27.9. Although GaAs and InP possess similar band gaps (see *Section 5.8*) to Si, they exhibit higher electron mobilities making them of great commercial value for high-speed computer circuitry. Ternary materials are also important, e.g. $GaAs_{1-x}P_x$ is the semiconductor of choice in LEDs in pocket calculator, digital watch and similar displays, the colour of the emitted light depending on the band gap (see Table 27.4). In such devices, the semiconductor converts electrical energy into optical energy.

Thin films of GaAs are deposited commercially using CVD techniques by reactions such as 27.13; slow hydrolysis

Table 27.4 The dependence of the wavelength, λ, of the emitted radiation from $GaAs_{1-x}P_x$ on the composition of the material.

x in $GaAs_{1-x}P_x$	Substrate	λ / nm	Observed colour or region of spectrum
0.10	GaAs	780	Infrared
0.39	GaAs	660	Red
0.55	GaP	650	Red
0.65	GaP	630	Orange
0.75	GaP	610	Orange
0.85	GaP	590	Yellow

of GaAs in moist air means that films must be protectively coated.

$$Me_3Ga + AsH_3 \xrightarrow{900\,K} GaAs + 3CH_4 \qquad (27.13)$$

The commercial production of $GaAs_{1-x}P_x$ requires the epitaxal growth of the crystalline material on a substrate.

Epitaxal growth of a crystal on a substrate crystal is such that the growth follows the crystal axis of the substrate.

Figure 27.10 gives a representation of the apparatus used to deposit $GaAs_{1-x}P_x$; the operating temperature is typically

Fig. 27.10 Schematic representation of the CVD assembly used for the epitaxal growth of $GaAs_{1-x}P_x$; H_2 is the carrier gas.

n-InGaAs contact ⎫
n-GaAs contact ⎬ Emitter layers
n-InGaP emitter ⎭
p-GaAs base layer
n-GaAs collecting layers
GaAs substrate

Fig. 27.11 Typical components in a multilayer heterojunction bipolar transistor wafer, each deposited by CVD.

≈ 1050 K and H_2 is used as a carrier gas. Gallium (mp 303 K, bp 2477 K) is held in a vessel within the reactor and reacts with the incoming dry HCl to give GaCl which then disproportionates (scheme 27.14) providing Ga at the substrate.

$$\left. \begin{array}{l} 2Ga + 2HCl \longrightarrow 2GaCl + H_2 \\ 3GaCl \longrightarrow 2Ga + GaCl_3 \end{array} \right\} \qquad (27.14)$$

The proportions of the group 15 hydrides entering the reactor can be varied as required; they thermally decompose by reaction 27.15 giving elemental components for the ternary semiconductor at the substrate surface. High purity reagents are essential for the deposition of films that are of acceptable commercial grade.

$$2EH_3 \longrightarrow 2E + 3H_2 \qquad E = As \text{ or } P \qquad (27.15)$$

Table 27.4 illustrates how the variation in semiconductor composition affects the colour of light emitted from an $GaAs_{1-x}P_x$-containing LED. Dopants can be added to the semiconductor by injecting a volatile dopant-precursor into the PH_3 and AsH_3 gas inflow. For an n-type semiconductor, H_2S or Et_2Te may be used, providing S or Te atom dopants.

Mobile telephones incorporate multilayer III–V epitaxal heterojunction bipolar transistor wafers such as that illustrated in Figure 27.11. The p–n junctions on either side of the base layer are a crucial feature of semiconductor devices, and in the wafer shown in Figure 27.11 (and in other similar wafers), the p-type base layer must be highly doped to provide high frequency performance. Choice of dopant is critical, e.g. use of a Zn dopant (see below) results in its diffusion into the emitting n-type layers. This problem has been overcome by doping with C which exhibits a low diffusion coefficient; C-doped wafers have been used commercially since the early 1990s.

Metal deposition

The use of volatile molecular, often organometallic, precursors for the deposition of thin films of metals for contacts and wiring in electrical devices (i.e. semiconductor–metal connections) and as sources of dopants in semiconductors is an important part of modern manufacturing processes. The general strategy is to choose a volatile organometallic complex which can be thermally decomposed on the substrate, depositing the metal film and liberating organic products which can be removed in the exhaust gases. The use of methyl derivatives as precursors often leads to

higher than acceptable carbon contamination of the deposited metal film, and for this reason other substituents tend to be preferred.

Aluminium is deposited by MOCVD using R_3Al (e.g. R = Et) despite the fact that these compounds are pyrophoric. Vanadium films can be deposited by reaction 27.16.

$$VCl_4 + 2H_2 \xrightarrow{1450\,K} V + 4HCl \qquad (27.16)$$

Nickel films can be deposited from $Ni(CO)_4$, but temperature control is important since above 470 K, there is a high tendency for the deposition of carbon impurities. Other suitable precursors include $(\eta^5\text{-}Cp)_2Ni$ and $Ni(acac)_2$.

Gallium arsenide can be doped with Sn by using tin(IV) alkyl derivatives such as Me_4Sn and Bu_4Sn, although the former tends to result in carbon contamination. Zinc is added as a dopant to, for example, AlGaAs (to give a p-type semiconductor) and can be introduced by adding appropriate amounts of Et_2Zn to the volatile precursors for the ternary semiconductor (Me_3Al, Me_3Ga and AsH_3). Silicon, GaAs and InP may be doped with Er, and Cp_3Er is a suitable precursor; similarly, Cp_3Yb is used to dope InP with Yb.

Ceramic coatings

The development of appropriate CVD techniques has enabled rapid progress to be made in the commercialization of applying ceramic coatings to carbide tools used for cutting steel. Wear-resistant coatings of thickness ≈ 5–$10\,\mu m$ are now usually added to heavy-duty cutting tools to prolong their lifetime and allow the tools to operate at significantly higher cutting speeds. Multilayers can readily be applied using CVD, and the method is amenable to coating non-uniform surfaces.

A coating of Al_2O_3 provides resistance against abrasion and oxidation, and can be deposited by the reaction at a substrate (≈ 1200–1500 K) of $AlCl_3$, CO_2 and H_2. Abrasion resistance is also provided by TiC, while TiN gives a barrier against friction; the volatile precursors used for TiC are $TiCl_4$, CH_4 and H_2, and TiN is deposited using $TiCl_4$, N_2 and H_2, both at temperatures >1000 K. In general, nitride layers can be deposited using volatile metal chlorides, H_2 and N_2 as the molecular precursors; of particular importance for wear-resistant coatings are nitrides of Ti, Zr and Hf.

Perovskites and cuprate superconductors

Table 27.5 lists applications of some of the most commercially important mixed metal, perovskite-type oxides, and illustrates that it is the dielectric, ferroelectric, piezoelectric (see *Section 13.9*) and pyroelectric properties of these materials that are exploited in the electronics industry.

Ferroelectric means the spontaneous alignment of electric dipoles caused by interactions between them; domains form in an analogous manner to the domains of magnetic dipoles in a *ferromagnetic* material (see *Figure 20.25* and related discussion).

Table 27.5 Electronic applications of selected perovskite-type mixed metal oxides.

Mixed metal oxide	Properties of the material	Electronic applications
$BaTiO_3$	Dielectric	Sensors; dielectric amplifiers; memory devices
$Pb(Zr,Ti)O_3$	Dielectric; pyroelectric; piezoelectric	Memory devices; acoustic devices
La doped $Pb(Zr,Ti)O_3$	Electrooptic	Optical memory displays
$LiNbO_3$	Piezoelectric; electrooptic	Optical memory displays; acoustic devices; wave guides; lasers; holography
$K(Ta,Nb)O_3$	Pyroelectric; electrooptic	Pyrodetector; wave guides; frequency doubling.

The industrial fabrication of electronic devices containing perovskite-type metal oxides traditionally involves the preparation of powdered materials which are then cast as required. However, there is great interest at the research level in developing techniques for thin film deposition and in this section we consider the use of CVD methods.

Reaction 27.17 is one conventional method of preparing $BaTiO_3$. A second route (used industrially) involves the preparation of $BaTiO(ox)_2 \cdot 4H_2O$ (ox = oxalate) from $BaCl_2$, $TiCl_4$, H_2O and H_2ox, followed by thermal decomposition (scheme 27.18).

$$TiO_2 + BaCO_3 \xrightarrow{\Delta} BaTiO_3 + CO_2 \qquad (27.17)$$

$$\left. \begin{array}{c} BaTiO(ox)_2 \cdot 4H_2O \xrightarrow[\text{dehydrate}]{400\,K} BaTiO(ox)_2 \\ \Big\downarrow \; 600\,K, \; -CO, \; -CO_2 \\ \tfrac{1}{2}BaTi_2O_5 + \tfrac{1}{2}BaCO_3 \\ \Big\downarrow \; 900\,K, \; -CO_2 \\ BaTiO_3 \end{array} \right\} \qquad (27.18)$$

(27.6)

It is also possible to deposit $BaTiO_3$ by using CVD (Figure 27.12), the source of Ti being the alkoxide $Ti(O^iPr)_4$ and of Ba, a β-ketonate complex such as BaL_2 where $L^- = $ **27.6**. A typical reactor temperature for $BaTiO_3$ deposition is $\approx 500\,K$, and substrates that have been used include Pt, MgO, Si and Al_2O_3. Although often formulated as 'BaL_2', the precursor is not so simple and its exact formulation depends on its method of preparation, e.g. adducts such as $BaL_2 \cdot (MeOH)_3$ and $[BaL_2(OEt_2)]_2$, the tetramer Ba_4L_8, and the species $Ba_5L_9(H_2O)_3(OH)$. Any increased degree of oligomerization is accompanied by a decrease in volatility, a fact that militates against the use of the precursor in CVD. Complexes containing fluorinated β-ketonate ligands such as **27.7** possess higher volatilities than related species containing non-fluorinated ligands, but unfortunately, their use in CVD experiments leads to thin films of $BaTiO_3$ contaminated with fluoride.

Fig. 27.12 Schematic representation of a CVD setup used for the deposition of the perovskite $BaTiO_3$.

(27.7)

So far, we have illustrated the formation of binary (e.g. GaAs, TiC) and ternary (e.g. $GaAs_{1-x}P_x$, $BaTiO_3$) systems through the combination of two or three volatile precursors in the CVD reactor. A problem that may be encountered is how to control the stoichiometry of the deposited material; in some cases, controlling the ratios of the precursors works satisfactorily, but in other cases, better control is achieved by trying to find a suitable *single* precursor. There is active research in this area and is illustrated by the formation of $LiNbO_3$ from the alkoxide precursor $LiNb(OEt)_6$. The ceramic $LiNbO_3$ is used commercially for a range of electronic purposes (Table 27.5) and is conventionally prepared by reaction 27.19 or 27.20.

$$Li_2CO_3 + Nb_2O_5 \xrightarrow{\Delta} 2LiNbO_3 + CO_2 \qquad (27.19)$$

$$Li_2O + Nb_2O_5 \xrightarrow{fuse} 2LiNbO_3 \qquad (27.20)$$

In order to develop an appropriate CVD method for depositing $LiNbO_3$ from $LiNb(OEt)_6$, one major problem has to be overcome: the volatility of bulk $LiNb(OEt)_6$ is low, and hence an aerosol-type system is used to introduce the molecular precursor into the CVD reactor. Solid $LiNb(OEt)_6$ is dissolved in toluene and the solution converted into a fine mist using ultrasonic radiation. In the first part of the reactor (550 K), the mist volatilizes and is transported in a flow of the carrier gas into a higher temperature region containing the substrate on which thermal decomposition of $LiNb(OEt)_6$ occurs to give $LiNbO_3$. Such results for the formation of ternary (or more complex) ceramic materials and the development of *aerosol-assisted CVD* may have a potential for commercial application in the future.

The explosion of interest in cuprate superconductors (see *Section 27.4*) over the past decade has lead to active research interest into ways of depositing these materials as thin films. For example, CVD precursors and conditions for the deposition of $YBa_2Cu_3O_7$ have included BaL_2, CuL_2 and YL_3 ($L^- = \mathbf{27.6}$) with He/O_2 carrier gas, and an $LaAlO_3$ substrate at 970 K. Progress to date, however, has not reached a level that makes CVD commercially viable.

27.7 Inorganic fibres

A *fibre* (inorganic or organic) usually has a diameter <0.25 mm, a length-to-diameter ratio ≥10:1, and a cross-sectional area $<5 \times 10^{-3}$ mm^2; *whiskers* are included in this category.

Fibrous asbestos (a layer silicate which occurs naturally, see *Section 13.9*) was used for much of the 20th century as an insulating material. It is now well recognized that exposure to asbestos fibres causes lung damage (*asbestosis*) and alternative insulating materials have entered the commercial market. Certain forms of asbestos that do not utilize fibres of length 5–20 μm remain in use, e.g. in brake linings. Glass fibres have a wide range of applications, two of the major ones being insulation and reinforcement of other materials such as plastics. Aluminoborosilicate glass fibres are the most commonly employed; alumino-lime silicate glass fibres are suited to acid-resistant needs, and when a high tensile strength material is required, aluminosilicate glass is generally appropriate. While the use of glass fibres for insulation is widespread, high temperature working requires materials such as Al_2O_3 or ZrO_2.

We limit our main discussion in this section to B, C, SiC and Al_2O_3 fibres which can be employed for high temperature (>1300 K) operations. Much of today's fibre technology stems from the development of new, low density, high tensile strength materials for air and space travel. Boron fibres were among the first to be developed, with carbon and silicon carbide fibres entering and dominating the market in more recent years. Silicon carbide has the advantage over both B and C fibres in that it is resistant to oxidation at high temperatures, oxidizing in air only above ≈1250 K.

Boron fibres

Boron fibres can be manufactured by CVD, boron being deposited on a heated tungsten substrate (1550 K) by reaction 27.21; the reactor is schematically represented in Figure 27.13. The

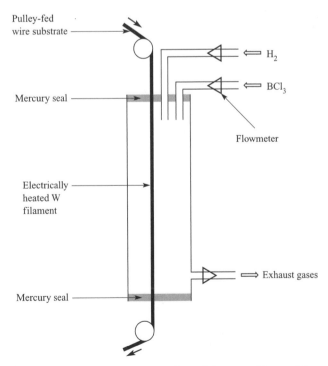

Fig. 27.13 Schematic representation of the assembly used for the manufacture of boron fibres by CVD using a tungsten substrate.

tungsten substrate is drawn through the reactor making the boron fibre production a continuous process. The proportion of H_2 and BCl_3 that interacts in the reactor is low and unchanged gases are recycled after first separating them from HCl.

$$2BCl_3 + 3H_2 \xrightarrow{\Delta} 2B + 6HCl \qquad (27.21)$$

A final step in manufacture is to coat the fibre with SiC or B_4C; this provides protection against reactions with other elements at high operating temperatures and ensures that the fibre retains its tensile strength at elevated temperature. Typically, the W wire substrate has an 8 μm diameter, the diameter of the boron fibre ≈ 150 μm, and the SiC or B_4C coating is ≈ 4 μm thick.

Carbon fibres

Since 1970, the commercial production of carbon fibres has risen dramatically. Where the low weight of a construction material is crucial, carbon-fibre reinforced polymers are now dominating the market. Body-parts for modern military aircraft contain $\leq 50\%$ by weight of carbon-fibre reinforced composites in place of aluminium. This trend is also being followed in modern commercial aircraft design. The performance of Formula 1 racing cars has been greatly enhanced by turning to body parts constructed from carbon-fibre reinforced materials. Carbon fibres are characterized by being stiff but brittle, and have a low density and high tensile strength; the high resistance to thermal shock arises from a high thermal conductivity but low coefficient of thermal expansion.

A number of different grades of carbon fibre are manufactured, but all are made by the thermal degradation of an appropriate polymeric organic precursor. For example, PAN-based carbon fibres are manufactured from polyacrylonitrile and may, depending on grade, retain a low nitrogen content. Their production is shown in scheme 27.22 in which atom X represents an arbitrary N content. Carbon fibres usually require a protective coating to provide resistance to reaction with other elements at elevated temperature.

(27.22)

X = N or C

The importance of carbon-fibre composite materials in the development of the space shuttle cannot be overemphasized. Reinforced *carbon–carbon composites* are used in the nose cone and wing leading edges to provide the resistance to thermal shock and stress required for re-entry into the Earth's atmosphere. Carbon–carbon composites are a particular group of carbon-fibre reinforced materials in which both the bulk material and fibres are carbon. The manufacturing process for the space shuttle's carbon–carbon composites starts by impregnating a graphitized rayon fabric with a phenolic resin and then subjecting the material to heat treatment to convert the phenolic resin to carbon. The next stage is impregnation with furfuryl alcohol (**27.8**) followed by heat treatment to convert this component to carbon. Three cycles of this process result in the desired composite material. The composite must be coated with SiC to render it resistant to oxidation. This coating is generated by heating the composite in contact with a mixture of Al_2O_3, SiO_2 and SiC in a furnace. Final impregnation with tetraethyl orthosilicate seals any surface imperfections.

(27.8)

Silicon carbide fibres

The resistance of SiC to high temperature working and oxidation makes it a valuable advanced material. Fibres of β-SiC are produced by CVD (see *Section 27.6*) using $R_{4-x}SiCl_x$ precursors or an alkane and chlorosilane in a reactor similar to that in Figure 27.13. Fibres marketed under the tradename of *Nicalon* are produced by a melt-spinning process. This begins with reaction 27.23, the product of which is pyrolyzed to give a carbosilane polymer.

$$n Me_2SiCl_2 \xrightarrow{Na} (Me_2Si)_n \qquad (27.23)$$

The polymer is heated until molten, forced through an appropriately sized aperture, and solidified as it is drawn into a fibre; the commercial process produces multifilament fibres rather than single strands and cross-linking of the filaments adds strength to the final product. The composition of Nicalon fibres is not as simple as this description implies: carbon and silica are present in addition to β-SiC.

Alumina fibres

Alumina fibres (often with silica content) are produced commercially on a large scale. Their high tensile strength, flexibility and inertness make them valuable in, for example, rope, thread (suitable for cloth manufacture), insulating material and electrical-cable coverings. A number of different manufacturing methods are in operation for the production of alumina–silica fibres, depending on the type of fibre and also the manufacturer. Polycrystalline Al_2O_3 fibres can be formed by extruding hydrated alumina slurries through suitable nozzles and then heating the extruded material. As an example of a fibre with silica content, continuous fibres containing 15% SiO_2 by weight are manufactured starting from Et_3Al. This is subject to partial hydrolysis to give a polymeric material which is dissolved along with an alkyl silicate in a suitable solvent. The viscous solution is amenable to fibre production by gel-spinning; the fibres so-formed are heated (*calcined*) to convert the material into alumina–silica. Further heating results in the formation of a polycrystalline material.

Summary

This chapter has been concerned with selected aspects of solid state inorganic chemistry. After brief treatments of the defect solid state, conductivity and superconductivity, we have concentrated on topics that illustrate the preparation and commercial applications of solid state materials and related research efforts. Many of the materials described are small-scale products for the electronics industry, but the preparations of ceramics and inorganic fibres of boron, carbon, silicon carbide and alumina are major industrial processes.

The following terms have been introduced in this chapter. Do you know what they mean?

- ❑ intrinsic defect
- ❑ extrinsic defect
- ❑ colour centre or F-centre
- ❑ cation or anion conductor
- ❑ superconductor
- ❑ ceramic material
- ❑ opacifier
- ❑ chemical vapour deposition (CVD)
- ❑ metal–organic chemical vapour deposition (MOCVD)
- ❑ epitaxal growth
- ❑ ferroelectric
- ❑ inorganic fibre

Suggested further reading

General and introductory texts

M. Ladd (1994) *Chemical Bonding in Solids and Fluids*, Ellis Horwood, Chichester.

L. Smart and E. Moore (1992) *Solid State Chemistry: An Introduction*, Chapman and Hall, London.

A.R. West (1999) *Basic Solid State Chemistry*, 2nd edn, Wiley-VCH, Weinheim – An introductory text which includes structures and bonding in solids, and electrical, magnetic and optical properties.

More specialized articles

A.K. Cheetham and P. Day, eds (1992) *Solid State Chemistry*, Clarendon Press, Oxford – Two volumes covering techniques (vol. 1) and compounds (vol. 2) in detail.

R.A. Eppler (1998) 'Ceramic colorants' in *Ullman's Encyclopedia of Industrial Inorganic Chemicals and Products*, Wiley-VCH, Weinheim, vol. 2, p. 1069 – Describes types and applications of ceramic pigments.

R.B. King, ed. (1994) *Encyclopedia of Inorganic Chemistry*, Wiley, Chichester – Contains the following detailed reviews: J.K. Burdett, 'Superconductivity', vol. 7, p. 4039; M. Greenblatt, 'Ionic conductors', vol. 3, p. 1584; W.H. McCarroll, 'Oxides: Solid state chemistry', vol. 6, p. 2903; R.J.D. Tilley, 'Defects in solids', vol. 2, p. 966.

T.T. Kodas and M. Hampden-Smith, eds (1994) *The Chemistry of Metal CVD*, VCH, Weinheim – Covers the deposition of a range of metals from organometallic precursors.

C.H. Winter and D.M. Hoffman, eds (1999) *Inorganic Materials Synthesis*, Oxford University Press, Oxford – A detailed coverage which includes inorganic thin films.

Problems

27.1 Explain what is meant by (a) a Schottky defect in $CaCl_2$, and (b) a Frenkel defect in AgBr. (c) Suggest what effect doping crystals of AgCl with $CdCl_2$ might have on the AgCl lattice structure.

27.2 Why are *d*-block metal oxides much more frequently non-stoichiometric than are non-*d*-block metal oxides?

27.3 When nickel(II) oxide is heated in O_2, some of the cations are oxidized and vacant cation sites are formed according to the equation:

$$4Ni^{2+}(s) + O_2(g) \rightleftharpoons 4Ni^{3+}(s) + 2\square_+ + 2O^{2-}(s)$$

where $\square_+$ denotes a vacant cation site and (s) denotes an ion in the solid. Account for the fact that the conductivity of the product is, for small deviations from stoichiometry, proportional to the sixth root of the pressure of O_2.

27.4 Comment on each of the following: (a) the difference between extrinsic and intrinsic defects; (b) why CaO is added to ZrO_2 used in refractory materials; (c) the formation of solid solutions of Al_2O_3 and Cr_2O_3.

27.5 Suggest why doping NiO with Li_2O in air (or the presence of O_2) leads to an increase in electrical conductivity, and comment on the dependence of this increase on the amount of lithium dopant.

27.6 Comment on the structural and compositional implications of (a) the Fe-deficiency of iron(II) oxide, and (b) the anion-excess nature of uranium(IV) oxide.

27.7 If Ag electrodes are placed in contact with and on either side of a piece of bulk AgI (mp 831 K) heated at 450 K, and current is passed through the cell for a given period, it is found that one electrode gains mass and the other loses mass. Rationalize these observations.

27.8 Comment on the following values of electrical conductivities: Na β-alumina, $3 \times 10^{-2}\,\Omega^{-1}\,cm^{-1}$ (at 298 K); Li_3N, $5 \times 10^{-3}\,\Omega^{-1}\,cm^{-1}$ (at 298 K); NaCl, $10^{-15}\,\Omega^{-1}\,cm^{-1}$ (at 300 K). Would you expect these values to be direction-independent with respect to the crystal under study?

27.9 A recently developed solid state battery consists of lithium and V_6O_{13} electrodes separated by a solid polymer electrolyte. Suggest how this battery might operate.

27.10 Discuss the variation in electrical conductivities along the series TiO, VO, MnO, FeO, CoO and NiO.

27.11 (a) The structure of $YBa_2Cu_3O_7$ can be described as consisting of rock salt and perovskite layers. Describe the origin of this description.
(b) Why is the potential replacement of NbTi by high-temperature superconducting components in MRI equipment of commercial interest?

27.12 Explain what is meant by 'doping' using as your examples (a) MgO doping of ZrO_2, (b) LaF_3 doping of CaF_2, (c) B doping of Si, and (c) As doping of Si.

27.13 Suggest likely products in the following reactions; (the reactions as shown are not necessarily balanced):
(a) $x\text{LiI} + V_2O_5 \xrightarrow{\Delta}$
(b) $CaO + WO_3 \xrightarrow{\Delta}$
(c) $SrO + Fe_2O_3 \xrightarrow{\Delta,\ \text{in presence of } O_2}$

27.14 Suggest possible solid state precursors for the formation of the following compounds by pyrolysis reactions:
(a) $BiCaVO_5$; (b) the Mo(VI) oxide $CuMo_2YO_8$;
(c) Li_3InO_3; (d) $Ru_2Y_2O_7$.

27.15 Give a brief outline of a typical CVD process and give examples of its use in the semiconductor industry.

27.16 Briefly discuss each of the following.

(a) Precursors for, and composition and uses of, CVD wear-resistant coatings.
(b) The production of GaAs thin films.
(c) The advantages of using LEDs over traditional glass-reflector Cat's eyes for road-lane markings.
(d) Problems in developing CVD methods for the deposition of perovskite and cuprate superconductors.

Chapter 28   The trace metals of life

28.1 Introduction

When one considers the chemistry of biological processes, the boundary between inorganic and organic chemistry is blurred. The *bulk biological* elements that are essential to all life include C, H, N, O (the four most abundant elements in biological systems) along with Na, K, Mg, Ca, P, S and Cl. The fundamental elements that make up the building blocks of biomolecules (e.g. amino acids, peptides, carbohydrates, proteins, lipids and nucleic acids) are C, H, N and O, with P playing its part in, for example, ATP and DNA (see *Box 14.10*) and S being the key to the coordinating abilities of cysteine residues in proteins. The roles of the less abundant, but nonetheless essential, elements include osmotic control and nerve action (Na, K and Cl), Mg^{2+} in chlorophyll (see *Section 11.8*), Mg^{2+}-containing enzymes involved in phosphate hydrolysis, structural functions of Ca^{2+} (e.g. bones, teeth, shells) and triggering actions of Ca^{2+} (e.g. in muscles). The *trace metals* are V, Cr, Mn, Fe, Co, Ni, Cu, Zn and Mo, while *trace non-metals* comprise B, Si, Se, F and I; their essentiality to life can be summarized as follows:

● V: accumulated by a few organisms (see Box 28.1), and has been shown to be essential for growth in rats and chicks;
● Cr: essential (see Table 28.1);
● Mn, Fe, Cu, Ni, Zn: essential to all organisms (see Table 28.1);
● Co: essential to mammals and many other organisms (see Table 28.1);
● Mo: essential to all organisms (see Table 28.1) although green algae may be an exception;
● B: essential to green algae and higher plants, but its role is unknown;
● Si: exoskeletons of marine diatoms composed of hydrated silica, but its role in other biological systems is less well defined;[†]

● Se: essential to mammals and some higher plants;
● F: its role is not fully established but its deficiency causes dental caries;
● I: essential to many organisms.

Despite their crucial role in life, the trace metals make up only a tiny fraction of the human body-weight (Table 28.1). In this chapter we look at the ways in which living systems store metals, and the manner in which trace metal ions take part in the transport of molecules such as O_2, electron transfer processes and catalysis. It is assumed that the reader has already studied Chapters 19 and 20, and is familiar with the general principles of *d*-block coordination chemistry: a study of the trace metals in biological systems is *applied coordination chemistry*.

Research progress in bioinorganic chemistry has been greatly assisted in recent years by the development of methods to solve protein structures using X-ray diffraction and NMR spectroscopy. Readers are encouraged to make use of the Protein Data Bank (PDB) to update the information given in this chapter; information is available using the worldwide web (http://www/rcsb.org/pdb).[‡]

Amino acids, peptides and proteins: some terminology

In this chapter, we refer to polypeptides and proteins, and we now give a brief résumé of some of the terminology needed.[§]

A *polypeptide* in Nature is formed by the condensation, in varying sequences, of the 20 naturally occurring α-amino acids. Structure **28.1** gives the general formula of an amino

---

[†] See: J.D. Birchall (1995) *Chemical Society Reviews*, vol. 24, p. 351 – 'The essentiality of silicon in biology'.
[‡] Protein structures in this chapter have been drawn using atomic coordinates from the Protein Data Bank.
[§] For a more detailed account, see for example: J. McMurry (1996) *Organic Chemistry*, 4th edn, Brooks/Cole, Pacific Grove, Chapter 27.

Table 28.1 Mass of each trace metal present in an average 70 kg human, and a summary of where the trace metals are found and their biological roles.

Metal	Mass / mg	Biological roles
V	0.11	Enzymes (nitrogenases, haloperoxidases)
Cr	14	Claimed (not yet proven) to be essential in glucose metabolism in higher mammals
Mn	12	Enzymes (phosphatase, mitochondrial superoxide dismutase, glycosyl transferase); photoredox activity in Photosystem II (see *equation 21.52* and discussion)
Fe	4200	Electron transfer systems (Fe–S proteins, cytochromes); O_2 storage and transport (haemoglobin, myoglobin, haemerythrin); Fe storage (ferritin, transferritin); Fe transport proteins (siderophores); in enzymes (e.g. nitrogenases, hydrogenases, oxidases, reductases)
Co	3	Vitamin B_{12} coenzyme
Ni	15	Enzymes (urease, some hydrogenases)
Cu	72	Electron transfer systems (blue copper proteins); O_2 storage and transport (haemocyanin); Cu transport proteins (ceruloplasmin)
Zn	2300	Acts as a Lewis acid (e.g. in hydrolysis processes involving carboxypeptidase, carbonic anhydrase, alcohol dehydrogenase); structural roles
Mo	5	Enzymes (nitrogenases, reductases, hydroxylases)

acid and **28.2** shows a peptide link formed after the condensation of two amino acid residues. A peptide chain has an *N-terminus* (corresponding to an NH_2 group) and a *C-terminus* (corresponding to a CO_2H group). The names, abbreviations and structures of those amino acids that are mentioned in this chapter are listed in Table 28.2. All but glycine are chiral, but Nature is specific in the enantiomers that it uses.

Proteins are high molecular mass polypeptides with complex structures. The sequence of amino acids gives the primary structure of the protein, while the secondary and tertiary structures reveal the spatial properties of the peptide chain. Haemoglobin, myoglobin and most metalloenzymes are *globular proteins* in which the polypeptide chains are coiled into near-spherical structures. The *prosthetic group* in a protein is an additional, non-amino acid component of a protein, which is essential for the biological activity of the protein. We shall be concerned with prosthetic groups containing metal centres, e.g. haem is the prosthetic group

Table 28.2 Some of the 20 naturally occurring α-amino acids.

Name of amino acid	Abbreviation for amino acid residue (alternative abbreviation used in sequence specification)	R group in:
L-Arginine	Arg (R)	$CH_2CH_2CH_2NHC(=NH)NH_2$
L-Aspartic acid	Asp (D)	CH_2CO_2H
L-Cysteine	Cys (C)	CH_2SH
L-Glutamic acid	Glu (E)	$CH_2CH_2CO_2H$
Glycine	Gly (G)	H
L-Histidine	His (H)	
L-Leucine	Leu (L)	CH_2CHMe_2
L-Lysine	Lys (K)	$CH_2CH_2CH_2CH_2NH_2$
L-Methionine	Met (M)	CH_2CH_2SMe
L-Serine	Ser (S)	CH_2OH
L-Threonine	Thr (T)	$CH(OH)Me$
L-Tyrosine	Tyr (Y)	

(28.1)   (28.2)

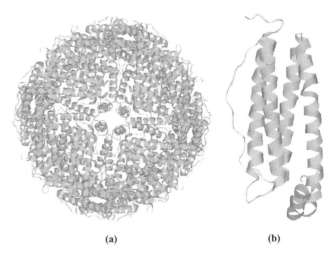

in haemoglobin and myoglobin. The proteins that we discuss contain metals (*metalloproteins*) and the form of the protein with the metal removed is called the *apoprotein*; the prefix *apo*- before a particular protein (e.g. ferritin and apoferritin) signifies the metal-free species. The difference between a protein and the corresponding apoprotein is analogous to that between a metal complex and the corresponding free ligand.

28.2 Metal storage and transport: Fe, Cu, Zn and V

Living organisms require ways of storing and transporting trace metals, and storing the metal in a non-toxic form is clearly critical. Consider Fe, the most important trace metal in humans. Table 28.1 gives the average mass of Fe present in a 70 kg human, and this level needs to be maintained through a dietary intake (typically 6–40 mg per day) offsetting loss through, for example, bleeding. There is no excretory loss of Fe, a phenomenon not shared by other metals present in the body. The amount of Fe stored in the body far exceeds that taken in per day, but only a very small fraction of the iron in the body is actually in use at any one time; the mammalian system is very effective at recycling Fe. Whereas we can discuss the storage and transport of Fe in some detail, less information is presently available about the storage and transport of other trace metals.

Iron storage and transport

In mammals, the task of transferring iron from dietary sources to haemoglobin (see *Section 28.3*) initially involves the absorption of Fe(II) after passage through the stomach, followed by uptake into the blood in the form of the Fe(III)-containing metalloproteins *transferrins*. Iron is transported as transferrin to protein 'storage vessels' until it is required for incorporation into haemoglobin. In mammals, iron is stored mainly in the liver (typically 250–1400 ppm of Fe is

Fig. 28.1 (a) The structure of the protein shell in ferritin (isolated from the bull frog) which shows the polypeptide chains in 'ribbon' representation. (b) One of the 24 equivalent units that are present in the protein shell of ferritin.

present), bone marrow and spleen in the form of *ferritin*, a water-soluble metalloprotein. *Apoferritin* has been isolated from, for example, horse spleen and has a molecular weight of ≈445 000. X-ray diffraction studies confirm that it consists of 24 equivalent units (each with 163 amino acid residues) arranged so as to form a hollow shell (Figure 28.1), the cavity of which has a diameter of ≈8000 pm. In *ferritin*, this cavity contains up to 4500 high-spin Fe^{3+} centres in the form of a *microcrystalline* oxohydroxophosphate of composition $(FeO \cdot OH)_8(FeO \cdot H_2PO_4)$. Results of an EXAFS (see *Box 26.2*) study indicate that this core comprises double layers of approximately close-packed O^{2-} and $[OH]^-$ ions, with interstitial sites between the layers occupied by Fe(III) centres. Adjacent [OFeO]-triple layer blocks are only weakly associated with each other. The phosphate groups in the iron-containing core appear to function as terminators and linking groups to the protein shell.

While the structures of apoferritin and ferritin are fairly well established, the manner in which iron is transported in and out of the protein cavity is still under investigation. It is proposed that iron enters as Fe^{2+} and is oxidized once inside the protein. The formation of the crystalline core is an example of *biomineralization* and it is a remarkable achievement of evolution that iron can be stored in mammals effectively as hydrated iron(III) oxide, i.e. a form closely related to rust!

As we illustrate throughout this chapter, studying appropriate model compounds gives valuable insight into related, but more complicated, bioinorganic systems. The synthesis of large iron-oxo clusters from mono- and dinuclear precursors is of research interest in relation to modelling the formation of the core of ferritin, and reactions 28.1 and 28.2 give two examples. The Fe_6O_{14}-core of the product of reaction 28.2 is shown in Figure 28.2.

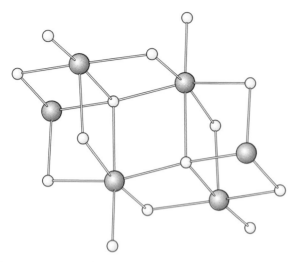

Fig. 28.2 A model for the biomineralization of ferritin. The Fe_6O_{14}-core of $Fe_6(OMe)_4(\mu\text{-}OMe)_8(\mu_4\text{-}O)_2L_2$ ($L = N(CH_2CH_2NH_2)_3$) determined by X-ray diffraction [V.S.Nair *et al.* (1992) *Inorg. Chem.*, vol. 31, p. 4048]. Colour code: Fe, blue; O, pale grey.

$$Fe(OAc)_2 + LiOMe$$

in presence of O_2 | in MeOH

$$Fe_{12}(OAc)_3(\mu\text{-}OAc)_3(MeOH)_4(\mu\text{-}OMe)_8(\mu_3\text{-}OMe)_{10}(\mu_6\text{-}O)_2$$

(28.1)

$$Fe(O_3SCF_3)_2 + L \xrightarrow{\text{in MeOH}} Fe_6(OMe)_4(\mu\text{-}OMe)_8(\mu_4\text{-}O)_2L_2$$

(28.2)

where $L = N(CH_2CH_2NH_2)_3$.

The *transferrins* are *glycoproteins* (i.e. compounds of proteins and carbohydrates) and include *serum transferrin*, *lactoferrin* (present in milk) and *ovotransferrin* (present in egg white).

In humans, serum transferrin transports $\approx 40\,mg$ of iron per day to the bone marrow. It contains a single polypeptide chain (molecular weight of $\approx 80\,000$) coiled in such a way as to contain two pockets suitable for binding Fe^{3+}. Each pocket presents hard *N*- and *O*-donor atoms to the metal centre (Figure 28.3), but the presence of a $[CO_3]^{2-}$ or $[HCO_3]^-$ ligand is also essential. The stability constant for the Fe^{3+} complex is very high ($\log \beta = 28$ at pH 7.4), making transferrin extremely efficient as an iron transporting and scavenging agent in the body. The exact mechanism by which the Fe^{3+} enters and leaves the cavity has not been elucidated, but it seems reasonable that a change in conformation of the polypeptide chain facilitates the process.

Aerobic microorganisms also require iron, but cannot simply absorb it from their aqueous environment since Fe^{3+} is precipitated as $Fe(OH)_3$ ($K_{sp} = 2.64 \times 10^{-39}$). Evolution has provided these organisms with *O*-donor polydentate ligands called *siderophores* which scavenge for iron. Examples of siderophores are the anions derived from *enterobactin* (Figure 28.4a), *desferrichrome* (Figure 28.5a) and *desferrioxamine* (Figure 28.5b). Enterobactin, H_6L, possesses three catechol groups and in the deprotonated form binds Fe^{3+} to give the complex $[FeL]^{3-}$ in which Fe^{3+} is in an octahedral (or close to octahedral) environment. The ligand itself is chiral and spectroscopic data (electronic and circular dichroism spectra) have shown that the Λ-complex is formed diastereoselectively (see *Box 19.2*). The crystal structure of iron(III) enterobactin has not been determined, but studies on model compounds provide relevant information. The model ligand **28.3** is closely related to enterobactin and gives a complex with Fe^{3+} for which $\log \beta$ is close to the value for iron(III) enterobactin. The V(IV) complex of enterobactin (reaction 28.3) has been structurally characterized by X-ray diffraction, and although the radius of a V(IV) centre (58 nm) is smaller than that of Fe(III) (65 nm), the

Fig. 28.3 Schematic representation of an Fe^{3+} binding site in transferrin; the coordinated $[CO_3]^{2-}$ points towards the positively charged Arg residue and the *N*-terminus of a helix. The binding site in human lactoferrin has been determined by protein X-ray crystallography.

Fig. 28.4 (a) The structure of the siderophore enterobactin, H_6L, showing the donor atoms; OH groups are deprotonated before coordination to Fe^{3+}. (b) The structure of the vanadium(IV) complex $[VL]^{2-}$ (H_6L = enterobactin) determined by X-ray diffraction of the K^+ salt [T.B. Karpishin *et al.* (1993) *J. Am. Chem. Soc.*, vol. 115, p. 1842]. Hydrogen atoms are omitted for clarity; colour code: V, blue; C, grey; O, white; N, black.

Fig. 28.5 The structures of the siderophores (a) desferrichrome and (b) desferrioxamine, showing the donor atoms; OH groups are deprotonated before coordination to Fe^{3+}.

gross structural features of the Fe(III) and V(IV) complexes should be similar. The three 'arms' of the ligand lie above the central macrocycle allowing each arm to act as an O,O'-donor (Figure 28.4b). The 6-coordinate V(IV) centre is in an environment described as trigonal prismatic with a twist angle of 28° (see structures **19.9**, **19.10** and **19.12**).

$$[V(O)(acac)_2] + \underset{\text{enterobactin}}{\underset{\text{see 21.7}}{H_6L}} + 4KOH$$

$$\xrightarrow{\text{MeOH}} K_2[VL] + 2K[acac] + 5H_2O \qquad (28.3)$$

(28.3)

High-spin Fe^{3+} complexes of the siderophores are kinetically labile (see *Section 25.2*). If Fe^{3+} is exchanged for Cr^{3+}, kinetically inert complexes are obtained which can be studied in solution as models for the Fe^{3+} complexes.

The complexes that transport iron in mammals and micro-organisms have very high overall stability constants (see above) and, although exact mechanisms have not been elucidated, it is reasonable to propose that reduction to Fe^{2+} is required since the stability constant for the Fe^{2+} complex is orders of magnitude lower than that for the Fe^{3+} complex.

Metallothioneins: transporting some toxic metals

Transporting soft metal centres is important in protection against toxic metals such as Cd^{2+} and Hg^{2+}. Complexation requires soft ligands which are provided by Nature in the form of cysteine residues (Table 28.2) in *thioneins*, the metal complexes of which are *metallothioneins*. Thioneins also bind Cu^+ and Zn^{2+}, but their active role in transporting and storing these metals in mammals has not been confirmed. Thioneins are small proteins containing ≈ 62 amino acids, about one-third of which is cysteine. The Cys residues are either adjacent to each other or separated by one other amino acid residue, thus providing pockets

of *S*-donor sites ideally suited to scavenging soft metal ions. Both Cd and Hg have NMR active nuclei (the most important are ^{113}Cd, 12% abundance, $I = \frac{1}{2}$; ^{199}Hg, 17% abundance, $I = \frac{1}{2}$) and the application of NMR spectroscopy to probe the coordination sites in Cd- and Hg-containing metallothioneins has greatly aided structural determination.

The presence of Hg^{2+}, Cd^{2+}, Cu^+ and Zn^{2+} induces the production of thioneins in the liver and kidneys of mammals. Between 4 and 12 metal centres can be bound by one thionein; Zn^{2+} Hg^{2+}, Cd^{2+} centres are likely to be in tetrahedral environments, while Cu^+ may be 3-coordinate. The structure of the Cd/Zn-containing metallothionein isoform II from rat liver has been determined by X-ray diffraction, and Figure 28.6a illustrates the folded protein chain consisting of 61 amino acid residues of which 20 are Cys groups. One Cd^{2+} and two Zn^{2+} centres are bound in one pocket of the folded chain, and four Cd^{2+} in the other (Figures 28.6b and 28.6c).

Thiolate and related complexes are studied as models for metallothioneins. For example, the Cu(I)-containing metallothionein in yeast has been modelled by $[Cu_4(SPh)_6]^{2-}$ (**28.4**), while model studies on canine liver cuprothionein have utilized complex **28.5** in which the Cys residues are 'replaced' by thiourea ligands. Among Cd_xS_y-containing clusters studied as models for Cd^{2+}-containing metallothioneins is $[Cd_3(SC_6H_2{}^iPr_3)_7]^-$ (**28.6**).

(28.4)

(28.5)

BOX 28.1   **RESOURCES, ENVIRONMENTAL AND BIOLOGICAL**

The specialists: organisms that store vanadium

Storage and transport of vanadium is a specialized affair. Just why certain organisms accumulate high levels of vanadium is unknown and the biological functions of this trace metal have yet to be established.

The fungus *Amanita muscari* (the deadly poisonous *fly agaric* toadstool) contains ≤400 times more vanadium than is typical of plants, and the amount present is independent of the vanadium content of the soil in which the fungus grows. *Amanita muscari* takes up the metal by using the conjugate base of (*S,S*)-2,2'-(hydroxyimino)dipropionic acid (H$_3$L) to transport and store the trace metal as the V(IV) complex [VL$_2$]$^{2-}$, *amavadin*.

(*S,S*)-2,2'-(hydroxyimino)dipropionic acid

The formation of a complex of 'naked' V(IV) is in contrast to the more common occurrence of complexes containing [VO]$^{2+}$ (see *Section 21.6*). The structure of the amavadin derivative Λ-[V(HL)$_2$]·H$_3$PO$_4$·H$_2$O has recently been solved. The complex contains five chiral centres, one of which is the V(IV) centre. The latter is 8-coordinate and each HL^{2-} ligand acts as an *N,O,O',O''*-donor; the N−O-unit coordinates in a side-on (η^2) manner. *Amanita muscari* contains a 1:1 mixture of the Λ- and Δ-forms of amavadin. Amavadin undergoes a reversible one-electron oxidation without a change in structure, and this observation may be significant in view of a possible role in electron transfer.

The levels of vanadium present in some ocean-dwelling ascidians such as the sea squirt *Ascidia nigra* are extraordinarily

high, up to 10^7 times greater than in the surrounding water. The metal is taken up from seawater (where it is typically present ≈1.1–1.8 × 10^{-3} ppm) in the form of [VO$_4$]$^{3-}$ and is stored in vacuoles in specialized blood cells called *vanadocytes*. Here it is reduced to V^{3+} or [VO]$^{2+}$ by the polyphenolic blood pigment *tunichrome*. (Note the structural relationship between tunichrome and L-DOPA, **26.2**.) Storage of vanadium must involve the formation of V^{3+} or [VO]$^{2+}$ complexes, but the nature of these species is not known.

Tunichrome

Further information

R.E. Berry, E.M. Armstrong, R.L. Beddoes, D. Collison, S.N. Ertok, M. Helliwell and C.D. Garner (1999) *Angewandte Chemie, International Edition in English*, vol. 38 p. 795 – 'The structural characterization of amavadin'.

D. Rehder (1991) *Angewandte Chemie, International Edition in English*, vol. 30, p. 148 – 'The bioinorganic chemistry of vanadium'.

(28.6)

28.3 Dealing with O$_2$

Haemoglobin and myoglobin

In mammals, O$_2$ (taken in by respiration) is carried in the bloodstream by *haemoglobin* and is stored in the tissues in *myoglobin*. Both haemoglobin and myoglobin are *haem-iron proteins*. Myoglobin has a molecular weight of ≈17 000 and is a monomer with a protein chain consisting of 153 amino acid residues. Haemoglobin has a molecular weight of ≈64 500 and is a tetramer, one of the four units of which is shown in Figure 28.7a. The protein chain in myoglobin and in each chain of haemoglobin contains a protoporphyrin IX group (see *Figure 11.8a* for porphyrin) which, together with

(a)

(b)   **(c)**

Fig. 28.6 (a) The backbone (folded to give two pockets) of the polypeptide chain in metallothionein isoform II from rat liver. Each pocket contains a multinuclear metal unit (Zn, blue; Cd, black) coordinated by cysteine residues; the composition and structures of these two units are (b) $CdZn_2S_8$ and (c) Cd_4S_{10}.

a histidine residue tethered to the protein backbone, contains an Fe centre. A porphyrin ring containing an Fe centre is called a *haem group* and the one present in haemoglobin is highlighted in Figure 28.7a; the enlarged view in Figure 28.7b shows that the Fe(II) centre is in a square-based pyramidal environment when in its 'rest state', also referred to as the deoxy-form. When O_2 binds to the haem group, it enters *trans* to the His residue to give an octahedral species (**28.7**); we return to details of this structure later. Although each of the four units in haemoglobin contains a haem group, the four groups do *not* operate independently of each other: the binding (and release) of O_2 is a *cooperative* process. As the tetramer binds successive O_2 molecules, the affinity of the 'vacant' haem groups for O_2 *increases* such that the affinity for the fourth site is ≈ 300 times that of the first haem unit. The 'cooperativity' can be rationalized in terms of communication between the haem groups arising from conformational changes in the protein chains. Consider the haem group in its rest state in Figure 28.7b: it contains high-spin Fe(II) lying ≈ 40 pm out of the plane of the N,N',N'',N'''-donor set of the porphyrin group and drawn towards the His residue; the high-spin Fe(II) centre is apparently too large to fit within the plane of the four N-donor atoms. When O_2 enters the sixth coordination site, the iron centre (now low-spin Fe^{3+},

see below) moves into the plane of the porphyrin ring and pulls the His residue with it. This in turn perturbs not only the protein chain to which the His group is attached, but also the other three protein subunits, and a cooperative process triggers the other haem units to successively bind O_2 more avidly. When O_2 is released from haemoglobin to myoglobin, the loss of the first O_2 molecule triggers the release of the remaining three. Myoglobin does not exhibit this cooperative effect since it comprises only one protein chain. When bound in either haemoglobin or myoglobin, the O_2 molecule resides in a sterically protected cavity; the importance of this becomes clear when we look at model compounds.

$$
\begin{array}{c}
\text{N(His)} \\
| \\
\text{(Porph)N}\,/\!/\!/\!/\!\!\!\underset{\displaystyle |}{\text{Fe(III)}}\!\!\!\cdots\text{N(Porph)} \\
\text{(Porph)N}\quad\quad\text{N(Porph)} \\
| \\
\text{O} \\
\text{O}
\end{array}
$$

(28.7)

Many man- and woman-years of research activity have gone into reaching our current level of understanding of O_2 uptake by myoglobin and haemoglobin, and various proposals have been put forward to describe the nature of the iron centre and of the O_2 species in the oxy-forms of these proteins. Some model studies have involved the reactions of O_2 with certain Co(II) Schiff base[†] complexes. Reactions such as that represented in equation 28.4 yield Co(III) compounds in which the O_2 molecule is bound 'end-on' to the metal centre; the Co−O−O bond angle is $\approx 125°$ and the O−O bond length ≈ 126 pm (compare values of 121 pm in O_2 and 134 pm in $[O_2]^-$, see *Box 15.2*).

Co(II)   Co(III)

L = monodentate ligand, e.g. py

(28.4)

The Co(III) complex formed in reaction 28.4 can be considered to contain coordinated $[O_2]^-$, but the presence of the axial base, L, is crucial to the formation of the monomeric product. In its absence, a dicobalt species with a Co−O−O−Co peroxo-bridge (i.e. analogous to those discussed in *Section 21.10*) is formed.

A logical ligand to model the active sites in myoglobin and haemoglobin is one derived from porphyrin. Tetraphenyl-porphyrin (H_2tpp, **28.8**) is readily available, but the reaction

---

[†] A Schiff base is an imine formed by condensation of a primary amine and a carbonyl.

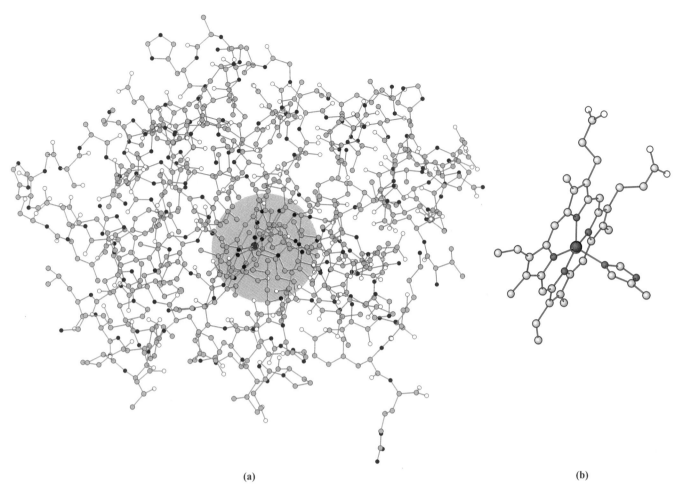

Fig. 28.7 (a) One of the chains of the mammalian protein haemoglobin; H atoms have been omitted for clarity. The haem unit is highlighted by the blue shaded circle. (b) Enlargement of the haem unit from diagram (a) showing the rest state in which Fe(II) is in a 5-coordinate environment consisting of a protoporphyrin IX ligand (see also structure **28.12**) and a histidine residue; hydrogen atoms are omitted. Colour code in (b): Fe, blue; N, black; O, white; C, grey; S, pale blue.

of the Fe(II) complex Fe(tpp)$_2$ with O$_2$ leads to a peroxo-bridged Fe(III) complex (equation 28.5).

(28.8)

(28.9)

Fe(II)

(28.5)

Fe(III)   Fe(III)

Interaction with the second iron centre can be prevented by using a porphyrin ligand with bulky substituents; an example is ligand **28.9**, a so-called 'picket-fence' porphyrin. The four substituents in ligand **28.9** form a cavity, and reaction 28.6 shows the binding of O_2 within this cavity. The axial ligand is 1-methylimidazole which is structurally similar to a His residue; the system clearly resembles the iron environment in haemoglobin (compare Figure 28.7).

(28.6)

The solid state structure of the product of reaction 28.6 has been determined by X-ray diffraction and confirms an end-on, bent coordination mode of the O_2 group; the O–O bond length is 125 pm and the Fe–O–O bond angle is 136°. The vibrational spectrum of the complex exhibits an absorption at $1159 \, cm^{-1}$ assigned to $\nu(O-O)$, and, when compared with values of $\nu(O-O)$ of $1560 \, cm^{-1}$ for O_2, $\approx 1140 \, cm^{-1}$ for $[O_2]^-$ and $\approx 800 \, cm^{-1}$ for $[O_2]^{2-}$, it suggests

the presence of an $[O_2]^-$ ligand. Oxyhaemoglobin and oxymyoglobin are characterized by values of $\nu(O-O) = 1107$ and $1103 \, cm^{-1}$, respectively. The current model for O_2 binding to the low-spin Fe(II) centre in haemoglobin and myoglobin is that coordination is accompanied by electron transfer, oxidizing high-spin Fe(II) to low-spin Fe(III) and reducing O_2 to $[O_2]^-$. Both low-spin Fe(III) (d^5) and $[O_2]^-$ contain an unpaired electron, and the fact that the oxyforms of the proteins are diamagnetic can be understood in terms of antiferromagnetic coupling between the Fe(III) centre and $[O_2]^-$ ligand (see *Section 20.8*).

In binding to a haem-group, O_2 acts as a π-acceptor ligand (see *Section 20.4*). It is not surprising, therefore, that other π-acceptor ligands can take the place of O_2 in haemoglobin or myoglobin, and this is the basis of the toxicity of CO. Cyanide, however, although a π-acceptor ligand, favours higher oxidation state metal centres and binds to Fe(III) in *cytochromes* (see *Section 28.4*); $[CN]^-$ poisoning is *not* caused by $[CN]^-$ blocking the O_2-binding sites in haemoglobin.

Haemocyanin

Haemocyanins are O_2-carrying copper-containing proteins in molluscs (e.g. whelks, snails, squid) and arthropods (e.g. lobsters, crabs, shrimps, horseshoe crabs, scorpions), and although the name suggests the presence of a haem-group, haemocyanins are *not* haem-proteins. Haemocyanins isolated from arthropods and molluscs are hexameric (M_r per unit $\approx 75\,000$) while those from molluscs possess 10 or 20 subunits, each with $M_r \approx 350\,000$ to $450\,000$. The deoxyform of a haemocyanin is colourless and contains Cu(I), while O_2-binding results in the blue Cu(II) form. Only recently have the structures of a deoxyhaemocyanin (isolated from the spiny lobster) and oxyhaemocyanin (isolated from the Atlantic horseshoe crab) been confirmed. The folded protein chain of one subunit of the deoxy-form is shown in Figure 28.8a. Buried within the metalloprotein are two adjacent Cu(I) centres (Cu⋯Cu = 354 pm, i.e. non-bonded), each of which is bound by three histidine residues (Figure 28.8b, and structure **28.10**).

(28.10)

The active site of the structurally characterized oxyhaemocyanin is shown in Figure 28.8c; the $Cu_2(His)_6$-unit (Cu⋯Cu = 360 pm) resembles that in the deoxy-form. The O_2 unit is bound in a bridging mode with an O–O bond length of 140 pm, typical of that found in peroxide complexes. The O_2-binding site is formulated as $Cu(II)-[O_2]^{2-}-Cu(II)$, i.e. electron transfer accompanies O_2 binding. Resonance Raman spectroscopic data are consistent with this

BOX 28.2   RESOURCES, ENVIRONMENTAL AND BIOLOGICAL

The specialists: how the blood-sucking *Rhodnius prolixus* utilizes NO

Nitrophorins are haem proteins which are present in the salivary glands of the blood-sucking insect *Rhodnius prolixus*. Binding of NO to the Fe(III) centre in nitrophorin (NP1) is reversible, and is dependent on pH. Crucial to the process of blood-sucking by *Rhodnius prolixus* is the fact that NO binds 10 times more tightly at pH 5 (i.e. the pH of the saliva within the insect) than at pH 7 (i.e. the physiological pH of the victim). Once insect saliva is released into the victim, NO is released causing expansion of the blood vessels (vasodilation) and inhibiting blood clotting. In response to being bitten, the victim releases histamine to aid healing of the wound.

NO bound to NP1 (NP1-NO)   Histamine

Studying the NO complex is difficult because of ready oxidation of the NO ligand. However, the crystal structure of a [CN]⁻-containing analogue of NP1–NO was determined in 1998, confirming the binding of the [CN]⁻ ligand (and by analogy, the NO ligand) to the haem Fe centre (angle Fe−C−N = 173°). The cyano group is lodged in a pocket of the protein chain between two leucine residues (see *Table 28.2*). Structural data for the histamine complex show that this same protein pocket hosts the histamine ligand, indicating that NO and histamine compete for the same binding site. At physiological pH, the haem unit in NP1 binds histamine ≈100 times more strongly than NO; this should both aid the dissociation of NO and inhibit the role of histamine, both of which work in favour of the attacking *Rhodnius prolixus*.

Further information

A. Weichsel, J.F. Andersen, D.E. Champagne, F.A. Walker and W.R. Montfort (1998) *Nature Structural Biology*, vol. 5, p. 304 – 'Crystal structures of a nitric oxide transport protein from a blood-sucking insect'.

formulation: $\nu(O-O) \approx 750 \, cm^{-1}$ compared with $\approx 800 \, cm^{-1}$ for $[O_2]^{2-}$. The Cu(II) centres are strongly antiferromagnetically coupled, with the μ-$[O_2]^{2-}$ ligand being involved in a *superexchange* mechanism (see *Section 20.8*).

Many model compounds have been studied in attempts to understand the binding of O_2 in haemocyanin, and often involve imidazole or pyrazole derivatives to represent His residues. In the light of the recent crystallographic data, the model that most closely resembles oxyhaemocyanin is the peroxo dicopper(II) complex (**28.11**) in which each Cu(II) centre is coordinated by a isopropyl-derivatized tris-pyrazolylborate ligand. Like oxyhaemocyanin, complex **28.11** is diamagnetic as a result of antiferromagnetically coupled Cu(II) centres; the Raman spectrum shows an absorption at $741 \, cm^{-1}$ assigned to $\nu(O-O)$ which agrees well with the value for oxyhaemocyanin.

(28.11)

Haemerythrin

In marine invertebrates such as annelids (segmented earthworms), molluscs and arthropods (see above), O_2 is

Fig. 28.8 The structure of deoxyhaemocyanin from the spiny lobster (*Panulirus interruptus*): (a) the backbone of the protein chain and the positions of the two Cu(I) centres shown in blue, and (b) the active site in which the two Cu(I) centres are bound by histidine residues. (c) The O_2-binding site in oxyhaemocyanin from the Atlantic horseshoe crab (*Limulus polyphemus*). Hydrogen atoms are omitted; colour code: Cu, blue; C, grey; O, white; N, black.

transported by *haemerythrin*, a non-haem Fe-containing protein. In the blood, the metalloprotein ($M_r \approx 108\,000$) consists of eight subunits, each with 113 amino acid residues and a diiron-active site; in tissues, fewer subunits make up the metalloprotein. Unlike haemoglobin, haemerythrin exhibits no cooperativity between the subunits during O_2 binding.

The structures of the deoxy- and oxy-forms of haemerythrin have been determined crystallographically (Figure 28.9). In the deoxy-form, a hydroxy-bridged $[Fe(II)]_2$ unit is present as shown in structure **28.12** (see *Table 28.2*); the dotted lines represent connections into the protein backbone. The two Fe(II) centres in deoxyhaemerythrin are strongly antiferromagnetically coupled through the Fe–O–Fe bridge.

The left-hand Fe(II) centre in **28.12** is coordinatively unsaturated and adds O_2 to give oxyhaemerythrin (Figure 28.9c). The hydroxyl H atom in **28.12** participates in

O_2-binding, becoming part of an $[HO_2]^-$ ligand, but remaining associated with the μ-oxo group by hydrogen bond formation (equation 28.7).

(28.12)

Fig. 28.9 The structure of deoxyhaemerythrin from the sipunculid worm (*Themiste dyscrita*): (a) the backbone of the protein chain showing the position of the Fe–O–Fe unit and (b) the active site in which the two Fe(II) centres are bound by histidine, glutamate and aspartate residues. (c) The O₂-binding site in oxyhaemerythrin from *Themiste dyscrita*. Hydrogen atoms are omitted; colour code: Fe, blue; C, grey; O, white; N, black.

(28.7)

(28.13)

Many model studies have focused on *methaemerythrin*, i.e. the oxidized Fe(III)–Fe(III) form of haemerythrin which contains an oxo (rather than hydroxy) bridge. Methaemerythrin does not bind O₂, but does interact with ligands such as [N₃]⁻

and $[SCN]^-$. Reaction 28.8 makes use of the trispyrazolylborate ligand, $[HBpz_3]^-$, to model three His residues; the product (**28.13**) contains antiferromagnetically coupled Fe(III) centres.

$$Fe(ClO_4)_3 + Na[O_2CMe] + K[HBpz_3]$$

$$\longrightarrow [Fe_2(HBpz_3)_2(\mu\text{-}O_2CMe)_2(\mu\text{-}O)] \quad (28.8)$$

Cytochromes P-450

Oxygenases are enzymes that insert oxygen into other molecules; a *monooxygenase* inserts one oxygen atom, and a *dioxygenase* inserts two.

The cytochromes P-450 are metalloenzymes which function as monooxygenases, catalyzing the insertion of oxygen into a C−H bond of an aromatic or aliphatic hydrocarbon, i.e. the conversion of RH to ROH. Two examples of the biological utilization of this reaction are in drug metabolism and steroid synthesis. The oxygen atom originates from O_2: one O atom is inserted into the organic substrate and one atom is reduced to H_2O.

$$\begin{array}{c} S(Cys) \\ | \\ (Porph)N_{////}\!\!\!\!\!\!\!\!\!\!\;—Fe(III)—\;\!\!\!\!\!\!\!\!\!_{\backslash\backslash}N(Porph) \\ (Porph)N\!\!\!\diagdown\qquad\diagdown N(Porph) \end{array}$$

(28.14)

The active site in a cytochrome P-450 is a haem unit, and structural data for cytochrome P-450 complexed with (1*S*)-camphor have confirmed the presence of an axially coordinated cysteine residue. The active site contains a 5-coordinate high-spin Fe(III) centre, schematically represented by structure **28.14**. Carbon monoxide adducts of cytochromes P-450 absorb at 450 nm and this is the origin of the name of the enzyme. The catalytic cycle for the conversion of RH to ROH follows the sequence of steps:

- binding of the organic substrate RH to the Fe(III) centre;
- one-electron reduction of the Fe(III) to high-spin Fe(II);
- binding of O_2 to give a diamagnetic complex, the bonding in which can be described in a similar way to oxyhaemoglobin and oxymyoglobin described earlier;
- reduction and protonation of one O atom to give H_2O, leaving an iron-oxo species, from which the O atom is transferred to RH to give ROH.

The iron-oxo species may contain Fe(IV) or Fe(V), and this is reduced back to Fe(III) upon transfer of the O atom. The insertion of O into the C−H bond in RH probably involves a radical pathway.

28.4 Biological redox processes

In this section we look at ways in which Nature carries out redox chemistry with reference to blue copper proteins,

iron–sulfur proteins and cytochromes; the redox steps in Photosystem II were outlined in the discussion accompanying equation 21.52. We have already discussed two topics of prime importance to electron transfer in Nature. The first is the way in which the reduction potential of a metal redox couple such as Fe^{3+}/Fe^{2+} can be tuned by altering the ligands coordinated to the metal centre; look back at the values of E^o for Fe^{3+}/Fe^{2+} redox couples listed in *Table 7.1*. The second is the discussion of *Marcus–Hush theory* in *Section 25.5*; this theory applies to electron transfer in bioinorganic systems where communication between redox active metal centres may be over relatively long distances as we shall illustrate in the following examples.

Blue copper proteins

There are three classes of copper centres in blue copper proteins:

- A Type 1 centre is characterized by an intense absorption in the electronic spectrum with $\lambda_{max} \approx 600$ nm, and $\varepsilon_{max} \approx 100$ times greater than that of aqueous Cu^{2+}; the absorption is assigned to charge transfer from a cysteine ligand to Cu^{2+}. In the EPR spectrum (Cu^{2+} has one unpaired electron), narrow hyperfine splitting is observed.[†]
- A Type 2 centre exhibits electronic spectroscopic characteristics typical of Cu^{2+}, and the EPR spectrum is typical of a Cu^{2+} centre in a simple coordination complex.
- A Type 3 centre exhibits an absorption with $\lambda_{max} \approx 330$ nm and exists as a pair of Cu(II) centres which are antiferromagnetically coupled to give a diamagnetic system; hence, there is no EPR spectroscopic signature. The Cu_2-unit can function as a two-electron transfer centre and is involved in the reduction of O_2.

Blue copper proteins contain a minimum of one Type 1 Cu centre, and those in this class include *plastocyanins* and *azurins*. Plastocyanins are present in higher plants and blue-green algae where they transport electrons between Photosystems I and II (see above). The protein chain in a plastocyanin comprises between 97 and 104 amino acid residues (most typically 99) and has $M_r \approx 10\,500$. Azurins occur in some bacteria and are involved in electron transport in the conversion of $[NO_3]^-$ to N_2. Typically, the protein chain contains 128 or 129 amino acid residues ($M_r \approx 14\,600$).

Single crystal structural data have provided valuable information about blue copper proteins containing Type 1 Cu centres. Figure 28.10a shows a representation of the folded protein chain of spinach plastocyanin. The Cu(II) centre lies within a pocket in the chain, bound by a Cys, a Met and two His residues (Figure 28.10b); the S(Met) atom is significantly further away from the Cu(II) centre than is S(Cys).

† For details of EPR spectroscopy, see for example: R.V. Parish (1990), *NMR, NQR, EPR and Mössbauer Spectroscopy in Inorganic Chemistry*, Ellis Horwood, Chichester.

Fig. 28.10 The structure of spinach plastocyanin: (a) the backbone of the protein chain showing the position of the Cu(II) centre (in blue) and (b) the coordination sphere of the Cu(II) centre, consisting of one methionine, one cysteine and two histidine residues. The structure of azurin from *Pseudomonas putida*: (c) the backbone of the protein chain showing the position of the Cu(II) centre (in blue) and (d) the Cu(II) centre, coordinated by a methionine, a cysteine and two histidine residues; one O atom from the glycine residue adjacent to one of the histidines interacts weakly with the metal centre. Hydrogen atoms are omitted; colour code: Cu, dark blue; S, pale blue; C, grey; N, black; O, white.

Figure 28.10c shows the backbone of the protein chain in azurin isolated from the bacterium *Pseudomonas putida*. The coordination environment of the Cu(II) centre resembles that in plastocyanin with Cu−S(Met) > Cu−S(Cys), but in addition, an O atom from an adjacent Gly residue is involved in a weak coordinate interaction (Figure 28.10d). Structural studies have also been carried out on the reduced forms of plastocyanin and azurin. In each case, the coordination sphere remains the same except for changes in the Cu−L bond lengths; typically, the bonds lengthen by 5–10 pm on going from Cu(II) to Cu(I). The observed coordination spheres can be considered as suiting *both* Cu(I) and Cu(II) (see *Section 21.12*) and thus facilitate rapid electron transfer. It should be noted, however, that in each structure discussed above, *three* donor atoms are more closely bound than the remaining donors and this indicates that binding of Cu(I) is more favourable than that of Cu(II). This is supported by

the high reduction potentials (measured at pH 7) of plasto-cyanin (+370 mV) and azurin (+308 mV).

Oxidases are enzymes that use O_2 as an electron acceptor.

Multicopper blue copper proteins include *ascorbate oxidase* and *laccase*. These are metalloenzymes that catalyze the reduction of O_2 to H_2O (equation 28.9) and, at the same time, an organic substrate (e.g. a phenol) undergoes a one-electron oxidation. The overall scheme can be written in the form of equation 28.10; $R^•$ undergoes polymerization.

$$O_2 + 4H^+ + 4e^- \rightleftharpoons 2H_2O \qquad (28.9)$$

$$4RH + O_2 \longrightarrow 4R^• + 2H_2O \qquad (28.10)$$

Spectroscopic data are consistent with the presence of all three types of copper site in ascorbate oxidase and laccase,

Fig. 28.11 The structure of ascorbate oxidase isolated from courgettes (zucchini; *Cucurbita pepo medullosa*). The right-hand diagrams show the coordination environments around the Cu centres (see text). Colour coding: Cu, dark blue; S, pale blue; N, black; O, white; C, grey.

and this was confirmed crystallographically in 1992 for ascorbate oxidase, isolated from courgettes (zucchini; *Cucurbita pepo medullosa*). Figure 28.11 shows one unit of ascorbate oxidase in which four Cu(II) centres are accommodated within the folds of the protein chain. Three Cu centres form a triangular array (non-bonded Cu⋯⋯Cu separations) while the fourth Cu atom (a Type 1 centre) is a significant distance away (>1200 pm), but indirectly connected to the Cu₃ unit by the protein chain. The coordination sphere of the Type 1 centre is similar to that in the oxidized form of plastocyanin (compare Figure 28.11 with Figure 28.10b) with the metal bound by one Met residue (Cu−S = 290 pm), one Cys residue (Cu−S = 213 pm) and two His groups. The Cu₃-unit lies within eight His residues (Figure 28.11), and can be subdivided into Type 2 and Type 3 Cu centres. The Type 2 centre is coordinated by two His groups and either an H₂O or [OH]⁻ ligand (the experimental data cannot distinguish between them). The Type 3 centre consists of two Cu atoms bridged by either an O²⁻ or [OH]⁻ ligand; magnetic data show these Cu centres to be antiferromagnetically coupled.

Reduction of O₂ occurs at a Type 2/Type 3 Cu₃ site, with the remote Type 1 Cu centre acting as the main electron acceptor, removing electrons from the organic substrate; details of the mechanism are not understood.

Laccase has been isolated from lacquer trees (e.g. *Rhus vernifera*) and from various fungi. The crystal structure of laccase obtained from the ink cap fungus *Coprinus cinereus* was reported in 1998, but the metalloenzyme was serendipitously isolated in a Type 2 Cu-depleted form, i.e. the coordination site for the type 2 Cu centre can be located in the protein structure, but is vacant. Despite the fact that the loss of the Type 2 Cu affects the local structure of the adjacent Type 3 Cu₂-site, the crystallographic data available for laccase have confirmed that a model based on the known structural features of ascorbate oxidase is essentially correct. One interesting difference involves the Type 1 Cu site. In ascorbate oxidase, this is pseudo-tetrahedral (Figure 28.11) with Cu−S(Met) > Cu−S(Cys). In laccase, the coordination sphere of the Type 1 Cu is trigonal planar, consisting of one Cys and two His residues.

The mitochondrial electron-transfer chain

Mitochondria are the sites in cells where raw, biological fuels are converted into energy.

Before continuing the discussion of specific electron-transfer systems, we take a look at the *mitochondrial electron-transfer chain*, i.e. the chain of redox reactions that occurs in living cells. This allows us to appreciate how the different systems discussed later fit together. Each system transfers one or more electrons and operates within a small range of reduction potentials as illustrated in Figure 28.12; diagrams **28.15** and **28.16** show the structures of the coenzymes [NAD]$^+$ and FAD, respectively.

(28.15)

(28.16)

At one end of the chain in Figure 28.12, *cytochrome c oxidase* catalyzes the reduction of O_2 to H_2O (equation 28.9 for which $E' = +815$ mV). The E' scale (applicable to measurements at pH 7) in Figure 28.12 extends to -414 mV which corresponds to reaction 28.11 at pH 7, and this range of potentials corresponds to those accessible under physiological conditions.

$$2H^+ + 2e^- \rightleftharpoons H_2 \qquad (28.11)$$

Most redox reactions involving organic molecules occur in the range 0 mV $> E' > -400$ mV. The oxidation of a biological 'fuel' (e.g. carbohydrate) involves reactions in which electrons are passed through members of the electron transport chain until eventually H_2 and the electrons enter the [NAD]$^+$/NADH couple. Electron transfer in *steps* utilizing redox couples provided by the metal centres in metalloproteins is an essential feature of biological systems. There is a mismatch, however: oxidations and reductions of organic molecules typically involve two-electron processes, whereas redox changes at metal centres involve one-electron steps. The mediators in the electron transport chain are *quinones*, organic molecules which can undergo *both* one- and two-electron processes (equation 28.12).

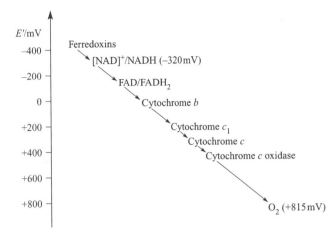

Fig. 28.12 A schematic representation of part of the mitochondrial electron-transfer chain; reduction potentials, E', are measured at physiological pH 7 and are with respect to the standard hydrogen electrode. Reduction potentials quoted in this chapter are with respect to the standard hydrogen electrode.

Quinone

Semiquinone

Hydroquinone

(28.12)

At several points in the mitochondrial electron-transfer chain, the release of energy is coupled to the synthesis of ATP from ADP (see *Box 14.10*), and this provides a means of storing energy in living cells.

Iron–sulfur proteins

The existence of iron–sulfur proteins in our present *oxidizing* environment has to be attributed to the fact that, during a stage in evolution, the environment was a *reducing* one.[†] Iron–sulfur proteins are of relatively low molecular weight and contain high-spin Fe(II) or Fe(III) coordinated tetrahedrally by four *S*-donors. The latter are either S^{2-} (i.e. discrete sulfide ions) or Cys residues attached to the protein backbone;

† For a fuller discussion, see J.J.R. Fraústo da Silva and R.J.P. Williams (1991) *The Biological Chemistry of the Elements*, Oxford University Press, Oxford, p. 331.

the sulfide (but not the cysteine) sulfur can be liberated as H_2S by the action of dilute acid. The FeS_4 centres occur singly in *rubredoxins*, but are combined into di-, tri- or tetrairon units in *ferredoxins*. The biological functions of iron–sulfur proteins include electron transfer processes, nitrogen fixation, catalytic sites in hydrogenases, and oxidation of NADH to $[NAD]^+$ in mitochondria (Figure 28.12).

> *Hydrogenases* are enzymes that catalyze the reaction:
> $2H^+ + 2e^- \longrightarrow H_2$.

The simplest iron–sulfur proteins are *rubredoxins* ($M_r \approx 6000$) which are present in bacteria and contain single FeS_4 centres in which all the *S*-donors are from Cys residues. Figure 28.13 shows the structure of rubredoxin isolated from the bacterium *Clostridium pasteurianum*. The metal site lies in a pocket of the folded protein chain; the

four Fe–S(Cys) bonds are of similar length (227–235 pm) and the S–Fe–S bond angles lie in the range 103–113°. The reduction potential for the Fe^{3+}/Fe^{2+} couple is sensitive to the conformation of the protein chain forming the pocket in which the FeS_4-unit lies. Consequently, a range of reduction potentials has been observed depending on the exact origin of the rubredoxin, but all are close to 0 V, e.g. −58 mV for rubredoxin from *Clostridium pasteurianum*. Rubredoxins function as one-electron transfer sites, with the iron centre shuttling between Fe(II) and Fe(III). Upon oxidation, the Fe–S bond lengths shorten by ≈5 pm.

Ferredoxins occur in bacteria, plants and animals and are of several types:

- [2Fe–2S] ferredoxins contain two Fe centres, bridged by two S^{2-} ligands with the tetrahedral coordination sphere of each metal completed by two Cys residues (Figure 28.14a);

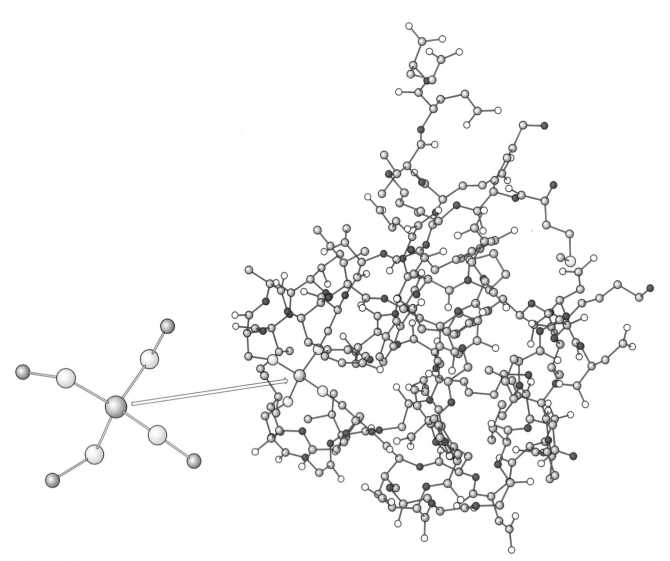

Fig. 28.13 The structure of rubredoxin from the bacterium *Clostridium pasteurianum*. The Fe(Cys)₄ site is enlarged to show the tetrahedral coordination environment. Hydrogen atoms are omitted; colour code: Fe, dark blue; S, pale blue; C, grey; O, white; N, black.

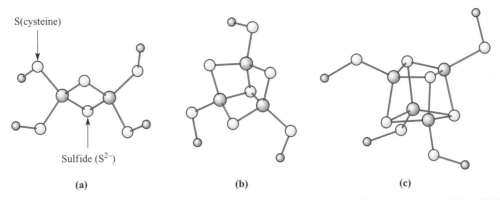

S(cysteine)

Sulfide (S^{2-})

(a)   (b)   (c)

Fig. 28.14 The iron–sulfur units from ferredoxins, structurally characterized by X-ray diffraction: (a) the [2Fe–2S] ferredoxin from spinach (*Spinacia oleracea*), (b) the [3Fe–4S] ferredoxin from the bacterium *Azotobacter vinelandii*, and (c) the [4Fe–4S] ferredoxin from the bacterium *Chromatium vinosum*. Hydrogen atoms are omitted; colour code: Fe, dark blue; S, pale blue; C, grey.

- [3Fe–4S] ferredoxins contain three Fe and four S^{2-} centres arranged in an approximately cubic framework with one corner vacant; this unit is connected to the protein backbone by Cys residues (Figure 28.14b);
- [4Fe–4S] resemble [3Fe–4S] ferredoxins, but contain an additional FeS(Cys) group which completes the approximately cubic cluster core (Figure 28.14c).

The advantage of ferredoxins over rubredoxins in terms of redox chemistry is that by combining several Fe centres in close proximity, it is possible to access a greater range of reduction potentials. Different conformations of the protein pockets which surround the Fe_xS_y clusters affect the detailed structural features of the cluster cores and, thus, their reduction potentials, e.g. −420 mV for spinach [2Fe–2S] ferredoxin, and −270 mV for adrenal [2Fe–2S] ferredoxin. A [2Fe–2S] ferredoxin acts as a one-electron transfer centre, going from an Fe(II)/Fe(II) state in the reduced form to an Fe(II)/Fe(III) state when oxidized and vice versa. Evidence for the localized, mixed valence species comes from EPR spectroscopic data.

A [4Fe–4S] ferredoxin also transfers one electron, and typical reduction potentials lie around −300 to −450 mV corresponding to the half-reaction 28.13. Note that a [4Fe–4S] ferredoxin containing *four* Fe(II) centres is never accessed in biology.

$$2Fe(III)\cdot2Fe(II) + e^- \rightleftharpoons Fe(III)\cdot3Fe(II) \qquad (28.13)$$

The two species represented in equation 28.13 do not actually possess localized Fe(II) and Fe(III) centres, rather the electrons are delocalized over the cluster core. One could envisage further oxidation to species that are formally 3Fe(III)·Fe(II) and 4Fe(III). Whereas the latter is never accessed under physiological conditions, 3Fe(III)·Fe(II) is the oxidized form of HIPIP (*high potential protein*), i.e. 2Fe(III)·2Fe(II) is the reduced form of HIPIP or the oxidized form of ferredoxin. In contrast to the reduction potentials of ferredoxins, those of HIPIPs are *positive*, e.g. +360 mV for HIPIP isolated from the bacterium *Chromatium vinosum*. Within a given metalloprotein, redox reactions involving

two electrons which effectively convert a ferredoxin into HIPIP do *not* occur.

Although we have focused on individual structural units in rubredoxins, ferredoxins and HIPIPs, we should note that some metalloproteins contain more than one Fe_xS_y unit. For example, the ferredoxin isolated from *Azotobacter vinelandii* contains both [4Fe–4S] and [3Fe–4S] units, with the closest Fe····Fe separation between units being ≈930 pm.

Oxygenic photosynthesis involves the cytochrome b_6f complex which is made up of subunits including cytochrome f containing one c haem, cytochrome b_6 with two b haems, and Rieske protein which is a high potential protein containing a [2Fe–2S] cluster. The latter is distinguished from a [2Fe–2S] ferredoxin by having one Fe centre bound by two His (rather than Cys) residues (Figure 28.15). Rieske protein is the electron transfer site in the oxidation of plastoquinol (a hydroquinone) to plastosemiquinone, during which protons are released. Rieske protein has a *positive* reduction potential (+290 mV) for that isolated from spinach chloroplasts, contrasting with *negative* values for [2Fe–2S] ferredoxins. The difference must be attributed to the His versus Cys coordination of one Fe centre.

Iron-only (as distinct from FeNi-containing) *hydrogenases* are present in a small number of microbes, and function at the end of the electron chain, catalyzing the reduction of H^+ to H_2. The crystal structure of the Fe-only hydrogenase isolated from the bacterium *Clostridium pasteurianum* provides an example of a remarkable iron–sulfur protein. The protein contains one [2Fe–2S] group, three [4Fe–4S] units and the so-called 'hydrogen cluster' shown in Figure 28.16 which has an unprecedented structure for a biological metalloprotein. It consists of an [4Fe–4S] unit connected to a diiron subcluster by a bridging Cys residue; in the diiron unit, the Fe atoms are bridged by two S atoms, themselves bridged by an atom or group of uncertain identity, possibly H_2O. Figure 28.16 shows the presence of one bridging and four terminal CO ligands, but it is not possible from the structural data to rule out the possibility that these are $[CN]^-$ ligands; IR spectroscopic data are consistent with

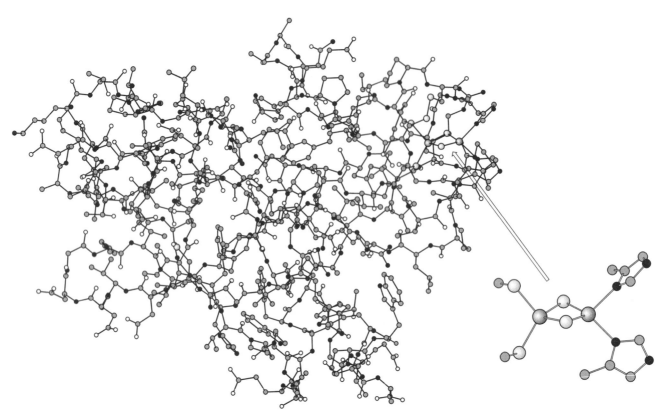

Fig. 28.15 The structure of Rieske protein from spinach (*Spinacia oleracea*) chloroplast. The $Fe_2(Cys)_2(Hist)_2$ site is enlarged. Hydrogen atoms are omitted; colour code: Fe, dark blue; S, pale blue; C, grey; O, white; N, black.

coordinated CO or $[CN]^-$. The final ligand on one of the Fe centres is assumed to be H_2O. Despite these uncertainties, the structural characterization of this hydrogenase has provided important insight into a possible mechanism for the reduction of H^+ to H_2. The protein chain conformation is such that a Cys residue (non-coordinated) is situated close to the terminally coordinated H_2O ligand. It is proposed that the one-electron reduction of H^+ is accompanied by the displacement of the terminal H_2O ligand and formation

of an Fe–H interaction, i.e. an H ligand. The adjacent Cys residue (for which $pK_a \approx 8$) provides H^+, leading to the release of H_2 by reaction 28.14. Water then refills the vacant site on the Fe centre.

$$[Fe]-H + H^+ \longrightarrow [Fe] + H_2 \qquad (28.14)$$

When the metalloenzyme is treated with CO, it becomes inactive, and this can readily be rationalized in terms of the CO irreversibly displacing the coordinated H_2O ligand.

Nitrogen fixation by bacteria involves the reduction of N_2 to NH_3 catalyzed by *nitrogenases*; concomitant with this process is the hydrolysis of ATP. Recent studies of nitrogenase proteins from the bacteria *Azobacter vinelandii* and *Clostridium pasteurianum* have provided structural details of the proteins involved. Two metalloproteins make up the nitrogenase system: an Fe protein which couples the hydrolysis of ATP to electron transfer, and an FeMo protein which is responsible for binding N_2. The dual role of these proteins can be summarized in three steps:

- reduction of Fe protein;
- one-electron transfer from the Fe protein to FeMo protein in a process which also involves ATP hydrolysis;
- electron and H^+ transfer to N_2.

The Fe protein is a dimer and contains one [4Fe–4S] ferredoxin unit held by Cys residues between the two halves of the protein, in a site which is relatively exposed on the surface of the protein. The FeMo protein contains two cluster

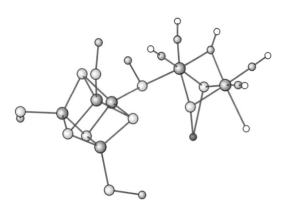

Fig. 28.16 The structure of the 'hydrogen cluster' in the Fe-only hydrogenase isolated from *Clostridium pasteurianum*. Colour code: Fe, dark blue; S, pale blue; O, white; C, grey; the identity of the black atom is uncertain, and it is not possible from the crystallographic data to unambiguously assign the CO ligands (see text).

(a)

(b)

Fig. 28.17 The structures of the two types of cluster units present in the nitrogenase molybdenum–iron protein isolated from *Clostridium pasteurianum*: (a) the cysteine-bridged double [4Fe–4S] cluster, and (b) the FeMo cofactor in which the identity of atom X has not been verified. Hydrogen atoms are omitted; colour code: Fe, dark blue; S, pale blue; C, grey; O, white; N and Mo, black.

sites, both buried within the protein and exhibiting intriguing structures. The first cluster (referred to as the 'P-cluster', Figure 28.17a) consists of two Cys-bridged [4Fe–4S] units which differ from ferredoxins not only in having two shared Cys residues, but in the fact that one Fe centre is bound by the oxygen atom of a serine residue (see *Table 28.2*). The second cluster (Figure 28.17b) is called the FeMo cofactor, and consists of a [4Fe–3S] unit connected by three bridging atoms to a [3Fe–1Mo–3S] unit; two of the bridging atoms are sulfurs but the identity of the third (labelled X in Figure 28.17b) is not known with certainty. Unlike other Fe–S clusters discussed so far, all but one of the Fe centres in the FeMo cofactor are 3-coordinate. The Mo centre is approximately octahedral; it is bound to the protein backbone by a His residue and is also coordinated by a didentate homocitrate ligand. The closest distance between metal centres in the two metal clusters in the FeMo protein is $\approx 1400\,\text{pm}$, a separation which is amenable to electron transfer (see *Section 25.5*). The way in which the Fe and FeMo proteins act together to catalyze the conversion of N_2 to NH_3 has yet to be established. Proposals that Mo is the N_2-binding site have now been replaced by the view that one of the Fe sites in the FeMo cofactor is responsible; the structural data confirm that the Mo centre is coordinately saturated while six Fe centres are coordinately unsaturated.

Before leaving iron–sulfur proteins, we must mention the important contributions that model studies have made, in particular before protein X-ray structural data were available. For discrete clusters of the type formed by reaction 28.15 and shown in diagram **28.17**, it is possible to investigate magnetic, electronic spectroscopic and electrochemical properties, record [57]Fe Mössbauer spectra (see *Section 2.12*) and determine accurate structural data by X-ray diffraction. Working with metalloproteins is, of course, far more difficult.

$$FeCl_3 + NaOMe + NaHS + PhCH_2SH$$
$$\longrightarrow Na_2[Fe_4S_4(SCH_2Ph)_4] \qquad (28.15)$$

(28.17)

Model compound **28.17** (and related complexes) contains high-spin Fe centres; formally there are two Fe(II) and two

----- to protein backbone

(a) **(b)**

Fig. 28.18 (a) The backbone of the protein chain of horse heart cytochrome c, showing the position of the haem unit (shown in blue). (b) An enlargement of the coordination sphere of the iron site showing the residues which are covalently linked to the protein chain. Hydrogen atoms have been omitted; colour code: Fe, dark blue; S, pale blue; N, black; C, grey; O, white.

Fe(III), but spectroscopic data are consistent with four equivalent metal centres and, therefore, delocalization of electrons within the cage.

Cytochromes

Figure 28.12 showed *cytochromes* to be vital members of the mitochondrial electron-transfer chain; they are also essential components in plant chloroplasts for photosynthesis. Cytochromes are haem proteins, and the ability of the iron centre to undergo reversible Fe(III) $\rightleftharpoons$ Fe(II) changes allows them to act as one-electron transfer centres. Many different cytochromes are known with the reduction potential for the Fe^{3+}/Fe^{2+} couple being tuned by the surrounding protein environment. Cytochromes belong to various families, e.g. cytochromes a, cytochromes b and cytochromes c, which are denoted according to the substituents on the haem group. We saw in Section 28.3 that in the O_2-carrying haem proteins the 'rest state' contains a 5-coordinate Fe(II) centre which becomes 6-coordinate after O_2 uptake. In contrast, the electron-transfer cytochromes b and c contain 6-coordinate Fe which is present as either Fe(II) or Fe(III); there is little change in ligand conformation as the redox change occurs. Figure 28.18 shows the structure of cytochrome c isolated from horse heart; compare the haem structure with that in haemoglobin (Figure 28.7). In cytochrome c, the haem unit is bound to the protein backbone through axial His and Met residues, and through two Met residues which are covalently linked to the porphyrin ring.

In the mitochondrial electron-transfer chain, cytochrome c accepts an electron from cytochrome c_1 and then transfers it to cytochrome c oxidase (equation 28.16); ultimately, the electron is used in the four-electron reduction of O_2 (see

below). The oxidized forms of the cytochromes in equation 28.16 contain Fe(III), and the reduced forms, Fe(II).

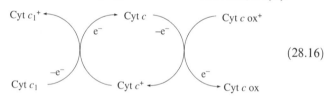

(28.16)

It is proposed that an electron is transferred by tunnelling through one of the exposed edges of the haem unit (recall that the porphyrin ring is conjugated). In relation to this, it is instructive to look at the arrangement of the haem units in cytochrome $c554$, a tetrahaem protein isolated from the bacterium *Nitrosomonas europaea* and essential to the nitrification pathway: NH_3 is converted to NH_2OH (catalyzed by *ammonia monooxygenase*) which is then oxidized to $[NO_3]^-$ (catalyzed by *hydroxylamine oxidoreductase*). The role of cytochrome $c554$ is to accept pairs of electrons from hydroxylamine oxidoreductase and transfer them, via cytochrome $c552$, to terminal oxidases. The crystal structure of cytochrome $c554$ shows that the four haem units are arranged in pairs such that the porphyrin rings are approximately parallel, and have overlapping edges. Adjacent pairs are then approximately perpendicular to each other (Figure 28.19). Such arrangements have been observed in other multi-haem cytochromes and are presumably set up to provide efficient electron transfer pathways between the edges of the haem groups.

The exact nature of the metal sites in cytochrome c oxidase was finally resolved in 1995. This terminal member of the mitochondrial electron transfer chain catalyzes the reduction of O_2 to H_2O (equation 28.9), and contains four active metal centres (Cu_A, Cu_B, haem a and haem a_3) which couple electron

Fig. 28.19 Cytochrome *c*554 isolated from *Nitrosomonas europaea*: the backbone of the protein chain and the four haem units (shown in blue). The Fe····Fe distances between haem units are ≈950 pm, 1220 pm and 920 pm.

transfer to proton pumping. Electron transfer involves the Cu_A and haem *a* sites, electrons being transferred from cytochrome *c* (equation 28.16) to Cu_A and then to haem *a*. Haem a_3 and Cu_B provide the site for O_2 binding and O_2 to H_2O conversion, and are involved in pumping H^+ (four per O_2 molecule) across the mitochondrial inner membrane. Until 1995, proposals for the nature of the metal sites were based largely on spectroscopic data and the fact that the Cu_B····Fe(haem a_3) centres were strongly antiferromagnetically coupled; the latter suggested the possible presence of a bridging ligand. Crystallographic data have now cleared the uncertainty, revealing the following structural features:

- Fe(haem *a*) is 6-coordinate with His residues in the axial sites;

- Cu_A is a dicopper site bridged by Cys residues, with a Cu_2S_2 core that is not unlike that in a [2Fe–2S] ferredoxin;
- the 3-coordinate Cu_B and 5-coordinate Fe(haem a_3) lie ≈450 pm apart and are *not* connected by a bridging ligand.

Figure 28.20 shows the active metal sites in the oxidized form of cytochrome *c* oxidase and the spatial relationship between them; they lie within a protein which has $M_r \approx 20\,000$ and is made up of 13 different polypeptide subunits. Detailed structural studies of the protein chains have shown that a hydrogen-bonded system which incorporates residues in the protein backbone, haem propanoate side chains, and a His reside bound to Cu_A may provide an electron-transfer 'highway' between Cu_A and haem *a*.

Many model systems have been developed to aid our understanding of electron transfer and O_2 binding by cytochromes. Of particular relevance is the Co(II)/Cu(I) heterometallic complex **28.18** which is designed to model the haem a_3/Cu_B site in cytochrome *c* oxidase (Figure 28.20). Although compound **28.18** contains a Co(II) rather than Fe(II) porphyrin, the ability of Co to bind and activate O_2 is similar to that of Fe. Both the Co(II) and Cu(I) centres can undergo one-electron oxidations. One equivalent of **28.18** strongly binds one equivalent of O_2, and a peroxide complex is proposed; the latter exhibits an absorption in the IR spectrum at 804 cm^{-1} assigned to ν(O–O), and this compares with ν(O–O) ≈ 800 cm^{-1} for $[O_2]^{2-}$. It has also been shown that O_2 bound to complex **28.18** undergoes a four-electron reduction to H_2O at physiological pH, thus mimicking the biological function of cytochrome *c* oxidase.

Finally in this section, we should note that it is the strong binding of $[CN]^-$ to Fe(III) in cytochromes that renders cyanide toxic.

(28.18)

Fig. 28.20 The Cu_A, Cu_B, haem a and haem a_3 sites in cytochrome c oxidase extracted from bovine (*Bos taurus*) heart muscle. The lower right-hand diagram shows the relative positions and orientations of the metal sites within the protein; an enlargement of each site shows details of the ligand spheres. Hydrogen atoms have been omitted; colour code: Cu, white; Fe, dark blue; S, pale blue; N, black; C, grey; O, white.

28.5 The Zn^{2+} ion: Nature's Lewis acid

In this section we focus on the Zn(II)-containing enzymes *carbonic anhydrase II* and *carboxypeptidases A and G2*. These are somewhat different from other systems so far described. Zinc(II) is not a redox active centre, and so cannot take part in electron-transfer processes. It is, however, a hard metal centre (see *Table 6.9*) and is ideally suited to coordination by *N*- and *O*-donors. It is also highly polarizing, and the activity of Zn(II)-containing metalloenzymes depends on the Lewis acidity of the metal centre.

Carbonic anhydrase II

Human carbonic anhydrase II (CAII) is present in red blood cells and catalyzes the reversible hydration of CO_2 (reaction 28.17). This process in slow ($k = 0.037\,s^{-1}$) but is fundamental to the removal of CO_2 from actively metabolizing

sites; CAII increases the rate of hydrolysis by a factor of $\approx 10^7$ at physiological pH.

$$H_2O + CO_2 \rightleftharpoons [HCO_3]^- + H^+ \qquad (28.17)$$

The metalloprotein consists of 260 amino acids and contains a Zn^{2+} ion bound by three His residues in a pocket $\approx 1500\,pm$ deep; the tetrahedral coordination sphere is completed by a hydroxide ion (Figure 28.21a). The peptide chain environment around the active site is crucial to the catalytic activity of the site: the $[OH]^-$ ligand is hydrogen bonded to an adjacent glutamic acid residue, and to the OH group of an adjacent threonine residue (Table 28.2). Next to the Zn^{2+} centre lies a hydrophobic pocket which 'captures' CO_2. The catalytic cycle by which CO_2 is hydrolyzed is shown in Figure 28.21b. After release of $[HCO_3]^-$, the coordinated H_2O ligand must be deprotonated in order to regenerate the active site, and the proton is transferred via a hydrogen-bonded network to a His residue (*non*-coordinated to Zn^{2+}) within the catalytic pocket.

Fig. 28.21 (a) Schematic representation of the active site in human carbonic anhydrase II (CAII). (b) The catalytic cycle for the hydration of CO_2 catalyzed by CAII.

Carboxypeptidase A

Carboxypeptidase A (CPA) is a pancreatic metalloenzyme which catalyzes the cleavage of a peptide link in a polypeptide chain. The site of cleavage is specific in two ways: it occurs at the C-terminal amino acid (equation 28.18), and it exhibits a high selectivity for substrates in which the C-terminal amino acid contains a large aliphatic or Ph substituent. The latter arises from the presence, near to the active site, of a hydrophobic pocket in the protein which is compatible with the accommodation of, for example, a Ph group (see below).

(28.18)

Carboxypeptidase A is monomeric ($M_r \approx 34\,500$) and exists in three forms (α, β and γ) which contain 307, 305 and 300 amino acids respectively. Near the surface of the protein lies a pocket in which a Zn^{2+} ion is bound to the protein backbone by one didentate Glu and two His residues. A 5-coordinate coordination sphere is completed by a water molecule (Figure 28.23a).

The mechanism by which the CPA-catalyzed peptide-link cleavage occurs has drawn much research attention, and the pathway that is currently favoured is illustrated in a schematic form in Figure 28.22. In the first step, the peptide to be cleaved is 'maneouvered' into position close to the Zn^{2+} site; the dominant substrate–protein interactions involved at this stage (Figure 28.22a) are:

- salt-bridge formation between the C-terminal carboxylate group of the substrate and residue Arg-145[†] which is positively charged;
- intermolecular interactions between the non-polar group R' and residues in a hydrophobic pocket of the protein chain.

These interactions may be supplemented by hydrogen bond formation (shown in Figure 28.22a) between the OH group of Tyr-248 and the amine group indicated in the figure, and between Arg-127 and the C=O group adjacent to the peptide cleavage site. This latter interaction polarizes the carbonyl group, activating it towards nucleophilic attack. The nucleophile is the H_2O ligand coordinated to Zn^{2+}; the Lewis acidity of the metal ion polarizes the O–H bonds, and (although this is not a unique proposal) it is likely that the carboxylate group of Glu-270 assists in the process by removing H^+ from the H_2O ligand (Figure 28.22b). Figure 28.22c shows the next step in the proposed mechanism: the cleavage of the peptide C–N bond for

---

[†] We have not previously included residue numbers, but do so in this discussion for the sake of clarity. Residues are numbered sequentially along the protein chain.

Fig. 28.22 Schematic representation of the generally accepted mechanism for the CPA-catalyzed cleavage of a *C*-terminal peptide-link; see *Figure 28.22a* for a more detailed diagram of the coordination sphere of the Zn^{2+} ion. The blue line represents the protein chain; only residues mentioned in the discussion are shown.

Cleavage of the C–N bond

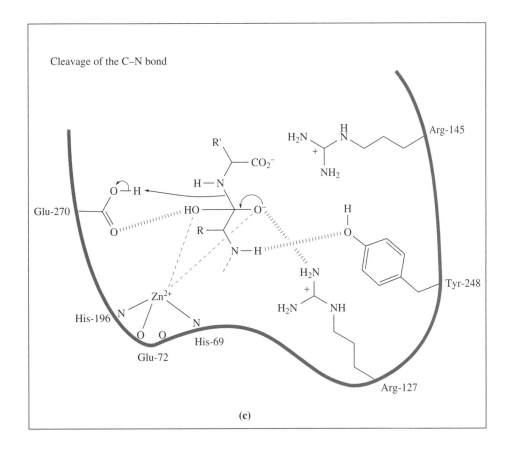

(c)

Proton transfer to give NH₃⁺ group and carboxylate

(d)

Fig. 28.22 continued

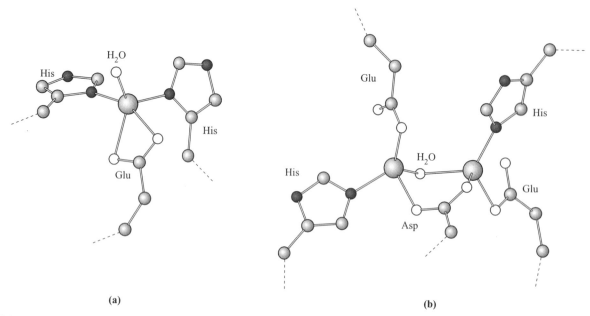

Fig. 28.23 The structures of the active sites in (a) α-carboxypeptidase A (CPA) isolated from bovine (*Bos taurus*) pancreas, and (b) carboxypeptidase G2 (CPG2) isolated from *Pseudomonas* sp.; see *Table 28.2* for amino acid abbreviations. Colour code: Zn, blue; C, grey; O, white; N, black.

which H$^+$ is probably provided by Glu-270. It appears likely that the second H$^+$ required for the formation of the NH$_3^+$ group on the departing terminal amino acid comes from the terminal CO$_2$H group of the remaining portion of the substrate (Figure 28.22d). Figure 28.22c shows Glu-72 bound in a monodentate manner to the Zn^{2+} centre, whereas in the rest state, a didentate mode has been confirmed (Figure 28.22a). A change from a di- to monodentate coordination appears to be associated with the formation of the Zn^{2+}····O····H(Arg-127) interaction illustrated in Figure 28.22c, the Zn^{2+} ion being able to move towards Arg-127 as the interaction develops. To complete the catalytic cycle, an H$_2$O ligand refills the vacant site on the Zn^{2+} centre. Details of this mechanism are based upon a range of data including kinetic and molecular mechanics studies and investigations of Co^{2+} substituted species (see below).

Carboxypeptidase G2

The carboxypeptidase family of enzymes also includes carboxypeptidase G2 (CPG2) which catalyzes the cleavage of *C*-terminal glutamate from folate (**28.19**) and related compounds such as methotrexate (in which NH$_2$ replaces the OH group in the pterin group, and NMe replaces NH in the 4-amino benzoic acid unit).

Folic acid is required for growth, and the growth of tumours can be inhibited by using cancer treatment drugs which reduce the levels of folates. Structural data for the enzyme CPG2 have recently provided valuable information which should assist design of such drugs. Carboxypeptidase G2 (isolated from bacteria of *Pseudomonas* sp.) is a dimeric protein with $M_r \approx 41\,800$ per unit; each monomer contains two domains,

one containing the active site and one intimately involved in dimerization. Unlike carboxypeptidase A, the active site of CPG2 contains two Zn(II) centres, separated by 330 pm and bridged by an Asp residue and a water molecule (Figure 28.23b). Each Zn^{2+} ion is further coordinated by His and Glu residues of the protein chain to give a tetrahedral environment. The pocket containing the Zn$_2$-unit also contains arginine and lysine residues (Table 28.2) which may be involved in binding the substrate molecule, positioning it correctly for interaction with the catalytic site.

(28.19)

Cobalt-for-zinc ion substitution

A practical disadvantage of working with metalloproteins containing Zn^{2+} is the d^{10} configuration of the ion: the metal site cannot be probed by using UV-VIS or EPR spectroscopies or by magnetic measurements. Studies involving Co^{2+}-for-Zn^{2+} substitution provide a metal centre' that is amenable to investigation by spectroscopic and magnetic techniques (Co^{2+} is a d^7 ion), the choice of Co^{2+} being because:

- the ionic radii of Co^{2+} and Zn^{2+} are about the same;
- Co^{2+} can tolerate similar coordination environments to Zn^{2+};
- it is often possible to replace Zn^{2+} in a protein by Co^{2+} without greatly perturbing the protein conformation.

A typical method of metal ion substitution is shown in scheme 28.19 in which the ligand L removes Zn^{2+} by complexation.

$$[PZn^{2+}] \xrightarrow{L} [AP] \xrightarrow{Co^{2+}} [PCo^{2+}] \qquad (28.19)$$

P = protein in the metalloprotein; AP = inactive apoprotein

$$(28.20) \qquad\qquad (28.21)$$

For example, treatment of carbonic anhydrase with **28.20** (or its conjugate base) results in the removal of Zn^{2+} and the formation of the catalytically inactive apoprotein. Reaction of the apoprotein with Co^{2+} gives a cobalt-substituted enzyme, $[PCo^{2+}]$, which catalyzes the hydration of CO_2. Similarly, the Zn^{2+} ion can be removed from carboxypeptidase A by treatment with bpy (**28.21**), and after insertion of Co^{2+}, the model metalloenzyme $[PCo^{2+}]$ is found to be active (actually more so than native carboxypeptidase A) with respect to peptide cleavage. Investigations can be carried out with $[PCo^{2+}]$ which are impossible with native zinc enzymes, e.g. electronic spectroscopic data provide insight into coordination geometries, and monitoring the electronic spectrum as a function of pH indicates whether ligands such as H_2O are deprotonated or not.

Summary

In this chapter, we have described the essential roles of some trace metal ions in living organisms: metal storage, O_2 utilization, redox processes and catalysis. We have also illustrated the importance of using model compounds to gain insight into the operation of far more complex metalloproteins. Recent advances in structural characterization of proteins, during the last decade of the 20th century added greatly to our knowledge of bioinorganic chemistry and this trend is bound to continue. Our discussion has in no way been comprehensive, but has introduced some of the main types of metalloprotein which are essential to life.

The following terms have been introduced in this chapter. Do you know what they mean?

- ❑ trace metals
- ❑ polypeptide
- ❑ protein
- ❑ metalloprotein
- ❑ apoprotein
- ❑ ferritin
- ❑ transferrin
- ❑ siderophore
- ❑ metallothionein
- ❑ haem-protein
- ❑ haemoglobin
- ❑ myoglobin
- ❑ haemocyanin
- ❑ haemerythrin
- ❑ blue copper proteins
- ❑ oxidase
- ❑ hydrogenase
- ❑ plastocyanin
- ❑ azurin
- ❑ ascorbate oxidase
- ❑ laccase
- ❑ mitochondrial electron-transfer chain
- ❑ rubredoxin
- ❑ ferredoxins
- ❑ nitrogenase
- ❑ cytochrome
- ❑ cytochrome *c*
- ❑ cytochrome *c* oxidase
- ❑ carbonic anhydrase II
- ❑ carboxypeptidase A
- ❑ carboxypeptidase G2

Suggested further reading

Bioinorganic chemistry is a fast-moving area and readers interested in the area are advised to update the following reading list by consulting major chemical journals, in particular *Angewandte Chemie*, *Chemical Communications*, *Journal of the American Chemical Society*, *Nature*, *Science*, *Nature Structural Biology* and *Structure*.

General sources

I. Bertini, H.B. Gray, S.J. Lippard and J.S. Valentine (1994) *Bioinorganic Chemistry*, University Science Books, Mill Valley – An excellent and detailed text, one of the best currently available.

J.A. Cowan (1997) *Inorganic Biochemistry: An Introduction*, 2nd edn, Wiley-VCH, New York – An up-to-date text covering a wider range of topics than in this chapter and including case studies.

D.E. Fenton (1995) *Biocoordination Chemistry*, Oxford University Press, Oxford – A clearly written, introductory text.

J.J.R. Fraústo da Silva and R.J.P. Williams (1991) *The Biological Chemistry of the Elements*, Oxford University Press, Oxford – An excellent, detailed text.

W. Kaim and B. Schwederski (1994) *Bioinorganic Chemistry: Inorganic Elements in the Chemistry of Life*, Wiley-VCH, Weinheim – A detailed text covering the roles of inorganic elements in living organisms, as well as applications in chemotherapy.

S.J. Lippard and J.M. Berg (1994) *Principles of Bioinorganic Chemistry*, University Science Books, Mill Valley – One of the primary modern texts dealing with bioinorganic chemistry.

More-specialized articles including model compounds

D.W. Christianson and C.A. Fierke (1996) *Accounts of Chemical Research*, vol. 29, p. 331 – 'Carbonic anhydrase: Evolution of the zinc binding site by Nature and by design'.

D.E. Fenton (1999) *Chemical Society Reviews* vol. 28, p. 159 – 'Metallobiosites and their synthetic analogues – a belief in synergism'.

R.B. King, ed. (1994) *Encyclopedia of Inorganic Chemistry*, Wiley, Chichester: volumes 2 and 4 contain detailed reviews on copper and iron proteins, respectively.

J.G. Leigh, G.R. Moore and M.T. Wilson (1993) 'Biological iron' in *Chemistry of iron*, ed. J. Silver, Blackie, London, p. 181.

A.K. Powell (1993) 'Models for iron biomolecules' in *Chemistry of iron*, ed. J. Silver, Blackie, London, p. 244.

P.C. Wilkins and R.G. Wilkins (1987) *Coordination Chemistry Reviews*, vol. 79, p. 195 – 'The coordination chemistry of the binuclear iron site in hemerythrin'.

Problems

28.1 Give brief descriptions of the following: (a) peptide; (b) naturally occurring amino acids; (c) metalloprotein; (d) apoprotein; (e) haem unit.

28.2 Give an account of the storage and transport of metalloproteins in mammals. How does the uptake of iron by aerobic microorganisms differ from that in mammals?

28.3 The complex $[CrL_3]^{3-}$ where $H_2L = 1,2-(HO)_2C_6H_4$ is a model complex for enterobactin. How is the model related to enterobactin, and what is the reason for chromium-for-iron substitution?

28.4 Comment on the following observations:

(a) Thioneins bind, for example, Cd^{2+} in cysteine-rich pockets.
(b) $[Cu_4(SPh)_6]^{2-}$ is a model for the Cu-containing metallothionein in yeast.
(c) Imidazole and trispyrazolylborate derivatives are often used to model histidine-binding sites.

28.5 (a) Briefly describe the mode of binding of O_2 to the iron centre in one haem unit of haemoglobin. (b) What are 'picket fence' porphyrins and why are they used in model studies of O_2 binding to myoglobin or haemoglobin? (c) The binding of O_2 to haemoglobin exhibits a 'cooperativity' effect. What is meant by this statement? (d) Why is the change from deoxyhaemoglobin to the oxy-form accompanied by a decrease in the observed magnetic moment?

28.6 Compare the modes of binding of O_2 to the metal centres in (a) myoglobin, (b) haemerythrin and (c) haemocyanin. Indicate what supporting experimental evidence is available for the structures you describe.

28.7 Differentiate between Type 1, Type 2 and Type 3 copper centres in blue copper proteins, giving both experimental and structural distinctions.

28.8 Describe the structure of the copper site in plastocyanin and discuss the features of both the metal centre and metal-binding site that allow it to function as an electron transfer site.

28.9 Ascorbate oxidase contains four copper centres. Discuss their coordination environments, and classify the centres as Type 1, 2 or 3. What is the function of ascorbate oxidase and how do the copper centres facilitate this function?

28.10 Comment on the following observations:

(a) 'Blue copper proteins' are not always blue.
(b) Two different metalloproteins, both containing [4Fe–4S] ferredoxins bound to the protein chain by Cys ligands, exhibit reduction potentials of +350 and +490 mV.
(c) The toxicity of CO is associated with binding to haemoglobin, but that of $[CN]^-$ is not.

28.11 What is the mitochondrial electron-transfer chain, and what role do quinones play in the chain?

28.12 Model compounds are often used to model iron–sulfur proteins. Comment on the applicability of the following models, and on the data given.

(a) $[Fe(SPh)_4]^{2-}$ as a model for rubredoxin; observed values of μ_{eff} are $5.85\,\mu_B$ for the oxidized form of the model compound and $5.05\,\mu_B$ for the reduced form.
(b) $[Fe_2(\mu\text{-}S)_2(SPh)_4]^{2-}$ as a model for the active site in spinach ferredoxin;
(c) compound **28.22** as a model for part of the active sites in nitrogenase; the Mössbauer spectrum of **28.22** is consistent with equivalent Fe centres, each with an oxidation state of 2.67.

(28.22)

28.13 For a [4Fe–4S] protein, the following series of redox reactions are possible, in which each step is a one-electron reduction or oxidation:

$$4Fe(III) \rightleftharpoons 3Fe(III) \cdot Fe(II) \rightleftharpoons 2Fe(III) \cdot 2Fe(II)$$
$$\rightleftharpoons Fe(III) \cdot 3Fe(II) \rightleftharpoons 4Fe(II)$$

(a) Which of these couples are accessible under physiological conditions? (b) Which couple represents the HIPIP system? (c) How do the redox potentials of the HIPIP and [4Fe–4S] ferredoxin system differ and how does this affect their roles in the mitochondrial electron-transfer chain?

28.14 Comment on the similarities and differences between a [2Fe–2S] ferredoxin and Rieske protein, both in terms of structure and function.

28.15 (a) Outline the similarities and differences between the haem units in deoxymyoglobin and cytochrome *c*. (b) What function does cytochrome *c* perform in mammals?

28.16 (a) What is the function of cytochrome *c* oxidase? (b) Describe the four active metal-containing sites in cytochrome *c* oxidase and the proposed way in which they work together to fulfil the role of the metalloprotein.

28.17 Give an explanation for the following observations (part d assumes Box 28.2 has been studied):

(a) both haemoglobin and cytochromes contain haem-iron;

(b) cytochrome *c* oxidase contains more than one metal centre;

(c) each subunit in deoxyhaemoglobin contains 5-coordinate Fe(II), but in cytochrome *c*, the Fe centre is always 6-coordinate;

(d) nitrophorin NP1 reversibly binds NO.

28.18 Discuss the role of Zn^{2+} as an example of a Lewis acid at work in a biological system.

28.19 The hydrolysis of the acid anhydride **28.23** by $[OH]^-$ is catalyzed by Zn^{2+} ions. The rate equation is of the form:

$$Rate = k[\mathbf{28.23}][Zn^{2+}][OH^-]$$

It is also known that the addition of Zn^{2+} does not accelerate hydrolysis by H_2O or attack by other nucleophiles. Suggest a mechanism for this reaction.

(28.23)

28.20 Why is metal substitution used to investigate the metal binding site in carbonic anhydrase? Discuss the type of information that might be forthcoming from such a study.

Appendices

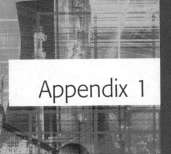

Appendix 1 Greek letters with pronunciations

Upper case letter	Lower case letter	Pronounced
A	α	alpha
B	β	beta
Γ	γ	gamma
Δ	δ	delta
E	ε	epsilon
Z	ζ	zeta
H	η	eta
Θ	θ	theta
I	ι	iota
K	κ	kappa
Λ	λ	lambda
M	μ	mu
N	ν	nu
Ξ	ξ	xi
O	ο	omicron
Π	π	pi
P	ρ	rho
Σ	σ	sigma
T	τ	tau
Υ	υ	upsilon
Φ	φ	phi
X	χ	chi
Ψ	ψ	psi
Ω	ω	omega

For ligand structures, see Table 6.7. Where a symbol has more than one meaning, the context of its use should make the meaning clear. For further information on SI symbols and names of units, see: *Quantities, Units and Symbols in Physical Chemistry* (1993) IUPAC, 2nd edn, Blackwell Science, Oxford.

a	cross-sectional area		ccp	cubic close-packed
a_i	relative activity of a component i		CFC	chlorofluorocarbon
a_0	Bohr radius of the H atom		CFSE	crystal field stabilization energy
A	ampere (unit of current)		cm	centimetre (unit of length)
A	absorbance		cm^3	cubic centimetre (unit of volume)
A	frequency factor (in Arrhenius equation)		cm^{-1}	reciprocal centimetre (wavenumber)
			conc	concentrated
A	Madelung constant		Cp	cyclopentadienyl
A	mass number (of an atom)		cr	crystal
A_r	relative atomic mass		CT	charge transfer
$A(\theta,\phi)$	angular wavefunction		CVD	chemical vapour deposition
A mechanism	associative mechanism		Cys	cysteine
Å	ångstrom (non-SI unit of length, used for bond distances)		d	bond distance or internuclear separation
acacH	acetylacetone		d-	dextro- (see Box 19.2)
ADP	adenosine diphosphate		d	day (non-SI unit of time)
aq	aqueous		D	bond dissociation enthalpy
Arg	arginine		$\bar{D}$	average bond dissociation enthalpy
Asp	aspartic acid		D mechanism	dissociative mechanism
atm	atmosphere (non-SI unit of pressure)		D	debye (non-SI unit of electric dipole moment)
ATP	adenosine triphosphate			
ax	axial		Dcb mechanism	conjugate–base mechanism
B	magnetic field strength		dec	decomposition
B	Racah parameter		DHA	9,10-dihydroanthracene
bar	bar (unit of pressure)		dien	1,4,7-triazaheptane
9-BBN	9-borabicyclo[3.3.1]nonane		dil	dilute
bcc	body-centred cubic		dm^3	cubic decimetre (unit of volume)
bp	boiling point		DME	dimethoxyethane
bpy	2,2'-bipyridine		DMF	N,N-dimethylformamide
Bq	becquerel (unit of radioactivity)		dmgH$_2$	dimethylglyoxime
^{n}Bu	n-butyl		DMSO	dimethylsulfoxide
^{t}Bu	*tert*-butyl		DNA	deoxyribonucleic acid
c	coefficient (in wavefunctions)		E	energy
c	concentration (of solution)		E	identity operator
c	speed of light		E	bond enthalpy term
c-C_6H_{11}	cyclohexyl		e	charge on the electron
C	Curie constant		e^-	electron
C	coulomb (unit of charge)		EA	electron affinity
Ci	curie (non-SI unit of radioactivity)		E_a	activation energy
C_n	n-fold rotation axis		E_{cell}	electrochemical cell potential

E^o	standard reduction potential		K_p	equilibrium constant expressed in terms of partial pressures
EDTAH$_4$	N,N,N',N'-ethylenediaminetetraacetic acid (see Table 6.7)		K_{self}	self-ionization constant
en	1,2-diaminoethane		K_{sp}	solubility product constant
EPR	electron paramagnetic resonance		K_w	self-ionization constant of water
eq	equatorial		kg	kilogram (unit of mass)
ESR	electron spin resonance		kJ	kilojoule (unit of energy)
Et	ethyl		kPa	kilopascal (unit of pressure)
eV	electron volt			
EXAFS	extended X-ray absorption fine structure		L	Avogadro's number
			L	total (resultant) orbital quantum number
F	Faraday constant		L	ligand
FAD	flavin adenine dinucleotide		l	liquid
fcc	face-centred cubic		l	length
FID	free induction decay		l	orbital quantum number
FT	Fourier transform		l-	laevo- (see Box 19.2)
			ℓ	path length
G	Gibbs free energy		LCAO	linear combination of atomic orbitals
g	gas		LED	light-emitting diode
g	gram (unit of mass)		Leu	leucine
Glu	glutamic acid		LFER	linear free energy relationship
Gly	glycine		LFSE	ligand field stabilization energy
			LGO	ligand group orbital
H	enthalpy		LMCT	ligand-to-metal charge transfer
h	Planck constant		Ln	lanthanoid
h	hour (non-SI unit of time)		LUMO	lowest unoccupied molecular orbital
$[HBpz_3]^-$	trispyrazolylborate		Lys	lysine
hcp	hexagonal close-packed			
HIPIP	high potential protein		M	molarity
His	histidine		m	mass
HMPA	hexamethylphosphoramide (see structure **10.1**)		m	metre (unit of length)
HOMO	highest occupied molecular orbital		m^3	cubic metre (unit of volume)
Hz	hertz (unit of frequency)		m_e	electron rest mass
$h\nu$	high frequency radiation (for a photolysis reaction)		m_i	molality
			m_i^o	standard state molality
			m_l	magnetic quantum number
I	nuclear spin quantum number		M_L	total (resultant) orbital magnetic quantum number
i	centre of inversion			
I_a mechanism	associative interchange mechanism		m_s	magnetic spin quantum number
I_d mechanism	dissociative interchange mechanism		M_S	magnetic spin quantum number for the multi-electron system
IE	ionization energy		M_r	relative molecular mass
IR	infrared		Me	methyl
IUPAC	International Union of Pure and Applied Chemistry		Mes	mesityl (2,4,6-Me$_3$C$_6$H$_2$)
			Met	methionine
j	inner quantum number		min	minute (non-SI unit of time)
J	joule (unit of energy)		MLCT	metal-to-ligand charge transfer
J	spin–spin coupling constant		MO	molecular orbital
J	total (resultant) inner quantum number		MOCVD	metal–organic chemical vapour deposition
			mol	mole (unit of quantity)
k	force constant		mp	melting point
k	rate constant		Mt	megatonne
k	Boltzmann constant			
K	kelvin (unit of temperature)		N	normalization factor
K	equilibrium constant		N	number of nuclides
K_a	acid dissociation constant		n	neutron
K_b	base dissociation constant		n	Born exponent
K_c	equilibrium constant expressed in terms of concentrations		n	number of (e.g. moles)
			n	principal quantum number

n	nucleophilicity parameter		T	temperature
$[NAD]^+$	nicotinamide adenine dinucleotide		T_c	critical temperature of a superconductor
NASICON	Na super ionic conductor		T_C	Curie temperature
nm	nanometre (unit of length)		T_N	Néel temperature
NMR	nuclear magnetic resonance		t	tonne (metric)
			t	time
oxH_2	oxalic acid		$t_{\frac{1}{2}}$	half-life
			^tBu	tert-butyl
P	pressure		THF	tetrahydrofuran
Pa	pascal (unit of pressure)		Thr	threonine
PES	photoelectron spectroscopy		TMEDA	N,N,N',N'-tetramethylethylenediamine
Ph	phenyl		TMS	tetramethylsilane
phen	1,10-phenanthroline		TOF	catalytic turnover frequency
pK_a	$-\log K_a$		TON	catalytic turnover number
pm	picometre (unit of length)		$tppH_2$	tetraphenylporphyrin
ppm	part per million		tpy	2,2':6',2''-terpyridine
ppt	precipitate		trien	1,4,7,10-tetraazadecane (see Table 6.7)
Pr	propyl		Tyr	tyrosine
^iPr	iso-propyl			
PVC	polyvinylchloride		U	internal energy
py	pyridine		u	atomic mass unit
pzH	pyrazole		UV	ultraviolet
			UV-VIS	ultraviolet-visible
q	point charge			
Q	reaction quotient		V	potential difference
			V	volume
R	general alkyl group		V	volt (unit of potential difference)
R	molar gas constant		v	vapour
R	Rydberg constant		v	velocity
R	resistance		VB	valence bond
R-	Cahn–Ingold–Prelog notation for an enantiomer (see Box 19.2)		ve	valence electrons (in electron counting)
r	radial distance		VIS	visible
r	radius		VSEPR	valence-shell electron-pair repulsion
$R(r)$	radial wavefunction			
r_{cov}	covalent radius		[X]	concentration of X
r_{ion}	ionic radius			
r_v	van der Waals radius		yr	year (non-SI unit of time)
r_{metal}	metallic radius			
RDS	rate determining step		z	number of moles of electrons transferred in an electrochemical cell
RF	radiofrequency		Z	atomic number
			Z	effective collision frequency in solution
S	entropy		Z_{eff}	effective nuclear charge
S	overlap integral		$|z_-|$	modulus of the negative charge
S	total spin quantum number		$|z_+|$	modulus of the positive charge
S	screening (or shielding) constant		ZSM-5	a type of zeolite (see Section 26.6)
S-	Cahn–Ingold–Prelog notation for an enantiomer (see Box 19.2)			
s	second (unit of time)		α	polarizability of an atom or ion
s	solid		$[\alpha]$	specific rotation
s	spin quantum number			
s	nucleophilicity discrimination factor		β	stability constant
			β^-	beta-particle
S_n	n-fold improper rotation axis		β^+	positron
S_N1cb mechanism	conjugate-base mechanism			
Ser	serine		δ	chemical shift
soln	solution		δ-	label for an enantiomer (see Box 19.2)
solv	solvated; solvent		δ^-	partial negative charge
SQUID	superconducting quantum interference device		δ^+	partial positive charge
			Δ	change in

Δ-	label for enantiomer with right-handedness (see Box 19.2)	τ_1	spin relaxation time (in NMR spectroscopy)
Δ	heat (in a pyrolysis reaction)	ν	total number of particles produced per molecule of solute
Δ_{oct}	octahedral crystal field splitting energy		
Δ_{tet}	tetrahedral crystal field splitting energy	ν	frequency
ΔH°	standard enthalpy change	$\bar{\nu}$	wavenumber
$\Delta H^{\ddagger}$	enthalpy change of activation	ν_e	neutrino
$\Delta_a H$	enthalpy change of atomization		
$\Delta_c H$	enthalpy change of combustion	χ	magnetic susceptibility
$\Delta_{EA} H$	enthalpy change associated with the gain of an electron	χ_m	molar magnetic susceptibility
		χ	electronegativity
$\Delta_f H$	enthalpy change of formation	χ^{AR}	Allred–Rochow electronegativity
$\Delta_{fus} H$	enthalpy change of fusion	χ^M	Mulliken electronegativity
$\Delta_{hyd} H$	enthalpy change of hydration	χ^P	Pauling electronegativity
$\Delta_{lattice} H$	enthalpy change for the formation of an ionic lattice		
		ψ	wavefunction
$\Delta_r H$	enthalpy change of reaction		
$\Delta_{sol} H$	enthalpy change of solution	Ω	ohm (unit of resistance)
$\Delta_{solv} H$	enthalpy change of solvation		
$\Delta_{vap} H$	enthalpy change of vaporization	2c-2e	two-centre two-electron
ΔG°	standard free energy change	3c-2e	three-centre two-electron
$\Delta G^{\ddagger}$	free energy of activation		
$\Delta_f G$	free energy change of formation	$(+)$-	label for specific rotation of an enantiomer (see Box 19.2)
$\Delta_r G$	free energy change of reaction		
ΔS	entropy change	$(-)$-	label for specific rotation of an enantiomer (see Box 19.2)
ΔS°	standard entropy change		
$\Delta S^{\ddagger}$	entropy change of activation	$^{\circ}$ or $^{\ominus}$	standard state
$\Delta U(0\,K)$	internal energy change at $0\,K$	$\ddagger$	(called a 'double dagger') activated complex; transition state
$\Delta V^{\ddagger}$	volume of activation		
		$^{\circ}$	degree
ε	molar extinction (or absorption) coefficient	$>$	is greater than
ε_{max}	molar extinction coefficient corresponding to an absorption maximum (in an electronic spectrum)	$\gg$	is much greater than
		$<$	is less than
		$\ll$	is much less than
ε_0	permittivity of a vacuum	$\geq$	is greater than or equal to
ε_r	relative permittivity (dielectric constant)	$\leq$	is less than or equal to
		$\approx$	is approximately equal to
η	hapticity of a ligand (see Box 18.1)	$=$	is equal to
		$\neq$	is not equal to
λ-	label for an enantiomer (see Box 19.2)	$\rightleftharpoons$	equilibrium
λ	spin-orbit coupling constant	$\propto$	is proportional to
λ	wavelength	$\times$	multiplied by
λ_{max}	wavelength corresponding to an absorption maximum (in an electronic spectrum)	∞	infinity
		$\pm$	plus or minus
Λ-	label for enantiomer with left-handedness (see Box 19.2)	$\sqrt{}$	square root of
		$\sqrt[3]{}$	cube root of
		$\lvert x \rvert$	modulus of x
μ	electric dipole moment	$\sum$	summation of
μ	reduced mass	Δ	change in (for example, ΔH is 'change in enthalpy')
μ	refractive index		
μ(spin only)	spin only magnetic moment	$\angle$	angle
μ_B	Bohr magneton	log	logarithm to base 10 ($\log_{10}$)
μ_{eff}	effective magnetic moment	ln	natural logarithm, *i.e.* logarithm to base e ($\log_e$)
μ_i	chemical potential of component i		
μ_i°	standard chemical potential of i	$\int$	integral of
μ-	bridging ligand		
		$\dfrac{d}{dx}$	differential with respect to x
ρ	density		
		$\dfrac{\partial}{\partial x}$	partial differential with respect to x
σ	mirror plane		

The character tables given in this appendix are for some commonly encountered point groups. Complete tables are available in many physical and theoretical chemistry texts, e.g. see Chapter 3 reading list.

C_1	E
A	1

C_s	E	σ_h		
A'	1	1	x, y, R_z	x^2, y^2, z^2, xy
A''	1	-1	z, R_x, R_y	yz, xz

C_2	E	C_2		
A	1	1	z, R_z	x^2, y^2, z^2, xy
B	1	-1	x, y, R_x, R_y	yz, xz

C_{2v}	E	C_2	$\sigma_v(xz)$	$\sigma_v'(yz)$		
A_1	1	1	1	1	z	x^2, y^2, z^2
A_2	1	1	-1	-1	R_z	xy
B_1	1	-1	1	-1	x, R_y	xz
B_2	1	-1	-1	1	y, R_x	yz

C_{3v}	E	$2C_3$	$3\sigma_v$		
A_1	1	1	1	z	$x^2 + y^2, z^2$
A_2	1	1	-1	R_z	
E	2	-1	0	$(x, y) (R_x, R_y)$	$(x^2 - y^2, xy) (xz, yz)$

D_3	E	$2C_3$	$3C_2$		
A_1	1	1	1		$x^2 + y^2, z^2$
A_2	1	1	-1	z, R_z	
E	2	-1	0	$(x, y) (R_x, R_y)$	$(x^2 - y^2, xy) (xz, yz)$

D_{3h}	E	$2C_3$	$3C_2$	σ_h	$2S_3$	$3\sigma_v$		
A_1'	1	1	1	1	1	1		$x^2 + y^2, z^2$
A_2'	1	1	-1	1	1	-1	R_z	
E'	2	-1	0	2	-1	0	(x, y)	$(x^2 - y^2, xy)$
A_1''	1	1	1	-1	-1	-1		
A_2''	1	1	-1	-1	-1	1	z	
E''	2	-1	0	-2	1	0	(R_x, R_y)	(xz, yz)

D_{4h}	E	$2C_4$	C_2	$2C_2'$	$2C_2''$	i	$2S_4$	σ_h	$2\sigma_v$	$2\sigma_d$		
A_{1g}	1	1	1	1	1	1	1	1	1	1		$x^2 + y^2, z^2$
A_{2g}	1	1	1	-1	-1	1	1	1	-1	-1	R_z	
B_{1g}	1	-1	1	1	-1	1	-1	1	1	-1		$x^2 - y^2$
B_{2g}	1	-1	1	-1	1	1	-1	1	-1	1		xy
E_g	2	0	-2	0	0	2	0	-2	0	0	(R_x, R_y)	(xz, yz)
A_{1u}	1	1	1	1	1	-1	-1	-1	-1	-1		
A_{2u}	1	1	1	-1	-1	-1	-1	-1	1	1	z	
B_{1u}	1	-1	1	1	-1	-1	1	-1	-1	1		
B_{2u}	1	-1	1	-1	1	-1	1	-1	1	-1		
E_u	2	0	-2	0	0	-2	0	2	0	0	(x, y)	

T_d	E	$8C_3$	$3C_2$	$6S_4$	$6\sigma_d$		
A_1	1	1	1	1	1		$x^2 + y^2 + z^2$
A_2	1	1	1	-1	-1		
E	2	-1	2	0	0		$(2z^2 - x^2 - y^2, x^2 - y^2)$
T_1	3	0	-1	1	-1	(R_x, R_y, R_z)	
T_2	3	0	-1	-1	1	(x, y, z)	(xy, xz, yz)

The frequency of electromagnetic radiation is related to its wavelength by the equation:

$$\text{Wavelength } (\lambda) = \frac{\text{Speed of light } (c)}{\text{Frequency } (\nu)}$$

where $c = 3.0 \times 10^8 \, \text{m s}^{-1}$.

$$\text{Wavenumber } (\bar{\nu}) = \frac{1}{\text{Wavelength}}$$

with units in cm^{-1} (pronounced 'reciprocal centimetre')

Energy $(E) = $ Planck's constant $(h) \times$ Frequency (ν) where $h = 6.626 \times 10^{-34} \, \text{J s}$

The energy given in the last column is measured per mole of photons.

Frequency $\nu\,/\,\mathrm{Hz}$	Wavelength $\lambda\,/\,\mathrm{m}$	Wavenumber $\bar{\nu}\,/\,\mathrm{cm}^{-1}$	Type of radiation	Energy $E\,/\,\mathrm{kJ\,mol}^{-1}$
10^{21}	10^{-13}	10^{11}		10^{9}
10^{20}	10^{-12}	10^{10}	γ-ray	10^{8}
10^{19}	10^{-11}	10^{9}		10^{7}
10^{18}	10^{-10}	10^{8}	X-ray	10^{6}
10^{17}	10^{-9}	10^{7}		10^{5}
10^{16}	10^{-8}	10^{6}	Vacuum ultraviolet	10^{4}
10^{15}	10^{-7}	10^{5}	Ultraviolet	10^{3}
10^{14}	10^{-6}	10^{4}	Visible	10^{2}
10^{13}	10^{-5}	10^{3}	Near infrared	10^{1}
10^{12}	10^{-4}	10^{2}	Far infrared	$10^{0} = 1$
10^{11}	10^{-3}	10^{1}		10^{-1}
10^{10}	10^{-2}	$10^{0} = 1$	Microwave	10^{-2}
10^{9}	10^{-1}	10^{-1}		10^{-3}
10^{8}	$10^{0} = 1$	10^{-2}		10^{-4}
10^{7}	10^{1}	10^{-3}		10^{-5}
10^{6}	10^{2}	10^{-4}		10^{-6}
10^{5}	10^{3}	10^{-5}	Radiowave	10^{-7}
10^{4}	10^{4}	10^{-6}		10^{-8}
10^{3}	10^{5}	10^{-7}		10^{-9}

Visible region:
Violet ≈ 400 nm
Blue
Green
Yellow
Orange
Red ≈ 700 nm

Appendix 5 Naturally occurring isotopes and their abundances

Data from *WebElements* by Mark Winter. Further information on radiocative nuclides can be found using the Web link accompanying this appendix. ✳

Element	Symbol	Atomic number, Z	Mass number of isotope (% abundance)
Actinium	Ac	89	artificial isotopes only; mass number range 224–229
Aluminium	Al	13	27(100)
Americium	Am	95	artificial isotopes only; mass number range 237–245
Antimony	Sb	51	121(57.3), 123(42.7)
Argon	Ar	18	36(0.34), 38(0.06), 40(99.6)
Astatine	At	85	artificial isotopes only; mass number range 205–211
Arsenic	As	33	75(100)
Barium	Ba	56	130(0.11), 132(0.10), 134(2.42), 135(6.59), 136(7.85), 137(11.23), 138(71.70)
Berkelium	Bk	97	artificial isotopes only; mass number range 243–250
Beryllium	Be	4	9(100)
Bismuth	Bi	83	209(100)
Boron	B	5	10(19.9), 11(80.1)
Bromine	Br	35	79(50.69), 81(49.31)
Cadmium	Cd	48	106(1.25), 108(0.89), 110(12.49), 111(12.80), 112(24.13), 113(12.22), 114(28.73), 116(7.49)
Caesium	Cs	55	133(100)
Calcium	Ca	20	40(96.94), 42(0.65), 43(0.13), 44(2.09), 48(0.19)
Californium	Cf	98	artificial isotopes only; mass number range 246–255
Carbon	C	6	12(98.9), 13(1.1)
Cerium	Ce	58	136(0.19), 138(0.25), 140(88.48), 142(11.08)
Chlorine	Cl	17	35(75.77), 37(24.23)
Chromium	Cr	24	50(4.345), 52(83.79), 53(9.50), 54(2.365)
Cobalt	Co	27	59(100)
Copper	Cu	29	63(69.2), 65(30.8)
Curium	Cm	96	artificial isotopes only; mass number range 240–250
Dysprosium	Dy	66	156(0.06), 158(0.10), 160(2.34), 161(18.9), 162(25.5), 163(24.9), 164(28.2)
Einsteinium	Es	99	artificial isotopes only; mass number range 249–256
Erbium	Er	68	162(0.14), 164(1.61), 166(33.6), 167(22.95), 168(26.8), 170(14.9)
Europium	Eu	63	151(47.8), 153(52.2)
Fermium	Fm	100	artificial isotopes only; mass number range 251–257
Fluorine	F	9	19(100)
Francium	Fr	87	artificial isotopes only; mass number range 210–227
Gadolinium	Gd	64	152(0.20), 154(2.18), 155(14.80), 156(20.47), 157(15.65), 158(24.84), 160(21.86)
Gallium	Ga	31	69(60.1), 71(39.9)
Germanium	Ge	32	70(20.5), 72(27.4), 73(7.8), 74(36.5), 76(7.8)
Gold	Au	79	197(100)
Hafnium	Hf	72	174(0.16), 176(5.20), 177(18.61), 178(27.30), 179(13.63), 180(35.10)
Helium	He	2	3(<0.001), 4(>99.999)
Holmium	Ho	67	165(100)
Hydrogen	H	1	1(99.985), 2(0.015)
Indium	In	49	113(4.3), 115(95.7)
Iodine	I	53	127(100)
Iridium	Ir	77	191(37.3), 193(62.7)
Iron	Fe	26	54(5.8), 56(91.7), 57(2.2), 58(0.3)
Krypton	Kr	36	78(0.35), 80(2.25), 82(11.6), 83(11.5), 84(57.0), 86(17.3)
Lanthanum	La	57	138(0.09), 139(99.91)

Element	Symbol	Atomic number, Z	Mass number of isotope (% abundance)
Lawrencium	Lr	103	artificial isotopes only; mass number range 253–262
Lead	Pb	82	204(1.4), 206(24.1), 207(22.1), 208(52.4)
Lithium	Li	3	6(7.5), 7(92.5)
Lutetium	Lu	71	175(97.41), 176(2.59)
Magnesium	Mg	12	24(78.99), 25(10.00), 26(11.01)
Manganese	Mn	25	55 (100)
Mendelevium	Md	101	artificial isotopes only; mass number range 247–260
Mercury	Hg	80	196(0.14), 198(10.02), 199(16.84), 200(23.13), 201(13.22), 202(29.80), 204(6.85)
Molybdenum	Mo	42	92(14.84), 94(9.25), 95(15.92), 96(16.68), 97(9.55), 98(24.13), 100(9.63)
Neodymium	Nd	60	142(27.13), 143(12.18), 144(23.80), 145(8.30), 146(17.19), 148(5.76), 150(5.64)
Neon	Ne	10	20(90.48), 21(0.27), 22(9.25)
Neptunium	Np	93	artificial isotopes only; mass number range 234–240
Nickel	Ni	28	58(68.27), 60(26.10), 61(1.13), 62(3.59), 64(0.91)
Niobium	Nb	41	93(100)
Nitrogen	N	7	14(99.63), 15(0.37)
Nobelium	No	102	artificial isotopes only; mass number range 250–262
Osmium	Os	76	184(0.02), 186(1.58), 187(1.6), 188(13.3), 189(16.1), 190(26.4), 192(41.0)
Oxygen	O	8	16(99.76), 17(0.04), 18(0.20)
Palladium	Pd	46	102(1.02), 104(11.14), 105(22.33), 106(27.33), 108(26.46), 110(11.72)
Phosphorus	P	15	31(100)
Platinum	Pt	78	190(0.01), 192(0.79), 194(32.9), 195(33.8), 196(25.3), 198(7.2)
Plutonium	Pu	94	artificial isotopes only; mass number range 234–246
Polonium	Po	84	artificial isotopes only; mass number range 204–210
Potassium	K	19	39(93.26), 40(0.01), 41(6.73)
Praseodymium	Pr	59	141(100)
Promethium	Pm	61	artificial isotopes only; mass number range 141–151
Protactinium	Pa	91	artificial isotopes only; mass number range 228–234
Radium	Ra	88	artificial isotopes only; mass number range 223–230
Radon	Rn	86	artificial isotopes only; mass number range 208–224
Rhenium	Re	75	185(37.40), 187(62.60)
Rhodium	Rh	45	103(100)
Rubidium	Rb	37	85(72.16), 87(27.84)
Ruthenium	Ru	44	96(5.52), 98(1.88), 99(12.7), 100(12.6), 101(17.0), 102(31.6), 104(18.7)
Samarium	Sm	62	144(3.1), 147(15.0), 148(11.3), 149(13.8), 150(7.4), 152(26.7), 154(22.7)
Scandium	Sc	21	45(100)
Selenium	Se	34	74(0.9), 76(9.2), 77(7.6), 78(23.6), 80(49.7), 82(9.0)
Silicon	Si	14	28(92.23), 29(4.67), 30(3.10)
Silver	Ag	47	107(51.84), 109(48.16)
Sodium	Na	11	23(100)
Strontium	Sr	38	84(0.56), 86(9.86), 87(7.00), 88(82.58)
Sulfur	S	16	32(95.02), 33(0.75), 34(4.21), 36(0.02)
Tantalum	Ta	73	180(0.01), 181(99.99)
Technetium	Tc	43	artificial isotopes only; mass number range 95–99
Tellurium	Te	52	120(0.09), 122(2.60), 123(0.91), 124(4.82), 125(7.14), 126(18.95), 128(31.69), 130(33.80)
Terbium	Tb	65	159(100)
Thallium	Tl	81	203(29.52), 205(70.48)
Thorium	Th	90	232(100)
Thulium	Tm	69	169(100)
Tin	Sn	50	112(0.97), 114(0.65), 115(0.36), 116(14.53), 117(7.68), 118(24.22), 119(8.58), 120(32.59), 122(4.63), 124(5.79)
Titanium	Ti	22	46(8.0), 47(7.3), 48(73.8), 49(5.5), 50(5.4)
Tungsten	W	74	180(0.13), 182(26.3), 183(14.3), 184(30.67), 186(28.6)
Uranium	U	92	234(0.005), 235(0.72), 236(99.275)
Vanadium	V	23	50(0.25), 51(99.75)
Xenon	Xe	54	124(0.10), 126(0.09), 128(1.91), 129(26.4), 130(4.1), 131(21.2), 132(26.9), 134(10.4), 136(8.9)
Ytterbium	Yb	70	168(0.13), 170(3.05), 171(14.3), 172(21.9), 173(16.12), 174(31.8), 176(12.7)
Yttrium	Y	39	89(100)
Zinc	Zn	30	64(48.6), 66(27.9), 67(4.1), 68(18.8), 70(0.6)
Zirconium	Zr	40	90(51.45), 91(11.22), 92(17.15), 94(17.38), 96(2.8)

Van der Waals, metallic, covalent and ionic radii for the s-, p- and first row d-block elements

The ionic radius varies with the charge and coordination number of the ion; a coordination number of 6 refers to octahedral coordination, and of 4 refers to tetrahedral unless otherwise specified. Data for the heavier d-block metals and the lanthanoids and actinoids are listed in Tables 22.1 and 24.1.

	Element	Van der Waals radius, r_v / pm	Metallic radius for 12-coordinate metal, r_{metal} / pm	Covalent radius, r_{cov} / pm	Ionic radius		
					Ionic radius, r_{ion} / pm	Charge on ion	Coordination number of the ion
Hydrogen	H	120		37[‡]			
Group 1	Li		157		76	1+	6
	Na		191		102	1+	6
	K		235		138	1+	6
	Rb		250		149	1+	6
	Cs		272		170	1+	6
Group 2	Be		112		27	2+	4
	Mg		160		72	2+	6
	Ca		197		100	2+	6
	Sr		215		126	2+	8
	Ba		224		142	2+	8
Group 13	B	208		88			
	Al		143	130	54	3+	6
	Ga		153	122	62	3+	6
	In		167	150	80	3+	6
	Tl		171	155	89	3+	6
					159	1+	8
Group 14	C	185		77			
	Si	210		118			
	Ge			122	53	4+	6
	Sn		158	140	74	4+	6
	Pb		175	154	119	2+	6
					65	4+	4
					78	4+	6
Group 15	N	154		75	171	3−	6
	P	190		110			
	As	200		122			
	Sb	220		143			
	Bi	240	182	152	103	3+	6
					76	5+	6
Group 16	O	140		73	140	2−	6
	S	185		103	184	2−	6
	Se	200		117	198	2−	6
	Te	220		135	211	2−	6

[‡] Sometimes it is more appropriate to use a value of 30 pm in organic compounds.

	Element	Van der Waals radius, r_v / pm	Metallic radius for 12-coordinate metal, r_{metal} / pm	Covalent radius, r_{cov} / pm	Ionic radius		
					Ionic radius, r_{ion} / pm	Charge on ion	Coordination number of the ion
Group 17	F	135		71	133	1−	6
	Cl	180		99	181	1−	6
	Br	195		114	196	1−	6
	I	215		133	220	1−	6
Group 18	He	99					
	Ne	160					
	Ar	191					
	Kr	197					
	Xe	214					
First row *d*-block elements	Sc		164		75	3+	6
	Ti		147		86	2+	6
					67	3+	6
					61	4+	6
	V		135		79	2+	6
					64	3+	6
					58	4+	6
					53	4+	5
					54	5+	6
					46	5+	5
	Cr		129		73	2+	6 (low-spin)
					80	2+	6 (high-spin)
					62	3+	6
	Mn		137		67	2+	6 (low-spin)
					83	2+	6 (high-spin)
					58	3+	6 (low-spin)
					65	3+	6 (high-spin)
					39	4+	4
					53	4+	6
	Fe		126		61	2+	6 (low-spin)
					78	2+	6 (high-spin)
					55	3+	6 (low-spin)
					65	3+	6 (high-spin)
	Co		125		65	2+	6 (low-spin)
					75	2+	6 (high-spin)
					55	3+	6 (low-spin)
					61	3+	6 (high-spin)
	Ni		125		55	2+	4
					44	2+	4 (square planar)
					69	2+	6
					56	3+	6 (low-spin)
					60	3+	6 (high-spin)
	Cu		128		46	1+	2
					60	1+	4
					57	2+	4 (square planar)
					73	2+	6
	Zn		137		60	2+	4
					74	2+	6

Pauling electronegativity values (χ^P) for selected elements of the periodic table

Values are dependent on oxidation state.

Group 1	Group 2		Group 13	Group 14	Group 15	Group 16	Group 17
H 2.2							
Li 1.0	Be 1.6		B 2.0	C 2.6	N 3.0	O 3.4	F 4.0
Na 0.9	Mg 1.3		Al(III) 1.6	Si 1.9	P 2.2	S 2.6	Cl 3.2
K 0.8	Ca 1.0		Ga(III) 1.8	Ge(IV) 2.0	As(III) 2.2	Se 2.6	Br 3.0
Rb 0.8	Sr 0.9	(*d*-block elements)	In(III) 1.8	Sn(II) 1.8 Sn(IV) 2.0	Sb 2.1	Te 2.1	I 2.7
Cs 0.8	Ba 0.9		Tl(I) 1.6 Tl(III) 2.0	Pb(II) 1.9 Pb(IV) 2.3	Bi 2.0	Po 2.0	At 2.2

Ground state electronic configurations of the elements and ionization energies for the first five ionizations[‡]

$IE(n)$ in kJ mol^{-1} for the processes:

$IE(1)$	$M(g) \longrightarrow M^+(g)$
$IE(2)$	$M^+(g) \longrightarrow M^{2+}(g)$
$IE(3)$	$M^{2+}(g) \longrightarrow M^{3+}(g)$
$IE(4)$	$M^{3+}(g) \longrightarrow M^{4+}(g)$
$IE(5)$	$M^{4+}(g) \longrightarrow M^{5+}(g)$

Atomic number, Z	Element	Ground state electronic configuration	$IE(1)$	$IE(2)$	$IE(3)$	$IE(4)$	$IE(5)$
1	H	$1s^1$	1312				
2	He	$1s^2 = $ [He]	2372	5250			
3	Li	[He]$2s^1$	520.2	7298	11820		
4	Be	[He]$2s^2$	899.5	1757	14850	21010	
5	B	[He]$2s^2 2p^1$	800.6	2427	3660	25030	32830
6	C	[He]$2s^2 2p^2$	1086	2353	4620	6223	37830
7	N	[He]$2s^2 2p^3$	1402	2856	4578	7475	9445
8	O	[He]$2s^2 2p^4$	1314	3388	5300	7469	10990
9	F	[He]$2s^2 2p^5$	1681	3375	6050	8408	11020
10	Ne	[He]$2s^2 2p^6 = $ [Ne]	2081	3952	6122	9371	12180
11	Na	[Ne]$3s^1$	495.8	4562	6910	9543	13350
12	Mg	[Ne]$3s^2$	737.7	1451	7733	10540	13630
13	Al	[Ne]$3s^2 3p^1$	577.5	1817	2745	11580	14840
14	Si	[Ne]$3s^2 3p^2$	786.5	1577	3232	4356	16090
15	P	[Ne]$3s^2 3p^3$	1012	1907	2914	4964	6274
16	S	[Ne]$3s^2 3p^4$	999.6	2252	3357	4556	7004
17	Cl	[Ne]$3s^2 3p^5$	1251	2298	3822	5159	6540
18	Ar	[Ne]$3s^2 3p^6 = $ [Ar]	1521	2666	3931	5771	7238
19	K	[Ar]$4s^1$	418.8	3052	4420	5877	7975
20	Ca	[Ar]$4s^2$	589.8	1145	4912	6491	8153
21	Sc	[Ar]$4s^2 3d^1$	633.1	1235	2389	7091	8843
22	Ti	[Ar]$4s^2 3d^2$	658.8	1310	2653	4175	9581
23	V	[Ar]$4s^2 3d^3$	650.9	1414	2828	4507	6299
24	Cr	[Ar]$4s^1 3d^5$	652.9	1591	2987	4743	6702
25	Mn	[Ar]$4s^2 3d^5$	717.3	1509	3248	4940	6990
26	Fe	[Ar]$4s^2 3d^6$	762.5	1562	2957	5290	7240
27	Co	[Ar]$4s^2 3d^7$	760.4	1648	3232	4950	7670
28	Ni	[Ar]$4s^2 3d^8$	737.1	1753	3395	5300	7339
29	Cu	[Ar]$4s^1 3d^{10}$	745.5	1958	3555	5536	7700
30	Zn	[Ar]$4s^2 3d^{10}$	906.4	1733	3833	5730	7970
31	Ga	[Ar]$4s^2 3d^{10} 4p^1$	578.8	1979	2963	6200	

[‡] Values are from several sources, but mostly from the *Handbook of Chemistry and Physics* (1993) 74th edn, CRC Press, Boca Raton, and from the NIST Physics Laboratory, Physical Reference Data (see Web link below). The values in kJ mol^{-1} are quoted to 4 significant figures or less depending upon the accuracy of the original data in eV. A conversion factor of 1 eV $= 96.485$ kJ mol^{-1} has been applied. For updates to values, use the Web link accompanying this appendix. ✱

Atomic number, Z	Element	Ground state electronic configuration	$IE(1)$	$IE(2)$	$IE(3)$	$IE(4)$	$IE(5)$
32	Ge	$[Ar]4s^2 3d^{10} 4p^2$	762.2	1537	3302	4411	9020
33	As	$[Ar]4s^2 3d^{10} 4p^3$	947.0	1798	2735	4837	6043
34	Se	$[Ar]4s^2 3d^{10} 4p^4$	941.0	2045	2974	4144	6590
35	Br	$[Ar]4s^2 3d^{10} 4p^5$	1140	2100	3500	4560	5760
36	Kr	$[Ar]4s^2 3d^{10} 4p^6 = [Kr]$	1351	2350	3565	5070	6240
37	Rb	$[Kr]5s^1$	403.0	2633	3900	5080	6850
38	Sr	$[Kr]5s^2$	549.5	1064	4138	5500	6910
39	Y	$[Kr]5s^2 4d^1$	599.8	1181	1980	5847	7430
40	Zr	$[Kr]5s^2 4d^2$	640.1	1267	2218	3313	7752
41	Nb	$[Kr]5s^1 4d^4$	652.1	1382	2416	3700	4877
42	Mo	$[Kr]5s^1 4d^5$	684.3	1559	2618	4480	5257
43	Tc	$[Kr]5s^2 4d^5$	702	1472	2850		
44	Ru	$[Kr]5s^1 4d^7$	710.2	1617	2747		
45	Rh	$[Kr]5s^1 4d^8$	719.7	1744	2997		
46	Pd	$[Kr]5s^0 4d^{10}$	804.4	1875	3177		
47	Ag	$[Kr]5s^1 4d^{10}$	731.0	2073	3361		
48	Cd	$[Kr]5s^2 4d^{10}$	867.8	1631	3616		
49	In	$[Kr]5s^2 4d^{10} 5p^1$	558.3	1821	2704	5200	
50	Sn	$[Kr]5s^2 4d^{10} 5p^2$	708.6	1412	2943	3930	6974
51	Sb	$[Kr]5s^2 4d^{10} 5p^3$	830.6	1595	2440	4260	5400
52	Te	$[Kr]5s^2 4d^{10} 5p^4$	869.3	1790	2698	3610	5668
53	I	$[Kr]5s^2 4d^{10} 5p^5$	1008	1846	3200		
54	Xe	$[Kr]5s^2 4d^{10} 5p^6 = [Xe]$	1170	2046	3099		
55	Cs	$[Xe]6s^1$	375.7	2234	3400		
56	Ba	$[Xe]6s^2$	502.8	965.2	3619		
57	La	$[Xe]6s^2 5d^1$	538.1	1067	1850	4819	5940
58	Ce	$[Xe]4f^1 6s^2 5d^1$	534.4	1047	1949	3546	6325
59	Pr	$[Xe]4f^3 6s^2$	527.2	1018	2086	3761	5551
60	Nd	$[Xe]4f^4 6s^2$	533.1	1035	2130	3898	
61	Pm	$[Xe]4f^5 6s^2$	538.8	1052	2150	3970	
62	Sm	$[Xe]4f^6 6s^2$	544.5	1068	2260	3990	
63	Eu	$[Xe]4f^7 6s^2$	547.1	1085	2404	4120	
64	Gd	$[Xe]4f^7 6s^2 5d^1$	593.4	1167	1990	4245	
65	Tb	$[Xe]4f^9 6s^2$	565.8	1112	2114	3839	
66	Dy	$[Xe]4f^{10} 6s^2$	573.0	1126	2200	3990	
67	Ho	$[Xe]4f^{11} 6s^2$	581.0	1139	2204	4100	
68	Er	$[Xe]4f^{12} 6s^2$	589.3	1151	2194	4120	
69	Tm	$[Xe]4f^{13} 6s^2$	596.7	1163	2285	4120	
70	Yb	$[Xe]4f^{14} 6s^2$	603.4	1175	2417	4203	
71	Lu	$[Xe]4f^{14} 6s^2 5d^1$	523.5	1340	2022	4366	
72	Hf	$[Xe]4f^{14} 6s^2 5d^2$	658.5	1440	2250	3216	
73	Ta	$[Xe]4f^{14} 6s^2 5d^3$	728.4	1500	2100		
74	W	$[Xe]4f^{14} 6s^2 5d^4$	758.8	1700	2300		
75	Re	$[Xe]4f^{14} 6s^2 5d^5$	755.8	1260	2510		
76	Os	$[Xe]4f^{14} 6s^2 5d^6$	814.2	1600	2400		
77	Ir	$[Xe]4f^{14} 6s^2 5d^7$	865.2	1680	2600		
78	Pt	$[Xe]4f^{14} 6s^1 5d^9$	864.4	1791	2800		
79	Au	$[Xe]4f^{14} 6s^1 5d^{10}$	890.1	1980	2900		
80	Hg	$[Xe]4f^{14} 6s^2 5d^{10}$	1007	1810	3300		
81	Tl	$[Xe]4f^{14} 6s^2 5d^{10} 6p^1$	589.4	1971	2878	4900	
82	Pb	$[Xe]4f^{14} 6s^2 5d^{10} 6p^2$	715.6	1450	3081	4083	6640
83	Bi	$[Xe]4f^{14} 6s^2 5d^{10} 6p^3$	703.3	1610	2466	4370	5400
84	Po	$[Xe]4f^{14} 6s^2 5d^{10} 6p^4$	812.1	1800	2700		
85	At	$[Xe]4f^{14} 6s^2 5d^{10} 6p^5$	930	1600	2900		
86	Rn	$[Xe]4f^{14} 6s^2 5d^{10} 6p^6 = [Rn]$	1037				
87	Fr	$[Rn]7s^1$	393.0	2100	3100		
88	Ra	$[Rn]7s^2$	509.3	979.0	3300		

Atomic number, Z	Element	Ground state electronic configuration	$IE(1)$	$IE(2)$	$IE(3)$	$IE(4)$	$IE(5)$
89	Ac	$[Rn]6d^1 7s^2$	499	1170	1900		
90	Th	$[Rn]6d^2 7s^2$	608.5	1110	1930	2780	
91	Pa	$[Rn]5f^2 7s^2 6d^1$	568	1130	1810		
92	U	$[Rn]5f^3 7s^2 6d^1$	597.6	1440	1840		
93	Np	$[Rn]5f^4 7s^2 6d^1$	604.5	1130	1880		
94	Pu	$[Rn]5f^6 7s^2$	581.4	1130	2100		
95	Am	$[Rn]5f^7 7s^2$	576.4	1160	2160		
96	Cm	$[Rn]5f^7 7s^2 6d^1$	578.0	1200	2050		
97	Bk	$[Rn]5f^9 7s^2$	598.0	1190	2150		
98	Cf	$[Rn]5f^{10} 7s^2$	606.1	1210	2280		
99	Es	$[Rn]5f^{11} 7s^2$	619	1220	2330		
100	Fm	$[Rn]5f^{12} 7s^2$	627	1230	2350		
101	Md	$[Rn]5f^{13} 7s^2$	635	1240	2450		
102	No	$[Rn]5f^{14} 7s^2$	642	1250	2600		
103	Lr	$[Rn]5f^{14} 7s^2 6d^1$	440 (?)				

Approximate enthalpy changes, $\Delta_{EA}H(298\ K)$ associated with the gain of one electron by a gaseous atom or anion. A negative enthalpy (ΔH), but a positive electron affinity (EA), corresponds to an exothermic process, (see Section 1.10).

$\Delta_{EA}H(298\ K) \approx \Delta U(0\ K) = -EA$

	Process	$\approx \Delta_{EA}H$ / kJ mol^{-1}
Hydrogen	$H(g) + e^- \longrightarrow H^-(g)$	−73
Group 1	$Li(g) + e^- \longrightarrow Li^-(g)$	−60
	$Na(g) + e^- \longrightarrow Na^-(g)$	−53
	$K(g) + e^- \longrightarrow K^-(g)$	−48
	$Rb(g) + e^- \longrightarrow Rb^-(g)$	−47
	$Cs(g) + e^- \longrightarrow Cs^-(g)$	−45
Group 15	$N(g) + e^- \longrightarrow N^-(g)$	≈ 0
	$P(g) + e^- \longrightarrow P^-(g)$	−72
	$As(g) + e^- \longrightarrow As^-(g)$	−78
	$Sb(g) + e^- \longrightarrow Sb^-(g)$	−103
	$Bi(g) + e^- \longrightarrow Bi^-(g)$	−91
Group 16	$O(g) + e^- \longrightarrow O^-(g)$	−141
	$O^-(g) + e^- \longrightarrow O^{2-}(g)$	+798
	$S(g) + e^- \longrightarrow S^-(g)$	−201
	$S^-(g) + e^- \longrightarrow S^{2-}(g)$	+640
	$Se(g) + e^- \longrightarrow Se^-(g)$	−195
	$Te(g) + e^- \longrightarrow Te^-(g)$	−190
Group 17	$F(g) + e^- \longrightarrow F^-(g)$	−328
	$Cl(g) + e^- \longrightarrow Cl^-(g)$	−349
	$Br(g) + e^- \longrightarrow Br^-(g)$	−325
	$I(g) + e^- \longrightarrow I^-(g)$	−295

Enthalpies are given in kJ mol^{-1} for the process:

$$\frac{1}{n}E_n(\text{standard state}) \longrightarrow E(g)$$

Elements (E) are arranged according to their position in the periodic table. The lanthanoids and actinoids are excluded. The noble gases are omitted because they are monatomic at 298 K.

1	2	3	4	5	6	7	8	9	10	11	12	13	14	15	16	17
H 218																
Li 161	**Be** 324											**B** 582	**C** 717	**N** 473	**O** 249	**F** 79
Na 108	**Mg** 146											**Al** 330	**Si** 456	**P** 315	**S** 277	**Cl** 121
K 90	**Ca** 178	**Sc** 378	**Ti** 470	**V** 514	**Cr** 397	**Mn** 283	**Fe** 418	**Co** 428	**Ni** 430	**Cu** 338	**Zn** 130	**Ga** 277	**Ge** 375	**As** 302	**Se** 227	**Br** 112
Rb 82	**Sr** 164	**Y** 423	**Zr** 609	**Nb** 721	**Mo** 658	**Tc** 677	**Ru** 651	**Rh** 556	**Pd** 377	**Ag** 285	**Cd** 112	**In** 243	**Sn** 302	**Sb** 264	**Te** 197	**I** 107
Cs 78	**Ba** 178	**La** 423	**Hf** 619	**Ta** 782	**W** 850	**Re** 774	**Os** 787	**Ir** 669	**Pt** 566	**Au** 368	**Hg** 61	**Tl** 182	**Pb** 195	**Bi** 210	**Po** $\approx$146	**At** 92

Selected standard reduction potentials (298 K)

The concentration of each aqueous solution is $1\,mol\,dm^{-3}$ and the pressure of a gaseous component is $1\,bar$ $(10^5\,Pa)$. (Changing the standard pressure to $1\,atm$ $(101\,300\,Pa)$ makes no difference to the values of E° at this level of accuracy.) Each half-cell listed contains the specified solution species at a concentration of $1\,mol\,dm^{-3}$; where the half-cell contains $[OH]^-$, the value of E° refers to $[OH^-] = 1\,mol\,dm^{-3}$, hence the notation $E^{\circ}_{[OH^-]=1}$ (see Box 7.1).

Reduction half-equation	E° or $E^{\circ}_{[OH^-]=1}$ / V
$Li^+(aq) + e^- \rightleftharpoons Li(s)$	-3.04
$Cs^+(aq) + e^- \rightleftharpoons Cs(s)$	-3.03
$Rb^+(aq) + e^- \rightleftharpoons Rb(s)$	-2.98
$K^+(aq) + e^- \rightleftharpoons K(s)$	-2.93
$Ca^{2+}(aq) + 2e^- \rightleftharpoons Ca(s)$	-2.87
$Na^+(aq) + e^- \rightleftharpoons Na(s)$	-2.71
$La^{3+}(aq) + 3e^- \rightleftharpoons La(s)$	-2.38
$Mg^{2+}(aq) + 2e^- \rightleftharpoons Mg(s)$	-2.37
$Y^{3+}(aq) + 3e^- \rightleftharpoons Y(s)$	-2.37
$Sc^{3+}(aq) + 3e^- \rightleftharpoons Sc(s)$	-2.03
$Al^{3+}(aq) + 3e^- \rightleftharpoons Al(s)$	-1.66
$[HPO_3]^{2-} + 2H_2O + 2e^- \rightleftharpoons [H_2PO_2]^- + 3[OH]^-$	-1.65
$Ti^{2+}(aq) + 2e^- \rightleftharpoons Ti(s)$	-1.63
$Mn^{2+}(aq) + 2e^- \rightleftharpoons Mn(s)$	-1.19
$V^{2+}(aq) + 2e^- \rightleftharpoons V(s)$	-1.18
$Te(s) + 2e^- \rightleftharpoons Te^{2-}(aq)$	-1.14
$2[SO_3]^{2-} + 2H_2O + 2e^- \rightleftharpoons 4[OH]^- + [S_2O_4]^{2-}$	-1.12
$[SO_4]^{2-}(aq) + H_2O(l) + 2e^- \rightleftharpoons [SO_3]^{2-}(aq) + 2[OH]^-(aq)$	-0.93
$Se(s) + 2e^- \rightleftharpoons Se^{2-}(aq)$	-0.92
$Cr^{2+}(aq) + 2e^- \rightleftharpoons Cr(s)$	-0.91
$2[NO_3]^-(aq) + 2H_2O(l) + 2e^- \rightleftharpoons N_2O_4(g) + 4[OH]^-(aq)$	-0.85
$2H_2O(l) + 2e^- \rightleftharpoons H_2(g) + 2[OH]^-(aq)$	-0.82
$Zn^{2+}(aq) + 2e^- \rightleftharpoons Zn(s)$	-0.76
$Cr^{3+}(aq) + 3e^- \rightleftharpoons Cr(s)$	-0.74
$S(s) + 2e^- \rightleftharpoons S^{2-}(aq)$	-0.48
$[NO_2]^-(aq) + H_2O(l) + e^- \rightleftharpoons NO(g) + 2[OH]^-(aq)$	-0.46
$Fe^{2+}(aq) + 2e^- \rightleftharpoons Fe(s)$	-0.44
$Cr^{3+}(aq) + e^- \rightleftharpoons Cr^{2+}(aq)$	-0.41
$Ti^{3+}(aq) + e^- \rightleftharpoons Ti^{2+}(aq)$	-0.37
$PbSO_4(s) + 2e^- \rightleftharpoons Pb(s) + [SO_4]^{2-}(aq)$	-0.36
$Tl^+(aq) + e^- \rightleftharpoons Tl(s)$	-0.34
$Co^{2+}(aq) + 2e^- \rightleftharpoons Co(s)$	-0.28
$H_3PO_4(aq) + 2H^+(aq) + 2e^- \rightleftharpoons H_3PO_3(aq) + H_2O(l)$	-0.28
$V^{3+}(aq) + e^- \rightleftharpoons V^{2+}(aq)$	-0.26
$Ni^{2+}(aq) + 2e^- \rightleftharpoons Ni(s)$	-0.25
$2[SO_4]^{2-} + 4H^+ + 2e^- \rightleftharpoons [S_2O_6]^{2-} + 2H_2O$	-0.22

Reduction half-equation	E° or $E^{\circ}_{[OH^-]=1}$ / V
$O_2 + 2H_2O + 2e^- \rightleftharpoons H_2O_2 + 2[OH]^-$	-0.15
$Sn^{2+}(aq) + 2e^- \rightleftharpoons Sn(s)$	-0.14
$Pb^{2+}(aq) + 2e^- \rightleftharpoons Pb(s)$	-0.13
$Fe^{3+}(aq) + 3e^- \rightleftharpoons Fe(s)$	-0.04
$2H^+(aq, 1\,mol\,dm^{-3}) + 2e^- \rightleftharpoons H_2(g, 1\,bar)$	0
$[NO_3]^-(aq) + H_2O(l) + 2e^- \rightleftharpoons [NO_2]^-(aq) + 2[OH]^-(aq)$	$+0.01$
$[S_4O_6]^{2-}(aq) + 2e^- \rightleftharpoons 2[S_2O_3]^{2-}(aq)$	$+0.08$
$S(s) + 2H^+(aq) + 2e^- \rightleftharpoons H_2S(aq)$	$+0.14$
$2[NO_2]^-(aq) + 3H_2O(l) + 4e^- \rightleftharpoons N_2O(g) + 6[OH]^-(aq)$	$+0.15$
$Cu^{2+}(aq) + e^- \rightleftharpoons Cu^+(aq)$	$+0.15$
$Sn^{4+}(aq) + 2e^- \rightleftharpoons Sn^{2+}(aq)$	$+0.15$
$[SO_4]^{2-}(aq) + 4H^+(aq) + 2e^- \rightleftharpoons H_2SO_3(aq) + H_2O(l)$	$+0.17$
$AgCl(s) + e^- \rightleftharpoons Ag(s) + Cl^-(aq)$	$+0.22$
$Cu^{2+}(aq) + 2e^- \rightleftharpoons Cu(s)$	$+0.34$
$[VO]^{2+}(aq) + 2H^+(aq) + e^- \rightleftharpoons V^{3+}(aq) + H_2O(l)$	$+0.34$
$[ClO_4]^-(aq) + H_2O(l) + 2e^- \rightleftharpoons [ClO_3]^-(aq) + 2[OH]^-(aq)$	$+0.36$
$[Fe(CN)_6]^{3-}(aq) + e^- \rightleftharpoons [Fe(CN)_6]^{4-}(aq)$	$+0.36$
$O_2(g) + 2H_2O(l) + 4e^- \rightleftharpoons 4[OH]^-(aq)$	$+0.40$
$Cu^+(aq) + e^- \rightleftharpoons Cu(s)$	$+0.52$
$I_2(aq) + 2e^- \rightleftharpoons 2I^-(aq)$	$+0.54$
$H_3AsO_4(aq) + 2H^+(aq) + 2e^- \rightleftharpoons HAsO_2(aq) + 2H_2O(l)$	$+0.56$
$[MnO_4]^-(aq) + e^- \rightleftharpoons [MnO_4]^{2-}(aq)$	$+0.56$
$[MnO_4]^-(aq) + 2H_2O(aq) + 3e^- \rightleftharpoons MnO_2(s) + 4[OH]^-(aq)$	$+0.59$
$[BrO_3]^-(aq) + 3H_2O(l) + 6e^- \rightleftharpoons Br^-(aq) + 6[OH]^-(aq)$	$+0.61$
$O_2(g) + 2H^+(aq) + 2e^- \rightleftharpoons H_2O_2(aq)$	$+0.70$
$[BrO]^-(aq) + H_2O(l) + 2e^- \rightleftharpoons Br^-(aq) + 2[OH]^-(aq)$	$+0.76$
$Fe^{3+}(aq) + e^- \rightleftharpoons Fe^{2+}(aq)$	$+0.77$
$Ag^+(aq) + e^- \rightleftharpoons Ag(s)$	$+0.80$
$[ClO]^-(aq) + H_2O(l) + 2e^- \rightleftharpoons Cl^-(aq) + 2[OH]^-(aq)$	$+0.84$
$2HNO_2(aq) + 4H^+(aq) + 4e^- \rightleftharpoons H_2N_2O_2(aq) + 2H_2O(l)$	$+0.86$
$[NO_3]^-(aq) + 3H^+(aq) + 2e^- \rightleftharpoons HNO_2(aq) + H_2O(l)$	$+0.93$
$Pd^{2+} + 2e^- \rightleftharpoons Pd$	$+0.95$
$[NO_3]^-(aq) + 4H^+(aq) + 3e^- \rightleftharpoons NO(g) + 2H_2O(l)$	$+0.96$
$HNO_2(aq) + H^+(aq) + e^- \rightleftharpoons NO(g) + H_2O(l)$	$+0.98$
$[VO_2]^+(aq) + 2H^+(aq) + e^- \rightleftharpoons [VO]^{2+}(aq) + H_2O(l)$	$+0.99$
$[Fe(bpy)_3]^{3+}(aq) + e^- \rightleftharpoons [Fe(bpy)_3]^{2+}(aq)$	$+1.03$
$[IO_3]^-(aq) + 6H^+(aq) + 6e^- \rightleftharpoons I^-(aq) + 3H_2O(l)$	$+1.09$
$Br_2(aq) + 2e^- \rightleftharpoons 2Br^-(aq)$	$+1.09$
$[Fe(phen)_3]^{3+}(aq) + e^- \rightleftharpoons [Fe(phen)_3]^{2+}(aq)$	$+1.12$
$Pt^{2+} + 2e^- \rightleftharpoons Pt$	$+1.18$
$2[IO_3]^-(aq) + 12H^+(aq) + 10e^- \rightleftharpoons I_2(aq) + 6H_2O(l)$	$+1.20$
$O_2(g) + 4H^+(aq) + 4e^- \rightleftharpoons 2H_2O(l)$	$+1.23$
$MnO_2(s) + 4H^+(aq) + 2e^- \rightleftharpoons Mn^{2+}(aq) + 2H_2O$	$+1.23$
$Tl^{3+}(aq) + 2e^- \rightleftharpoons Tl^+(aq)$	$+1.25$
$2HNO_2(aq) + 4H^+(aq) + 4e^- \rightleftharpoons N_2O(g) + 3H_2O(l)$	$+1.30$
$[Cr_2O_7]^{2-}(aq) + 14H^+(aq) + 6e^- \rightleftharpoons 2Cr^{3+}(aq) + 7H_2O(l)$	$+1.33$
$Cl_2(aq) + 2e^- \rightleftharpoons 2Cl^-(aq)$	$+1.36$
$2[ClO_4]^-(aq) + 16H^+(aq) + 14e^- \rightleftharpoons Cl_2(aq) + 8H_2O(l)$	$+1.39$
$[ClO_4]^-(aq) + 8H^+(aq) + 8e^- \rightleftharpoons Cl^-(aq) + 4H_2O(l)$	$+1.39$
$[BrO_3]^-(aq) + 6H^+(aq) + 6e^- \rightleftharpoons Br^-(aq) + 3H_2O(l)$	$+1.42$
$[ClO_3]^-(aq) + 6H^+(aq) + 6e^- \rightleftharpoons Cl^-(aq) + 3H_2O(l)$	$+1.45$
$2[ClO_3]^-(aq) + 12H^+(aq) + 10e^- \rightleftharpoons Cl_2(aq) + 6H_2O(l)$	$+1.47$

Reduction half-equation	E° or $E^{\circ}_{[OH^-]=1}$ / V
$2[BrO_3]^-(aq) + 12H^+(aq) + 10e^- \rightleftharpoons Br_2(aq) + 6H_2O(l)$	+1.48
$HOCl(aq) + H^+(aq) + 2e^- \rightleftharpoons Cl^-(aq) + H_2O(l)$	+1.48
$[MnO_4]^-(aq) + 8H^+(aq) + 5e^- \rightleftharpoons Mn^{2+}(aq) + 4H_2O(l)$	+1.51
$Mn^{3+}(aq) + e^- \rightleftharpoons Mn^{2+}(aq)$	+1.54
$2HOCl(aq) + 2H^+(aq) + 2e^- \rightleftharpoons Cl_2(aq) + 2H_2O(l)$	+1.61
$[MnO_4]^-(aq) + 4H^+(aq) + 3e^- \rightleftharpoons MnO_2(s) + 2H_2O(l)$	+1.69
$PbO_2(s) + 4H^+(aq) + [SO_4]^{2-}(aq) + 2e^- \rightleftharpoons PbSO_4(s) + 2H_2O(l)$	+1.69
$Ce^{4+}(aq) + e^- \rightleftharpoons Ce^{3+}(aq)$	+1.72
$H_2O_2(aq) + 2H^+(aq) + 2e^- \rightleftharpoons 2H_2O(l)$	+1.78
$Co^{3+}(aq) + e^- \rightleftharpoons Co^{2+}(aq)$	+1.92
$[S_2O_8]^{2-}(aq) + 2e^- \rightleftharpoons 2[SO_4]^{2-}(aq)$	+2.01
$XeO_3(aq) + 6H^+(aq) + 6e^- \rightleftharpoons Xe(g) + 3H_2O(l)$	+2.10
$[FeO_4]^{2-} + 8H^+ + 3e^- \rightleftharpoons Fe^{3+} + 4H_2O$	+2.20
$H_4XeO_6(aq) + 2H^+(aq) + 2e^- \rightleftharpoons XeO_3(aq) + 3H_2O(l)$	+2.42
$F_2(aq) + 2e^- \rightleftharpoons 2F^-(aq)$	+2.87

Answers to non-descriptive problems

Full methods of working for all problems are given in the accompanying *Solutions Manual*. Where *no* answer is given below, guidelines are given in the *Solutions Manual*.

Chapter 1

1.1 (a) $^{27}_{13}Al$, 13 p, 13 e, 14 n; (b) $^{79}_{35}Br$, 35 p, 35 e, 44 n; $^{81}_{35}Br$, 35 p, 35 e, 46 n; (c) $^{54}_{26}Fe$, 26 p, 26 e, 28 n; $^{56}_{26}Fe$, 26 p, 26 e, 30 n; $^{57}_{26}Fe$, 26 p, 26 e, 31 n; $^{58}_{26}Fe$, 26 p, 26 e, 32 n.

1.2 Assume ^3H can be ignored; % $^1H = 99.20$, % $^2H = 0.80$.

1.4 Conversion from nm to cm^{-1}: 1 nm = 10^{-9} m = 10^{-7} cm.

1.5 For $n = 2$, $r = 211.7$ pm; for $n = 3$, $r = 476.4$ pm.

1.6 (a) $n = 1$, $l = 0$, $m_l = 0$; (b) $n = 4$, $l = 0$, $m_l = 0$; (c) $n = 5$, $l = 0$, $m_l = 0$.

1.7 $n = 3$, $l = 1$, $m_l = -1$; $n = 3$, $l = 1$, $m_l = 0$; $n = 3$, $l = 1$, $m_l = 1$.

1.8 7; $4f$; $n = 4$, $l = 3$, $m_l = -3$; $n = 4$, $l = 3$, $m_l = -2$; $n = 4$, $l = 3$, $m_l = -1$; $n = 4$, $l = 3$, $m_l = 0$; $n = 4$, $l = 3$, $m_l = 1$; $n = 4$, $l = 3$, $m_l = 2$; $n = 4$, $l = 3$, $m_l = 3$.

1.9 (b); (e).

1.11 $n = 1$, $E = -1312$; $n = 2$, $E = -328$; $n = 3$, $E = -145.8$; $n = 4$, $E = -82$; $n = 5$, $E = -52.5$ kJ mol^{-1}; the larger is the value of n, the higher (less negative) the energy level; the energy levels get closer together as n increases.

1.13 Energy level diagrams similar to Figure 1.14 showing the following configurations: (a) $2s^2 2p^5$; (b) $3s^2 3p^1$; (c) $3s^2$.

1.14 (a) $Sn(g) \longrightarrow Sn^{4+}(g) + e^-$; endothermic; (b) $Al(g) \longrightarrow Al^{3+}(g) + 3e^-$.

1.15 Group 1.

1.18 (a) $+657$ kJ mol^{-1}.

1.20 (a) single; (b) single; (c) double; (d) single.

1.22 (b) VB theory predicts all to be diamagnetic.

1.23 (a) Single; (b) single; (c) double; (d) triple; (e) single.

1.25 (a) $\frac{1}{2}$, 1; (b) yes (H_2 and $[He_2]^{2+}$ are isoelectronic).

1.26 (b) The bond distances can be correlated with the bond orders calculated using equation 1.32: O_2, 2.0; $[O_2]^+$, 2.5; $[O_2]^-$, 1.5; $[O_2]^{2-}$, 1.0. (c) O_2, $[O_2]^+$ and $[O_2]^-$.

1.27 (a) Polar, $N^{\delta-}-H^{\delta+}$; (b) polar, $F^{\delta-}-Br^{\delta+}$; (c) slightly polar, $C^{\delta-}-H^{\delta+}$; (d) polar, $P^{\delta+}-Cl^{\delta-}$; (e) non-polar.

1.28 HF and $[OH]^-$; CO_2 and $[NO_2]^+$; NH_3 and $[H_3O]^+$; $SiCl_4$ and $[AlCl_4]^-$.

1.29 (a) Bent; (b) tetrahedral; (c) trigonal pyramidal; (d) trigonal bipyramidal; (e) trigonal pyramidal; (f) pentagonal bipyramidal; (g) linear; (h) bent; (i) trigonal planar.

1.30 Five points of negative charge, all associated with bonds gives a trigonal bipyramidal arrangement with the S=O (greatest electron density in the bonding region) lying in the equatorial plane; (a) single; (b) double.

1.31 (a) Bent, polar; (b) linear, non-polar; (c) bent, polar; (d) trigonal planar, non-polar; (e) trigonal bipyramidal, non-polar; (f) planar, polar; (g) planar, non-polar; (h) linear, polar.

1.32 (a) Trigonal planar; no isomers; (b) tetrahedral; no isomers; (c) trigonal bipyramidal; Me, axially or equatorially sited; (d) octahedral; *cis* or *trans*.

Chapter 2

2.1 (a) 9p, 9e, 10n; (b) 27p, 27e, 32n; (c) 92p, 92e, 143n.

2.4 5.98×10^{23} mol^{-1}.

2.5 $k = 0.0039$ s^{-1}; $t_{\frac{1}{2}} = 180$ s.

2.6 7.55×10^{-10} s^{-1}.

2.7 Reading across each row of table:
α-particle: -2; -2; -4; yes
β-particle: $+1$; -1; 0; yes
positron: -1; $+1$; 0; yes
(n,γ) reaction: 0; $+1$; $+1$; no

2.8 Refer to Table 2.1 to check answers.

2.9 (a) $^{58}_{26}Fe + 2n \longrightarrow {}^{60}_{27}Co + \beta^-$;
(b) $^{55}_{25}Mn + n \longrightarrow {}^{56}_{25}Mn + \gamma$;

(c) $^{32}_{16}S + n \longrightarrow ^{32}_{15}P + p$;

(d) $^{23}_{11}Na + \gamma \longrightarrow ^{20}_{11}Na + 3n$.

2.10 (a) $^{92}_{36}Kr$; (b) $^{97}_{40}Zr$.

2.11 (a) Fast; (b) slow; (c) slow.

2.12 $t_{\frac{1}{2}} = 300$ days.

2.13 Shift to $2120\,cm^{-1}$.

2.16 $2.01 \times 10^{-7}\,mol\,dm^{-3}$.

2.18 J_{PF} and $J_{PH} \gg J_{HH}$ (for directly attached pairs of nuclei).

2.19 2 ^{13}C environments; each ^{13}C couples to 3 equivalent ^{19}F; larger value of J_{CF} is due to ^{19}F directly attached to ^{13}C and smaller J_{CF} is long range coupling.

2.20 A doublet for Ph_2PH with large J_{PH}; a singlet for PPh_3.

2.21 (a) Coupling of ^{31}P to 9 equivalent ^1H; (b) doublet $J_{PH} = 2.7\,Hz$.

2.22 (a) Coupling to 2 equivalent ^1H gives triplet; (b) only 4.7% of the terminal ^1H are attached to ^{29}Si; observe singlet with overlapping doublet ($J_{SiH} = 194\,Hz$); relative intensities of 3 lines 2.35:95.3:2.35.

2.23 (a) Binomial quartet; coupling to 3 equivalent ^1H; (b) doublet of quartets; coupling to one ^{31}P (gives doublet) and to 3 equivalent ^1H (gives quartet).

2.24 (a) Disphenoidal; (b) static structure at 175 K contains 2 equatorial and 2 axial F giving 2 triplets (J_{FF}); at 298 K, a fluxional process renders all ^{19}F equivalent.

2.25 Consistent for all except (b); VSEPR predicts PF_5 to be trigonal bipyramidal with 2 F environments in a ratio 2:3.

Chapter 3

3.1 (a) Trigonal planar; non-polar; (b) bent; polar; (c) trigonal pyramidal; polar; (d) linear; non-polar; (e) tetrahedral; polar.

3.3 (a) 8-fold; (b) 2-fold; (c) 5-fold; (d) 3-fold *or* (a) C_8; (b) C_2; (c) C_5; (d) C_3.

3.4 Bent; E, C_2, σ_v and σ_v'.

3.5 C_2 axis bisecting the O–O bond; see Chapter 3 of the *Solutions Manual*.

3.6 Labels are C_3, C_2 ($\times 3$), σ_h and σ_v ($\times 3$).

3.7 (a) Lose C_3 axis, two C_2 axes, two σ_v planes; (b) lose C_2 axis, σ_v plane; (c) σ_h plane.

3.8 (a) NH_3, PBr_3, $[SO_4]^{2-}$; (b) the planar species SO_3; $AlCl_3$; $[NO_3]^-$.

3.9 $[ICl_4]^-$; XeF_4.

3.10 (a) 2 (disphenoidal); (b) 2 (bent); (c) 9 (octahedral); (d) 2 (disphenoidal); (e) 2 (bent); (f) 4 (trigonal planar).

3.11 (a) Ethane-like; (b) staggered conformation; (c) yes, at the midpoint of the Si–Si bond; (d) eclipsed conformation; (e) no.

3.12 (a) No (trigonal planar); (b) no (tetrahedral); (c) yes (square planar); (d) no (trigonal bipyramidal); (e) no (pentagonal planar); (f) yes (octahedral); (g) yes (planar); (h) no (*hint:* think about π-bonding and the relative orientations of the C–H bonds).

3.14 C_{3v} (trigonal pyramidal).

3.15 Linear.

3.16 C_{4v}.

3.17 Structure is T-shaped.

3.19 (a) and (e) T_d; (b) and (d) C_{3v}; (c) C_{2v}.

3.20 (a) C_{2v}; (b) yes.

3.21 I_h.

3.22 (a) 3; (b) 9; (c) 4; (d) 3; (e) 6.

3.23 (a) 3; (b) 2; (c) 4; (d) 3; (e) 2; (f) 3.

Chapter 4

4.1 (c) From $\psi_{sp\,hybrid} = c_1\psi_{2s} + c_2\psi_{2p_x}$ and $\psi_{sp\,hybrid} = c_3\psi_{2s} - c_4\psi_{2p_x}$; for 2s, $c_1 = c_3$ and normalization means that for 2s: $c_1{}^2 + c_3{}^2 = 1$ giving $c_1 = c_3 = {}^1/_{\sqrt{2}}$.

4.2 (b) Start with 3 equations with 9 coefficients (compare above); $c_1 = c_4 = c_7$ and normalization means $c_1{}^2 + c_4{}^2 + c_7{}^2 = 1$, giving $c_1 = c_4 = c_7 = {}^1/_{\sqrt{3}}$. Other values of c_n determined likewise (full working in the accompanying *Solutions Manual*).

4.4 (a) Assuming shaded lobes of p_x and $d_{x^2-y^2}$ point along the $+x$ axis, and shaded lobe of p_y points along the $+y$ axis (see Figures 1.10 and 1.11), diagrams should show the combinations: $(s + p_x + d_{x^2-y^2})$; $(s - p_x + d_{x^2-y^2})$; $(s + p_y - d_{x^2-y^2})$; $(s - p_y - d_{x^2-y^2})$; (b) each is 25% s, 50% p, 25% d.

4.5 (a) Tetrahedral, sp^3; (b) square planar, 2 lone pairs, sp^3d^2; (c) trigonal pyramidal, 1 lone pair, sp^3; (d) bent, 2 lone pairs, sp^3; (e) trigonal bipyramidal, sp^3d; (f) octahedral, sp^3d^2; (g) T-shaped, 2 lone pairs, sp^3d; (h) trigonal planar, sp^2.

4.6 (a) sp^2; (b) sp^3.

4.7 (a) Trigonal bipyramidal (Figure 1.33); (b) sp^3d for σ-bonding; S–O π-bond formation between an O $2p$ and S $3d$ atomic orbital.

4.8 $[CO_3]^{2-}$ is isoelectronic and isostructural with $[NO_3]^-$; answer should resemble worked example 4.2.

4.9 (a) Linear; (b) sp; (c) σ-bond formation using C sp and O sp^2; leaves two orthogonal $2p$ atomic orbitals on C; form a π-bond using a $2p$ atomic orbital on each O; (d) 2; (e) see **4A**; yes.

$$\ddot{O} = C = \ddot{O}$$

(4A)

4.15 $[NH_4]^+$ is isoelectronic with CH_4; the description of bonding in $[NH_4]^+$ is essentially the same as that for CH_4.

4.16 (a) Ignoring the lone pairs, see **4B**; no, all 2c-2e bonds; (b) from MO diagrams: bond order in $I_2 = 1$; bond order in $[I_3]^+ = 1$ (MO diagram is similar to that for H_2O); bond order in $[I_3]^- = \frac{1}{2}$ (MO diagram similar to that for XeF_2). Lower bond order is consistent with longer bonds.

$$I—I \qquad \left[\begin{array}{c} I \\ I \diagup \diagdown I \end{array} \right]^+ \qquad \left[I—I—I \right]^-$$

(4B)

4.20 (a) The MO derived from LGO(5) has terminal H and some B–B bonding character; (b) no.

4.21 (a) 14.

Chapter 5

5.2 (a) 12; (b) 12; (c) 8; (d) 12 (same as ccp); (e) 6.

5.3 (a) Higher temp. form is the bcc lattice; polymorphism; (b) see text for $\beta \longrightarrow \alpha$-Sn transition.

5.4 (a) $^1/_n Co_n(s) \longrightarrow Co(g)$.

5.14 (b) $-662\,\text{kJ}\,\text{mol}^{-1}$.

5.15 $\Delta_{\text{lattice}}H^\circ(298\,\text{K}) = -2050\,\text{kJ}\,\text{mol}^{-1} \approx \Delta U(0\,\text{K})$.

5.16 (a) $609\,\text{kJ}\,\text{mol}^{-1}$; (b) $657\,\text{kJ}\,\text{mol}^{-1}$.

5.18 (a) $-621.2\,\text{kJ}\,\text{mol}^{-1}$; (b) $-632.2\,\text{kJ}\,\text{mol}^{-1}$.

Chapter 6

6.1 (a) 0.18; (b) 3.24×10^{-7}.

6.2 Largest K_a (smallest pK_a) refers to loss of first proton and so on.

6.4 (b) $pK_b(1) = 3.29$; $pK_b(2) = 6.44$.

6.9 (a) Basic; (b) amphoteric; (c) acidic; (d) acidic; (e) amphoteric; (f) acidic; (g) amphoteric; (h) amphoteric.

6.11 (a) $[Ag^+][Cl^-]$; (b) $[Ca^{2+}][CO_3^{2-}]$; (c) $[Ca^{2+}][F^-]^2$.

6.12 (a) $\sqrt{K_{sp}}$; (b) $\sqrt{K_{sp}}$; (c) $\sqrt[3]{\dfrac{K_{sp}}{4}}$.

6.13 $2.40 \times 10^{-4}\,\text{g}$.

6.15 (a) $\Delta_f G^\circ(K^+,\text{aq}) = -282.7\,\text{kJ}\,\text{mol}^{-1}$; $\Delta_f G^\circ(F^-,\text{aq}) = -276.9\,\text{kJ}\,\text{mol}^{-1}$; (b) $-21.8\,\text{kJ}\,\text{mol}^{-1}$; (c) $\Delta_{\text{sol}}G^\circ$ is significantly negative, and so the solubility of KF in water is relatively high.

6.16 6.0×10^{-29}.

6.20 (a) $1.37 \times 10^{-5}\,\text{g}$ per $100\,\text{g}$ H_2O; (b) $2.01 \times 10^{-11}\,\text{g}$ per $100\,\text{g}$ solution.

6.25 (a) $K_2 = \dfrac{[M(H_2O)_4L_2^{z+}]}{[M(H_2O)_5L^{z+}][L]}$; $K_4 = \dfrac{[M(H_2O)_2L_4^{z+}]}{[M(H_2O)_3L_3^{z+}][L]}$

(b) $\beta_2 = \dfrac{[M(H_2O)_4L_2^{z+}]}{[M(H_2O)_6^{z+}][L]^2}$; $\beta_4 = \dfrac{[M(H_2O)_2L_4^{z+}]}{[M(H_2O)_6^{z+}][L]^4}$

6.26 (b) -50; -46; $-34\,\text{kJ}\,\text{mol}^{-1}$.

6.27 (a) 3; (b) 3; (c) 3; (d) 4; (e) 6.

Chapter 7

7.1 (a) Ca, +2; O, −2; (b) H, +1; O, −2; (c) H, +1; F, −1; (d) Fe, +2; Cl, −1; (e) Xe, +6; F, −1; (f) Os, +8; O, −2; (g) Na, +1; S, +6; O, −2; (h) P, +5; O, −2; (i) Pd, +2; Cl, −1; (j) Cl, +7; O, −2; (k) Cr, +3; H, +1; O, −2.

7.2 (a) Cr, +6 to +3; (b) K, 0 to +1; (c) Fe, +3 to 0; Al, 0 to +3; (d) Mn, +7 to +4.

7.3 All redox reactions *except* for (c), (e) and (h); for redox, red = reduced, ox = oxidized: (a) N, red; Mg, ox; (b) N, ox; O, red; (d) Sb (ox); F in F_2 (red); (f) C (ox); O in O_2, (red); (g) Mn, (red); Cl, two atoms only (ox).

7.4 Changes are: (a) N, $2 \times (-3)$; Mg, $3 \times (+2)$; (b) N, $2 \times (+2)$; O, $2 \times (-2)$; (d) Sb, +2; F, $2 \times (-1)$; (f) C, $2 \times (+2)$; O, $2 \times (-2)$; (g) Mn, −2; Cl, $2 \times (+1)$.

7.5 (a) $2Ag^+(aq) + Zn(s) \longrightarrow 2Ag(s) + Zn^{2+}(aq)$; $E^\circ_{\text{cell}} = 1.56\,\text{V}$; $\Delta G^\circ = -301\,\text{kJ}$ per mole of reaction; (b) $Cl_2(aq) + 2Br^-(aq) \longrightarrow 2Cl^-(aq) + Br_2(aq)$; $E^\circ_{\text{cell}} = 0.27\,\text{V}$; $\Delta G^\circ = -52.1\,\text{kJ}$ per mole of reaction; (c) $[Cr_2O_7]^{2-}(aq) + 14H^+(aq) + 6Fe^{2+}(aq) \longrightarrow 2Cr^{3+}(aq) + 7H_2O(l) + 6Fe^{3+}(aq)$; $E^\circ_{\text{cell}} = 0.56\,\text{V}$; $\Delta G^\circ = -324\,\text{kJ}$ per mole of reaction.

7.7 (a) +1.48; (b) +1.34; (c) +1.20 V.

7.8 (a) 1.08 V; (b) $-208\,\text{kJ}\,\text{mol}^{-1}$; (c) kinetically stable; additives act as catalysts.

7.9 0.34 V.

7.10 (a) +0.74 V; (b) less easily (ΔG° is less negative).

7.11 $-0.15\,\mathrm{V}$.

7.13 $K \approx 10^{42}$.

7.14 (c).

7.15 $\Delta G^{\circ}(298\,\mathrm{K}) = 41.5\,\mathrm{kJ}$ per mole of reaction; disproportionation of precipitated CuCl is thermodynamically unfavourable.

7.18 (a)

$$[\mathrm{VO_2}]^{+} \xrightarrow{+0.99} [\mathrm{VO}]^{2+} \xrightarrow{+0.34} \mathrm{V}^{3+} \xrightarrow{-0.26} \mathrm{V}^{2+} \xrightarrow{-1.18} \mathrm{V}$$

(b) no species disproportionates.

7.20 (a) $1.22\,\mathrm{V}$.

Chapter 8

8.3 Polar: (a); (b); (c); (d); (e); (f); (h); (i); (j); non-polar: (g); (k).

8.4 (a) $2\mathrm{KI} + \mathrm{Zn(NH_2)_2}$; (b) $\mathrm{K_2[Zn(NH_2)_4]}$; (c) $\mathrm{GeH_4} + 2\mathrm{MgBr_2} + 4\mathrm{NH_3}$; (d) $[\mathrm{NH_4}]^{+} + [\mathrm{CH_3CO_2}]^{-}$; (e) $\mathrm{Na_2O_2}$; $\mathrm{NaO_2}$; (f) $\mathrm{K[HC\equiv C]} + \mathrm{NH_3}$; in aqu. sol., $\mathrm{CH_3CO_2H}$ only *partially* dissociates.

8.5 (a) $\mathrm{Zn} + 2\mathrm{NaNH_2} + 2\mathrm{NH_3} \longrightarrow \mathrm{Na_2[Zn(NH_2)_4]} + \mathrm{H_2}$
$[\mathrm{Zn(NH_2)_4}]^{2-} + 2[\mathrm{NH_4}]^{+} \longrightarrow \mathrm{Zn(NH_2)_2} + 4\mathrm{NH_3}$
$\mathrm{Zn(NH_2)_2} + 2\mathrm{NH_4I} \longrightarrow [\mathrm{Zn(NH_3)_4}]\mathrm{I_2}$
(b) In water: $2\mathrm{K} + 2\mathrm{H_2O} \longrightarrow 2\mathrm{KOH} + \mathrm{H_2}$; in liquid $\mathrm{NH_3}$, at low concentrations: form $\mathrm{K}^{+}(\mathrm{NH_3}) + \mathrm{e}^{-}(\mathrm{NH_3})$; on standing, $2\mathrm{NH_3} + 2\mathrm{e}^{-} \longrightarrow 2[\mathrm{NH_2}]^{-} + \mathrm{H_2}$.

8.6 (a) $\mathrm{H_2NNH_2}$; (b) $\mathrm{Hg_3N_2}$; (c) $\mathrm{O_2NNH_2}$; (d) $\mathrm{MeNH_2}$; (e) $\mathrm{OC(NH_2)_2}$; (f) $[\mathrm{Cr(NH_3)_6}]\mathrm{Cl_3}$.

8.7 $\mathrm{AlF_3} + \mathrm{NaF} \longrightarrow \mathrm{Na[AlF_4]}$ (soluble in liquid HF)
$\mathrm{Na[AlF_4]} + \mathrm{BF_3} \longrightarrow \mathrm{AlF_3}$ (precipitate) $+ \mathrm{Na[BF_4]}$.

8.8 Species formed: (a) $[\mathrm{ClF_2}]^{+} + [\mathrm{HF_2}]^{-}$; (b) $[\mathrm{MeOH_2}]^{+} + [\mathrm{HF_2}]^{-}$; (c) $[\mathrm{Et_2OH}]^{+} + [\mathrm{HF_2}]^{-}$; (d) $\mathrm{Cs}^{+} + [\mathrm{HF_2}]^{-}$; (e) $\mathrm{Sr}^{2+} + 2[\mathrm{HF_2}]^{-}$; (f) $[\mathrm{H_2F}]^{+} + [\mathrm{ClO_4}]^{-}$.

8.9 (a) $\mathrm{H_2S_2O_7} + \mathrm{H_2SO_4} \longrightarrow [\mathrm{H_3SO_4}]^{+} + [\mathrm{HS_2O_7}]^{-}$;
(b) relatively strong acid.

8.10 (a) Basic; $\mathrm{H_2O} + \mathrm{H_2SO_4} \longrightarrow [\mathrm{H_3O}]^{+} + [\mathrm{HSO_4}]^{-}$;
(b) Basic; $\mathrm{NH_3} + \mathrm{H_2SO_4} \longrightarrow [\mathrm{NH_4}]^{+} + [\mathrm{HSO_4}]^{-}$;
(c) $\mathrm{HCO_2H} + \mathrm{H_2SO_4} \longrightarrow \mathrm{CO} + [\mathrm{H_3O}]^{+} + [\mathrm{HSO_4}]^{-}$;
(d) Basic; $\mathrm{H_3PO_4} + \mathrm{H_2SO_4} \longrightarrow [\mathrm{H_4PO_4}]^{+} + [\mathrm{HSO_4}]^{-}$;
(e) Basic;
$\mathrm{HCl} + 2\mathrm{H_2SO_4} \longrightarrow \mathrm{HOSO_2Cl} + [\mathrm{H_3O}]^{+} + [\mathrm{HSO_4}]^{-}$

8.12 (a) $\mathrm{Ph_2C{=}CH_2} + \mathrm{HCl} \rightleftharpoons [\mathrm{Ph_2CCH_3}]^{+} + \mathrm{Cl}^{-}$; equilibrium then upset by: $\mathrm{Cl}^{-} + \mathrm{BCl_3} \longrightarrow [\mathrm{BCl_4}]^{-}$ with an increase in conductivity but further addition of $\mathrm{BCl_3}$ has no effect.
(b) $\mathrm{N_2O_4} \rightleftharpoons [\mathrm{NO}]^{+} + [\mathrm{NO_3}]^{-}$
$[\mathrm{NO_3}]^{-} + \mathrm{H_2SO_4} \rightleftharpoons [\mathrm{NO_2}]^{+} + [\mathrm{HSO_4}]^{-} + [\mathrm{OH}]^{-}$

$[\mathrm{OH}]^{-} + 2\mathrm{H_2SO_4} \rightleftharpoons [\mathrm{H_3O}]^{+} + 2[\mathrm{HSO_4}]^{-}$
Overall: $\mathrm{N_2O_4} + 3\mathrm{H_2SO_4} \rightleftharpoons$
$\qquad [\mathrm{NO}]^{+} + [\mathrm{NO_2}]^{+} + [\mathrm{H_3O}]^{+} + 3[\mathrm{HSO_4}]^{-}$

8.15 (a) Terminal and bridge Al–Cl are 2c-2e bonds; localized bonding; the bridge involves use of lone pair from Cl.
(b) $[\mathrm{Al_2Cl_7}]^{-} + \mathrm{AlCl_3} \rightleftharpoons [\mathrm{Al_3Cl_{10}}]^{-}$.

Chapter 9

9.2 (b) See **9A** and **9B**.

(9A)

(9B)

9.3 1:1:1 three-line signal.

9.4 Sample contains small amounts of $\mathrm{CD_2HCN}$; $^{1}\mathrm{H}$–$^{2}\mathrm{H}$ spin–spin coupling gives 1:2:3:2:1 signal; on probability grounds, $\mathrm{CDH_2CN}$ and $\mathrm{CH_3CN}$ present in negligible amounts.

9.5 React $\mathrm{D_2O} + \mathrm{AlCl_3}$ to prepare DCl; then $\mathrm{Li[AlH_4]} + \mathrm{DCl}$; accurate measurement of M_{r}, or of density of water formed on combustion.

9.6 In dilute solutions, *tert*-BuOH $\approx$ monomeric; $3610\,\mathrm{cm}^{-1}$ due to $\nu(\mathrm{OH})$; in more concentrated solutions, hydrogen-bonded association weakens covalent O–H bond; band (broad) is shifted to lower frequency.

9.7 $\mathrm{MCl} + \mathrm{HCl} \rightleftharpoons \mathrm{M[HCl_2]}$ equilibrium position is governed by relative lattice energies of MCl and $\mathrm{M[HCl_2]}$.

9.10 (a) $\mathrm{KH} + \mathrm{NH_3} \longrightarrow \mathrm{KNH_2} + \mathrm{H_2}$;
$\mathrm{KH} + \mathrm{EtOH} \longrightarrow \mathrm{KOEt} + \mathrm{H_2}$.

9.11 (a) $2\mathrm{H_2O} \longrightarrow 2\mathrm{H_2} + \mathrm{O_2}$;
(b) $2\mathrm{LiH} \longrightarrow 2\mathrm{Li} + \mathrm{H_2}$;
(c) $\mathrm{CaH_2} + \mathrm{H_2O} \longrightarrow \mathrm{Ca(OH)_2} + \mathrm{H_2}$;
(d) $\mathrm{Mg} + 2\mathrm{HNO_3} \longrightarrow \mathrm{Mg(NO_3)_2} + \mathrm{H_2}$
(e) $2\mathrm{H_2} + \mathrm{O_2} \longrightarrow 2\mathrm{H_2O}$;
(f) $\mathrm{CuO} + \mathrm{H_2} \xrightarrow{\Delta} \mathrm{Cu} + \mathrm{H_2O}$.

9.12 $\mathrm{H_2O_2}$ is kinetically stable.

9.13 (b) Mg: 6-coordinate; octahedral; H: 3-coordinate; trigonal planar.

9.14 Ratio coordination numbers for Al:$\mathrm{H} = 6$:2; stoichiometry is 1:3.

Chapter 10

10.1 (b) ns^1.

10.6 (a) $^{40}_{19}K \longrightarrow ^{40}_{18}Ar + \beta^+$; (b) $0.57\,dm^3$.

10.8 Gives LiF and NaI.

10.9 Halide exchange between $[PtCl_4]^{2-}$ and KBr or KI.

10.11 Phase of the solid in equilibrium with the dissolved salt must alter at 305 K.

10.12 (a) N^{3-} wholly in unit, Li^+ per unit $= 6 \times \frac{1}{3} = 2$; (b) consider both layers 1 and 2 to obtain Li_3N.

10.14 Disproportionation.

10.15 (a) $[O_2]^-$; (b) $[O_2]^{2-}$; (c) $[O_3]^-$; (d) $[N_3]^-$; (e) N^{3-}; (f) Na^-.

10.17 (a) $[C\equiv N]^-$ isoelectronic with CO; bonding as in CO (Section 1.17); (b) as for KOH (Section 10.6).

10.19 (a) $NaH + H_2O \longrightarrow NaOH + H_2$;
(b) $KOH + CH_3CO_2H \longrightarrow [CH_3CO_2]K + H_2O$;
(c) $2NaN_3 \longrightarrow 2Na + 3N_2$;
(d) $K_2O_2 + 2H_2O \longrightarrow 2KOH + H_2O_2 \longrightarrow 2KOH + H_2O + \frac{1}{2}O_2$;
(e) $NaF + BF_3 \longrightarrow Na[BF_4]$;
(f) Cathode: $K^+ + e^- \longrightarrow K$; anode: $2Br^- \longrightarrow Br_2 + 2e^-$;
(g) Cathode: $2H_2O + 2e^- \longrightarrow 2[OH]^- + H_2$; anode: $2Cl^- \longrightarrow Cl_2 + 2e^-$.

Chapter 11

11.2 $Ca(OH)_2 = 1.05 \times 10^{-2}\,mol\,dm^{-3}$;
$Mg(OH)_2 = 1.12 \times 10^{-4}\,mol\,dm^{-3}$; relative solubilities $= 94:1$.

11.3 (a) $3Mg + N_2 \xrightarrow{\Delta} Mg_3N_2$;
(b) $Mg_3N_2 + 6H_2O \longrightarrow 2NH_3 + 3Mg(OH)_2$.

11.4 (a) Mg^{2+} replace Na^+ ions, and $[C\equiv C]^{2-}$ replace Cl^- in NaCl lattice; $[C\equiv C]^{2-}$ is not spherical, so elongation along one axis; (b) free rotation of $[CN]^-$ in NaCN means $[CN]^-$ ion is pseudo-spherical.

11.5 (a) $[NH_4]_2[BeF_4] \xrightarrow{\Delta} BeF_2 + 2NH_4F$
(b) $2NaCl + BeCl_2 \longrightarrow Na_2[BeCl_4]$
(c) $BeF_2 \xrightarrow{water} [Be(H_2O)_4]^{2+} + 2F^-$

11.6 (a) See **11A**; sp^2; (b) See **11B**.

(11A)   (11B)

11.7 (a) See Figure 5.21; (b) per unit cell, 2 Mg^{2+} and 4 F^- ions, giving 1:2 $Mg^{2+}:F^-$ ratio.

11.9 (a) $CaCl_2$ forms a hydrate;
$CaH_2 + H_2O \longrightarrow Ca(OH)_2 + H_2$.

11.10 (a) See discussion of disproportionation of CaF in Section 5.16; (b) dissolve each in dilute HCl, measure $\Delta_r H^\circ$, and apply Hess cycle.

11.11 (a) SrO_2 and H_2O_2, conjugate base and acid respectively; HCl and $SrCl_2$, conjugate acid and base respectively; (b) base + weak acid: $BaO_2 + 2H_2O \longrightarrow Ba(OH)_2 + H_2O_2$.

11.12 (a) $MO + H_2O \longrightarrow M(OH)_2$: Sr, $\Delta_r H^\circ = -81.5$; Ba, $\Delta_r H^\circ = -105.7\,kJ\,mol^{-1}$.

11.13 (a) Bubble CO_2 through limewater;
(b) $Ca(OH)_2(aq) + CO_2(g) \longrightarrow CaCO_3(s) + H_2O(l)$;
(c) white precipitate, 'milky' appearance.

11.16 A complex such as $[MgOMg]^{2+}$ or its hydrate formed in $MgCl_2(aq)$.

Chapter 12

12.2 See **12A**.

(12A)

12.4 (a) $B_2O_3(s) + 3Mg(s) \xrightarrow{\Delta} 2B(s) + 3MgO(s)$
(b) Al_2O_3 is amphoteric, Fe_2O_3 is basic; only Al_2O_3 reacts, leaving solid Fe_2O_3:
$Al_2O_3(s) + 3H_2O(l) + 2NaOH(aq) \longrightarrow 2Na[Al(OH)_4](aq)$;
(c) $2Na[Al(OH)_4](aq) + CO_2(g) \longrightarrow Al_2O_3 \cdot 3H_2O(s) + Na_2CO_3(aq) + H_2O(l)$

12.5 (a) See Figure 2.10; (b) 1:1:1:1 multiplet; (c) doublet $[J(^{11}B-^{31}P)]$ of quartets $[J(^{11}B-^{1}H)]$; (d) singlet.

12.6 $\Delta_r H^\circ = -851.5\,kJ$ per mole of Fe_2O_3 (or Al_2O_3); enough energy released to melt the iron formed.

12.9 (a) $Me_3N \cdot BH_3$ forms; ^{11}B NMR spectrum of $THF \cdot BH_3$ and $Me_3N \cdot BH_3$ shows 1:3:3:1 quartet, but at different chemical shifts; (b) no; no change in ^{11}B or ^{31}P NMR spectra; (c) yes; monitor solution by ^{11}B NMR spectroscopy; (d) formation of complex through one or two $P \longrightarrow B$ bonds; use ^{31}P or ^{11}B NMR spectroscopy.

12.10 (a) Attack by H_2O on Al (but not B) possible; expansion of valence shell octet for Al; (b) reaction steps are
(i) $B_2H_6 \underset{}{\overset{fast}{\rightleftharpoons}} 2BH_3$, (ii) $BH_3 + H_2O \xrightarrow{slow}$ products;
(c) $B(OH)_3 + 2[HF_2]^- \longrightarrow [BF_4]^- + 2H_2O + [OH]^-$.

12.11 (a) $B(OEt)_3 + 3HCl$; (b) $EtOH \cdot BF_3$;
(c) $B(NHPh)_3 + 3HCl$; (d) KBF_4 (ionic salt).

12.12 (a) $Na_3[AlF_6]$; (b) $CaTiO_3$; (c) rewrite $Na_3[AlF_6]$ as
$Na_2[NaAlF_6] \equiv NaXF_3$; cryolite has perovskite lattice
with $\frac{2}{3}$Na in Ca sites (Figure 5.23), and Al $+\frac{1}{3}$Na in Ti
sites.

12.13 (a) $[MBr_6]^{3-}$, octahedral; $[MCl_5]^{2-}$, trigonal
bipyramidal; $[MBr_4]^-$, tetrahedral; (b) crystal packing
effects; (c) $TlCl_3 + H_2N(CH_2)_5NH_2 + 2HCl$;
$TlCl_3 + 3CsCl$; (d) monomeric $GaCl_2$ would be
paramagnetic; $Ga[GaCl_4]$ contains diamagnetic Ga^+ and
$[GaCl_4]^-$ ions.

12.14 (a) $AlF_3 + 3F^- \longrightarrow [AlF_6]^{3-}$; on adding BF_3, formation
of $[BF_4]^-$ causes displacement and precipitation of AlF_3.
(b) Data indicate common species for $GaCl_2$ and
$GaCl_3/HCl$; i.e. $[GaCl_4]^-$. (c) Solid TlI_3 is $Tl^+[I_3]^-$;
hydrated Tl_2O_3 is insoluble, and oxidation of $Tl^+(aq)$ to
solid Tl_2O_3 is very much easier than to $Tl^{3+}(aq)$; I_2 is the
oxidizing agent.

12.15 (a) At 298 K, terminal and bridging Hs involved in
dynamic process; process persists at 203 K; (b) all ^{11}B
nuclei equivalent; quintet due to coupling of ^{11}B nucleus
to 4, equivalent ^1H nuclei, indicating that dynamic
process involves exchange of terminal and bridging H
about a single B atom; (c) the *IR timescale $\neq$ NMR
timescale*.

12.17 Use localized 2c-2e bonds; coordinate $N \longrightarrow Al$ bonds.

12.18 B_5H_9, *nido*-cage, square-based pyramid with 4 bridging
H; $[B_8H_8]^{2-}$, *closo*-dodecahedron; $C_2B_{10}H_{12}$, *closo*-
icosahedron; *nido*-$[B_6H_9]^-$, pentagonal pyramid with 3
bridging H atoms; $C_2B_{10}H_{12}$ could have C atoms
adjacent (1,2-isomer), or apart (1,7- and 1,12-isomers).

12.19 (a) From Wade's rules, adding 2 electrons means parent
deltahedron changes from $n = 6$ (for B_5H_9) to $n = 7$
(for B_5H_{11}); predict a change from *nido* to *arachno*.
(b) Anion must be undergoing dynamic process in
solution, with all 8 H equivalent and 'seeing' every B.
(c) Intermolecular elimination of H_2, suggests cage-
coupling by formation of inter-cage B–B bond; coupling
could be B_{apical}–B_{apical}, B_{apical}–B_{basal} or B_{basal}–B_{basal}.

12.20 (a) 1-BrB_5H_8, isomerizing to 2-BrB_5H_8;
(b) $B_4H_8(PF_3) + H_2$; (c) $K[1-BrB_5H_7] + H_2$;
(d) $4B(OR)_3 + MeB(OR)_2 + 11H_2$ (the B–C bond is not
hydrolyzed).

Chapter 13

13.4 $[C_{60}]^{n-}$ occupy 8 corner sites and 6 face sites $=$
$(8 \times \frac{1}{8}) + (6 \times \frac{1}{2}) = 4$; K^+ occupy 9 sites inside the unit

cell and 12 edge sites $= 9 + (12 \times \frac{1}{4}) = 12$; ratio
$[C_{60}]^{n-}:K^+ = 1:3$.

13.6 (a) Mg_2C_3 and CaC_2 contain $[C=C=C]^{4-}$ and $[C\equiv C]^{2-}$
ions respectively, ThC_2 contains $[C_2]^{4-}$, and TiC is an
interstitial carbide; (b) $[NH_4]Br$ acts as an acid in liquid
NH_3; (c) Si–H (or Si–D) is not broken in rate
determining step, and presumably $[OH]^-$ attacks Si
centre.

13.8 (a) Linear; (b) linear; (c); trigonal pyramidal; (d) trigonal
bipyramidal; (e) tetrahedral at each Si $(Cl_3Si–O–SiCl_3)$;
(f) octahedral; (g) octahedral; (h) tetrahedral.

13.9 (a) $[Sn_9Tl]^{3-}$ possesses 11 pairs of electrons for cluster
bonding; *closo* cage; (b) 2 isomers because Tl could
occupy one of the two different sites.

13.11 (a) $GeCl_4 + 2H_2O \longrightarrow GeO_2 + 4HCl$; GeO_2 has infinite
lattice structure, quartz form has tetrahedral GeO_4-units;
(b) $SiCl_4 + 4NaOH \longrightarrow Na_4SiO_4 + 4HCl$; discrete
$[SiO_4]^{4-}$ ion not present, polymeric species;
(c) $CsF + GeF_2 \longrightarrow Cs[GeF_3]$; trigonal pyramidal
$[GeF_3]^-$ ions;
(d) $2SiH_3Cl + H_2O \longrightarrow (H_3Si)_2O + 2HCl$;
(e) $2SiF_4 + 4H_2O \longrightarrow SiO_2 + 2[H_3O]^+ + [SiF_6]^{2-} + 2HF$;
octahedral $[SiF_6]^{2-}$;
(f) $2[Bu_4P]Cl + SnCl_4 \longrightarrow [Bu_4P]_2[SiCl_6]$; octahedral
$[SiCl_6]^{2-}$.

13.12 Each spectrum is ^{119}Sn *observed*, so all splittings are due
to ^{119}Sn–^{19}F spin–spin couplings; each species is
octahedral: (a) single F environment, so doublet;
(b) *trans*- and *cis*-isomers both have 2 equivalent F sites,
giving a triplet; (c) A $=$ *mer*-isomer with 2 F sites (1:2);
B $=$ *fac*-isomer with 3 equivalent F nuclei; (d) A $=$ *trans*-
Cl atoms giving 4 equivalent F sites; B $=$ *cis*-Cl atoms,
giving 2 F environments (2:2); (e) 2 F environments (1:4);
(f) 6 F atoms equivalent.

13.13 (a) $[Sn(OH)_6]^{2-} + H_2$; (b) $PbSO_4$; (c) Na_2CS_3;
(d) –SiH_2O– polymers; (e) $ClCH_2SiH_3$.

13.14 (a) Dissolve each in conc $HF(aq)$, measure $\Delta_r H^\circ$, and
apply Hess Law; (b) Si–Si and Si–H bond energies from
$\Delta_c H^\circ$ for Si_2H_6 and SiH_4 and other appropriate data;
apply Pauling relationship; (c) determine Pb(IV) by
allowing it to oxidize I^- and titrating I_2 formed with
thiosulfate (or heat with HCl, pass Cl_2 into KI(aq), and
titrate I_2 formed against thiosulfate).

13.15 At 1000 K, CO is more thermodynamically stable than
SnO_2; so C reduces SnO_2 to Sn at 1000 K (but not at 500
or 750 K).

13.16 (a) Fe^{2+} replaces Mg^{2+} with no structural change (r_{ion},
see Appendix 6); (b) see Figure 13.14 to see that Al^{3+} can
replace Si^{4+} with electrical neutrality conserved by Ca^{2+}
replacing Na^+; (c) Al^{3+} can replace Si^{4+} in silica structure
with interstitial Li^+ maintaining electrical neutrality.

13.17 I = $(CN)_2$; II = CS_2; III = CO_2; all linear ($D_{\infty h}$).

13.18 KCN(aq) is very alkaline owing to hydrolysis; $[CN]^-$ competes unsuccessfully with $[OH]^-$ for Al^{3+}.

13.19 (a) $NH_3 + H_2CO_3$; then forms $CO_2 + H_2O$; (b) same as (a); (c) $NH_3 + H_2CO_2S$; then forms $OCS + H_2O$.

Chapter 14

14.1 (a) 0; (b) +5; (c) +3; (d) +4; (e) +2; (f) −3; (g) −1; (h) 0; (i) +5; (j) +3; (k) +5.

14.2 (a) +932; (b) −274; (c) −450 kJ per mole of reaction.

14.4 (a) $Ca_3P_2 + 6H_2O \longrightarrow 3Ca(OH)_2 + 2PH_3$;
(b) $NaOH + NH_4Cl \longrightarrow NaCl + NH_3 + H_2O$;
(c) $Mg(NO_3)_2 + 2NH_3 + 2H_2O \longrightarrow$
$$Mg(OH)_2(s) + 2NH_4NO_3;$$
(d) $AsH_3 + 4I_2 + 4H_2O \longrightarrow H_3AsO_4 + 8HI$;
(e) $PH_3 + KNH_2 \xrightarrow{\text{liquid } NH_3} KPH_2 + NH_3$.

14.5 (a) HCl(aq) is fully ionized; solutions of NH_3 contain dissolved NH_3; (b) $[NH_4][NH_2CO_2]$ is salt of very weak acid.

14.7 $HNO_3(aq) + 6H^+ + 6e^- \rightleftharpoons NH_2OH(aq) + 2H_2O(l)$
$[BrO_3]^-(aq) + 6H^+(aq) + 6e^- \rightleftharpoons Br^-(aq) + 3H_2O(l)$
$NH_2OH(aq) + [BrO_3]^-(aq) \longrightarrow$
$$HNO_3(aq) + Br^-(aq) + H_2O(l)$$

14.8 (a) $3NaNH_2 + NaNO_3 \longrightarrow NaN_3 + 3NaOH + NH_3$;
(b) Na with liquid NH_3;
(c) $2NaN_3 + Pb(NO_3)_2 \longrightarrow Pb(N_3)_2 + 2NaNO_3$.

14.9 (a) Species include $[CN_2]^{2-}$, $[NO_2]^+$, $[NCO]^-$; (b) bonding scheme similar to that for CO_2.

14.10 (b) Unit cell contains 2 complete As, and $[(4 \times \frac{1}{4}) + (8 \times \frac{1}{8})]$ Ni = 2 Ni; i.e. 1:1.

14.11 (a) Electron diffraction, or vibrational spectroscopy; (b) Raman (not IR) spectroscopy.

14.12 (b) Assume *spherical* $[PCl_4]^+$ and $[PCl_6]^-$ ions.

14.13 Each ion contains equivalent F centres: (a) doublet (coupling to ^{31}P); (b) 1:1:1:1:1:1 signal (coupling to ^{121}Sb) superimposed on a 1:1:1:1:1:1:1:1 signal (coupling to ^{123}Sb), relative abundances ^{121}Sb:$^{123}Sb \approx 1$:1.

14.14 (a) $[PCl_4]^+[SbCl_6]^-$; (b) $K^+[AsF_6]^-$; (c) $[NO]^+[SbF_6]^-$; (d) $[H_2F]^+[SbF_6]^-$; tendency for Sb–F–Sb bridge formation may give $[Sb_2F_{11}]^-$ or higher association.

14.15 (a) See Figure 14.9b and **14.27**; $[Sb_2F_{11}]^-$, no lone pairs, 12 electrons in valence shell of each Sb(V) centre; $[Sb_2F_7]^-$, one lone pair and four bonding pairs per Sb(III) gives trigonal bipyramidal arrangements with lone pairs in equatorial sites; (b) both chains with octahedral Bi(III).

14.16 $[NO]^+$ is isoelectronic with CO and the MO diagram is similar; NO has one more electron than $[NO]^+$ and this occupies a π^* MO; frequency of vibration depends on the force constant which increases as bond strengthens.

14.17 **B** = N_2O.

14.19 (a) Triple-rutile lattice; (b) need three rutile-type unit cells to give unambiguous description of lattice; (c) O: 3-coordinate (trigonal planar); Fe: 6-coordinate (octahedral); Sb: 6-coordinate (octahedral); (d) Fe: one central + eight corners = 2 Fe; Sb: two central + eight edge = 4 Sb; O: ten central + four face = 12 O; stoichiometry = 2:4:12 = 1:2:6.

14.20 (a) $[P_3O_{10}]^{5-}$ gives two signals in ^{31}P NMR spectrum (rel. integrals 2:1); $[P_4O_{13}]^{6-}$ gives two signals of equal integral. (b) AsF_5 is isostructural with PF_5, **14.23**; two ^{19}F peaks of relative integrals 3:2 (eq:ax) will coalesce at a higher temperature if rapid exchange occurs. (c) Refer to structure in Figure 14.16a, replacing three Cl by NMe_2 groups; there are 3 possible isomers giving one, two or three signals in the ^1H NMR spectrum; or use ^{31}P NMR.

14.21 (a) If $2Ti(III) \longrightarrow 2Ti(IV)$, change in oxidation state for N is −1 to −3; product is NH_3. (b) If $2Ag(I) \longrightarrow 2Ag(0)$, change in oxidation state for P is +3 to +5; product is $[PO_4]^{3-}$. (c) If $2I(0) \longrightarrow I(-1)$ twice, change in oxidation state for P is +1 to +3 to +5, i.e. $H_3PO_2 \longrightarrow H_3PO_3 \longrightarrow H_3PO_4$.

14.22 (a) Tetrahedral; (b) planar; (c) trigonal pyramidal at N, bent at O; (d) tetrahedral; (e) trigonal bipyramidal with axial F atoms.

14.23 (a) $K^{15}NO_3 + Al$, NaOH(aq) $\longrightarrow$ $^{15}NH_3$; pass over Na; (b) Oxidize $^{15}NH_3$ [see part (a)] with CuO or NaOCl; (c) $K^{15}NO_3 + Hg$, $H_2SO_4 \longrightarrow$ ^{15}NO; combine with Cl_2, $AlCl_3$.

14.24 (a) Reduce to ^{32}P; treat with NaOH(aq); (b) ^{32}P [see part (a)] + limited Cl_2; hydrolyze the product; (c) ^{32}P [see part (a)] + excess S to $^{32}P_4S_{10}$; treat with Na_2S.

14.25 **D** = N_2.

14.26 Combination of Al + P is isoelectronic with Si + Si.

Chapter 15

15.1 (b) ns^2np^4.

15.2 $^{209}_{83}Bi(n,\gamma)^{210}_{83}Bi \xrightarrow{\beta^-} {}^{210}_{84}Po$.

15.3 Anode: $4[OH]^-(aq) \longrightarrow O_2(g) + 2H_2O(l) + 4e^-$; cathode $2H^+(aq) + 2e^- \longrightarrow H_2(g)$.

15.4 $8E \longrightarrow 4E_2$: $\Delta_r H^\circ = -[4 \times E(E{=}E)] = -1992\,kJ\,mol^{-1}$ for $E = O$, and $-1708\,kJ\,mol^{-1}$ for $E = S$; $8E \longrightarrow E_8$: $\Delta_r H^\circ = -[8 \times E(E{-}E)] = -1168\,kJ\,mol^{-1}$ for $E = O$, and $-2128\,kJ\,mol^{-1}$ for $E = S$.

15.5 (a) $E^\circ_{cell} = 1.08\,V$, so $\Delta_r G^\circ$ is negative. (b) $59.9\,g\,dm^{-3}$.

15.6 (a) $2Ce^{4+} + H_2O_2 \longrightarrow 2Ce^{3+} + O_2 + 2H^+$; (b) $2I^- + H_2O_2 + 2H^+ \longrightarrow I_2 + 2H_2O$.

15.7 (a) $Mn(OH)_2 + H_2O_2 \longrightarrow MnO_2 + 2H_2O$; (b) MnO_2 will catalyze decomposition of H_2O_2: $2H_2O_2 \longrightarrow 2H_2O + O_2$.

15.8 (a) Bent; (b) trigonal pyramidal; (c) bent; (d) disphenoidal; (e) octahedral; (f) bent at each S (but 2 isomers, see text).

15.9 (a) SF_4 is an F^- donor or acceptor; BF_3 is an F^- acceptor; CsF is source of F^-; (b) gives RCF_3.

15.12 $[TeF_7]^-$ is pentagonal bipyramidal; binomial octet in ^{125}Te NMR spectrum means it is fluxional on NMR timescale; ^{19}F NMR spectrum, singlet (equivalent due to fluxionality) attached to non spin active Te; 0.9% ^{123}Te and 7.0% ^{125}Te couple to give two doublets, i.e. satellites.

15.13 (a) all isoelectronic and isostructural; (b) isoelectronic: CO_2, SiO_2 and $[NO_2]^+$; isostructural CO_2 and $[NO_2]^+$; isoelectronic: SO_2 and TeO_2, but not isostructural; (c) all isoelectronic, but only SO_3 and $[PO_3]^-$ are isostructural; (d) all isoelectronic and isostructural.

15.14 (a) SO_3, trigonal planar; $[SO_3]^{2-}$, trigonal pyramidal.

15.16 (a) Reaction required is:
$[SO_4]^{2-} + 8H^+ + 8e^- \rightleftharpoons S^{2-} + 2H_2O$; this is assisted by very high $[H^+]$ and very low solubility of CuS. (b) Expected from VSEPR. (c) White precipitate is $Ag_2S_2O_3$, dissolves forming $[Ag(S_2O_3)_3]^{5-}$; disproportionation of $[S_2O_3]^{2-}$

$$[S_2O_3]^{2-} + H_2O \longrightarrow S^{2-} + [SO_4]^{2-} + 2H^+$$

brought about by removal of S^{2-} as insoluble Ag_2S.

15.17 (a) $[S_2O_4]^{2-} + 2Ag^+ + H_2O \longrightarrow [S_2O_5]^{2-} + 2Ag + 2H^+$; (b) $[S_2O_4]^{2-} + 3I_2 + 4H_2O \longrightarrow 2[SO_4]^{2-} + 6I^- + 8H^+$.

15.18 See **15A**.

$$NH_2$$
$$|$$
$$S$$
$$O{=} \quad \diagdown OH$$
$$O$$

(15A)

15.20 S_2O, **15.32**; $[S_2O_3]^{2-}$, **15.45**; NSF, **15.51**; NSF_3, **15.52**; $[NS_2]^+$, **15.58**; S_2N_2, **15.56**.

Chapter 16

16.2 (a) $2X^- + Cl_2 \longrightarrow X_2 + 2Cl^-$ (X = Br or I); (b) see reaction 10.1; to prevent recombination of Na and Cl_2 to NaCl; (c) $F_2 + H_2 \longrightarrow 2HF$; explosive radical chain reaction.

16.3 Lone pair–lone pair repulsions between lone pairs on O and F weaken bond.

16.5 ClF, 170; BrF, 185; BrCl, 213; ICl, 232; IBr, 247 pm; agreement with Table 16.3 good where $[\chi^P(Y) - \chi^P(X)]$ is small.

16.6 (a) $2AgCl + 2ClF_3 \longrightarrow 2AgF_2 + Cl_2 + 2ClF$ (AgF_2 rather than AgF because ClF_3 is a very strong oxidizing agent);
(b) $2ClF + BF_3 \longrightarrow [Cl_2F]^+[BF_4]^-$;
(c) $CsF + IF_5 \longrightarrow Cs^+[IF_6]^-$;
(d) $SbF_5 + ClF_5 \longrightarrow [ClF_4]^+[SbF_6]^-$ or $2SbF_5 + ClF_5 \longrightarrow [ClF_4]^+[Sb_2F_{11}]^-$;
(e) $Me_4NF + IF_7 \longrightarrow [Me_4N]^+[IF_8]^-$;
(f) $K[BrF_4] \xrightarrow{\Delta} KF + BrF_3$

16.8 (a) Square planar; (b) bent; (c) disphenoidal (based on trigonal bipyramid); (d) pentagonal bipyramidal; (e) planar (see **16.7**); (f) octahedral; (g) square-based pyramidal.

16.9 (a) BrF_5: doublet and quintet (J_{FF}), rel. int. 4:1; $[IF_6]^+$: singlet; (b) BrF_5 likely to be fluxional, high temperature limiting spectrum is singlet; $[IF_6]^+$: singlet at all temperatures.

16.10 (a) Valence shell of Al has 8 electrons, I has 12 electrons; use VSEPR. (b) Of three pairs of possible products, the three *salts* would have approximately the same lattice energies, so bond strength of the covalent product is decisive factor. (c) Solution in *n*-hexane contains uncomplexed I_2; charge transfer complexes formed in other solvents; pyridine complex is most stable.

16.11 (a) Disphenoidal; (b) see **16.25**, (c) bent, (d) square-based pyramid.

16.12 (a) In cold alkali:
$Cl_2 + 2NaOH \longrightarrow NaCl + NaOCl + H_2O$;
in hot alkali:
$3Cl_2 + 6NaOH \longrightarrow NaClO_3 + 5NaCl + 3H_2O$;
(b) $[IO_4]^- + 2I^- + H_2O \longrightarrow [IO_3]^- + I_2 + 2[OH]^-$;
$[IO_3]^- + 5I^- + 6H^+ \longrightarrow 3I_2 + 3H_2O$;
(c) $[IO_4]^-$.

16.14 (a) $[ClO_3]^- + 6Fe^{2+} + 6H^+ \longrightarrow Cl^- + 6Fe^{3+} + 3H_2O$;
(b) $[IO_3]^- + 3[SO_3]^{2-} \longrightarrow I^- + 3[SO_4]^{2-}$
(partial reduction also possible);
(c) $[IO_3]^- + 5Br^- + 6H^+ \longrightarrow 2Br_2 + IBr + 3H_2O$.

16.15 (a) Determine total chlorine by addition of excess of I^- and titration with thiosulfate; only HCl is a strong acid so

concentration can be determined by pH measurement. $\Delta_r H^\circ$ found by measuring K at different temps.
(b) Neutralize solution of weighed amount of oxide with $NaHCO_3$ and titrate I_2 against thiosulfate; add excess dilute HCl and titrate again.
(c) Raman spectroscopy to find stretching frequency, that of $[Cl_2]^- < Cl_2$.

16.16 (a) HF vapour is polymeric, hydrogen bonds not broken on vaporization; those in H_2O are. (b) Iodide complex with Ag^+ must be more stable than chloride complex.

16.17 (a) N–H····F hydrogen bond formation; structure similar to that of ice. (b) For the product HX, HI has weakest bond.

Chapter 17

17.2 Electronic configurations: He_2, $\sigma(1s)^2\sigma^*(1s)^2$; $[He_2]^+$, $\sigma(1s)^2\sigma^*(1s)^1$.

17.3 Linear XeF_2; square planar XeF_4; distorted octahedral XeF_6.

17.4 8 bonding pairs and one lone pair; stereochemically inactive lone pair.

17.5 (a) From hydrolysis of XeF_2; $\Delta_f H^\circ$(HF, 298 K) is known. (b) Use thermochemical cycle relating $[XeF_2(s)]$, $[XeF_2(g)]$, $[Xe(g) + 2F(g)]$, $[Xe(g) + F_2(g)]$.

17.6 Consider $Xe + Cl_2 \longrightarrow XeCl_2$ versus F analogue; weaker Xe–Cl than Xe–F bond; stronger Cl–Cl than F–F bond.

17.7 From Born–Haber cycle assuming lattice energies of XeF and CsF $\approx$ equal.

17.8 $[XeO_6]^{4-}$, octahedral; $XeOF_2$, T-shaped; $XeOF_4$, square pyramidal (O in apical site); XeO_2F_2, disphenoidal; XeO_2F_4, octahedral; XeO_3F_2, trigonal bipyramidal (axial F).

17.9 (a) $CsF + XeF_4 \longrightarrow Cs[XeF_5]$;
(b) $SiO_2 + 2XeOF_4 \longrightarrow SiF_4 + 2XeO_2F_2$ or:
$SiO_2 + XeOF_4 \longrightarrow SiF_4 + XeO_3$;
(c) $XeF_2 + SbF_5 \longrightarrow [XeF][SbF_6]$
or: $2XeF_2 + SbF_5 \longrightarrow [Xe_2F_3][SbF_6]$; or:
$XeF_2 + 2SbF_5 \longrightarrow [XeF][Sb_2F_{11}]$
(d) $2XeF_6 + 16[OH]^- \longrightarrow [XeO_6]^{4-} + Xe + O_2 +$
$8H_2O + 12F^-$;
(e) $2KrF_2 + 2H_2O \longrightarrow 2Kr + O_2 + 4HF$.

Chapter 18

18.1 (a) $MeBr + 2Li \xrightarrow{Et_2O} MeLi + LiBr$;

(b) $Na + (C_6H_5)_2 \xrightarrow{THF} Na^+[(C_6H_5)_2]^-$;
(c) $^nBuLi + H_2O \longrightarrow ^nBuH + LiOH$;
(d) $Na + C_5H_6 \longrightarrow Na^+[C_5H_5]^-$; i.e. Na[Cp].

18.4 (a) $Mg + 2C_5H_6 \longrightarrow (\eta^5\text{-}C_5H_5)_2Mg$ (i.e. Cp_2Mg);
(b) $MgCl_2 + LiR \longrightarrow RMgCl + LiCl$ or
$MgCl_2 + 2LiR \longrightarrow R_2Mg + 2LiCl$;
(c) $RBeCl \xrightarrow{LiAlH_4} RBeH$.

18.5 To make each Mg centre 4-coordinate, $n = 4$.

18.6 (a) Smaller K when steric demands of R are smaller, meaning that the dimer is favoured.

18.7 (a) $Al_2Me_6 + 6H_2O \longrightarrow 2Al(OH)_3 + 6CH_4$
(b) $nAlR_3 + nR'NH_2 \longrightarrow (RAlNR')_n + 2nRH$
(e.g. $n = 2$);
(c) $Me_3SiCl + Na[C_5H_5] \longrightarrow Me_3Si(\eta^1\text{-}C_5H_5) + NaCl$;
(d) $2Me_2SiCl_2 + Li[AlH_4] \longrightarrow$
$2Me_2SiH_2 + LiCl + AlCl_3$.

18.9 Anthracene (**L**) and K give $K^+[(L)]^-$; the radical anion acts as a reducing agent, $Sn(IV) \longrightarrow Sn(II)$ (regenerating anthracene); KBr is second product.

18.10 (a) Chain similar to **18.18**; octahedral; (b) chain; trigonal bipyramidal; (c) monomeric; tetrahedral; (d) monomeric; octahedral.

18.11 (a) Et_3SnOH or $(Et_3Sn)_2O$; (b) $(\eta^1\text{-}Cp)Et_3Sn$;
(c) $(Et_3Sn)_2S$; (d) Et_3PhSn; (e) $Et_3SnSnEt_3$.

18.12 (a) Tilt angle α (Figure 18.9) increases as the steric demands of R increase; tilting rationalized in terms of presence of lone pair of electrons.

18.13 $\mathbf{A} = Br_2InCHBr_2 \cdot C_4H_8O_2$; $\mathbf{B} = [Ph_4P]^+_3[HC(InBr_3)_3]^{3-}$.

18.15 (a) $Me_3Sb \cdot BH_3$; (b) Me_3SbO; (c) Me_3SbBr_2;
(d) Me_3SbCl_2; $[Me_6Sb]^-$; (e) Me_4SbI; (f) Me_3SbBr_2;
$Me_3Sb(OEt)_2$.

Chapter 19

19.3 Trend in E° values irregular across period; variation in ionization energies is not enough to account for variation in E°.

19.6 (a) Ions generally too small; (b) charge distribution;
(c) oxidizing power of O and F; (apply electroneutrality principle in b and c).

19.7 (a) +2; d^5; (b) +2; d^6; (c) +3; d^6; (d) +7; d^0; (e) +2; d^8;
(f) +3; d^1; (g) +3; d^2; (h) +3; d^3.

19.8 (a) Linear; (b) trigonal planar; (c) tetrahedral; (d) trigonal bipyramidal *or* square-based pyramidal; (e) octahedral.

19.10 (a) Two, axial (2 C) and equatorial (3 C); (b) low energy fluxional process; Berry pseudo-rotation (see Section 2.11).

19.12 Tripodal ligand; trigonal bipyramidal with central N of ligand and Cl in axial sites.

19.13 (a) Aqueous solutions of $BaCl_2$ and $[Co(NH_3)_5Br][SO_4]$ give $BaSO_4$ ppt; aqueous solutions of $AgNO_3$ and $[Co(NH_3)_5(SO_4)]Br$ give AgBr ppt; only free ion can be precipitated; (b) needs quantitative precipitation of free Cl^- by $AgNO_3$; (c) Co(III) salts are ionization isomers; Cr(III) salts are hydration isomers; (d) *trans*- and *cis*-$[CrCl_2(H_2O)_4]$.

19.14 (a) $[Co(bpy)_2(CN)_2]^+[Fe(bpy)(CN)_4]^-$; $[Fe(bpy)_2(CN)_2]^+[Co(bpy)(CN)_4]^-$; $[Fe(bpy)_3]^{3+}[Co(CN)_6]^{3-}$; (b) *trans*- and *cis*-$[Co(bpy)_2(CN)_2]^+$, and *cis*-$[Co(bpy)_2(CN)_2]^+$ has optical isomers; similarly for $[Fe(bpy)_2(CN)_2]^+$; $[Fe(bpy)_3]^{3+}$ has optical isomers.

19.15 Ignoring conformations of the chelate rings: (a) 4 depending on orientations of the Me groups; (b) 2.

19.16 8; Δ metal configuration with $(\delta\delta\delta)$, $(\delta\delta\lambda)$, $(\delta\lambda\lambda)$ or $(\lambda\lambda\lambda)$; similarly for Λ. All are related as diastereomers except those in which every chiral centre has changed configuration, e.g. Δ-$(\delta\delta\lambda)$ and Λ-$(\lambda\lambda\delta)$.

19.17 (a) Optical; (b) geometrical (*cis* and *trans*), and the *cis*-isomer has optical isomers; (c) geometrical (*trans* and *cis*) as square planar; (d) no isomers; *cis* arrangement; (e) geometrical (*trans* and *cis*); *cis* isomer has optical isomers.

19.18 (a) IR spectroscopy; (b) as for (a); ^{195}Pt is NMR active and ^{31}P NMR spectra of the *cis*- and *trans*-isomers show satellites with J_{PtP} *cis* > *trans*; (c) ^{31}P NMR spectroscopy, *fac*-isomer has one P environment, *mer*-isomer has two; Rh is spin-active, observe doublet for *fac* (J_{RhP}); for *mer*-isomer, observe doublet of triplets (J_{RhP} and $J_{PP'}$) and doublet of doublets (J_{RhP} and $J_{PP'}$) with relative integrals 1:2.

19.19 All octahedral; (a) *mer* and *fac*; (b) *cis* and *trans*, plus enantiomers for *cis*-isomer; (c) only *mer*-isomer.

Chapter 20

20.2 Green is absorbed; appears purple.

20.3 (a) *N*-donors; didentate; may be monodentate; (b) *N*-donors; didentate; (c) *C*-donor; monodentate; may bridge; (d) *N*-donor; monodentate; may bridge; (e) *C*-donor; monodentate; (f) *N*-donors; didentate; (g) *O*-donors; didentate; (h) *N*- or *S*-donor; monodentate; (i) *P*-donor; monodentate.

20.4 $Br^- < F^- < [OH]^- < H_2O < NH_3 < [CN]^-$

20.5 (a) $[Cr(H_2O)_6]^{3+}$ (higher oxidation state); (b) $[Cr(NH_3)_6]^{3+}$ (stronger field ligand); (c) $[Fe(CN)_6]^{3-}$ (higher oxidation state); (d) $[Ni(en)_3]^{2+}$ (stronger field ligand); (e) $[ReF_6]^{2-}$ (third row metal); (f) $[Rh(en)_3]^{3+}$ (second row metal).

20.6 (a) No possibility in d^8 case of promoting an electron from a fully occupied t_{2g} orbital to an *empty* e_g orbital; (c) magnetic data (μ_{eff}).

20.8 (a) Octahedral, low-spin d^5; (b) octahedral, low-spin d^3; (d) octahedral, high-spin d^4; (e) octahedral, high-spin d^5; (f) square planar, d^8; (g) tetrahedral, d^7; (h) tetrahedral, d^8.

20.10 (b) $F^- < H_2O < NH_3 <$ en $< [CN]^- < I^-$.

20.11 (a) In Co^{2+}, t_2 orbitals all singly occupied; in tetrahedral Cu^{2+}, t_2 orbitals asymmetrically filled and as a consequence, complex suffers Jahn–Teller distortion; (b) Jahn–Teller effect in the excited state $t_{2g}{}^3 e_g{}^3$ arising when electron is promoted from ground state $t_{2g}{}^4 e_g{}^2$.

20.12 (a) See table below; E and T_2; (b) see Table 20.5; tetrahedral: A_2, T_2 and T_1; octahedral: A_{2g}, T_{2g} and T_{1g}.

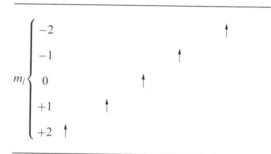

20.13 (a) $10\,000\,cm^{-1} = 1000$ nm; $30\,000\,cm^{-1} = 333$ nm; (b) 400–700 nm; $25\,000$–$14\,285\,cm^{-1}$; (c) $[Ni(H_2O)_6]^{2+}$: green; $[Ni(NH_3)_6]^{2+}$: purple; (d) H_2O weaker field ligand than NH_3; relative energies of transitions are estimated from Orgel diagram: $[Ni(H_2O)_6]^{2+} < [Ni(NH_3)_6]^{2+}$; $E \propto$ wavenumber or $E \propto 1/$wavelength.

20.14 (a) Cr(III) is d^3, so 3 bands; (b) *trans*-$[Co(en)_2F_2]^+$ has centre of symmetry, *cis* has not; charge transfer (CT) from Cl^- to Co^{3+} probably accounts for more intense colour of chloro complex; CT for F^- is most unlikely.

20.15 $x =$ (a) 4; (b) 3; (c) 2; assume one can ignore magnetic moment associated with orbital angular momentum.

20.17 (a) $1.73\,\mu_B$; (b) take into account spin–orbit coupling.

20.18 Octahedral Ni^{2+} (d^8) should have no orbital contribution; tetrahedral Ni^{2+} will have an orbital contribution, so $\mu_{eff} \neq \mu$(spin-only); all electrons paired in square planar Ni^{2+}.

20.20 Normal spinel would have tetrahedral Ni^{2+} with two octahedral Mn^{3+}; inverse spinel would have tetrahedral Mn^{3+}, octahedral Mn^{3+} and octahedral Ni^{2+}; compare LFSE values:

LFSE tet. Ni^{2+} + oct. Mn^{3+} = $-15\,622\,cm^{-1}$

LFSE oct. Ni^{2+} + tet. Mn^{3+} = $-13\,933\,cm^{-1}$

Predict normal spinel; factor not taken into account is Jahn–Teller effect for Mn^{3+}; although predict normal spinel by a small margin, the structure is actually an inverse spinel.

20.21 (a) Difference in LFSE on going from octahedral aqua ion to tetrahedral chloro complex much less for Co^{2+} (d^7) than Ni^{2+} (d^8); (b) indicates $H_4[Fe(CN)_6]$ is weak acid in respect of 4th dissociation constant; H^+ complexing of $[Fe(CN)_6]^{4-}$ makes reduction easier; (c) LFSE plays a *minor* part; there is a loss of LFSE on reduction of Mn^{3+}, gain on reduction of Fe^{3+} and loss on reduction of Cr^{3+}; the decisive factor is large value of IE_3 for Mn.

Chapter 21

21.3 Ether is chelating ligand through two O atoms; $[BH_4]^-$ ligand may be mono-, di- or tridentate; suggest three didentate $[BH_4]^-$ ligands.

21.4 (a) Li_2TiO_3 must have NaCl structure, i.e. $[Li^+]_2Ti^{4+}[O^{2-}]_3$; Li^+, Ti^{4+} and Mg^{2+} are about the same size; electrical neutrality must be maintained; (b) E° for $Ti^{4+} + e^- \rightleftharpoons Ti^{3+}$ is $+0.1\,V$ at pH 0, so one might think that in alkali no reaction with Ti^{3+}; but TiO_2 extremely insoluble and H_2 also upsets the equilibrium.

21.5 Yellow ammonium vanadate in acidic solution contains $[VO_2]^+$; reduction by SO_2 gives blue $[VO]^{2+}$; Zn reduction to purple V^{2+}.

21.6 $2VBr_3 \xrightarrow{\Delta} VBr_4 + VBr_2$; $2VBr_4 \xrightarrow{\Delta} 2VBr_3 + Br_2$; with removal of Br_2, VBr_2 is final product.

21.7 Compound is an alum containing $[V(H_2O)_6]^{3+}$, octahedral d^2; μ(spin-only) = $2.83\,\mu_B$; three bands for d^2 ion.

21.8 [Cr(**21.63**)]; hexadentate N,N',N'',O,O',O''; *fac*-isomer only imposed by *fac*-coordination of macrocycle.

21.9 Cr should be oxidized to Cr^{3+} but air should have no further action.

21.10 (a) Colorimetry (for $[MnO_4]^-$) or gas evolution (CO_2); (b) autocatalysis.

21.12 (a) Mössbauer spectrum; (b) show Fe^{3+}(aq) changes colour at high $[Cl^-]$ and also changes colour if Cl^- displaced by F^-; (c) treat ppt with acid to give MnO_2 and $[MnO_4]^-$, determine both with oxalic acid in strongly acidic solution.

21.13 (a) $2Fe + 3Cl_2 \longrightarrow 2FeCl_3$;
(b) $Fe + I_2 \longrightarrow FeI_2$;
(c) $2FeSO_4 + 2H_2SO_4 \longrightarrow Fe_2(SO_4)_3 + SO_2 + 2H_2O$;
(d) $[Fe(H_2O)_6]^{3+} + [SCN]^- \longrightarrow$
$[Fe(H_2O)_5(SCN\text{-}N)]^{2+} + H_2O$;
(e) $[Fe(H_2O)_6]^{3+} + 3[C_2O_4]^{2-} \longrightarrow [Fe(C_2O_4)_3]^{3-} + 6H_2O$; on standing, the Fe(III) oxidizes oxalate;
(f) $FeO + H_2SO_4 \longrightarrow FeSO_4 + H_2O$;
(g) $FeSO_4 + 2NaOH \longrightarrow$
$Fe(OH)_2$(precipitate) + Na_2SO_4.

21.14 (a) Compare lattice energy value determined from Born cycle with that interpolated from values for MnF_2 and ZnF_2; (b) $K \approx 10^{35}$.

21.15 $Co^{II}Co^{III}_2O_4$: in *normal* spinel the Co^{3+} ions occupy octahedral sites, favoured for low-spin d^6 configuration (LFSE).

21.16 (a) $[Co(en)_2Cl_2]^+$ is low-spin d^6 so diamagnetic; $[CoCl_4]^{2-}$ is d^7, tetrahedral, $e^4t_2^3$, no orbital contribution expected; μ(spin-only) = $3.87\,\mu_B$; here, spin–orbit coupling appears not to be important; (b) values $> \mu$(spin-only); due to spin–orbit coupling; μ_{eff} *inversely* related to ligand field strength.

21.17 (a) Green ppt is hydrated $Ni(CN)_2$; yellow solution contains $[Ni(CN)_4]^{2-}$, and red $[Ni(CN)_5]^{3-}$; (b) $K_2[Ni(CN)_4]$ reduced to give $K_4[Ni_2(CN)_6]$ (see **21.41**) or $K_4[Ni(CN)_4]$.

21.18 Gives octahedral *trans*-$[Ni(L)_2(H_2O)_2]$ (paramagnetic) then square planar $[Ni(L)_2]$ (diamagnetic compound); isomerism involves relative orientations of the Ph groups of ligand, **L**.

21.19 (a) $CuSO_4 + 2NaOH \longrightarrow Cu(OH)_2(s) + Na_2SO_4$;
(b) $CuO + Cu + 2HCl \longrightarrow 2CuCl + H_2O$;
(c) $Cu + 4HNO_3$(conc) $\longrightarrow Cu(NO_3)_2 + 2H_2O + 2NO_2$;
(d) $Cu(OH)_2 + 4NH_3 \longrightarrow [Cu(NH_3)_4]^{2+} + 2[OH]^-$;
(e) $ZnSO_4 + 2NaOH \longrightarrow Zn(OH)_2(s) + Na_2SO_4$;
$Zn(OH)_2(s) + 2NaOH \longrightarrow Na_2[Zn(OH)_4]$;
(f) $ZnS + 2HCl \longrightarrow H_2S + ZnCl_2$.

21.20 (b) $[Pd(Hdmg)_2]$ analogous to $[Ni(Hdmg)_2]$.

21.21 HCl can act in two ways: preferential complexing of Cu^{2+} by Cl^-, and diminution of reducing power of SO_2 because of $[H^+]$ in equilibrium:

$$[SO_4]^{2-} + 4H^+ + 2e^- \rightleftharpoons SO_2 + 2H_2O$$

Try effect of replacing HCl by (a) saturated LiCl or

another very soluble chloride; (b) $HClO_4$ or another very strong acid which is not easily reduced.

21.22 (a) Square planar; (b) tetrahedral; (c) tetrahedral. Distinguish by magnetic data.

21.23 (a) $[MnO_4]^-$; (b) $[MnO_4]^{2-}$; (c) $[Cr_2O_7]^{2-}$; (d) $[VO]^{2+}$; (e) $[VO_4]^{3-}$ (*ortho*), $[VO_3]^-$ (*meta*); (f) $[Fe(CN)_6]^{3-}$. Permanganate.

21.26 $X = K_3[Fe(ox)_3] \cdot 3H_2O$; analysis gives $ox^{2-}:Fe = 3:1$, hence $3K^+$ needed, and $3H_2O$ to make 100%.

$$[Fe(ox)_3]^{3-} + 3[OH]^- \longrightarrow$$
$$\qquad Fe(OH)_3 \text{ (or } Fe_2O_3 \cdot 3H_2O) + 3ox^{2-}$$

$$2K_3[Fe(ox)_3] \longrightarrow 2Fe[ox] + 3K_2[ox] + 2CO_2$$

$[Fe(ox)_3]^{3-}$ is chiral but reaction with $[OH]^-$ suggests anion may be too labile to be resolved into enantiomers.

21.27 $A = [Co(DMSO)_6][ClO_4]_2$; $B = [Co(DMSO)_6][CoCl_4]$.

21.28 $Cu^{2+} + H_2S \longrightarrow CuS + 2H^+$; very low solubility product of CuS allows its precipitation in acid solution. Reduction is:

$$[SO_4]^{2-} + 4H^+ + 2e^- \longrightarrow SO_2 + 2H_2O$$

but also: $[SO_4]^{2-} + 8H^+ + 8e^- \longrightarrow S^{2-} + 4H_2O$ with the very high $[H^+]$ and insolubility of CuS combining to bring about the second reaction.

Chapter 22

22.3 (a) Assume $CrCl_2$ and WCl_2 have same structure; calculate $\Delta_{lattice}H^\circ(CrCl_2)$ and estimate $\Delta_{lattice}H^\circ(WCl_2)$ using $\Delta U \propto 1/r$; $\Delta_{lattice}H^\circ(WCl_2) \approx -2450$ to $-2500 \, kJ \, mol^{-1}$; Born–Haber cycle gives $\Delta_f H^\circ(WCl_2) \approx +353$ to $+403 \, kJ \, mol^{-1}$.

22.4 (a) Same lattice structure and same unit cell size but A_r Hf $\gg$ Zr; (b) Nb(IV) is d^1; NbF_4 has no Nb–Nb interactions, but $NbCl_4$ and $NbBr_4$ contain pairs of Nb atoms (see **22.10**).

22.5 (a) $Cs[NbBr_6]$; (b) $K_2[TaF_7]$ or $K_3[TaF_8]$ more likely than $K[TaF_6]$ under conditions given; (c) $[Nb(bpy)F_5]$ is one possible product; (d) MF_5 (M = Nb, Ta), tetramer **22.4**; $NbBr_5$, dimer **22.5**; $[NbBr_6]^-$, octahedral; $[TaF_7]^{2-}$, monocapped octahedron; $[TaF_8]^{3-}$, square antiprism; $[Nb(bpy)F_5]$, 7-coordinate pentagonal bipyramid possible.

22.6 $[Cl_3M(\mu\text{-}Cl)_3MCl_3]^{3-}$; no Cr–Cr bonding (Cr(III) is d^3); W≡W bond pairs up metal electrons.

22.7 (a) $[Mo_6Cl_8]Cl_2Cl_{4/2} = [Mo_6(\mu_3\text{-}Cl)_8]^{4+}$ with two extra terminal Cl *trans* to each other, and four equatorial Cls involved in bridging; $[Mo_6Cl_8]Cl_2Cl_{4/2} = [Mo_6Cl_8]Cl_{2+2} = Mo_6Cl_{12} = MoCl_2$; (b) W = s^2d^4;

valence electrons $= 36 + 8 - 4 = 40$; 16 used for 8 M–Cl; 24 left for 12 W–W single bonds.

22.10 Re≡Re bond, eclipsed ligands; description as for Cr≡Cr.

22.11 $ReCl_4$ (**22.31**), Re–Re; Re_3Cl_9, Re=Re; $[Re_2Cl_8]^{2-}$, Re≡Re; $[Re_2Cl_9]^-$, Re–Re; $[Re_2Cl_4(\mu\text{-}Ph_2PCH_2CH_2PPh_2)_2]$, Re≡Re.

22.14 (b) *fac-* and *mer*-isomers; ^{31}P NMR spectroscopy is diagnostic; the ^1H decoupled spectrum of *fac*-isomer exhibits a singlet; for *mer*-isomer, triplet and doublet (J_{PP}). [Hydride signals in ^1H NMR spectra also diagnostic.]

22.15 IR spectroscopic data show H or D is present:

$$[RhBr_3(AsMePh_2)_3] \underset{Br_2}{\overset{H_2PO_2}{\rightleftharpoons}} [RhBr_2H(AsMePh_2)_3]$$

22.16 (a) β-$PdCl_2$ (**22.63**) related to $[Nb_6Cl_{12}]^{2+}$ (Figure 22.6b) but no M–M bonding in the Pd_6 core.

22.17 (a) X-ray diffraction definitive; in its absence, *cis-* and *trans*-$[PtCl_2(NH_3)_2]$ distinguished by dipole moments and IR spectroscopy; $[Pt(NH_3)_4][PtCl_4]$ is a 1:1 electrolyte; (b) $[(H_3N)_2Pt(\mu\text{-}Cl)_2Pt(NH_3)_2]Cl_2$ is 1:2 electrolyte; no absorptions in IR spectrum due to $v(Pt\text{-}Cl)_{terminal}$.

22.18 (a) $K_2[PtI_4]$, square planar anion; (b) *cis*-$[PtCl_2(NH_3)_2]$, square planar; (c) $[PtCl_2(phen)]$, square planar and didentate ligand, so *cis*; (d) $[PtCl(tpy)]Cl$, square planar, tridentate tpy; (e) $K_2[Pt(CN)_4]$, square planar anion, stacked in solid state.

22.19 For *trans*-$[PdCl_2(R_2P(CH_2)_nPR_2)]$ to form, $(CH_2)_n$ chain must be long; smaller chains give *cis*-monomer. Dimer will be of type **22A** with *trans*-arrangement most likely.

(22A)

22.20 (a) Bulky $EtNH_2$ ligands prevent cation–anion stacking, so discrete ions; (b) complex $[Ag_2I]^+$ more stable than $[Ag_2Cl]^+$ (see Table 6.9); (c) equilibrium involved is: $Hg^{2+} + Hg \rightleftharpoons [Hg_2]^{2+}$ rather than: $Hg^{2+} + Hg \rightleftharpoons 2Hg^+$.

Chapter 23

23.3 (a) $[V(CO)_6]^-$ and $Cr(CO)_6$ isoelectronic; greater negative charge leads to more back-donation (b) 4-Me group does not affect cone angle but in 2-position, makes ligand more bulky; (c) $Me_3NO + CO \longrightarrow Me_3N + CO_2$; MeCN occupies vacant site but easily replaced by PPh_3; (d) free $HC{\equiv}CH$ is linear; back-donation from Os reduces C–C bond order, making C atoms more sp^2-like.

23.4 (b) Shift consistent with metal–hydride; ^1H nucleus of bridging H couples to 4 equivalent ^{31}P nuclei (100%, $I = \frac{1}{2}$) to give binomial quintet.

23.5 (a) Significant population of π^* MO causes C–C bond to lengthen; (b) replacement of THF ligand by PPh_3; (c) in $Fe(CO)_5$, 2025 and 2000 cm^{-1} due to ν_{CO}; PPh_3 poorer π-acceptor than CO.

23.8 $Fe(CO)_5 + Na_2[Fe(CO)_4] \longrightarrow CO + Na_2[Fe_2(CO)_8]$; $[(OC)_4Fe{-}Fe(CO)_4]^{2-}$ isoelectronic and isostructural with *solution* structure of $Co_2(CO)_8$.

23.10 (a) Give bonding description to account for occupation of π^*-MO of alkene; (b) fluxionality, change in hapticity.

23.11 $Os_7(CO)_{21}$: capped octahedral; $[Os_8(CO)_{22})]^{2-}$: bicapped octahedral.

23.12 Electrons counts: (a) 86; (b) 60; (c) 72; (d) 64; (e) 48; (f) 48; (g) 86; (h) 48; (i) 60.

23.13 (a) $Os_5(CO)_{18}$ has 76 electrons; 3 edge-sharing triangles $= (3 \times 48) - (2 \times 34) = 76$; (b) write as $[\{Ir_4(CO)_{11}\}_2]^{2-}$; Ir–Ir bond *between* clusters is 2c-2e localized, giving two 60-electron tetrahedra.

23.14 (a) $Fe(CO)_4(\eta^2\text{-}C_2H_4)$, trigonal bipyramidal, equatorial C_2H_4; (b) $Na[Re(CO)_5]$; anion trigonal bipyramidal; (c) $Mn(CO)_4(NO)$; trigonal bipyramidal (2 isomers possible); (d) $HMn(CO)_5$; octahedral; (e) $Ni(CO)_3(PPh_3)$ or $Ni(CO)_2(PPh_3)_2$; tetrahedral.

23.15 For CO insertion, 25% product is $Mn(CO)_5Me$ (no ^{13}CO) and 75% is $Mn(^{13}CO)(CO)_4Me$ with ^{13}CO *cis* to Me.

23.20 (a) $[(\eta^5\text{-}Cp)_2Fe]^+[FeCl_4]^-$; (b) $(\eta^5\text{-}Cp)Fe(\eta^5\text{-}C_5H_4C(O)Ph)$ and $(\eta^5\text{-}C_5H_4C(O)Ph)_2Fe$; (c) $[(\eta^5\text{-}Cp)Fe(\eta^6\text{-}C_6H_5Me)]^+[AlCl_4]^-$; (d) $NaCl + (\eta^5\text{-}Cp)FeCo(CO)_6$.

23.21 ^1H NMR spectroscopy; η^5-Cp gives singlet, $\eta^5\text{-}C_5H_4C(O)Me$ gives singlet (Me) and two multiplets. Could also use ^{13}C NMR spectroscopy.

23.22 Use 18-electron rule:

A

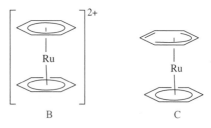

B   C

23.23 (a) 18-electron rule suggests L acts as 4-electron donor. (b) L becomes η^6 (see equation 23.102, left side); (c) $[Ph_3C]^+$ abstracts H^-.

Chapter 24

24.2 $^2F_{5/2}$; 2.54 μ_B.

24.3 Consider cycle for: $3LnX_2 \longrightarrow 2LnX_3 + Ln$; for a given Ln, difference in lattice energy between $3LnX_2$ and $2LnX_3$ is the governing factor, and is least when X is largest.

24.4 Determine electrical conductivity.

24.5 (a) Near constancy originates in small variation in metal ion size which affects interactions with H_2O and $[EDTA]^{4-}$ similarly; hexadentate $[EDTA]^{4-}$ has 4 O-donors and so ΔH^o for replacement of H_2O is small. (b) Complex formation by anions: $Cl^- > [SO_4]^{2-} > [NO_3]^- > [ClO_4]^-$. (c) $BaCeO_3$ is a mixed oxide.

24.8 Hard Ln^{3+} suggests $[NCS]^-$ N-bonded; $[Ln(NCS)_6]^{3-}$, octahedral; 8-coordinate $[Ln(NCS)_7(H_2O)]^{4-}$ could be dodecahedral, square antiprismatic, cubic or distorted variants (hexagonal bipyramidal less likely); $[Ln(NCS)_7]^{4-}$ could be pentagonal bipyramidal, capped octahedral, or distorted variants.

24.9 (b) Sandwich complexes $[(\eta^8\text{-}C_8H_8)_2Sm]^-$ (K^+salt) and $[(\eta^8\text{-}C_8H_8)_2Sm]^{2-}$.

24.10 (b) Zn amalgam should reduce Np(VI) to Np(III); O_2 at pH 0 should oxidize Np(III) to $[NpO_2]^+$ and some $[NpO_2]^{2+}$ (oxidation might be slow).

24.11 U(VI) $\longrightarrow$ U(IV) after aeration; UF_4 formed and then: $2UF_4 + O_2 \longrightarrow UF_6 + UO_2F_2$.

24.12 (a) UF_6; (b) $PaCl_5$, then $PaCl_4$; (c) UO_2; (d) $UCl_4 + UCl_6$; (e) $U(OC_6H_2\text{-}2,4,6\text{-}Me_3)_3$.

24.14 (a) **A**, ^{239}U; **B**, ^{239}Np; **C**, ^{239}Pu; (b) **D**, ^{241}Pu; **E**, ^{241}Am; **F**, ^{242}Cm.

24.15 (a) **A**, $^{253}_{99}Es$; **B**, $^{244}_{94}Pu$; **C**, $^{249}_{98}Cf$; (b) **D**, $^{248}_{96}Cm$; **E**, $^{249}_{98}Cf$.

24.16 (a) All Th(IV) compounds: $Th^{4+}(I^-)_2(e^-)_2$, $Th^{4+}(I^-)_3(e^-)$ and ThI_4; (b) solid state salts contain a linear UO_2 *unit* (formally $[UO_2]^{2+}$) with other ligands in equatorial plane, e.g. $[UO_2(H_2O)_5]^{2+}$; (c) monomer only if R is very bulky, e.g. $R = 2,6\text{-}^tBu_2C_6H_3$.

24.17 (a) $(\eta^5\text{-}Cp)_3ThRu(CO)_2(\eta^5\text{-}Cp)$; $(\eta^5\text{-}Cp)_3ThCHMeEt$; $(\eta^5\text{-}Cp)_3ThCH_2Ph$; (b) bulkier organic ligand hinders redistribution reaction (see equation 24.45); (c) to give $(\eta^5\text{-}C_5Me_5)(\eta^8\text{-}C_8H_8)U(THF)_x$ (in practice, $x = 1$).

24.18 (a) $UCl_4 + 4(\eta^3\text{-}C_3H_5)MgCl$ in Et_2O;
(b) $U(\eta^3\text{-}C_3H_5)_4 + HCl \longrightarrow$
$\qquad U(\eta^3\text{-}C_3H_5)_3Cl + CH_3CH{=}CH_2$.

Chapter 25

25.4 Consider usual square planar rate law, equation 25.9 with k_{obs} given by equation 25.11; suggest pathways are:

$[Rh(cod)(PPh_3)_2]^+ + py \xrightarrow{k_2} [Rh(cod)(PPh_3)(py)]^+ + PPh_3$

competes with:

$[Rh(cod)(PPh_3)_2]^+ + S \xrightarrow{k_1} [Rh(cod)(PPh_3)S]^+ + PPh_3$

$[Rh(cod)(PPh_3)S]^+ + py \xrightarrow{fast} [Rh(cod)(PPh_3)(py)]^+ + S$

Plot of k_{obs} *vs* [py] is linear; gradient $= k_2 = 322\,dm\,mol^{-1}\,s^{-1}$; intercept $= k_1 = 25\,s^{-1}$.

25.5 (b) *trans*-$[PtCl_2(PEt_3)_2]$ and Cl^-.

25.6 (a) As Figure 25.3 with $L^1 = L^3 = L$, and $L^2 = X = Cl$; (b) rearrangement of 5-coordinate intermediate may be possible, giving *cis* + *trans*-$[PtL_2ClY]^+$.

25.7 Positive $\Delta V^{\ddagger}$ suggests dissociative (D or I_d); the rate law suggests associative mechanism; apply Eigen–Wilkins mechanism to account for apparent second order kinetics.

25.8 (a) Step 1, only one product possible; *trans*-effect of $Cl^- > H_2O$, so specific isomer formation observed; (b) $[RhCl_5(H_2O)]^{2-}$ from *trans*-$[RhCl_4(H_2O)_2]^- + Cl^-$, or from $[RhCl_6]^{3-} + H_2O$; *cis*-$[RhCl_4(H_2O)_2]^-$ from $[RhCl_5(H_2O)]^{2-} + H_2O$ (*trans*-effect of Cl^-); *fac*-$[RhCl_3(H_2O)_3]$ from *cis*-$[RhCl_4(H_2O)_2]^- + H_2O$ (*trans*-effect of Cl^-).

25.9 All group 9, d^6; magnitude of Δ_{oct} increases down group.

25.11 Inversion at N; simple amines cannot be resolved.

25.12 These are $acac^-$ type ligands; common mechanism involving dissociation of one end of chelate and reformation of Co–O bond; this may exchange $C(O)CH_3$ and $C(O)CD_3$ groups.

25.13 $-\dfrac{d[SCN^-]}{dt} = \left(k_1 + \dfrac{k_2K_1}{[H^+]}\right)[Fe][SCN^-]$
$\qquad\qquad - \left(k_{-1} + \dfrac{k_{-2}K_2}{[H^+]}\right)[Fe(SCN)]$

where $[Fe] = [Fe(H_2O)_6{}^{3+}]$ and
$[Fe(SCN)] = [Fe(H_2O)_5(SCN)^{2+}]$.

25.14 First step involves breaking one Co–O bond in carbonato chelate ring; $H_2(^{18}O)$ fills vacant site; protonation of pendant carbonate-O atom.

25.16 Both sets of data are the same within experimental error; (a) $\Delta H^{\ddagger} = 128\,kJ\,mol^{-1}$; $\Delta S^{\ddagger} = 95\,J\,K^{-1}\,mol^{-1}$; (b) data consistent with racemization being by a dissociative mechanism.

25.17 *Dcb* mechanism; $[NH_2]^-$ in NH_3 is analogous to $[OH]^-$ in H_2O.

25.18 'Three hump' profiles, the highest E_A is (a) bridge formation, (b) electron transfer, (c) bridge breaking; (b) is most common.

25.20 **I**: both low-spin, similar Rh–N bond lengths; **II**: $[Co(NH_3)_6]^{3+}$ is low-spin, becomes high-spin (and has longer Co–N) after reduction; Δ_rG° helps reaction; **III**: see text discussion, Section 25.5; $\Delta G^\circ = 0$ for self-exchanges **I** and **III**.

25.21 (a) **A**: $[Fe^{III}(CN)_6]^{3-} + [Fe^{II}(CN)_6]^{4-} \longrightarrow$
$\qquad\qquad\qquad [Fe^{II}(CN)_6]^{4-} + [Fe^{III}(CN)_6]^{3-}$;
B: $[Mo^V(CN)_8]^{3-} + [Mo^{IV}(CN)_8]^{4-} \longrightarrow$
$\qquad\qquad\qquad [Mo^{IV}(CN)_8]^{4-} + [Mo^V(CN)_8]^{3-}$;
C: $[Mo^V(CN)_8]^{3-} + [Fe^{II}(CN)_6]^{4-} \longrightarrow$
$\qquad\qquad\qquad [Mo^{IV}(CN)_8]^{4-} + [Fe^{III}(CN)_6]^{3-}$;
(b) $4.3 \times 10^4\,dm^3\,mol^{-1}\,s^{-1}$; agrees well with experiment implying outer-sphere mechanism.

25.22 (a) $2[Fe(CN)_6]^{3-} + H_2A \rightleftharpoons 2[Fe(CN)_6]^{4-} + A + 2H^+$; (b) $K_{12} = 4.1 \times 10^{-5}$; $k_{12} = 530\,dm^3\,mol^{-1}\,s^{-1}$ in good agreement (same order of magnitude) as observed k; outer-sphere mechanism.

Chapter 26

26.1 (a) First, formation of active catalytic species; step 1 = oxidative addition; step 2 = alkene insertion; step 3 = β-elimination; step 4 = elimination of HX; (b) no β-H present.

26.4 (a) $PhMeC{=}CHPh$; $H_2C{=}C(CO_2H)(NHC(O)Me)$; (b) $\approx 8\%$ *S* and 92% *R*.

26.5 (a) Base cycle on inner part of Figure 26.6; (b) regioselectivity is *n:i* ratio; greater selectivity to linear aldehyde at lower temperature.

26.6

(I)   **(II)**   **(III)**

(**I**) highest yield, by alkene isomerization and then as in Figure 26.6; (**III**) lowest yield (sterically hindered); (**II**) formed as secondary product with both (**I**) and (**III**), see Figure 26.6.

26.7 Steps in cycle: loss of PPh_3 from $HRh(CO)(PPh_3)_3$ to give *trans*-$HRh(CO)(PPh_3)_2$; alkene addition gives 18e Rh which undergoes H migration to give σ-alkyl, 16e-complex; oxidative addition of H_2, then H migration, elimination of RCH_2CH_3 to regenerate catalyst.

26.8 (a) Similar IR absorptions indicate similar amounts of back donation to CO ligands and so similar charge distribution in complexes; (b) complex needs to be water-soluble, so Na^+ salt best choice.

26.10 (a) Active 16e-complex is NiL_3, so dissociation step is important; K depends on steric factors;

(b) $NiL_3 \xrightarrow{HCN} Ni(H)(CN)L_3 \xrightarrow{-L} Ni(H)(CN)L_2$

$\xrightarrow{CH_2=CHCH=CH_2} Ni(\eta^3\text{-}C_4H_7)(CN)L_2$;

(c) transfer of CN to give either **26A** or **26B**; the linear alkene is needed for the commercial process.

(26A)   (26B)

26.11 (a) 46 electron count, so unsaturated; (b) addition of alkene to an $Os(CO)_3$ vertex; transfer of one cluster H to give σ-bonded alkyl bound to Os at C(2); β-elimination gives alkene, *E*- or *Z*-isomer.

26.14 (a) Increases yield of SO_3; (b) reduces yield.

26.15 (b) V: strong chemisorption of N, nitride formation; Pt: high $\Delta G^{\ddagger}$ for N_2 adsorption; Os: rare and expensive compared to catalyst used (Fe_3O_4).

26.18 Discuss (a) shape selectivity; (b) acid catalyst.

Chapter 27

27.1 (a) Vacant Ca^{2+} and Cl^- sites must be in 1:2 ratio; (b) see Figure 5.27; (c) Ag^+ and Cd^{2+} similar size; replacement

of Ag^+ by Cd^{2+} gives charge imbalance countered by an extra Ag^+ vacancy.

27.3 For small deviations from stoichiometry, $[Ni^{2+}]$ and $[O^{2-}]$ are nearly constant, so $K = [Ni^{3+}]^4[\square_+]^2/p(O_2)$. Since $[\square_+] = \frac{1}{2}[Ni^{3+}]$, $K \propto [Ni^{3+}]^6/p(O_2)$ with conductivity $\propto [Ni^{3+}]$ and hence $\propto p(O_2)^{1/6}$.

27.4 (c) Cations of same charge and approximately same size.

27.5 Creation of positive hole as electron hops from Ni^{2+} to Ni^{3+}; as more Li^+ incorporated, more Ni^{3+} sites created, and conductivity increases.

27.7 AgI is a solid Ag^+ ion conductor; passage of Ag^+ (not e^-) occurs through solid electrolyte.

27.9 V_6O_{13} reversibly intercalates Li^+;

$$x Li^+ + V_6O_{13} + x e^- \underset{\text{charge}}{\overset{\text{discharge}}{\rightleftharpoons}} Li_x V_6 O_{13}$$

27.11 (b) Cost of cooling to T_c; $T_c = 9.5\,K$ for NbTi requiring He coolant; $T_c > 77\,K$ could utilize liquid N_2.

27.13 (a) $Li_xV_2O_5 + I_2$; (b) $CaWO_4$; (c) Sr_2FeO_4 (or $SrFeO_3$).

27.14 (a) Bi_2O_3, V_2O_5, CaO; (b) Cu_2O, MoO_3, Y_2O_3; (c) Li_2O, In_2O_3; (d) RuO_2, Y_2O_3.

Chapter 28

28.3 Octahedral complex with 3 catecholate ligands; the Cr^{3+} complex (d^3) is kinetically inert, so solution studies practicable.

28.4 (a) Soft *S*-donors compatible with soft metal ion; (b) protein binding sites coordinate several metals in cluster units; (c) similar C_3N_2 heterocyclic rings present in each.

28.10 (a) Cu^{2+} state blue; Cu^+, colourless; (b) changes in conformation of metal-binding pocket alters coordination environment and also reduction potential; (c) CO blocks O_2 binding site by coordinating to Fe^{2+}, but $[CN]^-$ favours Fe^{3+} and binds tightly to cytochrome haem.

28.12 (a) $[Fe(SPh)_4]^{2-}$ models $Fe\{S(Cys)\}_4$-site; for Fe^{2+} and Fe^{3+}, μ(spin-only) values are 4.90 and 5.92 μ_B; (b) spinach ferredoxin is a [2Fe–2S] system with an $Fe_2(\mu\text{-}S)_2\{S(Cys)\}_4$ core; (c) **28.22** models half of FeMo cofactor but fails to include coordinatively unsaturated Fe centres; Mössbauer data consistent with delocalization of charge.

28.13 (a) Middle two; 4Fe(III) and 4Fe(II) states are not accessed; (b) 3Fe(III)·Fe(II) $\rightleftharpoons$ 2Fe(III)·2Fe(II).

28.19 Rate data show substrate, Zn^{2+} and $[OH]^-$ all involved in rate-determining step; suggest complex of type **28A** formed (X could be H_2O or a substrate O); possible mechanism (with charges left off coordinated $[OH]^-$ and CO_2^-) is shown in **28A**. Degree of protonation depends on pH.

(28A)

Index

Note: (B) indicates text in a Box, (F) a Figure, (N) a footnote, (T) a Table and (WE) a Worked Example.